Encyclopedia of Database Systems

Ling Liu • M. Tamer Özsu
Editors

Encyclopedia of Database Systems

Second Edition

Volume 4

M–P

With 1374 Figures and 143 Tables

 Springer

Editors
Ling Liu
Georgia Institute of Technology College
of Computing
Atlanta, GA, USA

M. Tamer Özsu
University of Waterloo School of Computer Science
Waterloo, ON, Canada

ISBN 978-1-4614-8266-6   ISBN 978-1-4614-8265-9 (eBook)
ISBN 978-1-4614-8264-2 (print and electronic bundle)
https://doi.org/10.1007/978-1-4614-8265-9

Library of Congress Control Number: 2018938558

1st edition: © Springer Science+Business Media, LLC 2009 (USA)
© Springer Science+Business Media, LLC, part of Springer Nature 2018

Printed on acid-free paper

This Springer imprint is published by the registered company Springer Science+Business
Media, LLC part of Springer Nature.
The registered company address is: 233 Spring Street, New York, NY 10013, U.S.A.

Preface to the Second Edition

Since the release of the first volume of this Encyclopedia, big data has emerged as a central feature of information technology innovation in many business, science, and engineering fields. Databases are one of the fundamental core technologies for big data systems and big data analytics. In order to extract features and derive values from big data, it must be stored, processed, and analyzed in a timely manner. Not surprisingly, big data not only fuels the development and deployment of database systems and database technologies, it also opens doors to new opportunities and new challenges in the field of databases. As data grows in volume, velocity, variety, and with the attendant veracity issues, there is a growing demand for volume-scalable databases, velocity-adaptive databases, and variety-capable databases that can handle data quality issues properly. As machine learning and artificial intelligence renew their momentum with the power of big data, there is an increasing demand for new generation of database systems that are built for extracting features from databases as efficient and effective as conventional database systems are capable of for querying databases.

The first edition of the *Encyclopedia of Database Systems* is a comprehensive, multivolume collection of over 1,250 in-depth entries (3,067 including synonyms), covering important concepts on all aspects of database systems, including areas of current interest and research results of historical significance. This second edition of *Encyclopedia of Database Systems* expands the first edition by enriching the content of existing entries, expanding existing topic areas with new entries, adding a set of cutting-edge topic areas, including cloud data management, crowdsourcing, data analytics, data provenance management, graph data management, social networks, and uncertain data management to name a few. The new entries and the new topic areas were determined through discussions and consultations with the Advisory Board of the *Encyclopedia of Database Systems*. Each of the new topic areas was managed by a new Area Editor who, together with the editor-in-chief, further developed the content for each area, soliciting experts in the field as contributors to write the entries, and performed the necessary technical editing. We also reviewed the entries from the first edition and revised them as needed to bring them up-to-date.

We would like to thank the members of the Advisory Board, the Editorial Board, and all of the authors for their contributions to this second edition. We would also like to thank Springer's editors and staff, including Susan Lagerstrom-Fife, Michael Hermann, and Sonja Peterson for their assistance and support throughout the project, and Annalea Manalili for her involvement in the early period of this project.

In closing, we trust the Encyclopedia can serve as a valuable source for students, researchers, and practitioners who need a quick and authoritative reference to the subject on database systems. Suggestions and feedbacks to further improve the Encyclopedia are welcome from readers and from the community.

Preface to the First Edition

We are in an information era where generating and storing large amounts of data are commonplace. A growing number of organizations routinely handle terabytes and exabytes of data, and individual digital data collections easily reach multiple gigabytes. Along with the increases in volume, the modality of digitized data that requires efficient management and the access modes to these data have become more varied. It is increasingly common for business and personal data collections to include images, video, voice, and unstructured text; the retrieval of these data comprises various forms, including structured queries, keyword search, and visual access. Data have become a highly valued asset for governments, industries and individuals, and the management of these data collections remains a critical technical challenge.

Database technology has matured over the past four decades and is now quite ubiquitous in many applications that deal with more traditional business data. The challenges of expanding data management to include other data modalities while maintaining the fundamental tenets of database management (data independence, data integrity, data consistency, etc.) are issues that the community continues to work on. The links between database management and other fields such as information retrieval, multimedia retrieval, and data visualization are increasingly blurred.

This multi-volume *Encyclopedia of Database Systems* provides easy access to important concepts on all aspects of database systems, including areas of current interest and research results of historical significance. It is a comprehensive collection of over 1,250 in-depth entries (3,067 including synonyms) that present coverage of the important concepts, issues, emerging technology and future trends in the field of database technologies, systems, and applications. The content of the *Encyclopedia* was determined through wide consultations. We were assisted by an Advisory Board in coming up with the overall structure and content. Each of these areas were put under the control of Area Editors (70 in total) who further developed the content for each area, soliciting experts in the field as contributors to write the entries, and performed the necessary technical editing. Some of them even wrote entries themselves. Nearly 1,000 authors were involved in writing entries.

The intended audience for the *Encyclopedia* is technically broad and diverse. It includes anyone concerned with database system technology and its applications. Specifically, the *Encyclopedia* can serve as a valuable and authoritative reference for students, researchers and practitioners who need a quick and authoritative reference to the subject of databases, data management, and database systems. We anticipate that many people will benefit from this reference work, including database specialists, software developers, scientists and engineers who need to deal with (structured, semi-structured or unstructured) large datasets. In addition, database and data mining researchers and scholars in the many areas that apply database technologies, such as artificial intelligence, software engineering, robotics and computer vision, machine learning, finance and marketing are expected to benefit from the *Encyclopedia*.

We would like to thank the members of the Advisory Board, the Editorial Board, and the individual contributors for their help in creating this *Encyclopedia*. The success of the *Encyclopedia* could not have been achieved without the expertise and the effort of the many contributors. Our sincere thanks also go to Springer's editors and staff, including Jennifer Carlson, Susan Lagerstrom-Fife, Oona Schmid, and Susan Bednarczyk for their support throughout the project.

Finally, we would very much like to hear from readers for any suggestions regarding the *Encyclopedia's* content. With a project of this size and scope, it is quite possible that we may have missed some concepts. It is also possible that some entries may benefit from revisions and clarifications. We are committed to issuing periodic updates and we look forward to the feedback from the community to improve the *Encyclopedia*.

Ling Liu
M. Tamer Özsu

List of Topics

Replica Control
Replica Freshness
Replicated Data Types
Replication Based on Group Communication
Replication Based on Paxos
Replication for Availability and Fault Tolerance
Replication for Scalability
Strong Consistency Models for Replicated Data
WAN Data Replication
Weak Consistency Models for Replicated Data

Storage Structures and Systems

Section Editor: *Masaru Kitsuregawa*

Active Storage
Backup and Restore
Checksum and Cyclic Redundancy Check
    Mechanism
Continuous Data Protection
Database Machine
Deduplication
Direct Attached Storage
Disaster Recovery
Disk Power Saving
Information Lifecycle Management
Intelligent Storage Systems
IP Storage
Logical Volume Manager
Massive Array of Idle Disks
Multitier Storage Systems
Network Attached Secure Device
Network Attached Storage
Point-in-Time Copy
Redundant Arrays of Independent Disks
Replication
SAN File System
Storage Area Network
Storage Consolidation
Storage Management Initiative Specification
Storage Management
Storage Network Architectures
Storage Networking Industry Association
Storage Power Management
Storage Protection
Storage Resource Management
Storage Virtualization
Write Once Read Many

Views and View Management

Section Editor: *Yannis Kotidis*

Answering Queries Using Views
Incremental Maintenance of Views with
    Aggregates
Maintenance of Materialized Views with
    Outer-Joins
Maintenance of Recursive Views
Self-Maintenance of Views
Side-Effect-Free View Updates
Updates Through Views
View Adaptation
View Definition
View Maintenance Aspects
View Maintenance
Views
Web Views

Structured Text Retrieval

Section Editor: *Jaap Kamps*

Aggregation-Based Structured Text Retrieval
Content-and-Structure Query
Content-Only Query
Contextualization in Structured Text Retrieval
Entity Retrieval
Evaluation Metrics for Structured Text Retrieval
Indexing Units of Structured Text Retrieval
Initiative for the Evaluation of XML Retrieval
Integrated DB and IR Approaches
Logical Document Structure
Managing Compressed Structured Text
Narrowed Extended XPath I
Presenting Structured Text Retrieval Results
Processing Overlaps in Structured Text Retrieval
Processing Structural Constraints
Profiles and Context for Structured Text
    Retrieval
Propagation-Based Structured Text Retrieval
Relationships in Structured Text Retrieval
Relevance
Specificity
Structure Weight
Structured Document Retrieval
Structured Text Retrieval Models
Term Statistics for Structured Text Retrieval
XML Retrieval

Association Rule Mining

Section Editor: *Jian Pei*

Workflow Management

Section Editor: *Barbara Pernici*

Stream Mining

Section Editor: *Divesh Srivastava*

Distributed Database Systems

Section Editor: *Kian-Lee Tan*

About the Editors

Ling Liu Georgia Institute of Technology College of Computing, Atlanta, GA, USA

Ling Liu is Professor of Computer Science in the College of Computing at Georgia Institute of Technology. She holds a Ph.D. (1992) in Computer and Information Science from Tilburg University, The Netherlands. Dr. Liu directs the research programs in the Distributed Data Intensive Systems Lab (DiSL), examining various aspects of data intensive systems, ranging from big data systems, cloud computing, databases, Internet and mobile systems and services, machine learning, to social and crowd computing, with the focus on performance, availability, security, privacy, and trust. Prof. Liu is an elected IEEE Fellow and a recipient of IEEE Computer Society Technical Achievement Award (2012). She has published over 300 international journal and conference articles and is a recipient of the best paper award from numerous top venues, including ICDCS, WWW, IEEE Cloud, IEEE ICWS, and ACM/IEEE CCGrid. In addition to serving as general chair and PC chairs of numerous IEEE and ACM conferences in big data, distributed computing, cloud computing, data engineering, and very large databases fields, Prof. Liu served as the Editor-in-Chief of *IEEE Transactions on Service Computing* (2013–2016) and also served on editorial boards of over a dozen international journals. Her current research is sponsored primarily by NSF and IBM.

M. Tamer Özsu University of Waterloo School of Computer Science, Waterloo, ON, Canada

M. Tamer Özsu is Professor of Computer Science at the David R. Cheriton School of Computer Science and the Associate Dean of Research of the Faculty of Mathematics at the University of Waterloo. He was the Director of the Cheriton School of Computer Science from January 2007 to June 2010.

His research is in data management focusing on large-scale data distribution and management of nontraditional data, currently focusing on graph and RDF data. His publications include the book *Principles of Distributed Database Systems* (with Patrick Valduriez), which is now in its third edition. He was the Founding Series Editor of *Synthesis Lectures on Data Management* (Morgan & Claypool) and is now the Editor-in-Chief of *ACM Books*. He serves on the editorial boards of three journals and two book series.

He is a Fellow of the Royal Society of Canada, American Association for the Advancement of Science (AAAS), Association for Computing Machinery (ACM), and the Institute of Electrical and Electronics Engineers (IEEE). He is an elected member of the Science Academy, Turkey, and a member of Sigma Xi. He was awarded the ACM SIGMOD Test-of-Time Award in 2015, the ACM SIGMOD Contributions Award in 2006, and the Ohio State University College of Engineering Distinguished Alumnus Award in 2008.

Advisory Board

Ramesh Jain Department of Computer Science, School of Information and Computer Sciences, University of California Irvine, Irvine, CA, USA

Peter MG Apers Centre for Telematics and Information Technology, University of Twente, Enschede, The Netherlands

Timos Sellis Data Science Research Institute, Swinburne University of Technology, Hawthorn, VIC, Australia

Matthias Jarke Informatik 5 Information Systems, RWTH-Aachen University, Aachen, Germany

Ricardo Baeza-Yate Department of Information and Communication Technologies, University of Pompeu Fabra, Barcelona, Spain

Jai Menon Cloudistics, Reston, VA, USA

Beng Chin Ooi School of Computing, National University of Singapore, Singapore, Singapore

Elisa Bertino Department of Computer Science, Purdue University, West Lafeyette, IN, USA

Erhard Rahm Fakultät für Mathematik und Informatik, Institut für Informatik, Universität Leipzig, Leipzig, Germany

Gerhard Weikum Department 5: Databases and Information Systems, Max-Planck-Institut für Informatik, Saarbrücken, Germany

Stefano Ceri Department of Electronics, Information and Bioengineering, Politecnico di Milano, Milano, Italy

Asuman Dogac SRDC Software Research and Development and Consultancy Ltd., Cankaya/Ankara, Turkey

Hans-Joerg Schek Department of Computer Science, ETH Zürich, Zürich, Switzerland

Alon Halevy Recruit Institute of Technology, Mountain View, CA, USA

Jennifer Widom Frederick Emmons Terman School of Engineering, Stanford University, Stanford, CA, USA

John Mylopoulos Department of Computer Science, University of Toronto, Toronto, ON, Canada

Jiawei Han Department of Computer Science, University of Illinois at Urbana-Champaign, Urbana, IL, USA

Lizhu Zhou Department of Computer Science and Technology, Tsinghua University, Beijing, China

Theo Härder Department of Computer Science, University of Kaiserslautern, Kaiserslautern, Germany

Serge Abiteboul INRIA and ENS, Paris, France

Frank Tompa David R. Cheriton School of Computer Science, University of Waterloo, Waterloo, ON, Canada

Patrick Valduriez INRIA and LIRMM, Montpellier, France

Gustavo Alonso Department of Computer Science, ETH Zürich, Zürich, Switzerland

Krithi Ramamritham Department of Computer Science and Engineering, Indian Institute of Technology Bombay, Mumbai, India

Area Editors

Peer-to-Peer Data Management

Karl Aberer Department of Computer Science, École Polytechnique Fédérale de Lausanne (EPFL), Lausanne, Switzerland

Database Management System Architectures

Anastasia Ailamaki Department of Computer Science, Ecole Polytechnique Fédérale de Lausanne, Lausanne, Switzerland

Information Retrieval Models

Giambattista Amati Fondazione Ugo Bordoni, Rome, Italy

XML Data Management

Sihem Amer-Yahia CNRS, University Grenoble Alpes, Saint Martin D'Hères, France

Database Middleware

Cristiana Amza Electrical and Computer Engineering, University of Toronto, Toronto, ON, Canada

Database Management Utilities

Philippe Bonnet Department of Computer Science, IT University of Copenhagen, Copenhagen, Denmark

Visual Interfaces

Tiziana Catarci Department of Computer Engineering, Automation and Management, Sapienza – Università di Roma, Rome, Italy

Stream Data Management

Ugur Cetintemel Department of Computer Science, Brown University, Providence, RI, USA

Querying over Data Integration Systems

Kevin Chang Department of Computer Science, University of Illinois at Urbana-Champaign, Urbana-Champaign, IL, USA

Self Management

Surajit Chaudhuri Microsoft Corporation, Redmond, CA, USA

Text Mining

Zheng Chen Microsoft Corporation, Beijing, China

Extended Transaction Models

Panos K. Chrysanthis Department of Computer Science, School of Computing and Information, University of Pittsburgh, Pittsburgh, PA, USA

Privacy-Preserving Data Mining

Chris Clifton Department of Computer Science, Purdue University, West Lafayette, IN, USA

Digital Libraries

Amr El Abbadi Department of Computer Science, UC Santa Barbara, Santa Barbara, CA, USA

Data Models

David W. Embley Department of Computer Science, Brigham Young University, Provo, UT, USA

Complex Event Processing

Opher Etzion Department of Information Systems, Yezreel Valley College, Jezreel Valley, Israel

Database Security and Privacy

Elena Ferrari Department of Computer Science, Università degli Studi dell'Insubria, Varese, Italy

Semantic Web and Ontologies

Avigdor Gal Industrial Engineering and Management, Technion – Israel Institute of Technology, Haifa, Israel

Data Cleaning

Venkatesh Ganti Alation, Redwood City, CA, USA

Web Data Extraction

Georg Gottlob Computing Lab, Oxford University, Oxford, UK

Sensor Networks

Le Gruenwald School of Computer Science, University of Oklahoma, Norman, OK, USA

Data Clustering

Dimitrios Gunopulos Department of Informatics and Telecommunications, National and Kapodistrian University of Athens, Athens, Greece

Scientific Databases

Amarnath Gupta San Diego Supercomputer Center, University of California San Diego, La Jolla, CA, USA

Geographical Information Systems

Ralf Hartmut Güting Department of Computer Science, FernUniversität in Hagen, Hagen, Germany

Data Visualization

Hans Hinterberger Department of Computer Science, ETH Zurich, Zurich, Switzerland

Web Services and Service Oriented Architectures

Hans-Arno Jacobsen Department of Electrical and Computer Engineering, University of Toronto, Toronto, ON, Canada

Metadata Management

Manfred Jeusfeld IIT, University of Skövde, Skövde, Sweden

Health Informatics Databases

Vipul Kashyap CIGNA Healthcare, Bloomfield, CT, USA

Visual Data Mining

Daniel A. Keim Computer Science Department, University of Konstanz, Konstanz, Germany

Data Replication

Bettina Kemme School of Computer Science, McGill University, Montreal, QC, Canada

Storage Structures and Systems

Masaru Kitsuregawa Institute of Industrial Science, University of Tokyo, Tokyo, Japan

Views and View Management

Yannis Kotidis Department of Informatics, Athens University of Economics and Business, Athens, Greece

Structured Text Retrieval

Jaap Kamps Faculty of Humanities, University of Amsterdam, Amsterdam, The Netherlands

Information Quality

Yang W. Lee School of Business, Northeastern University, Boston, MA, USA

Relational Theory

Leonid Libkin School of Informatics, University of Edinburgh, Edinburgh, UK

Information Retrieval Evaluation Measures

Weiyi Meng Department of Computer Science, State University of New York at Binghamton, Binghamton, NY, USA

Logical Data Integration

Renée J. Miller Department of Computer Science, University of Toronto, Toronto, ON, Canada

Database Design

Alexander Borgida Department of Computer Science, Rutgers University, New Brunswick, NJ, USA

Text Indexing Techniques

Mario A. Nascimento Department of Computing Science, University of Alberta, Edmonton, AB, Canada

Data Quality

Felix Naumann Hasso Plattner Institute, University of Potsdam, Potsdam, Germany

Web Search and Crawl

Cong Yu Google Research, New York, NY, USA

Multimedia Databases

Vincent Oria Department of Computer Science, New Jersey Institute of Technology, Newark, NJ, USA

Shin'ichi Satoh Digital Content and Media Sciences ReseaMultimedia Information Research Division, National Institute of Informatics, Tokyo, Japan

Active Databases

M. Tamer Özsu Cheriton School of Computer Science, University of Waterloo, Waterloo, ON, Canada

Spatial, Spatiotemporal, and Multidimensional Databases

Dimitris Papadias Department of Computer Science and Engineering, Hong Kong University of Science and Technology, Kowloon, China

Data Warehouse

Torben Bach Pedersen Department of Computer Science, Aalborg University, Aalborg, Denmark

Stefano Rizzi DISI – University of Bologna, Bologna, Italy

Association Rule Mining

Jian Pei School of Computing Science, Simon Fraser University, Burnaby, BC, Canada

Workflow Management

Barbara Pernici Department di Elettronica e Informazione, Politecnico di Milano, Milan, Italy

Query Processing and Optimization

Evaggelia Pitoura Department of Computer Science and Engineering, University of Ioannina, Ioannina, Greece

Data Management for the Life Sciences

Louiqa Raschid Robert H. Smith School of Business, University of Maryland, College Park, MD, USA

Information Retrieval Operations

Edie Rasmussen Library, Archival and Information Studies, The University of British Columbia, VC, Canada

Query Languages

Tore Risch Department of Information Technology, Uppsala University, Uppsala, Sweden

Database Tuning and Performance

Dennis Shasha Department of Computer Science, New York University, New York, NY, USA

Classification and Decision Trees

Kyuseok Shim School of Electrical Engineering and Computer Science, Seoul National University, Seoul, Republic of Korea

Temporal Databases

Christian S. Jensen Department of Computer Science, Aalborg University, Aalborg, Denmark

Richard T. Snodgrass Department of Computer Science, University of Arizona, Tucson, AZ, USA

Stream Mining

Divesh Srivastava AT&T Labs-Research, Bedminster, NJ, USA

Distributed Database Systems

Kian-Lee Tan Department of Computer Science, National University of Singapore, Singapore, Singapore

Logics and Databases

Val Tannen Department of Computer and Information Science, University of Pennsylvania, Philadelphia, PA, USA

Structured and Semi-structured Document Databases

Frank Tompa David R. Cheriton School of Computer Science, University of Waterloo, Waterloo, ON, Canada

Indexing

Vassilis J. Tsotras Department of Computer Science and Engineering, University of California-Riverside, Riverside, CA, USA

Parallel Database Systems

Patrick Valduriez INRIA and LIRMM, Montpellier, France

Advanced Storage Systems

Kaladhar Voruganti Equinix, San Francisco, CA, USA

Transaction Management

Gottfried Vossen Department of Information Systems, Westfälische
Wilhelms-Universität, Münster, Germany

Mobile and Ubiquitous Data Management

Ouri Wolfson Department of Computer Science, University of Illinois at Chicago, Chicago, IL, USA

Multimedia Information Retrieval

Jeffrey Xu Yu Department of Systems Engineering and Engineering Management, The Chinese University of Hong Kong, Hong Kong, China

Approximation and Data Reduction Techniques

Xiaofang Zhou School of Information Technology and Electrical Engineering, University of Queensland, Brisbane, Australia

Social Networks

Nick Koudas Department of Computer Science, University of Toronto, Toronto, ON, Canada

Cloud Data Management

Amr El Abbadi Department of Computer Science, UC Santa Barbara, Santa Barbara, CA, USA

Data Analytics

Fatma Özcan IBM Research – Almaden, San Jose, CA, USA

Data Management Fundamentals

Ramez Elmasri Department of Computer Science and Engineering, The University of Texas at Arlington, Arlington, TX, USA

NoSQL Databases

M. Tamer Özsu Cheriton School of Computer Science, University of Waterloo, Waterloo, ON, Canada

Ling Liu College of Computing, Georgia Institute of Technology, Atlanta, GA, USA

Graph Data Management

Lei Chen Department of Computer Science and Engineering, The Hong Kong University of Science and Technology, Hong Kong, China

Data Provenance Management

Juliana Freire Computer Science and Engineering, New York University, New York, NY, USA

Ranking Queries

Ihab F. Ilyas Cheriton School of Computer Science, University of Waterloo, Waterloo, ON, Canada

Uncertain Data Management

Minos Garofalakis Technical University of Crete, Chania, Greece

Crowd Sourcing

Reynold Cheng Computer Science, The University of Hong Kong, Hong Kong, China

List of Contributors

Daniel Abadi Yale University, New Haven, CT, USA

Sofiane Abbar Qatar Computing Research Institute, Doha, Qatar

Alberto Abelló Polytechnic University of Catalonia, Barcelona, Spain

Serge Abiteboul Inria, Paris, France

Maribel Acosta Institute AIFB, Karlsruhe Institute of Technology, Karlsruhe, Germany

Ioannis Aekaterinidis University of Patras, Rio, Patras, Greece

Nitin Agarwal University of Arkansas, Little Rock, AR, USA

Charu C. Aggarwal IBM T. J. Watson Research Center, Yorktown Heights, NY, USA

Lalitha Agnihotri McGraw-Hill Education, New York, NY, USA

Marcos K. Aguilera VMware Research, Palo Alto, CA, USA

Yanif Ahmad Department of Computer Science, Brown University, Providence, RI, USA

Gail-Joon Ahn Arizona State University, Tempe, AZ, USA

Anastasia Ailamaki Informatique et Communications, Ecole Polytechnique Fédérale de Lausanne, Lausanne, Switzerland

Ablimit Aji Analytics Lab, Hewlett Packard, Palo Alto, CA, USA

Alexander Alexandrov Database and Information Management (DIMA), Institute of Software Engineering and Theoretical Computer Science, Berlin, Germany

Yousef J. Al-Houmaily Institute of Public Administration, Riyadh, Saudi Arabia

Mohammed Eunus Ali Department of Computer Science and Engineering, Bangladesh University of Engineering and Technology (BUET), Dhaka, Bangladesh

Robert B. Allen Drexel University, Philadelphia, PA, USA

Gustavo Alonso ETH Zürich, Zurich, Switzerland

Omar Alonso Microsoft Silicon Valley, Mountain View, CA, USA

Bernd Amann Pierre & Marie Curie University (UPMC), Paris, France

Giambattista Amati Fondazione Ugo Bordoni, Rome, Italy

Sihem Amer-Yahia CNRS, Univ. Grenoble Alps, Grenoble, France

Laboratoire d'Informatique de Grenoble, CNRS-LIG, Saint Martin-d'Hères, Grenoble, France

Rainer von Ammon Center for Information Technology Transfer GmbH (CITT), Regensburg, Germany

Robert A. Amsler CSC, Falls Church, VA, USA

Yael Amsterdamer Department of Computer Science, Bar Ilan University, Ramat Gan, Israel

Cristiana Amza Department of Electrical and Computer Engineering, University of Toronto, Toronto, ON, Canada

George Anadiotis VU University Amsterdam, Amsterdam, The Netherlands

Mihael Ankerst Ludwig-Maximilians-Universität München, Munich, Germany

Sameer Antani National Institutes of Health, Bethesda, MD, USA

Grigoris Antoniou Foundation for Research and Technology-Hellas (FORTH), Heraklion, Greece

Arvind Arasu Microsoft Research, Redmond, WA, USA

Danilo Ardagna Politechnico di Milano University, Milan, Italy

Walid G. Aref Purdue University, West Lafayette, IN, USA

Marcelo Arenas Pontifical Catholic University of Chile, Santiago, Chile

Nikos Armenatzoglou Department of Computer Science and Engineering, Hong Kong University of Science and Technology, Kowloon, Hong Kong, Hong Kong

Samuel Aronson Harvard Medical School – Partners Healthcare Center for Genetics and Genomics, Boston, MA, USA

Paavo Arvola University of Tampere, Tampere, Finland

Colin Atkinson Software Engineering, University of Mannheim, Mannheim, Germany

Noboru Babaguchi Osaka University, Osaka, Japan

Shivnath Babu Duke University, Durham, NC, USA

Nathan Backman Computer Science, Buena Vista University, Storm Lake, IA, USA

Kenneth Paul Baclawski Northeastern University, Boston, MA, USA

Ricardo Baeza-Yates NTENT, USA - Univ. Pompeu Fabra, Spain - Univ. de Chile, Chile

James Bailey University of Melbourne, Melbourne, VIC, Australia

Peter Bailis Department of Computer Science, Stanford University, Palo Alto, CA, USA

Sumeet Bajaj Stony Brook University, Stony Brook, NY, USA

Peter Bak IBM Watson Health, Foundational Innovation, Haifa, Israel

Magdalena Balazinska University of Washington, Seattle, WA, USA

Krisztian Balog University of Stavanger, Stavanger, Norway

Farnoush Banaei-Kashani Computer Science and Engineering, University of Colorado Denver, Denver, CO, USA

Jie Bao Data Management, Analytics and Services (DMAS) and Ubiquitous Computing Group (Ubicomp), Microsoft Research Asia, Beijing, China

Stefano Baraldi University of Florence, Florence, Italy

Mauro Barbieri Phillips Research Europe, Eindhoven, The Netherlands

Denilson Barbosa University of Alberta, Edmonton, AL, Canada

Pablo Barceló University of Chile, Santiago, Chile

Luciano Baresi Dipartimento di Elettronica, Informazione e Bioingegneria – Politecnico di Milano, Milano, Italy

Ilaria Bartolini Department of Computer Science and Engineering (DISI), University of Bologna, Bologna, Italy

Saleh Basalamah Computer Science, Umm Al-Qura University, Mecca, Makkah Province, Saudi Arabia

Sugato Basu Google Inc, Mountain View, CA, USA

Carlo Batini University of Milano-Bicocca, Milan, Italy

Michal Batko Masaryk University, Brno, Czech Republic

Peter Baumann Jacobs University, Bremen, Germany

Robert Baumgartner Vienna University of Technology, Vienna, Austria

Sean Bechhofer University of Manchester, Manchester, UK

Steven M. Beitzel Telcordia Technologies, Piscataway, NJ, USA

Ladjel Bellatreche LIAS/ISAE-ENSMA, Poitiers University, Futuroscope, France

Omar Benjelloun Google Inc., New York, NY, USA

Véronique Benzaken University Paris 11, Orsay Cedex, France

Rafael Berlanga Department of Computer Languages and Systems, Universitat Jaume I, Castellón, Spain

Mikael Berndtsson University of Skövde, The Informatics Research Centre, Skövde, Sweden

University of Skövde, School of Informatics, Skövde, Sweden

Philip A. Bernstein Microsoft Corporation, Redmond, WA, USA

Damon Andrew Berry University of Massachusetts, Lowell, MA, USA

Leopoldo Bertossi Carleton University, Ottawa, ON, Canada

Claudio Bettini Dipartimento di Informatica, Università degli Studi di Milano, Milan, Italy

Nigel Bevan Professional Usability Services, London, UK

Bharat Bhargava Purdue University, West Lafayette, IN, USA

Arnab Bhattacharya Indian Institute of Technology, Kanpur, India

Ernst Biersack Eurecom, Sophia Antipolis, France

Alberto Del Bimbo University of Florence, Florence, Italy

Carsten Binnig Computer Science-Database Systems, Brown University, Providence, RI, USA

Christian Bizer Web-based Systems Group, University of Mannheim, Mannheim, Germany

Alan F. Blackwell University of Cambridge, Cambridge, UK

Carlos Blanco GSyA and ISTR Research Groups, Department of Computer Science and Electronics, Faculty of Sciences, University of Cantabria, Santander, Spain

Marina Blanton University of Notre Dame, Notre Dame, IN, USA

Toine Bogers Department of Communication and Psychology, Aalborg University Copenhagen, Copenhagen, Denmark

Philip Bohannon Yahoo! Research, Santa Clara, CA, USA

Michael H. Böhlen Free University of Bozen-Bolzano, Bozen-Bolzano, Italy

University of Zurich, Zürich, Switzerland

Christian Böhm University of Munich, Munich, Germany

Peter Boncz CWI, Amsterdam, The Netherlands

Philippe Bonnet Department of Computer Science, IT University of Copenhagen, Copenhagen, Denmark

Alexander Borgida Rutgers University, New Brunswick, NJ, USA

Vineyak Borkar CTO and VP of Engineering, X15 Software, San Francisco, CA, USA

Chavdar Botev Yahoo Research!, Cornell University, Ithaca, NY, USA

Sara Bouchenak University of Grenoble I – INRIA, Grenoble, France

Luc Bouganim INRIA Saclay and UVSQ, Le Chesnay, France

Nozha Boujemaa INRIA Paris-Rocquencourt, Le Chesnay, France

Shawn Bowers University of California-Davis, Davis, CA, USA

Stéphane Bressan National University of Singapore, School of Computing, Department of Computer Science, Singapore, Singapore

Martin Breunig University of Osnabrueck, Osnabrueck, Germany

Scott A. Bridwell University of Utah, Salt Lake City, UT, USA

Thomas Brinkhoff Institute for Applied Photogrammetry and Geoinformatics (IAPG), Oldenburg, Germany

Nieves R. Brisaboa Database Laboratory, Department of Computer Science, University of A Coruña, A Coruña, Spain

Andrei Broder Yahoo! Research, Santa Clara, CA, USA

Nicolas Bruno Microsoft Corporation, Redmond, WA, USA

François Bry University of Munich, Munich, Germany

Yingyi Bu Chinese University of Hong Kong, Hong Kong, China

Alejandro Buchmann Darmstadt University of Technology, Darmstadt, Germany

Thilina Buddhika Colorado State University, Fort Collins, CO, USA

Chiranjeeb Buragohain Amazon.com, Seattle, WA, USA

Thorsten Büring Ludwig-Maximilians-University Munich, Munich, Germany

Benjamin Bustos Department of Computer Science, University of Chile, Santiago, Chile

David J. Buttler Lawrence Livermore National Laboratory, Livermore, CA, USA

Yanli Cai Shanghai Jiao Tong University, Shanghai, China

Diego Calvanese Research Centre for Knowledge and Data (KRDB), Free University of Bozen-Bolzano, Bolzano, Italy

Guadalupe Canahuate The Ohio State University, Columbus, OH, USA

K. Selcuk Candan Arizona State University, Tempe, AZ, USA

Turkmen Canli University of Illinois at Chicago, Chicago, IL, USA

Alan Cannon Napier University, Edinburgh, UK

Cornelia Caragea Computer Science and Engineering, University of North Texas, Denton, TX, USA

Barbara Carminati Department of Theoretical and Applied Science, University of Insubria, Varese, Italy

Sheelagh Carpendale University of Calgary, Calgary, AB, Canada

Michael W. Carroll Villanova University School of Law, Villanova, PA, USA

Ben Carterette University of Massachusetts Amherst, Amherst, MA, USA

Marco A. Casanova Pontifical Catholic University of Rio de Janeiro, Rio de Janeiro, Brazil

Giuseppe Castagna C.N.R.S. and University Paris 7, Paris, France

Tiziana Catarci Dipartimento di Ingegneria Informatica, Automatica e Gestionale "A.Ruberti", Sapienza – Università di Roma, Rome, Italy

James Caverlee Department of Computer Science, Texas A&M University, College Station, TX, USA

Emmanuel Cecchet EPFL, Lausanne, Switzerland

Wojciech Cellary Department of Information Technology, Poznan University of Economics, Poznan, Poland

Ana Cerdeira-Pena Database Laboratory, Department of Computer Science, University of A Coruña, A Coruña, Spain

Michal Ceresna Lixto Software GmbH, Vienna, Austria

Ugur Cetintemel Department of Computer Science, Brown University, Providence, RI, USA

Soumen Chakrabarti Indian Institute of Technology of Bombay, Mumbai, India

Don Chamberlin IBM Almaden Research Center, San Jose, CA, USA

Allen Chan IBM Toronto Software Lab, Markham, ON, Canada

Chee-Yong Chan National University of Singapore, Singapore, Singapore

K. Mani Chandy California Institute of Technology, Pasadena, CA, USA

Edward Y. Chang Google Research, Mountain View, CA, USA

Kevin Chang Department of Computer Science, University of Illinois at Urbana-Champaign, Urbana, IL, USA

Adriane Chapman University of Southampton, Southampton, UK

Surajit Chaudhuri Microsoft Research, Microsoft Corporation, Redmond, WA, USA

Elizabeth S. Chen Partners HealthCare System, Boston, MA, USA

James L. Chen University of Chicago, Chicago, IL, USA

Jin Chen Computer Engineering Research Group, University of Toronto, Toronto, ON, Canada

Jinjun Chen Swinburne University of Technology, Melbourne, VIC, Australia

Jinchuan Chen Key Laboratory of Data Engineering and Knowledge Engineering, Ministry of Education, Renmin University of China, Beijing

Lei Chen Hong Kong University of Science and Technology, Hong Kong, China

Peter P. Chen Louisiana State University, Baton Rouge, LA, USA

James Cheney University of Edinburgh, Edinburgh, UK

Hong Cheng Department of Systems Engineering and Engineering Management, The Chinese University of Hong Kong, Hong Kong, China

Reynold Cheng Computer Science, The University of Hong Kong, Hong Kong, China

Vivying S. Y. Cheng Hong Kong University of Science and Technology, Hong Kong, China

InduShobha N. Chengalur-Smith University at Albany – SUNY, Albany, NY, USA

Mitch Cherniack Brandeis University, Wattham, MA, USA

Yun Chi NEC Laboratories America, Cupertino, CA, USA

Fernando Chirigati NYU Tandon School of Engineering, Brooklyn, NY, USA

Rada Chirkova North Carolina State University, Raleigh, NC, USA

Laura Chiticariu Scalable Natural Language Processing, IBM Research – Almaden, San Jose, CA, USA

Jan Chomicki Department of Computer Science and Engineering, State University of New York at Buffalo, Buffalo, NY, USA

Fred Chong Computer Science, University of Chicago, Chicago, IL, USA

Stephanie Chow University of Ontario Institute of Technology, Oshawa, ON, Canada

Peter Christen Research School of Computer Science, The Australian National University, Canberra, Australia

Vassilis Christophides INRIA Paris-Roquencourt, Paris, France

Panos K. Chrysanthis Department of Computer Science, University of Pittsburgh, Pittsburgh, PA, USA

Paolo Ciaccia Computer Science and Engineering, University of Bologna, Bologna, Italy

John Cieslewicz Google Inc., Mountain View, CA, USA

Gianluigi Ciocca University of Milano-Bicocca, Milan, Italy

Eugene Clark Harvard Medical School – Partners Healthcare Center for Genetics and Genomics, Boston, MA, USA

Charles L. A. Clarke University of Waterloo, Waterloo, ON, Canada

William R. Claycomb CERT Insider Threat Center, Software Engineering Institute, Carnegie Mellon University, Pittsburgh, PA, USA

Eliseo Clementini University of L'Aguila, L'Aguila, Italy

Chris Clifton Department of Computer Science, Purdue University, West Lafayette, IN, USA

Edith Cohen AT&T Labs-Research, Florham Park, NJ, USA

Sara Cohen The Rachel and Selim Benin School of Computer Science and Engineering, The Hebrew University of Jerusalem, Jerusalem, Israel

Sarah Cohen-Boulakia University Paris-Sud, Orsay Cedex, France

Carlo Combi Department of Computer Science, University of Verona, Verona, VR, Italy

Mariano P. Consens University of Toronto, Toronto, ON, Canada

Dianne Cook Iowa State University, Ames, IA, USA

Graham Cormode Computer Science, University of Warwick, Warwick, UK

Antonio Corral University of Almeria, Almeria, Spain

Maria Francesca Costabile Department of Computer Science, University of Bari, Bari, Italy

Nick Craswell Microsoft Research Cambridge, Cambridge, UK

Fabio Crestani University of Lugano, Lugano, Switzerland

Marco Antonio Cristo FUCAPI, Manaus, Brazil

Maxime Crochemore King's College London, London, UK

Université Paris-Est, Paris, France

Andrew Crotty Database Group, Brown University, Providence, RI, USA

Matthew G. Crowson University of Chicago, Chicago, IL, USA

Michel Crucianu Conservatoire National des Arts et Métiers, Paris, France

Philippe Cudré-Mauroux Massachusetts Institute of Technology, Cambridge, MA, USA

Sonia Leila Da Silva Cerveteri, Italy

Peter Dadam University of Ulm, Ulm, Germany

Mehmet M. Dalkiliç Indiana University, Bloomington, IN, USA

Nilesh Dalvi Airbnb, San Francisco, CA, USA

Marina Danilevsky IBM Almaden Research Center, San Jose, CA, USA

Minh Dao-Tran Institute of Information Systems, Vienna University of Technology, Vienna, Austria

Gautam Das Department of Computer Science and Engineering, University of Texas at Arlington, Arlington, TX, USA

Mahashweta Das Visa Research, Palo Alto, CA, USA

Sudipto Das Microsoft Research, Redmond, WA, USA

Manoranjan Dash Nanyang Technological University, Singapore, Singapore

Anupam Datta Computer Science Department and Electrical and Computer Engineering Department, Carnegie Mellon University, Pittsburgh, PA, USA

Anwitaman Datta Nanyang Technological University, Singapore, Singapore

Ian Davidson University of California-Davis, Davis, CA, USA

Susan B. Davidson Department of Computer and Information Science, University of Pennsylvania, Philadelphia, PA, USA

Todd Davis Department of Computer Science and Software Engineering, Concordia University, Montreal, QC, Canada

Maria De Marsico Sapienza University of Rome, Rome, Italy

Edleno Silva De Moura Federal University of Amazonas, Manaus, Brazil

Antonios Deligiannakis University of Athens, Athens, Greece

Alex Delis University of Athens, Athens, Greece

Alan Demers Cornell University, Ithaca, NY, USA

Jennifer Dempsey University of Arizona, Tucson, AZ, USA

Raytheon Missile Systems, Tucson, AZ, USA

Ke Deng University of Queensland, Brisbane, QLD, Australia

Amol Deshpande University of Maryland, College Park, MD, USA

Zoran Despotovic NTT DoCoMo Communications Laboratories Europe, Munich, Germany

Alin Deutsch University of California-San Diego, La Jolla, CA, USA

Yanlei Diao University of Massachusetts Amherst, Amherst, MA, USA

Suzanne W. Dietrich Arizona State University, Phoenix, AZ, USA

Nevenka Dimitrova Philips Research, Briarcliff Manor, New York, USA

Bolin Ding University of Illinois at Urbana-Champaign, Urbana, IL, USA

Chris Ding University of Texas at Arlington, Arlington, TX, USA

Alan Dix Lancaster University, Lancaster, UK

Belayadi Djahida National High School for Computer Science (ESI), Algiers, Algeria

Hong-Hai Do SAP AG, Dresden, Germany

Gillian Dobbie University of Auckland, Auckland, New Zealand

Alin Dobra University of Florida, Gainesville, FL, USA

Vlastislav Dohnal Masaryk University, Brno, Czech Republic

Mario Döller University of Applied Science Kufstein, Kufstein, Austria

Carlotta Domeniconi George Mason University, Fairfax, VA, USA

Josep Domingo-Ferrer Universitat Rovira i Virgili, Tarragona, Catalonia, Spain

Guozhu Dong Wright State University, Dayton, OH, USA

Xin Luna Dong Amazon, Seattle, WA, USA

Chitra Dorai IBM T. J. Watson Research Center, Hawthorne, NY, USA

Zhicheng Dou Nankai University, Tianjin, China

Ahlame Douzal CNRS, Univ. Grenoble Alps, Grenoble, France

Yang Du Northeastern University, Boston, MA, USA

Susan Dumais Microsoft Research, Redmond, WA, USA

Marlon Dumas University of Tartu, Tartu, Estonia

Schahram Dustdar Technical University of Vienna, Vienna, Austria

Curtis E. Dyreson Utah State University, Logan, UT, USA

Johann Eder Department of Informatics-Systems, Alpen-Adria-Universität Klagenfurt, Klagenfurt, Austria

Milad Eftekhar University of Toronto, Toronto, ON, Canada

Thomas Eiter Institute of Information Systems, Vienna University of Technology, Vienna, Austria

Ibrahim Abu El-Khair Information Science Department, School of Social Sciences, Umm Al-Qura University, Mecca, Saudi Arabia

Ahmed K. Elmagarmid Purdue University, West Lafayette, IN, USA

Qatar Computing Research Institute, HBKU, Doha, Qatar

Ramez Elmasri Computer Science, The University of Texas at Arlington, Arlington, TX, USA

Aaron J. Elmore Department of Computer Science, University of Chicago, Chicago, IL, USA

Sameh Elnikety Microsoft Research, Redmond, WA, USA

David W. Embley Brigham Young University, Provo, UT, USA

Vincent Englebert University of Namur, Namur, Belgium

AnnMarie Ericsson University of Skövde, Skövde, Sweden

Martin Ester Simon Fraser University, Burnaby, BC, Canada

Opher Etzion IBM Software Group, IBM Haifa Labs, Haifa University Campus, Haifa, Israel

Patrick Eugster Purdue University, West Lafayette, IN, USA

Ronald Fagin IBM Almaden Research Center, San Jose, CA, USA

Ju Fan DEKE Lab and School of Information, Renmin University of China, Beijing, China

Wei Fan IBM T.J. Watson Research, Hawthorne, NY, USA

Wenfei Fan University of Edinburgh, Edinburgh, UK

Beihang University, Beijing, China

Hui Fang University of Delaware, Newark, DE, USA

Alan Fekete University of Sydney, Sydney, NSW, Australia

Jean-Daniel Fekete INRIA, LRI University Paris Sud, Orsay Cedex, France

Pascal Felber University of Neuchatel, Neuchatel, Switzerland

Paolino Di Felice University of L'Aguila, L'Aguila, Italy

Hakan Ferhatosmanoglu The Ohio State University, Columbus, OH, USA

Eduardo B. Fernandez Florida Atlantic University, Boca Raton, FL, USA

Eduardo Fernández-Medina GSyA Research Group, Department of Information Technologies and Systems, Institute of Information Technologies and Systems, Escuela Superior de Informática, University of Castilla-La Mancha, Ciudad Real, Spain

Paolo Ferragina Department of Computer Science, University of Pisa, Pisa, Italy

Elena Ferrari DiSTA, University of Insubria, Varese, Italy

Dennis Fetterly Google, Inc., Mountain View, CA, USA

Stephen E. Fienberg Carnegie Mellon University, Pittsburgh, PA, USA

Michael Fink Institute of Information Systems, Vienna University of Technology, Vienna, Austria

Peter M. Fischer Computer Science Department, University of Freiburg, Freiburg, Germany

Simone Fischer-Hübner Karlstad University, Karlstad, Sweden

Fabian Flöck GESIS – Leibniz Institute for the Social Sciences, Köln, Germany

Avrilia Floratou Microsoft, Sunnyvale, CA, USA

Leila De Floriani University of Genova, Genoa, Italy

Christian Fluhr CEA LIST, Fontenay-aux, Roses, France

Greg Flurry IBM SOA Advanced Technology, Armonk, NY, USA

Edward A. Fox Virginia Tech, Blacksburg, VA, USA

Chiara Francalanci Politecnico di Milano University, Milan, Italy

Andrew U. Frank Vienna University of Technology, Vienna, Austria

Michael J. Franklin University of California-Berkeley, Berkeley, CA, USA

Keir Fraser University of Cambridge, Cambridge, UK

Juliana Freire NYU Tandon School of Engineering, Brooklyn, NY, USA

NYU Center for Data Science, New York, NY, USA

New York University, New York, NY, USA

Elias Frentzos University of Piraeus, Piraeus, Greece

Johann-Christoph Freytag Humboldt University of Berlin, Berlin, Germany

Ophir Frieder Georgetown University, Washington, DC, USA

Oliver Frölich Lixto Software GmbH, Vienna, Austria

Ada Wai-Chee Fu Chinese University of Hong Kong, Hong Kong, China

Xiang Fu University of Southern California, Los Angeles, CA, USA

Kazuhisa Fujimoto Hitachi Ltd., Tokyo, Japan

Tim Furche University of Munich, Munich, Germany

Ariel Fuxman Microsoft Research, Mountain View, CA, USA

Silvia Gabrielli Bruno Kessler Foundation, Trento, Italy

Isabella Gagliardi National Research Council (CNR), Milan, Italy

Avigdor Gal Faculty of Industrial Engineering and Management, Technion–Israel Institute of Technology, Haifa, Israel

Alex Galakatos Database Group, Brown University, Providence, RI, USA

Department of Computer Science, Brown University, Providence, RI, USA

Wojciech Galuba EPFL, Lausanne, Switzerland

Johann Gamper Free University of Bozen-Bolzano, Bolzano, Italy

Weihao Gan University of Southern California, Los Angeles, CA, USA

Vijay Gandhi University of Minnesota, Minneapolis, MN, USA

Venkatesh Ganti Microsoft Research, Microsoft Corporation, Redmond, WA, USA

Dengfeng Gao IBM Silicon Valley Lab, San Jose, CA, USA

Like Gao Teradata Corporation, San Diego, CA, USA

Wei Gao Qatar Computing Research Institute, Doha, Qatar

Minos Garofalakis Technical University of Crete, Chania, Greece

Wolfgang Gatterbauer University of Washington, Seattle, WA, USA

Buğra Gedik Department of Computer Engineering, Bilkent University, Ankara, Turkey

IBM T.J. Watson Research Center, Hawthorne, NY, USA

Floris Geerts University of Antwerp, Antwerp, Belgium

Johannes Gehrke Cornell University, Ithaca, NY, USA

Betsy George Oracle (America), Nashua, NH, USA

Lawrence Gerstley PSMI Consulting, San Francisco, CA, USA

Michael Gertz Heidelberg University, Heidelberg, Germany

Giorgio Ghelli Dipartimento di Informatica, Università di Pisa, Pisa, Italy

Gabriel Ghinita National University of Singapore, Singapore, Singapore

Giuseppe De Giacomo Dip. di Ingegneria Informatica Automatica e Gestionale Antonio Ruberti, Sapienza Università di Roma, Rome, Italy

Phillip B. Gibbons Computer Science Department and the Electrical and Computer Engineering Department, Carnegie Mellon University, Pittsburgh, PA, USA

Sarunas Girdzijauskas EPFL, Lausanne, Switzerland

Fausto Giunchiglia University of Trento, Trento, Italy

Kazuo Goda The University of Tokyo, Tokyo, Japan

Max Goebel Vienna University of Technology, Vienna, Austria

Bart Goethals University of Antwerp, Antwerp, Belgium

Martin Gogolla University of Bremen, Bremen, Germany

Aniruddha Gokhale Vanderbilt University, Nashville, TN, USA

Lukasz Golab University of Waterloo, Waterloo, ON, Canada

Matteo Golfarelli DISI – University of Bologna, Bologna, Italy

Arturo González-Ferrer Innovation Unit, Instituto de Investigación Sanitaria del Hospital Clínico San Carlos (IdISSC), Madrid, Spain

Michael F. Goodchild University of California-Santa Barbara, Santa Barbara, CA, USA

Georg Gottlob Computing Laboratory, Oxford University, Oxford, UK

Valerie Gouet-Brunet CNAM Paris, Paris, France

Ramesh Govindan University of Southern California, Los Angeles, CA, USA

Tyrone Gradison Proficiency Labs, Ashland, OR, USA

Goetz Graefe Google, Inc., Mountain View, CA, USA

Gösta Grahne Concordia University, Montreal, QC, Canada

Fabio Grandi Alma Mater Studiorum Università di Bologna, Bologna, Italy

Tyrone Grandison Proficiency Labs, Ashland, OR, USA

Peter M. D. Gray University of Aberdeen, Aberdeen, UK

Todd J. Green University of Pennsylvania, Philadelphia, PA, USA

Georges Grinstein University of Massachusetts, Lowell, MA, USA

Tom Gruber RealTravel, Emerald Hills, CA, USA

Le Gruenwald School of Computer Science, University of Oklahoma, Norman, OK, USA

Torsten Grust University of Tübingen, Tübingen, Germany

Dirk Van Gucht Indiana University, Bloomington, IN, USA

Carlos Guestrin Carnegie Mellon University, Pittsburgh, PA, USA

Dimitrios Gunopulos Department of Computer Science and Engineering, The University of California at Riverside, Bourns College of Engineering, Riverside, CA, USA

Amarnath Gupta San Diego Supercomputer Center, University of California San Diego, La Jolla, CA, USA

Himanshu Gupta Stony Brook University, Stony Brook, NY, USA

Cathal Gurrin Dublin City University, Dublin, Ireland

Ralf Hartmut Güting Fakultät für Mathematik und Informatik, Fernuniversität Hagen, Hagen, Germany

Computer Science, University of Hagen, Hagen, Germany

Marc Gyssens Hasselt University, Hasselt, Belgium

Peter J. Haas IBM Almaden Research Center, San Jose, CA, USA

Karl Hahn BMW AG, Munich, Germany

Jean-Luc Hainaut University of Namur, Namur, Belgium

Alon Halevy The Recruit Institute of Technology, Mountain View, CA, USA

Google Inc., Mountain View, CA, USA

Maria Halkidi University of Piraeus, Piraeus, Greece

Terry Halpin Neumont University, South Jordan, UT, USA

Jiawei Han University of Illinois at Urbana-Champaign, Urbana, IL, USA

Alan Hanjalic Delft University of Technology, Delft, The Netherlands

David Hansen The Australian e-Health Research Centre, Brisbane, QLD, Australia

Jörgen Hansson University of Skövde, Skövde, Sweden

Nikos Hardavellas Carnegie Mellon University, Pittsburgh, PA, USA

Theo Härder University of Kaiserslautern, Kaiserslautern, Germany

David Harel The Weizmann Institute of Science, Rehovot, Israel

Jayant R. Haritsa Indian Institute of Science, Bangalore, India

Stavros Harizopoulos HP Labs, Palo Alto, CA, USA

Per F. V. Hasle Royal School of Library and Information Science, University of Copenhagen, Copenhagen S, Denmark

Jordan T. Hastings Department of Geography, University of California-Santa Barbara, Santa Barbara, CA, USA

Alexander Hauptmann Carnegie Mellon University, Pittsburgh, PA, USA

Helwig Hauser University of Bergen, Bergen, Norway

Manfred Hauswirth Open Distributed Systems, Technical University of Berlin, Berlin, Germany

Fraunhofer FOKUS, Galway, Germany

Ben He University of Glasgow, Glasgow, UK

Thomas Heinis Imperial College London, London, UK

Pat Helland Microsoft Corporation, Redmond, WA, USA

Joseph M. Hellerstein University of California-Berkeley, Berkeley, CA, USA

Jean Henrard University of Namur, Namur, Belgium

John Herring Oracle USA Inc, Nashua, NH, USA

Nicolas Hervé INRIA Paris-Rocquencourt, Le Chesnay, France

Marcus Herzog Vienna University of Technology, Vienna, Austria

Jean-Marc Hick University of Namur, Namur, Belgium

Jan Hidders University of Antwerp, Antwerpen, Belgium

Djoerd Hiemstra University of Twente, Enschede, The Netherlands

Linda L. Hill University of California-Santa Barbara, Santa Barbara, CA, USA

Alexander Hinneburg Institute of Computer Science, Martin-Luther-University Halle-Wittenberg, Halle/Saale, Germany

Hans Hinterberger Department of Computer Science, ETH Zurich, Zurich, Switzerland

Howard Ho IBM Almaden Research Center, San Jose, CA, USA

Erik Hoel Environmental Systems Research Institute, Redlands, CA, USA

Vasant Honavar Iowa State University, Ames, IA, USA

Mingsheng Hong Cornell University, Ithaca, NY, USA

Katja Hose Department of Computer Science, Aalborg University, Aalborg, Denmark

Haruo Hosoya The University of Tokyo, Tokyo, Japan

Vagelis Hristidis Department of Computer Science and Engineering, University of California, Riverside, Riverside, CA, USA

Wynne Hsu National University of Singapore, Singapore, Singapore

Yu-Ling Hsueh Computer Science and Information Engineering Department, National Chung Cheng University, Taiwan, Republic of China

Jian Hu Microsoft Research Asia, Haidian, China

Kien A. Hua University of Central Florida, Orlando, FL, USA

Xian-Sheng Hua Microsoft Research Asia, Beijing, China

Jun Huan University of Kansas, Lawrence, KS, USA

Haoda Huang Microsoft Research Asia, Beijing, China

Michael Huggett University of British Columbia, Vancouver, BC, Canada

Patrick C. K. Hung University of Ontario Institute of Technology, Oshawa, ON, Canada

Jeong-Hyon Hwang Department of Computer Science, University at Albany – State University of New York, Albany, NY, USA

Noha Ibrahim Grenoble Informatics Laboratory (LIG), Grenoble, France

Ichiro Ide Graduate School of Informatics, Nagoya University, Nagoya, Aichi, Japan

Sergio Ilarri University of Zaragoza, Zaragoza, Spain

Ihab F. Ilyas Cheriton School of Computer Science, University of Waterloo, Waterloo, ON, Canada

Alfred Inselberg Tel Aviv University, Tel Aviv, Israel

Yannis Ioannidis University of Athens, Athens, Greece

Ekaterini Ioannou Faculty of Pure and Applied Sciences, Open University of Cyprus, Nicosia, Cyprus

Panagiotis G. Ipeirotis New York University, New York, NY, USA

Zachary G. Ives Computer and Information Science Department, University of Pennsylvania, Philadelphia, PA, USA

Hans-Arno Jacobsen Department of Electrical and Computer Engineering, University of Toronto, Toronto, ON, Canada

H. V. Jagadish University of Michigan, Ann Arbor, MI, USA

Alejandro Jaimes Telefonica R&D, Madrid, Spain

Ramesh Jain University of California, Irvine, CA, USA

Sushil Jajodia George Mason University, Fairfax, VA, USA

Greg Janée University of California-Santa Barbara, Santa Barbara, CA, USA

Kalervo Järvelin University of Tampere, Tampere, Finland

Christian S. Jensen Department of Computer Science, Aalborg University, Aalborg, Denmark

Eric C. Jensen Twitter, Inc., San Francisco, CA, USA

Manfred Jeusfeld IIT, University of Skövde, Skövde, Sweden

Aura Frames, New York City, NY, USA

Heng Ji New York University, New York, NY, USA

Zhe Jiang University of Alabama, Tuscaloosa, AL, USA

Ricardo Jiménez-Peris Distributed Systems Lab, Universidad Politecnica de Madrid, Madrid, Spain

Hai Jin Service Computing Technology and System Lab, Cluster and Grid Computing Lab, School of Computer Science and Technology, Huazhong University of Science and Technology, Wuhan, China

Jiashun Jin Carnegie Mellon University, Pittsburgh, PA, USA

Ruoming Jin Department of Computer Science, Kent State University, Kent, OH, USA

Ryan Johnson Carnegie Mellon University, Pittsburg, PA, USA

Theodore Johnson AT&T Labs – Research, Florham Park, NJ, USA

Christopher B. Jones Cardiff University, Cardiff, UK

Rosie Jones Yahoo! Research, Burbank, CA, USA

James B. D. Joshi University of Pittsburgh, Pittsburgh, PA, USA

Vanja Josifovski Uppsala University, Uppsala, Sweden

Marko Junkkari University of Tampere, Tampere, Finland

Jan Jurjens The Open University, Buckinghamshire, UK

Mouna Kacimi Max-Planck Institute for Informatics, Saarbrücken, Germany

Tamer Kahveci University of Florida, Gainesville, FL, USA

Panos Kalnis National University of Singapore, Singapore, Singapore

Jaap Kamps University of Amsterdam, Amsterdam, The Netherlands

James Kang University of Minnesota, Minneapolis, MN, USA

Carl-Christian Kanne University of Mannheim, Mannheim, Germany

Aman Kansal Microsoft Research, Redmond, WA, USA

Murat Kantarcıoğlu University of Texas at Dallas, Richardson, TX, USA

Ben Kao Department of Computer Science, The University of Hong Kong, Hong Kong, China

George Karabatis University of Maryland, Baltimore Country (UMBC), Baltimore, MD, USA

Grigoris Karvounarakis LogicBlox, Atlanta, GA, USA

George Karypis University of Minnesota, Minneapolis, MN, USA

Vipul Kashyap Clinical Programs, CIGNA Healthcare, Bloomfield, CT, USA

Yannis Katsis University of California-San Diego, La Jolla, CA, USA

Raghav Kaushik Microsoft Research, Redmond, WA, USA

Gabriella Kazai Microsoft Research Cambridge, Cambridge, UK

Daniel A. Keim Computer Science Department, University of Konstanz, Konstanz, Germany

Jaana Kekäläinen University of Tampere, Tampere, Finland

Anastasios Kementsietsidis IBM T.J. Watson Research Center, Hawthorne, NY, USA

Bettina Kemme School of Computer Science, McGill University, Montreal, QC, Canada

Jessie Kennedy Napier University, Edinburgh, UK

Vijay Khatri Operations and Decision Technologies Department, Kelley School of Business, Indiana University, Bloomington, IN, USA

Ashfaq Khokhar University of Illinois at Chicago, Chicago, IL, USA

Daniel Kifer Yahoo! Research, Santa Clara, CA, USA

Stephen Kimani Director ICSIT, Jomo Kenyatta University of Agriculture and Technology (JKUAT), Juja, Kenya

Sofia Kleisarchaki CNRS, Univ. Grenoble Alps, Grenoble, France

Craig A. Knoblock University of Southern California, Marina del Rey, Los Angeles, CA, USA

Christoph Koch Cornell University, Ithaca, New York, NY, USA

EPFL, Lausanne, Switzerland

Solmaz Kolahi University of British Columbia, Vancouver, BC, Canada

George Kollios Boston University, Boston, MA, USA

Christian Koncilia Institute of Informatics-Systems, University of Klagenfurt, Klagenfurt, Austria

Roberto Konow Department of Computer Science, University of Chile, Santiago, Chile

Marijn Koolen Research and Development, Huygens ING, Royal Netherlands Academy of Arts and Sciences, Amsterdam, The Netherlands

David Koop University of Massachusetts Dartmouth, Dartmouth, MA, USA

Poon Wei Koot Nanyang Technological University, Singapore, Singapore

Julius Köpke Department of Informatics-Systems, Alpen-Adria-Universität Klagenfurt, Klagenfurt, Austria

Flip R. Korn AT&T Labs–Research, Florham Park, NJ, USA

Harald Kosch University of Passau, Passau, Germany

Cartik R. Kothari Biomedical Informatics, Ohio State University, College of Medicine, Columbus, OH, USA

Yannis Kotidis Department of Informatics, Athens University of Economics and Business, Athens, Greece

Spyros Kotoulas VU University Amsterdam, Amsterdam, The Netherlands

Manolis Koubarakis University of Athens, Athens, Greece

Konstantinos Koutroumbas Institute for Space Applications and Remote Sensing, Athens, Greece

Bernd J. Krämer University of Hagen, Hagen, Germany

Tim Kraska Department of Computer Science, Brown University, Providence, RI, USA

Werner Kriechbaum IBM Development Lab, Böblingen, Germany

Hans-Peter Kriegel Ludwig-Maximilians-University, Munich, Germany

Chandra Krintz Department of Computer Science, University of California, Santa Barbara, CA, USA

Rajasekar Krishnamurthy IBM Almaden Research Center, San Jose, CA, USA

Peer Kröger Ludwig-Maximilians-Universität München, Munich, Germany

Thomas Kühne School of Engineering and Computer Science, Victoria University of Wellington, Wellington, New Zealand

Krishna Kulkarni Independent Consultant, San Jose, CA, USA

Ravi Kumar Yahoo Research, Santa Clara, CA, USA

Nicholas Kushmerick VMWare, Seattle, WA, USA

Alan G. Labouseur School of Computer Science and Mathematics, Marist College, Poughkeepsie, NY, USA

Alexandros Labrinidis Department of Computer Science, University of Pittsburgh, Pittsburgh, PA, USA

Zoé Lacroix Arizona State University, Tempe, AZ, USA

Alberto H. F. Laender Federal University of Minas Gerais, Belo Horizonte, Brazil

Bibudh Lahiri Iowa State University, Ames, IA, USA

Laks V. S. Lakshmanan University of British Columbia, Vancouver, BC, Canada

Mounia Lalmas Yahoo! Inc., London, UK

Lea Landucci University of Florence, Florence, Italy

Birger Larsen Royal School of Library and Information Science, Copenhagen, Denmark

Mary Lynette Larsgaard University of California-Santa Barbara, Santa Barbara, CA, USA

Per-Åke Larson Microsoft Corporation, Redmond, WA, USA

Robert Laurini INSA-Lyon, University of Lyon, Lyon, France

LIRIS, INSA-Lyon, Lyon, France

Georg Lausen University of Freiburg, Freiburg, Germany

Jens Lechtenbörger University of Münster, Münster, Germany

Thierry Lecroq Université de Rouen, Rouen, France

Dongwon Lee The Pennsylvania State University, Park, PA, USA

Victor E. Lee John Carroll University, University Heights, OH, USA

Yang W. Lee College of Business Administration, Northeastern University, Boston, MA, USA

Pieter De Leenheer Vrije Universiteit Brussel, Collibra NV, Brussels, Belgium

Wolfgang Lehner Dresden University of Technology, Dresden, Germany

Domenico Lembo Dip. di Ingegneria Informatica Automatica e Gestionale Antonio Ruberti, Sapienza Università di Roma, Rome, Italy

Ronny Lempel Yahoo! Research, Haifa, Israel

Maurizio Lenzerini Dip. di Ingegneria Informatica Automatica e Gestionale Antonio Ruberti, Sapienza Università di Roma, Rome, Italy

Kristina Lerman University of Southern California, Marina del Rey, Los Angeles, CA, USA

Ulf Leser Humboldt University of Berlin, Berlin, Germany

Carson Kai-Sang Leung Department of Computer Science, University of Manitoba, Winnipeg, MB, Canada

Mariano Leva Dipartimento di Ingegneria Informatica, Automatica e Gestionale "A.Ruberti", Sapienza – Università di Roma, Roma, Italy

Stefano Levialdi Sapienza University of Rome, Rome, Italy

Brian Levine University of Massachusetts, Amherst, MA, USA

Changqing Li Duke University, Durham, NC, USA

Chen Li University of California – Irvine, School of Information and Computer Sciences, Irvine, CA, USA

Chengkai Li University of Texas at Arlington, Arlington, TX, USA

Hua Li Microsoft Research Asia, Beijing, China

Jinyan Li Nanyang Technological University, Singapore, Singapore

Ninghui Li Purdue University, West Lafayette, IN, USA

Ping Li Cornell University, Ithaca, NY, USA

Qing Li City University of Hong Kong, Hong Kong, China

Xue Li The University of Queensland, Brisbane, QLD, Australia

Yunyao Li IBM Almaden Research Center, San Jose, CA, USA

Ying Li Cognitive People Solutions, IBM Human Resources, Armonk, NY, USA

Xiang Lian Department of Computer Science, Kent State University, Kent, OH, USA

Leonid Libkin School of Informatics, University of Edinburgh, Edinburgh, Scotland, UK

Sam S. Lightstone IBM Canada Ltd, Markham, ON, Canada

Jimmy Lin University of Maryland, College Park, MD, USA

Tsau Young Lin Department of Computer Science, San Jose State University, San Jose, CA, USA

Xuemin Lin University of New South Wales, Sydney, NSW, Australia

Tok Wang Ling National University of Singapore, Singapore, Singapore

Bing Liu University of Illinois at Chicago, Chicago, IL, USA

Danzhou Liu University of Central Florida, Orlando, FL, USA

Guimei Liu Institute for Infocomm Research, Singapore, Singapore

Huan Liu Data Mining and Machine Learning Lab, School of Computing, Informatics, and Decision Systems Engineering, Arizona State University, Tempe, AZ, USA

Jinze Liu University of Kentucky, Lexington, KY, USA

Lin Liu Department of Computer Science, Kent State University, Kent, OH, USA

Ning Liu Microsoft Research Asia, Beijing, China

Qing Liu CSIRO, Hobart, TAS, Australia

Xiangyu Liu Xiamen University, Xiamen, China

Vebjorn Ljosa Broad Institute of MIT and Harvard, Cambridge, MA, USA

David Lomet Microsoft Research, Redmond, WA, USA

Cheng Long School of Electronics, Electrical Engineering and Computer Science, Queen's University Belfast, Kowloon, Hong Kong

Boon Thau Loo ETH Zurich, Zurich, Switzerland

Phillip Lord Newcastle University, Newcastle-Upon-Tyne, UK

Nikos A. Lorentzos Informatics Laboratory, Department of Agricultural Economics and Rural Development, Agricultural University of Athens, Athens, Greece

Lie Lu Microsoft Research Asia, Beijing, China

Bertram Ludäscher University of California-Davis, Davis, CA, USA

Yan Luo University of Illinois at Chicago, Chicago, IL, USA

Yves A. Lussier University of Chicago, Chicago, IL, USA

Ioanna Lykourentzou CRP Henri Tudor, Esch-sur-Alzette, Luxembourg

Craig MacDonald University of Glasgow, Glasgow, UK

Ashwin Machanavajjhala Cornell University, Ithaca, NY, USA

Samuel Madden Massachusetts Institute of Technology, Cambridge, MA, USA

Paola Magillo University of Genova, Genoa, Italy

Ahmed R. Mahmood Computer Science, Purdue University, West Lafayette, IN, USA

David Maier Portland State University, Portland, OR, USA

Ratul kr. Majumdar Department of Computer Science and Engineering, Indian Institute of Technology Bombay, Mumbai, India

Jan Małuszyński Linköping University, Linköping, Sweden

Nikos Mamoulis University of Hong Kong, Hong Kong, China

Stefan Manegold CWI, Amsterdam, The Netherlands

Murali Mani Worcester Polytechnic, Worcester, MA, USA

Serge Mankovski CA Labs, CA Inc., Thornhill, ON, Canada

Ioana Manolescu INRIA Saclay–Île de France, Orsay, France

Yannis Manolopoulos Aristotle University of Thessaloniki, Thessaloniki, Greece

Florian Mansmann University of Konstanz, Konstanz, Germany

Svetlana Mansmann University of Konstanz, Konstanz, Germany

Shahar Maoz The Weizmann Institute of Science, Rehovot, Israel

Patrick Marcel Département Informatique, Laboratoire d'Informatique, Université François Rabelais Tours, Blois, France

Amélie Marian Computer Science Department, Rutgers University, New Brunswick, NJ, USA

Volker Markl IBM Almaden Research Center, San Jose, CA, USA

David Martin Nuance Communications, Sunnyvale, CA, USA

Maria Vanina Martinez University of Maryland, College Park, MD, USA

Maristella Matera Politecnico di Milano, Milan, Italy

Michael Mathioudakis Université de Lyon, CNRS, INSA-Lyon, LIRIS, UMR5205, F-69621, France

Marta Mattoso Federal University of Rio de Janeiro, Rio de Janeiro, Brazil

Andrea Maurino University of Milano-Bicocca, Milan, Italy

Jose-Norberto Mazón University of Alicante, Alicante, Spain

John McCloud CERT Insider Threat Center, Software Engineering Institute, Carnegie Mellon University, Pittsburgh, PA, USA

Kevin S. McCurley Google Research, Mountain View, CA, USA

Andrew McGregor Microsoft Research, Silicon Valley, Mountain View, CA, USA

Timothy McPhillips University of California-Davis, Davis, CA, USA

Massimo Mecella Dipartimento di Ingegneria Informatica, Automatica e Gestionale "A.Ruberti", Sapienza – Università di Roma, Roma, Italy

Brahim Medjahed The University of Michigan–Dearborn, Dearborn, MI, USA

Carlo Meghini The Italian National Research Council, Pisa, Italy

Tao Mei Microsoft Research Asia, Beijing, China

Jonas Mellin University of Skövde, The Informatics Research Centre, Skövde, Sweden

University of Skövde, School of Informatics, Skövde, Sweden

Massimo Melucci University of Padua, Padua, Italy

Niccolò Meneghetti Computer Science and Engineering Department, University at Buffalo, Buffalo, NY, USA

Weiyi Meng Department of Computer Science, State University of New York at Binghamton, Binghamton, NY, USA

Ahmed Metwally LinkedIn Corp., Mountain View, CA, USA

Jan Michels Oracle Corporation, Redwood Shores, CA, USA

Gerome Miklau University of Massachusetts, Amherst, MA, USA

Alessandra Mileo Insight Centre for Data Analytics, Dublin City University, Dublin, Ireland

Harvey J. Miller University of Utah, Salt Lake City, UT, USA

Renée J. Miller Department of Computer Science, University of Toronto, Toronto, ON, Canada

Tova Milo School of Computer Science, Tel Aviv University, Tel Aviv, Israel

Umar Farooq Minhas Microsoft Research, Redmond, WA, USA

Paolo Missier School of Computing Science, Newcastle University, Newcastle upon Tyne, UK

Prasenjit Mitra The Pennsylvania State University, University Park, PA, USA

Michael Mitzenmacher Harvard University, Boston, MA, USA

Mukesh Mohania IBM Research, Melbourne, VIC, Australia

Mohamed F. Mokbel Department of Computer Science and Engineering, University of Minnesota-Twin Cities, Minneapolis, MN, USA

Angelo Montanari University of Udine, Udine, Italy

Reagan W. Moore School of Information and Library Science, University of North Carolina at Chapel Hill, Chapel Hill, NC, USA

Konstantinos Morfonios Oracle, Redwood City, CA, USA

Peter Mork The MITRE Corporation, McLean, VA, USA

Mirella M. Moro Departamento de Ciencia da Computaçao, Universidade Federal de Minas Gerais – UFMG, Belo Horizonte, MG, Brazil

Kyriakos Mouratidis Singapore Management University, Singapore, Singapore

Kamesh Munagala Duke University, Durham, NC, USA

Ethan V. Munson Department of EECS, University of Wisconsin-Milwaukee, Milwaukee, WI, USA

Shawn Murphy Massachusetts General Hospital, Boston, MA, USA

John Mylopoulos Department of Computer Science, University of Toronto, Toronto, ON, Canada

Marta Patiño-Martínez Distributed Systems Lab, Universidad Politecnica de Madrid, Madrid, Spain

ETSI Informáticos, Universidad Politécnica de Madrid (UPM), Madrid, Spain

Frank Nack University of Amsterdam, Amsterdam, The Netherlands

Marc Najork Google, Inc., Mountain View, CA, USA

Ullas Nambiar Zensar Technologies Ltd, Pune, India

Alexandros Nanopoulos Aristotle University, Thessaloniki, Greece

Vivek Narasayya Microsoft Corporation, Redmond, WA, USA

Mario A. Nascimento Department of Computing Science, University of Alberta, Edmonton, AB, Canada

Alan Nash Aleph One LLC, La Jolla, CA, USA

Harald Naumann Vienna University of Technology, Vienna, Austria

Gonzalo Navarro Department of Computer Science, University of Chile, Santiago, Chile

Wolfgang Nejdl L3S Research Center, University of Hannover, Hannover, Germany

Thomas Neumann Max-Planck Institute for Informatics, Saarbrücken, Germany

Bernd Neumayr Department for Business Informatics – Data and Knowledge Engineering, Johannes Kepler University Linz, Linz, Austria

Frank Neven Hasselt University and Transnational University of Limburg, Diepenbeek, Belgium

Chong-Wah Ngo City University of Hong Kong, Hong Kong, China

Peter Niblett IBM United Kingdom Limited, Winchester, UK

Naoko Nitta Osaka University, Osaka, Japan

Igor Nitto Department of Computer Science, University of Pisa, Pisa, Italy

Cheng Niu Microsoft Research Asia, Beijing, China

Vilém Novák Institute for Research and Applications of Fuzzy Modeling, University of Ostrava, Ostrava, Czech Republic

Chimezie Ogbuji Cleveland Clinic Foundation, Cleveland, OH, USA

Peter Øhrstrøm Aalborg University, Aalborg, Denmark

Christine M. O'Keefe CSIRO Preventative Health National Research Flagship, Acton, ACT, Australia

Paul W. Olsen Department of Computer Science, The College of Saint Rose, Albany, NY, USA

Dan Olteanu Department of Computer Science, University of Oxford, Oxford, UK

Behrooz Omidvar-Tehrani Interactive Data Systems Group, Ohio State University, Columbus, OH, USA

Patrick O'Neil University of Massachusetts, Boston, MA, USA

Beng Chin Ooi School of Computing, National University of Singapore, Singapore, Singapore

Iadh Ounis University of Glasgow, Glasgow, UK

Mourad Ouzzani Qatar Computing Research Institute, HBKU, Doha, Qatar

Fatma Özcan IBM Research – Almaden, San Jose, CA, USA

M. Tamer Özsu Cheriton School of Computer Science, University of Waterloo, Waterloo, ON, Canada

Esther Pacitti INRIA and LINA, University of Nantes, Nantes, France

Chris D. Paice Lancaster University, Lancaster, UK

Noël de Palma INPG – INRIA, Grenoble, France

Nathaniel Palmer Workflow Management Coalition, Hingham, MA, USA

Themis Palpanas Paris Descartes University, Paris, France

Biswanath Panda Cornell University, Ithaca, NY, USA

Ippokratis Pandis Carnegie Mellon University, Pittsburgh, PA, USA

Amazon Web Services, Seattle, WA, USA

Dimitris Papadias Department of Computer Science and Engineering, Hong Kong University of Science and Technology, Kowloon, Hong Kong, Hong Kong

Spiros Papadimitriou IBM T.J. Watson Research Center, Hawthorne, NY, USA

Apostolos N. Papadopoulos Aristotle University of Thessaloniki, Thessaloniki, Greece

Yannis Papakonstantinou University of California-San Diego, La Jolla, CA, USA

Jan Paredaens University of Antwerp, Antwerpen, Belgium

Christine Parent University of Lausanne, Lausanne, Switzerland

Josiane Xavier Parreira Siemens AG, Galway, Austria

Gabriella Pasi Department of Informatics, Systems and Communication, University of Milano-Bicocca, Milan, Italy

Chintan Patel Columbia University, New York, NY, USA

Jignesh M. Patel University of Wisconsin-Madison, Madison, WI, USA

Norman W. Paton University of Manchester, Manchester, UK

Cesare Pautasso University of Lugano, Lugano, Switzerland

Torben Bach Pedersen Department of Computer Science, Aalborg University, Aalborg, Denmark

Fernando Pedone Università della Svizzera Italiana (USI), Lugano, Switzerland

Jovan Pehcevski INRIA Paris-Rocquencourt, Le Chesnay Cedex, France

Jian Pei School of Computing Science, Simon Fraser University, Burnaby, BC, Canada

Ronald Peikert ETH Zurich, Zurich, Switzerland

Mor Peleg Department of Information Systems, University of Haifa, Haifa, Israel

Fuchun Peng Yahoo! Inc., Sunnyvale, CA, USA

Peng Peng Alibaba, Yu Hang District, Hangzhou, China

Liam Peyton University of Ottawa, Ottawa, ON, Canada

Dieter Pfoser Department of Geography and Geoinformation Science, George Mason University, Fairfax, VA, USA

Danh Le Phuoc Open Distributed Systems, Technical University of Berlin, Berlin, Germany

Mario Piattini University of Castilla-La Mancha, Ciudad Real, Spain

Benjamin C. Pierce University of Pennsylvania, Philadelphia, PA, USA

Karen Pinel-Sauvagnat IRIT laboratory, University of Toulouse, Toulouse, France

Leo L. Pipino University of Massachusetts, Lowell, MA, USA

Peter Pirolli Palo Alto Research Center, Palo Alto, CA, USA

Evaggelia Pitoura Department of Computer Science and Engineering, University of Ioannina, Ioannina, Greece

Benjamin Piwowarski University of Glasgow, Glasgow, UK

Vassilis Plachouras Yahoo! Research, Barcelona, Spain

Catherine Plaisant University of Maryland, College Park, MD, USA

Claudia Plant University of Vienna, Vienna, Austria

Christian Platzer Technical University of Vienna, Vienna, Austria

Dimitris Plexousakis Foundation for Research and Technology-Hellas (FORTH), Heraklion, Greece

Neoklis Polyzotis University of California Santa Cruz, Santa Cruz, CA, USA

Raymond K. Pon University of California, Los Angeles, CA, USA

Lucian Popa IBM Almaden Research Center, San Jose, CA, USA

Alexandra Poulovassilis University of London, London, UK

Sunil Prabhakar Purdue University, West Lafayette, IN, USA

Cecilia M. Procopiuc AT&T Labs, Florham Park, NJ, USA

Enrico Puppo Department of Informatics, Bioengineering, Robotics and Systems Engineering, University of Genova, Genoa, Italy

Ross S. Purves University of Zurich, Zurich, Switzerland

Vivien Quéma CNRS, INRIA, Saint-Ismier Cedex, France

Christoph Quix RWTH Aachen University, Aachen, Germany

Sriram Raghavan IBM Almaden Research Center, San Jose, CA, USA

Erhard Rahm University of Leipzig, Leipzig, Germany

Habibur Rahman Department of Computer Science and Engineering, University of Texas at Arlington, Arlington, TX, USA

Krithi Ramamritham Department of Computer Science and Engineering, Indian Institute of Technology Bombay, Mumbai, India

Maya Ramanath Max-Planck Institute for Informatics, Saarbrücken, Germany

Georgina Ramírez Yahoo! Research Barcelona, Barcelona, Spain

Edie Rasmussen Library, Archival and Information Studies, The University of British Columbia, Vancouver, BC, Canada

Indrakshi Ray Colorado State University, Fort Collins, CO, USA

Colin R. Reeves Coventry University, Coventry, UK

Payam Refaeilzadeh Google Inc., Los Angeles, CA, USA

D. R. Reforgiato University of Maryland, College Park, MD, USA

Bernd Reiner Technical University of Munich, Munich, Germany

Frederick Reiss IBM Almaden Research Center, San Jose, CA, USA

Harald Reiterer University of Konstanz, Constance, Germany

Matthias Renz Ludwig-Maximilians-Universität München, Munich, Germany

Andreas Reuter Heidelberg Laureate Forum Foundation, Schloss-Wolfsbrunnenweg 33, Heidelberg, Germany

Peter Revesz University of Nebraska-Lincoln, Lincoln, NE, USA

Mirek Riedewald Cornell University, Ithaca, NY, USA

Rami Rifaieh University of California-San Diego, San Diego, CA, USA

Stefanie Rinderle-Ma University of Vienna, Vienna, Austria

Tore Risch Department of Information Technology, Uppsala University, Uppsala, Sweden

Thomas Rist University of Applied Sciences, Augsburg, Germany

Stefano Rizzi DISI, University of Bologna, Bologna, Italy

Stephen Robertson Microsoft Research Cambridge, Cambridge, UK

Roberto A. Rocha Partners eCare, Partners HealthCare System, Wellesley, MA, USA

John F. Roddick Flinders University, Adelaide, SA, Australia

Thomas Roelleke Queen Mary University of London, London, UK

Didier Roland University of Namur, Namur, Belgium

Oscar Romero Polytechnic University of Catalonia, Barcelona, Spain

Rafael Romero University of Alicante, Alicante, Spain

Riccardo Rosati Dip. di Ingegneria Informatica Automatica e Gestionale Antonio Ruberti, Sapienza Università di Roma, Rome, Italy

Timothy Roscoe ETH Zurich, Zurich, Switzerland

Kenneth A. Ross Columbia University, New York, NY, USA

Prasan Roy Sclera, Inc., Walnut, CA, USA

Senjuti Basu Roy Department of Computer Science, New Jersey Institute of Technology, Tacoma, WA, USA

Sudeepa Roy Department of Computer Science, Duke University, Durham, NC, USA

Yong Rui Microsoft China R&D Group, Redmond, WA, USA

Dan Russler Oracle Health Sciences, Redwood Shores, CA, USA

Georgia Tech Research Institute, Atlanta, Georgia, USA

Michael Rys Microsoft Corporation, Sammamish, WA, USA

Giovanni Maria Sacco Dipartimento di Informatica, Università di Torino, Torino, Italy

Tetsuya Sakai Waseda University, Tokyo, Japan

Kenneth Salem University of Waterloo, Waterloo, ON, Canada

Simonas Šaltenis Aalborg University, Aalborg, Denmark

George Samaras University of Cyprus, Nicosia, Cyprus

Giuseppe Santucci University of Rome, Rome, Italy

Maria Luisa Sapino University of Turin, Turin, Italy

Sunita Sarawagi IIT Bombay, Mumbai, India

Anatol Sargin University of Augsburg, Augsburg, Germany

Mohamed Sarwat School of Computing, Informatics, and Decision Systems Engineering, Arizona State University, Tempe, AZ, USA

Kai-Uwe Sattler Technische Universität Ilmenau, Ilmenau, Germany

Monica Scannapieco University of Rome, Rome, Italy

Matthias Schäfer University of Konstanz, Konstanz, Germany

Sebastian Schaffert Salzburg Research, Salzburg, Austria

Ralf Schenkel Campus II Department IV – Computer Science, Professorship for databases and information systems, University of Trier, Trier, Germany

Raimondo Schettini University of Milano-Bicocca, Milan, Italy

Peter Scheuermann Department of ECpE, Iowa State University, Ames, IA, USA

Ulrich Schiel Federal University of Campina Grande, Campina Grande, Brazil

Markus Schneider University of Florida, Gainesville, FL, USA

Marc H. Scholl University of Konstanz, Konstanz, Germany

Michel Scholl Cedric-CNAM, Paris, France

Tobias Schreck Department of Computer Science and Biomedical Engineering, Institute of Computer Graphics and Knowledge Visualization, Graz University of Technology, Graz, Austria

Michael Schrefl University of Linz, Linz, Austria

Erich Schubert Heidelberg University, Heidelberg, Germany

Matthias Schubert Ludwig-Maximilians-University, Munich, Germany

Christoph G. Schuetz Department for Business Informatics – Data and Knowledge Engineering, Johannes Kepler University Linz, Linz, Austria

Heiko Schuldt Department of Mathematics and Computer Science, Databases and Information Systems Research Group, University of Basel, Basel, Switzerland

Heidrun Schumann University of Rostock, Rostock, Germany

Felix Schwagereit University of Koblenz-Landau, Koblenz, Germany

Nicole Schweikardt Johann Wolfgang Goethe-University, Frankfurt am Main, Frankfurt, Germany

Fabrizio Sebastiani Qatar Computing Research Institute, Doha, Qatar

Nicu Sebe University of Amsterdam, Amsterdam, Netherlands

Monica Sebillo University of Salerno, Salerno, Italy

Thomas Seidl RWTH Aachen University, Aachen, Germany

Manuel Serrano University of Alicante, Alicante, Spain

Amnon Shabo (Shvo) University of Haifa, Haifa, Israel

Mehul A. Shah Amazon Web Services (AWS), Seattle, WA, USA

Nigam Shah Stanford University, Stanford, CA, USA

Cyrus Shahabi University of Southern California, Los Angeles, CA, USA

Jayavel Shanmugasundaram Yahoo Research!, Santa Clara, NY, USA

Marc Shapiro Inria Paris, Paris, France

Sorbonne-Universités-UPMC-LIP6, Paris, France

Mohamed Sharaf Electrical and Computer Engineering, University of Toronto, Toronto, ON, Canada

Mehdi Sharifzadeh Google, Santa Monica, CA, USA

Jayant Sharma Oracle USA Inc, Nashua, NH, USA

Guy Sharon IBM Research Labs-Haifa, Haifa, Israel

Dennis Shasha Department of Computer Science, New York University, New York, NY, USA

Shashi Shekhar Department of Computer Science, University of Minnesota, Minneapolis, MN, USA

Jialie Shen Singapore Management University, Singapore, Singapore

Xuehua Shen Google, Inc., Mountain View, CA, USA

Dou Shen Microsoft Corporation, Redmond, WA, USA

Baidu, Inc., Beijing City, China

Heng Tao Shen School of Information Technology and Electrical Engineering, The University of Queensland, Brisbane, QLD, Australia

University of Electronic Science and Technology of China, Chengdu, Sichuan Sheng, China

Rao Shen Yahoo!, Sunnyvale, CA, USA

Frank Y. Shih New Jersey Institute of Technology, Newark, NJ, USA

Arie Shoshani Lawrence Berkeley National Laboratory, Berkeley, CA, USA

Pavel Shvaiko University of Trento, Trento, Italy

Wolf Siberski L3S Research Center, University of Hannover, Hannover, Germany

Ronny Siebes VU University Amsterdam, Amsterdam, The Netherlands

Laurynas Šikšnys Department of Computer Science, Aalborg University, Aalborg, Denmark

Adam Silberstein Yahoo! Research Silicon Valley, Santa Clara, CA, USA

Fabrizio Silvestri Yahoo Inc, London, UK

Alkis Simitsis HP Labs, Palo Alto, CA, USA

Simeon J. Simoff University of Western Sydney, Sydney, NSW, Australia

Elena Simperl Electronics and Computer Science, University of Southampton, Southampton, UK

Radu Sion Stony Brook University, Stony Brook, NY, USA

Mike Sips Stanford University, Stanford, CA, USA

Cristina Sirangelo IRIF, Paris Diderot University, Paris, France

Yannis Sismanis IBM Almaden Research Center, Almaden, CA, USA

Hala Skaf-Molli Computer Science, University of Nantes, Nantes, France

Spiros Skiadopoulos University of Peloponnese, Tripoli, Greece

Richard T. Snodgrass Department of Computer Science, University of Arizona, Tucson, AZ, USA

Dataware Ventures, Tucson, AZ, USA

Cees Snoek University of Amsterdam, Amsterdam, The Netherlands

Mohamed A. Soliman Datometry Inc., San Francisco, CA, USA

Il-Yeol Song College of Computing and Informatics, Drexel University, Philadelphia, PA, USA

Ruihua Song Microsoft Research Asia, Beijing, China

Jingkuan Song Columbia University, New York, NY, USA

Stefano Spaccapietra EPFL, Lausanne, Switzerland

Greg Speegle Department of Computer Science, Baylor University, Waco, TX, USA

Padmini Srinivasan The University of Iowa, Iowa City, IA, USA

Venkat Srinivasan Virginia Tech, Blacksburg, VA, USA

Divesh Srivastava AT&T Labs – Research, AT&T, Bedminster, NJ, USA

Steffen Staab Institute for Web Science and Technologies – WeST, University of Koblenz-Landau, Koblenz, Germany

Constantine Stephanidis Foundation for Research and Technology-Hellas (FORTH), Heraklion, Greece

University of Crete, Heraklion, Greece

Robert Stevens University of Manchester, Manchester, UK

Andreas Stoffel University of Konstanz, Konstanz, Germany

Michael Stonebraker Massachusetts Institute of Technology, Cambridge, MA, USA

Umberto Straccia The Italian National Research Council, Pisa, Italy

Martin J. Strauss University of Michigan, Ann Arbor, MI, USA

Diane M. Strong Worcester Polytechnic Institute, Worcester, MA, USA

Jianwen Su University of California-Santa Barbara, Santa Barbara, CA, USA

Kazimierz Subieta Polish-Japanese Institute of Information Technology, Warsaw, Poland

V. S. Subrahmanian University of Maryland, College Park, MD, USA

Dan Suciu University of Washington, Seattle, WA, USA

S. Sudarshan Indian Institute of Technology, Bombay, India

Torsten Suel Yahoo! Research, Sunnyvale, CA, USA

Jian-Tao Sun Microsoft Research Asia, Beijing, China

Subhash Suri University of California-Santa Barbara, Santa Barbara, CA, USA

Jaroslaw Szlichta University of Ontario Institute of Technology, Oshawa, ON, Canada

Stefan Tai University of Karlsruhe, Karlsruhe, Germany

Kian-Lee Tan Department of Computer Science, National University of Singapore, Singapore, Singapore

Pang-Ning Tan Michigan State University, East Lansing, MI, USA

Wang-Chiew Tan University of California-Santa Cruz, Santa Cruz, CA, USA

Letizia Tanca Computer Science, Politecnico di Milano, Milan, Italy

Lei Tang Chief Data Scientist, Clari Inc., Sunnyvale, CA, USA

Wei Tang Teradata Corporation, El Segundo, CA, USA

Egemen Tanin Computing and Information Systems, University of Melbourne, Melbourne, VIC, Australia

Val Tannen Department of Computer and Information Science, University of Pennsylvania, Philadelphia, PA, USA

Abdullah Uz Tansel Baruch College, CUNY, New York, NY, USA

Yufei Tao Chinese University of Hong Kong, Hong Kong, China

Sandeep Tata IBM Almaden Research Center, San Jose, CA, USA

Nesime Tatbul Intel Labs and MIT, Cambridge, MA, USA

Christophe Taton INPG – INRIA, Grenoble, France

Behrooz Omidvar Tehrani Laboratoire d'Informatique de Grenoble, Saint-Martin d'Hères, France

Paolo Terenziani Dipartimento di Scienze e Innovazione Tecnologica (DiSIT), Università del Piemonte Orientale "Amedeo Avogadro", Alessandria, Italy

Alexandre Termier LIG (Laboratoire d'Informatique de Grenoble), HADAS team, Université Joseph Fourier, Saint Martin d'Hères, France

Evimaria Terzi Computer Science Department, Boston University, Boston, MA, USA

IBM Almaden Research Center, San Jose, CA, USA

Bernhard Thalheim Christian-Albrechts University, Kiel, Germany

Martin Theobald Institute of Databases and Information Systems (DBIS), Ulm University, Ulm, Germany

Stanford University, Stanford, CA, USA

Sergios Theodoridis University of Athens, Athens, Greece

Yannis Theodoridis University of Piraeus, Piraeus, Greece

Saravanan Thirumuruganathan Department of Computer Science and Engineering, University of Texas at Arlington, Arlington, TX, USA

Qatar Computing Research Institute, Hamad Bin Khalifa University, Doha, Qatar

Stephen W. Thomas Dataware Ventures, Kingston, ON, Canada

Alexander Thomasian Thomasian and Associates, Pleasantville, NY, USA

Christian Thomsen Department of Computer Science, Aalborg University, Aalborg, Denmark

Bhavani Thuraisingham The University of Texas at Dallas, Richardson, TX, USA

Srikanta Tirthapura Iowa State University, Ames, IA, USA

Wee Hyong Tok National University of Singapore, Singapore, Singapore

David Toman University of Waterloo, Waterloo, ON, Canada

Frank Tompa David R. Cheriton School of Computer Science, University of Waterloo, Waterloo, ON, Canada

Alejandro Z. Tomsic Sorbonne Universités-UPMC-LIP6, Paris, France

Inria Paris, Paris, France

Rodney Topor Griffith University, Nathan, Australia

Riccardo Torlone University of Rome, Rome, Italy

Kristian Torp Aalborg University, Aalborg, Denmark

Nicola Torpei University of Florence, Florence, Italy

Nerius Tradišauskas Aalborg University, Aalborg, Denmark

Goce Trajcevski Department of ECpE, Iowa State University, Ames, IA, USA

Peter Triantafillou University of Patras, Rio, Patras, Greece

Silke Trißl Humboldt University of Berlin, Berlin, Germany

Andrew Trotman University of Otago, Dunedin, New Zealand

Juan Trujillo Lucentia Research Group, Department of Information Languages and Systems, Facultad de Informática, University of Alicante, Alicante, Spain

Beth Trushkowsky Department of Computer Science, Harvey Mudd College, Claremont, CA, USA

Panayiotis Tsaparas Department of Computer Science and Engineering, University of Ioannina, Ioannina, Greece

Theodora Tsikrika Center for Mathematics and Computer Science, Amsterdam, The Netherlands

Vassilis J. Tsotras University of California-Riverside, Riverside, CA, USA

Mikalai Tsytsarau University of Trento, Povo, Italy

Peter A. Tucker Whitworth University, Spokane, WA, USA

Anthony K. H. Tung National University of Singapore, Singapore, Singapore

Deepak Turaga IBM Research, San Francisco, CA, USA

Theodoros Tzouramanis University of the Aegean, Samos, Greece

Antti Ukkonen Helsinki University of Technology, Helsinki, Finland

Mollie Ullman-Cullere Harvard Medical School – Partners Healthcare Center for Genetics and Genomics, Boston, MA, USA

Ali Ünlü University of Augsburg, Augsburg, Germany

Antony Unwin Augsburg University, Augsburg, Germany

Susan D. Urban Arizona State University, Phoenix, AZ, USA

Jaideep Vaidya Rutgers University, Newark, NJ, USA

Alejandro A. Vaisman Instituto Tecnológico de Buenos Aires, Buenos Aires, Argentina

Shivakumar Vaithyanathan IBM Almaden Research Center, San Jose, CA, USA

Athena Vakali Aristotle University, Thessaloniki, Greece

Patrick Valduriez INRIA, LINA, Nantes, France

Maarten van Steen VU University, Amsterdam, The Netherlands

W. M. P. van der Aalst Eindhoven University of Technology, Eindhoven, The Netherlands

Christelle Vangenot EPFL, Lausanne, Switzerland

Stijn Vansummeren Hasselt University and Transnational University of Limburg, Diepenbeek, Belgium

Vasilis Vassalos Athens University of Economics and Business, Athens, Greece

Michael Vassilakopoulos University of Thessaly, Volos, Greece

Panos Vassiliadis University of Ioannina, Ioannina, Greece

Michalis Vazirgiannis Athens University of Economics and Business, Athens, Greece

Olga Vechtomova University of Waterloo, Waterloo, ON, Canada

Erik Vee Yahoo! Research, Silicon Valley, CA, USA

Jari Veijalainen University of Jyvaskyla, Jyvaskyla, Finland

Yannis Velegrakis Department of Information Engineering and Computer Science, University of Trento, Trento, Italy

Suresh Venkatasubramanian University of Utah, Salt Lake City, UT, USA

Rossano Venturini Department of Computer Science, University of Pisa, Pisa, Italy

Victor Vianu University of California-San Diego, La Jolla, CA, USA

Maria-Esther Vidal Computer Science, Universidad Simon Bolivar, Caracas, Venezuela

Millist Vincent University of South Australia, Adelaide, SA, Australia

Giuliana Vitiello University of Salerno, Salerno, Italy

Michail Vlachos IBM T.J. Watson Research Center, Hawthorne, NY, USA

Akrivi Vlachou Athena Research and Innovation Center, Institute for the Management of Information Systems, Athens, Greece

Hoang Vo Computer Science, Stony Brook University, Stony Brook, NY, USA

Hoang Tam Vo IBM Research, Melbourne, VIC, Australia

Agnès Voisard Fraunhofer Institute for Software and Systems Engineering (ISST), Berlin, Germany

Kaladhar Voruganti Advanced Development Group, Network Appliance, Sunnyvale, CA, USA

Gottfried Vossen Department of Information Systems, Westfälische Wilhelms-Universität, Münster, Germany

Daisy Zhe Wang Computer and Information Science and Engineering (CISE), University of Florida, Gainesville, FL, USA

Feng Wang City University of Hong Kong, Hong Kong, China

Fusheng Wang Stony Brook University, Stony Brook, NY, USA

Jianyong Wang Tsinghua University, Beijing, China

Jun Wang Queen Mary University of London, London, UK

Meng Wang Microsoft Research Asia, Beijing, China

X. Sean Wang School of Computer Science, Fudan University, Shanghai, China

Xin-Jing Wang Microsoft Research Asia, Beijing, China

Micros Facebook, CA, USA

Zhengkui Wang InfoComm Technology, Singapore Institute of Technology, Singapore, Singapore

Matthew O. Ward Worcester Polytechnic Institute, Worcester, MA, USA

Segev Wasserkrug IBM Research Labs-Haifa, Haifa, Israel

Hans Weda Phillips Research Europe, Eindhoven, The Netherlands

Gerhard Weikum Department 5: Databases and Information Systems, Max-Planck-Institut für Informatik, Saarbrücken, Germany

Michael Weiner Regenstrief Institute, Inc., Indiana University School of Medicine, Indianapolis, IN, USA

Michael Weiss Carleton University, Ottawa, ON, Canada

Ji-Rong Wen Microsoft Research Asia, Beijing, China

Chunhua Weng Columbia University, New York, NY, USA

Mathias Weske University of Potsdam, Potsdam, Germany

Thijs Westerveld Teezir Search Solutions, Ede, Netherlands

Till Westmann Oracle Labs, Redwood City, CA, USA

Karl Wiggisser Institute of Informatics-Systems, University of Klagenfurt, Klagenfurt, Austria

Jef Wijsen University of Mons, Mons, Belgium

Mark D. Wilkinson University of British Columbia, Vancouver, BC, Canada

Graham Wills SPSS Inc., Chicago, IL, USA

Ian H. Witten University of Waikato, Hamilton, New Zealand

Kent Wittenburg Mitsubishi Electric Research Laboratories, Inc., Cambridge, MA, USA

Eric Wohlstadter University of British Columbia, Vancouver, BC, Canada

Dietmar Wolfram University of Wisconsin-Milwaukee, Milwaukee, WI, USA

Ouri Wolfson Mobile Information Systems Center (MOBIS), The University of Illinois at Chicago, Chicago, IL, USA

Department of CS, University of Illinois at Chicago, Chicago, IL, USA

Janette Wong IBM Canada Ltd, Markham, ON, Canada

Raymond Chi-Wing Wong Department of Computer Science and Engineering, The Hong Kong University of Science and Technology, Clear Water Bay, Kowloon, Hong Kong

Peter T. Wood Birkbeck, University of London, London, UK

David Woodruff IBM Almaden Research Center, San Jose, CA, USA

Marcel Worring University of Amsterdam, Amsterdam, The Netherlands

Adam Wright Partners HealthCare, Boston, MA, USA

Sai Wu Zhejiang University, Hangzhou, Zhejiang, People's Republic of China

Yuqing Wu Indiana University, Bloomington, IN, USA

Alex Wun University of Toronto, Toronto, ON, Canada

Ming Xiong Bell Labs, Murray Hill, NJ, USA

Google, Inc., New York, NY, USA

Guandong Xu University of Technology Sydney, Sydney, Australia

Hua Xu Columbia University, New York, NY, USA

Jun Yan Microsoft Research Asia, Haidian, China

Xifeng Yan IBM T. J. Watson Research Center, Hawthorne, NY, USA

Jun Yang Duke University, Durham, NC, USA

Li Yang Western Michigan University, Kalamazoo, MI, USA

Ming-Hsuan Yang University of California at Merced, Merced, CA, USA

Seungwon Yang Virginia Tech, Blacksburg, VA, USA

Yang Yang Center for Future Media and School of Computer Science and Engineering, University of Electronic Science and Technology of China, Chengdu, Sichuan, China

Yun Yang Swinburne University of Technology, Melbourne, VIC, Australia

Yu Yang City University of Hong Kong, Hong Kong, China

Yong Yao Cornell University, Ithaca, NY, USA

Mikalai Yatskevich University of Trento, Trento, Italy

Xun Yi Computer Science and Info Tech, RMIT University, Melbourne, VIC, Australia

Hiroshi Yoshida VLSI Design and Education Center, University of Tokyo, Tokyo, Japan

Fujitsu Limited, Yokohama, Japan

Masatoshi Yoshikawa University of Kyoto, Kyoto, Japan

Matthew Young-Lai Sybase iAnywhere, Waterloo, ON, Canada

Google, Inc., Mountain View, CA, USA

Hwanjo Yu University of Iowa, Iowa City, IA, USA

Ting Yu North Carolina State University, Raleigh, NC, USA

Cong Yu Google Research, New York, NY, USA

Philip S. Yu Computer Science Department, University of Illinois at Chicago, Chicago, IL, USA

Jeffrey Xu Yu Department of Systems Engineering and Engineering Management, The Chinese University of Hong Kong, Hong Kong, China

Pingpeng Yuan Service Computing Technology and System Lab, Cluster and Grid Computing Lab, School of Computer Science and Technology, Huazhong University of Science and Technology, Wuhan, China

Vladimir Zadorozhny University of Pittsburgh, Pittsburgh, PA, USA

Matei Zaharia Douglas T. Ross Career Development Professor of Software Technology, MIT CSAIL, Cambridge, MA, USA

Ilya Zaihrayeu University of Trento, Trento, Italy

Mohammed J. Zaki Rensselaer Polytechnic Institute, Troy, NY, USA

Carlo Zaniolo University of California-Los Angeles, Los Angeles, CA, USA

Hugo Zaragoza Yahoo! Research, Barcelona, Spain

Stan Zdonik Brown University, Providence, RI, USA

Demetrios Zeinalipour-Yazti Department of Computer Science, Nicosia, Cyprus

Hans Zeller Hewlett-Packard Laboratories, Palo Alto, CA, USA

Pavel Zezula Masaryk University, Brno, Czech Republic

Cheng Xiang Zhai University of Illinois at Urbana-Champaign, Urbana, IL, USA

Aidong Zhang State University of New York, Buffalo, NY, USA

Benyu Zhang Microsoft Research Asia, Beijing, China

Donghui Zhang Paradigm4, Inc., Waltham, MA, USA

Dongxiang Zhang School of Computer Science and Engineering, University of Electronic Science and Technology of China, Sichuan, China

Ethan Zhang University of California, Santa Cruz, CA, USA

Jin Zhang University of Wisconsin Milwaukee, Milwaukee, WI, USA

Kun Zhang Xavier University of Louisiana, New Orleans, LA, USA

Lei Zhang Microsoft Research Asia, Beijing, China

Lei Zhang Microsoft Research, Redmond, WA, USA

Li Zhang Peking University, Beijing, China

Meihui Zhang Information Systems Technology and Design, Singapore University of Technology and Design, Singapore, Singapore

Qing Zhang The Australian e-health Research Center, Brisbane, Australia

Rui Zhang University of Melbourne, Melbourne, VIC, Australia

Dataware Ventures, Tucson, AZ, USA

Dataware Ventures, Redondo Beach, CA, USA

Yanchun Zhang Victoria University, Melbourne, VIC, Australia

Yi Zhang Yahoo! Inc., Santa Clara, CA, USA

Yue Zhang University of Pittsburgh, Pittsburgh, PA, USA

Zhen Zhang University of Illinois at Urbana-Champaign, Urbana, IL, USA

Feng Zhao Microsoft Research, Redmond, WA, USA

Ying Zhao Tsinghua University, Beijing, China

Baihua Zheng Singapore Management University, Singapore, Singapore

Yi Zheng University of Ontario Institute of Technology, Oshawa, ON, Canada

Yu Zheng Data Management, Analytics and Services (DMAS) and Ubiquitous Computing Group (Ubicomp), Microsoft Research Asia, Beijing, China

Zhi-Hua Zhou National Key Lab for Novel Software Technology, Nanjing University, Nanjing, China

Jingren Zhou Alibaba Group, Hangzhou, China

Li Zhou Partners HealthCare System Inc., Boston, MA, USA

Xiaofang Zhou School of Information Technology and Electrical Engineering, University of Queensland, Brisbane, QLD, Australia

Huaiyu Zhu IBM Almaden Research Center, San Jose, CA, USA

Xiaofeng Zhu Guangxi Normal University, Guilin, Guangxi, People's Republic of China

Xingquan Zhu Florida Atlantic University, Boca Raton, FL, USA

Cai-Nicolas Ziegler Siemens AG, Munich, Germany

Hartmut Ziegler University of Konstanz, Konstanz, Germany

Esteban Zimányi CoDE, Université Libre de Bruxelles, Brussels, Belgium

Arthur   Zimek Ludwig-Maximilians-Universität   München,   Munich, Germany

Department of Mathematics and Computer Science, University of Southern Denmark, Odense, Denmark

Roger Zimmermann Department of Computer Science, School of Computing, National University of Singapore, Singapore, Republic of Singapore

Lei Zou Institute of Computer Science and Technology, Peking University, Beijing, China

M

Machine Learning in Computational Biology

Cornelia Caragea[1] and Vasant Honavar[2]
[1]Computer Science and Engineering, University of North Texas, Denton, TX, USA
[2]Iowa State University, Ames, IA, USA

Synonyms

Data mining in bioinformatics; Data mining in computational biology; Data mining in systems biology; Machine learning in bioinformatics; Machine learning in systems biology

Definition

Advances in high throughput sequencing and "omics" technologies and the resulting exponential growth in the amount of macromolecular sequence, structure, gene expression measurements, have unleashed a transformation of biology from a data-poor science into an increasingly data-rich science. Despite these advances, biology today, much like physics was before Newton and Leibnitz, has remained a largely descriptive science. Machine learning [6] currently offers some of the most cost-effective tools for building predictive models from biological data, e.g., for annotating new genomic sequences, for predicting macromolecular function, for identifying functionally important sites in proteins, for identifying genetic markers of diseases, and for discovering the networks of genetic interactions that orchestrate important biological processes [3]. Advances in machine learning e.g., improved methods for learning from highly unbalanced datasets, for learning complex structures of class labels (e.g., labels linked by directed acyclic graphs as opposed to one of several mutually exclusive labels) from richly structured data such as macromolecular sequences, three-dimensional molecular structures, and reliable methods for assessing the performance of the resulting models, are critical to the transformation of biology from a descriptive science into a predictive science.

Historical Background

Large scale genome sequencing efforts have resulted in the availability of hundreds of complete genome sequences. More importantly, the GenBank repository of nucleic acid sequences is doubling in size every 18 months [4]. Similarly, structural genomics efforts have led to a corresponding increase in the number of macromolecular (e.g., protein) structures [5]. At present, there are over a thousand databases of interest to biologists [16]. The emergence of high-throughput "omics" techniques, e.g., for measuring the expression of thousands of genes under different perturbations, has made possible system-wide measurements of biological variables [8]. Consequently, discoveries in biological sciences are increasingly enabled by machine learning.

Some representative applications of machine learning in computational and systems biology include: identifying the protein-coding genes (including gene boundaries, intron-exon structure) from genomic DNA sequences; predicting the function(s) of a protein from its primary (amino acid) sequence (and when available, structure and its interacting partners); identifying functionally important sites (e.g., protein-protein, protein-DNA, protein-RNA binding sites, post-translational modification sites) from the protein's amino acid sequence and, when available, from the protein's structure; classifying protein sequences (and structures) into structural classes; Identifying functional modules (subsets of genes that function together) and genetic networks from gene expression data.

These applications collectively span the entire spectrum of machine learning problems including supervised learning, unsupervised learning (or cluster analysis), and system identification. For example, protein function prediction can be formulated as a supervised learning problem: given a dataset of protein sequences with experimentally determined function labels, induce a classifier that correctly labels a novel protein sequence. The problem of identifying functional modules from gene expression data can be formulated as an unsupervised learning problem: given expression measurements of a set of genes under different conditions (e.g., perturbations, time points), and a distance metric for measuring the similarity or distance between expression profiles of a pair of genes, identify clusters of genes that are co-expressed (and hence are likely to be co-regulated). The problem of constructing gene networks from gene expression data can be formulated as a system identification problem: given expression measurements of a set of genes under different conditions (e.g., perturbations, time points), and available background knowledge or assumptions, construct a model (e.g., a boolean network, a bayesian network) that explains the observed gene expression measurements and predicts the effects of experimental perturbations (e.g., gene knockouts).

Foundations

Challenges presented by computational and systems biology applications have driven, and in turn benefited from, advances in machine learning. Some of these developments are described below.

Multi-Label Classification: In the traditional classification problem, an instance x_i, $i = 1, \ldots, n$, is associated with a single class label y_j from a finite, disjoint set of class labels Y, $j = 1, \ldots, k$, $k = |Y|$ (*single-label classification problem*). If the set Y has only two elements, then the problem is referred to as the *binary classification problem*. Otherwise, if Y has more than two elements, then it is referred to as *multi-class classification problem*. However, in many biological applications, an instance $\mathbf{x}_i$ is associated with a subset of, not necessarily disjoint, class labels in Y (*multi-label classification problem*). For example, many genes and proteins are multifunctional. Most of the existing algorithms cannot simultaneously label a gene or protein with several, not necessarily mutually exclusive functions. Each instance is then assigned to a subset of nodes in the hierarchy, yielding a *hierarchical multi-label classification problem* or a *structured output classification problem*. The most common approach to dealing with *multi-label classification problem* [7] is to transform the problem into k binary classification problems, one for each different label $y_j \in Y$, $j = 1, \ldots, k$. The transformation consists of constructing k datasets, D_j, each containing all instances of the original dataset, such that an instance in D_j, $j = 1, \ldots, k$, is labeled with 1 if it has label y_j in the original dataset, and 0 otherwise. During classification, for a new unlabeled instance $\mathbf{x}_{test}$, each individual classifier C_j, $j = 1, \ldots, k$, returns a prediction that $\mathbf{x}_{test}$ belongs to the class label y_j or not. However, the transformed datasets that result from this approach are highly unbalanced, typically, with the number of positively labeled instances being significantly smaller than the number of negatively labeled instances, requiring the use of methods that can cope with unbalanced data. Alternative evaluation metrics need to be developed for assessing the performance of multi-label classifiers. This task is complicated by correlations among the class labels.

Learning from Unbalanced Data: Many of the macromolecular sequence classification problems present the problem of learning from highly *unbalanced* data. For example, only a small fraction of amino acids in an RNA-binding protein binds to RNAs. Classifiers that are trained to optimize accuracy generally perform rather poorly on the minority class. Hence, if accurate classification of instances from the minority class is important (or equivalently, the false positives and false negatives have unequal costs or risks associated with them), it is necessary to change the distribution of positive and negative instances *during training* by randomly selecting a subset of the training data for the majority class, or alternatively, assigning different *weights* to positive and negative samples (and learn from the resulting weighted samples). More recently, *ensemble classifiers* [11] have been shown to improve the performance of sequence classifiers on unbalanced datasets. Unbalanced datasets also complicate both the training and the assessment of the predictive performance of classifiers. *Accuracy* is not a useful performance measure in such scenarios. Indeed, no single performance measure provides a complete picture of the classifier's performance. Hence, it is much more useful to examine ROC (Receiver Operating Characteristic) or precision-recall curves [3]. Of particular interest are methods that can directly optimize alternative performance measures that take into account the unbalanced nature of the dataset and user-specified tradeoff between false positive and false negative rates.

Data Representation: Many computational and systems biology applications of machine learning present challenges in data representation. Consider for example, the problem of identifying functionally important sites (e.g., RNA-binding residues) from amino acid sequences. In this case, given an amino acid sequence, the classifier needs to assign a binary label (1 for an RNA-binding residue and 0 for a non RNA-binding residue) to each letter of the sequence. To solve this problem using standard machine learning algorithms that work with a fixed number of input features, it is fairly common to use a *sliding window* approach [12] to generate a collection of fixed length windows, where each window corresponds to the target amino acid and an equal number of its sequence neighbors on each side. The classifier is trained to label the target residue. Similarly, identifying binding sites from a three-dimensional structure of the protein requires transforming the problem into one that can be handled by a traditional machine learning method. Such transformations, while they allow the use of existing machine learning methods on macromolecular sequence and structure labeling problems, complicate the task of assessing the performance of the resulting classifier (see below).

Performance Assessment: Standard approaches to assessing the performance of classifiers rely on k-fold cross-validation wherein a dataset is partitioned into k disjoint subsets (folds). The performance measure of interest is estimated by averaging the measured performance of the classifier on k runs of a cross-validation experiment, each using a different choice of the k − 1 subsets for training and the remaining subset for testing the classifier. The fixed length window representation described above complicates this procedure on macromolecular sequence labeling problems. The training and test sets obtained by random partitioning of the dataset of labeled windows can contain windows that originate from the same sequence, thereby violating a critical requirement for cross-validation, namely, that the training and test data be disjoint. The resulting overlap between training and test data can yield overly optimistic estimates of performance of the classifier. A better alternative is to perform sequence-based (as opposed to window-based) cross-validation by partitioning the set of sequences (instead of windows) into disjoint folds. This procedure guarantees that training and test sets are indeed disjoint [9]. Obtaining realistic estimates of performance in sequence classification and sequence labeling problems also requires the use of *non-redundant* datasets [13].

M

Learning from Sparse Datasets: In gene expression datasets, the number of genes is typically in the hundreds or thousands, whereas the number of measurements (conditions, perturbations) is typically fewer than ten. This presents significant challenges in inferring genetic network models from gene expression data because the number of variables (genes) far exceeds the number of observations or data samples. Approaches to dealing with this challenge require reducing the effective number of variables via variable selection [17] or abstraction i.e., by grouping variables into clusters that behave similarly under the observed conditions. Another approach to dealing with sparsity of data in such settings is to incorporate information from multiple datasets [18].

Key Applications

Protein Function Prediction: Proteins are the principal catalytic agents, structural elements, signal transmitters, transporters and molecular machines in cells. Understanding protein function is critical to understanding diseases and ultimately in designing new drugs. Until recently, the primary source of information about protein function has come from biochemical, structural, or genetic experiments on individual proteins. However, with the rapid increase in number of genome sequences, and the corresponding growth in the number of protein sequences, the numbers of experimentally determined structures and functional annotations has significantly lagged the number of protein sequences. With the availability of datasets of protein sequences with experimentally determined functions, there is increasing use of sequence or structural homology-based transfer of annotation from already annotated sequences to new protein sequences. However, the effectiveness of such homology-based methods drops dramatically when the sequence similarity between the target sequence and the reference sequence falls below 30%. In many instances, the function of a protein is determined by conserved local sequence motifs. However, approaches that assign function to a protein based on the presence of a single motif (the so-called characteristic motif) fail to take advantage of multiple sequence motifs that are correlated with critical structural features (e.g., binding pockets) that play a critical role in protein function. Against this background, machine learning methods offer an attractive approach to training classifiers to assign putative functions to protein sequences. Machine learning methods have been applied, with varying degrees of success, to the problem of protein function prediction. Several studies have demonstrated that machine learning methods, used in conjunction with traditional sequence or structural homology based techniques and sequence motif-based methods outperform the latter in terms of accuracy of function prediction (based on cross-validation experiments). However, the efficacy of alternative approaches in genome-wide prediction of functions of protein-coding sequences from newly sequenced genomes remains to be established. There is also significant room for improving current methods for protein function prediction.

Identification of Potential Functional Annotation Errors in Genes and Proteins: As noted above, to close the sequence-function gap, there is an increasing reliance on automated methods in large-scale genome-wide annotation efforts. Such efforts often rely on transfer of annotations from previously annotated proteins, based on sequence or structural similarity. Consequently, they are susceptible to several sources of error including errors in the original annotations from which new annotations are inferred, errors in the algorithms, bugs in the software used to process the data, and clerical errors on the part of human curators. The effect of such errors can be magnified because they can propagate from one set of annotated sequences to another. Because of the increasing reliance of biologists on reliable functional annotations for formulation of hypotheses, design of experiments, and interpretation of results, incorrect annotations can lead to wasted effort and erroneous conclusions. Hence, there is an urgent need for computational methods for checking consistency of such annotations against independent sources of evidence and detecting potential annotation errors. A recent study has demonstrated the usefulness of machine learning methods to *identify and correct* potential annotation errors [1].

Identification of Functionally Important Sites in Proteins: Protein-protein, protein-DNA, and protein-RNA interactions play a pivotal role in protein function. Reliable identification of such interaction sites from protein sequences has broad applications ranging from rational drug design to the analysis of metabolic and signal transduction networks. Experimental detection of interaction sites must come from determination of the structure of protein-protein, protein-DNA and protein-RNA complexes. However, experimental determination of such complexes lags far behind the number of known protein sequences. Hence, there is a need for development of reliable computational methods for identifying functionally important sites from a protein sequence (and when available, its structure, but not the complex). This problem can be formulated as a sequence (or structure) labeling problem. Several groups have developed and applied, with varying degrees of success, machine learning methods for identification of functionally important sites in proteins (see [14, 21, 22] for some examples). However, there is significant room for improving such methods.

Discovery and Analysis of Gene and Protein Networks: Understanding how the parts of biological systems (e.g., genes, proteins, metabolites) work together to form dynamic functional units, e.g., how genetic interactions and environmental factors orchestrate development, aging, and response to disease, is one of the major foci of the rapidly emerging field of systems biology [8]. Some of the key challenges include the following: uncovering the biophysical basis and essential macromolecular sequence and structural features of macromolecular interactions; comprehending how temporal and spatial clusters of genes, proteins, and signaling agents correspond to genetic, developmental and regulatory networks [10]; discovering topological and other characteristics of these networks [19]; and explaining the emergence of systems-level properties of networks from the interactions

among their parts. Machine learning methods have been developed and applied, with varying degrees of success, in learning predictive models including boolean networks [20] and bayesian networks [15] from gene expression data. However, there is significant room for improving the accuracy and robustness of such algorithms by taking advantage of multiple types of data and by using active learning.

Future Directions

Although many machine learning algorithms have had significant success in computational biology, several challenges remain. These include the development of: efficient algorithms for learning predictive models from distributed data; cumulative learning algorithms that can efficiently update a learned model to accommodate changes in the underlying data used to train the model; effective methods for learning from sparse, noisy, high-dimensional data; and effective approaches to make use of the large amounts of unlabeled or partially labeled data; algorithms for learning predictive models from disparate types of data: macromolecular sequence, structure, expression, interaction, and dynamics; and algorithms that leverage optimal experiment design with active learning in settings where data is expensive to obtain.

Cross-References

▶ Biological Networks
▶ Biostatistics and Data Analysis
▶ Classification
▶ Data Mining
▶ Graph Database

Recommended Reading

1. Andorf C, Dobbs D, Honavar V. Exploring inconsistencies in genome-wide protein function annotations: a machine learning approach. BMC Bioinform. 2007;8(1):284.
2. Ashburner M, Ball CA, Blake JA, Botstein D, Butler H, Cherry JM, Davis AP, Dolinski K, Dwight SS, Ep-

M

pig JT, Harris MA, Hill DP, Issel-Tarver L, Kasarskis A, Lewis S, Matese JC, Richardson JE, Ringwald M, Rubin GM, Sherlock G. Gene ontology: tool for the unification of biology. Nat Gene. 2000;25(1): 25–9.

3. Baldi P, Brunak S. Bioinformatics: the machine learning approach. Cambridge, MA: MIT; 2001.

4. Benson DA, Karsch-Mizrachi I, Lipman DJ, Ostell J, Wheeler DL. Genbank. Nucleic Acids Res. 2007;35D(Database issue):21–D25.

5. Berman HM, Westbrook J, Feng Z, Gilliland G, Bhat TN, Weissig H, Shindyalov IN, Bourne PE. The protein data bank. Nucleic Acids Res. 2000;28(1): 235–42.

6. Bishop CM. Pattern recognition and machine learning. Berlin: Springer; 2006.

7. Boutell MR, Luo J, Shen X, Brown CM. Learning multi-label scene classification. Pattern Recogn. 2004;37(9):1757–71.

8. Bruggeman FJ, Westerhoff HV. The nature of systems biology. Trends Microbiol. 2007;15(1):15–50.

9. Caragea C, Sinapov J, Dobbs D, and Honavar V. Assessing the performance of macromolecular sequence classifiers. In: Proceedings of the IEEE 7th International Symposium on Bioinformatics and Bioengineering; 2007. p. 320–6.

10. de Jong H. Modeling and simulation of genetic regulatory systems: a literature review. J Comput Biol. 2002;9(1):67–103.

11. Diettrich TG. Ensemble methods in machine learning. Springer, Berlin. In: Proceedings of the 1st International Workshop on Multiple Classifier Systems; 2000. p. 1–15.

12. Diettrich TG. Machine learning for sequential data: a review. In: Proceedings of the Joint IAPR International Workshop on Structural, Syntactic, and Statistical Pattern Recognition; 2002. p. 15–30.

13. El-Manzalawy Y, Dobbs D, Honavar V. On evaluating MHC-II binding peptide prediction methods. PLoS One. 2008;3(9):e3268.

14. El-Manzalawy Y., Dobbs D., Honavar V. Predicting linear B-cell epitopes using string kernels. J Mole Recogn. 2008; 21(4):243–255.

15. Friedman N, Linial M, Nachman I, Pe'er D. Using bayesian networks to analyze expression data. J Comput Biol. 2000;7(3–4):601–20.

16. Galperin MY. The molecular biology database collection: 2008 update. Nucleic Acids Res. 2008;36(Database issue):D2–4.

17. Guyon I, Elisseeff A. An introduction to variable and feature selection. J Mach Learn Res. 2003;3(7–8):1157–82.

18. Hecker L, Alcon T, Honavar V, Greenlee H. Querying multiple large-scale gene expression datasets from the developing retina using a seed network to prioritize experimental targets. Bioinform Biol Insights. 2008;2:91–102.

19. Jeong H, Tombor B, Albert R, Oltvai ZN, Barabasi A-L. The large-scale organization of metabolic networks. Nature. 1987;407(6804):651–4.

20. Lahdesmaki H, Shmulevich I, Yli-Harja O. On learning gene regulatory networks under the boolean network model. Mach Learn. 2007;52(1–2): 147–67.

21. Terribilini M, Lee J-H, Yan C, Jernigan RL, Honavar V, Dobbs D. Predicting RNA-binding sites from amino acid sequence. RNA J. 2006;12(8): 1450–62.

22. Yan C, Terribilini M, Wu F, Jernigan RL, Dobbs D, Honavar V. Predicting DNA-binding sites of proteins from amino acid sequence. BMC Bioinform. 2006;7:262.

Main Memory

Peter Boncz
CWI, Amsterdam, The Netherlands

Synonyms

Primary memory; Random access memory (RAM)

Definition

Primary storage, presently known as main memory, is the largest memory directly accessible to the CPU in the prevalent Von Neumann model and stores both data and instructions (program code). The CPU continuously reads instructions stored there and executes them. Also called Random Access Memory (RAM), to indicate that load/store instructions can access data at any location at the same cost, it is usually implemented using DRAM chips, which are connected to the CPU and other peripherals (disk drive, network) via a bus.

Key Points

The earliest computers used tubes, then transistors and since the 1970s in integrated circuits. RAM chips generally store a bit of data in either the state of a flip-flop, as in SRAM (static RAM), or as a charge in a capacitor (or transistor gate), as in DRAM (dynamic RAM). Some types have circuitry to detect and/or correct random faults called memory errors in the stored data, using parity bits or error correction codes (ECC). RAM of the read-only type, ROM, instead uses a metal mask to permanently enable/disable selected transistors, instead of storing a charge in them.

The main memory available to a program in most operating systems, while primarily relying on RAM, can be increased by disk memory. That is, the memory access instructions supported by a CPU work on so-called virtual memory, where an abstract virtual memory space is divided into pages. At any time, a page either resides in a swap-file on disk or in RAM, where it must be in order for the CPU to access it. When memory is accessed, the Memory Management Unit (MMU) of the CPU transparently translates the virtual address into its current physical address. If the memory page is not in RAM, it generates a page fault, to be a handled by the OS which then has to perform I/O to the swap file. If a high percentage of the memory access generates a page fault, this is called thrashing, and severely lowers performance.

Over the past decades, the density of RAM chips has increased, following a planned evolution of finer chip production process sizes, popularly known as "Moore's Law." This has led to an increase in RAM capacity as well as bandwidth. Access latency has also decreased, however, the physical distance on the motherboard between DRAM chips and CPU results in a minimum access latency of around 50 ns (real RAM latencies are often higher). In current multi-GHz CPUs this means that a memory access instruction takes hundreds of cycles to execute. Typically, a high percentage of instructions in a program can be memory access instructions (up to 33%) and the RAM latency can seriously impact performance. This problem is known as the "memory wall."

To counter the performance problems of the memory wall, modern computer architecture now features a memory hierarchy that besides DRAM also includes SRAM cache memories, typically located on the CPU chip. Memory access instructions transfer memory in units of cachelines, typically 64 bytes at a time (this cache line size is also related to the width of the memory bus). Memory access instruction first checks whether the accessed cache line is in the highest (fastest/smallest) L1 cache. This takes just a few CPU cycles. If a cache miss occurs, the memory access instruction checks the next cache level. Only if no cache contains the cache line, memory access is performed. Therefore, like virtual memory page thrashing, the CPU cache hit ratio achieved by a program now materially affects performance.

While in the past access to the DRAM chips over the bus was typically performed by a chipset, in between CPU and memory, some modern CPU architectures have moved the memory controller logic onto the CPU chip itself, which tends to reduce access latency. Also, to better serve the memory bandwidth requirement multi-CPU systems, modern architectures often have a dedicated memory bus between the CPU and DRAM. In a Symmetric Multi-Processing (SMP) this leads to a so-called Non-Uniform Memory Access architecture (NUMA), where access to the memory directly connected to a CPU is faster than access to the memory connected to another CPU.

While database systems traditionally focus on the disk access pattern (i.e., I/O), modern database systems, as well as main-memory database systems (that do not rely on I/O in the first place) now must carefully plan the in-memory data storage format used as well as the memory access patterns caused by query processing algorithms, in order to optimize the use of the CPU caches and avoid high cache miss ratios. The increased RAM sizes as well as the increased impact of I/O latency also leads to a trend to rely more on main memory as the preferred storage medium in database processing.

Cross-References

▶ Disk

Main Memory DBMS

Peter Boncz
CWI, Amsterdam, The Netherlands

Synonyms

In-memory DBMS; MMDBMS

Definition

A main memory database system is a DBMS that primarily relies on main memory for computer data storage. In contrast, conventional database management systems typically employ hard disk based persistent storage.

Key Points

The main advantage of MMDBMS over normal DBMS technology is superior performance, as I/O cost is no more a performance cost factor. With I/O as main optimization focus eliminated, the architecture of main memory database systems typically aims at optimizing CPU cost and CPU cache usage, leading to different data layout strategies (avoiding complex tuple representations) as well as indexing structures (e.g., B-trees with lower-fan-outs with nodes of one or a few CPU cache lines).

While built on top of volatile storage, most MMDB products offer ACID properties, via the following mechanisms: (i) Transaction Logging, which records changes to the database in a journal file and facilitates automatic recovery of an in-memory database, (ii) Non-volatile RAM, usually in the form of static RAM backed up with battery power (battery RAM), or an electrically erasable programmable ROM (EEPROM). With this storage, the MMDB system can recover the data store from its last consistent state upon reboot, (iii) High availability implementations that rely on database replication, with automatic failover to an identical standby database in the event of primary database failure.

Main-memory database systems were originally popular in real-time systems (used in e.g., telecommunications) for their fast and more predictable performance, and this continues to be the case. However, with increasing RAM sizes allowing more problems to be addressed using a MMDBMS, this technology is proliferating into many other areas, such as on-line transaction systems, and recently in decision support. Main memory database systems are also deployed as drop-in systems that intercept read-only queries on cached data from an existing disk-based DBMS, thus reducing its workload and providing fast answers to a large percentage of the workload.

Examples of main-memory database systems are MonetDB, SolidDB, TimesTen and DataBlitz. MySQL offers a main-memory backend based on Heap tables. The MySQL Cluster product is a parallel main memory system that offers ACID properties through high availability.

Cross-References

▶ Disk
▶ Main Memory
▶ Processor Cache

Recommended Reading

1. Bohannon P, Lieuwen DF, Rastogi R, Silberschatz A, Seshadri S, Sudarshan S. The architecture of the dalí main-memory storage manager. Multimedia Tools Appl. 1997;4(2):115–51.
2. Boncz PA, Kersten ML. MIL primitives for querying a fragmented world. VLDB J. 1999;8(2):101–19.
3. DeWitt DJ, Katz RH, Olken F, Shapiro LD, Stonebraker M, Wood DA. Implementation techniques for main memory database systems. In: Proceedings of the ACM SIGMOD International Conference on Management of Data; 1984. p. 1–8.
4. Hvasshovd S-O, Torbjørnsen Ø, Bratsberg SE, Holager P. The ClustRa telecom database: high availability, high throughput, and real-time response. In: Proceedings of the 21th International Conference on Very Large Data Bases; 1995. p. 469–77.

Maintenance of Materialized Views with Outer-Joins

Per-Åke Larson
Microsoft Corporation, Redmond, WA, USA

Definition

An materialized outer-join view is a materialized view whose defining expression contains at least one outer join. View maintenance refers to the process of bringing the view up to date after one or more of the underlying base tables has been updated. View maintenance can always be done by recomputing the result, known as a full refresh, but this is usually prohibitively expensive. Incremental view maintenance, that is, only applying the minimal changes required to bring the view up to date, is normally more efficient.

Historical Background

Full outer join (called generalized join) was proposed by Lacroix and Pirotte in 1976 [4]. During the 1980s, there was considerable discussion in the research literature about the use and power of outer joins. Commercial systems began supporting outer joins in the late 1980s and at the time of writing (2007) all major commercial systems do. Optimization of outer-join queries was an active research area during the 1990s. Outer join was first included in the 1992 SQL standard. The first view matching algorithm for outer-join views was published by Larson and Zhou [5] in 2005.

In 1998 Griffin and Kumar [2] published the first paper covering incremental maintenance of materialized outer-join views. A paper 2006 by Gupta and Mumick [3] described a more efficient procedure but, unfortunately, it does not always produce the correct result. In 2007, Larson and Zhou [6] introduced an method for efficient incremental maintenance of outer-join views. At the time of writing, only Oracle allows (a limited form of) materialized outer-join views.

Foundations

Larson and Zhou [6] showed that incremental maintenance of an outer-join view can be divided into two steps: computing and applying a *primary delta* and a *secondary delta*. The first step is very similar to maintaining an inner-join view while the second step is a "clean-up" step.

This entry describes Larson's and Zhou's maintenance procedure for a view without aggregation when the update consists of insertions into one of its base tables. The reader is referred to the original paper [6] for a more complete description of how to handle deletions, views with aggregation, and how to exploit foreign-key constraints to simplify maintenance. Examples illustrating the procedure use a database consisting of the following three tables. Primary keys are underlined.

O(<u>Okey</u>, Odate, Ocustomer),
L(<u>okey, pkey</u>, Qty, Price),
P(<u>Pkey</u>, Pname).

The following materialized view consisting of two full outer joins will be used as a running example.

$$ MV = \left(L \bowtie \genfrac{}{}{0pt}{}{fo}{p(l,p)} P \right) \bowtie \genfrac{}{}{0pt}{}{fo}{p(l,o)} O $$

where the join predicates are defined as $p(l,p) \equiv (l.pkey = p.pkey)$ and $p(l,o) \equiv (l.okey = o.okey)$.

Join-Disjunctive Normal Form

The view maintenance procedure builds on the join-disjunctive normal form for outer-join expressions introduced by Galindo-Legaria [6]. The normal form is described in this section by an example; more details can be found in [6, 1].

Let T_1 and T_2 be tables with schemas S_1 and S_2, respectively. The *outer union*, denoted by $T_1 \uplus T_2$, first null-extends (pads with nulls) the tuples of each operand to schema $S_1 \cup S_2$ and then takes the union of the results (without duplicate elimination).

Let t_1 and t_2 be tuples with the same schema. Tuple t_1 is said to *subsume* tuple t_2 if t_1 agrees with t_2 on all columns where they both are non-null and t_1 contains fewer null values than t_2. The operator *removal of subsumed tuples* of T, denoted by $T\downarrow$, returns the tuples of T that are not subsumed by any other tuple in T.

The *minimum union* of tables T_1 and T_2 is defined as $T_1 \oplus T_2 = (T_1 \uplus T_1)\downarrow$. Minimum union is both commutative and associative.

Left outer join can be rewritten as $T_1 \bowtie_p^{lo} T_2 = T_1 \bowtie_p T_2 \oplus T_1$ and right outer join as $T_1 \bowtie_p^{ro} T_2 = T_1 \bowtie_p T_2 \oplus T_2$. Full outer join can be rewritten as $T_1 \bowtie_p^{fo} T_2 = T_1 \bowtie_p T_2 \oplus T_1 \oplus T_2$.

The example view was defined as

$$MV = \left(L \bowtie_{p(l,p)}^{fo} P \right) \bowtie_{p(l,p)}^{fo} O.$$

Conversion to normal form is done bottom up by applying the rewrite rules above. First rewrite the join between L and P in terms of inner joins and minimum union, which yields

$$MV = \left(\sigma_{p(l,p)} (L \times P) \oplus L \oplus P \right) \bowtie_{p(l,o)}^{fo} O.$$

Then apply the same rewrite to the second outer join, which produces

$$MV = \left(\left(\sigma_{p(l,p)} (L \times P) \oplus L \oplus P \right) \bowtie_{(l,o)} O \right)$$
$$\oplus \left(\sigma_{p(l,p)} (L \times P) \oplus L \oplus P \right) \oplus O.$$

Inner join distributes over minimum union in the same way as over regular union. Applying this transformation to the join with O produces

$$MV = \sigma_{p(l,p) \wedge p(l,o)} (O \times L \times P)$$
$$\oplus \sigma_{p(l,o)} (O \times L) \oplus \sigma_{p(l,o)} (O \times P).$$
$$\oplus \sigma_{p(l,p)} (L \times P) \oplus L \oplus P \oplus O$$

The view expression is now in join-disjunctive form but it can be further simplified. The term $\sigma_{p(l,o)}(O \times P)$ can be eliminated because the join predicate will never be satisfied.

$$MV = \sigma_{P(l,p) \wedge P(l,o)} (O \times L \times P) \oplus \sigma_{p(l,o)} (O \times L)$$
$$\oplus \sigma_{p(l,p)} (L \times P) \oplus L \oplus P \oplus O.$$

The normal form shows what form of tuples are found in MV. For example, it "contains" all tuples in the join of O and L. Most such tuples are represented implicitly by being included in a wider tuple composed of tuples from O, L and P; only the non-subsumed tuples are stored explicitly in the view.

As illustrated by this example, an outer-join expression E over a set of tables $\mathcal{U}$ can be converted to a normal form consisting of the minimum union of terms composed from selections and inner joins (but no outer joins). More formally, the join-disjunctive normal form of E equals

$$E = E_1 \oplus E_2 \oplus \cdots \oplus E_n$$

where each term E_i is of the form $E_i = \sigma_{pi}(T_{i1} \times T_{i2} \times \ldots \times T_{im})$. $\mathcal{T}_i = \{T_{i1}, T_{i2} \ldots T_{im}\}$ is a (unique) subset of the tables in $\mathcal{U}$. Predicate p_i is the conjunction of a subset of the selection and join predicates found in the original form of the query.

The Subsumption Graph

Every term in the normal form of the view has a unique set of source tables drawn from $\mathcal{U}$ and is null-extended on all other tables in the view. The set of source tables of term E_i is denoted by $\mathcal{T}_i$ and the set of tables on which it is null-extended by S_i, $S_i = \mathcal{U} - \mathcal{T}_i$.

A tuple produced by a term with source tables $\mathcal{T}_i$ can only be subsumed by tuples produced by terms whose source set is a superset of $\mathcal{T}_i$. The subsumption relationships among terms can be modeled by a DAG called the subsumption graph.

The *subsumption graph* of E contains a node n_i for each term E_i in the normal form and the node is labeled with the source table set $\mathcal{T}_i$ of E_i. There is an edge from a node n_i to a node n_j, if $\mathcal{T}_i$ is a minimal superset of $\mathcal{T}_i$. $\mathcal{T}_i$ is a minimal superset of $\mathcal{T}_j$ if there does not exist a node n_k in the graph such that $\mathcal{T}_j \subset \mathcal{T}_k \subset \mathcal{T}_i$.

The subsumption graph for view MV is shown to the left in Fig. 1. The importance of the subsumption graph lies in the following observation: when checking whether a tuple of a term is sub-

Maintenance of Materialized Views with Outer-Joins, Fig. 1 Subsumption graph and maintenance graph for view *MV*

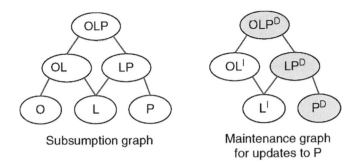

Subsumption graph

Maintenance graph for updates to P

sumed, it is sufficient to check against tuples in the term's immediate parent terms. For example, to determine whether a *P* tuple p_1 is subsumed, all that is needed is to check whether it joins with an *L* tuple. If it does, the resulting tuple, which subsumes p_1, is included in the *LP* term.

The result of an outer-join expression is represented in a minimal form. Only the non-subsumed tuples produced by a term E_i in the normal form are explicitly represented. A subsumed tuple is represented implicitly by being included in a subsuming tuple. The *net contribution* of a term, denoted by D_i, is the set of non-subsumed tuples of term E_i in the normal form of expression *E*. Then *E* can then be written in the form

$$E = D_1 \uplus D_2 \uplus \cdots \uplus D_n.$$

Consider a view *V* and suppose one of the its base tables *T* is modified. This may affect the net contribution of a term D_i in one of three ways:

1. Directly, which occurs if *T* is among the tables in $\mathcal{T}_i$;
2. Indirectly, which occurs if *T* is not among the tables in $\mathcal{T}_i$ but it is among the source tables of at least one of its parent nodes;
3. No effect, otherwise.

Based on this classification of how terms are affected, a *view maintenance graph* is created as follows.

1. Eliminate from the subsumption graph all nodes that are unaffected by the update of *T*.

2. Mark the remaining nodes by *D* or *I* depending on whether the node is affected directly or indirectly.

The maintenance graph for view when updating *P* is shown to the right in Fig. 1. The maintenance graph is used primarily to identify which terms of a view are indirectly affected and thus may require maintenance.

Maintenance Procedure

Suppose table *T* has been updated. If so, any view *V* that references *T* needs to be maintained. The first step is to compute the view's maintenance graph and classify the terms as directly affected, indirectly affected, and unaffected. Without loss of generality, assume that the view has *n* terms, of which terms $1, 2, \ldots, k$ are directly affected, terms $k+1, k+2, \ldots, k+m$ are indirectly affected, and terms $k+m+1, k+m+2, \ldots, n$ are not affected. The view expression can then be rewritten in the form

$$V = V^D \uplus V^I \uplus V^U \text{ where}$$

$$V^D = \uplus_{i=1}^{k} D_i \quad V^I = \uplus_{i=k+1}^{k+m} D_i,$$

$$V^U = \uplus_{i=k+m+1}^{n} D_i.$$

From this form of the expression, one can see that to update the view, two delta expressions must be evaluated and applied to the view

$$\Delta V^D = \uplus_{i=1}^{k} \Delta D_i, \quad \Delta V^I = \uplus_{i=k+1}^{k+m} \Delta D_i.$$

ΔV^D is called the *primary delta* and ΔV^I the *secondary delta*. In summary, maintenance of a view V after insertions into one of its underlying base tables can be performed in two steps.

1. Compute the primary delta ΔV^D and insert the resulting tuples into the view.
2. If there are indirectly affected terms, compute the secondary delta ΔV^I and delete the resulting tuples from the view.

Computing the Primary Delta

An expression that computes the primary delta, ΔV^D, can be constructed by the following simple algorithm.

1. Traverse the operator tree for V along the path from T to the root. On any join operator encountered, apply commutativity rules to ensure that the input referencing T is on the left.
2. Traverse the path from T to the root of V. Convert any full outer join operator encountered to a left outer join and any right outer join operator to an inner join.
3. Substitute T by ΔT.

Step 1 is a normal rewrite of the view expression and does not change the result. Step 2 modifies the expression so that it computes only V^D. After Step 2, the operators on the path from T to the root consists only of selects, inner joins, and left outer joins and the delta expression is always the left input.

Figure 2 illustrates the transformation process for the example view MV when table P is updated. The resulting expression for computing the primary delta is

$$\Delta MV^D = \left(\Delta P \bowtie^{lo}_{p(l,p)} L\right) \bowtie^{lo}_{p(l,o)} O$$

Computing the Secondary Delta

The secondary delta can be computed efficiently from the primary delta and either the view or base tables. Only the case when using the view is described here. Recall that the base tables have already been updated and the primary delta has been applied to the view.

The primary delta ΔV^D contains the union of the deltas for all directly affected terms. However, deltas for individual terms are needed to compute the secondary delta. Each term is defined over a unique set of tables and null extended on all others so tuples from a particular term are easily identified and can be extracted from ΔV^D by simple selection predicates.

Let $null(T)$ denote a predicate that evaluates to true if a tuple is null-extended on table T. $null(T)$ can be implemented in SQL as "$T.c$ is null" where c is any column of T that does not contain nulls, for example, a column of a key. When applying $null$ and $\neg null$ to a set of tables $T = \{T_1, T_2 \ldots, T_n\}$, the shorthand notations $n(\mathcal{T}) = \vee_{T_i \in T} null(T_i)$ and $nn(\mathcal{T}) = \vee_{T_i \in T} \neg null(T_i)$ are used.

For the example view, MV, the primary delta contains deltas of three directly affected terms, see Fig. 1. Non-subsumed tuples from, for example, the LP-term are uniquely identified by the fact that they are composed of a real tuple from L and from P but are null extended on O. Hence, ΔD_{LP} can be extracted from ΔV^D as follows:

Maintenance of Materialized Views with Outer-Joins, Fig. 2 Constructing primary-delta expression for insertions into table P

$$\Delta MV^D = \left(\Delta P \bowtie^{lo}_{p(l,p)} L\right) \bowtie^{lo}_{p(l,o)} O$$

$$\Delta D_{LP} = \pi_{(LP).*}\sigma_{nn(LP)\wedge n(o)}\Delta M V^D$$

where $nn(LP) = \neg\, null(L) \wedge \neg\, null(P)$ and $n(O) = null(O)$.

ΔD_{LP} contains only the delta of the net contribution of the term. ΔE_{LP} contains the complete delta of the term, including both subsumed and non-subsumed tuples. Tuples in ΔE_{LP} are composed of real tuples from L, and from P, and may or may not be null extended on O. Hence, ΔE_{LP} can be extracted from ΔV^D as follows:

$$\Delta E_{LP} = \delta\pi_{(LPo).*}\sigma_{nn}(LP)\Delta M V^D.$$

The duplicate elimination (δ) is necessary because an LP tuple may have joined with multiple O tuples.

Continuing with the running example, the secondary delta consists of ΔD_{OL} and ΔD_L. ΔD_{OL} is null extended on P and the OLP-term is its only parent so it can be computed as:

$$\Delta D_{OL} = \sigma_{nn(OL)\wedge n(P)}\left(M V + \Delta M V^D\right) \\ \bowtie_{eq(OL)}^{ls}\sigma_{nn(OLP)}\Delta M V^D.$$

This expression makes sense intuitively. The first part selects from the view all orphaned (non-subsumed) tuples of term E_{OL} contained in the view after the primary delta has been applied. The second part extracts from the primary delta all tuples added to the parent term E_{OLP}. The join is a left semijoin and outputs every tuple from the left operand that joins with one or more tuples in the right operand. The complete expression thus amounts to finding all currently orphaned tuples of the term and retaining those that cease to be orphans because of the insert. Those tuples should be deleted from the view.

D_L is null extended on O, and P and has one directly affected parent, the LP-term. ΔD_L can be computed as:

$$\Delta D_L = \sigma_{nn(L)\wedge n(OP)}\left(M V + \Delta M V^D\right) \\ \bowtie_{eq(L)}^{ls}\sigma_{nn}(LP)\Delta M V^D.$$

Summary

In summary, after insertion into table P of a set of tuples ΔP, the example view MV can be brought up to date as follows. First compute the primary delta

$$\Delta M V^D = \left(\Delta P \bowtie_{p(l,p)}^{lo} L\right) \bowtie_{p(l,o)}^{lo} O$$

and insert the resulting tuples into the view, resulting in $MV + \Delta MV^D$. Then compute the secondary delta

$$\Delta M V^I = \Delta D_{OL} \uplus \Delta D_L \\ = \sigma_{nn(OL)\wedge n(P)}\left(M V + \Delta M V^D\right) \\ \bowtie_{eq(OL)\,\sigma_{nn}(OLP)}^{ls}\Delta M V^D \uplus \\ \sigma_{nn(L)\wedge n(OP)}\left(M V + \Delta M V^D\right) \\ \bowtie_{eq(L)\,\sigma_{nn}(LP)}^{ls}\Delta M V^D$$

and delete the resulting tuples from the view.

Key Applications

Queries containing outer joins are often used in analysis queries over large tables in data warehouses. Materialized outer-join views, especially when aggregated, can be very beneficial in such scenarios.

Cross-References

▸ Views

Recommended Reading

1. Galindo-Legaria C. Outerjoins as disjunctions. In: Proceedings of the ACM SIGMOD International Conference on Management of Data; 1994. p. 348–58.
2. Griffin T, Kumar B. Algebraic change propagation for semijoin and outerjoin queries. ACM SIGMOD Rec. 1998;27(3):22–7.

M

3. Gupta H, Mumick IS. Incremental maintenance of aggregate and outerjoin expressions. Inf Syst. 2006;31(6):435–64.
4. Lacroix M, Pirotte A. Generalized joins. ACM SIGMOD Rec. 1976;8(3):14–5.
5. Larson P, Zhou J. View matching for outer-join views. In: Proceedings of the 31st International Conference on Very Large Data Bases; 2005. p. 445–56.
6. Larson P, Zhou J. Efficient maintenance of materialized outer-join views. In: Proceedings of the 23rd International Conference on Data Engineering; 2007. p. 56–65.

Maintenance of Recursive Views

Suzanne W. Dietrich
Arizona State University, Phoenix, AZ, USA

Synonyms

Incremental maintenance of recursive views; Recursive view maintenance

Definition

A view is a derived or virtual table that is typically defined by a query, providing an abstraction or an alternate perspective of the data that allows for more intuitive query specifications using these views. Each reference to the view name results in the retrieval of the view definition and the recomputation of the view to answer the query in which the view was referenced. When views are materialized, the tuples of the computed view are stored in the database with appropriate index structures so that subsequent access to the view can efficiently retrieve tuples to avoid the cost of recomputing the entire view on subsequent references to the view. However, the materialized view must be updated if any relation that it depends on has changed. Rather than recomputing the entire view on a change, an incremental view maintenance algorithm uses the change to incrementally compute updates to the materialized view in response to that change. A recursive view is a virtual table definition that depends on itself. A canonical example of a recursive view is the transitive closure of a relationship stored in the database that can be modeled as directed edges in a graph. The transitive closure essentially determines the reachability relationship between the nodes in the graph. Typical examples of transitive closure include common hierarchies such as employee-supervisor, bill of materials (parts-subparts), ancestor, and course prerequisites. The incremental view maintenance algorithms for the maintenance of recursive views have additional challenges posed by the recursive nature of the view definition.

Historical Background

A view definition relates a view name to a query defined in the query language of the database. Initially, incremental view maintenance algorithms were explored in the context of non-recursive view definitions involving select, project, and join query expressions, known as SPJ expressions in the literature. The power of recursive views was first introduced in the *Datalog* query language, which is a declarative logic programming language established as the database query language for deductive databases in the 1980s. Deductive databases assume the theoretical foundations of relational data but use Datalog as the query language. Since its relational foundations assume first normal form, Datalog looks like a subset of the Prolog programming language without function symbols. However, Datalog does not assume Prolog's top-down left-to-right programming language evaluation strategy. The evaluation of Datalog needed to be founded on the fundamentals of database query optimization. In a database system, a user needs to only specify a correct declarative query, and it is the responsibility of the database system to efficiently execute that specification. The evaluation of Datalog was further complicated by the fact that Datalog allows for relational views that include union and recursion in the presence of negation. Therefore, the view definitions in Datalog were more expressive than the traditional select-project-join views available in relational databases at that time. Therefore, the incremental view maintenance algorithms for recursive views in the early 1990s are typically formulated in the context of the evaluation of Datalog. The power to define a recursive union in SQL was added in the SQL:1999 standard.

Historically, it is important to note that the incremental maintenance of recursive views is related to the areas of integrity constraint checking and condition monitoring in active databases. These three areas were being explored in the research literature at around the same time. In integrity constraint checking, the database is assumed to be in a consistent state, and when a change occurs in the database, it needs to incrementally determine whether the database is still in a consistent state. In active databases, the database is responsible for actively checking whether a condition that it is responsible for monitoring is now satisfied by incrementally evaluating condition specifications affected by changes to the database. Although closely related, there are differences in the underlying assumptions for these problems.

Foundations

Recursive View Definition

A canonical example of a recursive view definition is the reachability of nodes in a directed graph. In Datalog, the reach view consists of two rules. The first non-recursive rule serves as the base or seed case and indicates that if the stored or base table edge defines a directed edge from the source node to the destination node, then the destination can be reached from the source. The second rule is recursive. If the source node can reach some intermediate node and there is an edge from that intermediate node to a destination node, then the source can reach the destination:

reach(Source, Destination) :-
   edge(Source, Destination).
reach(Source, Destination) :-
   reach(Source, Intermediate),
   edge(Intermediate, Destination).

Intuitively, one can think of the recursive rule as an unfolding of the joins required to compute the reachability of paths of length two, then paths of length three, and so on until the data of the underlying graph is exhausted.

In SQL, this recursive view is defined with the following recursive query expression:

with recursive reach(source, destination) as
(select E.source, E.destination
 from edge E)
union
(select S.source, D.destination
 from reach S, edge D
 where S.destination = D.source)

SQL limits recursive queries to linear recursions, which means that there is at most one direct invocation of a recursive item. The specification of the reach view above is an example of a linear recursion. There is another linear recursive specification of reach where the direct recursive call appears on the right side of the join versus the left side of the join:

reach(Source, Destination) :-
   edge(Source, Intermediate),
   reach(Intermediate, Destination).

However, there is a logically equivalent specification of reach that is nonlinear:

reach(Source, Destination) :-
   reach(Source, Intermediate),
   reach(Intermediate, Destination).

The goal of Datalog evaluation is to allow the user to specify the recursive view declaratively in a logically correct way, and it is the system's responsibility to optimize the evaluation of the query.

SQL also restricts recursions to those defined in deductive databases as stratified *Datalog with negation*. Without negation, a recursive Datalog program has a unique solution that corresponds to the theoretical fixpoint semantics or meaning of the logical specification. In the computation of the reach view, each unfolding of the recursion joins the current instance of the recursive view with the edge relation until no new tuples can be added. The view instance has reached a fixed point and will not change. When negation is introduced, the interaction of recursion and negation must be considered. The concept of stratified negation means that there can be no negation through a recursive computation, i.e., a view cannot be defined in terms of its own negation. Recursive views can contain negation, but the negation must be in the context of relations that are either stored or completely computed before the application of the negation. This imposed level of evaluation with respect to negation and recursion is called strata. For stratified Datalog with negation, there also exists a theoretical fixpoint that represents the intuitive meaning of the program.

Consider an example of a view defining a peer as two employees that are not related in the employee-supervisor hierarchy:

peer(A, B):-
    employee(A, . . .), employee(B, . . .),
    not (supervisor(A,B)), not(supervisor(B,A)).
supervisor(Emp, Sup)
    immediateSupervisor(Emp,Sup).
supervisor(Emp, Sup) :-
    supervisor(Emp, S) :-
    immediateSupervisor(S, Sup).

Since peer depends on having supervisor materialized for the negation, peer is in a higher stratum than supervisor. Therefore, the strata provide the levels in which the database system needs to compute views to answer a query.

Evaluation of Recursive Queries

Initial research in the area emphasized the efficient and complete evaluation of recursive queries. The intuitive evaluation of the recursive view that unions the join of the current view instance with the base data at each unfolding is known as a naïve bottom-up algorithm. In a bottom-up approach to evaluating a rule, the known collection of facts is used to satisfy the subgoals on the right-hand side of the rule, generating new facts for the relation on the left-hand side of the rule. To improve the efficiency of the naïve algorithm, a semi-naïve approach can be taken that only uses the new tuples for the recursive view from the last join to use in the join at the next iteration. A disadvantage of this bottom-up approach for evaluating a query is that the entire view is computed even when a query may be asking for a small subset of the data. This eager approach is not an issue in the context of materializing an entire view.

Another recursive query evaluation approach considered a top-down strategy as in Prolog's evaluation. In a top-down approach to evaluation, the evaluation starts with the query and works toward the collection of facts in the database. In the context of the reach recursive view, the reach query is unified with the left-hand side of the non-recursive rule and rewritten as a query involving edge. The edge facts are then matched to provide answers. The second recursive rule is then used to rewrite the reach query with the query consisting of the goals on the right-hand side of the rule. This evaluation process continues, satisfying the goals with facts or rewriting the goals using the rules. The unification of a goal with the left-hand side of a rule naturally filters the evaluation by binding variables in the rule to constants that appear in the query. However, the evaluation of a left-recursive query using Prolog's evaluation strategy enters an infinite loop on cyclic data by attempting to prove the same query over and over again. A logic programmer would not write a logic program that enters an infinite loop, but the deductive database community was interested in the evaluation of truly declarative query specifications.

The resulting evaluation approaches combine the best of top-down filtering with bottom-up materialization. The magic sets technique added top-down filtering by cleverly rewriting the original rules so that a bottom-up evaluation would take advantage of constants appearing in the query

[1]. Memoing was added to a top-down evaluation strategy to achieve the duplicate elimination feature that is inherent in a bottom-up evaluation of sets of tuples [3]. This duplicate elimination feature avoids the infinite loops on cyclic data. Top-down memoing is complete for subsets of Datalog on certain types of queries [4]. For stratified Datalog with negation, top-down memoing still requires iteration to guarantee complete evaluation. Further research explored additional optimizations as well as implementations of deductive database systems [12] and led to research in active databases and materialized view maintenance.

Incremental Evaluation of Recursive Views

A view maintenance algorithm uses the change to incrementally determine updates to the view. Consider a change in the underlying graph for the transitive closure example. If a new edge is inserted, this edge may result in a change to the materialized reach view by adding a connection between two nodes that did not exist before. However, another possibility is that the new edge added another path between two nodes that were already in the materialized view. A similar situation applies on the removal of an edge. The deletion could result in a change in the reachability between nodes, or it could result in the removal of a path, but the nodes are still connected via another route. In addition, in the general case, a view may depend on many relations including other (recursive) views in the presence of negation. Therefore, the approaches for the incremental maintenance of recursive views typically involve a propagation or derivation phase that determines an approximation or overestimate of the changes and a filtering or rederivation phase that checks whether the potential change represents a change to the view. There are differences in the underlying details of how these phases are performed.

The two incremental view maintenance algorithms that will be presented by example are the DRed algorithm [7] and the PF Algorithm [8]. Both the DRed and PF algorithms handle recursive stratified Datalog programs with negation.

There are other algorithms developed for special cases of Datalog programs and queries, such as the counting technique for non-recursive programs, but this exposition will explore these more general approaches for incremental view maintenance. Historically, the PF algorithm was developed in the context of top-down memoing, whereas DRed assumes a bottom-up semi-naïve evaluation. To assist with the comparison of the approaches, the notation introduced for the DRed algorithm [7] will be used to present both algorithms in the context of the transitive closure motivational example.

Figure 1 provides a graphical representation of an edge relation. Assume that the view for reach is materialized and the edge(e,f) is deleted from the graph. The potential deletions or overestimates for reach, denoted by δ^-(reach), are computed by creating Δ^- rules for each rule computing reach. Each reach rule has k Δ^- rules where k corresponds to the number of subgoals in the body of the rule. The ith Δ^- rule uses the current estimate of deleted tuples (δ^-) for the ith subgoal. For the non-recursive rule, there is only one subgoal. Therefore, there is only one Δ^- rule indicating that potential edge deletions generate potential deletions to the reach view.

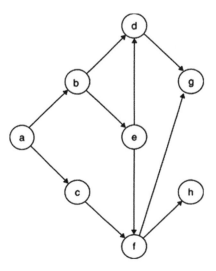

Maintenance of Recursive Views, Fig. 1 Sample graph

Δ^-(r1): δ^-(reach(S, D)):- δ^-(edge(S, D)).

Since the recursive rule has two subgoals, there are two Δ^- rules:

Δ^-(r21): δ^-(reach(S, D)):- δ^-(reach(S, I)), edge(I, D).
Δ^-(r22): δ^-(reach(S, D)):- reach(S, I), δ^-(edge(I, D)).

Potential deletions to the reach view as well as the edge relation can generate potential deletions to the view.

These potential deletions need to be filtered by determining whether there exist alternative derivations or paths between the nodes computed in the potential deletion. There is a Δ^r rule defined for each reach rule that determines the rederivation of the potential deletions, which is denoted by δ^+(reach):

Δ^r(r1): δ^+(reach(S, D)):- δ^- (reach(S, D)), edge^v(S, D).
Δ^r(r2): δ^+(reach(S, D)):- δ^-(reach(S, D)), reach^v(S, I), edge^v(I, D).

The superscript v on the subgoals in the rule indicates the use of the current instance of the relation corresponding to the subgoal. If the potential deletion is still reachable in the new database instance, then there exists another route between the source and destination, and it should not be removed from the materialized view. The actual removals to reach, indicated by Δ^-(reach), are

the set of potential deletions minus the set of alternative derivations:

Δ^-(reach) = δ^-(reach) - δ^+ (reach)

Table 1 illustrates the evaluation of the DRed algorithm for incrementally maintaining the reach view on the deletion of edge(e,f) from Fig. 1. The DRed algorithm uses a bottom-up evaluation of the given rules, starting with the deletion δ^-(edge(e,f)). In the first step, the Δ^- rules compute the overestimate of the deletions to reach. The result of the Δ^- rules is shown in the right column, which indicates the potential deletions to reach as δ^-(reach). The second step uses the Δ^r rules to filter the potential deletions. The right column illustrates the source destination pairs that are still reachable after the deletion of edge(e,f) as δ^+(reach). The tuples that must be removed from the materialized view are indicated by Δ^-(reach):{(e,f) (e,h) (b,f) (b,h)}.

The PF (propagate filter) algorithm on the same example is shown in Table 2. PF starts by propagating the edge deletion using the non-recursive rule, which generates a potential deletion of reach(e,f). This approximation is immediately filtered to determine whether there exists another path between e and f. Since there is no alternate route, the tuple (e,f) is identified as an actual change and is then propagated. The propagation of Δ^-(reach):{(e,f)} identifies (e,g) and (e,h) as potential deletions. However, the

Maintenance of Recursive Views, Table 1 DRed algorithm on deletion of edge(e,f) on materialized reach view

DRed algorithm		
Step 1	Compute overestimate of potential deletions	δ^-(reach)
	Δ^-(r1): δ^-(reach(S, D)):- δ^-(edge(S, D))	(e,f)
	Δ^-(r21): δ^-(reach(S, D)):- δ^-(reach(S, I)), edge(I, D)	(e,g) (e,h)
	Δ^-(r22): δ^-(reach(S, D)):- reach(S, I), δ^-(edge(I, D))	(a,f) (b,f)
	Repeat until no change: No new tuples for Δ^-(r1) and Δ^-(r22)	
	Δ^-(r21): δ^-(reach(S, D)):- δ^-(reach(S, I)), edge(I, D)	(a,g) (a,h) (b,g) (b,h)
	Last iteration does not generate any new tuples	
Step 2	Find alternative derivations to remove potential deletions	δ^+(reach)
	Δ^r(r1): δ^+ (reach(S, D)):- δ^-(reach(S, D)), edge(S, D)	
	Δ^r(r2): δ^+ (reach(S, D)):- δ^-(reach(S, D)), reach^v(S, I), edge^v(I, D)	(e,g) (a,f) (a,g) (a,h) (b,g)
Step 3	Compute actual changes to reach	Δ^-(reach)
	Δ^-(reach) = δ^-(reach) - δ^+(reach)	(e,f) (e,h) (b,f) (b,h)

Maintenance of Recursive Views, Table 2 PF algorithm on deletion of edge(e,f) on materialized reach view

PF algorithm					
Propagate			Filter		
	Rule	δ^-(reach)	δ^+(reach)	Δ^-(reach)	
δ^-(edge):{(e,f)}	Δ^-(r1)	(e,f)	{}	(e,f)	
Δ^-(reach):{(e,f)}	Δ^-(r21)	(e,g) (e,h)	(e,g)	(e,h)	
Δ^-(reach):{(e,h)}	Δ^-(r21)	{}		{}	
δ^-(edge):{(e,f)}	Δ^-(r22)	(a,f) (b,f)	(a,f)	(b,f)	
Δ^-(reach):{(b,f)}	Δ^-(r21)	(b,g) (b,h)	(b,g)	(b,h)	
Δ^-(reach):{(b,h)}	Δ^-(r21)	{}		{}	

filtering phase identifies that there is still a path from e to g, so (e,h) is identified as a removal to reach. The propagation of (e,h) does not identify any potential deletions. The propagation of the initial edge deletion δ^-(edge):{(e,f)} must be propagated through the recursive rule for reach using Δ^-(r22). The potential deletions are immediately filtered, and only actual changes are propagated. The PF algorithm also identifies the tuples {(e,f) (e,h) (b,f) (b,h)} to be removed from the materialized view.

As shown in the above deletion example, the DRed and PF algorithms both compute overestimates or approximations of tuples to be deleted from the recursive materialized view. The PF algorithm eagerly filters the potential deletions before propagating them. The DRed algorithm propagates the potential deletions within a stratum but filters the overestimates before propagating them to the next stratum. There are scenarios in which the DRed algorithm outperforms the PF algorithm and others in which the PF algorithm outperforms the DRed algorithm.

For the case of insertions, the PF algorithm operates in a manner similar to deletions, by approximating the tuples to be added and filtering the potential additions by determining whether the tuple was provable in the old database state. However, the DRed algorithm uses the bottom-up semi-naïve algorithm for Datalog evaluation to provide an inherent mechanism for determining insertions to the materialized view. In semi-naïve evaluation, the original rules are executed once to provide the seed or base answers. Then incremental versions of the rules are executed until a fixpoint is reached. The

incremental rules are formed by creating k rules associated with a rule where k corresponds to the number of subgoals in the right-hand side of the rule. The ith incremental rule uses only the new tuples from the last iteration for the ith subgoal. However, when the ith subgoal is a stored relation, then the corresponding incremental rules are removed since they will not contribute to the incremental evaluation. For the motivational example, the incremental rule for reach is

$$\Delta reach(S,I), edge(I, D)$$

where Δreach represents the new reach tuples computed on the previous iteration. Since a set of tuples is being computed, duplicate proofs are automatically filtered and are not considered new tuples. This is the inherent memoing in bottom-up evaluation that handles cycles in the underlying data.

Key Applications

Query optimization; Condition monitoring; Integrity constraint checking; Data warehousing; Data mining; Network management; Mobile systems

Cross-References

▶ Datalog
▶ Incremental Maintenance of Views with Aggregates

► Maintenance of Materialized Views with Outer-Joins

► View Maintenance

Recommended Reading

1. Bancilhon F, Maier D, Sagiv Y, Ullman J. Magic sets and other strange ways to implement logic programs. In: Proceedings of the 5th ACM SIGACT-SIGMOD Symposium on Principles of Database Systems; 1986. p. 1–5.
2. Ceri S, Widom J. Deriving production rules for incremental view maintenance. In: Proceedings of the 17th International Conference on Very Large Data Bases; 1991. p. 577–89.
3. Dietrich SW. Extension tables: memo relations in logic programming. In: Proceedings of the IEEE 4th Symposium on Logic Programming; 1987. p. 264–72.
4. Dietrich SW, Fan C. On the completeness of naive memoing in prolog. New Gener Comput. 1997;15(2):141–62.
5. Dong G, Su J. Incremental maintenance of recursive views using relational calculus/SQL. ACM SIGMOD Rec. 2000;29(1):44–51.
6. Gupta A, Mumick IS. Materialized views: techniques, implementations, and applications. Cambridge: The MIT Press; 1999.
7. Gupta A, Mumick IS, Subrahmanian VS. Maintaining views incrementally. In: Proceedings of the ACM SIGMOD International Conference on Management of Data; 1993. p. 157–66.
8. Harrison JV, Dietrich SW. Maintenance of materialized views in a deductive database: an update propagation approach. In: Proceedings of the Workshop on Deductive Databases; 1992. p. 56–65.
9. Küchenhoff V. On the efficient computation of the difference between consecutive database states. In: Proceedings of the 2nd International Conference on Deductive and Object-Oriented Databases; 1991. p. 478–502.
10. Martinenghi D, Christiansen H. Efficient integrity constraint checking for databases with recursive views. In: Proceedings of the 9th East European Conference on Advances in Databases and Information Systems; 2005. p. 109–24.
11. Ramakrishnan R. Applications of logic databases. Norwell: Kluwer; 1995.
12. Ramakrishnan R, Ullman D. A survey of deductive database systems. J Logic Programming. 1995;23(2):125–49.
13. Ullman J. Principles of database and knowledge base systems, vol. 1, 2. Rockville: Computer Science Press; 1989.
14. Urpí T, Olivé A. A method for change computation in deductive databases. In: Proceedings of the 18th International Conference on Very Large Data Bases; 1992. p. 225–37.

Managing Compressed Structured Text

Nieves R. Brisaboa[1], Ana Cerdeira-Pena[1], and Gonzalo Navarro[2]
[1]Database Laboratory, Department of Computer Science, University of A Coruña, A Coruña, Spain
[2]Department of Computer Science, University of Chile, Santiago, Chile

Synonyms

Compressing XML; Searching compressed XML

Definition

Compressing structured text is the problem of creating a reduced-space representation from which the original data can be re-created exactly. Compared to plain text compression, the goal is to take advantage of the structural properties of the data. A more ambitious goal is that of being able of manipulating this text in compressed form, without decompressing it. This entry focuses on compressing, navigating, and searching structured text, as those are the areas where more advances have been made.

Historical Background

Modeling data using structured text has been a topic of interest at least since the 1980s, with a significant burst of activity in the 1990s [3]. Since then, the widespread adoption of XML (appearing in 1998, see the current version at http://www.w3.org/TR/xml) as the standard to represent structured text has unified the efforts of the community around this particular format. Very early, however, the same features that made XML particularly appealing for both human and machine processing were pointed out as significant sources of redundancy and wasting of storage space and bandwidth. This was especially relevant for wireless transmission and triggered the proposal of the WAP Binary XML Content Format (WBXML) as early as 1999 (see http://www.w3.org/TR/wbxml), where simple techniques to compress XML prior to its transmission were devised.

In parallel, there has been a growing interest in not only compressing the data for storage or transmission, but in manipulating it in compressed form. The reason is the widening gaps in the memory hierarchy. A more compact data representation has the potential of fitting in a faster memory, where manipulating it can be orders of magnitudes faster, even if it requires more operations, than a naive representation fitting only in a slower memory. Moreover, reducing the space usage may be key to meet the requirements of memory-limited devices (such as mobile devices). Regarding distributed scenarios, compact data representations are particularly appealing to minimize the number of machines used, their energy consumption, and the overall communication costs.

Scientific Fundamentals

For concreteness, the entry will focus on the de facto standard XML, where the structure is a tree or a forest marked with beginning and ending tags in the text. There can be free text between every consecutive pair of tags. In fact this encompasses many other structured text proposals; hence, most of the material of the entry applies to structured text in general, with minimal changes. In XML, the tags can have attributes and associated values, and there might be available grammar giving the permissible context-free syntax of the structured document (called the DTD).

Compression of Structured Text

An obvious approach to compressing structured text is to regard it as plain text and use any of the well-known text compression methods, leading to so-called *XML blind* compressors. Yet, considering the structure might yield improved compression performance ratios. Many *XML conscious compressors* have been proposed to exploit structure in different ways. Rather than fully describing each tool individually, the main principles behind them will be presented and then illustrated with a few examples:

- The structure can be regarded as a labeled tree, where the labels are the tag and attribute names (which we call collectively tags), and the content as free text. Attribute information can be handled as text as well or as special data attached to tree nodes.
- The structure and the content can be compressed separately, which has proved to give good results.
  - The structure can be compressed in several ways, which can range from a simple scheme of assigning numbers to the different tag names, to sophisticated tree grammar compression methods. The latter may take advantage of the DTD when it is available.
  - The text content can be compressed using any text compression method. Semi-static compressors permit accessing the content at random without decompressing all from the beginning, whereas adaptive compressors may achieve better compression ratios when the text data is heterogeneous. Splitting the text into blocks that are compressed adaptively yields trade-offs.
  - Structure can be used, in addition, to boost text compression. If the text contents are grouped according to the structural path toward the root, and each group is compressed separately, compression ratios improve noticeably. This can be as simple as grouping texts that are under the same tag (i.e., considering only the deepest tree node containing the text) or as sophisticated as considering the full path toward the root. Texts can also be separated by data type (e.g., dates, numbers, etc.)
- Even if encoding tags and contents separately, they can be stored in the file in their original order, so that the document can be handled as a plain uncompressed document. These are *homomorphic* compressors. Alternatively, *nonhomomorphic* compressors store structure and

content separately, with some pointer information to reconstruct the tree. In this case, the structure pointers may help to point out relevant content to scan in the querying process.

Most tools that compress XML run on diverse combinations of these principles. Some of the most prominent examples are as follows:

- *XMill* was the first compressor separating structure and content. It uses dictionary compression for the tags, while text content and attribute values are grouped based on the rooted data path and their data type and then compressed independently. The sequences are finally passed to a back-end general text compressor.
- *Millau* is another early XML compressor that generates separate streams for structure and content. The structural one is encoded simply using WBXML, but it is optimized by taking the DTD as the base grammar. The content stream is compressed with a general text compressor.
- *XMLPPM* encodes the tokens and passes them to one of four different PPM models, depending on their syntactic meaning. To exploit correlations between different syntactic classes, *XMLPPM* injects previous symbols into the multiplexed models to be used as context.
- *XGrind* is a homomorphic compressor. It compresses the tags using dictionary encoding and uses different Huffman coders for the data content associated to each tag and attribute name.
- *XPress* is also homomorphic and applies different compression schemes based on token types, but tags are encoded regarding their full path to the root. Paths are mapped to intervals of real numbers so that the interval of a suffix of a path contains the interval of a path.
- *AXECHOP* produces a context-free grammar of the document structure (after tokenizing it). The grammar is then compressed with an adaptive arithmetic coder. Text containers are separately created according to the tag enclosing the texts and compressed with *bzip2*.

- *LZCS* is aimed at trees with repetitive topology. It converts the tree into a directed acyclic graph, by factoring common subtrees.
- *TinyT* uses a more powerful tree grammar compression, more specifically TreeRepair [12], which leads to very small representations on repetitive topologies.
- *XBzip* does not explicitly separate structure from content, but its construction automatically leads to a division based on tree paths. The compressor is based on the XBW transform [9], which succinctly represents labeled trees.

Other well-known XML compressors are *XCQ, SCM, XQueC, SCA, XComp, RNGzip, XWRT, QXT, XQZip, XSeq*, and *TREECHOP*. The appropriate references, as well as a more exhaustive coverage, can be found in recent surveys [7, 17].

Navigating and Searching in Compressed Form

The most popular retrieval operations on structured text are related to *navigating* the tree and to *searching* it. Compressors providing some kind of support of these operations are usually referred as *queryable* methods, in contrast with the *non-queryable* ones, that just aim to reduce the amount of space used. Navigating means moving from a node to its children, parent, and siblings. Searching means various *path-matching* operations such as finding all the paths where a node labeled A is the parent of another labeled B and that one is the ancestor of another labeled C, which in turn contains text where word W appears. A popular language combining navigation and searching operations is XPath (see http:// www.w3.org/TR/xpath20).

Several of the schemes above permit accessing and decompressing any part of the text at random positions. This is because they retain the original order of the components of the document and compress using a semi-static model (e.g., *XGrind* and *XPress*). Those compression methods are transparent, in the sense that the classical techniques to navigate and search XML data, sequentially or using indexes, can be used almost directly over this compressed representation. Other techniques, such as *XCQ*, allow random access under a slightly more complex scheme, because some work is needed in order to start decompression at a specific point. Finally, techniques based on adaptive compression (such as *XMLPPM*) usually achieve better compression ratios but need to decompress the whole data before they can operate on it. These are considered non-queryable representations. Other non-queryable compressors are *XMill*, *AXECHOP*, *XComp*, and *XWRT*.

Among the *queryable* solutions, some techniques take advantage of the separation between structure and content in order to run queries faster than scanning all the data. This is the case of *XCQ*, where the table that points from each different tree path to all the contents compressed under the corresponding model is useful to avoid traversing those contents if the path does not match a path-matching query. *XQueC* and *QXT* also work on a similar basic idea, but from different points of view. While *XQueC* focuses on query speed and query extent rather than compression efficiency, by creating several auxiliary data structures and indices, *QXT* aims at effective compression and does not keep any index; thus, it offers a more limited query support. Another example is *XPress*, which encodes paths in a way that the codes themselves permit checking containment between two paths and encodes numerical values in a way that allows directly performing range queries in compressed form.

Tree grammars may also support tree traversals without decompression. For instance, *LZCS* is aimed at compressing highly structured data, by replacing identical subtrees by a pointer to their first occurrence. The structure can be navigated almost transparently, and path-matching operations can be sped up by factoring out the work done on repeated substructures. *TinyT* uses a more general tree grammar compressor, and it efficiently handles navigation and some structural queries. Many such queries

can be solved by running a *tree automaton* on the tree [11].

The de facto XPath query language, however, requires much more than handling just traversals and a few path queries. In the rest of this entry, we will describe two recent approaches to support XPath functionality over succinct representations of XML data.

Succinct Encodings for Labeled Trees

Succinct representations of labeled trees are an algorithmic development that finds applications in navigating structured text in compressed form. In its simplest form, a general labeled tree of n nodes can be represented using a sequence P of $2n$ balanced parentheses and a sequence L of n labels (which correspond to tag names and will be regarded as atomic for simplicity). This is obtained by traversing the tree in preorder (i.e., first the current node and then recursively each of its children). As the tree is traversed, an opening parenthesis is added to P each time one goes down to a child and a closing parenthesis when going up back to the parent. That is called the *balanced parentheses* representation of a tree. In L, the labels are added in preorder.

Figure 1 shows an example representation of a labeled tree as a sequence of parentheses and labels in preorder. It is not hard to rebuild the tree from this representation. However, what is really challenging is to navigate the tree directly in this representation (where a node is represented by the position of its opening parenthesis).

An essential operation to achieve efficient navigation in compressed form is the *rank* operation on bitmaps: $rank(P, i)$ is the number of 1s (here representing opening parentheses) in $P[1, i]$. One immediate application of *rank* is to obtain the label of a given node i, as $L[rank(P, i)]$. For example, consider the second child of the root in Fig. 1. It is represented by the opening parenthesis at position 8 in the sequence. Its label is therefore $L[rank("(()())(()(()))", 8)] = L[5] = "C"$. Another application of *rank* is to compute the depth of a node i. This is the number of opening minus closing

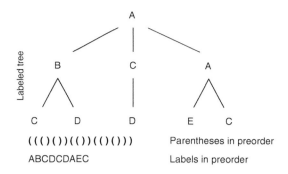

((()())(())(()()))   Parentheses in preorder

ABCDCDAEC   Labels in preorder

Managing Compressed Structured Text, Fig. 1
Succinct representation of a labeled tree, based on
parentheses and labels sequences

parentheses in $P[1, i]$, that is, $rank(P, i) - (i - rank(P, i)) = 2 \cdot rank(P, i) - i$. For example, the depth of the second child of the root is $2 \cdot rank(\text{``}((()())(())(()())\text{''}, 8) - 8 = 2 \cdot 5 - 8 = 2$. The dual of operation $rank$ is $select(P, i)$, which gives the position of the i-th 1 in P. This yields the tree node with preorder i or the tree node corresponding to label $L[i]$.

A large number of tree traversal and query operations can be supported with $rank$, $select$, and a few extra primitives: Operation $close(i)$ gives the position of the parenthesis that closes i (i.e., the next parenthesis with the same depth of i). Operation $enclose(i)$ gives the lowest parenthesis that contains i (i.e., the preceding parenthesis with depth smaller than that of i). All those operations can be solved efficiently with little extra space on top of P [1].

With these two operations, one can navigate the tree as follows. The next sibling of i is $close(i) + 1$ (unless it is a ')', in which case i is the last child of its parent). The first child of i is $i + 1$ unless $P[i + 1]$ is a ')', in which case i is a leaf and hence has no children. The parent of i is $enclose(i)$. The size of the subtree rooted at i is $(close(i) - i + 1)/2$. For example, consider the first child of the root in Fig. 1, such that $i = 2$. It finishes at $close(i) = 7$. Its next sibling is $close(i) + 1 = 8$, the node of the previous examples. Its first child is $i + 1 = 3$, the leftmost tree leaf. Its parent is $enclose(i) = 1$, the root. The size of its subtree is $(close(i) - i + 1)/2 = (7 - 2 + 1)/2 = 3$.

In order to enrich the navigation using the labels, sequence $L[1, n]$ may also be processed for symbol $rank$ and $select$ operations, where $rank_c(L, i)$ is the number of occurrences of c in $L[1, i]$ and $select_c(L, j)$ is the position of the j-th occurrence of c in L. Different sequence representations supporting this functionality exist. We refer the reader to a recent practical development [4].

For example, the following procedure finds all the descendants of node i which are labeled c: (i) Find the position $j = rank(P, i)$ of node i in the sequence of labels. (ii) Compute $k = rank_c(L, j - 1)$, the number of occurrences of c prior to j. (iii) Find the positions $p_r = select_c(L, k + r)$ of c from j onward, for successive r values until $select(P, p_r) > close(i)$, that is, until the answers are not anymore descendants of i. For example, consider again the first child of the root in Fig. 1, where $i = 2$ and $close(i) = 7$, and find its descendants labeled "D". The first step is to compute $j = rank(P, 2) = 2$, the position of its label in L. Now, $k = rank_{\text{``}D\text{''}}(L, 1) = 0$ tells that there are zero occurrences of "$D$" before $L[2]$. Now the next occurrences of "$D$" in $L$ are found as $select_{\text{``}D\text{''}}(L, 1) = 4$, $select_{\text{``}D\text{''}}(L, 2) = 6$, ... The first such occurrence is mapped to the tree node $select(P, 4) = 5$ (the second tree leaf), which is within the subtree of i because $i \leq 5 \leq close(i)$. The second occurrence of "D" is already outside the tree because $select(P, 6) = 9$ exceeds $close(i) = 7$.

The XPath language integrates path matching and content queries. *SXSI* [2] is a recent system that integrates the encoding of labeled trees just described with a compressed *self-indexed* representation of the collection of text nodes in the XML document. A self-index [14] is a representation of a text that not only allows accessing arbitrary text passages but also supports efficient substring searches on it. SXSI addresses a relevant subset of XPath by combining queries on the parentheses sequence, $rank/select$ operations on the sequence of labels, and pattern matching queries on the text. Queries are solved by using a tree automaton that traverses the XML structure using the described representations to direct the navigation toward just the relevant parts of the structure.

Different realizations of SXSI fit different scenarios. In the original proposal [2], they aim at speed; thus, they use the so-called "fully functional" parentheses representation [1], an FM-index [8] for the text contents, and a plain representation of the tag sequence L for fast access, plus one bitmap B_c for each tag c, with $B_c[i] = 1$ iff $L[i] = c$; thus, $rank_c(L, i) = rank(B_c, i)$ and $select_c(L, i) = select(B_c, i)$ are solved efficiently. The resulting index uses nearly the same space of the original XML data, and its query efficiency is competitive with non-compressed representations, such as that of *MonetDB*.

Another realization [16] aims at greatly reducing space on repetitive XML collections (e.g., versioned collections of XML data), where grammar compression stands out. Tree-grammar-based representations for the structure are not powerful enough to support the XPath operations. Instead, they use a grammar-compressed representation of the parentheses sequence, enriched with data that supports the described navigation operations [15]. This is complemented with a new grammar-compressed representation for L that supports *rank/select* operations. Finally, a self-index aimed at repetitive text collections, called *run-length compressed suffix array* [13], is used. On real-life highly repetitive collections (e.g., versioned software repositories), the final representation is much smaller than the original data (e.g., 25%, dominated by the text component). In exchange, XPath queries are noticeably slower.

Integrating Indexing and Compression

We have mentioned self-indexing as the concept of representing text data in compressed form so that the representation itself is queryable, for pattern-matching queries in that case. This concept can be translated into structured text compression, so that the representation of the XML data is itself an index that supports fast XPath queries. The first representation that built on this idea [9] was the basis of *XBzip*, but its query support is limited to very simple path-matching queries. In this final section, we describe a more

recent development, *XXS* [6], which supports a large subset of XPath, yet it is limited to XML collections containing natural language. *XXS* represents the XML collections in 35%–50% of their original space and is competitive in time efficiency with *SXSI* and *MonetDB*, which use much more space.

XXS represents the XML collection using a data structure called the XML Wavelet Tree (XWT). The XWT representation of a document separates the tokens into four categories: (i) start/end tags, (ii) attribute names, (iii) comments and processing instructions, and (iv) text content and attribute values. The words of each vocabulary are statistically encoded using a byte-wise representation called (s, c)-DC [5], which is almost as good as byte-wise Huffman codes but more flexible. This encoding is tailored to make XWT suitable for querying: the codewords of the vocabularies (i), (ii), and (iii), called *special*, are forced to start with specific byte values.

Instead of writing down the codes one after the other, the bytes of all the codewords are reordered and arranged into the XWT nodes. The root of the tree (i.e., the first level) contains the first bytes of the codewords, in the same order as in the XML collection. Then, each node Bx in the second level stores the second bytes of the codewords whose first byte is b_x (preserving again the text order). That is, the second byte corresponding to the j-th occurrence of byte b_x in the root is located at position j in node Bx, and so on. Operations *rank/select* on the byte sequences are sufficient to efficiently access and query the XWT.

Figure 2 shows an example XWT built from an XML document. Note that the combination of the XWT arrangement, combined with the usage of specific first bytes to encode the special vocabularies, isolates those special words below some XWT nodes. This yields several benefits. For instance, attribute isolation provides direct access to this type of tokens during query evaluation, while keeping comments and processing instructions separated allows skipping those fragments in general text searches. Yet, even more important is the implicit structural isolation. Note that the subtree below node $B3$ in Fig. 2 stores only start/end tags and that they follow the document

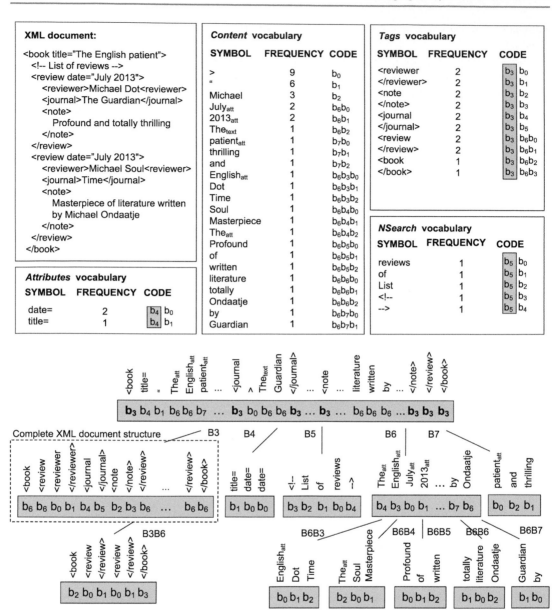

Managing Compressed Structured Text, Fig. 2 Example of an XWT structure built from an XML document

order. Hence, the root of this subtree (i.e., node *B*3) actually matches a balanced parentheses representation of the XML document structure. Both structures are used in conjunction by *XXS* to solve queries.

Instead of tree automata, *XXS* uses the so-called *bottom-up* query evaluation [10], where the leaves of the query syntax tree are solved first, and they feed the data to the higher nodes.

The evaluation process is built on the following principles: (i) map subtrees to segments in a line to facilitate structural comparisons, (ii) use *lazy* evaluation to produce only the necessary data, and (iii) a *skipping* strategy that propagates restrictions from the top query nodes to the bottom nodes, thus avoiding processing unnecessary parts of the tree. The XWT representation is used to implement these tasks efficiently.

Key Applications

Any application managing structured text, particularly if it has to transmit it over slow channels or operate within limited fast memory, even if there is an unlimited supply of slower memory, benefits from these techniques.

Future Directions

The future of the area is likely to be in manipulating XML in compressed form, and in this aspect, much more than XPath is needed. Note that the compressed representations we have considered are static. Languages like XQuery (www.w3.org/TR/xquery), which query but also transform the XML data, are much more demanding than XPath, in particular because manipulating XML requires generating new data as the result of queries.

Experimental Results

Experiments can be found in the cited papers. The most recent ones are those of *SXSI* [2] and *XXS* [6].

Url to Code

Several public XML compressors are available, for example, *XMill* (https://sourceforge.net/projects/xmill), *XMLPPM* (https://sourceforge.net/projects/xmlppm), *LZCS* (http://www.infor.uva.es/~jadiego/download.php), *XGrind* (https://sourceforge.net/projects/xgrind), *XWRT* (http://xwrt.sourceforge.net/), *XBzipIndex* (http://pages.di.unipi.it/ferragina/software.html), *SXSI* (http://fclaude.recoded.cl/archives/193), and *XXS* (http://vios.dc.fi.udc.es/xxs/).

Cross-References

► Semi-structured Data
► XML
► XPath/XQuery

Recommended Reading

1. Arroyuelo D, Cánovas R, Navarro G, Sadakane K. Succinct trees in practice. In: Proceedings of the 11th Workshop on Algorithm Engineering and Experiments; 2009. p. 84–97.
2. Arroyuelo D, Claude F, Maneth S, Mäkinen V, Navarro G, Nguyen K, Sirén J, Välimäki N. Fast in-memory XPath search using compressed indexes. In: Proceedings of the 26th International Conference on Data Engineering; 2010. p. 417–28.
3. Baeza-Yates R, Navarro G. Integrating contents and structure in text retrieval. ACM SIGMOD Rec. 1996;25(1):67–79.
4. Barbay J, Claude F, Gagie T, Navarro G, Nekrich Y. Efficient fully-compressed sequence representations. Algorithmica. 2014;69(1):232–68.
5. Brisaboa NR, Fariña A, Navarro G, Paramá JR. Lightweight natural language text compression. Inf Retr. 2007;10(1):1–33.
6. Brisaboa NR, Cerdeira-Pena A, Navarro G. XXS: efficient XPath evaluation on compressed XML documents. ACM Trans Inf Syst. 2014; 32(3):13.
7. Cerdeira-Pena A. Compressed self-indexed XML representation with efficient XPath evaluation. PhD thesis, Department of Computer Science, University of A Coruña, 2013.
8. Ferragina P, Manzini G. Indexing compressed text. J ACM. 2005;52(4):552–81.
9. Ferragina P, Luccio F, Manzini G, Muthukrishnan S. Compressing and indexing labeled trees, with applications. J ACM. 2009;57(1):4:1–4:33.
10. Gottlob G, Koch C, Pichler R. Efficient algorithms for processing XPath queries. ACM Trans Database Syst. 2005;30(2):444–91.
11. Lohrey M, Maneth S, Mennicke R. The complexity of tree automata and XPath on grammar-compressed trees. Theor Comput Sci. 2006;363(2):196–210.
12. Lohrey M, Maneth S, Mennicke R. XML tree structure compression using RePair. Inf Syst. 2013;38(8):1150–67.
13. Mäkinen V, Navarro G, Sirén J, Välimäki N. Storage and retrieval of highly repetitive sequence collections. J Comput Biol. 2010;17(3):281–308.
14. Navarro G, Mäkinen V. Compressed full-text indexes. ACM Comput Surv. 2007;39(1):2.
15. Navarro G, Ordóñez A. Faster compressed suffix trees for repetitive text collections. In: Proceedings of the 13th International Symposium on Experimental Algorithms; 2014. p. 424–35.
16. Navarro G, Ordóñez A. Grammar compressed sequences with rank/select support. In: Proceedings of the 21st International Symposium on String Processing and Information Retrieval; 2014.
17. Sakr S. XML compression techniques: a survey and comparison. J Comput Syst Sci. 2009;75(5):303–22.

M

Managing Data Integration Uncertainty

Xin Luna Dong[1] and Alon Halevy[2,3]
[1]Amazon, Seattle, WA, USA
[2]The Recruit Institute of Technology, Mountain View, CA, USA
[3]Google Inc., Mountain View, CA, USA

Synonyms

Probabilistic schema alignment

Definition

Consider a set of source schemas $S = \{S_1, \ldots, S_n\}$ in the same domain, where different schemas may describe the domain in different ways. An important component in data integration is *schema alignment*, including three steps: (1) creating a *mediated schema M* that provides a unified and virtual view of the disparate sources and captures the salient aspects of the domain being considered, (2) generating *attribute matching* that matches attributes in each source schema $S_i, i \in [1, n]$, to the corresponding attributes in the mediated schema M, and (3) building a *schema mapping* between each source schema S_i and the mediated schema M to specify the semantic relationships between the contents of the source and that of the mediated data. The result schema mappings are used to reformulate a user query into a set of queries on the underlying data sources for query answering. Uncertainty can arise in every step of schema alignment: it may not be clear what is the best way to model a domain; the semantics of an attribute from a source may be fuzzy; the semantic mapping between sources may be unclear. Managing the uncertainty requires modeling uncertainty as a first-class citizen in schema mappings and incorporating the uncertainty in query answers returned to the users.

Historical Background

A data integration system relies on the schema mappings between the data sources and the mediated schema for query reformulation. However, it is well known that creating and maintaining such mappings is nontrivial and requires significant resources, upfront effort, and technical expertise. From its early days, the goal of data integration has often been to integrate tens to hundreds of data sources created independently in an organization. Semiautomatic tools have been created to simplify schema alignment, but domain experts still often need to get heavily involved in refining the automatically generated mappings. As a result, schema alignment is one of the major bottlenecks in building a data integration system.

More recently, data integration projects involve integrating structured data from the Web, either in the form of Deep Web data or in the form of Web tables or Web lists. A huge number of data sources exist, and the schemas of the data can keep changing. As a result, having perfect schema mappings and keeping them in sync with the constantly evolving source schemas are infeasible. It is thus critical to manage uncertainty as an integral part of data integration.

Managing data integration uncertainty is based on the idea of providing a *pay-as-you-go* data management system (often referred as a *Dataspace support platform*) that takes a data *co-existence* approach and emphasizes best-effort results. The system provides some services from the outset and evolves the schema mappings between the different sources on an as-needed basis [1]. Given a query, such a platform generates best-effort (or, approximate) answers from data sources where perfect schema mappings do not exist. When it discovers a large number of sophisticated queries or data mining tasks over certain sources, it will guide the users to make additional efforts to integrate those sources more closely. Handling uncertainty is a core part for providing best-effort services on dataspaces.

Data integration uncertainties are addressed in two ways. First, a *probabilistic mediated schema* is built to capture the uncertainty on how to model the domain. Each possible schema in the probabilistic mediated schema represents one way of clustering the source attributes, where attributes in the same cluster is considered as having the same semantic meaning [2].

Second, a *probabilistic schema mapping* is built between each source schema and each possible mediated schema in the probabilistic mediated schema to capture the uncertainty on the semantics of the attributes and of the schema. Each possible mapping in the probabilistic schema mapping represents one way of matching source attributes and the attribute clusters in the mediated schema [3].

Scientific Fundamentals

Consider the setting in which each of the sources is a single relational table; as a result, the schema mapping can be easily inferred from the attribute matchings. Figure 1 shows the three major components in managing uncertainty in data integration: probabilistic mediated schema, probabilistic schema mappings, and query answering in this context. Hereafter, each component is briefly summarized for this simplified setting.

Probabilistic mediated schema: Consider automatically inferring a mediated schema from a set of data sources; in the context where each schema contains a single relational table, the mediated schema can be thought of as a *clustering* of source attributes, where similar attributes are grouped into the same cluster. Formally, denote the attributes in schema $S_i, i \in [1, n]$, by $\mathbf{A}(S_i)$, and the set of all source attributes as $\mathcal{A} = \mathbf{A}(S_1) \cup \cdots \cup \mathbf{A}(S_n)$. The mediated schema for the set of sources $\{S_1, \ldots, S_n\}$ can be denoted by $M = \{\mathbf{A}_1, \ldots, \mathbf{A}_m\}$, where each $\mathbf{A}_i, i \in [1, m]$, is called a *mediated attribute*. The mediated attributes are *sets* of attributes from the sources, i.e., $\mathbf{A}_i \subseteq \mathcal{A}$, and $\mathbf{A}_i \cap \mathbf{A}_j = \emptyset$ for each $i, j \in [1, m], i \neq j$. If a query contains an attribute $a \in \mathbf{A}_i, i \in [1, m]$, then when answering the query, a is replaced everywhere with $\mathbf{A}_i$.

Because of the heterogeneity of the data sources being integrated, uncertainty exists on the semantics of the source attributes and in turn on the clustering. A *probabilistic mediated schema (p-med-schema)* describes a probability distribution on a set of possible mediated schemas, where the probability of a mediated schema indicates the likelihood that the schema correctly describes the domain of the sources. Formally, the p-med-schema can be denoted by

$$\mathbf{M} = \{(M_1, Pr(M_1)), \ldots, (M_l, Pr(M_l))\}, \text{ where } Pr(M_i) \in (0, 1] \text{ and } \Sigma_{i=1}^{l} Pr(M_i) = 1.$$

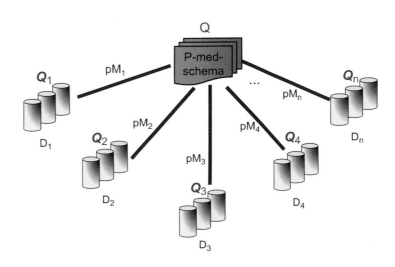

Managing Data Integration Uncertainty, Fig. 1 Query answering in a probabilistic data integration system

Probabilistic schema mapping: The goal of a schema mapping is to specify the semantic relationships between a *source schema* $S = \langle s_1, \ldots, s_m \rangle$ and a *target schema* $T = \langle t_1, \ldots, t_n \rangle$ (in the context of data integration, the mediated schema can be considered as the target schema). Consider the special case where the schema mapping contains only one-to-one matching, denoted by m, between attributes in S and attributes in T.

A *probabilistic schema mapping (p-mapping)* describes a probability distribution on a set of *possible* schema mappings between a source schema and a target schema. Formally, a p-mapping can be denoted by $pM = (S, T, \mathbf{m})$, where $\mathbf{m} = \{(m_1, Pr(m_1)), \ldots, (m_l, Pr(m_l))\}$, such that $Pr(m_i) \in (0, 1]$ and $\sum_{i=1}^{l} Pr(m_i) = 1$.

In addition to probabilistic schema mappings as stated previously, other probabilistic models have been proposed for managing uncertainty in schema semantics, such as assigning a probability for matching of every pair of source and target attributes [4].

Query answering: With a traditional schema mapping m, for each source instance D_S, there can be an infinite number of target instances that are consistent with D_S and m. Let $\mathbf{T}_m(D_S)$ be the set of all such target instances. The set of answers to a query Q is the intersection of the answers on all instances in $\mathbf{T}_m(D_S)$, which are called *certain answers* of Q.

Intuitively, a probabilistic schema mapping models the uncertainty about which of the mappings in pM is the correct one. To incorporate the uncertainty in answers, a probability is assigned to every answer tuple to represent its likelihood of being a certain answer. When a schema alignment system produces a set of candidate schema mappings, there are two ways to interpret the uncertainty: (1) a single mapping in pM is the correct one and it applies to all the data in D_S, or (2) several mappings are partially correct and each is suitable for a subset of tuples in D_S, though it is not known which mapping is the right one for a specific tuple. The semantics of query answering has been defined under both interpretations. The first interpretation is referred to as the *by-table*

semantics, and the second one is referred to as the *by-tuple* semantics of probabilistic mappings. Note that one cannot argue for one interpretation over the other; the needs of the application should dictate the appropriate semantics.

Specifically, a p-mapping pM describes a set of possible worlds, each with a possible mapping $m \in pM$. In by-table semantics, a source table can fall in one of the possible worlds; that is, the possible mapping associated with that possible world applies to the whole source table. Certain answers are generated with respect to each possible mapping; the probability of an answer t is the sum of the probabilities of the mappings w.r.t. which t is deemed to be a certain answer.

In by-tuple semantics, different tuples in a source table can fall in different possible worlds; that is, different possible mappings associated with those possible worlds can apply to different source tuples. Thus, a consistent target instance is defined by a *mapping sequence* that assigns a (possibly different) mapping in $\mathbf{m}$ to each source tuple in D_S (without losing generality, in order to compare between such sequences, some order is assigned to the tuples in the instance). Certain answers are generated with respect to each mapping sequence; the probability of an answer t is the sum of the probabilities of the mapping sequence w.r.t. which t is deemed to be a certain answer.

Answering queries under by-table semantics is conceptually simple: one can compute the certain answers of a query Q under each of the mappings $m \in \mathbf{m}$, attach the probability $Pr(m)$ to every certain answer under m, and add up the probabilities of the different mappings for each answer tuple t. Answering queries under by-tuple semantics requires computing the certain answers for every *mapping sequence* generated by pM. However, the number of such mapping sequences is exponential in the size of the input data. It has been proved that in general, answering SPJ queries in by-table semantics is in PTIME in the size of the data and the mapping, whereas in by-tuple semantics, the problem is #P-complete in the size of the data and in PTIME in the size of the mapping. Later work has discussed complexity of answering queries with aggregate operators [5].

Finally, when a p-med-schema exists and for each possible mediated schema there exists a p-mapping, query answering proceeds in two steps. First, it answers the query with respect to each possible mediated schema. Second, for each answer tuple, it computes the final probability as the sum of its probabilities under each possible mediated schema, weighted by the probabilities of the mediated schemas.

Key Applications

Dataspace Systems

The key idea of a dataspace systems is to provide best-effort services on heterogeneous data sets, and this requires handling uncertainty at various levels. Das Sarma et al. [2] shows empirically that using p-med-schemas and p-mappings allows quickly building a data integration system and providing high-quality answers.

Probabilistic Data Warehousing

P-med-schemas and p-mappings can be applied in building a data warehouse from a large number of heterogeneous data sources. The result is a probabilistic database where the probabilities represent uncertainty on the semantics of the data.

Probabilistic Knowledge Base Construction

Knowledge base construction involves knowledge extraction, schema alignment, entity reconciliation, and knowledge cleaning, so it shares a lot of common tasks with data integration. Being able to manage the many types of uncertainties is at the core of constructing a knowledge base that balances quality and coverage. Recent efforts of building a probabilistic knowledge base [6] share many insights with probabilistic data integration.

Cross-References

▶ Probabilistic Databases

Recommended Reading

1. Franklin M, Halevy AY, Maier D. From databases to dataspaces: a new abstraction for information management. Sigmod Rec. 2005;34(4):27–33.
2. Sarma AD, Dong XL, Halevy A. Bootstrapping pay-as-you-go data integration systems. In: Proceedings of the ACM SIGMOD International Conference on Management of Data; 2008. p. 861–74.
3. Dong X, Halevy AY, Yu C. Data integration with uncertainties. In: Proceedings of the 33rd International Conference on Very Large Data Bases; 2007. p. 687–98.
4. Gal A, Anaby-Tavor A, Trombetta A, Montesi D. A framework for modeling and evaluating automatic semantic reconciliation. VLDB J. 2003;14:50–67.
5. Gal A, Martinez MV, Simari GI, Subrahmanian VS. Aggregate query answering under uncertain schema mappings. In: Proceedings of the 25th International Conference on Data Engineering; 2009. p. 940–51.
6. Dong XL, Gabrilovich E, Heitz G, Horn W, Lao N, Murphy K, et al. Knowledge vault: a web-scale approach to probabilistic knowledge fusion. In: Proceedings of the 20th ACM SIGKDD International Conference on Knowledge Discovery and Data Mining; 2014.

M

Managing Probabilistic Entity Extraction

Daisy Zhe Wang
Computer and Information Science and Engineering (CISE), University of Florida, Gainesville, FL, USA

Synonyms

Probabilistic databases; Probabilistic information extraction; Probabilistic knowledge bases

Definition

Entity extraction is the process of extracting structured entities with corresponding attributes from unstructured text data. For example, a structured paper entity can be extracted from a citation with corresponding author names, title, and journal names. Alternatively, a professor entity can be extracted from his or her homepage with corresponding job title, email, and research interests. The result of entity extraction is a set of structured entity records.

Probabilistic entity extractions are structured entity attributes and records extracted from text each associated with probability of correctness. The probability of correctness is usually generated from the state-of-the-art statistical information extraction models due to the imperfect nature of automatic entity extraction process.

The management of probabilistic entity extractions requires not only scalable execution of statistical information extraction from unstructured text data but also probabilistic entity resolution, inference, and querying. A management system over probabilistic entity extractions should support efficient processing and optimization over a probabilistic data model and statistical inference operators. Such systems are sometimes called probabilistic knowledge bases.

Historical Background

Automatically extracting structured entities from instructed text is a challenging problem and has a long history of attempts spanning early rule-based systems like Rapier and GATE to the newer statistical methods like conditional random fields and hidden Markov models implemented in open-source natural language processing packages such as NLTK, LingPipe, and Stanford parser [4].

In the database community, work on information extraction (IE) has centered on two major architectural themes. First, there has been interest in the design of declarative languages and systems to easily specify, optimize, and execute IE tasks [2, 5, 6]. Second, IE has been a primary motivating application for the groundswell of work on probabilistic database systems (PDBS) [7], which can model the uncertainty inherent in IE outputs and enable users to write declarative queries that deal with such uncertainty. The BayesStore [8] merged these two ideas into a single architecture and build a unified database system that provides a query-oriented language for specifying, optimizing,

and executing IE tasks, and that supports a principled probabilistic framework for querying the probabilistic entities generated from IE.

Foundations

Extracting Probabilistic Entities

Information extraction (IE) is one type of text analysis that extracts entities, such as person names, companies, and events, from the text of news articles, emails, and tweets. This is done by tagging the tokens in the text with labels (e.g., location, event), which are structural information (i.e., metadata) over the text. In other words, the goal of IE is to automatically extract structured information from unstructured text. For example, information extraction tools, such as Stanford parser, can extract the organization *NFL*, the event *Super Bowl*, the locations *Meadowlands Stadium*, *the Big Apple*, *New York*, and the sport teams *Jets* and *Giants* from the following piece of text.

NFL owners voted to put the 2014 Super Bowl in the new $1.6 billion Meadowlands Stadium in the Big Apple, home to the New York Jets and Giants.

In addition to entities, attributes and relationships can also be extracted using similar IE techniques. For example, a *cost* attribute can be extracted for entity *Meadowlands Stadium* with value $1.6 *billion*, and the *located in* relationship can be extracted between entities *Meadowlands Stadium* and *the Big Apple*.

State-of-the-art IE over text is performed using probabilistic models and probabilistic inference algorithms from the statistical IE literature [4]. A probabilistic model is learned from the training text corpus, which contains human-annotated labels of entities, attributes, and relationships. The probabilistic model encodes (1) a probability distribution of all possible extractions (i.e., label sequences) that could be given to a piece of text and (2) the statistical correlation between the token and the corresponding label assigned to it. For example, if the word is *NFL*, it is highly likely that it should be labeled as an organization. The model also encodes the correlation between adjacent tokens and labels in the text sequence. For example, given the following token *Bowl*, it is more likely that *Super* is part of the event name *Super Bowl*.

Probabilistic Data Model

A probabilistic data model treats uncertain text extractions (i.e., labels) and probabilistic model of uncertainty as first-class citizens in a probabilistic database. Abstractly, a probabilistic entity extraction database $\mathcal{DB}^p = <\mathcal{R}, F>$ consists of two key components: (1) a collection of *incomplete relations* $\mathcal{R}$ and (2) a *probability distribution function* F that quantifies the uncertainty associated with all incomplete relations in $\mathcal{R}$.

An incomplete relation $R \in \mathcal{R}$ is defined over a schema $\mathcal{A}^d \cup \mathcal{A}^p$ comprising a (nonempty) subset $\mathcal{A}^d$ of *deterministic attributes* (that includes all candidate and foreign key attributes in R) and a subset $\mathcal{A}^p$ of *probabilistic attributes*. Deterministic attributes have no uncertainty associated with any of their values – they always contain a legal value from their domain, or the traditional SQL *NULL*. On the other hand, the values of probabilistic attributes may be present or *missing* from R. Given a tuple $t \in R$, non-missing values for probabilistic attributes are considered *evidence*, representing our partial knowledge of t. Missing values for probabilistic attribute $A_i \in \mathcal{A}^p$ capture attribute-level uncertainty; formally, each such missing value is associated with an RV X_j ranging over $\mathsf{dom}(A_i)$ (the domain of attribute A_i). In addition, *tuple-existence uncertainty* can be represented using a probabilistic Boolean attribute $\mathtt{Exist}^p$ to capture the uncertainty of each tuple's *existence* in an incomplete relation. Existence uncertainty can also arise during query processing over probabilistic extraction tables.

The second component of a probabilistic entity extraction database $\mathcal{DB}^p = <\mathcal{R}, F>$ is a probability distribution function F that models the *joint distribution* of all missing value RVs in extraction relations in $\mathcal{R}$. Thus, assuming n such RVs, $X_1, \ldots, X_n$, F denotes the joint probability distribution $\mathsf{Pr}(X_1, \ldots, X_n)$. The association of F with conventional PDB *possible-worlds semantics* [1] is now straightforward:

every complete assignment of values to all X_i's maps to a *single* possible world for $\mathcal{DB}^p$. From our definition of F, it is evident that it is a potentially huge mathematical object – the straightforward method for representing and storing this n-dimensional distribution F is essentially linear in the number of possible worlds.

Data Model for Probabilistic Entity Extraction

Although IE has been cited as a key driving application for probabilistic database research, there has been little work on effectively supporting the probabilistic IE process, storing the probabilistic IE outputs, and executing queries over them in a PDB. Gupta and Sarawagi's work [3] presented a mapping from probabilistic IE outputs to a PDB, but it showed that with a probabilistic database, which requires independent assumptions or supports only limited probabilistic correlations, the probability distribution over the possible extractions from an IE model can only be stored approximately.

We describe an instance of the probabilistic data model described in the last section that can support a state-of-the-art statistical IE model – conditional random fields (CRF) by (1) storing text strings in an incomplete relation R as an inverted file with a probabilistic label^p attribute and (2) representing the CRF model that encodes the probability distribution over all possible extractions.

The token table TOKENTBL is an incomplete relation R, which stores text strings as relations in a database, in a manner akin to the inverted files commonly used in information retrieval. Each tuple in TOKENTBL records a unique occurrence of a token, which is identified by the text string ID (strID) and the position (pos) the token is taken from. A TOKENTBL has the following schema:

TOKENTBL (strID, pos, token, label^p)

The TOKENTBL contains one probabilistic attribute – label^p, which can contain missing values, whose probability distribution can be computed from the CRF model. The deterministic attributes of the TOKENTBL are populated by parsing the input text strings $\mathcal{D}$, with label values marked as missing by default. Figure 1a shows the CRF model over an address string which encodes the second component: probabilistic distribution F over R in $\mathcal{DB}^p$.

Probabilistic Entity Table

The statistical IE methods and probabilistic extractions generate probabilistic entities. Depending on the type of text, the format of the probabilistic entity extractions is different. One type is text strings which usually exist in text fields in a database, each representing a cohesive entity with different attributes. Examples include address, product description, and bibliography strings. Entity extraction over each such text string can generate a unique entity with fixed number of attributes. Each text string is parsed into one probabilistic record in an *entity table*.

An entity table contains a set of probabilistic attributes, one for each possible label. In this case, each tuple in the entity table has an independent distribution, defined by the CRF model over the corresponding text string. Figure 2 shows the maximum-likelihood (ML) view of the entity table with three address strings.

The entity table is defined and generated by a *pivot operation* over the possible labelings in the text token sequence and their distribution. For each possible labeling of a text string, the pivot operation generates one possible world of the corresponding record in the entity table. For example, the labeling corresponding to label-sequence $y1$ in Fig. 1b generates the first tuple in the ML view of the probabilistic entity table in Fig. 2.

The second and more general type of text is sentences exist in large amounts of articles, emails, webpages, and social media. Entity extraction over such text usually generate triples with (subject, verb, object), where each triple represent an extracted attribute of a specific entity of relationship between two entities. For example, *Meadowlands Stadium costs 1.6 billion* is an extracted attribute and *New York Jets is located in the Big Apple* is an extracted relationship, respectively.

In this case, each text document is parsed into a set of probabilistic facts in a probabilistic *triple table*. Figure 3 shows a snippet of a triple table that contains all the facts (i.e., attributes and relationships) extracted from the Web. The triple table is a document-oriented view of the probabilistic extractions. A pivot operation can

Managing Probabilistic Entity Extraction, Fig. 1
(**a**) Example CRF model for an address string; (**b**) two possible entity extractions in the form of label sequences $y1$, $y2$

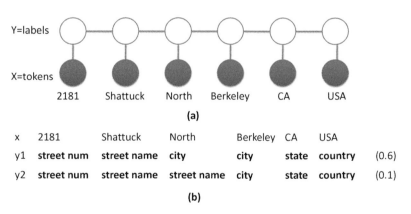

x	2181	Shattuck	North	Berkeley	CA	USA	
y1	**street num**	**street name**	**city**	**city**	**state**	**country**	(0.6)
y2	**street num**	**street name**	**street name**	**city**	**state**	**country**	(0.1)

(b)

strID	apt. num^p	street num^p	street name^p	city^p	state^p	country^p
1	null	2181	Shattuck	North Berkeley	CA	USA
2	12B	331	Fillmore St.	Seattle	WA	USA
3	224B	null	Ford South St.	Louis	MO	USA

Managing Probabilistic Entity Extraction, Fig. 2 An example maximum-likelihood view of the probabilistic entity table extracted from address text strings

Managing Probabilistic Entity Extraction, Fig. 3 Sample triples of extracted facts from the Web

subject	verb	attrib/object
Barack Obama	wasBornIn	Hawaii
Barack Obama	isMarriedTo	Michelle Obama
Barack Obama	isLeaderOf	United States
Barack Obama	isLeaderOf	New York Air National Guard
Stephen Harper	wasBornIn	Toronto
Stephen Harper	isMarriedTo	Laureen Harper
Stephen Harper	isLeaderOf	Canada
Stephen Harper	isLeaderOf	Calgary

entity	wasBornIn	isMarriedTo	isLeaderOf
Barack Obama	Hawaii	Michelle Obama	{United States, New York Air National Guard, ...}
Stephen Harper	Toronto	Laureen Harper	{Canada, Calgary}
Larry Page	Lansing, Michigan	Lucinda Southworth	{Google, Google Search, ...}
Bernie Machen	Greenwood, Mississippi	Chris Ackerman	University of Florida

Managing Probabilistic Entity Extraction, Fig. 4 An entity table generated from sample triples

be performed over the triple table to generate an *entity table* with one record for each entity with a fixed set of attributes and relationships. Figure 4 shows an example probabilistic entity table generated from the triple fact table via pivoting and aggregation operations.

Querying Probabilistic Entities via Inference

There are two families of queries we consider computing on the extracted data in the entity tables. Given that the ML view of an entity table can be defined as

```
CREATE VIEW entityTbl1-ML as
    SELECT *, rank() OVER
        (ORDER BY prob(*)
            DESC) r
    FROM    entityTbl1
    WHERE   r = 1;
```

one family is the set of deterministic SPJ queries over the ML views of the entity tables. An example of a selection query over the ML view of an entity table is to find all the *Address* tuples whose *streetname* contains "Sacramento" in the ML extraction:

```
SELECT *
FROM Address-ML
WHERE streetname like
    '%Sacramento%'
```

Inference operators computing maximum-likelihood values, such as Viterbi algorithm, can be used to answer the above queries along with relational operators. In some cases, pushing selection and join conditions before or into the inference operators can achieve better performance.

The other family computes the top-k results of probabilistic SPJ queries over entity tables using the set of possible worlds (i.e., segmentations) induced by the CRF distribution. Optionally, a prescribed threshold can be applied to the result probabilities. The queries in this family have the following general form, where SQLQuery is a view using a standard SPJ query:

```
SELECT *, rank() OVER
    (ORDER BY prob(*) DESC) r
FROM    SQLQuery
WHERE   r <= k [ AND prob(*)
    > threshold ]
```

The following query computes the top-1 result of probabilistic selection with condition *streetname contains 'Sacramento'* over the probabilistic *Address* entity table:

```
SELECT *, rank() OVER
    (ORDER BY prob(*) DESC) r
FROM ( SELECT *
        FROM Address
        WHERE streetname
```

M

```
                like '%Sacramento%'
) as Address
WHERE r = 1 AND prob(*)
   > threshold
```

As another example, the following query computes the top-1 (ML) result of a probabilistic join operation over the probabilistic entity tables *Address* and *Company*, with an equality join condition on *city*.

```
SELECT *, rank() OVER
   (ORDER BY prob(*) DESC) r
FROM (  SELECT *
        FROM Address A,
          Company C
        WHERE A.city =
          C.city
) as AddrComp   WHERE r = 1
```

Inference operators computing top-K and marginal probability distribution such as Markov-chain Monte Carlo (MCMC) algorithms can be used to answer the above queries. The relational operators add additional factors in the original probabilistic model representing F in $\mathcal{DB}^p$ [8]. Further optimizations can be achieved by using incremental inference algorithm and by choosing the most efficient inference algorithm based on data and model statistics.

Key Applications

First application domain is automatic knowledge base construction, which can be either general knowledge bases such as DBpedia and YAGO or domain-specific knowledge bases such as UMLS in the biomedical domain. Such knowledge bases can support more sophisticated information retrieval and integration.

Another application domain is situational awareness during different situation including disasters and protest events. Different from the first application where the entities are extracted from a relatively static or slow changing text corpus, this application requires extraction and management systems over streaming data.

Cross-References

▶ Probabilistic Databases

Recommended Reading

1. Dalvi N, Suciu D. Efficient Query Evaluation on Probabilistic Databases. In: Proceedings of the 30th International Conference on Very Large Data Bases; 2004.
2. Doan A, Ramakrishnan R, Chen F, DeRose P, Lee Y, McCann R, Sayyadian M, Shen W. Community information management. 2006.
3. Gupta R, Sarawagi S. Curating probabilistic databases from information extraction models. In: Proceedings of the 32nd International Conference on Very Large Data Bases; 2006.
4. Manning CD, Schütze H. Foundations of statistical natural language processing. Cambridge, MA: MIT Press; 1999.
5. Reiss F, Raghavan S, Krishnamurthy R, Zhu H, Vaithyanathan S. An algebraic approach to rule-based information extraction. In: Proceedings of the 24th International Conference on Data Engineering; 2008.
6. Shen W, Doan A, Naughton J, Ramakrishnan R. Declarative information extraction using datalog with embedded extraction predicates. In: Proceedings of the 33rd International Conference on Very Large Data Bases; 2007.
7. Suciu D, Olteanu D, Ré C, Koch C. Probabilistic databases, synthesis lectures on data management. San Rafael: Morgan and Claypool; 2011.
8. Wang D, Michelakis E, Garofalakis M, Hellerstein J. BayesStore: managing large, uncertain data repositories with probabilistic graphical models. In: Proceedings of the 33rd International Conference on Very Large Data Bases; 2008.

Mandatory Access Control

Bhavani Thuraisingham
The University of Texas at Dallas, Richardson, TX, USA

Synonyms

Multilevel security

Definition

As stated in [1], *"in computer security, 'mandatory access control (MAC)' refers to a kind of access control defined by the National Computer Security Center's Trusted Computer System Evaluation Criteria (TCSEC) as a means of restricting access to objects based on the sensitivity (as represented by a label) of the information contained in the objects and the formal authorization (i.e., clearance) of subjects to access information of such sensitivity."* With operating systems, the subjects are processes and objects are files. The goal is to ensure that when a subject accesses a file, no unauthorized information is leaked.

Key Point

MAC Models: MAC models were developed initially for secure operating systems mainly in the 1970s and early 1980s, and started with the Bell and La Padula security model. This model has two properties: the simple security property and the *-property (pronounced the star property). The simple security property states that a subject has read access to an object if the subject's security level dominated the level of the object. The *-property states that a subject has write access to an object if the subject's security level is dominated by that of the object [2]. Since then, variations of this model as well as a popular model called the noninterference model [3] have been proposed. The noninterference model is essentially about higher-level processes not interfering with lower level processes. Note that with the Bell and La Padula model, a higher level process can covertly send information to a lower level process by manipulating the file locks, even though there can be no write down due to the star property. The noninterference model prevents such covert communication.

MAC for Database Systems: While Database Management Systems (DBMS) must deal with many of the same security concerns as operating systems (identification and authentication, access control, auditing), there are characteristics of DBMSs that introduce additional security challenges. For example, objects in DBMSs tend to be of varying sizes and can be of fine granularity such as relations, attributes and elements. This contrasts with operating systems where the granularity tends to be coarse such as files or segments. Because of the fine granularity in database systems the objects on which MAC is performed may differ. In operating systems MAC is usually performed on the same object such as a file whereas in DBMSs it could be on relations and attributes. The simple security and * property are both applicable for database systems. However many of the database systems have modified, the *-property to read as follows: *A subject has write access to an object if the subject's level is that of the object.* This means a subject can modify relations at its level. Various commercial secure DBMS products have emerged. These products have been evaluated using the Trusted Database Interpretation which interprets the TCSEC for database systems.

MAC for Networks: For applications in defense and intelligence multilevel secure networks are essential. The idea here is for the network protocols such as a TCP/IP (Transmission Control Protocol/Internet Protocol) protocols operate at multiple security levels. The Bell and La Padula model has been extended for networks. Furthermore, the commercial multilevel networks have been evaluated using the Trusted Network Interpretation that interprets the TCSEC for networks.

Cross-References

▶ Multilevel Secure Database Management System

Recommended Reading

1. http://en.wikipedia.org/wiki/Mandatory_access_control
2. Bell D, LaPadula L. "Secure Computer Systems: Mathematical Foundations and Model," M74-244. Bedford: The MITRE Corporation; 1973.
3. Goguen J, Meseguer J. Security policies and security models. In: Proceedings of the IEEE Symposium on Security and Privacy; 1982. p. 11–20.

M

MANET Databases

Yan Luo[1] and Ouri Wolfson[2,3]
[1]University of Illinois at Chicago, Chicago, IL, USA
[2] Mobile Information Systems Center (MOBIS), The University of Illinois at Chicago, Chicago, IL, USA
[3]Department of CS, University of Illinois at Chicago, Chicago, IL, USA

Synonyms

Mobile ad hoc network databases

Definition

A mobile ad hoc network (MANET) database is a database that is stored in the peers of a MANET. The network is composed by a finite set of mobile peers that communicate with each other via short range wireless protocols, such as IEEE 802.11, Bluetooth, Zigbee, or Ultra Wide Band (UWB). These protocols provide broadband (typically tens of Mbps) but short-range (typically 10-100 m) wireless communication. On each mobile peer there is a local database that stores and manages a collection of data items, or reports. A report is a set of values sensed or entered by the user at a particular time, or otherwise obtained by a mobile peer. Often a report describes a physical resource such as an available parking slot. All the local databases maintained by the mobile peers form the MANET database. The peers communicate reports and queries to neighbors directly, and the reports and queries propagate by transitive multi-hop transmissions. Figure 1 below illustrates the definition.

MANET databases enable matchmaking or resource discovery services in many application domains, including social networks, transportation, mobile electronic commerce, emergency response, and homeland security.

Communication is often restricted by bandwidth and power constraints on the mobile peers. Furthermore, reports need to be stored and later forwarded, thus memory constraints on the mobile devices constitute a problem as well. Thus, careful and efficient utilization of scarce peer resources (specifically bandwidth, power, and memory) are an important challenge for MANET databases.

Historical Background

Consider mobile users that search for local resources. Assuming that the information about the existence and location of such a resource resides on a server, a communication infrastructure is necessary to access the server. Such an infrastructure may not be available in military/combat situations, disaster recovery, in a commercial flight, etc. Even if the infrastructure and a server are both available, a user may not be willing to pay the dollar-cost that is usually involved in accessing the server through the cellular infrastructure. Furthermore, cellular bandwidth is limited (e.g., 130 character text messages). In other words, a client-server approach may have accessibility problems.

Currently, Google and local.com provide static local information (e.g., the location of a restaurant, pharmacy, etc.), but not dynamic information such as the location of a taxi cab, a nearby person of interest, or an available parking slot. These dynamic resources are temporary in nature, and thus require timely, real-time update rates. Such rates are unlikely to be provided for the country or the world by a centralized server farm, e.g., Google. Thus, dynamic local resources may require local servers, each dedicated to a limited geographic area. However, for many areas such a local server may not exist due to lack of a profitable business model, and if it exists it may be unavailable (such servers are unlikely to have the reliability of global sites such as Google). Furthermore, the data on the server may be unavailable due to propagation delays (think of sudden-brake information that needs to be propagated to a server and from there to the trailing vehicles), or due to device limitations (e.g., a cab customer's cell-phone may have Bluetooth but not internet

MANET Databases, Fig. 1
A MANET database

access to update the server), or due to the fact that updates from mobile devices may involve a communication cost that nobody is willing to pay, or due to the fact that the local server (e.g., of Starbucks) may accept only updates from certain users or certain applications but not others. In short, a client-(local)-server may have both accessibility and availability problems.

Thus, a MANET database can substitute or augment the client-(local)-server approach. Communication in the MANET is free since it uses the unlicensed spectrum, and larger in bandwidth than the cellular infrastructure, thus can provide media rich information, such as maps, menus, and even video. A mobile user may search the MANET database only, or combine it with a client-server search.

Currently, there are quite a few experimental projects in MANET databases. These can be roughly classified into pedestrians and vehicular projects. Vehicular projects deal with high mobility and high communication topology change-rates, whereas pedestrians projects have a strong concern with power issues. The following are several active experimental MANET database projects for pedestrians and vehicles:

Pedestrians Projects
- *7DS* – Columbia University
  - http://www.cs.unc.edu/~maria/7ds/
  - Focuses on accessing web pages in environments where only some peers have access to the fixed infrastructure.
- *iClouds* – Darmstadt University

- http://iclouds.tk.informatik.tu-darmstadt. de/
- Focuses on the provision of incentives to brokers (intermediaries) to participate in MANET databases.
- *MoGATU* – University of Maryland, Baltimore County
  - http://mogatu.umbc.edu/
  - Focuses on the processing of complex data management operations, such as joins, in a collaborative fashion.
- *PeopleNet* – National University of Singapore
  - http://www.ece.nus.edu.sg/research/projects/ abstract.asp? Prj=101
  - Proposes the concept of information Bazaars, each of which specializes in a particular type of information; reports and queries are propagated to the appropriate bazaar by the fixed infrastructure.
- *MoB* – University of Wisconsin and Cambridge University
  - http://www.cs.wisc.edu/~suman/projects/ agora/
  - Focuses on incentives and the sharing among peers of virtual information resources such as bandwidth.
- *Mobi-Dik* – University of Illinois at Chicago
  - http://www.cs.uic.edu/~wolfson/html/p2p. html
  - Focuses on information representing physical resources, and proposes stateless algorithms for query processing, with particular concerns for power, bandwidth, and memory constraints.

Vehicular Projects

- *CarTALK 2000* – A European project
  - http://www.cartalk2000.net/
  - Develops a co-operative driver assistance system based upon inter-vehicle communication and MANET databases via self-organizing vehicular ad hoc networks.
- *FleetNet* – Internet on the Road Project
  - http://www.ccrle.nec.de/Projects/fleetnet.htm
  - Develops a wireless multi-hop ad hoc network for intervehicle communication to improve the driver's and passenger's safety and comfort. A data dissemination method called "contention-based forwarding" (CBF) is proposed in which the next hop in the forwarding process is selected through a distributed contention mechanism based on the current positions of neighbors.
- *VII* – Vehicle Infrastructure Integration, a US DOT project
  - http://www.its.dot.gov/vii/
  - The objective of the project is to deploy advanced vehicle-to-vehicle and vehicle-to-infrastructure communications that could keep vehicles from leaving the road and enhance their safe movement through intersections.
- *Grassroots, Trafficview* – Rutgers University

TrafficInfo – University of Illinois at Chicago

- http://paul.rutgers.edu/~gsamir/dataspace/grassroots.html
- http://discolab.rutgers.edu/traffic/veh_apps.htm
- http://cts.cs.uic.edu/
- These projects develop an environment in which each vehicle contributes a small piece of traffic information (its current speed and location) to the network, using the P2P paradigm, and each vehicle aggregates the pieces into a useful picture of the local traffic.

Foundations

There are two main paradigms for answering queries in MANET databases, one is report pulling and the other one is report pushing.

Report pulling means that a mobile peer issues a query which is flooded in the whole network, and the answer-reports will be pulled from the mobile peers that have them (see e.g., [2]). Report pulling is widely used in resource discovery, such as route discovery in mobile ad hoc networks and file discovery by query flooding in wired P2P networks like Gnutella. Flooding in a wireless network is in fact relatively efficient as compared to wired networks because of the wireless broadcast advantage, but there are also disadvantages which will be explained below.

Another possible approach for data dissemination is report pushing. Report pushing is the dual problem of report pulling; reports are flooded, and consumed by peers whose query is answered by received reports. So far there exist mechanisms to broadcast information in the complete network, or in a specific geographic area (geocast), apart from to any one specific mobile node (unicast/mobile ad hoc routing) or any one arbitrary node (anycast). Report pushing paradigm can be further divided into stateful methods and stateless methods. Most stateful methods are topology-based, i.e., they impose a structure of links in the network, and maintain states of data dissemination. PStree [4], which organizes the peers as a tree, is an example of topology based methods.

Another group of stateful methods is cluster- or hierarchy-based method, such as [14], in which moving peers are grouped into some clusters or hierarchies and the cluster heads are randomly selected. Reports are disseminated through the network in a cluster or hierarchy manner, which means that reports are first disseminated to every cluster head, and each cluster head then broadcasts the reports to the member peers in its group. Although cluster- or hierarchy-based methods can minimize the energy dissipation in moving peers, these methods will fail or cost more energy in highly mobile environments since they have to maintain a hierarchy structure and frequently reselect cluster heads.

Another stateful paradigm consists of location-based methods (see [9]). In location-based methods, each moving peer knows the location of itself and its neighbors through some localization techniques, such as GPS or Atomic Multilateration (see [9]).

The simplest location-based data dissemination is Greedy Forwarding, in which each moving peer transmits a report to a neighbor that is closer to the destination than itself. However, Greedy Forwarding can fail in some cases, such as when a report is stuck in local minima, which means that the report stays in a mobile peer whose neighbors are all further from the destination. Therefore, some recovery strategies are proposed, such as GPSR (Greedy Perimeter Stateless Routing [6]). Other location-based methods, such as GAF (Geographic Adaptive Fidelity [17]) and GEAR (Geographical and Energy Aware Routing [18]), take advantage of knowledge about both location and energy to disseminate information and resources more efficiently.

In stateless methods, the most basic and simplest one is flooding-based method, such as [11]. In flooding-based methods, mobile peers simply propagate received reports to all neighboring mobile peers until the destination or maximum hop is reached. Each report is propagated as soon as it is received. Flooding-based methods have many advantages, such as no state maintenance, no route discovery, and easy deployment. However, they inherently cannot overcome several problems, such as implosion, overlap, and resource blindness. Implosion refers to the waste of resources taking place when a node forwards a message to a neighbor although the latter may have already received it from another source. Overlap occurs when two nodes read the same report, and thus push into the network the same information. Resource blindness denotes the inability of the protocol to adapt the node's behavior to its current availability of resources, mainly power [12]. Therefore, other stateless methods are proposed, such as gossiping-based methods and negotiation-based methods.

Gossiping-based methods, such as [3], improve flooding by transmitting received reports to a subset of randomly selected neighbors; another option is to have some neighbors simply drop the report. For example, the neighbors that are not themselves interested in the report drop it. The advantages of gossiping-based methods include reducing the implosion and lowering the system overhead. However, dissemination, and thus performance, is reduced compared to pure flooding.

Negotiation-based methods solve the implosion and overlap problem by transmitting first the id's of reports; the reports themselves are transmitted only when requested (see [7]). Thus, some extra data transmission is involved, which costs more memory, bandwidth, and energy. In addition, in negotiation-based methods, moving peers have to generate meta-data or a signature for every report so that negotiation can be carried out, which will increase the system overhead and decrease the efficiency.

Another important stateless paradigm for data dissemination in MANET databases is store-and-forward. In contrast to flooding, store-and-forward does not propagate reports as soon as they are received; rather they are stored and rebroadcast later. This obviously introduces storage and bandwidth problems, if too many reports need to be saved and rebroadcast at the same time. To address these, methods such as [5] rank all the reports in a peer's database in terms of their relevance (or expected utility), and then the reports are communicated and saved in the order of their relevance. Or, the reports requested and communicated are the ones with the relevance above a certain threshold. The notion of relevance quantifies the importance or the expected utility of a report to a peer at a particular time and at a particular location. Other store-and-forward methods include PeopleNet [10] and 7DS [13].

In summary, the paradigms for data dissemination in MANET databases are summarized in Fig. 2 below.

M

MANET Databases, Fig. 2
Query answering methods
in MANET databases

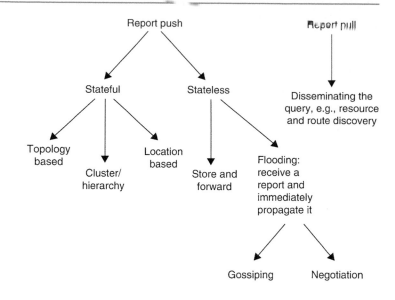

Key Applications

MANET databases provide mobile users a search
engine for transient and highly dynamic information in a local geospatial environment. MANET
databases employ a unified model for both the
cellular infrastructure and the mobile ad hoc environments. When the infrastructure is available,
it can be augmented by the MANET database
approach.

Consider a MANET database platform, i.e., a
set of software services for data management in a
MANET environment; it is similar to a regular
Database Management System, but geared to
mobile P2P interactions. Such a platform will enable quick building of matchmaking or resource
discovery services in many application domains,
including social networks, emergency response
and homeland security, the military, airport applications, mobile e-commerce, and transportation.

Social Networks
In a large professional, political, or social gathering, MANET databases are useful to automatically facilitate a face-to-face meeting based on
matching profiles. For example, in a professional
gathering, MANET databases enable attendees
to specify queries (interest profiles) and resource
descriptions (expertise) to facilitate face-to-face
meetings, when mutual interest is detected. Thus,

the individual's profile that is stored in MANET
databases will serve as a "wearable web-site."
Similarly, MANET databases can facilitate face-
to-face meetings for singles matchmaking.

Emergency Response, Homeland Security, and the Military
MANET databases offer the capability to ex-
tend decision-making and coordination capability. Consider workers in disaster areas, soldiers
and military personnel operating in environments
where the wireless fixed infrastructure is significantly degraded or non-existent. As mobile
users involved in an emergency response naturally cluster around the location of interest, a
self-forming, high-bandwidth network that allows database search without the need of potentially compromised infrastructure could be of
great benefit. For instance, the search could specify a picture of a wanted person.

Airport Applications
A potential opportunity that will benefit both
the consumer and the airport operations is
the dissemination and querying of real-time
information regarding flight changes, delays,
queue length, parking information, special
security alerts and procedures, and baggage
information. This can augment the present audio
announcements that often cannot be heard in

nearby restaurants, stores, or restrooms, and augment the limited number of displays.

Mobile E-Commerce

Consider short-range wireless broadcast and mobile P2P dissemination of a merchant's sale and inventory information. It will enable a customer (whose cell phone is query-capable) who enters a mall to locate a desired product at the best price. When a significant percentage of people have mobile devices that can query retail data, merchants will be motivated to provide inventory/sale/coupons information electronically to nearby potential customers. The information will be disseminated and queried in a P2P fashion (in, say, a mall or airport) by the MANET database.

Transportation Safety and Efficiency

MANET databases can improve safety and mobility by enabling travelers to cooperate intelligently and automatically. A vehicle will be able to automatically and transitively communicate to trailing vehicles its "slow speed" message when it encounters an accident, congestion, or dangerous road surface conditions. This will allow other drivers to make decisions such as finding alternative roads. Also, early warning messages may allow a following vehicle to anticipate sudden braking, or a malfunctioning brake light, and thus prevent pile-ups in some situations. Similarly, other resource information, such as ridesharing opportunities, transfer protection (transfer bus requested to wait for passengers), will be propagated transitively, improving the efficiency of the transportation system.

Future Directions

Further work is necessary on data models for mobile P2P search applications. Work on sensor databases (e.g., Tinydb [8]) addresses datamodels and languages for sensors, but considers query processing in an environment of static peers (see e.g., POS [1]). Cartel [5] addresses the translation of these abstractions to an environment in which cars transfer collected data to a central database via fixed access points. Work

on MANET protocols deals mainly with routing and multicasting. In this landscape there is a gap, namely general query-processing in MANET's; such processing needs to be cognizant of many issues related to peer-mobility. For example, existing mobile P2P query processing methods deal with simple queries, e.g., selections; each query is satisfied by one or more reports. However, in many application classes one may be interested in more sophisticated queries, e.g., aggregation. For instance, in mobile electronic commerce a user may be interested in the minimum gas price within the next 30 miles on the highway. Processing of such P2P queries may present interesting optimization opportunities.

After information about a mobile resource is found, localization is often critical for finding the physical resource. However, existing (self-)localization techniques are insufficient. For example, GPS is not available indoors and the accuracy of GPS is not reliable. Thus, furthering the state of the art on localization is important for mobile P2P search.

As discussed above, MANET databases do not guarantee answer completeness. In this sense, the integration with an available infrastructure such as the internet or a cellular network may improve performance significantly. This integration has two aspects. First, using the communication infrastructure in order to process queries more efficiently; and second, using data on the fixed network in order to provide better and more answers to a query. The seamless integration of MANET databases and infrastructure databases introduces important research challenges.

Other important research directions include: incentives for broker participation in query processing (see [16]), and transactions/atomicity/recovery issues in databases distributed over mobile peers (virtual currency must be transferred from one peer to another in an atomic fashion, otherwise may be lost).

Of course, work on efficient resource utilization in mobile peers, and coping with sparse networks and dynamic topologies is still very important for mobile P2P search.

M

Recommended Reading

1. Cox L, Castro M, Rowstron A. POS: Practical Order Statistics for wireless sensor networks. In: Proceedings of the 23rd IEEE International Conference on Distributed Computing Systems; 2003. p. 52.
2. Das SM, Pucha H, Hu YC. Ekta: an efficient DHT substrate for distributed applications in Mobile Ad hoc Networks. In: Proceedings of the 6th IEEE Workshop on Mobile Computing Systems and Applications; 2004. p. 163–73.
3. Datta A, Quarteroni S, Aberer K. Autonomous gossiping: a self-organizing epidemic algorithm for selective information dissemination in wireless Mobile Ad-Hoc Networks. In: Proceedings of the International Conference on Semantics of a Networked World; 2004. p. 126–43.
4. Huang Y, Molina HG. Publish/subscribe in a mobile environment. In: Proceedings of the 2nd ACM International Workshop on Data Engineering for Wireless and Mobile Access; 2001. p. 27–34.
5. Hull B, et al. CarTel: a distributed mobile sensor computing system. In: Proceedings of the 4th International Conference on Embedded Networked Sensor Systems; 2006. p. 125–38.
6. Karp B, Kung HT. GPSR: Greedy Perimeter Stateless Routing for wireless sensor networks. In: Proceedings of the 6th Annual International Conference on Mobile Computing and Networking; 2000. p. 243–54.
7. Kulik J, Heinzelman W, Balakrishnan H. Negotiation-based protocols for disseminating information in wireless sensor networks. Wireless Netw. 2002;8(2–3):169–85.
8. Madden SR, Franklin MJ, Hellerstein JM, Hong W. Tiny DB: an acquisitional query processing system for sensor networks. ACM Trans Database Syst. 2005;30(1):122–73.
9. Mauve M, Widmer A, Hartenstein H. A survey on position-based routing in Mobile Ad-Hoc networks. IEEE Netw. 2001;15(6):30–9.
10. Motani M, Srinivasan V, Nuggehalli P. PeopleNet: engineering a wireless virtual social network. In: Proceedings of the 11th Annual International Conference on Mobile Computing and Networking; 2005. p. 243–57.
11. Oliveira R, Bernardo L, Pinto P. Flooding techniques for resource discovery on high mobility MANETs. In: Proceedings of the International Workshop on Wireless Ad-hoc Networks; 2005.
12. Papadopoulos AA, McCann JA. Towards the design of an energy-efficient, location-aware routing protocol for mobile, Ad-hoc Sensor Networks. In: Proceedings of the 15th International Workshop on Database and Expert Systems Applications; 2004. p. 705–9.
13. Papadopouli M, Schulzrinne H. Design and implementation of a P2P Data dissemination and prefetching tool for mobile users. In: Proceedings of the 1st New York Metro Area Networking Workshop; 2001.
14. Visvanathan A, Youn JH, Deogun J. Hierarchical data dissemination scheme for large scale sensor networks. In: Proceedings of the IEEE International Conference on Communication; 2005. p. 3030–6.
15. Wolfson O, Xu B, Yin HB, Cao H. Search-and-discover in Mobile P2P Network Databases. In: Proceedings of the 23rd International Conference on Distributed Computing Systems; 2006. p. 65.
16. Xu B, Wolfson O, Rishe N. Benefit and pricing of spatio-temporal information in Mobile Peer-to-Peer Networks. In: Proceedings of the 39th Annual Hawaii International Conference on System Sciences; 2006. p. 2236.
17. Xu Y, Heidemann J, Estrin D. Geography-informed energy conservation for Ad hoc Routing. In: Proceedings of the 7th Annual International Conference on Mobile Computing and Networking; 2001. p. 70–84.
18. Yu Y, Govindan R, Estrin D. Geographical and energy aware routing: a recursive data dissemination protocol for wireless sensor networks. Technical Report UCLA/CSD-TR-01-0023, UCLA. May 2001.

MAP

Steven M. Beitzel[1], Eric C. Jensen[2], and Ophir Frieder[3]
[1]Telcordia Technologies, Piscataway, NJ, USA
[2]Twitter, Inc., San Francisco, CA, USA
[3]Georgetown University, Washington, DC, USA

Synonyms

Mean average precision

Definition

The Mean Average Precision (MAP) is the arithmetic mean of the average precision values for an information retrieval system over a set of n query topics. It can be expressed as follows:

$$\mathrm{MAP} = \frac{1}{n}\sum_n \mathrm{AP}_n$$

where AP represents the Average Precision value for a given topic from the evaluation set of n topics.

Key Points

The Mean Average Precision evaluation metric has long been used as the de facto "gold standard" for information retrieval system evaluation at the NIST Text Retrieval Conference (TREC) [1]. Many TREC tracks over the years have evaluated run submissions using the *trec_eval* program, which calculates Mean Average Precision, along with several other evaluation metrics. Much of the published research in the information retrieval field over the last 25 years relies on observed difference in MAP to draw conclusions about the effectiveness of a studied technique or system relative to a baseline.

Recently, the explosive growth of the World Wide Web and the corresponding difficulty of creating test collections that are representative, robust, and of appropriate scale has created new challenges for the research community. One such challenge is how to best evaluate systems in cases of incomplete relevance information. It has been shown that ranking systems by their MAP scores when relevance information is incomplete does not correlate highly with their rankings with complete judgments. This is a key weakness of MAP as a metric. In response to this problem, new metrics (such as BPref, for example) have been proposed that attempt to compensate for often incomplete relevance information [2].

Cross-References

▶ Average Precision
▶ BPref
▶ Chart
▶ Effectiveness Involving Multiple Queries

Recommended Reading

1. National Institute of Standards and Technology. TREC-2004 common evaluation measures. Available online at: http://trec.nist.gov/pubs/trec14/appendices/CE.MEASURES05.pdf. Retrieved on 27 Aug 2007.

2. Sakai T. Alternatives to BPref. In: Proceedings of the 30th Annual International ACM SIGIR Conference on Research and Development in Information Retrieval; 2007. p. 71–8.

Map Matching

Christian S. Jensen[1] and Nerius Tradišauskas[2]
[1]Department of Computer Science, Aalborg University, Aalborg, Denmark
[2]Aalborg University, Aalborg, Denmark

Synonyms

Position snapping

Definition

Map matching denotes a procedure that assigns geographical objects to locations on a digital map. The most typical geographical objects are point positions obtained from a positioning system, often a GPS receiver. In typical uses, the GPS positions derive from a receiver located in a vehicle or other moving object traveling in a road network, and the digital map models the embedding into geographical space of the roads by means of polylines that approximate the center lines of the roads. The GPS positions generally do not intersect with the polylines, due to inaccuracies. The aim of map matching is then to place the GPS positions at their "right" locations on the polylines in the map.

Map matching is useful for a number of purposes. Map matching is used when a navigation system displays the vehicle's location on a map. In many applications, information such as speed limits are assigned to the representations of roads in a digital map-map matching offers a means of relating such information to moving objects. Map matching may also be used for the representation of a route of a vehicle by means of the (sub-) polylines in the digital map.

M

Two general types of map matching exist, namely on-line map matching and off-line map matching. With on-line map matching, the map location of an object's current position needs to be determined in real time. Only past, but not future, positions are available. Vehicle navigation systems exemplify this type of map matching. In off-line map matching, a static data set of positions is given, meaning that all future positions are available when map matching a position. Thus better map matching may result when compared to on-line map matching. For example, off-line map matching may be used for billing in pay-per-use scenarios (insurance, road pricing).

Historical Background

One of the earliest map matching algorithms found in the literature dates back to 1971 and is due to R.L. French (see the overview in reference [1]). A 1996 paper by Berstein and Kornhauser [2] offers a brief introduction to the map matching problem and its variations.

The scientific literature contains a range of papers that address different aspects of the map matching problem. White et al. [3] study techniques that pay special attention to intersection areas, where map matching can be particularly challenging. Taylor et al. [4] propose a map matching technique that uses differential corrections and height, which leads to improved performance. Quddus et al. [5] provide a summary of different on-line and off-line map matching algorithms and describe advantages and disadvantages of these. Quddus et al. [6] have most recently proposed a map matching algorithm that utilizes techniques from fuzzy logic. This technique shows improved accuracy of polyline identification and the positioning on polylines. A complex off-line map matching algorithm was recently developed by Bratkatsoulas et al. [7] that uses the Fréchet distance to map match GPS position samples recorded only every 30 seconds. (While GPS receivers typically output a position every second, it may be that only some of these are saved for use in subsequent off-line map matching).

Foundations

Basics The most common use of map matching occurs in transportation where the GPS positions obtained from a GPS receiver in a vehicle are map matched to a digital representation of a road network. An example of GPS positions from a vehicle (dots) and a digital road network are shown in Fig. 1. The vehicle's trip started on road #2 and continued along road #1. In Fig. 2a the start of the trip is enlarged. The dots represent the vehicle's GPS locations, and the triangles represent the corresponding positions map matched onto the road network. The road network locations are typically expressed by using linear referencing, which is a standard means of indicating such locations. With linear referencing, a tuple (#2,5.2,+1) captures the road that the vehicle is driving on (#2), a position on that road measured as a distance from the beginning of the road (5.2 distance units). The third element captures a perpendicular distance from the road location given by the first two elements. In the tuple, the displacement is one distance unit to the left. Because a GPS position is typically mapped to the closest location in the road digital network (i.e., a perpendicular

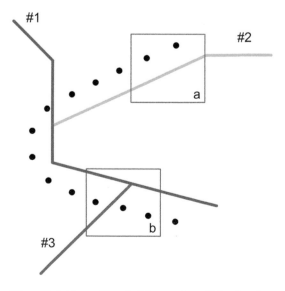

Map Matching, Fig. 1 Map and vehicle location example

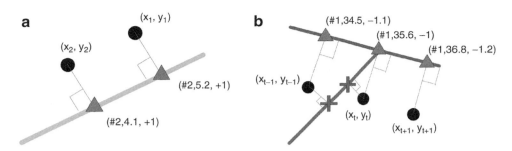

Map Matching, Fig. 2 Map matching example

projection is used), the third element is capable of capturing the GPS position that was map matched to the road network location. Linear referencing is supported by, e.g., the Oracle DBMS [6].

In Fig. 2b, a place where map matching is challenging is enlarged. The crosses represent two instance of wrong map matching to the nearest road, and the triangles represent the correct map matching.

Categorization of Map Matching Algorithms
Map matching can be divided into on-line and off-line map matching. On-line map matching occurs in real-time. Here, the map matching algorithm tries to identify the network location of a GPS position every time a new position is received. The algorithm has available the current position as well as information about the map matching of previous positions.

This contrasts off-line map matching, which occurs after a trip is over and all the positions from the start to the end are known. Off-line map matching is more accurate than on-line map matching, as more information (i.e., future positions) is available. An off-line algorithm does not provide a map matching result until the entire trip has been map matched.

On-line map matching is mostly used in vehicle navigation, tracking, and other applications that need the most recent network location of a vehicle. Off-line map matching is mostly used to determine as accurately as possible which route a vehicle was driving. Off-line map matching may also be used in scenarios where GPS data are received in batches from content providers and where the purpose is to build speed maps

that capture the expected travel speeds for different road segments and time intervals. Network locations of vehicles are essential in road load analysis, road pricing, and similar applications.

Map matching algorithms can also be divided into point-to-point, point-to-polyline, and polyline-to-polyline approaches. In point-to-point map matching, a point out of a point set is identified as a match for the given position. In point-to-polyline matching, a polyline in a polyline set is identified, and a point on the polyline is identified that represents the given point as a polyline-set location. In the typical scenario, the polyline set represents a road network, and the position is a GPS position. Finally, in polyline-to-polyline map matching, polylines are identified from the polylines in a road network that best match a given polyline that is usually constructed from point positions.

Map Matching Principles This section follows the explanation of the basic principles of map matching by considering in some detail the commonly used point-to-polyline map matching for the on-line case.

Map matching algorithms often consist of two overall steps (some algorithms skip the first of these two steps or include an extra step).

In the start-up step, the polyline in the digital road network on which the vehicle is initially located is found. Specifically, map matching is done for the first GPS position received. The correct outcome of this step is very important, as the map matching algorithms use the connections between roads, or the road network topology, to determine the ensuing polylines of the vehicle's movement path. Therefore, algorithms often perform special operations to determine a reliable match for the first GPS position.

In the steady-state step, the subsequent polylines in the digital road network are identified to form the route along which the vehicle is moving. This subsequent map matching follows a standard pattern:

1. Extract the relevant information from the record received from the GPS receiver.
2. Select candidate polylines from the digital road network.
3. Use algorithm-specific heuristics to determine the most suitable polyline among the candidate polylines.
4. Determine the vehicle position on the selected polyline.

First, information such as latitude, longitude, speed, and heading is extracted from the record obtained from the positioning unit and is converted into an appropriate format (a unified coordinate and metric system consistent with the digital road network).

Second, the candidate polylines are selected. Usually the polylines that are within a certain threshold distance of the GPS position are selected. An alternative approach is to select the n polylines nearest to the GPS position. The polylines found make up the candidate polyline set.

Third, specific heuristics are used to select the best polyline among the candidates. A common approach is to assign weights to the candidate polylines according to different criteria. The polyline with the highest sum of weights is then chosen. Some algorithms reduce the set of candidate polylines prior to assigning weights. For example, polylines that are perpendicular to the vehicle's heading may be disregarded, as may polylines that are not connected with the polyline currently being considered.

Fourth, the vehicle position on the selected polyline is found. The usual approach is to select the location on the polyline that is closest to the vehicle's position. This is a point-to-polyline projection. The projection of the position may be an end point of a line segment in the polyline, or it may be in the interior of a line segment. Quddus et al. [5] propose a more sophisticated approach that uses both the distance traveled since the last map matching and the "raw" projection of the GPS position onto the polyline.

Dead Reckoning In certain regions, the signals emitted by the GPS satellites may be very weak due to obstacles. This in turn degrades the accuracies of the positions produced by GPS receivers in those regions, and in some cases no GPS position may be produced. In such cases, both online and off-line map matching algorithms may utilize dead reckoning to estimate the movement of a vehicle in the road network. The use of dead reckoning is particularly attractive when the average speed of the vehicle is known (preferably every second) and when the road on which the vehicle is driving neither splits nor has intersections. This occurs when the road is inside a tunnel with no exits.

Heuristics for the Selection of a Polyline Different algorithms use different heuristics and weights when attempting to identify the best polyline among the candidate polylines. Each candidate polyline is assigned a weight for each criterion considered, and the polyline with the highest sum of weights is then selected. Common weighting criteria include the following:

• Weight for the proximity of a polyline. A polyline is assigned a weight according to its proximity to the GPS position being map matched. It is natural to assume that the closer a polyline is to the position, the better a candidate the polyline is.
• Weight for the continuity of a polyline. This weight is assigned to each polyline for being a continuation of the previously map matched polyline. This weight represents the reasoning that vehicles tend to drive on the same road most of the time.
• Weight for direction similarity. Polylines whose bearing is similar to the vehicle's heading are assigned higher values.

- Weight for topology. Higher weights are added to polylines that are connected to the polyline currently being map matched to.

Algorithms may also include weights for speed limit changes, shortest distance, road category, one-way streets, etc. Off-line map matching algorithms may use fewer, but more robust weights that are not suitable for the on-line map matching algorithms.

Execution Time Constraints and Accuracy In on-line map matching, the algorithms must keep up with the GPS device that usually emits one position per second. With current on-board computing units, it is usually not a problem for on-line map matching algorithms map match a GPS position within 1 second. For off-line map matching, there are no real-time constraints.

Key Applications

Map matching is essential for applications that rely on the positioning of a user within a road network.

This occurs in vehicle navigation where the user's position is to be displayed so that it coincides with the road network. When GPS data is used for the construction of speed maps, map matching is used. Such speed maps may be used for travel time prediction, route construction, and capacity planning. In metered services such as insurance and road pricing, map matching is also used.

Further, map matching is used in location-based services where the content to be retrieved is positioned within the road network. Services that offer network-context awareness utilize map matching. These including current network location context awareness [9], and route context awareness [10].

Tracking also plays a key role in intelligent speed adaptation [11] where drivers are alerted when they exceed the current speed limit. The speed limits are attached to the digital road network, so map matching is needed to identify the current speed limit. Finally, map matching is important for a range of other application within intelligent transportation systems.

Future Directions

In the near future, digital road networks will have accurate lane information embedded, and positioning technologies will be accurate enough to enable lane-level positioning. This will necessitate the extension of map matching techniques to function at the lane level.

Cross-References

▶ Compression of Mobile Location Data
▶ Location-Based Services
▶ Mobile Database
▶ Mobile Sensor Network Data Management
▶ Road Networks
▶ Spatial Network Databases
▶ Spatial Operations and Map Operations
▶ Spatiotemporal Interpolation Algorithms
▶ Spatiotemporal Trajectories

M

Recommended Reading

1. Bernstein D and Kornhauser A. An introduction to map matching for personal navigation assistants. New Jersey TIDE Center. 1996. http://www.njtide.org/reports/mapmatchintro.pdf.
2. Brakatsoulas S, Pfoser D, Salas R, Wenk C. On map-matching vehicle tracking data. In: Proceedings of the 31st International Conference on Very Large Data Bases; 2005. p. 853–64.
3. Brilingaitė A, Jensen CS. Enabling routes of road network constrained movements as mobile service context. Geoinformatica. 2007;11(1):55–102.
4. Civilis A, Jensen CS, Pakalnis S. Techniques for efficient road-network-based tracking of moving objects. IEEE Trans Knowl Data Eng. 2005;17(5):698–712.
5. French R.L. Historical overview of automobile navigation technology. In: Proceedings of the 36th IEEE Vehicular Technology conference; 1986. p. 350–8.
6. Oracle Corporation. Oracle spatial and oracle locator. http://www.oracle.com/technology/products/spatial/index.html.
7. Quddus M, Ochieng W, Zhao L, Noland R. A general map matching algorithm for transport telematics applications. GPS Solutions. 2003;7(3):157–67.

8. Quddus MA, Noland RB, Ochieng WY. A high accuracy fuzzy logic based map matching algorithm for road transport. J Intell Transp Syst. 2006;10(3):103–15.
9. Taylor G, Blewitt G, Steup D, Corbett S, Car A. Road reduction filtering for GPS-GIS navigation. Trans GIS. 2001;5(3):193–207.
10. Tradisauskas N, Juhl J, Lahrmann H, Jensen CS. Map matching for intelligent speed adaptation. In: Proceedings of the 6th European Congress on Intelligent Transport Systems and Services; 2007.
11. White CE, Bernstein D, Kornhauser AL. Some map matching algorithms for personal navigation assistants. Transp Res C. 2000;8(1–6):91–108.

MapReduce

Sai Wu
Zhejiang University, Hangzhou, Zhejiang, People's Republic of China

Scientific Fundamentals

MapReduce refers to both a programming model and the corresponding distributed framework. Its model is composed of two phases, map and reduce, which manipulate data formated as key-value pairs. Map phase splits and sorts data on keys, whereas reduce phase applies user-defined function to process data with the same key. In this way, MapReduce is a typical divide-and-conquer framework that is designed to handle embarrassingly parallel problems, namely problems that can be split into sub-tasks with little or no synchronization costs.

Definition

MapReduce is a programming framework that allows users to process large-scaled data by leveraging the parallelism among a cluster of nodes. It is also used to refer to the distributed engine which splits and disseminates users' jobs and monitors their processing in the cluster. MapReduce is a typical divide-and-conquer framework, since it transforms the user code into an embarrassingly parallel job, where little or no effort is required to synchronize subtasks.

Historical Background

To compute PageRank values and build inverted indexes for their large repository of Web pages, Google designed and implemented MapReduce [1], a parallel programming framework that supports processing data with thousands of compute nodes concurrently. Google revealed MapReduce in their 2004 OSDI paper. However, it is commonly believed that MapReduce had already been successfully deployed inside Google for years.

Google did not give too many details about their MapReduce implementations. However, as a general programming framework, MapReduce has many third-party open-source implementations, among which the most famous one is Hadoop [2]. Hadoop includes two parts, the HDFS (Hadoop Distributed File System) and MapReduce core. HDFS follows the design philosophy of Google File System [3] by splitting large files into equal-size chunks. Each mapper processes one specific data chunk. Dittrich et al. showed the basic workflow of a MapReduce process in Hadoop [4]. Users can extend functions like map, reduce, combine, and split to implement their own processing logics and optimization techniques.

Hadoop is now growing up as a full-fledged ecosystem for big data processing. On top of the basic Hadoop, the community has developed HBase [5] (a key-value store), Hive [6] (a data warehousing system), Mahout [7] (a machine learning toolkit), and Pig Latin [8] (a data flow language). It has been successfully deployed on Yahoo's cluster consisting of 4,500 compute nodes to sort 1T data within 1 min [9].

Besides the unstructured and semi-structured data, MapReduce is also employed to process relational data which is normally handled by the database systems. In their famous blog (MapReduce: A major step backwards), database researchers, David DeWitt and Michael

Stonebreaker, argued that MapReduce does not support schema and index, making it inefficient in processing relational data. They further compared MapReduce and the database system using a customized benchmark [10]. The conclusion is that database system is more efficient in processing complex analytic jobs for its support of schema and index, while MapReduce supports ad-hoc one-time analysis much easier. Initially, MapReduce is not designed as a database system, but its flexibility allows more sophisticated optimization techniques to be applied to reduce the performance gap. Specifically, popular database techniques, such as indexes [11], query optimization [12], replicated join [13], and shared scan [14], are being introduced to MapReduce systems and show significant improvements over the original MapReduce framework. Feng Li et al. summarized previous work on using MapReduce to process relational data in their survey paper [15].

MapReduce systems enhanced with those database techniques are very similar to a conventional distributed database system. On the other hand, conventional DBMSs, such as Oracle and Aster, implement the MapReduce workflow as a user-defined function. The two paradigms, MapReduce and DBMS, start to merge together. One representative system is HadoopDB [16] which uses Hadoop to process joins and pushes single table operators (e.g., select and project) to the DBMS. It achieves a better performance than pure MapReduce system while still providing the flexibility.

In mid-2014, Google's senior VP, Urs Hölzle, announced that MapReduce was replaced by a new system, Dataflow. He explained that MapReduce is no longer sufficient for the scale of data analytics the company needed. Its batch processing model is also too limited for applications that require real-time response.

On the other hand, MapReduce is also challenged by the new programming frameworks designed by the research community. For instance, GraphLab [18] simulates Google's Pregel [19] engine to support efficient graph data processing. Spark [20] is said to be 100 times faster than Hadoop when handling the same data analytic

job. epiC [20] uses more flexible actor model to reduce the synchronization costs in the iterative jobs.

Foundations

MapReduce is both a programming framework and a distributed processing engine. Figure 1 shows the architecture of a MapReduce engine (Hadoop implementation).

The engine includes one master node (a secondary master node can be created as a backup) and multiple worker nodes. The master node runs a scheduler process, responsible for disseminating tasks to different map and reducing processes and monitoring their processing. Each worker node can run one or more map/reduce processes. The map process includes four modules. The input module reads in a data chunk from the DFS (distributed file system) and splits it into key-value pairs. The map module applies the user-defined map function to process the key-value pairs from the input module which are forwarded to the combine module. Combine module is an optional module which applies the same reduce function as defined in the reduce process to compute a pre-aggregation result. This is to reduce the network communication overhead between map and reduce processes. Finally, the partition model splits the intermediate results of map processes for different reduce processes.

The reduce process has three core modules. The group module organizes the key-value pairs received from map processes via their keys. The reduce model applies the user-defined reduce function to process the pairs with the same key. The output model writes back the results to the DFS.

MapReduce provides two primitive functions, map and reduce, to manipulate data formatted as key-value pairs. Each input key-value pair will be transformed into one or more key-value pairs by map processes, and a merge-sort algorithm is applied to group data via keys. The intermediate results of map processes are first buffered in local

M

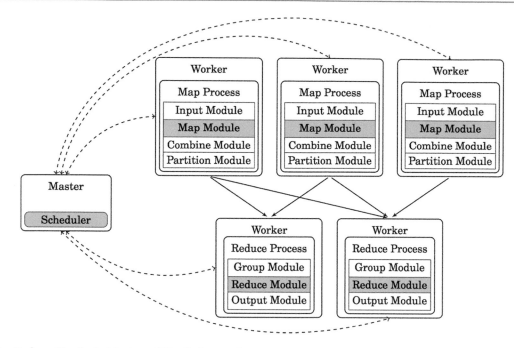

MapReduce, Fig. 1 Architecture of MapReduce engine

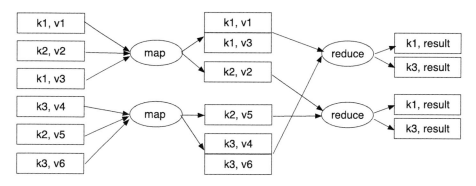

MapReduce, Fig. 2 MapReduce data flow

file system and then shuffled to reduce processes based on the partition module (normally a hash function on keys is applied).

Figure 2 illustrates the data flow of a MapReduce job. The input of map function is a stream of key-value pairs from the DFS. By default, each map process will handle key-value pairs in one specific data chunk. The size of data chunk is tunable and may significantly affect the performance. For each key-value pair, map process applies the used-defined function to generate one or more new key-value pairs, which are further sorted via keys using the merge-sort algorithm.

The intermediate results (key-value pairs) are partitioned using a hash function on keys (e.g., we use mod function in the figure). Key-value pairs with the same hash key are shuffled to the same reduce process, where pairs with the same keys are grouped together and the user-defined function is applied. The final results are written back to the DFS as a temporary file. So the consecutive MapReduce jobs can load them to continue the processing.

MapReduce framework supports auto-failure recovery. If a map process fails, the scheduler assigns its data chunk to a new map process and restarts the processing. If a reduce process fails, a new one is created to pull the data from the corresponding map processes to redo the processing. In real deployment, the scheduler normally assigns one specific task to multiple map/reduce processes. So if one map/reduce process completes its task, those responsible for the same task will be terminated.

Key Applications

MapReduce has been widely adopted to support various applications that need to handle huge volumes of data.

It was originally designed to build inverted index and compute PageRank values. This is also the most popular use case of MapReduce, namely, processing and transforming data in batch. Some data-importing tools and ETL tools leverage MapReduce to provide a scalable service. Example systems, such as SciHadoop and SciMate, process extremely large scientific datasets on MapReduce.

Due to its flexibility, many data mining and machine learning algorithms can be re-implemented on MapReduce and the training cost can be effectively reduced. For instance, Mahout is an Apache project on top of Hadoop which includes the implementations of clustering algorithms, classification algorithms, regression algorithms, and neural network algorithms. Behemoth is a NLP tool that can be integrated with Mahout. Weka, a famous machine learning toolkit, can work with Hadoop to facilitate the training process.

Recently, BI (business intelligent) systems start using MapReduce as their backend processing engine. They use MapReduce-based data warehouse systems (e.g., Hive and Cheetah) to maintain data and build cubes. They support R languages, so users' analytic queries can be automatically transformed into MapReduce jobs. They also provide various visualization interfaces to allow users to explore the data and detect the hidden patterns.

URL to Code

Hadoop and its ecosystem (HDFS, Hive, HBase, Pig, Manhout) can be found and downloaded from the apache website: https://hadoop.apache.org/. Cloudera's integrated Hadoop installation package, CDH, can be obtained at http://www.cloudera.com. Users can also deploy a customized MapReduce cluster using Amazon's Web service (http://aws.amazon.com/elasticmapreduce/).

Recommended Reading

1. Dean J, Ghemawat S. MapReduce: simplified data processing on large clusters. In: Proceedings of the 6th USENIX Symposium on Operating System Design and Implementation; 2004. p. 137–50.
2. https://hadoop.apache.org/
3. Sanjay Ghemawat, Howard Gobioff, Shun-Tak Leung. The Google file system. In: Proceedings of the 19th ACM Symposium on Operating System Principles; 2003. p. 29–43.
4. Dittrich J, Quiané-Ruiz J-A, Jindal A, Kargin Y, Setty V, Schad J. Hadoop++: making a yellow elephant run like a cheetah (without It even noticing). Proc VLDB Endow. 2010;3(1):518–29.
5. http://hbase.apache.org
6. Thusoo A, Sarma JS, Jain N, Shao Z, Chakka P, Anthony S, Liu H, Wyckoff P, Murthy R. Hive – a warehousing solution over a map-reduce framework. Proc VLDB Endow. 2009;2(2):1626–9.
7. http://mahout.apache.org
8. Olston C, Reed B, Srivastava U, Kumar R, Tomkins A. Pig latin: a not-so-foreign language for data processing. In: Proceedings of the ACM SIGMOD International Conference on Management of Data; 2008. p. 1099–110.
9. https://developer.yahoo.com/blogs/hadoop/
10. Pavlo A, Paulson E, Rasin A, Abadi DJ, DeWitt DJ, Madden S, Stonebraker M. A comparison of approaches to large-scale data analysis. In: Proceedings of the ACM SIGMOD International Conference on Management of Data; 2009. p. 165–78.
11. Jiang D, Ooi BC, Shi L, Wu S. The performance of MapReduce: an in-depth study. Proc VLDB Endow. 2010;3(1):472–83.
12. Sai Wu, Feng Li, Sharad Mehrotra, Beng Chin Ooi. Query optimization for massively parallel data processing. In: Proceedings of the 2nd ACM Symposium on Cloud Computing; 2011. p. 12.
13. Afrati FN, Das Sarma A, Menestrina D, Parameswaran AG, Ullman JD. Fuzzy joins using MapReduce. In: Proceedings of the 28th International Conference on Data Engineering; 2012. p. 498–509.

M

14. Nykiel T, Potamias M, Mishra C, Kollios G, Koudas N. MRShare: sharing across multiple queries in MapReduce. Proc VLDB Endow. 2010;3(1):494–505.
15. Li F, Ooi BC, Tamer Özsu M, Wu S. Distributed data management using MapReduce. ACM Comput Surv. 2014;46(3):31:1–31:42.
16. Abouzeid A, Bajda-Pawlikowski K, Abadi DJ, Rasin A, Silberschatz A. HadoopDB: an architectural hybrid of MapReduce and DBMS technologies for analytical workloads. Proc VLDB Endow. 2009;2(1):922–33.
17. http://www.informationweek.com/cloud/software-as-a-service/google-i-o-hello-dataflow-goodbye-mapreduce/d/d-id/1278917
18. Low Y, Gonzalez J, Kyrola A, Bickson D, Guestrin C, Hellerstein JM. Distributed GraphLab: a framework for machine learning in the cloud. Proc VLDB Endow. 2012;5(8):716–27.
19. Malewicz G, Austern MH, Bik AJC, Dehnert JC, Horn I, Leiser N, Czajkowski G. Pregel: a system for large-scale graph processing. In: Proceedings of the ACM SIGMOD International Conference on Management of Data; 2010. p. 135–46.
20. Jiang D, Chen G, Ooi BC, Tan K-L, Wu S. epiC: an extensible and scalable system for processing big data. Proc VLDB Endow. 2014;7(7):541–52.

Markup Language

Ethan V. Munson
Department of EECS, University of Wisconsin-Milwaukee, Milwaukee, WI, USA

Definition

A markup language is specification language that annotates content through the insertion of *marks* into the content itself. Markup languages differ from programming languages in that they treat data, rather than commands or declarations, as the primary element in the language.

Key Points

Markup languages were initially developed for text document formatting systems, though they are not limited to text. In fact, the term *markup* was taken directly from the jargon of the publishing business, where editors and typographers would "mark up" draft documents to indicate corrections or printing effects. Markup languages are generally quite declarative and have little, if any, computational semantics. The marks inserted into the content are often called "tags" because that term is used by XML.

Cross-References

▶ Document
▶ Document Representations (Inclusive Native and Relational)

MashUp

Alex Wun
University of Toronto, Toronto, ON, Canada

Definition

A MashUp is a web application that combines data from multiple sources, creating a new hybrid web application with functionality unavailable in the original individual applications that sourced the data.

Key Points

An emerging trend in web applications is to provide public APIs for accessing data that has traditionally been used only internally by those applications. The main purpose of providing access to traditionally private web application data is to encourage user-driven development. In other words, consumers are expected to take that public data and build custom applications for other consumers – thereby adding value to the original data sources. MashUps are web applications that take advantage of these publicly accessible data sources by correlating the data obtained from different sources and deriving some novel functionality. A simple and common example is correlating a data source that has location information (such as wireless hotspot locations) with cartographic data (from Google or Yahoo maps for example) to produce a graphical map of wireless hotspots.

MashUps are conceptually related to portals, which also collect data from multiple sources for presentation. However, portals perform server-side aggregation whereas MashUps can also perform this aggregation on the client-side (i.e., correlation can occur in the scripts of a web page). Additionally, portals present data collected from disparate sources together but without interaction between data sets. In contrast, MashUps focus heavily on merging disparate data sets into one unified representation. For example, a news portal would simply present a set of interesting articles gathered from various sources on a single page while a news MashUp would correlate textual news with related images and multimedia as well as automatically linking related articles in a single view.

While similar to MashUps, service composition is a more generic concept that focuses on orchestrating web service calls as part of some higher level application logic. The coordination of web service calls in a service composition are often more process-centric rather than data-centric. For example, a flight-booking composite service would query the flight reservation services of different airlines to book a flight based on customer requirements, while a flight-booking MashUp would gather flight data from various airlines and present the data to customers in a single unified view – likely correlated with other useful data such as weather.

There is currently no standardization of technologies or tools used to develop MashUps. In fact, many industry leaders such as Google, Microsoft, and Yahoo are pushing their own MashUp development tools. In particular, many of these tools are targeting non-programmers in hopes of expanding the base of users capable of contributing to and developing MashUps.

Cross-References

▶ AJAX
▶ Web 2.0/3.0

Massive Array of Idle Disks

Kazuo Goda
The University of Tokyo, Tokyo, Japan

Synonyms

MAID

Definition

The term Massive Array of Idle Disks refers to an energy-efficient disk array which has the capability of changing its disk drives into a low-power mode when the disk drives are not busy. Disk drives of the array may be controlled individually or in a group. The term Massive Array of Idle Disks is often abbreviated to MAID. A MAID disk array may have additional functions such as data migration/replication and access prediction to improve the energy saving.

Key Points

The basic idea of MAID is to save energy by exploiting storage access locality. That is, some disk drives which are installed in a disk array are frequently accessed whereas the others are rarely busy. MAID tries to spin down or power off such "low-temperature" disk drives to decrease the total energy consumption. The original papers [1, 2] which introduced MAID in 2002 studied different design choices and configurations. MAID disk arrays are mainly used for archival storage (replacing conventional tape libraries) or nearline storage (which falls between online storage and archival storage).

Cross-References

▶ Storage Power Management

Recommended Reading

1. Colarelli D, Grunwald D. Massive Arrays of Idle Disks for Storage Archives. In: Proceedings of the 2002 ACM/IEEE Conference on Supercomputing; 2002. p. 1–11.
2. Colarelli D, Grunwald D, Neufeld M. The case for massive arrays of idle disks (MAID). In: Proceedings of the 1st USENIX Conference on File and Storage Technologies Work-in Progress Reports; 2002.
3. Storage Network Industry Association. The dictionary of storage networking terminology. Also available at: http://www.snia.org/.

Matrix Masking

Stephen E. Fienberg and Jiashun Jin
Carnegie Mellon University, Pittsburgh, PA, USA

Synonyms

Adding noise; Data perturbation; Recodings; Sampling; Synthetic data

Definition

Matrix Masking refers to a class of statistical disclosure limitation (SDL) methods used to protect confidentiality of statistical data, transforming an $n \times p$ (cases by variables) data matrix Z through pre- and post-multiplication and the possible addition of noise.

Key Points

Duncan and Pearson [3] and many others subsequently categorize the methodology used for SDL in terms of transformations of an $n \times p$ (cases by variables) data matrix Z of the form

$$Z \rightarrow AZB + C, \qquad (1)$$

where A is a matrix that operates on the n cases, B is a matrix that operates on the p variables, and C is a matrix that adds perturbations or noise.

Matrix masking includes a wide variety of standard approaches to SDL: (i) adding noise, i.e., the C in matrix masking transformation of equation [1]; (ii) releasing a subset of observations (delete rows from Z), i.e., sampling; (iii) cell suppression for cross-classifications; (iv) including simulated data (add rows to Z); (v) releasing a subset of variables (delete columns from Z); (vi) switching selected column values for pairs of rows (data swapping). Even when one has applied a mask to a data set, the possibilities of both identity and attribute disclosure remain, although the risks may be substantially diminished.

Cross-References

▶ Individually Identifiable Data
▶ Inference Control in Statistical Databases
▶ Privacy
▶ Randomization Methods to Ensure Data Privacy
▶ Statistical Disclosure Limitation for Data Access

Recommended Reading

1. Doyle P, Lane JI, Theeuwes JJM, Zayatz L, editors. Confidentiality, disclosure and data access: theory and practical application for statistical agencies. New York: Elsevier; 2001.
2. Duncan GT, Jabine TB, De Wolf VA, editors. Private lives and public policies. Report of the Committee on National Statistics' panel on confidentiality and data access. Washington, DC: National Academy Press; 1993.
3. Duncan GT, Pearson RB. Enhancing access to microdata while protecting confidentiality: prospects for the future (with discussion). Stat Sci. 1991;6(3):219–39.
4. Federal Committee on Statistical Methodology. Report on statistical disclosure limitation methodology, Statistical policy working paper 22. Washington, DC: U.S. Office of Management and Budget; 1994.

Max-Pattern Mining

Guimei Liu
Institute for Infocomm Research, Singapore,
Singapore

Synonyms

Maximal itemset mining

Definition

Let $I = \{i_1, i_2 \ldots, i_n\}$ be a set of items and $D = \{t_1, t_2 \ldots, t_N\}$ be a transaction database, where $t_i (i \in [1, N])$ is a transaction and $t_i \subseteq I$. Every subset of I is called an *itemset*. If an itemset contains k items, then it is called a k-itemset. The support of an itemset X in D is defined as the percentage of transactions in D containing X, that is, $\sup(X) = |\{t | t \in D \wedge X \subseteq t\}| / |D|$. If the support of an itemset exceeds a user-specified minimum support threshold, then the itemset is called a *frequent itemset* or a *frequent pattern*. If an itemset is frequent but none of its supersets is frequent, then the itemset is called a *maximal pattern*. The task of maximal pattern mining is given a minimum support threshold, to enumerate all the maximal patterns from a given transaction database.

The concept of maximal patterns can be and has already been extended to more complex patterns, such as sequential patterns, frequent subtrees, and frequent subgraphs. For each type of patterns, a pattern is maximal if it satisfies the given constraints but none of its super-patterns satisfies the given constraints.

Historical Background

If a k-itemset is frequent, then all of its subsets are frequent and the number of them is $2^k - 1$. Datasets collected from some domains can be very dense and contain very long patterns. Any algorithm which produces the complete set of frequent itemsets suffers from generating numerous short patterns on these datasets, and most of the short patterns may be useless. Some researchers have noticed the long pattern problem and suggested mining only maximal frequent patterns [1, 3, 8]. The set of maximal patterns provides a concise view of the frequent patterns, and it can be orders of magnitude smaller than the complete set of frequent patterns. The complete set of frequent patterns can be derived from the set of maximal patterns, but the support information is lost.

A different concept called *closed itemset* or closed pattern has also been proposed to reduce result size. A pattern is closed if none of its supersets has the same support as it. Closed patterns retain the support information of frequent patterns. The complete set of frequent patterns can be derived from the set of frequent closed patterns without information loss. A maximal pattern must be a closed pattern, but not vice versa.

Given a set of items I, the search space of the *frequent itemset* mining problem is the power set of I, and it can be represented as a set-enumeration tree given a specific order of I [9]. Figure 1 shows the search space tree for $I = \{a, b, c, d, e\}$, and the items are sorted lexicographically. Every node in the search space tree represents an itemset. For every itemset X in the tree, only the items after the last item of X can be appended to X to form a longer itemset. These items are called *candidate extensions* of X, denoted as *cand_ext(X)*. For example, items d and e are candidate extensions of ac, while item b is not a candidate extension of ac because b is before c in lexicographic order. Mining maximal patterns can be viewed as finding a border through the search space tree such that all the nodes below the border are infrequent and all the nodes above the border are frequent. As shown in Fig. 1, the dotted line represents the border. All the nodes above the dotted line are frequent and all the nodes below the dotted line are infrequent. Among all the nodes above the border, only leaf nodes can be maximal, but not all the leaf nodes are maximal; every internal node has at least one frequent child (superset) thus cannot be maximal. The goal of maximal pattern mining is to find the border by counting support for as less as possible itemsets. Most, if not all, existing maximal pattern mining

M

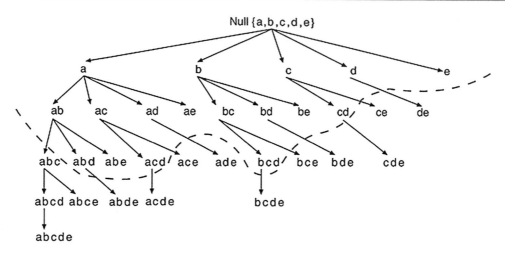

Max-Pattern Mining, Fig. 1 Search space tree for $I = \{a, b, c, d, e\}$

algorithms try to find some frequent long patterns first, and then use these long patterns to prune non-maximal patterns.

The first attempt at mining maximal patterns is made by Gunopoulos et al. [6], and an algorithm called *dualize and advance* is proposed. The algorithm is based on the observation that given a set of maximal patterns, any other maximal pattern not in the set must contain at least one common item with the complement of every maximal pattern in the set, where the complement of an itemset X is defined as $I - X$. The algorithm works as follows. It first uses a greedy search to generate some maximal patterns, denoted as H, and then finds the minimal patterns that contain at least one common item with the complement of every maximal pattern in H. Here a pattern is minimal if none of its subsets satisfies the condition. These minimal patterns are called minimal transversals of H. If all the minimal transversals of H are infrequent, it means that all the maximal patterns are all in H already. Otherwise, there exists some minimal transversal X of H such that X is frequent; the algorithm then finds a maximal superset of X, denoted as Y; and Y must be a maximal pattern. The algorithm puts Y in H and then generates the minimal transversals of the updated H. This process is repeated until no frequent minimal transversals of HH exists. The upper bound for the time complexity of this algorithm is sub-exponential to the output size.

Pincer-search [7] combines the bottom-up and top-down search strategy and approaches the border from both directions. The bottom-up search is similar to the a priori algorithm [2], and the top-down search is implemented by maintaining a set called maximum-frequent-candidate set (MFCS). MFCS is a minimum cardinality set of itemsets such that the union of all the subsets of its elements contains all the frequent itemsets that have been discovered so far but does not contain any itemsets that have been determined to be infrequent. Pincer-search uses MFCS to prune those candidate itemsets that have a frequent superset in MFCS to reduce the database scan times and the support counting cost.

Both the dualize and advance algorithm and the Pincer-search algorithm maintain the set of candidate maximal patterns during the mining process and use them to prune short non-maximal patterns. The main difference between the two algorithms is that Pincer-search considers large sets that may be frequent first and then shrinks them to find the real maximal patterns, while the dualize and advance algorithm starts from some seed maximal patterns and uses them to find other maximal patterns. Both algorithms can prune non-maximal patterns effectively, but maintaining and manipulating the set of candidate maximal patterns can be very costly.

Zaki et al. propose two maximal pattern mining algorithms MaxClique and MaxEclat [13]. Both algorithms rely on a preprocessing step to cluster itemsets and then use a hybrid bottom-up and top-down approach to find maximal patterns from each cluster with a vertical data representation. The two algorithms differ in how the itemsets are clustered. The purpose of the clustering step is to find some potential maximal patterns, and the two algorithms use these potential maximal patterns to restrict the search space. However, the cost of the clustering step can be very high.

Max-Miner [3] is a very practical algorithm for mining maximal patterns. It uses the bottom-up search strategy to traverse the search space as the a priori algorithm [2], but it always attempts to look ahead in order to quickly identify long patterns. By identifying a long pattern first, Max-Miner can prune all its subsets from consideration. Two pruning techniques are proposed in the Max-Miner algorithm: pruning based on superset frequency and the dynamic reordering technique. These two pruning techniques are very effective in removing non-maximal patterns and have been adopted by all later maximal pattern mining algorithms:

1. *Superset frequency pruning*. This technique is also called the lookahead technique. It is based on the observation that the itemsets in the sub search space tree rooted at X are subsets of $X \cup cand\_ext(X)$. Therefore, if $X \cup cand\_ext(X)$ is frequent, then none of the itemsets in the subtree rooted at X can be maximal and the whole branch can be pruned. There are two ways to check whether $X \cup cand\_ext(X)$ is frequent. One way is to check whether $X \cup cand\_ext(X)$ is a subset of some maximal pattern that has already been discovered. The other way is to look at the support of $X \cup cand\_ext(X)$ in the database, which can be done when counting the support of X's immediate supersets.

2. *Dynamic item reordering*. At every node X, Max-Miner sorts the items in $cand\_ext(X)$ in ascending frequency order. The candidate extensions of an item include all the items that are after it in the ascending frequency order. Let i_1 and i_2 be two items in $cand\_ext(X)$. Item i_1 is before i_2 in the ordering if $sup(X \cup \{i_1\})$ is smaller than $sup(X \cup \{i_2\})$, and item i_2 is a candidate extension of $X \cup \{i_1\}$. The motivation behind dynamic item reordering is to increase the effectiveness of the superset frequency pruning technique. The superset frequency pruning can be applied when $X \cup cand\_ext(X)$ is frequent. It is therefore desirable to make many Xs satisfy this condition. A good heuristic for accomplishing this is to force the most frequent items to be the candidate extensions of all other items because items with high frequency are more likely to be part of long frequent itemsets.

Besides the above two pruning techniques, Max-Miner also uses a technique that can often determine when a new candidate itemset is frequent before accessing the database. The idea is to use information gathered during previous database passes to compute a good lower bound on the number of transactions that contain the itemset.

The Max-Miner algorithm uses the breadth-first search order to explore the search space, which makes it not very efficient on dense datasets. DepthProject [1], MAFIA [4], GenMax [5], and AFOPT-Max [8] use the depth-first search strategy to traverse the search space. The depth-first search strategy is capable of finding long patterns first, which makes the superset frequency pruning technique more effective. These algorithms differ mainly in their support counting technique. Both DepthProject and MAFIA assume that the dataset fits in the main memory. At any point in the search, DepthProject maintains the projected transaction sets for some of the nodes on the path from the root to the node currently being processed, where a projected transaction of a node contains only the candidate extensions of the node. It is possible that a projected transaction is empty; in this case, the projected transaction is discarded. Since the projected database is substantially smaller than the original database both in terms of the number of transactions and the number of items, the process of finding the support counts is speeded up substantially. DepthProject also uses a bucketing technique to speed up the support counting. MAFIA and GenMax use the vertical mining technique, that is, each itemset is associated with a transaction id (tid) bitmap, and support counting is performed by tid bitmap join. MAFIA uses another pruning technique called parent equivalence pruning (PEP), which is essentially to remove frequent non-closed itemsets. GenMax [5] proposes a progressive focusing technique to improve the efficiency of superset searching. AFOPT-Max uses a prefix-tree structure to store projected transactions, which can make the lookahead pruning technique be performed more efficiently.

The concept of maximal patterns has been extended to other similar data mining problems dealing with complex data structures. Yang et al. [12] devise an algorithm that combines statistical sampling and a technique called border collapsing to discover long sequential patterns with sufficiently high confidence in a noisy environment. Xiao et al. [10] propose an algorithm to mine maximal frequent subtrees from a database of unordered labeled trees.

Scientific Fundamentals

In practice, mining maximal patterns is much cheaper than mining the complete set of frequent itemsets. However, the worst-case time complexity of maximal pattern mining is the same as mining all frequent patterns. Yang [11] studies the complexity of the maximal pattern mining problem and proves that the problem of counting the number of distinct maximal frequent itemsets in a transaction database, given an arbitrary support threshold, is #P-complete.

Key Applications

Maximal pattern mining is applicable to dense domains where extracting all frequent patterns is not feasible. It can also be used as a preprocessing step to improve the efficiency of frequent pattern mining and to decide appropriate thresholds for frequent pattern mining and association rule mining.

Experimental Results

Each introduced method has an accompanying experimental evaluation in the corresponding reference. A comprehensive comparison of different algorithms can be found at http://fimi. cs.helsinki.fi/experiments/.

Data Sets

A collection of datasets commonly used for experiments can be found at http://fimi.cs.helsinki. fi/data/.

URL to Code

http://fimi.cs.helsinki.fi/src/.

Cross-References

▶ Closed Itemset Mining and Nonredundant Association Rule Mining
▶ Frequent Itemsets and Association Rules

Recommended Reading

1. Agrawal R, Srikant R. Fast algorithms for mining association rules in large databases. In: Proceedings of the 20th International Conference on Very Large Data Bases; 1994. p. 487–99.
2. Agarwal RC, Aggarwal CC, Prasad VVV. Depth first generation of long patterns. In: Proceedings of the 6th ACM SIGKDD International Conference on Knowledge Discovery and Data Mining; 2000. p. 108–18.
3. Bayardo RJ Jr. Efficiently mining long patterns from databases. In: Proceedings of the ACM SIGMOD International Conference on Management of Data; 1998. p. 85–93.
4. Burdick D, Calimlim M, Gehrke J. Mafia: a maximal frequent itemset algorithm for transactional databases. In: Proceedings of the 17th International Conference on Data Engineering; 2001. p. 443–52.
5. Gouda K, Zaki MJ. GenMax: efficiently mining maximal frequent itemsets. In: Proceedings of the 1st IEEE International Conference on Data Mining; 2001. p. 163–70.
6. Gunopulos D, Mannila H, Saluja S. Discovering all most specific sentences by randomized algorithms. In: Proceedings of the 6th International Conference on Database Theory; 1997. p. 215–29.
7. Lin DI, Kedem ZM. Pincer search: a new algorithm for discovering the maximum frequent set. In: Advances in Database Technology, Proceedings of the 1st International Conference on Extending Database Technology; 1998. p. 105–19.
8. Liu G, Lu H, Lou W, Xu Y, Yu JX. Efficient mining of frequent patterns using ascending frequency ordered prefix-tree. Data Min Knowl Disc. 2004;9(3):249–74.

9. Rymon R. Search through systematic set enumeration. In: Proceedings of the 3rd International Conference on Principles of Knowledge Representation and Reasoning; 1992. p. 268–75.
10. Xiao Y, Yao JF, Li Z, Dunham MH. Efficient data mining for maximal frequent subtrees. In: Proceedings of the 3rd IEEE International Conference on Data Mining; 2003. p. 379–86.
11. Yang G. The complexity of mining maximal frequent itemsets and maximal frequent patterns. In: Proceedings of the 10th ACM SIGKDD International Conference on Knowledge Discovery and Data Mining; 2004. p. 344–53.
12. Yang J, Wang W, Yu PS, Han J. Mining long sequential patterns in a noisy environment. In: Proceedings of the ACM SIGMOD International Conference on Management of Data; 2002. p. 406–17.
13. Zaki MJ, Parthasarathy S, Ogihara M, Li W. New algorithms for fast discovery of association rules. In: Proceedings of the 3rd International Conference on Knowledge Discovery and Data Mining; 1997. p. 283–6.

Mean Reciprocal Rank

Nick Craswell
Microsoft Research Cambridge, Cambridge, UK

Synonyms

Mean reciprocal rank of the first relevant document; MRR; MRR1

Definition

The Reciprocal Rank (RR) information retrieval measure calculates the reciprocal of the rank at which the first relevant document was retrieved. RR is 1 if a relevant document was retrieved at rank 1, if not it is 0.5 if a relevant document was retrieved at rank 2 and so on. When averaged across queries, the measure is called the Mean Reciprocal Rank (MRR).

Key Points

Mean Reciprocal Rank is associated with a user model where the user only wishes to see one relevant document. Assuming that the user will look down the ranking until a relevant document is found, and that document is at rank n, then the precision of the set they view is 1/n, which is also the reciprocal rank measure. For this reason, MRR is equivalent to Mean Average Precision in cases where each query has precisely one relevant document. MRR is not a shallow measure, in that its value changes whenever the required document is moved, although the change is much larger when moving from rank 1 to rank 2 (change is 0.5) compared to moving from rank 100 to 1,000 (change of 0.009).

MRR is an appropriate measure for known item search, where the user is trying to find a document that he either has seen before or knows to exist. This is called *navigational search* in the case of web search. In a case where there are multiple copies of the required document, or otherwise a set of relevant documents that are substitutes, MRR can still be applied based on the first copy.

Cross-References

▶ Average Precision
▶ Precision-Oriented Effectiveness Measures

M

Measure

Torben Bach Pedersen
Department of Computer Science, Aalborg University, Aalborg, Denmark

Synonyms

Numerical fact

Definition

A *measure* is a numerical property of a multidimensional *cube*, e.g., sales price, coupled with an aggregation formula, e.g., SUM. It captures numerical information to be used for aggregate computations.

Key Points

As an example, a three-dimensional cube for capturing sales may have a Product *dimension* P, a Time dimension T, and a Store dimension S, capturing the product sold, the time of sale, and the store it was sold in, for each sale, respectively. The cube has two measures: DollarSales and ItemSales, capturing the sales price and the number of items sold, respectively. Item-Sales can be viewed as a function: ItemSales: $Dom(P) \times Dom(T) \times Dom(S) \mapsto \mathbb{N}_0$ that given a certain combination of dimension values returns the total number of items sold for that combination. If a dimension value corresponds to a higher level in the dimension *hierarchy*, e.g., a product group or even *all* products, the result is an aggregation of several lower-level measure values [2].

In a multidimensional database, measures generally represent the properties of the chosen facts that the users want to study, e.g., with the purpose of optimizing them. Measures then take on different values for different combinations of dimension values. The property and aggregation formula are chosen such that the value of a measure is meaningful for all combinations of aggregation levels [2]. The formula is defined in the metadata and thus not stored redundantly with the data. Although most multidimensional data models have measures, some do not. In these, dimension values are also used for computations, thus obviating the need for measures, but at the expense of some user-friendliness [4].

It is important to distinguish three classes of measures, namely *additive*, *semi-additive*, and *non-additive* measures, as these behave quite differently in computations. Additive measure values can be summed meaningfully along any dimension. For example, it makes sense to add the total sales over Product, Store, and Time, as this causes no overlap among the real-world phenomena that caused the individual values. Here, the so-called *type* of the measure [2, 3] is *Flow*. Semi-additive measure values cannot be summed along one or more of the dimensions, most often the Time dimension. For these dimensions, alternative aggregation functions such as AVG or MAX are sometimes meaningful.

Semi-additive measures generally occur for so-called "snapshot facts", where the type of the measure is *Stock*. For example, it does not make sense to sum inventory levels across time, as the same inventory item, e.g., a specific product item, may be counted several times, but it is meaningful to sum inventory levels across products and stores. Non-additive measure values cannot be summed along any dimension, usually because the type of the measure is *Value-Per-Unit*, e.g., an item unit price or a discount percentage. Again, alternative aggregation functions may sometimes be used.

A similar, but different, classification applies to the aggregation formula function, which can be either: *distributive,* where higher-level aggregate values can be computed directly from lower-level aggregate values, e.g., SUM; *algebraic,* where higher-level aggregate values can be computed from aggregate values of other, distributive measures, e.g., AVG, or *holistic,* where aggregate values at all levels have to be computed directly from the base data, e.g., MEDIAN [1]. Typically, only measures with distributive aggregation functions are stored in the data cube, while measures with algebraic and holistic aggregation functions are computed on the fly.

In OLAP systems for BI, measures are often used as the building blocks for complex formulas that capture important business characteristics, so-called *Key Performance Indicators (KPIs)* [6]. Although most measures have simple data types, usually numeric, more complex types of measures exist for non-conventional data, e.g., spatial (geo objects such as polygons) and text data (set of frequent terms) [5, 6].

Cross-References

► Cube
► Data Warehousing on Nonconventional Data
► Dimension
► Hierarchy
► Multidimensional Modeling
► Online Analytical Processing
► Summarizability

Recommended Reading

1. Gray J, Chaudhuri S, Bosworth A, Layman A, Reichart D, Venkatrao M, Pellow F, Pirahesh H. Data cube: a relational aggregation operator generalizing group-by, cross-tab, and sub totals. Data Min Knowl Discov. 1997;1(1):29–53.
2. Jensen CS, Pedersen TB, Thomsen C. Multidimensional databases and data warehousing, Synthesis lectures on data management. San Rafael: Morgan Claypool; 2010.
3. Kimball R, Ross M, Thornthwaite W, Mundy J, Becker B. The data warehouse lifecycle toolkit. 2nd ed. Indianapolis: Wiley; 2008.
4. Pedersen TB, Jensen CS, Dyreson CE. A foundation for capturing and querying complex multidimensional data. Inf Syst. 2001;26(5):383–423.
5. Pedersen TB. Managing complex multidimensional data. In: Aufaure M-A, Zimányi E, editors. Business intelligence – second European summer school, eBISS 2012. Brussels: Springer LNBIB; 2013. July 15–21, 2012, Tutorial Lectures.
6. Vaisman A, Zimányi E. Data warehouse systems – design and implementation. Berlin: Springer; 2014.

Mediation

Cesare Pautasso
University of Lugano, Lugano, Switzerland

Synonyms

Adaptation; Bridging; Mapping; Transformation

Definition

Mediation is the process of reconciling differences to reach an agreement between different parties. In databases, the goal of mediation is to compute a common view over multiple, distinct, and heterogeneous sources of data. In software architecture, a component plays the role of mediator if it achieves interoperability by decoupling heterogeneous component having mismatching interfaces. Protocol mediation enables the exchange of information between autonomous endpoints that use incompatible communication protocols.

Mediation middleware helps applications deal with heterogeneity. By hiding the multiplicity and the complexity of the underlying systems, it transforms a one-to-many interaction (the application communicating with multiple data sources) into a simpler, one-to-one interaction (the application communicates with the mediator) and shifts the complexity of handling the communication with multiple, heterogeneous parties into a reusable component: the mediator.

Historical Background

In the context of information systems, the concept of mediation has been introduced by Gio Wiederhold in 1992 as the organizing principle for the interoperation of heterogeneous software components and data sources [8]. Mediation is performed by *a layer of intelligent middleware services in information systems, linking data resources and application programs* [9]. Programs need to consume information of multiple data resources and mediators offer them a solution to deal with representation and abstraction problems and exchange objects across multiple, heterogeneous information sources [5]. In the original vision, mediators were seen as independently developed, reusable software modules exploiting expert knowledge about the data to create aggregated information for application programs found at a higher level of abstraction. As shown in Fig. 1, queries from applications are translated by mediators into sub-queries sent out to the different data sources. The various sub-answers are then collected, integrated and returned by the mediator as a single answer to the application.

Foundations

Mediation techniques help to deal with the integration of heterogeneous and incompatible systems. A good understanding of the nature of the problem of dealing with heterogeneity helps to

Mediation, Fig. 1
Mediation architecture

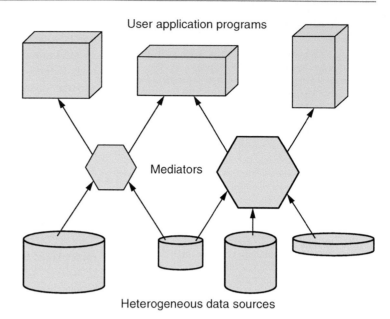

determine when and how mediation should be applied [6]. Conflicts requiring mediation may be found at the level of specific data elements (or messages) to be exchanged or at the level of schema (or interface metadata) definitions. Data values may be stored with multiple representations (syntactic mismatches) or the same representation may be interpreted in different ways (semantic mismatches). Data units (e.g., centimeters versus inches) are also a significant source of conflicts and potential bugs if they are not correctly accounted for when performing data fusion. Identification mismatches are due to the use of locally unique key identifiers that need to be reconciled when tuples from different databases are joined. At the schema level, conflicts involve mismatches in the naming of corresponding elements ("Customer" versus "Client" table); conflicts in the structure ("Street, Number, Zip, City" versus "Address") and granularity ("Purchase Order" versus "Order Line Item") of corresponding data types.

It is important to address such mismatches and conflicts at both data and schema levels. To do so, typical mediation middleware tools provide support for operations such as:

1. Selection, filtering, and aggregation of data
2. Data type, encoding and format conversion

3. Join, comparison, and fusion of data originating from multiple data sources
4. Resolution of conflicts between inconsistent sources
5. Multiplexing and demultiplexing over different channels
6. Lookup-based data translation
7. View materialization and caching of intermediate results

Mediator components can be developed using both imperative and declarative programming languages. Automatic schema matching techniques and algorithms that can be applied to support the development of mediators [7].

Mediation can be provided according to two styles:

- Standard-Based Mediation. The mediator transforms each of the incompatible interfaces to comply with a standardized interface that features the least common denominator between all representations. In order for two parties to communicate, this style requires performing two back-to-back transformations (to the standard representation and from the standard representation), thus introduces a larger performance overhead. However, in terms of development cost and maintainability of the mediator, providing support for a new interface only requires introducing one additional transformation in the mediator (assuming that the new interface can be funneled through the standard one).

- Point-to-Point Mediation. The mediator directly maps each pair of incompatible interfaces. This way, the performance overhead is minimized as only one transformation is required as data is exchanged between two parties. Also, there is no need to define a standardized representation, which may be a difficult task in some cases. However, this style should be applied to mediate between a limited (and fixed) set of interfaces only. The complexity of maintaining this kind of mediator does not scale: adding support for a new interface requires to introduce n additional mappings in the mediator (one for each existing interface).

Key Applications

Mediation plays a prominent role in applications that require integrating data and functionality originally provided by separate systems (especially in database federation and data integration [10]). For example, a mediator performing currency conversion makes it possible to compare prices of products on sale in different markets. Likewise, mediation is needed to build a country-wide census database out of local databases maintained by different cities, if these have been created independently based on different data models and schemas.

In the context of Service-Oriented Architectures, the role of mediator is associated with the communication bus (or Enterprise Service Bus). Through the bus, services exchange messages without being aware that messages may be transformed while in transit to reconcile differences between service interfaces. Such transformations may be based on context and semantics metadata [4]. Mediation is one of the main features of the Web Services Modeling Framework [1], which applies it to address heterogeneity going beyond traditional data mappings. Also mediators for

Business Logic (to compensate between different message ordering constraints found in the public processes of service) and Message Exchange Protocols (for translation between different transport protocols, e.g., to provide reliable and secure message delivery) are introduced in the framework.

In object-oriented design, the *mediator behavioral pattern* has been applied to decouple multiple collaborating objects ([3], p. 273). Instead of having the objects depend on each other and partitioning the interaction logic among them, a mediator object is introduced. It contains and centralizes some potentially complex interaction logic. All objects only refer to the mediator and interact through it. This pattern has also been named *mapper* in the context of enterprise application architecture [2]. Similar to the other applications of mediation, also in this case the interfaces and the overall interactions are simplified thanks to the mediator. However, the price of employing an additional layer of indirection between the elements of a system must be paid.

Cross-References

- ► Enterprise Service Bus
- ► Event Transformation
- ► Schema Mapping
- ► View Adaptation

Recommended Reading

1. Fensel D, Bussler C. The web service modeling framework WSMF. Electron Commer Res Appl. 2002;1(1):113–37.
2. Fowler M. Patterns of enterprise application architecture. Reading: Addison-Wesley; 2002.
3. Gamma E, Helm R, Johnson R, Vlissides J. Design patterns: elements of reusable software. Reading: Addison-Wesley; 1995.
4. Mrissa M, Ghedira C, Benslimane D, Maamar Z, Rosenberg F, Dustdar S. A context-based mediation approach to compose semantic web services. ACM Trans Internet Technol. 2007;8(1):4.
5. Papakonstantinou Y, Garcia-Molina H, Widom J. Object exchange across heterogeneous information sources. In: Proceedings of the 11th International Conference on Data Engineering; 1995. p. 251–60.

6. Park J, Ram S. Information systems interoperability: what lies beneath? ACM Trans Inf Syst. 2004;22(4):595–632.
7. Rahm E, Bernstein PA. A survey of approaches to automatic schema matching. Int VLDB J. 2001;10(4):334–50.
8. Wiederhold G. Mediators in the architecture of future information systems. Computer. 1992;25(4):38–49.
9. Wiederhold G, Genesereth MR. The conceptual basis for mediation services. IEEE Expert. 1997;12(5):38–47.
10. Ziegler P. Data integration project worldwide (2008). http://www.ifi.unizh.ch/~pziegler/IntegrationProjects.html

Membership Query

Mirella M. Moro
Departamento de Ciencia da Computaçao,
Universidade Federal de Minas Gerais – UFMG,
Belo Horizonte, MG, Brazil

Synonyms

Equality query; Equality selection

Definition

Consider a relation R whose schema contains some attribute A taking values over a domain D. A *membership query* retrieves all tuples in R with $A = x$ ($x \in D$).

Key Points

A membership query effectively checks membership in a set (relation). As such, it can be implemented using either a hash-based index (built on the attribute(s) involved in the query) or a B+-tree. If a hashing scheme is used, each indexed value is placed on an appropriate hash bucket. Then all records that satisfy $A = x$ are located on the bucket responsible for value x. If A is a numeric attribute (and can thus be indexed using an order-preserving access method like a B+-tree), the membership query is a special case of a range query where the range interval *[low, high]* is reduced to a single value (*low = high = x*).

Cross-References

▶ B+-Tree

Memory Hierarchy

Stefan Manegold
CWI, Amsterdam, The Netherlands

Synonyms

Hierarchical memory system

Definition

A Hierarchical Memory System – or Memory Hierarchy for short – is an economical solution to provide computer programs with (virtually) unlimited fast memory, taking advantage of locality and cost-performance of memory technology. Computer storage and memory hardware – from disk drives to DRAM main memory to SRAM CPU caches – shares the limitation that as they became faster, they become more expensive (per capacity), and thus smaller. Consequently, memory hierarchies are organized into several levels, starting from huge and inexpensive but slow disk systems to DRAM main memory and SRAM CPU caches (both off and on chip) to registers in the CPU core. Each level closer to the CPU is faster but smaller than the next level one step down in the hierarchy. Memory hierarchies exploit the principle of locality, i.e., the property that computer programs do not access all their code and data uniformly, but rather focus on referencing only small fractions for given periods of time. Consequently, during each period of time, only the fraction currently referenced – also called hot-set, locality-set, or working-set – needs to be present in the fastest memory level, while the remaining data and code can stay in slower levels. In general, all data in one level is also found in all (slower but larger) memory levels below it.

Historical Background

A three-level memory hierarchy, consisting of (i) the CPU's registers, (ii) DRAM main-memory as primary storage and (iii) secondary storage, has been in use since the introduction of drums and then disk drives as secondary storage. In this memory hierarchy, the decision of which data are loaded into a higher level at what time (and written back in case it is modified) is completely under software control. Application programs determine when to load data from secondary to primary memory, and compilers determine when to load data from main memory into CPU registers.

With the introduction of virtual memory around 1960 [10] application software – and in particular their programmers – are provided with uniformly addressable memory larger than physical memory. The operating system takes care of loading the references portion of data into main memory, automating page transfers, and hence relieving programmers from this task. In the late 1960s, the discovery of the locality principle [7] led to the invention of working sets [6] that exploit locality properties to predict data references, and enabled the design of page replacement algorithms that make virtual memory work efficiently and reliably in multiprogramming environments. Since then, virtual memory has become an inherent feature of operating systems. In contrast to many other application programs, a database management system usually does not rely on the operating system's generic virtual memory management, only. Instead, it implements its own buffer pool, exploiting domain specific knowledge to implement application-specific replacement algorithms.

Until the 1980s, the three-level memory hierarchy has been the state-of-the art, mainly because main memory is considered fast enough to serve the CPU – or better, the CPU were "slow enough." Since then, a continuously growing performance gap develops between CPU and main memory. With the chip integration technology following Moore's Law [12], i.e., doubling the number of transistors per chip area roughly every 1.5 years, the performance has grown exponentially, due to increasing clock-speeds, increasing inherent parallelism, or both. Advanced manufacturing techniques has grown the capacity and – thanks to even wider and faster system buses – their data transfer bandwidth of main memory similarly. Memory access latency, however, has lagged behind, demonstrating at most a slight linear improvement.

To bridge this performance gap, in the 1980's hardware designers started to extend the memory hierarchy by adding small (and expensive) but fast SRAM cache memories between the CPU and main memory. Initially, a single cache level is added, either located on the system board between CPU and main memory, or integrated on the CPU chip. With advancing integration and manufacturing techniques, more levels are added. Nowadays, two to three cache levels integrated on the CPU chip represent the most common configuration.

The main difference between cache memories and the original three levels of the memory hierarchy is that their contents are completely controlled by hardware logic. Relying on the locality principle, carefully turned replacement algorithms (usually variations of LRU), decide when data are loaded into or evicted from which cache level. While ensuring transparent use and easy portability, this approach leaves programs virtually without means to explicitly control the content of CPU caches. In modern systems, software prefetching commands have been introduced to provide programs with limited control to (pre-)load data into caches without actually accessing it.

The scientific computing and algorithm communities – both usually focusing on compute-intensive tasks on memory-based data

M

sets – quickly realized that they have to make the algorithms and data structures *cache-conscious* to exploit the performance potentials of ever faster CPU and CPU caches effectively and efficiently.

The database world initially ignored the new hardware developments, assuming that optimizing disk access (I/O) is still the key to high performance execution of data-intensive tasks on large disk-based data sets. First proposals of cache-conscious database algorithms occurred in the mid 1990s [14]. It was not until the end of the twentieth century that the database community realized that memory access has become a severe bottleneck also for database query processing performance, taking up to 90 % of the execution time [4, 5]. In the last decade, the development of hardware-aware database technology from system architectures over data structures to query processing algorithms has become a very active and recognized research area [1–3, 13].

Foundations

Memory- and Cache-Architectures

Modern computer architectures have a *hierarchical memory system*, as depicted in Fig. 1. The main memory on the system board consists of *DRAM* (*Dynamic Random Access Memory*)

chips. While CPU speeds are increasing rapidly, DRAM access latency has hardly progressed over time. To narrow the exponentially growing performance gap between CPU speed and memory latency (cf., Fig. 2), *cache memories* have been introduced, consisting of fast but expensive *SRAM* (*Static Random Access Memory*) chips. SRAM cells are usually made-up of six transistors per memory bit, and hence, they consume a rather large area on the chips. DRAM cells require a single transistor and a small capacitor to store a single bit. Thus, DRAMs can store much more data than SRAMs of equal (physical) size. But due to some current leakage, the capacitor in DRAMs get discharged over time, and have to be recharged (*refreshed*) periodically to keep their information. These refreshes slow down access.

The fundamental principle of all cache architectures is "*reference locality*," i.e., the assumption that at any time the CPU, thus the program, repeatedly accesses only a limited amount of data (i.e., memory) that fits in the cache. Only the first access is "slow," as the data have to be loaded from main memory. This is called a *compulsory cache miss* (see below). Subsequent accesses (to the same data or memory addresses) are then "fast", as the data are then available in the cache. This is called a *cache hit*. The fraction of memory accesses that can be fulfilled from the cache is

Memory Hierarchy, Fig. 1
Hierarchical memory architecture

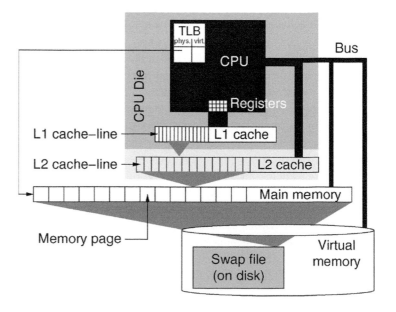

Memory Hierarchy, Fig. 2
Trends in CPU and DRAM
speed

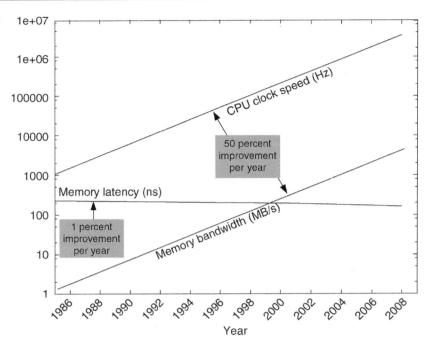

called *cache hit rate*; analogously, the fraction of memory accesses that cannot be fulfilled from the cache is called *cache miss rate*.

Cache memories are often organized in *multiple cascading levels* between the main memory and the CPU. As they become faster, but smaller, the closer they are to the CPU. Originally, there was one level (typically 64–512 KB) of cache memory located on the system board. As the chip manufacturing processes improved, a small cache of about 4–16 KB was integrated on the CPU's die itself, allowing much faster access. The on-board cache is typically not replaced by the on-chip cache, but rather both make up a cache hierarchy, with the one on chip called *first level* (*L1*) cache and the one on board called *second level* (*L2*) cache. Over time, the L2 cache has also been integrated on the CPU's die (e.g., with Intel's Pentium III "Coppermine," or AMD's Athlon "Thunderbird"). On PC systems, the on-board cache has since disappeared, keeping two cache levels. On other platforms, e.g., workstations based on Compaq's (formerly DEC's) Alpha CPU, the on-board cache is kept as *third level* (*L3*) cache, next to the two levels on the die. Most recent multi-core CPUs usually have a private L1 cache of 16–64 KB per core. The

L2 cache is also integrated on the CPU's die. L2 configuration vary between one private L2 per core to one single L2 that is shared among all cores. On Intel's Core 2 Quad, for instance, each of the two L2 caches is shared by two of the four cores. Typical L2 sizes are in the order of 1–8 MB.

To simplify presentation, the remainder of this entry assumes a typical system with two cache levels (L1 and L2). However, the discussion can easily be generalized to an arbitrary number of cascading cache levels in a straightforward way.

In practice, caches memories do not only cache the data used by an application, but also the program itself, more accurately, the instructions that are currently being executed. With respect to caching, there is one major difference between data and program. Usually, a program must not be modified while it is running, i.e., the caches may be read-only. Data, however, requires caches that also allow modification of the cached data. Therefore, almost all systems nowadays implement two separate L1 caches, a read-only one for instructions and a read-write one for data. The L2 cache, however, is usually a single "unified" read-write cache used for both instructions and data.

M

- Caches are characterized by three major parameters: Capacity (C), Line Size (Z), and Associativity (A):
  - **Capacity** (C) A cache's capacity defines its total size in bytes. Typical cache sizes range from 8 KB to 8 MB.
  - **Line Size** (Z) Caches are organized in *cache lines*, which represent the smallest unit of transfer between adjacent cache levels. Whenever a cache miss occurs, a complete cache line (i.e., multiple consecutive words) is loaded from the next cache level or from main memory, transferring all bits in the cache line in parallel over a wide bus. This exploits spatial locality, increasing the chances of cache hits for future references to data that are "close to" the reference that caused a cache miss. Typical cache line sizes range from 16 to 256 bytes. Dividing the cache capacity by the cache line size yields the *number of available cache lines* in the cache: $\# = C/Z$.
  - **Associativity** (A) To which cache line the memory is loaded depends on the memory address and on the cache's *associativity*. An *A-way set associative* cache allows loading a line in A different positions. If $A > 1$, some *cache replacement* policy chooses one from among the A candidates. *Least Recently Used* (*LRU*) is the most common replacement algorithm. In case $A = 1$, the cache is called *directly-mapped*. This organization causes the least (virtually no) overhead in determining the cache line candidate. However, it also offers the least flexibility and may cause a lot of *conflict misses* (see below). The other extreme case is a *fully associative* cache. Here, each memory address can be loaded to any line in the cache ($A = \#$). This avoids conflict misses, and only *capacity misses* (see below) occur as the cache capacity is exceeded. However, determining the cache line candidate in this strategy causes a relatively high overhead that increases with the cache size. Hence, it is feasible for only smaller caches. Current PCs

and workstations typically implement two-way to eight-way set associative caches.

With multiple cache levels, two types are distinguished: inclusive and exclusive caches. With *inclusive caches*, all data stored in L1 is also stored in L2. As data are loaded from memory, they are stored in all cache levels. Whenever a cache line needs to be replaced in L1 (because a mapping conflict occurs or as the capacity is exceeded), its original content can simply be discarded as another copy of that data still remains in the (usually larger) L2. The new content is then loaded from where it is found (either L2 or main memory). The total capacity of an inclusive cache hierarchy is hence determined by the largest level. With *exclusive caches*, all cached data are stored in exactly one cache level. As data are loaded from memory, they get stored only in the L1 cache. When a cache lines needs to be replaced in L1, its original content is first written back to L2. If the new content is then found in L2, it is moved from L2 to L1, otherwise, it is copied from main memory to L1. Compared to inclusive cache hierarchies, exclusive cache hierarchies virtually extend the cache size, as the total capacity becomes the sum of all levels. However, the "swap" of cache lines between adjacent cache levels in case of a cache miss also causes more "traffic" on the bus and hence increases the cache miss latency.

- Cache misses can be classified into the following disjoint types [9]:
  - **Compulsory** The very first reference to a cache line always causes a cache miss, which is hence classified as a compulsory miss, i.e., an unavoidable miss (even) in an infinite cache. The number of compulsory misses obviously depends only on the data volume and the cache line size.
  - **Capacity** A reference that misses in a fully associative cache is classified as a capacity miss because the finite sized cache is unable to hold all the referenced data. Capacity misses can be minimized by increasing the temporal and spatial locality of references in the algorithm. Increasing

cache size also reduces the capacity misses because it captures more locality.

- **Conflict** A reference that hits in a fully associative cache but misses in an A-way set associative cache is classified as a conflict miss. This is because even though the cache is large enough to hold all the recently accessed data, its associativity constraints force some of the required data out of the cache prematurely. For instance, alternately accessing just two memory addresses that "happen to be" mapped to the same cache line will cause a conflict cache miss with each access. Conflict misses are the hardest to remove because they occur due to address conflicts in the data structure layout and are specific to a cache size and associativity. Data structures would, in general, have to be remapped to minimize conflicting addresses. Increasing the associativity of a cache will decrease the conflict misses.
- **Coherence** Only in case of multi-processor or multi-core systems with private per processor/core high-level caches but shared lower-level caches and/or main memory, the following can occur. If two or more cores access the same data item, it will be loaded in each private cache. If one core than modifies this data item in its private cache, the other copies are invalidated and cannot server futures references. Instead, to ensure cache coherence, a cache miss occurs, and the modified data item has to be loaded from the cache that holds the most up-to-date copy.

Memory Access Costs

In general, memory access costs are characterized by the following three aspects:

Latency

Latency is the time span that passes after issuing a data access request until the requested data is available in the CPU. In hierarchical memory systems, the latency increases with the distance from the CPU. Accessing data that are already available in the L1 cache causes *L1 access latency* (λ_{L1}), which is typically rather small (one or two CPU cycles). In case the requested data are not found in L1, an *L1 miss* occurs, additionally delaying the data access by *L2 access latency* (λ_{L2}) for accessing the L2 cache. Analogously, if the data are not yet available in L2, an *L2 miss* occurs, further delaying the access by *memory access latency* (λ_{Mem}) to finally load the data from main memory. Hence, the total latency to access data that are in neither cache is $\lambda_{Mem} + \lambda_{L2} + \lambda_{L1}$. As L1 accesses cannot be avoided, L1 access latency is often assumed to be included in the pure CPU costs, leaving only memory access latency and L2 access latency as explicit memory access costs. As mentioned above, all current hardware actually transfers multiple consecutive words, i.e., a complete cache line, during this time.

When a CPU requests data from a certain memory address, modern DRAM chips supply not only the requested data, but also the data from subsequent addresses. The data are then available without additional address request. This feature is called *Extended Data Output* (EDO). Anticipating sequential memory access, EDO reduces the effective latency. Hence, two types of latency for memory access need to be distinguished. *Sequential access latency* (λ^s) occurs with sequential memory access, exploiting the EDO feature. With random memory access, EDO does not speed up memory access. Thus, *random access latency* (λ^r) is usually higher than sequential access latency.

Bandwidth

Bandwidth is a metric for the data volume (in megabytes) that can be transferred between CPU and main memory per second. Bandwidth usually decreases with the distance from the CPU, i.e., between L1 and L2 more data can be transferred per time than between L2 and main memory. The different bandwidths are referred to as *L2 access bandwidth* (β_{L2}) and *memory access bandwidth* (β_{Mem}), respectively. In conventional hardware, the memory bandwidth used to be simply the cache line size divided by the memory latency. Modern multiprocessor systems typically provide excess bandwidth capacity $\beta' \geq \beta$. To exploit this, caches need to be *non-blocking*, i.e., they need to allow more than one outstanding memory load at a time, and the CPU has to be able to issue subsequent load requests while waiting for the first one(s) to be resolved. Further, the access pattern needs to be sequential, in order to exploit the EDO feature as described above.

Indicating its dependency on sequential access, the excess bandwidth is referred to as *sequential access bandwidth* ($\beta^s = \beta'$). The respective *sequential access latency* is defined as $\lambda^s = Z/\beta^s$. For *random access latency* as described above, the respective *random access bandwidth* is defined as $\beta^r = Z/\lambda^r$.

On some architectures, there is a difference between read and write bandwidth, but this difference tends to be small.

Address Translation

For data access, logical virtual memory addresses used by application code have to be translated to physical page addresses in the main memory of the computer. In modern CPUs, a *Translation Lookaside Buffer* (*TLB*) is used as a cache for physical page addresses, holding the translation for the most recently used pages (typically 64). If a logical address is found in the TLB, the translation has no additional costs. Otherwise, a *TLB miss* occurs. The more pages an application uses (which also depends on the often configurable size of the memory pages), the higher the probability of TLB misses.

The actual *TLB miss latency* (l_{TLB}) depends on whether a system handles a TLB miss in hardware or in software. With software-handled TLB, TLB miss latency can be up to an order of magnitude larger than with hardware-handled TLB. Hardware-handled TLB fetches the translation from a fixed memory structure that is just filled by the operating system. Software-handled TLB leaves the translation method entirely to the operating system, but requires trapping to a routine in the operating system kernel on each TLB miss. Depending on the implementation and hardware architecture, TLB misses can therefore be more costly than a main-memory access. Moreover, as address translation often requires

accessing some memory structure, this can in turn trigger additional memory cache misses.

TLBs can be treated similar to memory caches, using the memory page size as their cache line size, and calculating their (virtual) capacity as *number_of_entries* × *page_size*. TLBs are usually fully associative. Like caches, TLBs can be organized in multiple cascading levels.

For TLBs, there is no difference between sequential and random access latency. Further, bandwidth is irrelevant for TLBs, because a TLB miss does not cause any data transfer.

Unified Hardware Model

Summarizing the above discussion, one can describe a computer's memory hardware as a cascading hierarchy of N levels of caches (including TLBs) [11]. An index $i \in \{1, \ldots, N\}$ added to the parameters described above identifies to the respective value of a specific level. The relation between access latency and access bandwidth then becomes $\lambda_{i+1} = Z_i/\beta_{i+1}$. Exploiting the dualism that an access to level $i + 1$ is caused a miss on level i allows some simplification of the notation. Introducing the *miss latency* $l_i = \lambda_{i+1}$ and the respective *miss bandwidth* $b_i = \beta_{i+1}$ yields $l_i = Z_i/b_i$. Each cache level is characterized by the parameters given in Table 1. Costs for L1

Memory Hierarchy, Table 1 Characteristic parameters per cache level ($i \in \{1, \ldots, N\}$)[2]

Description	Unit	Symbol
Cache name (level)	–	L_i
Cache capacity	[bytes]	C_i
Cache block size	[bytes]	Z_i
Number of cache lines	–	$\#_i = C_i/Z_i$
Cache associativity	–	A_i
Sequential access		
Access bandwidth	[bytes/ns]	β^S_{i+1}
Access latency	[ns]	$\lambda^s_{i+1} = Z_i/\beta^S_{i+1}$
Miss latency	[ns]	$l_i^s = \lambda^S_{i+1}$
Miss bandwidth	[bytes/ns]	$b_i^s = \beta^S_{i+1}$
Random access		
Access latency	[ns]	λ^r_{i+1}
Access bandwidth	[bytes/ns]	$\beta^r_{i+1} = Z_i/\lambda^r_{i+1}$
Miss bandwidth	[bytes/ns]	$b_i^r = \beta^r_{i+1}$
Miss latency	[ns]	$l_i^r = \lambda^r_{i+1}$

cache accesses are assumed to be included in the CPU costs, i.e., λ_1 and β_1 are not used and hence undefined.

Manegold developed a system independent C program called *Calibrator* to measure these parameters on any computer hardware.

Key Applications

In the last decade, the database community has done much research on modifying existing and developing new database technology (system architecture, data structures, query processing algorithms) to exploit the characteristics of the extended memory hierarchy efficiently and effectively, improving query evaluation performance up to orders of magnitude.

URL to Code

Manegold's cache-memory and TLB calibration tool *Calibrator* is available at http://homepages.cwi.nl/~manegold/Calibrator/calibrator.shtml

Cross-References

▶ Buffer Management
▶ Buffer Manager
▶ Buffer Pool
▶ Cache-Conscious Query Processing
▶ Locality
▶ Main Memory
▶ Main Memory DBMS
▶ Processor Cache

Recommended Reading

1. Ailamaki A, Boncz PA, Manegold S, editors. In: Proceedings of the 1st Workshop on Data Management on New Hardware; 2005.
2. Ailamaki A, Boncz PA, Manegold S, editors. In: Proceedings of the 2nd Workshop on Data Management on New Hardware; 2006.
3. Ailamaki A, Luo Q, editors. In: Proceedings of the 3rd Workshop on Data Management on New Hardware; 2007.
4. Ailamaki AG, DeWitt DJ., Hill MD, Wood DA. DBMSs on a modern processor: where does time go? In: Proceedings of the 25th International Conference on Very Large Data Bases; 1999. p. 266–77.
5. Boncz PA, Manegold S, Kersten M. Database architecture optimized for the New Bottleneck: memory access. In: Proceedings of the 25th International Conference on Very Large Data Bases; 1999. p. 54–65.
6. Denning PJ. The working set model for program behaviour. Commun ACM. 1968;11(5):323–33.
7. Denning PJ. The locality principle. Commun ACM. 2005;48(7):19–24.
8. Hennessy JL, Patterson DA. Computer architecture – a quantitative approach. 3rd ed. San Mateo: Morgan Kaufmann; 2003.
9. Hill MD, Smith AJ. Evaluating associativity in CPU caches. IEEE Trans Comput. 1989;38(12):1612–30.
10. Kilburn T, Edwards DBC, Lanigan MI, Sumner FH. One-level storage system. IRE Trans Electronic Comput. 1962;2(11):223–35.
11. Manegold S. Understanding, modeling, and improving main-memory database performance. PhD thesis, Universiteit van Amsterdam, Amsterdam, The Netherlands; 2002.
12. Moore GE. Cramming more components onto integrated circuits. Electronics. 1965;38(8):114–7.
13. Ross K, Luo Q, editors. In: Proceedings of the 3rd Workshop on Data Management on New Hardware; 2007.
14. Shatdal A, Kant C, Naughton J. Cache conscious algorithms for relational query processing. In: Proceedings of the 20th International Conference on Very Large Data Bases; 1994. p. 510–2.

M

Memory Locality

Stefan Manegold
CWI, Amsterdam, The Netherlands

Synonyms

Locality of reference; Locality principle; Principle of locality

Definition

Locality refers to the phenomenon that computer programs – or computational processes in general – do not access all of their data items uniformly and independently, but rather in a clustered and/or dependent/correlated manner. Some data items are accessed more often than others, repeated accesses to the same data item occur in bursts, and related items are usually accessed together, concurrently or within a short time interval.

There are two types of locality:

1. *Temporal locality* means that accesses to the same data item are grouped in time, i.e., multiple accesses to the same data item occur in rather short time intervals compared to rather long time periods where the same data item is not accessed. Hence, temporal locality is the concept that a data item that is referenced by a program at one point in time will be referenced again sometime in the near future.
2. *Spatial locality* means that data items that are stored physically close to each other tend to be accessed together. Hence spatial locality is the concept that likelihood of referencing a data item by a program is higher if a data item near it has been referenced recently.

Locality belongs to the most fundamental principles of computer science.

Key Point

The discovery of the locality principle dates back to the 1960s. Denning's pioneering work on working-set memory management exploits the locality principle to avoid thrashing of virtual memory systems high levels of multiprogramming [2, 3]. Today, this is the key to making virtual memory systems work reliably and efficiently.

In particular, with modern hierarchical memory architectures, exploiting and increasing locality in algorithms and data structures is the key to achieving high performance.

Increased temporal locality ensures that once a data item is referenced, and hence loaded into a fast high-level memory (e.g., cache), all subsequent references occur in short succession while the data item is still available in the cache. Ideally,

this data item will not be required, again, once it is evicted from the fast high-level memory.

Data transfer between adjacent levels of hierarchical memory systems does not happen per byte but with larger granularities, e.g., pages of multiple KB or even MB at a time between disk and main memory, cache lines of tens to hundreds of bytes between main memory and CPU cache. *Increased spatial locality* ensures that all data bytes/items that are loaded with each transfer are indeed useful for the program.

Database system architecture exploits and increases locality in numerous ways. Key examples for increased temporal locality are, for instance partitioned join algorithms, where iterating over small partition of the outer relation increases temporal locality of repeated access to the inner relation [4, 5]. In fact, smaller partitions of the inner relation also increased spatial locality. Examples of spatial locality range from clustered indices over tuned page layouts such as PAX [1] to the decision between column-stores and row-stores to optimal support of column-major (OLAP) of row-major (OLTP) workloads.

Cross-References

▶ Buffer Management
▶ Buffer Manager
▶ Buffer Pool
▶ Cache-Conscious Query Processing
▶ Main Memory
▶ Main Memory DBMS
▶ Memory Hierarchy
▶ Processor Cache

Recommended Reading

1. Ailamaki AG, DeWitt DJ, Hill MD, Skounakis M. Weaving relations for cache performance. In: Proceedings of the 27th International Conference on Very Large Data Bases; 2001. p. 169–180.
2. Denning PJ. The working set model for program behaviour. Commun ACM. 1968;11(5):323–33.
3. Denning PJ. The locality principle. Commun ACM. 2005;48(7):19–24.
4. Manegold S, Boncz PA, Kersten ML. Optimizing main-memory join on modern hardware. IEEE Trans Knowl Data Eng. July 2002;14(4):709–30.

5. Shatdal A, Kant C, Naughton J. Cache conscious algorithms for relational query processing. In: Proceedings of the 20th International Conference on Very Large Data Bases; 1994. p. 510–12.

Merkle Trees

Barbara Carminati
Department of Theoretical and Applied Science, University of Insubria, Varese, Italy

Synonyms

Authentication trees; Hash trees; Merkle hash trees

Definition

Merkle trees are data structures devised to authenticate, with a unique signature, a set of messages, by at the same time making an intended verifier able to verify authenticity of a single message without the disclosure of the other messages. In particular, given a set of messages $M = \{m_1, \ldots, m_n\}$, the Merkle tree build on them is a binary tree whose leaves contain the hash value of each message m in M, whereas internal nodes contain the concatenation of the hash values corresponding to its children.

Key Point

A Merkle tree is a data structure introduced by Merkle in 1979 [1] to improve the Lamport-Diffie one-time signature scheme [2]. In this digital signature scheme, keys can be used to sign, at most, one message. This implies that for each signed message a new public key has to be generated and published. As consequence, Lamport-Diffie one-time digital signature scheme requires publishing a large amount of public keys. To overcome this drawback, in [1] Merkle proposed a tree-structure, called *authentication tree*, with the aim of authenticating a large number of public keys to be used in one-time signature scheme.

In general, Merkle trees can be exploited to authenticate with a unique signature a set of messages by, at the same time, making an intended verifier able to authenticate a single message without the disclosure of the other messages. Given a set of messages $M = \{m_1, \ldots, m_n\}$, the corresponding Merkle tree is computed by means of the following bottom-up recursive computation: at the beginning, for each different message $m \in M$, a different leaf containing the hash value of m is inserted into the tree; then, for each internal node, the associated value is equal to $h(h_l \| h_r)$, where $h_l \| h_r$ denotes the concatenation of the hash values corresponding to the left and right children nodes, and $h()$ is an hash function. The root node of the resulting binary hash tree represents the digest of all messages, and thus it can be digitally signed by using a standard signature technique. The main benefit of this method is that a user is able to validate the signature by having a subset of messages, providing him/her with a set of additional hash values corresponding to missing messages. Indeed, these additional hash values, together with the provided set of original messages, make a user able to locally rebuild the binary tree and, therefore, to validate the signature generated on its root. Consider, for instance, the following set of messages $M = \{m_1, m_2, m_3, m_4\}$. The Merkle tree created with them is a complete binary tree with height of two. More precisely, according to the recursive computation, the root value of the Merkle tree is equal to $h(hr_l \| hr_r)$, where hr_l is its left children with value $h(h(m_1) \| h(m_2))$, whereas hr_r is its right children with value $h(h(m_3) \| h(m_4))$. Assume, now, that a user receives only messages m_1 and m_2. To make him/her able to validate the signature, he/she must be provided also with hash value hr_r. Indeed, by having m_1 and m_2 messages, the user is able to calculate hr_l. Then, using hr_r and hr_l he/she can compute the hash value of the root, and thus verify the signature.

Cross-References

▸ Digital Signatures
▸ Secure Data Outsourcing

Recommended Reading

1. Merkle R. Secrecy, Authentication, and public key systems. Electrical Engineering, PhD Thesis, Stanford University, 1979.
2. Lamport L. Constructing digital signatures from a one-way function. Technical Report CSL-98, SRI International, Palo Alto, 1979.

Message Authentication Codes

Marina Blanton
University of Notre Dame, Notre Dame, IN, USA

Synonyms

MAC; Message integrity codes

Definition

A *message authentication code* (MAC) is a short fixed-length value which is used to authenticate a message. A MAC algorithm can roughly be viewed as a hash function that takes as input two functionally distinct values: a secret key and a message. The output of a MAC algorithm is a short string computed in such a way that it is infeasible to produce the correct output on a message without the knowledge of the key. Thus, the MAC value protects both the *integrity* and *authenticity* of a message by allowing the entity in possession of the secret key to detect any changes to the message content.

Key Points

While MAC functions have conceptual similarities to keyed cryptographic hash functions, they have specific security requirements for authentication purposes that make them a distinct security tool. More precisely, an attacker who does not have access to the secret key and has not seen the MAC value for a specific message before should not be able to compute that value. MAC functions use symmetric techniques (i.e., the same key is used to create and verify a MAC) and thus are different from digital signatures where the signing and verification keys differ. Practical MAC algorithms can be constructed from cryptographic hash functions (for example, HMAC) or from block ciphers (for example, CBC-MAC and others).

Cross-References

▸ Authentication
▸ Digital Signatures
▸ Hash Functions
▸ Symmetric Encryption

Recommended Reading

1. Krawczyk H, Bellare M, Canetti R. HMAC: keyed-hashing for message authentication, RFC 2104. Internet Engineering Task Force (IETF), 1997.
2. Stallings W. Cryptography and network security: principles and practices. 4th ed. Upper Saddle River: Pearson-Prentice Hall; 2006.

Message Queuing Systems

Sara Bouchenak[1] and Noël de Palma[2]
[1]University of Grenoble I – INRIA, Grenoble, France
[2]INPG – INRIA, Grenoble, France

Synonyms

Message-oriented middleware (MOM); Message-oriented systems; Messaging systems; Queuing systems

Definition

A message is an information sent by a sender process to a receiver process. A message queue is a mechanism that allows a sender process and a receiver process to exchange messages. The sender posts a message in the queue, and the receiver retrieves the message from the queue. A message queuing system provides a means to build distributed systems, where distributed processes communicate through messages exchanged via queues.

Key Points

A message queuing system provides several facilities, such as creating messages, creating queues, initializing sender and receiver processes, and providing a means to send and receive messages.

First of all, a message queuing system provides a facility to build a message and fill it with data. Properties may be associated with a message, such as the message size, the message expiration time and the message priority.

A message queuing system also provides facilities to create a queue and, optionally, to associate parameters with a queue, such as the queue length (i.e., the maximum number of messages a queue may hold), the queue topics (i.e., the types of messages the queues may contain), etc.

The senders and receivers of messages may communicate in a synchronous way or in an asynchronous way. With a synchronous communication protocol, a receiver waits for a message from a sender, i.e., it blocks until the message arrives. Whereas with an asynchronous communication protocol, the receiver continues executing and is notified of the reception of a message when this one arrives.

Furthermore, the destination of a message may be specified either explicitly or implicitly. In the explicit mode, the sender specifies the queue to which the message is sent. While in the implicit mode, the sender specifies a topic to which a message is sent, and the message queuing system is responsible of automatically finding the queues that correspond to that topic before sending the message to these queues.

Several message queuing systems are proposed, some are proprietary and others are open source. Oracle proposes Advanced Queuing for Oracle databases [3], Skype has Skytools PgQ for PostgreSQL databases [5], IBM provides WebSphere MQ, Microsoft has MSMQ [1], and Sun Microsystems defines Java Message Service (JMS) as a specification of a Java standard for message queuing systems [6]. Open source message queuing systems include ActiveMQ [7], JBoss Messaging [2], and JORAM [4].

Cross-References

▶ Adaptive Middleware for Message Queuing Systems

Recommended Reading

1. IBM. WebSphere MQ. 2008. http://www-306.ibm.com/software/integration/wmq/
2. JBoss. JBoss messaging. 2008. http://labs.jboss.com/jbossmessaging/
3. Oracle. Oracle9i application developer's guide – advanced queuing. 2008. http://download.oracle.com/docs/cd/B10500_1/appdev.920/a96587/toc.htm
4. ScalAgent. JORAM: java open reliable asynchronous messaging. 2008. http://joram.objectweb.org/
5. Skype. SkyTools PgQ. 2008. https://developer.skype.com/SkypeGarage/DbProjects/SkyTools
6. Sun Microsystems. Java message service (JMS). 2008. http://java.sun.com/products/jms/
7. The Apache software foundation. Apache ActiveMQ. 2008. http://activemq.apache.org/

Meta Data Repository

Christoph Quix
RWTH Aachen University, Aachen, Germany

Synonyms

Meta data base; Meta data management system; Meta data manager; Metadata Registry, ISO/IEC 11179

Definition

A meta data repository (MDR) is a component which manages meta data. In the context of database systems, one example of meta data is information about the database schema, i.e., a description of the data. In addition, MDRs can manage information about the processes which create, use, or update the data, the hardware components that host these processes or the database system, or other (human) resources which make use of the data [6]. As meta data is also data, meta data repositories offer the same functionality for meta data as database management systems (DBMS) for data, e.g., queries, updates, transactions, access control.

Moreover, as meta data is semantically rich data, MDRs often employ object-oriented data models as the basis for the definition of meta data. MDRs should also offer predefined meta models for different types of meta data, so that the user is able to store meta data directly, without defining a meta model in advance.

Another task for a meta data repository is the integration of meta data from various sources into a comprehensive meta data model.

Historical Background

The first components that managed meta data in the context of database systems were dictionary systems which were integrated into the database management systems (DBMS) [12]. These dictionaries were already part of early DBMS products, such as IDMS or IBM IMS. For example, the integrated data dictionary (IDD) of IDMS was a separate database inside the IDMS which was used to maintain meta data of products in the IDMS family [11]. It could be extended also to maintain other types of meta data.

The relational database systems developed in the 1980s also had integrated dictionary systems (also known as system catalogs) to maintain definitions about tables, views, columns, etc. These dictionaries were mainly used by the DBMS itself, but they could be also queried by users and other applications to retrieve information about the contents of a database.

In the 1980s, ANSI started to develop a standard for Information Resource Dictionary Systems (IRDS) which was later adopted by ISO [6]. The standard defined the content, structure, and functionality of an IRDS. The main requirements stated by the IRDS standard are the availability of data modeling facilities, extensibility (i.e., the possibility to add new data types), and the provision of standard DBMS functionality such as query and reporting facilities, integrity and constraint management, and access control.

With the growing need for integrated information systems, stand-alone meta data repository systems became more popular in the 1990s. In contrast to the integrated repositories in DBMS products, a stand-alone MDR was able to manage meta data from different systems. The requirement for meta data integration was especially important for data warehouses, where data that was managed by independent, heterogeneous systems should be integrated into a common data store with a uniformed data model. The availability of meta data of the data sources was a prerequisite for data integration. In this context, some companies tried to build enterprise wide meta data repositories which were supposed to manage all meta data that is available in the enterprise. Such an ambitious goal was hard to achieve, and often, the return-of-investment of such a system was not as high as expected [5]. Therefore, the MDR market was significantly reduced at the end of the 1990s.

Since 2000, two trends for meta data repositories gained importance: community-focused repositories and federated repositories [5]. Communicty-focused repositories are employed in communities (within a company), which share a common interest, such as data warehousing or enterprise application integration. In these communities, the main problem of interoperability of meta data tools could be solved by dedicated bridging technologies, because of the limited scope of the meta data. With the rising importance of service oriented architecture (SOA), MDR needed again to address a broader scope of meta data. Therefore, federated solutions for MDRs are considered to be a solution for the meta data integration problem. In a federated MDR, there are still several MDRs for different communities but federated queries across several MDRs are enabled [5].

Foundations

Requirements

Requirements for MDRs were stated in the IRDS standard [6], in [2], and in [1]:

1. *Dynamic extensibility.* The MDR should provide easy functionalities for the extension of the built-in data models.
2. *Management of objects and relationships.* Objects and relationships between objects should be managed by the MDR.
3. *Notification.* An operation on a specific object in the MDR might trigger other operations on the same or different objects. Therefore, the MDR must be able to notify applications which are interested in certain events. In addition, the invocation of methods inside the MDR (based on other events) should be also possible.
4. *Version management.* Versioning of meta data is required to track the evolution of a meta data object. It is also important to know which versions of two objects were active at a specific time. It might be also necessary to maintain relationships between older and newer versions of an object.
5. *Configuration management.* A configuration is a set of meta data objects which belong together in respect of content, e.g., they all describe the state of one component. The MDR should be able to manage a configuration as one group. Configurations can be also versioned.
6. *Integrity constraints.* The MDR must provide a language for the definition of integrity constraints on meta data, and enforce the compliance of the meta data with these constraints.

7. *Query and reporting functionality.* To retrieve meta data from the repository, the MDR needs to offer a query language. In addition, user-configurable reports should be also supported.
8. *User access.* If the MDR can be accessed directly by end-users or administrators, the MDR needs to support: a browsing facility for meta data, so that users can navigate through the metadata; an access control, so that users see or update only meta data which they are allowed to; a sophisticated user interface if the users are also allowed to update the meta data, so that the integrity of the MDR is maintained.
9. *Interoperability.* To enable interoperability with other tools and repositories, the MDR should support standards for meta data exchange (such as XMI) and offer an API (application program interface).

Architecture

There are several MDRs already available (see "Systems" section below), each having its own unique architecture. However, by abstracting from these concrete architectures, several components which are common for all MDRs can be identified:

1. *Repository.* The repository component is the internal data store of the MDR and therefore the core of the MDR. As MDRs have to provide similar functionality for meta data as DBMS for data, the repository is often implemented on top of a DBMS.
2. *Meta data manager.* The meta data manager acts as the controller of the repository. As all accesses to the repository should go through the meta data manager, it provides an interface for external applications. Using this interface, applications can store, update, and query meta data.
3. *Models.* A MDR needs to come with already predefined meta models (or information models) which can be directly employed by the users of the MDR to store meta data. If the user has to define its own meta models, the effort to get the MDR running might be too high for the application. Nevertheless, it should be

M

possible to extend the existing models for the specific requirements of the applications that use the MDR.

4. *User interface.* As described above, a MDR can be also accessed by users, which are either end-users using the data or processes described in the MDR, or administrators controlling the system of which the MDR is a part. The user interface can consist of a query facility, a meta data browser, an administrator interface, and an interface to update the meta data.

As mentioned above, a current trend for MDRs is the idea of a federated MDR. This changes the standard architecture described before: the repository component in a federated MDR is not one single data store, the meta data can be distributed across several independent and heterogeneous components. In a federated MDR architecture, the meta data could be stored in files, databases, or managed by specific applications. This increases the complexity of the meta data manager significantly, as meta data queries have to be transformed into queries of the individual systems holding the meta data.

Systems

There are several MDRs available in the market. They can be classified as stand-alone MDRs, repositories integrated into larger software platforms, open source systems and research prototypes.

The market for stand-alone MDRs is changing frequently as companies specialized on MDRs are being acquired by other companies. The current products for separate meta data management solutions are, for example, *ASG Rochade*, *Adaptive Metadata Manager*, and *Advantage Repository*. These products came mainly from the data management area (especially used as MDRs in data warehouse systems), but are now also addressing other areas such as enterprise application integration and service-oriented architectures. Other systems, such as *Logidex* from *LogicLibrary* or *BEA AquaLogic Registry Repository*, have been originally developed

as meta data systems for service-oriented architectures.

As mentioned above, large software companies are also addressing the meta data challenges in their software or technology platforms. For example, *IBM* has an integrated MDR in their information integration framework.

There are also a few open source systems which can be used as MDR. Two examples are *Repository in a Box* and *XMDR*. In the research community, *ConceptBase* [9] is a MDR which has been used in several research projects. ConceptBase provides a very flexible data model which can be used for any kind of meta data structure.

Key Applications

There are various application areas for MDRs, basically in all areas in which the management of meta data is necessary. The most important applications for MDRs are situations in which meta data from different sources has to be integrated in one repository. This goes usually beyond the capabilities of builtin MDRs, i.e., repositories which are integrated with other software components.

Data integration in general is an application area in which MDRs play a central role. If data has to be integrated from heterogeneous systems, the description of this data is required to enable the integration. Data warehouse systems [8] are an example for an architecture of integrated data management in which the role of MDRs has been defined explicitly.

In the context of data warehouse systems, also the problem of data quality has been discussed [7, 10]. Meta data is often the basis for data quality measurements, e.g., meta data describes the provenance, the age, the semantics of data. Therefore, MDRs are important components for data quality projects.

A MDR can also be used as a resource for structured documentation about IT systems. In addition to the "usual" meta data artefacts such as models and mappings for data integration, also a documentation of the employed systems and their architecture in an organization is useful.

As discussed before, service-oriented architecture (SOA) are becoming an important concept for software development. As a system based on a SOA is a distributed and often heterogeneous system, the management of meta data in a SOA is also important. Therefore, a MDR is often also a component in a SOA.

Another application area for MDRs might the management of meta data on the web, such as RDF or OWL ontologies. However, the web is build on the idea of decentralized data management which is in conflict with the concept of a central, integrated repository for all kind of meta data. Nevertheless, MDRs can be useful to manage the meta data at a specific site, e.g., the ontologies which are offered by that site and their mappings to other ontologies.

Future Directions

The integration of meta data will remain to be a challenge, however the meta data will be integrated, either materialized in a repository, federated with a virtual integration system, or some combination of these. With the rising importance of web applications, service-oriented architectures and similar concepts, it can be expected that more loosely coupled MDRs with a federated integration become more successful. Existing or upcoming meta data standards, such as such as CWM (common warehouse metamodel) and XMI (XML metadata interchange), might simplify the task, but the integration of meta data will remain a problem. Another trend is the integration of MDR into larger software platforms as it is done (or planned) by the major software vendors.

Meta data integration and new architectures for MDRs are also interesting questions for the research community: How can such systems be built, that enable the integrated querying of various meta data sources? Which lessons can be applied for meta data that have already been learned at the data level? Other research questions for MDRs and meta data management are

addressed in model management [3, 4] which investigates formal methods for working with data models. The challenge for MDRs here is to provide generic structures for the representation of models and mappings.

Cross-References

▶ Data Warehouse Metadata
▶ Metadata Registry, ISO/IEC 11179
▶ Metamodel
▶ Meta Object Facility

Recommended Reading

1. Bauer A, Günzel H, editors. Data-Warehouse-Systeme: Architektur, Entwicklung, Anwendung. Heidelberg: dpunkt-Verlag; 2001.
2. Bernstein PA. Repositories and object oriented databases. ACM SIGMOD Rec. 1998;27(1):88–96.
3. Bernstein PA, Halevy AY, Pottinger R. A vision for management of complex models. ACM SIGMOD Rec. 2000;29(4):55–63.
4. Bernstein PA, Melnik S. Model management 2.0: manipulating richer mappings. In: Proceedings of the ACM SIGMOD International Conference on Management of Data; 2007. p. 1–12.
5. Blechar M. IT Metadata Repository Magic Quadrant Update 2002. Gartner, Inc., 2002.
6. ISO/IEC Information technology – Information Resource Dictionary System (IRDS) Framework. International Standard ISO/IEC 10027:1990, DIN Deutsches Institut für Normung, e.V., 1990.
7. Jarke M, Lenzerini M, Vassiliou Y, Vassiliadis P. Fundamentals of data warehouses. Berlin: Springer; 2000.
8. Jarke M, Vassiliou Y. Foundations of data warehouse quality – a review of the DWQ project. In: Proceedings of the 2nd International Conference Information Quality; 1997. p. 299–313.
9. Jeusfeld MA, Jarke M, Nissen HW, Staudt M. ConceptBase – Managing conceptual models about information systems. In: Bernus P, Mertins K, Schmidt G, editors. Handbook on architectures of information systems. Berlin: Springer; 1998. p. 265–85.
10. Tayi GK, Ballou DP. Examining data quality. Commun ACM. 1998;41(2):54–7.
11. Wikipedia – The Free Encyclopedia IDMS (Integrated Database Management System). Article in the encyclopedia, URL http://en.wikipedia.org/wiki/IDMS, 2008.
12. Wikipedia – The Free Encyclopedia. Metadata. Article in the encyclopedia, URL http://en.wikipedia.org/wiki/Metadata, 2008.

M

Meta Object Facility

Wei Tang
Teradata Corporation, El Segundo, CA, USA

Synonyms

MOF

Definition

The Meta Object Facility (MOF) is an OMG metamodeling and metadata repository standard. It is an extensible model driven integration framework for defining, manipulating and integrating metadata and data in a platform independent manner. MOF-based standards are in use for integrating tools, applications and data [1].

Key Points

MOF was developed as a response to a request for proposal (RFP), issued by the OMG Analysis and Design Task Force, for Metadata repository facility (http://www.omg.org/cgi-bin/doc?cf/96-05-02). The purpose of the facility was to support the creation, manipulation, and interchange of meta models.

MOF provides a metadata management framework, and a set of metadata services to enable the development and interoperability of model and metadata driven systems. The MOF metadata framework is typically depicted as a four-layer architecture as shown in Table 1:

The MOF specification has three core parts:

1. The specification of the MOF Model
   (a) The MOF's built-in meta-metamodel, the "abstract language" for defining MOF metamodels
2. The MOF IDL Mapping
   (a) A standard set of templates that map an MOF metamodel onto a corresponding set of CORBA IDL interfaces

Meta Object Facility, Table 1 OMG's metadata architecture

Meta-level	MOF terms	Examples
M3	Meta-metamodel	The "MOF Model"
M2	Metamodel, meta-metadata	UML Metamodel, CWM Metamodel
M1	Model, metadata	UML models, CWM metadata
M0	Object, data	Modeled systems, Warehouse data

3. The MOF's interfaces
   (a) The set of IDL interfaces for the CORBA objects that represent an MOF metamodel

The OMG adopted the MOF version 1.0 in November 1997. The most recent revision of MOF, 2.0, was adopted in January 2006 and based on the following OMG specifications:

- MOF 1.4 Specification – MOF 2.0 is a major revision of the MOF 1.4 Specification. MOF 2.0 addresses issues deferred to MOF 2.0 by the MOF 1.4 RTF.
- UML 2.0 Infrastructure Convenience Document: ptc/04-10-14 - MOF 2.0 reuses a subset of the UML 2.0 Infrastructure Library packages.
- MOF 2.0 XMI Convenience document: ptc/04-06-11 - Defines the XML mapping requirements for MOF 2.0 and UML 2.0.

The MOF 2 Model is made up of two main packages, Essential MOF (EMOF) and Complete MOF (CMOF).

1. The EMOF Model merges the Basic package from UML2 and merges the Reflection, Identifiers, and Extension capability packages to provide services for discovering, manipulating, identifying, and extending metadata.
2. The CMOF Model is the metamodel used to specify other metamodels such as UML2. It is built from EMOF and the Core:Constructs of UML 2. The Model package does not define any classes of its own. Rather, it merges packages with its extensions that together define basic metamodeling capabilities.

Examples of metadata driven systems that use MOF include modeling and development tools, data warehouse systems, metadata repositories etc. A number of technologies standardized by OMG, including UML, MOF, CWM, SPEM, XMI, and various UML profiles, use MOF and MOF derived technologies (specifically XMI and more recently JMI which are mappings of MOF to XML and Java respectively) for metadata-driven interchange and metadata manipulation. MOF mappings from MOF to W3C XML and XSD are specified in the XMI (ISO/IEC 19503) specification. Mappings from MOF to Java are in the JMI (Java Metadata Interchange) specification defined by the Java Community Process.

Note that MOF 2.0 is closely related to UML 2.0. MOF 2.0 specification integrates and reuses the complementary UML 2.0 Infrastructure submission to provide a more consistent modeling and metadata framework for OMG's Model Driven Architecture. UML 2.0 provides the modeling framework and notation, MOF 2.0 provides the metadata management framework and metadata services.

MOF was also incorporated into an ISO/IEC (the International Organization for Standardization/the International Electrotechnical Commission) standard (19,502:2005) in November 2005. The standard defines a metamodel (defined using the MOF), a set of interfaces (defined using ODP IDL - ITU-T Recommendation X.920 (1997) | ISO/IEC 14750:1999), that can be used to define and manipulate a set of interoperable metamodels and their corresponding models (including the Unified Modeling Language metamodel - ISO/IEC 19501:2005, the MOF meta-metamodel, as well as future standard technologies that will be specified using metamodels). It also defines the mapping from MOF to ODP IDL (ITU rec X920|ISO 14750).

In conclusion, the MOF provides the infrastructure for implementing design and reuse repositories, application development tool frameworks, etc. The MOF specifies precise mapping rules that enable the CORBA interfaces for metamodels to be generated automatically, thus encouraging consistency in manipulating metadata in all phases of the distributed application development cycle.

Cross-References

▸ Metadata
▸ Metamodel
▸ Unified Modeling Language
▸ XML Metadata Interchange

Recommended Reading

1. Common warehouse metamodel (CWM). Available at http://www.omg.org/technology/documents/formal/cwm.htm (accessed on September 22, 2008).
2. ISO/IEC standard 19502:2005 (Information Technology - Meta Object Facility). Available at http://www.iso.org/iso/iso_catalogue/catalogue_tc/catalogue_detail.htm?csnumber=32621
3. MOF Query/Views/Transformations. Available at http://www.omg.org/spec/QVT/ (current version 1.0)
4. MOF 2.0 versioning and development lifecycle. Available at http://www.omg.org/technology/documents/formal/MOF_version.htm
5. OMG's meta object facility. Available at http://www.omg.org/mof/ (current version 2.0)

M

Metadata

Manfred Jeusfeld
IIT, University of Skövde, Skövde, Sweden
Aura Frames, New York City, NY, USA

Definition

Metadata is data linked to some data item, i.e., metadata is data about data. The metadata of a data item specifies how the data item was created, in which context it can be used, how it was transformed, or how it can be interpreted or processed. The earliest use of metadata are bibliographic records about books, such as the author of the book. In principle, any type of data item can have metadata attached to it. The type of the data item itself can be metadata and determines which other metadata fields may be attached to the data item.

Key Points

The purpose of metadata is to provide contextual information for a data item. It may be used by humans to determine the usability of a data item. Likewise, computer programs can read the metadata in order to guide the processing of a data item. Metadata can be included in the data item (e.g., the date and location of a photography), or it may be stored apart of the data item. In the latter case, the data item requires being identifiable. In databases, metadata fields can be represented next to data fields, virtually blurring the distinction between metadata and data.

Metadata is mostly used in domains where the structure of the data item is rather complex. Metadata typically has a simple structure such as name/value pairs. Applications domains are word processing, multimedia processing, system design, data warehouses, data quality management, and others. The common characteristic of these domains is the presence of many data items of the same type, which need to be managed according to some criteria. Metadata allows providing the necessary information to check these criteria. In the semantic web, metadata can be represented in RDF and related formalisms such as the Dublin Core.

There is no limitation on the size of metadata attached to data items. It can be that the size of metadata exceeds the size of the data item itself. For example, the complete change history of a document is metadata of the document.

The schema of a database can be interpreted as metadata about the database. It specifies the type of the data items stored in the database. Likewise, a metamodel can be interpreted as metadata about schemas (or models).

Cross-References

▶ Database Schema
▶ Dublin Core
▶ Metamodel
▶ Resource Description Framework

Recommended Reading

1. Duval E, Hodgins W, Sutton SA, Weibel S. Metadata principles and practicalities. D-Lib Mag. 2002;8(4). https://doi.org/10.1045/april2002-weibel

Metadata Interchange Specification

Wei Tang
Teradata Corporation, El Segundo, CA, USA

Synonyms

MDIS

Definition

Metadata Interchange Specification (MDIS) is a standard proposed by the Metadata Coalition (MDC) for defining metadata.

Key Points

Metadata Coalition (MDC) was an organization of database and data warehouse venders founded in October 1995. Its aim was to define a tactical set of standard specifications for the access and interchange of meta-data between software tools.

In July 1996, Metadata Interchange Specification (MDIS) 1.0 was officially ratified by MDC.

The MDIS Version 1.0 specification represents Coalition member input and recommendations collected and synthesized by the Coalition's technical subcommittee which included representatives from Business Objects, ETI, IBM, Platinum Technology, Price Waterhouse, Prism Solutions, R&O and SAS Institute. The latest version of

MDIS is 1.1, which was published in August 1997.

The Metadata Interchange Specification draws a distinction between:

- The Application Metamodel – the tables, etc., used to "hold" the metadata for schemas, etc., for a particular application; for example, the set of tables used to store metadata in Composer may differ significantly from those used by the Bachman Data Analyst.
- The Metadata Metamodel – the set of objects that the MDIS can be used to describe. These represent the information that is common (i.e., represented) by one or more classes of tools, such as data discovery tools, data extraction tools, replication tools, user query tools, database servers, etc. The metadata metamodel should be:
  - Independent of any application metamodel.
  - Character-based so as to be hardware/platform-independent.
  - Fully qualified so that the definition of each object is uniquely identified.

There are two basic aspects of the specification:

1. Those that pertain to the semantics and syntax used to represent the metadata to be exchanged. These items are those that are typically found in a specifications document.
2. Those that pertain to some framework in which the specification will be used. This second set of items is two file-based semaphores that are used by the specification's import and export functions to help the user of the specification control consistency.

MDIS consists of a metamodel, which defines the syntax and semantics of the metadata to be exchanged, as well as the specification of a framework for supporting an actual MDIS implementation. The MDIS Metamodel is a hierarchically structured, semantic database model that's defined by a tag language. The metamodel consists of a number of generic, semantic constructs, such as Element, Record, View, Dimension, Level, and Subschema, plus a Relationship entity that can be used in the specification of associations between arbitrary source and target constructs. The MDIS metamodel may be extended through the use of named properties that are understood to be tool-specific and not defined within MDIS. Interchange is accomplished via an ASCII file representation of an instance of this metamodel. Although support for an API is mentioned in the specification, no API definition is provided.

The MDIS Access Framework specifies several fairly general mechanisms that support the interchange of metamodel instances. The Tool and Configuration Profiles define semaphores that ensure consistent, bidirectional metadata exchange between tools. The MDIS Profile defines a number of system parameters (environment variables) that would be necessary in the definition of an MDIS deployment. Finally, Import and Export functions are exposed by the framework as the primary file interchange mechanisms for use by tools.

MDIS 1.1 was planned to be incorporated with Microsoft's Open Information Model (OIM) when Microsoft joined the MDC in December 1998. MDC decided later that MDIS be superseded by OIM. In 2000, the Metadata Coalition merged with the Object Management Group (OMG). OMG has worked on integrating OIM into its Common Warehouse Model (CWM) in order to provide a single standard for modeling meta-data in data warehouses. MDC, MDIS, and OIM are no longer in existence today (as independent entities).

Cross-References

▶ Common Warehouse Metamodel

Recommended Reading

1. Metadata Interchange Specification (MDIS) Version 1.1. Available at: http://www.eda.org/rassp/documents/atl/MDIS-11.pdf.

Metadata Registry, ISO/IEC 11179

Raymond K. Pon[1] and David J. Buttler[2]
[1]University of California, Los Angeles, CA, USA
[2]Lawrence Livermore National Laboratory, Livermore, CA, USA

Synonyms

MDR; Metadata repository

Definition

ISO/IEC-11179 [10] is an international standard that documents the standardization and registration of metadata to make data understandable and shareable. This standardization and registration allows for easier locating, retrieving, and transmitting data from disparate databases. The standard defines the how metadata are conceptually modeled and how they are shared among parties, but does not define how data is physically represented as bits and bytes. The standard consists of six parts. Part 1 [5] provides a high-level overview of the standard and defines the basic element of a metadata registry – a data element. Part 2 [7] defines the procedures for registering classification schemes and classifying administered items in a metadata registry (MDR). Part 3 [4] specifies the structure of an MDR. Part 4 [6] specifies requirements and recommendations for constructing definitions for data and metadata. Part 5 [8] defines how administered items are named and identified. Part 6 [9] defines how administered items are registered and assigned an identifier.

Historical Background

The first edition of the standard was published by the Technical Committee ISO/IEC JTC1, Information Technology Subcommittee 32, Data Management and Interchange, starting in 1994 and completed in 2000. The second edition was started in 2004 and was completed in 2005. The second edition cancels and replaces the first edition of the standard.

Foundations

Metadata is data that describes other data. A metadata registry is a database of metadata. The database allows for the registration of metadata, which enables the identification, provenance tracking, and quality monitoring of metadata. Identification is accomplished by assigning a unique identifier to each object registered in the registry. Provenance details the source of the metadata and the object described. Monitoring quality ensures that the metadata accomplishes its designed task. An MDR also manages the semantics of data, so that data can be re-used and interchanged. An MDR is organized so that application designers can determine whether a suitable object described in the MDR already exists so that it may be reused instead of developing a new object.

Part 1: Framework

Part 1 introduces the building blocks of the MDR standard: data elements, value domains, data element concepts, conceptual domains, and classification schemes. An MDR is organized as a collection of concepts, which are mental constructs created by a unique combination of characteristics. A concept system is a set of concepts with relations among them. One such concept system that classifies objects is a classification scheme. A classification scheme is organized with some specified structure and is designed for assigning objects to concepts defined within it.

The basic construct in a metadata registry is the data element. A data element consists of a data element concept and a representation. A data element concept (DEC) is a concept that can be represented as a data element described independently of any particular representation. The representation of a data element consists of a value domain, a data-type, units of measure, and a representation class. A data element concept may consist of an object class, which is a set of abstractions in the real world that can be identified with explicit boundaries, and a property, which is a characteristic common to all members of an object class. A value domain is a set of permissible values. Each value domain is a member of the extension of a concept known as the conceptual domain. A conceptual domain is a set of value meanings, which are the associated meanings to values.

An MDR contains metadata describing data constructs. Registering a metadata item makes it a registry item. If the registry item is subject to administration, it is called an administered item. An ISO/IEC 11179 MDR consists of two levels: the conceptual level and the representational level. The conceptual level contains the classes for the data element concept and conceptual domain. The representational level contains classes for data element and value domain.

Part 2: Classification

Part 2 provides a conceptual model for managing concept systems used as classification schemes. Associating an object with a concept from a classification scheme provides additional understanding of the object, comparative information across similar objects, an understanding of an object within the context of a subject matter field, and the ability to identify differences of meaning between similar objects.

Classification schemes are registered in an MDR by recording their attributes, such as those regarding its designation, definition, classification scheme, administration record, reference document, submission, stewardship, registration authority, and registrar.

Part 2 also defines the mechanism for classifying an administered item, which is the assignment of a concept to an object. Objects can also be linked together by relationships linking concepts in the concept system.

Part 3: Registry Metamodel and Basic Attributes

Part 3 describes the basic attributes that are required to describe metadata items and the structure for a metadata registry. The standard uses a metamodel to describe the structure of

an MDR. A metamodel is a model that describes other models. The registry metamodel is specified as a conceptual data model, which describes how relevant information is structured in the real world, and is expressed in the Unified Modeling Language [13].

The registry model is divided into six regions:

- **The administration and identification region:** supports the administrative aspects of administered items in the MDR. This region manages the identification and registration of items submitted to the registry, organizations that have submitted and/or are responsible for items in the registry, supporting documentation, and relationships among administered items. An administered item can be a classification scheme, a conceptual domain, context for an administered item, a data element, a data element concept, an object class, a property, a representation class, and a value domain. An administered item is associated with an administration record, which records administrative information about the administered item in the registry.

- **The naming and definition region:** manages the names and definitions of administered items and the contexts for names. Each administered item is named and defined within one or more contexts. A context defines the scope within which the data has meaning, such as a business domain, a subject area, an information system, a data model, or standards document.

- **The classification region:** manages the registration and administration of classification schemes and their constituent classification scheme items. It is also used to classify administered items.

- **The data element concepts region:** maintains information on the concepts upon which the data elements are developed, primarily focusing on semantics.

- **The conceptual and value domain region:** administrates the conceptual domains and value domains.

- **The data element region:** administrates data elements, which provide the formal representations for some information (e.g., a fact, observation, etc.) about an object. Data elements are reusable and shareable representations of data element concepts.

Part 4: Formulation of Data Definitions

Part 4 specifies the requirements and recommendations for constructing data and metadata definitions. A data definition must be stated in the singular. It also must be a descriptive phrase, containing only commonly understood abbreviations, that state what the concept is (as opposed to what the concept is not). A data definition must also be expressed without embedding definitions of other data. The standard also recommends that a data definition should be concise, precise, and unambiguous when stating the essential meaning of the concept. Additionally, a data definition should be self-contained and be expressed without embedding rationale, functional usage, or procedural information, circular reasoning. Terminology and consistent logical structure for related definitions should also be used.

Part 5: Naming and Identification Principles

Part 5 defines the naming and identification of the data element concept, the conceptual domain, data element, and value domain. Each administered item has a unique data identifier within the register of a Registration Authority (RA), which is the organization responsible for an MDR. The international registration data identifier (IRDI) uniquely identifies an administered item globally and consists of a registration authority identifier (RAI), data identifier (DI), and version identifier (VI).

Each administered item has at least one name within a registry of an RA. Each name for an administered item is specified within a context. A naming convention can be used for formulating names. A naming convention may address the scope of the naming convention and the authority that establishes the name. A naming convention may additionally address semantic,

syntactic, lexical, and uniqueness rules. Semantic rules govern the existence of the source and content of the terms in a name. Syntactic rules govern the required term order. Lexical rules govern term lists, name length, character set, and language. Uniqueness rules determine whether or not names must be unique.

Part 6: Registration

Part 6 specifies how administered items are registered and assigned an IRDI. Metadata in the MDR is also associated with a registration status, which is a designation of the level of registration or quality of the administered item. There are two types of status categories: lifecycle and documentation. The lifecycle registration status categories address the development and progression of the metadata and the preferences of usage of the administered item. The documentation registration status categories are used when there is no further development in the quality of metadata or use of the administered item.

Each RA establishes its own procedures for the necessary activities of its MDR. Some activities include the submission, progression, harmonization, modification, retirement, and administration of administered items.

Key Applications

The standardization that ISO/IEC 11179 provides enables for the easy sharing of data. For example, many organizations exchange data between computer systems using data integration technologies. In data warehousing schemes, completed transactions must be regularly transferred to separate data warehouses. Exchanges of data can be accomplished more easily if data is defined precisely so that automatic methods can be employed. By having a repository of metadata that describes data, application designers can reuse and share data between computer systems, making the sharing of data easier. ISO/IEC 11179 also simplifies data manipulation by software by enabling the manipulation of data based on characteristics described by the metadata in the registry. This also allows for the development of a data representation model for CASE tools and repositories [3].

There are several organizations that have developed MDRs that comply with ISO/IEC 11179, such as the Australian Institute of Health and Welfare [1], the US Department of Justice [14], the US Environmental Protection Agency [15], the Minnesota Department of Education [11], and the Minnesota Department of Revenue [12]. Currently, there is also an MDR available developed by Data Foundations [2].

Cross-References

▶ Metadata

Recommended Reading

1. Australian Institute of Health and Welfare. Metadata Online Registry (METeOR). http://meteor.aihw.gov.au/, 2007.
2. Data Foundations. Metadata Registry. http://www.datafoundations.com/solutions/data_registries.shtml, 2007.
3. ISO/IEC JTC1 SC32. Part 1: Framework for the specification and standardization of data elements. Information Technology – Metadata registries (MDR). 1st edn. 1999.
4. ISO/IEC JTC1 SC32. Part 3: Registry metamodel and basic attributes. Information Technology – Metadata registries (MDR). 2nd edn. 2003.
5. ISO/IEC JTC1 SC32. Part 1: Framework. Information Technology – Metadata registries (MDR). 2nd edn. 2004.
6. ISO/IEC JTC1 SC32. Part 4: Formulation of data definitions. Information Technology – Metadata registries (MDR). 2nd edn. 2004.
7. ISO/IEC JTC1 SC32. Part 2: Classification. Information Technology – Metadata registries (MDR). 2nd edn. 2005.
8. ISO/IEC JTC1 SC32. Part 5: Naming and identification principles. Information Technology – Metadata registries (MDR). 2nd edn. 2005.
9. ISO/IEC JTC1 SC32. Part 6: Registration. Information Technology – Metadata registries (MDR). 2nd edn. 2005.
10. ISO/IEC JTC1 SC32. ISO/IEC 11179, Information Technology – Metadata registries (MDR). 2007.
11. Minnesota Department of Education. Metadata Registry (K-12 Data). http://education.state.mn.us/mde-dd, 2007.
12. Minnesota Department of Revenue. Property Taxation (Real Estate Transactions). http://proptax.mdor.state.mn.us/mdr, 2007.
13. Object Management Group. Unified Modeling Language. http://www.uml.org/, 2007.
14. US Department of Justice. Global Justice XML Data Model (GJXDM). http://justicexml.gtri.gatech.edu/, 2007.
15. US Environmental Protection Agency. Environmental Health Registry. http://www.epa.gov/edr/, 2007.

Metamodel

Manfred Jeusfeld
IIT, University of Skövde, Skövde, Sweden
Aura Frames, New York City, NY, USA

Synonyms

Metamodel

Definition

A metamodel is a model that consists of statements about models. Hence, a metamodel is also a model but its universe of discourse is a set of models, namely, those models that are of interest to the creator of the metamodel. In the context of information systems, a metamodel contains statements about the constructs used in models about information systems. The statements in a metamodel can define the constructs or can express true and desired properties of the constructs. Like models are abstractions of some reality, metamodels are abstractions of models. The continuation of the abstraction leads to metametamodels, being models of metamodels containing statements about metamodels. Metamodeling is the activity of designing metamodels (and metametamodels). Metamodeling is applied to design new modeling languages and to extend existing modeling languages.

A second sense of the term metamodel is the specification of the generation of mathematical models, in particular sets of mathematical equations that describe some reality.

M

Metamodel, Fig. 1
Abrial's definition of the
binary data model

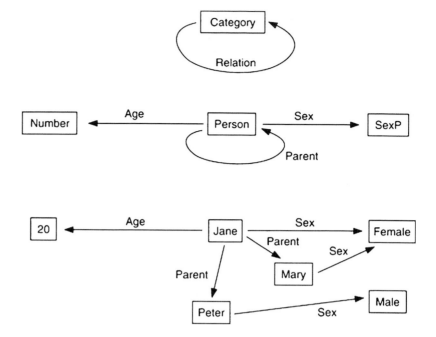

Historical Background

One of the earliest metamodels is the definition of the binary data model of Abrial [1]. Abrial distinguished three abstraction levels: the data level of a database, the schema of the database (model), and the category level (metamodel) (Fig. 1).

The metamodel defining the binary data model consists of the construct *Category* and the construct *relation*. Abrial interpreted the abstraction between the levels as classification. For example, *Jane* is classified to *Person* and *Person* is classified to *Category*. A similar classification holds for the relations.

In the 1980s, the use of metamodels became so widespread that an ISO standard, the Information Resource Dictionary Standard [2], was defined. It extended Abrial's view by a fourth level, i.e., by metametamodels. In the late 1990s, the Object Management Group (OMG) [5] consolidated and standardized the terminology of metamodels. They distinguished the levels M_0 (information), M_1 (model), M_2 (metamodel), and M_3 (metametamodel). The M_3 level is under the control of OMG. It defined four basic constructs (classes, associations, data types, and packages). The M_2

level is used to define modeling languages such as UML, IDL, and so forth.

Besides the standardization efforts, there were several metamodeling languages developed from the 1990s onward that mostly adopted the four-level approach. Examples are the Telos language and the GOPPR language of MetaEdit+ [9].

Foundations

Metamodels in computer science and related domains are mainly used to facilitate conceptual modeling, to define constructs of the conceptual modeling languages, to specify constraints on the use of constructs, and to encode the similarities of different models (and metamodels). As conceptual modeling is about representing concepts, an element of a metamodel is also a concept say *meta concept*, being interpreted by all entities that are defined or constrained by the meta concept. For example, the meta concept *EntityType* is a construct of the entity-relationship diagramming language. It is interpreted by all possible entity types of all possible entity relationship diagrams. Essentially, *EntityType* is the name of a set that has all possible entity types as elements. A problem with this set view is that it immediately introduces sets of sets (=concepts in metamodels) and sets of sets of sets (concepts of metametamodels). To overcome this complexity, metamodels were originally only investigated as level pairs: (metametamodel vs. metamodel), (metamodel vs. model), and (model vs. data). First a metametamodel is developed. Then, a metamodel or several metamodels are expressed as instances of the metametamodel, then models are expressed in terms of the metamodels. The lowest level (M_0 in MOF) is typically not expressed in conceptual modeling since it is about data or actual activities of some application domain. The pair-wise approach allows to keep the set-oriented semantics or other forms of semantics specification relying on distinguishing a concept from its instances.

The set-oriented semantics is mirrored by a logical interpretation, in which concepts are represented by unary predicates and relations and attributes are represented by binary predicates. For example, *EntityType(Employee)* is the fact expressing that *Employee* (model level) is classified to *EntityType* (metamodel level). A fact *Employee(Jane)* would then express that *Jane* (data level) is an instance of *Employee*. If one restricts to just two consecutive levels, predicate symbols can be distinguished from constant symbols. In other words, the underlying logic is a first-order logic. Scaling the semantics to more than two levels would move the logic to higher order.

One can avoid higher-order semantics by introducing a binary predicate *In(x,c)* where x is some concept of some modeling level M_i and c is a concept of the next higher modeling level M_{i+1}. This framework allows to represent the facts *In(Employee,EntityType)* and *In(Jane,EntityType)* without leaving first-order logic.

The specific choice of the underlying semantics for metamodels determines to which degree a metamodel can express the intended meaning of the concepts included in a metamodel. In UML, the semantics of the UML constructs are defined in a metamodel using OCL (object constraint language [6]). Current metamodeling tools dominantly use cardinality constraints as means to constrain the semantics of metamodel concepts. Constraints exceeding cardinalities have to be expressed in OCL or script languages.

The second sense of metamodels, the generation of mathematical equations to describe some reality, is, for example, used by Bailey and Basili [3] to develop a formal framework for understanding real-world phenomena in the domain of software engineering.

Key Applications

Metamodels became a popular technique at the end of the 1990s. The current specification of UML is supporting metamodeling in order to extend the capabilities of the language and to adapt it to specific modeling domains. Tools supporting metamodeling are among others MetaEdit+ [9], ConceptBase [4], and Aris [7]. The MetaEdit+ tool claims to accelerate system development by orders of magnitude since the concepts of a metamodel can be linked to parameterized program code.

M

Future Directions

An open problem of metamodels is their utility. If a metamodel is represented as a UML class diagram, then it does list the allowed constructs but it does not explain how to use it in a meaningful way, i.e., to represent models in terms of the metamodel. Conceptual modeling textbooks motivate constructs by examples and discuss scenarios in which certain constructs are usable. This pragmatic level is neglected by metamodels.

Metamodels should be seen as part of the larger model-driven architecture framework. That framework (also defined by OMG) is based on the assumption that system development is essentially a series of model transformations. The design of system development methods is then the combination of metamodeling and the specification of suitable model transformations.

The relationship between metamodels (or metametamodel) with ontologies is not yet well understood. Ontologies rely on two levels of abstraction: the concepts defined in the ontology and the real-world objects being the interpretations of the concepts. Apparently, an ontology makes no difference between a model level concept like *Employee* and a metamodel level concept like *EntityType*. See also [8] for a discussion.

Cross-References

▶ Telos

Recommended Reading

1. Abrial JR. Data semantics. In database management. In: Proceedings of the IFIP Working Conference on Database Management; 1974. p. 1–60.
2. American National Standard Institute. American National Standard X3.138–1988, Information Resource Dictionary System (IRDS). American National Standard Institute; 1989.
3. Bailey JW, Basili VR. A meta-model for software development resource expenditures. In: Proceedings of the 5th International Conference on Software Engineering; 1981. p. 107–16.
4. Jeusfeld MA, Jarke M, Nissen HW, Staudt M. Managing conceptual models about information systems. In: Bernus P, Mertins K, Schmidt G, editors. Handbook on architectures of information systems. 2nd ed. Berlin/Heidelberg/New York: Springer; 2006. p. 273–94.
5. Object Management Group. Meta Object Facility (MOF) Specification, Version 1.4; April 2002. Available at: http://www.omg.org/technology/documents/formal/mof.htm
6. Object Management Group. Object Constraint Language, OMG Available Specification Version 2.0; May 2006. Available at: http://www.omg.org/cgi-bin/doc?formal/2006-05-01
7. Scheer A-W, Schneider K. ARIS – architecture of integrated information systems. In: Bernus P, Mertins K, Schmidt G, editors. Handbook on architectures of information systems. 2nd ed. Berlin/Heidelberg/New York: Springer; 2006. p. 605–23.
8. Terrasse M-N, Savonnet M, Leclercq E, Grison T, Becker G. Do we need metamodels and ontologies for engineering platforms? In: Proceedings of the 2006 International Workshop on Global Integrated Model Management; 2006. p. 21–8.
9. Tolvanen J-P. MetaEdit+: integrated modeling and metamodeling environment for domain-specific languages. In: Proceedings of the 21st ACM SIGPLAN Conference on Object-Oriented Programming Systems, Languages & Applications; 2006. p. 690–1.

Metasearch Engines

Weiyi Meng
Department of Computer Science, State University of New York at Binghamton, Binghamton, NY, USA

Synonyms

Federated search engine

Definition

Metasearch is to utilize multiple other search systems (called *component search systems*) to perform simultaneous search. A metasearch engine is a search system that enables metasearch. To perform a basic metasearch, a user query is sent to multiple existing search engines by the metasearch engine; when the search results returned from the search engines are received by the metasearch engine, they are merged into a single ranked list and the merged list is presented to the user. Key issues include how to pass user queries to component search engines, how to extract correct search results from the result pages returned from component search engines, and how to merge the results returned from different component search sources. More sophisticated metasearch engines, especially those that have a large number of component search engines, also perform *search engine selection* (also referred to as *database selection*), i.e., identify the component search engines that are most appropriate for each given query and send the query to only these component search engines. To identify appropriate component search engines to use for a query requires the estimation of the usefulness of each component search engine with respect to the query based on some usefulness measure.

Historical Background

The earliest web-based metasearch engine is probably the MetaCrawler system [12] that became operational since June 1995. (The MetaCrawler's website (www.metacrawler. com) says the system was first developed in 1994.) Motivations for metasearch include (1) increased search coverage because a metasearch engine effectively combines the coverage of all component search engines, (2) improved convenience for users because a metasearch engine allows users to get information from multiple sources with one query submission and a metasearch engine hides the differences in query formats of different search engines from the users, and (3) better retrieval effectiveness because the result merging component can naturally incorporate the voting mechanism, i.e., results that are highly ranked by multiple component search engines are more likely to be relevant than those that are returned by only one of them. Over the last 20 years, many metasearch engines have been developed and deployed on the web. Most of them are built on top of a small number of popular general-purpose search engines, but there are also metasearch engines that are connected to more specialized search engines (e.g., medical/health search engines), and some are connected to over 1,000 search engines.

Even the earliest metasearch engines tackled the issues of search result extraction and result merging. Result merging is one of the most fundamental components in metasearch, and, as a result, it has received a lot of attention in the metasearch and distributed information retrieval (DIR) communities, and a wide range of solutions has been proposed to achieve effective result merging. Since different search engines may index a different set of web pages and some search engines are better than others for queries in different subject areas, it is important to identify the appropriate search engines for each user query. The importance of search engine selection was realized early in metasearch research, and many approaches have been proposed to address this issue. Many of the result merging and search engine selection techniques are covered in two recent works [10, 13].

Most metasearch engines are built on top of other search engines without explicit

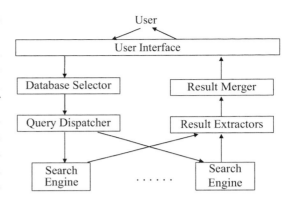

Metasearch Engines, Fig. 1 Metasearch engine component architecture

collaboration from these search engines. As a result, creating these metasearch engines requires a connection program and an extraction program (wrapper) for each component search engine. The former is needed to pass the query from the metasearch engine to the search engine and receive search results returned from the search engine, and the latter is used to extract the search result records from the result pages returned from the search engine. While the programs may not be difficult to produce by an experienced programmer, maintaining their validity can be a serious problem as they can become obsolete when the used search engines change their connection parameters and/or result display format. In addition, for applications that need to connect to hundreds or thousands of search engines, it can be very expensive and time-consuming to produce and maintain these programs. As a result, in recent years, automatic wrapper generation techniques have received much attention [10]. Figure 1 shows a basic architecture of a typical metasearch engine.

Scientific Fundamentals

Result Merging
Result merging is to combine the search results returned from multiple search engines into a single ranked list. Early search engines often associated a numerical matching score (similarity score) to each retrieved search result, and the

result merging algorithms at that time were designed to "normalize" the scores returned from different search engines into values within a common range with the goal of making them more comparable. Normalized scores will then be used to re-rank all the search results. When matching scores are not available, the ranks of the search results from component search engines can be aggregated using voting-based techniques (e.g., Borda count [1]). Score normalization and rank aggregation may also take into consideration the estimated usefulness of each selected search engine with respect to the query, which is obtained during the search engine selection step. For example, the normalized score of a result can be weighted by the usefulness score of the search engine that returned the result. This increases the chance for the results from more useful search engines to be ranked higher.

Another result merging technique is to download all returned documents from their local servers and compute their matching scores using a common similarity function employed by the metasearch engine. The results will then be ranked based on these scores. For example, the Inquirus metasearch engine employed this approach. The advantage of this approach is that it provides a uniform way to compute ranking scores so the resulted ranking makes more sense. Its main drawback is the longer response time due to the delay caused by downloading the documents and analyzing them on the fly. Most modern search engines display the title of each retrieved result together with a short summary called *snippet*. The title and snippet of a result can often provide good clues on whether or not the result is relevant to a query. As a result, result merging algorithms that rely on titles and snippets have been proposed (e.g., [9]). When titles and snippets are used to perform the merging, a matching score of each result with the query can be computed based on several factors such as the number of unique query terms that appear in the title/snippet and the proximity of the query terms in the title/snippet.

It is possible that the same result is retrieved from multiple search engines. Such results are more likely to be relevant to the query based on the observation that different ranking algorithms tend to retrieve the same set of relevant results but different sets of irrelevant results [6]. To help rank these results higher in the merged list, the ranking scores of these results from different search engines can be added up to produce the final score for the result. The search results are then ranked in descending order of the final scores.

Search Engine Selection

To enable search engine selection, some information that can represent the contents of the documents of each component search engine needs to be collected first. Such information for a search engine is called the *representative* of the search engine. The representatives of all component search engines used by the metasearch engine are collected in advance and are stored with the metasearch engine. During search engine selection for a given query, component search engines are ranked based on how well their representatives match with the query. A large number of search engine selection techniques have been proposed, and they often use different types of representatives. A simple representative of a search engine may contain only a few selected key words or a short description. This type of representative is usually produced manually by someone familiar with the contents of the search engine, but it can also be automatically generated. As this type of representatives provides only a general description of the contents of search engines, the accuracy of using such representatives for search engine selection is usually low. More elaborate representatives consist of detailed statistical information for each term in each search engine. In [2], the *document frequency* and *collection frequency* of each term (the latter is the number of component search engines that contain the term) are used to represent each search engine. In [11], the *adjusted maximum normalized weight* of each term across all documents in a search engine is used to represent a search engine. For a given term t and a search engine S, its adjusted maximum normalized weight is computed as follows: compute the normalized weight of t in every document (i.e., the term frequency weight of t divided by the length of the document) in S, find the maximum value among these weights, and multiply this maximum weight by the global idf weight of t across all component search engines. In [15], the notion of optimal search engine ranking is proposed based on the objective of retrieving the m most similar (relevant) documents with respect to a given query Q from across all component search engines: n search engines are said to be optimally ranked in order $[S_1, S_2, \ldots, S_n]$ if for any integer m, an integer k can be found such that the m most similar documents are contained in $[S_1, \ldots, S_k]$ and each of these k search engines contain at least one of the m most similar documents. It is shown in [15] that a necessary and sufficient condition for the component search engines to be optimally ranked is to order the search engines in descending order of the similarity of the most similar document with respect to Q in each search engine. Different techniques have been proposed to estimate the similarity of the most similar document with respect to a given query and a given search engine [11, 15]. Since it is impractical to find out all the terms that appear in some pages in a search engine, an approximate vocabulary of terms for a search engine can be used. Such an approximate vocabulary can be obtained from pages retrieved from the search engine using probe queries [3].

There are also techniques that create search engine representatives by learning from the search results of past queries. Essentially such type of representatives is the knowledge indicating the past performance of a search engine with respect to different queries. In the SavvySearch metasearch engine [5], for each component search engine S, a weight is maintained for every term that has appeared in previous queries. After each query Q is evaluated, the weight of each term in the representative that appears in Q is increased or decreased depending on whether or not S returns useful results. Over time, if a term for S has a large positive (negative) weight, then S is considered to have responded well (poorly) to the term in the past. For a new query received by the metasearch engine, the weights of the query terms in the representatives

of different search engines are aggregated to rank the search engines. In the ProFusion metasearch engine, training queries are used to find out how well each search engine responds to queries in different categories. The knowledge learned about each search engine from training queries is used to select search engines for each user query, and the knowledge is continuously updated based on the user's reaction to the search result, i.e., whether or not a particular retrieved result is clicked by the user.

Automatic Search Engine Connection

The search interfaces of most search engines are implemented using the HTML *form* tag with a query textbox. In most cases, the form tag of a search engine contains all information needed to make the connection to the search engine, i.e., sending queries and receiving search results, via a program. Such information includes the name and the location of the program (i.e., the search engine server) that evaluates user queries, the network connection method (i.e., the HTTP request method, usually GET or POST), and the name associated with the query textbox that is used to save the query string. The form tag of each component search engine interface is usually preprocessed to extract the information needed for program connection, and the extracted information is saved at the metasearch engine. The existence of JavaScript in the form tag usually makes extracting the connection information more difficult. After the metasearch engine receives a query and a particular search engine, among possibly other search engines, is selected to evaluate this query, the query is assigned to the name of the query textbox of the search engine and sent to the server of the search engine using the HTTP request method supported by the search engine. After the query is evaluated by the search engine, one or more result pages containing the search results are returned to the metasearch engine for further processing.

Automatic Search Result Extraction

A result page returned by a search engine is a dynamically generated HTML page. In addition to the search result records for a query, a result

M

page usually also contains some unwanted information/links such as advertisements and sponsored links. It is important to correctly extract the search result records on each result page. A typical search result record corresponds to a retrieved document, and it usually contains the URL and the title of the page as well as a short summary (snippet) of the document. Since different component search engines produce result pages in different format, a separate result extraction program (also called *extraction wrapper*) needs to be generated for each search engine. Automatic wrapper generation for search engines has received a lot of attention in recent years, and different techniques have been proposed. Most of them analyze the source HTML files of the result pages as text strings or tag trees (DOM trees) to find the repeating patterns of the search record records. Two surveys covering many extraction techniques can be found in [4, 14]. Some more recent works also utilize certain visual information on result pages to help identify result patterns (e.g., [8]).

Key Applications

The main application of metasearch is to support search. It can be an effective mechanism to search both Surface Web and Deep Web data sources. By providing a common search interface over multiple search engines, metasearch eliminates users' burden to search multiple sources separately. When a metasearch engine employs certain special component search engines, it can support interesting special applications. For example, for a large organization with many branches (e.g., a university system may have many campuses), if each branch has its own search engine, then a metasearch engine connecting to all branch search engines becomes an organization-wide search engine. As another example, if a metasearch engine is created over multiple e-commerce search engines selling the same type of product, then a comparison-shopping system can be created. Of course, for

comparison-shopping applications, a different type of result merging is needed, such as listing different search results that correspond to the same product in non-descending order of the prices.

Future Directions

Component search engines employed by a metasearch engine may change their connection parameters and result display format anytime. These changes can make the affected search engines unusable in the metasearch engine unless the corresponding connection programs and result extraction wrappers are changed accordingly. How to monitor the changes of search engines and make the corresponding changes in the metasearch engine automatically and timely is an area that needs urgent attention from metasearch engine researchers and developers.

Most of today's metasearch engines employ only a small number of general-purpose search engines. Building large-scale metasearch engines using numerous specialized search engines is another area that deserves more attention. The largest metasearch engine that had been reported was AllInOneNews [7]. This metasearch engine connected to more than 1,000 news search engines. Challenges arising from building very large-scale metasearch engines include automatic generation and maintenance of high-quality search engine representatives needed for efficient and effective search engine selection and highly automated techniques to add search engines into metasearch engines and to adapt to changes of search engines.

Cross-References

▶ Deep-Web Search
▶ Document Length Normalization
▶ Information Extraction
▶ Information Retrieval

- Inverse Document Frequency
- Personalized Web Search
- Search Engine Metrics
- Snippet
- Term Weighting
- Web Data Extraction System
- Web Information Retrieval Models
- Wrapper Induction
- Wrapper Maintenance
- Wrapper Stability

Recommended Reading

1. Aslam J, Montague M. Models for metasearch. In: Proceedings of the 24th Annual International ACM SIGIR Conference on Research and Development in Information Retrieval; 2001. p. 276–84.
2. Callan J, Lu Z, Croft, WB. Searching distributed collections with inference networks. In: Proceedings of the 18th Annual International ACM SIGIR Conference on Research and Development in Information Retrieval; 1995.
3. Callan J, Connell M, Du A. Automatic discovery of language models for text databases. In: Proceedings of the ACM SIGMOD International Conference on Management of Data; 1999.
4. Chang CH, Kayed M, Girgis MR, Shaalan KF. A survey of web information extraction systems. IEEE Trans Knowl Data Eng. 2006;18(10):1411–28.
5. Dreilinger D, Howe A. Experiences with selecting search engines using metasearch. ACM Trans Inf Sys. 1997;15(3):195–222.
6. Lee JH. Combining multiple evidence from different properties of weighting schemes. In: Proceedings of the 18th Annual International ACM SIGIR Conference on Research and Development in Information Retrieval; 1995. p. 180–8.
7. Liu KL, Meng W, Qiu J, Yu C, Raghavan V, Wu Z, Lu Y, He H, Zhao H. AllInOneNews: development and evaluation of a large-scale news metasearch engine. In: Proceedings of the ACM SIGMOD International Conference on Management of Data; 2007. p. 1017–28.
8. Liu W, Meng X, Meng W. ViDE: a vision-based approach for deep web data extraction. IEEE Trans Knowl Data Eng. 2010;22(3):447–60.
9. Lu Y, Meng W, Shu L, Yu C, Liu K. Evaluation of result merging strategies for metasearch engines. In: Proceedings of the 6th International Conference on Web Information Systems Engineering; 2005. p. 53–66.
10. Meng W, Yu C. Advanced metasearch engine technology. Morgan & Claypool Publishers; 2010.
11. Meng W, Wu Z, Yu C, Li Z. A highly scalable and effective method for metasearch. ACM Trans on Inf Sys. 2001;19(3):310–35.
12. Selberg E, Etzioni O. The MetaCrawler architecture for resource aggregation on the web. IEEE Expert. 1997;12(1):11–4.
13. Shokouhi M, Si L. Federated search. Found Trends Info Retrieval. 2011;5(1):1–102.
14. Sleiman HA, Corchuelo R. A survey on region extractors from web documents. IEEE Trans Knowl Data Eng. 2013;25(9):1960–81.
15. Yu C, Liu K, Meng W, Wu Z, Rishe N. A methodology to retrieve text documents from multiple databases. IEEE Trans Knowl Data Eng. 2002;14(6):1347–61.

Metric Space

Pavel Zezula, Michal Batko, and
Vlastislav Dohnal
Masaryk University, Brno, Czech Republic

Synonyms

Distance space

Definition

In mathematics, a metric space is a pair $M = (D, d)$, where D is a domain of objects (or objects'; *keys* or *indexed descriptors*) and d is a total (distance) function. The properties of the function $d: D \times D \mapsto R$, sometimes called the metric space postulates, are typically characterized as:

(p1)	$\forall x, y \in D, d(x, y) \geq 0$	Non-negativity,
(p2)	$\forall x, y \in D, d(x, y) = d(y, x)$	Symmetry,
(p3)	$\forall x \in D, d(x, x) = 0$	Reflexivity,
(p4)	$\forall x, y \in D, x \neq y \Rightarrow d(x, y) > 0$	Positiveness,
(p5)	$\forall x, y, z \in D, d(x, z) \leq d(x, y) + d(y, z)$	Triangle inequality

Key Points

Modifying or even abandoning some of the metric function properties leads to interesting concepts that can better suit the reality in many situations. A *pseudo-metric* function does not satisfy the positiveness property (p4), i.e., there can be pairs of different objects that have zero distance. However, these functions can be transformed to the standard metric by regarding any pair of objects with zero distance as a single object. If the symmetry property (p2) does not hold, the function is called a *quasi-metric*. For example, a car-driving distance in a city where one-way streets exist is a quasi-metric. The following equation allows to transform a quasi-metric into a standard metric: $d_{\text{sym}}(x, y) = d_{\text{asym}}(x, y) + d_{\text{asym}}(y, x)$. By tightening the triangle inequality property (p5) to $\forall x, y, z \in D, d(x, z) \leq \max\{d(x, y), d(y, z)\}$, an *ultra-metric* also called *super-metric* is obtained. The geometric characterization of the ultra-metric requires every triangle to have at least two sides of equal length, i.e., to be isosceles. A metric space M is bounded if there exists a number r, such that $d(x, y) \leq r$ for any $x, y \in D$. More details about metric functions can be found in [3].

Cross-References

▶ Closest-Pair Query
▶ Indexing Metric Spaces
▶ Information Retrieval
▶ Nearest Neighbor Query
▶ Spatial Indexing Techniques

Recommended Reading

1. Burago D, Burago YD, Ivanov S. A course in metric geometry. Providence: American Mathematical Society; 2001.
2. Bryant V. Metric spaces: iteration and application. New York: Cambridge University Press; 1985.
3. Zezula P, Amato G, Dohnal V, Batko M. Similarity search: the metric space approach. Berlin: Springer; 2006.

Microaggregation

Josep Domingo-Ferrer
Universitat Rovira i Virgili, Tarragona,
Catalonia, Spain

Definition

Microaggregation is a family of masking methods for statistical disclosure control of numerical microdata (although variants for categorical data exist). The rationale behind microaggregation is that confidentiality rules in use allow publication of microdata sets if records correspond to groups of k or more individuals, where no individual dominates (i.e., contributes too much to) the group and k is a threshold value. Strict application of such confidentiality rules leads to replacing individual values with values computed on small aggregates (microaggregates) prior to publication. This is the basic principle of microaggregation.

To obtain microaggregates in a microdata set with n records, these are combined to form g groups of size at least k. For each attribute, the average value over each group is computed and is used to replace each of the original averaged values. Groups are formed using a criterion of maximal similarity. Once the procedure has been completed, the resulting (modified) records can be published.

The optimal k-partition (from the information loss point of view) is defined to be the one that maximizes within-group homogeneity. The higher the within-group homogeneity, the lower the information loss, since microaggregation replaces values in a group by the group centroid. The sum of squares criterion is common to measure homogeneity in clustering. The within-groups sum of squares SSE is defined as

$$\text{SSE} = \sum_{i=1}^{g} \sum_{j=1}^{n_i} \left(x_{ij} - \overline{x}_i\right)' \left(x_{ij} - \overline{x}_i\right)$$

The lower SSE, the higher the within-group homogeneity. Thus, in terms of sums of squares, the optimal k-partition is the one that minimizes SSE.

Key Points

For a microdata set consisting of p attributes, these can be microaggregated together or partitioned into several groups of attributes. Also the way to form groups may vary. Several taxonomies are possible to classify the microaggregation algorithms in the literature: (i) fixed group size vs. variable group size; (ii) exact optimal (only for the univariate case vs. heuristic microaggregation; (iii) continuous vs. categorical microaggregation.

To illustrate, a heuristic algorithm called MDAV (maximum distance to average vector, by Domingo-Ferrer, Mateo-Sanz and Torra) is next given for multivariate fixed group size microaggregation on unprojected continuous data. MDAV has been implemented in the μ-Argus package:

1. Compute the average record $\overline{x}$ of all records in the dataset. Consider the most distant record x_r to the average record $\overline{x}$ (using the squared Euclidean distance).
2. Find the most distant record x_s from the record x_r considered in the previous step.
3. Form two groups around x_r and x_s, respectively. One group contains x_r and the $k-1$ records closest to x_r. The other group contains x_s and the $k-1$ records closest to x_s.
4. If there are at least $3k$ records which do not belong to any of the two groups formed in Step 3, go to Step 1 taking as new dataset the previous dataset minus the groups formed in the last instance of Step 3.
5. If there are between $3k-1$ and $2k$ records which do not belong to any of the two groups formed in Step 3: (i) compute the average record $\overline{x}$ of the remaining records; (ii) find the most distant record x_r from $\overline{x}$; (iii) form a group containing x_r and the $k-1$ records closest to x_r; (iv) form another group containing the rest of records. Exit the Algorithm.
6. If there are less than $2k$ records which do not belong to the groups formed in Step 3, form a new group with those records and exit the Algorithm.

The above algorithm can be applied independently to each group of attributes resulting from partitioning the set of attributes in the dataset. Microaggregation can be used to achieve k-anonymity.

Cross-References

▶ Inference Control in Statistical Databases
▶ k-Anonymity
▶ Microdata
▶ SDC Score

Recommended Reading

1. Domingo-Ferrer J, Mateo-Sanz JM. Practical data-oriented microaggregation for statistical disclosure control. IEEE Trans Knowl Data Eng. 2002;14(1):189–201.
2. Domingo-Ferrer J, Sebé F, Solanas A. A polynomial-time approximation to optimal multivariate microaggregation. Comput Math Appl. 2008;55(4):714–32.
3. Domingo-Ferrer J, Torra V. Ordinal, continuous and heterogenerous k-anonymity through microaggregation. Data Min Knowl Dis. 2005;11(2):195–212.
4. Hundepool A, Van de Wetering A, Ramaswamy R, Franconi L, Capobianchi A, DeWolf P-P, Domingo-Ferrer J, Torra V, Brand R, Giessing S. μ-ARGUS version 4.0 software and user's manual. Statistics Netherlands, Voorburg, May 2005. http://neon.vb.cbs.nl/casc

M

Microbenchmark

Denilson Barbosa[1], Ioana Manolescu[2], and Jeffrey Xu Yu[3]
[1]University of Alberta, Edmonton, AL, Canada
[2]INRIA Saclay–Ȋlle de France, Orsay, France
[3]Department of Systems Engineering and Engineering Management, The Chinese University of Hong Kong, Hong Kong, China

Definition

A micro-benchmark is an experimental tool that studies a given aspect (e.g., performance, resource consumption) of XML processing tool. The studied aspect is called the target of the micro-benchmark. A micro-benchmark includes a parametric measure and guidelines, explaining which data and/or operation parameters may impact the target, and suggesting value ranges for these parameters.

Key Points

Micro-benchmarks help capture the behavior of an XML processing system on a given operation, as a result of varying one given parameter. In other words, the goal of a micro-benchmark is to study the *precise* effect of a given system feature or aspect *in isolation*.

Micro-benchmarks were first introduced for object-oriented databases [2]. An XML benchmark sharing some micro-benchmark features is the Michigan benchmark [3]. The MemBeR project [1], developed jointly by researchers at INRIA Futurs, the University of Amsterdam, and University of Antwerpen provides a comprehensive repository of micro-benchmarks for XML.

Unlike application benchmarks, micro-benchmarks do not directly help determining which XML processing system is most appropriate for a given task. Rather, they are helpful in assessing particular modules, algorithms and techniques present inside an XML processing tool. Micro-benchmarks are therefore typically very useful to system developers.

Cross-References

▶ Application Benchmark
▶ XML Benchmarks

Recommended Reading

1. Afanasiev L, Manolescu I, Michiels P. MemBeR: a micro-benchmark repository for XQuery. In: Proceedings of the 3rd International XML Database Symposium on Database and XML Technologies; 2005, p. 144–61.
2. Carey MJ, DeWitt DJ, Naughton JF. The OO7 Benchmark. In: Proceedings of the ACM SIGMOD International Conference on Management of Data; 1993, p. 12–21.
3. Runapongsa K, Patel JM, Jagadish HV, Chen Y, Al-Khalifa S. The Michigan benchmark: towards XML query performance diagnostics. Inf Syst. 2006;31(2):73–97.

Microdata

Josep Domingo-Ferrer
Universitat Rovira i Virgili, Tarragona,
Catalonia, Spain

Synonyms

Individual data

Definition

A *microdata* file V with s respondents and t attributes is an $s \times t$ matrix where V_{ij} is the value of attribute j for respondent i. Attributes can be numerical (e.g., age, salary) or categorical (e.g., gender, job).

Key Points

The attributes in a microdata set can be classified in four categories which are not necessarily disjoint [1, 2]:

1. *Identifiers*. These are attributes that *unambiguously* identify the respondent. Examples are the passport number, social security number, name-surname, etc.
2. *Quasi-identifiers or key attributes*. These are attributes which identify the respondent with some degree of ambiguity. (Nonetheless, a combination of quasi-identifiers may provide unambiguous identification.) Examples are address, gender, age, telephone number, etc.

3. *Confidential outcome attributes*. These are attributes which contain sensitive information on the respondent. Examples are salary, religion, political affiliation, health condition, etc.
4. *Non-confidential outcome attributes*. Those attributes which do not fall in any of the categories above.

Cross-References

- ▸ k-Anonymity
- ▸ Data Rank/Swapping
- ▸ Inference Control in Statistical Databases
- ▸ Microdata Rounding
- ▸ Noise Addition
- ▸ Nonperturbative Masking Methods
- ▸ PRAM
- ▸ SDC Score
- ▸ Tabular Data

Recommended Reading

1. Dalenius T. The invasion of privacy problem and statistics production: an overview. Statistik Tidskrift. 1974;12:213–25.
2. Samarati P. Protecting respondents'; identities in microdata release. IEEE Trans Knowl Data Eng. 2001;13(6):1010–27.

Microdata Rounding

Josep Domingo-Ferrer
Universitat Rovira i Virgili, Tarragona,
Catalonia, Spain

Synonyms

Rounding

Definition

Microdata rounding is a family of masking methods for statistical disclosure control of numerical microdata; a similar principle can be used to protect tabular data. Rounding replaces original values of attributes with rounded values. For a given attribute X_i, rounded values are chosen among a set of rounding points defining a *rounding set* (often the multiples of a given base value).

Key Points

In a multivariate original dataset, rounding is usually performed one attribute at a time (*univariate* rounding); however, multivariate rounding is also possible [1, 2]. The operating principle of rounding makes it suitable for continuous data.

Cross-References

- ▸ Inference Control in Statistical Databases
- ▸ Microdata
- ▸ SDC Score

Recommended Reading

1. Cox LH, Kim JJ. Effects of rounding on the quality and confidentiality of statistical data. In: Domingo-Ferrer J, Franconi L, editors. Privacy in statistical databases. LNCS, vol. 4302. Heidelberg: Springer; 2006. p. 48–56.
2. Willenborg L, DeWaal T. Elements of statistical disclosure control. New York: Springer; 2001.

M

Middleware Support for Database Replication and Caching

Emmanuel Cecchet
EPFL, Lausanne, Switzerland

Definition

Database replication is a technique that aims at providing higher availability and performance than a single RDBMS. A database replication middleware implements a number of replication algorithms on top of existing RDBMS. Features provided by the replication middleware include load balancing, caching, and fault tolerance.

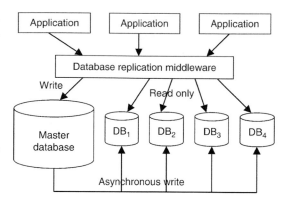

Middleware Support for Database Replication and Caching, Fig. 1 Master/slave scale-out scenario

Historical Background

Database replication is a well-known mechanism for performance scaling and availability of databases across a wide range of requirements. Limitations of 2-phase commit and synchronous replication have been pointed out early on by Gray et al. [7]. Since then, research on middleware-based replication addresses these issues and tries to provide solutions for better performance and availability while maintaining consistency guarantees for applications.

Foundations

Database replication is a wide area of research that encompasses multiple architectures and possible designs. This entry does not address in-core database replication, where the replication algorithms are implemented inside the database engine. Instead, it focuses on middleware-based replication, where the replication logic is implemented in a set of middleware components, outside the database engine.

Shared Disk Versus Shared Nothing

Two main architecture designs can be chosen for database replication. *Shared disk* replication is mostly used by in-core implementations, where replicas share the data storage, such as, a SAN (Storage Area Network). Middleware-based replication usually uses a *shared nothing* architecture, where each replica has its own local storage. This allows disk IOs to be distributed among replicas and prevents the storage from being a Single Point of Failure (SPOF).

Master/Slave Versus Multi-master

Database replication is often used as a way to scale up performance. Such efforts are typically targeted at increasing read performance or at increasing write performance; increasing both simultaneously is difficult.

Master-slave replication, depicted in Fig. 1, is popular because it improves read performance. This setup is frequently used in e-commerce applications, with slave databases dedicated to product catalog browsing, while the master performs all catalog updates.

In this scenario, read-only content is accessed on the slave nodes and updates are sent to the master. If the application can tolerate loose consistency, any data can be read at any time from the slaves given a freshness guarantee. As long as the master node can handle all updates, the system can scale linearly simply by adding slave nodes.

Multi-master replication, as shown on Fig. 2, allows each replica that owns a full copy of the database to serve both read and write requests. The replicated system then behaves as a centralized database which theoretically does not require any application modifications. However, replicas need to synchronize to agree on a serializable execution order of transactions so that each replica executes update transactions in the same order. Also, concurrent transactions might conflict leading to aborts and limiting the system scalability [7]. Even though real applications generally avoid conflicting transactions, significant efforts are spent to optimize for this problem in middleware replication research. However, the volume of update transactions remains the limiting performance factor for such systems.

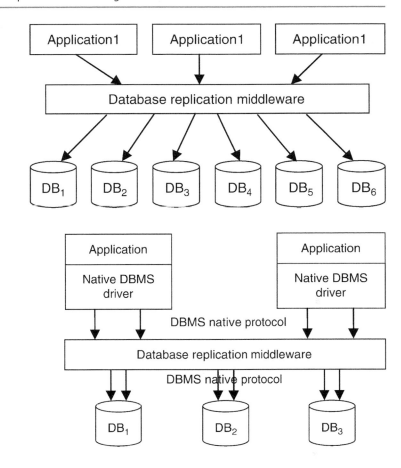

Middleware Support for Database Replication and Caching, Fig. 2
Multi-Master database replication

Middleware Support for Database Replication and Caching, Fig. 3
Query interception at the DBMS protocol level

Middleware Design

The replication middleware has to intercept client requests to process them and run them through the replication algorithm. A first technique consists of keeping existing database drivers and to intercept queries at the database protocol level as show on Fig. 3. This has the significant advantage to not have to re-implement database drivers but the database protocol specification must be available, which is not always the case for commercial databases. Moreover, this design requires multi-protocol implementations and bridges to support heterogeneous clusters.

Another technique provides the application with a replacement driver that can communicate with the replication middleware and that is API compatible with the original driver so that application changes are not required. Figure 4 shows an example of a middleware that intercepts queries at the JDBC level. The client application uses the middleware JDBC driver, and the

middleware uses the database native JDBC driver to access the replicas. The middleware driver typically adds functionality such as load balancing and failover that is usually not present in standalone database drivers. This is a popular approach introduced by C-JDBC [3] (now Sequoia [9]) and used in other prototypes like Tashkent [5] and Ganymed [8].

Concurrency Control

In replicated database systems, each replica runs Snapshot Isolation (SI) as its local concurrency control and the replicated system provides Generalized Snapshot Isolation (GSI) to the clients.

Snapshot isolation (SI) is a multi-version database concurrency control algorithm for centralized databases. In snapshot isolation, when a transaction begins it receives a logical copy, called snapshot, of the database for the duration of the transaction. This snapshot is the most recent version of the committed state of

**Middleware Support for
Database Replication
and Caching, Fig. 4**
Query interception in
JDBC-based replication

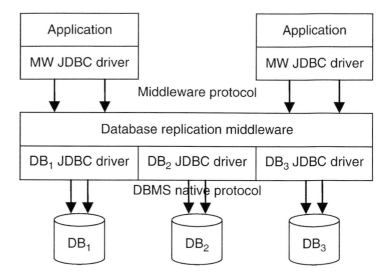

the database. Once assigned, the snapshot is unaffected by (i.e., isolated from) concurrently running transactions. When an update transaction commits, it produces a new version of the database.

Many database vendors use SI, e.g., PostgreSQL, Oracle and Microsoft SQL Server. SI is weaker than serializability but in practice many applications run serializably under SI including the widely used database benchmarks TPC-C and TPC-W. SI has attractive performance properties. Most notably, read-only transactions do not block or abort-they do not need read-locks, and they do not cause update transactions to block or abort.

Generalized Snapshot Isolation (GSI) extends SI to replicated databases such that the performance properties of SI in a centralized setting are maintained in a replicated setting. In addition, workloads that are serializable under SI are also serializable under GSI.

Informally, a replica using GSI works as follows. When a transaction starts, the replica assigns its latest snapshot to the transaction. All read and write operations of a transaction, e.g., the SELECT, UPDATE, INSERT and DELETE SQL statements, are executed locally on the replica. At commit, the replica extracts the transaction writeset. If the writeset is empty (i.e., it is a read-only transaction), the transaction commits immediately. Otherwise, certification is

performed to detect write-write conflicts among update transactions in the system. If no conflict is found, then the transaction commits, otherwise it aborts.

Certification results in total order on the commits of update transactions. Since committing an update transaction creates a new version (snapshot) of the database, the total order defines the sequence of snapshots the database goes through. Therefore, processing update transactions proceeds as follows: When a replica receives update transaction T, it executes T against a snapshot. At commit, the certification service receives the writeset of T and the version of the assigned snapshot. If certification is successful, the replica applies writesets of concurrent update transactions that committed before T in the order determined during certification and then commits T. Certification is a stateful service because it maintains recent committed writesets and their versions.

Statement Replication Versus Transaction Replication

Multi-master replication can be implemented either by multicasting every update statement (statement replication) or by capturing transaction writesets (the set of data W updated by a transaction T, such that applying W onto a replica is equivalent to executing T on it) and propagating them after certification (transaction replication). Both approaches have different performance/availability tradeoffs.

Statement-based replication requires that the execution of an update statement produces the same result on each replica. However, many SQL statements may produce different results on every replica if they are not processed before execution. This requires macros related to timing or random numbers to be preprocessed for a deterministic execution cluster-wide. Moreover, stored procedures or user-defined functions must have a deterministic behavior to prevent replicas from diverging in content. The advantage of statement-based replication is that it can replicate any kind of SQL statement including DDL (Data Definition Language) queries that alters the database schema or requests that modify non-persistent objects such as environment variables, sequences or temporary tables.

Transaction replication relies on writeset extraction that is usually implemented using triggers. This requires declaring additional triggers on every database table. This can become complex if the database already uses triggers, materialized views or temporary tables. Writeset extraction does not capture changes such as auto-incremented keys, sequence values or environment variable updates. Queries altering such database structures change the replica they execute on and can result in cluster divergence. Moreover, most of these data structures cannot be rolled back (for instance, an auto-incremented key or sequence number incremented in a transaction is not decremented at rollback time).

With statement replication, all replicas execute write transactions simultaneously in the same serializable order, whereas transaction replication executes update transaction at only one replica and propagates the transaction writeset to other replicas only after certification at commit time. Therefore, transaction replication usually offers better performance over statement replication as long as the writeset extraction and certification mechanisms are efficient. Statement replication offers a better infrastructure for failover during a transaction since each replica has a copy of every transactional context. With transaction replication, the failure of the replica executing the transaction will systematically abort the transaction and force the transaction to retry.

High Availability

High availability is often synonymous with little downtime. Such downtime can be either planned or unplanned, depending on whether it occurs under the control of the administrator or not. Planned downtime is incurred during most software and hardware maintenance operations, while unplanned downtime can strike at any time and is caused by foreseeable and unforeseeable failures (hardware failures, software bugs, human error, etc.).

A system's availability is the ratio of its uptime to total time. In practice, it is computed as the ratio between the expected time of continuous operation between failures to total time, or

$$\text{Availability} = \frac{MTTF}{MTTF + MTTR} \Rightarrow \text{Unavailability}$$
$$= \frac{MTTR}{MTTF + MTTR} \approx \frac{MTTR}{MTTF}$$

where MTTF is Mean Time To Failure and MTTR is Mean Time To Repair. Since MTTF >> MTTR, one can approximate unavailability (ratio of downtime to total time) as MTTR/MTTF.

The goal of replication together with failover/failback is to reduce MTTR, and thus reduce unavailability. Failover is the ability for users of a database node to be switched over to another database node containing a replica of the data whenever the node they were connected to has failed. Failback happens when the original replica comes back from its failure and users are reallocated to that replica.

A replicated database built for availability must eliminate any single point of failure (SPOF). This means that the middleware components (load balancer, certifier...) must also be replicated. Group communication libraries are used to synchronize the state of the different components. Total order is usually required by replication protocols to ensure a serializable execution order.

M

Several database drivers or connection pools offer automatic reconnection when a failure is detected. This technique only offers session failover but not failover of the transactional context. Sequoia [9] (the continuation of the C-JDBC project) provides transparent failover without losing transactional context. Failover code is available in the middleware to handle a database failure and additional code is available in the middleware driver running in the application to handle a middleware failure. A fully transparent failover requires consistently replicated state kept at all components, and is more easily achieved using statement-based rather than transaction-based replication. In the latter case, the transaction is only played at a single replica; if the replica fails, the entire transaction has to be replayed at another replica, which cannot succeed without the cooperation of the application.

Load Balancing

Load balancing aims at dispatching user requests or transactions to the replica that can provide consistent data with the lowest latency. Load balancer design is tightly coupled with the replication strategy implemented. Static strategies such as round-robin or even weighted-round-robin are usually not well adapted to the dynamic nature of transactional workloads. Algorithms taking into account replica resource usage such as LPRF (Least Pending Request First) perform much better. With additional information on transaction working set, it is also possible to optimize load balancing to improve in-memory request execution such as MALB (Memory-Aware Load Balancing) used in Tashkent+ [6]. More information on load balancing can be found in [1].

Caching

To reduce request execution time, the middleware can provide multiple caches. C-JDBC [3] provides three different caches. The parsing cache stores the results of query parsing so that a query that is executed several times is parsed only once. The metadata cache records all ResultSet metadata such as column names and types associated with a query result.

These caches work with query skeletons found in PreparedStatements used by application servers. A query skeleton is a query where all variable fields are replaced with question marks and filled at runtime with a specific API. An example of a query skeleton is "SELECT * FROM t WHERE x=?". In this example, a parsing or metadata cache hit will occur for any value of x.

The query result cache is used to store the ResultSet associated with each query. The query result cache reduces the request response time as well as the load on the database replicas. By default, the cache provides strong consistency. In other words, C-JDBC invalidates cache entries that may contain stale data as a result of an update query. Cache consistency may be relaxed using user-defined rules. The results of queries that can accept stale data can be kept in the cache for a time specified by a staleness limit, even though subsequent update queries may have rendered the cached entry inconsistent.

Different cache invalidation granularities are available ranging from database-wide invalidation to table-based or column-based invalidation. An extra optimization concerns queries that select a unique row based on a primary key. These queries are often issued by application servers using JDO (Java Data Objects) or EJB (Enterprise Java Beans) technologies. These entries are never invalidated on inserts since a newly inserted row will always have a different primary key value and therefore will not affect this kind of cache entries. Moreover, update or delete operations on these entries can be easily performed in the cache.

Key Applications

Middleware-based database replication is currently used in many e-Commerce production environments that require both high availability and performance scalability. Many open problems remain on the integration of databases with replication middleware, failure detection and transparent failover/failback, autonomic management and software upgrades.

Future Directions

Current research trends explore autonomic behavior for replicated databases [4] to automate all management operations such as provisioning, tuning, failure repair and recovery. Heterogeneous clustering is also used in the context of satellites databases [8] to scale legacy databases with open source databases. Partial replication is studied in conjunction with WAN (Wide Area Network) replication for global applications spanning over multiple datacenters distributed on different continents. A summary of the remaining gaps between the theory and practice of middleware-based replication can be found in [2].

URL to Code

The Sequoia source code is available from http://sequoia.continuent.org

Cross-References

► Autonomous Replication
► Consistency Models for Replicated Data
► Database Clusters
► Database Middleware
► Data Partitioning
► Data Replication
► Distributed Database Design
► Distributed Database Systems
► Distributed DBMS
► Distributed Deadlock Management
► Distributed Query Processing
► Distributed Recovery
► Extraction, Transformation, and Loading
► Inter-query Parallelism
► Middleware Support for Precise Failure Semantics
► One-Copy-Serializability
► Partial Replication
► Replication

► Replica Control
► Replica Freshness
► Replicated Database Concurrency Control
► Replication Based on Group Communication
► Replication for Scalability
► Replication in Multitier Architectures
► Shared-Disk Architecture
► Shared-Nothing Architecture
► Snapshot Isolation
► Strong Consistency Models for Replicated Data
► Transactional Middleware
► Weak Consistency Models for Replicated Data

Recommended Reading

1. Amza C, Cox A, Zwaenepoel W. A comparative evaluation of transparent scaling techniques for dynamic content servers. In: Proceedings of the 21st International Conference on Data Engineering; 2005. p. 230–41.
2. Cecchet E, Candea G, Ailamaki A. Middleware-based database replication: the gaps between theory and practice. In: Proceedings of the ACM SIGMOD International Conference on Management of Data; 2008. p. 739–52.
3. Cecchet E, Marguerite J, Zwaenepoel W. C-JDBC: flexible database clustering middleware. In: Proceedings of the USENIX 2004 Annual Technical Conference; 2004.
4. Chen J, Soundararajan G, Amza C. Autonomic provisioning of backend databases in dynamic content web servers. In: Proceedings of the 2006 IEEE International Conference on Autonomic Computing; 2006. p. 231–42.
5. Elnikety S, Dropsho S, Pedone F. Tashkent: uniting durability with transaction ordering for high-performance scalable database replication. In: Proceedings of the 1st ACM SIGOPS/EuroSys European Conference on Computer System; 2006. p. 117–30.
6. Elnikety S, Dropsho S, Zwaenepoel W. Tashkent+: memory-aware load balancing and update filtering in replicated databases. In: Proceedings of the 2nd ACM SIGOPS/EuroSys European Conference on Computer System; 2007. p. 399–412.
7. Gray JN, Helland P, O'Neil P, Shasha D. The dangers of replication and a solution. In: Proceedings of the ACM SIGMOD International Conference on Management of Data; 1996. p. 173–82.
8. Plattner C, Alonso G, Özsu MT. Extending DBMSs with satellite databases. VLDB J. 2008;17(4):657–82.
9. Sequoia project. Available at: http://sequoia.continuent.org

M

Middleware Support for Precise Failure Semantics

Vivien Quéma

CNRS, INRIA, Saint-Ismier Cedex, France

Definition

Providing support for precise failure semantics requires defining an appropriate correctness criterion for replicated action execution of a replication algorithm. Such a correctness criterion allows formally verifying that a sequence of actions is executed correctly. In the context of replication, a sequence of actions executed is correctly if their side-effect appears to have happened exactly-once.

Historical Background

Reasoning about the behavior of concurrent programs has been an active research area during the past decades. Of particular interest in this area are the works on linearizability, a consistency criterion for concurrent objects [4], and on serializability, a consistency criterion for concurrent transactions [5]. These two criteria facilitate certain kinds of formal reasoning by transforming assertions about complex concurrent behavior into assertions about simpler sequential behavior. Moreover, these consistency criteria are local properties: the correctness of individual objects or services is used to reason about system-level correctness. Recently, Frølund and Guerraoui introduced x-ability [2] (exactly-once ability), a correctness criterion for replicated services. X-ability is independent of particular replication algorithms. Although they facilitate reasoning in similar ways, there are fundamental differences between x-ability on one hand and serializability [5] and linearizability [4] on the other. X-ability has safety as well as liveness aspects to it whereas serializability and linearizability are safety conditions only. X-ability is a theory of distribution and partial failures where serializability and linearizability are theories of concurrency. X-ability does not specify correctness for concurrent invocations of a replicated service. More precisely, x-ability states constraints about the concurrency among replicas in the context of a given request (intra-request concurrency), but ignores the concurrency that originates from different requests (inter-request concurrency). This entry gives a precise description of the x-ability theory.

Foundations

Frølund and Guerraoui proposed x-ability [2] (exactly-once ability), a correctness criterion for replicated services. X-ability is independent of particular replication algorithms. The main idea behind x-ability is to consider a replicated service correct if it provides the illusion of a single, fault-tolerant entity. More precisely, an x-able service must satisfy a contract with its clients as well as a contract with third-party entities. In terms of clients, a service must provide idempotent, non-blocking request processing. Moreover, it must deliver replies that are consistent with its invocation history. The side-effect of a service, on third-party entities, must obey exactly-once semantics. X-ability is a local property: replicated services can be specified and implemented independently, and later composed in the implementation of more complex replicated services.

To model side-effects, x-ability is based on the notion of action execution. Actions are executed correctly (i.e., are *x-able*) if their side-effect *appears* to have happened *exactly-once*. The side-effect of actions can be the modification of a shared state or the invocation of another (replicated or non-replicated) service. X-ability represents the execution of actions as event histories and defines the notion of "appears to have happened exactly-once" in terms of history equivalence: an event history h is x-able if it is equivalent to a history h' obtained under failure-free conditions. Being defined relative to failure-free executions, x-ability encompasses both safety and liveness. It is a safety property because it states that certain partial histories must not occur. It is also a liveness property since it enforces guarantees about what must occur. History equivalence is defined relatively to the execution of two particular kinds of actions, namely *idempotent* and *undoable* actions:

- The side-effect of a history with n incarnations of an idempotent action is equivalent to a history with a single incarnation. For example, writing a particular value to a data object is an idempotent action.
- An undoable action is similar to a transaction: its side-effect can be cancelled up to a certain point (the commit point), after which the side-effect is permanent. Thus, the side-effect of a history with a cancelled action is equivalent to the side-effect of a history with no action at all.

System Model

The replicated service is implemented by a set of replicas. The functionality of the service is captured by a *state machine*. Each replica has its own copy of the state machine. A state machine exports a number of *actions*. An action takes an input value and produces an output value. In addition, an action may modify the internal state of its state machine and it may communicate with external entities. A client can invoke a replica's state machine by sending a request to the replica. A request contains the name of an action and an input value for the action. When it receives a request, a replica invokes its state machine based on the values in the request. If no failures occur, the replica returns the action's output value to the client. The execution of an action may fail or the replica executing the action may fail. If the action fails, it returns an exception (or error) value as the execution result. Formally speaking, action names are modeled as elements of a set *Action* (referred to using the letter a). The set *Value* contains the input and output values associated with actions. Furthermore, two sets, *Request* and *Result*, are defined as follows: $Request = (Action \times Value)$ and $Result = Value$. This signifies that a request is simply a pair that contains an action name and an input value (noted "(a, v)" for a request with action name a and value v).

The actions performed by state machines are represented by *events*. More precisely, the x-ability theory considers two kinds of events: start events to represent the invocation of a state-machine action by a process, and completion events to represent the successful completion of a state-machine action: a process receives a non-exception value back from the state machine. The causal and temporal relationship between action execution and event observation is subject to the following axioms: (i) an action's start event cannot be observed unless the action is invoked, (ii) an action's completion event cannot be observed before its start event, and (iii) if an action returns successfully, then its start and completion events have been observed. Events are modeled as elements of the set *Event*. Events are structured values with the following structure: $e ::= S(a, iv) \mid C(a, ov)$. The event $S(a, iv)$ captures the start of executing the action a with iv as argument. The event $C(a, ov)$ captures the completion of executing the action a, and ov is the output value produced by the action.

A sequence of events form a *history*. The notion of a sequence captures the total order in which events are observed. Histories are modeled as elements of the set *History*. Histories are structured values as defined by the following syntax: $h ::= \Lambda \mid e_1 \ldots e_n \mid h_1 \bullet \cdots \bullet h_n$. The symbol Λ denotes the empty history – a history with no events. The history $e_1 \ldots e_n$ contains the events e_1 through e_n. The history $h_1 \bullet \ldots \bullet h_n$ is the concatenation of histories h_1 through h_n. The semantics of concatenating histories is to concatenate the corresponding event sequences. An action a *appears* with input value iv in a history h (noted $(a, iv) \in h$) if h contains a start event produced by the execution of a on iv.

To capture structural properties of histories, the x-ability theory defines the notion of history *patterns*. Formally speaking, patterns are elements of the set *Pattern* (referred to using the letter p). The abstract syntax for patterns is depicted in Fig. 1. A simple pattern sp matches single-action histories. The pattern $[a, iv, ov]$ matches a history that contains the events from a failure-free execution of an action a. The value iv is the input to a and ov is the output from a. The pattern $?[a, iv, ov]$ matches a history in which a may have failed. A matching history may be the empty history, it may contain a start event only, or it may contain both the start and completion event of a. The pattern $|sp_1\|h\,sp_2$ matches a history h' that contains an interleaving of three sub-histories h_1, h_2, and h, where h_1 matches sp_1, h_2 matches sp_2, and h is an arbitrary history. The interleaving is constrained as follows: the first event in h_1 must also be the first event in h' and the last event in h_2 must also be the last event in h'.

X-ability defines *pattern matching* as a relation $\lhd$ between elements of the set *History* and elements of the set *Pattern*. In other words, $\lhd$ is a subset of *History* $\times$ *Pattern* (the set of all pairs from *History* and *Pattern*). Pattern matching rules are shown in Fig. 2. A history that matches a simple pattern contains at most two events. X-ability defines two operators on such histories: *first* and *second* (see Fig. 3). The *first* operator returns the first element in a history, if any, and Λ otherwise. The *second* operator returns the second element in a history of length two, the only element in a history of length one, and empty history otherwise.

$$sp ::= [a, iv, ov] \mid ?[a, iv, ov]$$
$$p ::= sp \mid sp_1\|h\,sp_2$$

Middleware Support for Precise Failure Semantics, Fig. 1 Abstract syntax for history patterns

X-Able Histories

To be fault-tolerant, a replicated service must be prepared to invoke the same action multiple times until the action executes successfully. To provide replication transparency, the service must have exactly-once semantics relative to its environment – the service must maintain the illusion that the action was executed once only. In short, an x-able history is a history that maintains the illusion of exactly-once but possibly contains multiple incarnations of the same action. The rest of this section describes how the x-ability theory defines the notion of x-able history.

The x-ability theory defines a *history reduction* relation, $\Rightarrow$, on histories as follows: if $h \Rightarrow h'$, then the execution that produced h has the same side-effect as an execution that produced h'. Essentially, a history is x-able if it can be reduced, under $\Rightarrow$, to a history that could arise from a system that does not fail. Two particular types of actions are considered: *idempotent* and *undoable*. The corresponding sets are called *Idempotent* and *Undoable*. The set *Idempotent* contains the names of idempotent actions. The notation a^i indicates that the action a is idempotent. The set *Undoable* contains names of undoable actions. The notation a^u indicates that an action a is undoable. An undoable action, a^u, has two associated actions: a cancellation action, a^{-1}, and a commit action, a^c. The commit and cancellation actions for an action a^u take the same arguments as a^u, and they

Middleware Support for Precise Failure Semantics, Fig. 2 Pattern matching rules

(1)
$$\rhd \subseteq (\text{History} \times \text{Pattern})$$

(2)
$$S(a, iv)C(a, ov) \rhd [a, iv, ov]$$

(3)
$$\Lambda \rhd ?[a, iv, ov]$$

(4)
$$S(a, iv) \rhd ?[a, iv, ov]$$

(5)
$$S(a, iv)C(a, ov) \rhd ?[a, iv, ov]$$

(6)
$$\frac{h_1 \rhd sp_1 \quad h_2 \rhd sp_2}{(h_1 \bullet h \bullet h_2) \rhd (sp_1\|h\,sp_2)}$$

(7)
$$\frac{h_1 \rhd sp_1 \quad h_2 \rhd sp_2}{(\text{first}(h_1) \bullet h_3 \bullet \text{second}(h_1) \bullet h_4 \bullet \text{first}(h_2) \bullet h_5 \bullet \text{second}(h_2)) \rhd (sp_1\|h_3 \bullet h_4 \bullet h_5\,sp_2)}$$

(8)
$$\frac{h_1 \rhd sp_1 \quad h_2 \rhd sp_2}{(\text{first}(h_1) \bullet h_3 \bullet \text{first}(h_2) \bullet h_4 \bullet \text{second}(h_1) \bullet h_5 \bullet \text{second}(h_2)) \rhd (sp_1\|h_3 \bullet h_4 \bullet h_5\,sp_2)}$$

$$
\begin{array}{llll}
(9) & \text{first}(\Lambda) &=& \Lambda & \text{first}(e_1 e_2) &=& e_1 \\
(10) & \text{first}(e) &=& e & \text{second}(\Lambda) &=& \Lambda \\
(11) & \text{second}(e) &=& e & \text{second}(e_1 e_2) &=& e_2
\end{array}
$$

Middleware Support for Precise Failure Semantics, Fig. 3 The definition of first and second

Middleware Support for Precise Failure Semantics, Fig. 4 Definition of history reduction

$$
(12) \qquad \Rightarrow\ \subseteq (\texttt{History} \times \texttt{History})
$$

$$
(13) \qquad \frac{h_1 \Rightarrow h_2 \quad h_2 \Rightarrow h_3}{h_1 \Rightarrow h_3}
$$

$$
(14) \qquad \frac{h \rhd (?[a^i, iv, ov]\ \|_{h'}\ [a^i, iv, ov])}{h_1 \bullet h \bullet h_2 \Rightarrow h_1 \bullet h' \bullet (S(a^i, iv)C(a^i, ov)) \bullet h_2}
$$

$$
(15) \qquad \frac{h \rhd (?[a^u, iv, ov]\ \|_{h'}\ [a^{-1}, iv, \text{nil}]) \quad (a^u, iv) \notin h_1 \quad (a^c, iv) \notin h'}{h_1 \bullet h \bullet h_2 \Rightarrow h_1 \bullet h' \bullet h_2}
$$

$$
(16) \qquad \frac{h \rhd (?[a^c, iv, \text{nil}]\ \|_{h'}\ [a^c, iv, \text{nil}]) \quad (a^u, iv) \notin h'}{h_1 \bullet h \bullet h_2 \Rightarrow h_1 \bullet h' \bullet (S(a^c, iv)C(a^c, \text{nil})) \bullet h_2}
$$

return the value *nil*. Moreover, cancellation and commit actions are idempotent.

Figure 4 defines the $\Rightarrow$ operator in terms of idempotent and undoable actions. The first inference rule (13) defines $\Rightarrow$ as a transitive relation. The second rule (14) captures the semantics of idempotent actions. If a history contains a successfully executed idempotent action a^i, then the events from a previous attempt to execute a^i can be removed. The third rule (15) is concerned with cancellation of undoable actions. Intuitively, if an undoable action is successfully cancelled, then its side-effect can be removed. The fourth rule (16) states that commit actions are idempotent.

The x-ability theory defines a *failure-free history* as a history that could have been produced by a failure-free execution of a single state-machine action. To define the notion of failure-free history, the x-ability theory relies on the definition of a function, called *eventsof*, which returns the failure-free history associated with an action and its values.

$$
\begin{aligned}
\text{eventsof}\,(a^u, iv, ov) &= S\,(a^u, iv)\,C\,(a^u, ov) \\
& \quad\ S\,(a^c, iv)\,C\,(a^c, \text{nil})
\end{aligned} \tag{17}
$$

$$
\text{eventsof}\,(a^i) = S\,(a^u, iv)\,C\,(a^i, ov)| \tag{18}
$$

Due to non-determinism, there are multiple failure-free histories which are possible for a given action a and a given input value iv. The set of all possible histories, $FailureFree_{(a, iv)}$, is defined as follows:

$$
\begin{aligned}
FailureFree_{(a,iv)} = \{ h \in \text{History} \,|\, \exists ov \in \} \\
\text{Result} : h \\
(a, iv, ov)\}
\end{aligned}
= \text{eventsof}
\tag{19}
$$

An *x-able history* is defined as a history that can be reduced to a failure-free history. Formally speaking, an x-able history is one that satisfies the predicate *x-able* on histories:

$$
\begin{aligned}
& X - 4\text{able}_{(a,iv)}(h) \\
&= \begin{cases} \text{true} & \text{if}\exists h' \in FailureFree_{(a,iv)} : h \Rightarrow h' \\ & \text{false} \quad \text{otherwise} \end{cases}
\end{aligned}
\tag{20}
$$

This definition of x-able histories applies to single-action histories, that is, a history that arises from a particular request. This reflects the fact that x-ability only specifies correctness relative to distribution and failures, it does not specify correctness for the concurrent processing of multiple requests from different clients.

Client-Service Consistency

The x-ability theory formalizes the relationship between clients and services. More precisely, the reply value given to a client in response to a request must be the value returned from the server-side state machine when the service processes the request. Moreover, the service is not allowed to invent requests. The server-side history is used to define the constraints for requests and replies. This history contains a request value as part of start events and reply values as part of completion events. The x-ability theory introduces the notion of *history signature*, which captures the client-side information (request and result) that is legal relative to a given server-side history. Because of non-determinism and server-side retry, a history can have multiple signatures. The set of signatures is defined by the following inference rules:

$$\frac{h \Rightarrow S\,(a^u, iv)\,C\,(a^u, ov)\,S\,(a^c, iv)\,C\,(a^c, \mathrm{nil})}{(a, iv, ov) \in \mathrm{signature}(h)} \tag{21}$$

$$\frac{h \Rightarrow S\,(a^u, iv)\,C\,(a^i, ov)}{(a, iv, ov) \in \mathrm{signature}(h)} \tag{22}$$

If a client submits a sequence of requests, one after the other, later requests should be processed in the context of earlier requests. To prevent a service from forgetting the effect of previous requests, the x-ability theory assumes the existence of a set *PossibleReply* that contains the possible reply values for a given request. To capture the history-sensitive nature of the set of possible replies, *PossibleReply* is defined in the context of a request sequence $R_1 \ldots R_n$. The interpretation of *PossibleReply* in the context of a sequence is the set of possible replies to request R_n after the state machine has executed the requests $R_1 \ldots R_{n-1}$ one after the other. Thus, the set is written as follows: $\mathrm{PossibleReply}_{(R_1 \ldots R_n)}$.

X-Able Services

The x-ability theory provides a formal specification of replication that is independent of a particular replication protocol. Formally speaking, a replicated service consists of a server-side state machine S and a client-side action *submit*. The state machine captures the functionality of the service. It is executed by a set of server processes $s_1 \ldots s_n$ that each have a copy of S. The action *submit* can be used by any process p to invoke the service. The action takes a value in the domain *Request* and, when executed, produces a value in the domain *Result*. Correctness is specified relative to a single client C. Thus, the considered system consists of the processes $s_1 \ldots s_n$ and C only. The client submits one request at a time, and the service is x-able if the following conditions hold:

- R1. The action *submit* is idempotent.
- R2. The client C will eventually be able to execute *submit* successfully.
- R3. If the client submits a request (a, iv), then the server-side history for (a, iv) is either empty or it satisfies x-able$_{(a, iv)}$.
- R4. If the client receives a reply ov in response to a request (a, iv), and if the server-side history for executing this request is h, then $(a, iv, ov) \in signature(h)$.
- R5. If the client successfully submits a sequence of requests, $R_1 \ldots R_n$, and receives the reply R' in response to R_n, then R' is in $\mathrm{PossibleReply}_{(R_1 \ldots R_n)}$.

The first two requirements (R1 and R2) are concerned with the contract between a service and its clients. Clients use the action *submit* to invoke the service. Because *submit* is idempotent, clients can repeatedly invoke the service without concern for duplicating side-effects. The second requirement (R2) is a liveness property. The action *submit* is not allowed to fail an infinite number of times. The requirement also makes a service non-blocking in the sense that *submit* is guaranteed to eventually return a value. The third requirement (R3) deals with the server-side side-effect of executing a request. The resulting server-side history must be x-able, that is, it must be equivalent (under history reduction) to a failure-free history. The fourth requirement (R4) forces an algorithm to preserve consistency between the client-side view (request and reply) and the server-side view (the side-effect). This requirement, prevents the *submit* action from inventing reply values. It also prevents the service from inventing request values. The fifth requirement (R5) forces the service to correctly maintain S's state, if any. The server-side history must be equivalent to a failure-free execution of the sequence $R_1 \ldots R_n$. But since R_1 may result in a transformation of S's state, the actions executed for R_2 may depend on this state transformation. So, a replication algorithm must ensure that the state resulting from R_1 is used as a context for executing R_2. The replication algorithm cannot assume that R_1 did not update the state of S, or that the state update is immaterial to the processing of R_2.

Key Applications

A key application of the x-ability theory is the design transactions protocols for three-tier applications. Such applications encompass three layers: human users interact with front-end clients (e.g., browsers), middle-tier application servers (e.g., Web servers) contain the business logic of the application, and perform transactions against back-end databases. Three-tier applications usually rely on replication and transaction-processing techniques. It has been defined in [1, 3] the notion of the Exactly-Once Transaction (e-Transaction) abstraction: an abstraction that encompasses both safety and liveness properties in three-tier environments and ensures end-to-end reliability.

Recommended Reading

1. Frølund S, Guerraoui R. Implementing e-transactions with asynchronous replication. IEEE Trans Parallel Distrib Syst. 2001;12(2):133–46.
2. Frølund S, Guerraoui R. X-ability: a theory of replication. Distrib Comput. 2001;14(4):231–49.
3. Frølund S, Guerraoui R. e-Transactions: end-to-end reliability for three-tier architectures. IEEE Trans Software Eng. 2002;28(4):378–95.
4. Herlihy M, Wing JM. Linearizability: a correctness condition for concurrent objects. ACM Trans Program Lang Syst. 1990;12(3):463–92.
5. Papadimitriou CH. The serializability of concurrent database updates. J ACM. 1979;26(4):631–53.

Mining of Chemical Data

Xifeng Yan
IBM T. J. Watson Research Center, Hawthorne, NY, USA

Definition

Given a set of chemical compounds, chemical data mining is to characterize the compounds present in the data set and apply a variety of mining methods to discover relationships between the compounds and their biological and chemical activities.

Historical Background

In 1969, Hansch [6] introduced quantitative structure-activity relationship (QSAR) analysis which attempts to correlate physicochemical or structural properties of compounds with biological and chemical activities. These physicochemical and structural properties are determined empirically or by computational methods. QSAR prefers vectorial mappings of compounds, which are usually coded by existing physicochemical and structural fingerprints. Dehaspe et al. [3] applied inductive logic programming to predict chemical carcinogenicity by mining frequent substructures in chemical datasets, which identifies new structural fingerprints so that QSAR could build comprehensive analytical models.

Foundations

Chemical compounds are unstructured data with noexplicit vector representation. For chemical compounds, various similarity measures are defined, which could be classified into three categories: (i) physicochemical property-based, e.g., toxicity and weight; (ii) structure-based; and (iii) feature-based. The structure-based similarity measure directly compares the topology of two chemical compounds, e.g., maximum common subgraph, graph edit distance, and graph kernel. As for the feature-based similarity measure, each graph is represented as a feature vector, $\mathbf{x} = [x_1, x_2, \ldots, x_n]$, where x_i is the value of feature f_i. A feature could be physicochemical or structural. The similarity between two graphs is measured by the similarity between their feature vectors. Bunke and Shearer [1] used maximum common subgraph to measure structure similarity. Given two graphs G and G', if P is the maximum common subgraph of G and G', then the structure similarity between G and G' is defined by

$$\frac{2\,|E(P)|}{E(G)\,|+|\,E\,(G')}\,,$$

where $E(G)$ is the edge set of G.

Kashima et al. [7] introduced marginalized kernels between labeled graphs,

$$K\left(G, G'\right) = \sum_h \sum_{h'} K_z\left(z, z'\right) p(h|G)\, p\left(h'|G'\right),$$

where $z = [G, h]$ and $K_z(z, z')$ is the joint kernel over z. The hidden variable h is a path generated by random walks and the joint kernel K_z is defined as a kernel between these paths. Other sophisticated graph kernels are also available. For example, Fröhlich et al. [5] proposed optimal assignment kernels for attributed molecular graphs, which compute an optimal assignment from the atoms of one molecule to those of another one, including local structures and neighborhood information.

The structure-based similarity measure can serve general chemical data mining such as chemical structure classification and clustering. The implicit definition of feature space makes it hard to interpret, and hard to adapt to many powerful data management and analytical tools such as R-tree and support vector machine. An alternative approach is to mine the most interesting features from chemical data directly, such as patterns that are discriminative between compounds with different chemical activities. Figure 1 depicts the pipeline of this approach built on features discovered by a mining process.

The feature-based mining framework includes three steps: (i) mine patterns/features from chemical data, (ii) select discriminative or significant features, and (iii) perform advanced mining. The first step is the process of finding and extracting useful features from raw datasets. One kind of features used in data mining is frequent substructures, the common structures that occur in many compounds. Formally, given a graph dataset $D = \{G_1, G_2, \ldots, G_n\}$ and a minimum frequency threshold θ, frequent substructures are subgraphs that are contained by at least $\theta|D|$ graphs in D. A set of graph pattern mining algorithms are available for mining frequent substructures, including SUBDUE, Warmr, AGM, gSpan, FSG, MoFa/MoSS, FFSM, Gaston, and so on. Generally, for mining graph patterns measured by an objective function F, there are two related mining tasks: (i) enumeration task, find all of subgraphs g such that $F(g)$ is no less than a threshold; and (ii) optimization task, find a subgraph g^* such that

Mining of Chemical Data, Fig. 1
Feature-based mining framework

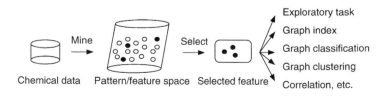

$$g^* = \operatorname{argmax}_g F(g).$$

The enumeration task might encounter the exponential number of patterns as the traditional frequent substructure mining does. To resolve this issue, one may rank patterns according to their objective score and select patterns with the highest value. The feature-based mining framework finds many key applications in chemical data mining including, but not limited to, chemical graph search, classification and clustering.

Chemical graph search aims to find graphs that contain a specific query structure. It is inefficient to scan the whole database and check each graph. Yan et al. [9] applied the feature-based mining framework to support fast search using frequent substructures selected by the following criterion. Let substructures $f_1, f_2, \ldots, f_n$ be selected features. Given a new substructure x, the selectivity power of x can be measured by

$$1 - \operatorname{Pr}\left(x \,|\, f_{\varphi_1}, \ldots, f_{\varphi_m}\right), f_{\varphi_i} \subseteq x, 1 \le \varphi_i \le n.$$

which shows the absence probability of x given the presence of $f_{\varphi_1}, \ldots, f_{\varphi_m}$ in a graph. When the selectivity is high, substructure x is a good candidate to index.

In addition to molecule search, chemical data classification and clustering could also benefit from graph patterns. A typical setting of molecule classification is to induce a mapping $h(\mathbf{g}) : G \rightarrow \{\pm 1\}$ from the training samples $D = \{\mathbf{g}_i, y_i\}^n_{i=1}$, where $\mathbf{g}_i \in G$ is a labeled graph and $y_i \in \{\pm 1\}$ is the class label. Feature-based classification models were proposed in [8, 4]. In these models, graphs are first transformed to vectors using discriminative substructures, which are then processed by standard classification methods.

Key Applications

Chemical Structure Search
   Chemical Classification
   Chemical Clustering
   Quantitative Structure-Activity Relationship Analysis

Experimental Results

PubChem (see dataset URL) provides information on the biological activities of small molecules, containing the bioassay records for anti-cancer screen tests with different cancer cell lines. Each dataset belongs to a certain type of cancer screen with the outcome active or inactive. These bioassays are experimented to identify the chemical compounds that display the desired and reproducible behavior against cancers. Chemical compound classification is to computationally predict the activity of untested compounds given the bioassay data. This process can replace or supplement the physical assay techniques. Furthermore, it could also identify substructures that are critical to specific biological or chemical activities.

The following experiment demonstrates the effectiveness of graph kernel method [7, 5] (optimal assignment kernel, OA) and pattern-based classification [8, 4] (PA), both of which show good accuracy. From the PubChem screen tests, 11 bioassay datasets are displayed. Since the active class is very rare (around 5%) in these datasets, 500 active compounds and 2000 inactive compounds are randomly sampled from each dataset for performance evaluation. The classification accuracy is evaluated with fivefold cross validation. For both methods, the same implementation of support vector machine, LIBSVM [2], with parameter C selected from $[2^{-5}, 2^5]$, is used. Table 1 shows AUC by OA and PA. The area under the ROC curve (AUC) is a measure of the model accuracy, in the range of [0, 1]. A perfect model will have an area of one. As shown in Table 1, PA achieves comparable results with OA. A detailed examination shows that PA is able to discover substructures that determine the activity of compounds, without domain knowledge.

Data Sets

PubChem provides bioassay records for anti-cancer screen tests with different cancer cell lines, available at http://pubchem.ncbi.nlm.nih. gov

M

Mining of Chemical Data, Table 1 Chemical compound classification

Dataset	OA	PA
MCF-7: Breast	0.68 ± 0.12	0.67 ± 0.10
MOLT-4: Leukemia	0.65 ± 0.06	0.66 ± 0.06
NCI-H23: Non-Small Cell Lung	0.79 ± 0.08	0.76 ± 0.09
OVCAR-8: Ovarian	0.67 ± 0.04	0.72 ± 0.06
P388: Leukemia	0.79 ± 0.07	0.82 ± 0.04
PC-3: Prostate	0.66 ± 0.09	0.69 ± 0.09
SF-295: Central Nerv Sys	0.75 ± 0.11	0.72 ± 0.12
SN12C: Renal	0.75 ± 0.08	0.75 ± 0.06
SW-620: Colon	0.70 ± 0.02	0.74 ± 0.06
UACC257: Melanoma	0.65 ± 0.05	0.64 ± 0.05
Yeast: Yeast anticancer	0.64 ± 0.04	0.71 ± 0.05
Average	0.70 ± 0.07	0.72 ± 0.07

Cross-References

▶ Frequent Graph Patterns
▶ Graph Database

Recommended Reading

1. Bunke H, Shearer K. A graph distance metric based on the maximal common subgraph. Pattern Recogn Lett. 1998;19(3):255–9.
2. Chang C-C, Lin C-J. LIBSVM: a library for support vector machines. 2001. Software available at http://www.csie.ntu.edu.tw/~cjlin/libsvm
3. Dehaspe L, Toivonen H, King R. Finding frequent substructures in chemical compounds. In: Proceedings of the 4th International Conference on Knowledge Discovery and Data Mining; 1998. p. 30–6.
4. Deshpande M, Kuramochi M, Wale N, Karypis G. Frequent substructure-based approaches for classifying chemical compounds. IEEE Trans Knowl Data Eng. 2005;17(8):1036–50.
5. Fröhlich H, Wegner J, Sieker F, Zell A. Optimal assignment kernels for attributed molecular graphs. In: Proceedings of the 22nd International Conference on Machine Learning; 2005. p. 225–32.
6. Hansch C. A quantitative approach to biochemical structure-activity relationships. Acc Chem Res. 1969;2(8):232–9.
7. Kashima H, Tsuda K, Inokuchi A. Marginalized kernels between labeled graphs. In: Proceedings of the 20th International Conference on Machine Learning; 2003. p. 321–28.
8. Kramer S, Raedt L, Helma C. Molecular feature mining in HIV data. In: Proceedings of the 7th ACM SIGKDD International Conference on Knowledge Discovery and Data Mining; 2001. p. 136–43.
9. Yan X, Yu PS, Han J. Graph indexing: a frequent structure-based approach. In: Proceedings of the ACM SIGMOD International Conference on Management of Data; 2004. p. 335–46.

Mobile Database

Ouri Wolfson
Mobile Information Systems Center (MOBIS), The University of Illinois at Chicago, Chicago, IL, USA
Department of CS, University of Illinois at Chicago, Chicago, IL, USA

Definition

A mobile database is a database that resides on a mobile device such as a PDA, a smart phone, or a laptop. Such devices are often limited in resources such as memory, computing power, and battery power.

Key Points

Due to device limitations, a mobile database is often much smaller than its counterpart residing on servers and mainframes. A mobile database is managed by a Database Management System (DBMS). Again, due to resource constraints, such a system often has limited functionality compared to a full blown database management system. For example, mobile databases are single user systems, and therefore a concurrency control mechanism is not required. Other DBMS components such as query processing and recovery may also be limited.

Queries to the mobile database are usually posed by the user of the mobile device. Updates of the database may originate from the user, or from a central server, or directly from other mobile devices. Updates from the server are communicated wirelessly. Such communication takes place either via a point-to-point connection between the mobile device (the client) and the server, or via broadcasting by the server [Acharya S, Franklin M, and Zdonik S (1995), Dissemination-based Data Delivery Using Broadcast Disks, Personal Communications]. Direct updates from other mobile devices may use short-range wireless communication protocols such as Bluetooth or Wifi [1].

Recommended Reading

1. Cao H, Wolfson O, Xu B, Yin H. MOBI-DIC: MOBIle DIscovery of loCal Resources in Peer-to-Peer Wireless Network. Bull Comput Soc Tech Comm Data Eng. 2005;28(3):11–18. (Special Issue on Database Issues for Location Data Management).

Mobile Interfaces

Giuseppe Santucci
University of Rome, Rome, Italy

Synonyms

Handhelds interfaces; Navigation system interfaces

Definition

Mobile interfaces are interfaces specifically designed for little portable electronic devices (handhelds), like cellular phones, personal digital assistants (PDAs), and pagers. Mobile interfaces provide means to execute complex activities on highly constrained devices, characterized by small screens, low computing power, and limited input/output (I/O) capabilities. While this term is mainly used to refer to interfaces for classical Web applications, like email and Web browsing it should be intended in a broader sense, encompassing all Web and non-Web applications that run on little mobile equipments (e.g., GPS navigation systems).

Historical Background

Mobile interfaces came up with the proliferation of cellular phones through the 1980s. At that time, the notion of mobile interface was very primitive and the offered services were very simple: handling lists of contacts, usually limited to <name, phone-number> pairs, starting/answering a phone call, writing short text, setting phone preferences (volume, light, ring tone, etc.). Even in this restricted scenario, the vendors had to deal with challenging issues, posed by the very limited early cellular phones capabilities, in terms of screen dimension and resolution, limited computing power, and I/O capabilities. No standards were around and different, unrelated solutions were adopted. The most used interaction strategy was to mimic the typical hierarchical computer menus with very poor results: the user was (and still is) forced to access functions and services through a series of boring menus. The main reason of failure is that while PCs can present to the user a complete list of all possible choices mobile phones can show a small number of options at time (usually one), forcing the user to remember the paths to the commands. That results in a very large number of key presses, errors, and mental overwhelming: many of the advantages of the conventional menus are lost [12].

For about 10 years the unique available portable devices were mainly cellular phones; from 1993 to 1998 Apple Computer (now Apple Inc.) marketed the "Apple Newton," the first line of PDAs, personal digital assistants, a term introduced on January 7, 1992 by John Sculley at the Consumer Electronics Show in Las Vegas, Nevada, referring to the Apple Newton. The Apple's official name for the PDA was "MessagePad" and it was based on the ARM 610 RISC processor using a dedicated operating system (Newton OS, an allusion to Isaac Newton's apple); however, the word Newton was popularly used to refer to the device and its software. The new device was characterized by a wider touchable monochromatic screen (366 × 240) with retro illumination (only for the top version). The increased computational power and the larger screen allowed for very innovative interaction techniques (for a portable device): icons, simulated touchable QWERTY keyboard, and handwriting recognition (called Calligrapher). Moreover, the user was allowed to turn the screen horizontally ("landscape") as well as vertically ("portrait") preserving the handwriting recognition functionality (see Fig. 1).

Even the data management was quite innovative: programs were able to convert and share data (e.g., the calendar was able to refer to names in the address book or to convert a note into an appointment). Finally all the devices were equipped with a built in infrared and an expansion port for connecting to a modem or Ethernet. Disregarding the color absence and the limited device connectivity, the Palm Newton functionalities were quite similar to the ones of modern PDAs!

In spite of the innovative interface and capabilities, the product was not a successful one. The main reasons were the poor handwriting recognition accuracy, the high price, and the non comfortable size (it did not fit in a regular pocket).

Still, the pioneering Apple ideas were captured by other vendors and many similar devices were around in the next years. Among them, the Palm series with the PalmOS operative system and the Graffiti handwriting recognition system conquered (in 1999) about 80% of the world market, mostly for the effective handwriting system and for the availability of a huge amount of third part software. On February 2000 Palm marketed the PalmIIIc, the first color PDA. Concerning the interface it is worth noting the different Graffiti philosophy: instead of allowing the user to write in his/her own style it forced the user to learn a set of pen strokes for each character (see Fig. 2). This narrowed the possibility for erroneous input, although memorization of the stroke patterns did increase the learning curve for the user. Some studies demonstrated that even if the error rate with Graffiti was higher than using the virtual

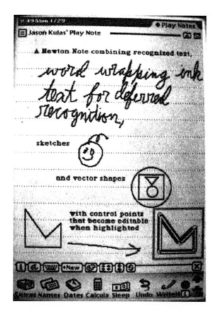

Mobile Interfaces, Fig. 1 A screenshot from Apple Newton

The graffiti® alphabet

Letter	Strokes	Letter	Strokes
A	∧	N	N
B	ℬ ℬ	O	○ ○
C	C	P	p p
D	Ɗ Ɖ	Q	ơ
E	Ɛ	R	ℛ ℛ
F	Γ Γ	S	S
G	G 6	T	⌐
H	h	U	U
I	ſ	V	⌄ U
J	J	W	W
K	⋋	X	X ✕
L	L	Y	4 ʏ
M	m m	Z	Z

Mobile Interfaces, Fig. 2 The PalmOS Graffiti

keyboard (19.3 vs. 4.1%), users prefer Graffiti, because its usage is more natural.

As soon as the hardware made it possible a deeper integration between PDA and cellular phones started: phones were equipped with larger screen and complex applications (smartphones), while PDA included phone hardware. Moreover, the so called pager devices, mainly intended for on the way email, came up. In 2004, smartphones were outselling PDAs for the first time. Nowadays, the difference between these three categories in term of capabilities and interface is very little. The main consequence of this integration is that all these devices allow for Internet access (through GPRS or WI-FI) and it is very likely that in the next future there will be more people accessing the Internet via mobile devices (phones, PDA, etc) than via conventional PCs. According to this issue many researchers are now dealing with the problem of designing friendly interface for accessing Internet through portable devices.

Foundations

The research effort concerning mobile interfaces followed two main paths: designing interface for a specific application running on a specific device (e.g., the agenda interface of a specific cellular phone) or designing web based applications that can be deployed on multiple devices (e.g., a web browser working on different PDAs and cellular phones).

Mobile Interface for Specific Devices

The problem of dealing with little screen is not new. Much literature from the 1980s and early 1990s, written even before the Web, deals with user interfaces on small screens. At that time the focus was on first generation cash dispensing ATMs, electronic typewriters and photocopiers. All of these systems could display only a limited number of text lines to communicate with the user. Research looked into the impact of reduced screen size on comprehension [3], reading rate [4], and interaction [12]. This research is still

valid and has new relevance to mobile Web devices [8].

The availability of graphic display and the diffusion of navigation system raised the issue of interacting with graphics and maps [6]. Moreover, the increasing usage of such systems while driving a car posed several new issues concerning distracting factors and security [1].

Multi Target Applications

In this case, research deals with techniques able to support the design of applications that run on several devices that present different interaction capabilities (e.g., small/large screens, keyboard-/keypad input, etc.). In order to address this issue, researchers have adopted two main approaches:

1. Designing applications in order to let them run on different platforms
2. Adapting existing systems in order to render them usable on platform that were not originally included as service provider (e.g., standard web sites)

In the first approach the solution is offered at design time, that is, the designer explicitly knows that the system will run on multiple devices. Some research exists that follows this approach and the key issues consist in giving models, methods, and tools to support the designer in the creation of multi-device applications and in offering techniques and algorithms to generate the final interface at run time for the specific device utilized. In [10], a tool for support is presented together with a well defined development life cycle: the designer starts from abstract tasks definition and upon it designs the interface in abstract. The final interfaces are statically generated at design time by following suggestions offered by the system and further refined in order to accommodate specific details. Puerta et al. in [5] propose a similar framework, defining a model to design user interfaces in abstract as well, but their work envisions also adaptive techniques to produce the final user interface. A central mediator agent is responsible for translating the abstract specification of the user interface, according to a description of device's characteristics, into the fi-

M

nal interface. Fundamental questions arising from this kind of research are what kind of models should be employed to specify abstract interaction elements and what strategies/techniques can permit an effective translation of abstract models into real interfaces.

The second approach is the one that attempts to bring existing systems, typically web sites, to small screen devices by means of some sort of adaptation/filtering. Among them, the WAP forum (www.wapforum.org) defined a protocol that enables Web-like services on little, portable devices. The interface uses a card metaphor and the designers claim that it is highly appropriate and usable. However, Nielson [11] raised doubts about its effectiveness. A similar proposal is i-mode, developed by the dominant Japanese carrier NTT DoCoMo, and widely used in Japan. Moreover, the World Wide Web Consortium (W3C) is also developing a framework that will enable Web content to be accessed on a diverse range of devices [13]. This framework will allow a document to exist in multiple variants, each variant specifying the type of support needed by the browser to display its contents. Device capabilities and user preferences will also be captured. The information about documents, devices and users will be used to automatically adapt a Web page to best suit the device and user.

Key Applications

Users of mobile interfaces are growing at high speed. According to IDC's Worldwide Quarterly Mobile Phone Tracker (www.idc.com), worldwide mobile phone shipments rose 19.1% year over year and increased sequentially 8.8% in 2005 to reach 208.3 million units. While unconnected PDA's are a declining segment, the emerging navigator market is still increasing about 10 million cars guided by navigation systems, 43 million dedicated portable navigators, and 16 millions smart phones based navigators forecasted for 2010 (www.strategyanalytics.net).

These figures make clear the importance of mobile interfaces in the coming years.

Future Directions

There are some emerging issues that will likely affect the behavior and structure of mobile interfaces:

- Search versus menu browsing. Several proposals relies on the idea that, in order to start an action, it is better to search it (i.e., to write down the command name) instead of browsing boring menus [7, 9].
- Multimodal interaction. When the hardware allows it, mobile interfaces can exploit multimodal input/output (e.g., voice commands, likely together with the aforementioned search strategy).
- Context and user modeling. Some proposal describe interfaces that can adapt themselves according to the user preferences and the context [2], e.g., an interface for browsing tourist information presents a user with list of vegetarian restaurants at walking distance, according to the user location, discovered through the integrated GPS device, knowing that the user is vegetarian and that s/he is currently walking.

These research issues could greatly improve the mobile interfaces in the next years.

Cross-References

▶ Multimodal Interfaces
▶ Visual Interfaces
▶ WIMP Interfaces

Recommended Reading

1. Commission of the European Communities. Commission recommendation of 22 December 2006 on safe and efficient in-vehicle information and communication systems: update of the European Statement of Principles on Human Machine Interface; 2006.
2. Coutaz J, Crowley J, Dobson S, Garlan D. Context is key. Commun ACM. 2005;48(3):49–53.
3. Dillon A, Richardson J, McKnight C. The effect of display size and text splitting on reading lengthy text from the screen. Behav Inform Technol. 1990;9(3):215–27.

4. Duchnicky RL, Kolers PA. Readability of text scrolled on visual display terminals as a function of window size. Hum Factors. 1983;25(6):683–92.
5. Eisenstein J, Vanderdonckt J, Puerta A. Applying model-based techniques to the development of UIs for mobile computers. In: Proceedings of the 6th International Conference on Intelligent User Interfaces; 2001. p. 69–76.
6. Frey PR, Rouse WB, Garris RD. Big graphics and little screens: designing graphical displays for maintenance tasks. IEEE Trans Syst Man Cybern. 1992;22(1):10–20.
7. Graf S, Spiessl W, Schmidt A, Winter A, Rigoll G. In-car interaction using search based user interfaces. In: Proceedings of the 26th Annual SIGCHI Conference on Human factors in Computing Systems; 2008. p. 1685–88.
8. Jones M, Marsden G, Mohd-Nasir N, Boone K, Buchanan G. Improving web interaction on small displays. In: Proceedings of the 8th International World Wide Web Conference, Toronto; 1999. Also reprinted in Int J Comput Telecommun Netw;31(11–16):1129–7, 1999.
9. Marsden G, Gillary P, Jones M, Thimbleby H. Successful user interface design from efficient computer algorithms. In: Proceedings of the ACM CHI 2000 Conference on Human Factors in Computing Systems; 2000. p. 181–2.
10. Mori G, Paternò F, Santoro C. Tool support for designing nomadic applications. In: Proceedings of the 2003 International Conference on Intelligent User Interfaces; 2003. p. 141–8.
11. Nielson J. Graceful degradation of scalable internet services. Available online at: http://www.useit.com/alertbox/991031.html. 1999.
12. Swierenga SJ. Menuing and scrolling as alternative information access techniques for computer systems: interfacing with the user. In: Proceedings of the 34th Annual Meeting of Human Factors Society; 1990. p. 356–9.
13. W3C Mobile Activity Statement. Available online at: http://www.w3.org/Mobile/Activity

Mobile Resource Search

Sergio Ilarri
University of Zaragoza, Zaragoza, Spain

Synonyms

Event discovery in mobile computing; Mobile discovery of local resources; Mobile local search; Mobile local search for spatiotemporal resources

Definition

In mobile computing environments, mobile users frequently need to access different types of resources to satisfy a certain need. For this purpose, an appropriate resource must first be identified and located. This is a difficult problem, due to several factors. First of all, as users move from one location to another, the appropriate resources may change continuously: they must be both accessible for the user (e.g., within a reachable range) and relevant for his/her intended purpose; this problem is further compounded when the resources are also mobile (e.g., in the scenario of a user searching for a taxi). Secondly, many resources are scarce and used in mutual exclusion (e.g., parking spaces, non-shareable taxi cabs, etc.), which may eventually generate a competition among mobile users to try to get the resource. Thirdly, the relevance of a resource for a mobile user must be determined, in order to direct him/her toward the resources that are more promising, and taking into account that the data available about them may be subject to some uncertainty. This entry describes the task of mobile resource search and briefly outlines the main challenges and application scenarios.

Historical Background

Mobile resource search is a problem that belongs to the field of mobile computing. Key initial research on mobile computing dates back from the 1990s [12], but technological advances in wireless communication technologies and mobile devices, along with their popularity and widespread use, have reinforced the relevance of mobile computing. It now includes new scenarios such as mobile peer-to-peer (mobile P2P) networks/mobile ad hoc networks (MANETs) [4], mobile cloud computing [6], and vehicular networks (VANETs) [15].

In all these scenarios, there has been a great interest in offering customized information and services that users may find suitable. The basic idea is to exploit context information (the time of the day, the direction and whereabouts of the user, the activity that he/she is performing, etc.) to offer *location-dependent data* [19] and *context-aware mobile services* [2] to the user. The data provided to the user is frequently information about *physical resources*, such as a parking slot, an available taxi cab, a certain type of person (e.g., a doctor), a hotel, or a friend. It may also be a *virtual resource* such as a service, or simply information about the environment (e.g., traffic information, information about an accident, information about an event in the city, etc.) [14].

Current research on mobile resource search has its roots in the field of *location-dependent query processing* [10], whose goal is to efficiently obtain up-to-date data that satisfy certain spatial constraints (such as information about nearby resources). Regarding the architecture required to support mobile resource search, we can distinguish three types of proposals (e.g., see [7, 10, 17]):

- *Centralized approaches.* The first solutions proposed were based on the use of a centralized server offering a directory service to locate the resources. That central server manages information about all the available resources and processes queries about them.
- *Distributed approaches based on a support infrastructure.* Due to scalability and reliability reasons, distributed solutions started to emerge, based on the use of several servers connected through a fixed network. Data management responsibilities are often assigned based on geographic criteria: each server manages data concerning a different geographic area.
- *Mobile peer-to-peer approaches.* Finally, the most recent trend is to try to benefit from direct interactions among nearby mobile devices to exchange information about resources in a peer-to-peer (P2P) way, opportunistically.

Mobile P2P approaches are based on the use of short-range wireless communication technologies. They are expected to minimize potential accessibility, scalability, and availability problems, as well as the deployment cost (a fixed support infrastructure is not needed). The fact that it is free to use encourages the cooperation among mobile devices to willingly share information about the resources. Due to these reasons, current research on mobile resource search usually considers either pure mobile P2P approaches or hybrid approaches that combine ad hoc interactions with traditional client/server. The recent emergence of VANETs [15], which are mobile P2P networks whose nodes are vehicles, have also attracted research on mobile resource search in that context.

Scientific Fundamentals

A suitable mobile resource search should consider the context of the user to direct him/her toward appropriate resources. One of the most important and exploited context dimensions is the location of the user, and so *location-based services* (*LBS*) [16] have attracted significant attention both in research and in industry. As an example, a great amount of research focuses on the problem of efficient processing of *location-dependent queries* [10], which are queries whose answer depends on the locations of certain moving objects (the mobile user or other objects he/she is asking about), such as "give me the taxi cabs nearby," "find petrol stations on my route," or "suggest me appropriate hotels located within 0.5 miles."

As opposed to traditional queries (also called instantaneous queries, snapshot queries, or one-shot queries), location-dependent queries must be processed as *continuous queries*, which means that the answer to the query must be kept active and must be continuously refreshed until explicitly canceled by the user. This is important because, due to the continuous mobility of objects, the answer may be changing all the time. For example, in the case of mobile resources, even if the set of relevant resources does not change, the locations of the resources will change along time. The general goal is to retrieve information about *local resources* [3] that are interesting for the user and reachable. Therefore, location-dependent operators such as *inside/range/within* operators (that retrieve the moving entities located within a certain spatial area around a given object or reference point) and *(k-)nearest neighbor* (kNN) operators (that retrieve the k nearest resources of a certain type to a given moving object) have been defined and studied.

The challenge is how to find and retrieve information about those resources in an efficient and effective way. First, techniques are needed to obtain only the interesting resources, thus avoiding overloading the user with information about resources that are not relevant. For example, the filtering of irrelevant resources may be based on the type of resource (e.g., ignore information about taxi cabs if the user is not interested in taxis at that moment), spatial criteria (e.g., retrieve only resources located within a certain radio, near the expected route or trajectory of the user, etc.), or temporal criteria (consider only resources that have been advertised or detected recently). Second, the search procedures should work in real time (i.e., exhibit a low latency), as most physical resources have a very temporary status and can appear and disappear at any time (e.g., a taxi cab and a parking spot can become available/unavailable in a very short time). Third, as the information about the resources is usually obtained through mobile crowdsourcing [11], the proposed solutions should consider the potential existing data uncertainty (e.g., see [8, 9]).

Key Applications

We can consider a number of applications of mobile resource search, depending on the specific scenario considered and the type of resource searched. There are both *competitive resources* (such as parking slots, cab customers, and cabs) and *noncompetitive resources* (such as information about traffic conditions and petrol stations, which are shareable). Moreover, according to the criteria that determine their relevance,

resources can also be classified into *temporal resources* (available during a limited period of time), *spatial resources* (attached to a location), and more generally *spatiotemporal resources* (events) [3, 18]. It can also be distinguishing between *static resources* (e.g., parking spaces or petrol stations that do not move) and *mobile resources* (e.g., taxi cabs). Finally, depending on the economic model supporting and encouraging the exchange of resource information in a mobile P2P network, a resource can be a *producer-paid resource* (the producer pays a fee for advertising the resource, such as in the case of a petrol station announcing its services in the vicinity) or a *consumer-paid resource* (the consumer pays a fee for receiving information about the resource, such as in the case of a vehicle searching for an available parking space) [18].

Many application domains of mobile resource search have been identified (e.g., see [3]), including social networks, transportation, mobile electronic commerce, emergency response, and homeland security. Some applications concern drivers (and/or passengers) of vehicles and others are useful for pedestrians. In the following, we discuss some key applications of mobile resource search.

Searching Available Parking Spaces

According to several reports, finding an available parking space is a time-consuming task with a negative impact in several dimensions. Besides leading to drivers' dissatisfaction due to the large amount of time required to find parking, it also causes traffic congestion, implies the consumption of a significant amount of fuel, and contributes to environmental pollution. Parking spaces are precious resources for drivers, and so this is a prominent application of mobile resource search.

However, there are several challenges. For example, it has been shown that providing deliberately information about an available parking space to all the drivers that may be interested in it may actually worsen the problem (i.e., increase the search time in comparison with a *blind search*, where the driver is not assisted by any service) [5, 13]. The reason is that it can lead to

M

additional competition (i.e., a herding effect [3]): several drivers receiving the information could try to reach the same parking space and only one of them will succeed. Several solutions to this problem have been proposed, such as using a reservation protocol that books the parking space and allocates it to a single driver, restricting the communication of the information about the parking space (e.g., ideally providing it only to a single interested vehicle), trying to maximize the probability that the driver will find a parking space available (e.g., the driver can be directed toward areas with good parking availability rather than to a specific parking spot), exploiting the gap between the system optimum (SO) and Nash equilibrium (NE) in matching vehicles and parking slots [1], or estimating the availability probability at the time of arrival to make the driver aware of the competition. Besides, it would be interesting to have applications able to search available parking spaces of different types (private parking and garages, parking lots, on-street parking, payment-based parking areas on the street, and even home parking), and considering different data collection mechanisms to accumulate information about these resources (e.g., sensors of different type, information directly provided by drivers who see or release a parking space, etc.).

Searching other competitive resources such as municipal bike stations (with at least one bike available in the case of a pedestrian searching for a bike, or a slot to leave a bike in the case of a bike driver trying to return it) would bring similar challenges.

Searching Taxi Cabs/Taxi Customers

A typical interesting application for pedestrians would be to find available taxi cabs in the vicinity. Depending on the city and the taxi company considered, a taxi cab may be a competitive resource or a resource that can be shared with other people, but in any case, the temporary availability status of taxis makes them a scarce resource in many scenarios. Traditional booking services for taxis based on phone reservation are subject to some limitations, as, for example, the customer has to select and call a specific taxi company and he/she

cannot obtain any information regarding nearby taxis. An approach where mobile users are able to dynamically obtain information in real time about nearby taxis and book them would be much more flexible and could lead to a better matchmaking among customers and taxis, by exploiting information about trajectories and intentions that a traditional customer service of the taxi company would probably not be able to take into account otherwise.

With a mobile resource search application, not only mobile users could be able to monitor and/or search for an appropriate taxi, but taxi drivers with space in the car could also search for potential customers nearby.

Searching Vehicles and People for Ridesharing

A mobile resource application quite related to the previous one is in the context of ridesharing (also popularly known as carpooling, car sharing, or lift sharing), where people could dynamically share space in private cars to go to certain places. In this case, private cars are shareable resources that could be searched by users needing a ride. Similarly, potential passengers that could share the transportation expenses are also resources from the point of view of the private cars. Providing a suitable matchmaking based on the preferences of the users and their planned routes is key for the success of this type of service. Moreover, there are also issues regarding the reliability and safety of the service, which may be encouraged by applying compensation policies in case of agreement violation and using rating systems that penalize bad behavior.

Searching Charging Stations for Electric Vehicles

Charging stations for electric vehicles could also be relevant resources for drivers in a near future. Despite advances in charging techniques, it is expected that charging electric vehicles will consume a considerable amount of time in comparison with traditional refueling. Therefore, in order to find appropriate charging stations, the ideal solution should incorporate effective planning/allocation procedures to try to maximize the throughput of the charging stations and at the same time minimize the waiting time for drivers. Drivers may also have a preference for charging stations that have other facilities nearby (restaurants, coffee shops, etc.).

Searching Competition-Free Resources

Finally, if we consider general resources that are not necessarily competitive, then the range of mobile resource search applications expands. Indeed, any MANET database application or LBS that provides information about entities could belong to this category (e.g., friend-finder applications, social networks allowing searching nearby people with certain features or expertise, mobile applications that recommend products or sites such as restaurants or shops, etc.).

Future Directions

Further research related to mobile resource search is required. The following aspects could be highlighted:

- *Mobile resource search in mobile P2P scenarios*. Many issues for future work are related to the specifics of the underlying architecture considered. Particularly, as mobile resource search in mobile P2P is a promising area, there are interesting open issues derived from the mobile P2P environment, such as the need to: develop effective incentive mechanisms to encourage sharing information about resources [18], ensure the reliability of the information shared, perform an efficient query processing and routing exploiting the capabilities available on the mobile devices, and appropriately handle the existing network conditions and mobility issues (possibility of sparse networks in some areas, disconnections and network partitioning, high mobility because the devices providing information as well as the device searching for resources could move at any time, etc.).
- *Fair and effective management of the competition*. Although several proposals have studied

the competition problem that may arise in the access to scarce resources, mainly in the scenario of parking spaces, the development of a general solution that is able to provide a fair and effective sharing in a variety of scenarios seems to be missing. For example, even if there were a mechanism to assign a given resource to a single searcher, several allocation policies could be applied, such as considering the user that has been searching for the longest amount of time, the user that has received less resources, the user that has the highest probability to successfully reach the resource, the user with the shortest travel time, etc.

- *Interoperability between different systems*. Mobile resource search usually assumes a homogeneous environment where both the clients searching for resources and the providers of information about resources represent the different types of resources in the same way. In these circumstances, sharing information about resources is only possible among equivalent applications. In other words, two different systems providing information about certain types of resources would not be able to interoperate. To enable a flexible and general mobile resource search where different services can cooperate, knowledge representation techniques (such as ontologies) to represent resources would be needed.
- *Detection of suitable resources*. Estimating the relevance of resources to support an efficient and effective mobile resource search is a key issue. Moreover, obtaining information about potential resources, either through push-based proactive dissemination of resource information or through pull-based querying, should be done with a minimum overhead and at the same time avoiding missing information about relevant resources. Although there are several studies and proposals (coming mainly from the field of MANETs and VANETs), this is still an active area of research.
- *Impact on users*. Studying the impact of mobile resource search applications on the behav-

ior of the users is also an interesting avenue for further research. For example, some studies in the context of driving alerts indicate that providing irrelevant information to the drivers may lead to their distrust in the system and their eventual desensitization; it could be interesting to extend these studies to the area of mobile resource search.

Finally, mobile resource search and recommendation systems are similar in spirit (e.g., in both cases the goal is to recommend relevant items to the user) but for the moment quite far from each other in practice. This is mainly explained by the fact that recommendation systems have mainly focused on static environments and tackled mainly the algorithmic problems rather than those related to data collection and data management; only very recently mobile recommendations and Context-Aware Recommendation Systems (CARS) have emerged. In the future, it could be interesting to try to bridge the gap between these two research fields, as they may benefit from each other.

Cross-References

▶ Continuous Monitoring of Spatial Queries
▶ Data Management for VANETs
▶ Location-Based Recommendation
▶ Location-Based Services
▶ MANET Databases
▶ Moving Object

Recommended Reading

1. Ayala D, Wolfson O, Xu B, Dasgupta B, Lin J. Parking slot assignment games. In: Agrawal D, Cruz I, Jensen CS, Ofek E, Tanin E, editors. Proceedings of the 19th ACM SIGSPATIAL International Conference on Advances in Geographic Information Systems; 2011. p. 299–308.
2. Baldauf M, Dustdar S, Rosenberg F. A survey on context-aware systems. Int J Ad Hoc Ubiquitous Comput. 2007;2(4):263–77.
3. Cao H, Wolfson O, Xu B, Yin H. MOBI-DIC: MOBIle DIscovery of loCal resources in peer-to-peer wireless network. IEEE Data Eng Bull. 2005;28(3):11–8.
4. Chlamtac I, Conti M, Liu JJN. Mobile ad hoc networking: imperatives and challenges. Ad Hoc Netw. 2003;1(1):13–64.
5. Delot T, Ilarri S, Lecomte S, Cenerario N. Sharing with caution: managing parking spaces in vehicular networks. Mob Inf Syst. 2013;9(1):69–98.
6. Dinh HT, Lee C, Niyato D, Wang P. A survey of mobile cloud computing: architecture, applications, and approaches. Wirel Commun Mob Comput. 2013;13(18):1587–611.
7. Edwards WK. Discovery systems in ubiquitous computing. IEEE Pervasive Comput. 2006;5(2):70–7.
8. Guo Q, Wolfson O. PRESENTs: probabilistic REsource-SEarch NeTworks. In: Proceedings of the 23rd ACM SIGSPATIAL International Conference on Advances in Geographic Information Systems; 2015.
9. Guo Q, Wolfson O. Spatio-temporal resource search with uncertain data. In: Chow C, Jayaraman P, Wu W, editors. Proceedings of the 17th IEEE International Conference on Mobile Data Management; 2016. p. 66–71.
10. Ilarri S, Mena E, Illarramendi A. Location-dependent query processing: where we are and where we are heading. ACM Comput Surv. 2010;42(3):12: 1–73.
11. Ilarri S, Wolfson O, Delot T. Collaborative sensing for urban transportation. IEEE Data Eng Bull. 2014;37(4):3–14.
12. Imielinski T, Korth HF, editors. Mobile computing. Norwell: Kluwer Academic Publishers; 1996.
13. Kokolaki E, Karaliopoulos M, Stavrakakis I. Opportunistically assisted parking service discovery: now it helps, now it does not. Pervasive Mob Comput. 2012;8(2):210–27.
14. Lee DL, Xu J, Zheng B, Lee WC. Data management in location-dependent information services. IEEE Pervasive Comput. 2002;1(3):65–72.
15. Olariu S, Weigle MC. Vehicular networks: from theory to practice. Boca Raton: Chapman & Hall/CRC; 2009.
16. Schiller J, Voisard A. Location based services. San Francisco: Morgan Kaufmann Publishers Inc.; 2004.
17. Ververidis CN, Polyzos GC. Service discovery for mobile ad hoc networks: a survey of issues and techniques. IEEE Commun Surv Tutor. 2008;10(3): 30–45.
18. Wolfson O, Xu B, Sistla AP. An economic model for resource exchange in mobile peer to peer networks. In: Hatzopoulos M, Manolopoulos Y, editors. Proceedings of the 16th International Conference on Scientific and Statistical Database Management; 2004. p. 235–44.
19. Zheng B, Xu J, Lee DL. Cache invalidation and replacement strategies for location-dependent data in mobile environments. IEEE Trans Comput. 2002;51(10):1141–53.

Mobile Sensor Network Data Management

Demetrios Zeinalipour-Yazti[1] and
Panos K. Chrysanthis[2]
[1]Department of Computer Science, Nicosia, Cyprus
[2]Department of Computer Science, University of Pittsburgh, Pittsburgh, PA, USA

Synonyms

Mobile wireless sensor network data management; MSN data management

Definition

Mobile Sensor Network (MSN) Data Management refers to a collection of centralized and distributed algorithms, architectures and systems to handle (store, process and analyze) the immense amount of spatio-temporal data that is cooperatively generated by collections of sensing devices that move in space over time.

Formally, given a set of n homogenous or heterogeneous mobile sensors $\{s_1, s_2, \ldots, s_n\}$ that are capable of acquiring m physical attributes $\{a_1, a_2, \ldots, a_m\}$ from their environment at every discrete time instance t (i.e., datahas a temporal dimension), an implicit or explicit mechanism that enables each s_i ($i \leq n$) to move in some multi-dimensional Euclidean space (i.e., data has one or more spatial dimensions), MSN Data Management provides the foundation to handle spatio-temporal data in the form (s_i, t, x, $[y, z,]a_1[, \ldots, a_m]$), where x, y, z defines three possible spatial dimensions and the bracket expression "[ ]" denotes the optional arguments in the tuple definition. In a more general perspective, MSN Data Management deals with algorithms, architectures and systems for in-network and out-of-network query processing, access methods, storage, data modeling, data warehousing, data movement and data mining.

Historical Background

The improvements in hardware design along with the wide availability of economically viable embedded sensor systems have enabled scientists to acquire environmental conditions at extremely high resolutions. Early approaches to monitor the physical world were primarily composed of passive sensing devices, such as those utilized in wired weather monitoring infrastructures, that could transmit their readings to more powerful processing units for storage and analysis. The evolution of passive sensing devices has been succeeded by the development of *Stationary Wireless Sensor Networks (Stationary WSNs)*. These are composed of many tiny computers, often no bigger than a coin or a credit card, that feature a low frequency processor, some flash memory for storage, a radio for short-range wireless communication, on-chip sensors and an energy source such as AA batteries or solar panels. Applications of stationary WSNs have emerged in many domains ranging from environmental monitoring [15] to seismic and structural monitoring as well as industry manufacturing.

The transfer of information in such networks is conducted without electrical conductors (i.e., wires) using technologies such as radio frequency (RF), infrared light, acoustic energy and others, as the mobility aspect inherently hinders the deployment of any technology that physically connects nodes with wires. Since communication is the most energy demanding factor in such networks, data management researchers have primarily focused on the development of energy-conscious algorithms and techniques.

In particular, declarative approaches such as TinyDB [9] and Cougar [16] perform a combination of in-network aggregation and filtering in order to reduce the energy consumption while conveying data to the *querying node (sink)*. Additionally, approaches such as TiNA [13] and MINT Views [17] take into account intelligent in-network data reduction techniques to further reduce the consumption of energy. *Data Centric Routing* approaches, such as directed diffusion [8], establish low-latency paths between the sink and the sensors in order to reduce the cost

M

of communication. *Data Centric Storage* [14] schemes organize data with the same attribute (e.g., humidity readings) on the same node in the network in order to offer efficient location and retrieval of sensor data.

The evolution of stationary WSNs in conjunction with the advances made by the distributed robotics and low power embedded systems communities have led to a new class of *Mobile (and Wireless) Sensor Networks (MSNs)* that can be utilized for land [3, 5, 10], ocean exploration [11], air monitoring [1], automobile applications [7, 6], Habitant Monitoring [12] and a wide range of other scenarios. MSNs have a similar architecture to their stationary counterparts, thus are governed by the same energy and processing limitations, but are supplemented with implicit or explicit mechanisms that enable these devices to move in space (e.g., motor or sea/air current) over time. Additionally, MSN devices might derive their coordinates through absolute (e.g., dedicated Geographic Positioning System hardware) or relative means (e.g., *localization techniques*, which enable sensing devices to derive their coordinates using the signal strength, time difference of arrival or angle of arrival). There are several classes of MSNs which can coarsely be structured into the following classes: (i) highly mobile, which contains scenarios in which devices move at high velocities such as cars, human with cell phones, airplanes, and others; (ii) mostly static, which contains scenarios in which devices move at low velocities such as monitoring sensors in a shop floor with moving robots; and (iii) hybrid, which contains both classes such as an airplane that has sensors installed on inside and outside.

Foundations

The unique characteristics of MSNs create novel data management opportunities and challenges that have not been addressed in other contexts including those of mobile databases and stationary WSNs. In order to realize the advantages of such networks, researchers have to re-examine existing data management and data processing approaches in order to consider sensor and user mobility; develop new approaches that consider the impact of mobility and capture its trade-offs. Finally, MSN data management researchers are challenged with structuring these networks as huge distributed databases whose edges consist of numerous "receptors" (e.g., RFID readers or sensor networks) and internal nodes form a pyramid scheme for (in-network) aggregation and (pipelined) data stream processing.

There are numerous advantages of MSNs over their stationary counterparts. In particular, MSNs offer: (i) *dynamic network coverage*, by reaching areas that have not been adequately sampled; (ii) *data routing repair*, by replacing failed routing nodes and by calibrating the operation of the network; (iii) *data muling*, by collecting and disseminating data/readings from stationary nodes out of range; (iv) *staged data stream processing*, by conducting in-network processing of continuous and ad-hoc queries; and (v) *user access points*, by enabling connection to handheld and other mobile devices that are out of range from the communication infrastructure.

These advantages enable a wide range of new applications whose data management requirements go beyond those of stationary WSNs. In particular, MSN system software is required to handle: (i) *the past*, by recording and providing access to history data; (ii) *the present*, by providing access to current readings of sensor data; (iii) *the future*, by generating predictions; (iv) *distributed spatio-temporal data*, by providing new means of distributed data storage, indexing and querying of spatio-temporal data repositories; (v) *data uncertainty*, by providing new means of handling real world signals that are inherently uncertain; (vi) *self-configurability*, by withstanding "harsh" real-life environments; and (vii) *data and service mash-ups*, by enabling other innovative applications that build on top of existing data and services.

In light of the above characteristics, the most predominant data management challenges that have prevailed in the context of MSNs include:

In-Network Storage The absence of a stationary network structure in MSNs makes continuous data acquisition to some sink point a non-intuitive task (e.g., mobile nodes might be out of communication range from the sink). In particular, the absence of an always accessible sink mandates that acquisition has to be succeeded by in-network storage of the acquired events so that these events can later be retrieved by the user. Mobile devices usually utilize flash memory as opposed to magnetic disks, which are not shock-resistant and thus are not appropriate for a mobile setting. Consequently, a major challenge in MSNs is to extend local storage structures and access methods in order to provide efficient access to the data stored on the local flash media of a sensor device while traditional database research has mainly focused on issues related to magnetic disks.

Flexible and Expressive Query Types In a traditional database management system, there is a single correct answer to a given query on a given database instance. When querying MSNs the situation is notably different as there are many more degrees of freedom and the underlying querying engine needs to be guided regarding which alternative execution strategy is the right one, typically on the basis of target answer quality and resource availability. In this context, there are additional relevant parameters that include: (i) *Resolution:* physical sensor data can be observed at multiple resolutions along space and time dimensions; (ii) *Confidence:* more often than not, correctness of query results can be specified only in probabilistic terms due to the inherent uncertainty in the sensor hardware and the modeling process; (iii) *Alternative models:* in some cases, several alternative models apply to a single scenario. Each alternative typically represents a different point in the efficiency (resource consumption) and effectiveness (result quality) spectrum, thereby allowing a tradeoff between these two metrics on the basis of application-level expectations. The prime challenge is to define new declarative query languages that make use of these new parameters while allowing a highly flexible and optimizable implementation. Additionally, approximate query processing with controlled result accuracy becomes vital for dynamic mobile environments with varying node veloci-

ties, changing data traffic patterns, information redundancy, uncertainty, and inevitable flexible load shedding techniques. Finally, in order to have an efficient and optimized implementation of query types, MSNs will need to consider cross-layer optimization since all layers of the data stack are involved in query execution.

Efficient Query Routing Trees Query routing and resolution in stationary WSNs is typically founded on some type of query routing tree that provides each sensor with a path over which answers can be transmitted to the sink. In a MSN, such a query routing tree can neither be constructed in an efficient manner nor be maintained efficiently as the network topology is transient. The dynamic nature of the underlying physical network tremendously complicates the interchange of information between nodes during the resolution of a query. In particular, it is known that sensing devices tend to power down their transceiver (transmitter-receiver) during periods of inactivity in order to conserve energy [2]. While stationary WSNs define transceiver scheduling approaches, such as those defined in TAG [9], Cougar [16] and MicroPulse [1], in order to enable accurate transceiver allocation schemes, such approaches are not suitable for mobile settings in which a sensor is not aware of its designated parent node in the query tree hierarchy. Consequently, nodes are not able to agree on rendezvous time-points on which data interchange can occur.

Purpose-Driven Data Reduction The amount of data generated from MSNs can be overwhelming. Consequently, a main challenge is to provide data reduction techniques which will be tuned to the semantics of the target application. Furthermore, data reduction must take into account the entire spectrum of uses, ranging from real-time to offline, supporting both snapshot and continuous queries that take advantage of designated optimization opportunities (e.g., multi-query) especially targeted for mobile environments. Finally, it must also consider the inherently dynamic aspects of these environments and the possibility of in-network data reduction (e.g., in-network aggregation).

Perimeter Construction and Swarm-Like Behavior In many types of MSNs, new events are more prevalent at the periphery of the network (e.g., water detection and contamination detection) rather than uniformly throughout the network (which is more typically for applications like fire detection). This creates the necessity to construct the perimeter of a MSN in an online and distributed manner. Additionally, many types of MSNs are expected to feature a swarm-like behavior (The term *Swarm (or Flock)* refers to a group of objects that exhibit a polarized, non-colliding and aggregate motion.). For instance, consider a MSN design that consists of several rovers that are deployed as a swarm in order to detect events of interest (e.g., the presence of water) [18]. The swarm might collaboratively collect spatio-temporal events of interest and store them in the swarm until an operator requests them. In order to increase the availability of the detected answers, in the presence of unpredictable failures, individual rovers can perform replication of detected events to neighboring nodes. That creates challenges in data aggregation, data fusion and data storage that have not yet been addressed.

Enforcement of Security, Privacy and Trust Frequent node migrations and disconnections in MSNs, as well as resource constraints raise severe concerns with respect to security, privacy and trust. Additionally, the cost of traditional secure data dissemination approaches (e.g., using encryption) may be prohibitively high in volatile mobile environments. As such, research on encryption-free data dissemination strategies becomes very relevant here. This includes strategies to deliver separate and under-defined data shares, secure multiparty computation and advanced information recovery techniques.

Context-Awareness and Self-Everything Providing a useful level of situational awareness in an unobtrusive way is crucial to the success of any application utilizing MSNs as this can be used to improve functionality by including preferences

from the users but can also be used to improve performance (e.g., better network routing decisions if the exact topology is known). Note that context is often obvious in stationary WSN deployments (i.e., a specific sensor is always in the same location) but in the context of a MSN additional data management measures need to be taken into account in order to enable this parameter. Additionally, it is crucial for them to be "plug-and-play" and self-everything (i.e., self-configurable and self-adaptive) as application deployment of sensors in the field is famously hard, even without the mobility aspect which is introducing additional challenges. Finally, a crucial parameter is that of being adaptive both in how to deal with the system issues (i.e., how to adapt from failures in network connectivity) and also with user-interface/application issues (i.e., how to adapt the application when the context changes).

Key Applications

MSN Data Management algorithms, architectures and systems will play a significant role in the development of future applications in a wide range of disciplines including the following:

Environmental and Habitant Monitoring A large class of MSN applications have already emerged in the context of environmental and habitant monitoring systems. Consider an ocean monitoring environment that consists of n independent surface drifters floating on the sea surface and equipped with either acoustic or radio communication capabilities. The operator of such a MSN might seek to answer queries of the type: *"Has the MSN identified an area of contamination and where exactly?"*. The MSN architecture circumvents the peculiarities of individual sensors, is less prone to failures and is potentially much cheaper. Similar applications have also emerged with MSNs of car robots, such as CotsBots [3], Robomotes [5] or Millibots [10], and MSNs of Unmanned Aerial Vehicles (UAVs), such as SensorFlock [1], in which devices can fly autonomously based on complex interactions with their peers. One final challenging application in this class is that of detecting a phenomenon that itself is mobile, for example a brush fire which is being carried around by high winds.

Intelligent Transportation Systems Sensing systems have been utilized over the years in order to better manage traffic with the ultimate goal of reducing accidents and minimizing the time and the energy (gasoline) wasted while staying idle in traffic. Since cars are already equipped with a wide range of sensors, the generated information can be shared in a vehicle-to-vehicle network. For example the ABS system can detect when the road is slippery or when the driver is hitting the brakes thus this information can be broadcasted to the surrounding cars but also to the many cars back and forth, as needed, in order to make sure that everybody can safely stop with current weather conditions and car speeds.

Medical Applications This class includes applications that monitor humans in order to improve living conditions and in order to define early warning systems that identify when human life is at risk. For instance, Nike+ is an example for monitoring the health of a group of runners that have simple sensing devices embedded in their running shoes. Such an application would require embedded storage and retrieval techniques in order to administer the local amounts of data. Applications in support of the elderly and those needing constant supervision (e.g., due to chronic diseases like diabetes, allergies, etc.) are another example in which MSN data management techniques will play an important role. Wellness applications could also be envisioned, where a health "dose" of exercise is administered according to ones needs and capabilities. Another area are systems to protect soldiers on the battlefield. SPARTNET has recently developed wearable physiological sensor systems that collect, organize and interpret data on the health status of soldiers in order to improve situational and medical awareness during field trainings. Such systems could be augmented with functionality of detecting and reporting threats that are either derived from

individual signals (e.g., when a soldiers personal health monitor shows erratic life-signals) and from correlated signals that are derived from multiple sensors/soldiers (e.g., by recognizing when a small group of soldiers is deviating away from the expected formation). Finally, disaster and emergency management are another prime area where MSN data management techniques will play a major impact.

Location-Based Services and the Sensor Web The last group of challenging motivating applications is that of real-time location-based services, for example a service that can report whether there are any available parking spaces or a service that can keep track of buses moving and report how delayed a certain bus is. Many of these services become more powerful with the integration of data from the *Sensor Web* (i.e., live sensor data) with the *Web* (i.e., static content available online) and the *Deep-web* (i.e., data that is stored in a database, but are accessible through a web page or a web service).

Cross-References

▶ Sensor Networks
▶ Spatial Network Databases

Recommended Reading

1. Allred J, Hasan AB, Panichsakul S, Pisano B, Gray P, Huang J-H, Han R, Lawrence D, and Mohseni K. SensorFlock: an airborne wireless sensor network of micro-air vehicles. In: Proceedings of the 5th International Conference on Embedded Networked Sensor Systems; 2007. p. 117–29.
2. Andreou P, Zeinalipour-Yazti D, Chrysanthis PK, Samaras G. Workload-aware optimization of query routing trees in wireless sensor networks. In: Proceedings of the 9th International Conference on Mobile Data Management; 2008. p. 189–96.
3. Bergbreiter S, Pister KSJ. CotsBots: an off-the-shelf platform for distributed robotics. In: Proceedings of the IEEE/RSJ International Conference on Intelligent Robots and Systems; 2003. p. 1632–7.
4. Chintalapudi K, Govindan R. Localized edge detection in sensor fields. Ad-hoc Networks. 2003;1(2–3):273–91.

5. Dantu K, Rahimi MH, Shah H, Babel S, Dhariwal A, Sukhatme GS. Robomote: enabling mobility in sensor networks. In: Proceedings of the 4th International Symposium on Information Processing in Sensor Networks; 2005.

6. Eriksson J, Girod L, Hull B, Newton R, Madden S, Balakrishnan H. The Pothole Patrol: using a mobile sensor network for road surface monitoring. In: Proceedings of the 6th International Conference on Mobile Systems, Applications and Services; 2008. p. 29–39.

7. Hull B, Bychkovsky V, Chen K, Goraczko M, Miu A, Shih E, Zhang Y, Balakrishnan H, Madden S. CarTel: a distributed mobile sensor computing system. In: Proceedings of the 4th International Conference on Embedded Networked Sensor Systems; 2006. p. 125–38.

8. Intanagonwiwat C, Govindan R, Estrin D. Directed diffusion: a scalable and robust communication paradigm for sensor networks. In: Proceedings of the 6th Annual International Conference on Mobile Computing and Networking; 2000. p. 56–67.

9. Madden SR, Franklin MJ, Hellerstein JM, Hong W. The design of an acquisitional query processor for sensor networks. In: Proceedings of the ACM SIGMOD International Conference on Management of Data; 2003.

10. Navarro-Serment LE, Grabowski R, Paredis CJJ, Khosla PK. Millibots: the development of a framework and algorithms for a distributed heterogeneous robot team. IEEE Robot Autom Mag. 2002;9(4).

11. Nittel S, Trigoni N, Ferentinos K, Neville F, Nural A, Pettigrew N. A drift-tolerant model for data management in ocean sensor networks. In: Proceedings of the 6th ACM International Workshop on Data Engineering for Wireless and Mobile Access; 2007. p. 49–58.

12. Sadler C, Zhang P, Martonosi M, Lyon S. Hardware design experiences in ZebraNet. In: Proceedings of the 2nd International Conference on Embedded Networked Sensor Systems; 2004. p. 227–38.

13. Sharaf M, Beaver J, Labrinidis A, Chrysanthis PK. Balancing energy efficiency and quality of aggregate data in sensor networks. VLDB J. 2004;13(4):384–403.

14. Shenker S, Ratnasamy S, Karp B, Govindan R, Estrin D. Data-centric storage in sensornets. SIGCOMM Comput Commun Rev. 2003;33(1):137–42.

15. Szewczyk R, Mainwaring A, Polastre J, Anderson J, Culler D. An analysis of a large scale habitat monitoring application. In: Proceedings of the 2nd International Conference on Embedded Networked Sensor Systems; 2004. p. 214–26.

16. Yao Y, Gehrke JE. The cougar approach to in-network query processing in sensor networks. ACM SIGMOD Rec. 2002;32(3):9–18.

17. Zeinalipour-Yazti D, Andreou P, Chrysanthis P, Samaras G. MINT views: materialized in-network top-k views in sensor networks. In: Proceedings of the 2007 International Conference on Mobile Data Management; 2007. p. 182–9.

18. Zeinalipour-Yazti D, Andreou P, Chrysanthis P, Samaras G. SenseSwarm: a perimeter-based data acquisition framework for mobile sensor networks. In: Proceedings of the VLDB Workshop on Data Management for Sensor Networks; 2007. p. 13–8.

Model Management

Christoph Quix
RWTH Aachen University, Aachen, Germany

Definition

Model management comprises technologies and mechanisms to support the integration, transformation, evolution, and matching of models. It aims at supporting metadata-intensive applications such as database design, data integration, and data warehousing. To achieve this goal, a model management system has to provide definitions for *models* (i.e., schemas represented in some metamodel), *mappings* (i.e., relationships between different models), and *operators* (i.e., operations that manipulate models and mappings). Model management has become more and more important, since the interoperability and/or integration of heterogeneous information systems is a frequent requirement of organizations. Some important operations in model management are *Merge* (integration of two models), *Match* (creating a mapping between two models), and *Model-Gen* (transforming a model given in one modeling language into a corresponding model in a different modeling language).

The current understanding of model management has been defined in [4] and focuses mainly on (but is not limited to) the management of *data* models. Most of the problems mentioned before have been already addressed separately and for specific applications. The goal now is to build a model management system (MMS) which unifies the previous approaches by providing a set of *generic* structures representing models and mappings, and the definition of *generic* operations on these structures. Such a system could then be used by an application to solve model management tasks.

Historical Background

Model management, as it is understood today, has been defined by Bernstein et al. [4]. In the 1980s, the term "model management" was used in the context of decision support systems, but this refered mainly to mathematical models. Dolk [7] stated first the requirement for a theory for models similar to the relational database theory. Such a theory should include formal definitions of models and operations on models and could be used as a basis for the implementation of a model management system. This work was based on a draft of the *Information Resource Dictionary System* (IRDS) standard (ISO/IEC 10027:1990) which was accepted in 1990. The IRDS standard clarified the terminology of modeling systems and defined a framework structure for such systems as a four-level hierarchy: at the lowest level reside data instances which are described by a model (or schema) on the next higher level. This model is expressed in some modeling language (or metamodel) which is located at the third level. The highest level contains a metametamodel which can be used to define metamodels.

The new definition of model management in 2000 [4] integrated the research efforts of several previously loosely coupled areas. Therefore, research in model management did not start from scratch, rather it could build already on many results such as schema integration [3], or model transformation [1]. The main contribution of [4] was the definition of operations (such as Match, Merge, Compose) which a model management system should offer. Furthermore, it was required that models and mappings are considered as first class objects and that operations should address them as a whole and not only one model element at a time.

Since the vision of model management has been stated, research diverted into the areas of schema matching [14], model transformation [1], generic metamodels [10], schema integration [3], and the definition and composition of mappings [8]. Each of these areas will be summarized briefly in the next section.

Recently, research on model management has been summarized in [5]. It was also emphasized that more expressive mapping languages are required than proposed in the original vision of model management. Model management systems should also include a component in which the mappings can be executed.

Foundations

Schema Matching

Schema matching is the task of identifying a set of correspondences (also called a morphism or a mapping) between schema elements. Many aspects have to be considered during the process of matching, such as data values, element names, constraint information, structure information, domain knowledge, cardinality relationships, and so on. All this information is useful in understanding the semantics of a schema, but it can be a very time consuming problem to collect this information. Therefore, automatic methods are required for schema matching.

A multitude of methods have been proposed for schema matching [14] using different types of information to identify similar elements. The following categories of schema matchers are frequently used:

- Element-Level Matchers take only the information of one schema element separately into account. These can either use linguistic information (name of the element) or constraint information (data type, key constraints).
- Structure-Level Matchers use graph matching approaches to measure the similarity of the structures implied the schema.
- Instance-Level Matchers use also data instances to match schema elements. If the instance sets of two elements are similar, or

M

have a similar value distribution, this might indicate a similarity of the schema elements.

- Machine-Learning Matchers use either instance data or previously identified matches as training data for a machine-learning system. Based on this training data, the system should then detect similar matches in new schema matching problems.

It has been agreed that no single method can solve the schema matching problem in general. Therefore, matching frameworks have been developed which are able to combine multiple individual matching methods to achieve a better result.

Model Transformation

Model transformation (also called *ModelGen*) is the task of transforming a given model M in some particular modeling language into a corresponding model M' in some other modeling language. A classical example for such a transformation is the transformation of an entity-relationship model into a relational database schema. In general, the transformation cannot guarantee that all semantic information of M is still present in the resulting model M' as the target modeling language might have limited expressivity.

Whereas transformation of models between different modeling languages is a frequent task, it has up to now mainly been addressed in specialized settings which map from one particular metamodel to another fixed metamodel. Recent approaches for generic model transformations are based on generic model representations and use a rule-based system to transform a generic representation of the original model into a different model in the generic metamodel. As a side effect, these approaches generate also instance-level mappings which are able to transform data conforming to the original model into data of the generated model.

Another important application area of model transformation is MDA (Model Driven Architecture). The MDA concept is based on the transfor-

mation of abstract, conceptual models into more concrete implementation-oriented models.

Generic Metamodel

A generic representation of models is a prerequisite for building a model management system. Without a generic representation, model operations would have to be implemented for each modeling language that should be supported by the system. Especially for the task of model transformation, a generic representation of models is advantageous as the necessary transformations have just to be implemented for the generic representation.

Such a generic representation is called a *generic metamodel*. A generic metamodel should be able to represent models originally represented in different metamodels (or modeling languages) in a generic way without loosing detailed information about the semantics of the model.

First implementations of model management systems used rather simple graph representations of models, e.g., Rondo [13]. Although the graph-based approach might allow an efficient implementation of operations which do not rely on a detailed representation of the models (such as schema matching), it is more difficult to implement more complex operations (such as model transformation or schema integration).

Schema integration approaches used rather abstract generic metamodels. More detailed generic metamodels have been used for model transformation. In [1], the authors describe a metamodel consisting of "superclasses" of the modeling constructs in the native metamodels. The transition between this internal representation and a native metamodel is described as a set of patterns. This induces the concept of a *supermodel* which is the union of patterns defined for any supported native metamodel.

A detailed generic metamodel called *GeRoMe* (Generic Role based Metamodel) is proposed in [10]. *GeRoMe* employs the *role based* modeling approach in which an object is regarded as playing roles in collaborations with other objects. This allows to describe the properties of model elements as accurately as possible while using only metaclasses and roles from a relatively small set. Therefore, *GeRoMe* provides a generic, yet detailed representation of data models originally represented in different metamodels.

Schema Integration

In model management, the *Merge* operator addresses the problem of schema integration, i.e., generating a merged model given two input models and a mapping between them. The merged model should contain all the information contained in the input models and the mapping. In this context, a mapping is not just a simple set of correspondences between model elements; it might have itself a complex structure and is therefore often regarded also as a *mapping model*. A mapping model is necessary because the models to be merged also have complex structures, which usually do not correspond to each other; the mapping model then acts as a "bridge" to connect these heterogeneous structures.

These structural heterogeneities are one class of conflicts which have to be solved in schema integration. Other types of conflicts are semantic conflicts (model elements describe overlapping sets of objects), descriptive conflicts (the same elements are described by different sets of properties; this includes also name conflicts), and heterogeneity conflicts (models are described in different modeling languages) [15]. The resolution of these conflicts is the main problem in schema integration.

The problem of schema integration has been addressed for various metamodels, such as variants of the ER metamodel [15] or generic metamodels.

Mappings

Depending on the application area, such as data translation, query translation or model merging, schema mappings come in different flavors. One can distinguish between correspondences, also called morphisms, extensional and intensional mappings. Correspondences usually do not have a formal semantics but only state informally that

the respective model elements are similar. Morphisms are often - as the result of a schema matching operation [14] - the starting point for specifying more formal mappings. Intensional mappings are usually used for schema integration (see above) and are based on the *possible* instances of a schema. In contrast to intensional mappings, extensional mappings refer to the actual instances of a schema.

Extensional mappings are defined as local-as-view (LAV), global-as-view (GAV), source-to-target tuple generating dependencies (s-t tgds), [12], second order tuple generating dependencies (SO tgds) [8], or similar formalisms. Each of these classes has certain advantages and disadvantages when it comes to properties such as composability, invertibility or execution of the mappings.

Composition is an important sub-problem when dealing with extensional mappings. In general, the problem of composing mappings has the following definition: given a mapping M_{12} from model S_1 to model S_2, and a mapping M_{23} from model S_2 to model S_3, derive a mapping M_{13} from model S_1 to model S_3 that is equivalent to the successive application of M_{12} and M_{23} [8].

Fagin et al. [8] explored the properties of the composition of schema mappings specified by a finite set of s-t tgds. They proved that the language of s-t tgds is not closed under composition. To ameliorate the problem, they introduced the class of SO tgds which are closed under composition.

Model Management Systems

In the recent years, several prototype systems related to model management have been developed. Many of them focus only on some particular aspects of model management whereas only a few try to address a broader range of model management operators. The systems Rondo [13] and GeRoMeSuite [11] aim at providing a complete set of model management operators and are not restricted to particular modeling languages. Clio [9] focuses especially on mappings between XML and relational

M

databases, the generation, and composition of these mappings. COMA++ [2] provides schema matching functionality.

Key Applications

Model management can be applied in scenarios, in which the management of complex data models is necessary. For example, data warehouses (DWs) are one application area for model management systems [6]. For integrating a new data source into the DW, a *Match* operator could be used to identify the similarities between the source schema and the DW schema. If the DW schema has not yet been created, a *Merge* operator can be used to generate it from several source schemas. The composition of mappings can be used after a source schema S has been evolved to a new version S': if a mapping from S' to S is known (either given by the schema evolution operation, or computed by a *Match* operator), one can compose this mapping with the original mapping between the source schema S and the integrated schema T to get a mapping from the new version of the source schema S' to the integrated schema T.

Such applications of model management operators are also possible for other integrated information systems. For example, web services or e-business systems often have to take a message in XML format and store it in some relational data store for further processing. To implement this data transformation, a mapping between the XML schema and the schema of the relational database has to be defined. Such tasks are already supported by commercial products (e.g., Microsoft BizTalk, Altova MapForce). These products already use some model management operators (e.g., simple *Match* operators to simplify the task of mapping definition), but could further benefit from more powerful techniques for mapping generation. Clio started as a research prototype for mapping generation between XML and relational schemas, results have now been integrated into the IBM Information Integration platform.

Another application area for model management is the design and development of software applications. Model transformation is an inherent problem in software development, models have to be transformed into new metamodels, and then interoperability between the original model and the transformed model has to be implemented. A classical example is the transformation of an object-oriented data model of an application to a relational database schema, for which data access objects later have to be implemented to enable the synchronization between the data in the database and in the application. Such tasks are supported by frameworks (e.g., ADO.NET Entity Framework or the Entity Beans in Java 2 Enterprise Edition).

Future Directions

Mappings will become more and more important for MMS in the future [5]. The initial vision of model management, which considered mappings as rather simple correspondences attached with some complex expressions which contain the semantics of the mapping, was too limited. Mappings are in practice very complex and therefore, a rich mapping language is required in a MMS. The MMS must also be able to reason about the mappings. Furthermore, the MMS should not only enable the definition of a mapping, but support also its application; for example, using it for the integration of two schemas or for data transformation between two data stores. In addition, complex processes involving mappings (such as ETL process in data warehouses) have also to be considered.

Cross-References

► Meta Data Repository
► Metamodel
► Schema Mapping
► Schema Mapping Composition
► Schema Matching

Recommended Reading

1. Atzeni P, Torlone R. Management of multiple models in an extensible database design tool. In: Advances in Database Technology, Proceedings of the 5th International Conference on Extending Database Technology; 1996. p. 79–95.
2. Aumueller D, Do HH, Massmann S, Rahm E. Schema and ontology matching with COMA++. In: Proceedings of the ACM SIGMOD International Conference on Management of Data; 2005. p. 906–8.
3. Batini C, Lenzerini M, Navathe SB. A comparative analysis of methodologies for database schema integration. ACM Comput Surv. 1986;18(4): 323–64.
4. Bernstein PA, Halevy AY, Pottinger R. A vision for management of complex models. ACM SIGMOD Rec. 2000;29(4):55–63.
5. Bernstein PA, Melnik S. Model management 2.0: Manipulating richer mappings. In: Proceedings of the ACM SIGMOD International Conference on Management of Data; 2007. p. 1–12.
6. Bernstein PA, Rahm E. Data warehousing scenarios for model management. In: Proceedings of the 19th International Conference on Conceptual Modeling; 2000. p. 1–15.
7. Dolk DR. Model management and structured modeling: the role of an information resource dictionary system. Commun ACM. 1988; 31(6); 704–18.
8. Fagin R, Kolaitis PG, Popa L, Tan WC. Composing schema mappings: second-order dependencies to the rescue. ACM Trans Database Syst. 2005;30(4): 994–1055.
9. Hernández MA, Miller RJ, Haas LM. Clio: a semiautomatic tool for schema mapping. In: Proceedings of the ACM SIGMOD International Conference on Management of Data; 2001. p. 607.
10. Kensche D, Quix C, Chatti MA, Jarke M. GeRoMe: a generic role based metamodel for model management. J Data Semant. 2007; VIII:82–117.
11. Kensche D, Quix C, Li X, Li Y. GeRoMeSuite: a system for holistic generic model management. In: Proceedings of the 33rd International Conference on Very Large Data Bases; 2007. p. 1322–5.
12. Lenzerini M. Data integration: a theoretical perspective. In: Proceedings of the 21st ACM SIGACT-SIGMOD-SIGART Symposium on Principles of Database Systems; 2002. p. 233–46.
13. Melnik S, Rahm E, Bernstein PA. Rondo: a programming platform for generic model management. In: Proceedings of the ACM SIGMOD International Conference on Management of Data; 2003. p. 193–204.
14. Rahm E, Bernstein PA. A survey of approaches to automatic schema matching. VLDB J. 2001;10(4):334–50.
15. Spaccapietra S, Parent C. View integration: a step forward in solving structural conflicts. IEEE Trans Knowl Data Eng. 1994;6(2):258–74.

Model-Based Querying in Sensor Networks

Amol Deshpande[1], Carlos Guestrin[2], and Samuel Madden[3]
[1]University of Maryland, College Park, MD, USA
[2]Carnegie Mellon University, Pittsburgh, PA, USA
[3]Massachusetts Institute of Technology, Cambridge, MA, USA

Synonyms

Approximate querying; Model-driven data acquisition

Definition

The data generated by sensor networks or other distributed measurement infrastructures is typically incomplete, imprecise, and often erroneous, such that it is not an accurate representation of physical reality. To map raw sensor readings onto physical reality, a mathematical description, a *model*, of the underlying system or process is required to complement the sensor data. Models can help provide more robust interpretations of sensor readings: by accounting for spatial or temporal biases in the observed data, by identifying sensors that are providing faulty data, by extrapolating the values of missing sensor data, or by inferring hidden variables that may not be directly observable. Models also offer a principled approach to predict future states of a system. Finally, since models incorporate spatio-temporal correlations in the environment (which tend to be very strong in many monitoring applications), they lead to significantly more energy-efficient query execution – by exploiting such attribute correlations, it is often possible to use a small set of observations to provide approximations of the values of a large number of attributes.

Model-based querying over a sensor network consists of two components: (i) identifying and/or building a model for a given sensor network, and (ii) executing declarative queries against a sensor network that has been augmented with such a model (these steps may happen serially or concurrently). The queries may be on future or hidden states of the system, and are posed in a declarative SQL-like language. Since the cost of acquiring sensor readings from the sensor nodes is the dominant cost in these scenarios, the optimization goal typically is to minimize the total data acquisition cost.

Historical Background

Statistical and probabilistic models have been a mainstay in the scientific and engineering communities for a long time, and are commonly used for a variety of reasons, from simple pre-processing tasks for removing noise (e.g., using Kalman Filters) to complex analysis tasks for prediction purposes (e.g., to predict weather or traffic flow). Standard books on machine learning and statistics should be consulted for more details (e.g., Cowell et al. [3], Russell and Norvig [14]).

The first work to combine models, declarative SQL-like queries and live data acquisition in sensor networks was the BBQ System [7, 6]. The authors proposed a general architecture for model-based querying, and posed the optimization problem of selecting the best sensor readings to acquire to satisfy a user query (which can be seen as a generalization of the *value of information problem* [14]). The authors proposed several algorithms for solving this optimization problem; they also evaluated the approach on a several real-world sensor network datasets, and demonstrated that model-based querying can provide high-fidelity representation of the real phenomena and leads to significant performance gains versus traditional data acquisition techniques. Several works since then have considerably expanded upon the basic idea, including development of sophisticated algorithms for data acquisition [12, 13], more

complex query types [15], and integration into a relational database system [8, 11].

The querying aspect of this problem has many similarities to the problem of approximate query processing in database systems, which often uses model-like *synopses*. For example, the AQUA project [1] proposed a number of sampling-based synopses that can provide approximate answers to a variety of queries using a fraction of the total data in a database. As with BBQ, such answers typically include tight bounds on the correctness of answers. AQUA, however, is designed to work in an environment where it is possible to generate an independent random sample of data (something that is quite tricky to do in sensor networks, as losses are correlated and communicating random samples may require the participation of a large part of the network). AQUA also does not exploit correlations, which means that it lacks the *predictive* power of representations based on probabilistic models. Deshpande et al. [4] and Getoor et al. [9] proposed exploiting data correlations through use of graphical modeling techniques for approximate query processing, but, unlike BBQ, neither provide any guarantees on the answers returned. Furthermore, the optimization goal of approximate query processing is typically not to minimize the data acquisition cost, rather it is minimizing the size of the synopsis, while maintaining reasonable accuracy.

Foundations

Figure 1 shows the most common architecture of a model-based querying system (adapted from the architecture of the BBQ system). The model itself is located at a centralized, Internet-connected basestation, which also interacts with the user. The user may issue either continuous or ad hoc queries against the sensor network, using a declarative SQL-like language. The key module in this architecture is the *query planner and model updater*, which is in charge of maintaining the model and answering the user queries (possibly by acquiring more data from the underlying sensor network). The following

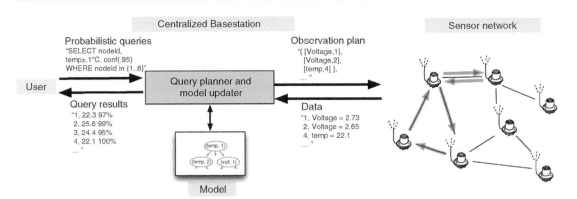

Model-Based Querying in Sensor Networks, Fig. 1 Architecture of a model-based querying system (Adapted from BBQ [7,6])

sections elaborate on the various components of such a system.

Model

A model is essentially a simplified representation of the underlying system or the process, and describes how various attributes of the system interact with each other, and how they evolve over time. Hence, the exact form of the model is heavily dependent on the system being modeled, and an astounding range of models have been developed over the years for different environments. For ease of exposition, the rest of this entry focuses on a dynamic model similar to the one used in the BBQ system.

Let $X_1, \ldots, X_n$ denote the (n) attributes of interest in the sensor network. Further, let X_i^t denote the value of X_i at time t (assuming that time is discrete). At any time t, a subset of these attributes may be observed and communicated to the basestation; here, the observations at time t are denoted by $\mathbf{o}^t$ (note that hidden variables can never be observed). The attributes typically correspond to the properties being monitored by the sensor nodes (e.g., *temperature* on sensor number 5, *voltage* on sensor number 8). However, more generally, they may be *hidden variables* that are of interest, but cannot be directly observed. For example, it may be useful to model and query a hidden boolean variable that denotes whether a sensor is faulty [11] – the value of this variable can be inferred using the model and the actual observations from the sensor.

The model encodes the spatial and temporal relationships between these attributes of interest. At any time t, the model provides us with a posterior *probability density function* (pdf), $p\left(X_1^t, \ldots, X_n^t | \mathbf{o}^{1 \ldots (t-1)}\right)$, assigning a probability for each possible assignment to the attributes at time t given the observations made so far. Such a joint distribution can capture all the spatial correlations between the attributes; more compact representations like Bayesian networks can be used instead as well.

To model the temporal correlations, it is common to make a *Markov* assumption; given the values of *all* attributes at time t, one assumes that the values of the attributes at time $t + 1$ are independent of those for any time earlier than t. This assumption leads to a simple model for a dynamic system where the dynamics are summarized by a conditional density called the *transition model*, $p\left(X_1^{t+1}, \ldots, X_n^{t+1} | X_1^t, \ldots, X_n^t\right)$. Using a transition model, one can compute $p\left(X_1^{t+1}, \ldots, X_n^{t+1} | \mathbf{o}^{1 \ldots t}\right)$ from $p\left(X_1^t, \ldots, X_n^t | \mathbf{o}^{1 \ldots t}\right)$ using standard probabilistic procedures. Different transition models may be used for different time periods (e.g., hour of day, day of week, season, etc.) to model the differences in the way the attributes evolve at different times.

Learning the model. Typically in probabilistic modeling, a *class* of models is chosen (usually with input from a *domain expert*), and learning techniques are then used to pick the best model in the class. Model parameters are typically learned from training data, but can also be directly inferred if the behavior of the underlying physical process is well-understood. In BBQ, the model was learned from historical data, which consisted of readings from all of the monitored attributes over some period of time.

Updating the model. Given the formulation above, model updates are fairly straightforward. When a new set of observations arrives (say $\mathbf{o}^t$), it can be incorporated into the model by conditioning on the observations to compute new distributions (i.e., by computing $p\left(X_1^t, \ldots, X_n^t | \mathbf{o}^{1 \ldots t}\right)$ from $p\left(X_1^t, \ldots, X_n^t | \mathbf{o}^{1 \ldots (t-1)}\right)$. Similarly, as time advances, the transition model is used to compute the new distribution for time $t + 1$ (i.e., $p\left(X_1^{t+1}, \ldots, X_n^{t+1} | \mathbf{o}^{1 \ldots t}\right)$ is computed from $p\left(X_1^t, \ldots, X_n^t | \mathbf{o}^{1 \ldots t}\right)$ and $p\left(X_1^{t+1}, \ldots, X_n^{t+1} | X_1^t \ldots X_n^t\right)$.

Query Planning and Execution

User queries are typically posed in a declarative SQL-like language that may be augmented with constructs that allow users to specify the approximation that the user is willing to tolerate, and the desired confidence in the answer. For example, the user may ask the system to report the temperature readings at all sensors within ± 0.5, with confidence 95%. For many applications (e.g., building temperature control), such approximate answers may be more than sufficient. Tolerance for such approximations along with the correlations encoded by the model can lead to significant energy savings in answering such queries.

Answering queries probabilistically based on a pdf over the query attributes is conceptually straightforward; to illustrate this process, consider two types of queries (here, assume that all queries are posed over attributes $X_1^t, \ldots, X_n^t$, and the corresponding pdf is given by $P\left(X_1^t, \ldots, X_n^t\right)$:

Value query. A value query [6] computes an approximation of the values of the attributes to within $\pm \varepsilon$ of the true value, with confidence at least $1 - \delta$. Answering such a query involves computing the expected value of each of the attribute, μ_i^t, using standard probability theory. These μ_i^t's will be the reported values. The pdf can then be used again to compute the probability that X_i^t is within ε from the

mean, $P(X_i^t \in [\mu_i^t - \varepsilon, \mu_i^t + \varepsilon])$. If all of these probabilities meet or exceed user specified confidence threshold, then the requested readings can be directly reported as the means μ_i^t. If the model's confidence is too low, additional readings must be acquired before answering the query (see below).

Max query. Consider an *entity* version of this query [2] where the user wants to know the identity of the sensor reporting the maximum value. A naive approach to answering this query is to compute, for each sensor, the probability that its value is the maximum. If the maximum of these probabilities is above $1 - \delta$, then an answer can be returned immediately; otherwise, more readings must be acquired. Although conceptually simple, computing the probability that a given sensor is reporting the maximum value is nontrivial, and requires complex integration that may be computationally infeasible [15].

If the model is not able to provide sufficient confidence in the answer, the system must acquire more readings from the sensor network, to bring the model's confidence up to the user specified threshold. Suppose the system observes a set of attributes $\mathcal{O} \subset \{X_1, \ldots, X_n\}$. After incorporating these observations into the model and recomputing the answer, typically the confidence in the (new) answer will be higher (this is not always true). The new confidence will typically depend on the actual observed values. Let $R(\mathcal{O})$ denote the *expected* confidence in the answer after observing $\mathcal{O}$. Then, the optimization problem of deciding which attributes to observe can be stated as follows:

$$\begin{aligned} \text{minimize}_{\mathcal{O} \subseteq \{1, \ldots, n\}} \quad & C(\mathcal{O}), \\ \text{such that} \quad & R(\mathcal{O}) \geq 1 - \delta. \end{aligned} \tag{1}$$

where $C(\mathcal{O})$ denotes the *data acquisition cost* of observing the values of attributes in $\mathcal{O}$.

This optimization problem combines three problems that are known to be intractable, making it very hard to solve it in general:

1. Answering queries using a pdf: as mentioned above, this can involve complex numerical

integration even for simple queries such as max.

2. hoosing the minimum set of sensor readings to acquire to satisfy the query (this is similar to the classic *value of information problem* [14, 12, 15, 10]).

3. Finding the optimal way to collect a required set of sensor readings from the sensor network that minimizes the total communication cost. Meliou et al. [13] present several approximation algorithms for this NP-Hard problem.

Example

Figure 2 illustrates the query answering process using a simple example, where the model takes the form of time-varying *bivariate Gaussian (normal)* distribution over two attributes, X_1 and X_2. This was the basic model used in the BBQ system. A bivariate Gaussian is the natural extension of the familiar unidimensional normal probability density function (pdf), known as the "bell curve". Just as with its one-dimensional counterpart, a bivariate Gaussian can be expressed as a function of two parameters: a length-2 vector of means, μ, and a 2×2 matrix of covariances, Σ. Figure 2a shows a three-dimensional rendering of a Gaussian over the two attributes at time t, X_1^t and X_2^t the z axis represents the *joint density* that $X_2^t = x$ and $X_1^t = y$).

Now, consider a *value query* over this model, posed at time t, which asks for the values of X_1^t and X_2^t, within $\pm \varepsilon$, with confidence $1 - \delta$. The reported values in this case would be the means (μ), and the confidence can be computed easily using Σ (details can be found in [6]). Considering the high initial covariance, it is unlikely that the Gaussian in Figure 2a can achieve the required confidence. Suppose the system decides to observe X_1^t. Figure 2b shows the result of incorporating this observation into the model. Note that not only does the spread of X_1^t reduces to near zero, because of the high correlation between X_1^t and X_2^t, the variance of X_2^t also reduces dramatically, allowing the system to answer the query with required confidence.

Then, after some time has passed, the belief about the values of X_1 and X_2 (at time $t' > t$) will be "spread out", again providing a high-variance Gaussian over two attributes, although both the mean and variance may have shifted from their initial values, as shown in Fig. 2c.

Key Applications

Model-based querying systems like BBQ, that exploit statistical modeling techniques and optimize the utilization of a network of resource constrained devices could have significant impact in a number of application domains, ranging from control and automation in buildings [16] to highway traffic monitoring. Integrating a model into the data acquisition process can significantly improve data quality and reduce data uncertainty. The ability to query over missing, future or hidden states of the system will prove essential in many applications where sensor failures are common (e.g., highway traffic monitoring) or where direct observation of the variables of interest is

a 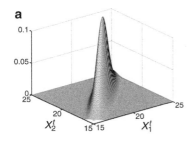 **b** 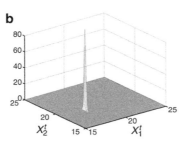 **c**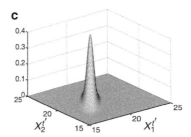

Model-Based Querying in Sensor Networks, Fig. 2
Example of Gaussians: (a) 3D plot of a 2D Gaussian with high covariance; (b) the resulting Gaussian after a particular value of X_1^t has been observed (because of measurement noise, there might still be some uncertainty about the true value of X_1^t); (c) the uncertainty about X_1 and X_2 increases as time advances to t'.

not feasible. Finally, model-based querying has the potential to significantly reduce the cost of data acquisition, and thus can improve the life of a resource-constrained measurement infrastructure (e.g., battery-powered wireless sensor networks) manyfold.

Future Directions

Model-based querying is a new and exciting research area with many open challenges that are bound to become more important with the increasingly widespread use of models for managing sensor data. The two most important challenges are dealing with a wide variety of models that may be used in practice, and designing algorithms for query processing and data acquisition; these are discussed briefly below:

Model selection and training. The choice of model affects many aspects of model-based querying, most importantly the accuracy of the answers and the confidence bounds that can be provided with them. The problem of selecting the right model class has been widely studied [3, 14] but can be difficult in some applications. Furthermore, developing a new system for each different model is not feasible. Ideally, using a new model should involve little to no effort on the part of user. Given a large variety of models that may be applicable in various different scenarios, this may turn out to be a tremendous challenge.

Algorithms for query answering and data acquisition. Irrespective of the model selected, when and how to acquire data in response to a user query raises many hard research challenges. As discussed above, this problem combines three very hard problems, and designing general-purpose algorithms that can work across the spectrum of different possible model remains an open problem.

See Deshpande et al. [5] for a more elaborate discussion of the challenges in model-based querying.

Cross-References

▶ Approximate Query Processing
▶ Continuous Queries in Sensor Networks
▶ Data Acquisition and Dissemination in Sensor Networks
▶ Data Uncertainty Management in Sensor Networks

Recommended Reading

1. Acharya S, Gibbons PB, Poosala V, Ramaswamy S. Join synopses for approximate query answering. In: Proceedings of the ACM SIGMOD International Conference on Management of Data; 1999. p. 275–86.
2. Cheng R, Kalashnikov DV, Prabhakar S. Evaluating probabilistic queries over imprecise data. In: Proceedings of the ACM SIGMOD International Conference on Management of Data; 2003. p. 551–62.
3. Cowell R, Dawid P, Lauritzen S, Spiegelhalter D. Probabilistic networks and expert systems. New York: Spinger; 1999.
4. Deshpande A, Garofalakis M, Rastogi R. Independence is good: dependency-based histogram synopses for high-dimensional data. In: Proceedings of the ACM SIGMOD International Conference on Management of Data; 2001. p. 199–210.
5. Deshpande A, Guestrin C, Madden S. Using probabilistic models for data management in acquisitional environments. In: Proceedings of the 2nd Biennial Conference on Innovative Data Systems Research; 2005. p. 317–28.
6. Deshpande A, Guestrin C, Madden S, Hellerstein J, Hong W. Model-driven approximate querying in sensor networks. VLDB J. 2005;14(4):417–43.
7. Deshpande A, Guestrin C, Madden S, Hellerstein JM, Hong W. Model-driven data acquisition in sensor networks. In: Proceedings of the 30th International Conference on Very Large Data Bases; 2004. p. 588–99.
8. Deshpande A, Madden S. MauveDB: supporting model-based user views in database systems. In: Proceedings of the ACM SIGMOD International Conference on Management of Data; 2006. p. 73–84.
9. Getoor L, Taskar B, Koller D. Selectivity estimation using probabilistic models. In: Proceedings of the ACM SIGMOD International Conference on Management of Data; 2001. p. 461–72.
10. Goel A, Guha S, Munagala K. Asking the right questions: model-driven optimization using probes. In: Proceedings of the 25th ACM SIGACT-SIGMOD-SIGART Symposium on Principles of Database Systems; 2006. p. 203–12.
11. Kanagal B, Deshpande A. Online filtering, smoothing and probabilistic modeling of streaming data. In: Proceedings of the 24th International Conference on Data Engineering; 2008. p. 1160–1169.
12. Krause A, Guestrin C, Gupta A, Kleinberg J. Near-optimal sensor placements: maximizing information while minimizing communication cost. In: Proceed-

ings of the 5th International Symposium on Information Processing in Sensor Networks; 2006. p. 2–10.

13. Meliou A, Chu D, Hellerstein J, Guestrin C, Hong W. Data gathering tours in sensor networks. In: Proceedings of the 5th International Symposium on Information Processing in Sensor Networks; 2006. p. 43–50.

14. Russell S, Norvig P. Artificial intelligence: a modern approach. Prentice Hall; 1994.

15. Silberstein A, Braynard R, Ellis C, Munagala K, Yang J. A sampling-based approach to optimizing top-k queries in sensor networks. In: Proceedings of the 22nd International Conference on Data Engineering; 2006. p. 68.

16. Singhvi V, Krause A, Guestrin C, Garrett Jr J, Matthews H. Intelligent light control using sensor networks. In: Proceedings of the 3rd International Conference on Embedded Networked Sensor Systems; 2005. p. 218–29.

Monotone Constraints

Carson Kai-Sang Leung
Department of Computer Science, University of Manitoba, Winnipeg, MB, Canada

Synonyms

Monotonic constraints

Definition

A constraint C is *monotone* if and only if for all itemsets S and S':

$$\text{if } S \supseteq S' \text{ and } S \text{ violates } C, \text{ then } S'$$
$$\text{violates } C.$$

Key Points

Monotone constraints [1–3] possess the following nice property. If an itemset S violates a monotone constraint C, then any of its subsets also violates C. Equivalently, all supersets of an itemset satisfying a monotone constraint C also satisfy C (i.e., C is upward closed). By exploiting this property, monotone constraints can be used for reducing computation in frequent itemset mining

with constraints. As frequent itemset mining with constraints aims to find frequent itemsets that satisfy the constraints, if an itemset S satisfies a monotone constraint C, no further constraint checking needs to be applied to any superset of S because all supersets of S are guaranteed to satisfy C. Examples of monotone constraints include $min\,(S.Price) \leq \$30$, which expresses that the minimum price of all items in an itemset S is at most \$30. Note that, if the minimum price of all items in S is at most \$30, adding more items to S would not increase its minimum price (i.e., supersets of S would also satisfy such a monotone constraint).

Cross-References

▶ Frequent Itemset Mining with Constraints

Recommended Reading

1. Brin S, Motwani R, Silverstein C. Beyond market baskets: generalizing association rules to correlations. In: Proceedings of the ACM SIGMOD International Conference on Management of Data; 1997. p. 265–76.

2. Grahne G, Lakshmanan LVS, Wang X. Efficient mining of constrained correlated sets. In: Proceedings of the 16th International Conference on Data Engineering; 2000. p. 512–21.

3. Pei J, Han J. Can we push more constraints into frequent pattern mining? In: Proceedings of the 6th ACM SIGKDD International Conference on Knowledge Discovery and Data Mining; p. 350–54.

M

Monte Carlo Methods for Uncertain Data

Peter J. Haas
IBM Almaden Research Center, San Jose, CA, USA

Synonyms

Sampling methods; Simulation methods; Stochastic methods

Definition

Uncertain datasets do not contain specific data values but rather representations of probability distributions over "possible worlds," that is, over possible realizations of the dataset. Queries over such datasets result in a probability distribution over possible answers, where each possible answer corresponds to the query result in one of the possible worlds. In this setting, the goal of query processing is to compute the query-result distribution or perhaps features of this distribution such as marginal probabilities, moments, modes, and quantiles. Basic Monte Carlo methods approximate the query-result distribution by, in essence, repeatedly generating possible-world instances and computing the query answer on each instance. The resulting samples from the query-result distribution are used to estimate quantities of interest using statistical methods. More sophisticated techniques try to improve on the efficiency of this procedure by exploiting problem structure or additional information to focus the sampling effort on the "most important" data values.

Historical Background

Interest in Monte Carlo methods for handling uncertain data arose in the database, machine learning, and networking communities as it became apparent that many important computations over uncertain data are intractable to exact solution. "Probabilistic database systems" such as MayBMS, MCDB, MystiQ, Orion, SPROUT, and TRIO—see [1] and references therein—attempt to extend relational database semantics with notions of probability, so as to exploit existing query processing technology. A typical goal of these systems is to compute, for a relational query Q over an uncertain database, the set of tuples that satisfy Q in at least one possible world and, for each such tuple t, the probability that t appears in the output. This latter appearance probability is the "count" (weighted by probability value) of the number of possi-

ble worlds in which t appears. Dalvi, Suciu, and colleagues, in a series of papers beginning with [2], established that many such analyses fall within the #P-complexity class of counting problems, rendering exact probability computations intractable for large databases. Starting with the work of Ulam in 1946, modern Monte Carlo methods have been recognized as powerful tools for obtaining approximate answers when exact computation is infeasible [3], and the authors of [2] proposed using *Karp-Luby sampling* [4] for approximate probability computation in #P-complex scenarios.

In parallel with these developments, probabilistic graphical models (PGMs)—such as Bayesian networks, Markov networks, and conditional random fields—were recognized by the machine learning community as powerful tools for concise representation of uncertain data [5]. The two research threads began to merge when a number of researchers used PGMs to represent uncertainty within a probabilistic database system. Examples of such systems include PrDB [6], BayesStore [7], and the system of Wick et al. [8]. In the PGM setting, each uncertain data item is viewed as a random variable, and the PGM concisely represents the joint probability distribution over the set of random variables. The key "inference" computations in this setting involve the computation of functionals over the joint distribution, often in the form of marginal probabilities for a specific set of variables of interest (perhaps conditioned on the values of some "observed" variables). Exact inference is known to be NP-hard in general, so Markov Chain Monte Carlo (MCMC) algorithms have been proposed for approximate inference.

Another important type of uncertain data arises in graph databases when the existence of edges is uncertain. For example, questions about the reliability of complex systems often reduce to the problem of computing the probability that such a random graph is connected. This computation is typically #P-complete [9], so again Monte Carlo algorithms are used to estimate probabilities of interest. Interest in Monte Carlo methods for uncertain graphs has grown as these graphs have been increasingly used for novel applications such as social and biological network analysis.

Scientific Fundamentals

The application of Monte Carlo methods to uncertain data depends strongly on how uncertainty is represented. Methods for the most common uncertainty representations are discussed below.

PC-Tables and Karp-Luby Sampling

Many schemes for representing probabilities in first-generation probabilistic relational databases are captured by "pc-tables" [1, Sec. 2.4]. The idea is to define a "relational skeleton" comprising a set of "possible tuples" along with a collection of random variables $X = \{X_1, X_2, \ldots, X_k\}$ and to associate with each possible tuple t a first-order logical expression $L_t[X]$ in the variables. To generate a realization d of the probabilistic database D, values for each of the variables are sampled from the variables' joint probability distribution P to obtain a realization $x = \{x_1, x_2, \ldots, x_k\}$. A possible tuple t is included in d if and only if $L_t[x] = 1$. (For boolean variables and expressions, denote *true* by 1 and *false* by 0.) Thus the a priori probability that t will appear in a realization is $P(L_t[X] = 1)$. Perhaps the simplest special case of this setup is a "tuple independent" (TID) database in which $X_1, X_2, \ldots, X_k$ are mutually independent boolean random variables, one per possible tuple, and $L_t[X] = X_t$. Figure 1 shows two base relations R and S in a TID database, each with three attributes. The boolean random variable associated with each tuple is displayed on the right. The column labeled "P" that has been appended to the tables gives for each tuple i the probability that $X_i = 1$ and hence the probability that tuple i will appear in a realization of the database. There are 32 possible worlds. For example, the probability of seeing a possible world with $R = \{(1, \text{Alice}, \text{SF}), (3, \text{Laura}, \text{NY})\}$ and $S = \{(1, \text{Sally}, \text{NY})\}$ is given by $P(X_1 = 1, X_2 = 0, X_3 = 1, X_4 = 0, X_5 = 1) = 0.6 \times (1 - 0.2) \times 0.5 \times (1 - 0.8) \times 0.7 = 0.0336$,

id	name	loc	P	
1	Alice	SF	0.6	X_1
2	Bob	NY	0.2	X_2
3	Laura	NY	0.5	X_3

R

id	name	loc	P	
4	Fred	SF	0.8	X_4
5	Sally	NY	0.7	X_5

S

name1	name2	P	
Alice	Fred	0.48	$X_1 \wedge X_4$
Bob	Sally	0.14	$X_2 \wedge X_5$
Laura	Sally	0.35	$X_3 \wedge X_5$

$$T = R \bowtie_{\text{loc}} S$$

Monte Carlo Methods for Uncertain Data, Fig. 1
Three relations in a TID database

where the individual probabilities have been multiplied together because of independence. Now suppose that a relation T is defined via an SQL query of the form SELECT R.name as name1, S.name as name2 FROM R, S WHERE R.loc = S.loc. Observe that the tuple (Alice, Fred) only appears in the result if Alice appears in R and Fred appears in S, that is, if $X_1 \wedge X_4 = 1$. The logical expression $X_1 \wedge X_2$ is called the "lineage" of the tuple (Alice, Fred) in T. The probability that this tuple appears in the answer is $P(X_1 \wedge X_4 = 1) = 0.6 \times 0.8 = 0.48$. Figure 1 displays the lineage for all possible tuples in T. The notion of lineage extends to more complex scenarios such as "block independent" (BID) databases. Here the tuples are partitioned into blocks, and the inclusion or exclusion of tuples in different blocks occurs independently. At most one tuple in each block is included, so that the appearance probabilities for the tuples in a block sum to at most 1. The BID model contains the TID model as a special case and also encompasses "attribute-level" uncertainty, where a probability distribution is specified over the possible value of each uncertain attribute [1, Sec. 1.2.3].

One of the most basic computations in a probabilistic database system is to determine the probability that a given tuple appears in a query answer. To do this, the standard "intensional" approach first computes the lineage of the tuple, which takes the form of a first-order logical expression $L_t[X]$, and then computes $p_t = P(L_t[X] = 1)$. Because this problem is

#P-hard, Monte Carlo approximation methods are of interest. A direct approach repeatedly generates independent random samples $X^{(1)}, X^{(2)}, \ldots, X^{(n)}$ from the distribution of X, where $X^{(i)} = \{X_1^{(i)}, X_2^{(i)}, \ldots, X_k^{(i)}\}$. The estimator of p_t is then $\hat{p}_{t,n} = (1/n) \sum_{i=1}^{n} L_t[X^{(i)}]$. Standard statistical theory shows that, for a sample size of $n = \lceil 4 \log(2/\delta)/(p_t \epsilon^2) \rceil$, one can estimate p_t to within $\pm 100\epsilon\%$ with probability at least $100(1 - \delta)\%$. A key difficulty is that the procedure can require huge sample sizes. For example, when $P(X_i = 0) = P(X_i = 1) = 1/2$ for each X_i, the probability p_t can be as small as 2^{-k} and k can be on the order of the number of possible tuples. The required sample size n can thus be exponentially large in the data size.

The *Karp-Luby sampling* scheme can often avoid the foregoing problem when $L_t[X]$ is in "disjunctive normal form" (DNF), that is, $L_t[X]$ has the special form $C_1 \vee C_2 \vee \cdots \vee C_m$, where each clause C_i is a conjunction of variables in $\{X_1, X_2, \ldots, X_k, \neg X_1, \neg X_2 \ldots, \neg X_k\}$. It is known, for example, that the time complexity for computing a DNF representation of the lineage for a query result is polynomial in the data size when the database is TID or BID and the query is a "union of conjunctive queries" (UCQ)—essentially meaning that the query has no logical negations [1, Sec. 5.3.2]. The key idea behind Karp-Luby is to sample not from the set of all possible realizations of X but only from those realizations that satisfy at least one of the clauses. It can be shown that, to estimate p_t to within $\pm 100\epsilon\%$ with probability at least $100(1 - \delta)\%$, the required sample size is $n = \lceil 3m \log(2/\delta)/\epsilon^2 \rceil$; see [1]. Thus the potential $O(2^k)$ complexity factor in the direct approach is replaced by $O(m)$, which often represents a significant reduction in complexity.

The foregoing discussion has focused on estimating the probability that a given tuple appears in a query result. In practice, it is desirable to report the set of most probable tuples. Ré et al. [10] use the Karp-Luby algorithm to return the top-k most probable query results, the idea being to simulate each candidate just enough to produce a correct ranking. The reported set of top-k results

is exact, but the reported probabilities for these results are Monte Carlo estimates.

PGMs and MCMC

One of the most general PGM models that has been proposed for probabilistic databases is the factor graph model; see [8]. The uncertain data items are represented as a collection $Y = \{Y_1, Y_2, \ldots, Y_k\}$ of random variables. There are also "observed" random variables $X = \{X_1, X_2, \ldots, X_m\}$ which are treated as constants. Both X and Y can include additional random variables that do not correspond to the value of an individual data item, but to a function of the data items such as a sum. The conditional probability over the uncertain data Y, given the observed data X, is concisely represented as $P(Y = y \mid X = x) = (1/Z_x) \prod_{i=1}^{l} \phi_i(y_{\{i\}}, x_{\{i\}})$, where $\phi_1, \ldots, \phi_l$ are a small set of *factors* and $Z_x = \sum_y \prod_{i=1}^{l} \phi_i(y_{\{i\}}, x_{\{i\}})$ is a normalization constant such that the probabilities over the possible values of Y add up to 1. For each i, the quantities $y_{\{i\}}$ and $x_{\{i\}}$ denote the (small) subsets of y and x that comprise the input arguments to ϕ_i. The range of ϕ_i comprises the nonnegative real numbers, and ϕ_i often has the form $\exp(\theta_i \cdot \psi(y_{\{i\}}, x_{\{i\}}))$, where θ_i is a real-valued parameter. Factors can also be binary valued, thereby enforcing constraints on the set of allowed possible worlds. This model contains Markov random fields and Bayesian networks as special cases.

For this class of probabilistic databases, the primary challenge to direct Monte Carlo methods is that the above normalization constant Z_x is #P-hard to compute in general, so that it is difficult to directly generate a realization of $W = Y \cup X$. The idea behind MCMC methods is to generate a Markov chain $W_0, W_1, \ldots$ such that the corresponding marginal distributions $\pi_0, \pi_1, \ldots$ converge to π, the distribution of W, i.e., the possible world distribution. Thus, after a "burn-in" period of length N, the random variables $\{Q(W_n)\}_{n \geq N}$ can be treated as (dependent) samples from the query-result distribution and used to estimate distribution features of interest. The algorithm proceeds by choosing an initial realization W_0. A *proposal* realization W_1' is generated according to a proposal distribution $q(\cdot|W_0)$. With probability $\alpha(W_1', W_0)$ the proposal is accepted and $W_1 \leftarrow W_1'$; otherwise, the proposal is rejected and $W_1 \leftarrow W_0$. Here

$$\alpha(w', w) = \min\left(1, \frac{\pi(w')q(w|w')}{\pi(w)q(w'|w)}\right). \quad (1)$$

This procedure is iterated to generate the Markov chain. A key advantage of this approach is that the normalization constant Z_x cancels out in (1) and hence does not need to be computed. Moreover, $q(w'|w)$ is defined such that w' is obtained from w by changing the values of a very small number of random variables (usually via simple procedures). Therefore, only factors whose argument variables are changed by q need to be recomputed. The authors in [8] use a standard database system to execute a query on a possible world and use an in-memory factor-graph processing system to carry out the MCMC updates. The system also exploits the extensive overlap between successive possible world realizations by using well-known view maintenance techniques. The system also exploits parallelization opportunities.

Ilyas and Soliman [11] discuss the use of MCMC techniques to solve top-k ranking queries in which the score used for ranking is uncertain because of uncertainty in the data attribute values. They also extend the MCMC technique to handle ranking queries where the tuples to be ranked are themselves the result of a join operation.

Generative Uncertainty Models: MCDB, PIP, and SimSQL

As can be seen from the above discussion, uncertain database systems represent uncertainty using random variables. These random variables either directly represent uncertain data values or appear in first-order logical expressions that determine the presence or absence of a tuple in a database realization. In virtually all of the systems described so far, the joint distribution of the random variables is specified explicitly, either by directly storing the possible values together with their probabilities or, more concisely, explicitly specifying joint or conditional distributions of subsets of the random variables along with

rules on how to combine these distributions to form the complete joint probability distribution. Systems based on factor graphs store distribution parameter values—e.g., the θ_i values in the exponential representation given above—together with factor functions. The ORION system [12] allows specification of certain continuous probability distributions, such as the normal and exponential distributions. These database systems can be viewed as using a simple stochastic model of uncertainty: an uncertain data value X is generated as a sample from an explicitly specified distribution.

The Monte Carlo Database System (MCDB) incorporates Monte Carlo techniques into the fabric of the system in order to encompass a broad class of stochastic models and to handle uncertainty in great generality; see [13] and references therein. MCDB allows an analyst to attach arbitrary stochastic models to a database, thereby specifying, in addition to the ordinary tables in the database, "stochastic" tables that contain uncertain data. The stochastic models are implemented as user- and system-defined libraries of external C++ programs called *Variable Generation functions*, or VG functions for short. A call to a VG function generates a pseudorandom sample from an underlying stochastic model; VG functions are usually parameterized on the current state of the non-random relations (e.g., tables of historical sales data or delivery times) via SQL queries that define the input relations to the VG function. Generating a sample of each uncertain data value creates a database realization. For example, an uncertain data value X might correspond to the next week's value of an exotic stock option. In this case the underlying model might comprise a stochastic differential equation that describes the evolution of the underlying stock price over time. There is typically no closed form expression for the distribution of X, and the VG function might internally simulate the stock price over a period of one week. As another example, a VG function might generate a sequence of tuples based on a random walk model; here the number of tuples generated might itself be random, so that there is no "deterministic relational skeleton" of possible tuple values as required for the pc-table uncertainty representation. As a final example, a VG function can internally execute an MCMC algorithm to generate samples from complex joint distributions over a collection of random variables.

MCDB runs an SQL query over each database realization, thereby generating samples from the query-result distribution. Features such as marginal distributions, moments, and quantiles can then be estimated from the samples, e.g., the expected value of a "total sales" query over uncertain customer data can be estimated by the average of the "total sales" values for the different database realizations. To ensure acceptable practical performance, MCDB employs novel query processing algorithms that execute a query plan only once, processing "tuple bundles" rather than ordinary tuples, and executing Monte Carlo operations as late as possible in the query plan. A tuple bundle encapsulates the instantiations of a tuple over a set of database instances. As mentioned in [13], MCDB has been extended to deal with risk analysis by efficiently estimating extreme quantiles using a sophisticated "Gibbs cloning" technique that systematically focuses the sampling effort on possible worlds that yield extreme query results. MCDB has also been extended to handle threshold queries of the form "Which regions will see more than a 2% decline in sales with at least 50% probability?" and to exploit MapReduce parallelism. The MCDB approach has also been extended to array databases with uncertain data [14]. The PIP system of Kennedy and Koch [15] combines the pc-table uncertainty representation and Monte Carlo techniques as in MCDB. The idea is to first generate the lineage and then sample from the random variables that appear in the first-order formulas. This approach can yield superior performance for certain MCDB-style queries that focus on expected values and have simple VG-function parameterizations.

SimSQL [16] is a re-implementation and extension of MCDB. SimSQL allows data in stochastic database tables to be used to parametrize the processes that stochastically generate the data in other stochastic database tables. Moreover, SimSQL allows both ver-

sioning and recursive definitions of stochastic database tables. Whereas MCDB merely allowed generation of sample realizations of a given stochastic database D—in other words, a static database-valued random variable—the foregoing extensions enable SimSQL to generate realizations of a database-valued Markov chain $D[0], D[1], D[2], \ldots$. That is, the stochastic mechanism that generates a realization of the ith database state $D[i]$ may explicitly depend on the prior state $D[i-1]$. As with MCDB, queries are expressed in SQL. Thus SimSQL can handle uncertain data that evolves over time and can execute scalable MCMC algorithms for Bayesian machine learning. SimSQL supports most of the declarative part of SQL, including deeply nested, correlated subqueries. SQL codes are compiled by the system into a series of Hadoop MapReduce jobs and hence can exploit parallelism.

Uncertain Graph Data and Related Models

Graph databases are becoming increasingly important, and so there has been increasing interest in queries over uncertain graph data. In a classical uncertain graph, an edge e is included in the graph with a probability $p(e)$, independently of the other edges. Note that the existence of a path between two points is equivalent to the truth of a DNF expression in the binary random variables X_{ij} that indicate whether an edge exists between nodes i and j. Not surprisingly, the Karp-Luby algorithm plays a role in Monte Carlo algorithms in this setting [17]. More recently, efficient Monte Carlo algorithms employing unequal-probability sampling have been proposed for answering "distance constraint reachability" queries—queries that determine the probability that the distance between a pair of nodes is at or below a threshold—as well as for shortest-path queries in uncertain graphs where X_{ij} variables for edges that share a vertex can be correlated; see [18] and references therein. Another means for increasing efficiency is stratified random sampling, which has been proposed for queries on uncertain graphs [19]. The idea is to group the possible worlds into disjoint strata such that the worlds within a given stratum are "similar" according to a problem-specific definition. A specified number of worlds are generated from each stratum, thereby ensuring that the different types of possible worlds are well represented in the overall sample of graph realizations. Emrich et al. [20] present a system for Monte-Carlo-based processing of a variety of uncertain graph queries such as shortest path, nearest neighbor, and reversed nearest neighbor queries and provide some pointers to the literature. Uncertain XML data can be viewed as a type of uncertain graph data; see [21] and references therein for a discussion and comparison of Monte Carlo methods for uncertain XML.

A variant of the k-nearest-neighbor (KNN) problem on an uncertain graph is the KNN problem for data points with uncertain positions in d-dimensional euclidean space, i.e., static uncertain objects. This problem is closely related to the top-k ranking queries discussed previously: the "score" is the distance from the query point. Zhang et al. [22] discuss Monte Carlo algorithms where the key idea is to index the possible worlds using an R-tree to increase query-answering efficiency. A related problem is that of querying uncertain moving object data. Recently, Monte Carlo algorithms for nearest-neighbor and reverse-nearest-neighbor queries have been developed that employ a Bayesian framework to combine first-order Markov models of object movement with a set of observations on the objects; see [23] and references therein.

Key Applications

Key applications involving uncertain data include sensor and RFID data, data integration over inconsistent sources, and information extraction from text and other media. Uncertain graphs arise in the analysis of social, telecommunication, and biological networks.

Future Directions

Future directions around Monte Carlo methods for probabilistic database systems encompass improvements to the usability of such systems as well as the application of such systems to real-world management of uncertain data. For example, to improve the usability of the SimSQL system for Bayesian machine learning, Gao et al. [24] have recently introduced BUDS, a declarative front end language for SimSQL that allows succinct and simple specification of large-scale Bayesian machine learning algorithms on a distributed computing platform. Other potential applications of Monte-Carlo-based database systems include massive scale stochastic agent-based simulations and applications in risk management. With respect to the latter, organizations such as Probability Management (www.probabilitymanagement.org) have been promoting standards for certifying and communicating uncertain data within and between organizations, in order to enable a coherent, consistent approach to risk management. The SIPmath standard, in particular, calls for the distribution of uncertain data items to be represented as vectors of sample realizations of the data values. This is essentially the same as the tuple bundles used in MCDB and SimSQL, and so such systems are ideally positioned to be part of a risk management ecosystem.

Url to Code

FACTORIE (PGM inference package used in [8]): https://github.com/factorie/factorie

MayBMS, SPROUT, and Pip: http://maybms.sourceforge.net

MystiQ: https://homes.cs.washington.edu/~suciu/project-mystiq.html

Orion: http://orion.cs.purdue.edu

SimSQL: http://cmj4.web.rice.edu/SimSQL/SimSQL.html

Trio: http://infolab.stanford.edu/trio

Cross-References

▶ Data Uncertainty Management in Sensor Networks
▶ Incomplete Information
▶ Inconsistent Databases
▶ Karp-Luby Sampling
▶ Nearest Neighbor Query
▶ Probabilistic Databases
▶ Uncertainty Management in Scientific Database Systems

Recommended Reading

1. Suciu D, Olteanu D, Ré C, Koch C. Probabilistic databases. Synthesis lectures on data management. San Rafael: Morgan & Claypool; 2011.
2. Dalvi NN, Suciu D. Efficient query evaluation on probabilistic databases. In: Proceedings of the 30th International Conference on Very Large Data Bases; 2004. p. 864–75.
3. Robert CP, Casella G. Monte Carlo statistical methods. 2nd ed. New York: Springer; 2010.
4. Karp RM, Luby M. Monte-Carlo algorithms for enumeration and reliability problems. In: Proceedings of the 24th Annual Symposium on Foundations of Computer Science; 1983. p. 56–64.
5. Koller D, Friedman N. Probabilistic graphical models: Principles and techniques. Cambridge: MIT Press; 2009.
6. Sen P, Deshpande A, Getoor L. PrDB: managing and exploiting rich correlations in probabilistic databases. VLDB J. 2009;18(5):1065–90.
7. Wang DZ, Michelakis E, Garofalakis MN, Hellerstein JM. BayesStore: managing large, uncertain data repositories with probabilistic graphical models. PVLDB. 2008;1(1):340–51.
8. Wick ML, McCallum A, Miklau G. Scalable probabilistic databases with factor graphs and MCMC. Proc VLDB Endow. 2010;3(1):794–804.
9. Provan JS, Ball MO. The complexity of counting cuts and of computing the probability that a graph is connected. SIAM J Comput. 1983;12(4): 777–88.
10. Ré C, Dalvi NN, Suciu D. Efficient top-k query evaluation on probabilistic data. In: Proceedings of the 23rd International Conference on Data Engineering; 2007. p. 886–95.
11. Ilyas IF, Soliman MA. Probabilistic ranking techniques in relational databases. San Rafael: Morgan & Claypool Publishers; 2011.
12. Singh S, Mayfield C, Mittal S, Prabhakar S, Hambrusch S, Shah R. Orion 2.0: native support for uncertain data. In: Proceedings of the ACM SIGMOD International Conference on Management of Data; 2008. p. 1239–42.
13. Jampani R, Xu F, Wu M, Perez LL, Jermaine C, Haas PJ. The Monte Carlo database system: stochastic analysis close to the data. ACM Trans. Database Syst. 2011;36(3):1–41.
14. Ge T, Grabiner D, Zdonik SB. Monte Carlo query processing of uncertain multidimensional array data. In: Proceedings of the 27th International Conference on Data Engineering; 2011. p. 936–47.
15. Kennedy O, Koch C. PIP: a database system for great and small expectations. In: Proceedings of the 26th International Conference on Data Engineering; 2010. p. 157–68.
16. Cai Z, Vagena Z, Perez LL, Arumugam S, Haas PJ, Jermaine CM. Simulation of database-valued Markov chains using SimSQL. In: Proceedings of the ACM SIGMOD International Conference on Management of Data; 2013. p. 637–48.
17. Zou L, Peng P, Zhao D. Top-K possible shortest path query over a large uncertain graph. In: Proceedings of the 12th International Conference on Web Information Systems Engineering; 2011. p. 72–86.
18. Cheng Y, Yuan Y, Wang G, Qiao B, Wang Z. Efficient sampling methods for shortest path query over uncertain graphs. In: Proceedings of the 19th International Conference on Database Systems for Advanced Applications; 2014. p. 124–40.
19. Li R, Yu JX, Mao R, Jin T. Efficient and accurate query evaluation on uncertain graphs via recursive stratified sampling. In: Proceedings of the 30th International Conference on Data Engineering; 2014. p. 892–903.
20. Emrich T, Kriegel H, Niedermayer J, Renz M, Suhartha A, Züfle A. Exploration of Monte Carlo based probabilistic query processing in uncertain graphs. In: Proceedings of the 21st ACM International Conference on Information and Knowledge Management; 2012. p. 2728–30.
21. Souihli A, Senellart P. Optimizing approximations of DNF query lineage in probabilistic XML. In: Proceedings of the 29th International Conference on Data Engineering; 2013. p. 721–32.
22. Zhang Y, Lin X, Zhu G, Zhang W, Lin Q. Efficient rank based KNN query processing over uncertain data. In: Proceedings of the 26th International Conference on Data Engineering; 2010. p. 28–39.
23. Emrich T, Kriegel H, Mamoulis N, Niedermayer J, Renz M, Züfle A. Reverse-nearest neighbor queries on uncertain moving object trajectories. In: Proceedings of the 19th International Conference on Database Systems for Advanced Applications; 2014. p. 92–107.
24. Gao ZJ, Luo S, Perez LL, Jermaine C. The BUDS language for distributed Bayesian machine learning. In: Proceedings of the ACM SIGMOD International Conference on Management of Data; 2017. p. 961–76.

Moving Object

Ralf Hartmut Güting
Fakultät für Mathematik und Informatik,
Fernuniversität Hagen, Hagen, Germany
Computer Science, University of Hagen, Hagen,
Germany

Synonyms

Time dependent geometry

Definition

A moving object is essentially a time dependent geometry. Moving objects are the entities represented and queried in moving objects databases.

Key Points

The term emphasizes the fact that geometries may change continuously (whereas earlier work on spatio-temporal databases allowed only discrete changes, e.g., of land parcels). One can distinguish between moving objects for which only the time dependent position is of interest and those for which also shape and extent are relevant and may change over time. The first can be characterized as *moving points*, the second as *moving regions*. For example, moving points could represent people, vehicles (such as cars, trucks, ships or air planes), or animals. Moving regions could be hurricanes, forest fires, spread of epidemic diseases etc. Moving point data may be captured by GPS devices or RFID tags; moving region data may result from processing sequences of satellite images, for example. Moving points and moving regions can be made available as data types in suitable type systems; such a design can be found in [1]. Such an environment may have further "moving" data types (e.g., *moving lines*).

Cross-References

▶ Moving Objects Databases and Tracking

Recommended Reading

1. Güting RH, Böhlen MH, Erwig M, Jensen CS, Lorentzos NA, Schneider M, Vazirgiannis M. A foundation for representing and querying moving objects in databases. ACM Trans Database Syst. 2000;25(1): 1–42.

Moving Objects Databases and Tracking

Ralf Hartmut Güting
Fakultät für Mathematik und Informatik,
Fernuniversität Hagen, Hagen, Germany
Computer Science, University of Hagen, Hagen,
Germany

Synonyms

Spatio-Temporal Databases; Trajectory Databases

Definition

Moving objects database systems provide concepts in their data model and data structures in the implementation to represent moving objects, i.e., continuously changing geometries. Two important abstractions are *moving point*, representing an entity for which only the time dependent position is of interest, and *moving region*, representing an entity for which also the time dependent shape and extent is relevant. Examples of moving points are cars, trucks, air planes, ships, mobile phone users, RFID equipped goods, or polar bears; examples of moving regions are forest fires, deforestation of the Amazon rain forest, oil spills in the sea, armies, epidemic diseases, hurricanes, and so forth.

There are two flavors of such databases. The first represents information about a set of currently moving objects. Basically, one is interested in efficiently maintaining their location information and asking queries about the current and expected near future positions and relationships between objects. In this case, no information about histories of movement is kept. This is sometimes also called a *tracking database*.

The second represents complete histories of movements. The goal in the design of query languages for moving objects is to be able to ask any kind of questions about such movements, perform analyses, and derive information, in a way as simple and elegant as possible. The underlying system must support efficient execution of such analyses. This view is associated with the term *moving objects database*, sometimes also called *trajectory database*.

Historical Background

The field of moving objects databases came into being in the late 1990s of the last century mainly by two parallel developments. First, a model was developed [15, 18] that allows one to keep track in a database of a set of time dependent locations, e.g., to represent vehicles. The authors observed that one should store in a database not the locations directly, which would require high update rates, but rather a motion vector, representing an object's expected position over time. An update to the database is needed only when the deviation between the expected position and the real position exceeds some threshold. At the same time this concept introduces an inherent, but bounded uncertainty about an object's real location. The model was formalized introducing the concept of a *dynamic attribute*. This is an attribute of a normal data type which changes implicitly over time. This implies that results of queries over such attributes also change implicitly over time. A related query language FTL (future temporal logic) was introduced that allows one to specify time dependent relationships between expected positions of moving objects.

Second, the European project CHORO-CHRONOS set out to integrate concepts from spatial and temporal databases. In this case, one represents in a database time dependent geometries of various kinds such as points, lines, or regions. Earlier work on spatio-temporal databases had generally admitted only discrete changes. This restriction was dropped and continuously changing geometries were considered. A model was developed based on the idea of *spatio-temporal data types* to represent histories of continuously changing geometries [3, 6]. The model offers data types such as *moving point* or *moving region* together with a comprehensive set of operations. For example, there are operations to compute the projection of a moving point into the plane, yielding a *line* value, or to compute the distance between a moving point and a moving region, returning a time dependent real number, or *moving real*, for short. Such data types can be embedded into a DBMS data model as attribute types and can be implemented as an extension package.

A second approach to data modeling for moving object histories was pursued in CHOROCHRONOS. Here, the constraint model was applied to the representation of moving objects [14], and a prototype called Dedale was implemented. Constraint databases can represent geometries in n-dimensional spaces; since moving objects exist in 3D (2D + time) or 4D (3D + time) spaces, they can be handled by this approach. Several researchers outside CHOROCHRONOS also contributed to the development of constraint-based models for moving objects. See also "▶ Constraint Query Languages."

Scientific Fundamentals

Modeling and Querying Current Movement (Tracking)

Consider first moving objects databases for current and near future movement, or tracking databases. Sets of moving entities might be taxicabs in a city, trucks of a logistics company, or military vehicles in a military application. Possible queries might be:

- Retrieve the three free cabs closest to Cottle Road 52 (a passenger request position).
- Which trucks are within 10 kms of truck T70 (which needs assistance)?
- Retrieve the friendly helicopters that will arrive in the valley within the next 15 min and then stay in the valley for at least 10 min.

Statically, the positions of a fleet of taxi-cabs, for example, could be easily represented in a relation

```
taxi-cabs(id: int, pos: point)
```

Unfortunately, this representation needs frequent updates to keep the deviation between real position and position in the database small. This is not feasible for large sets of moving objects.

The MOST (moving objects spatio-temporal) data model [15, 18], discussed in this section, stores instead of absolute positions a motion vector which represents a position as a linear function of time. This defines an expected position for a moving object. The distance between the expected position and the real position is called the *deviation*. Furthermore, a *distance threshold* is introduced and a kind of contract between a moving object and the database server managing its position is assumed. The contract requires that the moving object observes the deviation and sends an update to the server when it exceeds the threshold. Hence, the threshold establishes a bound on the *uncertainty* about an object's real position.

A fundamental new concept in the MOST model is that of a *dynamic attribute*. Each attribute of an object class is classified to be either static or dynamic. A dynamic attribute is of a standard data type (e.g., *int*, *real*) within the DBMS conceptual model, but changes its value automatically over time. This means that queries involving such attributes also have time dependent results, even if time is not mentioned in the query and no updates to the database occur.

The MOST model assumes that time advances in discrete steps, so-called clock ticks. Hence, time can be represented by integer values. For a data type to be eligible for use in a dynamic attribute, it is necessary that the type has a value 0 and an addition operation. This holds for numeric types but can be extended to types like *point*. A dynamic attribute A of type T is then internally represented by three subattributes $A.value$, $A.updatetime$, and $A.function$, where $A.value$ is of type T, $A.updatetime$ is a time value, and $A.function$ is a function $f: int \rightarrow T$ such that at time $t = 0$, $f(t) = 0$. The semantics of this representation is called the value of A at time t and defined as

$$value(A, t) = A.value +$$
$$A.function(t - A.updatetime)$$
$$\text{for } t \geq A.updatetime$$

When attribute A is mentioned in a query, its dynamic value $value(A, t)$ is meant.

With dynamic attributes, for each clock tick one obtains a new state of the database, even without explicit updates. Such a sequence of states is called a database history. With each explicit update, all subsequent states change so that one obtains a new database history. One can now define different types of queries:

- An *instantaneous query* issued at a time t_0 is evaluated once on the database history starting at time t_0.
- A *continuous query* issued at time t_0 is (conceptually) re-evaluated for each clock tick. Hence, it is evaluated once on the database history starting at time t_0, then on the history starting at t_1, then on ... t_2, and so forth.

Of course, reevaluating a continuous query on each clock tick is not feasible. Instead, the evaluation algorithm for such queries is executed only once and returns a time dependent result, in the form of a set of tuples with associated time stamps. A re-evaluation is only necessary when explicit updates occur.

The query language associated with the MOST model is called FTL (future temporal logic). Here are a few example queries formulated in FTL.

1. Which trucks are within 10 kms of truck T70?

```
RETRIEVE t
FROM trucks t, trucks s
WHERE s.id = 'T70' ∧ dist(s, t) <= 10
```

Here nothing special happens, yet, the result is time dependent.

2. Retrieve the helicopters that will arrive in the valley within the next 15 min and then stay in the valley for at least 10 min.

```
RETRIEVE h
FROM helicopters h
WHERE eventually_within_15 (inside(h,
Valley) ∧ always_for_10 (inside(h,
Valley))
```

Here *Valley* is a polygon object.
The general form of a query in FTL is

```
RETRIEVE <target-list> FROM <object
classes> WHERE <FTL-formula>
```

FTL formulas may contain special time dependent constructs, in particular:

- If f and g are formulas, then f **until** g and **nexttime** f are formulas

Informally, the meaning is that for a given database state s, f **until** g holds if there exists a future state s' on the database history such that g holds in state s' and for all states from s up to s', f holds. Similarly, **nexttime** f holds in state s_{i+1} if f holds in state s_i. Based on such temporal operators, one can define bounded temporal operators like **eventually_within_c** g or **always_for_c** g as they occur in the second example query.

Modeling and Querying History of Movement

Now, consider the problem of representing complete histories of movement in a database. The scope is also extended from point objects to more complex geometrical shapes.

The idea of the approach [3] presented in the following is to introduce "▶ Spatiotemporal Data Types" that encapsulate time dependent geometries with suitable operations. For moving objects, point and region appear to be most relevant, leading to data types *moving point* and *moving region*, respectively. The *moving point* type (*mpoint* for short) can represent entities such as vehicles, people, or animals moving around whereas the *moving region* type (*mregion*) can represent hurricanes, forest fires, armies, or flocks of animals, for example. Geometrically, values of spatio-temporal data types are embedded into a 3D space (2D + time) if objects move in the 2D plane, or in a 4D space if movement in the 3D space is modeled. Hence, a moving point and a moving region can be visualized as shown in Fig. 1.

Data types may be embedded in the role of attribute types into a DBMS data model. For example, in a relational setting, there may be relations to represent the movements of air planes or storms:

```
flight (id:string, from:string,
  to:string, route:mpoint)
weather (id: string, kind: string,
  area: mregion)
```

The data types include suitable operations such as:

intersection:	*mpoint* × *mregion*	→ *mpoint*
trajectory:	*mpoint*	→ *line*
deftime:	*mpoint*	→ *periods*
length:	*line*	→ *real*

One discovers quickly that in addition to the main types of interest, *mpoint* and *mregion*, related spatial and temporal as well as other time dependent types are needed. The operations

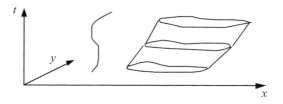

Moving Objects Databases and Tracking, Fig. 1
A moving point and a moving region

above have the following meaning: **Intersection** returns the part of a moving point whenever it lies inside a moving region, which is a moving point (*mpoint*) again. **Trajectory** projects a moving point into the plane, yielding a *line* value (a curve in the 2D space). **Deftime** returns the set of time intervals when a moving point is defined, of a data type called *periods*. **Length** returns the length of a *line* value.

Given such operations, one may formulate queries:

"Find all flights from Düsseldorf that are longer than 5000 kms."

```
select id
from flights
where from = 'DUS' and length
  (trajectory(route)) > 5000
```

"At what times was flight BA488 within the snow storm with id S16?"

```
select deftime(intersection(f.route, w.
  area))
from flights as f, weather as w
where f.id = 'BA488' and w.id = 'S16'
```

Reference [3] develops the basic idea, discusses the distinction between *abstract models* (using infinite sets, and describing e.g., a moving region as a function from time into region values) and *discrete models* (selecting a suitable finite representation, e.g., describing a moving region as a polyhedron in the 3D space), and clarifies several fundamental questions related to this approach. A system of related data types and operations for moving objects is carefully defined in [6], emphasizing genericity, closure, and consistency. The semantics of these types is defined at the abstract level.

Implementation is based on the discrete model which uses the so-called *sliced representation* as illustrated in Fig. 2. A temporal function value is

represented as a time-ordered sequence of *units* where each unit has an associated time interval and time intervals of different units are disjoint. Each unit is capable of representing a piece of the moving object by a "simple" function. Simple functions are linear functions for moving points or regions, and quadratic polynomials (or square roots of such) for moving reals, for example.

Within a database system, an extension module (data blade, cartridge, extender, etc.) can be provided offering implementations of such types and operations. The sliced representation is basically stored in an array of units. (It is a bit more complicated in case of variable size units as for a moving region, for example.) Because values of moving object types can be large and complex, the DBMS must provide suitable storage techniques for managing large objects. Large parts of this design have been implemented prototypically in the SECONDO extensible DBMS [1] and in Hermes [10]. Both systems are freely available for download, see URLs below.

Related Issues

In this short closing section, some issues related to moving objects databases are briefly discussed. Many related issues are treated in other entries of this encyclopedia; see the cross references.

Uncertainty

Locations of moving objects are most often captured using GPS devices at certain instants of time. This introduces an inherent uncertainty already for the sampled positions (due to some inaccuracy of the GPS device) and in particular for the periods of time between measurements [12]. Bounded uncertainty is also introduced due to a contract between location server and moving

Moving Objects Databases and Tracking, Fig. 2 Sliced representation of a *moving*(*real*) and a *moving*(*points*) value

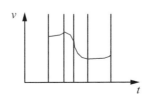
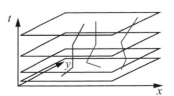

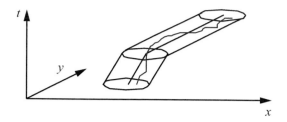

Moving Objects Databases and Tracking, Fig. 3
Geometry of an uncertain trajectory

object, as discussed above. The MOST model includes concepts to deal with this uncertainty in querying [18]. For history of movement, one can consider uncertain trajectories based on an uncertainty threshold, resulting in a shape of a kind of slanted cylinder (Fig. 3). It is only known that the real position is somewhere inside this volume. Based on this model, Trajcevski et al. [17] have defined a set of predicates between a trajectory and a region in space taking uncertainty and aggregation over time into account.

Movement in Networks
Whereas the basic case is free movement in the Euclidean plane, it is obvious that vehicles usually move on transport networks. There is a branch of research on network-constrained movement (e.g., [5, 16]); there is also work on indexing network based movements. For network-based movement, captured GPS positions have to be mapped to the transportation network; this is called "▶ Map Matching."

Spatio-Temporal Indexing
A lot of research exists on indexing movement, both for expected near future movement (see "▶ Indexing of the Current and Near-Future Positions of Moving Objects") and for history of movement (see "▶ Indexing Historical Spatiotemporal Data").

Query Processing for Continuous/Location-Based Queries
Continuous queries for moving objects have been studied in depth, for example, maintaining the result of nearest neighbor or range queries both for moving query and moving data objects (see "▶ Continuous Monitoring of Spatial Queries").

Spatiotemporal Aggregation and Selectivity Estimation
Another subfield of research in moving objects databases considers the problem of computing precisely or estimating the numbers of moving objects within certain areas in space and time – hence, of computing aggregates. For example, various index structures have been proposed to compute efficiently such aggregates. This is also related to the problem of performing selectivity estimation for spatio-temporal query processing. See "▶ Spatiotemporal Data Warehouses."

Key Applications

Databases for querying current and near future movement like the MOST model described are the foundation for "▶ Location-Based Services." Service providers can keep track of the positions of mobile users and notify them of upcoming service offers even some time ahead. For example, gas stations, hotels, shopping centers, sightseeing spots, or hospitals in case of an emergency might be interesting services for car travelers.

Several applications need to keep track of the current positions of a large collection of moving objects, for example, logistics companies, parcel delivery services, taxi fleet management, public transport systems, air traffic control. Marine mammals or other animals are traced in biological applications. Obviously, the military is also interested in keeping track of fighting units in battlefield management.

Database systems for querying history of movement are needed for more complex analyses of recorded movements. For example, in air traffic control one may go back in time to any particular instant or period to analyze dangerous situations or even accidents. Logistics companies may analyze the paths taken by their delivery vehicles to determine whether optimizations are possible. Public transport systems in a city may be analyzed to understand reachability of any place in the city at different periods of the day. Movements of animals may be analyzed in biological studies. Historical modeling may represent movements of people or tribes and actually animate and query such movements over the centuries.

The book [13] presents a set of chapters on applications such as car traffic monitoring, maritime monitoring, air traffic analysis, animal movement, and person monitoring with Bluetooth tracking.

Future Directions

Some research trends that have already been pursued for several years are data mining on large sets of trajectories, closely related to the issue of privacy for human movements [4]. Visual analytics are another approach to make sense of large trajectory data sets (e.g., [13, Chap. 8]). Another trend is to move from geometric to "semantic" trajectories, that is, to describe movement in terms of meaningful locations or properties [9]. This can also be viewed as studying geometric trajectories together with one or multiple annotations. Yet another direction is to study scalable trajectory data management using distributed query processing on large clusters [8].

Text books covering the topics presented in this entry in more detail are [7, 11, 13].

Experimental Results

Running times for queries of the BerlinMOD benchmark (see below), evaluated in the SECONDO system, can be found in [2].

Data Sets

A collection of links to data sets with real spatio-temporal data can be found at

http://chorochronos.datastories.org/

Within the GeoLife project of Microsoft Asia, GPS trajectories of 182 people have been col-

lected in a period of more than 3 years. The dataset contains about 18,000 trajectories covering a total distance of about 1.2 million kms and more than 48,000 h of traveling. This data set is available at

http://research.microsoft.com/en-us/projects/
    geolife/

A scalable benchmark data set is available, the so-called BerlinMOD benchmark [2]. It is based on a simulation of the movements of 2000 people's vehicles in the city of Berlin, observed over 1 month (at scale factor 1). The benchmark contains a number of test queries. Test data are generated by the SECONDO system. The benchmark can be found and the relevant resources downloaded at

http://dna.fernuni-hagen.de/Secondo.html/Berlin
    MOD/BerlinMOD.html

See also "► Real and Synthetic Test Datasets"; this entry should include links to further test data generators.

URL to Code

The two prototypical moving objects database systems SECONDO and Hermes (for histories of movement, or trajectories) are freely available for download at

http://dna.fernuni-hagen.de/Secondo.html
http://infolab.cs.unipi.gr/hermes

Hermes is in fact available in two versions sitting on top of Oracle and Postgres, respectively.

Cross-References

► Constraint Query Languages
► Continuous Monitoring of Spatial Queries
► Indexing Historical Spatiotemporal Data
► Indexing of the Current and Near-Future Positions of Moving Objects
► Location-Based Services
► Map Matching
► Real and Synthetic Test Datasets

▶ Spatiotemporal Data Mining
▶ Spatiotemporal Data Types
▶ Spatiotemporal Data Warehouses
▶ Spatial and Spatiotemporal Data Models and Languages

Recommended Reading

1. Almeida VT, Güting RH, Behr T. Querying moving objects in SECONDO. In: Proceedings of the International Conference Mobile Data Management; 2006. p. 47–51.
2. Düntgen C, Behr T, Güting RH. BerlinMOD: a benchmark for moving object databases. VLDB J. 2009;18(6):1335–68.
3. Erwig M, Güting RH, Schneider M, Vazirgiannis M. Spatio-temporal data types: an approach to modeling and querying moving objects in databases. GeoInformatica. 1999;3(3):265–91.
4. Giannotti F, Pedreschi D, editors. Mobility, data mining and privacy. Berlin: Springer; 2008.
5. Güting RH, de Almeida VT, Ding Z. Modeling and querying moving objects in networks. VLDB J. 2006;15(2):165–90.
6. Güting RH, Böhlen MH, Erwig M, Jensen CS, Lorentzos NA, Schneider M, Vazirgiannis M. A foundation for representing and querying moving objects in databases. ACM Trans Database Syst. 2000;25(1):1–42.
7. Güting RH, Schneider M. Moving objects databases. Amsterdam: Morgan Kaufmann Publishers; 2005.
8. Lu J, Güting RH. Parallel SECONDO: a practical system for large-scale processing of moving objects. Proceedings of the 30th International Conference on Data Engineering; 2014. p. 1190–3.
9. Parent C, Spaccapietra S, Renso C, Andrienko GL, Andrienko NV, Bogorny V, Damiani ML, Gkoulalas-Divanis A, de Macêdo JAF, Pelekis N, Theodoridis Y, Yan Z. Semantic trajectories modeling and analysis. ACM Comput Surv. 2013;45(4):42.
10. Pelekis N, Frentzos E, Giatrakos N, Theodoridis Y. HERMES: a trajectory DB engine for mobility-centric applications. Int J Knowl-Based Organ. 2015;5(2):19–41.
11. Pelekis N, Theodoridis Y. Mobility data management and exploration. New York: Springer; 2014.
12. Pfoser D, Jensen CS. Capturing the uncertainty of moving-object representations. In: Proceedings of the 6th International Symposium on Spatial Databases; 1999. p. 111–31.
13. Renso C, Spaccapietra S, Zimányi E. Mobility data: modeling, management, and understanding. Cambridge, UK: Cambridge University Press; 2013.
14. Rigaux P, Scholl M, Segoufin L, Grumbach S. Building a constraint-based spatial database system: model, languages, and implementation. Inf Syst. 2003;28(6):563–95.
15. Sistla AP, Wolfson O, Chamberlain S, Dao S. Modeling and querying moving objects. In: Proceedings of the 13th International Conference on Data Engineering; 1997. p. 422–32.
16. Speicys L, Jensen CS, Kligys A. Computational data modeling for network-constrained moving objects. In: Proceedings of the 11th ACM Symposium on Advances in Geographic Information Systems; 2003. p. 118–25.
17. Trajcevski G, Wolfson O, Hinrichs K, Chamberlain S. Managing uncertainty in moving objects databases. ACM Trans Database Syst. 2004;29(3):463–507.
18. Wolfson O, Chamberlain S, Dao S, Jiang L, Mendez G. Cost and imprecision in modeling the position of moving objects. In: Proceedings of the 14th International Conference on Data Engineering; 1998. p. 588–96.

Multi-data Center Replication Protocols

Marcos K. Aguilera
VMware Research, Palo Alto, CA, USA

Synonyms

Geo-distributed replication protocols; Geo-replication protocols

Definition

Multi-data center replication protocols serve to coordinate access to data that is replicated across data centers. The data centers are often separated by large distances, causing significant delays in communication and occasional network outages. The protocols ensure that the replicas remain identical or sufficiently close, so that data accesses satisfy a consistency guarantee suited to a particular application (Consistency Properties).

Historical Background

Multi-data center replication protocols originate from replication protocols in database systems, distributed file systems, and mobile systems (► Data Replication; Replication for High Availability). The desire to replicate data across data centers has increased in the past decade, as cloud-based Web applications have grown considerably. Applications such as Web mail, e-commerce, Web search, and social networks now include hundreds of millions of users or more globally. Many of these applications are deployed across several data centers around the world, to provide closer access to users, to protect user data against disasters, and to scale beyond the capacity of a single data center. Most systems have used asynchronous replication protocols, but some recent systems have started to use synchronous replication enabled by the faster dedicated network connections between data centers.

Scientific Fundamentals

The circumference of the Earth is around 40,000 km. To cover this distance, it takes light 133 ms, which is the theoretical minimum network round-trip time between two antipodal locations. In practice, network latencies are much larger than the theoretical minimum, due to network switching latencies and the lack of direct links between locations. With data centers separated by thousands of km or more, actual network round-trip latencies are in the tens to hundreds of milliseconds. In addition, spikes of traffic between data centers may exceed the available network bandwidth, resulting in queuing delays and momentarily increasing network latencies to seconds or more. These latencies are orders of magnitude higher than latencies within a data center, which are in the tens to hundreds of microseconds. These large latencies are an important consideration for multi-data center replication protocols, which for efficiency must avoid too many rounds of communication across data centers.

A second important consideration for replication protocols is when to acknowledge completion of updates to clients. Because remote communication is slow, some protocols acknowledge an update before remote replicas have seen it;

other protocols wait for the remote replicas. The former is faster but creates complications when replicas fail or clients issue conflicting updates, as discussed below.

A third important consideration is how to deal with *conflicting updates* at different replicas. Two updates are conflicting if they affect the same data item in incompatible ways – say, both updates overwrite the data item. Conflicting updates can be problematic if they propagate concurrently; they could cause different replicas to receive and apply the updates in different orders, causing the presumed replicas to diverge – an undesirable outcome.

A fourth important consideration is who takes on the role of the *coordinator* of the protocol. The coordinator is the process that actively drives the protocol for a given client request, while other processes typically have a passive role of responding to messages. In some protocols, the coordinator must be one of the replicas; the client first submits its request to one of the replicas, and that replica becomes the coordinator. In other protocols, the client itself can be the coordinator, which can save communication and avoid bottlenecks at the replicas. Unless otherwise stated, in the protocols below the client can be the coordinator.

To discuss the replication protocols, we classify them alongside two dimensions. The first is whether updates are applied synchronously or asynchronously. With *synchronous replication*, the protocol waits for updates to be propagated to remote replicas before acknowledging to clients. Because of high network latencies, the waiting takes significant time. With *asynchronous replication*, the updates are propagated in the background, that is, the protocol acknowledges the completion of the update, while the update may still be propagating to remote replicas. This is faster than synchronous replication but creates complications.

The second dimension is what type of service is offered by the storage system. Replication protocols are closely intertwined with the underlying storage service; for example, when clients update data, the replication machinery must propagate the update in accordance with the semantics of the storage service; similarly, when clients read data, the storage service must take into consideration where the data has been replicated. In fact, replication and the storage service are typically fused into a single monolithic piece, making it difficult to separate the two. For these reasons, replication protocols often depend on the storage service.

Storage services fall into three broad categories: *read-write*, *state machine*, and *transaction*. The *read-write service* supports two operations: one to read and one to write data items. The replication protocol must implement each of these operations. Examples of this storage service include key-value storage systems and block storage systems.

The *state machine* service supports *read-modify-write* operations – operations that modify a data item relative to its current state (increment value, conditionally update value, etc.). To handle such operations, one can model each data item as a state machine and each operation as a function that maps a state to a new state and some result. The available functions depend on the storage service. In some systems, the functions are fixed; in others, they are extensible. Examples of this storage service include file systems, objects storage systems, and table storage systems.

The *transactional service* allows applications to start, execute, and commit transactions over multiple data items. This service is broader than the state machine service because operations of the latter involve one data item at a time. The replication protocol specifies how to execute and commit transactions. Examples of this storage service include transactional key-value storage systems and database systems.

The combination of two choices for one dimension and three for the other yields six possibilities, shown as quadrants in Table 1. The quadrants are populated with various replication protocols, described below.

Synchronous Replication

Synchronous replication protocols are simpler than asynchronous ones, but they are also slower because update operations wait for at least one round-trip across data centers. Some synchronous protocols perform reads at any data center, so only updates are delayed; others delay both reads and updates. Next is a description of the protocols in the *synchronous replication* row of Table 1.

Synchronous replication with read-write service. A replication protocol specifies how each operation gets executed. The read-write service has only two operations, read and write. The simplest replication protocol is *write-all*, where a client executes writes at all replicas (waiting for them to complete) and reads at one replica – say, the closest one. With concurrent writes, replicas must apply them in the same order. To do that, one could designate a replica to order writes. This can be inefficient as the designated replica becomes a bottleneck. A better scheme assigns unique timestamps to values when they are written; replicas then order the writes by timestamp.

Multi-data Center Replication Protocols, Table 1 Classification of multi-data center replication protocols

	Read-write service	State machine service	Transactional service
Synchronous replication	Write-all	State machine approach	2PL+2PC
	Write-quorum		Transaction total ordering
			Consensus+2PL+2PC
Asynchronous replication	Primary copy	Primary copy	Primary copy
	Last timestamp wins	Commutative ops	Preferred copy
	History merge	History merge	Message futures
		History reorder	Commutative ops
			History merge
			History reorder

Write-all has a problem: write performance is limited by the slowest replica. Furthermore, if that replica is unresponsive, writes block until it recovers or is removed from the system. The *write-quorum* protocol addresses these problems. A client submits writes to all replicas but waits only for a majority to complete. If a replica is slow, it does not affect the performance. Differently from write-all, the client submits a read to all replicas; it waits for a majority to complete, and chooses the value with the highest timestamp. Because two majorities of replicas always intersect at one replica at least, the protocol ensures that reads see the latest written value.

Synchronous replication with state machine service. Under the *state machine approach* [17], each replica is an identical deterministic state machine that starts from the same state, and replicas agree on the same sequence of inputs. Since the inputs are identical, replicas go through the same state sequence. The inputs are the storage operations that clients want to execute. To agree on the input sequence, replicas use instances of a *consensus protocol*, such as Paxos [10], Chandra-Toueg's $\diamond S$ protocol [4], Viewstamped Replication [15], and the PBFT protocol [3].

Synchronous replication with transactional service. To provide a replicated transactional service, a simple approach is to treat the replicas of an item as individual items that are updated together in a transaction. Specifically, transactions write to all replicas and read from any replica. Transactions ensure that replicas remain identical. Transactions are implemented using known techniques (*distributed two-phase commit*, *two-phase locking*, etc.).

Another approach is *transaction total ordering*. Each transaction executes entirely at a single replica. For read-only transactions, no further action is needed. Update transactions are committed by a mechanism that totally orders all transactions. This mechanism could be *atomic broadcast* [8], as in the *deferred update replication* approach. Or the mechanism could be a log implemented with consensus, as in Megatore [2] and in [16]. Either way, all replicas learn the transaction ordering and update their local state accordingly. In one scheme, the system orders transactions before transaction validation; then, each replica validates the ordered transactions using some deterministic rule so that all replicas agree on whether each transaction aborts or commits. In another scheme, the system validates before ordering; specifically, the system validates a transaction for a specific position in the ordering, and, if validation succeeds, the system tries to order the transaction in that position.

A variant of transaction total ordering permits the storage state to be partitioned across servers. The system orders all transactions before they execute. Once ordered, transactions execute at each data center using two-phase locking. To ensure replicas converge, transaction execution must be deterministic. This scheme is used in Calvin [20] in synchronous replication mode.

Another replication scheme combines consensus, two-phase locking, and two-phase commit; it also permits partitioning the storage state across servers. Consensus replicates each partition across data centers using the state machine approach, and two-phase locking and two-phase commit provide transactions across these replicated partitions. More precisely, consensus provides state machines whose inputs are the basic actions in two-phase locking and two-phase commit (acquire lock, release lock, prepare, commit). There is a state machine for each storage partition, and the state machine replicas are placed in different data centers. A coordinator then drives two-phase locking and two-phase commit to execute and commit the transaction, by issuing inputs to the state machines. This scheme is used in Spanner [6] and MDCC [9]. Spanner uses Paxos [10] for consensus. MDCC is based on similar principles but reduces data center round-trips through careful design and the use of faster consensus protocols.

Asynchronous Replication

With asynchronous replication, updates are acknowledged quickly to clients, and they are propagated to remote replicas in the background. This feature, however, brings two issues. First, if a disaster strikes a data center, acknowledged updates

M

could be lost, as they have not been replicated yet. This problem is mitigated by bounding the time to propagate an update (e.g., 30 s), so that only recent updates are lost. To bound the delay, one throttles the updates when too many updates queue up for propagation.

Another issue with asynchronous replication is conflicting updates: an update may be acknowledged to a client, while a conflicting update is concurrently acknowledged at another data center. This situation leads to lost updates or undesirable semantics. Each protocol deals with this problem in a different manner.

Next is a description of these protocols, in the *asynchronous replication* row of Table 1.

Asynchronous replication with read-write service. One way to address conflicting updates is to serialize updates at one replica, called the *primary copy* or *primary*. With a read-write service, updates are write operations. The primary serializes writes to the same data item, so replicas apply the writes in the same order. Reads are performed at any replica. At the primary, a read returns the most recent value; at other replicas, the read may return stale values, which is acceptable to some applications. This protocol is very simple. However, it has two drawbacks. First, clients must wait for the primary to perform writes, and the wait is significant if the primary is in a separate data center. Second, the primary becomes a write bottleneck.

Instead of using a primary copy, another approach is *last timestamp wins*. The system assigns a timestamp to each write, and replicas apply writes in increasing timestamp order. If a replica receives a write with a timestamp smaller than the timestamp of its data item, the replica ignores the write. This idea is known as the *Thomas write rule*. Reads are executed at any replica. This protocol is simple; however, writes may be lost, as a write with a higher timestamp obliterates a write with a smaller timestamp.

This problem can be addressed with the *history merge* protocol. The protocol tracks the causal relation of operations and applies the writes to the replicas in causal order; if two writes are not causally related, the system merges their effect. More precisely, an operation op1 *causally*

precedes op2 if either (a) op1 executes before op2 at the same client, (b) op1 writes a value that op2 reads (For simplicity, this definition assumes that a given value is written at most once, but the definition can be extended to the general case.), or (c) for some op3, op1 precedes op3, and op3 precedes op2. To track this precedence relation, the system can use *vector clocks*, as in Dynamo [7]; or it can attach to each operation a list of recent operations that precede it, as in COPS [12]. When writes are not causally related, replicas combine them using an application-specific merge function that takes the current value of the data item and merges it with a new value. Merge functions must ensure that replicas converge. A simple merge function is to assign timestamps to written values and then choose the value with highest timestamp. This merge function causes lost updates. A better merge function should produce values that combine the writes according to application semantics. The benefit of this scheme is that updates need not wait for remote replicas, and concurrent updates are not lost given a suitable merge function. However, the drawback is that applications must provide reasonable merge functions, which can be hard or impossible. Because history merge applies writes in causal order, it provides a property called *causal consistency*.

Asynchronous replication with state machine service. With a state machine service, one can use *primary copy* or *history merge*. They work as in the read-write service, except that, instead of writes, updates are functions that modify a data item. For *history merge*, the merge function must handle such updates, which complicates it further. Primary copy is used in both PNUTS [5] and Azure Table storage. History merge is used in Eiger [13].

The state machine service allows the technique of *commutative operations* [11, 21]. Applications carefully choose update functions that commute. Then, there is no need for a primary copy or a merge function: each replica can simply apply the updates as they arrive. Because operations commute, the replicas converge. The advantage of this scheme is that updates are fast and concurrent updates are not lost. The drawback is that applications must use commutative updates, which may be infeasible.

Another technique is *history reorder*, proposed in Bayou [19]. Basically, replicas revise the history of updates so that, after revision, replicas apply the updates in the same order. More precisely, each update operation is assigned a *tentative ordering* when it is received from the client by a replica. The update is quickly acknowledged to the client, without waiting for other replicas. After the update is propagated to all replicas, it is assigned a final *stable ordering*, and the replicas reevaluate the effect of the update if its ordering changed. Doing this requires undoing previous updates using a log at the replicas. The advantage of this scheme is that updates are not lost and they are fast, but there is a caveat: the effect of updates may subsequently change as the update is reordered. If the application cannot tolerate this change, it must wait until the stable ordering, which is slow.

Asynchronous replication with transactional service. The techniques for state machines also apply to the transactional service. With primary copy, update transactions execute at the primary; implicit is the assumption that data items have the same primary, so that update transactions can execute at one server. A simple generalization is to have a *primary data center*; data items can have different primaries, but all primaries are in one data center. Update transactions, then, are distributed transactions in the primary data center. Read-only transactions execute at any data center, but transactions outside the primary may observe stale data. To obtain transaction isolation, known techniques are used. For example, if replicas keep multiple versions, transactions can use multi-version concurrency control to read from snapshots. Primary data center is used in Calvin [20] in asynchronous replication mode.

With *preferred copy*, each data item has a *preferred replica* or, more generally, a *preferred data center*. Unlike with primary copy, the preferred replica or data center can vary per item, and data items can be updated at any data center. Updates are faster in the preferred data center as the commit protocol contacts the preferred data center of each item updated in the transaction.

The advantage of this technique is that updates are fast for data in its preferred data center. The drawback is that the commit protocol provides an isolation property weaker than serializability, called *parallel snapshot isolation*. This scheme is used in Walter [18].

With *message futures* [14], replicas maintain a log with pending and committed local transactions and periodically exchange their logs. A replica can commit its local transaction tx when (1) it knows that all replicas are aware of its log up to a certain log entry, namely, the latest entry when the replica last sent its log prior to tx's start, and (2) items read by tx have not changed. This scheme permits optimizing the system for one chosen replica: if a replica sends its log at a much lower frequency than all others, conditions (1) and (2) are often met immediately for new transactions at that replica; these transactions commit without remote communication. However, the optimization comes at the expense of the other replicas, which must wait for the infrequent transmissions of the chosen replica. This protocol has the benefit that it guarantees serializability; however, only transactions of a chosen data center commit quickly.

The technique of *commutative operations* works with a transactional service. Transactions must be composed of commutative operations on abstract data types [21], instead of reads and writes, so that the entire transaction is commutative.

History merge requires a merge function for transactions. While this can be efficient, such merge functions are even more complicated than for read-write or state machines.

With *history reorder*, instead of reordering operations, entire transactions are reordered. The caveat is that, when a transaction is reordered, the data it reads may change and the application must be designed to handle these changes.

Key Applications

Multi-data center replication is used for four reasons. First, it allows applications to grow beyond the capacity of a data center. Second, it protects the data of applications against disasters that destroy a data center. Third, it improves data availability despite outages of the network connecting the data centers. Fourth, it allows applications to run at data centers closer to users, providing faster responses. The replication techniques described above are used in production systems in large data centers running global applications at Amazon, Facebook, Google, Microsoft, and others.

Cross-References

▶ Data Replication

Recommended Reading

1. Attiya H, Bar-Noy A, Dolev D. Sharing memory robustly in message-passing systems. J ACM. 1995;42(1):124–42.
2. Baker J, et al. Megastore: providing scalable, highly available storage for interactive services. In: Proceedings of the 5th Biennial Conference on Innovative Data Systems Research; 2011. p. 223–34.
3. Castro M, Liskov B. Practical Byzantine fault tolerance and proactive recovery. ACM Trans Comput Syst. 2002;20(4):398–461.
4. Chandra TD, Toueg S. Unreliable failure detectors for reliable distributed systems. J ACM. 1996;43(2): 225–67.
5. Cooper BF, et al. PNUTS: Yahoo!'s hosted data serving platform. Proc VLDB Endowment. 2008;1(2):1277–88.
6. Corbett JC, et al. Spanner: Google's globally-distributed database. In: Proceedings of the 10th USENIX Symposium on Operating System Design and Implementation; 2012. p. 251–64.
7. DeCandia G, et al. Dynamo: Amazon's highly available key-value store. In: Proceedings of the 21st ACM Symposium on Operating System Principles; 2007. p. 205–20.
8. Hadzilacos V, Toueg S. A modular approach to fault-tolerant broadcasts and related problems. Technical report 94-1425. Department of Computer Science, Cornell University, Ithaca, NY; 1994.
9. Kraska T, Pang G, Franklin M, Madden S, Fekete A. MDCC: multi-data center consistency. In: Proceedings of the 8th ACM SIGOPS/EuroSys European Conference on Computer Systems; 2013. p. 113–26.
10. Lamport L. The part-time parliament. ACM Trans Comput Syst. 1998;16(2):133–69.
11. Letia M, Preguiça N, Shapiro M. Consistency without concurrency control in large, dynamic systems. In: Proceedings of the International Workshop on Large Scale Distributed Systems and Middleware; 2009.
12. Lloyd W, Freedman M, Kaminsky M, Andersen D. Don't settle for eventual: stronger consistency for wide-area storage with cops. In: Proceedings of the ACM Symposium on Operating Systems Principles; 2011. p. 401–16.
13. Lloyd W, Freedman M, Kaminsky M, Andersen D. Stronger semantics for low-latency geo-replicated storage. In: Proceedings of the 10th USENIX Symposium on Networked Systems Design & Implementation; 2013. p. 313–28.
14. Nawab F, Agrawal D, Abbadi AE. Message futures: fast commitment of transactions in multi-datacenter environments. In: Proceedings of the 6th Biennial Conference on Innovative Data Systems Research; 2013.
15. Oki BM, Liskov BH. Viewstamped replication: a new primary copy method to support highly-available distributed systems. In: Proceedings of the ACM Symposium on Principles of Distributed Computing; 1988. p. 8–17.
16. Patterson S, Elmore AJ, Nawab F, Agrawal D, Abbadi AE. Serializability, not serial: concurrency control and availability in multi-datacenter datastores. Proc VLDB Endowment. 2012;5(11):1459–70.
17. Schneider F. Implementing fault-tolerant services using the state machine approach: a tutorial. ACM Comput Surv. 1990;22(4):299–319.
18. Sovran Y, Power R, Aguilera MK, Li J. Transactional storage for geo-replicated systems. In: Proceedings of the ACM Symposium on Operating Systems Principles; 2011. p. 385–400.
19. Terry DB, et al. Managing update conflicts in Bayou, a weakly connected replicated storage system. In: Proceedings of the ACM Symposium on Operating Systems Principles; 1995. p. 172–83.
20. Thomson A, Diamond T, Weng S-C, Ren K, Shao P, Abadi DJ. Calvin: fast distributed transactions for partitioned database systems. In: Proceedings of the International Conference on Management of Data; 2012. p. 1–2.
21. Weihl W. Commutativity-based concurrency control for abstract data types. IEEE Trans Comput. 1988;37(12).

Multi-datacenter Consistency Properties

Peter Bailis
Department of Computer Science, Stanford University, Palo Alto, CA, USA

Synonyms

Consistency; Geo-replication

Definition

Multi-datacenter consistency refers to the integrity of application data that is stored in multiple, possibly geographically distant locations. Due to large communication delays between sites, traditional protocols for enforcing integrity despite concurrent operation on separate copies of data may be prohibitively expensive. As a result, multi-datacenter storage systems increasingly implement a range of techniques that avoid coordination. The most basic strategy foregoes all coordination and only provides users with the guarantee that eventually all replicas agree (i.e., *eventual consistency*). However, many systems avoid coordination while still preserving various integrity criteria. Generally, the more application semantics that are made available to a storage engine, the more that coordination can be safely avoided without compromising application integrity. A growing set of abstract data type implementations (e.g., counters) designed to avoid data loss despite concurrent updates are reusable but require care in composition. In contrast, techniques that assume greater access to program text and invariants can enforce consistency with a minimum of coordination but typically require either use of restricted languages or increased programmer overhead.

Historical Background

While application architects have long made use of off-site disaster recovery (e.g., off-site tape backups, passive secondary datacenters), the rise of large-scale Internet services and cloud computing in the 1990s and 2000s spurred the widespread adoption of elastic, scale-out computing infrastructure. The availability of cheap, commodity, and multi-datacenter computing lowered the bar for an increasing number of developers to adopt distributed database designs and consider geo-replication as a feature in their applications. In many cases demanding "always on" availability and low latency execution at global scale, multi-datacenter deployment became an inevitability. Thus, while distribution and replication are perennial concerns in data management, the scale and popularity of distribution at large and geo-replicated scale currently enjoys unprecedented popularity.

Scientific Fundamentals

Coordination and Serializability

The maintenance of application data is a core concern in data management and is at the heart of transaction processing and concurrency control. Specifically, during concurrent access, database systems typically ensure that application *consistency* or integrity criteria (e.g., usernames are unique) are upheld on behalf of applications [19]. The classic mechanism for enforcing consistency is to bundle ordered groups of operations into *transactions* and use *serializable isolation* to ensure that the result of executing the transactions is equivalent to some serial execution of the transactions [10]. Serializable isolation is remarkably convenient for programmers as it obviates the need to reason about concurrency: insofar as each transaction leaves the database in a consistent state (i.e., integrity criteria hold), their serializable composition will leave the database in a consistent state.

However, since the earliest days of transaction processing, system implementers realized that serializability incurs costly coordination penalties across concurrently executing processes [18]. That is, for arbitrary read-write transactions over arbitrary data items, concurrent transactions cannot make progress independently while guaranteeing serializability [14]. On a single node, these *coordination* penalties can be severe, resulting in order-of-magnitude slowdowns compared to coordination-free execution. However, in the distributed environment, they are especially severe.

In particular, communication networks powering distributed systems incur latencies between a few microseconds (via cutting-edge networking hardware such as RDMA and InfiniBand) and hundreds of milliseconds (via cross-continental interconnects). Given that modern networks can-

M

not communicate faster than the speed of light, this places an upper bound on transaction performance: on Earth, a message sent across the equator requires around 67 ms delay. Thus, a serializable database simultaneously executing transactions on two replicas on opposite sides of the Earth can achieve throughput no greater than 15 transactions per second. In practice, networks are considerably slower than the speed of light, incurring even greater coordination penalties [5].

In summary, there is a fundamental tension between classical approaches to maintaining application consistency and desires for coordination-free execution in a wide-area distributed environment. Serializable transactions are at odds with the availability, latency, and scalability benefits offered by a coordination-free system architecture in which any server can respond. While these challenges are endemic to all distributed systems, this trade-off is especially important for database systems operating from multiple datacenters.

Beyond Serializability

Given demands for "always on" availability, latency, and scalability, database system architects developed a range of designs that sought to avoid coordination by providing non-serializable semantics [16]. Serializability takes a myopic approach to maintaining application integrity; in contrast, these systems provide a range of alternative, application-specific guarantees ranging from few, if any, guarantees to enforcement of arbitrary application integrity criteria. As Kung and Papadimitriou demonstrate [22], without knowledge of application semantics, non-serializable isolation can compromise application integrity. As a result, generally, the more semantics that are made available to a storage engine, the more that coordination can be safely avoided without compromising application integrity. Some semantics fundamentally require coordination for enforcement, while others do not. We survey several popular approaches below.

i.) Eventual Consistency

One of the most common and most basic strategies for ensuring coordination-free execution is

to only guarantee eventual consistency or the property that, the absence of new updates to the database, all reads will eventually return the same value for each item [7]. This allows servers to perform operations entirely locally and asynchronously propagate writes among replicas as they are received. While eventual consistency guarantees *liveness* (something good eventually happens: all reads return the same value for each item), it lacks any form of *safety* (nothing bad happens: any value can be returned in the meantime and any value can eventually be chosen) [1]. As a result, eventual consistency is remarkably difficult to program.

In practice, eventually consistent systems often return nontrivial results: for example, in Facebook's eventually consistent TAO data store, 99.9996% of reads returned the last written value [25]; other studies of popular services have demonstrated similar behavior [8, 30]. However, given millions to billions of requests per second, this small number of stale reads can still cause problems, especially for "hot" items.

In summary, eventual consistency is useful to system builders but can be remarkably challenging to reason about. Eventual consistency obviates reasoning about (or writing code to handle) many corner-case scenarios, including node and network failures, communication delays, and service unavailability: an eventually consistent database can indefinitely return any value without violating its semantics. Many first-generation Internet-scale databases, including Amazon's Dynamo and S3 databases and Google's BigTable, provided this property. While eventual consistency dates to at least the 1970s [21], these applications were some of the first at scale.

ii.) Leverage Full Application Semantics

At the opposite extreme, several approaches to maintaining distributed consistency assume knowledge of the entire application semantics, from program text to application invariants. This allows an inherently precise notion of coordination requirements but is also the most heavyweight among non-serializable transaction mechanisms.

As an example, invariant confluence [5] is a necessary and sufficient criterion for maintaining invariants during concurrent execution of transactions on multiple replicas; it states that the set of invariant-preserving states reachable by executing program transactions must be closed (i.e., must also preserve invariants) under the application of a user- or database-specified "merge" operator that reconciles these states. Under set-based merge, a set of transactions that assigns user IDs is not invariant confluent with respect to the invariant that user IDs are unique (requires coordination to prevent collisions), but the same set of transactions is invariant confluent with respect to the invariant that user IDs are nonnegative (as merging any two databases with nonnegative IDs will not produce negative IDs).

Once it is established that coordination can be safely avoided, the database must also determine *how* to avoid coordination. Traditionally, this required new protocol design. However, a range of automated techniques such as the Homeostasis Protocol [28] and Sieve [24] automate this process using program analysis, in effect, performing query optimization for coordination. In this model, perhaps surprisingly, many program invariants do not require coordination; for example, in a recent study of 67 popular open-source applications, only approximately 13% of invariants were not invariant confluent [4]. As a result, there is a wide opportunity for coordination avoidance that is missed by serializable protocols.

An additional caveat with the above techniques is that they typically assume that, in the absence of an invariant, concurrent execution is safe. That is, if the program supplies no invariants about a data item, the item can assume any value under concurrent execution. As a result, it is imperative that invariants/integrity criteria are provided *in full* to the database engine and/or analysis routine, lest coordination-free execution inadvertently compromise some unknown consistency criteria. This is a direct corollary of Kung and Papadimitriou's seminal results.

Currently, these techniques are not in widespread use in practice. However, their basic principles can be applied to the design of coordination-avoiding systems and applications, and they provide a precise guideline as to when coordination can be avoided and when applications must pay the price (or alternatively adjust their semantics to a cheaper model).

iii.) Exploit More Limited Application Information

A compromise with the above techniques is to seek and exploit more limited application semantics. There are several ways to accomplish this, ranging from leveraging programmer annotations to exploiting more conservative properties.

Annotations: A range of systems dating to SDD-1 [11] leverage programmer annotations that indicate which sets of transactions (and which operations within transactions) require coordination (i.e., may conflict). These techniques effectively delegate the task of determining which operations require coordination to the programmer—a pragmatic compromise given limited ability to automatically analyze program text and invariants.

Commutativity: If the return value of transactions is identical independent of order of execution, transactions can execute on different replicas in different orders [31]. If programmers "push down" their program logic into the database (e.g., via language closures), transactions can be re-executed on separate servers [23]. This is sufficient but not necessary to maintain consistency. In practice, few commercial database engines allow this behavior. It is also possible to provide similar behavior by determining a total order on transactions via background consensus; in this case, transactions may be reordered following local execution [15]. There is considerable work on exploiting commutativity both for abstract data types and in the programming language literature.

Monotonicity: If program logic is monotone, meaning facts that a program emits are never retracted later (e.g., no negation), the program outcomes are guaranteed to be deterministic (i.e., confluent) despite message reordering [2]. This is a useful property, although the deterministic outcome of the program is not guaranteed to satisfy arbitrary (non-monotonic) program invariants. Like commutativity, it is possible to perform program analysis for monotonicity (e.g., searching for instances of negation).

Limited APIs: It is possible to implement a restricted API that minimizes coordination in implementation. This is particularly promising for high-value use cases, where many transactions fit a given pattern (e.g., graph queries in TAO). This optimization process is analogous to the design of concurrent data structures: the semantics of the API dictate the degree of parallelism achievable; for example, in database engines, indices are maintained using non-serializable transactions. Systems can adopt similar techniques in the distributed, multi-datacenter setting. For example, Commutative Replicated Data Types (CRDTs) [29] provide a library of abstract data types with the guarantee that all successful operation invocations will be reflected in the final database state. The use of CRDTs is not by itself sufficient to guarantee integrity for all applications but nevertheless obviates a large number of common problems, such as lost updates to a shared counter. As another example, RAMP transactions [6] provide support for distributed, coordination-free implementation of several kinds of distributed materialized views. A suite of protocols such as the Escrow transaction method [26] and the demarcation protocol [9] amortize the overhead of coordination across multiple operations; for example, we could split the remaining balance in a bank account across five datacenters, and datacenters need only to communicate when one datacenter runs out of allocated funds. Several of these techniques are adopted in commercial and open-source databases.

Compensation: A large class of techniques in the literature adopt an alternative programming model in which users specify compensating actions (e.g., apologies [20] and sagas [17]) in the event of conflicts. In the event that two users reserve the same user ID, the system could run a prespecified compensating action to reassign one of the conflicting IDs. This is similar in spirit to the merge procedure described above, but the programming model differs slightly (i.e., merge reconciles two divergent database states, while a compensating action is explicitly triggered in the event of a conflict).

The key challenges in adopting compensating actions are twofold. First, it is difficult to enumerate all possible conflicts and even more difficult to completely specify the appropriate conflict resolution policy. This is analogous to specifying a possibly huge number of exception handlers for a commensurately large set of (possibly unknown) exception types. Second, while compensating actions *eventually* resolve all conflicts, they may alter the ultimate values of data that has already been returned to a client. For example, if a passenger prints an airplane boarding pass for a seat that is ultimately reassigned via a compensating action, the boarding pass will still exist in physical form; in cases such as this, it is impossible to revoke the effects of the operation outside the system boundaries. Compensating actions were a highly popular research area in the 1980s and 1990s but have seen limited adoption today. However, for particularly high-value use cases such as banking and retail, limited forms of compensating actions are common. For example, if concurrent ATM withdrawals result in negative balances, a bank will likely issue an overdraft fee to the customer; in the event of discrepancies between an inventory system and an order placed on an e-commerce site, the site may provide the customer with a compensatory coupon. When revenue and customer retention are at stake, application writers (in this case, business operators) are often more willing to reason about common, critical conflict modes.

WAN Serializability
A final option is to simply forego coordination-free execution and opt for serializability. To circumvent throughput restrictions due to coordination, system designers often optimize for particular workload mixes. For example, Google Spanner, which uses two-phase locking with two-phase commit, faces considerable throughput

restrictions on read-write transactions. However, the design is optimized for a read-mostly workload (99%+), and read-only transactions can proceed with locking.

In addition, this final approach is often coupled with considerable effort in data modeling. A rough design principle is to minimize the scope of coordination, and, in the serializable context, data modeling (e.g., restricting transactions to a single server) can dramatically reduce coordination costs. Systems such as G-Store [13] automate this process by automatically migrating data items accessed by concurrent transactions.

Key Characteristics

There are several distinguishing characteristics between multi-datacenter transaction engines and single-node transaction engines. Network costs dominate most all other costs; network delay is frequently more expensive than reading from disk and orders of magnitude more expensive than reading from DRAM or processor cache. As a result, the relative overhead of concurrency control procedures (e.g., calculating deltas, performing garbage collection) is limited compared to a single, main-memory database. In addition, because updates are sent over the network, there is greater leeway in terms of reconciling concurrent updates. In a shared memory environment, the memory controller will choose a single "winning" write to a given memory cell; in a distributed environment, incorporating custom conflict resolution ("merge") policies incurs limited performance overhead.

Application-level consistency criteria are often confused with distributed "consistency" models such as linearizability. Both are semantic criteria regarding operations in a distributed environment, and distributed consistency models are useful building blocks for enforcing application-level criteria. However, the key distinguishing characteristic between application-level consistency criteria and distributed consistency models is that the former pertain to application-level properties (i.e., relate to application constructs such as users, usernames, status updates), while the latter pertain to standardized abstractions

such as distributed read-write registers and consensus objects. Application consistency criteria are, by definition, not standardized as they are an inherent property of particular applications. However, many applications share correctness criteria such as referential integrity and uniqueness constraints; in these cases, databases often provide native support for enforcing them. Distributed consistency models are closer in spirit to isolation guarantees than application consistency criteria: both capture (in)admissible executions at a low level (typically using a read-write abstraction). Like isolation guarantees, distributed consistency models are often difficult to reason about; moreover, the literature contains an enormous number of models, many of which are incomparable. As a result, it is difficult to ascertain what distributed consistency (or weak isolation) guarantee an application requires for correctness. Instead, it is often much easier for users to reason about the application correctness criteria directly. Fortunately, it is possible to maintain many application criteria without coordination, while, in contrast, (and, like serializability) other traditional distributed consistency models fundamentally require coordination. This is one of the most significant practical implications of focusing on the application instead of generic abstractions such as read-write registers and is a common source of confusion in distributed protocol design. For example, Brewer's CAP Theorem [12] is often erroneously cited as a rationale for foregoing application consistency in coordination-free system design, but, in fact, as the results above demonstrate, this is not the case: CAP only applies to a narrow distributed consistency criteria (linearizability) and not *arbitrary* correctness criteria [3].

Key Applications

There are few problems in database system design with such firm and difficult trade-offs as in multi-datacenter consistency. Faced with requirements for guaranteed, "always on" availability, low latency, and scalable execution, application writers must (provably) seek alternative means of enforcing correctness criteria. This trade-off is not abstract: every major Internet service, including Amazon, Google, Facebook, Twitter, and LinkedIn, and an increasing number of cloud customers, must grapple with these issues when designing and deploying their applications. Barring radical advances in quantum communication technologies, multi-datacenter consistency will become increasingly pervasive.

Future Directions

Multi-datacenter consistency and protocol design ("Multi-Datacenter Replication Protocols" by Marcos K. Aguilera.) is a rich area of ongoing research and development. A range of techniques is seeing widespread renewed interest in practice, particularly pertaining to providing advanced functionality such as counters and indices without the overhead of traditional engines. The feature gap between coordination-free systems and traditional single-node counterparts is quickly shrinking. This has been spurred by a combination of practitioner interest and commensurate attention from the research community. While more speculative proposal, such as whole-language design, have yet to see major adoption, they are considerably more usable and advanced than approaches in prior decades, which relied much more heavily on programmer annotation.

The continued evolution of multi-datacenter consistency will be driven by the gradual absorption of common programming patterns into common APIs. For example, Facebook's adoption of a graph-based data model (and similar projects in industry) hint that graph-based semantics (e.g., multi-get, multi-put) may be a "sweet spot" for future coordination-avoiding systems. In addition, the use of large-scale distributed machine learning models has spurred a renaissance in interest in numerically oriented consistency guarantees (e.g., return the current value of this variable within 5%). While these models date to the early 2000s [32], they saw limited adoption until recently. Even now, many distributed ML

training procedures such as HogWild! [27] opt for an eventually consistent approach; the value of bounded staleness guarantees remains to be seen.

Recommended Reading

1. Alpern B, Schneider FB. Defining liveness. Inf Process Lett. 1985;21(4):181–5.
2. Ameloot TJ, Neven F, Van Den Bussche J. Relational transducers for declarative networking. J ACM. 2013;60(2):15:1–15:38.
3. Bailis P, Davidson A, Fekete A, Ghodsi A, Hellerstein JM, Stoica I. Highly available transactions: virtues and limitations. In: Proceedings of the 40th International Conference on Very Large Data Bases; 2014.
4. Bailis P, Fekete A, Franklin MJ, Ghodsi A, Hellerstein JM, Stoica I. Feral concurrency control: an empirical investigation of modern application integrity. In: Proceedings of the ACM SIGMOD International Conference on Management of Data; 2015.
5. Bailis P, Fekete A, Franklin MJ, Hellerstein JM, Ghodsi A, Stoica I. Coordination avoidance in database systems. In: Proceedings of the 41st International Conference on Very Large Data Bases; 2015.
6. Bailis P, Fekete A, Ghodsi A, Hellerstein JM, Stoica I. Scalable atomic visibility with RAMP transactions. In: Proceedings of the ACM SIGMOD International Conference on Management of Data; 2014.
7. Bailis P, Ghodsi A. Eventual consistency today: limitations, extensions, and beyond. ACM Queue. 2013;11(3):20–32.
8. Bailis P, Venkataraman S, Franklin MJ, Hellerstein JM, Stoica I. VLDB J. 2014;23(2):279–302
9. Barbará-Millá D, Garcia-Molina H. The demarcation protocol: a technique for maintaining constraints in distributed database systems. VLDB J. 1994;3(3):325–53.
10. Bernstein PA, Hadzilacos V, Goodman N. Concurrency control and recovery in database systems, vol. 370. New York: Addison-wesley; 1987.
11. Bernstein PA, Shipman DW, Rothnie JB Jr. Concurrency control in a system for distributed databases (SDD-1). ACM Trans Database Syst. 1980;5(1): 18–51
12. Brewer E. CAP twelve years later: how the "rules" have changed. Computer 2012;45(2):23–9.
13. Das S, Agrawal D, El Abbadi A. G-store: a scalable data store for transactional multi key access in the cloud. In: Proceedings of the 1st ACM Symposium on Cloud Computing; 2010.
14. Davidson SB, Garcia-Molina H, Skeen D. Consistency in partitioned networks. ACM Comput Surv. 1985;17(3):341–70.
15. Fekete A, Gupta D, Luchangco V, Lynch N, Shvartsman A. Eventually-serializable data services. In: Proceedings of the ACM SIGACT-SIGOPS 15th Sym-

posium on the Principles of Distributed Computing; 1996. p. 300–9.

16. Fox A, Gribble SD, Chawathe Y, Brewer EA, Gauthier P. Cluster-based scalable network services. ACM SIGOPS Oper Syst Rev. 1997;31(5):78–91.

17. Garcia-Molina H, Salem K. Sagas. In: Proceedings of the ACM SIGMOD International Conference on Management of Data; 1987.

18. Gray JN, Lorie RA, Putzolu GR, Traiger IL. Granularity of locks and degrees of consistency in a shared data base. Technical report, IBM, 1976.

19. Haerder T, Reuter A. Principles of transaction-oriented database recovery. ACM Comput Surv. 1983;15(4):287–317.

20. Helland P, Campbell D. Building on quicksand. In: Proceedings of the 4th Biennial Conference on Innovative Data Systems Research; 2009.

21. Johnson PR, Thomas RH. Rfc 667: the maintenance of duplicate databases. Technical report, 1 1975.

22. Kung H-T, Papadimitriou CH. An optimality theory of concurrency control for databases. In: Proceedings of the ACM SIGMOD International Conference on Management of Data; 1979.

23. Li C, Porto D, Clement A, Gehrke J, Preguiça N, Rodrigues R. Making geo-replicated systems fast as possible, consistent when necessary. In: Proceedings of the 10th USENIX Symposium on Operating System Design and Implementation; 2012.

24. Li C, Leitao J, Clement A, Preguiça N, Rodrigues R et al. Automating the choice of consistency levels in replicated systems. In: Proceedings of the USENIX 2014 Annual Technical Conference; 2014.

25. Lu H, Veeraraghavan K, Ajoux P, Hunt J, Song YJ, Tobagus W, Kumar S, Lloyd W. Existential consistency: measuring and understanding consistency at facebook. In: Proceedings of the 25th ACM Symposium on Operating System Principles; 2015.

26. O'Neil PE. The Escrow transactional method. ACM Trans Database Syst. 1986;11(4):405–30.

27. Recht B, Ré C, Wright S, Niu F. Hogwild: a lock-free approach to parallelizing stochastic gradient descent. In: Advances in Neural Information Proceedings of the Systems 24, Proceedings of the 25th Annual Conference on Neural Information Proceedings of the Systems; 2011.

28. Roy S, Kot L, Bender G, Ding B, Hojjat H, Koch C, Foster N, Gehrke J. The homeostasis protocol: avoiding transaction coordination through program analysis. In: Proceedings of the ACM SIGMOD International Conference on Management of Data; 2015.

29. Shapiro M, Preguiça N, Baquero C, Zawirski M. A comprehensive study of convergent and commutative replicated data types. INRIA TR 7506. 2011.

30. Wada H, Fekete A, Zhao L, Lee K, Liu A. Data consistency properties and the trade-offs in commercial cloud storage: the consumers' perspective. In: Proceedings of the 5th Biennial Conference on Innovative Data Systems Research; 2011.

31. Weihl W. Specification and implementation of atomic data types. PhD thesis, Massachusetts Institute of Technology, 1984.

32. Yu H, Vahdat A. Design and evaluation of a conit-based continuous consistency model for replicated services. ACM Trans Comput Syst. 2002;20(3):239–82.

Multidimensional Data Formats

Amarnath Gupta
San Diego Supercomputer Center, University of California San Diego, La Jolla, CA, USA

Definition

The term "multidimensional data" is used in two different ways in data management. In the first sense, it refers to data aggregates created by different groupings of relational data for online analytical processing. In the second sense, the term refers to data that can be described as arrays over heterogeneous data types together with metadata to describe them.

Example: HDF (Hierarchical Data Format) and NetCDF (network Common Data Form) are well known multidimensional data formats used in scientific applications.

Key Points

The goal of a multidimensional data format is to enable random access to very large, very complex, heterogeneous data, such that the data is self describing, sharable, compact, extendible, and archivable. For example, a composite of 900 files from a seismic simulation has been organized in HDF5 format to create a terabyte-sized dataset. One can mix tables, images, small metadata, streams of data from instruments, and structured grids all in the same HDF file. While multidimensional file formats are very flexible, they present the challenge of storing such large datasets and providing concurrent, random access to any part of the data required by user queries. Design of novel index structures over these formats is an area of active research.

Cross-References

▶ Bitmap-based Index Structures
▶ Query Evaluation Techniques for Multidimensional Data
▶ Storage of Large Scale Multidimensional Data

Recommended Reading

1. Home page of the HDF group. Available at: http://hdf.ncsa.uiuc.edu/.
2. Home page of the NetCDF group. Available at: http://www.unidata.ucar.edu/software/netcdf/.
3. Wu K, Otoo EJ, Shoshani A. "An efficient compression scheme for bitmap indices". Technical Report LBNL 49626, Lawrence Berkeley National Laboratory, Berkeley, 2002.

Multidimensional Modeling

Torben Bach Pedersen
Department of Computer Science, Aalborg University, Aalborg, Denmark

Synonyms

Dimensional modeling; Star schema modeling

Definition

Multidimensional modeling is the process of modeling the data in a universe of discourse using the modeling constructs provided by the multidimensional data model. Briefly, the multidimensional data model categorizes data as being either *facts* with associated numerical *measures* or as being *dimensions* that characterize the facts and are mostly textual. For example, in a retail business, *products* are sold to *customers* at certain *times* in certain *amounts* and at certain *prices*. A typical fact would be a *purchase*. Typical measures would be the amount and price of the purchase. Typical dimensions would be the location of the purchase, the type of product being purchased, and the time of the purchase. Queries then aggregate measure values over ranges of dimension values to produce results such as the total sales per month and product type.

More precisely, a number of different formal multidimensional data models have been proposed at the conceptual and logical level. While none of these are considered as *the* standard model and they differ in the details, all models agree on the major concepts of the "canonical" multidimensional data model, as described above.

Historical Background

Multidimensional databases do not have their origin in database technology but stem from multidimensional matrix algebra, which has been used for (manual) data analyses since the late nineteenth century. During the late 1960s, two companies, IRI and Comshare, independently began the development of systems that later turned into multidimensional database systems. The IRI Express tool became very popular in the marketing analysis area in the late 1970s and early 1980s; it later turned into a market-leading OLAP tool and was acquired by Oracle. Concurrently, the Comshare system developed into System W, which was heavily used for financial planning, analysis, and reporting during the 1980s.

A concurrent development started in the early 1980s in the area of the so-called statistical data management which focused on modeling and managing statistical data [1], initially within social science contexts such as census data. Many important concepts of multidimensional modeling such as *summarizability* (ensuring correct aggregate query results for complex data) have their roots in this area. An overview is found in [15].

In 1991, Arbor was formed with the specific purpose of creating "a multiuser, multidimensional database server," which resulted in the Essbase system. Arbor, now Hyperion, later licensed a basic version of Essbase to IBM for integration into DB2. It was Arbor and Codd who in 1993 coined the term *online analytical processing* (OLAP) [2].

Another significant development in the early 1990s was the advent of large *data warehouses* [5, 6] for storing and analyzing massive amounts of enterprise data. Data warehouses are typically based on relational *star schemas* or *snowflake schemas*, an approach to implementing multidimensional databases using relational database technology. The 1996 version of Kimball and Ross [6] popularized the use of star schema modeling for data warehouses.

From the mid 1990s and beyond, the introduction of the "data cube" operator [4] sparked a considerable research interest in the field of modeling multidimensional databases for use in data warehouses and OLAP.

In 1998, Microsoft shipped its MS OLAP Server, the first multidimensional system aimed at the mass market. This has lead to the current situation where multidimensional systems are commodity products that are shipped at no extra cost together with leading relational database systems.

A more in-depth coverage of the history of multidimensional databases is available in the literature [17]. Surveys of multidimensional data models can also be found in the literature [13, 19].

Scientific Fundamentals

First, an overview of the concept of a multidimensional *cube* is given, then dimensions, facts, and measures are covered in turn.

Data Cubes

Data cubes provide true multidimensionality. They generalize spreadsheets to any number of dimensions. In addition, hierarchies in dimensions and formulas are first-class, built-in

concepts, meaning that these are supported without duplicating their definitions. A collection of related cubes is commonly referred to as a *multidimensional database* or a *multidimensional data warehouse*.

A dimensional cube, e.g., CD sales, can be obtained by including additional dimensions apart from just the album and the city where the album was sold. The most pertinent example of an additional dimension is a time dimension, but it is also possible to include other dimensions, e.g., an artist dimension that describes the artists associated with albums. In a cube, the combinations of a dimension value from each dimension define the *cells* of the cube. The actual sales counts are stored in the corresponding cells.

In a cube, dimensions are first-class concepts with associated domains, meaning that the addition of new dimension values is easily handled. Although the term "cube" implies three dimensions, a cube can have any number of dimensions. It turns out that most real-world cubes have 4–12 dimensions [5, 6, 17]. Although there is no theoretical limit to the number of dimensions, current tools often experience performance problems when the number of dimensions is more than 10–15. To better suggest the high number of dimensions, the term "hypercube" is often used instead of "cube."

Figure 1 illustrates a three-dimensional cube based on the number of CD sales of two particular albums in Aalborg, Denmark, and New York, USA, for 2006 and 2007. The cube then contains sales counts for two cities, two albums, and 2 years. Depending on the specific application, a highly varying percentage of the cells in a cube are nonempty, meaning that cubes range from *sparse* to *dense*. Cubes tend to become increasingly sparse with increasing dimensionality and with increasingly finer granularities of the dimension values.

A nonempty cell is called a *fact*. The example has a fact for each combination of time, album, and city where at least one sale was made. A fact has associated with it a number of *measures*. These are numerical values that "live" within the cells. In our case, there is only one measure, the sales count.

M

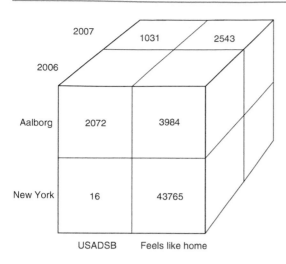

Multidimensional Modeling, Fig. 1 Sales data cube

Generally, only two or three dimensions may be viewed at the same time, although for low-cardinality dimensions, up to four dimensions can be shown by nesting one dimension within another on the axes. Thus, the dimensionality of a cube is reduced at query time by *projecting* it down to two or three dimensions via *aggregation* of the measure values across the projected-out dimensions. For example, if the user wants to view just sales by city and time, she aggregates over the entire dimension that characterizes the sales by album for each combination of city and time.

An important goal of multidimensional modeling is to "provide as much context as possible for the facts" [6]. The concept of *dimension* is the central means of providing this context. One consequence of this is a different view on *data redundancy* than in relational databases. In multidimensional databases, controlled redundancy is generally considered appropriate, as long as it considerably increases the information value of the data. One reason to allow redundancy is that multidimensional data is often *derived* from other data sources, e.g., data from a transactional relational system, rather than being "born" as multidimensional data, meaning that updates can more easily be handled [6]. However, there is usually no redundancy in the facts, only in the dimensions.

Having introduced the cube, its principal elements, dimensions, facts, and measures, are now described in more detail.

Dimensions

The notion of a dimension is an essential and distinguishing concept for multidimensional databases. Dimensions are used for two purposes: the *selection* of data and the *grouping* of data at a desired level of detail.

A dimension is organized into a containment-like *hierarchy* composed of a number of *levels*, each of which represents a level of detail that is of interest to the analyses to be performed. The instances of the dimension are typically called *dimension values*. Each such value belongs to a particular level.

In some cases, it is advantageous for a dimension to have *multiple hierarchies* defined on it [18]. For example, a time dimension may have hierarchies for both *Fiscal Year* and *Calendar Year* defined on it. Multiple hierarchies share one or more common lowest level(s), e.g., day and month, and then group these into multiple levels higher up, e.g., Fiscal Quarter and Calendar Quarter to allow for easy reference to several ways of grouping. Most multidimensional models allow multiple hierarchies. A dimension hierarchy is defined in the metadata of the cube or the metadata of the multidimensional database, if dimensions can be shared among different cubes.

In Fig. 2, the schema and instances of a sample *location* dimension capturing the cities where CDs are sold are shown. The location dimension has three levels, the city level being the lowest. City-level values are grouped into *country*-level values, i.e., countries. For example, Aalborg is in Denmark. The ⊤ ("top") level represents *all* of the dimension, i.e., every dimension value is part of the ⊤ ("top") value.

In some multidimensional models, a level may have associated with it a number of *level properties* that are used to hold simple, nonhierarchical information [5, 18]. For example, the duration of an album can be a level property in the album level of the music dimension. This information could also be captured using an extra duration

Multidimensional Modeling, Fig. 2 Schema and instance for the location dimension

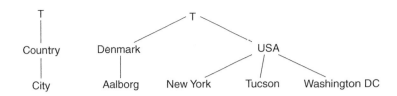

dimension. Using the level property has the effect of not increasing the dimensionality of the cube.

Unlike the linear spaces used in matrix algebra, there is typically no ordering and/or distance metric on the dimension values in multidimensional models. Rather, the only ordering is the containment of lower-level values in higher-level values. However, for some dimensions, e.g., the time dimension, an ordering of the dimension values is available and is used for calculating cumulative information such as "total sales in year to date."

Most models require dimension hierarchies to form *balanced trees* [5]. This means that the dimension hierarchy must have uniform height everywhere, e.g., all departments, even small ones, must be subdivided into project groups. Additionally, direct links between dimension values can only go between immediate parent-child levels and not jump two or more levels. For example, all cities are first grouped into states and then into countries; cities cannot be grouped directly under countries (as is the case in Denmark which has no states). Finally, each non-top value has precisely one parent, e.g., a product must belong to exactly one product group. Below, the relaxation of these constraints is discussed.

Facts

Facts are the objects that represent the *subject* of the desired analyses, i.e., the interesting "thing," event, or process that is to be analyzed to better understand its behavior.

In most multidimensional data models, the facts are *implicitly* defined by their combination of dimension values. If a nonempty cell exists for a particular combination, a fact exists; otherwise, no fact exists. Some other models treat facts as first-class objects with a separate identity [13]. Next, most multidimensional models require that each fact be mapped to precisely one dimension value at the lowest level in each dimension. Other models relax this requirement [11].

A fact has a certain *granularity*, determined by the levels from which its combination of dimension values are drawn. For example, the fact granularity in our example cube is "year by album by city." Granularities consisting of higher-level or lower-level dimension levels than a given granularity, e.g., "year by album genre by city" or "day by album by city," are said to be *coarser* or *finer* than the given granularity, respectively.

It is commonplace to distinguish among three kinds of facts: *event* facts, *state* facts, and *cumulative snapshot* facts [5, 6]. Event facts typically model *events in the real world*. The real-world events can be captured as facts either 1-to-1, e.g., every single sale becomes a unique fact, or many-to-1, e.g., all sales sharing the same time, store, and product, are combined into a single, compound fact. Examples of event facts include sales, clicks on web pages, and movement of goods in and out of (real) warehouses (flow).

A snapshot fact models the *state* of a given process at a given point in time. Typical examples of snapshot facts include the inventory levels of different stockkeeping units (SKUs) in different stores and warehouses and the number of users using a web site. This means that snapshot facts should be interpreted with care, since the underlying physical instances may "overlap" and appear in more than one fact. For example, if the inventory levels (facts) for canned beans are two in January 2015 and two in February 2015, this may be the same two cans of beans, or three or four different ones.

Cumulative snapshot facts are used to handle information about *a process up to a certain point in time*. For example, the total sales in the year to date may be considered as a fact. Then the total sales up to and including the current month this

M

year can be easily compared to the figure for the corresponding month last year.

Often, all three types of facts can be found in a given data warehouse, as they support complementary classes of analyses. Indeed, the same base data, e.g., the movement of goods in a (real) warehouse, may often find its way into three cubes of different types, e.g., warehouse flow, warehouse inventory, and warehouse flow, in year to date.

Measures

A *measure* has two components: a *numerical property* of a fact, e.g., the sales price or profit, and a *formula* (most often a simple aggregation function such as SUM) that can be used to combine several measure values into one. In a multidimensional database, measures generally represent the properties of the chosen facts that the users want to study, e.g., with the purpose of optimizing them.

Measures then take on different values for different facts. The property and formula are chosen such that the value of a measure is meaningful for all combinations of aggregation levels. The formula is defined in the metadata and thus not replicated as in the spreadsheet example. Most multidimensional data models provide the built-in concept of measures, but a few models do not. In these models, dimension values are used for computations instead [13].

It is important to distinguish among three classes of measures, namely, *additive*, *semi-additive*, and *nonadditive* measures, as these behave quite differently in computations.

Additive measure values can be summed meaningfully along any dimension. For example, it makes sense to add the total sales over album, location, and time, as this causes no overlap among the real-world phenomena that caused the individual values. Additive measures occur for any kind of fact.

Semi-additive measure values cannot be summed along one or more of the dimensions, most often the time dimension. Semi-additive measures generally occur when the fact is of type snapshot or cumulative snapshot. For example, it does not make sense to sum inventory levels across time, as the same inventory item, e.g., a specific physical instance of an album, may be counted several times, but it is meaningful to sum inventory levels across albums and stores.

Nonadditive measure values cannot (meaningfully) be summed along any dimension, usually because the measure is a so-called unit measure that is computed as a relative level. For example, the *gross margin* for a particular cell in a sales cube is the gross profit (sales price minus cost) divided by the cost and cannot be added up to find the gross margins at higher levels. Nonadditive measures can occur for any kind of fact.

A similar, but different, classification divides measures based on their aggregation formula into being either *distributive*, where higher-level aggregate values can be computed directly from lower-level aggregate values, e.g., SUM; *algebraic*, where higher-level aggregate values can be computed from aggregate values of other, distributive measures, e.g., AVG; or *holistic*, where aggregate values at all levels have to be computed directly from the base data, e.g., MEDIAN [18]. Distributive measures are typically *stored* in the data cube, where values of lower-level cells, e.g., monthly sales, can be used to compute value of higher-level cells, e.g., quarterly sales. Algebraic and holistic measure values can also be stored in the cube but cannot be directly reused as for distributive measures. Thus, underlying distributive *support measures* are often stored instead for algebraic measures, e.g., for an algebraic average measure, the distributive measures of sum and count are stored and used to compute the averages on demand. Holistic measures have to be either stored at all needed levels or computed on the fly from base data.

The Modeling Process

Now, the process to be carried out when doing multidimensional modeling is covered. One difference from "ordinary" data modeling is that the multidimensional modeler should not try to include all the available data and all their relationships in the model but only those parts which are essential "drivers" of the business. Another difference is that redundancy may be okay (in a few, well-chosen places) if introducing redundancy makes the model more intuitive for the user. For example, time-related information may be stored in both a calendar year time dimension and a fiscal year time dimension, or specific customer info may be present both in a person-oriented customer dimension or a group-oriented demographics dimension.

As a simple starting point, Kimball [6, 7] advocates a four-step process when doing multidimensional modeling:

1. Choose the business process(es) to model.
2. Choose the grain of the business process.
3. Choose the dimensions.
4. Choose the measures.

Step 1 refers to the facts that not all business processes may be equally important for the business. For example, in a supermarket, there are business processes for *sales* and *purchases*, but the sales process is probably the one with the largest potential for increasing profits and should thus be prioritized. Step 2 says that data should be captured at the right grain, or granularity, compared to the analysis needs. For example, "individual sales items" may be captured, or perhaps (slightly aggregated) "total sales per product per store per day" may be precise enough, enabling performance and storage gains. Step 3 then goes on to refine the schema of each part of the grain into a complete dimension with levels and attributes. For the example above, a store, a product, and a time dimension are specified. Finally, Step 4 chooses the numerical measures to capture for each combination of dimension values, for example, dollar sales, unit sales, dollar cost, profit, etc.

When doing multidimensional modeling "in the large" for many types of data (many cubes) and several user groups, the most important task is to ensure that analysis results are comparable across cubes, i.e., that the cubes are somehow "compatible." This is ensured by (as far as possible) picking dimensions and measures from a set of common so-called "conformed" dimensions and measures [6, 7] rather than "redefining"

the same concept, e.g., product, each time it occurs in a new context. New cubes can then be put onto the common "DW bus" [7] and used together. This sounds easier than it is, since it often requires quite a struggle with different parts of an organization to define, for example, a common product dimension that can be used by everyone. In the literature, several methods for multidimensional modeling have been proposed that are more effective and efficient than Kimball's; please see the cross-reference "▶ Data Warehouse Life Cycle and Design" for details.

Complex Multidimensional Modeling

Multidimensional data modeling is not always as simple as described above. A complexity that is almost always present is that of handling *change* in the dimension values. Kimball [6, 7] calls this the problem of *slowly changing dimensions*. For example, customer addresses, product category names, and the way products are categorized may change over time. This must be handled to ensure correct results both for current and historical data. Kimball originally suggested three types of slowly changing dimensions: Type 1 (overwrite previous value with current value), Type 2 (keep versions of dimension rows), and Type 3 (keep previous and current value in different columns). These were later supplemented with Type 0 (keep original value), Type 4 (separate tables for current and historical data), and Type 6 (combining Type $1 + 2 + 3$). Finally, the concept of *minidimensions* [6, 7] advocates the separation of relatively static information (customer name, gender, etc.) and dynamic information (income, number of kids, etc.) into separate dimensions.

The traditional multidimensional data models and implementation techniques assume that the data being modeled is quite regular. Specifically, it is typically assumed that all facts map (directly) to dimension values at the lowest levels of the dimensions and only to one value in each dimension. Further, it is assumed that the dimension hierarchies are simply balanced trees. In many cases, this is adequate to support the desired applications satisfactorily. However, situations occur where these assumptions fail.

M

In such situations, the support offered by "standard" multidimensional models and systems is inadequate, and more advanced concepts and techniques are called for. Now, the impact of irregular hierarchies on the performance-enhancing technique known as partial, or practical, precomputation is reviewed.

Complex multidimensional data are problematic as they are not summarizable. Intuitively, data is *summarizable* if the results of higher-level aggregates can be derived from the results of lower-level aggregates. Without summarizability, users will either get wrong query results, if they base them on lower-level results, or the system cannot use precomputed lower-level results to compute higher-level results. When it is no longer possible to precompute, store, and subsequently reuse lower-level results for the computation of higher-level results, aggregates must instead be calculated directly from base data, which leads to considerable increases in computational costs.

It has been shown that summarizability requires that aggregate functions be distributive and that the ordering of dimension values be *strict*, *onto*, and *covering* [8, 13]. Informally, a dimension hierarchy is *strict* if no dimension value has more than one (direct) parent, *onto* if the hierarchy is balanced, and *covering* if no containment path skips a level. Intuitively, this means that dimension hierarchies must be balanced trees. If this is not the case, some lower-level values will be either double-counted or not counted when reusing intermediate query results.

Figure 3 contains two dimension hierarchies: a location hierarchy including a state level and the hierarchy for the organization dimension for some company. The hierarchy to the left is *non-*

covering, as Denmark has no states. If aggregates at the state level are precomputed, there will be no values for Aalborg and Copenhagen, meaning that facts mapped to these cities will not be counted when computing country totals.

To the right in Fig. 3, the hierarchy is non-onto because the research department has no further subdivision. If aggregates are materialized at the lowest level, facts mapping directly to the research department will not be counted. The hierarchy is also non-strict as the TestCenter is shared between finance and logistics. If aggregates are materialized at the middle level, data for TestCenter will be counted twice, for both finance and logistics, which is, in fact, what is desired at this level. However, this means that data will be double-counted if these aggregates are then combined into the grand total.

Several design solutions exist that aim to solve the problems associated with irregular hierarchies by altering the dimension schemas or hierarchies [3, 9, 12, 18].

Although not strictly multidimensional, the so-called data vault modeling [20] is becoming more popular for modeling *enterprise data warehouses* (rather than data marts, which are almost always multidimensional). The key idea of data vault modeling is to divide data into *hubs* (collections of *stable* business keys), *links* (associations/transactions between business keys), *satellites* (temporal and descriptive attributes), or *reference tables* (lookup tables for reference/classification data). Data vault modeling represents a compromise between multidimensional modeling and the traditional 3NF modeling of enterprise data warehouses suggested by Inmon.

Multidimensional Modeling, Fig. 3
Irregular dimensions

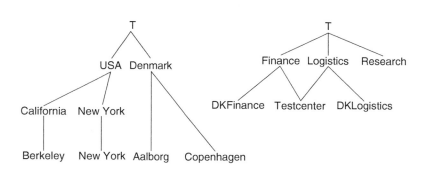

Key Applications

Multidimensional data models have three important application areas within data analysis. First, multidimensional models are used in *data warehousing* [5]. Briefly, a data warehouse is a large repository of integrated data obtained from several sources in an enterprise for the specific purpose of data analysis. Typically, this data is modeled as being multidimensional, as this offers good support for data analyses.

Second, multidimensional models lie at the core of *online analytical processing* (OLAP) systems. Such systems provide fast answers to queries that aggregate large amounts of the so-called detail data to find overall trends, and they present the results in a multidimensional fashion. Consequently, a multidimensional data organization has proven to be particularly well suited for OLAP. The acknowledged "OLAP Report" (now "The BI Verdict") [16] provided an "acid test" for OLAP by defining OLAP as "fast analysis of shared multidimensional information" (FASMI). In this definition, "fast" refers to the expectation of response times that are within a few seconds, "analysis" refers to the need for easy-to-use support for business logic and statistical analyses, "shared" suggests a need for security mechanisms and concurrency control for multiple users, "multidimensional" refers to the expectation that a data model with hierarchical dimensions is used, and "information" suggests that the system must be able to manage all the required data and derived information.

Third, multidimensional data are increasingly becoming the basis for *data mining*, where the aim is to (semi-)automatically discover unknown knowledge in large databases. Indeed, it turns out that multidimensionally organized data are also particularly well suited for the queries posed by data mining tools, so-called online analytical mining (OLAM).

Future Directions

A pressing need for multidimensional modeling is the aspect of standardization, i.e., agreeing on a common data model, a graphical notation for it, and support by tools. Also, better integration between ordinary "operational modeling" and multidimensional modeling is needed. Another future research line is the modeling of important system aspects such as security, quality, requirements, evolution, and interoperability [14]. This will be extended to also cover the modeling of business intelligence applications such as data mining, patterns, *extraction-transformation-loading* (ETL), *what-if analysis*, and business process modeling [14].

Another important line of research covers modeling of more (complex) types of data, including integrating multidimensional data with complex and nonconventional types of data, such as spatial/spatiotemporal/mobile, temporal, text, multimedia, (semantic) Web, networks, and graphs [10, 11, 18].

Emerging applications of multidimensional databases include analytics applications trying to predict the future (*predictive analytics*) and further using the predictions to suggest (prescribe) optimal decisions/actions in certain scenarios (*prescriptive analytics*). Another trend is that multidimensional databases form the cornerstone of "big data" business intelligence using cloud-based technology for scalability and collaboration, the so-called cloud intelligence.

Cross-References

- ▶ Business Intelligence
- ▶ Cloud Intelligence
- ▶ Cube
- ▶ Data Warehouse
- ▶ Data Warehouse Life Cycle and Design
- ▶ Data Warehouse Maintenance, Evolution, and Versioning
- ▶ Data Warehousing in Cloud Environments
- ▶ Data Warehousing on Nonconventional Data
- ▶ Data Warehousing Systems: Foundations and Architectures
- ▶ Dimension
- ▶ Hierarchy
- ▶ Measure
- ▶ OLAM

▶ Online Analytical Processing

▶ Predictive Analytics

▶ Prescriptive Analytics

▶ Statistical Data Management

▶ Summarizability

▶ What-If Analysis

Recommended Reading

1. Chan P, Shoshani A. SUBJECT: a directory driven system for organizing and accessing large statistical databases. In: Proceedings of the 9th International Conference on Very Data Bases; 1983. p. 553–63.
2. Codd EF. Providing OLAP (On-line Analytical Processing) to user-analysts: an IT mandate. E.F. Codd and Assoc., 1993.
3. Dyreson CE, Pedersen TB, Jensen CS. Incomplete information in multidimensional databases. In: Rafanelli M, editor. Multidimensional databases: problems and solutions. Hershey: Idea Group Publishing; 2003.
4. Gray J, Chaudhuri S, Bosworth A, Layman A, Venkatrao M, Reichart D, Pellow F, Pirahesh H. Data cube: a relational aggregation operator generalizing groupby, cross-tab and sub-totals. Data Min Knowl Disc. 1997;1(1):29–54.
5. Jensen CS, Pedersen TB, Thomsen C. Multidimensional databases and data warehousing. Synthesis lectures on data management. San Rafael: Morgan Claypool; 2010.
6. Kimball R, Ross M. The data warehouse toolkit: the complete guide to dimensional modeling. 2nd ed. New York: Wiley; 2002.
7. Kimball R, Ross M, Thornthwaite W, Mundy J, Becker B. The data warehouse lifecycle toolkit. 2nd ed. New York: Wiley; 2008.
8. Lenz H, Shoshani A. Summarizability in OLAP and statistical data bases. In: Proceedings of the 9th International Conference on Scientific and Statistical Database Management; 1997. p. 39–48.
9. Niemi T, Nummenmaa J, Thanisch P. Logical multidimensional database design for ragged and unbalanced aggregation. In: Proceedings of the 3rd International Workshop on Design and Management of Data Warehouses; 2001. Paper 7.
10. Pedersen TB. Warehousing the world: a vision for data warehouse research. New trends in data warehousing and data analysis: 1–17, Springer Annals of Information Systems, 2009.
11. Pedersen T B. Managing complex multidimensional data. In: Aufaure M-A., Zimányi E, editors. Business intelligence – second European Summer School, eBISS 2012, Brussels, Belgium, July 15–21, 2012, Tutorial Lectures. Springer LNBIB, 2013.
12. Pedersen TB, Jensen CS, Dyreson CE. Extending practical pre-aggregation in on-line analytical processing. In: Proceedings of the 25th International Conference on Very Large Data Bases; 1999. p. 663–74.
13. Pedersen TB, Jensen CS, Dyreson CE. A foundation for capturing and querying complex multidimensional data. Inf Syst. 2001;26(5):383–423.
14. Rizzi S, Abello A, Lechtenbörger J, and Trujillo J. Research in data warehouse modeling and design: dead or alive? In: Proceedings of the ACM 9th International Workshop on Data Warehousing and OLAP; 2006. p. 3–10.
15. Shoshani A. OLAP and statistical databases: similarities and differences. In: Proceedings of the 16th ACM SIGACT-SIGMOD-SIGART Symposium on Principles of Database Systems; 1997. p. 185–96.
16. The BI Verdict http://barc-research.com/bi-verdict/. Current as of August 21, 2014.
17. Thomsen E. OLAP solutions: building multidimensional information systems. New York: Wiley; 1997.
18. Vaisman A, Zimányi E. Data warehouse systems – design and implementation. Berlin: Springer; 2014.
19. Vassiliadis P, Sellis TK. A survey of logical models for OLAP databases. ACM SIGMOD Rec. 1999;28(4):64–9.
20. Wikipedia, data vault modeling. Available at http://en.wikipedia.org/wiki/Data_Vault_Modeling. Accessed on 14 Aug 2014.

Multidimensional Scaling

Heng Tao Shen
School of Information Technology and Electrical Engineering, The University of Queensland, Brisbane, QLD, Australia
University of Electronic Science and Technology of China, Chengdu, Sichuan Sheng, China

Synonyms

MDS

Definition

Multidimensional scaling (MDS) is a mathematical dimension reduction technique that best preserves the interpoint distances by analyzing gram matrix. Given any two points p_i and p_j in a dataset P, MSD aims to minimize the following objective function:

$$\text{Error} = \sum \left[d\,(pi, pj) - d'\,(pi, pj) \right] 2$$

where $d(pi,\ pj)d(pi,\ pj)$ and $d'(pi,\ pj)d'(pi,\ pj)$ represent the distance between points p_i and p_j in original space and the lower dimensional subspace, respectively.

Key Points

Multidimensional scaling (MDS) is a set of related statistical techniques often used in data visualization and analysis for exploring similarities or dissimilarities in data. An MDS algorithm starts with a matrix of point-point (dis)similarities and then assigns a location of each point in a low-dimensional space. The points are arranged in this subspace so that the distances between pairs of points have their original distance maximally retained. MDS is a generic term that includes many different specific types. These types can be classified according to whether the data are qualitative (called nonmetric MDS) or quantitative (metric MDS). The number of (dis)similarity matrices and the nature of the MDS model can also classify MDS types. This classification yields classical MDS (one matrix, unweighted model), replicated MDS (several matrices, unweighted model), and weighted MDS (several matrices, weighted model) [1, 2]. MDS is different from many other existing dimensionality reduction techniques, such as PCA and CCA [3].

MDS applications include scientific visualization and data mining in fields such as cognitive science, information science, psychophysics, psychometrics, marketing, and ecology [2].

Cross-References

► Dimensionality Reduction

► Discrete Wavelet Transform and Wavelet Synopses

► Principal Component Analysis

Recommended Reading

1. Cox MF, Cox MAA. Multidimensional scaling. Boca Raton: Chapman and Hall; 2001.
2. Young FW, Hamer RM. Multidimensional scaling: history, theory and applications. New York: Erlbaum; 1987.
3. Zhu X, Huang Z, Shen HT, Cheng J, Xu C. Dimensionality reduction by mixed kernel canonical correlation analysis. Pattern Recog (PR). 2012;45(8): 3003–16.

Multilevel Modeling

Bernd Neumayr[1] and Christoph G. Schuetz[2]
Department for Business Informatics – Data and Knowledge Engineering, Johannes Kepler University Linz, Linz, Austria

M

Synonyms

Deep metamodeling; Deep modeling; Multilevel metamodeling

Definition

Multilevel modeling extends object-oriented modeling with multiple levels of instantiation as well as deep characterization. As opposed to traditional two-level modeling, multilevel modeling overcomes the strict separation of class and object. The *clabject*, with class facet and object facet, becomes the central modeling element. Multilevel modeling arranges clabjects in arbitrary-depth hierarchies combining aspects of instantiation and specialization. A clabject not only specifies the schema of its members at the instantiation level immediately below but may also specify the schema of the members of its members, and so forth, at arbitrary instantiation levels below, which is referred to as *deep characterization*.

Historical Background

In object-oriented modeling, a *class* describes the common attributes of its many instances. An instance of a class is also referred to as *object*. A class, however, may itself be seen as an object with a distinct set of attributes that describe the class rather than its instances. The class then becomes instance of another class, the *metaclass*. A self-describing object, on the other hand, defines its class itself. The appellation of such a class varies between modeling approaches (e.g., singleton class, eigenclass, own type).

In multilevel modeling, the class-object or *clabject* – a term coined by Atkinson for a modeling primitive otherwise called "two-faceted construct" [13] with class facet and object facet alike – allows for unbounded meta-modeling with metaclasses, metaclasses of metaclasses, and so forth. The distinction between object, class, metaclass, meta-metaclass, and so forth becomes relative. To its instances an object is a class. To instances of its instances an object is a metaclass. *Deep characterization* allows an object to define schema elements of members at arbitrary instantiation levels below. An object may specify distinct sets of attributes to be instantiated by instances, instances of instances, and so forth.

In the late 1980s, Telos (see cross-reference) and its implementation ConceptBase, which employs the O-Telos dialect of Telos, were among the first approaches to support unbounded meta-modeling for data and knowledge engineering, with an arbitrary number of instantiation levels. In Telos, everything is an object and an object may be instance of any other object. Thus, the clabject effectively becomes the main modeling primitive in Telos. Telos, however, offers no direct support for deep characterization.

Among the first systems to introduce a kind of deep characterization was VODAK [7] in the early 1990s, albeit limited to three levels. The VODAK system realizes deep characterization through the organization of objects into types and, in parallel, into classes and metaclasses. Metaclasses, classes, and individual objects are uniformly treated as objects. Every object specifies an own type that describes its specific structure and behavior. Metaclasses and classes additionally specify an instance type describing structure and behavior of its instances. The own type of an object specializes the instance type of the object's class. Metaclasses additionally specify an instance-instance type describing structure and behavior of the instances of its instances. The instance type of a class specializes the instance-instance type of its class, which is a metaclass. In this regard, the instance-of relationship between individual object and class, as well as between class and metaclass, combines aspects of instantiation and specialization. In addition to instance-of relationships, an individual object may generalize another individual object and a class may generalize another class.

Potency-based deep instantiation [1], from the late 1990s onwards, has gone beyond the VODAK approach by allowing arbitrary levels of instantiation. A clabject may define properties with different potencies. Relating potency-based deep instantiation to the VODAK approach, potency-0 properties of a clabject define its own type, potency-1 properties define its instance type, potency-2 properties define its instance-instance type, potency-3 properties define the instance-instance-instance type, and so forth.

The distinction between linguistic and ontological metamodeling [1, 2], also referred to as orthogonal classification architecture (OCA), has since become a central aspect of multilevel modeling. Linguistic metamodeling refers to instantiation of primitives from the modeling language, e.g., CarModel and Porsche911GT3 are linguistic instances of Element. Ontological metamodeling, on the other hand, refers to instantiation relationships based on semantic criteria, e.g., Porsche911GT3 is ontological instance of CarModel. Multilevel modeling approaches commonly are multilevel with respect to ontological metamodeling but remain two-level with respect to linguistic metamodeling.

In the middle of the 1990s, the introduction of materialization [13] as an abstraction pattern brought a new type of relationship with both specialization and instantiation semantics. A materialization relationship associates a class of more abstract objects, e.g., product categories, with a class of more concrete objects, e.g., product models. Clabjects (then referred to as "two-faceted constructs") instantiate the former and specialize the latter. Cascaded use of materialization allows for unbounded levels of classification. Materialization relationships propagate attribute values from abstract to concrete objects either as attribute values or, depending on the propagation mechanism, promoting attribute values from the object facet to attributes of the class facet of a clabject.

The 1990s and 2000s also saw the emergence of power types [6, 12], an expressive pattern for grouping objects. A power type is "a type whose instances are subtypes of another type" [12]. The power type, e.g., CarModel, is always applied in conjunction with a partitioned type, e.g., CarIndividual. The instances of the power type are specializations of the partitioned type. Furthermore, the instances of the power type are typically modeled as clabjects.

More recently, m-objects and m-relationships [10] have combined aspects of deep instantiation, materialization, and power types. An m-object, e.g., Product, has a set of hierarchically ordered, named levels, e.g., Catalog, Category, Model, and Individual. For each level, an m-object defines a class; an m-object instantiates its top-level class, i.e., its own type. The relationships between these classes are kind of materialization. A class at one level is power type of the class at the level immediately below and partitioned type of the class at the level immediately above. An m-object, e.g., Car, may concretize another m-object, e.g., Product. The concretization relationship between m-objects has an instantiation and specialization facet. The concretizing m-object instantiates the concretized m-object's second-level class and specializes the concretized m-object's other classes. M-relationships relate m-objects at various levels.

A recent study [9] has shown the practical relevance of multilevel modeling by studying the occurrences of multilevel modeling patterns (independent of using a multilevel modeling approach) in existing meta-models from different domains. These multilevel modeling patterns are ubiquitous especially in software architecture and enterprise/process modeling, leading to the conclusion that many modeling problems are intrinsically multilevel. Extensions to deep instantiation and implementations of multilevel modeling environments (e.g., [8]) further improve applicability of multilevel modeling to practical problems.

The discussion [3,5] about the ontological and pragmatic adequacy of modeling with clabjects is ongoing. It has been argued [4, 15] that the compactness gained from representing different semantic relationships (such as instantiation and specialization) by a single relationship – referred to as ontological instance-of [2], materialization [13], or concretization [10] – leads to models that are more difficult to understand since "semantic clarity is traded for reduction of model size" [4]. The MLT multilevel theory [4] fosters semantic clarity by relating multilevel modeling with the ontological foundations of conceptual modeling.

Scientific Fundamentals

Multilevel modeling is an extension of traditional two-level modeling. The core patterns of multilevel modeling comprise class/object, property range/value, extension, and extension by auxiliary class. Among the additional patterns are clabject generalization, property specialization, heterogeneous level hierarchies, and level-crossing relationships. Multilevel modeling approaches aim at the creation of concise models with clear semantics as well as support for flexible querying.

Background from Two-Level Modeling

A database typically consists of database schema and database instance. A conceptual data model or semantic data model (see Cross-Reference) consists of an intensional definition of the database schema as well as an extensional definition of the database instance. Using terms from UML (see Cross-Reference), a conceptual data model describes the database schema using classes, associations, and attributes; attributes and associations are also referred to as properties. A conceptual data model describes the database instance using objects, links, and attribute values. Links and attribute values, also referred to as property values, are collected into slots of objects.

In a database schema, classes serve a dual purpose. First, classes collect objects into sets, i.e., each class is the set of its member objects, thereby providing an entry point for queries. Second, classes provide the structure of the data, i.e., each class defines the common attributes of its member objects. Whereas the database instance typically is considered dynamic, i.e., updated by users and applications, the database schema typically is considered static, i.e., fixed a priori and not changed by users or applications.

In many application domains, e.g., enterprise databases, the strong assumption of fixed database schema with a static set of classes versus dynamic database instance with changing objects is problematic. For example, a company may

organize its product catalog in a database with product categories collected into the Product-Category class, product models collected into the ProductModel class, and individual products collected into the ProductIndividual class (Fig. 1). The addition of a new product category, e.g., "Car," would see an instance of ProductCategory, e.g., an object Car, added to the database instance. In order to represent the particularities of a newly added product category, the database schema may be extended with subclasses of ProductModel and ProductIndividual. For example, the CarModel class introduces the maximum speed as attribute and an association to the EngineModel class with the specified horsepower as attribute; each CarModel object links to an EngineModel object. The CarIndividual class introduces an association to the EngineIndividual class with the engine identity number as attribute; each CarIndividual object links to an Engine-Individual object. Likewise, the addition of a new product model, e.g., "Porsche 911 GT3," would see an instance of ProductModel (or one of its subclasses), e.g., Porsche911GT3, added to the database instance. Furthermore, individual physical entities of the newly added product model may have their properties

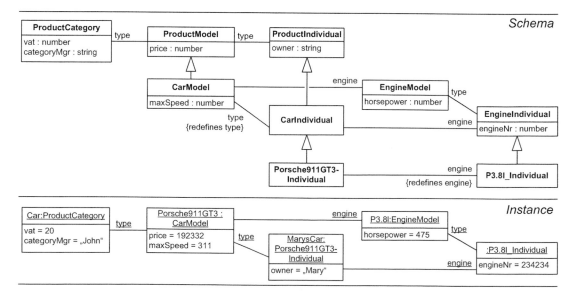

Multilevel Modeling, Fig. 1 Two-level representation (in UML) of a multilevel product catalog

Multilevel Modeling,
Fig. 2 Multilevel
representation (using deep
instantiation and named
levels) of a multilevel
product catalog

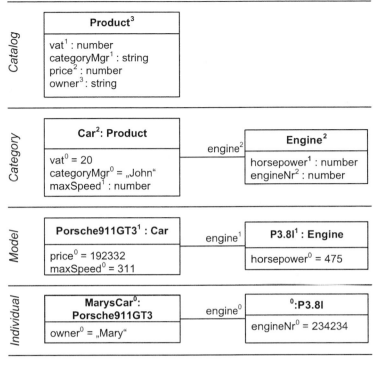

restricted, e.g., Porsche911GT3Individual objects have a P3.8l_Individual engine. Thus, the manipulation of the database instance may also entail introduction of classes in the database schema.

Multilevel modeling approaches aim for a better support of modeling situations with fluid schema/instance boundaries as well as avoidance of problems associated with the redundant representation of classes and objects in database schema and instance. Different approaches to multilevel modeling all have in common their reliance on a modeling primitive that embodies characteristics of class and object alike, usually referred to as *clabject*. For example, a product category "Car" may be represented by the Car clabject that instantiates – also referred to as concretization, ontological instantiation, or materialization in other multilevel modeling approaches – a ProductCategory class and serves as class of car models such as "Porsche 911 GT3." The clabject is believed to avoid accidental complexity in modeling situations lacking clear separation of schema and instance data.

Core Multilevel Modeling Patterns

In the remainder of this section, Fig. 2 serves to illustrate frequent patterns of multilevel modeling, based on the patterns identified by de Lara et al. [9]. The example shows a multilevel representation of a product catalog with different product categories, product models, and product individuals. The multilevel representation relies on deep instantiation as modeling approach together with named levels to increase its semantic clarity. The core patterns also translate to other multilevel modeling approaches.

The *class/object pattern* (cf. type-object [9]) is arguably the central pattern of multilevel modeling. The class/object pattern consists of two model elements, a class and an object. Many multilevel modeling approaches allow representation of the class/object pattern as a single model element with class facet and object facet. Consider, for example, the Porsche911GT3 clabject in Fig. 2 which is first an object instantiating the Car clabject with a value of 311 for the maxSpeed attribute. The Porsche911GT3 clabject also acts as class collecting instances such as MarysCar.

The class/object pattern may be cascaded and combined with deep characterization. A clabject then not only characterizes its immediate instances but also instances of its instances (which we refer to as deep instances) at arbitrary instantiation levels below. For example, in Fig. 2, Product represents the class of product categories (including the *Car* category) with properties vat and categoryMgr (of potency 1), the class of product models (including the *Porsche 911 GT3* model) with a price property (of potency 2) as well as the class of individual physical entities (including *Mary's Car*) with the owner property (of potency 3). The Car clabject is immediate instance of Product assigning a value to the vat and categoryMgr properties, the Porsche911GT3 clabject is deep instance of Product two levels below assigning a value to the price property, and the MarysCar clabject is deep instance of Product three instantiation levels below assigning a value to the owner property.

The *property range/value pattern* (cf. relation configuration [9]) allows for the redefinition of the range of a property according to property values of a clabject. For example, clabject Porsche911GT3 has P3.8l as value of the engine property. This property value acts as refinement of the range of the engine property, i.e., instances of Porsche911GT3 may only be related via the engine property to instances of P3.8l.

The class/object pattern and the property range/value pattern are two forms of the *schema/instance duality* which is central to multilevel modeling. Objects at higher levels act as specialized classes at lower levels. Property values at higher levels act as refined property ranges at lower levels.

The *extension* (cf. dynamic feature [9]) pattern allows for dynamic additions to the database schema. For example, the Car clabject in Fig. 2 extends the schema of its instances with respect to the 2-potency properties defined by the Product clabject. Consider the Product clabject which defines the price attribute (of potency 2) for deep instances two levels below. The Car clabject extends this schema with a maxSpeed attribute which is to be instantiated by instances of Car, i.e., deep instances of Product two instantiation levels below.

The *extension by auxiliary class* (cf. dynamic auxiliary domain concept [9]) pattern, in combination with the extension pattern, permits the dynamic addition of auxiliary classes that act as range of dynamically added properties. For example, Car extends the schema of its instances and of the instances of its instances with the engine property and introduces the Engine auxiliary class with the horsepower attribute. The P3.8l clabject instantiates Engine, defining a value of 475 for the horsepower property and being linked to Porsche911GT3, an instance of Car.

Class and Metaclass Facets of Clabjects

In multilevel clabject hierarchies with deep characterization, a clabject becomes a multifaceted construct with multiple class and metaclass facets. Likewise the instantiation relationship between two clabjects becomes a multifaceted relationship combining multiple instantiation and specialization relationships. Level names together with procedures for the construction of qualified names from clabject names and level names help to make more explicit these different facets and allow to directly refer to them and to describe the different relationships between facets of clabjects.

A qualified name combining a clabject name and a level name refers to a class facet of a clabject (see Def. 5 in [10], with clabjects referred to as m-objects). Qualified names act as substitutes for explicitly modeled classes, e.g., ProductModel and CarModel of Fig. 1, and serve as predefined entry points for querying. For example, clabject names Product and Car are combined with level name Model to construct qualified names Product⟨Model⟩ and Car⟨Model⟩ to refer to class facets of clabjects Product and Car that collect clabjects at level Model, such as Porsche911GT3, as members. The Porsche911GT3 clabject is member of Product⟨Model⟩ and Car⟨Model⟩. The Car⟨Model⟩ class facet is a subclass of the Product⟨Model⟩ class facet.

Level names and clabject names can further be combined to qualified names that refer to different metaclass facets of clabjects [10]. For example, Product⟨Category⟨Model⟩⟩ refers to the metaclass facet of clabject Product that has class facets such as Car⟨Model⟩ and Motorcycle⟨Model⟩ as members. Likewise, Product⟨Category⟨Individual⟩⟩ refers to the metaclass facet of the Product clabject which has class facets such as Car⟨Individual⟩ and Motorcycle⟨Individual⟩ as members. The Car⟨Model⟨Individual⟩⟩ metaclass facet, with members such as the Porsche911GT3 ⟨Individual⟩ class facet, is a subclass of the Product⟨Model⟨Individual⟩⟩ metaclass facet.

Additional Multilevel Modeling Patterns

Clabject generalization (cf. element classification [9]) allows for the arrangement in inheritance hierarchies of clabjects at the same level. For example, after introducing the additional Motorcycle product category, which also has a maxSpeed property, one could generalize the Motorcycle and Car clabjects to the generalized Vehicle clabject and move the maxSpeed property to Vehicle.

Multilevel property specialization allows for the specialization of properties to more specific properties. For example, a component property introduced at the Product clabject may be specialized as engine, transmission, and others by the Car clabject. Property specialization may be dependent or independent from the stepwise instantiation of clabjects. Materialization [13] comes with a slightly different approach, referred to as type 3 mechanism, where the values of a clabject's type 3 property become properties of the instance of the clabject.

Heterogeneous level hierarchies allow for relating clabjects in hierarchies with different numbers of levels. For example, in the product catalog example in Fig. 2, products are described at three levels of instantiation, namely, product category, product model, and product individual. Other kinds of entities may come with a different number of instantiation levels. For example, persons may be organized at a single level of instantiation, i.e., a Person class may be specified as a simple class with individual persons such as John and Mary as instances.

Heterogeneous level hierarchies are naturally supported by powertype-based approaches including materialization. The deep instantiation approach has been extended in this direction by the proposal of m-objects and m-relationships [10], dual deep instantiation [11], and by de Lara et al. [9]. M-objects additionally come with the possibility to dynamically introduce additional levels in sub-hierarchies, for example, an additional *car brand* level may be introduced for the *car* product category above the *car model* level.

Level-crossing relationships transcend strict metamodeling. The original approach to deep modeling [3] adheres to the rules of strict metamodeling, i.e., every element at some level is an instance of an element at the next higher level, and the instance-of relationship is the only type of relationship that crosses level boundaries. It has been observed by many authors that especially the latter restriction can hardly be maintained in many situations. For example, when modeling persons by a simple class Person which then acts as range of the categoryMgr and owner properties of the Product clabject, both the categoryMgr and the owner property crosses level boundaries. Level crossing relationships are naturally supported by powertype-based approaches. M-objects and m-relationships [10] and dual deep instantiation [11] support level-crossing relationships, the latter using dual potencies. The approach by de Lara et al. [8] supports level-crossing relationships using deep references.

Key Applications

In model-driven software engineering, multilevel modeling is applied to the specification of domain-specific languages [9]. In multilevel business process modeling [14], process instances (i.e., object life cycles) at higher levels impose dynamic constraints on the life cycles of their members at lower levels. In application integration, a multilevel model acts as the conceptual model of an integrated

M

enterprise database [11] or a heterogeneous IT ecosystem [15].

Cross-References

► Deep Instantiation
► Metamodel
► Semantic Data Model
► Telos
► Unified Modeling Language

Recommended Reading

1. Atkinson C, Kühne T. The essence of multilevel metamodeling. In: Gogolla M, Kobryn C, editors. UML 2001. LNCS, vol. 2185. Springer; 2001. p. 19–33.
2. Atkinson C, Kühne T. Model-driven development: a metamodeling foundation. IEEE Softw. 2003;20(5):36–41.
3. Atkinson C, Kühne T. In defence of deep modelling. Inf Softw Technol. 2015;64(C):36–51.
4. Carvalho VA, Almeida JPA, Fonseca CM, Guizzardi G. Extending the foundations of ontology-based conceptual modeling with a multi-level theory. In: Johannesson P, Lee M, Liddle SW, Opdahl AL, López OP, editors. ER 2015. LNCS, vol. 9381. Springer; 2015. p. 119–33.
5. Eriksson O, Henderson-Sellers B, Ågerfalk PJ. Ontological and linguistic metamodelling revisited: a language use approach. Inf Softw Technol. 2013;55(12):2099–124.
6. Gonzalez-Perez C, Henderson-Sellers B. A powertype-based metamodelling framework. Softw Syst Model. 2006;5(1):72–90.
7. Klas W, Neuhold EJ, Schrefl M. Metaclasses in VODAK and their application in database integration. GMD Technical Report (Arbeitspapiere der GMD); 1990.
8. de Lara J, Guerra E, Cobos R, Moreno-Llorena J. Extending deep meta-modelling for practical model-driven engineering. Comput J. 2014;57(1):36–58.
9. de Lara J, Guerra E, Cuadrado JS. When and how to use multilevel modelling. ACM Trans Softw Eng Methodol. 2014;24(2):12:1–12:46.
10. Neumayr B, Grün K, Schrefl M. Multi-level domain modeling with m-objects and m-relationships. In: Link S, Kirchberg M, editors. Proceedings of the 6th Asia-Pacific Conference on Conceptual Modeling; 2009. p. 107–16.
11. Neumayr B, Jeusfeld MA, Schrefl M, Schütz C. Dual deep instantiation and its ConceptBase implementation. In: Jarke M, Mylopoulos J, Quix C, Rolland C, Manolopoulos Y, Mouratidis H, Horkoff J, editors. CAiSE 2014. LNCS, vol. 8484. Springer; 2014. p. 503–17.
12. Odell J. Power types. J Object-Oriented Prog. 1994;7(2):8–12.
13. Pirotte A, Zimányi E, Massart D, Yakusheva T. Materialization: a powerful and ubiquitous abstraction pattern. In: Proceedings of the 20th International Conference on Very Large Data Bases; 1994. p. 630–641.
14. Schuetz CG. Multilevel business processes – modeling and data analysis. Springer Vieweg Wiesbaden; 2015.
15. Selway M, Stumptner M, Mayer W, Jordan A, Grossmann G, Schrefl M. A conceptual framework for large-scale ecosystem interoperability. In: Johannesson P, Lee M, Liddle SW, Opdahl AL, López OP, editors. ER 2015. LNCS, vol. 9381. Springer; 2015. p. 287–301.

Multilevel Recovery and the ARIES Algorithm

Gerhard Weikum
Department 5: Databases and Information Systems, Max-Planck-Institut für Informatik, Saarbrücken, Germany

Definition

In contrast to basic database recovery with page-level logging and redo/undo passes, *multi-level recovery* is needed whenever the database system uses fine-grained concurrency control, such as index-key locking or operation-based "semantic" conflict testing, or when log records describe composite operations that are not guaranteed to be atomic by single page writes (as a consequence of concurrency control or for other reasons). Advanced methods perform logging and recovery at multiple levels like pages and data objects (records, index entries, etc.). Page-level recovery is needed to ensure the atomicity and applicability of higher-level operations, and also for efficient redo. Higher-level recovery is needed to perform correct undo for composite operations of loser transactions. In addition, logged actions at all levels must be testable at recovery-time, by embedding extra information in database pages, typically using log sequence numbers (LSNs), and appropriate logging of recovery steps. A highly optimized instantiation of these principles is the *ARIES algorithm* (Algorithms for Recovery and Isolation Exploiting Semantics) [1, 2], the de facto standard solution for industrial-strength database systems. Its salient features are: very fast, potentially parallelizable or selective, redo for high availability; use of LSNs and compensation log records (CLRs) for tracking recovery progress; full-fledged support for crash and media recovery; suitability for all kinds of semantic concurrency control methods.

Historical Background

Crash recovery for composite operations on database records and indexes has first been addressed by [3], but that solution required heavy-weight check pointing (based on shadow storage) and was fairly inefficient. Some commercial engines developed various techniques to overcome these problems while supporting fine-grained concurrency control, but there is very little public literature on such system internals [4–6]. The ARIES algorithm was the first comprehensive solution [2] and has become the state-of-the-art method for industrial-strength recovery [1]. In parallel to and independently of the ARIES papers, research on multi-level recovery developed general principles and a systematic framework [7–9]. The textbook [10] discusses both the general framework and the ARIES algorithm in great detail. A correctness proof for the ARIES algorithm is given in [11].

Foundations

Basic database recovery methods log page modifications during normal operation. During the restart after a system crash (i.e., soft failure of the server with all disks intact), a three-phase recovery procedure is performed with an analysis pass, a redo pass, and an undo pass over the log file. This is appropriate when the database system uses page locking or some other page-granularity concurrency control. Such locking guarantees that all updates of loser transactions (i.e., uncommitted transactions that require undo) that are in conflict with updates of winner transactions (i.e., committed transactions that may require redo) follow the winners' updates in the log file. However, with fine-grained locking or some form of semantic concurrency control (e.g., exploiting commutative operations on hot-spot objects), this invariant does no longer hold and thus necessitates more advanced recovery algorithms.

As an example, consider the following execution history of two transactions t1 and t2 that insert various database records, with time proceeding from left to right and c2 denoting a successful commit of t2:

$$\text{insert1}(x) \quad \text{insert1}(y) \quad \text{crash} \quad \text{insert2}(z) \quad c2$$

The insertions may appear to be single operations, but they are not atomic from the system viewpoint. In fact, each of them may require multiple page writes to maintain indexes and other storage structures. For example, the following history of page writes may result from the above execution:

$$w1(p) \; w1(q) \; w1(s) \; w1(r) \; w1(p) \; w1(q) \; \text{crash}$$
$$w2(p) \; w2(s) \; c2$$

Here, p would be a data page into which all three records x, y, z are inserted, and q and s may be leaf pages of the same B + -tree index. For simplicity, the history does not show any read accesses like accesses for descending the tree. It may happen that the index update on behalf of record x triggers a leaf split of page q with a newly allocated page s and a corresponding update to the parent page r. The subsequent record operations may then access the new pages or the old page q, depending on where their corresponding keys now reside. This is a very normal situation for a database system, and it is perfectly admissible from a concurrency control viewpoint, because there are no (high-level) conflicts between the three insert operations. The system may have to use additional short-term locks or latches on pages to implement a multi-level concurrency control method, but this is very normal as well.

If the standard page-level recovery were applied to this situation, it would first redo (if necessary) all page writes of winner transaction t2 and then undo all page writes of loser transaction t1, using before-images of pages or byte-range-oriented log records. This would lead to two kinds of problems:

- Redoing the page write $w2(s)$ may be logically flawed and lead to an inconsistent index state if the effects of the preceding leaf split are not properly reflected in page s. But it is indeed possible that the loser updates $w1(q)$ and $w1(s)$ were not written to disk before the crash, and no redo would be performed for them.
- If all index updates were fully recovered and correctly captured in the database by the time the undo recovery for t1 takes place, undoing the write $w1(s)$ of page s would restore the page as of the time before t1 started, thus accidentally – and incorrectly – eliminating also the index update of the winner transaction t2.

If, on the other hand, the logging and recovery procedures were changed so that only record- and index-level operations are captured and redone or undone, one would run into a third problem:

- If the system crashed in the middle of a high-level operation, say in between the $w1(q)$ and the $w1(s)$ steps, the database may be left in an inconsistent state with partial effects of an operation. Such a database would not be recoverable as all logged operations could face a state that they cannot properly interpret.

These problems show the need for multi-level recovery; the solution must meet the following requirements:

- *Operation atomicity*: High-level operations that comprise multiple page writes must be guaranteed to appear atomic (i.e., have an all-or-nothing impact on the database).

- *High-level undo*: Operations of loser transactions must, in full generality, be undone by means of inverse operations at the same level of abstraction. For example, the insertion of an index key must be undone by performing a delete operation on that key, not by restoring the before-images of the underlying pages (and neither by corresponding byte-range modifications).
- *Testable operations*: Before invoking a high-level operation for undoing or redoing the effect of a prior operation, it must be tested whether the effects of the prior operation are indeed present (as they may have been lost by the crash or already undone/redone in a recovery procedure that was interrupted by another crash). This testability is crucial for handling non-idempotent operations.
- *Efficient redo*: As the restart time and thus the unavailability of the system is usually dominated by the redo pass, it is crucial that the redo actions are performed as efficiently as possible. This strongly suggests performing redo in terms of page writes rather than re-executing high-level operations.

Multi-level recovery methods address these requirements in the following way:

- For *proper undo*, both page-level writes and higher-level operations are logged. The page-level log records guarantee that high-level operations can always be made to appear atomic. The high-level log records guarantee that undo can be performed by means of inverse operations. During the undo of a high-level operation, page-level logging is again enabled. This way, the first two requirements are satisfied.
- For *testable operations*, when the recovery procedure undoes a (high-level) operation, both the resulting page writes and a marker for the inverse operation itself are logged. The latter kind of log record is referred to as a compensation log record (CLR). In addition, the standard technique of maintaining log sequence numbers (LSNs) in the headers of modified pages, as a form of virtual time

stamping, is used to be able to compare a log record to the state of a page and decide whether the logged action should be undone/redone or disregarded.

- For *efficient redo*, although redoing high-level operations may add to the repertoire of recovery actions, it is much more desirable to perform all redo steps in terms of page writes. This can leverage all kinds of acceleration techniques that have been developed for more conventional, page-level recovery like asynchronous check pointing and dirty-pages bookkeeping, smart scheduling of page reads from the database disk, parallelized per-page redo, and selective redo for pages with very high availability demands.
- Further considerations on the redo pass lead to the *repeating-history principle* [2]: rather than aiming to redo only winner updates or at least as few loser updates as possible, it is much simpler to redo all logged page writes regardless of their transaction status, thus effectively reconstructing the database as of the time of the crash.

All these principles together result in the following algorithmic template for multi-level recovery:

- Analysis pass for determining loser transactions.
- Redo pass by repeating history in terms of logged page writes.
- Undo pass for loser transactions, with page-level undo for incomplete high-level operations and high-level undo for complete (and possibly just redone) high-level operations. Logging at both levels is enabled during the undo pass, thus creating new log records: page-write log records during the operation's execution, and undo information for the entire operation at the very end, thus also marking the completion of the operation.

For all steps, *idempotence* is ensured by two means: for page writes the standard comparison of page-header LSN versus log-record LSN is performed; for high-level operations, only undo idempotence is a potential issue, and this is guaranteed by the fact that the preceding redo pass always repeats history so that all completely repeated operations need subsequently be undone by definition.

The *ARIES algorithm* is an integrated and highly optimized instantiation of these principles, with various additional features. Its recovery procedure performs three passes over the log: analysis, repeating-history redo, and undo. The analysis pass mostly follows standard recovery methods; the redo pass has been discussed above; the undo pass uses additional techniques based on the use of *compensation log records (CLRs)*. The following undo-relevant log records are produced by ARIES:

- During normal operation, page writes are logged in a way that they can be redone or undone (whichever is needed later), and each high-level operation is logged for undo purposes following all page-write log records that were produced during the operation's execution. The high-level log records have a backward pointer that points to the preceding high-level action of the same transaction, thus allowing the recovery manager to skip the operation's logged page writes.
- During the undo pass, when undoing a page write, a CLR is written with a backward pointer to the log record that precedes the undo page write within the same transaction. When undoing a high-level operation, normal page-write log records are written during the execution of the inverse operation, and a CLR for the entire high-level operation is written at the end. That CLR again has a backward pointer to its preceding high-level action, skipping its own page writes.

With these preparations, the undo procedure itself is rather straightforward. For each loser transaction, it locates the most recent log record and then follows the backward chain of log records. Whenever a CLR is encountered, this tells the recovery manager that the undo of the corresponding action is already completed (either already during normal operation or by the preceding redo pass) and the log record should thus be disregarded. Page-write log records are relevant for incomplete high-level operations; otherwise high-level log records determine the undo logic.

This undo procedure of the ARIES algorithm has a number of great benefits:

- It handles high-level undo in a correct and efficient way, thus allowing *fine-grained and semantic concurrency control*.
- It handles *nested rollbacks* in a correct and efficient way. These are situations where a transaction rollback is interrupted by a crash and later considered for undo or when the undo pass after a server crash is interrupted by a second crash. In all these situations, it is guaranteed that the amount of recovery work stays bounded, regardless of how many "nested" crashes might occur during recovery. This is important for high availability.
- For *media recovery*, restoring the database after disk failures, an analogous but even more severe situation arises. As media recovery always starts with a backup copy of the database and then repeats the history of a potentially very long archive log, rollbacks or undo steps for (soft) system crashes that happened long ago would interfere with log truncation and become performance showstoppers with pre-ARIES recovery methods. The way ARIES generates redo log records for undo actions and CLRs for progress tracking, media recovery is as fast as possible, which is crucial for availability.

For the example scenario given above, ARIES would create the following log records during normal operation, denoted in the form *LSN:action*. Note that log records 5 and 8 will only be used for undo purposes (if necessary). Further note that the example happens to show page-level log records for the second insert operation of t1 but no high-level log record. This may occur because of the crash happening

before the high-level log record was flushed to the log disk.

$$1 : w1(p)\ 2 : w1(q)\ 3 : w1(s)\ 4 : w1(r)$$
$$5 : \text{insert}1(x)\ 6 : w2(p)\ 7 : w2(s)\ 8 : \text{insert}2(z)$$
$$9 : w1(p)\ 10 : w1(q)\ 11 : c2$$

During recovery, the redo pass processes log records 1, 2, 3, 4, 6, 7, 9, and 10. Subsequently the undo pass processes log records 10, 9, and 5 (in this – chronologically reverse – order). As it does so, it will create the following new log records, with CLRs denoted in the form *LSN:action →
UndoNextLSN* with UndoNextLSN being the LSN of the log record to which the CLR has a backward pointer.

$$12 : w1(q) \rightarrow 9\ 13 : w1(p) \rightarrow 5\ 14 : w1(s)$$
$$15 : w1(p)\ 16 : \text{delete}1(x) \rightarrow 0$$

If the system crashed again immediately after the completion of the delete1(x) undo step, the redo pass would repeat the page writes with LSNs 12, 13, 14, and 15 (in addition to all writes with LSNs 1 through 11 that may need redo again). This means that all effects of t1 have been properly removed. The subsequent undo pass would then encounter the CLR 16, but its backward pointer immediately tells the recovery manager that it can skip all log records of transaction t1 as t1 had already been completely undone before the second crash.

If the system crashed again after the action with LSN 14 (a page write issued on behalf of the high-level undo of insert1(x)), the redo pass would repeat the page writes with LSNs 12, 13, and 14, thus effectively removing all effects of insert1(y) but only some partial effects of insert1(x). The subsequent undo pass would start with LSN 14, undo it and create a new CLR, and then encounter LSN 13, which points to LSN 5 which in turn is the next logged action to undo.

In general, ARIES can be implemented with very low overhead, and it is compatible with other optimizations in the storage engine of a database system: flexible free space management, flexible buffer management, acceleration techniques for the redo pass, and many more. For *high availability*, the redo pass of both crash and media recovery can be parallelized or performed selectively for most important page sets; media recovery efficiently works also with fuzzy backups without ever quiescing the system. For *index management*, extensions of ARIES have been developed that optimize the locking, logging, and recovery of index keys in B + -trees. Finally, there are also extensions of ARIES for the special requirements of *shared-disk clusters* with automated fail-over procedures and very high availability.

Future Directions

The ARIES algorithm is a mature and comprehensive solution that can be readily adopted for most data management systems. A salient property of the multi-level recovery framework is that it can be generalized to arbitrary kinds of composite operations (with deeper and flexible nestings). All the ARIES techniques for efficient repeating-history redo are directly applicable, and the undo procedures need to be extended to handle a conceptual stack of undo log records and corresponding CLRs – with the stack actually being embedded in the linear log. This generalization is of potential interest for modern applications like composite Web services or enterprise-level middleware with integrated recovery.

Cross-References

▶ Atomicity
▶ Transaction

Recommended Reading

1. Mohan C. Repeating history beyond ARIES. In: Proceedings of the 25th the International Conference on Very Large Data Bases; 1999. p. 1–17.
2. Mohan C, Haderle DJ, Lindsay BG, Pirahesh H, Schwarz PM. ARIES: a transaction recovery method supporting fine-granularity locking and partial rollbacks using write-ahead logging. ACM Trans Database Syst. 1992;17(1):94–162.
3. Gray J, McJones PR, Blasgen MW, Lindsay BG, Lorie RA, Price TG, Putzolu GR, Traiger IL. The recovery manager of the system R database manager. ACM Comput Surv. 1981;13(2):223–43.
4. Borr AJ. Robustness to crash in a distributed database: a non shared-memory multi-processor approach. In: Proceedings of the 10th International Conference on Very Large Data Bases; 1984. p. 445–53.
5. Crus RA. Data recovery in IBM database 2. IBM Syst J. 1984;23(2):178–88.
6. Gray J, Reuter A. Transaction processing: concepts and techniques. San Francisco: Morgan Kaufmann; 1993.
7. Lomet DB. MLR: a recovery method for multi-level systems. In: Proceedings of the ACM SIGMOD International Conference on Management of Data; 1992. p. 185–94.
8. Moss JEB, Griffeth ND, Graham MH. Abstraction in recovery management. In: Proceedings of the ACM SIGMOD International Conference on Management of Data; 1986. p. 72–83.
9. Weikum G, Hasse C, Brössler P, Muth P. Multi-level recovery. In: Proceedings of the 9th ACM SIGACT-SIGMOD-SIGART Symposium on Principles of Database Systems; 1990. p. 109–23.
10. Weikum G, Vossen G. Transactional information systems: theory, algorithms, and the practice of concurrency control and recovery. San Francisco: Morgan Kaufmann; 2001.
11. Kuo D. Model and verification of a data manager based on ARIES. ACM Trans Database Syst. 1996;21(4):427–79.

M

Multilevel Secure Database Management System

Bhavani Thuraisingham
The University of Texas at Dallas, Richardson, TX, USA

Synonyms

Secure database systems; Trusted database systems

Definition

Many of the developments in the 1980s and 1990s in database security were on multilevel secure database management systems (MLS/DBMS). These systems were also called trusted database management systems (TDBMS). In a MLS/DBMS, users are cleared at different clearance levels such as Unclassified, Confidential, Secret and TopSecret. Data is assigned different sensitivity levels such as Unclassified, Confidential, Secret, and TopSecret. It is generally assumed that these security levels form a partially ordered lattice. For example, Unclassified < Confidential < Secret < TopSecret. Partial ordering comes from having different compartments. For example, Secret Compartment A may be incomparable to Secret Compartment B.

Historical Background

MLS/DBMSs have evolved from the developments in multilevel secure operating systems such as MULTICS and SCOMP (see for example [4]) and the developments in database systems. Few developments were reported in the late 1970s on MLS/DBMSs. However, during this time there were many developments in discretionary security, such as access control for System R and INGRES as well as many efforts on statistical database security. Then there was a major initiative by the Air Force and a summer study was convened. This summer study marks a significant milestone in the development of MLS/DBMSs [2].

The early developments in MLS/DBMSs influenced the Air Force Summer Study a great deal. Notable among these efforts are the Hinke-Schaefer approach to operating system providing mandatory security, the Ph.D. Thesis of Deborah Downs at UCLA (University of California at Los Angeles), the IP Sharp Model developed in Canada and the Naval Surveillance Model developed at the MITRE Corporation. The Hinke Schaefer approach [3] essentially developed a way to host MLS/DBMSs on top of the MULTICS MLS operating system. The system was based on the relational system and the idea was to partition the relation based on attributes and store the attributes in different files at differ-

ent levels. The operating system would then control access to the files. The early efforts showed a lot of promise to designing and developing MLS/DBMSs. As a result, the Air Force started a major initiative, which resulted in the summer study of 1982.

Since the summer study, several efforts were reported throughout the 1980s. Many of the efforts were based on the relational data model. At the end of that decade, the National Computer Security Center started a major effort to interpret the Trusted Computer Systems Evaluation Criteria for database systems [7]. This interpretation was called the Trusted Database Interpretation [8]. In the 1990s research focused on non-relational systems including MLS object database systems and deductive database systems. Work was also carried out on multilevel secure distributed database systems. Challenging research problems such as multilevel data models, inference problem and secure transaction processing were being investigated. Several commercial products began to emerge. Since the late 1990s, while the interest in MLS/DBMSs began to decline a little, efforts are still under way to examine multilevel security for emerging data management technologies. A detailed discussion of many of the developments with significant references are given in [5].

Foundations

Many of the developments were based on the relational model. The early systems were based on the Integrity Lock approach developed at the MITRE Corporation. Two prototypes were designed and developed. One used the MISTRESS relational database system and the other used the INGRES relational database system. Around 1985 TRW designed and developed a MLS/DBMS called ASD and this system was designed to be hosted on ASOS (the Army Secure Operating System). The approaches were based on the Trusted Subject based architecture. Later on TRW developed some extensions to ASD and the system was called ASD Views where access was granted on views (GARV88). Two of the notable systems designed in the late 1980s were the SeaView system by SRI International and LOCK Data Views system by Honeywell. These two efforts were funded by the then Rome Air Development Center and the goal was to focus on the longer term approaches proposed by the Summer Study. Both efforts influenced the commercial developments a great deal. Three other efforts worth mentioning are the SINTRA system developed by the Naval Research Laboratory, the SWORD system developed by the then Defense Research Agency and funded by the Ministry of Defense in the United Kingdom and the SDDBMS effort by Unisys. The SINTRA system was based on the distributed architecture proposed by the Air Force Summer Study. The SWORD system proposed some alternatives to the SeaView and LOCK Data Views data models. While the initial planning for these systems began in the late 1980s, the designs were actually developed in the early 1990s. The SDDBMS effort was funded by the Air Force Rome Laboratory and investigated both the partitioned and replicated approaches to designing an MLS/DBMS.

Around 1987, the Rome Air Development Center (now known as Air Force Research Laboratory in Rome) funded an effort to design an MLS/DBMS based on the Entity Relationship (ER) model. The ER model was initially developed in 1976 by Peter Chen and since then it has been used extensively to model applications. The goal of the security effort carried out by Gajnak and his colleagues was to explore security properties for the ER model as well as to explore the use of secure ER models to design DBMSs. The effort produced MLS ER models that have since been used to model secure applications. Furthermore variations of this model have been used to explore the inference problem by Burns, Thuraisingham and Smith. However, there does not appear to have been any efforts undertaken on designing MLS/DBMSs based on the ER model. In summary, the ER approach has contributed extensively toward designing MLS applications.

During the late 1980s, efforts began on designing MLS/DBMSs based on object models. Notable among these efforts is the one by Keefe, Tsai and Thuraisingham who designed the SODA model by Keefe and his colleagues. Later Thuraisingham designed the SORION and SO2 models. These models extended models such as ORION and O2 with security properties. Around 1990 Millen and Lunt produced an object model for secure knowledge base systems. Jajodia and Kogan developed a message-passing model in 1990. Finally MITRE designed a model called UFOS. Designs of MLS/DBMSs were also produced based on the various models. The designs essentially followed the designs proposed for MLS/DBMSs based on the relational model. However with the object model, one had to secure complex objects as well as handle secure method execution. While research progressed on designing MLS/DBMSs based on objects, there were also efforts on using object models for designing secure applications. Notable efforts were those by Sell and Thuraisingham. Today with the development of UML (Unified Modeling Language) there are efforts to design secure applications based on UML.

Around 1989 work began at MITRE on the design and development of multilevel secure distributed database systems (MLS/DDBMS). Prototypes connecting MLS/DBMSs at different sites were also developed. Work was then directed toward designing and developing MLS heterogeneous distributed database systems. These efforts focused on connecting multiple MLS/DBMSs, which are heterogeneous in nature. Research was also carried out on MLS federated databases by Thuraisingham and Rubinovitz.

In the late 1970 and throughout the 1980s there were many efforts on designing and developing logic-based database systems. These systems were called deductive databases. While investigating the inference problem, multilevel secure deductive database systems were designed. These systems were based on a logic called NTML (Non monotonic Typed Multilevel Logic) designed by Thuraisingham at MITRE. NTML essentially provides the reasoning capability across security levels, which are non-monotonic in nature. Essentially, it incorporates constructs to reason about the applications at different security levels. A

Prolog language based on NTML, which is called NTML-Prolog, was also designed. Both reasoning with the Closed World Assumption as well as with the Open World Assumption were investigated. Due to the fact that there was limited success with logic programming and the Japanese Fifth Generation Project, deductive systems are being used only for a few applications. If such applications are to be multilevel secure, then systems such as those based on NTML will be needed. Nevertheless there is use for NTML on handling problems such as the inference problem. Note that presently the integration of NTML-like logic with descriptive logics for secure semantic webs is being explored.

Researchers have identified several hard problems. The most notable hard problem is the Inference problem. Inference problem is the process of posing queries and deducing sensitive information form the legitimate responses received. Many efforts have been discussed in the literature to handle the inference problem First of all, Thuraisingham proved that the general inference problem was unsolvable [6] and this effort was stated by Dr. John Campbell of the National Security Agency as one of the significant developments in database security in [1]. Then Thuraisingham explored the use of security constraints and conceptual structures to handle various types of inferences. Note that the aggregation problem is a special case of the inference problem where collections of data elements are sensitive while the individual data elements are Unclassified. Another hard problem is secure transaction processing. Many efforts have been reported on reducing covert channels when processing transactions in MLS/DBMSs including the work of Jajodia, Bertino and Atluri among others. A third challenging problem is developing a multilevel secure relational data model. Various proposals have been developed including those by Jajodia and Sandhu, the Sea View model by Denning and her colleagues and the LOCK Data Views model by Honeywell. SWORD developed by Wiseman also proposed its own model. The problem is due to the fact that different users have different views of the same element. If multiple values are used to represent the same entity then the integrity of databases is violated. However, if what is called polyinstantiation is not enforced, then there is a potential for signaling channels. This is still an open problem.

Key Applications

The department of defense was the major funding agency for multilevel secure database management systems. The applications are mainly in the defense and intelligence area. However many of the concepts can be used to design systems that have multiple labels, privacy levels or roles. Therefore these systems can also be used to a limited extent for non-defense applications including healthcare and financial applications.

Future Directions

As technologies emerge, one can examine multilevel security issues for these emerging technologies. For example, as object database systems emerged in the 1980s, multilevel security for object databases began to be explored. Today there are many new technologies including data warehousing, e-commerce systems, multimedia systems, real-time systems and the web and digital libraries. Only a limited number of efforts have been reported on investigating multilevel security for the emerging data management systems. This is partly due to the fact the even for relational systems, there are hard problems to solve with respect to multilevel security. As the system becomes more complex, developing high assurance multilevel systems becomes an enormous challenge. For example, how can one develop usable multilevel secure systems say for digital libraries and e-commerce systems? How can one get acceptable performance? How does one verify huge systems such as the World Wide Web? At present, there is still a lot to do with respect to discretionary security for such emerging systems. As progress is made with assurance technologies and if there is a need for multilevel security for such emerging technologies, then research initiatives will commence for these areas.

Cross-References

▶ Database Security
▶ Mandatory Access Control
▶ Role-Based Access Control

Recommended Reading

1. Campbell J. A year of progress in database security. In: Proceedings of the National Computer Security Conference; 1990.
2. Committee on Multilevel Data Management Security. Air Force studies board. Multilevel Data Management Security. Washington, DC: National Academy Press; 1983.
3. Hinke T, Schaefer M. Secure data management system. System Development Corp., Technical Report RADC-TR-75-266, Nov 1975.
4. IEEE Computer Magazine, vol. 16, #7; 1983.
5. Thuraisingham B. Database and applications security: integrating data management and information security. Boca Raton: CRC Press; 2005.
6. Thuraisingham B. Recursion theoretic properties of the inference problem. Presented at the IEEE Computer Security Foundations Workshop, Franconia, Jun 1990 (also available as MITRE technical Paper MTP291, Jun 1990).
7. Trusted Computer Systems Evaluation Criteria, National Computer Security Center, MD; 1985.
8. Trusted Database Interpretation. National Computer Security Center, MD; 1991.

Multilevel Transactions and Object-Model Transactions

Gerhard Weikum
Department 5: Databases and Information Systems, Max-Planck-Institut für Informatik, Saarbrücken, Germany

Synonyms

Layered transactions; Open nested transactions

Definition

Multilevel transactions are a variant of nested transactions where nodes in a transaction tree correspond to executions of operations at particular levels of abstraction in a layered system architecture. The edges in a tree represent the implementation of an operation by a sequence (or partial ordering) or operations at the next lower level. An example instantiation of this model are transactions with record and index-key accesses as high-level operations which are in turn implemented by reads and writes of database pages as low-level operations. The model allows reasoning about the correctness of concurrent executions at different levels, aiming for serializability at the top level: equivalence to a sequential execution of the transaction roots. This way, semantic properties of operations, like different forms of commutativity, can be exploited for higher concurrency, and correctness proofs for the corresponding protocols can be derived. Likewise, multilevel transactions provide a framework for structuring recovery methods and reasoning about their correctness.

Multilevel transactions have wide applications outside of database engines as well. For example, transactional properties can be provided by middleware application servers, layered on top of a database system. A generalization of this approach is the notion of object-model transactions, also known as open nested transactions (trees where the nodes correspond to arbitrary method invocations). In contrast to multilevel transactions, there is no layering constraint anymore, and arbitrary caller-callee relations among objects of abstract data types can be expressed. This provides a model for reasoning about transactional guarantees in composite web services, and for structuring the design and run-time architecture of web-service-based applications.

Historical Background

Object-model transactions have been around for 30 years, going back to the work of Bjork and Davies on "spheres of control" [2]. The first work that made these concepts explicit and gave formal definitions is by Beeri et al. [1]. Parallel work on the important special case of multilevel transactions has been done by Moss et al. [7] and Weikum et al. [11, 13]. The textbook [14]

M

Multilevel Transactions and Object-Model Transactions, Fig. 1
Example of a multilevel transaction

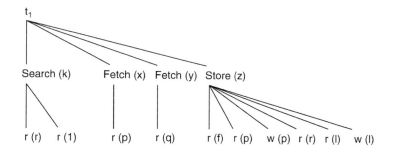

gives a detailed account of both conceptual and practical aspects. A broader perspective of extended transaction models is given by [9]. More recently, the concept of object-model transactions has received considerable attention also for long-running workflows (e.g., [10, 15]) and transactional memory (e.g., [8]).

On the system side, multilevel transaction protocols have been employed for concurrency control and recovery in various products and prototypes [3, 5, 6, 12]. Typically, the layered structure of the protocols is only implicit; a suite of additional smart implementation techniques is used for integrated, highly efficient code. An example for concurrency control is transaction-duration locking for index-manager operations combined with operation-duration latching. Examples for recovery are the ARIES family of algorithms by Mohan et al. [6] and the MLR algorithm by Lomet [5].

Foundations

Multilevel and object-model transactions are best understood in an object model where operations are invoked on arbitrary objects. This allows exploiting "semantic" properties of the invoked operations for the sake of improved performance. This model also captures situations where an operation on an object invokes other operations on the same or other objects. Often the implementation of an object and its operations requires calling operations of some lower-level types of objects Fig. 1.

For example, operations at the access layer of a database system, such as index searches, need to invoke page oriented operations at the storage

layer underneath. Similar invocation hierarchies may exist among a collection of business objects that are made available as abstract data type (ADT) instances within a data server or an application server, e.g., a "shopping cart" or a "bank account" object type along with operations like deposit, withdraw, get_balance, get_history, compute_interests, etc. The following figure depicts an example of a transaction execution against an object model scenario that refers to the internal layers of a database system Fig. 2.

The figure shows a transaction, labeled t1, which performs, during its execution, (i) an SQL Select command to retrieve all records from a database that satisfy a certain attribute-value condition, and, after inspecting the result set, (ii) an SQL command to insert a record for a new record with this attribute value. Since SQL commands are translated into query execution plans already at compile time, the operations invoked at run time refer to an internal level of index and record accesses. The Select command is executed by first issuing a Search operation with some key k on an index that returns the RIDs (i.e., addresses) of the result records. Next, these records, referred to as x and y in the figure, are fetched by dereferencing their RIDs. The Search operation in turn invokes operations at the underlying storage layer: read and write operations on pages. First the root of a B+ tree is read, labeled as page r in the figure, which points to a leaf page, labeled l, that contains the relevant RID list for key k. The subsequent Fetch operations to access the two result records x and y by their RIDs, require only one page access each to pages p and q, respectively. Finally, the SQL Insert command is executed as a Store operation, storing the new record and also maintaining the index. This in-

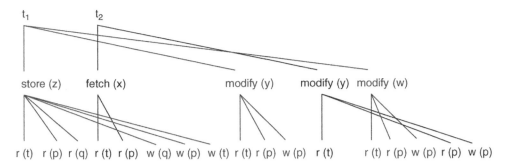

Multilevel Transactions and Object-Model Transactions, Fig. 2 Example of a multilevel schedule

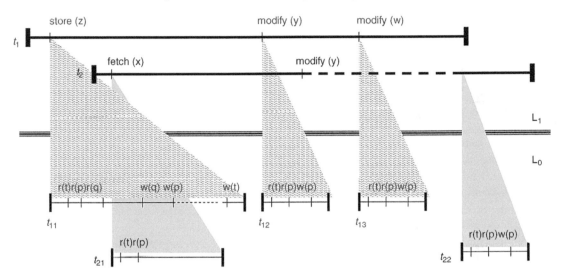

Multilevel Transactions and Object-Model Transactions, Fig. 3 Concurrent execution of multilevel schedules with a multilevel locking protocol

volves first reading a metadata page, labeled f that holds free space information in order to find a page p with sufficient empty space. Then that page is read and subsequently written after the new record z has been placed in the page. Finally, the RID of the new record z is added to the RID list of the key k in the index. This requires reading the B+ tree root page r, reading the proper leaf page l, and finally writing page l after the addition of the new RID.

This entire execution is represented in a graphical, compact form, by connecting the calling operation and the called operation with an arc when operations invoke other operations. As a convention, the caller is always placed closer to the top of the picture than the callee. Furthermore, the order in which operations are invoked is

represented by placing them in "chronological" order from left to right, which suffices for illustration purposes Fig. 3.

More formally, a **multilevel transaction tree** is defined as a partially ordered labeled tree with node labels being operation invocations and the leaf nodes denoting elementary (i.e., indivisible) read and write operations. Moreover, the tree must be perfectly balanced with all leaves having the same distance from the root; this constraint is dropped for the more general case of **object-model transactions (open nested transactions)**. Finally, a constraint is imposed on conflicting leaf nodes to be totally ordered so that no concurrent write-write or read-write pair of operations is possible on the same elementary object; for all other cases partial orders are allowed. It is important to note that transaction trees model executions and not programs. Thus, node labels are method names along with concrete input parameter values (and possibly even output parameter values if these are exploited in the reasoning about concurrency); edges denote the dynamic calling structure and not a static hierarchy.

A *concurrent execution of several transaction trees*, referred to as a multilevel schedule, is essentially an interleaved forest of the individual transaction trees. This is illustrated in the figure below, for two transactions with record-level and page-level operations, in the same spirit as the previous example but with some simplifications. Like before, the ordering of operations is indicated by drawing the leaf nodes in their execution order from left to right (assuming total ordering of leaves for simplicity). As the caller-callee relationship in transaction trees is captured by vertical or diagonal arcs, the crossing of such arcs indicates that two (non-leaf) operations are concurrent. In the figure the two transactions t1 and t2 are concurrent, and also the store and fetch operations execute concurrently and the same holds for the last two modify operations.

To reason about the correctness of such interleavings, it is first necessary to define the ordering of non-leaf operations: node o1 precedes o2 in the execution, o1 < o2, if all leaf-level descendants of o1 precede all leaf-level descendants of o2. The forest of labeled trees and this execution order < define a *multilevel schedule*, more generally, a schedule of transaction trees.

Following the standard argumentation about serializability for conventional, "flat" transactions, the goal for a correct schedule is to show that the execution is equivalent to a sequential one based on a notion of *conflicting versus non-conflicting operations*. Usually, *commutativity* properties of operations are the basis for defining conflict relations. In the example, this suggests that store and fetch operations on different objects as well as pairs of modify operations on different objects are non-conflicting (even if their implementations write the same page). Thus, their observed execution order could be changed, by swapping adjacent operations, without changing the overall effect of the schedule. But applying this principle to, for example, the store and fetch operations in the above schedule does not work because these two nodes are composite operations (i.e., non-leaf nodes) and executed concurrently among themselves. To disentangle the concurrency between these operations, one needs to reason about the execution ordering of their children, and in an actual system, one would need a *lower-level concurrency control* mechanism that treats the two operations as *subtransactions*. The goal of this disentangling, the counterpart to serial schedules in conventional concurrency control theory, are *isolated subtrees* for the two operations. A subtree rooted at node o is isolated if there is a total ordering among all its leaf-level descendants and o either precedes or follows all other operations o' that are not among its descendants (o < o' or o' < o). Once a subtree is isolated, the fact that its root is a composite operation is no longer important, and it is possible, for reasoning about equivalent executions, to *reduce an isolated subtree* to its root alone. This argument abstracts from the lower-level executions, as they are now (shown to be equivalent to) sequential.

Putting everything together, the above considerations lead to three rules for transforming a multilevel schedule into equivalent and abstracted executions, ideally leading to a sequential execution of the transaction roots:

- *Commutativity rule:* The order of two ordered leaf operations p and q with, say, the order p < q, can be reversed provided that
  - both are isolated, adjacent in that there is no other operation r with p < r < q, and commutative,
  - the operations belong to different transactions, and
  - the operations p and q do not have ancestors, say p' and q', respectively, which are non-commutative and totally ordered (in the order p' < q').
- *Ordering rule:* Two unordered leaf operations p and q can be (arbitrarily) ordered, i.e., as-

suming either p < q or q < p, if they are commutative.

- *Tree pruning rule:* An isolated subtree can be pruned and replaced by its root.

A schedule that, by applying the above rules, can be transformed into a sequential execution of the transaction roots is called **tree-reducible** or **multilevel serializable**. Note that this notion of multilevel serializability is much more liberal than the conventional notion of read-write-oriented serializability. The example schedule shown above is not serializable at the leaf level of read and write operations (i.e., if one ignored the level of search, fetch, store, and modify operations and simply connected all leaves directly to the roots), but these seemingly non-serializable effects on the low-level storage structures are irrelevant as long as they are properly handled within the scope of their parent operations and new transactions access the data through the higher-level operations like search, fetch, store, and modify.

The example schedule depicted above is multilevel serializable. It can be reduced as follows. First the two reads of the fetch(x) operation are commuted with their left-hand neighbors so that fetch(x) completely precedes store(z); analogously the r(t) step of modify(y) is commuted with its right-hand neighbors, the children of modify(w), so that modify(w) completely precedes modify(y). This establishes a serial order of the record-level operations, all of them now being isolated subtrees. This enables the application of the pruning rule to remove all page-level operations. Next, the fetch(x) operation of t2 is commuted with t1's store(z), modify(y), and modify(w) operations all the way to the right, producing an order where all of t1's operations precede all of t2's operations. This turns t1 and t2 into isolated subtrees. Finally, pruning the operations of t1 and t2 produces the sequential order of the transaction roots: t1 < t2.

The transformation rules do not directly lead to an efficient concurrency control protocol. Rather their purpose is to prove the correctness of protocols. But for the case of a layered system,

the way the example was handled points towards a practically viable protocol. The key is to consider pairs of adjacent levels and apply the transformation rules in a bottom-up manner. So first, the commutativity and ordering rules are used to establish a sequential execution of the parent nodes of the leaf-level nodes, then these isolated parents are reduced. Then, with the lowest level removed, this procedure is iterated through the levels until the roots of the entire transaction trees are isolated. This proof strategy can be directly turned into a protocol by enforcing conventional order-preserving conflict-serializability (OPCSR) for each pair of adjacent levels. Any protocol for OPCSR can be used, and even different protocols for different level pairs are possible. The most widely used protocol, two-phase locking, is often a natural choice, and then forms the following **multilevel locking protocol**:

- *Lock acquisition rule:* When an operation f(x) is issued, an f-mode lock on x needs to be acquired before the operation can start its execution.
- *Lock release rule:* Once a lock originally acquired by an operation f(x) with parent o (an operation at the next higher level) is released, no other descendant of o is allowed to acquire any locks.
- *Subtransaction rule:* At the termination of an operation o, all locks that have been acquired for descendants of o are released, thus treating o as a committed subtransaction. Note that the o-lock for o itself is still kept – until the parent of o terminates. The releasing of lower-level locks at the end of a subtransaction is the origin of the name "open nested transaction".

A possible execution of the example schedule under this multilevel locking protocol is shown in the figure below (with levels L1 and L0 referring to the record and page layer in a database engine, and t_{ij} denoting the jth subtransaction of transaction t_i).

The example shows that, despite many page-level conflicts, high concurrency is possible by exploiting the finer granularity and richer semantics of record-level operations. These benefits are even more pronounced for index-key operations. For this case, highly optimized special-purpose protocols like ARIES Key-Value Locking have been developed. One important optimization for both record and index operations is that the sub-transactions may use light-weight latching instead of full-fledged locks.

Another use case with wide applicability are operations on counters, such as increment and decrement or conditional variants on lower-bounded or upper-bounded counters. Such objects and operations are common in reservation systems, inventory control, financial trading, and so on. The relaxed (but not universal) commutativity properties of the operations can be leveraged for very high concurrency even if operations access the same object. Again, special implementation techniques like escrow locking have been developed for these settings. When counter operations have a composite nature, e.g., by automatically triggering updates on other objects, then the special commutativity techniques need to be embedded in a multilevel transaction framework.

A complication that arises from all these high-concurrency settings is that undo recovery (for transaction abort and to wipe out effects of incomplete transactions after a crash) can no longer be implemented merely by restoring prior page versions. Instead, adequately implemented forms of inverse operations need to be executed. Together with the composite nature of operations, this necessitates a form of multilevel recovery.

Most of the outlined principles and algorithms apply to the general case of object-model trans-actions as well. However, the absence of a layer-ing does incur some extra difficulties, which are beyond the scope of this entry. The algorithms for fully general object-model transactions are explained in detail in the textbook [14].

Future Directions

Multilevel transactions have originally been de-veloped in the database system context, but their usage and potential benefits are by no means limited to database management. So not surpris-ingly, object-model transactions and related con-cepts are being explored in the operating systems and programming languages community. Recent trends include, for example, enhancing the Java language with a notion of atomic blocks that can be defined for methods of arbitrary classes. This could largely simplify the management of concurrent threads with shared objects, and po-tentially also the handling of failures and other exceptions. The run-time environment could be based on an extended form of software transac-tional memory [8].

Another important trend is to enhance com-posite web services with transactional properties. Again, object-model transactions is a particularly intriguing paradigm because of its flexibility in allowing application-specific methods for provid-ing atomicity, isolation, and persistence. Adapt-ing and extending transactional concepts for web services and combining them with other aspects of service-oriented computing is the subject of ongoing research [15].

Cross-References

▶ Atomicity
▶ Escrow Transactions
▶ Multilevel Recovery and the ARIES Algorithm
▶ Nested Transaction Models
▶ Transaction
▶ Transaction Management

Recommended Reading

1. Beeri C, Bernstein PA, Goodman N. A model for concurrency in nested transactions systems. J ACM. 1989;36(2):230–69.
2. Davies CT, Davies Jr CT. Data processing spheres of control. IBM Syst J. 1978;17(2):179–98.
3. Gray J, Reuter A. Transaction processing: concepts and techniques. Los Altos: Morgan Kaufmann; 1993.
4. Greenfield P, Fekete A, Jang J, Kuo D, Nepal S. Isolation support for service-based applications: a position paper. In: Proceedings of the 3rd Biennial Conference on Innovative Data Systems Research; 2007. p. 314–23.

5. Lomet DB. MLR: a recovery method for multi-level systems. In: Proceedings of the ACM SIGMOD International Conference on Management of Data; 1992. p. 185–94.
6. Mohan C, Haderle DJ, Lindsay BG, Pirahesh H, Schwarz PM. ARIES: a transaction recovery method supporting fine-granularity locking and partial rollbacks using write-ahead logging. ACM Trans Database Syst. 1992;17(1):94–162.
7. Moss JEB, Griffeth ND, Graham MH. Abstraction in recovery management. In: Proceedings of the ACM SIGMOD International Conference on Management of Data; 1986. p. 72–83.
8. Ni Y, Menon V. Adl-Tabatabai A-R, Hosking AL, Hudson RL, Moss JEB, Saha B, Shpeisman T. Open nesting in software transactional memory. In: Proceedings of the 12th ACM SIGPLAN Symposium on Principles and Practice of Parallel Programming; 2007. p. 68–78.
9. Ramamritham K, Chrysanthis PK. A taxonomy of correctness criteria in database applications. VLDB J. 1996;5(1):85–97.
10. Schuldt H, Alonso G, Beeri C, Schek H-J. Atomicity and isolation for transactional processes. ACM Trans Database Syst. 2002;27(1):63–116.
11. Weikum G. Principles and realization strategies of multilevel transaction management. ACM Trans Database Syst. 1991;16(1):132–80.
12. Weikum G, Hasse C. Multi-level transaction management for complex objects: implementation, performance, parallelism. VLDB J. 1993;2(4): 407–53.
13. Weikum G, Schek H-J. Architectural issues of transaction management in multi-layered systems. In: Proceedings of the 10th International Conference on Very Large Data Bases; 1984. p. 454–65.
14. Weikum G, Vossen G. Transactional information systems: theory, algorithms, and the practice of concurrency control and recovery. Los Altos: Morgan Kaufmann; 2001.
15. Zimmermann O, Grundler J, Tai S, Leymann F. Architectural decisions and patterns for transactional workflows in SOA. In: Proceedings of the 5th International Conference on Service-Oriented Computing; 2007. p. 81–93.

Multimedia Data

Ramesh Jain
University of California, Irvine, CA, USA

Synonyms

Multimodal data

Definition

Multimedia in principle means data of more than one medium. It usually refers to data representing multiple types of medium to capture information and experiences related to objects and events. Commonly used forms of data are numbers, alphanumeric, text, images, audio, and video. In common usage, people refer a data set as multimedia only when time-dependent data such as audio and video are involved.

Historical Background

In early stages of computing, the major applications were scientific computations. In these applications, computers dealt with numbers and were programmed to carry out a sequence of calculations to solve a scientific problem. As people realized power of computing, new applications started emerging. Alphanumeric data was the next type of data to be used in different applications. In early days, these applications were mostly related to businesses. These applications were mostly to store large volumes of data to find desired information from this data set. These applications were the motivation of the development of current database technology.

Text is a special case of alphanumeric data. In text, there is a large string of alphanumeric data that humans associate with written language. Text has been the basis of written human communication and has become one of the most common data form. Most of the information and communication among humans takes place in text.

Next data type to start appearing on computers was images. Images started in many applications where they needed to be analyzed as well as in applications where computers were used to create and display images. Image processing and computer vision emerged as fields dealing with image analysis and understanding while computer graphics emerged as a field dealing with creation and display of images. Images were initially represented in two ways: a list of lines (called vectors) and a 2-dimensional array of intensity values. The second method has now become the most common method of representing images. Images represent a more complex data type because people perceive not the data, which is really a large collection of intensity values, but what the data represents. In computer generated images, the semantics of the pixels is determined and is known at the creation time. In all other images, the semantics must be determined. Computer vision researchers have been developing tools to automatically determine this semantics and have made progress. However, segmenting an image to determine objects in it has been a difficult problem and in general remains an unsolved problem.

Audio data represents variation of a signal over time. Signal processing deals with many types of signals, but due to its closeness to human perception, audio became an important signal type. Unlike regular numerical, alphanumerical, and image data, audio is time-varying or time dependent data. Like images, the numbers have semantics only when they are rendered, in this case using a speaker, to a human. Both images and audio are a collection of numbers that have strong semantics associated with them. This semantics can be associated only by segmenting the data and identifying each segment. Video is next in this sequence of semantic richness. Video is a time-dependent sequence of images synchronized with audio. This means that it brings with it enormous volume of data and richness of semantics.

In late 1980s, people started using the term multimedia to denote combination of text, audio, and video. This gained popularity because the technology had advanced enough to combine these media to articulate thoughts, messages, and stories using appropriate combination of these components and present them easily on computers, save them on CDs, and transmit and receive them using compression/decompression and streaming technologies. By the year 2000, multimedia had become a common data form on computers and Internet.

Foundations

Multimedia data is fundamentally different than the data traditional databases normally manage.

Some fundamental differences in multimedia data are discussed here by considering several aspects.

Types and Semantics

The data in early generation databases was either a number or a string of characters. Each data item usually represented value of an attribute. These attribute had clear and explicit semantics in the applications that used the data.

Multimedia data may be considered to be composed of numbers or strings. So an image may be viewed as a two dimensional array of integers. In multimedia applications, however, the semantics is not defined and used at the level of such basic types as in traditional applications. An image is usually considered an image that contains certain objects that are characterized by regions in the image. The relationships among these regions should also be captured. Depending on the context and an application, the semantics associated with an image may change and may need to be represented differently. Similarly an audio file may be viewed as a collection of phonemes rather than just integer values at a time instant representing sound energy. Video is a synchronized combination of audio and images. But if a video is considered just a combination of separate sound energy and images, then the semantics of video is lost. The semantics of video is due to synchronized combination of its components rather than individual elements.

Multimedia types cannot be considered simply by considering its atomic components. One must consider whole data. The data types and the semantics of multimedia data are the result of the "multi" and are not present in single (mono) medium that may be part of the whole data.

Gestalt philosophy is in action in multimedia data: the whole is bigger than the sum of its parts.

Sequence and Order

Many components of multimedia data are measurements using some sensors. These sensors measure some attribute of physical world. These measurements represent the attribute at a point in space at a particular time. The semantics of the data is intimately tied to the space and time underlying the data. The data is usually organized in the time sequence as it is acquired over some pre-defined spatial ordering of its acquisition using multiple sensors covering the space of interest.

Multimedia data could be archived data or live data. Archived data is the one that was acquired and stored and hence comes from a server. Live data is presented as it is being acquired. Live data is increasingly being used in many applications.

Size

Multimedia data is voluminous. Audio, Images, and video are much larger in size than alphanumeric data and text. Usually the size of traditional data can be measured in bytes to Kilobytes. Images usually, even the regular amateur photographs run into Megabytes and video easily runs into Gigabytes to Terabytes. Due to the size of the multimedia data, it is usually stored and transmitted in compressed form. For analysis and use of the data, it must be usually decompressed.

Meta data plays a significant role in the analysis of multimedia data and is commonly stored as part of the dataset. Metadata can be of two types: about context or about content. Contextual metadata is about the situation of the real world and the parameters of devices used in acquiring the data. Content related metadata is obtained either thru analysis of the data or by human annotation or interpretation of data.

Many different standards have evolved for compression of multimedia data and association and storage of metadata. Usually these standards related to the medium and are developed by international standards body. Some commonly used standards are JPEG for images, MP3 for audio, and MPEG for video.

Accessing Multimedia Data

Each multimedia data is usually large and represents measurements acquired using a sensor over space and time. Even an image is acquired at a location at a particular time and also contains measurements performed in space using an array of pixel. Each pixel represents measurements related to a particular point in three-dimensional space. Each image or audio video is usually represented as a separate file. This file may contain raw measurements in original form or in compressed form and may also contain associated metadata such as in EXIF data for photos acquired using digital cameras.

Multimedia data representing a measurement is usually represented as one file. In databases such data is usually represented as a pointer to the file, as a BLOB, or the name of the file.

In most current applications, multimedia data is accessed based on the metadata. All queries are formed based on metadata and then the correct file is retrieved and presented. The granularity at which multimedia data is accessed is at the level of file. Text search became so useful when it was applied to documents by analyzing and indexing all areas of a document. This content analysis and indexing based on content within a file will be very useful in multimedia data also. Research in content analysis of multimedia documents for content-based retrieval is an active research area currently.

Presentation

Multimedia data must be presented to a user by sending it to appropriate devices. Audio must be sent to speakers and images and video should also be displayed using special display programs. Displaying raw data in a file is not useful to users. In most cases, before displaying the data, it must be decompressed.

Considering large files and copyright issues, many times multimedia data is not transferred to users for storage, users are allowed to see or listen it only once each time a display request is made. Such playback of data is commonly called streaming of data and is commonly used with video. In streaming, the data from server is sent to a client only for displaying it once. This is also used in the context of live data also.

Key Applications

Computing at one time was mostly numeric, then it became alphanumeric. Now it is multimedia. Almost all applications in computing now deal with multimedia data. In a sense, the term multimedia was a good term to use in the last decade, but now it is a redundant term. In early days of computing there were two types of computing: analog and digital. Slowly all computing became

M

digital. Now no body normally uses the term digital computing because all computing is digital. In the same way all computing ranging from scientific to entertainment will use multimedia data and hence the term multimedia data or multimedia computing will shed "multimedia" and simply become data and computing.

Future Directions

Multimedia data has already become ubiquitous. With the increasing popularity of mobile phones with camera, digital cameras, and falling prices of sensors of different kinds multimedia data is becoming as widespread as alphanumeric data. Considering the current trend and human dependence on sensory data, it is likely that soon multimedia data will become more common than the traditional alphanumeric data. In terms of volume, multimedia data already may be far ahead of alphanumeric data.

Most of the current techniques which deal with multimedia data have two major limitations: first they mostly rely on metadata for access and they treat each type, such as images, audio, and video, as a separate type and hence create silos. What is required is dealing with all data, alphanumeric as well as different types of multimedia, as the data related to some physical objects or situations. This unified approach will treat all data in a unified manner and will not distinguish between media. Each media will be considered only as a source helping understand an object or a situation.

Cross-References

► Multimedia Databases

Recommended Reading

1. Jain R. Experiential computing. Commun ACM. 2003;46(7):48–55.
2. Kankanhalli MS, Wang J, Jain R. Experiential sampling in multimedia systems. IEEE Trans Multimed. 2006;8(5):937–46.
3. Rowe L, Jain R. ACM SIGMM retreat report on future directions in multimedia research. ACM Trans Multimed Comput Commun Appl. 2005;1(1):3–13.
4. Steinmetz R, Nahrstedt K. Multimedia fundamentals: media coding and content processing (IMSC Press Multimedia series). Prentice Hall; 2002.

Multimedia Data Buffering

Jeffrey Xu Yu
Department of Systems Engineering and Engineering Management, The Chinese University of Hong Kong, Hong Kong, China

Definition

Multimedia data are large in size and reside on disks. When users retrieve large multimedia data, in-memory buffers are used to reduce the number of disk I/Os, since memory is significantly faster than disk. The problem to be studied is to efficiently make use of buffers in the multimedia system to reduce the number of I/Os in order to get a better performance when multiple users are retrieving multiple multimedia data simultaneously. Existing works on multimedia data buffering focus on either the replacement algorithms to lower the number of cache misses or the buffer sharing algorithms when many simultaneous clients reference the same data item in memory.

Historical Background

Early works on multimedia data buffering focus on replacement algorithms to reduce the number of cache misses. Although in the traditional database systems, a number of different buffer replacement algorithms, such as the least recently used (LRU) and most recently used (MRU) algorithms are used to approximate the performance behavior of the optimal buffer replacement algorithm [1, 2, 6, 8, 15]. They do not reduce disk I/O significantly when they are used in a multimedia database system. Many new buffer replacement algorithms are proposed to save as much of the reserved disk bandwidth for continuous media data as possible. In [5], the effects of various buffer replacement algorithms on the number of glitches experienced by clients are studied. In [9], the authors introduce two buffer replacement algorithms, namely, the basic replacement algorithm (BASIC) and the distance-based replacement algorithm (DISTANCE), for multimedia database systems, which have a much better performance in comparison with LRU and MRU schema.

In terms of buffer sharing, a simple buffer replacement strategy may miss some opportunities to share memory buffers [14]. A straightforward use of LRU or LRU-k [6, 8] is shown to be inadequate [7]. Bridging [3, 4, 10–12] as a new technique is studied to facilitate data sharing in memory, but it can degrade the system performance. In [13], the authors observe that an uncontrolled buffer sharing scheme may reduce system performance, and introduce the Controlled Buffer Sharing (CBS), which can trade memory for disk bandwidth in order to minimize cost per stream.

Foundations

Buffer Replacement

Assume that each buffer in the buffer space of a system is of the same size and is tagged as either free or used. All the free buffers are kept in a free buffer pool. In order to meet the rate requirement for clients, the system must prefetch the required data block from disks into the buffer space, so that the required piece of data is already in the buffer space before being read. In each service cycle, the system first moves the buffers containing data blocks that were already consumed in the last service cycle to the free buffer pool, then determines which data block need to be pre-fetched from disk to the buffer next. If the block is not in the buffer space, then it allocates buffers, from the set of free buffers for the block and issues disk I/O to retrieve the needed data block from disk into the allocated buffers. The algorithm to decide which of the buffers should be allocated is referred to as the buffer replacement algorithm. Several general replacement algorithms are listed below, which are widely used in database management systems. (i) LRU: when a buffer is to be allocated, the buffer containing the block that is used least recently is selected. (ii) MRU: when a buffer is to be allocated, the buffer containing the block that is used most recently is selected. (iii) Optimal: when a buffer is to be allocated, the buffer containing the block that will not be referenced for the longest period of time is selected. Since arrival, pause, resume and jump time when playing an object are unknown in advance, the optimal algorithm can only be implemented for simulation studies.

For the multimedia database systems, the commonly used LRU and MRU algorithms may not reduce disk I/O significantly. Two buffer replacement algorithms are proposed. They are the basic replacement algorithm (BASIC) and the distance-based replacement algorithm (DISTANCE).

The BASIC Buffer Replacement Algorithm

The main idea behind the BASIC buffer replacement algorithm [9] is as follows. It is possible to estimate the duration by assuming each client will remain its consumption rate for a long period, even though it is difficult to decide which block will not be referenced for the longest period of time. It assumes that clients continue to consume data at the specified rate they are accessing the blocks. When there is a new request to allocate a buffer, the BASIC algorithm selects the buffer containing the block that will not be accessed for the longest period of time. If there are several such buffers, the algorithm will select the block with the highest offset-rate ratio (the ratio of offset/rate) to be replaced. The BASIC algorithm may reduce the miss ratio to nearly optimal, but it requires to sort clients and free buffers in the increasing order of their offset, which make the overhead of the BASIC algorithm very high. The DISTANCE algorithm is proposed to handle the overhead.

M

The DISTANCE Buffer Replacement Algorithm

The main idea behind the DISTANCE buffer replacement algorithm is based on distance between clients [9]. Suppose that there are clients, c_1, c_2, ..., accessing the same media data, M. Assume that each client, c_i, is accessing the M at a certain position of M, denoted as $p_i (M)$, and the data block on disk starting from $p_i (M)$ is kept in a buffer, B_i. Let all clients that are accessing the same media data M be sorted in order, c_1, c_2, ... Here, c_i is accessing M ahead of c_j if $i < j$, or in other words, $p_i (M) > p_j (M)$, because c_i has already accessed $p_j (M)$ and is now accessing $p_i (M)$. The distance between c_i and its next c_{i+1} is denoted as $dist_i$ which is equal to $p_i (M) - p_{i+1}(M)$. Note that the distance d_i is a value associated with the client c_i. Suppose all clients c_1, c_2, ..., are accessing their blocks in the buffers in the current cycle. They all need to move ahead and access the next data blocks. The question becomes which buffer they are accessing in the current cycle needs to be freed if the buffer is full. In brief, the buffers consumed by a client, c_i, will be kept longer if the next client, c_{i+1}, will need them shortly (small distance $dist_i$). The buffers consumed by a client, c_j, will be freed earlier if the next client, c_{j+1}, does not need to access the data block $p_j (M)$, that c_j has just accessed, shortly (large distance $dist_j$). The DISTANCE algorithm frees buffers consumed by clients in the previous cycle in the decreasing order of clients' $dist_i$. In other words, when a new buffer needs to be allocated and there are no free buffers, a buffer consumed by a client, which will not be accessed by its next client shortly, based on the distance between clients, will be selected as a victim to be freed.

The DISTANCE algorithm can be implemented by dynamically maintaining a client list which is ordered in the decreasing order of clients' $dist_i$. The overhead is lower than the BASIC algorithm.

Table 1 shows the comparison of overhead and cache misses of different buffer replacement algorithms, n_B is the number of buffers used.

Buffer Sharing

Consider buffer sharing, where cached data can be shared among all the clients. A naive approach is to use LRU or LRU-k, which is shown to be inadequate to efficiently share data. An example is given in Fig. 1. There are two displays, D_1 and D_2, and both reference different blocks of

Multimedia Data Buffering, Table 1 Overhead and cache misses of different buffer replacement algorithms (Table 1 in [9]).

n_B	LRU	MRU	BASIC	DISTANCE	# of refs
300	13:48 s/670,080	13:44 s/668,974	33:30 s/638,974	11:89 s/641,274	670,080
600	13:32 s/670,080	13:18 s/665,748	1:12 min/595,416	10:96 s/599,214	670,080
1200	16:80 s/665,934	16:27 s/657,634	3:41 min/549,570	12:28 s/554,640	670,080
2400	13:22 s/654,240	12:32 s/642,914	5:31 min/480,068	8:64 s/481,364	670,080

Multimedia Data Buffering, Fig. 1 Two displays may compete for buffer frames with LRU (Fig. 1 in [13])

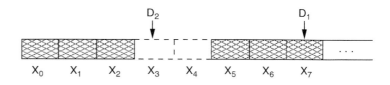

X_0 X_1 X_2 X_3 X_4 X_5 X_6 X_7

: Data block available when needed for display

: Data block not available when needed for display

the same clip. With LRU as a global buffer pool replacement policy, the blocks accessed by D_1 may be discarded before D_2 needs to access.

Bridging [3, 4, 10–12] as a technique is to form a bridge between the data blocks staged by two different clients referencing the same clip, which enables them to share memory and use one disk stream. As shown in Fig. 2, two displays D_1 and D_2 are supported using a single disk stream. The distance between D_1 and D_2 is 5. With the bridging technique, it holds the intermediate data pages between D_1 and D_2 in the buffer pool, and does not swap these pages out from the buffer pool. However, as analyzed in [11, 12], a potential problem is that a simple bridging may possibly exhaust the available buffer space, which will have great impacts on the system performance.

A Controlled Buffer Sharing (CBS) technique is proposed in [13], which increases disk bandwidth using memory in order to achieve two objectives, namely, minimization of cost per simultaneous stream, and balancing memory and disk utilization. The latter considers that unlimited memory consumption may in fact degrade the system performance. The framework of CBS is shown in Fig. 3. The framework consists of three components: a configuration planner, a system generator, and a buffer management technique. The configuration planner determines the amount of required buffer and disk bandwidth in support of a pre-specified performance objective. The system generator simply acts as a multiplier. The first two components are applied off-line to determine the system size. The buffer management technique controls the memory consumption at run time.

In the CBS framework, a distance threshold, d_t, is used to capture the cost of memory and disk bandwidth and control the number of pinned buffer blocks between two adjacent displays that access the same clip. As shown in Fig. 4, suppose $d_t = 5$, D_1 and D_2 can share one disk stream because their distance is below the specified threshold, while D_3 and D_4 cannot share one disk stream because their distance exceeds the threshold.

a

Current cycle:

D_2 D_1

... X_3 X_4 X_5 X_6 X_7 X_8 ...

b

Next cycle:

D_2 D_1

... X_4 X_5 X_6 X_7 X_8 X_9 ...

Multimedia Data Buffering, Fig. 2 Bridging (Fig. 2 in [13])

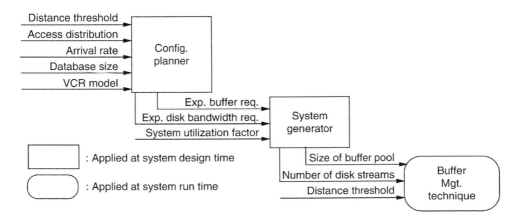

Multimedia Data Buffering, Fig. 3 The CBS Scheme (Fig. 4 in [13])

Multimedia Data Buffering, Fig. 4 The effectiveness of distance threshold ($d_t = 5$) (Fig. 5 in [13])

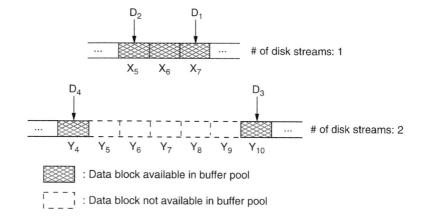

Key Applications

Buffering is widely used in retrieving and playing multimedia data, especially for network continuous media applications, where multiple users may need to display multiple medias simultaneously.

Cross-References

▶ Buffer Management
▶ Buffer Manager
▶ Continuous Multimedia Data Retrieval
▶ I/O Model of Computation
▶ Multimedia Data Buffering
▶ Multimedia Data Storage
▶ Multimedia Resource Scheduling

Recommended Reading

1. Chew KM, Reddy J, Romer TH, Silberschatz A. Kernel support for recoverable-persistent virtual memory. In: Proceedings of the USENIX MACH III Symposium; 1993. p. 215–34.
2. Chou HT, DeWitt DJ. An evaluation of buffer management strategies for relational database systems. In: Proceedings of the 11th International Conference on Very Large Data Bases; 1985. p. 127–41.
3. Dan A, Dias DM, Mukherjee R, Sitaram D, Tewari R. Buffering and caching in large-scale video servers. In: Digest of papers – COMPCON; 1995. p. 217–24.
4. Dan A, Sitaram D. Buffer management policy for an on-demand video server. IBM Research Report RC 19347.
5. Freedman CS, DeWitt DJ. The SPIFFI scalable video-on-demand system. ACM SIGMOD Rec. 1995;24(2):352–63.
6. Lee D, Choi J, Kim JH, Noh SH, Min SL, Cho Y, Kim CS. On the existence of a spectrum of policies that subsumes the least recently used (LRU) and least frequently used (IFU) policies. SIGMETRICS Perform Eval Rev. 1999;27(1):134–43.
7. Martin C. Demand paging for video-on-demand servers. In: Proceedings of the International Conference on Multimedia Computing and Systems; 1995. p. 264–72.
8. O'Neil EJ, O'Neil PE, Weikum G. The LRU-K page replacement algorithm for database disk buffering. In: Proceedings of the ACM SIGMOD International Conference on Management of Data; 1993. p. 297–306.
9. Özden B, Rastogi R, Silberschatz A. Multimedia information storage and management, chap. 7: buffer replacement algorithms for multimedia storage systems. Kluwer Academic. 1996.
10. Rotem D, Zhao JL. Buffer management for video database systems. In: Proceedings of the 11th International Conference on Data Engineering; 1995. p. 439–48.
11. Shi W, Ghandeharizadeh S. Buffer sharing in video-on-demand servers. SIGMETRICS Perform Eval Rev. 1997;25(2):13–20.
12. Shi W, Ghandeharizadeh S. Trading memory for disk bandwidth in video-on-demand servers. In: Proceedings of the 1998 ACM Symposium on Applied Computing; 1998. p. 505–12.
13. Shi W, Ghandeharizadeh S. Controlled buffer sharing in continuous media servers. Multimedia Tools Appl. 2004;23(2):131–59.
14. Christodoulakis S, Ailamaki N, Fragonikolakis Y, Koveos L, Leonidas K. An object oriented architecture for multimedia information systems. Data Eng. 1991;14(3):4–15.
15. Stonebraker M. Operating system support for database management. Readings in database systems. 3rd ed. San Francisco: Morgan Kaufmann; 1998. p. 83–9.

Multimedia Data Indexing

Paolo Ciaccia
Computer Science and Engineering, University of Bologna, Bologna, Italy

Synonyms

MM indexing

Definition

Multimedia (MM) data indexing refers to the problem of preprocessing a database of MM objects so that they can be efficiently searched for on the basis of their content. Due to the nature of MM data, indexing solutions are needed to efficiently support *similarity queries*, where the similarity of two objects is usually defined by some expert of the domain and can vary depending on the specific application. Peculiar features of MM indexing are the intrinsic high-dimensional nature of the data to be organized and the complexity of similarity criteria that are used to compare objects. Both aspects are therefore to be considered for designing efficient indexing solutions.

Historical Background

Earlier approaches to the problem of MM data indexing date back to the beginning of 1990s, when it became apparent the need of efficiently supporting queries on large collections of non-standard data types, such as images and time series. Representing the content of such data is typically done by automatically extracting some low-level features (e.g., the color distribution of a still image), so that the problem of finding objects similar to a given reference is transformed into the one looking for similar features. Although, at that time, many solutions from the pattern recognition field were available for this problem,

they were mainly concerned with the *effectiveness* issue (which features to consider and how to compare them), thus almost disregarding *efficiency* aspects.

The issue of making similarity query processing scalable to large databases was first considered in systems like QBIC [6] for the indexing of color images and by more focused approaches such as the one described by Jagadish in [8] for indexing shapes. Not surprisingly, these solution adopted index methods available at that time that had been developed for the case of low-dimensional spatial databases, such as R-trees and grid files. The peculiarity of MM data then originated a flourishing brand new stream of research, which resulted in many indexes explicitly addressing the problems of high-dimensional features and complex similarity criteria.

Foundations

Figure 1 illustrates the typical scenario to be dealt with for indexing multimedia data. The first step, *feature extraction*, is concerned with the problem of highlighting those relevant features, f_i, of an object o_i on which content-based search wants to be performed. In the figure, this is the shape of the image subject (a cheetah). The second step, *feature approximation*, is optional and aims to obtain a more compact representation, af_i, of f_i that can be inserted into a suitable index structure (third step). It has to be remarked that, while feature extraction is needed to define which are the relevant aspects of objects on which the search has to focus on, feature approximation is mainly motivated by feasibility and efficiency reasons. This is because it might not be possible to directly index non-approximate features and/or indexing approximate features might result in a better performance of the search algorithms.

Consider a collection $O = \{o_1, o_2, \ldots, o_n\}$ of MM objects with corresponding features $F = \{f_1, f_2, \ldots, f_n\}$ and approximate features $AF = \{af_1, af_2, \ldots af_n\}$. In order to compare

M

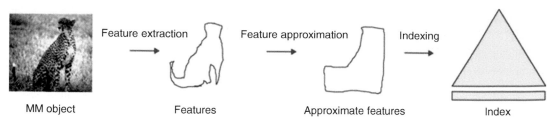

Multimedia Data Indexing, Fig. 1 The multimedia data indexing scenario

features, a *distance function d* is typically set up, where $d(f_i, f_j)$ measures how dissimilar are the feature values of objects o_i and o_j. Given a reference object q (the *query point*), a range query with radius $\in$, also called an $\in-$ similarity query, will return all the objects $o_i \in O$ such that $d(f_i, f(q)) \leq \in$, whereas a k-nearest neighbor query (k-NN) will return the k objects in O whose features are closest to those of q.

A simple yet remarkable result due to Agrawal, Faloutsos, and Swami [1], and now popularly known as the lower-bounding lemma, provides the basis for exactly solving queries by means of an index that organizes approximate features.

The lower-bounding lemma. Let I be an index that organizes the set of approximate features $AF = \{af_1, af_2, \ldots, af_n\}$ and that compares such features using an *approximate distance* d_{appr}. If, for any pair of objects, it is $d_{\mathrm{appr}}(af_i, af_j) \leq d(f_i, f_j)$, then the result of a range query obtained from I is guaranteed to contain the exact result, i.e., no false dismissals are present.

The result easily follows from the observation that, since d_{appr} lower bounds d by hypothesis, $d(f_i, f(q)) \leq \in$ implies $d_{\mathrm{appr}}(af_i, af(q)) \leq \in$. The lower-bounding lemma guarantees that querying the index with a search radius equal to $\in$ will return a result set that contains all the objects whose non-approximate features satisfy the query constraint.

Filter and Refine

When indexing is based on an approximate distance, a two-step *filter and refine* process is therefore needed, in which the role of the index is to filter out many irrelevant objects. The so-resulting candidate objects then need to be verified by using the actual distance d. The lower-bounding lemma is also the key for solving k-NN queries using a multistep query processing approach.

The effectiveness of the filter and refine approach depends on two contrasting requirements:

1. The approximate distance function d_{appr} should be a tight approximation of d, in order to minimize the number of false hits, i.e., those objects that do not satisfy the query constraint yet the index is not able to discard them. These are exactly those objects o_i for which both $d_{\mathrm{appr}}(af_i, af(q)) \leq \in$ and $d(f_i, f(q)) > \in$ hold.

2. At the same time, d_{appr} should be relatively cheap to compute as compared to d, in order to avoid wasting much time in the filter phase.

The literature on MM indexing abounds of examples showing how to derive effective approximations for complex distance functions. For instance, the QBIC system compares color images using a quadratic form distance function, $d_A^2(f_i, f_j) = (f_i - f_j)A(f_i - f_j)^T = \sum_{k=1}^{D} \sum_{l=1}^{D} a_{k,l}(f_{i,k} - f_{j,k})(f_{i,l} - f_{j,l})$, where $A = (a_{k,l})$ is a color-to-color similarity matrix and features are color histograms. Evaluating d_A has complexity $O(D^2)$, which becomes too costly even for moderately large values of D, the number of bins in the color histograms. In [6] it is demonstrated that using as approximate features the average RGB color of an image, which is a three-dimensional vector, and comparing average colors using the Euclidean distance, i.e., $d_{\mathrm{avg}}^2(af_i, af_j) = \sum_{k \in \{R, G, B\}}(af_{i,k} - af_{j,k})^2$, leads to derive that $d_{\mathrm{avg}}^2 \leq d_A^2/\lambda_1$, where λ_1

is the smallest eigenvalue of matrix A. Then, the lower-bounding lemma guarantees that querying an index built on average colors with a range query of radius $\epsilon \sqrt{\lambda_1}$ will not lead to any false dismissal.

As another relevant example, consider the problem of comparing feature vectors that represent time-varying signals. A distance function more robust than the Euclidean one to misalignments on the time domain is the dynamic time warping (DTW) distance. However, evaluating DTW has a complexity $O(D^2)$, which is untenable for long time series. In [9] Keogh introduces an effective lower-bounding function for the DTW distance. In essence, the idea is to construct an *envelope*, *Env(q)*, around the query time series q, after that an Euclidean-like distance between *Env(q)* and a stored sequence f_i can be computed in $O(D)$ time.

The Need for Approximate Features

As anticipated, there are several reasons for which approximate features might have to be considered. First, in many relevant cases the features of a MM object are represented through a high-dimensional vector, $f_i = (f_{i,1}, f_{i,2}, \dots, f_{i,D})$, with D of the order of the hundreds or even thousands. At such high dimensions, it is known that the performance of multidimensional indexes rapidly deteriorates, becoming either comparable to, or even worse than, that of a sequential scan. This phenomenon, known as the dimensionality curse, inhibits any approach based on a direct indexing of feature values. Besides ad hoc solutions, such as those above described, one might consider using some dimensionality reduction technique that projects feature vectors onto a (much) lower D'-dimensional space, $D' \ll D$, and then indexing the so-obtained D'-dimensional feature vectors. The effectiveness of such techniques however is highly variable, being dependent on the actual data distribution.

Another practical reason that could motivate the use of approximate features is the mismatch between the type of the features and the one natively supported by the index. As a simple example, consider an index implementation that only manages entries of an arbitrary, but fixed, size, and that objects to be indexed are regions of pixels described by their boundaries. Clearly, boundary descriptions have different sizes, depending on the shape of the region. In this case a possible solution would be to use a *conservative approximation* of boundaries, like minimum bounding rectangles.

Finally, it might also be the case that, although in principle actual feature values could be stored in the index, the distance function to be used on them cannot be supported by the index organization. A remarkable example is the DTW distance: since DTW is *not* a true metric, in that it does not satisfy the triangle inequality, no multidimensional index can directly process queries with such a distance function.

Metric Indexing

When features are not vectors and/or the distance function is not the Euclidean distance or some other (possibly weighted) L_p norms, coordinate-based spatial indexes cannot be used. There are many cases in which this situation shows up. For instance, in region-based image retrieval (RBIR), each image is first automatically segmented into a set of homogeneous regions, each of them being represented by a vector of low-level features (usually encoding color and texture information). Thus, each f_i is a *set of vectors* and as such cannot be indexed by a spatial index. As a further example, *graphs* representing, say, spatially located objects with their relationships cannot be directly supported by a coordinate-based index. In cases like these, one could consider using a metric index, such as the M-tree [5]. A metric index just requires the distance function d used to compare feature values to be a metric, i.e., a positive and symmetric function that also satisfies the triangle inequality: $d(f_i,f_j) \le d(f_i,f_k) + d(f_k,f_j) \; \forall f_i, f_j, f_k$. Although there is nowadays a large number of metric indexes available [14], as demonstrated in [3], all of them are based on the common principle of organizing the indexed features into a set of equivalence classes and then discarding some of these classes by exploiting the triangle inequality. For instance, in the case of the M-tree, each class corresponds to the set of feature values stored into a same leaf of the tree. Triangle inequality can also be applied to save some distance computations while searching the index, which turns out to be particularly relevant in the case of computationally demanding distance functions (a common case with MM data). This was first shown for the M-tree, in which distances between each feature value and its parent in the index tree are precomputed and stored in the tree. The idea is quite general and effective, an obvious trade-off existing between the amount of extra information stored in the tree and the benefit this has on pruning the search space. Along this direction, Skopal and Hoksza propose the M*-tree [12], a variant of the M-tree in which each entry in a node also includes its NN in that node, i.e., the *NN-graph* of the features in each node is maintained.

A common objection to metric indexes is that they are bound to use only a specific distance function, namely, the one with which the index is built. Along the direction of increasing flexibility, Ciaccia and Patella [4] introduce the QIC-M-tree, which is an extension of the M-tree able to support queries with any distance function d_Q from the same "family" of the distance d_I used to build the tree. On the condition that there exists a *scaling factor* $S_{d_I \to d_Q}$ such that $d_I(f_i, f_j) \leq S_{d_I \to d_Q} d_Q(f_i, f_j)$ holds (i.e., d_I lower bounds d_Q up to a constant factor), the lower-bounding lemma applies, and the index can answer queries based on d_Q. A similar idea allows the QIC-M-tree to use also a "cheap" approximate distance d_C as a filter before computing the "costly" d_I and d_Q functions.

Ad Hoc Solutions

The availability of general purpose metric indexes does not rule out the possibility of deriving better, more specialized solutions for the problem at hand. For instance, the STRG-Index [10] is a specialized structure for indexing spatiotemporal graphs arising from the modeling of video sequences. Consider a video segment with N frames. Each frame is first segmented into a set of homogeneous color regions, each of which becomes a node in the region adjacency graph

(RAG) of that frame, with edges connecting spatially adjacent regions. Node attributes (such as size, color, and location) are then defined, and the same is done for edges (in which case attributes such as the distance and the orientation between the centroids of connected regions can be used). Since a node representing a region can span multiple frames, nodes in consecutive RAGs can be connected to represent temporal aspects. The resulting graph is called spatiotemporal region graph (STRG). The STRG is then decomposed into a set of object graphs (OGs) and background graphs (BGs), and clusters of OGs are obtained for the purpose of indexing. Since the distance function used for comparing OGs (the so-called extended graph edit distance (EGED)) is a metric, any metric index could be used. The ad hoc STRG-Index proposed in [10] is a three-level metric tree, where the root node contains entries for the BGs, the intermediate level stores clusters of OGs, and individual OGs are inserted into the leaf level.

Extensions of available indexes might be also required as a consequence of feature approximation. An example is found in [13], where the problem of providing rotation-invariant retrieval of shapes under the Euclidean (L_2) distance is considered. After converting a shape boundary into a time series $f_i = (f_{i,1}, f_{i,2}, \ldots, f_{i,D})$ (this is quite a common way to represent shapes; see, e.g., [2]), a discrete Fourier transform (DFT) is applied to obtain a representation of $f\,i$ in the frequency domain. Due to Parseval's theorem, the DFT transformation preserves the Euclidean distance [1]. To obtain invariance to rotation, only the magnitude of DFT coefficient is retained. The so-resulting vectors F_i are then compressed by keeping only the k ($k \ll D$) coefficients with the highest magnitude (together with their position in the original vector) plus an error term ϵF_i given by the square root of the sum of the squares of dropped coefficients. This information allows a tight lower bound to be derived on the actual rotation-invariant Euclidean distance between f_i and a query shape q. For indexing, a variation of the VP-tree is introduced, which allows compressed features to be stored and searched.

Key Applications

Any application dealing with massive amounts of multimedia data requires effective indexing solutions for efficiently supporting similarity queries. This is further motivated by the complexity of distance functions that are of interest for multimedia data.

Future Directions

All the above indexing techniques and methods assume (at least) that the distance function is a metric. An interesting problem is to devise indexing methods for nonmetric distance functions that do not rely on the lower-bounding lemma. The work of Skopal [11] on *semimetrics* appears to be a relevant step on this direction. In the same spirit, Goial, Lifshits, and Schütse [7] study how to avoid turning the *similarity* search problem into a *distance*-based one, which in several cases might not yield a metric. Working directly with similarities is however more complex, since there is no analogue of the triangle inequality property for similarity values. Let $rank_y(x)$ be the rank of object x with respect to object y (i.e., x is the NN of y if $rank_y(x) = 1$). Then, [7] introduces the concept of *disorder constant DC*, the smallest value for which the *disorder inequality $rank_y(x) \leq DC(rank_z(x) + rank_z(y))$* holds $\forall x, y, z$ in the given dataset, and describes algorithms for NN search based on this idea. Making this approach practical for large MM databases remains an open problem.

Cross-References

▶ Curse of Dimensionality
▶ Dimensionality Reduction
▶ High-Dimensional Indexing
▶ Indexing and Similarity Search
▶ Indexing Metric Spaces
▶ Multimedia Data Querying
▶ Spatial Indexing Techniques

Recommended Reading

1. Agrawal R, Faloutsos C, Swami A. Efficient similarity search in sequence databases. In: Proceedings of the 4th International Conference on Foundations of Data Organizations and Algorithms; 1993. p. 69–84.
2. Bartolini I, Ciaccia P, Patella M. WARP: accurate retrieval of shapes using phase of Fourier descriptors and time warping distance. IEEE Trans Pattern Anal Machine Intell. 2005;27(1):142–7.
3. Chávez E, Navarro G, Baeza-Yates R, Marroquín JS. Proximity searching in metric spaces. ACM Comput Surv. 2001;33(3):273–321.
4. Ciaccia P, Patella M. Searching in metric spaces with user-defined and approximate distances. ACM Trans Database Syst. 2002;27(4):398–437.
5. Ciaccia P, Patella M, Zezula P. M-tree: an efficient access method for similarity search in metric spaces. In: Proceedings of the 23rd International Conference on Very Large Data Bases; 2007. p. 426–35.
6. Faloutsos C, Barber R, Flickner M, Hafner J, Niblack W, Petkovic D, Equitz W. Efficient and effective querying by image content. J Intell Inf Sys. 1994;3(3/4):231–62.
7. Goyal N, Lifshits Y, Schütse H. Disorder inequality: a combinatorial approach to nearest neighbor search. In: Proceedings of the 1st ACM International Conference on Web Search and Data Mining; 2008. p. 25–32.
8. Jagadish HV. A retrieval technique for similar shapes. In: Proceedings of the ACM SIGMOD International Conference on Management of Data; 1991. p. 208–17.
9. Keogh E. Exact indexing of dynamic time warping. In: Proceedings of the 28th International Conference on Very Large Data Bases; 2002. p. 406–17.
10. Lee J, Oh JH, Hwang S. STRG-index: spatiotemporal region graph indexing for large video databases. In: Proceedings of the ACM SIGMOD International Conference on Management of Data; 2005. p. 718–29.
11. Skopal T. On fast non-metric similarity search by metric access methods. In: Advances in Database Technology, Proceedings of the 10th International Conference on Extending Database Technology; 2006. p. 718–36.
12. Skopal T, Hoksza D. Improving the performance of M-tree family by nearest-neighbor graphs. In: Proceedings of the 11th East European Conference Advances in Databases and Information Systems; 2007. p. 172–88.
13. Vlachos M, Vagena Z, Yu PS, Athitsos V. Rotation invariant indexing of shapes and line drawings. In: Proceedings of the ACM International Conference on Information and Knowledge Management; 2005. p. 131–8.
14. Zezula P, Amato G, Dohnal V, Batko M. Similarity search: the metric space approach. Berlin: Springer; 2005.

M

Multimedia Data Querying

K. Selcuk Candan[1] and Maria Luisa Sapino[2]
[1]Arizona State University, Tempe, AZ, USA
[2]University of Turin, Turin, Italy

Definition

By its very nature, multimedia data querying shares the three V challenges ([V]olume, [V]elocity, and [V]ariety) of the so called "Big Data" applications. Systems supporting multimedia data querying, however, must tackle additional, more specific, challenges, including those posed by the *[H]igh-dimensional, [M]ulti-modal* (temporal, spatial, hierarchical, and graph-structured), and *inter-[L]inked* nature of most multimedia data as well as the *[I]mprecision* of the media features and *[S]parsity* of the observations in the real-world.

Moreover, since the end-users for most multimedia data querying tasks are us (i.e., humans), we need to consider fundamental constraints posed by *[H]uman* beings, from the difficulties they face in providing unambiguous specifications of interest or preference, subjectivity in their interpretations of results, and their limitations in perception and memory. Last, but not the least, since a large portion of multimedia data is human-centered, we also need to account for the users' (and others') needs for *[P]rivacy*.

Variety/heterogeneity in multimedia data is due to three factors: (a) the semantics of the data in different forms can be drastically different from each other, (b) the resource and processing requirements of various media differ substantially, and (c) the user needs and context have significant impacts on what is relevant. User dependence necessitates a process that can take low-level features that are available from the media and map to the high-level semantic features that require external knowledge. This is commonly referred to as the *semantic gap*.

While graph structured models are commonly used for representing interrelationship among media elements and associated semantic con-cepts, the models that can capture the imprecise and statistical nature of multimedia data and query processing are fuzzy and probabilistic in nature. Therefore multimedia data query evaluation requires fuzzy and probabilistic data and query models as well as appropriate query processing mechanisms. Probabilistic models rely on the premise that the sources of imprecision in data and query processing are inherently statistical and thus they commit onto probabilistic evaluation. Fuzzy models are more flexible and allow various different semantics, each applicable under different system requirements to be selected for query evaluation.

Historical Background

Multimedia querying is an inherently dynamic process, and systems for multimedia querying and exploration must be able to support, efficiently and effectively, a continuous exploration cycle involving four key steps:

- **Sense and integrate**: the system takes as inputs and integrates data, media, and models of the application space and continuously sensed real-time media data,
- **Filter, rank and recommend**: the system provides support for context-aware access to integrated media data sets,
- **Visualize and feedback**: the system acquires accurate user feedback through an intuitive data and result representation, and
- **Act and adapt**: the system provides continuous adaptation of models of data, context, and user preference based on user feedback.

Due to the possibly redundant ways to sense the environment, the alternative ways to process, filter, and fuse multimedia data, and the subjectivity involved in the interpretation of data and query results, multimedia data quality is inherently imprecise [6]:

- *Feature extraction algorithms that form the basis of content-based multimedia data query-*

ing are generally imprecise. For example, high error rate is encountered in motion capture data due to the multitude of environmental factors involved, including camera and object speed. Especially for video/audio/motion streams, data extracted through feature extraction modules are only *statistically* accurate and may be based on the frame rate or the position of the video camera related to the observed object.

• *It is rare that a multimedia querying system relies on exact object matching.* Instead, in many cases, multimedia databases leverage similarities between feature vectors to identify data objects that are similar to the query. In many cases, it is also necessary to account for *semantic similarities* between associated annotations and *partial matches,* where objects in the result satisfy some of the requirements in the query, but fail to satisfy all query conditions.

• *Imprecision can also be due to the available index structures which are imperfect.* Due to the sheer size of the data, many systems rely on clustering and classification algorithms for pruning during query processing.

• *Query formulation methods are not able to capture user's subjective intention perfectly.* For example, in Query by Example (QBE), which features, feature value ranges, feature combinations, or which similarity notions are to be used for processing is left to the system to figure out through feature significance analysis, user preferences, relevance feedback (Fig. 1), and/or collaborative filtering techniques, which are largely statistical and probabilistic in nature.

In many multimedia querying systems, more than one of these reasons coexist and, consequently, the system must take them into consideration collectively. Figure 2 provides an example

M

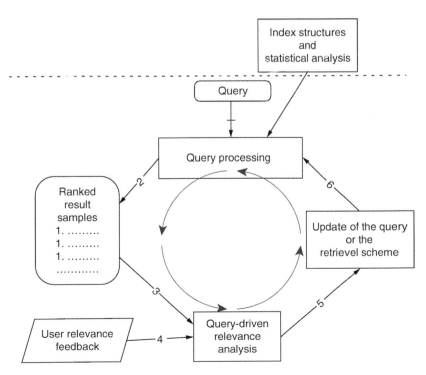

Multimedia Data Querying, Fig. 1 Multimedia query processing usually requires the semantic gap between what is stored in the database and how the user interprets the query and the data to be bridged through a relevance feedback cycle. This process itself is usually statistical in nature and, consequently, introduces probabilistic imprecision in the results

Multimedia Data Querying, Fig. 2 A sample multimedia query with imprecise and exact predicates

> *select image P, imageobject object*
> *where contains (P, object) and*
> *   semantically_similar(P.semanticannotation, "travel") and*
> *   visually_similar(object.imageproperties, "Fujimountain.jpg") and*
> *   after(P.date, 2007).*

query (in an SQL-like syntax used by the SEM-COG system [16]) which brings together imprecise and exact predicates. Processing this query requires assessment of different sources of imprecision and merging them into a single value. Traditional databases are not able to deal with imprecision since they are based on Boolean logic: predicates are treated as propositional functions, which return either *true* or *false*. A naive way to process queries is to transform imprecision into *true* or *false* by mapping values less than a cut-off to *false* and the remainder to *true*. With this naïve approach, partial results can be quickly refuted or validated based on their relationships to the cut-off. User provided cut-offs can also be leveraged for *filtering*, while maintaining the imprecision value of the results for further processing. In general, however, cut-off based early pruning leads to misses of relevant results. This leads to the need for data models and query evaluation mechanisms, which can take into account imprecision in the evaluation of the query criteria. In particular, the data and query models cannot be propositional in nature.

Foundations

Feature Representations

Features of interest of multimedia data can be diverse, including low-level content-based features, such as color, to higher-level semantic features that require external knowledge and semantic representations. Despite this diversity of features and feature models, however, in general, we can classify the representations common to many media features into five general classes [6]:

- Vectors: Given a set of independent properties of interest to describe multimedia objects, the vector model associates a muti-dimensional vector space, where the i-th dimension corresponds to the i-th property. Intuitively, the vector describes the composition of a given multimedia data object in terms of its quantifiable properties. Histograms for example are good candidates for being represented in the form of vectors.

- Sets: Set based models are necessary when a given media object can be partitioned into multiple components, each requiring a separate feature representation, for example as a vector. Note that sets of vectors are often represented as matrices and when media properties are not scalarly valued, but are vectors themselves, this leads to tensor-based representations.

- Strings/Sequences: Many multimedia data objects, such as text documents, audio files, or DNA sequences, are essentially sequences of symbols from a base alphabet. In fact, strings and sequences can even be used to represent more complex data, such as spacial distribution of features, in a more compact manner.

- Graphs/Trees: Most complex media objects, especially those that involve spatio-temporal structures, object composition hierarchies, or object references and interaction pathways (such as hyperlinks) can be modeled as trees or graphs.

- Fuzzy and Probabilistic representations: Vectors, strings/sequences, and graphs/trees all assume that the media data has an underlying simple structure that can be used as the common basis of representation. Many times, however, the underlying regularity may be too complex to be represented in the forms of vectors, strings, or graphs. In such a case, a richer (Boolean, fuzzy, or probabilistic) logic-based model may be more suitable.

Fuzzy and Probabilistic Models

Assessments of the degrees of imprecisions in multimedia data can take different forms. For example, if the data is generated through a sensor/operator with a quantifiable quality rate (for instance a function of the available sensor power), then a scalar-valued assessment of imprecision may be applicable. This is similar to the (so called type-1) fuzzy predicates, which (unlike propositional functions which return *true* or *false*) return a membership value to a fuzzy set. *In this simplest case, the quality assessment of a given object, o, is modeled as a value* $0 \le qa(o) \le 1$. A more general quality assessment model would take into account the uncertainties in the assessments themselves. These type of predicates, where sets have grades of membership that are themselves fuzzy, are referred to as type-2 fuzzy predicates. For example the assessment of a given object o can be modeled as a normal distribution of qualities, $qa(o) = No(qo, \xi o)$, where qo is the expected quality and ξo is the variance. Although the type-2 model can be more general and use different probability distributions, this specific model (using the normal distribution) is a generally applicable sampling-related imprecision as it relies on the well-known *central limit theorem*, which states that the average of the samples tends to be normally distributed, even when the distribution from which the average is computed is not normally distributed. Note that, in general, such complex statistical assessments of data precision can be hard to obtain. A compromise between the above two models represents the range of possible qualities of an object with a lower- and an upper-bound. In this case, *given an object o, its quality assessment, qa(o) is modeled as a pair* $< qo_{low}, qo_{high} >$, *where* $0 \le qo_{low} \le qo_{high} \le 1$.

Fuzzy data and query models for multimedia querying are based on the fuzzy set theory and fuzzy logic introduced by Zadeh in mid 1960s [21]. A fuzzy set, F, with domain D is defined using a membership function, $F: D \to [0,1]$. A fuzzy predicate, then, corresponds to a fuzzy set: instead of returning *true(1)* or *false(0)* values as in propositional functions, fuzzy predicates return the corresponding membership values (or scores). Fuzzy clauses combine fuzzy predicates and fuzzy logical operators into complex fuzzy statements. Like the predicates, the fuzzy clauses also have associated scores. The meaning of a fuzzy clause (i.e., the score it has, given the constituent predicate scores) depends on the semantics chosen for the fuzzy logical operators, *not* $(\neg)$, *and* $(\wedge)$, and *or* $(\vee)$.

Table 1 shows popular *min* and *product* fuzzy semantics used in multimedia querying. These two semantics (along with some others) have the property that binary conjunction and disjunction operators are triangular-norms (t-norms) and triangular-conorms (t-conorms). Intuitively, t-norm functions reflect the (boundary, commutativity, monotonicity, and associativity) properties of the corresponding Boolean operations. Although the property of capturing Boolean semantics is desirable in many applications of fuzzy logic, for multimedia querying, this is not always the case [5]. For instance, the partial match requirements invalidate the boundary conditions. Monotonicity can be too weak a condition for multimedia query processing. In many cases, according to real-world and artificial nearest-neighbor workloads, the highest-scoring predicates are interesting and the rest is not interesting. This implies that the *min* semantics, which gives the highest

Multimedia Data Querying, Table 1 Fuzzy *min* and *product* semantics for logic operators: $\mu_i(x)$ stands for the score of the predicate P_i on x

Min semantics	Product semantics
$\mu_{P_i \wedge P_j}(x) = min\{\mu_i(x), \mu_j(x)\}$	$\mu_{P_i \wedge P_j}(x) = \frac{\mu_i(x) \times \mu_j(x)}{max(\mu_i(x), \mu_j(x), \alpha)} \alpha \in [0, 1]$
$\mu_{P_i \wedge P_j}(x) = max\{\mu_i(x), \mu_j(x)\}$	$\mu_{P_i \vee P_j}(x) = $ $\frac{\mu_i(x) + \mu_j(x) - \mu_i(x) \times \mu_j(x) - min\{\mu_i(x), \mu_j(x), 1-\alpha\}}{max\{1-\mu_i(x), 1-\mu_j(x), \alpha\}}$
$\mu_{\neg P_i}(x) = 1 - \mu_i(x)$	$\mu_{\neg P_i}(x) = 1 - \mu_i(x)$

importance on the lowest scoring predicate, may not be suitable for real workloads. Other fuzzy semantics used in multimedia systems include arithmetic and geometric average semantics. Figure 3 visualizes the behavior of the fuzzy conjunction operator under different fuzzy semantics. It is well established that the only fuzzy semantics which preserves logical equivalence of statements (involving conjunction and disjunction) and is also monotonic is the *min* semantics. This, and the query processing efficiency it enables due to its simplicity, make it a popular choice despite its shortcomings.

Processing multimedia queries, like the one depicted in Fig. 2, under a fuzzy system requires extending query languages and query processors with fuzzy semantics. Many commercial database management systems include fuzzy extensions that are suitable for multimedia applications. Relational databases can be extended to capture fuzzy data in various different ways. In *tuple-level approaches,* the schema of each fuzzy relation is extended to include one or more attributes, each representing the degrees of imprecision of the tuples in the relation with respect to a different interpretation of the tuples. In these systems, the relational algebra operators (such as select, project, join, union, difference) are also extended to apply the selected fuzzy semantics to the tuple scores. In the *attribute-level approaches*, the degrees of uncertainty are associated individually to the attribute values. Especially when the imprecisions in the various attributes of a multimedia object are due to different reasons, attribute level approaches are more applicable due to their finer granularity. Furthermore, since each attribute can be treated as a fuzzy predicate on the multimedia object, query evaluation within these models can benefit more naturally from fuzzy logic evaluation schemes.

Processing these queries, on the other hand, requires significant extensions to the underlying database engines. For example, the underlying relational concepts, such as functional dependencies and normalization, need to be extended to cope with fuzziness in the stored data. In particular, in multimedia databases, users are usually interested in a result set which is ranked according to a ranking criterion which is generally user dependent (Fig. 1). Adali et al. [1] introduces a similarity algebra which brings together relational operators and results of multiple similarity implementations in a uniform language. Other algebraic treatments of fuzzy multimedia queries, relying on finer granularity attribute-based models, include the FNF^2 *algebra* [9]. When the requirement for exact matches is removed, the result space becomes significantly large, and thus, the query engine cannot rely on any processing scheme which would need to touch or enumerate all solutions. Consequently, query processing schemes would need to generate results as progressively (in decreasing order of relevance) as possible. Fagin [11] proposes ranked query evaluation algorithms, which assume that individual sources can progressively output sorted results and also enable random access. These algorithms also assume that the query has a monotone com-

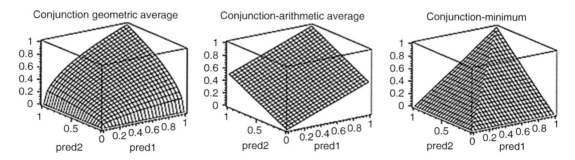

Multimedia Data Querying, Fig. 3 Visual representations of various binary fuzzy conjunction semantics: The horizontal axes correspond to the values between 0 and 1 for the two input conjuncts and the vertical axis represents the resulting scores according to the corresponding function

bined scoring function. Candan et al. [7] presents *approximate* ranked query processing techniques for cases where not all sub-queries are able to return ordered results. In turn, Fagin et al. [12] recognizes that there may be cases where random accesses are impossible and presents algorithms under monotonicity assumption to enumerate top-k objects without accessing all the data. These augment *monotonicity* with an *upper bound* principle, which enables bounding of the maximum possible score of a partial result. Qi et al. [19] establishes an alternative, *sum-max monotonicity* property and shows how to leverage this for developing a self-punctuating, horizon-based ranked join (HR-Join) operator for cases when the more strict monotonicity property does not hold. Top-k querying can also be viewed as a k-constrained optimization problem, where the goal function includes both a Boolean constraint characterizing the data of interest and a quantifying function which acts as the numeric optimization target [22]. Adali et al. [2] and Li et al. [17] extend the relational algebra to support ranking as a first-class construct. Li et al. [17] also presents a pipelined and incremental execution model of ranking query plans.

A particular challenge in multimedia querying is that (as shown in Fig. 1) the underlying query processing scheme needs to adapt to the specific needs and preferences of individual users. Due to its flexibility, the fuzzy model enables var-

ious mechanisms of adaptation. First of all, if user's feedback focuses on a particular attribute in the query, the way the fuzzy score of the corresponding predicate is computed can change based on the feedback. Secondly, the semantics of the fuzzy logic operator can be adapted based on the feedback of the user. A third mechanism through which user's feedback can be taken into account is to enrich the merge function, used for merging the fuzzy scores, with weights that regulate the impacts of the individual predicates. Fagin proposed a generic weighting mechanism that can be used for any fuzzy merge function [11]. The mechanism ensures that (a) the result is a continuous function of the weights (as long as the original merge function is continuous), (b) sub-queries with zero weight can be dropped without affecting the rest of the query, and (c) if all weights are equal, then the result is equal to the original, not-weighted merge function. Candan and Li [5], on the other hand, argued that the relative importance of predicates in a merge function should be measured in terms of the overall impacts changes in the scores that the individual predicates would have on the overall score (Fig. 4). Consequently, the relative importance of predicates can vary based on the scores the individual predicates take and the corresponding partial derivatives. A more direct mechanism to capture the user feedback is to modify the partial derivatives of the scoring functions appropriately.

Multimedia Data Querying, Fig. 4 The relative impact of the predicates in a scoring function can vary based on the scores of the individual predicates

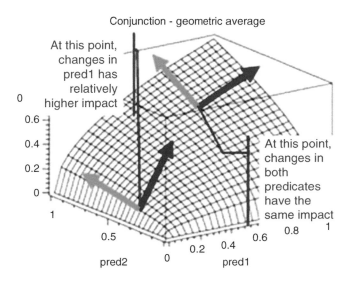

While the generic scheme presented by Fagin [11] would satisfy this for some merge functions, such as the arithmetic average, it would fail to capture this requirement for others, such as the commonly used product semantics.

Unlike the fuzzy models, which can capture a large spectrum of application requirements, probabilistic approaches to data and query modeling are applicable only to those cases where the source of imprecision is of statistical nature. These cases include probabilistic noise in data collection, sampling (over time, space, or population members) during data capture or processing, randomized and probabilistic algorithms (such Markov chains and Bayesian networks) used in media processing and pattern detection, and probabilistic consideration of relevance feedback. Dalvi and Suciu [10], for example, associates a value between 0 and 1 to each tuple in a given relation: the value expresses to probability with which a given tuple belongs to the relation. By extending SQL and the underlying relational algebra with probabilistic semantics and a theory of belief, the authors provide a probabilistic semantics for query processing with uncertain matches.

A general simplifying assumption in many probabilistic models is that the individual attributes (and the corresponding predicates) are independent of each other: consequently, the probability of a conjunction can be computed as the product of the probabilities of the conjuncts; i.e., under these conditions, the probabilistic model corresponds to the fuzzy product semantics. However, the independence assumption does not always hold (in fact, it rarely holds). Lakshmanan et al. [15] presents a probabilistic relational data model, an algebra, and aggregate operators that capture various types of interdependencies, including independence, mutual exclusion, as well as positive, negative, and conditional correlation.

While the simplest probabilistic models associate a single value between 0 and 1 to each attribute or tuple, more complete models represent the score in the form of an interval of possible values or more generally in terms of a probability distribution describing the possible values for the attribute or the tuple. Consequently, these models

are able to capture more realistic scenarios, where the imprecision in data collection and processing prevents the system to compute the exact *quality* of the individual media objects, but (based on the domain knowledge) can associate probability distributions to them. Note that relaxing the independence assumption or extending the model to capture non-singular probability distributions both necessitate changes in the underlying rank evaluation algorithms.

Other non-relational probabilistic models for multimedia querying includes Markov chains and Bayesian networks. A stochastic process is said to be Markovian if the conditional probability distribution of the future states depends only on the present. A Markov chain is a discrete-time stochastic process which is conditionally independent of the past states. A random walk on a graph, $G(V,E)$, is a Markov chain whose state at any time is described by a vertex of G and the transition probability is distributed equally among all outgoing edges. The transition probability distribution in the corresponding Markov model can be represented as a matrix, where the (i, j)'th element of this matrix, T_{ij}, describes the probability that, given that the current state is i, the process will be in state j in the next time unit; i.e., the n-step transition probabilities can be computed as the n'th power of the transition matrix. Markovian models are used heavily for linkage analysis in supporting queries over web, multimedia, and social network data with graphical representations.

A Bayesian network is another graphical probabilistic model used especially for representing probabilistic relationships between variables (e.g., objects, properties of the objects, or beliefs about the properties of the objects) [18]. In a Bayesian network nodes represent variables and edges between the nodes represent the relationships between the probability distributions of the corresponding variables. Consequently, once they are fully specified, Bayesian networks can be used for answering probabilistic queries given certain observations. However, in many cases, both the structure as well as the parameters of the network have to be learned through iterative and sampling-based heuristics, such as expectation maximization

(EM), and Markov Chain Monte Carlo (MCMC) algorithms. Hidden Markov models (HMMs), where some of the states are hidden (i.e., unknown), but variables that depend on these states are observable, are Bayesian networks used commonly in many machine-learning based multimedia pattern recognition applications. This involves training (i.e., given a sequence of observations, learning the parameters of the underlying HMM) and pattern recognition (i.e., given the parameters of an HMM, finding the most likely sequence of states that would produce a given output).

Dimensionality Reduction and Feature Selection

For most media types, there are multiple features that one can use for indexing and retrieval. In practice indexing more features (or having more feature dimensions to represent the data) is not always an effective way of managing a database: A fundamental problem with having to deal with a large number of dimensions is that searches in high dimensional vector spaces suffer from a dimensionality curse: range and nearest neighbor searches in high dimensional spaces fail to benefit from available multi-dimensional index structures [6], Ref: "Indexing" Section] and searches deteriorate to sequential scans of the entire database. This is known as the "*dimensionality curse*" in multimedia querying. While approximate index structures, such as Locality Sensitive Hashing (LSH [3]) help aleviate the problem to some degree, dimensionality reduction and feature selection techniques aim to tackle the dimensionality curse directly by reducing the number of aspects that need to be considered.

A feature can be selected or dropped based on (a) application semantics, (b) human perception, (c) discrimination power in the database, (d) object description effectiveness, or (e) query relevance [6]. Principal component analysis (PCA), also known as the Karhunen-Loeve, KL, transform) is a linear transform, which optimally decorrelates the input data. In other words, given a data set described in a vector space, PCA identifies a set of alternative bases for the space, along which the spread is maximized. The resulting

bases vectors have the greatest discriminatory power and thus are more important for indexing. Singular Value Decomposition (SVD), on the other hand, identifies a transformation which takes media data, described in terms of an m dimensional vector space, and maps them into a vector space defined by k (<=m) orthogonal basis vectors (also known as latent semantics) each with a score denoting its contributions in the given data set. Intuitively, SVD finds an optimal embedding (with minimal error) of a given data set into a space with smaller number of dimensions.

Other dimensionality reduction techniques include Probabilistic latent semantic indexing [14] and Latent Dirichlet Allocation [4].

Tensors and Tensor Decomposition

The tensor model maps each attribute of a given data to a mode in an n-dimensional array where each possible tuple is a cell, the existence (absence) of a particular tuple in a database instance can be denoted as 1 (0) in the cell; similarly, the model can also represent fuzzy or probabilistic tuples by filling the cells with values between 0 and 1. Intuitively, tensors are generalizations of matrices: while a matrix is essentially a two dimensional (or two-mode) array, a tensor is an array of arbitrary dimensions/modes.

As described above, a set of vectors (described as a matrix) is often analyzed for its latent semantics and indexed for search using a matrix decomposition operation known as the singular value decomposition (SVD). The more general analysis operation which applies to tensors with more than two modes is known as the tensor decomposition [6]. The two most popular tensor decompositions are the CANDECOMP/PARAFAC [8, 14] and the Tucker [20] decompositions. CANDECOMP [8] and PARAFAC [14] decompositions (together known as the CP decomposition) decompose the input tensor into a sum of component rank-one tensors. Tucker decomposition [20], on the other hand, decomposes a given tensor into a core tensor multiplied by a matrix along each mode. Intuitively, the CP decomposition can be represented in the form of a diagonal core tensor (describing the strengths of the clusters of data) and one factor matrix per mode (describing the contribution of the data/features to each cluster), the Tucker decomposition results in a dense core tensor (describing the relationships among the clusters of data) in addition to the factor matrices.

Multi-Dimensional Indexing

Vectors play an important role in representing and managing multimedia data. Once relevant features are identified and quantified, most media data can be mapped onto a multi-dimensional feature space and queries can be formulated as range searches or nearest-neighbor searches in this space. A naive way of executing these queries is to have a lookup file containing the vector representations of all the objects in the database and scan this file for the required matches, pruning those objects that do not satisfy the search condition. While this approach might be feasible for small databases where all objects fit into the main memory, for large databases, a full-scan of the database quickly becomes infeasible. Instead multimedia databases use specialized indexing techniques to help speed up search by pruning the irrelevant portions of the space and focusing on the parts which are likely to satisfy the search predicate. Commonly used multi-dimensional index structures include, quadtree variants, R-tree variants, SS- and SR-trees, and M-trees [6, Ref: "Indexing" Section].

Unfortunately, most multi-dimensional data indexing schemes fail to scale with the number of dimensions necessary to describe the multimedia data. A number of schemes, such as TV-trees, X-trees, Pyramid trees, VA-files, and approximate indexing techniques, such as Locality Sensitive Hashing [6, Ref: "Indexing" Section], have been developed to address this shortcoming.

Clustering

When the data cannot be mapped onto multi-dimensional space or when the space onto which the data is mapped is not suitable for efficient indexing, other techniques, such as clustering can be used for pruning the irrelevant data during query processing. Common clustering techniques include, graph partitioning (such as minimum cut or spectral graph partitioning), K-means, self-organizing maps (SOM), and confidence/perturbation-based clustering [6, Ref: "Data Clustering" Section]. As we have discussed above SVD and tensor decompositions can also be used for clustering data. While different clustering schemes operate differently, most first group related objects together and, then, select a single representative for the entire cluster. The query is then compared against the cluster representatives and only those clusters whose representatives match the query are further investigated. While this saves time by reducing the number of comparisons one would need to make with a naive scan of the entire database, it may result in misses if there are matching objects in clusters whose representatives are not sufficiently good match to the query. Thus, in general, clusters need to be compact and the representatives must be selected carefully.

Supervised Learning/Classification

Unlike indexing and clustering processes which aim to place similar media objects together for efficient access and pruning during retrieval, classification process aims to associate media objects into known semantically-meaningful categories. A classifier (such as decision tree, support-vector machines, or nearest-neighbor classifier [6, Ref: "▶ Decision Tree Classification" Section]) learns how to recognize from a given set of media-objects preclassified into a set of categories, what the critical features of these categories of interest are and, based on these, associates new media objects to these categories. The goal of the classification process is to analyze the input data and develop a description, or a model, for every class using the characteristics of the training data. Thus, the classification algorithms detect patterns, or sets of features, which define categories.

Relevance Feedback and Collaborative Filtering

Multimedia query processing often involves answering ill-posed questions: there may be multiple ways to interpret the query and data and the appropriate query processing strategy may be user- and use context-dependent. Especially when users are not sufficiently informed about the data (or sometimes of their interests) to formulate a precise query, feedback based data exploration can play a critical role in helping users find relevant information [6, Ref: "▶ Relevance Feedback for Content-Based Information Retrieval" Section].

After observing the initial set of results returned by the system, the user may be able to identify certain aspects of the objects or features that are critical, but not included in the original query. These explicit assertions of desirability or un-desirability are referred to as the hard feedback. Often, hard feedback is suitable for expert users who know what they are looking for but do not know the data to formulate "accurate" queries in advance.

When the user is exploring the data within the context of an initial query but does not have well-defined query criteria in mind yet, she may want the system to rank the results in the next iteration according to the positive or negative feedback she provides on the current results. To accommodate such declarations of preference, the system needs to support soft feedback. The soft feedback process is most suitable for users who are exploring the data.

In collaborative filtering schemes, on the other hand, analysis of similarities between different users' preferences are used for predicting whether a given user will find a given object relevant or not [Ref: "▶ Collaborative Filtering" Section].

Key Applications

Applications of multimedia querying include social media, personal and public photo/media collections, personal information management systems, digital libraries, on-line and print advertisement, digital entertainment, communications, long-distance collaborative systems, surveillance, security and alert detection, military, environmental monitoring, ambient and ubiquitous systems that provide real-time personalized services to humans, improved accessibility to blind and elderly, rehabilitation of patients through visual and haptic feedback, and interactive performing arts.

Future Directions

Naturally, each and every step of this multimedia querying and exploration cycle poses significant challenges. While it would be impossible to enumerate all the challenges, we can perhaps identify the following five "core" challenges that will help the multimedia querying and exploration process:

- Media annotation, summarization, and (dimensionality) reduction,
- User, community, context, preference modeling and feedback,
- Multi-modal and richly structured/linked data exploration,
- Dynamic/evolving multimedia data exploration, and of course
- Bridging the semantic gap in media exploration.

Unfortunately, while there is great amount of research in tackling these core challenges, we probably have to admit that we are still quite far from addressing any of these issues satisfactorily, especially within the context of large multimedia data collections. In fact, at least in the short- to medium-term future, these five issues will continue to form the core challenges in multimedia querying and exploration.

While most of the existing work in this area focused on content-based and object-based query processing, future directions in multimedia querying will involve understanding of how media objects affect users and how do they fit into users experiences in the real world. These require better understanding of underlying psychological and cognitive processes in human media processing. Ambient media-rich systems which collect and feed in diverse media from environmentally embedded sensors necessitate novel ways of continuous and distributed media processing and fusion schemes. Intelligent schemes to choose the right objects to process are needed to scale query processing workflows to the immense influx of real-time media data. In a similar manner, collaborative-filtering based query processing schemes that can help

M

overcoming the semantic gap between media and users' experiences will help the multimedia databases scale to Internet-scale media indexing and querying.

If there is one thing that is becoming more urgent, however, it is that the ever-increasing scale and the speed of the data implies that to support the above media querying/exploration cycle, our emphasis must shift towards development of integrated data platforms that can support, in an optimized and scalable manner, both media analysis (feature extraction, clustering, partitioning aggregation, summarization, classification, latent analysis) and data manipulation (filtering, integration, personalized and task-oriented retrieval) operations.

Cross-References

▶ Advanced Information Retrieval Measures
▶ Decision Tree Classification
▶ Fuzzy Models
▶ Graph Data Management in Scientific Applications
▶ Information Retrieval
▶ Information Retrieval Operations
▶ Multimedia Data
▶ Multimedia Data Indexing
▶ Multimedia Databases
▶ Multimedia Information Retrieval Model
▶ Multimedia Retrieval Evaluation
▶ Probabilistic Databases
▶ Probabilistic Retrieval Models and Binary Independence Retrieval (BIR) Model
▶ Social Networks
▶ Spatial and Spatiotemporal Data Models and Languages
▶ Stream Mining
▶ Temporal Database
▶ Top-K Selection Queries on Multimedia Datasets

Recommended Reading

1. Adali S, Bonatti PA, Sapino ML, Subrahmanian VS. A multi-similarity algebra. In: Proceedings of the ACM SIGMOD International Conference on Management of Data; 1998. p. 402–13.

2. Adali S, Bufi C, Sapino ML. Ranked relations: query languages and query processing methods for multimedia. Multimed Tools Appl. 2004;24(3): 197–214.

3. Andoni A, Indyk, P. Near-optimal hashing algorithms for approximate nearest neighbor in high dimensions. In: Proceedings of the 47th Annual IEEE Symposium on Foundations of Computer Science; 2006. p. 459–68.

4. Blei DM, Ng AY, Jordan MI. Latent Dirichlet allocation. J Mach Learn Res. 2003;3(4–5):993–1022.

5. Candan KS, Li W-S. On similarity measures for multimedia database applications. Knowl Inf Syst. 2001;3(1):30–51.

6. Candan KS, Sapino ML. Data management for multimedia retrieval. Cambridge University Press; 2010. ISBN-10:0521887399, ISBN-13: 978-0521887397.

7. Candan KS, Li W-S, Priya ML. Similarity-based ranking and query processing in multimedia databases. Data Knowl Eng. 2000;35(3):259–98.

8. Caroll JD, Chang JJ. Analysis of individual differences in multidimensional scaling via an n-way generalization of 'Eckart-Young' decomposition. Psychometrika. 1970;35(3):283–319.

9. Chianese A, Picariello A, Sansone L, Sapino ML. Managing uncertainties in image databases: a fuzzy approach. Multimed Tools Appl. 2004;23(3): 237–52.

10. Dalvi NN, Suciu D. Efficient query evaluation on probabilistic databases. In: Proceedings of the 30th International Conference on Very Large Data Bases; 2004. p. 864–75.

11. Fagin R. Fuzzy queries in multimedia database systems. In: Proceedings of the 17th ACM SIGACT-SIGMOD-SIGART Symposium on Principles of Database Systems; 1998. p. 1–10.

12. Fagin R, Lotem A, Naor M. Optimal aggregation algorithms for middleware. J Comput Syst Sci. 2003; 66(4):614–56.

13. Harshman RA. Foundations of the parafac procedure: models and conditions for an "explanatory" multimodal factor analysis. UCLA Work Papers Phon. 1970;16:1–84.

14. Hofmann T. Probabilistic latent semantic indexing. In: Proceedings of the 22nd Annual International ACM SIGIR Conference on Research and Development in Information Retrieval; 1999. p. 50–7.

15. Lakshmanan LV, Leone N, Ross R, Subrahmanian VS. ProbView: a flexible probabilistic database system. ACM Trans Database Syst. 1997;22(3): 419–69.

16. Li W-S, Candan KS. SEMCOG: a hybrid object-based image and video database system and its modeling, language, and query processing. Theory Pract Object Syst. 1999;5(3):163–80.

17. Li C, Chang KC-C, Ilyas IF, Song S. RankSQL: Query algebra and optimization for relational top-k queries. In: Proceedings of the ACM SIGMOD International Conference on Management of Data; 2005. p. 131–42.

18. Pearl J. Bayesian networks: a model of self-activated memory for evidential reasoning. In: Proceedings of the 7th Conference of the Cognitive Science Society; 1985. p. 329–34.
19. Qi Y, Candan KS, Sapino ML. Sum-Max monotonic ranked joins for evaluating top-K twig queries on weighted data graphs. In: Proceedings of the 33rd International Conference on Very Large Data Bases; 2007. p. 507–18.
20. Tucker LR. Some mathematical notes on three-mode factor analysis. Psychometrika. 1966;31(3):279–311.
21. Zadeh LA. Fuzzy sets. Inf Control. 1965;8(3): 338–53.
22. Zhang Z, Hwang S, Chang KC, Wang M, Lang CA, Chang Y. Boolean + Ranking: querying a database by K-constrained optimization. In: Proceedings of the ACM SIGMOD International Conference on Management of Data; n.d. p. 359–70.

Multimedia Data Storage

Jeffrey Xu Yu
Department of Systems Engineering and Engineering Management, The Chinese University of Hong Kong, Hong Kong, China

Definition

Data storage management, as one of the important functions in database management systems, is to manage data on disk in an efficient way to support data retrieval and data update. Multimedia data storage management is to manage continuous media data (audio/video) on disk. The uniqueness of multimedia data storage management is, in a multiuser environment, how to arrange the data storage to support a continuous retrieval of large continuous media data from disk to be displayed on screen, at a pre-specified rate, without any disruptions, which is also called hiccup-free display.

Historical Background

In a multimedia environment, the continuous media needs to be retrieved and displayed continuously. As magnetic disks are used as the mass storage device for multimedia data, zoning is one approach to increase the storage capacity of magnetic disks. Here, zones of a disk drive are different regions of the disk drive that usually have different transfer rates. A number of studies have investigated techniques to support a hiccup-free display of continuous media (video/audio) using magnetic disk drives with a single zone [1, 2, 9, 10] in the early 1990s. These studies assume a fixed transfer rate for a disk drive. These techniques can be possibly adopted to design a multi-zone disk system, but such a multi-zone disk system is then forced to use the minimum transfer rate of the zones for the entire disk, in order to guarantee a continuous display of continuous media objects. Such an approach is called Min-Z-tfr.

In the late 1990s and early 2000s, many new techniques are proposed to deal with the storage of continuous media in multi-zone disks [4, 5]. In [4], VARB and FIXB are proposed to place media objects on the multi-zone disks. VARB and FIXB techniques provide the average transfer rate of zones while ensuring a continuous display, compared with Min-Z-tfr, which is forced to use the minimum transfer rate of zones. VARB and FIXB increase the throughput of the system, while they also (i) increase startup latency, (ii) waste disk space, and (3) increase the amount of memory required to support more simultaneous displays. A configuration planner is proposed to decrease the drawbacks of VARB and FIXB, in order to meet the performance requirements of applications [4]. As VARB and FIXB [4] take account of a single media type only, RP, MTP, and MVP are proposed in [5] to support multiple media types with different bandwidth requirements. In [4, 5], the discussions focus on multimedia data placement across disk drives to support continuous display requirement.

Foundations

Hiccup-Free Display

The size of continuous media, especially videos, can be very large. The transmission of data must be just-in-time. In other words, data must be re-

M

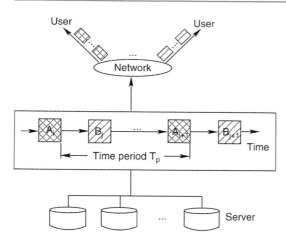

Multimedia Data Storage, Fig. 1 A hiccup free display (Fig. 3 in [5])

trieved from disks and transmitted to the display in a timely manner that prevents hiccups. A cycle-based data retrieval technique [6, 12] is designed to provide continuous display for multiple users. As shown in Fig. 1, it is a cycle-based data retrieval technique to support hiccup-free display [5]. Consider a constant-bit-rate (CBR) media. In order to guarantee a continuous display of a continuous media A, the system needs to retrieve the block of A_i before its immediate previous block A_{i-1} completes its display. For each block A_i, there are two tasks, namely, block retrieval and display initialization. This process of block retrieval and display initialization repeats in a cyclic manner until all A blocks have been displayed. If the time to retrieve a block, termed *block retrieval time*, T_p (as indicated in Fig. 1), is smaller than or equal to the time period to display a block, then the whole display process will be hiccup free [5]. The time interval between the time a request of A arrives and the time the display of A starts is called *startup latency*.

FIXB and VARB
Modern disk drivers are produced with multiple zones to meet the demands for a higher storage capacity [11]. A zone is a contiguous collection of disk cylinders where the tracks in the cylinders are supposed to have the same storage capacity. The outer zones have a higher transfer rate

in comparison with the inner zones. Two approaches, FIXB and VARB, are proposed in [4] to support a continuous display of continuous media using a single disk with multi-zones. Suppose that the disk consists of m zones, $Z_1, Z_2, \ldots, Z_m$, and the transfer rate of zone Z_i is R_i. Assume that each object X is partitioned into f blocks: $X_1, X_2, \ldots, X_f$.

With FIXB (Fixed Block Size), the blocks of an object X are rendered equi-sized, i.e., $X_i = X_j$, for any i and j. The system assigns the blocks of X to the zones in a round-robin manner. FIXB is designed to support a predetermined number of simultaneous displays (N). The retrieval process of this system is to scan the disk in one direction, for example, starting with the outermost zone moving inward, visiting one zone at a time and multiplexing the bandwidth of that zone among N block reads. A sweep is a scan of the zones. The time to perform one such a sweep is denoted as T_{Scan}. The time of reading N blocks from zone Z_i, denoted $T_{MUX}(Z_i)$, is dependent on the transfer rate of zone Z_i. As the transfer rate of zones varies, the time to read blocks from different zones also varies. To support hiccup-free displays, the system uses buffers to compensate for the low transfer rates of innermost zones.

VARB makes $T_{MUX}(Z_i)$ to be identical for all zones, using variable block sizes. The size of a block, $B(Z_i)$, is a function of the transfer rate of the zone Z_i. This results in an identical transfer time for all the blocks, T_{disk}, i.e., $T_{disk} = \frac{B(Z_i)}{R_i} = \frac{B(Z_j)}{R_j}$, for any i and j. Like FIXB, VARB assigns the blocks of an object to the zones in a round-robin manner. Unlike FIXB, with VARB, the blocks of an object X have different sizes depending on which zones the blocks are assigned to. Also, like FIXB, VARB employs memory to compensate for the low bandwidth of innermost zones.

With FIXB, the blocks of an object is equi-sized, whereas VARB renders the blocks in different sizes, which depends on the transfer rate of its assigned zone. FIXB is easy to be implemented in compared with VARB. But VARB requires a lower amount of memory and incurs a lower latency as compared to FIXB.

RP, MTP, and MVP

Based on zoning, there are different data transfer rates (R_i) to retrieve data from a disk. When a server is required to support multiple media types with different bandwidth requirements, the block reading time varies widely depending on the block size and its assigned zone. Suppose that there are n different media types to be supported. The block size of a media type i object is determined by $B_i = T_p \times D_i$ where D_i is the bandwidth requirement of the media type i, and T_p is a fixed time period which is set to be the same for all the media types. The transfer time (service time) to retrieve a block of a media type i object in zone Z_j is $s_{i,j} = \frac{B_i}{R_j}$. Suppose that there are b blocks, the average service time is computed as

$$\bar{s} = \sum_{i=1}^{b} F_i \sum_{j=1}^{n} P_{i,B_j} \sum_{k=1}^{m} \frac{P_{i,Z_k} B_j}{R_k}$$

where F_i is the access frequency of block i, for $1 \leq i \leq b$, P_{i,B_j} is the probability that the size of block i is B_j, and P_{i,Z_k} is the probability that this block is assigned to zone Z_k. The variance of service time is:

$$\sigma_s^2 = \sum_{i=1}^{b} F_i \sum_{j=1}^{n} \sum_{k=1}^{m} P_{i,B_j} P_{i,Z_k} (s_{j,k} - \bar{s})^2$$

Three approaches are proposed in [5]: RP, MTP, and MVP. RP (Random Placement) assigns blocks to the zones in a random manner. MTP (Maximizing Throughput Placement) sorts blocks based on their size and frequency of access $(F_i \times B_i)$. The blocks are assigned to the zones sequentially starting with the fastest zone, i.e., block i with the highest $F_i \times B_i$ value is assigned to the fastest zone. With MVP (Minimizing Variance Placement), a block of size B_i is placed on the zone Z_j (with R_j) which has the closest $\frac{B_i}{R_j}$ value to the average block reading time $(\overline{T}_B)$:

$$\overline{T}_B = \frac{average\ block\ size}{average\ transderrate} = \frac{\frac{1}{n}\sum_{i=1}^{n} B_i}{\frac{1}{m}\sum_{i=1}^{m} R_i}$$

Performance studies in [5] demonstrate that both MTP and MVP are superior to RP. MVP outperforms MTP regarding the average service time and/or variance of service time. One advantage of MVP is that it is not sensitive to the access frequency of objects.

Data Placement Across Disk Drivers

The bandwidth of a single disk is insufficient for the multimedia applications that strive to support thousands of simultaneous displays. One approach is to employ a multi-disk architecture. Assuming a system with D homogeneous disks, the data is striped across the disks in order to distribute the load of a display evenly across the disks [2, 3, 8].

The striping technique is as follows (Fig. 2). First, the disks are partitioned into k disk clusters where each cluster consists of d disks: $k = \frac{D}{d}$. An object X is partitioned into f blocks, X_1, $X_2, \ldots, X_f$, and the blocks of X are assigned to the k disk clusters in a round-robin manner, starting with an arbitrarily chosen disk cluster and zone, for example, zone Z_j in disk cluster C_i. In a disk cluster, each block of X, X_i, is declustered [7] into d fragments, $X_{i,j}$, where each fragment is assigned to a different disk in the disk cluster. As shown in Fig. 2, the X_0 block is assigned to the disk cluster C_0, and its two declustered fragments, $X_{0,0}$ and $X_{0,1}$ are assigned to the zone Z_0. Note that the fragments of a block need to be assigned to the same zone on the d disks in the disk cluster where the block is assigned to. In the retrieval of objects, one zone of all disks in a disk cluster is active per time period. To display object X of Fig. 2, it needs to access zone Z_0 in disk cluster C_0, when the disk cluster is idle, followed by accessing zone Z_1 in disk cluster C_1. This process repeats to retrieve/display all blocks of the object X.

Key Applications

Multimedia information systems have emerged as an essential component in many application domains ranging from library information systems to entertainment technology. The data storage

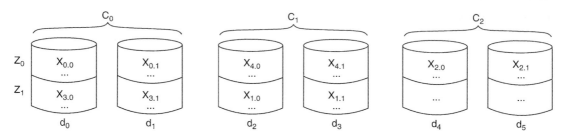

Multimedia Data Storage, Fig. 2   Three clusters with two logical zones per cluster (Fig. 12 in [4])

management is the basis to support a continuous display of multimedia objects.

Cross-References

▶ Continuous Multimedia Data Retrieval
▶ Multimedia Data Buffering
▶ Multimedia Resource Scheduling
▶ Storage Access Models
▶ Storage Devices
▶ Storage Management
▶ Storage Manager
▶ Storage Resource Management

Recommended Reading

1. Anderson DP, Homsy G. A continuous media I/O server and its synchronization mechanism. Comput. 1991;24(10):51–7.
2. Berson S, Ghandeharizadeh S, Muntz R, Ju X. Staggered striping in multimedia information systems. ACM SIGMOD Rec. 1994;23(2):79–90.
3. Ghandeharizadeh S, Kim S. Striping in multi-disk video servers. In: Proceedings of the SPIE High Density Data Recording and Retrieval Technology Conference; 1995.
4. Ghandeharizadeh S, Kim S, Shahabi C, Zimmermann R. Multimedia information storage and management. chap. 2: Placement of continuous media in multi-zone disks. Kluwer Academic; 1996.
5. Ghandeharizadeh S, Kim SH. Design of multi-user editing servers for continuous media. Multimedia Tools Appl. 2000;11(1):101–27.
6. Ghandeharizadeh S, Kim SH, Shi W, Zimmermann R. On minimizing startup latency in scalable continuous media servers. In: Proceedings of the SPIE Conference on Multimedia Computing and Networking; 1997.
7. Ghandeharizadeh S, Ramos L, Asad Z, Qureshi W. Object placement in parallel hypermedia systems. In: Proceedings of the 17th International Conference on Very Large Data Bases; 1991. p. 243–54.
8. Ozden B, Rastogi R, Silberschatz A. Disk striping in video server environments. In: Proceedings of the International Conference on Multimedia Computing and Systems; 1996. p. 580–9.
9. Rangan PV, Vin HM. Efficient storage techniques for digital continuous multimedia. IEEE Trans Knowl Data Eng. 1993;5(4):564–73.
10. Reddy ALN, Wyllie JC. I/O issues in a multimedia system. Comput. 1994;27(3):69–74.
11. Ruemmler C, Wilkes J. An introduction to disk drive modeling. Comput. 1994;27(3):17–28.
12. Tewari R, Mukherjee R, Dias DM, Vin HM. Design and performance tradeoffs in clustered video servers. In: Proceedings of the International Conference on Multimedia Computing and Systems; 1996. p. 144–50.

Multimedia Databases

Ramesh Jain
University of California, Irvine, CA, USA

Synonyms

Multimodal databases

Definition

Multimedia Databases are databases that contain and allow key data management operations with multimedia data. Traditional databases contained alphanumeric data and managed it for various applications. Increasingly, applications now contain multimedia data that requires defining additional types and requires development of operations for storage, management, access, and presentation of multimedia data. Multimedia databases must increasingly deal with issues related to managing multimedia data as well as the traditional data. Commonly, databases that manage images, audio, and video in addition to metadata related to these and other alphanumeric information are called multimedia databases. When databases contain only one of the images, audio, or video, they are called image databases, audio databases, and video databases, respectively. Considering the current trend, it is likely that most databases will slowly become multimedia databases.

Historical Background

The first in multimedia databases to appear were image databases that started appearing in late 1980s. Researchers in early image databases were more concerned with using databases for maintaining results of image processing operations to analyze and understand image analysis systems. Remote sensing and medical imaging produced images that needed to be saved and analyzed to extract information for various applications. In most of these applications, an environment to save images and processing results of these images were required.

The idea of making images an integral component of databases first started appearing in early 1990s. Relational data model had become the most common data model to deal with structured data and was used to store images as binary large objects (BLOBs) in these databases. To deal with images as first class data objects in images, a multilayered data model was proposed. This model considered image objects, and domain objects and suggested storage of those along with changes in relationships among objects. Some interesting developments in early systems evolved along two independent directions. In one direction [3], a user was considered an integral part of the query environment and feedback from user resulted in continuous refinement of queries leading to finding images that were required. In the other approach, many low level features

were computed and used for finding images using query by example approach. These two approaches adopted distinctly different directions, the first used domain knowledge and the second relied only on image features without any use of domain knowledge. The image features used commonly are different types and characteristics of color histograms and texture measures. These approaches are commonly called content-based retrieval, to differentiate them from metadata based retrieval. Commercially image database systems appeared in traditional database systems in mid 1990s. IBM used its QBIC technology in their DB2 database system and Oracle, Sybase, and Illustra used technology developed by a start-up company Virage. All search engines use image retrieval mostly based on the metadata that includes name of the file and text in the context of the image on a webpage.

Content based video retrieval result started in analyzing video into its constituent parts. At the lowest level is a frame, an individual image. Images are grouped into shots, shots into scenes, and scenes into episodes. All this data is extracted from video and stored in the database. Speech recognition techniques are used to prepare the transcript of the video and are also used in the database. Such systems found early use in TV program production and defense applications. Virage technology was used in these applications. Current search engines usually use metadata for searching video. Some specialized video search companies such as Blinkx (http://www.blinkx.com/) use predominantly text obtained using speech recognition or closed captions in television video. News videos have been one of the major application domains due to their applications as well as to good quality of audio available for these.

In audio databases, the signal is analyzed to detect characteristics that could be used in searching musical pieces that are similar to those. Such techniques, sometimes referred to as query by humming, were thought to be useful in finding music of interest.

Some effort has gone into analyzing CAD databases also for retrieving drawing and objects of interest.

M

Though some research has started in addressing multimedia data, rather than just one of the above multimedia types, most research is in either images, video, or audio. Text or metadata available in context of multimedia data is being considered in multimedia database research increasingly.

Foundations

The most fundamental difference in multimedia databases compared to the traditional databases is in the rich semantics of the data. Multimedia data, in addition to being lot more voluminous, is very rich in semantics. Many queries that users are interested require understanding of the semantics of the data. The problem becomes more complex because the semantics of multimedia is dependent not only on the data, but also on the specific user, the context in which the query is asked, and other sets of data available in the system.

Early approaches to multimedia databases did not consider the nature of multimedia data and just stored multimedia data either as BLOBs or links to files containing the data. These systems could allow only limited operations on multimedia data - usually limited to display or rendering of the data. No other operations or queries could be performed on this data.

With increasing use of multimedia data and applications that require use of multimedia data in many different ways, the nature of multimedia databases started changing. Currently, multimedia databases are still in their early stages. Many different concepts and approaches are being tried. Some of the important emerging ideas that are being tried in multimedia databases are discussed below. This area is currently a very active one and is likely to receive increasing attention both from academic and industrial research community.

A multimedia database system is considered to have the following four clear modules that need to work together to provide the functionality desired from them.

Data Analysis and Feature Extraction
A MMDB contains multimedia data but just storing the data as a BLOB does not allow any queries related to the content of the data. To solve this problem, data is analyzed to extract features from the data. These features can then be used to derive the required semantics. The features extracted depend on the nature of the type of the data and domain of application for the MMDB. These features could range from low-level features that are very general and do not depend on the application domain such as, color histograms and texture features for images to high level features directly tied to application domain such as shape of the tumor.

A significant amount of research related to MMDB is in specific media related research communities such as audio processing or computer vision. There is strong interest in finding efficient and effective features to interpret multimedia data in general as well as in specific applications.

Domain Knowledge and Interpretation
Multimedia data interpretation requires use of domain knowledge. Moreover, the knowledge required for interpretation of this data is not only the traditional domain knowledge represented using ontologies and similar techniques well developed for interpretation of text; but also media dependent models that require sophisticated classification approaches. In audio and video events must be detected in the data and that requires processing time dependent features.

There is a new emerging perspective that multimedia data should be considered evidence for real world events captured using such data. This requires modeling events and representing knowledge about domain events. This knowledge is then used in interpretation of multimedia data not in silos but together. Some progress is being made in representation of events.

Interaction and User Interface
Interaction environments used in traditional databases and search engines are not satisfactory in many applications of MMDB. Using keywords or names of objects, some limited searches can be performed, but many applications require concepts and ideas that require both continuous interactions and successive refinement of queries in what is called emergent semantics environment. Query by example including query

using sketches, humming, and some other non-textual approaches are being developed for some applications.

Presentation of results of queries also requires different techniques. Multimedia data is not very suited to list or record based presentations. Also, in many applications different media sources must be combined to create multimedia presentations with which a user can interact to refine and re-articulate their queries in the emergent semantics environment.

Storage, Matching, and Indexing

In most applications, the size of the multimedia data requires special attention. Commonly the video files can not be stored even using BLOBs. It is common to store file names of multimedia data and compute and store features from the data in the database. In most applications, for each file the number of features that should be stored becomes very large from hundreds to hundreds of thousands for each multimedia data item. These features are used in searching for correct results.

Unlike traditional databases, where records are searched based on exact matches, in MMDB search requires similarity matching. It is very rare to find a result using exact matching. Search in MMDB usually becomes finding data that has maximum similarity based on features. The similarity techniques [] requires comparing the features in queried data with all the data in a MMDB for evaluating similarity.

Indexing in MMDB for similarity computation requires representing features in a way that can allow fast computation for potentially similar objects. Many high dimensional techniques have been developed for organizing this data. The dimensionality of data, sometimes called curse of dimensionality, poses interesting challenges in such organization.

Key Applications

Multimedia data is becoming ubiquitous. Ranging from photos to videos, multimedia data is becoming part of all applications. Most emerging applications now have some kind of multimedia data that must be considered integral part of the databases. Moreover, emerging applications in all applications areas, ranging from homeland security to healthcare contain rich multimedia data. Based on current trend, it is safe to assume that in very near future, much of the data managed in databases will be multimedia. Some particular application domains where multimedia is natural and will continue dominating are entertainment, news, healthcare, and homeland security.

Future Directions

From structured data to semi-structured data and then to unstructured data, databases are being challenged to deal with increasingly semantic-rich environment. MMDB offer the biggest challenge to databases in terms of bridging the semantic gap.

Increasingly, applications are talking about situation modeling using real life sensor data. These applications combine live sensory data with other information to project current situation and also predict near future for users to take appropriate actions. These databases will require sophisticated tools to manage streaming multimedia data. Research efforts in these areas have already started and are likely to accelerate significantly in the near future.

Cross-References

▶ Multimedia Data

Recommended Reading

1. Bach J, Paul S, Jain R. An interactive image management system for face information retrieval. IEEE Trans Knowl Data Eng. Special Section on Multimedia Information Systems. 1993;5(4):619–28.
2. Gupta A, Weymouth T, Jain R. Semantic Queries with Pictures, The VIMSYS Model. In: Proceedings of the 17th International Conference on Very Large Data Bases; 1991. p. 3–6.
3. Jain R. Out of the Box Data Engineering Events in Heterogeneous Data (Keynote talk). In: Proceedings of the 19th International Conference on Data Engineering; 2003.
4. Jain R. Events and experiences in human centered computing. IEEE Comput. 2008;41(2):42–50.

M

5. Katayama N, Shin'ichi Satoh. The SR-tree: An Index Structure for High-Dimensional Nearest Neighbor Queries. In: Proceedings of the ACM SIGMOD International Conference on Management of Data; 1997. p. 369–80.

6. Lew M, Sebe N, Djerba C, Jain R. Content-based multimedia information retrieval: state of the art and challenges. ACM Trans Multimed Comput Commun Appl. 2006;2(1):1–19.

7. Santini S, Gupta A, Jain R. Emergent semantics through interaction in image databases. IEEE Trans Knowl Data Eng. 2001;13(3):337–51.

8. Santini S, Jain R. Similarity Measures. IEEE Trans Pattern Anal Mach Intell. 1999;21(9).

9. Smeulders A, Worring M, Santini S, Gupta A, Jain R. Image Databases at the end of the early years. IEEE Trans Pattern Anal Mach Intell. 2001;23(1).

Multimedia Information Retrieval Model

Carlo Meghini[1], Fabrizio Sebastiani[2], and Umberto Straccia[1]
[1]The Italian National Research Council, Pisa, Italy
[2]Qatar Computing Research Institute, Doha, Qatar

Synonyms

Content-based retrieval; Multimedia information discovery; Semantic-based retrieval

Definition

Given a collection of multimedia documents, the goal of multimedia information retrieval (MIR) is to find the documents that are relevant to a user information need. A multimedia document is a complex information object, with components of different kinds, such as text, images, video and sound, all in digital form.

Historical Background

The vast body of knowledge nowadays labeled as MIR, is the product of several streams of research, which have arisen independently of each others and proceeded largely in an autonomous way, until the beginning of 2000, when the difficulty of the problem and the lack of effective results made it evident that success could be achieved only through integration of methods. These streams can be grouped into three main areas:

The first area is that of *information retrieval* (IR) proper. The notion of IR attracted significant scientific interest from the late 1950s in the context of textual document retrieval. Early characterizations of IR simply relied on an "objective" notion of topic-relatedness (of a document to a query). Later, the essentially subjective concept of relevance gained ground, and eventually became the cornerstone of IR. Nowadays, IR is synonymous with "determination of relevance" [9].

Around the beginning of the 1980s, the area of multimedia documents came into existence and demanded an IR functionality that no classical method was able to answer, due to the *medium mismatch problem* (in the image database field, this is often called the *medium clash problem*). This problem refers to the fact that when documents and queries are expressed in different media, matching is difficult, as there is an inherent intermediate mapping process that needs to reformulate the concepts expressed in the medium used for queries (e.g., text) in terms of the other medium (e.g., images). In response to this demand, a wide range of methods for achieving IR on multimedia documents has been produced, mostly based on techniques developed in the areas of *signal processing* and *pattern matching*, initially foreign to the IR field. These methods are nowadays known as *similarity-based* methods, due to the fact that they use as queries an object of the same kind of the sought ones (e.g., a piece of text or an image) [6]. Originally, the term *content-based* was used to denote these methods, where the content in question was not the content of the multimedia object under study (e.g., the image) but that of the file that hosts it.

The last area is that of *semantic information processing* (SIP) which has developed across the information system and the artificial intelligence communities starting in the 1960s. The basic goal of SIP was the definition of artificial languages that could represent relevant aspects of a reality of interest (whence the appellation *semantic*), and of suitable operations on the ensuing representations that could support knowledge-intensive activities. Since the inception of the field, SIP methods are rooted in first-order mathematical logic, which offers the philosophically well-understood and computationally well-studied notions of syntax, semantics and inference as bases on which to build. Nowadays, SIP techniques are mostly employed in the context of Knowledge Organization Systems. In MIR, SIP methods have been used to develop sophisticate representations of the contents (in the sense of "semantics") of multimedia documents, in order to support the retrieval of these documents based on a logical model. According to this model, user's information needs are predicates expressed in the same language as that used for documents representations, and a document is retrieved if its representation logically implies the query. A wide range of logical models for IR have been proposed, corresponding to different ways of capturing the uncertainty inherent in IR, of expressing document contents, of achieving efficiency and effectiveness of retrieval [4].

To a lesser extent, the database area has also contributed to MIR, by providing indexing techniques for fast access to large collections of documents. Initially, typical structures such as inverted files and B-trees were employed. When similarity-based retrieval methods started to appear, novel structures, such as R- or M-trees were developed in order to support efficient processing of range and k nearest neighbors queries [15].

Foundations

MIR is a scientific discipline, endowed with many different approaches, each stemming from a different branch of the MIR history. All these approaches can be understood as addressing the same problem through a different aspect of multimedia documents.

Documents can be broadly divided from a user perspective into two main categories: simple and complex.

A document is *simple* if it cannot be further decomposed into other documents. Images and pieces of text are typically simple documents. A simple document is an arrangement of symbols that carry information via meaning, thus concurring in forming what is called the *content* of the document. In the case of text, the symbols are words (or their semantically significant fractions, such as stems, prefixes or suffixes), whereas for images the symbols are colors and textures. Simple documents can thus be characterized as having two parallel dimensions: that of *form* (or *syntax*, or *symbol*) and that of *content* (or *semantics*, or *meaning*). The form of a simple document is dependent on the medium that carries the document. On the contrary, the meaning of a simple document is the set of states of affairs (or "worlds") in which it is true, and is therefore medium-independent. For instance, the meaning of a piece of text is the set of (spatio-temporally determined) states of affairs in which the assertions made are true, and the meaning of an image is the set of such states of affairs in which the scene portrayed in the image indeed occurs.

Complex documents (or simply documents) are structured sets of simple documents. This leads to the identification of *structure* as the third dimension of documents. Document structure is typically a binary relation, whose graph is a tree rooted at the document and having the component simple documents as leaves. More complex structures may exist, for instance those requiring an ordering between the children of the same parent (such as between the chapters of a book), or those having an arity greater than 2 (such as synchronization amongst different streams of an audio-visual document).

Finally, documents, whether simple or complex, exist as independent entities characterized by (meta-)attributes (often called *metadata* in the digital libraries literature), which describe the relevant properties of such entities. The set of such attributes is usually called the *profile* of a document, and constitutes the fourth and last document dimension.

Corresponding to the four dimensions of documents just introduced, there can be four categories of retrieval, each one being a projection of the general problem of MIR onto a specific dimension. In addition, it is possible, and in some

M

cases desirable, to combine different kinds of retrieval within the same operation.

Retrieval based on document structure does not really lead to a genuine discovery, since the user must have already seen (or be otherwise aware of) the sought document(s) in order to be able to state a predicate on their structure. Retrieval based on document profile, from a purely logical point of view, is not different from content-based retrieval and in fact many metadata schema used for document description (notable, the Dublin Core Metadata Set) include attributes of both kinds.

Form-Based Multimedia Information Retrieval

The retrieval of information based on form addresses the syntactic properties of documents. In particular, form-based retrieval methods automatically create the document representations to be used in retrieval by extracting low-level features from documents, such as the number of occurrences of a certain word in a text, or the energy level in a certain region of an image. The resulting representations are abstractions which retain that part of the information originally present in the document that is considered sufficient to characterize the document for retrieval purposes. User queries to form-based retrieval engines may be documents themselves (this is especially true in the non-textual case, as this allows overcoming the medium mismatch problem), from which the system builds abstractions analogous to those of documents. Document and query abstractions are then compared by an appropriate function, aiming at assessing their degree of similarity. A document ranking results from these comparisons, in which the documents with the highest scores occur first.

In the case of text, form-based retrieval includes most of the traditional IR methods, ranging from simple string matching (as used in popular Web search engines) to the classical *tf-idf* term weighting method, to the most sophisticated algorithms for similarity measurement. Some of these methods make use of information structures, such as thesauri, for increasing retrieval effectiveness. However, what makes them form-based retrieval methods is their relying on a form-based document representation. Two categories of queries addressing text can be distinguished:

1. *Full-text* queries, each consisting of a *text pattern*, which denotes, in a deterministic way, a set of texts; when used as a query, the text pattern is supposed to retrieve any text layout belonging to its denotation.
2. *Similarity* queries, each consisting of a text, and aimed at retrieving those text layouts which are similar to the given text.

In a full-text query, the text pattern can be specified in many different ways, e.g., by enumeration, via a regular expression, or via ad hoc operators specific to text structure such as proximity, positional and inclusion operators [9].

Queries referring to the form dimension of images are called *visual* queries, and can be partitioned as follows:

1. *Concrete visual queries:* These consist of full-fledged images that are submitted to the system as a way to indicate a request to retrieve "similar" images; the addressed aspect of similarity may concern color [2, 7], texture [8, 14], appearance [12] or combination thereof [13].
2. *Abstract visual queries:* These are artificially constructed image elements (hence, "abstractions" of image layouts) that address specific aspects of image similarity; they can be further categorized into:
   (a) *Color queries:* specifications of color patches, used to indicate a request to retrieve those images in which a similar color patch occurs [6, 7].
   (b) *Shape queries:* specifications of one or more shapes (closed simple curves in the 2D space), used to indicate a request to retrieve those images in which the specified shapes occur as contours of significant objects [6, 11].
   (c) Combinations of the above [2].

Visual queries are processed by matching a vector of features extracted from the query image, with each of the homologous vectors extracted from the images candidate for retrieval. For concrete visual queries, the features are computed on the whole image, while for abstract visual queries only the features indicated in the query (such as shape or color) are represented in the vectors involved. For each of the above categories of visual queries, a number of different techniques have been proposed for performing image matching, depending on the features used to capture the aspect addressed by the category, or the method used to compute such features, or the function used to assess similarity.

Semantic Content-Based Multimedia Information Retrieval

On the contrary, semantic-based retrieval methods rely on symbolic representations of the meaning of documents, that is descriptions formulated in some suitable knowledge representation language, spelling out the truth conditions of the involved document. Various languages have been employed to this end, ranging from net-based to logical. Description Logics [1], or their Semantic Web syntactic forms such as OWL, are contractions of the Predicate Calculus that are most suitable candidates for this role, thanks to their being focused on the representation of concepts and to their computational amenability. Typically, meaning representations are constructed manually, perhaps with the assistance of some automatic tool; as a consequence, their usage on collections of remarkable size (text collections can reach nowadays up to millions of documents) is not viable. The social networking on which Web 2.0 is based may overcome this problem, as groups of up to thousands of users may get involved in the collaborative indexing process (flicker).

While semantic-based methods explicitly apply when a connection in meaning between documents and queries is sought, the status of form-based methods is, in this sense, ambiguous. On one hand, these methods may be viewed as pattern recognition tools that assist an information seeker by providing associative access to a collection of signals. On the other

hand, form-based methods may be viewed as an alternative way to approach the same problem addressed by semantic-based methods, that is deciding relevance, in the sense of connection in meaning, between documents and queries. This latter, much more ambitious view, can be justified only by relying on the assumption that there be a systematic correlation between "sameness" in low-level signal features and "sameness" in meaning. Establishing the systematic correlation between the expressions of a language and their meaning is precisely the goal of a *theory of meaning* (see, e.g., [5]), a subject of the philosophy of language that is still controversial, at least as far as the meaning of natural languages is concerned. So, pushed to its extreme consequences, the ambitious view of form-based retrieval leads to viewing a MIR system *as an algorithmic simulation of a theory of meaning*, in force of the fact that the sameness assumption is relied upon in every circumstance, not just in the few, happy cases in which everybody's intuition would bet on its truth. At present, this assumption seems more warranted in the case of text than in the case of non-textual media, as the representations employed by form-based textual retrieval methods (i.e., vectors of weighted words) come much closer to a semantic representation than the feature vectors employed by similarity-based image retrieval methods. Irrespectively of the tenability of the sameness assumption, the identification of the alleged syntactic-semantic correlation is at the moment a remote possibility, so the weaker view of form-based retrieval seems the only reasonable option.

Mixed Multimedia Information Retrieval

Suppose a user of a digital library is interested in retrieving all documents produced after January 2007, containing a critical review on a successful representation of a Mozart's opera, and with a picture showing Kiri in a blueish dress. This need addresses all dimensions of a document: it addresses structure because it states conditions on several parts of the desired documents; it addresses profile because it places a restriction on the production date; it addresses form- (in particular color-) and semantic-based image retrieval on a specific region of the involved image (the region must be blue and represent the singer Kiri) as well as on the whole image (must be a scene of a Mozart's opera); it addresses from-based text retrieval by requiring that the document contains a piece of text of a certain type and content. This is an example of mixed MIR, allowing the combination of different types of MIR in the context of the same query [10].

Emerging standards in multimedia document representation (notably, the ISO standard MPEG21) address all of the dimensions of a document. Consequently, their query languages support more and more mixed MIR.

Key Applications

Nowadays, MIR finds its natural context in *digital libraries,* a novel generation of information systems [3], born in the middle of the 1990s as a result of the First Digital Library Initiative. Digital Libraries are large collections of multimedia documents which are made on-line available on global infrastructures for discovery and access. MIR is a core service of any DL, addressing the discovery of multimedia documents.

Cross-References

▶ Information Retrieval

Recommended Reading

1. Baader F, Calvanese D, McGuiness D, Nardi D, Patel-Scheneider P, editors. The description logic handbook. Cambridge: Cambridge University Press; 2003.
2. Bach JR, Fuller C, Gupta A, Hampapur A, Horowitz B, Humphrey R, Jain R, Shu C.-F. The Virage image search engine: an open framework for image management. In: Proceedings of the 4th SPIE Conference on Storage and Retrieval for Still Images and Video Databases; 1996. p. 76–87.
3. Candela L, Castelli D, Pagano P, Thanos C Ioannidis Y, Koutrika G., Ross S, Schek HJ, Schuldt H. Setting the foundations of digital libraries. The DELOS manifesto. D-Lib Magazine. 13(3/4), March/April 2007.
4. Crestani F, Lalmas M, van Rijsbergen CJ,.editors. Logic and uncertainty in information retrieval: advanced models for the representation and retrieval of information. The Kluwer International Series On Information Retrieval 4. Boston: Kluwer Academic; 1998.
5. Davidson D. Truth and meaning. In: Inquiries into truth and interpretation. Oxford: Clarendon; 1991. p. 17–36.
6. Del Bimbo A. Visual information retrieval. Los Altos: Morgan Kaufmann; 1999.
7. Faloutsos C, Barber R, Flickner M, Hafner J, Niblack W. Efficient and effective querying by image content. J Intell Inform Syst. 1994;3(3–4):231–62.
8. Liu F, Picard RW. Periodicity, directionality, and randomness: wold features for image modelling and retrieval. IEEE Trans Pattern Analysis Machine Intell. 1996;18(7):722–33.
9. Manning CD, Raghavan P, Schütze H. An introduction to information retrieval. Cambridge: Cambridge University Press; 2007.
10. Meghini C, Sebastiani F, Straccia U. A model of multimedia information retrieval. J ACM. 2001;48(5):909–70.
11. Petrakis EG, Faloutsos C. Similarity searching in medical image databases. IEEE Trans Data Knowl Eng. 1997;9(3):435–47.
12. Ravela S, Manmatha R. Image retrieval by appearance. In: Proceedings of the 20th Annual International ACM SIGIR Conference on Research and Development in Information Retrieval; 1997. p. 278–85.
13. Rui Y, Huang TS, Ortega M, Mehrotra S. Relevance feedback: a power tool for interactive content-based image retrieval. IEEE Trans Circuits Syst Video Tech. 1998;8(5):644–55.
14. Smith JR, Chang S-F. Transform features for texture classification and discrimination in large image databases. In: Proceedings of the International Conference Image Processing; 1994. p. 407–11.
15. Zezula P, Amato G, Dohnal V, Batko M. Similarity search: the metric approach. Berlin: Springer; 2006.

Multimedia Metadata

Frank Nack
University of Amsterdam, Amsterdam,
The Netherlands

Synonyms

Hypermedia metadata; Meta-knowledge; Mixed-media; New media metadata

Definition

Multimedia is media that utilizes a combination of different content forms. In general, a multimedia asset includes a combination of at least two of the following: text, audio, still images, animation, video, and some sort of interactivity. There are two categories of multimedia content: linear and non-linear. Linear content progresses without any navigation control for the viewer, such as a cinema presentation. Non-linear content offers users interactivity to control progress. Examples are computer games or computer based training applications. Non-linear content is also known as hypermedia content.

Metadata is data about data of any sort in any media, describing an individual datum, content item, or a collection of data including multiple content items. In that way, metadata facilitates the understanding, characteristics, use and management of data.

Multimedia metadata is structured, encoded data that describes content and representation characteristics of information-bearing multimedia entities to facilitate the automatic or semiautomatic identification, discovery, assessment, and management of the described entities, as well as their generation, manipulation, and distribution.

Historical Background

The first appearance of the idea of mixed media was Vannevar Bush's "Memex" system ("memory extender"). In his article "As we may think" [1], published 1945 in the "The Atlantic Monthly," he proposed a device that is electronically linked to a library, able to display books and films from the library and automatically follow cross-references from one work to another. The Memex was the first simple and elegant approach towards multimedia information access. Further important steps toward the use of multimedia in digital systems in the 1960s were the introduction of "hyperlinks and hypermedia" by Ted Nelson [2], which allowed the generation of non-linear presentations, the development of the mouse by Douglas Engelbart [3], which supported the direct manipulation of objects on a computer screen, and the invention of the GUI, developed at Xerox Parc, which introduced media into computing.

In particular, it was the development of the web and digital consumer electronics that allowed computing to leave the realm of research or government organizations and enter society in large. The development of personal computers in the 1980s enabled desktop publishing, which further enhanced in the 1990s into adequate manipulation software for digital media, such as Adobe's Photoshop and Flash, or Appel's Final Cut Express. Yet, only the development of new distribution platforms in the 1990s, such as CD-ROM, DVD, and essentially the World Wide Web (Web) in combination with the rise of media technologies, such as the digital photo and video camera, or midi and MP3 player, stimulated the swift growth of digital media in mixed form. The growing amount of mixed media available to the public, and more recently provided by the public in form of user-generated content, requested for effective access mechanisms, which resulted in a steady development of content description technologies, mainly based on metadata.

Over the years, a number of metadata standards have been developed, which address various aspects of multimedia data, such as

- Low-level features: usually automatically extracted from the content.
- Semantic features: describing high-level concepts in form of key-word, ratings and links.
- Structure and identification: describing the spatial and temporal arrangement of one or more multimedia assets.
- Management: describing the information gathered during the life cycle of the multimedia asset, such as information about reuse, archiving, and rights management.

The major organizations who contributed to the development of standards are: the Dublin Core Metadata Initiative [4], the World Wide Web Consortium (W3C) [5], the Society for Motion Pictures and Television Engineering (SMPTE)

M

[6], the Moving Picture Expert Group (MPEG) [7], the TV-Anytime Consortium [8], and the International Press Telecommunications Council (IPTC) [9]. The common definition language between all theses languages is in one or the other way the Extensible Markup Language (XML) [10], defined by W3C.

The two major approaches towards open standards for multimedia metadata, with respect to content description and distribution for multimedia assets, were certainly provided by the W3C and MPEG, a working group of ISO/IEC charged with the development of video and audio encoding standards.

The W3C provided, besides the well known markup languages HTML, XHTML, CSS, SVG, in particular the Synchronized Multimedia Integration Language (SMIL). SMIL enables simple authoring of interactive audiovisual presentations, which integrate streaming audio and video with images, text or any other media type. SMIL [11] is an HTML-like language facilitating the authoring of a SMIL application by using a simple text-editor. SMIL 1.0 received recommendation status in 1998, SMIL 2.0 in 2005, and SMIL 3.0 is under development while this article is written.

Work in the Media Annotation Work has started in September 2008. This working group (http://www.w3.org/2008/01/media-annotations-wg.html) is chartered to provide a simple ontology to support cross-community data integration of information related to media objects on the Web, as well as an API to access the information. In addition there is the Media Fragments Working Group (http://www.w3.org/2008/WebVideo/Fragments/), which address temporal and spatial media fragments in the Web using Uniform Resource Identifiers (URI). Both groups should finish their work by June 2010.

MPEG's contribution to multimedia metadata are certainly the three standards MPEG-4 [12], MPEG-7 [13] and MPEG-21 [14].

With MPEG-4 the group entered the realm of media content, arisen due to the growing need for content manipulation and interaction, and expanded MPEG-1 to support video/audio "objects," 3D content, low bitrate encoding and support for Digital Rights Management. With respect to multimedia metadata MPEG4 also provides content authors with a textual syntax for the MPEG-4 Binary Format for Scenes (BIFS) to exchange their content with other authors, tools, or service providers. First, XMT is an XML-based abstraction of the object descriptor framework for BIFS animations. Moreover, it also respects existing practice for authoring content, such as SMIL, HTML, or Extensible 3D (X3D) by allowing the interchange of the format between a SMIL player, a Virtual Reality Modeling Language (VRML) player, and an MPEG player through using the relevant language representations such as XML Schema, MPEG-7 DDL, and VRML grammar. As such, the XMT serves as a unifying framework for representing multimedia content where otherwise fragmented technologies are integrated and the interoperability of the textual format between them is facilitated.

In the mid 1990s, the need for retrieving and manipulating digital media content form exploding digital libraries requested new ways of describing multimedia content on deeper semantic granularity, which resulted in MPEG-7, the multimedia content description standard. MPEG-7 provides a large set of descriptors and description schemata for video (part 3), audio (part 4) and multimedia content, including its presentation (part 5). All schemata are described in the Description Definition Language (DDL), which is modeled on XML-Schema, a schemata language developed by the WC3. Since MPEG-7 tries to establish the richest and most versatile set of audio-visual feature description structures by embracing standards such as SMPTE, or PTC, a 1:1 mapping to the text-oriented XML schema language could not be achieved (see [15] for a detailed description of existing DDL problems). However, the goal to be a highly interoperable standard among well-known industry standards and related standards of other domains – such as the area of digital libraries and ontologies using RDF or Dublin Core, was a useful exercise.

The beginning of the 21st century established an even faster exchange of multimedia data via the web, as higher bandwidth as well as access to high quality data became a commodity. The media businesses, such as the record or film industry, feared, due to peer-to-peer technology, for their markets and requested strict digital rights management enforcement. MPEG reacted with MPEG-21: MPEG describes this standard as a multimedia framework.

Both the W3C as well as MPEG provide open standards, which are able to describe adaptive content, represented in a single file that can be targeted to several platforms, such as mobile, broadband or the web. Both organizations compete with highly successful but rather proprietary industry standards, such as Adobe's Flash standard.

The latest trend on multimedia metadata also reflects the trend towards user-generated content, namely social tagging. A tag is a keyword or term associated with or assigned to a piece of information (a picture, a map, a video clip etc), which enables keyword-based classification and search. The advantage of tagging is its ease of use. This approach, though highly popular (e.g., in YouTube [16] and Flickr [17]), carries serious problems. Typically there is no information about the semantics of a tag, no matter if it is a single tag or a bag of tags. Additionally, different people may use drastically different terms to describe the same concept. This lack of semantic distinction can lead to inappropriate connections.

Foundations

The essential models describing the internal structures of multimedia compositions and the related production processes are: the Dexter Hypertext Reference Model [18], the Amsterdam Hypermedia Model [19], HyTime [20], the reference model for intelligent multimedia presentation systems (IMMPSs) [21], MPEG-4, SMIL, and the model of Canonical processes of media production [22].

The basic five functionalities that every multimedia system needs to address are: media content, layout, timing, linking and adaptivity.

Media Content

Readers who are interested in the fundamentals of the single media elements (i.e., their low-level as well as high-level feature descriptions) are referred to the articles on audio metadata, image metadata and video metadata in this encyclopedia.

Layout

Layout deals with the arrangement and style treatment of media on a screen. This means that layout determines how a particular media item is presented at a point in time and how it is rendered when activated. A multimedia presentation layout describes "the look and feel" of a composite of all of its media components. Thus, layout adds the semantic organization that enables a viewer to quickly and efficiently absorb the multiple content streams in a multimedia presentation.

There are in general three approaches towards multimedia layout, namely

1. *Embedded Layout,* where all layout decisions are resolved at media creation time and then performed by the presentation. Here the control lies ultimately by the designer of the presentation and there is no control, besides the required rendering, at the side of the media player.
2. *Dynamic Layout*, where all layout is dynamically determined by the user's media player, depending on the multimedia document structure or the timeline of the presentation. Here, the actual visual design is mainly in the hand of the media player rather than the presentation's designer.
3. *Compositing Layout*, which decouples media content from media placement. Here, a presentation is understood as a composite of relatively autonomous objects, where each uses embedded or dynamic layout models, which are then positioned into a arrangement by a presentation designer.

The essential classes and their attributes that describe a layout are:

• A root and several region elements, which establish the primary connection between media objects elements and the layout structure.

M

- Basic layout classes, such as referencing (region name and ID), scaling (z-index fit), positioning (width, height, top, left, bottom, right), background (back ground color, show back ground), and audio (sound level).

Timing

Timing deals with issues on how elements in a multimedia presentation get scheduled. Moreover, once an element is active, it needs to be determined how long it will be scheduled. The aim in a multimedia presentation is to go beyond the timing concepts known from audio and video objects.

Media timing: Media objects in multimedia presentations can either be *discrete* (e.g., text, with no implicit duration) or *continuous* (e.g., a music object, with an explicit duration defined within its encoding).

Presentation timing: A reference list to one or more media objects, describing timing primitives that determine the start and end time relative to one another.

There are basically 4 different ways of defining the active period of an object:

1. *Implicit duration*, as defined when the object was created (e.g., length of a video in sec).
2. *Explicit duration*, which describes the actual duration of the media object in the application (e.g., the actual duration might be shorter or longer than the implicit duration).
3. *Active duration*, which allows repetitions or other temporal manipulations of a media object.
4. *Rendered duration*, which describes the persistence of a media object at the end of its active duration.

There are various ways to describe time in values, such as full clock value (e.g., 7:45:23.76, where the last two items present ms), partial clock values (any sort of short base notation), time count values (numbers with a additional type string, e.g., 10S for "10 s"), and time context values, which are represented in three parts: a date field (YYY:MM:DD), a time field, and a time zone field.

Usually, a type of synchronization is required, as media objects start in relation to the container they belong to. A child element in a parallel container starts relative to the start time of the parent, whereas child elements in a sequential container are started relative to the end of their predecessor.

Linking

Linking defines and activates a non-linear navigation structure within and across documents.

The simplest form of a link is a pointer. The pointer defines an address of a document (e.g., a URI) and, optionally, an offset within the document. The element that identifies that a link exists is called a source anchor. If the anchor points to anything other than the beginning of a document, this anchor is called a destination anchor. The typical elements in HTML for linking are the <a> </a>element, to define the source anchor and link address, and the <area> </area> element, which is roughly the same, but it is applied only to a part of a media object. The basic linking attributes define the uri (href), the source and destination state (e.g., play or pause), the external or target state (e.g., true or false, the display environment) and the impact of link activation on the source and destination presentation (e.g., as non-negative percentage).

Both source and destination anchor need to express temporal moments, as their activation depends on the temporal behavior of the application. The three key temporal moments to be addressable are: the destination is already active, the destination is inactive, and the destination is inactive and the begin time is unresolved.

Finally, the link needs some attribute that describe geometry, as the linking into a region of a media item is possible. The core attributes are shape (values can be rectangle, circle or poly). The size and position of the anchor are defined via the origin of the coordinate space (the 0.0 point) and the resolution of the display device (support of rendering).

Adaptivity

The aim of multimedia presentations is usually to adapt them to the needs of the user, which might address either the runtime environment available to the user, or the personalized presentation wishes by the user.

There are four techniques to customize information in a presentation:

1. *Minimum set:* The multimedia presentation assumes a minimum set of performance, and device and user characteristics and the presentation document is designed based on this lowest denominator set (manageable solution but usually no compelling content)
2. *Multiple presentation set*: Each presentation represents a quality level, which the user selects on runtime (this one-size-fits-all approach is a dead end for the current trend towards portable and quality mix devices).
3. *Over-specified presentation*: All of the potential media items are available and it is the media player at the client side which makes a selection at run time (demands too much during the making phase).
4. *Control presentation*: The presentation contains pointers to all potential alternatives and only those used by the user would be sent from the server to the client (the advantage is that the presentation does not need to send copies of each of the various data streams across the network).

The essential elements a system would need to provide any of the above techniques are:

- Switch: which establishes a collection of alternatives for an interactive multimedia presentation;
- System control attributes, such as sys_language, sys_captions, sys_bitrate, sys_screensize, syt_cpu, etc.

Key Applications

Multimedia metadata is useful for the creation, manipulation, retrieval and distribution of mixed media sources within domains, such as

- The creative industries (e.g., fine arts, entertainment, commercial art, journalism, games, etc)
- The entertainment industries (e.g., special effects in movies and animations)
- Education (e.g., in computer based training courses and computer simulations, military or industrial training)
- Mathematical and scientific research (e.g., modeling and simulation)
- Medicine (e.g., virtual surgery or simulations of virus spread, etc)

Cross-References

- ▶ Audio Metadata
- ▶ Image Metadata
- ▶ Video Metadata

Recommended Reading

1. Bordegoni M, Faconti G, Maybury MT, Rist T, Ruggieri S, Trahanias P, Wilson M. A standard reference model for intelligent multimedia presentation systems. Comput Stand Interfaces. 1997;18(6–7):477. Available at http://kazan.cnuce.cnr.it/papers/abstracts/9708.IJCAI97.Immps.html.
2. Bush V. As we may think. Atl Mon. 1945;176(1):101–8.
3. Engelbart D. 1968. Available at http://sloan.stanford.edu/mousesite/1968Demo.html.
4. Flickr. Available at http://www.flickr.com/.
5. Gronbaek K, Trigg RH. Design issues for a dexter-based hypermedia system. Commun ACM. 1994;37(2):41–9.
6. Hardman L, Bulterman DCA, van Rossum G. February The amsterdam hypermedia model: adding time and context to the dexter model. Commun ACM. 1994;37(2):5062.
7. Hardman L, Obrenovic Z, Nack F, Kerherve B, Piersol K. Canonical processes of semantically annotated media production. Multimedia Syst. 2008;14(6):427–33.
8. Information processing – Hypermedia/Time-based structuring language (HyTime) – 2nd ed, ISO/IEC 10744:1997, WG8 PROJECT: JTC1.18.15.1. Available at http://www1.y12.doe.gov/capabilities/sgml/wg8/document/n1920/.
9. MPEG-4: ISO/IEC JTC1/SC29/WG11 N4668 2002. Available at http://www.chiariglione.org/mpeg/standards/mpeg-4/mpeg-4.htm.

M

10. MPEG-21: ISO/IEC JTC1/SC29/WG11/N5231 Shanghai, 2002. Available at http://www. chiariglione.org/mpeg/standards/mpeg-21/mpeg-21.htm.

11. MPEG-7: ISO/IEC JTC1/SC29/WG11N6828 Palma de Mallorca, 2004. Available at http://www. chiariglione.org/mpeg/standards/mpeg-7/mpeg-7. htm.

12. Nack F, van Ossenbruggen J, Hardman L. That obscure object of desire: multimedia metadata on the web (Part II). IEEE MultiMedia. 2005;12(1):54–63.

13. Nelson TH. A file structure for the complex, the changing, and the intermediate. In: The new media reader. Cambridge: Noha Wardrip-Fruin & Nick Montfort/The MIT Press; 2003. p. 133–46.

14. SMIL. Available at http://www.w3.org/AudioVideo/.

15. The dublin core metadata initiative. Available at http://www.dublincore.org/.

16. The extensible markup language (XML). Available at http://www.w3.org/XML/.

17. The international press telecommunications council [IPTC]. Available at http://www.iptc.org/pages/index.php.

18. The moving picture expert group (MPEG). Available at http://www.chiariglione.org/mpeg/.

19. The society for motion pictures and television engineering (SMPTE). Available at http://www.smpte. org/home/.

20. The TV-anytime consortium. Available at http:// www.tv-anytime.org/.

21. Youtube. Available at http://www.youtube.com/.

Multimedia Presentation Databases

V. S. Subrahmanian, Maria Vanina Martinez, and D. R. Reforgiato
University of Maryland, College Park, MD, USA

Synonyms

Multimedia presentation databases

Definition

A multimedia presentation consists of a set of media objects (such as images, text objects, video clips, and audio streams) presented in accordance with various temporal constraints specifying when the object should be presented, and spatial constraints specifying where the object should be presented on a screen. Today, multimedia presentations range from the millions of PowerPoint presentations users have created the world over, to more sophisticated presentations authored using tools such as Macromedia Director. Multimedia presentation databases provide the mechanisms needed to store, access, index, and query such collections of multimedia presentations.

Historical Background

Multimedia presentations have been in existence since the 1980s, when PowerPoint emerged as a presentation paradigm and animated computer video games started gaining popularity. Both of these paradigms allowed a multimedia presentation author to specify a set of objects (collections of images, video, audio clips, and text objects) and then specify how these objects should be presented. These objects could be presented in accordance with some temporal constraints that describe when and in conjunction with which other objects a given object should be presented. As an increasing number of authoring frameworks came into being, accompanied by an increasing need to create, collaborate on, and share presentations, the notion of multimedia presentations as a programming paradigm gradually came into existence.

Buchanan and Zellweger [6] were one of the first to recognize the need to treat multimedia presentations in a rigorous framework. They recognized that presentations consisted of a set of media objects, and they proposed presenting these objects in accordance with some very simple precedence constraints.

Later, Candan et al. [7, 8] extended the framework of Buchanan and Zellweger by describing a multimedia presentation as a set of media objects together with a very rich, but polynomially computable set of spatial and temporal constraints defining their presentation. They also showed how to help a presentation author identify when their presentation specifications were inconsistent and to minimally modify the presentation constraints so that consistency was restored. Related work by their co-authors showed how to deliver these presentations across a network in the presence of spatial and temporal constraints.

Adali et al. [1] introduced a relational model of data to support interactive multimedia presentations and define a variant of the relational algebra that allows users to dynamically query and create new presentations using parts of existing ones, generalizing select, project, and join operations. Lee et al. [11] present a graph data model for the specification of multimedia presentations, together with two icon-based based graphical query languages for multimedia presentations and the GCalculus (Graph Calculus), a relational calculus-style language that formalizes the use of temporal operators for querying presentation graphs that takes the content of a presentation into account [4, 5] propose methods to present the answer of a query to a multimedia database as a multimedia presentation [9] focuses on specializations and improvements of the above methods when querying databases consisting solely of PowerPoint information.

[10] treats multimedia presentations like a temporal database, and supports querying and reuse of parts of existing presentations. It considers algebraic operators such as insert, delete, and join, as well as user interface operations such as Fast Forward/Rewind, Skip, and links to other presentation databases, etc. [13] focuses on indexing and retrieving complex Flash movies. A generic framework called FLAME (FLash Access and Management Environment) based on a three-layer structure is presented to address this problem. This framework mines and understands the contents of the movies to address the representation, indexing and retrieval of the expressive movie elements, including heterogeneous components, dynamic effects, and the way in which the user can interact with the movie.

Foundations

A multimedia presentation consists of a set O of media objects and a set of constraints on the presentation of objects in O. A media-object o is a file such as an image file, a video file, a text file, or an audio file. Each type of file is assumed to have an associated player. For example, a video file may have QuickTime or the Windows Media players as its associated player.

The constraints associated with o fall into two categories: temporal and spatial constraints.

On the temporal side, each object o in O has two associated variables, $st(o)$ and $et(o)$ denoting, respectively, the start time and end time at which media object o is presented using its associated players. The *temporal specification* associated with a presentation is a set of constraints of the form

$$x - y <= c$$

where x, y are variables of the form $st(o_i)$, $et(o_j)$ and c is some constant. For example, if o is a video file, and there exists the constraint $et(o') - st(o) = 0$, then this means that the video object o should start playing as soon as object o' finishes being played out.

Likewise, on the spatial side, each object o in O has four variables $llx(o)$, $lly(o)$, $urx(y)$, $ury(o)$ denoting the x coordinate of the lower left corner of object o, the y coordinate of the lower left corner of object o, and likewise for the upper right corner's x and y coordinates [7] presents algorithms to check the consistency of spatial and temporal constraints, and to minimally modify the spatial/temporal constraints when they are inconsistent. In addition, each object o in O has an associated set of properties that can be stored (and queried) using any object oriented database management system.

A *multimedia presentation database* **M** consists of a set of multimedia objects (and their associated spatial and temporal constraints) [1] defines methods to query such multimedia presentation databases.

Consider the simplest operation: selection. Suppose the query is "*Select all objects o in **M** such that $C[o]$ holds and such that $st(o) > 10$.*" In this case, the goal is to look at each multimedia presentation m in **M**, and eliminate all objects o from m such that $st(o) < 10$. Also, it is necessary to eliminate all remaining objects such that $C[o]$ does not hold. The objects that survive

M

this elimination process must be presented in accordance with the original set of temporal and spatial constraints present in m. If m^* denotes the modification of m in this way, then $\sigma_C(M) = \{m^* \mid m \text{ in } M\}$.

In addition [1], defines other operations that allow videos to be concatenated together using various chromatic composition operators (such as smoothing, fading, etc.), methods to perform joins across videos, and methods to execute other kinds of relational style operations.

Key Applications

There are numerous possible applications for multimedia presentation databases. A simple application is an engine to query PowerPoint presentations, of which millions exist in the world today [9] proposes a PowerPoint database query algebra.

Another application is in the area of digital rights management. Suppose a major record company wants to identify all multimedia documents on the web that contain a clip of their copyrighted music. This corresponds to a select query on all multimedia documents on the web. The result would be a set of multimedia documents, some of which might infringe on the copyright holder's rights. The same might apply to online video on sources such as YouTube where it is not uncommon to find copyrighted material. The ability for an entertainment company to find gross violations of their copyright by searching through YouTube archives is critical.

Future Directions

As video games become ever more common, and as virtual worlds such as Second Life become increasingly popular with users, the ability to query games and *avatar* based systems will become increasingly important.

In addition, methods to index multimedia presentation databases are in their very infancy. Taking into account the graph based nature of presentation databases such as those in [1, 2, 11], it is important to note that methods to

index graphs may have a role to play. However, multimedia presentations are more complex than labeled directed graphs because presentation constraints can potentially be satisfied in many different ways.

Cross-References

▶ Temporal Constraints

Recommended Reading

1. Adali S, Sapino ML, Subrahmanian VS. A multimedia presentation algebra. In: Proceedings of the ACM SIGMOD International Conference on Management of Data; 1999. p. 121–32.
2. Adali S, Sapino ML, Subrahmanian VS. Interactive multimedia presentation databases, I: algebra and query equivalences. Multimed Syst. 2000;8(3): 212–30.
3. Bailey B, Konstan JA, Cooley R, Dejong M. Nsync – a toolkit for building interactive multimedia presentations. In: Proceedings of the 6th ACM International Conference on Multimedia; 1998. p. 257–66.
4. Baral C, Gonzalez G, Nandigam A. SQL+D: extended display capabilities for multimedia database queries. In: Proceedings of the 6th ACM International Conference on Multimedia; 1998. p. 109–14.
5. Baral C, Gonzalez G, Son TC. Design and implementation of display specification for multimedia answers. In: Proceedings of the 14th International Conference on Data Engineering; 1998. p. 558–65.
6. Buchanan MC, Zellweger P. Automatically generating consistent schedules for multimedia documents. Multimed Syst. 1993;1(2):55–67.
7. Candan KS, Prabhakaran B, Subrahmanian VS. CHIMP: a framework for supporting distributed multimedia document authoring and presentation. In: Proceedings of the 4th ACM International Conference on Multimedia; 1996. p. 329–40.
8. Candan K, Lemar E, Subrahmanian VS. View management in multimedia databases. VLDB J. 2000;9(2):131–53.
9. Fayzullin M, Subrahmanian VS. An algebra for powerpoint sources. Multimed Tools Appl J. 2004;24(3):273–301.
10. Jiao B. Multimedia presentation database system. In: Proceedings of the 8th ACM International Conference on Multimedia; 2000. p. 515–16.
11. Lee T, Sheng L, Bozkaya T, Balkir NH, Özsoyoglu ZM, Özsoyoglu G. Querying multimedia presentations based on content. In: Jeffay K, Zhang H, editors. Readings in multimedia computing and networking. San Francisco: Morgan Kaufmann Publishers; 2001. p. 413–37.

12. Wirag S. Modeling of adaptable multimedia documents. In: Proceedings of the 4th International Workshop on Interactive Distributed Multimedia Systems and Telecommunication Services; 1997. p. 420–9.
13. Yang J, Li Q, Wenyin L, Zhuang Y. Content-based retrieval of FlashTM movies: research issues, generic framework, and future directions. Multimed Tools Appl. 2007;34(1):1–23.

Multimedia Resource Scheduling

Jeffrey Xu Yu
Department of Systems Engineering and
Engineering Management, The Chinese
University of Hong Kong, Hong Kong, China

Definition

Multimedia information systems are different from the traditional information systems, where continuous media (audio/video) requests special storage and delivery requirements due to (i) the large transfer rate, (ii) the storage space required, and (iii) the real-time and continuous nature. Due to the special characteristic of continuous media, different types of scheduling are proposed, namely, the disk scheduling and stream scheduling. On one hand, the disk scheduling is to tackle both the large storage space and the corresponding large transfer rate requirements. On the other hand, the stream scheduling is to schedule requests from multiple clients, in order to minimize the delay in satisfying the requests. It attempts to support as many requests as possible, and at the same time, keep the real-time and continuous nature.

Historical Background

Continuous media adds additional requirements to the traditional information systems. In order to satisfy these new requirements, new scheduling algorithms are proposed.

Disk scheduling is designed to achieve the low service latency and high disk throughput requirements [14]. Continuous media, such as audio and video, adds additional real-time constraints to the disk scheduling problem. There are two categories of disk scheduling, namely, single disk scheduling and multiple disk scheduling. The disk scheduling algorithms, for continuous media, can possibly adapt one of the conventional disk scheduling strategies, which include round-robin, SCAN [14], and EDF [12]. For single disk scheduling for continuous media, there are SCAN-EDF [13], and sorting-set [8, 16] scheduling strategies. For multiple disk scheduling for continuous media, multiple disks can be accessed in parallel. The increased parallelism is mainly used to support more streams. There are two categories of multiple disk scheduling strategies, which are stripe data across the disks and replicate data across the disks. In additional, there are three schema for accessing striped data, which are called, striped retrieval, split-striped retrieval [15], and cyclic retrieval [2, 3].

Stream scheduling (or session scheduling) is designed to effectively allocate and share server and network resources. It aims at supporting as many clients as possible simultaneously, within the limited server and network resources, by assigning multiple client requests to a shared data stream. Three policies are proposed to effectively select the multimedia to be shared: FCFS, MQL, and FCFS-n [10]. In [6, 7], sharing is studied when a client pauses/resumes while viewing a long video. A two-level hierarchical scheduling policy is proposed to deal with time-varying load, such as the load at peek time and off-peek time, regarding the preservation of resources for the future requests [1]. In the two-level hierarchical scheduling, at the higher level, it focuses on channel allocation with consideration of how many requests will come in near future; at the lower level, it focuses on selecting clients to be served with consideration of the clients waiting time.

Foundations

There are mainly two types of resource scheduling in a multi-stream environment: (i) disk scheduling, and (ii) stream scheduling.

Disk Scheduling

In a multi-stream environment, multiple users are requesting to retrieve N continuous multimedia data streams in a similar time period. The system will serve each of the N data streams in rounds. In other words, in any i-th round, the system will serve all the requested data streams N by reading enough data for all the requests to be consumed until the next $i+1$-th round. Because there are differences between the transfer rate and the consumption rate, a scheduling strategy needs to deal with the differences between the transfer rate and the consumption rate.

Let r_c and r_t be the rate of display consumption of a data stream and the rate of data transfer from disk, respectively. And let p_{max} and d_{max} be the maximum time interval in any round and the maximum time interval between two consecutive reads for any data stream, respectively. Figure 1 illustrates the ideas on round-length and delay between two consecutive disk reads. In order to have enough data to be consumed by a display, at the consumption rate of r_c, in each round of the time interval at most p_{max} (to prevent starvation until the next round), the amount of data it needs is at least $r_c \cdot p_{max}$. The d_{max} shows the possible delay, called startup delay, to start consuming a data stream, which implies the time interval it needs to wait if it misses in the current round.

Consider the disk scheduling for multiple data streams on a single disk. The round-robin algorithm retrieves data for each data stream in a fixed order in every round. Due to the fixed order, the maximum startup delay (d_{max}) is similar to p_{max}. The SCAN algorithm [13] attempts to read more disk blocks while moving the disk head over the disk, and retrieve the requested blocks when the disk header passes over them [14]. The main advantages of the SCAN algorithm are: (i) it reduces the seek time to move the disk header from one location to another in order to read next disk block, and (ii) it maximizes the throughput of disk accesses. But, because the order of serving each data stream is not fixed, the startup delay for a given data stream can be larger up to $2p_{max}$. The SCAN-EDF algorithm [13] processes the data stream requests with the earliest deadline first, using the idea discussed in EDF [12] (Earliest-Deadline-First), and processes the requests, that be shared, using the SCAN algorithm. Note that the SCAN-EDF algorithm different from the other strategies, is not designed as a round-based algorithm. The sorting-set algorithm [8, 16] is designed in a way where round-robin and SCAN are treated as special cases of it. In brief, in the sorting-set algorithm, each round is further divided into several time slots. Each time slot is conceptually considered as a sorting set (or simply set). If a round is divided into m time slots, there are m time slots, and therefore there are m sorting sets, namely, set_1, set_2, . . . , and set_m. All the sorting sets are served in a fixed order in each run. When there is a request to retrieve a multimedia data stream, the requested data stream is assigned to a sorting set. If the entire round is treated as a single time slot, the sorting-set algorithm behaves like the SCAN algorithm. If the requested data stream is assigned to a unique sorting set, the sorting-set algorithm behaves like the round-robin algorithm.

For multiple disks accessing in parallel, there are two main categories of multiple disks scheduling strategies, namely, stripe data across the disks and replicate data across the disks. There are also different schema to retrieve the striped data: striped retrieval, split-striped retrieval, and cyclic retrieval. In the striped retrieval, entire stripes are retrieved in parallel. The split-stripe retrieval [15] retrieves some consecutive units of an entire stripe rather than the entire stripe at one time. The cyclic retrieval [2, 3] is designed to retrieve units of stripes for more than one data stream. The main idea behind it is to read a small portion of data for a data stream frequently. As comparison with the other striped approaches, cyclic retrieval does not need a large buffer space. Unlike the stripe based

Multimedia Resource Scheduling, Fig. 1 Round length and delay between reads (Fig. 3 in [9])

algorithms, data streams can be replicated across disks where each disk is treated individually and independently. If a data stream is requested frequently, it can be replicated in multiple disks.

Stream Scheduling

Continuous data stream retrieval needs to be guaranteed by reserving sufficient resources. The resources are referred to as logical channels (or simply channels). Stream scheduling policies need to increase the server capacity, or in other words, to increase the number of continuous data stream requests to be served with the limited number of channels. The stream scheduling needs to consider several facts regarding the possibility of sharing. There are popular videos which are viewed by many clients most of the time during a certain time period. Several clients may view the same video but start viewing at different times in a short time interval. A client may pause and then resume when viewing a long multimedia (video) at any time. The time interval between such pause and resume is not known, which can be short or long. A client may also change to another video after the pause.

Data stream sharing serves multiple clients by a single I/O stream, which is also referred to as batching of requests [4, 6]. A batching factor indicates how many clients can share a single I/O stream. In order to effectively support continuous multimedia requests, some requests need to be delayed, in order to be batched with other requests. There are three batching policies: FCFS, MQL, and FCFS-n [10]. With the FCFS (First Come First Served) policy, it queues all requests into a single queue. When there is a channel available, the FCFS policy selects the video, which is requested by the first client in the queue, to be served, in a first come first served fashion. If there are other requests in the queue that request the same video, they will be served by sharing the I/O stream. The MQL (Maximum Queue Length) policy maintains a queue for a requested video. If there are n videos to be requested, there will be n queues. The MQL policy chooses the video with the maximum queue length to be served, when a channel becomes available. With the MQL policy, the videos with a small number of requests (short queue) may not be served. Therefore, unlike FCFS which is a fair policy, the MQL policy is seen as unfair to the videos that are not requested by many clients. The FCFS-n policy is similar to the FCFS policy, except that the n hottest videos are assigned to dedicated streams. A dedicated stream will be served in a batching window in turn, and the remaining videos, that are not assigned to a dedicated stream, are served following the FCFS policy. With the policies, the batching factor can be increased, but the amount of waiting time for a client to wait may be increased as well.

Sharing a data stream is affected by the fact that a client may pause. A new channel may need to be allocated when the client resumes. Assume that a client may resume shortly after the pause, the system may maintain some channels (contingency channels), which can improve resource utilization. The system needs to guarantee that the delay between the receipt of a resume request from a client and playback is small in order to assure client satisfaction [4, 6]. Figure 2 illustrates the scheduling with contingency channels for VCR control operations (pause/resume) [6, 7]. The admission control policy determines the acceptance of a new request. The scheduling policy determines which request to be served on the

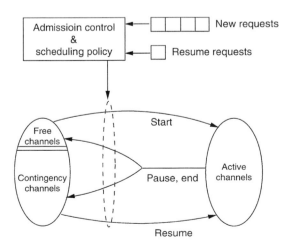

Multimedia Resource Scheduling, Fig. 2 Channel states under contingency policy (Fig. 2 in [5])

available channel, in order to maximize certain performance objectives [6]. When there is a pause request from a client, the admission control and scheduling policy determines if the client is the only client viewing the multimedia stream. If it is, the channel is freed and returned to either the free channel pool or the contingency pool. A certain number of contingency channels are maintained in the system. When there is a resume request from a client, the scheduling policy checks if there is another request being served within a predetermined time window, in order to maximize the possibility of sharing. If there is another request being served but is beyond the predetermined time window, the scheduling policy tries to use another contingency channel where possible. Otherwise, it will request a free channel with higher priority when such a free channel becomes available.

The arrival rate of requests to multimedia system may vary with the time of a day [11], peek time and off-peek time. The scheduling policy also needs to address the issues of time-varying load. Note: the optimal policy at the current may not be the optimal when the load changes shortly. A two-level hierarchical scheduling policy is proposed to deal with such load fluctuations [1]. The high-level scheduler controls channels allocation, whereas the low-level scheduler controls the clients selection to be served. Three high level policies are proposed to control channel allocation rate [1]: on-demand allocation, forced-wait, and pure rate control. The on-demand allocation allocates channels to requests when there are channels available. Under this policy, the waiting time for new requests may be long, which may cause clients to change from one video to another to view at high possibility. The forced-wait policy, as the name implies, forces the first request to a data stream to wait for up to a certain time (minimum wait time). The minimum wait time is a critical parameter, and can be difficult to be selected in a dynamic environment where the load changes dynamically. The pure rate control policy allocates channels uniformly in fixed time intervals (called measurement intervals) during such a time interval only a certain number of channels are used [1].

Key Applications

In a multimedia environment, multimedia objects are retrieved from either a digital library, or a video database, or an audio database, and are delivered to a large number of clients. The applications in such environments range from retrieving small multimedia objects (shopping, medias, education, etc.) to playback of large video objects (movie, entertainment, etc.).

Cross-References

▶ Continuous Multimedia Data Retrieval
▶ Multimedia Data Buffering
▶ Multimedia Data Storage
▶ Scheduler
▶ Storage Resource Management

Recommended Reading

1. Almeroth KC, Dan A, Sitaram D, Tetzlaff WH. Long term channel allocation strategies for video applications. IBM Research Report (RC 20249). 1995.
2. Berson S, Ghandeharizadeh S, Muntz R, Ju X. Staggered striping in multimedia information systems. ACM SIGMOD Rec. 1994;23(2):79–90.
3. Chen MS, Kandlur DD, Yu PS. Storage and retrieval methods to support fully interactive playout in a disk-array-based video server. Multimedia Syst. 1995;3(3):126–35.
4. Dan A, Shahabuddin P, Sitaram D, Towsley D. Channel allocation under batching and VCR control in video-on-demand systems. J Parallel Distrib Comput. 1995;30(2):168–79.
5. Dan A, Sitaram D. Session scheduling and resource sharing in multimedia systems. Chap. 11. In: Multimedia information storage and management. Boston: Kluwer Academic; 1996.
6. Dan A, Sitaram D, Shahabuddin P. Scheduling policies for an on-demand video server with batching. In: Proceedings of the 2nd ACM International Conference on Multimedia; 1994. p. 15–23.
7. Dey-Sircar JK, Salehi JD, Kurose JF, Towsley D. Providing VCR capabilities in large-scale video servers. In: Proceedings of the 2nd ACM International Conference on Multimedia; 1994. p. 25–32.
8. Gemmell DJ. Multimedia network file servers: multichannel delay sensitive data retrieval. In: Proceedings of the 1st ACM International Conference on Multimedia; 1993. p. 243–250.

9. Gemmell DJ. Disk scheduling for continuous media. Chap. 1. In: Multimedia information storage and management. Boston: Kluwer Academic; 1996.

10. Ghose D, Kim HJ. Scheduling video streams in video-on-demand systems: a survey. Multimed Tool Appl. 2000;11(2):167–95.

11. Little TDC, Venkatesh D. Prospects for interactive video-on-demand. IEEE Multimed. 1994;1(3):14–24.

12. Liu CL, Layland JW. Scheduling algorithms for multiprogramming in a hard-real-time environment. J ACM. 1973;20(1):46–61.

13. Reddy ALN, Wyllie JC. I/O issues in a multimedia system. Computer. 1994;27(3):69–74.

14. Teorey TJ, Pinkerton TB. A comparative analysis of disk scheduling policies. Commun ACM. 1972;15(3):177–84.

15. Tobagi FA, Pang J, Baird R, Gang M Streaming RAID: a disk array management system for video files. In: Proceedings of the 1st ACM International Conference on Multimedia; 1993. p. 393–400.

16. Yu PS, Chen MS, Kandlur DD. Grouped sweeping scheduling for DASD-based multimedia storage management. Multimedia Syst. 1993;1(3):99–109.

Multimedia Retrieval Evaluation

Thijs Westerveld
Teezir Search Solutions, Ede, Netherlands

Definition

Multimedia Retrieval Evaluation is the activity of measuring the effectiveness of one or more multimedia search techniques. A common way of evaluating multimedia retrieval systems is by comparing them to each other in community wide benchmarks. In such benchmarks participants are invited to submit their retrieval results for a given set of topics, the relevance of the submitted items is checked, and effectiveness measures for each of the submissions are reported. Multimedia retrieval evaluation measures the effectiveness of multimedia retrieval systems or techniques by looking at how well the information need as described by a topic is satisfied by the results retrieved by the system or technique. Efficiency of the techniques is typically not taken into account, but may be studied separately.

Historical Background

Until the mid-1990s, no commonly used evaluation methodology existed for multimedia retrieval. An important reason for this is that the field has merely been a showcase for computer vision techniques. Many papers in the field 'proved' the technical merits and usefulness of their approaches to image processing by showing a few well-chosen, and well-performing examples. Since 1996, the problem of systematically evaluating multimedia retrieval techniques has gained more and more interest. In that year, the *Mira* (Multimedia Information Retrieval Applications) working group was formed [2]. The group, consisting of people from the fields of information retrieval, digital libraries, and library science studied user behavior and information needs in multimedia retrieval situations. Based on their findings, they developed performance measures. Around the same time, in the multimedia community, the discussion on proper evaluation started, and Narasimhalu et al. [8] proposed measures for evaluating content-based information retrieval systems. These measures are based on comparing ranked lists of documents returned by a system to the perfect, or ideal, ranking. However, they do not specify how to obtain such a perfect ranking, nor do they propose a common test set. A year later, Smith [10] proposed to evaluate image retrieval using measures from the text retrieval community and in particular from *TREC*, the Text Retrieval Conference [12] for image retrieval evaluation. Again, no dataset was proposed. At the start of the twenty-first century, the evaluation problem gained more attention within the content-based image retrieval community, with the publication of three papers discussing benchmarking in visual retrieval [3, 6, 7]. These three papers call for a common test collection and evaluation methodology and a broader discussion on the topic. The *Benchathlon* network (http://www.benchathlon.net) was started to discuss the development of a benchmark for image retrieval. Then, in 2001, *TREC* started a video track [9] that evolved into the workshop now known as *TRECVID*.

Today, a variety of initiatives exists for evaluating the retrieval of different types of data in a variety of contexts, a list of these is provided below under data.

Foundations

Information retrieval is interactive. In web search, for example, queries are often changed or refined after an initial set of documents has been retrieved and inspected. In multimedia retrieval, where browsing is common, interactivity is perhaps even more important. Evaluation should take interactivity into account, and measure user satisfaction. The evaluation of a system as a whole in an interactive setting is often called an *operational test*. Such tests measure performance in a realistic situation. Designing such an operational test is difficult and expensive. Many users are needed to free the experiment of individual user effects, the experimental setup should not interfere with the user's natural behavior, and learning effects need to be minimized. Also, because there are many free variables, it is hard to attribute observations to particular causes. In contrast to these tests in fully operational environments, *laboratory tests* are defined as those tests in which possible sources of variability are controlled. Thus, laboratory tests can provide more specific information, even though they are further away from a realistic setting. Also, laboratory tests are cheaper to set up, because the interactive nature is ignored, and the user is removed from the loop. Laboratory tests measure the quality of the document ranking instead of user satisfaction. While some studies exist to evaluate multimedia retrieval systems in an operational setting by investigating user satisfaction, most approaches are studied in laboratory tests.

Current laboratory tests are based on the *Cranfield* paradigm [1]. In this paradigm, a test collection consists of a fixed set of documents, a fixed set of topics, and a fixed set of relevance judgements. Documents are the basic elements to retrieve, topics are descriptions of the information needs, and relevance judgements list the set of relevant documents for each topic.

The focus in laboratory tests is on comparative evaluation. Different approaches are tested, and their relative performance is measured. The process is as follows. Each approach produces a ranked lists of documents for each topic. The quality of the ranked lists is measured based on the positions of the relevant documents in the list. The results are averaged across all topics to obtain an overall quality measure.

For successful evaluation of retrieval techniques, two components are needed in addition to a test collection [4]: a measure that reflects the quality of the search and a statistical methodology for judging whether a measured difference between two techniques can be considered statistically significant. The measures used in multimedia retrieval evaluation are typically based on precision and recall, the fraction of retrieved documents that is relevant and the fraction of relevant documents that is retrieved. To measure recall, the complete set of relevant documents needs to be known. For larger collections this is impractical and a *pooling* method is used instead. With pooling, the assumption is that with a diverse set of techniques contributing runs to the evaluation, the probability that a relevant document is retrieved at a high rank by at least one of the approaches is high. A merged set of top ranked documents is assumed to contain most relevant documents and only this set of documents is manually judged for relevance. Documents that are not judged are assumed not relevant. In reality, some unjudged documents may certainly still be relevant, but it has been shown that this is of no influence to the comparative evaluation of search systems [11, 13, 14].

A number of aspects influence the reliability of evaluation results. First, a sufficiently large set of topics is needed. ?] suggest a minimum of 75. Second, the measures should be stable. This means it should not be influenced too much by chance effects. Clearly, measures based on few observations are less stable than measures based on many observations. For example, precision at rank 1 (is the first retrieved document relevant) is not a very stable measure. Third, there needs to be a reasonable difference between two approaches before deciding one approach is better than the other. Sparck Jones [5] suggests a 5% difference is noticeable, and a difference greater than 10% is material. Statistical significance tests take all these aspects into account and are useful in deciding whether an observed difference between two approaches is meaningful or simply due to chance.

Key Applications

Multimedia retrieval evaluation helps to better understand what works and what does not in the area of multimedia retrieval. Many practitioners in the field benefit from the area. It gives them the opportunity to test their ideas in a principled manner and allows them to build upon approaches that are known to be successful. More and more, the papers published in renowned journals and conferences demonstrate the usefulness of their techniques by a thorough evaluation on a well-known test collection.

Data Sets

Creating a test collection for multimedia retrieval systems takes a lot of effort. Especially the generation of ground truth data for a sufficient number of topics is something that a small company or research institution cannot manage on its own. Many workshops exist that solve this problem by sharing resources in a collaborative effort to create valuable and re-usable test collections. This approach was first taken in the Text Retrieval Conferences (TREC) [12], but many others followed. Below the main collections and evaluation platforms in multimedia are listed.

The *Corel* document set is a collection of stock photographs, which is divided into subsets each relating to a specific theme (e.g., *tigers*, *sunsets*, or *English pub signs*). The collection is often used in an evaluation setting by using the classification into themes as ground truth. Given a query image, all images from the same theme -and only those- are assumed relevant. Evaluation results based on Corel are highly sensitive to the exact themes used in the evaluation [7]. In addition, Corel

has a clear distinction between themes and an unusually high similarity within a theme because the photos in a theme often come from the same photographer or even the same location. This makes the collection more homogeneous than can be expected in a realistic setting.

TRECVID studies video retrieval. The data collections used have been dominated by broadcast news, but other raw and edited professional video footage is studied as well. TRECVID defines a number of tasks and provides test collections for each of them. In 2007, four main tasks existed:

Shot boundary detection: identify shot boundaries in the given video clips with their location and type (hard or soft transition).

High level feature extraction: For each of the pre-defined high-level features or concepts, detect the shots that contain the feature. Features that have been studied in the past include sky, road, face, vegetation, office and people marching.

Search: Given a textual description of an information need and/or one or more visual examples, find shots that satisfy this need.

Rushes summarization: Given a set of rushes, i.e., raw, unedited footage, provide a visual summary of this data that in a limited number of frames shows the key objects and events that are present in the footage.

As part of the Cross-Language Evaluation Forum (CLEF), *ImageCLEF* studies cross-language image retrieval. ImageCLEF concentrates on two main areas: retrieval of images from photographic collections and retrieval of images from medical collections.

The Initiative for the Evaluation of XML retrieval (INEX) aims to evaluate the effectiveness of XML retrieval systems. Within this initiative, the *INEX multimedia track* evaluates the retrieval of multimedia elements from a structured collection. The data collection used consists of wikipedia documents and the images contained in them. Both the retrieval of multimedia fragments (combinations of text and images) and the retrieval of images in isolation are studied.

ImagEVAL evaluates image processing technology for content-based image retrieval. The

M

assessment focuses on features relating to what collection holders (from defence, industry and cultural sectors) expect in terms of how images may be used. The tasks include recognizing transformed images, combined textual and visual search and object detection. The collections are a heterogeneous mix of professional images including stock photography, museum archives and industrial images.

The Music Information Retrieval Evaluation Exchange (*MIREX*) evaluates subtasks of music information retrieval. The datasets used by MIREX are cd-quality audio originating from (internet) record labels that allow tracks of their artists to be published. The tasks include genre classification, melody extraction, onset detection, tempo extraction and key finding.

Cross-References

▶ Advanced Information Retrieval Measures
▶ Multimedia Databases
▶ Multimedia Information Retrieval Model

Recommended Reading

1. Cleverdon CW. The cranfield tests on index languagr devices. Aslib Proc. 1967;19(6):173–92.
2. Draper SW, Dunlop MD, Ruthven I, van Rijsbergen CJ, editors. In: Proceedings of the MIRA 99: Evaluating Interactive Information Retrieval. eWiC, Electronic Workshops in Computing; 1999.
3. Gunther NJ, Beretta G. A benchmark for image retrieval using distributed systems over the internet: BIRDS-I. Technical report HPL-2000-162, HP laboratories. 2000.
4. Hull D. Using statistical testing in the evaluation of retrieval experiments. In: Proceedings of the 16th Annual International ACM SIGIR Conference on Research and Development in Information Retrieval; 1993. p. 329–38.
5. Jones KS. Automatic indexing. J Doc. 1974;30(4): 393–432.
6. Leung CHC, Ho-Shing IPH. Benchmarking for content-based visual information search. In: Advances in Visual Information Systems, 4th International Conference; 2000. p. 442–56.
7. Müller H, Müller W, McG Squire D, Marchand-Maillet S, Pun T. Performance evaluation in content-based image retrieval: overview and proposals. Pat-

tern Recogn Lett. (Special Issue on Image and Video Indexing). 2001;22(5):593–601.
8. Narasimhalu AD, Kankanhalli MS, Wu J. Benchmarking multimedia databases. Multimedia Tools Appl. 1997;4(3):333–56. ISSN 1380-7501
9. Smeaton AF, Over P, Costello CJ, de Vries AP, Doermann D, Hauptmann A, Rorvig ME, Smith JR, Wu L. The TREC-2001 video track: information retrieval on digital video information. In: Proceedings of the 6th European Conference on Research and Advanced Technology for Digital Libraries; 2002. p. 266–75.
10. Smith JR. Image retrieval evaluation. In: Proceedings of the IEEE Workshop on Content-Based Access of Image and Video Libraries; 1998. p. 112–3.
11. Voorhees EM, Harman DK. Overview of the eighth text retrieval conference (TREC-8). In: Proceedings of the 8th Text Retrieval Conference; 2000.
12. Voorhees EM, Harman DK. TREC: experiment and evaluation in information retrieval (digital libraries and electronic publishing). Cambridge: MIT; 2005. ISBN 0262220733.
13. Westerveld T. Trecvid as a re-usable test-collection for video retrieval. In: Proceedings of the Multimedia Information Retrieval Workshop; 2005.
14. Zobel J. How reliable are the results of large-scale information retrieval experiments? In: Proceedings of the 21st Annual International ACM SIGIR Conference on Research and Development in Information Retrieval; 1998. p. 307–14.

Multimedia Tagging

Xiaofeng Zhu
Guangxi Normal University, Guilin, Guangxi, People's Republic of China

Synonyms

High-level feature extraction; Interactive tagging; Multimedia annotation; Multimedia concept detection; Multimedia labeling; Tag location; Tag recommendation; Tag refinement

Definition

Multimedia data indicates large amounts of multi-/rich media data, such as text, graphics, images, music, video, and their combination. The basic elements of multimedia data are

Multimedia Tagging, Fig. 1 An illustration of video tagging on assigning single tag (e.g., gold fish) for a video

Multimedia Tagging, Fig. 2 An illustration of video tagging on assigning multiple tags (such as Gangnam Style and PSY) for a video

text, images, audio, animation, and video. Multimedia tagging is referred to as the process by which a computer system automatically assigns metadata in the form of captions or keywords to multimedia data for describing their content on semantic or syntactic levels. With such metadata, the management, summarization, and retrieval of multimedia content can be easily accomplished. The tags (i.e., metadata or captions or keywords) can be directly used to index multimedia data. According to the semantic and syntactic content, multimedia data can be assigned one tag or multiple tags. For example, the video on gold fish in Fig. 1 might be assigned a tag of "gold fish," while multiple tags can be added to a video, such as the tag "Gangnam Style" and "PSY" for a video in Fig. 2.

Historical Background

With the explosive growth of the study of multimedia data, multimedia tagging has attracted great interest of computer vision, multimedia, and information retrieval, in the past two decades. Now a number of techniques have been developed by combining humans and computers for more accurate and efficient video tagging in the Web 2.0 era.

Manual Multimedia Tagging Manual multimedia tagging is the most naive approach for multimedia tagging. Currently, manual multimedia tagging has been used for real applications, such as the collaborative nature of Web 2.0, Flickr (http://www.flickr.com), and YouTube (http://www.youtube.com). Moreover, these multimedia sharing websites not only allowed users to upload multimedia data but also encourage them to label the content with descriptive tags. Although manual multimedia tagging (even for large-scale multimedia data) becomes feasible, it still encounters the following problems [5] in real applications:

1. Expensive labor and time costs. Manual multimedia tagging is often constraint from the limited labor and time cost, even for small datasets.
2. Tagging quality. The tags assigned by non-professional users are usually noisy and incomplete, while professional users are always personal and will not achieve agreement for all tags. Low-quality tags will produce noise in many tag-based applications, such as multimedia retrieval.
3. Prohibiting for large-scale datasets. [7] shows that typically annotating 1 h of video with 100 concepts can take anywhere between 8 and 15 h, and the study in [6] shows that a user will need 6.8 s to tag an image on average. This indicates that a pure manual multimedia tagging encounters a number of problems for large-scale multimedia tagging.

Automatic Multimedia Tagging Because manual tagging is a time-consuming and labor-cost process, automatic multimedia tagging has attracted great research interest. The first group of automatic multimedia tagging methods is based on parametric topic models. The parametric topic models include probabilistic latent semantic analysis model, latent Dirichlet allocation model, joint distribution of the image features and the tags with mixture models, discriminative models (such as SVM, ranking SVM, and boosting), and so on [1]. While these methods achieve promising tagging results, their expensive training processes limit the application on large-scale datasets. Recently, the joint equal contribution model [3] and the TagProp model [2] rely on local nearest neighborhoods and work surprisingly well despite their simplicity. Despite the great progress made in the past years, the performance of automatic multimedia tagging is still far from satisfactory, such as low prediction accuracy.

Assistive Multimedia Tagging: Combining Humans and Comp Because manual multimedia tagging is with high accuracy and costs and automatic multimedia tagging is with low accuracy and costs, assistive tagging, combing humans and computers, complements manual methods and automatic methods. For example, active learning method for assistive tagging asked users to label samples selected by computers and then learned the concept models according to the labeled samples. The assistive tagging methods can categorized the techniques into the following three paradigms [5]:

1. Tagging with data selection and organization. The method only manually labels several representative sample paradigms that aim to reduce tagging costs or intelligently organize the to-be-tagged data for improving tagging efficiency.
2. Tag recommendation. The method first found a set of candidates, by which users directly selected the correct candidates. It can improve both the tagging efficiency and the quality of tags.
3. Tag processing. The method was proposed for improving the accuracy and completeness of tags.

Scientific Fundamentals

Because multimedia data include many kinds of data, for simplicity, we illustrate the general framework of video tagging in this section. The general framework of video tagging includes the following steps [4]:

Shot segmentation (step 1) and feature extraction (step 2) The first step in the analysis of a video is to partition it into its elementary subparts, called shots or keyframes, according to significant scene changes (cuts or progressive transitions). Usually scene change detection was used to split the video into structural elements like shots, scenes, etc. To perform scene change detection, a feature extraction from video sources should be developed. The popular trend on feature extraction is to concatenate all kinds of low-level features in a high-dimensional vector, such as edge distribution histograms, histograms of oriented gradient (HOG), color histograms, and Haar or Gabor wavelets. Another current trend on feature extraction is to build a bag of visual words (BoVW) from the original low-level feature vectors [6]. Because the before-mentioned features might not be discriminative enough for comparing videos with only the visual content, some frequent pattern mining techniques are usually designed because these approaches can select a compact set of relevant features according to class information.

Similarity calculation between videos (step 3) Although a video can be regarded as a sequence of images, similarity calculation between videos is always difficult due to the variations in either the lot of shots or the video duration. This leads to a number of methods for similarity calculation between videos: first, representing a whole video with a single description, consisting in the average of all frame histograms; second, computing the similarity between the two most similar frames of the videos. Note that such comparison uses a unique pair of frames but no sequential information; and third, making the best use of common identical frames (a.k.a., near duplicate [7]) to compare videos.

Tag propagation process (step 4) Usually, the tag propagation process will be ignored while learning multiple classifiers in the video auto tagging systems. However, the new tag propagation method should be designed in the near duplicate-based method.

Key Applications

VeoTag and Click.TV are two new contenders in the video tagging space. Their goal is to let users tag individual sections of a video so that others can skip straight to the juicy bits. Further down the line, the data could also be used to improve video search:

Veotag offers a very basic service, in which one can tag videos and explore the clip library. The design of Veotag is good and the tagging tool is easy to use.

Click.TV provides plenty of entertainment for the casual user. However, the site is focused on comments, rather than tags.

Cross-References

▶ Automatic Image Annotation
▶ Video Summarization

Recommended Reading

1. Chen M, Zheng A, Weinberger K. Fast image tagging. In: Proceedings of the 30th International Conference on Machine Learning; 2013. p. 1274–82.
2. Guillaumin M, Mensink T, Verbeek J, Schmid C. Tagprop: discriminative metric learning in nearest neighbor models for image auto-annotation. In: Proceedings of the 12th IEEE Conference on Computer Vision; 2009. p. 309–16.
3. Makadia A., Pavlovic V., Kumar S. A new baseline for image annotation. In: Proceedings of the 10th European Conference on Computer Vision; 2008. p. 316–29.
4. Tran HT, Fromont E, Jacquenet F, Jeudy B, Martins A, et al. Unsupervised video tag correction system. In: Extraction et gestion des connaissances; 2013. p. 461–6.
5. Wang M, Ni B, Hua XS, Chua TS. Assistive tagging: a survey of multimedia tagging with human-computer joint exploration. ACM Comput Surv. 2012;44(4):25.
6. Yang J, Jiang YG, Hauptmann AG, Ngo CW. Evaluating bag-of-visual-words representations in scene classification. In: Proceedings of the 9th ACM SIGMM International Workshop on Multimedia Information Retrieval; 2007. p. 197–206.
7. Zhao WL, Wu X, Ngo CW. On the annotation of web videos by efficient near-duplicate search. IEEE Trans Multimed. 2010;12(5):448–61.

Multimodal Interfaces

Monica Sebillo[1], Giuliana Vitiello[1], and Maria De Marsico[2]
[1]University of Salerno, Salerno, Italy
[2]Sapienza University of Rome, Rome, Italy

Definition

Multimodal interfaces are characterized by the (possibly simultaneous) use of multiple human sensory modalities and can support combined input/output modes.

The term *multimodal* recurs across several domains. Since its first use in the field of interface design, its affinity and derivation from the terms "mode" and "modality" were discussed. According to Merriam-Webster, one of the meanings for *mode* is "a possible, customary, or preferred way of doing something," whereas *modality* can be "one of the main avenues of sensation (as vision)." In *multimodal interfaces*, the former influences the way information is conveyed, the latter refers to the exploited communication channel. Both express peculiar aspects of a multimodal system, which is expected to provide users with flexibility and natural interaction.

Early multimodal interfaces simply combined display, keyboard, and mouse with voice (speech recognition/synthesis). Later, pen-based input, hand gestures, eye gaze, haptic input/output, head/body movements have been progressively used. As a result, modern multimodal interfaces aim at emulating the natural multi-sensorial forms of human-human dialogue, relying on the integration of advanced interaction modes.

M

To support the described variety of modes, the system underlying such an interface must include hardware to acquire and render multimodal expressions and must exploit recognition-based technologies to interpret them, with response times compatible with user's interaction pace.

Historical Background

Bolt's "Put That There" demonstration system [2] can be considered as one of the earliest multimodal systems. The underlying technology allowed a joint use of voice and gesture by processing speech in parallel with manual pointing within a virtual graphical space, where users could be made easily aware of the available facility and its usage. Spoken input was semantically processed, while deictic terms in the speech were resolved by processing the spatial coordinates derived from pointing. Since then, different proposals of multimodal interfaces can be found in literature [4, 5]. Many of them aim at integrating natural languages and direct manipulation in specific application domains, such as manufacturing environments and interactive maps.

The *CHI'90 Workshop on Multimedia and Multimodal Interface Design* represented a turning point [1]. The growing interest by the international community toward this research area induced the Program Committee to focus on both the integration of scientific knowledge about this discipline and the direction of future developments. Four main topics were identified for discussion, with respect to which interfaces could be designed and examined: structural principles for composition, media appropriateness, enabling technologies, and paradigms/metaphors/models/representations. Important issues emerged, from which some years later the first remarkable set of guidelines for the design of multimodal user interfaces stemmed [9].

Following the scientific attention to the multimodal paradigm, a variety of systems rapidly came out. Both hardware and software were enhanced to integrate parallel input streams, mostly acquired by speech and pen-based gestures. One of the first such prototypes was the QuickSet sys-

tem developed in 1994 [3]. It was an agent-based, collaborative, multimodal-multimedia map system, allowing the user to issue commands using a combination of speech and pen input. For illustration purposes, in [8] Oviatt provides a comparison among five different speech and gesture systems, which represented the most exemplifying research-level systems developed until the late 1990s. In her evaluation, the author takes into account some characteristics, such as the type of signal fusion and the sizes of gesture and speech vocabularies, on which integration and interpretation functionalities were based.

More recently, new combinations of speech and other modalities, such as lip movements and gaze, have been exploited thanks to advanced input/output technologies, whose effective integration into flexible yet reliable interfaces requires the underlying system robustness and performance stability. As a result, multimodal interfaces have pervaded new application domains, ranging from virtual reality systems, meant to support expert users in decision making and simulation of scenarios, to training systems and to person identification/verification systems for security purposes. Such systems may be further distinguished on the basis of their input modes, which can be either intentionally exploited by users (active input mode) such as speech and gestures, or captured by the system without an explicit request by users (passive input mode), such as facial expressions.

Foundations

In order to clarify the different aspects involved in a multimodal interface, the complexity of multimodal interaction can be described in terms of the well-known *execution-evaluation cycle* defined by Don Norman in 1988 [6].

In the case of multimodal interactive cycles, Norman's model of interaction may be reformulated as follows:

1. Establishing the goal.
2. Forming the intention.

3. Specifying the multimodal action sequence in terms of human output modalities.
4. Executing the multimodal action.
5. Perceiving the system state in terms of human input modalities.
6. Interpreting the system state.
7. Evaluating the system state with respect to the goals and the intentions.

- *Establishing the goal* is, as usual, the stage when the user determines what needs to be done in the given domain, in terms of a suitable task language.
- *Forming the intention*: at this stage the goal is translated into more precise user's intention, which will help the user determining the right sequence of actions that should be performed to achieve the goal
- *Specifying the multimodal action sequence.* The sequence of actions performed to accomplish the required task should be precisely stated at this stage. Here the complexity of multimodal interaction appears for the first time in the cycle. In fact, each multimodal action can be specified in terms of:
  1. Complementary human output sensory modalities (i.e., multiple utterances at once form the action) and/or
  2. Alternative human output sensory modalities (i.e., alternative, redundant utterances for the same action).

Examples of human output modalities include speaking, gesturing, gazing, touching, moving, facial expressions and some unintentional utterances such as blood pressure, temperature, heartbeat, excretion, etc. A user may, for instance, move an object in the interaction scene by speaking and pointing at (gesturing) the new object location (complementary modalities). Then, instead of gesturing (s)he may want to gaze at the new location on the interface where the object should be moved (alternative modality).

- *Executing each multimodal action.* At this stage, each human modality used to specify an action is translated into corresponding in-

teraction modes. Thus, each action is executed through

1. Complementary modes or
2. Alternative modes

Text, speech, Braille, mimicking, eye/motion capture, haptics, bio-electrical sensing are examples of modes used to translate human output modalities into the system input language.

When the execution of the whole sequence of multimodal actions is complete, the system reaches a new state and communicates it to the user again exploiting (possibly multiple) interaction modes, such as speech synthesis, display, haptic/tactile feedback, smell rendering and so on.

- *Perceiving the system state.* At this stage, the evaluation phase of the cycle begins. Depending on the combination of system output modes, the user may perceive the new state through multiple input sensory modalities, such as visual, auditory, tactile, and (in some revolutionary interfaces) even smelling and tasting.
- *Interpreting the system state.* Here the user is supposed to interpret the output of her/his sequence of actions to evaluate what has happened.
- *Evaluating the system state with respect to the goals and the intentions.* At the final stage, the user compares the new system state with her/his expectations, to evaluate if the initial goal has been effectively reached.

Of course a good mapping should be achieved between the execution and the evaluation phases in order to bridge what Norman calls the *gulf of execution* (i.e., the distance between user's specification of the action and the actions allowed by the system) and the *gulf of evaluation* (i.e., the distance between user's perception of the new state and her/his expectation). In multimodal interfaces the effective reduction of both gulfs crucially depends on the underlying interactive technology that must move as close as possible

M

towards human-human forms of interaction and communication. Thus, on the execution side, human output modalities should be supported by suitable computer input devices, e.g., by mapping gaze, speech, touch, gesturing and smelling onto cameras, microphones, keyboard, haptic sensors and the most recent olfactory sensors, respectively. From the evaluation perspective, computer output and its perception by user should be also tightly linked. This requires the adoption of output devices, like display, audio and haptic/olfactory rendering devices, able to quickly and effectively reach the human input sensory system, so that the user sees, hears, touches and, in general, feels the new system state as it is communicated by the multimodal interface.

As an example, the successful execution of direct manipulation tasks within an immersive virtual reality environment may critically depend on the abolition of any latency time between the moment an action is performed, e.g., by means of datagloves, and the moment the user recognizes the touch. In this way the user perceives gesturing as an act that can be directly realized at the interface. If this cannot be achieved, any usability benefits coming from the direct manipulation paradigm would be lost.

The ultimate advantage of multimodal interfaces is increased usability, in terms of both flexibility and robustness of the interaction when either redundant or complementary information is conveyed by modes. Higher flexibility is gained since multimodal interfaces can accommodate a wide range of users, tasks and environments for which each single mode may not be sufficient. Different types of information may be conveyed using the most appropriate or even less error prone modality, while alternation of different channels may prevent from fatigue in computer use intensive tasks. Redundancy of information through different communication channels is especially desirable when supporting accessibility, since users with different impairments may benefit from information and services otherwise difficult to obtain. As for the increased robustness of the interaction, the weaknesses of one modality may be offset by the strengths of another. More semantically rich input streams can support

mutual disambiguation for the execution phase. As in human-human communication, the correct decoding of transmitted messages requires interpreting the mix of audio-video signals.

An important research theme in multimodal interface design is how to integrate and synchronize different modes, taking into account that synchrony of different "tracks" of interaction in different modes, does not imply their simultaneity. At present, each unimodal technique is developed separately, with noticeable advances produced by improvements in both software recognition-based techniques and hardware input/output technologies. However, as pointed out in [9], an effective integration of the involved modal technologies requires a deep understanding of the "natural" integration patterns that characterize people's combined use of different communication modes, as widely studied by psychologists and cognitive experts. The issue of integration may become even more complex when a multimodal interface is designed to support collaborative work, namely the work by multiple users who may interact through the interface using several input/output modes, either synchronously or asynchronously, and either locally or remotely.

Key Applications

Recent advances in technology have been urging IT researchers to investigate innovative multimodal interfaces and interaction paradigms, able to exploit the increased technological power. The common key goal is to reproduce in the best possible way the interaction through different channels, typical of a human-human dialogue. As an example, real "physical" manipulation might be simulated even when the latter is not possible due to logistical problems (e.g., remote operation) or to dangerous settings (e.g., radioactive materials and areas), or when it is convenient to just simulate a real operation (e.g., for training purposes). A more natural and familiar way of managing objects and situations is also expected to improve global user performances and increase applications effectiveness.

Among the most recent efforts, research on haptic equipment deserves a mention. Related advances trigger new potentialities and convey novel features towards many domains, especially industrial, medical, and biotechnological. In the industrial world, the goal of improving competitiveness has led to the experimentation of haptic interfaces in fields like automotive and aerospace engineering, and texture manufacturing. In the medical domain, education and research activities are being increasingly improved by the adoption of haptic environments for virtual surgery simulation. Several new challenges are arising in the field of Biology/Biotechnology, where the adoption of visual interfaces connected to haptic devices is recognized as a powerful and straightforward mean to handle nano-objects, such as cells, and the possibility of force feedback offered by certain haptic systems, is envisioned as a considerable improvement of operator's perception. Last but not least, several multimodal interfaces enhanced with haptic feedback have been conceived to address major societal needs, e.g., by visually impaired people or wheelchair users.

In the following, a brief list of some further application domains is presented, where multimodal interfaces are presently investigated.

Interaction in Mobile Environments

The problem that has to be solved in applications designed for mobile environments is that hands, which are the usual interaction mediator for human-computer communication with traditional input devices, must be devoted to different crucial activities, e.g., controlling a steering wheel. In such situation, alternative modes should rather be exploited to interact with software applications such as a map browser. Moreover, user's visual attention must be focused on catching situations such as obstacles approach, so that relevant software events should be communicated for example through auditory signals, so as to relieve the user from continuously inspecting system state.

Geographic Information Systems

Multimodal interfaces are also being employed as a means to support decision makers in accessing and analyzing geospatial information in specific and critical scenarios, such as crisis management procedures and *what if* analyses. Some systems have been recently proposed, which rely on large screen displays and augmented reality tools for enhanced data visualization, as well as on collaborative advanced interfaces supporting speech and gesture recognition.

In these systems, multimodality becomes the way domain expert users can formulate appropriate requests to the underlying geographic information system and receive rapid responses, provided through different perspectives. Rapid feedback is in fact a crucial issue in situations when risk and vulnerability must be predicted as well as during exceptional events when recovery actions must be taken by users with complementary expertises.

Interaction in Adverse Settings

As discussed above, it is often necessary to substitute the human operator in adverse settings in a way that preserves both his/her health and the effectiveness of the interaction with environment objects. A much simpler case is when some communication channel might be hindered by disturbing conditions and the presence of other modes may provide possible missing information.

Multimodal Biometric Databases

Multimodal biometric databases are an example of tight integration of multiple input modes to achieve reliable person identification and verification. Fingerprints are the most well-known biometric method. More biometrics include hand conformation, iris scanning, features from face, ears or voice or handwriting. Despite noticeable progresses in biometrics research, no single bodily or behavioral trait satisfies acceptability, speed and reliability constraints of authentication in real applications. Single biometric systems are vulnerable to possible attacks, and may suffer from acquisition failures, or from the possible non-universality of the biometric feature, as in the case of deaf-mute subjects for voice recognition. The present trend is therefore towards multimodal systems, as flaws of an individual system

M

can be compensated by the availability of a higher number of alternative biometries. Integration of single responses is a crucial point, especially when different reliability degrees can be assigned to them due to input quality or effectiveness of recognition algorithms.

Interaction in Impairment Conditions

Accessibility is a transversal issue relating to different application domains. Physical impairments call for flexible system interfaces allowing universal access to services and information. What should be affected when designing for accessibility is the structure of both input and output for each application function. Functions need parameters including both data and events triggered by user's actions. In both cases, it is necessary to adapt the format manageable by a user possibly bearing a specific disability to the one acceptable by the functions. Such adaptation could be provided by special pieces of software (wrappers). A different wrapper is needed for each different disability situation. They would be connected to suitable interfaces allowing the user to issue commands and data according to his/her ability, and translating them for function call. Information and data returned by the system undergoes a symmetrical translation. In other words, multimodal input/output should be dynamically provided. The contribution of (disabled) accessibility experts to the overall design of wrappers is essential to obtain significant results. They can suggest the best suited interaction mechanisms and the best input/output modes to use.

Future Directions

The future challenge for multimodal interfaces is the ability to better and better mimic human-like sensory perception. Such interfaces will be able to reliably interpret continuous input from more different visual, auditory, and tactile sources, chosen according to the target users' tasks. More advanced recognition of users' natural communication modalities will be supported, and more sophisticated models of multimodal interaction

are expected to replace present bimodal systems. One of the problems to solve is to design and implement effective integration schemas among different modalities, based on available literature on human intersensory perception and on natural human-human multimodal interaction patterns. More research is required on human inclination to multimodal communication with applications, depending on different target tasks, and about integration and synchronization characteristics of multimodal input/output in different contexts and situations.

Cross-References

▶ Geographic Information System
▶ Visual Interfaces

Recommended Reading

1. Blattner MM, Dannenberg RB. CHI'90 Workshop on multimedia and multimodal interface design. SIGCHI Bull. 1990;22(2):54–8.
2. Bolt RA. Put that there: voice and gesture at the graphics interface. ACM Comput Graph. 1980;14(3):262–70.
3. Cohen PR, Johnston M, McGee DR, Oviatt SL, Pittman J, Smith I, Chen L, Clow J. QuickSet: multimodal interaction for distributed applications. In: Proceedings of the 5th ACM International Conference on Multimedia; 1997. p. 31–40.
4. European Telecommunications Standards Institute. Human Factors (HF); Multimodal interaction, communication and navigation guidelines ETSI EG 202 191 V1.1.1 (2003–08).
5. Jaimes A, Sebe N. Multimodal human-computer interaction: a survey. Comput Vis Image Underst. 2007;108(1–2):116–34.
6. Norman D. The design of everyday things. New York: Doubleday; 1988.
7. Oviatt S. Ten myths of multimodal interaction. Commun ACM. 1999;42(11):74–81.
8. Oviatt S. Multimodal interfaces. In: Jacko J, Sears A, editors. The human-computer interaction handbook: fundamentals, evolving technologies, and emerging applications. Mahwah: Lawrence Erlbaum; 2003.
9. Reeves LM, Lai J, Larson JA, Oviatt S, Balaji TS, Buisine S, Collings P, Cohen P, Kraal B, Martin JC, McTear M, Raman TV, Stanney KM, Su H, Wang QY. Guidelines formultimodal user interface design. Commun ACM. 2004;47(1):57–9.

10. Yuen PC, Tang YY, Wang PSP, editors. Multimodal interface for human-machine communication. River Edge: World Scientific; 2002.

Multi-pathing

Kaladhar Voruganti
Advanced Development Group, Network
Appliance, Sunnyvale, CA, USA

Definition

There can be multiple paths between a SCSI initiator and a SCSI target. Multiple paths between a host and a storage device are useful to provide more fault-tolerance as well as to improve system throughput. Multi-pathing software ensures that the same target volume is not seen as two separate LUNs by the host.

Key Points

The multiple paths can be configured in active-active or active-standby modes. In the active-active mode, both paths are actively transferring data. In the active-standby mode, the standby path does not actively transfer data. Some multi-pathing software allows for dynamic load balancing of traffic between the multiple paths. Typically, the storage controller vendor also provides the multi-pathing driver that runs on the host, and this software is usually limited to only operating with the vendor's storage devices. Software vendors are beginning to provide multi-pathing software that can interoperate with storage controllers from multiple storage vendors.

Cross-References

▶ Initiator
▶ Logical Unit Number
▶ Volume

Multiple Representation Modeling

Esteban Zimányi[1], Christine Parent[2], Stefano Spaccapietra[3], and Christelle Vangenot[3]
[1]CoDE, Université Libre de Bruxelles, Brussels, Belgium
[2]University of Lausanne, Lausanne, Switzerland
[3]EPFL, Lausanne, Switzerland

Synonyms

Multi-granularity modeling; Multi-resolution; Multi-scale

Definition

Geodata management systems (i.e., GIS and DBMS) are said to support *multiple representations* if they have the capability to record and manage multiple representations of the same real-world phenomena. For example, the same building may have two representations, one with administrative data (e.g., owner and address) and a geometry of type point, and the other one with technical information (e.g., material and height) and a geometry of type surface. Multirepresentation is essential to make a data repository suitable for use by various applications that focus on the same real world of interest, while each application has a specific perception matching its goals. Different perceptions translate into different requirements determining what information is kept and how it is structured, characterized, and valued. A typically used case is map agencies that edit a series of national maps at various scales and on various themes.

Factors that concur in generating different representations include the intended use of data and the level of detail matching the applications concerns. The former rules the choice of data structures (which objects, relationships, and attributes are relevant) and of value domains (e.g., whether the temperatures are stored in Celsius or Fahrenheit). The latter rules data resolution from

M

coarse to precise and impacts both the semantic and the spatial representation.

Multiple representation modeling is the activity of designing a data repository that consistently holds multiple representations for various perceptions of a given set of phenomena. It relies on a multirepresentation data model, i.e., a data model with constructs and rules to define and differentiate the various perceptions and for each perception the representations of the phenomena of the real world of interest.

Historical Background

Support for different requirements over the same data set has first been provided by using multiple data files, each one designed for a specific application. Aiming at consistency, databases looked instead for ways to gather data into a single repository, generating the need for multiple representations. First came facilities to define application-specific subschemas, as simple restrictions of the database schema. Later, more flexibility was achieved through the view mechanism. In object-oriented terms, views are virtual representations derived from existing data. They create either an alternative representation for existing objects (object preserving views) or new objects composed from existing objects (object generating views). Each view defines a single new representation. However, views do not provide multiple perceptions, i.e., there is no possibility to identify the collection of views that forms a consistent whole for an application.

Similarly, the concept of is-a link was borrowed from artificial intelligence to provide for various representations of the same object in a classification refinement hierarchy. However, the concept comes with a population inclusion constraint: The subtype population is included in the supertype population. This cannot cope with situations where the populations of two object types only overlap, i.e., the two populations may have specific objects not represented in the other population. Since then, the situation has not changed much. It is only in the last decade that the need for more flexible multirepresentation

and explicit support of various perceptions has been stressed by researchers.

Multirepresentation research in spatial databases can be traced back to the 1989 NCGIA program. In the early 1980's, maps began to be stored in geographic databases, and that opened up many new research tracks. The specification and implementation of cartographic generalization was one of them. It relied on the idea that cartographic generalization would allow the automatic derivation of maps at different scales from a single geographic database. Realizing that this full automation was not possible increased the focus on multiple representation databases [7, 9, 15]. Under the pressure of delivering maps at different scales, the first researches were, in the early 1990's, focusing on multi-scale data structures (i.e., data structures allowing to retrieve geometry of real-world objects for any given scale) and multiscale databases (i.e., databases in which the data for maps at different scales is stored and linked together). Later, the multiscale approach was refined and extended into the multirepresentation approach: scale indeed is not a concept relevant for databases and there is nowadays more to a geographical database than producing maps. The current scope of multirepresentation for GIS database comprises various techniques that can be used, within a single database, to automatically derive a coarser representation from another representation at finer resolution. These techniques aim at computing objects at coarser representations, for multiresolution analyses rather than for display. They include generic cartographic generalization operations (not driven by map display considerations) as well as operations performed on specific object types (e.g., selection of instances based on an ad-hoc predicate, aggregation of instances to create new objects). Some authors use the term model generalization to denote that the use of various techniques leads to the creation of a set of virtual databases for different resolution levels, and the mappings between their schemas (models in GIS terms). The automatic derivation rules (the mappings) allow update propagation from finer to coarser representations.

Currently, only a few simple multi-representation databases exist and are used to their expected potential.

Foundations

Early research on multi-representation in GIS was driven by cartographers's; requirements. This explains why the ability to draw maps at different scales has been for long the targeted objective, popularizing the concept of *multi-scale databases*. However, many other spatial applications that need to perform spatial data analysis require storing and managing specific representations where objects may have various geometries (derived one from another or not) and also show varying thematic characteristics (e.g., have different attributes and different relationships). Thus, the research domain evolved from multi-scale databases to *multi-representation databases*. Equally important is the capability to provide each application with a consistent set of data that corresponds to its own perception of the real world of interest, in short its own database. Therefore, support of multirepresentation should be complemented with support of multiperception.

Multiscale Databases

Multiresolution databases are still referred to by the GIS community as multiscale databases, despite the fact that scale does not apply to data storing. Scale is a concept related to the drawing of maps on paper or on screen. It is the ratio between measures on a map and the corresponding measures in the real world. Scale only characterizes an intended use of data. Instead, the level of resolution of a spatial database determines what geometries are stored. It defines a threshold such that only geometries beyond the threshold are captured and stored.

In early work by Timpf, the different map representations of the same real-world entities are interconnected using a directed acyclic graph data structure. The graph allows users to navigate among maps at different scales by zoom-in and zoom-out operations. This work later de-veloped into a more general Map Cube Model [14], supported by a theory on the structure of series of maps of the same region at different scales. Each map is described as a composition of four components: a set of lines representing transportation and hydrology networks, the set of areas, called containers, created by the lines of the trans-hydro network, areas that are a refinement of the container partition, and objects contained in these areas. The elements stored in each of the four components (e.g., streets, land-use areas, buildings) are then organized into aggregation and/or generalization hierarchies. This forms a graph for each component, where each level of the graph corresponds to a given scale.

Stell and Worboys have also proposed a solution to link a series of maps [12]. Their database organization is called a stratified map space. Each map gathers objects of a particular region that share the same semantic and spatial granularity. Maps are grouped by map spaces, i.e., sets of maps at the same granularity, describing various regions. The stratified map space is the set of all maps spaces organized according to a hierarchy based on different granularity levels. Transformation functions allow users to navigate in a stratified map space and propagate updates.

Multi-Representation Databases

Work on spatial multi-representation databases has followed two main tracks: either proposing new (conceptual) data models that include explicit description of multi-representation, or proposing frameworks that organize a set of existing classic (i.e., without multirepresentation) databases into a global multirepresentation repository.

Database Models for Multiple Representations

Several data models with specific concepts for multiple-representation modeling have been proposed. They range from simple solutions allowing users to associate various geometries to the same objects to more sophisticated solutions. They are discussed here according to the requirements for multiple-representation modeling.

A model for multirepresentation should allow one to characterize the same objects using different sets of attributes, and attributes with different values and different domains. This flexibility is supported by the MADS model [8], where multiple representations of a given phenomenon may be organized according to two strategies. In the first one, the various spatial and semantic descriptions of the same real-world phenomenon are merged into a single database construct. Each element of a description is qualified by a tag (called stamp) whose value identifies the perceptions for which it is relevant. Object and relationship types can thus have various sets of attributes depending on the perception. Attributes may bear various cardinalities or value domains according to the perception stamp; they can also contain a value that is a function of the perception stamp.

The Vuel approach [2] also offers the possibility of associating various semantic and spatial descriptions to the same real-world entities. In addition, various graphical representations, useful for drawing maps at different scales, can be defined. The data model is a snowflake model for spatial data warehousing. The fact table is composed of a specific kind of tuples, called vuels. A vuel fact is a particular representation of a real-world entity. It has three components: a geometry, a graphical description, and a semantic description. The vuel representation may vary according to these three dimensions. Moreover, the semantic dimension is a fact table itself with four dimensions: the class, the attribute, the domain of value, and the value dimensions. This allows the creation of various semantic descriptions by combining the dimensions (different classes, different sets of attributes, attributes with various domains and various values). OMT-G [3], a UML-based model, supports the modeling of multiple representations of data through a specific kind of relationship called conceptual generalization relationship. This relationship allows the definition of various views of the same real-world entities as subclasses of a shared super-class. The superclass describes the thematic attributes that are common to all the representations and it has no spatial representation. Each subclass describes its own view by specifying its own thematic and spatial

attributes. The subclasses inherit the common attributes from the superclass. A presentation diagram shows graphical representations that may be associated to a class and the operations to obtain them. MRSL [5], another UML-based model, supports multi-representation through the introduction of two concepts: representation objects (r-objects) and integration objects (i-objects). All r-objects corresponding to the same real-word phenomenon are linked by a monovalued or multivalued link to a single i-object, whose role is to ensure consistency among them. Each r-object specifies a specific set of attributes and values for the same real-world object.

Multiple representation modeling is not limited to associating multiple sets of attributes or values to one object. In particular, when changing the level of detail, objects may disappear, whereas others may be grouped. Thus, in addition, there is the need to put into correspondence one object with several objects or two different sets of objects.

In the second strategy of the MADS approach, the various descriptions of the same phenomena belong to separate object types. They can be linked by inter-representation links that are either traditional associations or multi-associations. Multi-associations are binary relationships that, contrarily to association relationships, do not link two objects but two groups of objects. A multi-association is needed whenever the real-world entities are not represented per se, but through two different decompositions; e.g., a decomposition of a road in segments according to the number of lanes, and one according to crossroads. The other modeling approaches only support correspondence links of kind association, and the supported cardinality of the link varies: in MRSL, a i-object can be linked to r-objects through 1:1 and 1:N links thus providing support for the 1:N and N:M correspondences. In Vuel, corresponding objects can also be linked through 1:1 or 1:N inter-representation links. However, there is no support for N:M inter-representation links. OMT-G does not support inter-representation associations between objects.

Not only do objects need multiple representations, but relationships also do. This is only supported by MADS. In MADS, all characteristics of a relationship may have various representations: its semantics (e.g., topological, aggregation, or plain), its roles, and its cardinalities. For instance, a relationship type can be a topological adjacency relationship in one description and a near relationship in another one with a more precise resolution.

In the spatial context, as data from one representation may often result from the derivation of the same data represented at another resolution, the representations of the same real-world entity are not independent and one may expect to be able to state constraints between these representations. Consistency constraints in databases are maintained through the definition of integrity constraints. Some constraints, such as cardinalities, are embedded in the concepts of the model - in particular some constraints are inherent to the multiple-representation concepts - while other constraints need to be defined in the application. MRSL is the only model proposing specific multirepresentation constraints: three kinds of rules can be associated to an i-object and its linked r-objects: consistency rules, which can be object or value correspondences, matching rules, and restoration rules. Matching rules specify how to match objects representing the same entity. They can be attribute comparison, spatial match operations, or global identifiers. Restoration rules are used to restore consistency between an i-object and its r-objects when needed.

Finally, considering that a multirepresentation database contains several representations of the same real-world phenomena, it is important to associate metadata to the representations to identify the application(s) they are relevant for, but also in order to know which representations together form a consistent whole for the application. This important requirement is fulfilled by MADS through the concept of perception stamp. In MADS, a perception stamp is a vector of values (e.g., a viewpoint and a resolution) that identifies a particular perception, and all elements of the database (types, properties, instances) are stamped for defining for which perception they are relevant. In Vuel, the designer can define views that are compositions of vuels. Each view defines a particular perception, thus providing a functionality similar to perception stamps.

Architectures for Distributed Representations

Instead of proposing new concepts allowing users to integrate multiple representations of the same real-world phenomena into a unique multirepresentation database, other proposals followed a less intrusive approach. Capitalizing on the fact that there already exist many spatial databases, these approaches create a multirepresentation framework out of a set of existing classic (i.e., describing a unique perception and resolution) databases. There are two main kinds of proposals: the first one focuses on the *definition of links between objects* in corresponding databases, the second one aims at building *federated database management systems*.

In the first category, the work of Kilpelainen [6] was one of the first proposals tackling multiple representations from a database point of view. It supports bi-directional links that allow one to propagate updates in both directions and perform reasoning processes in the form of generalization operators.

In federated spatial databases, users access a set of databases through a single integrated schema, which describes virtual multirepresentation objects. During query processing, multirepresentation objects are dynamically constructed by merging all the corresponding monorepresentation objects that exist in the various databases. There have been several proposals for spatial database integration [4]. Particularly interesting are those that build the integrated schema using multirepresentation concepts, e.g., [5], based on MRSL, and [11], based on MADS. Using MRSL, each r-object in the integrated schema holds an UML tag that identifies the corresponding source database. Using MADS, perception stamps can fulfill the same functionality.

Key Applications

Cartography

As they cannot automatically derive maps at different scales from a single detailed database, national map agencies have to create several databases, one per scale. For them, multirepresentation modeling is crucial for two main reasons:

1. To propagate updates [1]: The cost of updating can be lowered by entering updates only once in a database and propagating them, at least semi-automatically, to the other databases.
2. To enforce consistency [10]: Multi-representation databases play an important role in order to enforce consistency between the same data described at different levels of details. In addition, integrating existing databases to create a multi-representation database allows one to detect inconsistencies between the databases.

Multi-Scale Analysis

Multirepresentation databases can benefit many applications that need to analyze data at different levels of details or defined for different viewpoints. For example, a fire monitoring application may need very detailed data on current fires (to direct the action of fire brigades as precisely as possible), only need medium-level resolution data for records of past fires, and use low-level resolution data for generic organization of fire management activities.

Other candidate applications are those relying on spatial data warehouses, using spatial OLAP and spatial data cubes to perform multidimensional analysis. An example is traffic accident monitoring applications, e.g., for analysis of the number of deadly accidents according to multilevel criteria (by road, region, department, or state). Multirepresentation storage of spatial data is needed in order to drill-up and drill-down the cube [2].

Future Directions

Work in progress explores the use of multirepresentation capabilities in support of modularization of knowledge repositories. In particular, the semantic web community is developing various approaches to turn huge ontologies that are being built in several knowledge domains into smaller sets of more manageable ontological modules. Existing approaches follow both the integrated direction (a single ontology is modularized) and the distributed direction (various existing ontologies are interconnected within a global knowledge sharing system). A forthcoming book on Ontology Modularization [13] is due for publication in 2008.

Cross-References

▶ Database Design
▶ Distributed Spatial Databases
▶ Field-Based Spatial Modeling
▶ Geographic Information System
▶ Multidimensional Modeling
▶ Semantic Modeling for Geographic Information Systems
▶ Spatial and Spatiotemporal Data Models and Languages
▶ Spatial Data Types
▶ Topological Data Models
▶ Topological Relationships

Recommended Reading

1. Badard T, Lemarié C. Propagating updates between geographic databases with different scales. chapter 10. In: Atkinson P, Martin D, editors. Innovations in GIS 7: GIS and geo computation. London: Taylor and Francis; 2000. p. 135–46.
2. Bédard Y, Bernier E. Supporting multiple representations with spatial view management and the concept of VUEL. In: Proceedings of the Joint Workshop on Multiscale Representations of Spatial Data; 2002.
3. Borges K, Davis CA, Laender A. OMT-G: an object-oriented data model for geographic applications. Geo Informatica. 2001;5(3):221–60.
4. Devogele T, Parent C, Spaccapietra S. On spatial database integration. Int J Geogr Inf Syst. 1998;12(4):335–52.
5. Friis-Christensen A, Jensen CS, Nytun JP, Skogan D. A conceptual schema language for the management of multiple representations of geographic entities. Trans GIS. 2005;9(3):345–80.
6. Kilpeläinen T. Maintenance of topographic data by multiple representations. In: Proceedings of the Annual Conference and Exposition of GIS/LIS; 1998. p. 342–51.
7. Mustière S, Van Smaalen J. Database requirements for generalisation and multiple representations. In:

Mackaness WA, Ruas A, Sarjakoski T, editors. Generalisation of geographical information: cartographic modelling and applications. Amsterdam: Elsevier; 2007.

8. Parent C, Spaccapietra S, Zimányi E. Conceptual modeling for traditional and spatio-temporal applications:the MADS approach. Berlin: Springer; 2006.

9. Sarjakoski LT. Conceptual models of generalisation and multiple representation. In: Mackaness WA, Ruas A, Sarjakoski T, editors. Generalisation of geographical information: cartographic modelling and applications. Elsevier: Amsterdam; 2007. p. 11–36.

10. Sheeren D, Mustière S, Zucker JD. How to integrate heterogeneous spatial databases in a consistent way? In: Proceedings of the 8th East European Conference on Advances in Databases and Information Systems; 2004. p. 364–78.

11. Sotnykova A, Vangenot C, Cullot N, Bennacer N, Aufaure M-A. Semantic mappings in description logics for spatio-temporal database schema integration. In: Spaccapietra S, Zimanyi E, editors. Journal on Data Semantics III. Lecture notes in computer science, vol. 3534. Heidelberg: Springer. p. 143 67.

12. Stell JG, Worboys MF Stratified map spaces: a formal basis for multi-resolution spatial databases. In: Proceedings of the 8th International Symposium on Spatial Data Handling; 1998. p. 180–9.

13. Stuckenschmidt H, Parent C, Spaccapietra S. Modular ontologies. Berlin/New York: Springer LNCS; 2009.

14. Timpf S. Map cube model: a model for multi-scale data. In: Proceedings of the 8th International Symposium on Spatial Data Handling; 1998. p. 190 201.

15. Weibel R, Dutton G. Generalizing spatial data and dealing with multiple representations. In: Geographical information systems: principles, techniques, management and applications, 1, 2nd, P Longley, MF Goodchild, DJ Maguire, DW Rhind. New York/Chichester: Wiley; 1999. p. 125–155.

Multi-query Optimization

Prasan Roy[1] and S. Sudarshan[2]
[1]Sclera, Inc., Walnut, CA, USA
[2]Indian Institute of Technology, Bombay, India

Synonyms

Common subexpression elimination; Global query optimization; Multiple query optimization; Optimization of DAG-structured query evaluation plans

Definition

Multi-query optimization is the task of generating an optimal combined evaluation plan for a collection of multiple queries. Unlike traditional single-query optimization, multi-query optimization can exploit commonalities between queries, for example by computing common subexpressions (i.e., subexpressions that are shared by multiple queries) once and reusing them, or by sharing scans of relations from disk.

Historical Background

Early work on multi-query optimization includes work by Sellis [11], Park and Segev [7] and Rosenthal and Chakravarthy [9]. Shim et al. [12] consider heuristics to reduce the cost of multi-query optimization. However, even with heuristics, these approaches are extremely expensive for situations where each query may have a large number of alternative evaluation plans.

Subramanian and Venkataraman [13] consider sharing only among the best plans of the query; this approach can be implemented as an efficient, post-optimization phase in existing systems, but does not guarantee optimality. In fact, Roy et al. [10] show that it can be significantly suboptimal. Rao and Ross [8] address the problem of sharing common computation across multiple invocations of a subquery, which is a special case of multi-query optimization,

Roy et al. [10] address the problem of extending top-down cost-based query optimizers to support multi-query optimization, and present greedy heuristics, as well as implementation optimizations. Their techniques were shown to be practical and to give good results. Dalvi et al. [1] explores the possibility of sharing intermediate results by pipelining, avoiding unnecessary materializations. Diwan et al. [2] consider issues of scheduling and caching in multi-query optimization. Zhou et al. [14] discuss the implementation of multi-query optimization on a commercial query optimizer.

M

In addition to the motivation of optimizing a collection (batch) of queries, multi-query optimization has also been applied to other settings. For example, Mistry et al. [6] consider the issue of multi-query optimization in the context of view maintenance, while Fan et al. [3] point out the importance of multi-query processing in optimizing XPath queries.

Foundations

Multi-query optimization is more expensive than independent optimization of multiple queries, since a globally optimal plan may involve subplans that are sub-optimal for the individual queries.

Consider a batch consisting of two queries $(A \bowtie B \bowtie C)$ and $(B \bowtie C \bowtie D)$. A traditional system would evaluate each of these queries independently, using the individual best plans suggested by the query optimizer for each of these queries. Let these best plans be as shown in Fig. 1a. Suppose the base relations A, B, C and D each have a scan cost of 10 units (the actual unit of measure is not relevant to this example). Each of the joins have a cost of 100 units, giving a total evaluation cost of 460 units. On the other hand, in the plan shown in Fig. 1b, the common subexpression $(B \bowtie C)$ is first computed and materialized on the disk at a cost of 10. Then, it is scanned twice – the first time to join with A in order to compute $(A \bowtie B \bowtie C)$, and the second time to join it with D in order to compute $(B \bowtie C \bowtie D)$ – at a cost of 10 per scan. Each of these joins have a cost of 100 units. The total cost of this *consolidated* plan is thus 370 units, which is about 20% less

than the cost of the traditional plan of Fig. 1a. Although the benefit here is small, it could be significantly more for batches containing more queries.

The expression $(B \bowtie C)$ that is common between the two queries $(A \bowtie B \bowtie C)$ and $(B \bowtie C \bowtie D)$ in the above example is a *common subexpression* (CSE). A relation used in multiple queries can be thought of as a special case of a common subexpression. Although there is no need to compute and store it, a scan of the relation from disk can be shared by multiple queries.

A plan for a single complex query can have common subexpressions within itself. Traditional optimizers ignore the possibility of exploiting such a common subexpression, but some of the techniques for multi-query optimization, such as [10] can exploit such common subexpressions.

Challenges

The job of a multi-query optimizer can be broken into two parts: (i) recognize possibilities of shared computation by identifying CSEs, and (ii) find a globally optimal evaluation plan exploiting the CSEs identified.

Identifying CSEs

Each query can have a large number of alternative evaluation plans. Given a particular evaluation plan for each of a set of queries, it is straightforward to find common subexpressions amongst these plans. However, since the number of possible combinations of such plans is very large, enumerating them is not feasible.

Subexpressions that could be shared between some plans for two or more queries can however be identified without enumerating all possible plan combinations. The number of such

Multi-query Optimization, Fig. 1 Example illustrating benefits of sharing computation

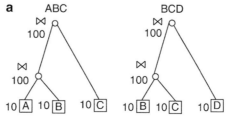

Traditional execution: No sharing
total cost = 460

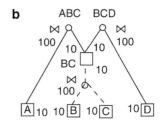

Execution with BC shared
total cost = 370

potentially common subexpressions is still very large, but smaller than the number of plan combinations.

Finding the Optimal Plan in Presence of CSEs

Traditional query optimizers use dynamic programming algorithms to find the best plan for an input query. These dynamic programming algorithms are applicable because, in the absence of sharing of common subexpressions, each subplan of the overall best plan is also the best plan for the subexpression it computes.

In the presence of sharing, such a property does not hold – as shown in the example above, a globally optimal plan can consist of subplans that are not globally optimal – and therefore a straightforward dynamic programming approach does not work. The problem of finding an optimal combined plan in presence of CSEs is therefore a strictly harder problem than traditional query optimization.

Engineering an Efficient Multi-query Optimizer

As mentioned earlier, a simple minded approach that iterates over all possible plans for each query and analyzes each combination of plans is very expensive, and infeasible for non-trivial queries. And conversely, a heuristic that only considers the individual best plan for each query does not work well, as mentioned earlier.

A more practical approach was presented in [10]. This approach efficiently finds the set of potentially common subexpressions for a set of queries, and then identifies the subset of CSEs to share, and the best resulting consolidated plan, using an iterative greedy heuristic.

Instead of enumerating the search space of possible plan combinations, the idea is to store all the plans across all the queries in a single compact data structure called the Logical Query DAG (LQDAG). The LQDAG is a refinement of the "memo" data structure used in transformational top-down optimizers, such as Volcano [4], to memorize the best plans of the intermediate results. (Such memorization, as done in top-down query optimizers such as Volcano, is equivalent to dynamic programming, as used in System R and other bottom-up query optimizers.)

Figure 2a shows a LQDAG for the query A ⋈ B ⋈ C; this LQDAG represents the three alternative plans for the query: (A ⋈ B) ⋈ C, A ⋈ (B ⋈ C) and B ⋈ (A ⋈ C). Each square node (equivalence node) in the LQDAG represents a distinct intermediate result, and each round node (operation node) below represents a distinct plan to compute the same from the underlying intermediate results. In general, a LQDAG can represent multiple queries in a consolidated manner, with a distinct root node for each distinct query. Figure 2b shows a consolidated LQDAG for the two example queries seen earlier, A ⋈ B ⋈ C, and B ⋈ C ⋈ D.

The CSEs for the given queries correspond to equivalence nodes in the LQDAG that are shared either within the same plan, or between plans for two distinct queries; [10] presents an efficient algorithm that identifies the set of all CSEs in a single bottom-up traversal of the LQDAG.

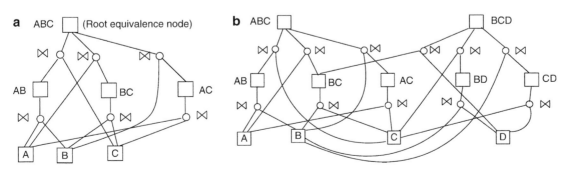

Multi-query Optimization, Fig. 2 (a) LQDAG for A ⋈ B ⋈ C, and (b) Combined LQDAG for A ⋈ B ⋈ C and B ⋈ C ⋈ D

After the CSEs are identified, the next task is to find the best consolidated plan for the queries exploiting these CSEs. When the number of CSEs is large, an exhaustive search is not feasible; a natural approach is then to use a greedy heuristic that iteratively picks the CSE with the greatest benefit (i.e., whose use would result in the greatest decrease in the overall evaluation cost), terminating when no further decrease is possible. This algorithm requires that the benefit of each candidate CSE be recomputed in each iteration – this involves finding the best plan that uses the candidate CSE, in addition to the CSEs selected in earlier iterations.

With multiple such optimization calls in each iteration, a naive implementation of the greedy heuristic would be too expensive to be practical. [10] shows how to make this approach practical by (i) incorporating additional heuristics to significantly reduce the number of benefit computations, and (ii) showing how to efficiently perform a benefit computation by exploiting the LQDAG representation of the plan space. Additional insights on the task of seamlessly incorporating multi-query optimization into the Microsoft SQL-Server query optimizer are presented by Zhou et al. [14].

The above approach assumes that CSEs are materialized and read back from disk when required. Dalvi et al. [1] shows how to schedule queries such that results can be pipelined to multiple uses, even with a limited buffer space, thereby minimizing IO. Diwan et al. [2] addresses the issue of caching results in limited memory, and scheduling queries to minimize cache usage.

Key Applications

The idea of sharing computation among different queries to save on time and resources is ubiquitous. As queries become increasingly expensive, the need for multi-query optimization to enable such savings is likely to increase as well. A few representative applications which motivate multi-query optimization are listed below.

- **On-Line Analytic Processing (OLAP) and Reporting**: A typical OLAP and reporting workload consists of queries with a significant amount of overlap. This overlap can occur for several reasons. For instance, queries might overlap in the kind of analysis they perform, or in the subset of data they are interested in; different queries could compute different aggregates over a join of the same set of tables. Alternatively, the queries could be against a virtual view; these queries clearly overlap at least in the computation of the result of the virtual view. Or else, the queries could involve common table expressions (specified using the WITH clause) that could be used in multiple places in the query; parts or whole of such common table expressions could be transiently materialized and reused [14]. Finally, the queries could involve correlated nested subqueries – invariant parts of these nested subqueries could be computed once and shared across invocations of the subquery [8].

- **Materialized View Maintenance**: Materialized views are supported by most major database systems today. Such materialized views must be updated when the underlying relations are updated. The maintenance plans for different views often share common computation. Mistry et al. [6] show how to exploit multi-query optimization to create an optimal view maintenance plan.

- **XML Query Processing**: In systems that store XML data in relational databases, the XPATH queries containing regular path expressions translate into a sequence of queries with significant overlap. Such queries are likely to benefit significantly from multi-query optimization [3].

- **Stream Query Processing**: Monitoring applications such as financial analysis and network intrusion detection often have to process multiple queries over a common stream of data. Such queries are likely to overlap significantly in the expressions they compute, and are likely to gain from multi-query optimization [5].

Cross-References

▶ Query Optimization

Recommended Reading

1. Dalvi NN, Sanghai SK, Roy P, Sudarshan S. Pipelining in multi-query optimization. J Comput Syst Sci. 2003;66(4):728–62.
2. Diwan AA, Sudarshan S, Thomas D. Scheduling and caching in multi-query optimization. In: Proceedings of the 13th International Conference on Management of Data; 2006.
3. Fan W, Yu JX, Lu H, Lu J, Rastogi R. Query translation from XPATH to SQL in the presence of recursive DTDs. In: Proceedings of the 31st International Conference on Very Large Data Bases; 2005. p. 337–48.
4. Graefe G, McKenna WJ. The volcano optimizer generator: extensibility and efficient search. In: Proceedings of the 9th International Conference on Data Engineering; 1993. p. 209–18.
5. Krishnamurthy S, Wu C, Franklin M. On-the-fly sharing for streamed aggregation. In: Proceedings of the ACM SIGMOD International Conference on Management of Data; 2006. p. 623–34.
6. Mistry H, Roy P, Sudarshan S, Ramamritham K. Materialized view selection and maintenance using multi-query optimization. In: Proceedings of the ACM SIGMOD International Conference on Management of Data; 2001. p. 307–18.
7. Park J, Segev A. Using common subexpressions to optimize multiple queries. In: Proceedings of the 4th International Conference on Data Engineering; 1988. p. 311–9.
8. Rao J, Ross KA. Reusing invariants: a new strategy for correlated queries. In: Proceedings of the ACM SIGMOD International Conference on Management of Data; 1998. p. 37–48.
9. Rosenthal A, Chakravarthy US. Anatomy of a modular multiple query optimizer. In: Proceedings of the 14th International Conference on Very Large Data Bases; 1988. p. 230–9.
10. Roy P, Seshadri S, Sudarshan S, Bhobe S. Efficient and extensible algorithms for multi query optimization. In: Proceedings of the ACM SIGMOD International Conference on Management of Data; 2000. p. 249–60.
11. Sellis TK. Multiple query optimization. ACM Trans Database Syst. 1988;13(1):23–52.
12. Shim K, Sellis T, Nau D. Improvements on a heuristic algorithm for multiple-query optimization. Data Knowl Eng. 1994;12(2):197–222.
13. Subramanian SN, Venkataraman S. Cost-based optimization of decision support queries using transient views. In: Proceedings of the ACM SIGMOD International Conference on Management of Data; 1998. p. 319–30.
14. Zhou J, Larson PÅ, Freytag JC, Lehner W. Efficient exploitation of similar subexpressions for query processing. In: Proceedings of the ACM SIGMOD International Conference on Management of Data; 2007. p. 533–44.

Multi-resolution Terrain Modeling

Enrico Puppo
Department of Informatics, Bioengineering, Robotics and Systems Engineering, University of Genova, Genoa, Italy

Synonyms

Level-of-detail (LOD) terrain modeling

Definition

Multi-resolution terrain models provide the capability of using different representations of terrain, at different levels of accuracy and complexity, depending on specific application needs. The major motivation behind multi-resolution is improving performance in processing and visualization. Given a terrain database, a multi-resolution model provides the mechanisms to answer queries that combine both spatial and resolution criteria. In the simplest case, one could ask for a representation of terrain on a given area and with a given accuracy in elevation. More sophisticated multi-resolution models support adaptive queries, also known as *selective refinement* queries, where resolution may vary smoothly on the extracted representation, according to some given criterion. For instance, one could ask for an accuracy of at least 10 m on a given range of elevations, smoothly degrading to say 100 m out of that range; similarly, high resolution could be focused in the proximity of a lineal feature (e.g., a road, a river); in view-dependent visualization, it is useful to have maximal resolution close to the viewpoint and degrade it smoothly according to distance from it; etc. Multi-resolution engines must be able to answer such queries in real time even on planetary-size databases. For instance, in view-dependent visualization, it may be necessary to change representation, hence answering a query, at each frame, i.e., 25–30 times per second.

Historical Background

The concept of multi-resolution has been known since the mid-1970s with seminal work by J. Clark [4]. Since then, many different proposals appeared in the literature. In the case of terrain data, the underlying representation, which is either a raster grid, or a triangulated irregular network (TIN), determines the different approaches to multi-resolution.

The design of multi-resolution terrain models is interrelated with *terrain generalization*, i.e., the problem of taking a representation of a terrain and generating another smaller representation of the same terrain at a lower accuracy. While just trivial solutions (subsampling, averaging) are possible on raster models, adaptivity of TINs can be exploited to seek for an optimal ratio between accuracy and size of representation. This problem has been shown to be NP-hard by Agarwal and Suri [2]. However, starting with seminal work by Fowler and Little in the late 1970s [5], many algorithms for terrain generalization have been proposed in the literature that achieve good results in practice. The first straightforward approach to multi-resolution terrain representation is given by layered models, which are built on both raster models and TINs, by applying generalization at different levels of resolution. A layered model thus provides just a collection of alternative representations of the same terrain at different resolutions.

Treelike models follow a hierarchical approach: a base model provides a coarse representation of terrain made of a small number of atomic cells; each such cell is refined by decomposition into smaller cells at the next level of resolution; refinement is repeated over several levels, and the model is maintained in a treelike data structure. A notable example is given by *restricted quadtrees*, introduced first by Von Herzen and Barr in 1987 [18] and widely developed later by several authors. This class of models is suited to efficiently manage data with a regular distribution; thus, it is usually built from raster data, even though the representation provided in output can sometimes be a TIN.

Later on, DAG-based models, aka CLOD (continuous level of detail) models, extended the hierarchical approach to manage irregular data while guaranteeing seamless representations across different levels of detail. Most of such models were developed during the 1990s, and they may all be seen as instances of a general framework introduced by Puppo in 1996 [15]: the basic elements of such a framework are local modification operations, which change the resolution of a representation locally, and the hierarchical organization on a directed acyclic graph allows to query them efficiently.

Several models developed during the first decade of the twenty-first century try to combine a regular hierarchical approach for space subdivision with a TIN representation inside space blocks, and they adopt mechanisms typical of CLOD models to obtain seamless representations. Some more recent approaches tend to go back to collections of raster models, which can be stored efficiently in graphics memory, while adaptive tessellations are generated on-the-fly by exploiting the computing power of GPUs. A seminal work in this direction is due to Losasso and Hoppe [11] and several other proposals followed in the last few years. GPU techniques are mainly devoted to terrain rendering, while their applicability to data analytics (e.g., within GIS) is still not clear.

Scientific Fundamentals

Regularity and adaptivity are somehow opposite features that characterize different approaches to multi-resolution.

The simplest way to perform terrain generalization and implicitly obtain a multi-resolution layered model from a database of regularly distributed data is based on subsampling. Given a raster grid of elevation data at high resolution, a coarser sub-grid is obtained by regularly sampling data along each axis with a fixed step. In this case, the term *resolution* is referred to the size of cells in the resulting grid or, in other terms, to the size of the step used to subsample. This method maintains the regular structure of data, but it provides no control on the loss of accuracy, and it is not adaptive. Thus, a large number of samples may be used even to represent terrain in flat areas, while vertical error could easily exceed the allowed tolerance on areas that contain sudden variations of altitude.

On the contrary, TIN representations are adaptive, so they can concentrate small triangles in areas where terrain has large variations, while large triangles can be used in relatively flat areas, while maintaining a nearly constant accuracy. Generalization algorithms, run with different thresholds on the same dataset at high resolution, provide a layered model consisting of a collection of TIN representations at different accuracies. Since TINs are inherently irregular, they need more complex and expensive algorithms and data structures to be maintained.

In both cases, layered models support simple queries to extract a representation at fixed resolution: it is sufficient to select the layer corresponding to the desired resolution and the region of interest within that layer. However, discrete models have several drawbacks: they usually provide only a small number of levels of detail; they cannot relate different representations of the same area at different resolutions; and they cannot combine data at different resolutions within a single representation.

In treelike models, each node represents exactly the same portion of terrain covered by its children. Having a straightforward hierarchical structure, these models can also act as spatial indexes. Most successful models have been developed for regularly distributed data (for instance, the multi-resolution model adopted in *Google Earth* falls in this category). The simplest example consists of a quadtree structure built over a pyramid of subsampled grids. Quadtrees need to be restricted and triangulated in order to obtain seamless representations at variable resolution (see Fig. 1).

Similar results are obtained by a hierarchical decomposition pattern based on triangle bisection (see Fig. 2), proposed in the mid-1990s by Lindstrom et al. [10] and later adopted in many variants by many authors. A square universe S is initially covered by two isosceles right triangles. The bisection rule subdivides a triangle into two similar triangles by splitting it at the midpoint v of its longest edge. A binary tree of right triangles is thus obtained. In order to extract seamless representations, such tree must be traversed in a proper way. In practice, adjacent triangles that are split by introducing a given vertex v will have to be split together during selective refinement. This scheme can be easily generalized to

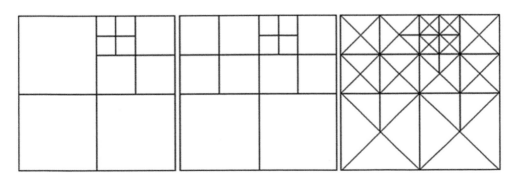

Multi-resolution Terrain Modeling, Fig. 1 In a restricted quadtree, cracks can be eliminated by balancing the level of adjacent quadrants and triangulating quadrants with suitable patterns

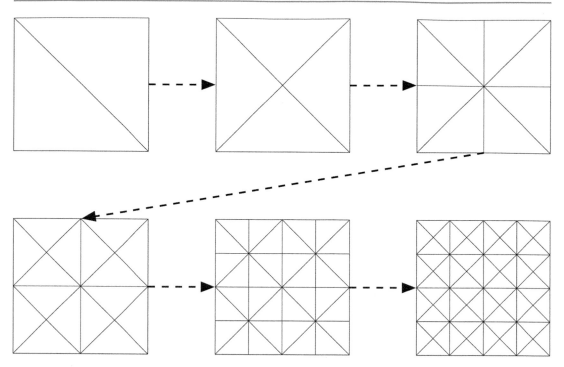

Multi-resolution Terrain Modeling, Fig. 2 Recursive triangle bisection generates a regular hierarchy based on the same triangles that appear in the restricted quadtree, but it exhibits a better flexibility

a spherical domain starting, e.g., from an octahedron. Efficient algorithms and data structures have been developed, which can support selective refinement efficiently even on planetary size databases.

The idea that a local portion of terrain can be represented with an either coarser or finer tessellation has been generalized in DAG hierarchies. The general framework for such hierarchies is described by the multi-triangulation (MT), as proposed by Puppo [15]: a hierarchy of local modifications is encoded in a directed acyclic graph, where each node represents a modification refining a given portion of terrain, and different nodes are linked in the DAG if they represent overlapping portions of terrain at different levels of detail. By traversing such graph in a proper order, selective refinement can be performed efficiently, and seamless representations can be built on-the-fly. Efficient implementations of this framework have been presented in the literature, which are based on atomic operations on TINs, such as the *edge*

collapse (see Fig. 3), which is at the basis of the *progressive meshes (PM)* introduced by Hoppe [8], and of many other models proposed in the literature.

DAG hierarchies based on irregular tessellations are not suitable for planet-size databases, and they are difficult to implement on secondary memory. On the other hand, regular hierarchies such as the one based on triangle bisection can be interpreted in the MT framework, thus providing a suitable trade-off between regularity and adaptivity. Cignoni et al. in 2003 proposed a model oriented to terrain rendering, the *BDAM*, which combines triangle bisection with adaptive schemes based on TINs [7]. The triangle bisection scheme is used as a spatial index, to obtain a coarse decomposition of the domain. Standard algorithms traverse such an index during selective refinement, and triangular blocks are collected from the proper levels of the tree. A TIN consisting of possibly a large number of triangles is associated to each triangular block in the spatial index and encoded with a compression

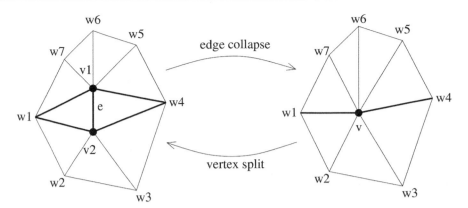

Multi-resolution Terrain Modeling, Fig. 3 *Edge collapse* is a local modification for iterative terrain generalization: edge e together with its endpoints v_1 and v_2 collapse to vertex v; triangles adjacent to e together with their other edges collapse to edges w_1v and w_4v, respectively. Edge collapse is inverted by a refinement operation called *vertex split*

scheme. In this way, the representation resulting from selective refinement is in fact given by the collection of TINs corresponding to triangular blocks extracted during traversal of the index. TINs can be maintained on efficient data structures, which may greatly improve performance. Other similar approaches have been proposed more recently, which leverage GPU computation and support high performance even on huge databases [1, 3, 9].

In the literature, also other kinds of multi-resolution models have been proposed, which follow a functional approach rather than a geometric one (see, e.g., [12]). The basic idea is that a function can be decomposed into a simpler part at low resolution, together with a collection of perturbations called wavelet coefficients which define its details at progressively finer levels of resolution. Wavelets have been widely used for multi-resolution representation and compression of signals and images, while their applications to terrain and surfaces are less popular. The discrete computation of wavelets requires a recursive subdivision of the domain into regular cells like equilateral triangles or squares. Therefore also these methods are suitable for rater data and resulting hierarchies correspond to either quaternary triangulations or quadtrees. In [17] a method is proposed, which exploits fast GPU wavelet mechanisms to support both ray-casting

rendering and interactive editing of huge terrain datasets.

Other recent approaches, which are mostly devoted to terrain rendering, rely on collections of subsampled grids and perform on-the-fly tessellation on the GPU. The geometry clipmaps proposed by Losasso and Hoppe [11] consist of a set of nested regular grids centered about the viewer, which are stored in the GPU memory, and used for rendering. As the viewpoint moves, the geometry clipmaps are updated in video memory by fetching from a high-resolution raster database. Tessellation is performed directly in the GPU. Very high frame rates can be obtained, even for huge datasets.

As already noticed, raster data allow for the design of implicit data structures that have a small overhead with respect to maintaining just the single resolution data at the highest available detail, and they are also suitable for implementation in secondary memory. Most efficient data structures, though, assume that the collection of all data in the database form a unique regular grid at a given (high) resolution. In most real cases, however, the database is rather a patchwork of partially overlapping grids at different resolutions. Gerstner in 2003 proposed a data structure for the scheme based on triangle bisection, which works also in the latter case [6]. More recently, Rocca et al. [16] addressed a similar problem

M

with a radically different approach. They propose a model that can encode a database consisting of a collection of geo-referenced, freely oriented, and freely aligned raster grids and that is able to support CLOD queries by sampling data on-the-fly. A collection of patches is generated by pre-proceeding raster data; patches are stored in a spatial index over a three-dimensional space, where two dimensions span the spatial domain and a third dimension an error axis. Each patch is regarded as a parallelepiped in such a space, covering its spatial domain and the range of accuracies for which the patch is relevant. A CLOD query is specified by a surface through this space-error space: patches intersecting the query surface can be extracted efficiently from the database and merged on-the-fly with a sampling algorithm.

For a more detailed treatment of multi-resolution terrain modeling, see, e.g., [13, 14, 19] and references therein.

Key Applications

Multi-resolution terrain modeling is essential to manage complexity in those applications that need to either analyze or visualize terrain data at different scales, such as planetary browsers, flight simulators, CAD tools for road design, and all intensive computational tasks related to terrain, such as drainage networks and visibility.

Cross-References

▶ Discrete Wavelet Transform and Wavelet Synopses
▶ Quadtrees (and Family)
▶ Simplicial Complex
▶ Triangulated Irregular Network

Recommended Reading

1. Adil Yalçin M, Weiss K, De Floriani L. GPU algorithms for diamond-based multiresolution terrain processing. In: Proceedings of the 11th Eurographics Conference on Parallel Graphics and Visualization; 2011. p. 121–30.

2. Agarwal PK, Suri S. Surface approximation and geometric partitions. In: Proceedings of the 5th ACM-SIAM Symposium On Discrete Algorithms; 1994. p. 24–33.

3. Bösch J, Goswami P, Pajarola R. Raster: simple and efficient terrain rendering on the GPU. In: Proceedings of the Eurographics 2009; 2009. p. 35–42.

4. Clark JH. Hierarchical geometric models for visible surface algorithms. Commun ACM. 1976;19(10):547–54.

5. Fowler RJ, Little JJ. Automatic extraction of irregular network digital terrain models. ACM Comput Graph. 1979;13(3):199–207.

6. Gerstner T. Multiresolution compression and visualization of global topographic data. Geoinformatica. 2003;7(1):7–32.

7. Gobbetti E, Marton F, Cignoni P, Di Benedetto M, Ganovelli F. C-BDAM – compressed batched dynamic adaptive meshes for terrain rendering. Comput Graph Forum. 2006;25(3):333–342.

8. Hoppe H. Progressive meshes. In: Proceedings of the 23rd Annual Conference on Computer Graphics and Interactive Techniques; 1996. p. 99–108.

9. Lindstrom P, Cohen JD. On-the-fly decompression and rendering of multiresolution terrain. In: Proceedings of the 2010 ACM SIGGRAPH Symposium on Interactive 3D Graphics and Games; 2010. p. 65–73.

10. Lindstrom P, Koller D, Ribarsky W, Hodges LF, Faust N, Turner GA. Real-time, continuous level of detail rendering of height fields. In: Proceedings of the 23rd Annual Conference on Computer Graphics and Interactive Techniques; 1996. p. 109–18.

11. Losasso F, Hoppe H. Geometry clipmaps: terrain rendering using nested regular grids. In: Proceedings of the 31st Annual Conference on Computer Graphics and Interactive Techniques; 2004. p. 769–76.

12. Lounsbery M, DeRose TD, Warren J. Multiresolution analysis for surfaces of arbitrary topological type. ACM Trans Graph. 1997;16(1):34–73.

13. Lübke D, Reddy M, Cohen JD, Varshney A, Watson B, Hübner R. Level Of detail for 3D graphics. San Francisco: Morgan Kaufmann; 2002.

14. Pajarola R, Gobbetti E. Survey of semi-regular multiresolution models for interactive terrain rendering. Vis Comput. 2007;23(8):583–605.

15. Puppo E. Variable resolution terrain surfaces. In: Proceedings of the 8th Canadian Conference on Computational Geometry; 1996. p. 202–10.

16. Rocca L, Panozzo D, Puppo E. Patchwork terrains: multi-resolution representation from arbitrary overlapping grids with dynamic update. In: Csurka G, Kraus M, Laramee R, Richard P, Braz J, editors. Computer vision, imaging and computer graphics. Theory and application. Communications in computer and information science, vol. 359. Berlin/Heidelberg: Springer; 2013. p. 48–66.

17. Treib M, Reichl F, Auer S, Westermann R. Interactive editing of gigasample terrain fields. Comput Graph Forum. 2012;31(2):383–92.

18. Von Herzen B, Barr AH. Accurate triangulations of deformed, intersecting surfaces. Comput Graph. 1987;21(4):103–10.
19. Weiss K, De Floriani L. Simplex and diamond hierarchies: models and applications. In: Hauser H, Reinhard E, editors. Eurographics 2010 – state of the art reports. Norrköping: Eurographics Association; 2010.

Multi step Query Processing

Peer Kröger and Matthias Renz
Ludwig-Maximilians-Universität, München

Synonyms

Filter/refinement query processing

Definition

A query on a database reports those objects which fulfill a given query predicate. A query processor has to evaluate the query predicate for each object in the database which is a candidate for the result set. Multi-step query processing (filter/refinement query processing) is a technique to speed up queries specifying query predicates that are complex and costly to evaluate. The idea is to save the costs of the evaluation of the complex query predicate by reducing the candidate set for which the query predicate has to be evaluated applying one or more filter steps. The aim of each filter step is to identify as many true hits (objects that truly fulfill the complex query predicate) and as many true drops (objects that truly do not fulfill the query predicate) as possible by applying a less costly query predicate. The remaining candidates that are not pruned as drops or reported as hits in one of the filter steps need to be tested in a refinement step where the exact (costly) query predicate is evaluated. Obviously, the less costly the filter predicates are and the smaller the number of candidates that need to be refined, the higher the performance gain of a multi-step query processing is over a single-step query processing.

In addition, if any of the applied filter steps is able to report true hits, first results can be reported to the user significantly sooner by a multi-step query processor compared to a single-step query processor.

Historical Background

In many database applications the management of complex objects is required. For example, the parts of a geographical map such as streets, lakes, forests – or generally *regions* – are stored as polylines or polygons. Queries on these complex objects usually involve complex query predicates that are costly to evaluate. For example, in order to retrieve all regions of a map that intersect with a given query window it is required to test the intersection of the query window and the database polygons which is computationally rather expensive. In such situations, the evaluation of the query predicate (e.g., "intersects the query window") becomes the bottleneck of query processing. Index structures are designed for shrinking down the search space of tentative hits in order to scale well for very large databases. Principally, the aim of index structures is the same as that of the filter-steps in multi-step query processing. However, index structures are only applicable for the first filter step. The reason is that index structures are designed to organize the entire database and cannot be applied to a reduced set of candidates.

To cope with complex data objects and costly query predicates, the paradigm of multi-step query processing (filter/refinement query processing) has been defined originally for spatial queries such as point queries and region queries on databases of spatial objects [2, 6]. This paradigm has been applied to similarity search in databases of complex objects performing general similarity queries such as distance range queries [1, 3] and k-nearest neighbor (kNN) queries [4] using costly distance functions. The key idea is to apply one or more filter steps each using cheaper query predicates (e.g., cheaper distance functions), the so-called *filter predicates*, in order to identify as many objects as possible as true

M

hits or true drops. For the remaining candidates, for which the query predicate cannot be decided using any of the filter steps, the exact (more costly) query predicate needs to be evaluated in a refinement step. To ensure correct results, the filter predicates are required to be based on *conservative approximations* of the exact objects. This ensures that if any object does not qualify for a filter predicate, it can also not qualify for the exact query predicate. For example, if the regions of a map are conservatively approximated by minimum bounding rectangles (MBRs) of the corresponding polygons, those regions whose corresponding MBRs do not intersect with the query window cannot intersect with the query window. This conservative property of the filter predicates enables discarding true drops. On the other hand, filter predicates that are based on progressive approximations of the exact objects can be used to identify true hits. For example, if the regions of a map are progressively approximated by an incircle of the corresponding polygons, those regions whose corresponding incircle intersect with the query window do also intersect with the query window.

Foundations

Multi-step query processing is usually used in applications where the objects in the database are complex and the queries launched on objects rely on costly predicates that cannot be evaluated efficiently. In such applications, the evaluation of the query predicate becomes the bottleneck in query execution.

General Schema of Multi-step Query Processing

Multi-step query processing is based on the following idea: design one or more filter predicates that can be evaluated much faster than the original query predicate and that can be used to shrink down the number of candidates for which it is unknown whether they qualify for the query predicate or not. The query processing starts with all database objects as candidates and applies the designed filters sequentially on the remaining candidates. Each filter ideally identifies true hits that can be added to the result set and true drops that can be pruned. The candidates that cannot be classified as true hits or true drops after all filter steps need to be refined by evaluating the (costly) original query predicate. This general schema is illustrated in Fig. 1. The order in which the single filter steps are applied usually depends on the cost of each filter step and on the *selectivity* of each filter step. The selectivity of a filter step determines the fraction of objects that are identified as true hit or true drop by the corresponding filter and do not need any further processing. In order to produce correct results, obviously, the filter steps must not produce false drops (i.e., drop objects that match the original query predicate

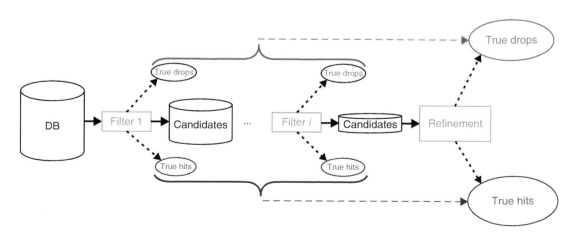

Multi step Query Processing, Fig. 1 General schema of multi-step query processing

according to a filter predicate) and false hits (i.e., report objects that do not match the original query predicate as hits according to a filter predicate).

In order to apply multi-step query processing, it is important to design appropriate filter predicates for the original query predicates of the given application. An appropriate filter predicate can usually be designed by designing a less complex representation that approximates the complex database objects. The evaluation of the query predicate on these less complex object approximations should be less costly than on the original object representation. In the following, special instances of multi-step query processing is discussed in more detail.

Example: Multi-step Query Processing of Similarity Queries

Usually, similarity between objects is expressed by means of a pair-wise distance function *dist*. A high distance between two objects denotes low similarity of these objects whereas a low distance implies high similarity. For example, if the database objects are points (of any dimensionality), *dist* could be the Euclidean distance, i.e., the vicinity of the corresponding points in the Euclidean space. If the database objects are sequences, *dist* could be the Edit distance. If the database objects are spatial regions (e.g., of a map), *dist* could be the smallest Euclidean distance between the corresponding polygons. The two most important and general types of similarity queries are distance range (DR) queries and *k*-nearest neighbor (*k*NN) queries.

A distance range query is a general query type in non-standard database systems such as spatial DBS, temporal DBS, and multi-media DBS. Given a query object q, a distance function $dist(.,.)$, and a distance threshold ε, a distance range query returns all database objects o that have a distance less or equal than ε to q, i.e., $dist(q, o) \leq \varepsilon$. They can be efficiently supported using index structures or multi-step query processing.

According to the above definition the query predicate of DR queries is given as follows: all

hits o must qualify the predicate $dist(q, o) \leq \varepsilon$, where q is the query object and ε is a distance threshold. The query predicate of *k*NN queries is quite similar to DR queries: all hits o must qualify the predicate $dist(q, o) \leq d(q, k)$, where q is the query object and $d(q, k)$ is the *k*-nearest neighbor distance. However, the big difference between DR queries and *k*NN queries is that the distance threshold ε is given in advance, whereas the value of $d(q, k)$ is usually not known at query time.

A filter predicate for identify true drops (conservative property) can be designed as follows. First, a less complex representations to conservatively approximate the exact objects should to be developed. Usually, this can only be implemented for spatial objects: the conservative approximation must completely contain the exact object, e.g., a minimum bounding box (MBR) is a conservative approximation of a polygon. A second step is essential: A (cheaper) distance function on the approximation must be designed that implements the *lower bounding property*. Let $dist(.,.)$ be the exact distance function on the exact database objects and $LB(.,.)$ the cheaper filter distance. The distance function $LB(.,.)$ lower bounds the exact distance $dist(.,.)$, if the following holds:

$$LB\,(x, y) \leq dist\,(x, y)$$

for all database objects x and y. Since the exact predicate of a similarity query usually determines the hits as those objects that have a distance less than a threshold ε to the query object q, all objects o with $\varepsilon < LB(q, o) \leq dist(q, o)$ can be excluded from the result set without further processing. In other words, the filter predicate is similar to the original query predicate, but uses LB instead of *dist*.

For example, if the database contains the regions of a map, and $dist(r_1, r_2)$ is the smallest Euclidean distance between the regions (polygons) r_1 and r_2, an appropriate filter can be designed as follows. The regions are approximated by MBRs and $LB(m_1, m_2)$ is defined as the smallest Euclidean distance between the MBRs m_1 and m_2 of r_1 and r_2, respectively. Obviously,

M

evaluating *LB* on the MBRs is usually much less complex and costly than evaluating *dist* on the polygons.

A filter predicate for identifying true hits can be designed analogously. First, a less complex representation to progressively approximate the exact objects should be developed. Again, this can usually be implemented only for spatial objects: the progressive approximation must be completely contained within the exact object, e.g., the maximal circle contained within a polygon (incircle) is a progressive approximation of that polygon. Again, a second step is essential: A (cheaper) distance function on the approximation must be designed that implements the upper bounding property. Again, let *dist*(.,.) be the exact distance function on the exact database objects and *UB*(.,.) the cheaper filter distance. The distance function *UB*(.,.) upper bounds the exact distance *dist*(.,.), if the following holds:

$$UB\,(x, y) \geq dist\,(x, y)$$

for all database objects *x* and *y*. All objects *o* with $\varepsilon > UB(q, o) \geq dist(q, o)$ can be added to the result set without further processing. In other words, the filter predicate is again similar to the original query predicate, but uses *UB* instead of *dist*.

Sometimes, an approximate representation of the database objects allows the definition of two distance functions, one lower bounding distance and one upper bounding distance. Filter predicates that do not use upper or lower distances cannot be applied to reduce the number of candidates.

Example: Algorithms for Multi-step Query Processing of Similarity Queries

The algorithm for multi-step distance range queries is rather easy. Since the distance threshold ε is known in advance, in each filter step, true hits and/or true drops are identified as described above, depending on the property of the distance used in the filter predicate.

On the other hand, a multi-step solution for *k*NN queries is not trivial, because in order to determine the exact value of $d(q, k)$ that can be used to identify objects based on any filter predicates as true hits or true drops, at least *k* objects need to be refined. Since the *k* nearest neighbors are not known in advance, the *k* objects that need to be refined to determine the exact value of $d(q, k)$ are not known. Obviously, this is a vicious circle.

The multi-step *k*NN query processing algorithm proposed in [4] tries to approximate $d(q, k)$ by refining any *k* objects and take the maximum value $d'(q, k)$ of these exact distances. then, a multi-step DR query with query object *q* and distance threshold $d'(q, k)$ is evaluated. The resulting (refined) objects are ranked in ascending exact distances to *q* and only the first *k* objects of this ranking are reported as final result.

In [7], the authors enhance this approach with an algorithm that minimizes the number of refinements. The basic assumption of this algorithm is that only a conservative filter is applied. In the case of only one filter step, the algorithm uses a ranking query in the filter step. Given a query object *q* and a distance function *dist*(.,.), a ranking query returns a sequence of the database objects in a database *D* sorted by ascending distances to *q*. A ranking query is a general query type in non-standard database systems such as spatial DBS, temporal DBS, and multi-media DBS and can be efficiently supported using index structures. In the context of multi-step query processing in the filter step a ranking query returns a ranking of the database objects sorted in ascending filter distances to the query object *q*. Initially, the first *k* objects of the ranking are refined and an approximation $d'(q, k)$ of the true value of $d(q, k)$ is determined from these refined distances as above. Then, in each iteration, the next object from the ranking is fetched as long as the filter distance of the next object in the ranking is greater than the current approximation $d'(q, k)$. As long as this is not the case, the currently fetched object is refined and $d'(q, k)$ is updated. This algorithmic schema can easily be extended to applying multiple filter steps. It can be shown that – if only a conservative filter is implemented – this algorithm is optimal with regard to the number of refinements.

Finally, the algorithm in [5] further enhances the preceding algorithms that take only a conservative filter into account, by additionally using a progressive filter. The algorithm is similar to that in [7] but determines $d'(q, k)$ from the progressive filter as long as this is possible rather than from exact distances. As a consequence, the proposed algorithm reduces the number of refinements significantly. It can be shown that – if both a conservative filter and a progressive filter are implemented – this algorithm is optimal with regard to the number of refinements.

Key Applications

More and more applications suffer from the increasing complexity of the objects and of the functions required to evaluate query predicates on such objects, e.g., complex distance functions or spatial intersections. In the meantime, the efficient support of multi-step query processing is essential for many application areas such as molecular biology, medical imaging, CAD systems, and multimedia databases.

In this context, one of the most important application where multi-step query processing is essential for efficient query processing is similarity search in time series databases. Time series may be very large. Typical similarity queries in time series databases are distance range queries and k-nearest neighbor queries. Due to the curse of dimensionality, similarity queries cannot efficiently be supported by indexing the time-series based on the raw data. A common method to overcome this problem is to reduce the dimensionality of the object descriptions and use this lower-dimensional feature space to index the time series. Similarity queries are then performed using the paradigm of multi-step query processing. In the filter step, approximated similarity distances are computed based on the dimensionality reduced representations, while the refinement step applies similarity distance functions based on the raw time series data. Usually, the filter step is conservative, i.e., the filter distances lower bound the exact distances.

Another important application which requires multi-step query processing is the support of proximity queries in spatial networks like road networks where point objects located within the road network that is represented by a graph are queried. Usually, the objects are positions of buildings or individuals like persons or cars that can have a static location or may move within the network. Example queries could be "retrieve all cars within the road network having a smaller distance to the fast-food restaurant Pinky than 5.0 km" or "give me the three filling stations having the smallest distance to my actual position". Since the motion of the objects is restricted by the network, i.e., objects can only move along a path in the network graph, the distance between two objects is not measured using the Euclidean distance. Rather, the length of the shortest path between two objects is used as distance measure. For each distance computation it is necessary to apply the Dijkstra algorithm which is too expensive to answer such proximity queries on large databases in real time. Therefore, distance approximations are needed, which can be computed more efficiently and can be used in the filter step of a multi-step query processing algorithm. The simplest road-network distance approximation that fulfills the lower bound criterion is the Euclidean distance. Another method to achieve suitable distance approximations is the pre-computation of distances based on certain landmarks (reference nodes). The distance approximation based on landmarks has the advantage that, in addition to the lower bounding distance approximation, it is possible to compute a distance approximation which fulfills the upper bounding property.

A further important application of multi-step query processing is the support of spatial queries in spatial databases, i.e., databases containing objects having a spatial extension. One of the most important query types in such databases is the point-in-polygon test. Given a database with two-dimensional polygon objects and a certain query point, retrieve all polygons that include the query point. Several filter steps can be applied for this problem to avoid unnecessary point-in-polygon-tests. For example, the polygons can be con-

M

servatively approximated by minimum bounding rectangles (MBRs). Obviously, MBRs that do not contain the query point can be discarded as true drops. On the other hand, progressive approximations of the polygons can be used to identify true hits.

Cross-References

▶ Closest-Pair Query
▶ High-Dimensional Indexing
▶ Indexing Metric Spaces
▶ Nearest Neighbor Query
▶ Spatial Indexing Techniques
▶ Spatial Join
▶ Spatiotemporal Data Mining

Recommended Reading

1. Agrawal R, Faloutsos C, Swami A. Efficient similarity search in sequence databases. In: Proceedings of the 4th International Conference on Foundations of Data Organization and Algorithms; 1993. p. 69–80.
2. Brinkhoff T, Horn H, Kriegel H-P, Schneider R. A storage and access architecture for efficient query processing in spatial database systems. In: Proceedings of the 3rd International Symposium on Advances in Spatial Databases; 1993. p. 357–76.
3. Faloutsos C, Ranganathan M, Manolopoulos Y. Fast subsequence matching in time series database. In: Proceedings of the ACM SIGMOD International Conference on Management of Data; 1994. p. 419.
4. Korn F, Sidiropoulos N, Faloutsos C, Siegel E, Protopapas Z. Fast nearest neighbor search in medical image databases. In: Proceedings of the 22nd International Conference on Very Large Data Bases; 1996. p. 215–26.
5. Kriegel H-P, Kröger P, Kunath P, Renz M. Generalizing the optimality of multi-step k-nearest neighbor query processing. In: Proceedings of the 10th International Symposium on Advances in Spatial and Temporal Databases; 2007. p. 75–9.
6. Orenstein J, Manola F. Probe spatial data modelling and query processing in an image database application. IEEE Trans Softw Eng. 1998;14(5):611–29.
7. Seidl T, Kriegel H-P. Optimal multi-step k-nearest neighbor search. In: Proceedings of the ACM SIGMOD International Conference on Management of Data; 1998. p. 154–16.

Multitenancy

Aaron J. Elmore
Department of Computer Science, University of Chicago, Chicago, IL, USA

Synonyms

Database-as-a-service; Database multi-tenancy; Multitenant databases

Definition

Database multitenancy is the ability for a database system to efficiently host multiple applications' data within a single logical database service. Here, each hosted application is referred to as a *tenant* in the system. As tenants are distinct applications, each tenant's data and performance must be isolated from each other. A multitenant database system is composed of one or more database servers that are multiplexed to host multiple tenants. Different techniques, or multitenancy models, exist to multiplex a database server and differ in how they isolate tenants from each other. Sample techniques include the use of multiple virtual machines, multiple schemas within a shared process, or collocating tenants' data within shared tables. Each multitenancy model exhibits trade-offs between tenant isolation, the maximum level of tenants per server (i.e., tenant consolidation), functionality, and development required.

Historical Background

Traditionally, database systems focus on hosting a single or limited number of applications within a database instance. However, organizations often must host a large number of small databases, which result in many servers dedicated to many databases. For instance, cloud computing service providers face the management of many database instances from various service offerings, including but not limited to database offerings. Large organizations, such as telecommunication companies and financial institutions, can also need to host many disjoint databases that arise from acquisitions, independent development teams, and different internal applications. Often these databases do not require the full resources of a dedicated server; therefore, hosting each database in an dedicated server results in wasted resources. With traditional database architectures lacking a focus on multitenant hosting, several efforts began to explore multitenant database solutions.

While the earliest research into building a *database-as-a-service* examined issues related to query privacy, interest grew in techniques to consolidate a large number of databases into a reduced number of servers. Several models for enabling multitenancy using existing database systems emerged. With virtualization technology maturing, the use of virtual machines (VMs) to multiplex a database server was quickly adopted. Virtualization provides several key features required for building a multitenant offering, including the ability to limit resource consumption and mechanisms for load-balancing tenants between servers. An alternative approach to enabling multitenancy relies on extending a standalone database system to more effectively host multiple tenants in separate schemas or catalogs. These extensions are implemented either as middleware for unmodified database systems or through direct modification of a database engine. For some environments, such as a software-as-a-service offering, there may be many hosted tenants with significant schema overlap. In this environment, using dedicated virtual machines (VMs) or a single database with dedicated catalogs may limit the number of tenants hosted by a server. To reduce the metadata overhead and uncoordinated database components, cloud service providers have explored multitenant solutions that collocate tenants' data within shared tables. The various multitenant models exhibit trade-offs between consolidation level, functionality, development required, and resource isolation between tenants. Research in database multitenancy explores various system architectures and challenges in efficient hosting and server orchestration.

Scientific Fundamentals

Enabling multitenancy in a database system raises challenges in providing effective consolidation – or the number of tenants that can be hosted – while isolating tenants' resource utilization and performance from each other. Limitations on consolidations often arise from tenant metadata overhead, redundant components wasting finite resources, or uncoordinated resource utilization (e.g., multiple transaction managers writing to independent write ahead logs) [1, 4]. Tenant isolation is the level that a tenant's performance can be impacted by collocated tenant workloads. If a system provides strong isolation, then resource-intensive tenants will likely have some form of limitation enforced to prevent resource starvation for other tenants. Systems with soft isolation will require load balancing when a tenant is not receiving adequate resources due to contention. The design of a multitenant system explores trade-offs in consolidation and isolation.

Many methods exist for enabling multitenancy, and each model differs in how tenants are isolated from each other. While a large number of approaches have been proposed, only a few multitenancy models are typically found. In the first common model, *shared hardware*, virtualization technology is used to separate tenants. In this model, each tenant has an independent database and operating system hosted in a virtual machine. The virtualization hypervisor is responsible for enforcing resource consumption limits for tenants. The shared hardware model provides strong resource isolation through use of mature virtualization technology and many advanced features, such as migration or high availability. Therefore, virtualization provides a rich environment where resource utilization can be controlled and tenants are highly portable. However, in certain scenarios it has been shown that this approach can limit the amount of tenant consolidation –

largely due to redundant system components and uncoordinated resource utilization between tenants [4].

With the second common model, *shared process*, many tenants are hosted within a single database process. Database authentication, table space, catalogs, or schemas can be used to isolate tenants. In this model, tenants share database resources, such as a buffer pool or write ahead log. Which components are shared is dependent on the system implementation. This model can provide improved consolidation over the shared hardware approach; however vast majority of database management systems (DBMSs) are not designed for resource isolation between tenants. Careful planning and resource utilization models must be utilized for planning which tenants to collocate in such a system. Regardless, a workload change can result in suboptimal tenant placement strategy and poor performance due to resource starvation for tenants in systems that lack strong resource allocation controls. Additionally, the resources needed for storing catalog information or dedicated client connections can limit the number of tenants hosted on a single server. Therefore, environments with a very large number of tenants may require higher levels of consolidation.

The *shared table* model emerged to host large numbers of tenants by collocating tenants' data within shared tables. While several techniques exist to provide a shared table system, the *pivot table* approach developed by Salesforce is the standard example [8]. Here, a heap table stores all data, with each row containing a tenant id, table id, column id, and a value stored in a generic format (e.g., bytes or a string). Additional metadata tables store information about tenants, table definitions, column types, and the columns composing a table. Such an approach allows each tenant to have a distinct table schema while limiting dedicated client connections or the amount of dedicated system metadata required for each tenant. Since data is mixed within large heap tables, resource and data isolation – or privacy – must be enforced through an external access layer. Additionally, query rewriting is required as data is not stored in a standard relational format. For these reasons, a shared table approach requires extensive software development and can experience performance limitations for complex workloads.

Beyond design considerations, research has explored techniques required to manage a large number of databases in a limited number of servers. For multitenant systems built on a shared hardware model, research has explored how to configure the available resources for each tenant's VM to achieve desired performance objectives [9]. These objectives typically include query latencies or workload throughput requirements; the agreement between the service provider and tenant is referred to as a *service level objective (SLO)* or a *service level agreement (SLAs)*. Beyond resource allocation configuration, research on shared hardware models has explored leveraging virtualization technology to provide high availability for database services. For the shared process model, research has focused on building a database service with minimal modifications to existing systems and extending database kernels to support resource allocation controls, similar to the role of a virtualization hypervisor. Sample systems that are built on top of existing DBMSs with minimal modifications include Relational Cloud [4], Delphi [5], and SQLAzure [2]. As these systems lack strong resource isolation, efforts have focused on placement to ensure tenants have adequate resources available. These projects have explored problems for systems with soft isolation in understanding resource requirements, modeling resource consumption, initial tenant consolidation, tenant load balancing, modeling the impact of colocation, tenant replica placement, enforcing SLAs, and live database migration [3–5, 7, 10]. SQLVM [6] is a project that seeks to embed concepts of virtualization into a database engine by enforcing limitations, or reservations, on resource consumption by tenants. Lastly, research on the shared table model examined performance trade-offs of various design alternatives for collocating tenants' data within shared tables. Examples of these approaches include using sparse columns or pivot tables to allow for flexible schemas on data stored in heap tables [1].

Key Applications

Several environments face the challenge of hosting many small databases. Cloud service providers that offer hosted database services host many small databases of varying resource requirements for both storage and workload processing. Public cloud providers have developed a variety of multitenant systems built upon the shared hardware, process, and table models. Enterprise organizations also face managing many databases that arise from acquisitions and disjoint development teams. In both of these environments, gross over-provisioning of servers is often used to simplify administration and to maintain performance SLAs. This over-provisioning results in wasting resources due to purchasing servers, datacenter space, and utilities required to power and cool servers. Research into multitenant systems seeks to alleviate wasted resources through more efficient and automated hosting of tenants on a reduced number of servers.

Cross-References

- ▶ Cloud Computing
- ▶ Data Partitioning
- ▶ Data Migration Management

Recommended Reading

1. Aulbach S, Grust T, Jacobs D, Kemper A, Rittinger J. Multi-tenant databases for software as a service: schema-mapping techniques. In: Proceedings of the ACM SIGMOD International Conference on Management of Data; 2008. p. 1195–206.
2. Bernstein PA, Cseri I, Dani N, Ellis N, Kalhan A, Kakivaya G, Lomet DB, Manne R, Novik L, Talius T. Adapting microsoft SQL server for cloud computing. In: Proceedings of the 27th International Conference on Data Engineering; 2011. p. 1255–63.
3. Curino C, Jones EPC, Madden S, Balakrishnan H. Workload-aware database monitoring and consolidation. In: Proceedings of the ACM SIGMOD International Conference on Management of Data; 2011. p. 313–24.
4. Curino C, Jones EPC, Popa RA, Malviya N, Wu E, Madden S, Balakrishnan H, Zeldovich N. Relational cloud: a database service for the cloud. In: Proceedings of the 5th Biennial Conference on Innovative Data Systems Research; 2011. p. 235–40.
5. Elmore AJ, Das S, Pucher A, Agrawal D, El Abbadi A, Yan X. Characterizing tenant behavior for placement and crisis mitigation in multitenant DBMSs. In: Proceedings of the ACM SIGMOD International Conference on Management of Data; 2013. p. 517–28.
6. Narasayya VR, Das S, Syamala M, Chandramouli B, Chaudhuri S. SQLVM: performance isolation in multi-tenant relational database-as-a-service. In: Proceedings of the 6th Biennial Conference on Innovative Data Systems Research; 2013.
7. Schaffner J, Januschowski T, Kercher M, Kraska T, Plattner H, Franklin MJ, Jacobs D. RTP: robust tenant placement for elastic in-memory database clusters. In: Proceedings of the ACM SIGMOD International Conference on Management of Data; 2013. p. 773–84.
8. Weissman CD, Bobrowski S. The design of the force.com multitenant internet application development platform. In: Proceedings of the ACM SIGMOD International Conference on Management of Data; 2009. p. 889–96.
9. Xiong P, Chi Y, Zhu S, Moon HJ, Pu C, Hacigümüs H. Intelligent management of virtualized resources for database systems in cloud environment. In: Proceedings of the 27th International Conference on Data Engineering; 2011. p. 87–98.
10. Yang F, Shanmugasundaram J, Yerneni R. A scalable data platform for a large number of small applications. In: Proceedings of the 4th Biennial Conference on Innovative Data Systems Research; 2009.

Multitier Architecture

Heiko Schuldt
Department of Mathematics and Computer Science, Databases and Information Systems Research Group, University of Basel, Basel, Switzerland

Synonyms

Multi-layered architecture; n-tier architecture

Definition

A *Multi-tier Architecture* is a software architecture in which different software components, organized in tiers (layers), provide dedicated functionality. The most common occurrence of a multi-tier architecture is a three-tier system consisting of a data management tier (mostly encompassing one or several database servers), an application tier (business logic) and a client tier (interface functionality). Novel deployments come with additional tiers. Web information systems, for instance, encompass a dedicated tier (web tier) between client and application layer.

Conceptually, a multi-tier architecture results from a repeated application of the client/server paradigm. A component in one of the middle tiers is client to the next lower tier and at the same time acts as server to the next higher tier.

Historical Background

Early generation software systems have been built in a monolithic way. This means that all the different tasks for implementing a particular application and presenting the results to a user are provided by a single dedicated software component. With the advent of client/server architectures in the 1980s, different tasks could be separated and possibly even be distributed across network boundaries. In a client/server architecture (two tier architecture), the client is responsible for presenting the application to the user while the server is in charge of data management. For the provision of business logic, two alternatives have emerged. First, in so-called fat client/thin server architectures, the client also provides business logic, in addition to presentation and user interfaces. This can be realized by using SQL against the underlying database server in the application program run by the client, either by embedding SQL into a higher programming language or by using the database server's call level interface (e.g., JDBC, ODBC). Second, in thin client/fat server architectures, the database server also provides business logic while the client solely focuses on presentation issues. Fat servers can be realized by using persistent stored modules or stored procedures inside the database server. In the case of evolving business logic, fat client architectures, although

being the most common variant of client/server systems,impose quite some challenges when new client releases need to be distributed in large deployments. In addition, a fat client architecture usually comes along with a high network load since data is completely processed at the client side. Fat servers, in contrast, impose a single point of failure and a potential performance bottleneck.

Three-tier architectures thus are the next step in the evolution of client/server architectures where both client and database server are freed from providing business logic. This task is taken over by an application layer (business tier) between client and database server. In multi-tier architectures, additional tiers are introduced, such as for instance a web tier between client and application layer.

Foundations

Multi-tier systems follow an architectural paradigm that is based on separation of concerns. The architecture considers a vertical decomposition of functionality into a stack of dedicated software layers. Between each pair of consecutive layers, a client/server style of interaction is applied, i.e., the lower layer acts as server for the next higher layer (see Fig. 1). Typical tiers in a three-tier architecture are *data management*, *business* and *client* tier. Multi-tier architectures consider additional layers, such as a web tier which hosts servlet containers and a web server and which is located between client tier and application tier.

In addition to vertical decomposition and distribution across tiers, in many cases multi-tier architectures also leverage horizontal distribution within tiers. For the data management tier, this means that several distributed database servers can be used. Most commonly, horizontal distribution is applied at the business tier, i.e., providing several application server instances [7].

The main benefit of multi-tier applications is that each tier can be deployed on different heterogeneous and distributed platforms. Load balancing within tiers, especially for the application tier,

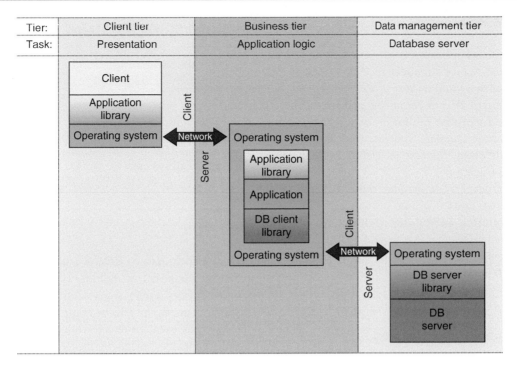

Multitier Architecture, Fig. 1 Structure of a three tier architecture

is supported by distributing requests across the different application server instances. This can be implemented by a dispatcher which accepts calls from the next higher layer and distributes them accordingly (this is done, for instance, in TP Monitors which allow to distribute requests among application processes at the middle tier in a three-tier architecture).

When multi-tier architectures are used in a business context, they have to support transactional interactions. Due to the inherent distribution of software components across layers and potentially even within layers, distributed transactions are needed. This is usually implemented by a two-phase commit protocol (2PC) [5] (depending on the application server and the middleware used, this can be done, for instance, via CORBA OTS, the Java Transaction Service JTS, etc.). While 2PC provides support for atomicity in distributed transactions, it does not take into account the layered architecture where transactions at one layer are implemented by using services and operations of the next lower layer. Multi-level transactions [10] take this structure

into account. SAP ERP [4], for instance, applies multi-level transactions by jointly considering the application server and data management tier. Asynchronous interactions between components in a multi-tier architecture require a message-oriented middleware (MOM). In this case, transactional semantics can be supported by persistent queues and queued transactions [1].

In order to increase the performance of multi-tier systems and to improve response times, caching is used at the application tier. For this, different database technologies such as replication, materialized views, etc. can be applied outside the DBMS [6].

Key Applications

Due to the proliferation of both commercial and open source application servers, multi-tier architectures can be found in a very large variety of different domains. Applications include, but are not limited to, distributed information systems, Web information systems, e-Commerce, etc.

M

Cross-References

▶ Application Server
▶ Client-Server Architecture
▶ Database Middleware
▶ Distributed Transaction Management
▶ Java Enterprise Edition
▶ Message Queuing Systems
▶ Middleware Support for Database Replication and Caching
▶ Multilevel Transactions and Object-Model Transactions
▶ Replication in Multitier Architectures
▶ Service-Oriented Architecture
▶ Transactional Middleware
▶ Web Services
▶ Web Transactions

Recommended Reading

1. Bernstein P, Newcomer E. Principles of transaction processing. 2nd ed. Los Altos: Morgan Kaufmann; 2009.
2. Birman K. Guide to reliable distributed systems: building high-assurance applications and cloud-hosted services. Berlin: Springer; 2012.
3. Britton C, Bye P. IT architectures and middleware. 2nd ed. Reading: Addison Wesley; 2004.
4. Buck-Emden R, Galimow J. SAP R/3 system: a client/server technology. Reading: Addison-Wesley; 1996.
5. Lindsay B, Selinger P, Galtieri C, Gray J, Lorie R, Price T, Putzolu F, Wade B. Notes on distributed databases. IBM Research Report RJ2571, San Jose; 1979.
6. Mohan C.. Tutorial: caching technologies for web applications. In: Proceedings of the 27th International Conference on Very Large Data Bases; 2001.
7. Mohan C.. Tutorial: application servers and associated technologies. In: Proceedings of the 28th International Conference on Very Large Data Bases; 2002.
8. Myerson J. The complete book of middleware. Philadelphia: Auerbach; 2002.
9. Orfali R, Harkey D, Edwards J. Client/server survival guide. 3rd ed. New York: Wiley; 1999.
10. Weikum G, Schek HJ. Concepts and applications of multilevel transactions and open nested transactions. In: Elmagarmid K, editor. Database transaction models for advanced applications. Los Altos: Morgan Kaufmann; 1992. p. 515–53.
11. Weikum G, Vossen G. Transactional information systems: theory, algorithms, and the practice of concurrency control. Los Altos: Morgan Kaufmann; 2001.

Multitier Storage Systems

Kazuo Goda
The University of Tokyo, Tokyo, Japan

Synonyms

Tiered storage systems

Definition

A Multitier Storage System is a storage system which is composed of multiple types (tiers) of storage devices and/or subsystems, each having distinct characteristics from other tiers. Typical technologies used for an individual storage tier are flash memory, magnetic disks, and tapes. A Multitier Storage System is usually designed to meet system requirements that any single tier would not be able to meet alone. To ease the administration, a Multitier Storage System often has the capability of automatically and dynamically moving data across tiers in accordance with the requirements. Another term Tiered Storage System is also utilized to refer to a Multitier Storage System.

Key Points

Individual storage technologies have unique characteristics. Combining different technologies to build a storage system is a natural solution in order to satisfy a design goal that any single technology cannot meet by itself. This idea can be seen as a conceptual extension of memory hierarchy management.

Early implementation was conducted for building cost-efficient filesystems for IBM mainframes in 1970s. Magnetic disks were so expensive for many users that they could not afford enough capacity to store all the data. IBM then provided a solution called Hierarchical

Storage Management (HSM). Its basic function is to place only online data (active data) on magnetic disks and to store all the other data (inactive data) on tapes such that users can utilize much larger capacity at less cost by allowing longer retrieval latency for the inactive data stored in tapes. The interesting property of HSM is that it is potentially transparent to host users. Data staging and destaging between magnetic disks and tapes are automatically controlled mainly based on data inactiveness and user-specified thresholds. Later, HSM was eventually ported to other operating systems and dedicated storage products. Recently, HSM is widely deployed for archiving storage solutions that are designed for long-term retention of inactive data.

The storage industry exhibited two technical changes since the 2000s. Magnetic disk manufacturers started to ship a new class of disk drive products, often called nearline disk drives, which could perform at lower speed but afford more capacity. Although traditional HSM was practiced in a two-tier design, this new class successfully created a new three-tier practice (comprising high-speed enterprise disks, capacity nearline disks, and tapes), which became widely deployed into enterprise systems.

Coincidentally, emerging storage networking technologies provided opportunities to sophisticate data and storage management. Traditional HSM makes migration decisions based on user-specified configuration parameters regarding file ages and file types. The industry proposed a more intelligent framework to realize automatic migration decisions in accordance with business-level policies that describe companies' business goals and compliance requirements. Technically this framework is a class of HSM, but it is often referred to as Information Lifecycle Management.

A recent hot topic is how to incorporate flash memory into the storage hierarchy where data are efficiently moved across the tiers in an automated and dynamic fashion. In contrast to traditional HSM that focused on long-term archival use, this new technology is expected to provide a new opportunity to accommodate wider spectrum of data including fresh online data and long-term inactive data in a single system. Current storage hierarchy is much deeper and more complicated. The term Multitier Storage Systems is more often used to highlight such recent technology trends, although the basic concept of Multitier Storage Systems is identical to that of the original HSM.

Cross-References

▶ Information Lifecycle Management

Recommended Readings

1. Freeman L. What's old is new again – Storage Tiering. SNIA tutorial; 2012.
2. Keith Winnard et al. IBM z/OS DFSMShsm primer. IBM Redbooks; 2015.
3. Troppens U, Erkens R, Müller W. Storage networks explained. London: Wiley; 2004.
4. Storage Network Industry Association. The dictionary of storage networking terminology. Available at: http://www.snia.org/

M

Multivalued Dependency

Solmaz Kolahi
University of British Columbia, Vancouver, BC, Canada

Synonyms

MVD

Definition

A *multivalued dependency (MVD)* over a relation schema $R[U]$, is an expression of the form

$X \twoheadrightarrow Y$, where $X, Y \subseteq U$. An instance I of $R[U]$ satisfies $X \twoheadrightarrow Y$, denoted by $I \models X \twoheadrightarrow Y$, if for every two tuples t_1, t_2 in I such that $t_1[X] = t_2[X]$, there is another tuple t_3 in I such that $t_3[X] = t_1[X] = t_2[X]$, $t_3[Y] = t_1[Y]$, and $t_3[Z] = t_2[Z]$, where $Z = U - XY$ (XY represents $X \cup Y$). In other words, for every value of X, the value of attributes in Y is independent of the value of attributes in Z. A multivalued dependency $X \twoheadrightarrow Y$ is a special case of a *join dependency* expressed as $\bowtie[XY, X(U - XY)]$, which specifies that the decomposition of any instance I satisfying $\bowtie [XY, X(U - XY)]$ into $\pi_{XY}(I)$ and $\pi_{X(U-XY)}(I)$ is lossless, i.e., $I = \pi_{XY}(I) \bowtie \pi_{X(U-XY)}(I)$.

Key Points

Multivalued dependencies, like functional dependencies, can cause redundancy in relational databases. For instance, in the following table, each director of the movie *The Matrix* is recorded once per actor of the movie, and this is because the instance satisfies the MVD *title* $\twoheadrightarrow$ *director*.

Multivalued dependencies have been considered in the normalization techniques that try to improve the schema of a database by disallowing redundancies. The most common normal form that takes MVDs into account is the Fourth Normal Form (4NF). The implication problem for MVDs can be solved in polynomial time. That is, given a set Σ of MVDs, it is possible to check whether an MVD $X \twoheadrightarrow Y$ is logically implied by Σ, denoted by $\Sigma \models X \twoheadrightarrow Y$, in the time that is polynomial in the size of Σ and $X \twoheadrightarrow Y$. Multivalued dependencies are usually considered together with functional dependencies (FDs) in the normalization of relational data. There is a sound and complete set of rules (axioms) that can be used to infer new dependencies from a set of MVDs and FDs defined over a relation $R[U]$:

MVD0 (complementation): If $X \twoheadrightarrow Y$, then $X \twoheadrightarrow (U - X)$.

Movies

Title	Director	Actor	Year
Pulp Fiction	Quentin Tarantino	John Travolta	1994
Pulp Fiction	Quentin Tarantino	Samuel L. Jackson	1994
The Matrix	Andy Wachowski	Keanu Reeves	1999
The Matrix	Andy Wachowski	Laurence Fishburne	1999
The Matrix	Larry Wachowski	Keanu Reeves	1999
The Matrix	Larry Wachowski	Laurence Fishburne	1999

MVD1 (reflexivity): If $Y \subseteq X$, then $X \twoheadrightarrow Y$.

MVD2 (augmentation): If $X \twoheadrightarrow Y$, then $XZ \twoheadrightarrow YZ$.

MVD3 (transitivity): If $X \twoheadrightarrow Y$ and $Y \twoheadrightarrow Z$, then $X \twoheadrightarrow (Z - Y)$.

FMVD1 (conversion): If $X \rightarrow Y$, then $X \twoheadrightarrow Y$.

FMVD2 (interaction): If $X \twoheadrightarrow Y$ and $XY \rightarrow Z$, then $X \rightarrow (Z - Y)$.

It is also known that the set $\{MVD0, \ldots, MVD3\}$ is an axiomatization for MVDs considered alone.

Cross-References

▶ Fourth Normal Form
▶ Functional Dependency
▶ Join
▶ Join Dependency
▶ Normal Forms and Normalization
▶ Projection

Recommended Reading

1. Abiteboul S, Hull R, Vianu V. Foundations of databases. Reading: Addison-Wesley; 1995.

Multivariate Visualization Methods

Antony Unwin
Augsburg University, Augsburg, Germany

Synonyms

Graphical displays of many variables

Definition

Multivariate datasets contain much information. One- and two-dimensional displays can reveal some of this, but complex pieces of information need more sophisticated displays that visualize several dimensions of the data simultaneously. Usually several displays are needed.

Historical Background

Graphical displays have been used for presenting and analysing data for many years. Playfair [10] produced some fine work over 200 years ago. Minard prepared what Tufte has called "the finest graphic ever drawn" in the middle of the nineteenth century, showing Napoleon's advance on and retreat from Moscow, including information on the size of the army and the temperature at the time. Neugebaur introduced many innovative ideas in the 1920s and 1930s. Most of these graphics are primarily one- or two-dimensional. Techniques for displaying higher dimensional data have mainly been suggested more recently.

Foundations

There are two quite different aims of data display: analysis and presentation. Graphics aid analysts in understanding data and in determining structure. Graphics are good for identifying outliers, for picking out local patterns, and for recognizing global features. Graphics are also valuable for conveying that information to others. Wilkinson's book [13] defines a formal structure. Unwin et al. [12] discuss graphics for large datasets. Theus et al. [11] present interactive graphics for exploring data. The Handbook of Data Visualization [2] provides an overview of the current state of play.

For displaying multivariate data, indeed for displaying data in general, it is important to distinguish between different data types. Variables may be categorical, ordinal, continuous, temporal, spatial or logical (and other specialist types could be added as well). In data analysis the most common types are continuous and categorical. Ordinal may sometimes be treated as categorical (when there are only a few distinct values) and sometimes as continuous (when there are many).

No printed graphic can display more than two dimensions fully at once. Multivariate graphics use projections, conditioning, and linking to capture higher dimensional information. Some displays can deal with very large numbers of cases (area displays such as mosaic plots), and some can potentially handle very many variables (parallel coordinate plots). Most displays are limited in both dimensions. One strategy is to use small multiples, multiple versions of the same graphic restricted to subsets of the data. Trellis plots [1] are the most important example of this approach. Whether many small displays are used or a range of large displays (which might be referred to as large multiples), more than one display will always be necessary to reveal the information in the data.

It is essential to bear in mind that to have enough evidence to confirm complex relationships lots of data are needed. Think of determining the effects of all the influences on car insurance premiums or of estimating the effects of factors in health studies (for instance breast cancer risk). Graphical methods must be able to deal with large datasets to be fully useful [12].

Multivariate Continuous Data

For multivariate continuous data the most popular graphic solution is parallel coordinate plots [8]. Further approaches include scatterplot matrices (sploms), showing all scatterplots of two variables at a time, and trellis plots, which display the data in subsets defined by conditioning variables. Glyphs, individual images for each case whose form depends on the separate variable values, can be an interesting possibility for smaller datasets (at most a few hundred cases). Matrix visualizations [2] are also interesting for smaller datasets and display each value by a color coding, with cases in the rows and variables in the columns. Including options for ordering both rows and columns is essential. Microarray data are often displayed in this way.

M

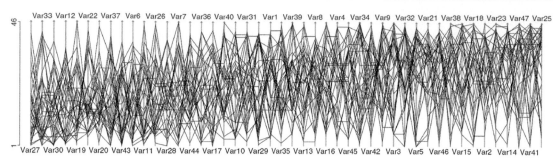

Multivariate Visualization Methods, Fig. 1 A parallel coordinate plot of the ratings of 46 wines by 32 judges. The axes are common scaled and have been ordered by mean values

Other alternatives, known under the common heading of dimension reduction plots, display two-dimensional approximations of multivariate data, e.g., multi-dimensional scaling (MDS) and biplots. Dynamic methods include the grand tour and projection pursuit [3], both of which work by moving smoothly through two dimensional projections of the data.

Figure 1 shows a parallel coordinate plot of 32 judges' rankings of 46 American and French Cabernet wines from a 1999 tasting. The axes (one for each wine) have been given a common scale and sorted by their mean ratings from left to right, so that the highest ranked wine is on the far left and the lowest on the far right. What is striking is the lack of agreement amongst the judges. While there is a discernible trend, it is obscured by the high variability of the ratings. Most wines were ranked best by at least one judge and worst by at least one other. So the main message of this plot is that while the league table of results and statistical tests (whether the ordering was significantly non-random) imply a consensus ranking of the wines, the data convey otherwise. Parallel coordinate plots, like all high-dimensional plots, require fine-tuning to reveal information. In this case, common scaling and sorting were important tools. As parallel coordinate plots are covered in another entry, they are not discussed in detail here. One interesting new variant is represented by textile plots [9] in which the axes are rescaled to make the individual lines linking cases as horizontal as possible.

In MDS [4] the attempt is made to find a low-dimensional (almost always two dimensional) approximation to high dimensional data by positioning points so that the distances between them in the low dimensional display are close to the distances between them in the original dimension. For a high number of dimensions this is unlikely to be effective, but it often produces interesting views. The MDS display depends on the criterion used to match the distances (e.g., emphasising the absolute differences or the relative differences). Since all possible pairs must be considered, it is not efficient for large datasets. MDS displays are not unique for two reasons: an optimal solution in terms of the criterion will not necessarily be found; any solution is rotation invariant. Figure 2 shows an MDS plot of five-dimensional data on 381 cars sold in Germany. Each case is represented by a circular-based glyph using these five dimensions and two additional ones. The selected group at the top of the display seem relatively well separated in this view. They are all midsize luxury cars.

Biplots were developed by Gabriel [5]. He pointed out that both cases and variables could be plotted on the same approximating low dimensional plot. The two axes are usually chosen to be the first two principal components. Lines representing variables which are well approximated appear longer than those which represent variables badly approximated and the angles between the lines reflect the correlations between the variables in the low dimensional hyperplane. More complex biplots are also possible [6]. Like MDS displays, biplots will not work well in general, neither for many cases nor many variables, but

**Multivariate
Visualization Methods,
Fig. 2** An MDS display
based on five variables for
cars sold in Germany. Each
car is represented by a
circular-based glyph

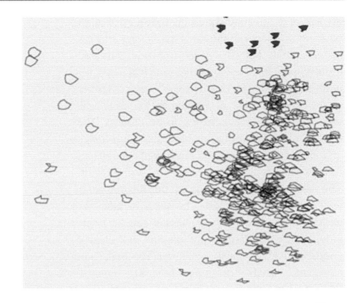

**Multivariate
Visualization Methods,
Fig. 3** A mosaicplot of the
numbers who sailed on the
Titanic with the survivors
selected. Women are to the
left and men to the right,
adults are below and
children above. Within
these groups the classes are
ordered first, second, third,
crew

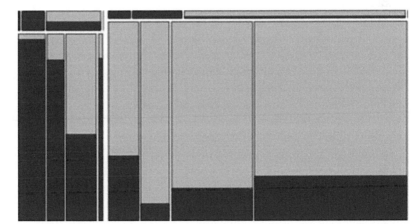

M

often the two-dimensional projections produced
can offer insightful views of the data.

Multivariate Categorical Data

Continuous data can always be sensibly binned
and compressed, while retaining the option of
zooming in to reveal the full level of detail.
This does not hold for categorical data, where
it may not be possible to combine any of the
individual categories with others. While displays
of single categorical variables are simple, the
number of combinations rises exponentially with
the number of variables. One binary variable
can be displayed in a barchart of two columns.
Twenty binary variables would give rise to 2^{20}
combinations, a little over a million, many of

which are likely to be empty, even for extremely
large datasets.

Classic mosaicplots were suggested by Hartigan [7] for displaying a small number of categorical variables in a multivariate way. Other variations (multiple barcharts, fluctuation diagrams,
equal binsize plots, doubledecker plots) [12] are
often more useful. All depend very much on a
careful choice of the ordering of variables and
on an informative choice of size and aspect ratio. The ordering of variables determines which
comparisons can be made, while the aspect ratio
influences how well the comparison can be made.

Figure 3 shows a mosaic plot of the Titanic
data with the order of variables, gender, age,
class. The block of four equally tall columns at

the left of the display shows the numbers of adult women in each of the three passenger classes and the crew, with the proportion who survived highlighted. It is obvious that survival rates for adult women declined across the three passenger classes (the number of women in the crew was too small for any conclusion to be drawn). The next block of four columns relates to adult males and shows that the second class adult males had the lowest survival rate, a rather surprising result. The smaller bars at the top of the display refer to the children on board. The survival rates for males and females within classes can be compared approximately in this display, but clearly another display would be better for that, one using the variable ordering class, age, gender. Even in this dataset with only four variables, one plot is not enough.

The main idea underlying all mosaicplots is that each combination of variable values is displayed by a rectangle whose area is proportional to the number of cases with that combination. The layout of the combinations is key in determining the interpretation, which can be difficult at the best of times, and is eased by providing interactive tools to query and adjust the graphic. Multiple barcharts are for comparing distributions of subsets (and are therefore related to trellis plots). Fluctuation diagrams are best for larger numbers of combinations to identify which are most common. Equal binsize plots and doubledecker plots are for comparing highlighted proportions.

Figure 4 shows a fluctuation diagram of a dataset from the Pakistan Labour Force Survey. Five variables are considered (the numbers of categories, including missings, are in brackets): gender [2], relation to head of household [9], marital status [4], literacy [3], and urban/rural [2], making 432 possible combinations in all. Although there are just under 140,000 cases, many of the combinations are empty or rare (e.g., few women and few single men are heads of households). The biggest single combination (male, son in household, never married, literate, living in a rural area) is highlighted and includes 9,678 cases or 7% of the dataset. Using interactive querying and animating the construction of the plot, one variable at a time, aids interpretation

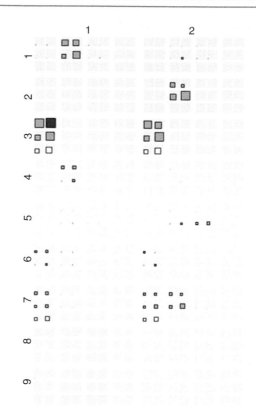

Multivariate Visualization Methods, Fig. 4 A fluctuation diagram of five variables from the Pakistani Labour Force Survey: gender, relation to head of household, marital status, literacy, urban/rural. The biggest single combination is highlighted

considerably. No display of several categorical variables at once can either be easy to grasp immediately or convey all the potentially available information.

Interactive Graphics and Multivariate Graphics

Although both parallel coordinate plots and mosaicplots can be used for static plots, they are much more effective when used interactively. Their necessarily complex nature (after all, they have to display multivariate structure) demands careful scrutiny to grasp the information in them to the full, and the gain of understanding can be considerably enhanced when these graphics are empowered with interactive tools.

Interactive graphics may also be used to gain insight into multivariate datasets using one- or two-dimensional displays. Multiple linked simple displays of the same dataset can be easier to interpret than complex multivariate plots.

Key Applications

Descriptive statistics and Exploratory Data Analysis.

Future Directions

Many other more or less esoteric multivariate visualizations have been proposed. None should be dismissed out of hand, every visualization is probably ideal for some particular dataset. Nevertheless any successful graphic should satisfy a number of criteria: it should be based on a readily recognizable and interpretable concept; it should be flexible and capable of being made interactive; it should be able to handle more than just three or four dimensions.

Displaying and interpreting even four-dimensional data is tricky. Dominating features can usually be seen, more subtle effects cannot. In higher dimensions the difficulties become much greater. At the moment it is impossible to visualize large numbers of categorical variables and although several hundred continuous variables can readily be displayed in parallel coordinate plots, the chances of identifying important features are slim. Nevertheless, graphics displays are useful for checking results found analytically and this can be very valuable. Visualizing multivariate data is new and progress is to be expected.

Visualization is currently mainly used for presentation of data, rather than for exploration of data. A single graphic can only display a limited number of aspects of a multivariate dataset and many are needed to convey all information available. The development of multivariate graphics should consider the design of sets of graphics rather than more elaborate versions of single ones. More interactive tools will have be developed. Sorting, rescaling, and querying are just some of the basics required.

Visualization is an important component of data analysis. It provides a complementary approach to analytic modeling and is much more suited to carrying out exploratory data analysis. Results found by models should be checked with graphics and ideas generated with graphics should be investigated analytically. The tighter integration of analytic and graphical methods would be of great advantage.

Cross-References

▶ Data Visualization
▶ Dynamic Graphics
▶ Parallel Coordinates
▶ Visual Data Mining
▶ Visualizing Categorical Data
▶ Visualizing Quantitative Data

Recommended Reading

1. Becker R, Cleveland W, Shyu MJ. The visual design and control of trellis display. J Comput Graph Stat. 1996;5(2):123–55.
2. Chen CH, Haerdle W, Unwin A. Handbook of data visualization. Berlin: Springer; 2007.
3. Cook D, Swayne D. Interactive and dynamic graphics for data analysis. New York: Springer; 2007.
4. Cox M, Cox M. Multidimensional scaling. London: Chapman and Hall; 2001.
5. Gabriel K. The biplot – graphic display of matrices with application to principal component analysis. Biometrika. 1971;58(3):453–67.
6. Gower J, Hand D. Biplots. London: Chapman & Hall; 1996.
7. Hartigan JA, Kleiner B. Mosaics for contingency tables. In: Proceedings of the 13th Symposium on the Interface; 1981. p. 268–73.
8. Inselberg A. Parallel coordinates. New York: Springer; 2008.
9. Kumasaka N, Shibata R. High dimesional data visualisation: the textile plot. Comput Stat Data Anal. 2008;52(7):3616–44.
10. Playfair W. Playfair's commercial and political atlas and statistical breviary. London: Cambridge University Press; 2005.
11. Theus M, Urbanek S. Interactive graphics for data analysis. London: CRC Press; 2008.
12. Unwin AR, Theus M, Hofmann H. Graphics of large datasets. New York: Springer; 2006.
13. Wilkinson L. The grammar of graphics. 2nd ed. New York: Springer; 2005.

M

Multiversion Serializability and Concurrency Control

Wojciech Cellary
Department of Information Technology, Poznan
University of Economics, Poznan, Poland

Synonyms

Multiversion concurrency control; Multiversion
concurrency control algorithms; Multiversion
databases

Definition

Given a multiversion database, where each data
item is a sequence of its versions, the number of
versions of a data item may be limited or not. If it
is unlimited, then each update of a data item over
the limit gives rise to its next version. If it is limited, than each update of a data item replaces its
oldest version. In case of limited number of versions, a database is called a K-version database.
In multiversion databases any read operation of
a data item, subsequent to a write operation of
this data item, may access any of its currently
existing versions. Thus, a multiversion schedule
of a transaction set differs from the ordinary,
mono-version schedule by a mapping of the data
item read operations into the data item version
read operations. Multiversion serializability plays
the same role for the multiversion databases,
as serializability for the ordinary, mono-version
ones. Multiversion serializability is used to prove
correctness of a concurrent execution of a set
of transactions, whose read and write operations
interleave, and, moreover, read operations may
access one of many available versions of a data
item.

Historical Background

Multiversion serializability problem was a hot
research topic in the mid-1980s. First works were
published by P.A. Bernstein and N. Goodman [2,
3] in 1983. Research was continued by G. Lausen
[7], next by S. Muro, T. Kameda, and T. Minoura
[8]. The next group of researchers involved was
composed of C.H. Papadimitriou, P.C. Kanellakis, and T. Hadzilacos [6, 9]. There is a lot of
work devoted to different variants of multiversion
concurrency control algorithms. A comprehensive background may be found in [5].

Foundations

Definition of a multiversion schedule. A *multiversion schedule mvs* of a set of transactions τ
is a triple mvs $(\tau) = (T(\tau), h, <_{\text{mvs}})$, where (i)
$T(\tau)$ is the set of all database operations involved
in the transactions of the set τ extended by the
database operations of two hypothetical initial
and final transactions and which, respectively,
write the initial state of the database and read
the final state of the database; (ii) h is a function
which maps each read operation $r_{ij}(x) \in T(\tau)$
into a write operation $w_{kl}(x) \in T(\tau)$; and (iii)
$<_{\text{mvs}} = \cup_i < T_i$ is a partial order relation over
$T(\tau)$ such that if $T_{ij} <_{T_i} T_{ik}$ then $T_{ij} <_{\text{mvs}} T_{ik}$ and
if $h(r_{ij}(x)) = w_{kl}(x)$ then $w_{kl}(x) <_{\text{mvs}} r_{ij}(x)$.
Function h defined above maps a read operation
of a data item into the write operation of a version
of this data item – more precisely – into the
write operation which creates the version of the
data item read. Relation $<_{mvs}$ is defined by two
conditions. The first one states that $<_{mvs}$ honors
all orderings stipulated by transactions of the
set τ. The second one states that a transaction
cannot read a version of a data item until it has
been created. A multiversion schedule is *serial*
if no two transactions are executed concurrently;
otherwise, it is *concurrent*.

Multiversion schedule equivalence. Two multiversion schedules are equivalent if they are *view*
and *state equivalent*. Two multiversion schedules
mvs $(\tau) = (T(\tau), h, <_{\text{mvs}})$ and $mvs'(\tau) =
(T(\tau), h', <_{mvs'})$ of the set τ are *view equivalent*
if and only if $h = h'$. If the transactions of
two multiversion schedules $mvs(\tau)$ and $mvs'(\tau)$
receive an identical view of the database, i.e., if
both multiversion schedules are view equivalent,

then all the write operations issued by transactions in both schedules are the same. Two multiversion schedules mvs $(\tau) = (T(\tau), h, <_{\mathrm{mvs}})$ and mvs$'(\tau) = (T(\tau), h', <_{\mathrm{mvs}\prime})$ of the set of transactions τ are *final-state equivalent* if and only if for every initial state of the database and any computations performed by the transactions contained in τ the final states of the database reached as the result of schedules $mvs(\tau)$ and $mvs'(\tau)$ are identical.

Standard serial multiversion schedule. A serial multiversion schedule mvs $(\tau) = (T(\tau), h, <_{\mathrm{mvs}})$ is *standard* if each read operation $r_{ij}(x) \in T(\tau)$ accesses the version of a data item x created by the last write operation $w_{kl}(x) \in T(\tau)$ preceding $r_{ij}(x)$. Since in a serial schedule, for every two transactions T_i and T_k, either all database operations of T_i precede all database operations of T_k or vice versa, the last write operation preceding a read operation is well defined. Note that a standard serial multiversion schedule in multiversion databases corresponds to a serial mono-version schedule in mono-version databases. From the consistency property of each transaction, i.e., from the assumption that each transaction separately preserves database consistency, it follows that a standard serial multiversion schedule must also preserve database consistency. On the basis of the above observation, it is possible to define the multiversion serializability criterion [3].

Multiversion serializability criterion. A multiversion schedule $mvs(\tau)$ is *correct* if it is equivalent to any standard serial multiversion schedule of the set τ. Intuitively, the above criterion can be interpreted as follows. A concurrent schedule of a set of transactions in a multiversion database is correct if it is equivalent to a serial schedule of the transactions in which data replication over versions is transparent.

Key Applications

Multiversion serializability is used to prove correctness of concurrency control algorithms devoted to multiversion databases. As an example, consider a multiversion two-phase locking

algorithm, called *WAB* [1, 3, 4], devoted to K-version databases. The concept of multiversion two-phase locking is broader than the concept of mono-version two-phase locking (cf. section on two-phase locking). An algorithm is a *multiversion two-phase locking algorithm* if it satisfies the following conditions:

1. There are two phases of transaction execution: the *locking phase* and the *unlocking phase*. During the locking phase, a transaction must obtain all locks it requests. The moment when all locks are granted, which is equivalent to the end of the locking phase and the beginning of the unlocking phase, is called the *commit point* of a transaction. New versions of the data items prepared in the transaction's private workspace are written to the database during the unlocking phase.
2. The execution order of a set of transactions τ is determined by the order of transaction commit points.
3. The execution of any transaction $T \in (\tau)$ does not require locking data items that T does not access.

The concepts of locking and unlocking phases do not have exactly the same meaning as the similar notions used in the mono-version two-phase locking algorithm. In the *WAB* algorithm, the process of setting the so-called certify lock is two phase, but not as in the two-phase locking algorithm, the process of accessing data. In the *WAB* algorithm, each transaction initiated in the database and each version of a data item is *certified* or *uncertified*. When a transaction begins, it is uncertified. Similarly, each new version of a data item prepared in the transaction's workspace is uncertified. A *certify operation* is introduced, denoted by $c(w_{ij}(x))$, where $w_{ij}(x)$ is a T_i's write operation and a new lock mode – the *certify lock* denoted by $CL(x)$. Certify locks are mutually incompatible. The algorithm requires that all certify and read operations of a data item x be $<_{mvs}$ related. Similarly, all certify operations must be $<_{mvs}$ related. The execution order of the certify operations determines a precedence relation $\ll_w$ defined on the set τ. The precedence relation $\ll_w$

specifies the order of transaction executions as follows: $T_i \ll_w T_k$ if and only if there exist such certify operations $c(w_{ij}(x))$ and $c(w_{kl}(x))$ that $c\left(w_{ij}(x)\right) \ll_{mvs} c\left(w_{kl}(x)\right)$.

According to the *WAB* algorithm, any read operation $T_{ij}(x)$ concerns the last certified version of a data item x or any uncertified version of this data item. The version selected depends on a particular implementation of the *WAB* algorithm. Any write operation $w_{ij}(x)$ prepares a new version of a data item x in the workspace of transaction T_i (the version prepared is uncertified). At the end of transaction execution, the transaction and the new versions of the data items it prepared are being certified. The T_i's certification is a two-phase locking procedure. It consists of certify-locking all data items that the transaction T_i accessed to write. The T_i's certification is completed, when all certify locks are set and the following conditions are satisfied:

1. At the moment of T_i's certification, the versions of all data items read by T_i are certified.
2. For each data item x that T_i wrote, all transactions that read certified versions of x are certified.

To satisfy condition (ii), a *certify token* is allocated to each data item x to forbid reading certified versions of x other than the last one. On the other hand, all uncertified versions of x are allowed to be read. When the transaction T_i's certification is completed (the commit point), the procedure for certifying the versions of data items prepared by T_i is initiated. It was proved in [3] that the *WAB* algorithm is correct in the sense that any schedule produced by it is multiversion serializable. The main drawback of this algorithm is a possibility of a deadlock.

Future Directions

In database systems, multiple versions of data items are necessary to ensure transaction atomicity and to recover from crashes. The original idea of multiversion concurrency control based on multiversion serializability was to use those versions also to increase the degree of transaction concurrency and as a result to improve database performance. However, such double use of versions decreases database reliability, because of the complexity of multiversion concurrency control. For practice, reliability of databases is of ultimate importance. This is why the concept of multiversion concurrency control was not well accepted in practice, except some implementations of two-version concurrency control concerning two values of each data item: before and after write operations. The concept of multiversion concurrency control may find attention in database systems applied in areas where ACID properties may be relaxed.

Cross-References

► Atomicity
► Concurrency Control: Traditional Approaches
► Replicated Database Concurrency Control
► Serializability
► Transaction
► Transaction Management
► Transaction Models: The Read/Write Approach

Recommended Reading

1. Bernstein PA, Goodman N. A sophisticate's introduction to distributed database concurrency control. In: Proceedings of the 8th International Conference on Very Data Bases; 1982. p. 62–76.
2. Bernstein PA, Goodman N. Concurrency control and recovery for replicated distributed databases. Technical Report TR-20/83, Harvard University; 1983.
3. Bernstein PA, Goodman N. Multiversion concurrency control – theory and algorithms. ACM Trans Database Syst. 1983;8(4):465–83.
4. Bernstein PA, Hadzilacos V, Goodman N. Concurrency control and recovery in database systems. Reading: Addison-Wesley; 1987.
5. Cellary W, Gelenbe E, Morzy T. Concurrency control in distributed database systems. North-Holland: Elsevier Science; 1988.
6. Hadzilacos T, Papadimitriou CH. Algorithmic aspects of multiversion concurrency control. J Comput Syst Sci. 1986;33(2):297–310.

7. Lausen G. Formal aspects of optimistic concurrency control in a multiple version database system. Inf Syst. 1983;8(4):291–301.

8. Muro S, Kameda T, Minoura T. Multi-version concurrency control scheme for database system. J Comput Syst Sci. 1984;29(2):207–24.

9. Papadimitriou CH, Kanellakis PC. On concurrency control by multiple versions. ACM Trans Database Syst. 1984;9(1):89–99.

M

Naive Tables

Gösta Grahne
Concordia University, Montreal, QC, Canada

Synonyms

Extended relations; Relations with marked nulls

Definition

The simplest way to incorporate unknown values into the relational model, is to allow *variables*, in addition to *constants*, as entries in the columns of relations. Such constructs are called *tables*, instead of *relations*. A table is an incomplete database, and *represents a set of complete databases*, each obtained by substituting all variables with constants. Different occurrences of the same variable (marked null) are substituted with the same constant. The substitution is thus a function from the variables and constants, to the constants, such that the function is identity on the constants. A table T then represents the set of relations, denoted $rep(T)$, defined as $\{v(T) : v \text{ is a valuation}\}$. Then the *certain answer* to a query q on a table T, denoted $sure(q, T)$ is the set of tuples that occur in every answer obtained by applying the query to every database in $rep(T)$. In other words, the certain answer to q on T is $sure(q, T) = \cap q(rep(T))$.

Key Points

To illustrate the above concepts, let tables T_1 and T_2 be as below, and let q be the relational expression $\sigma_{A=a \vee A=c}(\pi_{AC}(R_1 \bowtie R_2))$. (The schema of T_i is that of R_i, $i = 1,2$). Then applying q to T_1, T_2, which is denoted $\sigma_{A=a \vee A=c}(\pi_{AC}(T_1 \bowtie T_2))$, yields table $q(T_1, T_2)$ below.

T_1	A	B
	a	X
	Y	b
	c	b

T_2	B	C
	X	d
	b	Z

$q(T_1, T_2)$	A	C
	a	d
	c	Z

The variables/null-values are written in uppercase, to clearly distinguish them from the (lowercase) constants. Note, however, that $q(T_1, T_2)$ is not (necessarily) yet the certain answer. How was $q(T_1, T_2)$ derived from q and T_1, T_2, and how is the certain answer $sure(q,(T_1, T_2))$ obtained from $q(T_1, T_2)$? The answer to the second question is very simple: just drop all tuples containing variables from $q(T_1, T_2)$. The remaining tuples form the certain answer. In the example, the certain answer consists of tuple (a, c) only. The answer to the first question is not much more complicated: evaluate q on the tables, treating variables as "constants," pairwise distinct, and distinct from all "real" constants. This is also known as the Naive evaluation of q on T [2, 3]. In the example above, tuple (a, X) is joined with tuple (X, c) since they have the same value, represented by X, in the join column. This is done even though the "actual" value of X is not known, since in any valuation v the two occurrences of X are mapped to the same value. On the other hand, when performing the selection $\sigma_{A=a \vee A=c}$ the tuple (Y, Z) is not picked, since there is at least one valuation v, for which both $v(Y) \neq a$ and $v(Y) \neq c$. A characterization of the correctness of the Naive evaluation is given below.

Before going to the characterization, note that it is not always ideal to only return the certain answer. Namely, if the answer to q is to be materialized as a view for further querying, essential information is lost if the tuples with variables are dropped. For a simple example, a evaluating π_A on $sure(q, (T_1, T_2))$ gives tuple (a) as a sure answer, whereas evaluating $\pi_A(q(T_1, T_2))$, puts tuples (a) and (c) in the sure answer. As a consequence, query evaluation would not be compositional, unless $q(T_1, T_2)$ is stored as an "intermediate" answer. This "intermediate" answer is called the *exact answer* in [1], where the theory of query rewriting in information integration systems is extended to use the exact answer, instead of the certain one.

The correctness and completeness criteria for tables and query-evaluation is formalized using the notion of the representation system [2]. Here an alternative, equivalent formulation given in [3] is used: Consider a class of tables T, and a query language $\mathcal{Q}$. A triple (T, *rep*, $\mathcal{Q}$) is said to be a *representation system* if for every table $T \in$ T, and for every applicable $q \in \mathcal{Q}$, there exists a function (here also named) q, such that

$$\cap rep\,(q(T)) = \cap q\,(rep(T))\,, \text{and} \quad (1)$$

$$q' \circ q(T) = q'\,(q(T)) \quad (2)$$

for all applicable $q' \in \mathcal{Q}$.

Condition (1) says that the system can correctly compute the certain answer, and condition (2) states that the computation has to be uniformly recursive, following the structure of q. The important result is that, the class of Naive tables, and the class of all negation-free relational algebra expressions, together with Naive evaluation, form such a representation system.

And this result comes without any computational penalty.

Cross-References

▶ Certain (and Possible) Answers
▶ Incomplete Information
▶ Naive Tables

Recommended Reading

1. Grahne G, Kiricenko V. Towards an algebraic theory of information integration. Inf Comput. 2004;194(2):79–100.
2. Imielinski T, Lipski Jr W. Incomplete information in relational databases. J ACM. 1984;31(4):761–91.
3. Lipski W Jr. On relational algebra with marked nulls. In: Proceedings of the 4th ACM SIGACT-SIGMOD Symposium on Principles of Database systems: 1985. p. 201–3.

Narrowed Extended XPath I

Andrew Trotman
University of Otago, Dunedin, New Zealand

Synonyms

NEXI

Definition

NEXI is an information retrieval (IR) query language for searching structured and semi-structured document collections. The language was first introduced for searching XML documents at the annual INEX [3] evaluation forum in 2004, and it was used extensively by INEX for its element retrieval and XML-IR experiments.

Designed as the simplest query language that could possibly work, the language is a tiny subset of XPath 1.0 [1] with an added *about*() function for identifying elements about some given topic. The language has extensions for question answering, multimedia searching, and searching heterogeneous document collections. NEXI is a language with a strict syntax defined in YACC but it has no semantics; the interpretation of the query is the task of the search engine.

Historical Background

A common information retrieval query language for searching collections of XML documents was needed for specifying the queries at the first INEX in 2002. There, XML markup was chosen as the method of identifying keywords and the elements in which they should appear. It was also chosen as the method of identifying the preferred XML element to return to the user (the target element). The INEX 2002 query from topic 05 is given in Fig. 1. In this example, QBIC should occur in a bibl element, image retrieval may appear anywhere in the document, and the user is interested in a list of tig elements as the result of the query.

Two problems with this format were identified: first it allowed the specification of queries that could be resolved by a simple mechanical process; second the language was not sufficiently expressive for information retrieval queries.

A modified XPath 1.0 [1] was used at INEX 2003. In this variant the *contains*() function that required an element to contain the given content was replaced by an *about*() function that required an element to be about the content. Changing XPath in this way allowed fuzzy IR queries to be specified using a highly expressive language. However, an analysis of the queries showed high syntactic and semantic error rates [6].

O'Keefe and Trotman [6] proposed using the simplest query language that could possibly work

```
<title>
        <te>tig</te>
        <cw>QBIC</cw><ce>bibl</ce>
        <cw>image retrieval</cw>
</title>
```

Narrowed Extended XPath I, Fig. 1 INEX topic 05 in the 2002 XML format

and a novel syntax. The INEX Queries Working Group [7] rejected the syntax but embraced the philosophy. It identified the minimum requirements of an IR query language for information retrieval queries containing structural constraints. This language, although at the time without syntax or semantics, was to be used at INEX for evaluation purposes.

Trotman and Sigurbjörnsson [9] proposed the Narrowed Extended XPath I (NEXI) language based on the working group report. It was narrowed in so far as only the descendant axis was supported, and extended in so far as the *about*() function was added, all other functions and axis were dropped. A formal grammar and parser were published, and an online syntax checker was hosted by the authors.

The decision to reduce XPath resulted in fewer errors because it reduced the chance of making mistakes. NEXI has a precise mathematical formulation which matches intuitive user profiles [4]. For both naïve users with knowledge of just the tag names, and for more advanced users with additional knowledge of the inter-relationships of those tags, the language is safe and complete. That is, the user cannot make semantic mistakes, and can express every information need they have.

Foundations

Web queries typically contain between 2 and 3 terms per query [8]. Formal query languages for semi-structured data tend to be comprehensive. This mismatch became apparent at INEX 2003 where XPath was chosen as the preferred language for information retrieval experts to specify relatively simple queries, but where they were unable to write syntactically and semantically correct queries. Just as SQL is not an end-user query language, neither, it turned out, was XPath.

Requirements

After 2 years of experimentation with XML query languages at INEX, the needs of such a language became apparent. The INEX, Queries Working Group [7] specified that the language should:

- Be compatible with existing syntax for specifying content only (keyword) queries.
- Be based on XPath as that language was already well understood, but:
- Remove all unnecessary XPath axis used for describing paths. Limit to just the descendant axis was suggested. The child operator was considered particularly problematic as it was open to misinterpretation.
- Drop exact match of strings, and inequality of numbers. XPath path filtering remained, however all strings were expressed as aboutness.
- Support multiple data types including numeric and string.
- Be open for extensions for new data types (including names, locations, dates, etc.).
- Not include tag instancing (for example author [1], the first author).
- Have vague semantics open to interpretation by the search engine.
- Loosen the meaning of the Boolean operators AND and OR.
- Disallow multiple target elements. Although not explicit in the requirements, the implication is that the target element must be about the final clause in the query. It is a simple mechanical process to add non-target elements that are not about the query to the result – such as the author, title, source details to sections about something.
- Allow queries in which the target element was not specified and in which the search engine identified the ideal element.

Content Only (CO) Queries

NEXI addresses two kinds of queries on semi-structured and structured data: Content Only (CO) and Content And Structure (CAS) queries.

Content Only (CO) queries are the traditional IR query containing only keywords and phrases. No XML restrictions are seen and no mention is

given of a preferred result (target) element. For these the NEXI syntax is derived from popular search engines: search terms can be keywords, numbers, or phrases (delineated with quotes). Term restrictions can be specified using plus and minus.

Information Retrieval queries are by their very nature fuzzy. A user has an information need and from that need they express a query. There are many different queries they might specify from the same need, some of which might be more precise than the others. If a document in the document collection satisfies the user's information need, that document is relevant regardless of the query. That is, no query term might appear in a relevant document, or all the query terms might appear, either way the document is relevant. When specifying an IR query language it is important to avoid specifying semantics that violate this principle of relevance. The semantics of the terms with and without restriction in NEXI is, for example, specified this way:

- *"The '+' signifies the user expects the word will appear in a relevant element. The user will be surprised if a '-' word is found, but this will not prevent the document from being relevant. Words without a sign are specified because the user anticipates such terms will help the search engine to find relevant elements. As restrictions are only hints, it is entirely possible for the most relevant element to contain none of the query terms, or for that matter only the '-' terms."*

Or, in other words, it is the task of the search engine to identify relevant documents even if this involves ignoring the query.

In INEX topic 210 the author states:

- *"I'm developing a new lecture for the Master course 'Content Design' and want to discuss the topic "Multimedia document models and authoring". Therefore I want to do a quick background search to collect relevant articles in a reader. I expect to find information in abstracts or sections of articles. Multimedia content is an essential component of my lecture, thus for fragments to be relevant they should address document models of content authoring approaches for multimedia content. I'm not interested in single media approaches or issues that discuss storing multimedia objects."*

The query they give is

- +multimedia "document models" "content authoring"

in which "`document models`" is a phrase and +`multimedia` is a term-restricted search term (is positively selected for by the user).

Content and Structure (CAS) Queries

The second kind of query addressed by NEXI is the Content and Structure (CAS) query. These queries contain not only keywords but also structural constraints know as *structural hints*. Just as the keywords are hints passed to the search engine in an effort to help with the identification of relevant documents, so too are structural hints. CAS queries contain two kinds of structural hints, where to look (support elements), and what to return to the user (target elements).

Formally, queries may take one of the forms in Table 1:

A and C are paths and B and D are filters. Other forms could easily be added, but since NEXI was originally designed to address the INEX query problem, they are not formally included.

Paths (A and C in Table 1) are specified as a list of descendants separated by the descendant axis //. Formally, a path is an ordered sequence

of nodes $//E_1...//E_n$ starting with E_1 and finishing at E_n, and for all $e \in n$, E_e is an ancestor of E_{e+1}. An attribute node is indicated by the prefix @. Alternative paths are specified $(E_{na}|E_{nb})$. The wildcard * is used as a place holder.

For example, the path:

`//article//*//(sec|section)//@author`

describes an `author` attribute beneath either a `sec` or `section` element beneath something beneath an `article` element. The interpretation by the search engine is, of course, loose.

Filters (B and D in Table 1) can be either arithmetic or string. Arithmetic filters are specified as arithmetic comparisons (>, <, =, >=, <=) of numbers to relative-paths, for example:`.//year >= 2000`. String filters take the form `about(relative-path, COquery)`. Filters can be combined using the Boolean operators and, and or. Paths and filters are all considered hints and there is no requirement for the search engine to distinguish between the Boolean operators.

The target elements for the forms given in Table 1 are specified in column 2. Target elements, like support elements, are also hints. If, for example, the user specified paragraphs a subsection element might fulfill the user's information need.

An example of a valid NEXI CAS query (again, from INEX topic 230) is:

```
//article[about(.//bdy, "artificial
intelligence") and.//yr <= 2000]//
bdy[about(., chess) and about
(., algorithm)]
```

in which the target element is `//article//bdy`. The user has specified an arithmetic filter.`//yr. <= 2000`. Several string filters are used including `about(.//bdy, "artificial intelligence")`. A Boo-

Narrowed Extended XPath I, Table 1 Valid forms of NEXI CAS queries

Form	Target element	Meaning
//A[B]	A	Return A tags about B
//A[B]//C	A//C	Return C descendants of A where A is about B
//A[B]//C[D]	A//C	Return C descendants of A where A is about B and a C descendant of A are about D

lean operator is also used to separate two filters `about(., chess)` and `about(., algorithm)`.

The NEXI CAS query from INEX topic 210 is an alternative expression of the information need given in the previous section. That query is:

```
//article//(abs|sec)[about(., +
multimedia "document models"
"content authoring")]
```

in which the target element is either `//article //abs` or `//article//sec`. The same documents and elements are relevant to both queries, as relevance is with respect to the information need and not the specific query.

Key Applications

Information retrieval from structured and semi-structured document collections.

Future Directions

Although proposed as an XML query language for use in an evaluation forum, there is evidence it may also be an effective end-user language. Van Zwol et al. [13] compared NEXI to a graphical query language called Bricks. They found that a graphical query language reduced the time needed to find information, but that users were more satisfied with NEXI. Inherent in text query languages is the problem that users are required to know the structure (the DTD or schema) of the documents. In a heterogeneous environment this may not be possible, especially if new and different forms of data are constantly being added. Graphical query languages that translate into an intermediary text-based query language are one solution. This solution is seen with graphical user interfaces to relational databases.

Woodley et al. [12] further the model of NEXI as an intermediate language and compare NLPX (a natural language to NEXI translator) to that of Bricks (a graphic to NEXI translator). They show that users prefer a natural language interface, and that the performance of the two is comparable.

Ogilvie [10] examined the use of NEXI for question answering and proposed extensions to the language for this purpose. Dignum and van Zwol [2] proposed extensions for heterogeneous searching. Trotman and Sigurbjörnsson [10] unified these proposals and formally extended the language to include both – however, these extensions are not considered core to the language (any language extension philosophically deviates from the principle of "simplest that could possibly work"). Multimedia extensions to the language have also been used at INEX [11], again the extensions are not considered core to the language.

Experimental Results

The analysis of XPath queries used at INEX 2003 showed 63% of queries containing either syntactic or semantic errors [3]. An analysis of the errors in NEXI queries used at INEX 2004 showed that only 12% contained errors [12]. NEXI has been in use at INEX ever since.

Data Sets

NEXI queries from INEX 2004 to INEX 2009 can be downloaded from the INEX web site of the time: http://inex.otago.ac.nz. INEX queries from 2010 onwards are at the current INEX web site: https://inex.mmci.uni-saarland.de

INEX queries for 2003 and 2002 were translated into NEXI (where possible) and can be downloaded from the NEXI web page hosted by the University of Otago: http://metis.otago.ac.nz/abin/nexi.cgi

URL to Code

An online syntax checker, lex and yacc scripts, and a command line syntax checker can be downloaded from the NEXI web page hosted by the University of Otago: http://metis.otago.ac.nz/abin/nexi.cgi

Cross-References

▶ Content-and-Structure Query
▶ Content-Only Query
▶ INitiative for the Evaluation of XML Retrieval
▶ Processing Structural Constraints
▶ Query by Humming
▶ Query Languages for the Life Sciences
▶ Semi-Structured Query Languages
▶ Temporal Query Languages
▶ XML
▶ XPath/XQuery
▶ XQuery Full-Text
▶ XSL/XSLT

Recommended Reading

1. Clark J, DeRose S. XML path language (XPath) 1.0, W3C recommendation. The World Wide Web Consortium. Available at: http://www.w3.org/TR/xpath 1999.
2. Dignum V, van Zwol R. Guidelines for topic development in heterogeneous collections. In: Proceedings of the 3rd International Workshop of the Initiative for the Evaluation of XML Retrieval; 2004.
3. Fuhr N, Gövert N, Kazai G, Lalmas M. INEX: initiative for the evaluation of XML retrieval. In: Proceedings of the 25th Annual International ACM SIGIR Conference on Research and Development in Information Retrieval; 2002.
4. Kamps J, Marx M, Rijke M, Sigurbjörnsson B. Articulating information needs in XML query languages. Trans Inf Syst. 2006;24(4):407–36.
5. Ogilvie P. Retrieval using structure for question answering. In: Proceedings 1st Twente Data Management Workshop – XML Databases and Information Retrieval; 2004. p. 15–23.
6. O'Keefe RA, Trotman A. The simplest query language that could possibly work. In: Proceedings of the 2nd International Workshop of the Initiative for the Evaluation of XML Retrieval; 2003.
7. Sigurbjörnsson B, Trotman A. Queries: INEX 2003 working group report. In: Proceedings of the 2nd International Workshop of the Initiative for the Evaluation of XML Retrieval; 2003.
8. Spink A, Wolfram D, Jansen BJ, Saracevic T. Searching the web: the public and their queries. J Am Soc Inf Sci Tech. 2001;53(2):226–34.
9. Trotman A, Sigurbjörnsson B. Narrowed extended XPath I (NEXI). In: Proceedings of the 3rd International Workshop of the Initiative for the Evaluation of XML Retrieval; 2004. p. 16–40.
10. Trotman A, Sigurbjörnsson B. NEXI, now and next. In: Proceedings of the 3rd International Workshop of the Initiative for the Evaluation of XML Retrieval; 2004. p. 41–53.
11. Westerveld T, van Zwol R. Multimedia retrieval at INEX 2006. SIGIR Forum. 2007;41(1):58–63.
12. Woodley A, Geva S, Edwards SL. Comparing XML-IR query formation interfaces. Aust J Intell Inf Process Syst. 2007;9(2):64–71.
13. van Zwol R, Baas J, van Oostendorp H, Wiering F. Bricks: the building blocks to tackle query formulation in structured document retrieval. In: Proceedings of the 28th European Conference on IR Research; 2006. p. 314–25.

Natural Interaction

Stefano Baraldi, Alberto Del Bimbo,
Lea Landucci, and Nicola Torpei
University of Florence, Florence, Italy

Synonyms

Natural human-computer interaction; NHCI

Definition

The aim of Natural Human Computer Interaction (NHCI) research is to create new interactive frameworks that integrate human language and behaviour into tech applications, focusing on the way people live, work, play and interact with each other. Such frameworks have to be easy to use, intuitive, entertaining and non-intrusive.

The design of *natural interaction* systems is focused on recognizing innate and instinctive *human expressions* in relation to some object, and return to the user a corresponding feedback that has the characteristics of being both *expected* and *inspiring*. All of the technology and the intelligence is built inside the digital artifacts and the user is not asked to use external devices, wear anything, or learn any commands or procedures. An interesting challenge for NHCI is therefore to make systems self-explanatory by working on their "affordance" [11] and introducing simple and intuitive interaction languages.

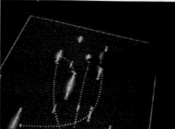

Natural Interaction, Fig. 1 Examples of natural interfaces for accessing digital contents and structures

The human expressions that can be utilized are those considered innate, meaning that they don't have to be learned. This includes vocal expressions and all the gestures used by humans to explore the nearby space or the immediate surroundings with their bodies, like: touching, pointing, stepping into zones, grabbing and manipulating objects (see Fig. 1). These direct actions express a clear sign of interest and necessitate a sudden reaction from the system.

The application fields range from browsing multimedia contents to exploration of knowledge structures; the scenarios involved can be either task-specific (in office or research contexts) or experiences created for casual interaction between the visitors and the media contents (like in museum exhibits and creative installations).

Natural interaction interfaces are very interesting to access, and explore large data sets like the ones contained in multimedia or geo-referenced databases. Data mining applications, which usually ask the user to enter complex search criteria and tweak the parameters to obtain the wanted data, can be designed with a visual analytic interface following these design guidelines. Queries are expressed visually, selecting interactive objects that embody selection criteria, creating clusters of them. Result sets provided by the database back-end can be iteratively shaped by directly manipulating the visual elements mapping the data, including other sets or reducing them.

Historical Background

At the beginning of the 1990s, Human-Computer Interaction [17] was completely integrated into the Computer Science field, its quick growth led to the development of technologies able to go over the standard limiting user-computer communication paradigm, approaching a kind of natural interaction.

Alex Pentland in his "The Dance of Bits and Atoms" (1996, [13]) said that "There is a deep divide between the world of bits, and the world of atoms. Current machines are blind and deaf; they are unaware of us or our desires unless we explicitly instruct them. Consequently, only experts use most machines, and even they must spend most of their time battling arcane languages and strange, clunky interface devices."

The key point is that technologies must be designed and developed adapting them to the users, not the contrary: this is the foundation of the "human centered design" [11, 12].

The goal, finally, becomes a technology at one user's service, suitable to the task requested and characterized by the complexity naturally embedded in the task itself [11, 12].

Foundations

Interactive artifacts, augmented spaces, ambient technology and ubiquitous computing are all fields of study in HCI, which concentrate on both the technological aspects and the modalities in which users can perform activities. The techniques proposed are often referred to as *multi-modal interaction*, focusing on how machines can understand commands coming from different channels of human communication [10]. Among those: speech recognition, natural language understanding and gesture recognition, have been applied in different mixtures with roles ranging from active (the system observes the user and is pro-active in interacting with him) to passive (the system expects some kind of command, often through a device that is worn by the user). In the last years, studies have been following the concept of natural interaction as a conjunction of different technologies and design principles, with a more radical view about the user freedom in using interactive artifacts.

Natural interaction systems can be modelled as the sum of different modules: the *sensing* subsystem, which gathers sensor data about user expressions and behaviour, and the *presentation* module, which realizes the dialogue with the user orchestrating the output of different kind of actuators (graphics display, audio, haptics).

Sensing

The initial expression of interest towards the digital artefact is distracted from its main role: transferring information and stimuli. For this reason, the artefact should hide all the needs for an external controlling device, and be able to *sense* what the user is trying to do.

From a technological point of view, sensing involves the use of different sensors that provide data about some physical dimension in the surroundings of the artefact. There are a great number of electronic sensors that have been used in many industrial fields for robotics, automation and automatic inspection. Ranging from video cameras and image processing algorithms, to capacitive sensors for touch and pressure sensing or accelerometers for gesture recognition and body articulation.

Every sensor can have very different ways of providing this kind of data in terms of resolution, range, tolerances and errors. Some of them are able to provide discrete data with a high certainty, while others (like video cameras) just provide a great amount of data that has to be processed and interpreted by algorithms in order to extract useful information.

Considering the sensing architecture as a whole, a processing logic must be applied on top of it to abstract all the singularities of the sensors and create a homogeneous model of events for the later stages of interaction. Discrete events arriving from the sensors can be considered as belonging to three different categories of data:

- *Presence*: this data usually comes with a high signal-to-noise ratio, meaning that it cannot be directly interpreted as a human expression. Nonetheless, it reports a general activity in a spatial area.
- *Behaviour*: as the certainty level of some event increases (thanks to the combination of different sensors data), data can be studied for a certain period of time and provide information about the behaviour of single users or groups.
- *Activity*: this kind of data are considered "certain" and interpreted as a clear user intention to be used for a direct control of the interface. Instead of providing just the bi-dimensional screen coordinates as, it happens with traditional touch-screens, advanced sensing can interpret speed or pressure as well as distance of the hands as they approach a surface.

Sensors can be displaced in the environment, embedded in the artefact and also worn by the user but for natural interaction systems the last option is not suggested. For indoor use, the most meaningful displacement of sensors is inside the artefact itself. In this way, space can remain flexible and structures can be moved around. Such a thing could not be done if the whole environment was disseminated with.

Intelligence and Presentation

The intelligence substratum is what orchestrates the signals coming from sensors and generates output for presentation.

The devices embedded in the artefacts that manage the output phase are called *actuators* and can range from visual displays, to audio systems, surfaces with tactile feedback, holograms and many others. In scenarios like museums or art installations, even simple forms of stimuli can be used and mixed to convey a sensation or to catch the attention of the user. In an information space like an office or a data-intensive environment, the main channel of parallel communication remains the visual one, and the actuator is the graphical display.

On traditional GUIs (Graphical User Interfaces), what happens "inside the screen" can be subject to inference and prediction. Modern interfaces gather data about their use and are predictive in searching and suggesting data to the user in a light and "polite" way. Natural interfaces can extend this reasoning to a lot more data coming from the sensors, because the interactive artefact itself tries to *sense* beyond the screen. Also, unlike traditional interfaces whose informational layout is totally independent from the physical place in which the device is installed and used, digital artefacts can consider their position in the environment and what is happening around them.

In natural graphical interfaces, the basic assumption is that digital elements are not just a representation of data, but a part of the environment, and they need to follow aesthetics rules too. As a first principle: digital elements should behave like objects in physical systems. Following this *analogy* the graphical display becomes a real *space*, and therefore has to provide the same affordance of a normal surface. If something changes on the visual interface it should happen in a smooth way, without sudden jumps, similar to what happens with hyperlinks on web pages. The models of movement and forces used in transitions have to mimic those of one real world, like: gravity, accelerating forces, momentum and friction. The user has the chance to understand what is going on without feeling disoriented, and thanks to these visual cues he can also expect what is going to happen.

The anticipation is more important because in shared natural interfaces the model of control of traditional interfaces, modelled around a question to confirm every action, cannot be used. A popup message box would block the interface occupying space and this is not feasible in a shared user environment. Instead, every action is directly executed and, in the time in which this happens, the visual features of objects influenced by the action gradually change. Another guideline for the visual interface design is *lightness*. On every graphical display, there is a conflict between the

level of contents belonging to the domain of the application (diversified media), and the level of widget elements (labels, buttons, etc.) that explain the interface to the user and give hints about the options. In natural interfaces this is even truer. Basically, the space is for contents, so any other element is stealing some attention from the user. For this reason, the widget level is kept to a minimal amount of symbols and there is little or no use of "global" interface elements.

Handling Complexity with TUIs

A challenge in natural interaction systems is tackling the complexity with simplicity. While the category of exploring applications can be realized using only innate gestures, in the case of complex applications (featuring multiple options and actions) simple and spontaneous hand gestures turn out to be not enough, and methods like speech understanding do not adapt well to noisy environments or to scenarios in which multiple people are interacting simultaneously.

The solutions could be:

1. Enriching the interaction language by adding new complex gestures that map to actions. This could distort the naturalness of interaction forcing users to learn unnatural gestures.
2. Introducing an intermediate visual level using interface elements (such as menus, icons etc.). This would reduce the interaction directness causing a conflict between digital contents and interface elements, both sharing the same visualization area.

The result is that such solutions could increase the user cognitive load. Tangible user interfaces (TUIs [8]) can be an alternative solution to those mentioned. They introduce physical, tangible objects that the system interprets as embodiment [6] of the elements of the interaction language. Users, manipulating those objects, inspired by their physical affordance, can have a more direct access to functions mapped to different objects [3].

There is a broad literature about TUIs which dates back to the first experiments of Hiroshi et al at the MIT, where a set of normal objects where

Natural Interaction, Table 1 Tangible interaction themes

Theme	Features
Tangible manipulation	Physical manipulation of tactile qualities
Spatial interaction	Movement in space
Embodied facilitation	Configuration of objects affecting group behavior
Expressive representation	Focus on digital and physical expressiveness

"sensed" by a system who could recognize and track some of their features, like the position in space. Many other systems have been developed using this paradigm, and today there is a growing familiarity with taxonomies that try to define the different styles and scenarios in which it can be used [7]. Table 1 illustrates the main different branches.

The physical objects, called *tangibles*, could address some of the issues related to complexity. They can become the *embodiment* of some aspects of the interaction between the user and the domain of multi-media contents handled by the application. In particular three possible roles can be distinguished.

- *The tangible as a simulacrum.* The physical object can be used as a representative of a single digital object or a collection. This means either a uniquely identified static element with a one-to-one mapping between the physical and the digital world, or a re-assignable element that can be associated with different data (like a container).
- *The tangible as a manipulator.* In this case the tangible represents a *function* that can be applied to a digital object. The closest metaphor is that of a tool in order to change one of its aspects, reveal other connected contents etc. This modality can be also extended to global functions whose target is the entire viewport or collection of digital objects.
- *The tangible as an avatar of the user.* In this case, every user would have a personal object that represents himself, to move across the contexts. An avatar object can be used in

conjunction with the others when the application needs an authentication or when the activity proposed requires the user to express a preference.

While the manipulator role is specific to the nature and local interaction with the artefact (e.g., a digital tabletop), the simulacrum and avatar roles exactly provide the abstraction that is needed in order to expand the affordance of the environment, providing a natural way to transport the productions across the artifacts. In this fashion, activities can be initiated on an artefact and continued somewhere else, like in a laboratory, different places provide different contexts and options for the same data.

Key Applications

Multimedia Browsing

Usually they are easy-to use systems where people can interact simultaneously with multimedia contents through their own bare-hand gesture.

This kind of application offers an intuitive approach to various multimedia objects, they don't require any kind of training or instructions.

Knowledge Exploration and Building

Interactive workspace featuring vision-based gesture recognition that allows multiple users to collaborate [2] in order to realize face-to-face contexts, designing a common workspace where users can build knowledge (activities like brainstorming or problem solving sessions), exploiting the useful scenario-specific characteristics.

Interactive Museum and Cultural Exhibits

Museums and exhibitions are often just a collection of objects, standing deaf in front of visitors. In many cases, objects are accompanied by textual descriptions, usually too short or long to be useful for the visitor. In the last decade, progress in multimedia has allowed for new, experimental forms of communication (using computer technologies) in public spaces [1].

N

Interactive Music Systems

Usually built over a tabletop tangible user interface, Interactive Music Systems allow several simultaneous performers to share complete control over the instrument by moving physical artefacts on the table surface while constructing different audio topologies in a kind of tangible modular synthesizer. The reacTable [9], a clear example, is a novel multi-user electro-acoustic musical instrument with a tabletop tangible user interface.

Cross-References

▶ Human-Computer Interaction
▶ Multimedia Databases
▶ Multimodal Interfaces
▶ Object Recognition
▶ Visual Interfaces
▶ Visual Perception
▶ Visual Representation

Recommended Reading

1. Alisi TM, Del Bimbo A, Valli A. Natural interfaces to enhance visitors' experiences. IEEE Multimed. 2005;12(3):80–5.
2. Baraldi S, Del Bimbo A, Landucci L, Valli A. wikiTable: finger driven interaction for collaborative knowledge-building workspaces. In: Proceedings of the 2006 IEEE International Conference on Computer Vision and Pattern Recognition Workshop; 2006. p. 144.
3. Baraldi S, Del Bimbo A, Landucci L, Torpei N, Cafini O, Farella E, Pieracci A, Benini L. Introducing TANGerINE: a tangible interactive natural environment. In: Proceedings of the 5th ACM International Conference on Multimedia; 2007. p. 831–4.
4. Colombo C, Del Bimbo A, Valli A. Visual capture and understanding of hand pointing actions in a 3-D environment. IEEE Trans Syst Man Cybern B Cybern. 2003;33(4):677–86.
5. Dietz P, Leigh D. DiamondTouch: a multi-user touch technology. In: Proceedings of the 14th Annual ACM Symposium on User Interface Software and Technology; 2001. p. 219–26.
6. Fishkin KP. A taxonomy for and analysis of tangible interfaces. Pers Ubiquit Comput. 2004;8(5):347–58.
7. Hornecker E, Buur J. Getting a grip on tangible interaction: a framework on physical space and social interaction. In: Proceedings of the SIGCHI Conference on Human Factors in Computing Systems; 2006. p. 437–46.
8. Ishii H, Ullmer B. Tangible bits: towards seamless interfaces between people, bits and atoms. In: Proceedings of the SIGCHI Conference on Human Factors in Computing Systems; 1997. p. 234–41.
9. Kaltenbrunner M, Jordà S, Geiger G, Alonso M. The reactable: a collaborative musical instrument. In: Proceedings of the Workshop on Tangible Interaction in Collaborative Environments; 2006.
10. Marsic I, Medl A, Flanagan J. Natural communication with information systems. Proc IEEE. 2000;88(8):1354–66.
11. Norman DA. The design of everyday things. Cambridge, MA: MIT Press; 1998.
12. Norman DA. The invisible computer. Cambridge, MA: MIT Press; 1999.
13. Pentland A. Smart rooms. Sci Am. 1996;274(4):54–62.
14. Prante T, Streitz NA, Tandler P. Roomware: computers disappear and interaction evolves. IEEE Comput. 2004;37(12):47–54.
15. Ulmer B, Ishii H. Emerging frameworks for tangible user interfaces. IBM Syst J. 2000;39(3–4):915–31.
16. Valli A. Notes on natural interaction http://naturalinteraction.org. 2005.
17. Wania CE, Atwood ME, McCain KW. How do design and evaluation interrelate in HCI research? In: Proceedings of the 6th Conference on Designing Interactive Systems; 2006. p. 90–8.

Near-Duplicate Retrieval

Heng Tao Shen
School of Information Technology and Electrical Engineering, The University of Queensland, Brisbane, QLD, Australia
University of Electronic Science and Technology of China, Chengdu, Sichuan Sheng, China

Synonyms

NDVR

Definition

Being a relatively new topic in the research society, near-duplicate video retrieval (NDVR) has a variety of definitions on near-duplicate videos (NDVs). Representative definitions include those defined in [1, 2]. In [1], NDVs are defined as those videos which have the same semantics, "but appear differently due to various changes introduced during capturing time (camera view point and setting, lighting condition, etc.), transformations (video format, frame rate, resize, crop, contrast, brightness, etc.), and editing operations (frame insertion, deletion, and content modification)." In [2], while the human perception of NDV matches many of the features presented in its technical definitions with respect to manipulations of visual content in [1], similar videos differing in overlaid or added visual content with additional information were not perceived as near-duplicates. Conversely, two different videos with distinct people and scenarios were considered to be NDVs because they shared the same semantics, provided that none of near-duplicates has additional information. It is evidenced that users perceive as NDVs those which are both visually similar and semantically

identical. This clearly differs itself from conventional Content-Based Video Retrieval (CBVR), because for the latter the content similarity is the only factor it needs to consider, while for the former both content and semantics need to be examined. Therefore, NDVR can be regarded as the bridge between traditional content-based retrieval (i.e., videos should have similar visual content regardless of semantics) and semantic-based retrieval (i.e., videos should have relevant semantics regardless of visual content) [3].

Main Text

The exponential growth of online videos, along with increasing user involvement in video-related activities, has been observed as a constant phenomenon during the last decade. User's time

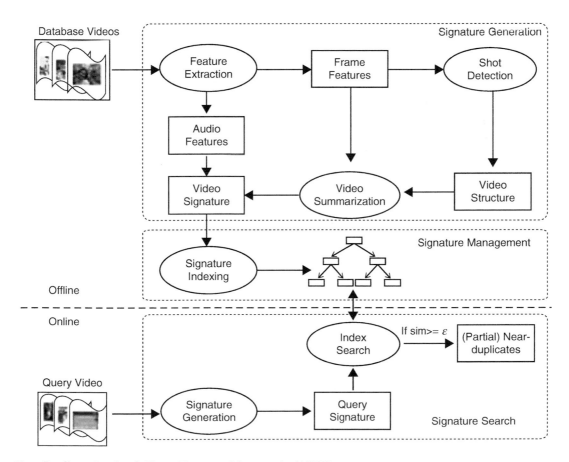

Near-Duplicate Retrieval, Fig. 1 The general framework of NDVR

spent on video capturing, editing, uploading, searching, and viewing has boosted to an unprecedented level. The massive publishing and sharing of videos has given rise to the existence of large amount of near-duplicate content. This imposes urgent demands on near-duplicate video retrieval (NDVR) as a key role in novel tasks such as video search, video copyright protection, video recommendation, and many more. Driven by its significance, near-duplicate video retrieval has recently attracted a lot of attention [3].

Figure 1 depicts the general framework for an NDVR system, including three major components, signature generation, signature management, and signature search, for a typical NDVR task. Normally, given a database video, in the signature generation component, compact and distinctive video signatures are firstly generated to facilitate retrieval. Video signatures can also take structural information into consideration. To achieve fast retrieval, in the signature management component, the generated video signatures are organized by effective indexing structures to avoid extensive data accesses. Given a query video, in the signature search component, its signature is first generated like that in the signature generation component. The obtained query signature is then searched in the database index to find NDVs, based on signature matching. If the similarity between a database video's signature and the query signature is greater than or equal to a predefined similarity threshold – ε, the database video is regarded as an NDV of the query. The above procedure describes the principles of an NDVR task. In real life, the existence of partial NDVs imposes additional requirements to the procedure. With the assistance of some specific techniques, such as fusing frame-level or segment-level video relevance to enable subsequence retrieval, the framework can be capable of detecting partial NDVs as well.

Cross-References

▶ Video Sequence Indexing

Recommended Reading

1. Huang Z, Shen HT, Shao J, Zhou X, Cui B. Bounded coordinate system indexing for real-time video clip search. ACM Trans Inf Syst. 2009;27(3):17: 1–17:33.
2. Cherubini M, de Oliveira R, Oliver N. Understanding near-duplicate videos: a user-centric approach. In: Proceedings of the 17th ACM International Conference on Multimedia; 2009, p. 35–44.
3. Liu J, Huang Z, Cai HY, Shen HT, Ngo C-W, Wang W. Near-duplicate video retrieval: current research and future trends. ACM Comput Surv. 2013;45(4):44.

Nearest Neighbor Classification

Thomas Seidl
RWTH Aachen University, Aachen, Germany

Synonyms

k-nearest neighbor classification; k-NN classification; NN classification

Definition

Nearest neighbor classification is a machine learning method that aims at labeling previously unseen query objects while distinguishing two or more destination classes. As any classifier, in general, it requires some training data with given labels and, thus, is an instance of supervised learning. In the simplest variant, the query object inherits the label from the closest sample object in the training set. Common variants extend the decision set from the single nearest neighbor within the training data to the set of k nearest neighbors for any $k > 1$. The decision rule combines the labels from these k decision objects, either by simple majority voting or by any distance-based or frequency-based weighting scheme, to decide the predicted label for the query object. Mean-based nearest neighbor classifiers group the training data and work on the means of classes rather than on the individual training objects. As nearest neighbor classifiers only require distance computation of objects as a basic operation, their applicability exceeds the domain of vector spaces and attribute-based tuple data, and includes metric data spaces even if there are no attributes or dimensions available. Nearest neighbor classification counts for being highly accurate but inefficient; the latter assessment is to be invalidated by using appropriate indexing structures that support fast k-nearest neighbor retrieval. The abbreviation "NN classification" tends to cause confusion with the totally different concept of neural network classification.

Historical Background

Nearest neighbor classification is one of the earliest classification schemata. Nevertheless, it is one of the most highly accurate ones, whilst obeying a very broad range of applicability. The following aspects distinguish it from other classification methods.

Multiple class labels. Nearest neighbor classification is not restricted to two-class problems where classification decisions have to distinguish two classes only. Moreover, a high number of different classes does not deteriorate the efficiency in contrast to many other classifiers which create individual models for all the classes in the training set.

Applicability to general metric data spaces. Nearest neighbor classification is applicable to data without any structured attribute representation, i.e., to non-vectorial data. As distance computations are the only required basic operation, the applicability includes complex multimedia data, time series and measurement series data, sequences and structural data to name just a few supported domains. Among the competing approaches in the field, only support vector machines equipped with respective kernel functions share this advantage. Other classifiers including decision trees, neural networks and Bayes density models rely on the attribute structure of the data.

Instance-based learning. Nearest neighbor classification derives decisions close to the individual training instances as any other method. No model is created for the training data and, thus, it is called lazy evaluation. Most of the competing classification approaches follow the eager evaluation paradigm which spends significant training effort in the determination of models for the training data. They produce decision tree structures, synaptic weights in neural networks, support vector weights, or density probability functions, respectively. The lazy model of nearest neighbor classification nevertheless allows for fast class decisions when spending (training) effort in the creation of an index on the training data.

Training data that changes dynamically. If the classifier has to follow training data that changes over time, without or even with respect to its statistical characteristics, a nearest neighbor classifier is superior to almost all of the competing classifiers. No model parameters such as decision tree structures or density weights have to be recomputed where the training data has changed. The lazy evaluation strategy ensures that the results always rely on the current training data status. The efficiency depends on how the underlying index structure supports the dynamic data changes.

Intuitive explanation component. Nearest neighbor classifiers provide illustrative explanations of their decisions by revealing the closest neighbors of the query object from the training data. Moreover, the corresponding similarity scores, i.e., the distances of the training items to the query object, provide some insight into the decision process, and the users may derive their confidence in the final decision result. In comparison, support vector machines analogously give illustrative weights to individual training objects, Bayes classifiers have a similar explanation power by providing probability values for each target class, and decision trees are slightly more intuitive by presenting the rules that produced the decision result. On the other hand, neural network classifiers lack illustrative explanations of their decisions.

Foundations

Classification Model

Nearest neighbor classifiers are a common classification model for which several variants exist. Along with the simple nearest neighbor model, k-nearest neighbor classification uses a set of k neighbors and the mean-based nearest neighbor model where individual training objects are generalized uses group representatives.

Simple nearest neighbor classification. The simplest variant of nearest neighbor classification may be formalized as follows: For any metric object space O, let $TS \subseteq O$ denote the set of labeled training data and $d : O \times O \rightarrow \Re_0^+$ is the chosen and thus distance function that reflects the dissimilarity of any two objects from O. Then, for any query object $q \in O$, the classifier evaluates the function $C_{\text{preliminary}}(q) = \{label(o) | o \in TS, \forall p \in TS - \{o\} : d(o,q) < d(p,q)\}$.

Unfortunately, this formalization is not valid for cases where several objects $o \in TS$ share the same minimal distance $d(o,q)$ to the query object, i.e., when the query hits perpendicular bisectors of these objects. Though in practice, these ties occur very rarely, a formalization that is equivalent to the previous one in other respects but in which ambiguities may be broken simply in a nondeterministic way is preferred (Fig. 1):

$$C_{nn}(q) = \left\{ label(o) \middle| o \in TS, \forall p \in TS : d(o,q) \leq d(p,q) \right\}$$

$$(1)$$

The decision model of nearest neighbor classification allows for a particularly broad range of applications. Its applicability covers vector data which are characterized by numerical attributes and includes general relational data where the tuples are represented by both numerical and categorical data as long as some metrics are defined on the categories. Moreover, if training and test data are neither taken from a vector space nor carry any categorical attributes but are only compared in terms of a distance function that reflects the dissimilarity of objects, nearest neighbor classifiers are still applicable. Examples of such applications include sequence data or structural data which may be compared by the edit distance, time series and measurement series compared by dynamic time warping, text and document data compared by the cosine distance, or multimedia data compared by complex similarity models.

Mean-based nearest neighbor classification. Nearest neighbor classifiers yield high quality decisions in terms of classification accuracy, and they are quite robust even for small training sets. Nevertheless, they tend to suffer from overfitting since the decisions are made closer at the training data than it is the case for any other method. As a consequence, erroneous training data, noise

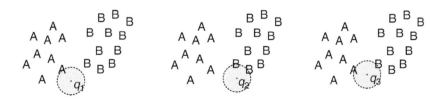

Nearest Neighbor Classification, Fig. 1 Ambiguity of nearest neighbors for a sample training data set with two classes, A and B. Query object q_1 has a unique nearest neighbor and will be assigned to class A. For query object q_2, two neighbors share the smallest distance but there is no ambiguity with respect to the class decision since both neighbors are in class B. Query q_3 yields a conflict as it has two nearest neighbors from different classes, namely A and B, respectively. In practice, these ties occur very rarely at least for numerical reasons but may be solved nondeterministically and outliers may badly affect the decisions. In order to increase the generalization power and to approach the overfitting problem, mean-based nearest neighbor classifiers and k-nearest neighbor classifiers have been developed as common variants of the simple model.

For mean-based nearest neighbor classification, the objects in the training data set are grouped into one or more clusters per class. These clusters are represented by their means, and nearest neighbors are then selected among the means rather than among the original individual training objects. The means inherit the class label of their cluster members, and the decision rule (1) from simple nearest neighbor classification is easily adopted to the mean model just by replacing the training data set by the set of means:

$$C_{m-nn}(q) = \begin{cases} label(m) \big| m \in means(TS), \\ \forall n \in means(TS) : \\ d\,(m,q) \le d\,(n,q) \end{cases} \tag{2}$$

Note that the computation of mean values requires the object space O to be a vector space, since the objects of a cluster need to be summed up followed by a scalar multiplication by the reciprocal value of the group's cardinality. Thus, the mean-based model does not apply to non-vectorial metric spaces which, in contrast, are supported by k-nearest neighbor classifiers.

k-nearest neighbor classification. Beside mean-based nearest neighbor classifiers, k-nearest neighbor classification is another quite

common approach to increase the generalization power and to decrease overfitting effects. Instead of looking at a single object to the query among the training data, a set of k nearest neighbors, $k > 1$, is taken into account when making the decision. One starts by defining the decision set to be a subset of the training data TS that contains the k objects closest to the query object q. Again, ties may be broken by a nondeterministic choice among equidistant neighbors:

$$NN(q,k) = \big\{ o \big| o \in TS, \forall p \in TS-NN(q,k) : \\ d\,(o,q) \le d\,(p,q) \}, \quad |NN\,(q,k)| = k \tag{3}$$

The decision rule based on majority vote now looks as follows (Fig. 2):

$$C_{majority}(q) = \arg\ \max_{l \in label(NN(q,k))} \\ \big\{ card\,\big\{ o \in NN\,(q,k) \big| label(o) = l \big\} \big\}. \tag{4}$$

Weighted k-nearest neighbor classification. For k-nearest neighbor classifiers, different weighting schemata have been developed which introduce distances or frequencies into the decision rule. Conceptually, majority voting represents a weighting schema with unit weights. A quite common variant is to use the squared distances of the decision objects to the query object as reciprocal weights. This way, the desired effect is obtained that the closer an object is to the query, the higher is its influence to the classification decision. The corresponding decision rule is as follows:

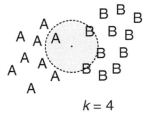 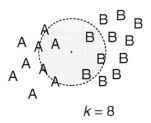

Nearest Neighbor Classification, Fig. 2 Dependency of the classification decision on the number of nearest neighbors. In the examples from left to right, the same query object is considered. A choice of $k = 1$ yields the class label A. For $k = 4$, class B holds the majority in the nearest neighbor set, and for $k = 8$, again label A is assigned to the query object

$$C_{dist}(q) = \text{argmax}_{l \in label(NN(q,k))} \left\{ \sum_{\substack{o \in NN(q,k) \\ label(o) = l}} \frac{1}{d(o,q)^2} \right\}. \tag{5}$$

An alternative weighting schema takes the relative frequency of the classes into account. Naturally, rare classes are represented only by a small number of instances in the training data. If at decision time, a nearest neighbor set contains a few of these rare instances, this occurrence is honored by the frequency weighting schema. This way, rare classes may outvote highly frequent classes even if these dominate the decision set by their number. Formally speaking, decisions are based on the fraction of training instances a class contributes to the decision set rather than on the absolute number of instances among the nearest neighbors (Fig. 3).

$$C_{freq}(q) = \text{argmax}_{l \in label(NN(q,k))} \left\{ \frac{card\left\{ o \middle| o \in NN(q,k), label(o) = l \right\}}{card\left\{ o \middle| o \in TD, label(o) = l \right\}} \right\} \tag{6}$$

Lazy Evaluation Model

Nearest neighbor classification follows the model of lazy evaluation. This means that there is no computation or estimation of model parameters at training time, but all calculations to make decisions are deferred to query time. Lazy evaluation is in contrast to the eager evaluation paradigm which requires some effort at training time to build an appropriate model. Examples of eager classifiers include decision trees, Bayes classifiers, neural networks, or support vector machines for which attribute-oriented decision flow diagrams, probability (mixture) density models, synaptic weights, or support vector weights are computed at training time, respectively. Lazy evaluation does not build models for the training data but derives its decisions directly from the training objects and, therefore, is called

instance-based learning. Nearest neighbor classification is a prominent example of lazy evaluation, and the simple 1-nn and extended k-nn variants purely reflect the instance-based paradigm. The mean-based variant, however, deviates a little from this idea since means of classes are computed at training time, and decisions are based on the means rather than on the individual instances. For lazy evaluation in general, questions about decision efficiency, dynamic changes of training data, and explanations of decisions arise; these issues are discussed in the following subsections. Note that mean-based nearest neighbor classification is not a pure lazy variant since means of classes are computed at training time.

Decision efficiency. By following the lazy evaluation model, nearest neighbor classifiers defer all data analysis to query time. So query processing tends to be computationally more expensive than with eager classifiers. Obviously, the most complex task is to determine the decision set, that is the set of (k) nearest neighbors for a query object; the subsequent weighting and voting is negligibly done in $O(k)$. Retrieving the k nearest neighbors in a set of n training objects depends on the data organization. Whereas a sequential scan requires $O(n)$ operations, the use of multidimensional or metric indexing structures may significantly speedup the retrieval of the decision set depending on the dimensionality and on the statistical characteristics of the data. For high-dimensional data, techniques for reduction of dimensionality and multi-step query processing help to keep k-nearest neighbor retrieval efficient. As strict lazy evaluation does not waste any training time to create a model, it is nevertheless highly recommended to spend some effort for the creation of an appropriate index on

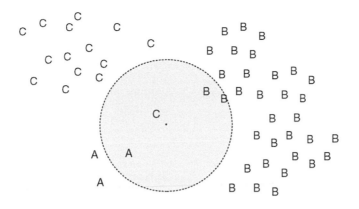

Nearest Neighbor Classification, Fig. 3 Dependency of class decision on the weighting schema. In the example, a 4-nearest neighbor set is selected for the query object. Pure majority voting decides for class B as the decision set includes two instances of class B. Distance weighting decides for class C as the close neighbor from class C dominates the farer neighbor from class A and even the two neighbors from class B. Frequency-based weighting yields class A since A contributes the highest fraction of its overall occurrences in the training data to the nearest neighbor set

the training data. In short, to train an efficient nearest neighbor classifier means to instantiate model creation by index creation.

Dynamically changing training data. As for lazy evaluation, no model is created at training time, a lazy classifier in general allows for dynamic changes of training data particularly well. For a ncarest neighbor classifier, efficiency depends on how the underlying index structure maintains the dynamic data set. This represents a common requirement for indexing and is supported well by almost all multidimensional or metric access methods. As training data changes, new examples are inserted into the index and outdated training instances are removed. This way, incremental training for low change rates and also online training with potentially high rates of incoming training data are enabled. Nearest neighbor classification, thus, always produces a result that relies on the current training data status.

Decision explanation. Lazy classifiers do not create models for the training data and, as an inherent limitation, no explicit knowledge is extracted which reflects the structure of the data. Though nearest neighbor classifiers do not provide intuitive models of the data characteristics, they nevertheless explain their individual decisions illustratively by revealing the closest neighbors of the query object within the labeled training data to the user. Additional information is provided by the similarity scores, i.e., the distances of the training items to the query object, thus yielding some insight into the decision process from which the users may derive their confidence in the final decision result.

Key Applications

Key applications of nearest neighbor classification include all areas where classification problems occur. Particularly well suited are nearest neighbor classifiers in the following cases:

1. Many classes need to be distinguished
2. Objects are represented by general metric data without an attribute structure
3. Classifier has to follow training data that change dynamically
4. Users require intuitive explanations of the system's decisions

Real applications are found in all areas of multimedia data exploration and cognitive systems including computer vision, speech recognition, medical imaging, robot motion planning, or sensor-based object recognition in general to name just a few.

Cross-References

Recommended Reading

1. Ankerst M, Kastenmuller G, Kriegel H-P, Seidl T. Nearest neighbor classification in 3D protein databases. In: Proceedings of the 7th International Conference on Intelligent Systems for Molecular Biology; 1999. p. 34–43.
2. Athistos V. Nearest neighbor retrieval and classification. 2007. Available at: http://cs-people.bu.edu/athitsos/nearest-neighbors/
3. Djouadi A, Bouktache E. A fast algorithm for the nearest-neighbor classifier. IEEE Trans Pattern Anal Mach Intell. 1997;19(3):277–82.
4. Duda RO, Hart PE, Stork DG. Pattern classification. 2nd ed. New York: Wiley; 2001.
5. Efros AA, Berg AC, Mori G, Malik J. Recognizing action at a distance. In: Proceedings of the 9th IEEE Conference on Computer Vision; 2003. p. 726–33.
6. Ghosh AK, Chaudhuri P, Murthy CA. On visualization and aggregation of nearest neighbor classifiers. IEEE Trans Pattern Anal Mach Intell. 2005;27(10):1592–602.
7. Han J, Kamber M. Data mining, concepts and techniques. 2nd ed. Amsterdam: Elsevier; 2006.
8. Hastie T, Tibshirami R, Friedman J. The elements of statistical learning: data mining. Inference and prediction. Springer series in statistics. New York: Springer; 2001.
9. Kriegel H-P, Pryakhin A, Schubert M. Multi-represented kNN-classification for large class sets. In: Proceedings of the 10th International Conference on Database Systems for Advanced Applications; 2005. p. 511–22.
10. Li Y, Yang J, Han J. 1Continuous K-nearest neighbor search for moving objects. In: Proceedings of the 16th International Conference on Scientific and Statistical Database Management; 2004. p. 123–6.
11. Shibata T, Kato T, Wada TK-D. Decision tree: an accelerated and memory efficient nearest neighbor classifier. In: Proceedings of the 2003 IEEE International Conference on Data Mining; 2003. p. 641–4.
12. Veenman CJ, Reinders MJT. The nearest subclass classifier: a compromise between the nearest mean and nearest neighbor classifier. IEEE Trans Pattern Anal Mach Intell. 2005;27(9):1417–29.

Nearest Neighbor Query

Dimitris Papadias
Department of Computer Science and
Engineering, Hong Kong University of Science
and Technology, Kowloon, Hong Kong, Hong
Kong

Synonyms

kNN query; NN query

Definition

Given a dataset P and a query point q, a k *nearest neighbor* (kNN) query returns the k closest data points $p_1, p_2, \ldots, p_k$ to q. For any point $p \in P - \{p_1, p_2, \ldots, p_k\}$ $dist(p_i, q) \leq dist(p, q) \ \forall \ 1 \leq i \leq k$, where $dist$ depends on the underlying distance metric. The usual metric is the L_x norm, or formally, given two points q, p whose coordinates on the i-th dimension ($1 \leq i \leq m$) are q_i and p_i respectively, their L_x distance is:

$$L_x(p, q) = \left(\sum_{i=1 \sim m} |p_i - q_i|^x \right)^{1/x}, \text{ for } x \neq \infty,$$

$$L_x(p, q) = \max_{i=1 \sim m} (|p_i - q_i|), \text{ for } x = \infty.$$

Most work on spatial and spatio-temporal databases focuses on Euclidean distance (i.e., L_2), but alternative definitions have been used in various domains (e.g., road networks, time series).

Key Points

Nearest neighbor search constitutes an important component for several tasks such as clustering, outlier detection, time series analysis, and image and document retrieval. For instance, finding the most similar series, image, or document to a given input is a NN query in a corresponding data space defined by the features of interest. Unlike spatial and spatio-temporal databases that usually involve two to three dimensions, these applications may lead to high-dimensional spaces. In order to avoid the high cost of distance computations in such cases, several methods follow a multi-step framework for processing NN queries [3]: (i) a dimensionality reduction technique is used to decrease the number of dimensions, (ii) the low-dimensional data are indexed, (iii) the index is used to efficiently retrieve a candidate NN (in low-dimensional space), (iv) the actual distance of the candidate (in the original space) is computed, and (v) steps (iii) and (iv) are repeated until no other candidate can lead to a better solution than the one already discovered.

In addition to conventional NN search, several alternative types of nearest neighbor queries have been proposed in the database literature. Given a multi-dimensional dataset P and a point q, a *reverse nearest neighbor* query retrieves all the points $p \in P$ that have q as their NN (The nearest neighbor relationship is not symmetric; i.e., the fact that q is the NN of p does not necessarily imply that p is the NN of q.) [1]. Given a set P of data points and a set Q of query points, an *aggregate nearest neighbor* query returns the data point p with the minimum *aggregate distance* [2]. The *aggregate distance* between a data point p and $Q = \{q_1, \ldots, q_n\}$ is defined as $f(dist(p,q_1), \ldots, dist(p,q_n))$. If, for instance, $f = sum$, the corresponding query reports the data point that minimizes the total distance from all query points.

Cross-References

▶ Nearest Neighbor Query in Spatiotemporal Databases
▶ Reverse Nearest Neighbor Query

Recommended Reading

1. Korn F, Muthukrishnan S. Influence sets based on reverse nearest neighbor queries. In: Proceeding of the ACM SIGMOD International Conference on Management of Data; 2000. p. 201–12.
2. Papadias D, Tao Y, Mouratidis K, Hui K. Aggregate nearest neighbor queries in spatial databases. ACM Trans Database Syst. 2005;30(2):529–76.
3. Seidl T, Kriegel H-P. Optimal multi-step k-nearest neighbor search. In: Proceeding of the ACM SIGMOD International Conference on Management of Data; 1998. p. 154–65.

Nearest Neighbor Query in Spatiotemporal Databases

Dimitris Papadias
Department of Computer Science and Engineering, Hong Kong University of Science and Technology, Kowloon, Hong Kong, Hong Kong

Synonyms

NN query; NN search

Definition

Given a set of points P in a multidimensional space, the nearest neighbor (NN) of a query point q is the point in P that is closest to q. Similarly, the k nearest neighbor (kNN) set of q consists of the k points in P with the smallest distances from q. In spatial and spatiotemporal databases, the distance is usually defined according to the Euclidean metric, and the dataset P is disk resident. Query algorithms aim at minimizing the processing cost. Other optimization criteria in the case of moving objects (or queries) include the network latency or the number of queries required for keeping the results up-to-date.

Historical Background

Nearest neighbor (NN) search is one of the oldest problems in computer science. Several algorithms and theoretical performance bounds have been devised for exact and approximate processing in main memory [1]. In spatial databases, existing algorithms assume that P is indexed by a spatial access method (usually an *R-tree* [2]) and utilize some pruning bounds to restrict the search space. Figure 1 shows an *R-tree* for point set $P = \{p_1, p_2, \ldots, p_{12}\}$ with a capacity of three entries per node (typically, the capacity is in the order of hundreds). Points that are close in space (e.g., p_1, p_2, p_3) are clustered in the same leaf node (N_3). Nodes are then recursively grouped together with the same principle until the top level, which consists of a single root. Given a node N and a query point q, the *mindist(N,q)* corresponds to the closest possible distance between q and any point in the subtree of node N.

The first NN algorithm for *R-trees* [8] searches the tree in a *depth-first* (DF) manner, recursively visiting the node with the minimum *mindist* from q, e.g., in Fig. 1, DF accesses the root, followed by N_1 and N_4, where the first potential nearest neighbor is found (p_5). Note that p_5 is not the actual NN (it is p_{11}), and the search continues. During backtracking to the upper level (node N_1), the algorithm prunes entries whose *mindist* is equal to or larger than the distance (*best_dist*) of the nearest neighbor already retrieved. In the example of Fig. 1, after discovering p_5, (i) *best_dist* is set to $dist(p_5, q)$, (ii) DF backtracks to the root level (without visiting N_3), and (iii) it follows the

path N_2, N_6 where the actual NN p_{11} is found. DF can be easily extended for the retrieval of $k > 1$ nearest neighbors: the k points discovered so far with the minimum overall distances are maintained in an ordered list of k pairs $< p, dist(p, q) >$ (sorted on ascending $dist(p, q)$ order) and *best_dist* equals the distance of the k-th NN. Whenever a better neighbor is found, it is inserted in the list, the last element is removed, and the value of *best_dist* is updated.

The DF algorithm is suboptimal, i.e., it accesses more nodes than necessary. Specifically, an optimal algorithm should visit only nodes intersecting the *search region*, i.e., a circle centered at the query point q with radius equal to the distance between q and its nearest neighbor [4]. In Fig. 1, for instance, an optimal algorithm should visit only the root, N_1, N_2, and N_6, whereas DF also visits N_4. The *best-first* (BF) algorithm of [5] achieves the optimal I/O performance by maintaining a heap H with the entries visited so far, sorted by their *mindist*. As with DF, BF starts from the root and inserts all the entries into H (together with their *mindist*), e.g., in Fig. 1, $H = \{<N_1, mindist(N_1, q)>, <N_2, mindist(N_2, q)>\}$. Then, at each step, BF visits the node in H with the smallest *mindist*. Continuing the example, the algorithm retrieves the content of N_1 and inserts all its entries in H, after which $H = \{<N_2, mindist(N_2, q)>, <N_4, mindist(N_4, q)>,$

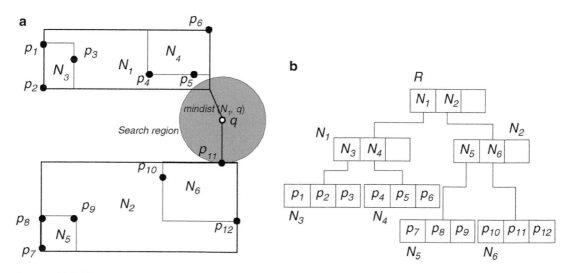

Nearest Neighbor Query in Spatiotemporal Databases, Fig. 1 Example of an R-tree and a point NN query

$<N_3, mindist(N_3, q)>\}$. The next two nodes accessed are N_2 and N_6 (inserted in H after visiting N_2), in which p_{11} is discovered as the current NN. At this time, the algorithm terminates (with p_{11} as the final result) since the next entry (N_4) in H is farther (from q) than p_{11}. Similarly to DF, BF can be easily extended to kNN queries ($k > 1$). In addition, BF is *incremental*, i.e., it can output the nearest neighbors in ascending order of their distance to the query without a predefined termination condition.

Foundations

In dynamic environments, the result of a conventional NN query may be immediately invalidated due to the query or data movement. Several techniques augment the result with some additional information regarding its validity. For instance, Zheng and Lee [14] assume an architecture where moving clients (e.g., mobile devices) send their queries to a server. The server precomputes and stores in an *R-tree* the *Voronoi diagram* of the (static) dataset. When an NN query q arrives at the server, the *Voronoi diagram* is used to efficiently compute its nearest neighbor, i.e., the point (o in Fig. 2), whose Voronoi cell covers q. In addition to the result, the server sends back to the client its *validity time T*, which is a conservative approximation assuming that the query speed is below a maximum value. In particular, T is the time that q will cross the closest boundary of the Voronoi cell of object o (in which case point a will become the nearest neighbor). Zhang et al. [13] propose the concept of *location-based* queries that return the *validity region* around the query point, where the result remains the same. For instance, in Fig. 2, a location-based query q will return object o and its Voronoi cell, which is computed on the fly using an *R-tree* on the data points.

For the same settings (moving query – static data objects), Song and Roussopoulos [10] reduce the number of queries required to keep the result up-to-date by introducing redundancy. In particular, when a kNN query arrives, the server returns to the client a number $m > k$ of neighbors. Let $dist(k)$ and $dist(m)$ be the distances of the kth and mth nearest neighbor from the query point q. If the client reissues the query at a new location q', it can be proven that the new k nearest neighbors will be among the m objects of the first query, provided that $2 \cdot dist(q', q) \leq dist(m) - dist(k)$. Figure 3 shows an example for a 2NN query at location q, where the server returns four results o, a, b, and c (the two nearest neighbors are o and a). When the client moves to a nearby location q', the 2 NN are o and b. If $2 \cdot dist(q,$

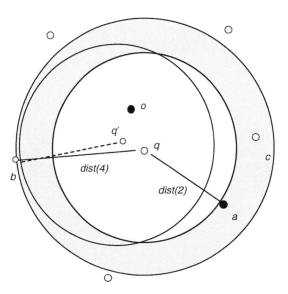

Nearest Neighbor Query in Spatiotemporal Databases, Fig. 2 Example of [6]

Nearest Neighbor Query in Spatiotemporal Databases, Fig. 3 Example of [8]

q') $\leq$ *dist(4)-dist(2)*, the client can determine this by computing the new distances (with respect to q') of the four objects, without having to issue a new query to the server.

Tao and Papadias [11] propose *time-parameterized* (TP) queries, assuming that the clients move with linear and known velocities. In addition to the current result R, the output of a TP query contains its *validity period T* and the *next change C* of the result (that will occur at the end of the validity period). Given the additional information (C and T), the client only needs to issue another TP query after the expiry of the current result. Figure 4 shows a TP NN, where the query point q is moving east with speed 1. Point d is the current nearest neighbor of q. The *influence time* $T_{INF}(p, q)$ of an object p is the time that p starts to get closer to q than the current nearest neighbor. For example, $T_{INF}(g, q) = 3$, because at this time g will come closer to q than d. The expiry time of the current result is the minimum $T_{INF}(o, q)$ of all objects, i.e., in Fig. 4, $T = 1.5$, at which point f will replace d as the NN of q. Based on the above observations, a TP NN query is processed in two steps: (i) a conventional algorithm (e.g., [5, 8]) retrieves the NN at the current time and (ii) a second pass computes the time-parameterized component (i.e., C and T) by applying again NN search and treating $T_{INF}(o, q)$ as the distance metric; the goal is to find the objects (C) with the minimum T_{INF}.

A *continuous* nearest neighbor (CNN) query [3, 11] retrieves the nearest neighbor (NN) of every point in a line segment $q = [s, e]$. In particular, the result contains a set of $<R, T>$ tuples, where R (for result) is a point of P, and T is the interval during which R is the NN of q. As an example consider Fig. 5, where $P = \{a, b, c, d, f, g, h\}$. The output of the query is $\{<a, [s, s_1]>, <c, [s_1, s_2]>, <f, [s_2, s_3]>, <h, [s_3, e]>\}$, meaning that point a is the NN for interval $[s, s_1]$, then at s_1, point c becomes the NN, etc. The points of the query segment (i.e., s_1, s_2, s_3) where there is a change of neighborhood are called *split points*. CNN algorithms use DF or BF traversal on R-trees to visit nodes and data points according to their proximity to the query segment. Visited data points introduce new split points, which are used to prune the search space. For instance, in Fig. 5, if a, c, f, and h have already been discovered, every node and data point (b, d, g) outside a circle defined by a split point (center) and its NN (radius) can be eliminated since they cannot affect the result.

Nearest Neighbor Query in Spatiotemporal Databases, Fig. 4 Example of TP NN query

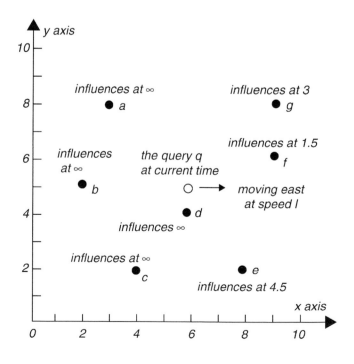

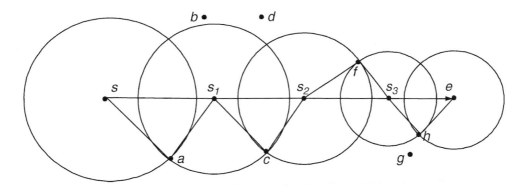

Nearest Neighbor Query in Spatiotemporal Databases, Fig. 5 Example CNN query

A *predictive* NN query retrieves the expected nearest neighbor of a (static or moving) query point (e.g., 30 s from now) based on the current motion patterns. Assuming that data points move linearly, they can be indexed by a TPR-tree [9]. The TPR-tree is similar to the *R-tree*, but takes into account both location and velocity in order to group data objects into nodes. Furthermore, each node is assigned a velocity vector so that its extent continuously encloses all the objects inside (i.e., a moving node that may grow with time). A predictive query is answered in the same way as in the *R-tree* (e.g., by adaptations of [5, 8]), except that the node extents at the (future) query time are dynamically computed (using the current location and velocity vector of the node). The concepts of TP and continuous queries also apply in this case. Tao et al. [12] propose an architecture and index for processing queries on objects moving with arbitrary motion patterns, unknown in advance.

Key Applications

Geographic Information Systems
Nearest neighbor search is one of the most common query types. Efficient algorithms are important for dealing with the large and ever-increasing amount of spatial data in several GIS applications.

Location-Based Services
Advanced NN algorithms will provide the means for enhanced location-based services based on

the proximity of mobile clients to potential facilities of interest.

Multi-criteria Decision Making
A number of decision support tasks can be modeled as nearest neighbor search. For instance, news WWW sites usually recommend to users the articles that are most similar to their previous choices. Nearest neighbor algorithms have also been used to process skyline queries.

Future Directions

All the above techniques take as input a single query and report its nearest neighbor set at the current time, possibly with some validity information (e.g., expiration time, Voronoi cell) or generate future results based on predictive features (e.g., velocity vectors of queries or data objects). On the other hand, *continuous monitoring of spatial queries* [6] (i) involves multiple, long-running queries (from geographically distributed clients), (ii) is concerned with both computing *and* keeping the results up-to-date, (iii) usually assumes main-memory processing to provide fast answers in an online fashion, and (iv) attempts to minimize factors such as the CPU or communication cost (as opposed to I/O overhead).

Another interesting problem concerns NN search in non-Euclidean spaces. For instance, in road networks, the distance between two points can be defined as the length of the shortest path connecting them or by the minimum time it takes to travel between them. In either case, the problem requires different algorithmic solutions [7].

Experimental Results

In general, for every presented method, there is an accompanying experimental evaluation in the corresponding reference. [5] compares BF and DF traversal for conventional NN search. [9] proposes cost models for TP NN and CNN queries and evaluates their accuracy, as well as the relative performance of the two query types for static (indexed by *R-trees*) and dynamic (indexed by *TPR-trees*) objects.

Datasets

A large collection of real datasets, commonly used for experiments, can be found at: http://www.rtreeportal.org/.

URL to Code

R-tree portal (see above) contains the code for most common spatial and spatiotemporal indexes, as well as data generators and several useful links for researchers and practitioners in spatiotemporal databases.

Cross-References

▶ Continuous Monitoring of Spatial Queries
▶ Nearest Neighbor Query
▶ R-Tree (and Family)
▶ Voronoi Diagrams

Recommended Reading

1. Arya S, Mount D, Netanyahu N, Silverman R, Wu A. An optimal algorithm for approximate nearest neighbor searching. J ACM. 1998;45(6):891–923.
2. Beckmann N, Kriegel H, Schneider R, Seeger B. The R*-tree: an efficient and robust access method for points and rectangles. In: Proceedings of the ACM SIGMOD International Conference on Management of Data; 1990. p. 322–31.
3. Benetis R, Jensen C, Karciauskas G, Saltenis S. Nearest neighbor and reverse nearest neighbor queries for moving objects. VLDB J. 2006;15(3):229–49.
4. Bohm C. A cost model for query processing in high dimensional data spaces. ACM Trans Database Syst. 2000;25(2):129–78.
5. Hjaltason G, Samet H. Distance browsing in spatial databases. ACM Trans Database Syst. 1999;24(2):265–318.
6. Mouratidis K, Hadjieleftheriou M, Papadias D. Conceptual partitioning: an efficient method for continuous nearest neighbor monitoring. In: Proceedings of the ACM SIGMOD International Conference on Management of Data; 2005. p. 634–45.
7. Papadias D, Zhang J, Mamoulis N, Tao Y. Query processing in spatial network databases. In: Proceedings of the 29th Conference on Very Large Data Bases; 2003. p. 790–801.
8. Roussopoulos N, Kelly S, Vincent F. Nearest neighbor queries. In: Proceedings of the ACM SIGMOD International Conference on Management of Data; 1995. p. 71–9.
9. Saltenis S, Jensen C, Leutenegger S, Lopez M. Indexing the positions of continuously moving objects. In: Proceedings of the ACM SIGMOD International Conference on Management of Data; 2000. p. 331–42.
10. Song Z, Roussopoulos N. K-nearest neighbor search for moving query point. In: Proceedings of the 7th International Symposium. Advances in Spatial and Temporal Databases; 2001. p. 79–96.
11. Tao Y, Papadias D. Spatial queries in dynamic environments. ACM Trans Database Syst. 2003;28(2):101–39.
12. Tao Y, Faloutsos C, Papadias D, Liu B. Prediction and indexing of moving objects with unknown motion patterns. In: Proceedings of the ACM SIGMOD International Conference on Management of Data; 2004. p. 611–22.
13. Zhang J, Zhu M, Papadias D, Tao Y, Lee D. Location-based spatial queries. In: Proceedings of the ACM SIGMOD International Conference on Management of Data; 2003. p. 443–54.
14. Zheng B, Lee D. Semantic caching in location-dependent query processing. In: Proceedings of the 7th International Symposium. Advances in Spatial and Temporal Databases; 2001. p. 97–116.

Nested Loop Join

Jingren Zhou
Alibaba Group, Hangzhou, China

Synonyms

Loop join; Nested loop join

Definition

The nested loop join is a common join algorithm in database systems using two nested loops. The algorithm starts with reading the *outer* relation R, and for each tuple $\mathcal{R} \in R$, the *inner* relation S is checked and matching tuples are added to the result.

```
Algorithm1: NestedLoopJoin: R⋈pred(r, s) S
foreach R ∈ R do
foreach S ∈ S do
if pred (R.r, S.s) then
add {R, S} to result
end
end
end
```

Key Points

One advantage of the nested loop join is that it can handle any kind of join predicates, unlike the sort-merge join and the hash join which mainly deal with an equality join predicate. An improvement over the simple nested loop join is the block nested loop join which effectively utilizes buffer pages and reduces disk I/Os.

Block Nested Loop Join

Suppose that the memory can hold B buffer pages. If there is enough memory to hold the smaller relation, say R, with at least two extra buffer pages left, the optimal approach is to read in the smaller relation R and to use one extra page as an input buffer to read in the larger relation S and the other extra buffer page as an output buffer.

If there is not enough memory to hold the smaller relation, the best approach is to break the outer relation R into *blocks* of $B - $ two pages each and scan the whole inner relation S for each block of R. As described before, one extra page is used as an input buffer and the other as an output buffer. In this case, the outer relation R is scanned only once while the inner relation S is scanned multiple times.

Cross-References

▸ Evaluation of Relational Operators

Recommended Reading

1. Mishra P, Eich MH. Join processing in relational databases. ACM Comput Surv. 1992;24(1):63–113.

Nested Transaction Models

George Karabatis
University of Maryland, Baltimore Country (UMBC), Baltimore, MD, USA

Definition

A *nested transaction model* as proposed by Moss is a generalization of the flat transaction model that allows nesting. A nested transaction forms a tree of transactions with the root being called a *top-level transaction* and all other nodes called *nested transactions* (*subtransactions*). Transactions having no subtransactions are called *leaf* transactions. Transactions with subtransactions are called *parents (ancestors)* and their subtransactions are called *children (descendants)*.

A subtransaction can commit or rollback by itself. However, the effects of the commit cannot take place unless the parent transaction also commits. Therefore, in order for any subtransaction to commit, the top-level transaction must commit. If a subtransaction aborts, all its children subtransactions (forming a subtree) are forced to abort even if they committed locally.

Historical Background

Nested transactions were introduced by Moss in 1981 [1] to overcome some of the limitations of the flat transaction model, as they allow a finer level of control in a transaction. For example, a failed operation in a flat transaction causes the entire transaction to be rolled back. On the contrary, a failed operation in a part of a nested transaction may be acceptable and the entire transaction may be allowed to commit, even if some of its operations failed. As an example, a trip consisting of a flight reservation, hotel accommodation and car rental can be implemented as a nested transaction with three subtransactions. If some of its subtransactions commit (e.g., flight and hotel), and some fail (e.g., car rental) the outcome may still be acceptable and allow the nested transaction to commit with partial results.

Nested transactions are based on the underlying concept of *spheres of control* developed by Bjork and Davies in 1973 [2, 3]. Davies describes the spheres of control in more detail in [4]; however, he portrays overall semantics of the spheres which convey a more abstract concept than nested transactions. Moss provides implementation details of nested transactions for a distributed environment. Prior to that, Reed in his dissertation [5] also describes an implementation of nested transactions in a distributed environment and one can see similarities and differences between the two approaches. For example, Moss uses locking whereas Reed uses timestamps. For a detailed comparison of the two approaches see [1].

Foundations

Structure of a Nested Transaction
The structure of a nested transaction is depicted by a transaction tree: its root represents the *top-level* transaction and its children represent *subtransactions* each one corresponding to a transactional unit. A subtransaction may be either a simple (*leaf*) transaction or a nested transaction, recursively expanding the structure to a hierarchy with multiple levels of transactions.

Leaf transactions are very similar to traditional ACID transactions, but they do not preserve the durability property of ACID transactions as will be explained in the Commit/Abort rules. Leaf transactions are the ones that actually manipulate the data in the database and perform the work. Intermediate level subtransactions and the top-level transaction operate on a higher level of abstraction: their purpose is to control when to create a new subtransaction. This hierarchy corresponds to the notion of nested spheres of control [3]. Figure 1 illustrates an example of a nested transaction and its corresponding transaction tree. It consists of a top-level transaction T_1 with two subtransactions T_2 and T_3; subtransaction T_2 is itself a nested transaction invoking subtransaction T_4. In this example, subtransactions T_3 and T_4 are leaf transactions performing the actual work. When multiple such nested transactions coexist their trees form a forest.

Synchronization of Nested Transactions
Subtransactions of nested transactions can execute concurrently and for proper synchronization Moss devised the following locking protocol: A subtransaction can either *hold* a lock or *retain* a lock. When a lock is held the transaction has exclusive rights to the object. After a subtransaction commits, its held or retained locks are *inherited* by the parent subtransaction. These inherited locks are not available to other subtransactions outside the subtree (sphere of control) of the holder; however, subtransactions within the subtree (descendants) can acquire a retained lock. The Locking Rules of nested transactions slightly paraphrased from Moss [1] are the following:

1. A transaction may hold a lock in write mode if no other transaction holds the lock (in any mode) and all retainers of the lock are ancestors of the requester.

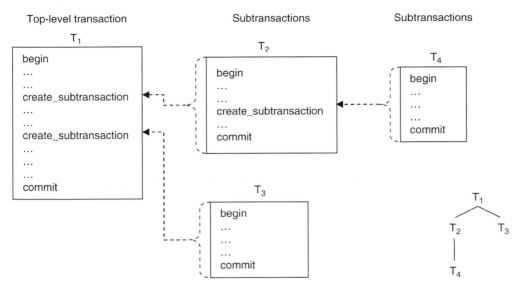

Nested Transaction Models, Fig. 1 Structure of an example nested transaction; to the *lower right* is a depiction of the corresponding nested transaction tree

2. A transaction may hold a lock in read mode if no other transaction holds the lock in write mode and all retainers of write locks are ancestors of the requester.
3. When a transaction commits, its parent inherits all locks the descendants held or retained in the same mode as the child held or retained them.
4. When a transaction aborts, all locks it holds or retains are released. Ancestors of the aborted transaction who were retaining the lock they continue to do so in the same mode as before the abort.

Commit/Abort of Nested Transactions

Moss describes a set of rules (state restoration algorithm in [1]) that dictate the behavior of nested transactions at commit/abort time. A subtransaction can unilaterally commit or abort independently of the others preserving atomicity, consistency and isolation from the ACID properties. The rules reworded from [1] are:

1. All children of a subtransaction must be resolved (committed or aborted) before it commits itself.
2. A subtransaction that unilaterally commits, first reaches its local commit point. However,

it cannot finally commit until its parent transaction commits. Consequently, a subtransaction will commit only if it commits locally and all its ancestors commit including the top-level transaction.
3. If one of the ancestors of a subtransaction aborts, then the subtransaction itself will have to abort. In other words, the abort of a subtransaction causes the abort of all its children. Even if the subtransaction commits, its actions must be undone due to the abort of an ancestor. This is the reason that nested subtransactions violate the durability of ACID properties.

A nested transaction at the top level (i.e., the entire nested transaction as a whole) does observe durability. Therefore, nested transactions as a whole obey the ACID properties, despite the fact that individual nested subtransactions may violate the durability property.

Advantages of Nested Transactions

1. Nested transactions allow for a simple composition of subtransactions improving modularity of the overall structure.
2. The concurrent execution of subtransactions that follow the prescribed rules allows for enhanced concurrency while preserving consis-

N

tency. This enhanced concurrency occurs because transactions can create subtransactions which execute concurrently.

3. The success or failure of a subtransaction is independent of the success of its siblings. When a subtransaction fails, its parent may try to execute it again, or ignore the failure, depending on the situation. Therefore, failures that are contained within the scope of a subtransaction may be tolerated and do not necessarily cause the failure of the entire nested transaction.

Gray and Reuter present a mechanism to emulate nested transactions using database savepoints [6]. This approach emulates recovery as accurately as prescribed in the nested transaction model; however, it emulates locking not as flexibly as described in nested transactions.

Key Applications

Several systems were influenced by the nested transactions model, and as such they either incorporated or adapted its concept. The implementation of nested transactions by Moss, and especially the locking mechanism was used in the Argus system in which Moss worked with Liskov [7]. Camelot is another system that used nested transactions. It is a transaction processing system developed at Carnegie Mellon University using the Avalon programming language [8]. Camelot was a precursor to the Encina distributed transaction processing monitor which uses the Transactional-C language. Encina was commercialized under the name Transarc Corporation, which was acquired by IBM in 1994.

Nested transactions have also been implemented in Berkeley DB – a database library with bindings for several programming languages, which originated at the University of California – Berkeley, distributed by Sleepycat Software, which was acquired by Oracle Corporation in 2006.

Several researchers have also proposed systems with transaction processing components that are designed for application domains such as manufacturing, CSCW, software engineering,

etc. where the ACID properties may be considered too strict; therefore, relaxed or advanced transaction models have been proposed to address the specific requirements that exist in these domains. Several of these transaction models appear in a book by Elmagarmid [9].

A notable example of an extension to the concept of nested transactions is the *multi-level transaction model* and its closely related *open nested transaction model* by Weikum and Schek [10, 11]. The multi-level transaction model is represented by a transaction tree like in nested transactions; however, unlike nested transactions the transaction tree is balanced. Multi-level concurrency control takes advantage of the semantics of conflicting operations in the same level of the tree, such as commutativity, in order to enhance concurrency. In open nested transactions, the subtransactions are not restricted to have the same depth and are allowed to commit their results without waiting for their ancestors to commit [12].

Cross-References

▶ ACID Properties
▶ Concurrency Control: Traditional Approaches
▶ Data Conflicts
▶ Database Management System
▶ Distributed DBMS
▶ Distributed Transaction Management
▶ Extended Transaction Models and the ACTA Framework
▶ Multilevel Transactions and Object-Model Transactions
▶ Open Nested Transaction Models
▶ Serializability
▶ Transaction
▶ Transaction Management
▶ Transaction Manager

Recommended Reading

1. Moss EB. Nested transactions: an approach to reliable distributed computing. Technical Report. PhD Thesis. UMI Order Number: TR-260: Massachusetts Institute of Technology; 1981. p. 178.

2. Bjork LA. Recovery scenario for a DB/DC system. In: Proceedings of the ACM Annual Conference; 1973. p. 142–6.
3. Davies CT. Recovery semantics for a DB/DC system. In: Proceedings of the ACM Annual Conference; 1973. p. 136–41.
4. Davies CT. Data processing spheres of control. IBM Syst J. 1978;17(2):179–98.
5. Reed DP. Naming and synchronization in a distributed computer system. Technical Report. PhD Thesis, UMI Order Number: TR-205: Massachusetts Institute of Technology; 1978. p. 181.
6. Gray J, Reuter A. Transaction processing: concepts and techniques. 1st ed. San Francisco: Morgan Kaufmann Publishers; 1992.
7. Liskov B. Distributed programming in Argus. Commun ACM. 1988;31(3):300–12.
8. Eppinger JL, Mummert LB, Spector AZ. Camelot and Avalon: a distributed transaction facility. San Mateo: Morgan Kaufmann Publishers; 1991.
9. Elmagarmid AK. Database transaction models for advanced applications. San Mateo: Morgan Kaufmann Publishers; 1992.
10. Weikum G. Principles and realization strategies of multilevel transaction management. ACM Trans Database Syst. 1991;16(1):132–80.
11. Weikum G, Schek H-J. Multi-level transactions and open nested transactions. Q Bull IEEE TC Data Eng. 1991;14(1):60–6.
12. Weikum G, Schek HJ. Concepts and applications of multilevel transactions and open nested transactions. In: Database transaction models for advanced applications. San Mateo: Morgan Kaufmann Publishers; 1992. p. 515–53.

Network Attached Secure Device

Kazuo Goda
The University of Tokyo, Tokyo, Japan

Synonyms

NASD; Object-based storage device; OSD

Definition

A Network Attached Secure Device (often abbreviated to NASD) is a networked storage device which offers object-level storage accesses. NASD was first explored in a research project of the same name, which started at Carnegie Mellon University in 1995. NASD is now more often called Object based Storage Device (shortened to OSD).

Key Points

NASD is an approach of decoupling only low-level functions of file systems from NAS devices. A NASD manages only stored objects, each of which is identified with a unique file descriptor, and their additional information such as metadata and attributes, whereas a central policy server manages higher-level information which is used for name space management and authorization. A bunch of NASDs, with the assistance of the central policy server, can together work as a file store. This may look similar to NAS at a glance, but the difference is that most commands and data are directly transferred between clients and NASDs. Conventional storage subsystems have a number of storage devices under a storage controller, which is thus likely to become a performance bottleneck. NASD has benefits of scaling aggregate bandwidth by spreading partial functions over a number of disk processors.

Object-level storage abstraction, introduced by NASD, is recognized to be the third alternative in addition to block-level storage abstraction typically seen in SAN storage environments and file-level storage abstraction in NAS storage environments. Content-Addressable Storage is another example which offers object-level storage accesses and is now widely deployed in enterprise systems.

The idea of NASD has been standardized as ANSI T10 SCSI OSD, which specifies an extension of the SCSI protocol for clients and devices to exchange objects and their related information. Thus, OSD is sometimes seen as a promising infrastructure of intelligent storage devices in which more intelligence will be incorporated.

N

Cross-References

▶ Intelligent Storage Systems

Recommended Reading

1. ANSI. Information Technology – SCSI Object-Based Storage Device Commands (OSD). Standard ANSI/INCITS 400-2004. 2004.
2. Gibson GA, Van Meter R. Network attached storage architecture. Commun ACM. 2000;43(11):37–45.
3. Gibson GA, Nagle DF, Amiri K, Chang FW, Feinberg EM, Gobioff H, Lee C, Ozceri B, Riedel E, Rochberg D, Zelenka J. File server scaling with network-attached secure disks. In: Proceedings of the 1997 ACM SIGMETRICS International Conference on Measurement and Modeling of Computer Systems; 1997. p. 272–84.

Network Attached Storage

Kazuo Goda
The University of Tokyo, Tokyo, Japan

Synonyms

NAS

Definition

Network attached storage is a storage device which is connected to a network and provides file access services. Network attached storage is often abbreviated to NAS. Although the term NAS refers originally to such a storage device, when the term is used in the context of explaining and comparing storage network architectures it sometimes refers to a storage network architecture in which NAS devices are mainly implemented.

Key Points

A NAS device is basically comprised of disk drives which store files and controllers which export access services to the files. A file sever which runs network file system (NFS) and/or common internet file system (CIFS) to export file sharing services is a type of NAS implementation, but recent NAS products are sometimes implemented using dedicated hardware and software to increase reliability and performance. A diskless NAS device which contains only controllers is sometimes referred to as a NAS gateway or a NAS head. A NAS gateway/head has two types of network ports: one is connected to disk storage devices over a SAN and the other is to NAS clients over IP networks. The clients are thus provided with access services towards files stored in the storage devices. That is, a NAS gateway/head can be seen as a service bridge between SAN and NAS systems.

Cross-References

▶ Direct Attached Storage
▶ Storage Area Network
▶ Storage Network Architectures

Recommended Reading

1. Storage Network Industry Association. The dictionary of storage networking terminology. Available at: http://www.snia.org/.
2. Troppens U, Erkens R, Müller W. Storage networks explained. New York: Wiley; 2004.

Network Data Model

Jean-Luc Hainaut
University of Namur, Namur, Belgium

Synonyms

CODASYL data model; DBTG data model

Definition

A database management system complies with the *network data model* when the data it manages are organized as data records connected through binary relationships. Data processing is based on navigational primitives according to which records are accessed and updated one at a time, as opposed to the set orientation of the relational query languages. Its most popular variant is the CODASYL DBTG data model that was first defined in the 1971 report from the CODASYL group, and that has been implemented into several major DBMSs. They were widely used in the seventies and eighties, but most of them are still active at the present time.

Historical Background

In 1962, C. Bachman of General Electric, New-York, started the development of a data management system according to which data records were interconnected via a network of relationships that could be navigated through [2]. Called Integrated Data Store (IDS), this disk-based system quickly became popular to support the storage, the management and the exploitation of corporate data.

IDS was the main basis of the work of the CODASYL Data Base Task Group (DBTG) that published its first major report in 1971 [3, 6], followed by a revision in 1973 [3, 9]. This report described a general architecture for DBMSs, where the respective roles of the operating system, the DBMS and application programs were clearly identified. It also provided a precise specification of languages for data structure definition (Data Description Language or Schema DDL), for data extraction and update (Data Manipulation Language or DML) and for defining interfaces for application programs through language-dependent views of data (Sub-schema DDL).

The 1978 report [7, 9] clarified the model. In particular physical specifications such as indexing structures and storage were removed from the DDL and collected into the Data Storage Description Language (DSDL), devoted to the physical schema description. In 1985, the X3H2 ANSI Database Standard committee issued standards for network database management systems, called NDL and based on the 1978 CODASYL report. However, due to the increasing dominance of the relational model these proposals have never been implemented nor updated afterwards.

Some of the most important implementations were Bull IDS/II (an upgrade of IDS), NCR DBS, Siemens UDS-1 and UDS-2, Digital DBMS-11, DBMS-10 and DBMS-20 (now distributed by Oracle Corp.), Data General DG/DBMS, Philips Phollas, Prime DBMS, Univac DMS 90 and DMS 1100 and Cullinane IDMS, a machine-independent rewriting of IDS (now distributed by Computer Associates). Other DBMSs have been developed, that follow more or less strictly the CODADYL specifications. Examples include Norsk-Data SYBAS, Burroughs DMS-2, CDC IMF, NCR IDM-9000, Cincom TOTAL and its clone HP IMAGE (which were said to define the *shallow* data model), and MDBS and Raima DbVista that both first appeared on MS-DOS PCs.

In the seventies, IBM IMS was the main competitor of CODASYL systems [11]. From the early eighties, they both had to face the increasing influence of relational DBMSs such as Oracle (from 1979) and IBM SQL/DS (from 1982 [8]). Nowadays, most CODASYL DBMSs provide an SQL interface, sometimes through an ODBC API. Though the use of CODASYL DBMSs is slowly decreasing, many large corporate databases are still managed by network DBMSs. This state of affairs will most probably last for the next decade. Network databases, as well as hierarchical databases, are most often qualified *legacy*, inasmuch as they are often expected to be replaced, sooner or later, by modern database engines.

Foundations

The presentation of the network model is based on the specifications published in the 1971 and 1973 reports, with which most CODASYL DBMSs comply.

The Languages

The data structures and the contents of a database can be created, updated and processed by means of four languages, namely the *Schema DDL* and *Sub-schema DDL*, through which the global schema and the sub-schemas of the database are declared, the *Data Storage Description Language* or *DSDL* (often named *DMCL*) that allows physical structures to be defined and tuned, and the *DML* through which application programs access and update the contents of the database.

Gross Architecture of a CODASYL DBMS

The CODASYL reports define the interactions between client application programs and the database. The resulting architecture actually laid down the principles of modern DBMSs. The DBMS includes (at least) three components, namely the DDL compiler, the DML compiler and the database control system (DBCS, or simply system). The DDL compiler translates the data description code into internal tables that are stored in the database, so that they can be exploited by the DML compiler at program compile time and by the DBCS at program run time. Either the DML compiler is integrated into the host language compiler (typically COBOL) or it acts as a precompiler. It parses applications programs and replaces DML statements with calls to DBMS procedures. The DBCS receives orders from the application programs and executes them. Each program includes a *user working area* (UWA) in which data to and from the database are stored. The UWA also comprises registers that inform the program on the status of the last operations, and in particular references to the last records accessed/updated in each record type, each area, each set type and globally for the current process. These references, called *currency indicators*, represent static predefined cursors that form the basis for the navigational facilities across the data.

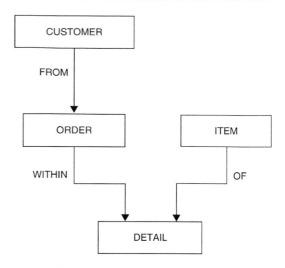

Network Data Model, Fig. 1 Diagram representation of the schema of a sample database

The Data Structures

The pictorial representation of a schema, or *Data Structure Diagram* [1], looks like a graph, where the nodes are the database *record types* and the edges are *set types*, that is, binary 1:N relationship types between record types, conventionally directed from the *one* side to the *many* side. Both record types and set types are named (Fig. 1). In this popular representation, record fields as well as various characteristics of the data structures are ignored.

Records and Record Types

A *record* is the data unit exchanged between the database and the application program. A program reads one record at a time from the database and stores one record at a time in the database. Records are classified into *record types* that define their common structure and default behavior. The intended goal of a record type is to represent a real world entity type. A database key, which is a database-wide system-controlled identifier, is associated with each record, acting as an object-id.

Each database includes the SYSTEM record type, with one occurrence only, that can be used to define access paths across user record types through SYSTEM-owned singular set types.

The schema specifies, via the record type *location mode*, how a record is stored and how it is retrieved when storing a dependent record. This feature is both very powerful and, when used at its full power, fairly complex. Two main variants are proposed:

- location mode *calc using field-list*: the record is stored according to a hashing technique (or later through B-tree techniques) applied to the record key, composed of one or several fields of the record type (*field-list*); at run time, the default way to access a record will be through this record key;
- location mode via *set type S*: the record is physically stored as close as possible to the current record of set type *S*; later on, the default way to access a record will be through an occurrence of *S* identified by its *set selection* mode.

Record Fields

A record type is composed of *fields*, the occurrences of which are *values*. Not surprisingly (CODASYL was also responsible for the COBOL specifications), their structure closely follow the record declaration of COBOL. The DDL offers the following field structures:

- *data item*: elementary piece of data of a certain type (arithmetic, string, implementor defined);
- *vector*: array of values of the same type; its size can be fixed or variable;
- *repeating group*: a somewhat misleading name for a possibly repeating aggregate of fields of any kind.

The fields of a record type can be atomic or compound, single-valued or multi-valued, mandatory or optional (through the null value); these three dimensions allow complex, multi-level field structures to be defined.

Sets and Set Types

Basically, a CODASYL *set* is a list of records made up of a head record (the *owner* of the set) followed by zero or more other records (the *members* of the set). A *set type* S is a schema construct defined by its name and comprising one owner record type and one or more member record type(s). Considering set type S with owner type A and member type B, any A record is the owner of one and only one occurrence of S and no B record can be a member of more than one occurrence of S. In other words, a set type materializes a 1:N relationship type. The owner and the members of a set type are distinct. This limitation has been dropped in the 1978 specifications, but has been kept in most implementations (exceptions: SYBAS and MDBS). Cyclic structures are allowed provided they include at least two record types. It must be noted that a set type can include more than one member record type.

The member records of S can be ordered (first, last, sorted, application-defined). This characteristic is static and cannot be changed at run-time as in SQL. The insertion of a member record in an occurrence of S can be performed at creation time (automatic insertion mode) or later by the application program (manual insertion mode). Once a record is a member of an occurrence of S, its status is governed by the retention mode; it can be removed at will (optional), it cannot be changed (fixed) or it can be moved from an occurrence to another but cannot be removed (mandatory).

The *set [occurrence] selection* of S defines the default way an occurrence of S is determined in certain DML operations such as storing records with automatic insertion mode.

Areas

An area is a named logical repository of records of one or several types. The records of a definite type can be distributed in more than one area. The intended goal is to offer a way to partition the set of the database records according to real world dimensions, such as geographic, organizational or temporal. However, since areas are mapped to physical devices, they are sometimes used to partition the data physically, e.g., across disk drives.

Schema and Sub-schemas

The data structures of each database are described by a schema expressed in the DDL. Though

DDL is host language independent, its syntax is reminiscent of COBOL. Views are defined by sub-schemas. Basically, a subschema is a host language dependent description of a subset of the data structures of a schema. Some slight variations are allowed, but they are less powerful than relational database capabilities. Figure 2 shows a fragment of the schema declaring the data structures of Fig. 1.

Data Manipulation

The DML allows application programs to ask the DBCS data retrieval and update services. The program accesses the data through a sub-schema that identifies the schema objects the instances of which can be retrieved and updated as well as their properties, such as the data type of each field. Exchange between the host language and the DBCS is performed via the UWA, a shared set of variables included in each running program. This set includes the currency indicators, the process status (e.g., the error indicators) and record variables in which the data to and from the database are temporarily stored. Many DML statements use the currency indicators as implicit arguments. Such is the case for set traversal and for record storing. Based on the currency indicators, on the *location mode* of record types and on the *set selection* option of set types, sophisticated positioning policies can be defined, leading to tight application code.

Data Retrieval

The primary aim of the *find* statement is to retrieve a definite record on the basis of its position in a specified collection and to make it the current of all the *communities* which it belongs to, that is, its database, its area, its record type and each of its set types. For instance, if an *ORDER* record is successfully retrieved, it becomes the current of the database for the running program (the *current of run unit*), the current of the *DOMESTIC* area, the current of the *ORDER* record type and the current of the *FROM* and *WITHIN* set types. The variants of the *find* statement allow the program to scan the records of an area, of a record type

```
schema name is ORDER-MANAGEMENT.

area name is DOMESTIC.
area name is FOREIGN.

record name is CUSTOMER within DOMESTIC, FOREIGN;
location mode is calc using CUST-NO duplicates not allowed.
2 CUST-NO type is character 10.
2 CUST-NAME type is character 45.
2 CUST-ADDRESS.
  4 STREET character 32.
  4 CITY character 20.
  4 PHONE type is character 15 occurs 3 times.

record name is ORDER within DOMESTIC;
location mode is via FROM set;
duplicates not allowed for ORD-NO.
  2 ORD-NO type is decimal 12.
  2 ORD-DATE type is date.

set name is FROM;
owner is CUSTOMER;
order is sorted duplicates not allowed;
member is ORDER mandatory automatic key is ascending ORD-NO;
set selection is through current of FROM.
```

Network Data Model, Fig. 2   Fragment of the DDL code defining the schema of Fig. 1

and of the members and the owner of a set. They also provide selective access among the members of a set.

The *get* statement transfers field values from a current record in the UWA, from which they can then be processed by the program.

Data Update

A record r is inserted in the database as follows: first, field values of r are stored in the UWA, then the current of each set in which r will be inserted is retrieved and finally a *store* instruction is issued. The *delete* instruction applies to the current record. For this operation, the DBCS enforces a *cascade* policy: if the record to be deleted is the owner of sets whose members have a mandatory or fixed retention mode, those members are deleted as well. The *modify* statement transfers in the current of a record type the new values that have been stored in the UWA. Insertion and removal of the current of a record type is performed by *insert* and *remove* instructions. Transferring a mandatory member from a set to another cannot be carried out by merely removing then inserting the record. A special case of the *modify* statement makes such a transfer possible. Later specifications as well as some implementations propose a specific statement for this operation.

Figure 3 shows, in an arbitrary procedural pseudo-code, a fragment that processes the orders of customer *C400* and another fragment that creates an *ORDER* record for the same customer.

Entity-Relationship to Network Mapping

Among the many DBMS data models that have been proposed since the late sixties, the network model is probably the closest to the Entity-Relationship model [5]. As a consequence, network database schemas tend to be more readable than those expressed in any other DBMS data model, at least for simple schemas. Each entity type is represented by a record type, each attribute by a field and each simple relationship type by a set type. Considering modern conceptual formalisms, the network model suffers from several deficiencies, notably the lack of generalization-specialization (is-a) hierarchies and the fact that relationship types are limited to the 1:N category. Translating an Entity-relationship schema into the network model requires the transformation of these missing constructs into standard structures.

Is-a hierarchies Three popular transformations can be applied to express this construct in standard data management systems, namely one record type per entity type, one record type per supertype and one record type per subtype. Representing each entity type by a distinct record type and forming a set type S with each supertype (as owner of S) and all its direct subtypes (as members of S) is an appropriate implementation of the first variant.

1:1 Relationship type This category is a special case of 1:N and can be expressed by a mere set type, together with dynamic restriction on the number of members in each set. However,

```
CUSTOMER.CUST-NO := "C400"           /* store search key value in the UWA
find CUSTOMER record                 /* find CUSTOMER record "C400"
find first ORDER record of FROM set   /* find first ORDER record owned by this CUSTOMER record
while ERROR-COUNT = 0
    get ORDER                        /* get item values of current ORDER record
    <process item values of current ORDER>
    find next ORDER record of FROM set /* find first ORDER record
end-while


CUSTOMER.CUST-NO := "C400"
find CUSTOMER record
if ERROR-COUNT = 0 then              /* make CUSTOMER record "C400" the current of FROM
    ORDER.ORD-NO := 30183            /* store value of ORD-NO in the UWA
    ORDER.CUST-DATE := "2008/08/19"  /* store value of ORD-DATE in the UWA
    store ORDER                      /* store ORDER record and insert it in the current FROM set
end-if
```

Network Data Model, Fig. 3 Two examples of data manipulation code

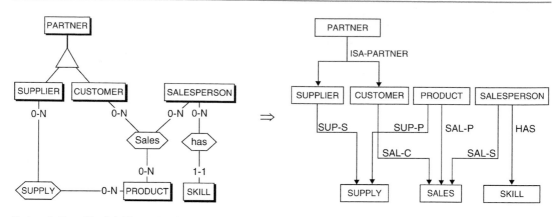

Network Data Model, Fig. 4 Partial translation of a representative Entity-relationship schema (*left*) into a network schema (*right*)

merging both record types when one of them depends on the other one (e.g., as an automatic, mandatory member) is also a common option.

Complex relationship type In most implementations, n-ary and N:N relationship types as well as those with attributes must be reduced to constructs based on 1:N relationship types only through standard transformations. A complex relationship type *R* is represented by a relationship record type *RT* and by as many set types as *R* has roles. The attributes of *R* are translated into fields of *RT*. Cyclic relationship types, if necessary, will be translated in the same way.

Figure 4 illustrates some of these principles.

Discussion

The network model offers a simple view of data that is close to semantic networks, a quality that accounts for much if its past success. The specifications published in the 1971 and 1973 reports exhibited a confusion between abstraction levels that it shared with most proposals of the seventies and that was clarified in later recommendations, notably the 1978 report and X3H2 NDL. In particular, the DDL includes aspects that pertain to logical, physical and procedural layers.

Though they were not implemented in most commercial DBMSs, the CODASYL recommendations included advanced features that are now usual in database technologies such as database procedures, derived fields, check and some kind of triggers.

Key Applications

CODASYL DBMSs have been widely used to manage large corporate databases submitted to both batch and OLTP (On-line Transaction Processing) applications. Compared with hierarchical and relational DBMS, their simple and intuitive though powerful model and languages made them very popular for the development of large and complex applications. However, their intrinsic lack of flexibility in rapidly evolving contexts and the absence of user-oriented interface made them less attractive for decisional applications, such as data warehouses.

Cross-References

▶ Database Management System
▶ Entity Relationship Model
▶ Hierarchical Data Model
▶ Relational Model

Recommended Reading

1. Bachman C. Data structure diagrams. ACM SIGMIS Database. 1969;1(2):4–9.
2. Bachman C. The programmer as navigator. Commun ACM. 1973;16(11):635–58.
3. DBTG C. CODASYL data base task group, April 1971 report. ACM: New York; 1971.
4. DDLC C. CODASYL data description language committee, CODASYL DDL. Journal of Development (June 1973), NBS Handbook 113 (Jan 1974). 1973.

5. Elmasri R, Navathe S. Fundamentals of database systems. 3rd ed. Addison-Wesley; 2000. (The appendix on the network data model has been removed from later editions but is now available on the authors' site).
6. Engels R.W. An analysis of the April 1971 DBTG report. In: Proceedings of the ACM SIGFIDET Workshop on Data Description, Access, and Control; 1971. p. 69–91.
7. Jones JL. Report on the CODASYL data description language committee. Inf Syst. 1978;3(4):247–320.
8. Michaels A, Mittman B, Carlson CA. Comparison of relational and CODASYL approaches to database management. ACM Comput. Surv. 1976;8(1):125–51.
9. Olle W. The CODASYL approach to data base management. New York: Wiley; 1978.
10. Taylor R, Frank R. CODASYL data-base management systems. ACM Comput Surv. 1976;8(1):67–103.
11. Tsichritzis D, Lochovsky F. Data base management systems. New York: Academic; 1977.

Neural Networks

Pang-Ning Tan
Michigan State University, East Lansing, MI, USA

Synonyms

Connectionist model; Parallel distributed processing

Definition

An artificial neural network (ANN) is an abstract computational model designed to solve a variety of supervised and unsupervised learning tasks. While the discussion in this chapter focuses only on supervised classification, readers who are interested in unsupervised learning using ANN may refer to the literature on vector quantization [1] and self-organizing maps [2]. An ANN consists of an assembly of simple processing units called neurons connected by a set of weighted edges (or synapses), as shown in Fig. 1. The neurons are often configured into a feed-forward multilayered topology, with outputs from one layer being fed into the next layer. The first layer, which is known as the input layer, encodes the attributes of the input data, while the last layer, known as the output layer, encodes the neural network's output. Hidden layers are the intermediary layers of neurons between the input and output layers. A feed-forward ANN without any hidden layer is called a perceptron [3]. Another common ANN architecture is the recurrent network, which allows a neuron to feed its output back into the inputs of other preceding neurons in the network. Such a network topology is useful for modeling temporal and sequential relationships in dynamical systems.

Historical Background

The design of an ANN was inspired by the desire to emulate how a human brain works. Interest in this field began to emerge following the seminal work of McCulloch and Pitts [4], who attempted to understand how complex patterns can be modeled in the brain using a large number of interconnected neurons. They presented a simplified model of a neuron and showed how a collection of these neurons could be used to represent logical propositions. Nevertheless, they did not provide an algorithm to estimate the weights of the network.

A major step forward in the study of ANN occurred when Hebb [5] formulated a postulate relating the cerebral activities of the brain to the synaptic connections between neurons. Hebb theorized that the process of learning takes place when a pair of neurons is activated repeatedly, thus making their synaptic connection stronger. By strengthening the connection, this enables the network to recognize the appropriate response when the same stimulus is re-applied. This idea forms the basis for what is now known as Hebbian learning.

The invention of the perceptron by Rosenblatt [3] marked another major development of the field. Rosenblatt demonstrated that a simple perceptron can be trained to recognize visual patterns by modifying its weights using an iter-

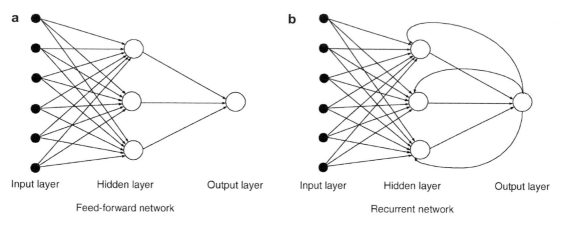

Neural Networks, Fig. 1 Architecture of artificial neural network

ative error correcting algorithm. A theorem was subsequently proposed to show the guaranteed convergence of the perceptron training rule in finite time – a result that subsequently triggered a wave of interest in the field. During this period, new perceptron training algorithms began to emerge, such as the Widrow-Hoff [6] and stochastic gradient descent [7] algorithms. Such interest, however, began to wane as Minsky and Papert [8] showed the practical limitations of perceptrons, particularly in terms of solving nonlinearly separable problems such as the exclusive-or (XOR) function. Though multilayer neural networks may overcome these limitations, there has yet to be any feasible learning algorithm that can automatically adjust the weights of the neurons in the hidden layer.

Interest in ANN was finally rekindled in the 1980s, fueled by the successful development of the Boltzmann machine [9] and the rediscovery of the backpropagation algorithm [10], both of which demonstrated the feasibility of training multilayer neural networks. Furthermore, as computers become cheaper, this permits more researchers to participate in the field and experiment with the capabilities of neural networks, unlike the situation in the 1960s. Research in ANN continued to flourish with the development of more complex networks such as radial basis function networks [11], Hopfield networks [12], Jordan networks [13], and Elman networks [14].

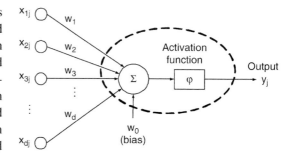

Neural Networks, Fig. 2 Structure of an artificial neuron

Foundations

Neurons are the elementary processing units of an ANN. The structure of a neuron is shown in Fig. 2. Each neuron consists of a set of weighted edges $\{w_1, w_2, \cdots, w_d\}$, a threshold w_0 (known as the bias), an adder Σ that computes the weighted sum of its input, and an activation function ϕ that transforms the weighted sum into its output value. The computation performed by a neuron can be expressed in the following form:

$$y_j = \varphi \left(\sum_{i=1}^{d} w_i x_{ij} + w_0 \right) \equiv \varphi \left(z_j \right) \quad (1)$$

The output of a neuron can be discrete or continuous, depending on the choice of activation function. For example, the Heaviside function can be used to produce a binary-valued output:

$$\varphi(z) = \begin{cases} 1, & z \geq 0 \\ 0, & \text{otherwise} \end{cases}$$

while the linear function $\phi(z) = z$ and sigmoid function

$$\varphi(z) = \frac{1}{1 + \exp(-z)}$$

produce a continuous-valued output.

An ANN must be trained to determine the appropriate set of weights for a given classification task. Training is accomplished by processing the training examples one at a time and comparing the predicted class of a training example to its actual class. The error in prediction is then used to modify the weights of the network. This process is repeated until a stopping criterion is satisfied.

Table 1 shows the activation functions and weight update formula employed by several perceptron learning algorithms. Notice that the amount of weight adjustment is proportional to the difference between the true output y_j and the predicted output $\phi(z_j)$ of a neuron. Computing the error term $[y_j - \phi(z_j)]$ in the weight update formula is straightforward for perceptrons, but is more challenging for multilayer networks because the true output of each hidden neuron is unknown. Furthermore, it is unclear the extent to which each of the neuron's weight contributes to the overall network error, an issue that is known as the credit assignment problem. The backpropagation algorithm attempts to overcome this problem by employing a differentiable activation function (such as the sigmoid function) so that a gradient descent strategy can be used to derive the weight update formula. In the backpropagation algorithm, the weight update step is decomposed into two phases. During the forward phase, the outputs of the neurons are computed one layer at a time, starting from neurons in the first hidden layer, followed by those in the next layer until the neurons in the output layer are computed. Next, during the backward phase, the errors are propagated in the reverse direction, starting from neurons in the output layer, followed by those in the preceding layer until the errors of neurons in the first hidden layer are computed.

There are several issues that must be considered when building an ANN model. First, the structure of the network must be appropriately chosen to avoid model overfitting. To determine the structure with the right model complexity, model selection approaches such as cross validation may be used. Regularization methods are also applicable to penalize network structures that are overly complex. Another possibility is to apply the Bayesian approach, which provides a principled framework for handling the model complexity problem. A review of the Bayesian approach for neural networks can be found in Ref. [3]. The cascade correlation algorithm [10] is another useful approach to grow the network dynamically from a structure with no hidden layers. Hidden nodes are then added one at a time until the residual error of the network falls within some acceptable level.

Since the error function of a multilayer neural network is not convex, convergence of the backpropagation algorithm to a local minimum is another potential issue to consider. One way to mitigate the problem is by adding a momentum term to the weight update formula to escape from the local minimum. Alternative strategies include repeating the network simulation with different initial weights and applying a stochastic gradient descent method to estimate the network weights.

Instead of learning a discriminant function to distinguish instances from different classes, an ANN can also be trained to produce posterior probabilities of their class memberships. This is

Neural Networks, Table 1 Perceptron training rules

Perceptron training rule	Activation function	Weight update formula
Rosenblatt's perceptron	Heaviside function	$w_i \leftarrow w_i + \eta \left[y_j - \phi(z_j) \right] x_{ij}$
Widrow-Hoff	Linear function	$w_i \leftarrow w_i + \eta \sum_j \left[y_j - \phi(z_j) \right] x_{ij}$
Stochastic gradient descent	Linear function	$w_i \leftarrow w_i + \eta \left[y_j - \phi(z_j) \right] x_{ij}$

desirable because a probabilistic framework provides a natural way to compensate for imbalanced class distributions, to incorporate the cost of misclassification into decision-making, and to fuse the outputs from multiple networks. An ANN can be designed to generate probabilistic outputs by training its weights to minimize the cross entropy error function [15] instead of the sum-of-square error function.

As ANN encodes a sophisticated mathematical function, the ability to explain how it makes its prediction is crucial for a number of applications, particularly in the medical and legal domains. Algorithms have been developed to extract symbolic representations or rules from a trained network [6]. Such algorithms generally fall into two categories. The first category generates rules based on features constructed from the weights of the network connections. The second category uses the trained network to label a data set and then applies rule extraction algorithms on the labeled data set.

In general, multilayer neural networks are universal approximators, allowing them to fit any type of function. Furthermore, since it is an abstract computational model, it is also applicable to a variety of supervised, unsupervised, and semi-supervised learning tasks.

Key Applications

ANN has been successfully applied to various applications including pattern recognition (face, handwriting, and speech recognition), finance (bankruptcy prediction, bond rating, and economic forecasting), manufacturing (process control, tool condition monitoring, and robot scheduling), and computer-aided diagnosis (arrhythmia identification and tumor detection in biological tissues).

Future Directions

The performance of supervised and unsupervised learning algorithms often depends on the quality of their input features. There have been considerable interests in recent years to design deep-layered network architectures that can automatically learn the appropriate feature representation for complex learning tasks [16]. The key challenge here is to develop efficient and scalable algorithms to learn the parameters of such architecture.

Experimental Results

The experimental evaluation often depends on the learning tasks. For supervised classification, some of the typical evaluation criteria include model accuracy, specificity, sensitivity, F-measure, and run-time for model building.

Data Sets

A large collection of data sets are available at the UCI data mining and machine learning repository. For a more comprehensive list of data sets and other information about neural networks, go to http://www.neural-forecasting.com/.

Url to Code

A comprehensive set of functions for implementing feed-forward and recurrent neural networks is available in MATLAB toolbox.

Cross-References

▶ Classification
▶ Decision Tree Classification

Recommended Reading

1. Haykin S. Neural networks – a comprehensive foundation. 2nd ed. Englewood: Prentice-Hall; 1998.
2. Kohonen T. Self-organizing maps. Berlin: Springer; 2001.
3. Rosenblatt F. Principles of neurodynamics. New York: Spartan Books; 1959.
4. McCulloch WS, Pitts W. A logical calculus of the ideas immanent in nervous activity. Bull Math Biophys. 1943;5(4):115–33.
5. Hebb D. The organization of behaviour. New York: Wiley; 1949.

6. Widrow B, Hoff ME Jr. Adaptive switching circuits. IRE WESCON convention record; 1960. p. 96–104.
7. Amari S. A theory of adaptive pattern classifiers. IEEE Trans Electron Comput. 1967;16(3):299–307.
8. Minsky M, Papert S. Perceptions: An introduction to computational geometry. Cambridge, MA: MIT; 1969.
9. Ackley DH, Hinton GE, Sejnowski TJ. A learning algorithm for Boltzmann machines. Cogn Sci. 1985;9(1):147–69.
10. Rumelhart DE, Hinton GE, Williams RJ. Learning representations by backpropagating errors. Nature. 1986;323(6088):533–6.
11. Powell M. Radial basis functions for multivariable interpolation: A review. In: Mason JC, Cox MG, editors. Algorithms for approximation. New York: Clarendon Press; 1987. p. 143–67.
12. Hopfield JJ. Neural networks and physical systems with emergent collective computational abilities. Proc Natl Acad Sci. 1982;79(8):2554–8.
13. Jordan MI. Attractor dynamics and parallelism in a connectionist sequential machine. In: Proceedings of the 8th Annual Conference of the Cognitive Science Society; 1986. p. 531–46.
14. Elman JL. Finding structure in time. Cogn Sci. 1990;14(2):179–211.
15. Lampinen J, Vehtari A. Bayesian approach for neural networks – review and case studies. Neural Netw. 2001;14(3):7–24.
16. Bengio Y, Courville A, Vincent P. Representation learning: A review and new perspectives. IEEE Trans Pattern Anal Mach Intell. 2003;35(8):1798–828.
17. Craven M, Shavlik JW. Learning symbolic rules using artificial neural networks. In: Proceedings of the 10th International Conference on Machine Learning; 1993. p. 73–80.
18. Fahlman SE, Lebiere C. The cascade correlation learning architecture. Adv Neural Inf Process Syst. 1989;2:524–32.
19. Hopfield JJ. Learning algorithms and probability distributions in feed-forward and feed-back networks. Proc Natl Acad Sci U S A. 1987;84(23):8429–33.

N-Gram Models

Djoerd Hiemstra
University of Twente, Enschede,
The Netherlands

Definition

In language modeling, n-gram models are probabilistic models of text that use some limited amount of history, or word dependencies, where n refers to the number of words that participate in the dependence relation.

Key Points

In automatic speech recognition, n-grams are important to model some of the structural usage of natural language, i.e., the model uses word dependencies to assign a higher probability to "how are you today" than to "are how today you," although both phrases contain the exact same words. If used in information retrieval, simple unigram language models (n-gram models with $n = 1$), i.e., models that do not use term dependencies, result in good quality retrieval in many studies. The use of bigram models (n-gram models with $n = 2$) would allow the system to model direct term dependencies, and treat the occurrence of "New York" differently from separate occurrences of "New" and "York," possibly improving retrieval performance. The use of trigram models would allow the system to find direct occurrences of "New York metro," etc. The following equations contain respectively (1) a unigram model, (2) a bigram model, and (3) a trigram model:

$$P(T_1, T_2, \ldots T_n | D) = P(T_1 | D) P(T_2 | D) \ldots$$
$$P(T_n | D)$$
$$(1)$$

$$P(T_1, T_2, \ldots T_n | D)$$
$$= P(T_1 | D) P(T_2 | T_1, D) \ldots P(T_n | T_{n-1}, D)$$
$$(2)$$

$$P(T_1, T_2, \ldots T_n | D)$$
$$= P(T_1 | D) P(T_2 | T_1, D) P(T_3 | T_1, T_2, D)$$
$$\ldots P(T_n | T_{n-2}, T_{n-1}, D)$$
$$(3)$$

The use of n-gram models increases the number of parameters to be estimated exponentially with n, so special care has to be taken to smooth the bigram or trigram probabilities. Several studies have shown small but significant improvements of using bigrams if smoothing parameters are properly tuned [2, 3]. Improvements of the use of n-grams and other term dependencies seem to be bigger on large data sets [1].

Cross-References

▶ Language Models
▶ Probability Smoothing

Recommended Reading

1. Metzler D, Bruce Croft W. A Markov random field model for term dependencies. In: Proceedings of the 31st Annual International ACM SIGIR Conference on Research and Development in Information Retrieval; 2005. p. 472–9.
2. Miller DRH, Leek T, Schwartz RM. A hidden Markov model information retrieval system. In: Proceedings of the 22nd Annual International ACM SIGIR Conference on Research and Development in Information Retrieval; 1999. p. 214–21.
3. Song F, Bruce Croft W. A general language model for information retrieval. In: Proceedings of the 22nd Annual International ACM SIGIR Conference on Research and Development in Information Retrieval; 1999. p. 4–9.

Noise Addition

Josep Domingo-Ferrer
Universitat Rovira i Virgili, Tarragona, Catalonia, Spain

Synonyms

Additive noise

Definition

Noise addition is a masking method for statistical disclosure control of numerical microdata that consists of adding random noise to original microdata.

Key Points

The noise addition algorithms in the literature are:

1. *Masking by uncorrelated noise addition.* The vector of observations x_j for the jth attribute of the original dataset X_j is replaced by a vector

$$z_j = x_j + \epsilon_j$$

where ε_j is a vector of normally distributed errors drawn from a random variable $\varepsilon_j \sim N\left(0, \sigma_{\varepsilon_j}^2\right)$, such that $Cov(\varepsilon_t, \varepsilon_l) = 0$ for all $t \neq l$. This does not preserve variances nor correlations.

2. *Masking by correlated noise addition.* Correlated noise addition also preserves means and additionally allows preservation of correlation coefficients. The difference with the previous method is that the covariance matrix of the errors is now proportional to the covariance matrix of the original data, i.e., $\varepsilon \sim N(0, \Sigma_\varepsilon)$, where $\Sigma_\varepsilon = \alpha \Sigma$.

3. *Masking by noise addition and linear transformation.* In Kim [1], a method is proposed that ensures by additional transformations that the sample covariance matrix of the masked attributes is an unbiased estimator for the covariance matrix of the original attributes.

4. *Masking by noise addition and nonlinear transformation.* An algorithm combining simple additive noise and nonlinear transformation is proposed in Sullivan [2]. The advantages of this proposal are that it can be applied to discrete attributes and that univariate distributions are preserved. Unfortunately, the application of this method is very time-consuming and requires expert knowledge on the data set and the algorithm.

See Brand [3] for more details on specific algorithms.

Cross-References

▶ Inference Control in Statistical Databases
▶ Microdata
▶ SDC Score

Recommended Reading

1. Kim JJ. A method for limiting disclosure in microdata based on random noise and transformation In: Proceedings of the Section on Survey Research Methods; 1986. p. 303–8.
2. Sullivan GR. The use of added error to avoid disclosure in microdata releases. PhD thesis, Iowa State University; 1989.
3. Brand R. Microdata protection through noise addition. In: Domingo-Ferrer J, editor. Inference control in statistical databases. LNCS, vol. 2316. Berlin: Springer; 2002. p. 97–116.

Nonparametric Data Reduction Techniques

Rui Zhang
University of Melbourne, Melbourne,
VIC, Australia
Dataware Ventures, Tucson, AZ, USA
Dataware Ventures, Redondo Beach, CA, USA

Definition

A nonparametric data reduction technique is a data reduction technique that does not assume any model for the data.

Key Points

Nonparametric data reduction (NDR) techniques is opposite to parametric data reduction (PDR) techniques. A PDR technique must assume a certain model for the data. Parameters of the model are determined before the data reduction is performed. A NDR technique does not assume any model and is applied to the data directly. The data reduction effectiveness of a PDR technique heavily depends on whether the model suits the data well. If well-suited, good accuracy as well as substantial data reduction can be achieved; otherwise, both cannot be achieved at the same time. A NDR technique yields more uniform effectiveness irrespective of the data, but it may not achieve as high data reduction as a well-suited PDR technique.

Popular NDR techniques include histograms, clustering and indexes. Histograms are used to approximate data distributions. An equidepth histogram can adjust itself to the data distribution and always gives good approximation of the data, no matter how the data are distributed. Clustering techniques also find the cluster centers automatically, irrespective of the data distribution. Although parameters are used in some clustering algorithms, such as the value of k in a k-means algorithm, k is a given parameter instead of a parameter determined by the actual data. Conceptually, indexes are similar to clustering and they also adjust themselves to the data distribution. No parameter needs to be determined from the data for indexes. A summary of data reduction techniques including NDR techniques can be found in [1].

Cross-References

▶ Histogram
▶ Parametric Data Reduction Techniques

Recommended Reading

1. Barbará D, DuMouchel W, Faloutsos C, Haas PJ, Hellerstein JM, Ioannidis YE, Jagadish HV, Johnson T, Ng RT, Poosala V, Ross KA, Sevcik KC. The New Jersey data reduction report. Q Bull IEEE TC Data Eng. 1997;20(4):3–45.

Nonperturbative Masking Methods

Josep Domingo-Ferrer
Universitat Rovira i Virgili, Tarragona,
Catalonia, Spain

Synonyms

Non-perturbative masking

Definition

Non-perturbative masking methods are SDC methods for microdata protection which do not alter data; rather, they produce partial suppressions or reductions of detail in the original dataset. Sampling, global recoding, top and bottom coding and local suppression are examples of non-perturbative masking methods.

Key Points

1. Sampling is a non-perturbative masking method for statistical disclosure control of microdata. Instead of publishing the original microdata file, what is published is a sample S of the original set of records. Sampling methods are suitable for categorical microdata, but for continuous microdata they should probably be combined with other masking methods. The reason is that sampling alone leaves a continuous attribute V_i unperturbed for all records in S. Thus, if attribute V_i is present in an external administrative public file, unique matches with the published sample are very likely: indeed, given a continuous attribute V_i and two respondents o_1 and o_2, it is highly unlikely that V_i will take the same value for both o_1 and o_2 unless $o_1 = o_2$ (this is true even if V_i has been truncated to represent it digitally).

2. Global recoding or generalization is a masking method for statistical disclosure control of microdata. For a categorical attribute V_i, several categories are combined to form new (less specific) categories, thus resulting in a new V'_i with $|D(V'_i)| < |D(V_i)|$ where $|\cdot|$ is the cardinality operator. For a numerical attribute, global recoding means replacing V_i by another attribute V'_i which is a discretized version of V_i. In other words, a potentially infinite range $D(V_i)$ is mapped onto a finite range $D(V'_i)$. This technique is more appropriate for categorical microdata, where it helps disguise records with strange combinations of categorical attributes. Global recoding is used heavily by statistical offices. Global recoding is implemented in the μ-Argus package. In combination with local suppression, it can be used to achieve k-anonymity.

3. Top coding and bottom coding are special cases of the global recoding masking method for statistical disclosure control of microdata. Their operating principle is that top values (those above a certain threshold), respectively bottom values (those below a certain threshold), are lumped together to form a new category. Top and bottom coding can be used on attributes that can be ranked, that is, numerical or categorical ordinal.

4. Local suppression or blanking is a masking method for statistical disclosure control of microdata. Certain values of individual attributes are suppressed with the aim of increasing the set of records agreeing on a combination of key values. Local suppression is implemented in the μ-Argus package. Ways to combine local suppression and global recoding are discussed in DeWaal and Willenborg [1]. In fact, the combination of both methods can be used to attain *k-anonymity*. If a numerical attribute V_i is part of a set of key attributes, then each combination of key values is probably unique. Since it does not make sense to systematically suppress the values of V_i, it can be asserted that local suppression is rather oriented to categorical attributes.

Cross-References

▶ Inference Control in Statistical Databases
▶ k-Anonymity
▶ Microdata

Recommended Reading

1. DeWaal AG, Willenborg LCRJ. Global recodings and local suppressions in microdata sets. In: Proceedings of the Statistics Canada Symposium; 1995. p. 121–32.

2. Hundepool A, Domingo-Ferrer J, Franconi L, Giessing S, Lenz R, Longhurst J, Schulte-Nordholt E, Seri G, DeWolf P-P. Handbook on statistical disclosure control (version 1.0). Eurostat (CENEX SDC Project Deliverable); 2006. http://neon.vb.cbs.nl/CENEX/
3. Hundepool A, Van de Wetering A, Ramaswamy R, Franconi F, Polettini S, Capobianchi A, DeWolf P-P, Domingo-Ferrer J, Torra V, Brand R, Giessing S. μ-ARGUS user's manual version 4.1, Feb 2007. http://neon.vb.cbs.nl/CASC
4. Willenborg L, DeWaal T. Elements of statistical disclosure control. New York: Springer; 2001.

Nonrelational Streams

Jeong-Hyon Hwang[1], Alan G. Labouseur[2], and Paul W. Olsen[3]
[1]Department of Computer Science, University at Albany – State University of New York, Albany, NY, USA
[2]School of Computer Science and Mathematics, Marist College, Poughkeepsie, NY, USA
[3]Department of Computer Science, The College of Saint Rose, Albany, NY, USA

Definition

A non-relational stream is a continuously generated, ordered collection of data items that are *not* relational tuples and therefore *not* readily processed by relational algebraic operators such as selection, projection, join, and aggregation. Each data item may be associated with a time stamp that represents the time when that data item was produced or received by a certain device or system. Many applications require highly efficient, low-latency, real-time processing techniques in order to keep up with high-volume data streams.

Main Text

Non-relational data streams have been studied in the following forms: graph streams, spatial streams, text streams, and XML streams.

Graph Streams

Each data item represents an insert, update, or delete operation on a vertex or an edge in a graph. Queries on these streams are concerned with estimating properties of the graph or finding patterns within that graph. See ▶ "Graph Mining on Streams".

Spatial Streams

Spatial data streams represent the motion of objects (e.g., people and taxicabs). Queries on these streams often take the form of range or nearest neighbor (NN) inquiries (e.g., locating the nearest taxi to the person requesting it). See ▶ "Continuous Monitoring of Spatial Queries".

Text Streams

Text streams (e.g., messages from Twitter, LinkedIn, Facebook, WeChat, and other social media) tend to be temporally ordered collections of text or text documents. Queries on text streams include text classification, topic detection and tracking (including event discovery from a stream of news stories), bursty event detection (e.g., reports of disease outbreaks), knowledge and opinion mining from blogs/chat transcripts, and search engine log files. See ▶ "Text Streaming Model".

XML Streams

XML streams are continuously generated series of XML documents. Queries on these streams may select only the XML documents that match certain criteria or may transform input streams into different output streams. See ▶ "XML Stream Processing".

Cross-References

- ▶ Continuous Monitoring of Spatial Queries
- ▶ Data Stream
- ▶ Graph Mining on Streams
- ▶ Text Streaming Model
- ▶ XML Stream Processing

Nonsequenced Semantics

Michael H. Böhlen[1,5], Christian S. Jensen[2], and
Richard T. Snodgrass[3,4]
[1]Free University of Bozen-Bolzano,
Bozen-Bolzano, Italy
[2]Department of Computer Science, Aalborg
University, Aalborg, Denmark
[3]Department of Computer Science, University of
Arizona, Tucson, AZ, USA
[4]Dataware Ventures, Tucson, AZ, USA
[5]University of Zurich, Zürich, Switzerland

Synonyms

Nontemporal semantics

Definition

Nonsequenced semantics guarantees that query
language statements can reference and manipulate the timestamps that capture the valid and
transaction time of data in temporal databases
as regular attribute values, with no built-in temporal semantics being enforced by the query
language.

Key Points

A temporal database generalizes a nontemporal
database and associates one or more timestamps
with database entities. Different authors have
suggested temporal query languages that provide
advanced support for formulating temporal statements. Results of these efforts include a variety of
temporal extensions of SQL, temporal algebras,
and temporal logics that simplify the management of temporal data.

Languages with built-in temporal support are
attractive because they offer convenient support
for formulating a wide range of common temporal statements. The classical example of advanced
built-in temporal support is *sequenced semantics*,
which makes it possible to conveniently interpret
a temporal database as a sequence of nontemporal
databases. To achieve built-in temporal support,
the timestamps are viewed as implicit attributes
that are given special semantics.

Built-in temporal support, however, may also
limit the expressiveness of the language when
compared to the original nontemporal language
where timestamps are explicit attributes. Nonsequenced semantics guarantees that statements
can manipulate timestamps as regular attribute
values with no built-in temporal semantics being
enforced. This ensures that the expressiveness of
the original language is preserved.

The availability of legacy statements with the
standard nontemporal semantics is also important
in the context of migration where users can be
expected to be well acquainted with the semantics
of their nontemporal language. Nonsequenced
semantics ensures that users are able to keep
using the paradigm they are familiar with and to
incrementally adopt the new features. Moreover,
from a theoretical perspective, any variant of temporal logic, a well-understood language that only
provides built-in temporal semantics, is strictly
less expressive than a first-order logic language
with explicit references to time [1, 4].

Each statement of the original language has
the potential to either be evaluated with temporal or nontemporal semantics. For example, a
count query can count the tuples at each time
instant (this would be temporal, i.e., sequenced,
semantics) or count the tuples actually stored in a
relation instance (this would be nontemporal, i.e.,
nonsequenced, semantics).

To distinguish the two semantics, different
approaches have been suggested. For instance,
TempSQL distinguishes between so-called current and classical users. ATSQL and SQL/Tem-
poral offer so-called statement modifiers that enable the users to choose between the two semantics at the granularity of statements. Below, statement modifiers are used for illustration.
Specifically, the ATSQL modifier *NSEQ VT* signals standard SQL semantics with full control
over the timestamp attributes of a valid-time
database.

The illustrations assume a database instance
with three relations:

Employee

ID	Name	VTIME
1	Bob	5–8
3	Pam	4–12
4	Sarah	1–5

Salary

ID	Amt	VTIME
1	20	4–10
3	20	6–9
4	20	6–9

Bonus

ID	Amt	VTIME
1	20	1–6
1	20	7–12
3	20	1–12

and the following queries:

NSEQ VT
*SELECT COUNT(*) FROM Bonus;*
NSEQ VT
SELECT E.ID FROM Employee AS E, Salary AS S
WHERE VTIME(E) PRECEDES VTIME(S) AND E.ID = S.ID;

Both queries are nonsequenced, i.e., the valid time is treated as a regular attribute without any special processing going on. The first query determines the number of bonuses that have been paid. It returns the number of tuples in the *Bonus* relation, which is equal to 3. Note that if the sequenced modifier was used (cf. *sequenced semantics*), then the time-varying count had been computed. With the given example, the count at each point in time would be 2. The second query joins *Employee* and *Salary*. The join is not performed at each snapshot (cf. *sequenced semantics*). Instead, it requires that the valid time of *Employee* precedes the valid time of *Salary*. The result is a nontemporal table.

Nonsequenced statements offer no built-in temporal support, but instead offer complete control. This is akin to programming in assembly language, where one can do everything, but everything is hard to do. The query language must provide a set of functions and predicates for expressing temporal relationships (e.g., *PRECEDES* [2])

and performing manipulations and computations on timestamps (e.g., *VTIME*). The resulting new query language constructs are relatively easy to implement because they only require changes at the level of built-in predicates and functions. Instead of using functions and predicates on timestamps, the use of temporal logic with temporal connectives has also been suggested.

Cross-References

▶ Allen's Relations
▶ Sequenced Semantics
▶ SQL-Based Temporal Query Languages
▶ Temporal Database
▶ Valid Time

Recommended Reading

1. Abiteboul S, Herr L, Van den Bussche J. Temporal versus first-order logic to query temporal databases. In: Proceedings of the 15th ACM SIGACT-SIGMOD-SIGART Symposium on Principles of Database Systems; 1996. p. 49–57.
2. Allen JF. Maintaining knowledge about temporal intervals. Commun ACM. 1983;16(11):832–43.
3. Böhlen MH, Jensen CS, Snodgrass RT. Temporal statement modifiers. ACM Trans Database Syst. 2000;25(4):48.
4. Toman D, Niwiński D. First-order queries over temporal databases inexpressible in temporal logic. In: Advances in Database Technology, Proceedings of the 5th International Conference on Extending Database Technology; 1996. p. 307–24.

Normal Form ORA-SS Schema Diagrams

Gillian Dobbie[1] and Tok Wang Ling[2]
[1]University of Auckland, Auckland, New Zealand
[2]National University of Singapore, Singapore, Singapore

Synonyms

NF-SS; Normalizing ORA-SS diagrams

Definition

Normal forms have been defined for data models, such as the relational data model, the nested relational data model, the object-oriented data model and more recently, the semi-structured data model, to recognize (and remove) certain kinds of redundant data. A normal form that has been defined for semi-structured data, based on the ORA-SS data model, is described.

Key Points

The definition of a normal form for ORA-SS (Object-Relationship-Attribute data model for Semi-structured data) diagrams is based on the definition of a normal form for the nested relational data model as described by Ling and Yan [1], recognizing the similarity between the nesting relationship in the nested relational data model and hierarchies in the semi-structured data model.

The definition for NF ORA-SS (normal form for the ORA-SS data model) can be broken into three main parts. The first part ensures that every object class is normalized. The second part ensures that every relationship type is normalized. The third part ensures that there is no data that can be derived from other data in the diagram. Examples of ORA-SS schemas that are not in NF ORA-SS are shown below, along with schemas that are in NF ORA-SS that capture the same information.

Figure 1a represents an object class *Course*. The attributes of course are *code, title, department* and *faculty*. This object class is not normalized since *code* determines *title* and *department*, but *department* determines *faculty*. What this means is that a department belongs to only one faculty, and this information needs to be stored once rather than once for every course. Figure 1b has object classes *Course, DepRef* and

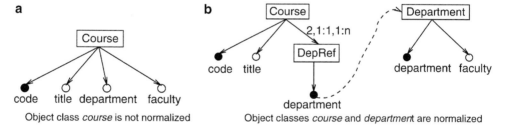

Normal Form ORA-SS Schema Diagrams, Fig. 1 Example of object class normal form

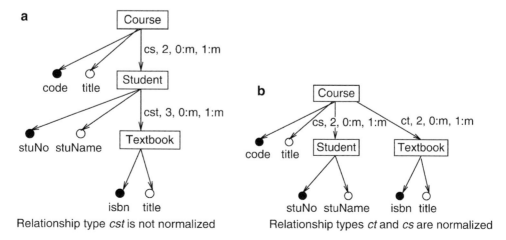

Normal Form ORA-SS Schema Diagrams, Fig. 2 Example of relationship type normal form

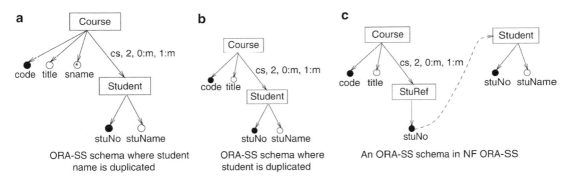

Normal Form ORA-SS Schema Diagrams, Fig. 3 Example of ORA-SS normal form (NF ORA-SS)

Department. Object class *DepRef* has a reference to object class *Department*, so the information about the faculty a department belongs to is stored only once. The object classes in Fig. 1b are normalized.

Figure 2a shows a relationship type *cst*, which is a ternary relationship type among the object classes *Course*, *Student*, and *Textbook*. Intuitively it makes sense because courses have students and textbooks. However, students and textbooks are independent of each other. A course has a certain set of students, irrespective of the textbooks for the course, and a course has a certain set of textbooks irrespective of the students taking the course. In the schema in Fig. 2a, the textbooks for a given course are repeated for each student taking the course. Figure 2b has two relationship types *cs* and *ct*, between *Course* and *Student*, and *Course* and *Textbook*, respectively. Both of these relationship types are normalized. However, the object classes *Student* and *Textbook* in Fig. 2b are still not normalized.

Figure 3a shows two object classes *Course* and *Student*. Object class *Course* has attributes *code, title* and *sname*. Attribute *sname* is a multivalued attribute and has the names of the students taking the course. This information is repeated in the attribute *stuName* that belongs to object class *Student*, so the ORA-SS schema in Fig. 3a is not in NF ORA-SS. If the student names are stored only in the attribute *stuName* of object class *Student*, as shown in Fig. 3b that redundancy is removed. Of course the many-to-many relationship type leads to the students name being repeated in every course the student takes. Using a reference to

represent this relationship removes this redundancy and the resulting ORA-SS diagram in Fig. 3c is in NF ORA-SS.

Cross-References

▶ Normal Forms and Normalization
▶ Object Relationship Attribute Data Model for Semistructured Data
▶ Semi-structured Database Design

Recommended Reading

1. Ling TW, Yan LL. NF-NR: a practical normal form for nested relations. J Syst Integr. 4(4):309–40.

Normal Forms and Normalization

Marcelo Arenas
Pontifical Catholic University of Chile, Santiago, Chile

Definition

A normal form defines a condition over a set of data dependencies, or semantic constraints, that has been specified as part of a database schema. This condition is used to check whether the design of the database has some desirable properties (for example, the database does not store redundant information), and if this is not the case, then it can also be used to convert the poorly designed database into an equivalent well-designed one.

Historical Background

Information is one of the most - if not the most - valuable assets of a company. Therefore, organizations need tools to allow them to structure, query and analyze their data, and, in particular, they need tools providing simple and fast access to their information.

During the last 40 years, relational databases have become the most popular computer application for storing and analyzing information. The simplicity and elegance of the relational model, where information is stored just as tables, has largely contributed to this success.

To use a relational database, a company first has to think of its data as organized in tables. How easy it is for a user to understand this organization and use the database depends on the design of these relations. If tables are not carefully selected, users can spend too much time executing simple operations, or may not be able to extract the desired information.

Since the beginnings of the relational model, it was clear for the research community that the process of designing a database is a nontrivial and time-consuming task. Even for simple application domains, there are many possible ways of storing the data of interest. Soon the difficulties in designing a database became clear for practitioners, and the design problem was recognized as one of the fundamental problems for the relational technology.

During the 70s and 80s, a lot of effort was put into developing methodologies to aid in the process of deciding how to store data in a relational database. The most prominent approaches developed at that time - which today are a standard part of the relational technology - were the entity-relationship and the normalization approaches. In the former approach, a diagram is used to specify the objects of an application domain and the relationships between them. The *schema* of the relational database, i.e., the set of tables and column names, is then automatically generated from the diagram. In the normalization approach, an already designed relational database is given as input, together with some semantic information provided by a user in the form of relationships between different parts of the database, called *data dependencies*. This semantic information is then used to check whether the design has some desirable properties, and if this is not the case, it can also be used to convert the poor design into an equivalent well-designed database.

Foundations

In the relational model, a database is viewed as a collection of relations or tables. For instance, a relational database storing information about courses in a university is shown in Fig. 1. This relation consists of a time-varying part, the data about courses, and a part considered to be time independent, the *schema* of the relation, which is given by the name of the relation (*Course*) and the names of the attributes of this relation.

Usually, the information contained in a database satisfies some *semantic* restrictions. For example, in the relation *Course* shown in Fig. 1, it is expected that only one title is associated with each course number. These restrictions are called *data dependencies*, and they are expressed by using suitable languages. For example, the previous constraint corresponds to a functional dependency, which is an expression of the form $X \rightarrow Y$, where X and Y are sets of attributes. A relation satisfies $X \rightarrow Y$ if for every pair of tuples t_1, t_2 in it, if t_1 and t_2 have the same values on X, then they have the same values on Y. Thus, for example, the relation shown in Fig. 1 satisfies functional dependency *Number* $\rightarrow$ Title since each course has only one title.

In [2], Codd showed that a database containing functional dependencies may exhibit some anomalies when the information is updated. For example, consider again the relational database shown in Fig. 1, which includes functional

Normal Forms and Normalization, Fig. 1 Example of a relational database

Course	Number	Title	Section	Room
	CSC258	Computer organization	1	LP266
	CSC258	Computer organization	2	GB258
	CSC258	Computer organization	3	LM161
	CSC258	Computer organization	3	GB248
	CSC434	Data management systems	1	GB248

CourseInfo	Number	Title
	CSC258	Computer organization
	CSC434	Data management systems

CourseTerm	Number	Section	Room
	CSC258	1	LP266
	CSC258	2	GB258
	CSC258	3	LM161
	CSC258	3	GB248
	CSC434	1	GB248

Normal Forms and Normalization, Fig. 2 Example of a normalized relational database

dependency $Number \rightarrow Title$. This database is prone to three different types of anomalies. First, if the name of the course with number CSC258 is changed to Computer Organization I, then four distinct cells need to be updated. If any of them is not updated, then the information in the database becomes inconsistent. This anomaly was called an *update anomaly* by Codd [2], and it arises because the instance is storing redundant information. Second, if the information is updated because a new semester is starting, and the course with number CSC434 is not given in that semester, then the last tuple of the instance is deleted and no information about CSC434 appears in the updated instance. This has the additional effect of deleting the title of the course, which will be the same the next time that CSC434 is offered. This anomaly was called a *deletion anomaly* by Codd [2], and it arises because the relation is storing information that is not directly related; the sections of a course vary from one term to another while its title is likely not to be changed from one semester to the next one. This can also lead to *insertion anomalies* [2]; if a new course (CSC336, Numerical Methods) is created, then it cannot be added to the database until at least one section and one room is assigned to the course.

To avoid updates anomalies, Codd introduced three increasingly restrictive *normal forms* [2, 3], which specify some syntactic properties that the set of functional dependencies in a database must satisfy. The most restrictive among them is known today as Boyce-Codd Normal Form (BCNF). Informally, a database schema S including a set Σ of functional dependencies is in BCNF, if there is no set of attributes $(X \cup \{A\} \cup \{B\})$ such that $A \notin X$, $B \notin X$, $X \rightarrow A$ holds in S but $X \rightarrow B$ does not hold in S, that is, if there are no attributes in S with different levels of association between them according to Σ (if $X \rightarrow A$ holds in S but $X \rightarrow B$ does not hold in S, then for every value of X, there exists only one value of A in the database but possibly many values of B.). For example, the database shown in Fig. 1 is not in BCNF since $Number \rightarrow Title$ holds in this database, while $Number \rightarrow Section$ does not hold.

In [2], Codd also introduced the first *normalization algorithm*, that is, a procedure that takes as input a relational schema that includes some data dependencies and does not satisfy some particular normal form, and produces a new schema that conforms to this normal form. For example, if S is the relational schema *Course* (*Number, Title, Section, Room*) and Σ is the set of functional dependencies {$Number \rightarrow Title$}, then the standard normalization algorithm for BCNF [1] produces a new schema where *Course* is split into two tables: *CourseInfo*(*Number, Title*) and *CourseTerm*(*Number, Section, Room*). In Fig. 2, it is shown how the information in the initial database in Fig. 1 is stored under the new schema. It should be noticed that the new schema is not prone to any of the anomalies mentioned at the beginning of this section:

- If the name of the course with number CSC258 is changed to Computer Organization I, then only one cell needs to be updated.
- If the information is updated because a new semester is starting, and the course with number CSC434 is not given in that semester, then the last tuple of relation *CourseTerm* is deleted. This does not have the additional effect of deleting the title of the course (this information is kept in the relation *CourseInfo*).
- If a new course (CSC336, Numerical Methods) is created, then it can be added to the database even if no section has been created for this course (tuple (CSC336, Numerical Methods) is included only in the relation *CourseInfo*).

A normalization algorithm takes as input a relational schema and generates a database schema in some particular normal form. It is desirable that these two are as similar as possible, that is, they should contain the same data and the same semantic information. These properties have been called *information losslessness* and *dependency preservation* in the literature, respectively.

Let S_1, S_2 be two database schemas. Intuitively, two instances I_1 of S_1 and I_2 of S_2 contain the same information if it is possible to retrieve the same information from them, that is, for every query Q_1 over I_1 there exists a query Q_2 over I_2 such that $Q_1(I_1) = Q_2(I_2)$, and vice versa. To formalize this notion, a query language has to be chosen. If this query language is relational algebra, then this notion is captured by the notion of calculously dominance introduced by Hull [7]. Schema S_2 *dominates* S_1 *calculously* if there exist relational algebra expressions Q over S_1 and Q' over S_2 satisfying the following property: For every instance I of S_1, there exists an instance I' of S_2 such that $Q(I) = I'$ and $Q'(I') = I$. Thus, every query Q_1 over I can be transformed into an equivalent query $Q_2 = Q_1 \circ Q'$ over I', since $Q_2(I') = Q_1(Q'(I')) = Q_1(I)$, and, analogously, every query Q_2 over I' can be transformed into

an equivalent query $Q_1 = Q_2 \circ Q$ over I, since $Q_1(I) = Q_2(Q(I)) = Q_2(I')$.

Normalization algorithms try to achieve the goal of information losslessness; if any of them transforms a database schema S into a database schema S', then S' should dominate S calculously. The standard normalization algorithm for BCNF uses only the projection operator to transform a schema [1] and, thus, calculously dominance is defined in terms of this operator and its inverse, the join operator. More precisely, the normalization algorithm mentioned in this section takes as input a relation R and a set of functional dependencies Σ, and uses the projection operator to transform it into a database schema in BCNF that is composed by some relations $R_1, \dots, R_n$. Then $R_1, \dots, R_n$ is a *lossless decomposition* of R if for every instance I of R, there is an instance I' of $R_1, \dots, R_n$ such that:

- For every $i \in \{1, \dots, n\}$, it holds that $I'_i = \pi_{U_i}(I)$, where I'_i is the R_i-relation of I' and U_i is the set of attribute of R_i, and
- $I = I'_1 \bowtie I'_2 \bowtie \dots \bowtie I'_n$, where each I'_i is the R_i-relation of I'.

That is, every instance I of S can be transformed into an instance I' of S' by using the projection operator, and I can be reconstructed from I' by using the join operator. For example, the relation *CourseInfo* in Fig. 2 can be obtained by projecting the relation *Course* in Fig. 1 over the set of attributes {*Number,Title*}, while the relation *CourseTerm* in Fig. 2 can be obtained by projecting the relation *Course* in Fig. 1 over the set of attributes {*Number, Section, Room*}. Moreover, the relation *Course* in Fig. 1 can be obtained by joining relations *CourseInfo* and *CourseTerm* in Fig. 2. Given that this holds for every instance I of the initial schema *Course(Number, Title, Section, Room)*, the new schema *CourseInfo(Number, Title)*, *CourseTerm(Number, Section, Room)* is said to be a lossless decomposition of the initial one.

Normalization algorithms also try to achieve the goal of dependency preservation; if any of them transforms a database schema S including a set Σ of data dependencies, into a database schema S' including a set Σ' of data dependencies, then Σ should be equivalent to Σ' (no semantic information is lost). In the running example, the standard normalization algorithm for BCNF produces a new schema *CourseInfo(Number,Title)* and *CourseTerm(Number, Section, Room)*, and also includes dependency *Number* → Title in the relation *CourseInfo*. Thus, the new schema is a dependency preserving decomposition of the initial one.

The normalization approach was proposed in the early 70s by Codd [2]. Since then, many researchers have studied the normalization problem for relational databases and other data models, and today it is possible to find normal forms for many different types of data dependencies: 3NF [2] and BCNF [3] for functional dependencies, 4NF [4] for multivalued dependencies, PJ/NF [5] and 5NFR [8] for join dependencies, and DK/NF [6] for general constraints. These normal forms, together with normalization algorithms for converting a poorly designed database into a well-designed database, can be found today in every database textbook.

Key Applications

Normal forms and normalization algorithms are essential to schema design, redundancy elimination, update anomaly prevention and efficient storage.

Cross-References

▶ Boyce-Codd Normal Form
▶ Fourth Normal Form
▶ Second Normal Form (2NF)
▶ Third Normal Form

Recommended Reading

1. Abiteboul S, Hull R, Vianu V. Foundations of databases. Reading: Addison-Wesley; 1995.
2. Codd EF. Further normalization of the data base relational model. In: Data base systems. Englewood Cliffs: Prentice-Hall; 1972. p. 33–64.
3. Codd EF. Recent investigations in relational data base systems. In: Proceedings of the IFIP Congress, Information Processing 74; 1974. p. 1017–21.
4. Fagin R. Multivalued dependencies and a new normal form for relational databases. ACM Trans Database Syst. 1977;2(3):262–78.
5. Fagin R. Normal forms and relational database operators. In: Proceedings of the ACM SIGMOD International Conferences on Management of Data; 1979. p. 153–60.
6. Fagin R. A normal form for relational databases that is based on domians and keys. ACM Trans Database Syst. 1981;6(3):387–415.
7. Hull R. Relative information capacity of simple relational database schemata. SIAM J Comput. 1986;15(3):856–86.
8. Vincent M. A corrected 5NF definition for relational database design. Theor Comput Sci. 1997;185(2):379–91.

NoSQL Stores

Sai Wu
Zhejiang University, Hangzhou, Zhejiang, People's Republic of China

Definition

NoSQL (originally referring to "non SQL") is a new type of data management system, which, different from the conventional database systems, does not model its data using the relational tabular model. To provide a highly scalable and available data access service, NoSQL systems may adopt various data models (e.g., key-value, graph, and document) based on the applications that they are designed for. The flexibility of NoSQL's data model makes it easier to scale to a large cluster. However, on the other hand, most NoSQL systems compromise the consistency for the scalability and availability (CAP theorem says we can only keep two features among consistency, availability, and partition tolerance). Many of them adopt the multi-version strategy and the eventual consistency model.

Applications can use the specific APIs (e.g., key based and vertex based) provided by the NoSQL system to access the data. Standard SQL is not supported, since most NoSQL systems are not designed to support relational operators. The atomic access is only limited to a single-key or document retrieval, and hence, supporting transaction over multiple keys or documents is very challenging.

Recently, the features of NoSQL and database systems are merged together. The result systems are called new SQL systems or NoSQL (not only SQL) systems. They support standard SQL and high-throughput transactions on top of a cluster.

Historical Background

Relational database management system (RDBMS) has dominated the data storage market for more than 30 years. Its success was attributed to some key features such as strong schema requirement, indexing technique, transaction support, concurrent control, and failure recovery. Those features become a double-edged sword when handling large-scale heterogeneous data.

With the popularity of smartphones, mobile and web applications start dominating the market, whereas as a data management system, RDBMS may also be too expensive for those applications. Many mobile applications will generate semi-structured (e.g., JSON document) or graph data (e.g., social graph). But before imported into the RDBMS, all data must be strictly formatted using the predefined schema. However, it is nontrivial to use a relational schema to handle semi-structured data and graph data. Moreover, based on the CAP theory, it is impossible (or too costly) to guarantee both strong consistency and availability when some network partition fails. As most applications require simple query interface and choose the availability over the consistency, a branch of new systems, such as BigTable [1], HBase, Redis, Dynamo [2], Cassandra [3], MongoDB, and Neo4J, is developed to serve different processing purposes. To distinguish with conventional database systems, those systems are called NoSQL systems, because they are not designed to support the relational model and its query language, SQL.

The simplest NoSQL systems are key-value stores. Every item in the store has two attributes. The key attribute is used to identify the item, while the value attribute maintains its data. The key-value model can be also extended to support relational data. In such systems (e.g., BigTable and HBase), the value attribute actually represents an arbitrary number of columns or column families. In principle, you can use the value attribute to maintain any complex data structures, but key-value store only allows you to query those structures via keys.

To address this problem, in document store such as MongoDB and CouchDB, each value is considered as a document that can contain complex data types like arrays, key-value pairs, or even nested arrays. Indexes are supported in the document store to facilitate the search on non-key values, and views are also used in some systems [4] to precompute the query results.

The key-value model is inefficient in processing graph data which actually can be represented as a sparse matrix. This requires a fundamental redesign for the data model and processing techniques. Example graph stores like Neo4J and HyperGraphDB support typical graph queries like retrieving nearest neighbors and searching isomorphic subgraphs. One interesting system is OrientDB which is designed as a document store but also maintains the relationships of items in a graph database. It can support efficient search for various types of data.

Compared to the full-fledged database systems, NoSQL systems are considered as lightweight systems, because they are normally sacrificing the features like strong consistency and transaction support for availability and scalability. For example, BigTable, HBase, Cassandra, and Dynamo only guarantee the atomicity for single-key operation and eventual consistency. It requires significant efforts to support transactions and standard SQL on NoSQL systems. However, those two features are becoming increasingly important for applications that need to process large-scale hybrid data (structured+unstructured data). This stimulates the design of new data manage systems, the NewSQL system or NoSQL (not only SQL) system.

Google's Spanner [5] is an early NewSQL system which extends the structure of BigTable to support transactions across multiple data centers. The idea of Spanner is re-implemented in an open-source project, Cockroach. Other popular NewSQL systems include FoundationDB, Clustrix, and OrientDB. The intuition of those systems is to provide the same scalable performance as previous NoSQL systems and support key features in RDBMS, such as transaction, index, and SQL. A detailed comparison between DBMSs and NoSQL/NewSQL systems can be found in [6].

To improve the search performance, key-value stores like Redis and Ramcloud maintain their data completely in memory, whereas MongoDB uses memory-mapped files to buffer a portion of data in memory. Hao Zhang et al. discussed how NoSQL systems leverage memory to support efficient and highly concurrent access [7].

Foundations

NoSQL systems adopt different architectures to process various types of data. For the simplest key-value model, the design goal is to efficiently locate the corresponding value for any input key.

This becomes a nontrivial task, when data are partitioned into multiple shards to achieve the scalability. Figure 1 illustrates the distributed architecture of HBase. Many other NoSQL systems share a similar design.

In HBase, there is a single master server in the cluster which maintains the metadata of the system, such as data type and statuses of servers in the cluster. Multiple region servers are established to help partition the key space. Each of them is responsible for one or more continuous key ranges. The region server maintains multiple data stores which refer to the formatted data stored in the DFS (Distributed File System). To speed up the data research and insertion, data store can buffer the most recently used data in memory. For search purpose, the master server records how key ranges are partitioned among the region servers. Master server and region servers synchronize their knowledge using a distributed service. In particular, HBase relies on the service of zookeepers for synchronization.

To reduce the overhead of master server, Cassandra and Dynamo adopt the DHT (Distributed Hash Table) to maintain the partitioning information of key space. Client can contact any region server to locate the position of a given key using the DHT protocol.

The key-value model is actually very flexible and can be extended to support relational

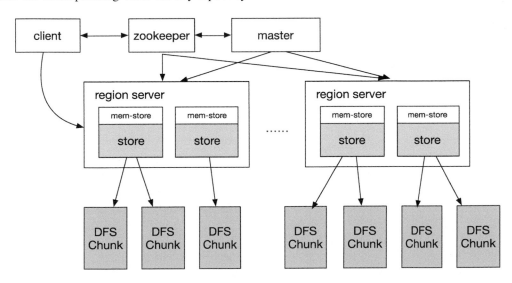

NoSQL Stores, Fig. 1 Architecture of HBase

NoSQL Stores, Fig. 2
Data model of
BigTable/HBase

key	value			
uid	column family: a		column family: b	
	name	age	salary	position
1001	Tom	23	12,300	Sales
1002	Jerry	25	45,600	Manager

NoSQL Stores, Fig. 3
Data model of document
store

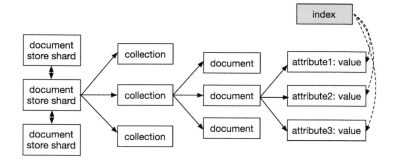

data. Figure 2 shows the data model used in BigTable and HBase. The column user ID is used as the key to address key-value pairs, while the remaining columns are stored as values. Columns can be also grouped into column families based on their semantic relationships. To reduce the storage and maintenance overhead, BigTable adopts the column-oriented storage model. Each column will be stored separately and clients can retrieve any specific column independently.

To provide a reliable data access service, multiple replicas are created for each key-value pair. Some systems adopt the versioning technique which allows users to retrieve stale data via version number. Early key-value system only supports atomic access for single-key-based retrieval with the eventual consistent model. New key-value system or NewSQL system can support DBMS-like transactions with strict consistency.

Document store adopts a slightly different design, where corresponding documents are grouped as a collection. Inside each document, user can define one or more attributes. For example, in MongoDB, each document is a JSON string that can have nested attributes. To facilitate the search, index is built for those attributes and keyword-based query can be efficiently answered (Fig. 3).

Key Applications

NoSQL systems have been widely used in the applications that need to process a huge number of concurrent accesses. For example, mobile game developers such as Zynga build their data management modules on top of NoSQL systems which can support millions of online gamers. E-commerce companies like the Hut Group leverage NoSQL techniques to help them handle the peak workload generated in holidays. Social websites such as Facebook and Twitter design their own NoSQL systems to maintain users' profiles and relationship. Foursquare migrates of venues and check-ins from its original relational architecture to MongoDB.

URL to Code

1. HBase: http://hbase.apache.org
2. CouchDB: http://couchdb.apache.org/
3. Cassandra: http://cassandra.apache.org/
4. CockroachDB: https://github.com/cockroach db/cockroach

Cross-References

▶ Self-Management Technology in Databases

Recommended Reading

1. Chang F, Dean J, Ghemawat S, Hsieh WC, Wallach DA, Burrows M, Chandra T, Fikes A, Gruber RE. Bigtable: a distributed storage system for structured data. ACM Trans Comput Syst. 2008;26(2):133.
2. DeCandia G, Hastorun D, Jampani M, Kakulapati G, Lakshman A, Pilchin A, Sivasubramanian S, Vosshall P, Vogels W. Dynamo: Amazon's highly available key-value store. In: Proceedings of the 21st ACM Symposium on Operating System Principles; 2007. p. 205–20.
3. Lakshman A, Malik P. Cassandra: structured storage system on a P2P network. In: Proceedings of the ACM SIGACT-SIGOPS Symposium on Principles of Distributed Computing; 2009. p. 5.
4. Iordanov B. HyperGraphDB: a generalized graph database. In: Proceedings of the International Conference on Web-Age Information Management; 2010. p. 25–36.
5. Corbett JC, Dean J, Epstein M, Fikes A, Frost C, Furman JJ, Ghemawat S, Gubarev A, Heiser C, Hochschild P, Hsieh WC, Kanthak S, Kogan E, Li H, Lloyd A, Melnik S, Mwaura D, Nagle D, Quinlan S, Rao R, Rolig L, Saito Y, Szymaniak M, Taylor C, Wang R, Woodford D. Spanner: Google's globally distributed database. ACM Trans Comput Syst. 2013;31(3):8.
6. Cattell R. Scalable SQL and NoSQL data stores. SIGMOD Record. 2010;39(4):12–27.
7. Zhang H, Chen G, Ooi BC, Tan KL, Zhang M. In-memory big data management and processing: a survey. IEEE Trans Knowl Data Eng. 2015;27(7):1920.

Now in Temporal Databases

Curtis E. Dyreson[1], Christian S. Jensen[2], and Richard T. Snodgrass[3,4]
[1] Utah State University, Logan, UT, USA
[2] Department of Computer Science, Aalborg University, Aalborg, Denmark
[3] Department of Computer Science, University of Arizona, Tucson, AZ, USA
[4] Dataware Ventures, Tucson, AZ, USA

Synonyms

Current date; Current time; Current timestamp; Until changed

Definition

The word *now* is a noun in the English language that means "at the present time." This notion appears in databases in three guises. The first use of *now* is as a function within queries, views, assertions, etc. For instance, in SQL, CURRENT_DATE within queries, etc., returns the current date as an SQL DATE value; CURRENT_TIME and CURRENT_TIMESTAMP are also available. These constructs are nullary functions.

In the context of a transaction that contains more than one occurrence of these functions, the issue of which time value(s) to return when these functions are invoked becomes important. When having these functions return the same (or consistent) value, it becomes a challenge to select this time and to synchronize it with the serialization time of the transaction containing the query.

The second use is as a database variable used extensively in temporal data model proposals, primarily as timestamp values associated with tuples or attribute values in temporal database instances. As an example, within transaction-time databases, the stop time of data that has not been logically deleted and thus is current is termed "until changed." A challenging aspect of supporting this notion of *now* has been to contend with instances that contain this variable when defining the semantics of queries and modification and when supporting queries and updates efficiently, e.g., with the aid of indices.

The third use of *now* is as a database variable with a specified offset (a "now-relative value") that can be stored within an implicit timestamp or as the value of an explicit attribute. Challenges include the specification of precise semantics for database instances that contain these variables and the indexing of such instances.

Historical Background

Time variables such as *now* are of interest and indeed are quite useful in databases, including databases managed by a conventional DBMS, that record time-varying information, the validity of which often depends on the current-time value. Such databases may be found in many application areas, such as banking, inventory management, and medical and personnel records.

The SQL nullary functions CURRENT_DATE, CURRENT_TIME, and CURRENT_TIMESTAMP have been present since SQL's precursor, SEQUEL 2 [1]. The transaction stop time of *until changed* has been present since the initial definition of transaction-time databases (e.g., in the early work of Ben-Zvi [2]).

The notion of *now* and the surprisingly subtle concerns with this ostensibly simple concept permeate the literature, and yield quite interesting solutions.

Foundations

The three notions of "now" in temporal databases are explored in more detail below.

SQL Nullary Functions
The following example illustrates the uses and limitations of *now*, specifically the SQL CURRENT_DATE, CURRENT_TIME, and CURRENT_TIMESTAMP, in conventional databases. Banking applications record when account balances for customers are valid. Examine the relation AccountBalance with attributes AccountNumber, Balance, FromDate, and ToDate. To determine the balance of account 12345 on January 1, 2007, one could use a simple SQL query.

```
SELECT Balance
FROM AccountBalance
WHERE Account = 12345
      AND FromDate <= DATE
      '2007-01-01'
      AND DATE '2007-01-01' < ToDate
```

To determine the balance of that account *today*, the nullary function CURRENT_DATE is available.

```
SELECT Balance
FROM AccountBalance
WHERE Account = 12345
      AND FromDate <= CURRENT_DATE
      AND CURRENT_DATE < ToDate
```

Interestingly, in the SQL standard, the semantics of the function are implementation-dependent, which opens it to various interpretations: "If an SQL-statement generally contains more than one reference to one or more <datetime value functions>s then all such references are effectively evaluated simultaneously. The time of evaluation of the <datetime value function> during the execution of the SQL-statement is implementation-dependent." (This is from the SQL:1999 standard.) So an implementation is afforded considerable freedom in choosing a definition of "current," including perhaps when the statement was presented to the system, or perhaps when the database was first defined.

Transactions take time to complete. If a transaction needs to insert or modify many tuples, it may take minutes. However, from the point of view of the user, the transaction is atomic (all or nothing) and serializable (placed in a total ordering). Ideally "current" should mean "when the transaction executed," that is, the instantaneous time the transaction logically executed, consistent with the serialization order.

Say a customer opens an account and deposits $200. This transaction results in a tuple being inserted into the AccountBalance table. The transaction started on January 14 at 11:49 P.M. and committed at 12:07 A.M. (that is, starting before midnight and completing afterwards). The tuple was inserted on 11:52 P.M. Another user also created a second account with an initial balance of $500 in a transaction that started on January 14 at 11:51 P.M. and committed on 11:59 P.M., inserting a tuple into the AccountBalance relation at 11:52 P.M. If the system uses the transaction start time, the following two tuples will be in the relation.

According to these two tuples, the sum of balances on January 14 is $700. But note that though both transactions began on January 14, only the second transaction committed by midnight (also note that the second transaction is earlier in the serialization order since it commits first). Hence, the actual aggregate balance on January 14 was never $700: the balance started at $0 (assuming there were initially no other tuples), then changed to $500. Only on January 15 did it increase to $700.

Suppose that instead of using the time at which the entire transaction began, the time of the actual insert statement (e.g., for the first transaction, January 14, 11:52 P.M.) is used; then the same problem occurs.

What is desired is for a time returned by CURRENT_DATE to be consistent with the serialization order and with the commit time, which unfortunately is not known when CURRENT_DATE is executed. Lomet et al. [3] showed how to utilize the lock manager to assign a commit time in such a way that it is consistent with the serialization order as well as with dates assigned as values to prior CURRENT_DATE invocations. Specifically, each use of CURRENT_DATE, etc. defines or narrows a *possible period* during which the commit time of the transaction must reside. For CURRENT_DATE this period is the particular 24 h of the day returned by this function. Read-write conflicts with other transactions further narrow the possible period. For example, if a particular transaction reads values written by another transaction, the transaction time of the reader must be later than that of the writer. At commit time, it is attractive to assign the earliest instant in the possible period to the transaction. Alternatively, if the possible period ever becomes empty, the transaction must be aborted. Lomet et al. outline important optimizations to render such calculations efficient.

Now in End or Stop Columns

The above discussion concerned how to determine what to store in the begin time of a tuple (e.g., in the FromDate column for the two example tuples in the table displayed above) and how to make this time consistent with that returned by CURRENT_DATE. We now consider what to store in the end time of a tuple (e.g., in the ToDate column). The validity of a tuple then starts when a deposit is made and extends until the current time, assuming no update transactions are committed. Thus, on January 16, the balance is valid from January 14 until January 16; on January 17, the balance is valid from January 14 until January 17; etc.

It is impractical to update the database each day (or millisecond) to correctly reflect the valid time of the balance. A more promising approach is to store a variable, such as *now*, in the ToDate field of a tuple, to indicate that the time when a balance is valid depends on the current time. In the example, it would be recorded on January 14 that the customer's balance of $200 is valid from January 14 through *now*. The CURRENT_DATE construct used in the above query cannot be stored in a column of an SQL table. All major commercial DBMSs have similar constructs, and impose this same restriction. The user is forced instead to store a specific time, which is cumbersome and inaccurate (Clifford et al. explain these difficulties in great detail [4]).

The solution is to allow one or more free, current-time variables to be stored in the database. Chief among these current-time variables is "*now*" (e.g., [5]), but a variety of other symbols have been used, including "-" [2], "∞" [6], "@" [7], and "*until-changed*" [8]. Such stored variables have advantages at both the semantic and implementation levels. They are expressive and space efficient and avoid the need for updates at every moment in time.

As an example, consider a variable database with the tuple ⟨*Jane, Assistant,* [*June 1, now*]⟩, with *now* being a variable. The query "List the faculty on June 15," evaluated on June 27, results in {??⟨*Jane, Assistant*⟩}.

Now-Relative Values

A now-relative instant generalizes and adds flexibility to the variable *now* by allowing an offset from this variable to be specified. Now-relative instants can be used to more accurately record the knowledge of Jane's employment. For example, it may be that hiring changes are recorded in the database only 3 days after they take effect. Assuming that Jane was hired on June 1, the definite knowledge of her employment is accurately captured in the tuple ⟨*Jane, Assistant,* [*June 1, now − 3 days*]⟩. This tuple states that Jane was an Assistant Professor from June 1 and until three days ago, but it contains no information about her employment as of, e.g., yesterday.

A now-relative instant thus includes a displacement, which is a signed duration, also termed a span, from *now*. In the example given above, the displacement is minus 3 days.

AccountNumber	Balance	FromDate	ToDate
121345	200	2007-01-14	...
543121	500	2007-01-14	...

Now-relative variables can be extended to be indeterminate [4], as can regular instants and the variable *now*.

The semantics of now variables has been formalized with an *extensionalization* that maps from a *variable database level* containing variables as values to an *extensional database level*. The extensional database exhibits three key differences when compared to the variable database level. First, no variables are allowed – the extensional level is fully ground. Second, timestamps are instants rather than intervals. Third, an extensional tuple has one additional temporal attribute, called a *reference time attribute*, which may be thought of as representing the time at which a meaning was given to the temporal variables in the original tuple. Whereas the variable-database level offers a convenient representation that end-users can understand and that is amenable to implementation, the mathematical simplicity of the extensional level supports a rigorous treatment of temporal databases in terms of first order logic.

When a variable database is queried, an additional problem surfaces: what to do when a variable is encountered during query evaluation. In the course of evaluating a user-level query, e.g., written in some dialect of SQL, it is common to transform it into an internal algebraic form that is suitable for subsequent rule or cost-based query optimization. As the query processor and optimizer are among the most complex components of a database management system, it is attractive if the added functionality of current-time-related timestamps necessitates only minimal changes to these components.

Perhaps the simplest approach to supporting querying is that, when a timestamp that contains a variable is used during query processing (e.g., in a test for overlap with another timestamp), a ground version of that timestamp is created and used instead. With this approach, only a new component that substitutes variable timestamps with ground timestamps has to be added, while existing components remain unchanged.

Put differently, a *bind* operator can be added to the set of operators already present. This operator is then utilized when user-level queries are mapped to the internal representation. The operator accepts any tuple with variables. It substitutes a ground value for each variable and thus returns a ground (but still variable-level) tuple. Intuitively, the *bind* operator sets the perspective of the observer, i.e., it sets the reference time. Existing query languages generally assume that the temporal perspective of an observer querying a database is the time when the observer initiates the query.

Torp et al. has shown how to implement such variables within a database system [9].

With *now* as a database variable, the temporal extent of a tuple becomes a non-constant function of time. As most indices assume that the extents being indexed are constant in-between updates, the indexing of the temporal extents of now-relative data poses new challenges.

For now-relative transaction-time data, one may index all data that is not now-relative (i.e., has a fixed end time) in one index, e.g., an R-tree, and all data that is now-relative (i.e., the end time is *now*) in another index where only the start time is indexed.

For bitemporal data, this approach can be generalized to one where tuples are distributed among four R-trees. The idea is again to overcome the inabilities of indices to cope with continuously evolving regions, by applying transformations to the growing bitemporal extents that render them stationary and thus amenable to indexing. Growing regions come in three shapes, each with its own transformation. These transformations are accompanied by matching query transformations [10].

In another approach, the R-tree is extended to store *now* for both valid and transaction time in the index. The resulting index, termed the GR-tree, thus accommodates bitemporal regions and uses minimum bounding regions that can be either static or growing and either rectangles or stair shapes. This approach has been extended to accommodate bitemporal data that also have spatial extents [11].

Key Applications

The notion of *now* as a nullary function representing the current time is common in database applications. For relational database management, SQL offers CURRENT_DATE, CURRENT_TIME, and CURRENT_ TIMESTAMP functions that return an appropriate SQL time value. In other kinds of database management systems, such as native XML database management systems, similar constructs can be found. For example, XQuery has a fn:current-time() function that returns a value of XML Schema's xs:time type. XQuery also has a fn:current-date() function.

Less common in existing database applications are the other notions of now: as a variable stored in a database to represent the ever-changing current time or as a time related to, but displaced from, the current time.

Future Directions

The convenience of using now variables poses challenges to the designers of database query languages. The user-defined time types available in SQL-92 can be extended to store now-relative variables as values in columns. The TSQL2, language [12] does so, and also supports those variables for valid and transaction time. In TSQL2, the "bind" operation is implicit; NOBIND is provided to store variables in the database. However, such variables have yet to be supported by commercial DBMSs. It may also be expected that at least one of the three uses of *now* will re-emerge as part of a temporal extension of XQuery or a language associated with the Semantic Web.

The impact of stored variables on database storage structures and access methods is a relatively unexplored area. Such stored variables may present optimization opportunities. For example, if the optimizer knows (through attribute statistics) that a large proportion of tuples has a "to" time of *now*, it may then decide that a sort-merge temporal join will be less effective. Finally, new kinds of variables, such as *here* for spatial and spatio-temporal databases, might offer an interesting extension of the framework discussed here.

Cross-References

► Bitemporal Indexing
► Bitemporal Interval
► Supporting Transaction Time Databases
► Temporal Query Languages
► Temporal Strata
► Temporal XML
► Time Period
► Transaction Time
► TSQL2
► Valid Time

Recommended Reading

1. Ben-Zvi J. The time relational model. PhD Dissertation, University of California, Los Angeles; 1982.
2. Bliujūtė R, Jensen CS, Šaltenis S, Slivinskas G. Light-weight indexing of bitemporal data. In: Proceedings of the 12th International Conference on Scientific and Statistical Database Management; 2000. p. 125–38.
3. Chamberlin DD, Astraham MM, Eswaran KP, Griffiths PP, Lorie RA, Mehl JW, Reisner P, Wade BW. SEQUEL 2: a unified approach to data definition, manipulation, and control. IBM J Res Dev. 1976;20(6):560–75.
4. Finger M. Handling database updates in two-dimensional temporal logic. J Appl Non Classical Logics. 1992;2(2):201–24.
5. Clifford J, Dyreson CE, Isakowitz T, Jensen CS, Snodgrass RT. On the semantics of "now." ACM Trans Database Syst. 1997;22(2):171–214.
6. Clifford J, Tansel AU. On an algebra for historical relational databases: two views. In: Proceedings of the ACM SIGMOD International Conference on Management of Data; 1985. p. 247–65.
7. Lomet D, Snodgrass RT, Jensen CS. Exploiting the lock manager for timestamping. In: Proceedings of the International Conference on Database Engineering and Applications; 2005. p. 357–68.
8. Lorentzos NA, Johnson RG. Extending relational algebra to manipulate temporal data. Inf Syst. 1988;13(3):289–96.

N

9. Montague R. Formal philosophy: selected papers of Richard Montague. New Haven: Yale University Press; 1974.
10. Šaltenis S, Jensen CS. Indexing of now-relative spatio-bitemporal data. VLDB J. 2002;11(1): 1–16.
11. Snodgrass RT. The temporal query language TQuel. ACM Trans Database Syst. 1988;12(2):247–98.
12. Snodgrass RT, editor. The TSQL2 Temporal Query Language. Norwell: Kluwer; 1995.
13. Torp K, Jensen CS, Snodgrass RT. Modification semantics in now-relative databases. Inf Syst. 2004;29(78):653–83.
14. Wiederhold G, Jajodia S, Litwin W. Integrating temporal data in a heterogeneous environment. In: Tansel A, Clifford J, Gadia SK, Jajodia S, Segev A, Snodgrass RT, editors. Temporal databases: theory, design, and implementation, Chapter 22. Redwood City: Benjamin/Cummings; 1993. p. 563–79.

Null Values

Leopoldo Bertossi
Carleton University, Ottawa, ON, Canada

Definition

Null values are used to represent *uncertain data values* in a database instance.

Key Points

Since the beginning of the relational data model, *null values* have been investigated, with the intention of capturing and representing data values that are uncertain. Depending on the intuitions and cases of uncertainty, different kinds of null values have been proposed, e.g., they may represent information that is withheld, inapplicable, missing, unknown, etc. Thus, in principle, it could be possible to find in a hypothetical database diverse classes of null values, and also several null values of the same class. However, in commercial relational DBMSs and in the SQL Standard, only a single constant, *NULL*, is used to represent the missing values.

Many semantic problems appear when null values are integrated with the rest of the relational data model, which essentially follows the semantics of predicate logic. Among them, (i) the interpretation of nulls values (for a particular intuition); (ii) the meaning of relational operations when applied to both null values and certain data values; and (iii) the characterization of consistency of databases containing null values.

Different formal semantics for null values have been proposed. A common and well-studied semantic for *incomplete databases* uses null values to represent unknown or missing values. Each null value in the database represents a whole set of possible values from the underlying data domain. The combination of concrete values that null values might take generates a class of alternative instances containing certain values. This *possible worlds semantics* makes true whatever is true in every alternative instance. However, the usage of null values in the SQL Standard and commercial DBMS still lacks a clear and complete formal semantic.

Cross-References

▶ Incomplete Information

Recommended Reading

1. Grahne G. The Problem of incomplete information in relational databases. LNCS, vol. 554. Secaucus: Springer; 1991.
2. Levene M, Loizou G. A guided tour of relational databases and beyond, Chapter 5. London: Springer; 1999.
3. Van der Meyden R. Logical approaches to incomplete information: a survey. In: Chomicki J, Saake G, editors. Logics for databases and information systems. Boston: Kluwer; 1998. p. 307–56.

O

OASIS

Serge Mankovski
CA Labs, CA Inc., Thornhill, ON, Canada

Synonyms

Organization for the advancement of structured information standards

Definition

OASIS is a non-for-profit consortium aiming at collaborative development and approval of open international, mainly XML-based, standards.

Key Points

OASIS was founded in 1993 under the name "SGML Open." The initial goal of the organization was to develop guidelines for interoperability among products using Standard Generalized Markup Language (SGML). In 1998 it changed name to OASIS to reflect on changing scope of its technical work.

OASIS consists of an open group of member organizations whose representatives work in committees developing standards, promoting standards adoption, product interoperability and standards conformance. In 2007 OASIS had 5000 participants representing 600 organizations and individual members in 100 countries. OASIS is governed by a member-elected Board in an annual election process. The board membership is based on the personal merits of Board nominees.

OASIS process allows participants to influence standards that affect their business, contribute to standards advancement and start new standards. The process is designed to promote industry consensus. OASIS strategy values creativity and consensus over conformity and control. It relies on the market to determine the particular approach taken in the development of sometimes overlapping standards.

OASIS maintains collaborative relationships with the International Electrotechnical Commission (IEC), International Organization for Standardization (ISO), International Telecommunication Union (ITU) and United Nations Electronic Commission for Europe (UN/ECE), and National Institute of Standards and Technology (NIST).

Among major accomplishments of the OASIS are such influential of standards as a group of ebXML standards, SAML, XACML, WSRP, WSDM, BPEL, OpenDocument, DITA, DocBook, LegalXML and others.

Cross-References

▶ Business Process Execution Language
▶ Service-Oriented Architecture

© Springer Science+Business Media, LLC, part of Springer Nature 2018
L. Liu, M. T. Özsu (eds.), *Encyclopedia of Database Systems*,
https://doi.org/10.1007/978-1-4614-8265-9

Recommended Reading

1. OASIS. Available at: http://www.oasis-open.org

Object Constraint Language

Martin Gogolla
University of Bremen, Bremen, Germany

Synonyms

OCL

Definition

The Unified Modeling Language (UML) includes a textual language called Object Constraint Language (OCL). OCL allows users to navigate class diagrams, to formulate queries, and to restrict class diagrams with integrity constraints. From a practical perspective, the OCL may be viewed as an object-oriented version of the Structured Query Language (SQL) originally developed for the relational data model. From a theoretical perspective, OCL may be viewed as a variant of first-order predicate logic with quantifiers on finite domains only. OCL has a well-defined syntax [1, 3] and semantics [2].

Key Points

The central language features in OCL are: navigation, logical connectives, collections and collection operations.

Navigation: The navigation features in OCL allow users to determine connected objects in the class diagram by using the dot operator ". ". Starting with an expression *expr* of start class *C*, one can apply a property *propC* of class *C* returning, for example, a collection of objects of class *D* by using the dot operator:

expr.propC. The expression *expr* could be a variable or a single object, for example. The navigation process can be repeated by writing *expr.propC.propD*, if *propD* is a property of class *D*.

Logical Connectives: OCL offers the usual logical connectives for conjunction (*and*), disjunction (*or*), and negation (*not*) as well as the implication (*implies*) and a binary exclusive (*xor*). An equality check (=) and a conditional (*if then else endif*) is provided on all types.

Collections: In OCL there are three kinds of collections: sets, bags, and sequences. A possible collection element can appear at most once in a set, and the insertion order in the set does not matter. An element can appear multiple times in a bag, and the order in the bag collection does not matter. An element can appear multiple times in a sequence in which the order is significant. Bags and sequences can be converted to sets with ->*asSet()*, sets and sequences to bags with ->*asBag()*, and sets and bags to sequences with ->*asSequence()*. The conversion to sequences assumes an order on the elements. The arrow notation is explained in more detail below.

Collection Operations: There is a large number of operations on collections in OCL. A lot of convenience and expressibility is based upon them. The most important operations on all collection types are the following: *forAll* realizes universal quantification, *exists* is existential quantification, *select* filters elements with a predicate, *collect* applies a term to each collection element, *size* determines the number of collection elements, *isEmpty* tests for emptiness, *includes* checks whether a possible element is included in the collection, and *including* builds a new collection *including* a new element.

In addition to the central language features, OCL also has special operations available only on particular collection, e.g., the operation *at* on sequences for retrieving an element by its position. All collection operations are applied with the arrow notation mentioned above. Roughly speaking, the dot notation is used when a property follows, i.e., an attribute or a role follows, and the arrow notation is employed when a collection operation follows.

Variables in collection operations: Most collection operations allow variables to be declared (possibly including a type specification), but the variable may be dropped if it is not needed.

Retrieving all Current Instances of a Class: Another important possibility is a feature to retrieve the finite set of all current instances of a class by appending *.allInstances* to the class name. In order to guarantee finite results *.allInstances* cannot be applied to data types like *String* or *Integer*.

Return types in collection operations: If the collection operations are applied to an argument of type *Set/Bag/Sequence(T)*, they behave as follows: *forAll* and exists returns a *Boolean, select* yields *Set/Bag/Sequence(T),* *collect* returns *Bag/Bag/Sequence(T';),* size gives back *Integer*, *isEmpty* yields *Boolean,* *includes* returns *Boolean*, and *including* gives back *Set/Bag/Sequence(T)*.

Most notably, the operation *collect(...)* changes the type of a *Set(T)* collection to a *Bag(T')* collection. The reason for this is that term inside the *collect* may evaluate to the same result for two different collection elements. In order to reflect that the result is captured for each collection element, the result appears as often as a respective collection element exists. This convention in OCL resembles the same approach in SQL: SQL queries with the additional keyword *distinct* return a set; plain SQL queries without *distinct* return a bag. In OCL, the convention is similar: OCL expressions using the additional conversion *asSet()* as in *collect(...)->asSet()* return a set; plain *collect(...)* expressions without *asSet()* return a bag.

Cross-References

▶ Unified Modeling Language

Recommended Reading

1. OMG (ed.). OMG object constraint language specification. OMG, 2007. www.omg.org.
2. Richters M, Gogolla M. On formalizing the UML object constraint language OCL. In: Proceedings of the 17th International Conference on Conceptual Modeling; 1998. p. 449–64.
3. Warmer J, Kleppe A. The object constraint language: getting your models ready for MDA. Boston: Addison-Wesley, Reading; 2003.

Object Data Models

Susan D. Urban and Suzanne W. Dietrich
Arizona State University, Phoenix, AZ, USA

Synonyms

ODB (object database); OODB (object-oriented database); ORDB (object-relational database)

Definition

An object data model provides support for objects as the basis for modeling in a database application. An object is an instance of a class, which is a complex type specification that defines both the state of its instance fields and the behavior provided by its methods. Object features also include a unique object identifier that can be used to refer to the object, as well as the organization of data into class hierarchies that support inheritance of state and behavior. The term object data model encompasses the data model for both object-oriented databases (OODBs) and object-relational databases (ORDBs). OODBs use an object-oriented programming language as the database language and provide inherent support for the persistence of objects with typical database functionality. ORDBs extend relational databases by providing additional support for objects.

Historical Background

The relational data model was developed in the 1970s, providing a way to organize data into tables with rows and columns [4]. Relationships between tables were defined by the concept of foreign keys, where a column (or multiple columns) in one table contained a reference to a primary key value (unique identifier) in another table. The simplicity of the relational data model was complemented by its formal foundation on set theory, thus providing powerful algebraic and calculus-based techniques for querying relational data.

Initially, relational data modeling concepts were used in business-oriented applications, where tables provided a natural structure for the organization of data. Users eventually began to experiment with the use of relational database concepts in new application domains, such as engineering design and geographic information systems. These new application areas required the use of complex data types that were not supported by the relational model. Furthermore, database designers were discovering that the process of normalizing data into table form was affecting performance for the retrieval of large, complex, and hierarchically structured data, requiring numerous join conditions to retrieve data from multiple tables. Around the same time, object-oriented programming languages (OOPLs) were also beginning to develop, defining the concept of user-defined classes, with instance fields, methods, and encapsulation for information hiding [14].

The OOPL approach of defining object structure together with object behavior eventually provided the basis for the development of Object-Oriented Database Systems (OODBs) in the mid-1980s. The *Object-Oriented Database System Manifesto*, written by leading researchers in the database field, was the first document to fully outline the characteristics of OODB technology [1]. OODBs provided a revolutionary concept for data modeling, with data objects organized as instances of user-defined classes. Classes were organized into class hierarchies, supporting inheritance of attributes and behavior. OODBs differed from relational technology through the use of internal object identifiers, rather than foreign keys, as a means for defining relationships between classes. OODBs also provided a more seamless integration of database and programming language technology, resolving the impedance mismatch problem that existed for relational database systems. The impedance mismatch problem refers to the disparity that exists between set-oriented relational database access and iterative one-record-at-a-time host language access. In the OODB paradigm, the OOPL provides a uniform, object-oriented view of data, with a single language for accessing the database and implementing the database application.

The relational database research community responded to the development of OODBs with the *Third Generation Database System Manifesto*, defining the manner in which relational technology can be extended to support object-oriented capabilities [13]. Rowe and Stonebraker developed Postgres as the first object-relational database system (ORDB), illustrating an evolutionary approach to integrating object-oriented and relational concepts [10]. ORDB concepts parallel those found in OODBs, with the notions of user-defined data types, object tables formed from user-defined types, hierarchies of user-defined types and object tables, rows of object tables with internal object identifiers, and relationships between object tables that use object identifiers as references.

Today, several OODB products exist in the market, and most relational database products provide some form of ORDB support. The following section elaborates on the common features of object data models and then differentiates between OODB and ORDB modeling concepts.

Foundations

Characteristics of Object Data Models

An object is one of the most fundamental concepts of the object data model, where an object represents an entity of interest in a specific application. An object has state, describing the specific structural properties of the object. An object also has behavior, defining the methods that are used to manipulate the object. Each method has a signature that includes the method name as well as the method parameters and types. The state and behavior of an object is expressed through an object type definition, which provides an interface for the object. Objects of the same interface are collected into a class, where each object is viewed as an instance of the class. A class definition supports the concept of encapsulation, separating the specification of a class from the implementation of its methods. The implementation of a method can therefore change without affecting the class interface and the way in which the interface is used in application code.

When an object of a class is instantiated, the object is assigned a unique, internal object identifier, or oid [6]. An oid is immutable, meaning that the value of the identifier cannot be changed. The state of an object, on the other hand, is mutable, meaning that the values of object properties can change. In an object data model, object identity is used as the basis for defining relationships between classes, instead of using object state, as in the relational model. As a result, the values of object properties can freely change without affecting the relationships that exist between objects. Object-based relationships between classes are referred to as object references.

Classes in an object model can be organized into class hierarchies, defining superclass and subclass relationships between classes. A class hierarchy provides for the inheritance of the state and behavior of a class, allowing subclasses to inherit the properties and methods of its superclasses while extending the subclass with additional properties or behavior that is specific to the subclass. Inheritance hierarchies provide a powerful mechanism to represent generalization/specialization relationships between classes, which simplify the specification of an object schema, as well as queries over the schema.

As an example of the above concepts, consider the Publisher application described in Fig. 1 using a Unified Modeling Language (UML) class diagram [11]. A Book is a class that is based on an object type that defines the state of a book (isbn, title, and listPrice), as well as the behavior of a book (the method calcBookSales for calculating the total sales of a book based on customer purchases). Publisher, Person, Author, and Customer are additional classes, also having state and behavior. Since authors and customers are specific types of people, the Author and Customer classes are defined to be subclasses of Person. Since personName and address are common to authors and customers, these attributes are defined at the Person level and inherited by instances of

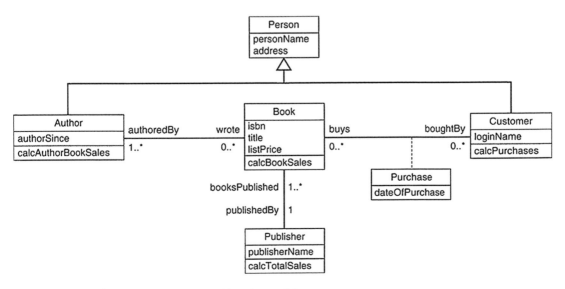

Object Data Models, Fig. 1 The publisher object data model

the Author and Customer classes. Furthermore, Author introduces additional state and behavior that is specific to authors, defining the date (authorSince) when an author first wrote a book as well as a method (calcAuthorBookSales) for calculating the total sales of the author's books. The Customer class similarly introduces state and behavior that is specific to a customer.

Relationships are also defined between the classes of the application:

- A book is authored by one or more authors; an author writes many books.
- A book is published by one publisher; a publisher publishes many books.
- A book is bought by many customers; a customer buys many books, also recording the date of each purchase.

For each relationship, specific instances of each class are related based on the object identity of each instance. For example, a book will establish a relationship to the publisher of the book using the oid of the publisher. In the relational model, the publisher name would be used as a foreign key to establish the relationship. If the publisher name changes, then the change in name must be propagated to the book that references the publisher. In the object data model, such changes in state do not affect relationships between objects since the relationship is based on an immutable, internal object identity.

A generic object model, such as the one shown in Fig. 1, can be mapped to either an object-oriented data model or an object-relational data model. The following subsections use the Publisher application in Fig. 1 to illustrate and explain OODB and ORDB approaches to object data modeling.

Object-Oriented Data Model

An object-oriented database (OODB) is a term typically used to refer to a database that uses objects as a building block and an object-oriented programming language as the database language. The database system supports the persistence of objects along with the features of concurrency and recovery control with efficient access and an ad hoc query language.

The Object Data Standard [2] developed as a standard to describe an object model, including a definition language for an object schema, and an ad-hoc query language. The object model supports the specification of classes having attributes and relationships between objects and the behavior of the class with methods. The Object Definition Language (ODL) provides a standard language for the specification of an object schema including properties and method signatures. A property is either an attribute, representing an instance field that describes the object, or a relationship, representing associations between objects. In ODL, relationships represent bidirectional associations with the database system being responsible for maintaining the integrity of the inverse association. An attribute can be used to define a unidirectional association. If needed, the association can be derived in the other direction using a method specification. The decision is based on trade-offs of storing and maintaining the association versus deriving the inverse direction on demand.

Figure 2 provides an ODL specification of the Publisher application. Each class has a named extent, which represents the set of objects of that type. The Author and Customer classes inherit from the Person class, extending each subclass with specialized attributes. The Book class has the isbn attribute that forms a key, being a unique value across all books. The association between Author and Book is represented as an inverse relationship, and the cardinality of the association is many-to-many since an author can write many books and a book can be written by multiple authors. The set collection type models multiple books and authors. Since the Purchase association class from Fig. 1 has an attribute describing the association, Purchase is modeled in ODL using reification, which is the process of transforming an abstract concept, such as an association, into a class. As a result, the Purchase class in Fig. 2 represents the many-to-many association between Book and Customer. Each instance of the Purchase class represents the

Object Data Models,
Fig. 2 ODL schema of the
publisher application

```
class Person
(extent people)
{attribute string personName,
attribute AddressType address
};

class Author extends Person
(extent authors)
{attribute Date authorSince,
 relationship set<Book> wrote
    inverse Book::authoredBy,
 public float calcAuthorBookSales();
};

class Customer extends Person
(extent customers)
{attribute string loginName,
relationship set<Purchase> buys
    inverse Purchase::purchasedBy,
public float calcPurchases();
};

class Purchase
(extent purchases)
{attribute Date dateOfPurchase,
relationship Book bookPurchased
    inverse Book::boughtBy,
relationship Customer purchasedBy
    inverse Customer::buys;
);
```

```
class Book
(extent books,
key isbn)
{attribute string isbn,
attribute string title,
attribute float listPrice,
relationship set<Author> authoredBy
    inverse Author::wrote,
relationship set<Purchase> boughtBy
    inverse Purchase::bookPurchased,
relationship Publisher publishedBy
    inverse Publisher::booksPublished,
public float calcBookSales();
};

class Publisher
(extent publishers)
{attribute string publisherName,
relationship set<Book> booksPublished
    inverse Book::publishedBy,
public float calcTotalSales();
};
```

purchase of a book by a customer. The Purchase class has the dateOfPurchase instance field, as well as two relationships indicating which Book (bookPurchased) and which Customer (purchasedBy) is involved in the purchase. The inverse relationships in Book (boughtBy) and Customer (buys) are related to instances of the Purchase class.

This ODL specification forms the basis of the definition of the object schema within the particular OOPL used with the OODB, such as C++, Java, and Smalltalk. The specification of the schema and the method implementation using a given OOPL is known as a language binding. In some OODB products, the ODL specification of the properties of the class are used to automatically generate the definition of the schema for the OOPL being used.

The standard also includes a declarative query language known as the Object Query Language (OQL). The OQL is based on the familiar select-from-where syntax of SQL. The select clause defines the structure of the result of the query.

The from clause specifies variables that range over collections within the schema, such as a class extent or a multivalued property. The where clause provides restrictions on the properties of the objects that are to be included in the result. Object references are traversed through the use of dot notation for single-valued properties and through the from clause for multivalued properties.

Consider a simple query that finds the name of a publisher of a book given its isbn:

```
select b.publishedBy.publisherName
from books b
where b.isbn = '978-1608454761';
```

This OQL query looks quite similar to SQL. In the from clause, the alias b ranges over the books extent. The where clause locates the book of interest. The select clause provides a path expression that navigates through the publishedBy single-valued property to return the name of the publisher.

Consider another query that finds the title and sales for books published by Springer:

```
select title: b.title,
       sales: b.calcBookSales()
from p in publishers,
     b in p.booksPublished
where p.publisherName = 'Springer';
```

This query illustrates the alternative syntax for the alias in the from clause, using the syntax "variable in collection." The alias p ranges over the publishers extent, whereas the alias b ranges over the multivalued relationship booksPublished of each publisher that satisfies the where condition. The select clause returns the name of each field and its value, where sales returns the results of a method call.

Object-Relational Data Model

An object-relational database (ORDB) refers to a relational database that has evolved by extending its data model to support user-defined types along with additional object features. An ORDB supports the traditional relational table in addition to introducing the concept of a typed table, which is similar to a class in an OODB. A typed table is created based on a user-defined type (UDT), which provides a way to define complex types with support for encapsulation. UDTs and their corresponding typed tables can be formed into class hierarchies with inheritance of state and behavior. The rows (or instances) of a typed table have object identifiers that are referred to as object references. Object references can be used to define relationships between tables that are based on object identity.

Figure 3 presents an ORDB schema of the Publisher application that is defined using the object-relational extensions to the SQL standard. The type personUdt is an example of specifying a UDT. The UDT defines the structure of the type by identifying attributes together with their type definitions. The phrase "instantiable not final ref is system generated" defines three properties of the type:

1. "Instantiable" indicates that the type supports a constructor function for the creation of in-

stances of the type. The phrase "not instantiable" can be used in the case where the type has a subtype and instances can only be created at the subtype level.
2. "Not final" indicates that the type can be specialized into a subtype. The phrase "final" can be used to indicate that a type cannot be further specialized.
3. "Ref is system generated" indicates that the database system is responsible for automatically generating an internal object identifier. The SQL standard supports other options for the generation of object identifiers, which include user-specified object-identifiers as well as identifiers that are derived from other attributes.

Definition of the personUdt type is followed by the specification of the person typed table, which is based on the personUdt type. The person-typed table automatically acquires columns for each of the attributes defined in personUdt. In addition, the person typed table has a column for an object identifier that is associated with every row in the table. The phrase "ref is personID" defines that the name of the object identifier column is personID. The definition of a typed table can add constraints to the columns that are defined in the type associated with the table. For example, personName is defined to be a primary key in the person typed table.

The authorUdt type is defined as a subtype of personUdt, as indicated by the "under personUdt" clause. In addition to defining the structure of the type, authorUdt also defines behavior with the definition of the calcAuthorBookSales method. Since authorUdt is a subtype of personUdt, authorUdt will inherit the object identifier (personID) defined in personUdt. For consistency, the author table is also defined to be a subtable of the person object table. The typed table hierarchy therefore parallels the UDT hierarchy. In a similar manner, customerUdt is defined to be a subtype of personUdt, and the customer typed table, based on customerUdt, is defined to be a subtable of the person table. UDTs and typed tables are also defined for the Book and Publisher classes from

```
create type personUdt as                          create table publisher of publisherUdt
(personName varchar(15),                          (primary key (publisherName),
address varchar(20))                               ref is publisherID system generated);
instantiable not final ref is system generated;
                                                  create type purchaseUdt as
create table person of personUdt                  (dateOfPurchase date,
(primary key (personName),                         purchasedBy ref (customerUdt) scope customer
ref is personID system generated);                     references are checked on delete cascade,
                                                  bookPurchased ref (bookUdt) scope book
create type authorUdt under personUdt as              references are checked on delete cascade)
(authorSince varchar(10),                          instantiable not final ref is system generated;
wrote ref (bookUdt) scope book array [10]
    references are checked on delete no action)   create table purchase of purchaseUdt
instantiable not final                            (ref is purchaseID system generated);
method calcAuthorBookSales() returns decimal;
                                                  create type bookUdt as
create table author of authorUdt under person;    (isbn varchar(30),
                                                  title varchar(50),
create type customerUdt under personUdt as        listPrice decimal,
(loginName varchar(10),                            authoredBy ref (authorUdt) scope author array[5]
buys ref (purchaseUdt) scope purchase array [50]      references are checked on delete no action,
    references are checked on delete no action)    boughtBy ref (purchaseUdt) scope purchase
instantiable not final                            array[1000]
method calcPurchases() returns decimal;               references are checked on delete set null,
                                                  publishedBy ref (publisherUdt) scope publisher
create table customer of customerUdt under person     references are checked on delete no action)
(unique (loginName));                              instantiable not final ref is system generated
                                                  method calcBookSales() returns decimal;
create type publisherUdt as
(publisherName varchar(30),                        create table book of bookUdt
booksPublished ref (bookUdt) scope book array [1000]  (primary key (isbn),
    references are checked on delete set null)     ref is bookID system generated);
instantiable not final ref is system generated
method calcTotalSales() returns decimal;
```

Object Data Models, Fig. 3 ORDB schema of publisher application

Fig. 1, as well as the (reified implementation of the) Purchase association class.

Figure 3 also illustrates the use of object references to represent identity-based relationships between UDTs. Recall from the object data model in Fig. 1 that a book is published by one publisher; a publisher publishes many books. In an ORDB, this relationship is established through the use of reference types. In the bookUdt, the publishedBy attribute has the type ref(publisherUdt), indicating that the value of publishedBy is a reference to the object identifier (publisherID) of a publisher. In the inverse direction, the type of booksPublished in the publisherUdt is an array of ref(bookUdt), indicating that booksPublished is an array of object references to books. Each attribute definition includes a scope clause and a "references are checked" clause. Since a UDT can be used to define multi-

ple tables, the scope clause defines the table of the object reference. The references clause specifies the same options for referential integrity of object references as originally defined for traditional relational tables.

To establish the fact that a book is published by a specific publisher, the object identifier of publisher is retrieved to create the relationship:

```
update book
      set publishedBy =
      (select publisherID
      from publisher
      where publisherName =
      'Morgan Claypool')
where isbn = '978-1608454761';
```

A similar update statement can be used to establish the relationship in the inverse direction by adding the book oid to the array of object references of the publisher.

References can be traversed to query information about relationships. For example, to return the name of the publisher of a specific book, the following query can be used:

```
select publishedBy.publisherName
from book
where isbn = '978-1608454761';
```

The dot notation in the select clause performs an implicit join between the book table and the publisher table, returning the name of the publisher. The deref() function can also be used to retrieve the entire structured type associated with a reference value. For example, the following query will return the full instance of the publisherUdt type, rather than just the publisherName:

```
select deref(publishedBy)
from book
where isbn = '978-1608454761';
```

In this case, the result of the query is a value of type publisherUdt, containing the publisher name and the array of references to books published by the publisher.

Key Applications

Computer-Aided Design, Geographic Information Systems, Computer-Aided Software Engineering, Embedded Systems, Real-time Control Systems.

Cross-References

► Conceptual Schema Design
► Database Design
► Extended Entity-Relationship Model
► OQL
► Relational Model
► Semantic Data Model
► Unified Modeling Language

Recommended Reading

1. Atkinson M, Bancilhon F, DeWitt D, Dittrich K, Maier D, Zdonik S. The object-oriented database system manifesto. In: Proceedings of the 1st International Conference on Deductive and Object-Oriented Databases; 1989. p. 223–40.
2. Cattell RGG, Barry DK, Berler M, Eastman J, Jordan D, Russell C, Schadow O, Stanienda T, Velez F, editors. The object data standard: ODMG 3.0. San Mateo: Morgan Kaufmann; 2000.
3. Chaudhri A, Zicari R, editors. Succeeding with object databases: a practical look at today's implementations with java and XML. New York: Wiley; 2000.
4. Codd EF. A relational model of data for large shared data banks. Commun ACM. 1970;13(6):377–87.
5. Dietrich SW, Urban SW. Fundamentals of object databases: object-oriented and object-relational design. San Rafeal: Morgan Claypool; 2011.
6. Koshafian S, Copeland G. Object identity. ACM SIGPLAN Not. 1986;21(11):406–16.
7. Loomis MES, Chaudhri A, editors. Object databases in practice. Upper Saddle River: Prentice Hall; 1997.
8. Melton J. Advanced SQL:1999: understanding object-relational and other advanced features. San Mateo: Morgan Kaufmann; 2002.
9. Object Databases. http://odbms.org/free-downloads-and-links/object-databases. Accessed 16 June 2014.
10. Rowe L, Stonebraker M. The postgres data model. In: Proceedings of the 13th International Conference on Very Large Data Bases; 1987. p. 83–96.
11. Rumbaugh J, Jacobson I, Booch G. The unified modeling language reference manual. Reading: Addison-Wesley; 1991.
12. Stonebraker M. Object-relational DBMSs: the next great wave. San Mateo: Morgan Kaufmann; 1995.
13. Stonebraker M, Rowe L, Lindsay B, Gray J, Carey M, Brodie M, Bernstein P, Beech D. Third generation database system manifesto: the committee for advanced DBMS function. ACM SIGMOD Rec. 1990;19(3):31–44.
14. Stroustrup B. The C++ programming language. 3rd ed. Reading: Addison-Wesley; 1997.
15. Zdonik SB, Maier D. Readings in object-oriented database systems. San Mateo: Morgan Kaufmann; 1990.

Object Identity

Susan D. Urban and Suzanne W. Dietrich
Arizona State University, Phoenix, AZ, USA

Synonyms

Object identifier; Object reference; OID

Definition

Object identity is a property of data that is created in the context of an object data model, where an object is assigned a unique internal object identifier, or OID. The object identifier is used to define associations between objects and to support retrieval and comparison of object-oriented data based on the internal identifier rather than the attribute values of an object.

Key Points

In an object data model, an object is created as an instance of a class. An object has an object identifier as well as a state. An object identifier is immutable, meaning that the value of the object identifier cannot change over the lifetime of the object. The state, on the other hand, is mutable, representing the attributes that describe the object and the relationships that define associations among objects. Relationships in an object data model are defined using object references based on internal object identifiers rather than attribute values as in a relational data model. As a result, the attribute values of an object can freely change without affecting identity-based relationships.

Variables that contain object references can be compared using either object identity or object equality. Two object references are identical if they contain the same object identifiers. In contrast, two object references, which possibly contain different object identifiers, are equal if the values of attributes and relationships in each object state are identical. Shallow equality is the process of comparing the immediate values of attributes and relationships. Deep equality involves the traversal of object references in the comparison process. Query languages for objects must incorporate operators to distinguish between object identity and object equality in the specification of object queries.

Cross-References

▶ Conceptual Schema Design
▶ Extended Entity-Relationship Model
▶ Object Data Models
▶ Object-Role Modeling
▶ Semantic Data Model
▶ Unified Modeling Language

Recommended Reading

1. Beeri C, Thalheim B. Identification as a primitive of database models. In: Proceedings of the 7th International Workshop on Foundations of Models and Languages for Data and Objects; 1998. p. 19–36.
2. Koshafian S, Copeland G. Object identity. ACM SIGPLAN Not. 1986;21(11):406–16.

Object Recognition

Ming-Hsuan Yang
University of California at Merced, Merced, CA, USA

Synonyms

Object identification; Object labeling

Definition

Object recognition is concerned with determining the identity of an object being observed in the image from a set of known labels. Oftentimes, it is assumed that the object being observed has been detected or there is a single object in the image.

Historical Background

As the holy grail of computer vision research is to tell a story from a single image or a sequence of images, object recognition has been studied for more than four decades [9, 22]. Significant efforts have been spent to develop representation schemes and algorithms aiming at recognizing generic objects in images taken under different imaging conditions (e.g., viewpoint, illumination, and occlusion). Within a limited scope of distinct objects, such as handwritten digits, fingerprints, faces, and road signs, substantial success has been achieved. Object recognition is also related to content-based image retrieval and multimedia indexing as a number of generic objects can be recognized. In addition, significant progress towards object categorization from images has been made in the recent years [17]. Note that object recognition has also been studied extensively in psychology, computational neuroscience and cognitive science [4, 9].

Foundations

Object recognition is one of the most fascinating abilities that humans easily possess since childhood. With a simple glance of an object, humans are able to tell its identity or category despite of the appearance variation due to change in pose, illumination, texture, deformation, and under occlusion. Furthermore, humans can easily generalize from observing a set of objects to recognizing objects that have never been seen before. For example, kids are able to generalize the concept of "chair" or "cup" after seeing just a few examples. Nevertheless, it is a daunting task to develop vision systems that match the cognitive capabilities of human beings, or systems that are able to tell the specific identity of an object being observed. The main reasons can be attributed to the following factors: relative pose of an object to a camera, lighting variation, and difficulty in generalizing across objects from a set of exemplar images. Central to object recognition systems are how the regularities of images, taken under different lighting and pose conditions, are extracted and recognized. In other words, all the algorithms adopt certain representations or models to capture these characteristics, thereby facilitating procedures to tell their identities. In addition, the representations can be either 2D or 3D geometric models. The recognition process, either generative or discriminative, is then carried out by matching the test image against the stored object representations or models.

Geometry-Based Approaches

Early attempts at object recognition were focused on using geometric models of objects to account for their appearance variation due to viewpoint and illumination change. The main idea is that the geometric description of a 3D object allows the projected shape to be accurately predicated in a 2D image under projective projection, thereby facilitating recognition process using edge or boundary information (which is invariant to certain illumination change). Much attention was made to extract geometric primitives (e.g., lines, circles, etc.) that are invariant to viewpoint change [13]. Nevertheless, it has been shown that such primitives can only be reliably extracted under limited conditions (controlled variation in lighting and viewpoint with certain occlusion). Mundy provides an excellent review on geometry-based object recognition research [12].

Appearance-Based Algorithms

In contrast to early efforts on geometry-based object recognition, most recent efforts have been centered on appearance-based techniques as advanced feature descriptors and pattern recognition algorithms are developed [8]. Most notably, the eigenface method has attracted much attention as it is one of the first face recognition systems that are computationally efficient and relatively accurate [21]. The underlying idea of this approach is to compute eigenvectors from a set of vectors where each one represents one face image as a raster scan vector of gray-scale pixel values. Each eigenvector, dubbed an eigenface, captures certain variance among all the vectors, and a small set of eigenvectors captures almost all the appearance variation of face images in the training set. Given a test image represented as a vector of gray-scale pixel values, its identity is determined by finding the nearest neighbor of this vector after being projected onto a subspace spanned by a set of eigenvectors. In other words, each face image can be represented by a linear combination of eigenfaces with minimum error (often in the L2 sense), and this linear combination constitutes a compact reorientation. The eigenface approach has been adopted in recognizing generic objects across different viewpoints [14] and modeling illumination variation [2].

As the goal of object recognition is to tell one object from the others, discriminative classifiers have been used to exploit the class specific information. Classifiers such as k-nearest neighbor, neural networks with radial basis function (RBF), dynamic link architecture, Fisher linear discriminant, support vector machines (SVM), sparse network of Winnows (SNoW), and boosting algorithms have been applied to recognize 3D objects from 2D images [16, 6, 1, 18, 19]. While appearance-based methods have shown promising results in object recognition under viewpoint and illumination change, they are less effective in handling occlusion. In addition, a large set of exemplars needs to be segmented from images for generative or discriminative methods to learn the appearance characteristics. These problems are partially addressed with parts-based representation schemes.

Feature-Based Algorithms

The central idea of feature-based object recognition algorithms lies in finding interest points, often occurred at intensity discontinuity, that are invariant to change due to scale, illumination and affine transformation (a brief review on interest point operators can be found in [8]). The scale-invariant feature transform (SIFT) descriptor is arguably one of the most widely used feature representation schemes for vision applications [8]. The SIFT approach uses extrema in scale space for automatic scale selection with a pyramid of difference of Gaussian filters, and keypoints with low contrast or poorly localized on an edge are removed. Next, a consistent orientation is assigned to each keypoint and its magnitude is computed based on the local image gradient histogram, thereby achieving invariance to image rotation. At each keypoint descriptor, the contribution of local image gradients are sampled and weighted by a Gaussian, and then represented by orientation histograms. For example, the 16×16 sample image region and 4×4 array of histograms with 8 orientation bins are often used, thereby providing a 128-dimensional feature vector for each keypoint. Objects can be indexed and recognized using the histograms of keypoints in images. Numerous applications have been developed using the SIFT descriptors, including object retrieval [15, 20], and object category discovery [5].

Although the SIFT approach is able to extract features that are insensitive to certain scale and illumination changes vision applications with large base line changes entail the need of affine invariant point and region operators [11]. A performance evaluation among various local descriptors can be found in [10], and a study on affine region detectors is presented in [11]. Finally, SIFT-based methods are expected to perform better for objects with rich texture information as sufficient number of keypoints can be extracted. On the other hand, they also require sophisticated indexing and matching algorithms for effective object recognition [8, 17].

Key Applications

Biometric recognition, and optical character/digit/document recognition are arguably the most widely used applications. In particular, face recognition has been studied extensively for decades and with large scale ongoing efforts [23]. On the other hand, biometric recognition systems based on iris or fingerprint as well as handwritten digit have become reliable technologies [3, 7]. Other object recognition applications include surveillance, industrial inspection, content-based image retrieval (CBIR), robotics, medical imaging, human computer interaction, and intelligent vehicle systems, to name a few.

Future Directions

With more reliable representation schemes and recognition algorithms being developed, tremendous progress has been made in the last decade towards recognizing objects under variation in viewpoint, illumination and under partial occlusion. Nevertheless, most working object recognition systems are still sensitive to large variation in illumination and heavy occlusion. In addition, most existing methods are developed to deal with rigid objects with limited intra-class variation. Future research will continue searching for robust representation schemes and recognition algorithms for recognizing generic objects.

Data Sets

Numerous face image sets are available on the web

- FERET face data set: http://www.itl.nist.gov/iad/humanid/feret/
- UMIST data set: http://images.ee.umist.ac.uk/danny/database.html
- Yale data set: http://cvc.yale.edu/projects/yalefacesB/yalefacesB.html
- AR data set: http://cobweb.ecn.purdue.edu/%7Ealeix/aleix_face_DB.html
- CMU PIE data set: http://www.ri.cmu.edu/projects/project_418.html

There are several large data sets for object recognition experiments,

- COIL data set: http://www1.cs.columbia.edu/CAVE/software/softlib/coil-100.php
- CalTech data sets: http://www.vision.caltech.edu/html-files/archive.html
- PASCAL visual object classes: http://www.pascal-network.org/challenges/VOC/

URL to Code

There are a few excellent short courses on object recognition in recent conferences available on the web.

- "Recognition and matching based on local invariant features" by Schmid and Lowe in IEEE Conference on Computer Vision and Pattern Recognition 2003: http://lear.inrialpes.fr/people/schmid/cvpr-tutorial03/

- "Learning and recognizing object categories" by Fei-Fei, Fergus and Torralba in IEEE International Conference on Computer Vision 2005:http://people.csail.mit.edu/torralba/shortCourseRLOC/
- "Recognizing and Learning Object Categories: Year 2007" by Fei-Fei, Fergus and Torralba in IEEE Conference on Computer Vision and Pattern Recognition 2005: http://people.csail.mit.edu/torralba/shortCourseRLOC/

Sample code for face recognition and SIFT descriptors:

- Face recognition: http://www.face-rec.org/
- Lowe's sample SIFT code: http://www.cs.ubc.ca/~lowe/keypoints/
- MATLAB implementation of SIFT descriptors by Vedaldi: http://vision.ucla.edu/~vedaldi/code/sift/sift.html
- libsift by Nowozin: http://user.cs.tu-berlin.de/~nowozin/libsift/

Grand challenge in object recognition:

- NIST face recognition grand challenge: http://www.frvt.org/FRGC/
- NIST multiple biometric grand challenge: http://face.nist.gov/mbgc/
- PASCAL visual object classes challenge 2007: http://www.pascal-network.org/challenges/VOC/voc2007/index.html

Cross-References

▶ Object Recognition

Recommended Reading

1. Belhumeur P, Hespanha J, Kriegman D. Eigenfaces vs. fisherfaces: recognition using class specific linear projection. IEEE Trans Pattern Anal Mach Intell. 1997;19(7):711–20.

2. Belhumeur P, Kriegman D. What is the set of images of an object under all possible illumination conditions. Int J Comput Vis. 1998;28(3):1–16.

3. Daugman J. Probing the uniqueness and randomness of iriscodes: results from 200 billion iris pair comparisons. Proc IEEE. 2006;94(11):1927–35.

4. Edelman S. Representation and recognition in vision. Cambridge: MIT; 1999.

5. Fergus R, Perona P, Zisserman A. Object class recognition by unsupervised scale-invariant learning. In: Proceedings of the IEEE International Conference on Computer Vision and Pattern Recognition; 2003. p. 264–71.

6. Lades M, Vorbrüggen JC, Buhmann J, Lange J, von der Malsburg C, Würtz RP, Konen W. Distortion invariant object recognition in the dynamic link architecture. IEEE Trans Comput. 1993;42(3):300–11.

7. Lecun Y, Bottou L, Bengio Y, Haffner P. Gradient-based learning applied to document recognition. Proc IEEE. 1998;86(11):2278–324.

8. Lowe D. Distinctive image features from scale-invariant keypoints. Int J Comput Vis. 2004;60(2):91–110.

9. Marr D. Vision. San Francisco: W.H. Freeman and Company; 1982.

10. Mikolajczyk K, Schmid C. A performance evaluation of local descriptors. IEEE Trans Pattern Analy Machine Intell. 2005;27(10):1615–30.

11. Mikolajczyk K, Tuytelaars T, Schmid C, Zisserman A, Matas J, Schaffalitzky F, Kadir T, Van Gool L. A comparison of affine region detectors. Int J Comput Vis. 2006;65(1/2):43–72.

12. Mundy J. Object recognition in the geometric era: a retrospective. In: Ponce J, Hebert M, Schmid C, Zisserman A, editors. Toward category-level object recognition. Springer: Berlin; 2006. p. 3–29.

13. Zisserman A, Mundy J. Geometric invariance in computer vision. Cambridge: MIT; 1992.

14. Murase H, Nayar SK. Visual learning and recognition of 3-D objects from appearance. Int J Comput Vis. 1995;14(1):5–24.

15. Nister D, Stewenius H. Scalable recognition with a vocabulary tree. In: Proceedings of the IEEE International Conference on Computer Vision and Pattern Recognition; 2006. p. 2161–8.

16. Poggio T, Edelman S. A network that learns to recognize 3D objects. Nature. 1990;343(6255):263–6.

17. Ponce J, Hebert M, Schmid C, Zisserman A. Toward category-level object recognition. Berlin: Springer; 2006.

18. Pontil M, Verri A. Support vector machines for 3D object recognition. IEEE Trans Pattern Anal Mach Intell. 1998;20(6):637–46.

19. Roth D, Yang M-H, Ahuja N. Learning to recognize objects. Neural Comput. 2002;14(5):1071–104.

20. Sivic J, Zisserman A. Video Google: a text retrieval approach to object matching in videos. In: Proceedings of the 9th IEEE Conference on Computer Vision; 2003. p. 1470–7.

21. Turk M, Pentland A. Eigenfaces for recognition. J Cogn Neurosci. 1991;3(1):71–86.

22. Ullman S. High-level vision: object recognition and visual recognition. Cambridge: MIT; 1996.

23. Zhao W, Chellappa R, Rosenfeld A, Phillips JP. Face recognition: a literature survey. ACM Comput Surv. 2003;35(4):399–458.

Object Relationship Attribute Data Model for Semistructured Data

Gillian Dobbie[1] and Tok Wang Ling[2]
[1]University of Auckland, Auckland, New Zealand
[2]National University of Singapore, Singapore, Singapore

Synonyms

ORA-SS data model; ORA-SS schema diagram

Definition

When a database schema is designed, a data model is initially used to model the real world constraints that are taken into account in the design of the schema. For semi-structured database design, it is necessary to capture the following constraints: object classes, n-ary relationship types, attributes of object classes, attributes of relationship types, cardinality, participation and uniqueness constraints, ordering, irregular and heterogeneous structures, for both data- and document-centric data.

Key Points

The ORA-SS (Object-Relationship-Attribute Data Model for Semi-structured Data) data model was designed [1] specifically to capture the constraints that are necessary for designing semi-structured databases, for normalization of schemas, and for defining views.

Figure 1 models the scenario where there is a department, with a name and many courses. A course has a unique code, a title, and many students, and a student has a unique student number, name, address, and many hobbies. For each course that a student takes, they have a grade. There is a tutor for each student in each course they take. A tutor has a unique staff number and a name, and each student can give feedback for the tutor they have in each course that they take.

A closer look is now taken at the notation used. Each of the rectangles represents an object class, the circles represent attributes and the labeled directed edges between object classes represent relationship types. A filled circle is an identifier, which is similar to a key in relational databases and an identifier of an object class. The "?" in the circle represents zero or one occurences of that attribute, while a "*" represents zero or more occurences. The default is one. An attribute in an ORA-SS diagram could be represented as an attribute or an element in an XML document. The label on the edge has *name, n, a:b, c:d*, where *name* is the name of the relationship type, *n* is the degree, *a:b* is the participation constraint on the parent and *c:d* is the participation constraint on the child. The participation constraint *a:b* indicates that the parent object participates in a minimum of *a* and a maximum of *b* relationships. Whereas, the participation constraint *c:d* indicates that the child object participates in a

minimum of *c* and a maximum of *d* relationships. A label on the edge between an object class and an attribute, *name*, indicates that the attribute belongs to relationship type *name*.

Consider the example in Fig. 1. There are object classes *Department, Course, Student* and *Tutor*. Object class *Department* has an identifier *name*, and is the parent object class in the relationship type between *Department* and *Course*. The relationship type is a binary relationship with name *dc*, and the participation constraint *1:m* indicates that a department has a minimum of *one* course and a maximum of *m* courses, where *m* means many, i.e., any number of courses. The participation constraint *1:1* indicates that each course must belong to *one* department and can belong to a maximum of *one* department. Each course has an identifier *code*, a required attribute *title*, and is the parent object class in the relationship type, *cs*, between *course* and *student*. Object class *Student* has an identifier *stuNo*, a required attribute *stuName*, an optional attribute *address*, and zero or more *hobby*. There is a binary relationship type, *cs*, between *Course* and *Student*, where a *Course* can have zero or more *Students*, and a *Student* takes one or more *Courses*. The attribute *grade* belongs to the relationship type, *cs*, that is it represents the grade a student scored in a particular course. There is a ternary relationship type among object classes *Course, Student* and *Tutor*. Each course-student pair can have zero to

Object Relationship Attribute Data Model for Semistructured Data, Fig. 1 An ORA-SS schema diagram

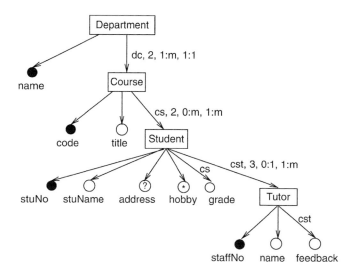

one tutor, and each tutor belongs to one or more course-student pairs. Each tutor has an identifier *staffNo*, a required attribute *name*, and there is an attribute *feedback* for each tutor from a particular student in a particular course.

Cross-References

▶ Entity Relationship Model
▶ Hierarchical Data Model
▶ Object Data Models
▶ Semi-structured Data Model
▶ Semi-structured Database Design
▶ XML Integrity Constraints
▶ XML Schema

Recommended Reading

1. Ling TW, Lee ML, Dobbie G. Semi-structured database design. Berlin/Heidelberg/New York: Springer; 2005.

Object Storage Protocol

Kaladhar Voruganti
Advanced Development Group, Network Appliance, Sunnyvale, CA, USA

Synonyms

Archive storage; Cloud storage; Key-value storage; Object storage

Definition

A RESTful protocol-based interface that is used to access storage in the form of objects.

Main Text

Traditional network attached protocols (NAS) provide support for POSIX semantics. Supporting POSIX makes the NAS protocols quite complex to both understand and implement. In comparison, the object protocol provides a RESTful interface (simple get and put operations) where typically one does not update an object. Instead, every update usually leads towards the creation of a new object. Object storage protocols do not adhere to POSIX semantics. Key-value interface is a specific form of object protocol where the value corresponds to an object. The underlying architecture for implementing a storage system that supports small objects (few kilobytes) is fundamentally very different than a storage system that has been designed for supporting large (mega or gigabytes) objects. Typically, a storage system that supports an object protocol maps the objects to the underlying blocks of a disk system. However, attempts have been made to design disks that natively understand objects but these systems have not gained much traction.

Scale-out systems that consist of a cluster of directly attached disk-based servers typically support an object storage protocol. It is important to note that a scale-out storage system can also support file or block based access protocols. These scale-out systems have a different design center in that they employ newer erasure coding algorithms for reliability instead of using the traditional RAID-5 algorithm. They also employ an eventual consistency model to keep the meta-data consistent across all of the nodes unlike a more traditional storage controller that employs strong consistency protocols to keep the meta-data consistent across the cluster of storage nodes.

In conclusion, storage industry is experiencing dramatic changes with respect to the new types of protocols being used to access storage (object, key-value store), new types of storage architectures (SSD-based performance optimized storage, and capacity optimized scale-out storage architectures), and the use of new types of persistent media (flash, PCM, STT-MRAM, NRAM).

Recommended Reading

1. Mesnier M, Ganger GR, Riedel E. Object-based storage. IEEE Commun Mag. 2013;41(8):84–91.
2. Ghemawat S, Gobioff H, Leung S-T. Google file system. In: Proceedings of the 19th ACM Symposium on Operating System Principles; 2003.
3. Mathew S. Overview of Amazon web services. Amazon White Paper. 2015 Dec.

Object-Role Modeling

Terry Halpin
Neumont University, South Jordan, UT, USA

Synonyms

Fact-oriented modeling; NIAM

Definition

Object-Role Modeling (ORM), also known as *fact-oriented modeling*, is a conceptual approach to modeling and querying the information semantics of business domains in terms of the underlying facts of interest, where all facts and rules may be verbalized in language readily understood by non-technical users of those business domains. Unlike Entity-Relationship (ER) modeling and Unified Modeling Language (UML) class diagrams, ORM treats all facts as relationships (unary, binary, ternary etc.). How facts are grouped into structures (e.g., attribute-based entity types, classes, relation schemes, XML schemas) is considered a design level, implementation issue that is irrelevant to the capturing of essential business semantics.

Avoiding attributes in the base model enhances semantic stability, populatability, and natural verbalization, facilitating communication with all stakeholders. For information modeling, fact-oriented graphical notations are typically far more expressive than those provided by other notations. Fact-oriented textual languages are based on formal subsets of native languages, so are easier to understand by business people than technical languages like UML's Object Constraint Language (OCL). Fact-oriented modeling includes procedures for mapping to attribute-based structures, so may also be used to front-end other approaches.

The fact-oriented modeling approach comprises a family of closely related "dialects", known variously as Object-Role Modeling (ORM), Natural Language Information Analysis Method (NIAM), and Fully-Communication Oriented Information Modeling (FCO-IM). While not adopting the ORM graphical notation, the Object-oriented Systems Model (OSM) [4] and the Semantics of Business Vocabulary and business Rules (SBVR) [13] initiative within the Object Management Group (OMG) are close relatives, with their attribute-free philosophy.

Historical Background

In 1973, Falkenberg generalized work by Abrial and Senko on binary relationships to *n*-ary relationships, and excluded attributes at the conceptual level to avoid "fuzzy" distinctions and to simplify schema evolution. Later, Falkenberg proposed the fundamental ORM framework, which he called the "object-role model" [5]. This framework allowed *n*-ary and nested relationships, but depicted roles with arrowed lines. Nijssen adapted this framework by introducing a circle-box notation for objects and roles, and adding a linguistic orientation and design procedure to provide a modeling method called ENALIM (Evolving NAtural Language Information Model) [12]. Nijssen's team of researchers at Control Data in Belgium developed the method further, including van Assche who classified object types into lexical object types (LOTs) and non-lexical object types (NOLOTs). Today, LOTs are commonly called "entity types" and NOLOTs are called "value types". Meersman added subtyping to the approach, and made major contributions to the RIDL query language [11] with Falkenberg and Nijssen. The method was renamed "aN Information Analysis Method" (NIAM). Later, the acronym "NIAM" was given different expansions, and is now known as "Natural language Information Analysis Method".

In the 1980s, Nijssen and Falkenberg worked on the design procedure and moved to the University of Queensland, where the method was further enhanced by Halpin, who provided the first full formalization, including schema equivalence proofs, and made several refinements and extensions. In 1989, Halpin and Nijssen co-authored a book on the approach, followed a year later by Wintraecken's book [16]. Today several books, including major works by Halpin [10], and Bakema et al. [1] expound on the approach.

Many researchers contributed to the fact-oriented approach over the years, and there is no space here to list them all. Today various versions exist, but all adhere to the fundamental object-role framework. Habrias developed an object-oriented version called MOON (Normalized Object-Oriented Method). The Predicator Set Model (PSM), developed mainly by ter Hofstede et al. [7], includes complex object constructors. De Troyer and Meersman developed a version with constructors called Natural Object-Relationship Model (NORM). Halpin developed an extended version simply called ORM, and with Bloesch and others developed an associated query language called ConQuer [2]. Bakema et al. [1] recast all entity types as nested relationships, to produce Fully Communication Oriented NIAM, which they later modified to Fully Communication Oriented Information Modeling (FCO-IM).

More recently, Meersman and others adapted ORM for ontology modeling, using a framework called DOGMA (Developing Ontology-Grounded Methodology and Applications) (http://www.starlab.vub.ac.be/website/). Nijssen and others extended NIAM to a version called NIAM2007. Halpin and others developed a second generation ORM (ORM 2), whose graphical notation is used in this article.

Foundations

ORM includes graphical and textual *languages* for modeling and querying information at the conceptual level, as well as *procedures* for designing conceptual models, transforming between different conceptual representations, forward engineering ORM schemas to implementation schemas (e.g., relational database schemas, object-oriented schemas, XML schemas, and external schemas) and reverse engineering implementation schemas to ORM schemas.

Attributes are not used as a base construct. Instead, all fact structures are expressed as *fact types* (relationship types). These may be unary (e.g., Person smokes), binary (e.g., Person was born on Date), ternary (e.g., Person visited Country in Year), and so on. This attribute-free nature has several advantages: *semantic stability* (minimize the impact of change caused by the need to record something about an attribute); *natural verbalization* (all facts and rules may be easily verbalized in sentences understandable to the domain expert); *populatability* (sample fact populations may be conveniently provided in fact tables); *null avoidance* (no nulls occur in populations of base fact types, which must be elementary or existential). Although attribute-free diagrams typically consume more space, this apparent disadvantage is easily overcome by using an ORM tool to automatically create attribute-based structures (e.g., ER, UML class, or relational schemas) as views of an ORM schema.

ORM's graphical language is far more expressive for data modeling purposes than that of UML or industrial versions of ER, as illustrated later. The *rich graphical notation* makes it easier to detect and express constraints, and to visually transform schemas into equivalent alternatives.

ORM includes *effective modeling procedures* for constructing and validating models. In step 1a of the Conceptual Schema Design Procedure (CSDP), the domain expert informally verbalizes facts of interest. In step 1b, the modeler formally rephrases the facts in natural yet unambiguous language, using standard reference patterns to ensure that entities are well identified. Verbalized fact instances are abstracted to fact types, which are then populated with sample instances. The constraints on the fact types are verbalized formally, a process that may be automated [8], and these verbalizations are checked with the domain expert, using positive populations to illustrate satisfaction of the constraints as well as counterexamples to illustrate what it means to violate a constraint. This approach to model *validation by verbalization and population* has proved extremely effective in industrial practice, with correct models typically obtained from the outset rather than going through unreliable iterative procedures.

Figure 1 lists the main graphical symbols in the ORM 2 notation [8], numbered for easy reference. An *entity type* (e.g., Person) is depicted as a named, soft rectangle (symbol 1), or alternatively an ellipse or hard rectangle. *Value type* (e.g., Person Name) shapes have dashed lines (symbol 2). Each entity type has a *reference scheme*, indicating how each instance may be mapped via predicates to a combination of one or more values. Injective (1:1 into) reference schemes

mapping entities (e.g., countries) to single values (e.g., country codes) may be abbreviated as in symbol 3 by displaying the *reference mode* in parentheses, e.g., Country (.code). The reference mode indicates how values relate to the entities. Values are constants with a known denotation, so require no reference scheme.

Relationships used for *preferred reference* are called *existential facts* (e.g., there exists a country that has the country code 'US'). The other relationships are *elementary facts* (e.g., The country with country code 'US' has a population of 301,000,000). The exclamation mark in symbol 4 declares that an object type is *independent* (instances may exist without participating in any elementary facts). Object types displayed in multiple places are shadowed (symbol 5).

A fact type results from applying a logical *predicate* to a sequence of one or more object types. Each predicate comprises a named sequence of one or more *roles* (parts played in the relationship). A predicate is sentence with object holes, one for each role, with each role depicted as a box and played by exactly one

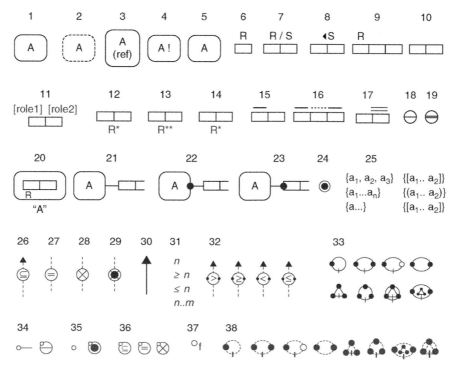

Object-Role Modeling, Fig. 1 Main ORM graphic symbols

object type. Symbol 6 shows a unary predicate (e.g., ... smokes), symbols 7 and 8 depict binary predicates (e.g., ... loves ...), and symbol 9 shows a ternary predicate. Predicates of higher *arity* (number of roles) are allowed. Each predicate has at least one *predicate reading*. ORM uses *mixfix* predicates, so objects may be placed at any position in the predicate (e.g., the fact type Person introduced Person to Person involves the predicate "... introduced ... to ... "). Mixfix predicates allow natural verbalization of *n*-ary relationships, as well as binary relationships where the verb is not in the infix position (e.g., in Japanese, verbs come at the end). By default, forward readings traverse the predicate from left to right (if displayed horizontally) or top to bottom (if displayed vertically). Other reading directions may be indicated by an arrow-tip (symbol 8). For binary predicates, forward and inverse readings may be separated by a slash (symbol 7). Duplicate predicate shapes are shadowed (symbol 10).

Roles may be given *role names*, displayed in square brackets (symbol 11). An asterisk indicates that the fact type is *derived* from one or more other fact types (symbol 12). If the fact type is derived and stored, a double asterisk is used (symbol 13). Fact types that are semi-derived are marked "+" (symbol 14). *Internal uniqueness constraints*, depicted as bars over one or more roles in a predicate, declare that instances for that role (combination) in the fact type population must be unique (e.g., symbols 15, 16). For example, a uniqueness constraint on the first role of Person was born in Country verbalizes as: **Each** person was born in **at most one** Country. If the constrained roles are not contiguous, a dotted line separates the constrained roles (symbol 16). A predicate may have many uniqueness constraints, at most one of which may be declared *preferred* by a double-bar (symbol 17). An *external uniqueness constraint* shown as a circled uniqueness bar (symbol 18) may be applied to two or more roles from different predicates by connecting to them with dotted lines. This indicates that instances of the role combination in the join of those predicates are unique. For example, if a state is identified by combining its state code and country, an external uniqueness constraint is added to the

roles played by Statecode and Country in: State has State code; State is in Country. Preferred external uniqueness constraints are depicted by a circled double-bar (symbol 19).

To talk about a relationship, one may *objectify* it (i.e., make an object out of it) so that it can play roles. Graphically, the objectified predicate (a.k.a. *nested* predicate) is enclosed in a soft rectangle, with its name in quotes (symbol 20). Roles are connected to their players by a line segment (symbol 21). A *mandatory role constraint* declares that every instance in the population of the role's object type must play that role. This is shown as a large dot placed at the object type end (symbol 22) or the role end (symbol 23). An *inclusive-or* (*disjunctive mandatory*) constraint applied to two or more roles indicates that all instances of the object type population must play at least one of those roles. This is shown by connecting the roles by dotted lines to a circled dot (symbol 24).

To restrict the population of an object type or role, the relevant values may be listed in braces (symbol 25). An ordered range may be declared separating end values by "..". For continuous ranges, a square/ round bracket indicates an end value is included/excluded. For example, "(0.10)" denotes the positive real numbers up to 10. These constraints are called *value constraints*.

Symbols 26–28 denote *set comparison constraints*, which apply only between compatible role sequences. A dotted arrow with a circled subset symbol depicts a *subset constraint*, restricting the population of the first sequence to be a subset of the second (symbol 26). A dotted line with a circled "=" symbol depicts an *equality constraint*, indicating the populations must be equal (symbol 27). A circled "X" (symbol 28) depicts an *exclusion constraint*, indicating the populations are mutually exclusive. Exclusion and equality constraints may be applied between two or more sequences. Combining an inclusive-or and exclusion constraint yields an *exclusive-or constraint* (symbol 29).

A solid arrow (symbol 30) from one object type to another indicates that the first is a (proper) *subtype* of the other (e.g., Woman is a subtype of Person). Mandatory (circled dot)

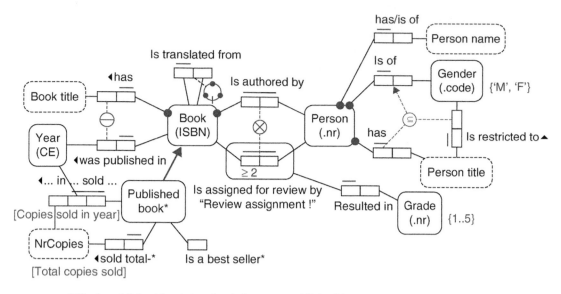

Object-Role Modeling, Fig. 2 An ORM schema for a book publishing domain

and exclusion (circled "X") constraints may be displayed between subtypes, but are implied by other constraints if the subtypes have formal definitions. Symbol 31 shows four kinds of *frequency constraint*. Applied to a role sequence, these indicate that instances that play those roles must do so *exactly n* times, *at least n* times, *at most n* times, or *at least n and at most m* times. Symbol 32 shows four varieties of *value-comparison constraint*. The arrow shows the direction in which to apply the circled operator between two instances of the same type (e.g., **For each** Employee, hiredate > birthdate).

Symbol 33 shows the main kinds of *ring constraint* that may apply to a pair of compatible roles. Read left to right and top row first, these indicate that the binary relation formed by the role population must respectively be ir-reflexive, asymmetric, antisymmetric, reflexive, intransitive, acyclic, intransitive and acyclic, or intransitive and asymmetric.

The previous constraints are *alethic* (necessary, so can't be violated) and are colored violet. ORM 2 also supports *deontic* rules (obligatory, but can be violated). These are colored blue, and either add an "o" for obligatory, or soften

lines to dashed lines. Displayed here are the deontic symbols for uniqueness (symbol 34), mandatory (symbol 35), set-comparison (symbol 36), frequency (symbol 37) and ring (symbol 38) constraints.

Figure 2 shows a sample ORM schema for a book publishing domain. A detailed discussion using the CSDP to develop this schema may be found elsewhere [9]. Each book is identified by an International Standard Book Number (ISBN), each person is identified by a person number, each grade is identified by a grade number in the range 1 through 5, each gender is identified by a code ('M' for male and 'F' for Female), and each year is identified by its common era (CE) number. Published Book is a derived subtype determined by the subtype definition shown at the bottom of the figure. Review Assignment objectifies the relationship Book is assigned for review by Person, and is independent since an instance of it may exist without playing any other role (one can known about a review assignment before knowing what grade will result from that assignment).

The internal uniqueness constraints (depicted as bars) and mandatory role constraints (solid dots) verbalize as follows: **Each** Book is translated from **at most one** Book; **Each** Book has **exactly one** Book Title; **Each** Book was published in **at most 1** Year; **For each** Published Book **and** Year, **that** Published Book in **that** Year sold **at most one** NrCopies; **Each** Published Book sold **at most one** total NrCopies; **It is possible that the same** Book is authored by **more than one** Person **and that more than one** Book is authored by **the same** Person; **Each** Book is authored by **some** Person; **It is possible that the same** Book is assigned for review by **more than one** Person **and that more than one** Book is assigned for review by **the same** Person; **Each** Review Assignment resulted in **at most one** Grade; **Each** Person has **exactly one** Person Name; **Each** Person has **at most one** Gender; **Each** Person has **at most one** Person Title; **Each** Person Title is restricted to **at most one** Gender.

The external uniqueness constraint (circled bar) indicates that the combination of BookTitle and Year applies to at most one Book. The acyclic ring constraint (circle with three dots and a bar) on the book translation predicate indicates that no book can be a translation of itself or any of its ancestor translation sources. The exclusion constraint (circled cross) indicates that no book can be assigned for review by one of its authors. The frequency constraint (≥ 2) indicates that each book that is assigned for review is assigned for review by at least two persons. The subset constraint (circled subset symbol) means that if a person has a title that is restricted to some gender, then the person must be of that gender. The first argument of this subset constraint is a person-gender role pair projected from a join path that performs a conceptual join on PersonTitle. The last two lines at the bottom of the schema declare two derivation rules, one specified in attribute-style using role names and the other in relational style using predicate readings.

Key Applications

ORM has been used productively in industry for over 30 years, in all kinds of business domains. *Commercial tools* supporting the fact-oriented approach include Microsoft's Visio for Enterprise Architects, and the FCO-IM tool CaseTalk (www.casetalk.com). CogNIAM, a tool supporting NIAM2007 is under development at PNA Active Media (http://cogniam.com/). *Free ORM tools* include VisioModeler and Infagon (www.mattic.com). Dogma Modeler (www.starlab.vub.ac.be) and T-Lex [15] are academic ORM-based tools for specifying ontologies. NORMA (http://sourceforge.net/projects/orm), an open-source plug-in to Microsoft® Visual Studio, is under development to provide deep support for ORM 2 [3].

Future Directions

Research in many countries is actively extending ORM in many areas (e.g., dynamic rules, ontology extensions, language extensions, process modeling). A detailed overview of this research may be found in [9]. General information about ORM, and links to other relevant sites, may be found at www.ORMFoundation.org and http://www.orm.net.

Cross-References

► Conceptual Schema Design
► Entity Relationship Model
► Unified Modeling Language

Recommended Reading

1. Bakema G, Zwart J, van der Lek H. Fully communication oriented information Modelling. The Netherlands: Ten Hagen Stam; 2000.
2. Bloesch A, Halpin T. Conceptual queries using ConQuer-II. In: Proceedings of the 16th International Conference on Conceptual Modeling; 1997. p. 113–26.
3. Curland M, Halpin T. Model driven development with NORMA. In: Proceedings of the 40th Annual Hawaii International Conference on System Sciences; 2007.
4. Embley D, Kurtz B, Woodfield S. Object-oriented systems analysis: a model-driven approach. Englewood Cliffs: Prentice Hall; 1992.
5. Falkenberg E. Concepts for modeling information. In: Proceedings of the IFIP Working Conference on Modelling in Data Base Management Systems; 1976. p. 95–109.

6. Halpin T. Comparing metamodels for ER, ORM and UML data models. In: Siau K, editor. Advanced topics in database research. Idea Publishing Group: Hershey; 2004. p. 23–44.

7. Halpin T. Fact-oriented modeling: past, present and future. In: Krogstie J, Opdahl A, Brinkkemper S, editors. Conceptual modelling in information systems engineering. Springer: Berlin Heidelberg; 2007. p. 19–38.

8. Halpin T, Curland M. Automated verbalization for ORM 2. In: On the move to meaningful internet systems 2006: OTM 2006 workshops. LNCS, vol. 4278. Heidelberg: Springer; 2006. p. 1181–90.

9. Halpin T, Evans K, Hallock P, MacLean W. Database modeling with Microsoft® Visio for enterprise architects. San Francisco: Morgan Kaufmann; 2003.

10. Halpin T, Morgan T. Information modeling and relational databases. 2nd ed. San Francisco: Morgan Kaufmann; 2008.

11. Meersman R. The RIDL conceptual language, research report. International Centre for Information Analysis Services, Control Data Belgium, Brussels; 1982.

12. Nijssen GM. Current issues in conceptual schema concepts. In: Proceedings of the IFIP Working Conference on Modelling in Data Base Management Systems; 1977. p. 31–66.

13. OMG. Semantics of usiness Vocabulary and Business Rules (SBVR). URL: http://www.omg.org/cgi-bin/doc?dtc/2006-08-05. 2007.

14. ter Hofstede AHM, Proper HA, van der Weide TP. Formal definition of a conceptual language for the description and manipulation of information models. Inf Syst. 1993;18(7):489–523.

15. Trog D, Vereecken J, Christiaens S, De Leenheer P, Meersman R. T-Lex: a role-based ontology engineering tool. In: On the move to meaningful internet systems 2006: OTM 2006 workshops. LNCS, vol. 4278. Heidelberg: Springer; 2006. p. 1191–200.

16. Wintraecken J. The NIAM information analysis method: theory and practice, vol. 1990. Deventer: Kluwer; 1990.

OLAM

Matteo Golfarelli
DISI – University of Bologna, Bologna, Italy

Synonyms

Data mining on top of data warehouse systems; OLAM

Definition

The term *Online Analytical Mining*, coined in 1997 by J. Han [9], refers to solutions that integrate online analytical processing (OLAP) with data mining functionalities so that mining can be performed in different portions of databases or data warehouses and at different levels of abstraction at the user's fingertips. In such a system, data mining techniques will beneficiate of a higher level of integration, consistency, and cleanness, and data warehouse users will be able to express more powerful queries directly from their user interface. Although no commercial tools make available a complete and integrated set of OLAM features, many data mining techniques have been extended to deal with specific data warehouse features, while new algorithms, that specifically address the OLAP user's advanced requirements, have been developed.

Historical Background

OLAM originated from the coupling of OLAP and data mining systems. OLAP is the main technique for carrying out analyses in the business intelligence area. The term refers to the possibility of carrying out complex queries issued by users with limited technical capabilities through a set of operators. OLAP relies on data organized according to the *multidimensional model*: a multidimensional cube hinges on a fact relevant to decision-making. It shows a set of events quantitatively described by a set of numeric measures. Each cube axis shows a possible analysis dimension. Each dimension can be analyzed at different detail levels specified by hierarchically structured attributes. Given an initial visualization of a cube, OLAP analysis proceeds through a set of operators that allow to navigate and change the data to be visualized by increasing/reducing the level of detail or by focusing on a set of facts through selections on hierarchies and measures.

An OLAP analysis is driven by the user, who progressively navigates the data cube using OLAP operators. Although simple, this approach represents a limit with respect to the possibilities provided by a data mining system where the complex searches through data are delegated to intelligent algorithms. In a data mining process, instead of manually specifying the next data to be retrieved, the users must specify the type of pattern she is looking for. A data mining algorithm will be in charge of instantiating the patterns through a "comprehensive" exploration of data. The main data mining pattern families [11] are *Classification*, *Clustering*, *Association*, *Time-series analysis*, and *Outlier detection*.

Running a data mining process on top of a data warehouse (DW) system ensures the algorithms can beneficiate of a higher level of integration, consistency, and cleanness and allows them to exploit the multidimensional nature of the cubes that make information available at different levels of detail.

Scientific Fundamentals

The goal of an OLAM solution is to enable a cube analysis based on a mix of OLAP operators and data mining algorithms. The analysis should be carried out through an easy-to-use interface, coupled with ad hoc visualization techniques. Let us initially list the set of desired functionalities for an OLAM solution as defined in [9].

- *Cubing then mining*: starting from a data cube, the user applies a set of OLAP operators first in order to select the aggregation level and the set of facts she is interested in and then applies on such data a data mining algorithm.
- *Mining then cubing*: an explorative OLAP analysis is carried out on the results of a data mining algorithm carried out on a data cube or on the result of a previous cubing step.
- *Cubing while mining*: it refers to the possibility of carrying out the same mining operation at different granularities or on different portions of the data cube.
- *Backtracking*: it refers to the possibility of carrying out alternative mining processes from the same initial data. This requires to backtrack to a preset marker and then launch an alternative data mining task.

- *Comparative mining*: it refers to the possibility of comparing the results of alternative data mining processes.

More recently several authors (e.g., [12, 14, 17]) exploited data mining algorithms during the DW design process to discover multidimensional schemata in a semi-automatic fashion. Although the OLAM concept has been sometimes adopted in this context too, we believe that OLAM must be strictly related to front-end functionalities made available to the users.

From an architectural point of view, an OLAM solution extends a DW one by adding an OLAM engine that runs in a combined manner with the OLAP one in order to extract useful knowledge from a collection of subject-oriented cubes (see Fig. 1). The user interface must be able to interactively and transparently apply OLAP operators and data mining algorithms.

Although the main commercial DBMS suites make data mining functionalities available (e.g., Oracle DBMS and R, Microsoft Analysis Services), no full OLAM architecture has been developed so far: data mining algorithms can be run on both source data and data cubes but no native, transparent, and interactive execution of OLAP and data mining is possible. The most interesting and complete framework born in the academic world is DBMiner [10], a data mining system for interactive mining of multiple-level knowledge in large relational databases and DWs. The system implements a wide spectrum of data mining functions, including generalization, characterization, discrimination, association, classification, and prediction. DBMiner performs interactive data mining at multiple concept levels on any user-specified set of data in a database using either an SQL-like Data Mining Query Language, DMQL, or a graphical user interface. Users may interactively set and adjust various thresholds, control a data mining process, perform roll-up or drill-down at multiple concept levels, and generate different types of outputs. Another framework supporting the cubing-then-mining paradigm is MRE-KDD [6] that includes two different layers: an OLAP one used for extracting some multidimensional views out a set of multidimensional

O

OLAM, Fig. 1 A reference architecture for an OLAM system

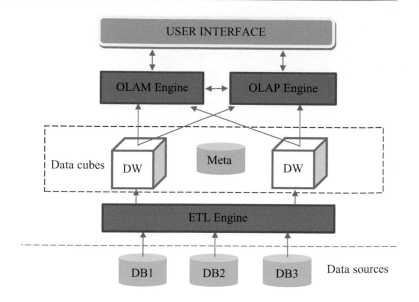

cubes, and an OLAM one that allows a set of data mining algorithms to be applied to such views.

The main idea behind the OLAM concept is the capability of mixing OLAP operations and data mining algorithms to multidimensional data in a transparent fashion. Researches in this area have been focused on either extending traditional data mining techniques to handle the specific features of multidimensional data or on creating new techniques specifically devised to meet the requirements of BI users.

Within the first group the two features to be specifically handled are *multi-granularity* of data and *data dimensionality*. A multidimensional database makes data available at several different granularities, all of them should be explored in order to extract all the interesting information. Let us consider, for example, association rules that may be true at a certain granularity but false at a finer one. In 2000 [16] studied multilevel association mining from a primitive data warehouse and proposed a mining algorithm. Since then, substantial works have been devoted to discovering multidimensional association rules from DWs [4, 13]. Data dimensionality and sparseness of data sets are intrinsic features of DWs since the number of facts stored in the DW is typically orders of magnitude lower than the number of cells defined by the cube dimensions.

High dimensionality (a multidimensional cube easily has from 10 to 20 dimensions) is recognized to be a negative feature of data sets since it limits performances and reduces the capability of identifying interesting patterns [8].

To cope with these problems, in [1], the authors propose an algorithm for predicting the measure values that minimize the effect of sparseness on the prediction process through the application of a cube transformation step, based on a dedicated aggregation technique. [15] proposes a framework based on the notion of information entropy that enables the identification of facts whose measures largely deviate from the underlying data distribution. In this case a preprocessing step is adopted to reduce the dimensionality of the search space.

As to the approaches that exploit data mining algorithms to solve an OLAP user requirement, we recall:

• MyOLAP [3] is an approach for expressing and evaluating OLAP preferences (i.e., desired characteristics of OLAP queries), devised by taking into account the specificities of OLAP queries. During an OLAP session, the user often does not exactly know what she is looking for. The reasons behind a specific phenomenon or trend may be hidden, and finding those reasons by manually applying different combinations of OLAP operators may be very frustrating. Preferences enable users to specify a pattern that describes the type of information she is searching for. Since preferences express soft constraints, when no data exactly match that pattern, the most similar data will be automatically searched through the data cube facts. From this point of view, preference queries can be regarded as an OLAM technique.

• Shrink [7], an OLAP operator aimed at balancing data precision with data size in cube visualization via pivot tables. Shrinking is aimed at solving the information flooding problem that may occur during an OLAP session when the user drills down her cube up to a very fine-grained level, because the huge number of facts returned makes it very hard to analyze them using a pivot table. The shrink operation fuses slices of similar data and replaces them with a single representative slice, respecting the constraints posed by dimension hierarchies, until the result is smaller than a given threshold. Shrink is implemented as a hierarchical clustering with constraint whose goal is the minimization of the approximation error.

Finally, a further research issue that couples data mining and OLAP is similarity. Finding the similarity between objects is the basis for many data mining techniques (e.g., clustering, outlier detection). In this direction many researches focused in defining a similarity criterion tailored on the specific features of OLAP queries [5] and OLAP sessions [2].

Key Applications

OLAM is a natural evolution of OLAP in response to OLAP users requiring more powerful analysis techniques. These requests are a consequence of an increased culture of data analysis of the company managers, and they become particularly relevant whenever the search space of the needed information is too large to be effectively explored through a manual explorative search.

Future Directions

The integration between OLAP and data mining is far to be completed since, as discussed so far, (1) no complete OLAM architectures exist yet; (2) adaptation of traditional data mining techniques to the multidimensional context is far from having reached a sufficient level of efficiency and effectiveness; (3) OLAP and business intelligence are continuously evolving disciplines, and thus new specific applications of data mining techniques arise every day. In particular, to solve points (1) and (2), research should work on the one hand toward implementing one of the most important features of an OLAM architecture, that is, the transparent and interactive application of OLAP operators and data mining algorithms. Furthermore, the ability to cope with the specific features of OLAP systems (i.e., data set size, dimensionality and sparseness; multi-granularity of data) must be largely improved.

Cross-References

▶ Association Rules
▶ Classification
▶ Dimension Reduction Techniques for Clustering
▶ Multidimensional Modeling
▶ Online Analytical Processing

Recommended Reading

1. Abdelbaki W, Yahia SB, Messaoud RB. NAP-SC: a neural approach for prediction over sparse cubes. In: Proceedings of the 8th International Conference on Advanced Data Mining and Applications; 2012. p. 340–52.
2. Aligon J, Golfarelli M, Marcel P, Rizzi S. Similarity measures for OLAP sessions. Int J Knowl Inf Syst. 2014;39(2):463–89.

3. Biondi P, Golfarelli M, Rizzi S. myOLAP: an approach to express and evaluate OLAP preferences. IEEE Trans Knowl Data Eng. 2011;23(7): 1050–64.
4. Chiang JK, Chu C. Multidimensional multi-granularities data mining for discover association rule. Trans Mach Learn Artif Intell. 2014;2(3): 73–89.
5. Chihcheng K, Li MZ. Techniques for finding similarity knowledge in OLAP reports. Expert Syst Appl. 2011;38(4):3743–56.
6. Cuzzocrea A. An OLAM-based framework for complex knowledge pattern discovery in distributed-and-heterogeneous-data-sources and cooperative information systems. In: Proceedings of the 9th International Conference on Data Warehousing and Knowledge Discovery; 2007. p. 181–98.
7. Golfarelli M, Rizzi S. Honey, I Shrunk the Cube. In: Proceedings of the 17th East European Conference on Advances in Databases and Information Systems; 2013. p. 176–89.
8. Golfarelli M, Turricchia E. A characterization of hierarchical computable distance functions for data warehouse systems. Decis Support Syst. 2014;62(June):144–57.
9. Han J. OLAP mining: an integration of OLAP with data mining, In: Proceedings of the 7th IFIP 2.6 Working Conference on Database Semantics; 1997. p. 3–20.
10. Han J, Fu Y, Wang W, Chiang J, et al. DBMiner: a system for mining knowledge in large relational databases. In: Proceedings of the 2nd International Conference on Knowledge Discovery and Data Mining; 1996. p. 250–55.
11. Han J, Kamber M, Pei J. Data mining: concepts and techniques. 3rd ed. Waltham: Morgan Kaufmann; 2011.
12. Jensen M, Holmgren T, Pedersen T. Discovering multidimensional structure in relational data. In: Proceedings of the 6th International Conference on Data Warehousing and Knowledge Discovery; 2004. p. 138–48.
13. Li Y, Wu J, Xu Y, Yang W. Granule mining oriented data warehousing model for representations of multidimensional association rules. Int J Intell Inf Database Syst. 2008;2(1):125–45.
14. Mansmann S, Rehman N, Weiler A, Scholl M. Discovering OLAP dimensions in semi-structured data. Inf Syst. 2014;44(Aug):120–33.
15. Palpanas T, Koudas N, Mendelzon A. Using datacube aggregates for approximate querying and deviation detection. IEEE Trans Knowl Data Eng. 2005;17(11):1465–77.
16. Psaila G, Lanzi P. Hierarchy-based mining of association rules in data warehouses. In: Proceedings of the 2000 ACM Symposium on Applied Computing; 2000. p. 307–12.
17. Usman M, Asghar S. An architecture for integrated online analytical mining. J Emerg Technol Web Intell. 2011;3(2):74–99.

OLAP Personalization and Recommendation

Patrick Marcel
Département Informatique, Laboratoire d'Informatique, Université François Rabelais Tours, Blois, France

Definition

Personalizing or recommending OLAP queries aims at making the OLAP user experience less disorientating when navigating huge amounts of multidimensional data (also called cubes). Such approaches allow coping with too many or too few query results, or suggesting new queries to pursue the navigation. Personalization allows adding preferences to a query for filtering out irrelevant results or ranking the results to focus on the most relevant first. It also allows turning selection predicates (hard constraints) into preferences (soft constraints) to favor nonempty answers. On the other end, recommendation allows to leverage the cube instance and/or past navigations on it to complement the current query result.

The general problem can be formally defined by given a sequence of queries $S = <q_1, \ldots, q_c>$ (a session from now on) over an instance I of a cube schema C, a user profile P (consisting of ordered multidimensional objects), and a set of past sessions L (a log from now on), generate a set of one or more queries $Q = \{q^p_1, \ldots q^p_n\}$ such that, typically:

- The queries in Q are sub-queries of q_c (personalization), in the classical sense of query inclusion, or none of the queries of Q are sub-queries of the queries of S (recommendation).
- The queries in Q maximize an interestingness score.

In this definition, S represents the current session, with q_c the last query of this session (the current query).

Historical Background

OLAP personalization and recommendation approaches are distant descendants of cooperative database [11] techniques aiming at enhancing database management systems with a cooperative behavior. Cooperation can be introduced at the different stages of the retrieval process, which is typically iterative. The purpose of the cooperation includes helping the user to formulate a query corresponding to an objective and acceptable by the database system, dealing with empty answers or too few results, or suggesting additional information and explaining the query result.

This retrieval process perfectly reflects the activity of OLAP users, who interactively analyze multidimensional data, often without exactly knowing what they are looking for. OLAP queries are normally formulated in the form of sequences called OLAP sessions, by using basic operations to transform one OLAP query into another, so that the new query gives a better understanding of the information retrieved so far. The huge number of possible aggregations and selections that can be operated on data may make the user experience disorientating, and OLAP sessions mainly include extemporary queries that may easily either return huge volumes of data (if their group-by sets are too fine), or little or even no information.

To facilitate this navigation, discovery-driven analysis of OLAP cubes [13, 14] was introduced as the definition of two kinds of advanced OLAP operators to guide the user toward interesting regions of the cube by automatically navigating the cube instance. The first kind tries to explain an unexpected significant difference observed in a query result by either looking for more detailed data contributing to the difference [13], or looking for less detailed data confirming an observed tendency. The second kind proposes to the user unexpected data in the cube based on the data she has already observed, by adapting the Maximum Entropy Principle [14].

It was also observed that past navigations, recorded in a (potentially multiusers) query log, could be used for anticipating the next user query. The works presented in [12] propose to pre-fetch cube data by analyzing the OLAP query log and using it to find the query that is the most likely to appear after the current query of the current session. To this end, past queries are grouped by common projections and selections, and a Markov model represents the OLAP sessions.

Finally, giving to the user the possibility to cope with too many or too few query, answers emerged in the database community as preference modeling and query personalization [15]. A first type of approaches extends relational query languages with operators to declare preferences (Preference SQL, Skyline) and an operator to compute dominating tuples (Winnow). Another type of approaches expands regular database queries by incorporating elements from a user profile, usually resulting in another query that is a sub-query of the initial one.

Scientific Fundamentals

Personalization and recommendation approaches can be categorized using the following criteria:

- Proactiveness: This first criterion allows distinguishing between query recommendation (suggesting new queries), which is inherently proactive, and query personalization (changing the current query q_c or post processing its results).
- The type of information used to generate Q: The approach may use all or a subset of the set of parameters (the current session S, the instance I, the cube schema C, the profile P, the log L). In particular, we distinguish current-state approaches, exploiting the content and schema of the current query result and database instance, from history-based approaches, exploiting the query log. We call collaborative approaches those approaches leveraging a multiuser log. Notably, queries can be treated either as simple expressions in a formal language or as the results of the partial or full evaluation of these expressions over the instance I. The full evaluation of an OLAP query is the set of facts (tuples) retrieved

by evaluating the query over the instance. The partial evaluation of an OLAP query is defined as the set of references (i.e., positions in a data cube) to be retrieved from the cube to answer the query, which requires only the instances of the dimensions.

- Prescriptiveness: Prescriptive approaches use profile elements as hard constraints that are added to a query (typically q_c), while non-prescriptive approaches use them as soft ones; tuples that satisfy as much profile criteria as possible are returned even if no tuples satisfy all of them.

Non-proactive approaches are based on the two types of personalization approaches found in the database literature that essentially differ in terms of prescriptiveness. The first approach [5] borrows from the query expansion paradigm, where an OLAP query is transformed into another query by rewriting the former to incorporate elements of the profile, while the second work [9] is inspired by the use of explicit preference constructors for expressing complex preferences directly within the query, à la Preference SQL.

The approach proposed in [5] defines the user profile P as a set of preferences over multidimensional objects and a visualization constraint. The preferences are defined as orders over dimensions and, for each dimension, an order over members (instance of the dimensions at various level of details). These preferences allow defining an order over the set of partially evaluated queries that can be expressed over the instance I. The visualization constraint is defined as an anti-monotone Boolean function over the set of partially evaluated queries. It can, for instance, be used to indicate the maximum number of references for displaying the query answer. The personalization $Q = \{q^p_I\}$ of query q_c consists of prescriptively expanding q_c with preferences of P, guaranteeing that (i) q^p_I is included in q_c, (ii) q^p_I only fetches preferred facts with respect to P, and (iii) q^p_I respects the visualization constraint. q^p_I is generated by starting from the set of most preferred references and iteratively adding less preferred references while the visualization constraint is satisfied.

The work of [9] proposes that elements of the profile P are written with each query. It introduces an algebra to annotate OLAP queries with preference expressions, for defining a strict partial order on the instance I. The algebra consists of a set of base constructors on attributes, measures, and hierarchies, composed by the Pareto (giving the same importance to two base preferences) or prioritization (giving priority to one of the base preferences) operators. This allows defining preferences on the schema, i.e., on the space of hierarchies, which are used to induce preferences on the space of data, thereby allowing defining preferences over group-by sets (aggregated data). A specific implementation is developed for evaluating preference queries expressed in this language, in order to calculate the personalized query q^p_I, without having to compute all the aggregations. In [2], it is proposed that preference constructors are automatically added to a current query q_c by mining a query log to identify which preferences could fit q_c.

If they also build upon the previous works (especially [12, 13, 14]), proactive approaches are more diverse than non-proactive ones. They range from current state to collaborative, with a combination thereof; they can be similarity based, preference based, or stochastic and vary in how they treat sessions and queries and in how they generate recommendations.

A current-state, preference-based approach is proposed in [10], with a principle similar to that described in [5], the main differences being that the recommendation q^p_I is usually not a subquery of q_c and that q_c is fully evaluated. q^p_I is derived from q_c using elements extracted from the user profile P that consists of a set of preference predicates, each with a score of interest. The best preferences (according to the interest score) that are consistent with q_c are incorporated to it, resulting in q^p_I.

The work described in [8] is both collaborative and current state and uses the operators introduced in [13] to discover in L the (fully evaluated) queries that investigated the same phenomenon as the one shown by q_c. Sessions are associated with a goal, and recommendations are those queries from former sessions having the same goal as that of the current session. The approach is composed of two steps. In an offline step, a multiuser query log L is processed to detect discovery-driven analysis sessions, i.e., sessions investigating (either by rolling up or drilling down) a pair of facts that show a significant difference (like, e.g., a drop of sales from 1 year to the following year). Those pairs are then arranged into a specialization relation based on the cube hierarchies. A goal is created for each most general pair recording the pair and its descendants, together with the queries that contain them. In the second step, at query time, if q_c investigates a pair that is a descendant of a pair discovered in L, then the set Q of queries associated with the corresponding goal is recommended. The main difference with the approach of [13] is that only L and not I is searched for interesting data.

Another two-step approach is described in [4], where queries are recommended using a probabilistic model of former sessions, inspired by that of [12]. In an offline step, the former queries of L are grouped with a density-based clustering that uses a similarity measure tailored for the syntax of OLAP queries. The Markov model organizes the query clusters into series of states, with a transition score for each pair of clusters. At query time, the current query q_c is matched with the closest state of the Markov model, in the sense of the average similarity between it and each query of the cluster. Then, the most probable state is identified, and the query of this cluster that is the most similar to q_c is recommended.

The work of [6] introduces a generic framework for similarity-based collaborative query recommendations, with a three-step approach for generating recommendations. In the first step, the current session S is compared to the sessions of L, to find the ones that are the most similar to S, in the sense of a similarity between sessions. In the second step, candidate recommendations are extracted from the sessions resulting of the first phase. Finally, in the last step, these candidate recommendations are further processed to be presented to the user. [7] instantiates this framework with partially evaluated queries. It introduces an extension of the edit distance to compare sessions and uses the Hausdorff distance between sets of references to compare partially evaluated queries. These distances enable the definition of a similarity measure for sessions to be used during the first step. In the second step, the last queries of the sessions that are the most similar to S are extracted to form Q. In the last step, these queries are ranked according to how close they are from the current query q_c. Another instance of this framework is described in [1], where only the syntax of queries is considered. The similarity measure between sessions is an extension of the Smith-Waterman alignment algorithm whose goal is to efficiently find the best alignment between subsequences of two given sequences by ignoring their nonmatching parts. This extension uses a query similarity measure tailored for the syntax of OLAP queries that compares the three parts of queries (the group-by set, the selection predicate set, and the measure set) and averages the result of these comparisons. During the first step, log sessions in L are aligned with S, and portions of the most similar log sessions are identified as potential futures for S. In the second step, a subsequence of one of these futures is chosen as a base recommendation r, based on its similarity with S and its frequency in L. Finally, in the third step, r is adapted to S, by characterizing (i) the differences between S and its aligned counterpart in the log session l from which r is extracted and (ii) the user's behavior during S. These characterizations adapt the technique of [2] and consist of extracting association rules from S and l.

A study of similarity measures tailored for OLAP sessions is provided in [3], where various similarity measures are devised and tested using both subjective (i.e., user) and objective tests.

Key Applications

OLAP personalization and recommendation techniques can be incorporated into any OLAP front-end tool that allows the user to compose and evaluate OLAP queries.

Future Directions

Although a number of different personalization and recommendation approaches already exist, a comprehensive comparative study of these approaches is still missing. Objective quality criteria for recommended queries only start to emerge [1] and should be completed, and subjective, user-based tests are still to be conducted. A long-term objective is to define a benchmark allowing assessing the effectiveness of OLAP recommendations and, more generally, how successful OLAP sessions are.

Experimental Results

Approaches are usually evaluated from both the efficiency and effectiveness point of view. Efficiency is crucial in the sense that OLAP sessions are interactive by nature and personalized or recommended queries must be computed on the fly. Efficiency is measured as the time taken to obtain the personalized or recommended queries, varying the characteristics of the information used to obtain them. [7, 8] showed that recommending an OLAP query can be computed efficiently for logs of reasonable sizes. [2, 9] showed that personalization puts no significant overhead in the querying process and that personalized queries are evaluated faster than non-personalized ones.

Effectiveness is typically measured in terms of reduction of the answer set for personalization approaches [2], or in terms of prediction accuracy for proactive approaches. In this latter case, [1, 7, 8] report effectiveness in terms of precision and recall of the recommendations when recommending for a sub-session of the log while the technique is trained on other parts of this log. More effectiveness quality criteria, including coverage, novelty, or foresight of recommendation, are proposed and tested in [1].

Data Sets

URL to Code
The I3 project hosts the Java code of the operators described in [13, 14] and used in [8], distributed under the terms of the GPL: http://www.it.iitb.ac. in/~sunita/icube/

CubeLoad is a parametric generator of OLAP workloads written in Java (used in [1]) that can be used to generate a realistic profile-based workload in the form of sessions: http://big.csr.unibo. it/?q=node/371

Cross-References

▸ Collaborative Filtering
▸ Cube
▸ Dimension
▸ Hierarchy
▸ Measure
▸ Multidimensional Modeling
▸ Online Analytical Processing
▸ Preference Specification
▸ Recommender Systems
▸ Skyline Queries and Pareto Optimality

Recommended Readings

1. Aligon J, Similarity based recommendation of OLAP sessions, doctoral dissertation, Université François Rabelais Tours, France; 2013.
2. Aligon J, Golfarelli M, Marcel P, Rizzi S, Turricchia E. Mining preferences from OLAP query logs for proactive personalization. In: Proceedings of the 15th East European Conference on Advances in Databases and Information Systems; 2011. p. 84–97.
3. Aligon J, Golfarelli M, Marcel P, Rizzi S, Turricchia E. Similarity measures for OLAP sessions. Knowl Inf Syst. 2014;39(2):463–89.
4. Aufaure M-A, Beauger NK, Marcel P, Rizzi S, Vanrompay Y. Predicting your next OLAP query based on recent analytical sessions. In: Proceedings of the 15th International Conference on Data Warehousing and Knowledge Discovery; 2013. p. 134–45.
5. Bellatreche L, Giacometti A, Marcel P, Mouloudi H, Laurent D. A personalization framework for OLAP queries. In: Proceedings of the ACM 8th International Workshop on Data Warehousing and OLAP; 2005. p. 9–18.
6. Giacometti A, Marcel P, Negre E. A framework for recommending OLAP queries. In: Proceedings of the ACM 11th International Workshop on Data Warehousing and OLAP; 2008. p. 73–80.
7. Giacometti A, Marcel P, Negre E. Recommending multidimensional queries. In: Proceedings of the 10th International Conference on Data Warehousing and Knowledge Discovery; 2009. p. 453–66.

8. Giacometti A, Marcel P, Negre E, Soulet A. Query recommendations for OLAP discovery-driven analysis. Int J Data Warehouse Min. 2011;7(2):1–25.
9. Golfarelli M, Rizzi S, Biondi P. myOLAP: an approach to express and evaluate OLAP preferences. IEEE Trans Knowl Data Eng. 2011;23(7):1050–64.
10. Jerbi H, Ravat F, Teste O, Zurfluh G. Preference-based recommendations for OLAP analysis. In: Proceedings of the 10th International Conference on Data Warehousing and Knowledge Discovery; 2009. p. 467–78.
11. Motro H. Cooperative database systems. In: Encyclopedia of library and information science. vol. 66 Supp 29. New York: Marcel Dekker; 2000. p. 79–97.
12. Sapia C. PROMISE: predicting query behavior to enable predictive caching strategies for OLAP systems. In: Proceedings of the 2nd International Conference on Data Warehousing and Knowledge Discovery; 2000. p. 224–33.
13. Sarawagi S. iDiff: informative summarization of differences in multidimensional aggregates. Data Min Knowl Discov. 2001a;5(4):255–76.
14. Sarawagi S. User-cognizant multidimensional analysis. VLDB J. 2001b;10(2–3):224–39.
15. Stefanidis K, Koutrika G, Pitoura E. A survey on representation, composition and application of preferences in database systems. ACM Trans Database Syst. 2011;36(3):19.

One-Copy-Serializability

Bettina Kemme
School of Computer Science, McGill University, Montreal, QC, Canada

Synonyms

Transactional consistency in a replicated database

Definition

While transactions typically specify their read and write operations on logical data items, a replicated database has to execute them over the physical data copies. When transactions run concurrently in the system, their executions may interfere. The replicated database system has to isolate these transactions. The strongest, and most well-known correctness criterion for replicated databases is 1-copy-serializability. A concurrent execution of transactions in a replicated database is 1-copy-serializable if it is "equivalent" to a serial execution of these transactions over a single logical copy of the database.

Main Text

A transaction is a sequence of read and write operations on the data items of the database. A read operation of transaction T_i on data item x is denoted as $r_i(x)$, a write operation on x as $w_i(x)$. A transaction T_i either ends with a commit c_i (all operations succeed) or with an abort a_i (whereby all effects on the data are undone before the termination).

A replicated database consists of a set of database servers $A, B, \ldots$ and each logical data item x of the database has a set of physical copies $x^A, x^B, \ldots$ where the index refers to the database server on which the copy resides. *Replica Control* translates each operation $o_i(x), o_i \in \{r, w\}$ of a transaction T_i on logical data item x into physical operations $o_i(x^A), o_i(x^B)$ on physical data copies. Given a set of transactions $\mathcal{T}$, a replicated history RH describes the execution of the physical operations of transactions in the replicated database. For simplicity, the following discussion only considers histories where all transactions commit. A database server A executes the subset of physical operations of the transactions in $\mathcal{T}$ performed on copies residing on A. The local history RH^A describes the order in which these operations occur. For simplicity, a local history is assumed to be a total order. RH is the union of all local histories with some additional ordering. In particular, if a transaction T_i executes $o_i(x)$ on logical data item x before $o_i(y)$ on logical data item y, and RH^A contains physical operation $o_i(x^A)$ and RH^B contains $o_i(y^B)$, then $o_i(x^A) <_{RH} o_i(y^B)$.

As an example, given $T_1 = w_1(y)w_1(x)$ and $T_2 = r_2(y)w_2(x)$ on logical data items x, and database servers A and B, both having a copy of both x and y, the local histories could be:

$$RH^A : w_1(y^A)r_2(y^A)w_1(x^A)w_2(x^A)c_1c_2$$
$$RH^B : w_1(y^B)w_1(x^B)w_2(x^B)c_2c_1$$

The replicated history RH is the union of these two local histories plus the ordering of $r_2(y^A) <_{RH} w_2(x^B)$.

Using this notation, the following defines 1-copy-serializability for the case that replica control uses ROWA (read-one-write-all-approach), i.e., where each read operation is performed on one copy while write operations are performed on all copies of the data item. Failures are ignored. In this restricted case *conflict-equivalence* can be exploited. Two physical operations o_i and o_j conflict, if they are from two different transactions, access the same data copy, and at least one is a write operation.

Definition 1 A replicated history RH over a set of transactions T in a replicated system with servers $A, B, \ldots$ is 1-copy-serializable if it is conflict-equivalent to a serial history H of T over the logical data items. This means that if $o_i(x^A), o_j(x^A) \in RH$ and the operations conflict, then $o_i(x) <_H o_j(x) \in H$ if and only if $o_i(x^A)$ is executed before $o_j(x^A)$ at server A.

Using conflict-equivalence, one can easily determine whether RH is 1-copy-serializable. For each local history RH^A the serialization graph $SG(RH^A)$ has each committed transaction as node, and contains an edge from T_i to T_j if $o_i(x^A)$ is executed before $o_j(x^A)$ and the two operations conflict. The serialization graph $SG(RH)$ is then the union of the local serialization graphs.

Theorem 1 *A replicated history RH over a set of transactions T and database servers $A, B, \ldots$ following the ROWA strategy is 1-copy-serializable if and only if its serialization graph $SG(RH)$ is acyclic.*

The example history above is 1-copy-serializable because its serialization graph contains only an edge from T_1 to T_2, i.e., in all local histories, and for any conflict between T_1 and T_2, T_1's operation is ordered before T_2's operation.

As soon as node failures are considered or both read and write operations only access a subset of copies, conflict-equivalence is not appropriate anymore because it might miss catching conflicts at the logical level. For that purpose, one can define 1-copy-serializability based on view-equivalence which observes which data versions a read operation accesses and in which order write operations occur.

Cross-References

▶ Consistency Models for Replicated Data
▶ Replica Control
▶ Replicated Database Concurrency Control
▶ Concurrency Control for Replicated Databases

Recommended Readings

1. Bernstein PA, Hadzilacos V, Goodman N. Concurrency control and recovery in database systems. Reading: Addison Wesley; 1987.

One-Pass Algorithm

Nicole Schweikardt
Johann Wolfgang Goethe-University, Frankfurt am Main, Frankfurt, Germany

Synonyms

Data stream algorithm; One-pass algorithm; Streaming algorithm

Definition

A one-pass algorithm receives as input a list of data items $x_1, x_2, x_3, \ldots$. It can read these data items only once, from left to right, i.e., in increasing order of the indices $i = 1, 2, 3, \ldots$. Critical parameters of a one-pass algorithm are (1) the size of the memory used by the algorithm, and (2) the processing time per data item x_i. Typically, a one-pass algorithm is designed for answering one particular query against the input data. To this end, the algorithm stores and maintains a suitable data structure which, for each i, is updated when reading data item x_i.

The two parameters *processing time per data item* and *memory size* are usually measured as functions depending on the size N of the input (different measures of the *input size* are considered in the literature, among them, e.g., the number of data items occurring in the input, as well as the total number of bits needed for storing the entire input). The ultimate goal when designing a one-pass algorithm is to keep the processing time per data item and the memory size *sublinear*, preferably polylogarithmic, in N. In particular, one typically aims at algorithms whose memory size is far smaller than the size of the input.

Key Points

The design and study of one-pass algorithms has a long tradition in many areas of computer science. For example, they are used in the area of *data stream processing*, where streams of huge amounts of data have to be monitored *on-the-fly* without first storing the entire data. A deterministic finite automaton on words can be viewed as a (very simple) example of a one-pass algorithm whose memory size and processing time per data item is constant, i.e., does not depend on the input size. For most computational problems, however, the amount of memory necessary for solving the problem grows with increasing input size. *Lower bounds* on the memory size needed for solving a problem by a one-pass algorithm are usually obtained by applying methods from *communication complexity* (see, e.g., [1, 2] for typical examples).

For many concrete problems it is even known that the memory needed for solving the problem by a deterministic one-pass algorithm is at least linear in the size N of the input. For some of these problems, however, *randomized* one-pass algorithms can still compute good *approximate* answers while using memory of size sublinear in N (cf. [1–3]). Typically, such algorithms are based on *sampling*, i.e., only a "representative" portion of the data is taken into account, and *random projections*, i.e., only a rough "sketch" of the data is stored in memory (see [3] for a comprehensive survey of according algorithmic techniques).

In the context of database systems these techniques are relevant, for example, for maintaining information needed for cost-based query optimization, e.g., estimates for the number of distinct values of an attribute, or the self-join size of a database relation. Efficient one-pass algorithms for incrementally updating these estimates can be found in [1].

In some application areas, rather than just a single pass, a small number P of sequential passes over the data may be available; the resulting algorithms are called *multi-pass algorithms* (see e.g., [2] for an analysis of the trade-off between the memory size and the number of passes necessary for solving particular problems).

Cross-References

▶ Approximate Query Processing
▶ Clustering on Streams
▶ Data Sketch/Synopsis
▶ Data Stream
▶ Event and Pattern Detection over Streams
▶ Stream Processing
▶ XML Stream Processing

Recommended Reading

1. Alon N, Matias Y, Szegedy M. The space complexity of approximating the frequency moments. J Comput Syst Sci. 1999;58(1):137–47.
2. Henzinger M, Raghavan P, Rajagopalan S. Computing on data streams. In: External memory algorithms. DIMACS series in discrete mathematics and theoretical computer science 50. Boston: American Mathematical Society; 1999. p. 107–18.
3. Muthukrishnan S. Data streams: algorithms and applications. Found Trends Theor Comput Sci. 2005;1(2):117–236.

Online Analytical Processing

Alberto Abelló and Oscar Romero
Polytechnic University of Catalonia, Barcelona,
Spain

Synonyms

OLAP

Definition

On-line analytical processing (OLAP) describes an approach to decision support, which aims to extract knowledge from a data warehouse, or more specifically, from data marts. Its main idea is providing navigation through data to non-expert users, so that they are able to interactively generate ad hoc queries without the intervention of IT professionals. This name was introduced in contrast to on-line transactional processing (OLTP), so that it reflected the different requirements and characteristics between these classes of uses. The concept falls in the area of business intelligence.

Historical Background

From the beginning of computerized data management, the possibility of using computers in data analysis has been evident for companies. However, early analysis tools needed the involvement of the IT department to help decision makers to query data. They were not interactive at all and demanded specific knowledge in computer science. By the mid-1980s, executive information systems appeared introducing new graphical, keyboard-free interfaces (like touch screens). However, executives were still tied to IT professionals for the definition of ad hoc queries, and prices of software and hardware requirements where prohibitive for small companies. Eventually, cheaper and easy-to-use spreadsheets became very popular among decision makers, but soon it was clear that they were not appropriate for using and sharing huge amounts of data. Thus, it was in 1993 that Codd et al. [5], coined the term OLAP. In that report, the authors defined 12 rules for a tool to be considered OLAP. These rules caused heated controversy, and they did not succeed as Codd's earlier proposal for relational database management systems (RDBMS). Nevertheless, the name OLAP became very popular and is broadly used.

Although the name OLAP comes from 1993 and the idea behind them goes back to the 1980s, there is not a formal definition for this concept, yet. As proposed by Nigel Pendse [13], OLAP tools should pass the FASMI (fast analysis of shared multidimensional information) test. Thus, they should be fast enough to allow interactive queries; they should help analysis tasks by providing flexibility in the usage of statistical tools and what–if studies; they should provide security (both in the sense of confidentiality and integrity) mechanisms to allow sharing data; they should provide a multidimensional view so that the data cube metaphor can be used by users; and, finally, they should also be able to manage large volumes of data (gigabytes can be considered a lower bound for volumes of data in decision support) and metadata. However, there are not measures and thresholds for all these characteristics in order to be able to establish whether one of them is fulfilled or not, and therefore it is always arguable that a given tool fulfills them. Nevertheless, it is generally agreed that in order to be considered an OLAP tool, it must offer a multidimensional view of data.

Since their first days, OLAP tools have been losing weight and lowering prices, while at the same time, offering more functionality, better user interfaces and easier administration. Thus, time has come for small companies to use OLAP. They can afford it and they are willing to use it in their decision processes. Part of OLAP industry was associated into the OLAP Council (created in January 1995), whose aim was the promotion and standardization of OLAP terminology and technology. However, some major vendors never became members of this council, so eventually it disappeared (last news date from 1999). Nowadays, there is no standardization institution

Online Analytical Processing, Fig. 1 Comparing OLTP Versus OLAP

	OLTP	OLAP
Usage	Application specific	Decision support
Workload	Predefined	Unforeseeable
Access	Read/Write	Read-only
Query structure	Simple	Complex
Records per operation	Tens/Hundreds	Thousands/Millions
Number of users	Thousands/Millions	Tens/Hundreds

specifically devoted to OLAP. Therefore, it seems difficult to have a standard data model and query language in the near future, despite the fact that it is clearly desirable.

Market (billion US$)	ROLAP tools		MOLAP tools	
	Europe	USA	Europe	USA
2005	1	2	0.75	0.25
2006	1.5	2.5	1	0.5

Online Analytical Processing, Fig. 2 Example of cross-tab or statistical table representation of a $2 \times 2 \times 2$ data cube

Foundations

OLAP environments have completely different requirements, compared to OLTP. Figure 1 summarizes the main differences. Firstly, their usage is different. While OLTP systems are conceived to solve a concrete problem and are used in the daily work of companies, OLAP systems are used in decision support. Thus, in the first case, since the addressed problem can be completely specified, the workload of the system is clearly predefined. Conversely, a decision support system aims to solve new problems every day. Therefore, ad hoc queries are executed. OLTP systems read as well as write data, while OLAP systems are considered read-only, because decision makers do not directly modify data. Nevertheless, the queries in a decision support system are much more complex, since they usually include big volumes of information processed by joining several tables, grouping data and calculating functions. Queries in OLTP systems do not usually involve volumes of data of the same magnitude, neither as many tables, nor groupings or calculations. The number of records in OLTP operations can be estimated as tens or hundreds at most, while OLAP queries usually involve thousands or even millions of records. Finally, the number of users is also different in both kinds of systems. OLTP systems can have thousands or millions of users (like in the case of cash machines), while OLAP systems have tens or maybe hundreds of users.

The main characteristic of OLAP is multidimensionality. The data cube metaphor is used to make user interaction easier and closer to decision makers' way of thinking, who would probably find SQL or any other text-based query language hard to understand and error prone. Thus, it is much easier for them to think in terms of the multidimensional model, where a Fact is a subject of analysis and its Dimensions are the different points of view that analysts could use to study the Fact. In this way, the instances of a Fact are shown in an n-dimensional space usually called Cube or Hypercube.

In order to show n-dimensional Cubes in two-dimensional interfaces, Cross-tabs or Statistical Tables such as the one in Fig. 2 (its data is entirely fictitious) are used. While in relational tables it is found that fixed columns and different instances are shown in each row, in Cross-tabs both columns and rows are fixed and interchangeable. In this example, you see three dimensions (i.e., Product, Place, and Year) that show the different points of view to analyze the OLAP tools market.

Multidimensionality is based on this fact-dimension dichotomy. A Dimension is considered to contain a hierarchy of aggregation levels representing different granularities (or levels of detail) to study data, and an aggregation level to contain descriptive attributes. On the other hand, a Fact contains quantitative attributes that are called measures. Dimensions of analysis arrange the multidimensional space where the Fact of study is depicted. Each instance of data is identified (i.e., placed in the multidimensional space) by a point in each of its analysis dimensions. Two different instances of data cannot be spotted in the same point of the multidimensional space. Therefore, given a point in each of the analysis dimensions they only determine one, and just one, instance of factual data. Moreover, data summarization that is performed must be correct, i.e., aggregated categories must be a partition (complementary and disjoint) and the kind of measure, aggregation function, and the dimension along which data is aggregated must be compatible. For example, stock, sum and time are not compatible, since stock measures cannot be added along temporal dimensions.

Operations

Unfortunately, there is no consensus on the set of multidimensional operations and how to name them. However, [14] provides a comparison of algebraic proposals in the academic literature, as well as a set of operations subsuming all of them. A sequence of these operations is known as an OLAP session. An OLAP session allows transformation of a starting query into a new query. Figure 3 draws the transitions generated by each one of these operations (circles and triangles represent different measures for Fact instances):

1. *Selection or dice*. By means of a logic predicate over the dimension attributes, this operation allows users to choose the subset of points of interest out of the whole n-dimensional space (Fig. 3a).
2. *Roll-up*. Also called "Drill-up", it groups cells in a Cube based on an aggregation hierarchy. This operation modifies the granularity of data by means of a many-to-one relationship which relates instances of two aggregation levels in the same Dimension, corresponding to a part-whole relationship (Fig. 3b from left to right). For example, it is possible to roll-up monthly sales into yearly sales moving from "Month" to "Year" aggregation level along the temporal dimension.

3. *Drill-down*. This is the counterpart of Roll-up. Thus, it removes the effect of that operation by going down through an aggregation hierarchy, and showing more detailed data (Fig. 3b from right to left).
4. *ChangeBase*. This operation reallocates exactly the same instances of a Cube into a new n-dimensional space with exactly the same number of points (Fig. 3c). Actually, it allows two different kinds of changes in the space: rearranging the multidimensional space by reordering the Dimensions, interchanging rows and columns in the Cross-tab (this is also known as Pivoting), or adding/removing dimensions to/from the space.
5. *Drill-across*. This operation changes the subject of analysis of the Cube, by showing measures regarding a new Fact. The n-dimensional space remains exactly the same, only the data placed in it change so that new measures can be analyzed (Fig. 3d). For example, if the Cube contains data about sales, this operation can be used to analyze data regarding production using the same Dimensions.
6. *Projection*. It selects a subset of measures from those available in the Cube (Fig. 3e).
7. *Set operations*. These operations allow users to operate two Cubes defined over the same n-dimensional space. Usually, Union (Fig. 3f), Difference and Intersection are considered.

This set of algebraic operations is minimal in the sense that none of the operations can be expressed in terms of others, nor can any operation be dropped without affecting functionality (some tools consider that the set of measures of a Fact conform to an artificial analysis dimension, as well; if so, Projection should be removed from the set of operations in order to be considered minimal, since it would be done by Selection over this artificial Dimension). Thus, other operations can be derived by sequences of these. It is the case of Slice (which reduces the dimensionality of the original Cube by fixing a point in a Dimension) by means of Selection and ChangeBase operations. It is also common that OLAP implementations use the term Slice&Dice to refer to the

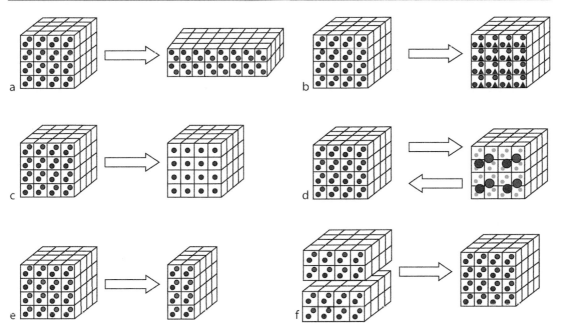

Online Analytical Processing, Fig. 3 Schema of operations on cubes

selection of fact instances, and some also introduce Drill-through to refer to directly accessing the data sources in order to lower the aggregation level below that in the OLAP repository or data mart.

Declarative Languages

There are some research proposals of declarative query languages for OLAP. Cabibbo and Torlone [4] propose a graphical query language, while Gyssens and Lakshmanan [9] propose a calculus. From the industry point of view, MDX (standing for multidimensional expressions) [12] is the de facto standard. It was introduced in 1997, and in spite of the specification being owned by Microsoft, it has been widely adopted. Its syntax resembles that of SQL:

```
[WITH <MeasureDefinition>+]
SELECT <DimensionSpecification>+
FROM <CubeName>
[WHERE <SlicerClause>]
```

However, its semantics are completely different. Roughly speaking, an MDX query gets the instances of a given Cube stated in the FROM clause and places them in the space defined by the

SELECT clause. Moreover, complex calculations can be defined in the WITH clause, and the dimensions not used in the SELECT clause can be sliced in the WHERE clause (if not explicitly sliced, it is assumed that dimensions that do not appear in the SELECT are sliced at the highest aggregation level: All).

```
WITH MEMBER [Measures].[pending] AS
    '[Measures].[Units Ordered]-[Measures].[Units
    Shipped]'
SELECT [Time].[2006].children ON COLUMNS,
[Warehouse].[Warehouse Name].members ON
    ROWS
FROM Inventory
WHERE ([Measures].[pending],[Trademark].
    [Acme]);
```

In the previous MDX query, an ad hoc measure "pending" is first defined as the difference between units ordered and shipped. Then, the children of the instance representing year 2006 (i.e., the 12 months of that year) are placed on columns, and the different members of the aggregation level "Warehouse Name" on rows. Now, this matrix is filled with the data in "Inventory" cube, showing the previously defined measure "pending" and slicing "Acme" trademark.

Key Applications

Managers are usually not trained to query databases by means of SQL. Moreover, if the query is relatively complex (several joins and subqueries, grouping, and functions) and the database schema is not small (with maybe hundreds of tables), using interactive SQL could be a nightmare even for SQL experts. Thus, OLAP is used to ease the tasks of these managers in extracting knowledge from the data warehouse by means of Drag&Drop, instead of typing SQL queries by hand. The primary idea behind OLAP is to be used to gain quick insight into data, whereas data mining is meant to thoroughly explore the correlations and hidden patterns in the data. Indeed, one naturally follows the other in most cases. In some tools, OLAP functionalities are intertwined with data mining functionalities (so called OLAM).

Some existing alternatives follow the same spirit as OLAP (i.e., quick analysis of data) and are sometimes incorrectly categorized as OLAP tools. This is the case of, for example, QlikView (http://www.qlik.com/es), which is based in associative rules. Thus, data is not arranged in a multidimensional fashion (most importantly, the concept of dimension hierarchies is not considered) and the potential analysis tasks enabled by QlikView substantially differ from those empowered by an OLAP tool (and viceversa).

Future Directions

Traditionally, operational data have been collected in the DW of the company by means of ETL flows, and deployed in Data Marts for later analysis with OLAP tools. However, not only real-time analytics, but also situational BI has been recognized as a real need in today world (see [11]). This entails the need of a much faster BI cycle, reducing the intervention of IT specialist at the same time that we integrate more heterogeneous (potentially providing lower data quality) sources.

Indeed, more and more data is available every day. Some come from public institutions (e.g.,

Open Data Portal (https://open-data.europa.eu) offered by the European Commission), and others from private companies like Facebook, Tweeter, etc. This phenomenon is fueling the Big Data business, which is directly related to analytics.

Some proposals, like [1], already appeared to fuse internal data cubes in the companies with external data in the Web. As explained in [2], to enable such possibility, semantics and reasoning are a must. Thus, we need to define the meaning of the data being offered to others. W3C already defined a vocabulary for the exchange of statistical data in [15]. Nevertheless, as outlined in [6], this is not enough and it must be enriched with OLAP metadata.

The role played by external data in current OLAP systems and the need to assist the user to explore these data repositories is addressed in [3]. There, the authors discuss how to capture the semantics of the queries posed by the users and exploit them to assist the user in her future analysis. Nevertheless, query recommendation should not be the only support provided by OLAP tools but also visualization support and self-tuning techniques according to the usage of the system (e.g., most used fact tables).

Also, OLAP has been traditionally related to the analysis of numerical data (e.g., sales, income, revenue, etc.), whereas new approaches are extending the multidimensional concept to any kind of data. For example, in [8] the authors propose to exploit the cube metaphor to analyze spatiotemporal data and highlight the relevance of designing dynamic dimensions (see for example [7]) and hierarchies (based on the available data) instead of design-time-based dimensions.

Other research directions in OLAP can be the improvement of user interaction and flexibility in the calculation of statistics (see Visual OLAP definitional entry), and the integration of what-if analysis (see What-if Analysis definitional entry). As proposed in [10], OLAP tools need to be extended with writing capabilities in order to provide planning functionalities.

Url to Code

Some OLAP vendors:

1. Microsoft Analysis Services: http://www.microsoft.com/en-us/server-cloud/products/analytics-platform-system
2. IBM Cognos: www.ibm.com/software/analytics/cognos
3. Oracle Business Intelligence: http://www.oracle.com/us/solutions/business-analytics/business-intelligence/overview/index.html
4. SAP Business Objects: http://www.sap.com/pc/analytics/business-intelligence/software/overview/bi-platform.html
5. MicroStrategy: http://www.microstrategy.com/us/platforms/analytics/self-service-analytics
6. Tableau: http://www.tableausoftware.com

Some open source OLAP tools:

1. Mondrian: http://mondrian.pentaho.org
2. Palo: http://www.palo.net

Cross-References

▶ Business Intelligence
▶ Cube
▶ Data Mart
▶ Data Mining
▶ Data Warehouse
▶ Dimension
▶ Hierarchy
▶ Hierarchical Data Summarization
▶ Measure
▶ Multidimensional Modeling
▶ OLAP Personalization and Recommendation
▶ Star Schema
▶ Summarizability
▶ Visual Online Analytical Processing (OLAP)
▶ What-If Analysis

Recommended Reading

1. Abelló A, Darmont J, Etcheverry L, Golfarelli M, Mazón J-N, Naumann F, Pedersen TB, Rizzi S, Trujillo J, Vassiliadis P, Vossen G. Fusion cubes: towards self-service business intelligence. Int J Data Warehouse Min. 2013;9(2):66–88.
2. Abelló A, Romero O, Pedersen TB, Berlanga R, Nebot V, Aramburu MJ, Simitsis A. Using semantic web technologies for exploratory OLAP: a survey. IEEE Trans Data Knowl Eng. 2014; PP(99):1. https://doi.org/10.1109/TKDE.2014.2330822.
3. Aufaure M-A, Cuzzocrea A, Favre C, Marcel P, Missaoui R. An envisioned approach for modeling and supporting user-centric query activities on data warehouses. Int J Data Warehouse Min. 2013;9(2):89–109.
4. Cabibbo L, Torlone R. From a procedural to a visual query language for OLAP. In: Proceedings of the 10th International Conference on Scientific and Statistical Database Management; 1998. p. 74–83.
5. Codd EF, Codd SB, Salley CT. Providing OLAP to user-analysts: an IT mandate. Technical report, E. F. Codd & Associates; 1993.
6. Etcheverry L, Vaisman A, Zimanyi E. Modeling and querying data warehouses on the semantic web using QB4OLAP. In: Proceedings of the 16th International Conference on Data Warehousing and Knowledge Discovery; 2014.
7. Golfarelli M, Graziani S, Rizzi S. Shrink: an OLAP operation for balancing precision and size of pivot tables. Data Knowl Eng. 2014;93(Sept): 19–41.
8. Gómez LI, Gómez SA, Vaisman AA. A generic data model and query language for spatiotemporal OLAP cube analysis. In: Proceedings of the 15th International Conference on Extending Database Technology; 2012. p. 300–11.
9. Gyssens M, Lakshmanan LVS. A foundation for multi-dimensional databases. In: Proceedings of the 23rd International Conference on Very Large Data Bases; 1997. p. 106–15.
10. Jaecksch B, Lehner W. The planning OLAP model – a multidimensional model with planning support. In: Proceedings of the 15th International Conference on Data Warehousing and Knowledge Discovery; 2013. p. 32–52.
11. Markl V. Situational business intelligence. In: Proceedings of the 2nd International Workshop on Business Intelligence for the Real Time Enterprise (in conjunction with the VLDB Conference); 2008.
12. Microsoft. Multidimensional expressions (MDX) reference; 2007. Available at http://msdn2.microsoft.com/en-us/library/ms145506.aspx. SQL Server books online.
13. Pendse N. The OLAP report – what is OLAP? 2007. Business Application Research Center.
14. Romero O, Abelló A. On the need of a reference algebra for OLAP. In: Proceedings of the 9th International Conference on Data Warehousing and Knowledge Discovery; 2007. p. 99–110.
15. W3C. The RDF data cube vocabulary; 2014. Available at http://www.w3.org/TR/vocab-data-cube. Recommendation.

O

Online Recovery in Parallel Database Systems

Ricardo Jiménez-Peris
Distributed Systems Lab, Universidad
Politecnica de Madrid,
Madrid, Spain

Synonyms

Continuous availability; High availability; 24×7 operation

Definition

Replication (also known as clustering) is a technique to provide high availability in parallel and distributed databases. High availability aims to provide continuous service operation. High availability has two faces. On one hand, it provides fault-tolerance by introducing redundancy in the form of replication, that is, having multiple copies or replicas of the data at different sites. On the other hand, since sites holding the replicas may crash and/or fail, in order to keep a given degree of availability, failed or new replicas should be reintroduced into the system. Introducing new replicas requires transferring to them the current state in a consistent fashion (known as *recovery*). A simple solution to this problem is *offline recovery*, that is, in order to obtain a quiescent state, request processing is suspended, then the state is transferred from a working replica (termed *recoverer replica*) to the new replica (*recovering replica*) and finally, request processing is resumed. Unfortunately, offline recovery results in a loss of availability, which defeats the original goal of replication, that is to provide high availability. The alternative is *online recovery*, in which transaction processing is not stopped while the recovery is performed. The main challenge for online recovery is to attain consistency, since the state to be transferred to the recovering replica(s) is a moving target. While the recovery

takes place, new transactions are processed and the state evolves during the recovery itself. Online recovery needs to be coordinated with the replica control protocol to enforce consistency.

Historical Background

Recovery is used in centralized databases to bring the database to a consistent state after a crash [1]. The consistency is attained by ensuring that the updates of committed transactions are reflected in the database and the updates of uncommitted (aborted) transactions are not reflected in it. In clustered databases, centralized recovery is used to bring a failed replica to a local consistent state, but then, since other working replicas may have processed transactions, centralized recovery is not sufficient and it has to be followed by a replica recovery [11]. During replica recovery, the failed replica (or a fresh new one) recovers the current state from the working replicas. What is meant by *current state* the state reflecting all the updates from transactions that will not be processed by the recovering replica, and not reflecting any of the updates from transactions that will be processed by the recovering replica after recovery. A failed replica only needs to recover the missed updates, while a new replica needs to recover the full database.

The seminal paper on online recovery for clustered databases is [8]. In this paper, a suite of protocols for online recovery is proposed. One of the protocols lies in a locking-based online recovery. The full database is locked atomically using recovery locks, a special kind of read lock. This guarantees a quiescent state of the database. Then, the recovery locks are released as data is transferred to the recovering replica. The atomic setting of the recovery locks acts as a synchronization point. All update requests processed before the recovery lock setting should be reflected in the transferred state. Requests submitted after the recovery locks are set should be processed by the recovering replica after recovery finishes. This protocol lies inbetween offline and online recovery. In the beginning, all the data are locked, and therefore the database is unavailable. This situation improves as recovery progresses, since recovery locks are released and the corresponding data become available. Another online recovery protocol proposed in [8] is a multi-round recovery. In this protocol, the state to be transferred (missed updates) to the recovering replica is sent in rounds. The first round would contain all the updates missed until the start of recovery. In the second round, the updates performed during the first round are sent, and so on. When the number of updates performed during the last round is small enough, a last round is run. During the last round, the recovering replica has to store all client requests to process them after finishing recovery. This recovery protocol if fully online and also significantly reduces the number of transactions to be stored during recovery, since it is only during the last round that the recovering replica has to store incoming transactions (typically only the resulting updates from them).

Another piece of work on online recovery was presented in [5]. This paper describes a log-based online recovery protocol in which a failed replica receives the prefix of the log corresponding to the update transactions it has missed. Since the log grows as the recovery progresses, the protocol has a special handshake protocol to finish recovery. The recoverer traverses the log from the first transaction the failed replica missed until it reaches the end of the log. At this point, the end recovery handshake protocol is started to determine which will be the last transaction to be sent as part of the recovery. The recovering replica starts storing requests that follow this transaction to process them after recovery finishes.

Online recovery has also been used for replicated data warehouses across the Internet [9]. In this work, each replica is located at an autonomous organization and exhibits an interface to execute queries. The online recovery protocol exploits the underlying architecture and performs recovery by issuing queries from the recovering replica to the working replicas. It also takes advantage of a facility for historical queries that enables executing a read only query providing a timestamp T. This historical query will return the same results as it happened at time T. The recovery protocol has 3 phases. In the first phase, the recovering site determines the latest time T for which it has all the committed updates. In the second phase, the recovering site runs historical queries at working replicas (that act as recoverer replicas) with a timestamp between the recovery point and a time closer to the present to catch up missed updates. Historical queries do not set read locks, and therefore do not block updates at the recoverer sites. In the final phase, a regular query (non-historical) is run to get the latest updates. In this case, read locks are set to guarantee the consistency.

Online recovery has also been studied in other contexts, such as diverse data replication and Byzantine data replication. In diverse data replication [4], each replica runs a database from a different vendor. This approach enables tolerating software failures since it has been observed that bugs in one product typically do not appear in a database from a different vendor [4]. In diverse data replication, recovery is slightly more complex, since it requires using a common abstract representation of the data. This might need some tuple translation during recovery in order to align fields with slightly different types.

Byzantine data replication [17] tolerates intrusions and provides continuous correct service despite them. An intrusion happens at a site when it is attacked. This site may behave arbitrarily to disrupt service provision. Intrusions are modeled as Byzantine (also known as arbitrary) failures. Byzantine replication typically resorts to diverse data replication to avoid common vulnerabilities. Typically tolerating f Byzantine failures requires $3 \cdot f + 1$ replicas. Byzantine data replication has a more involved recovery, since it should also mask Byzantine recoverers. This means that a sufficient number of recoverers is needed in order to mask Byzantine failures during recovery [2]. An additional issue in Byzantine fault tolerance is that once one replica has been successfully attacked, another one can be attacked, and so on. In order to reduce the window of vulnerability, proactive recovery has been proposed [2]. This approach recovers replicas proactively without waiting until they are crashed or attacked. During recovery, the recovering replica boots from read-only media and recovers the state from working replicas using a Byzantine recovery protocol. During recovery, the state is transferred from multiple working replicas to be able to tolerate Byzantine (attacked) recoverers. Replicas are recovered in a round-robin fashion forced by a reboot provoked by a hardware watchdog. In this way, even if a replica has been attacked silently (without the system and administrator noticing it), it will become operative again, thus, reducing the window of vulnerability of the system.

Foundations

High availability consists of two inseparable aspects. On one hand, it requires the ability to tolerate failures. This is typically achieved by introducing redundancy in the form of replication. However, in order to keep a given degree of availability, the ability to recover failed (or new) replicas is also necessary. Recovering failed replicas requires obtaining a quiescent state from a working replica (or *recoverer replica*) and transferring it to the new replica (or *recovering replica*). This quiescent state can be easily obtained by stopping transaction processing. This results in *offline recovery*. Although offline recovery guarantees the consistency of recovery, its major drawback is that it results in a loss of availability during the recovery process that defeats the original goal of replication, which is to provide high availability. The alternative is *online recovery*, that is, to transfer the state to a recovering replica without stopping transaction processing.

Replication can be used to provide scalability in addition to providing availability. In this case, availability results insufficient as a metrics to express the goodness of a recovery protocol [5]. Performability becomes a more appropriate metrics in this context. *Performability* is defined as the cumulative performance of a highly available system over a period of time in the presence of failures and recoveries. To understand why it is important, an extreme situation will be illustrated. A replicated system has a throughput of 1000 transactions per second (tps). During online recovery the system remains available, but its throughput decreases to 1 tps. Although this

system is available, it is clearly worse than a system that would deliver 990 tps during the online recovery (assuming that online recovery takes the same amount of time in both systems). However, when comparing the performability of both systems during online recovery, the former system would offer a very poor performability, while the second would offer a very high one close to the one in which the system is not recovering any replica.

Online recovery protocols typically consist of five phases:

1. *Local Recovery* brings the local state to a consistent state by means of centralized recovery.
2. *Find Last Committed Update Transaction* determines the last update transaction reflected in the local state.
3. *Global Recovery Start* initiates online recovery taking care of obtaining a quiescent state from one or more working replicas to be transferred to the recovering replica.
4. *Global Recovery* transfers a quiescent state from a working replica to the recovering replica.
5. *Global Recovery End* is the handshake protocol to determine the end of recovery.

The first and second phases depend on whether the recovering replica is a failed replica or a fresh new replica. For a failed replica, the first phase is typically performed automatically by the underlying database system upon recovery after a crash. For a failed replica, the second phase implies traversing the log or any other recovery information repository to find out which was the last committed update transaction. In most protocols, this information does not need to be precise. It can be enough to obtain the identifier of a committed transaction (e.g., a timestamp, log sequence number or transaction identifier, TID) close to the failure instant such that all previously committed transactions are also reflected in the local state. If some updates of latter transactions are reflected in the state, they will be rewritten during recovery. For a fresh new replica, the first phase is empty and the second phase first involves obtaining a checkpoint of the database and then performing the same processing as a failed replica. The checkpoint of the database also needs to be obtained in an online fashion using techniques such as point-in-time recovery [16] in order to keep the system available.

In the third phase, there is some communication between the recovering replica and the working replicas. This interaction has several purposes. First, the working replicas become aware of the new replica wanted to join the system. Second, the recovering replica informs the working replicas about the last known update. Third, a recoverer is elected to transfer the state to the recovering replica.

The fourth phase transfers the state from the recoverer to the recovering replica. The recovery process is synchronized with replica control to guarantee that a quiescent state is transferred to the recovering replica.

The fifth phase aims at finishing the recovery, which requires splitting the sequence of transactions into two disjoint sets (the ones whose state is reflected in the state transfer to the recovering replica, and the ones that should be processed by the recovering replica after finishing the recovery).

Online recovery depends on the specific features of the replica control protocol being used, such as eager vs. lazy, primary-backup vs. update-everywhere, kernel-based vs. middleware-based replication, etc. In eager replication, the coordination between replica control and online recovery is very tight to guarantee consistency [6, 7, 10, 13, 14]. In lazy replication, the coordination can be looser. For instance, in a freshness-based approach [3, 12], as far as the freshness requirement is satisfied the coordination can be more relaxed. In a primary-backup approach, the recovery of backups is simpler since they do not execute updates on their own, they just apply the updates coming from the primary [15]. In update-everywhere approaches [6, 7, 10, 13, 14] the recovery needs to be interleaved carefully with replica control to guarantee consistency [5, 8]. In kernel-based approaches, it is possible to use recovery protocols that use mechanisms within the kernel (such as locking) [8]. However, in middleware-based approaches,

the recovery can only use those mechanisms available at the database interface [5].

In this entry, two approaches will be examined in detail: an online recovery for kernel-based replication based on locking [8], and an online recovery for middleware-based replication based on logging [5].

First, a locking-based online recovery approach is studied. This approach is based on the most basic protocol from [8]. This basic version of the protocol transfers the full database. The recovery is coordinated with replica control to guarantee consistency. This coordination is materialized through locking. In what follows, the first generic phases of online recovery for this particular protocol are described. The first phase, local recovery, is orthogonal to online recovery it can just be ignored. The second phase consists in determining the last update known by the recovering replica. Since in this approach the full database is sent, this phase does not exist.

The third phase is recovery start. The recovering replica notifies to the working replicas that is willing to join the system and recover. In this protocol, in order to guarantee the quiescence of the transferred state, the recovery is started by initiating a transaction that sets atomically special read locks or recovery locks over all the tuples in the database. The quiescent state to be transferred corresponds to the state just after the atomic setting of the recovery locks. The recovery then takes place gradually. As soon as a recovery lock is granted, the tuple is read and sent to the recovering replica. Then, the lock is released. It should be noted that recovery locks are read locks and therefore do not delay read operations, only update ones. During the recovery, the recovering replica should store the incoming transactions (only those involving updates) to process them after recovery, since the associated updates are not incorporated in the state being transferred to it.

The end-of-recovery handshake is simple in this protocol. It is initiated after the sending of the last tuple, and the recovery message piggybacks is marked to indicate this fact. After receiving this message, the recovering replica starts to process all the stored transactions during the recovery. Depending on the replica control in place the recovering replica will execute the update transactions, or will receive the resulting updates from the other replicas (which is usually the case). In the latter case, incoming transactions do not need to be stored-just the update propagation messages from the other replicas.

There are many optimizations that can be performed over this basic protocol as described in [8]. Batching recovery messages limit the amount of data transferred during recovery by recording the updates performed during recovery, shortening the amount of updates to be stored during online recovery by using multiple phases, etc.

The second protocol that will be described is a log-based protocol [5]. This protocol is combined with a pessimistic replica control implemented as a middleware layer. A pessimistic replica control freely executes non-conflictive transactions in the replicated system, while conflictive ones are executed in the same relative order at all replicas. The replica control protocol over which online recovery is built is Nodo, described in [13]. Nodo is based on the notion of conflict classes. A transaction might access one or more conflict classes. Each conflict class has a master replica. Each combination of conflict classes also has a master replica, one of the master replicas of the individual conflict classes. Each conflict class has an associated transaction queue. Update transactions are sent to all replicas in the same order. Read-only transactions are only sent to one of the replicas. Transactions are queued in the conflict class queues relevant to them (the ones that they read or write). When a transaction is at the front of all the queues it has been enqueued, the master of the conflict class combination associated to the transaction executes it locally. If the transaction is read-only, then the results are returned directly to the client and removed from all the queues. If the transaction contains updates, the master extracts them from the database and sends them to all other replicas. Nodo keeps a log for each individual conflict class that is exploited by the online recovery protocol.

The online recovery protocol for Nodo guarantees consistent recovery by careful coordination with the replica control protocol of Nodo. Individual conflict classes can be recovered inde-

pendently, that is, each individual conflict class could use a different recoverer replica. The recovery of an individual conflict class is detailed in what follows. The first phase, local recovery, occurs when the replica recovers. The second phase, determining the last update reflected in the replica state, is done by looking at the local log. The recovering replica traverses the log to determine the identifier of the last transaction processed for the conflict class being recovered. This identifier is used in the fourth phase to initiate the recovery of the conflict class. One of the working replicas is elected as recoverer replica for this conflict class. In the fourth phase, the recoverer replica traverses the log of the conflict class being recovered. The recoverer replica continues processing updates involving this conflict class. This means that the log grows as is being traversed. The recovering replica just applies all the updates corresponding to the conflict class under recovery, but will discard them, since it is receiving them via the recovery. The recoverer replica will eventually reach the end of the log. At this point the fifth phase is performed to finish recovery. In the end-of-recovery handshake three different roles are involved, namely, the recoverer and recovering replicas, and the master replica of the conflict class under recovery. There is a race condition, since the master produces updates from locally executed transactions, while the recoverer needs to know the last update to be forwarded to the recovering replica and the recovering replica needs to know from which transaction it has to start to process update transactions. When the recoverer reaches the end of the log, it sends a request-end-of-recovery message to all replicas (only the master needs this message in the failure-free case). The master will then piggyback with the next update it forwards an end-of-recovery marker. This marker will indicate to the recoverer that this is the last update to be sent to the recovering replica. The marker will tell the recovering replica which is the last update part of the recovery, and therefore it will know from which point it has to apply updates received from the master replica.

The online recovery for Nodo, as stated before, allows recovering conflict classes independently. In fact, once a conflict class is recovered, the recovering replica could become master of that class. Additionally, several conflict classes can be recovered in parallel using different recoverers for each of them. The online recovery for Nodo has been designed to be adaptive. The goal is to obtain the highest performability. If the replicated system has a low load, the resources employed to recover the replica are increased (i.e., increasing the number of recoverers). If there is a peak load, during recovery, then the resources devoted to recovery can be decreased (i.e., decreasing the number of recoverers). In this way, performability is maximized. Nodo also enables dealing with overlapping recoveries in parallel. If a batch of sites is recovered in a time interval, they will start recovery at slightly different times. The recovery protocol takes care of recovering simultaneously conflict classes for all recovering replicas it knows. In this way, if one replica starts recovery after another one has recovered two conflict classes, they will perform simultaneously the recovery of the remaining N-2 conflict classes, being N the number of conflict classes, and alone the recovery of the first two conflict classes. This enables a more efficient dissemination of the recovery messages (e.g., exploiting multicast) and efficiently recovering batches of replicas.

Key Applications

Online recovery is a crucial technique to provide true high availability, that is, 24×7 operation. Its main potential users are all those organizations that provide services that should be continuously available. Among these potential users one can find enterprise data centers, software as a service (SaaS) platforms, services governed by service level agreements (SLAs), services for critical infrastructures (health-care, energy, police, etc.).

Cross-References

▶ Data Replication
▶ Replica Control

Recommended Reading

1. Bernstein PA, Hadzilacos V, Goodman N. Concurrency control and recovery in database systems. Reading: Addison Wesley; 1987.
2. Castro M, Liskov B. Practical byzantine fault tolerance and proactive recovery. ACM Trans Comput Syst. 2002;20(4):398–461.
3. Gançarski S, Naacke H, Pacitti E, Valduriez P. The leganet system: freshness-aware transaction routing in a database cluster. Inform Syst. 2007;32(2):320–43.
4. Gashi I, Popov P, Strigini L. Fault tolerance via diversity for off-the-shelf products: a study with SQL database servers. IEEE Trans Depend Secur Comput. 2007;4(4):280–94.
5. Jiménez-Peris R, Patiño-Martínez M, Alonso G. Non-intrusive, parallel recovery of replicated data. In: Proceedings of the 21st Symposium on Reliable Distributed Systems; 2002. p. 150–9.
6. Kemme B. and Alonso G. Don't be lazy, be consistent: Postgres-R, a new way to implement database replication. In: Proceedings of the 26th International Conference on Very Large Data Bases; 2000. p. 134–43.
7. Kemme B, Alonso G. A new approach to developing and implementing eager database replication protocols. ACM Trans Database Syst. 2000;25(3):333–79.
8. Kemme B, Bartoli A, Babaoglu O. Online reconfiguration in replicated databases based on group communication. In: Proceedings of the International Conference on Dependable Systems and Networks; 2001. p. 117–30.
9. Lau E Madden S. An integrated approach to recovery and high availability in an updatable, distributed data warehouse. In: Proceedings of the 32nd International Conference on Very Large Data Bases; 2006. p. 703–14.
10. Manassiev K, Amza C. Scaling and continuous availability in database server clusters through multiversion replication. In: Proceedings of the International Conference on Dependable Systems and Networks; 2007. p. 666–76.
11. Özsu MT, Valduriez P. Principles of distributed database systems. 2nd ed. Upper Saddle River: Prentice-Hall; 1999.
12. Pacitti E, Simon E. Update propagation strategies to improve freshness in lazy master replicated databases. VLDB J. 2000;8(3):305–18.
13. Patiño-Martínez M, Jiménez-Peris R, Kemme B, Alonso G. Middle-R: consistent database replication at the middleware level. ACM Trans Comput Syst. 2005;23(4):375–423.
14. Pedone F, Guerraoui R, Schiper A. The database state machine approach. Distrib Parallel Databases. 2003;14(1):71–98.
15. Plattner C, Alonso G. Ganymed: scalable replication for transactional web applications. In: Proceedings of the ACM/IFIP/USENIX 5th International Middleware Conference; 2004. p. 155–74.

O

16. PostgreSQL PostgreSQL Point in Time Recovery. http://www.postgresql.org/docs/8.0/interactive/backup-online.html.

17. Vandiver B, Balakrishnan H, Liskov B, Madden S. Tolerating Byzantine faults in database systems using commit barrier scheduling. In: Proceedings of the 21st ACM Symposium on Operating System Principles; 2007. p. 59–72.

Ontologies and Life Science Data Management

Robert Stevens[1] and Phillip Lord[2]
[1]University of Manchester, Manchester, UK
[2]Newcastle University, Newcastle-Upon-Tyne, UK

Synonyms

Knowledge management

Definition

Biology is a knowledge-rich discipline. Much of bioinformatics can, therefore, be characterized as *knowledge management*: organizing, storing and representing that knowledge to enable search, reuse and computation.

Most of the knowledge of biology is categorical; statements such as "fish gotta swim, birds gotta fly" cannot be easily represented as mathematical or statistical relationships. These statements can, however, be formalized using ontologies: a form of model which represents the key concepts of a domain.

Ontologies are now widely used in bioinformatics for a variety of tasks, enabling integration and management of multiple data or knowledge sources, and providing a structure for new knowledge as it is created.

Historical Background

Biological knowledge is highly complex. It is characterized not by the large size of the data sets that it uses, but by the large number of data types; from relatively simple data such as raw nucleotide sequence, through to anatomies, systems of interacting entities, to descriptions of phenotype.

In addition to its natural complexity, biology has traditionally operated as a "small science" - with a large number of individual, autonomous laboratories working independently. This has resulted in highly heterogeneous data; in addition to the natural complexity of the data, knowledge is often represented in many different ways [5]. There are, for example, at least twenty different file formats for representing DNA sequence.

Ontologies can be used to enable the knowledge management that overcomes these two forms of complexity. First, they can be used to represent complex, categorical knowledge of the sort common in biology. Second, by describing the heterogeneity of the representation of knowledge, they can provide a common, shared understanding that can be used to overcome this heterogeneity.

In post-genomic biology, the first of these has been the most common usage. Here, ontologies are used to generate a controlled vocabulary; for this, the Gene Ontology (GO) [14] provides the paragon for biological sciences. It represents three key aspects of genetics; the *molecular function* of a gene (product), the *biological process* in which the product is involved, and the *cellular component* in which it is located. GO has been used to annotate many genomes and has been used for annotations in UniProt and InterPro. Following on from the success of GO, the Open Biomedical Ontologies (OBO) now provides controlled vocabularies for describing many aspects of biological knowledge (http://obo.sf.net).

The second major use has been to enable access to or querying over multiple independent data sources. EcoCyc [8], for example, uses an ontology to provide a schema to integrate genome, proteome data and a number of pathway resources, while RiboWeb [1] was a similar style of ontology driven application for storing, managing and analyzing data from experiments on ribosomal structure. The TAMBIS system [7] used an ontology to mediate queries to a number of different data sources.

Most of these examples are *post-hoc* additions to existing systems; GO, for example, presents knowledge which is already present in other, less formal, representations. More recently, however, there has been a shift to the use of ontologies as a primary representation. The MGED Ontology (MO) [17] provides a vocabulary for reporting microarray experiments, while the Systems Biology Ontology (SBO) (http://www.ebi.ac.uk/sbo/) supports the representation of systems biology models.

Over the past decade, the use of ontologies has now become well-entrenched as a tool for organizing and structuring biomedical knowledge and, therefore, has become a key part of life science knowledge management [3].

Foundations

It was recognized early in bioinformatics that there is a massive problem with heterogeneous representations of data [5]. Such heterogeneities, particularly at the semantic level, exist both in the meanings of the structures that hold values (the schema) and in the meanings of the values themselves. To query or analyze meaningfully across data from different sources, therefore, there is a necessary reconciliation step to enable the data to be understood. A database, for example, might have either separate tables for substrate or product, or just one for reactant; as an orthogonal issue, a chemical might be "acetic acid" in one or "ethanoic acid" in another. Both types of mismatch need to be overcome.

This heterogeneity can occur both in descriptions of biology and also bioinformatics: it is possible to disagree on which genes are involved in a process, what those genes are called and what structure is used to hold information about those genes.

Ontologies provide a computable mechanism for working around these problems; they describe the entities and what is known about these entities within the data of biology, and provide a set of labels to describe these entities and their properties. As a result, an ontology can be used to describe the entities within a biological database. Terms

from the Gene Ontology, for example, are used to describe the major functional attributes of gene products in many databases; this, in turn, allows comparative studies of genes and their actions across species.

Ontologies need to be represented in a language; these are often called knowledge representation languages. They have a set of language elements for describing categories (also called classes, types, concepts or universals) of instances (also called objects, entities, individuals or particulars). The languages vary in their expressivity – that is, how much is it possible to say about what these elements mean. For example, if we state that ArB, where A and B are classes and r is a relationship, does this mean that every A has a B related to it by r; that for any A with a r relation, this r is to a B; to how many B's can an A have a relationship; that r has an inverse relationship, and so on. Some languages allow only trees of categories to be made, others more complex graphs. Finally, languages differ in their computational amenability [11].

Ontologies have found a variety of uses within bioinformatics:

Reference Ontology: An ontology can be used simply as a reference, encompassing what is understood about a domain with high-fidelity. Such an ontology is not skewed by any application bias, except that of correctness.

Controlled Vocabulary: The labels on the categories in an ontology provide a vocabulary with which discussion of those categories can be accomplished. By committing to use that vocabulary – controlling the words used in communication – a controlled vocabulary is established. This is the principal means of overcoming a large portion of heterogeneity.

Computational Component: An ontology can form a component in a software application. The strict semantics of its representation language can be used to make inferences from data described in terms of that ontology. This can simply be retrieving all the children of a given category (all instances of a child are also instances of the parent) to recognizing membership of a category based on facts known about an object.

There are many technical and social difficulties associated with using an ontology for data management. The choice of representation language can be key; an ontology for use as a computational component probably needs a more computationally amenable and expressive language than an ontology intended to provide a controlled vocabulary. It is often hard to engage with domain experts to ensure that the ontology reflects the domain, while maintaining ontological precision. There are a number of methodologies for ontology building [6], but the discipline is still nascent. Finally, adapting and updating the ontology when it is already in use can require rigorous, yet flexible policies.

Key Applications

Perhaps the best known ontology in biology is the Gene Ontology (GO) [14]. It started in 1998 with an aim of enabling queries across multiple databases for the key aspects or properties of the genes or proteins; it achieves this not by schema reconciliation but enabling the augmentation of existing knowledge; in short, it is one of the best examples of a reference ontology.

Another domain well served by reference models is that of anatomy. The Foundational Model of Anatomy, for example, aims to provide a "symbolic modeling of the structure of the human body in a computable form that is also understandable by humans" [12]. The aim is that this ontology provides a common representation into which others can be mapped.

One of the early uses for ontologies was enabling schema and value reconciliation. The TAMBIS [7] system was an early example. It uses an ontological representation of entities and their properties in biology expressed in a description logic [2]. These descriptions, that could be composed to more complex concept descriptions, were then transformed to queries against bioinformatics analysis services capable of retrieving instances of the concepts within the ontology. The ontology, then, could tell that the user that a *Protein* might have one or more *Homolog*s, while the system would understand that a BLAST search might reveal these *Homolog*s. A concept therefore also defined a query plan. More recently, the BioPAX ontology [10] provides a schema for representing biological pathway data; the different contributing databases could then release knowledge in the format. Both of these operate on the level of schema, but reconciliation to a common model also occurs at the level of the values held within a schema. The Gene Ontology, for example, in providing a controlled vocabulary for the functional attributes of gene products has allowed many genome resources to use common values within their schema.

Both BioPAX and TAMBIS enabled querying by assigning objects to categories in the ontology. Ontologies represent the properties by which objects can be recognized to be a member of a category. If these properties are recognizable computationally, then the ontology can be used to classify these members automatically. This approach has been used to classify the phosphatase proteins from three parasite genomes [4]. A final approach to ontological querying is to use the ontology as a basis for statistical analysis of individuals annotated with these ontologies. GO has been widely used for this purpose [9, 15].

Finally, ontologies have begun to be used for the representation of metadata about primary experimental data. The MGED society has led the way with the MGED Ontology (MO) which has been used for describing microarray data [17]. This ontology describes a number of aspects about an experiment including: the biological material used; experimental design and microarray equipment. Similar work is now underway for describing proteomics experiments [13]. These are coming together in the Ontology for Biomedical Investigations OBI that is providing a general framework for describing the protocols and analyses for many different kinds of experiment [16].

Future Directions

In the past decade ontologies have come to form a major aspect of information management in the life sciences. The field of ontology development in the life sciences now faces several challenges in the short and medium term.

- The wide scope of biology is a challenge; to describe many parts of it, also needs descriptions of closely related areas such as chemistry, geology and geography.
- Ontologies are starting to get very large. It is not clear whether current methodologies are scalable, both in terms of building, maintaining or using them. This has many implications for the formal expressive structures of the knowledge representation language, the ability to support modularity of these languages, and the social processes used to build ontologies.
- Dealing with change both as a result of the ontology development process and, perhaps more importantly, as a result of changes in knowledge itself. There are many different techniques for dealing with the former situation; there are many fewer for dealing with the latter. If datasets gathered over a long period of time are to be understood in the future, it may become as important to understand what was thought in the past as it is to manage the current state of knowledge.
- Currently many ontologies deal with a single level of granularity or the view point of a single discipline building sophisticated, computationally amenable ontologies necessitates crossing boundaries of granularity and discipline. It remains, however, unclear how to integrate these sorts of ontology.
- Ontologies currently fulfill the luxury end of the metadata market; they can be very expensive to build, maintain and deploy. Lower the cost is critical. Probably the best way to achieve this is to make them easier for domain scientists to build which leads to a second challenge; maintaining usability of ontologies and representation languages, while increasing their scale and computability.

It seems clear that ontologies will be in heavy use in the future within the life sciences. How well these challenges are answered will determine the uses to which they are put.

Cross-References

▶ Ontology

Recommended Reading

1. Altman R, Bada M, Chai X, Whirl CM, Chen R, Abernethy N. RiboWeb: an ontology-based system for collaborative molecular biology. IEEE Intell Syst. 1999;14(5):68–76.
2. Baader F, Calvanese D, McGuinness D, Nardi D, Patel-Schneider P, editors. The description logic handbook: theory, implementation and applications. Cambridge: Cambridge University Press; 2003.
3. Bodenreider O, Stevens R. Bio-ontologies: current trends and future directions. Brief Bioinform. 2006;7(3):256–74.
4. Brenchley R, Tariq H, McElhinney H, Szoor B, Stevens R, Matthews K, Tabernero L. The TriTryp Phosphatome analysis of the protein phosphatase catalytic domains. BMC Genome. 2007;8(1):434.
5. Davidson S, Overton C, Buneman P. Challenges in integrating biological data sources. J Comput Biol. 1995;2(4):557–72.
6. Fernàndez-Lòpez M, Gòmcz-Pèrez A. Overview and analysis of methodologies for building ontologies. Knowl Eng Rev. 2002;17(2):129–56.
7. Goble CA, Stevens R, Ng G, Bechhofer S, Paton NW, Baker P, Peim M, Brass A. Transparent access to multiple bioinformatics information sources. IBM Syst J, Special issue on deep computing for the life sciences. 2001;40(2e):532–52.
8. Karp P, Riley M, Saier M, Paulsen I, Paley S, Pellegrini-Toole A. The EcoCyc and metacyc databases. Nucleic Acids Res. 2000;28(1):56–9.
9. Lord PW, Stevens R, Brass A, Goble CA. Investigating semantic similarity measures across the gene ontology: the relationship between sequence and annotation. Bioinformatics. 2003;19(10):1275–83.
10. Luciano J. PAX of mind for pathway researchers. Drug Discov Today. 2005;10(13):937–42.
11. Ringland G, Duce D. Approaches to knowledge representation: an introduction knowledge-based and expert systems series. Chichester: Wiley; 1988.
12. Rosse C, Mejino JLV. A reference ontology for bioinformatics: the foundational model of anatomy. J Biomed Inform. 2003;36(6):478–500.
13. Taylor C, Paton N, Lilley K, Binz P, Julian Jr R, Jones A, Zhu W, Apweiler R, Aebersold R, Deutsch E, Dunn M, Heck A, Leitner A, Macht M, Mann M, Martens L, Neubert T, Patterson S, Ping P, Seymour S, Souda P, Tsugita A, Vandekerckhove J, Vondriska T, Whitelegge J, Wilkins M, Xenarios I, Yates 3rd

O

J, Hermjakob H. The Minimum Information About a Proteomics Experiment (MIAPE). Nat Biotechnol. 2007;25(8):887–93.

14. Ashburner M et al. The gene ontology consortium gene ontology: tool for the unification of biology. Nat Genet. 2000;25(1):25–9.

15. Wang H, Azuaje F, Bodenreider O, Dopazo J. Gene expression correlation and gene ontology-based similarity: an assessment of quantitative relationships. In: Proceedings of the IEEE Symposium on Computational Intelligence in Bioinformatics and Computational Biology; 2004. p. 25–31.

16. Whetzel P, Brinkman R, Causton H, Fan L, Field D, Fostel J, Fragaso G, Gray T, Heiskanen M, Hernandez-Boussard T, Morrison N, Parkinson H, Rocca-Serra P, Sansone SA, Schober D, Smith B, Stevens R, Stoeckert C, Taylor C, White J, Wood A. The FuGo working group development of FuGo: an ontology for functional genomics investigations. OMICS J Integrat Biol. 2006;10(2):199–204.

17. Whetzel PL, Parkinson H, Causton HC, Fan L, Fostel J, Fragoso G, Game L, Heiskanen M, Morrison N, Rocca-Serra P, Sansone SA, Taylor C, White J, Stoeckert CJ. The mged ontology: a resource for semantics-based description of microarray experiments. Bioinformatics. 2006;22(7):866–73.

Ontology

Tom Gruber
RealTravel, Emerald Hills, CA, USA

Synonyms

Computational ontology; Ontological engineering; Semantic data model

Definition

In the context of computer and information sciences, an ontology defines a set of representational primitives with which to model a domain of knowledge or discourse. The representational primitives are typically classes (or sets), attributes (or properties), and relationships (or relations among class members). The definitions of the representational primitives include information about their meaning and constraints on their logically consistent application. In the context of database systems, ontology can be viewed as a level of abstraction of data models, analogous to hierarchical and relational models, but intended for modeling knowledge about individuals, their attributes, and their relationships to other individuals. Ontologies are typically specified in languages that allow abstraction away from data structures and implementation strategies; in practice, the languages of ontologies are closer in expressive power to first-order logic than languages used to model databases. For this reason, ontologies are said to be at the "semantic" level, whereas database schema are models of data at the "logical" or "physical" level. Due to their independence from lower level data models, ontologies are used for integrating heterogeneous databases, enabling interoperability among disparate systems, and specifying interfaces to independent, knowledge-based services. In the technology stack of the Semantic Web standards [1], ontologies are called out as an explicit layer. There are now standard languages and a variety of commercial and open source tools for creating and working with ontologies.

Historical Background

The term "ontology" comes from the field of philosophy that is concerned with the study of being or existence. In philosophy, one can talk about an ontology as a theory of the nature of existence (e.g., Aristotle's ontology offers primitive categories, such as substance and quality, which were presumed to account for All That Is). In computer and information science, ontology is a technical term denoting an artifact that is *designed* for a purpose, which is to enable the modeling of knowledge about *some* domain, real or imagined.

The term had been adopted by early Artificial Intelligence (AI) researchers, who recognized the applicability of the work from mathematical logic [6] and argued that AI researchers could create new ontologies as computational models that enable certain kinds of automated reasoning [5]. In the 1980s the AI community came to use the term ontology to refer to both a theory of a modeled world (e.g., a Naïve Physics [5]) and a component of knowledge systems. Some researchers, drawing inspiration from philosophical ontologies, viewed computational ontology as a kind of applied philosophy [10].

In the early 1990s, an effort to create interoperability standards identified a technology stack that called out the ontology layer as a standard component of knowledge systems [8]. A widely cited web page and paper [3] associated with that effort is credited with a deliberate definition of ontology as a technical term in computer science. The paper defines ontology as an "explicit specification of a conceptualization," which is, in turn, "the objects, concepts, and other entities that are presumed to exist in some area of interest and the relationships that hold among them." While the terms specification and conceptualization have caused much debate, the essential points of this definition of ontology are:

- An ontology defines (specifies) the concepts, relationships, and other distinctions that are relevant for modeling a domain.
- The specification takes the form of the definitions of representational vocabulary (classes, relations, and so forth), which provide meanings for the vocabulary and formal constraints on its coherent use.

One objection to this definition is that it is overly broad, allowing for a range of specifications from simple glossaries to logical theories couched in predicate calculus [9]. But this holds true for data models of any complexity; for example, a relational database of a single table and column is still an instance of the relational data model. Taking a more pragmatic view, one can say that ontology is a tool and product of engineering and thereby defined by its use. From this perspective, what matters is the use of ontologies to provide the representational machinery with which to instantiate domain models in knowledge bases, make queries to knowledge-based services, and represent the results of calling such services. For example, an API to a search service might offer no more than a textual glossary of terms with which to formulate queries, and this would act as an ontology. On the other hand, today's W3C Semantic Web standard suggests a specific formalism for encoding ontologies (OWL), in several variants that vary in expressive power [7]. This reflects the intent that an ontology is a specification of an abstract data model (the domain conceptualization) that is independent of its particular form.

Foundations

Ontology is discussed here in the applied context of software and database engineering, yet it has a theoretical grounding as well. An ontology specifies a vocabulary with which to make assertions, which may be inputs or outputs of knowledge agents (such as a software program). As an *interface specification*, the ontology provides a language for communicating with the agent. An agent supporting this interface is not required to use the terms of the ontology as an *internal encoding* of its knowledge. Nonetheless, the definitions and formal constraints of the ontology do put restrictions on what can be *meaningfully* stated in this language. In essence, committing to an ontology (e.g., supporting an interface using the ontology's vocabulary) requires that statements that are asserted on inputs and outputs be *logically consistent* with the definitions and constraints of the ontology [3]. This is analogous to the requirement that rows of a database table (or insert statements in SQL) must be consistent with integrity constraints, which are stated declaratively and independently of internal data formats.

Similarly, while an ontology must be formulated in *some* representation language, it is intended to be a semantic level specification – that is, it is independent of data modeling strategy or implementation. For instance, a conventional database model may represent the identity of individuals using a primary key that assigns a unique identifier to each individual. However, the primary key identifier is an artifact of the modeling process and does not denote something in the domain. Ontologies are typically formulated in languages which are closer in expressive power to logical formalisms such as the predicate calculus. This allows the ontology designer to be able to state semantic constraints without forcing a particular encoding strategy. For example, in typical ontology formalisms one would be able to say that an individual was a member of class or has some attribute value without referring to any implementation patterns such as the use of primary key identifiers. Similarly, in an ontology one might represent constraints that hold across relations in a simple declaration (A is a subclass of B), which might be encoded as a join on foreign keys in the relational model.

Ontology engineering is concerned with making representational choices that capture the relevant distinctions of a domain at the highest level of abstraction while still being as clear as possible about the meanings of terms. As in other forms of data modeling, there is knowledge and skill required. The heritage of computational ontology in philosophical ontology is a rich body of theory about how to make ontological distinctions in a systematic and coherent manner. For example, many of the insights of "formal ontology" motivated by understanding "the real world" can be applied when building computational ontologies for worlds of data [4]. When ontologies are encoded in standard formalisms, it is also possible to reuse large, previously designed ontologies motivated by systematic accounts of human knowledge or language [11]. In this context, ontologies embody the results of academic research, and offer an operational method to put theory to practice in database systems.

Key Applications

Ontologies are part of the W3C standards stack for the Semantic Web, in which they are used to specify standard conceptual vocabularies in which to exchange data among systems, provide services for answering queries, publish reusable knowledge bases, and offer services to facilitate interoperability across multiple, heterogeneous systems and databases. The key role of ontologies with respect to database systems is to specify a data modeling representation at a level of abstraction above specific database designs (logical or physical), so that data can be exported, translated, queried, and unified across independently developed systems and services. Successful applications to date include database interoperability, cross database search, and the integration of web services.

Recommended Reading

1. Berners-Lee T, Hendler J, Lassila O. The semantic web. Scientific American; 2001.
2. Gruber TR. A translation approach to portable ontology specifications. Knowl Acquis. 1993;5(2):199–220.
3. Gruber TR. Toward principles for the design of ontologies used for knowledge sharing. Int J Hum Comput Stud. 1995;43(5–6):907–28.
4. Guarino N. Formal ontology, conceptual analysis and knowledge representation. Int J Hum Comput Stud. 1995;43(5–6):625–40.
5. Hayes PJ. The second naive physics manifesto. In: Moore RC, Hobbs J, editors. Formal theories of the common-sense world. Norwood: Ablex; 1985.
6. McCarthy J. Circumscription – a form of non-monotonic reasoning. Artif Intell. 1980;5(13):27–39.
7. McGuinness DL, van Harmelen F. OWL web ontology language. W3C Recommendation, February 10, 2004. Available online at: http://www.w3.org/TR/owl-features/.
8. Neches R, Fikes RE, Finin T, Gruber TR, Patil R, Senator T, Swartout WR. Enabling technology for knowledge sharing. AI Mag. 1991;12(3):16–36.
9. Smith B, Welty C. Ontology – towards a new synthesis. In: Proceedings of the International Conference on Formal Ontology in Information Systems; 2001.
10. Sowa JF. Conceptual structures: information processing in mind and machine. Reading: Addison Wesley; 1984.
11. Standard Upper Ontology Working Group (SUO). IEEE P1600.1. Available online at: http://suo.ieee.org/.

Ontology Elicitation

Pieter De Leenheer
Vrije Universiteit Brussel, Collibra NV, Brussels, Belgium

Synonyms

Knowledge creation; Ontology acquisition; Ontology argumentation; Ontology learning; Ontology negotiation

Definition

Ontology elicitation embraces the family of methods and techniques to explicate, negotiate, and ultimately agree on a partial account of the structure and semantics of a particular domain, as well as on the symbols used to represent and apply this semantics unambiguously.

Ontology elicitation only results in a *partial* account because the formal definition of an ontology cannot completely specify the intended structure and semantics of each concept in the domain, but at best can approximate it. Therefore, the key for scalability is to reach the appropriate amount of consensus on relevant ontological definitions through an effective meaning negotiation in an efficient manner.

Historical Background

Ontology elicitation is based on techniques of *knowledge acquisition*, a subfield of AI that is concerned with eliciting and representing knowledge of human experts so that it can later be used in some application. Two typical knowledge acquisition methods can be distinguished:

1. *Top-down* (deductive) *knowledge elicitation* techniques are used to acquire knowledge directly from human domain experts. Examples include interviewing, case study, and mind mapping techniques.
2. *Bottom-up* (inductive) *machine learning* techniques use different methods to infer knowledge (e.g., concepts and rules) patterns from sets of data. A well-known example is *formal concept analysis* [10].

More formal methods for top-down knowledge acquisition use *knowledge modeling* as a way of structuring projects, acquiring and validating and storing knowledge for future use. Knowledge models include: symbolic character-based languages (e.g., logic, OWL), diagrammatic representations (networks, ladders, taxonomies, concept maps), tabular representations (e.g., matrices), structured text (e.g., hypertext) [16], and conceptual modeling.

Conceptual Modeling

Certain methods and techniques from the database field for *conceptual modeling* (e.g., ER, UML, dataflow diagrams) have been proven useful for ontology elicitation. For example, in [13], ORM/NIAM has been adopted. Figure 1 shows an example of a minimal ORM diagram (on top) explicating the semantics that is implicit in the relational table schema and population (on the bottom). For example, this ORM diagram already reveals what the table cannot, the semantics of the relation of attribute "person" to attributes "city" and "country," and that "first name" and "last name" are both part of a "name." Furthermore, "city" and "country" appear not to be related at all.

A key characteristic of NIAM/ORM is that the analysis of information is based on natural language. This brings the advantage that the analysis can be done by the domain experts using their own vocabulary, and hence avoiding invalid interpretations. Furthermore, this attribute-free approach seen in the NIAM/ORM approach promotes semantic stability.

Data Schema Versus Ontology

Data models, such as data or XML schemas, typically specify the structure and integrity of data sets. Hence, building data schemas for an enterprise usually depends on the specific needs and tasks that have to be performed within this enterprise. *Data engineering* languages such as SQL aim to maintain the integrity of data sets and only use a typical set of language constructs to that aim, e.g., foreign keys. The schema vocabulary is basically to be understood intuitively (via the terms used) by the human database designer(s). The semantics of data schemas often constitute an informal agreement between the developers and an intended group of users of the data schema, and finds its way only in application programs that use the data schema instead of manifesting itself as an agreement that is shared amongst the community [20]. When new functional requirements pop up, the schema is updated on the fly. One designated individual usually controls this schema update process.

Ontology Elicitation,
Fig. 1 Illustration of a
minimal ORM diagram
explicating the implicit
semantics for a relational
table

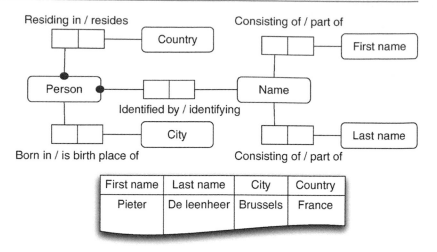

In (collaborative) ontology elicitation, however, absolute meaning is essential for all practical purposes, hence all elements in an ontology must ultimately be the result of agreements among human agents such as designers, domain experts, and users. In practice, correct and unambiguous reference to concepts or entities in the schema vocabulary is a real problem; often harder than agreeing about their properties, and obviously not solved by assigning system-owned identifiers.

Foundations

In collaborative ontology elicitation, multiple stakeholders have overlapping or contradicting perspectives about the intended structure, semantics, and vocabulary of the domain concepts [5]. This is principally caused by three facts: (i) no matter how expressive ontologies might be, they are all in fact lexical representations of concepts, relationships, and semantic constraints; (ii) linguistically, there is no bijective mapping between a concept and its lexical representation; and (iii) concepts can have different properties and values in different contexts of use. Hence, humans play an important role in the interpretation and negotiation of meaning during the elicitation and application of ontologies [7]. These principles can be illustrated by considering Stamper's *semiotic ladder* [21] that consists of six views or levels on signs from

the perspective of physics, empirics, syntactics, semantics, pragmatics and the social world, that together form a complex conceptual structure. In this article, we only consider syntactical or *lexical* level, *semantic* level, and *pragmatic* level (Fig. 2). Ontology elicitation can be considered as a process that gradually takes ontological elements through these levels.

Lexical Versus Semantic Level

At the start of the elicitation of an ontology, its basic terminology for labeling concepts and relationships are extracted from various resources such as a text corpus [3], existing schemas [16], from so-called *serious games* [19] or rashly formulated by human domain experts through, e.g., tagging [22]. Many ontology engineering approaches focus merely on the conceptual modeling task, hence the distinction between lexical level (term for a concept) and semantic level (the concept itself) is often weak or ignored. In order to represent concepts and relationships lexically, they usually are given a uniquely identifying term (or label). However, the meaning of a concept behind a lexical term is influenced by the *elicitation context*, which is the context of the resource the term was extracted from. When eliciting and unifying information from multiple sources, this can easily give rise to misunderstandings and ambiguities, therefore the meaning of all terms used for ontology representation purposes should be articulated appropriately.

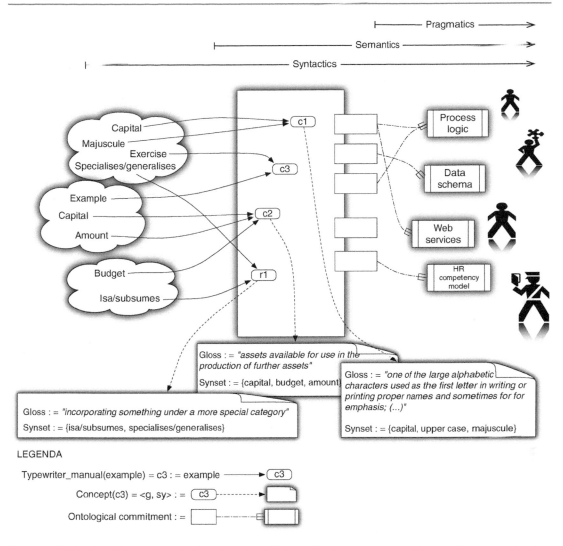

Ontology Elicitation, Fig. 2 Three levels of ontology elicitation: lexical level, semantic level, and pragmatic level

This is illustrated in Fig. 2: the full arrows denote the *meaning articulation* mappings between terms in organizational vocabularies (cloud on the left) and unique *concept identifiers* (e.g., c_1, r_1, etc.). The mapping of each unique concept identifiers to a particular explication of a meaning, i.e., a *concept definition* is defined by the dashed arrows.

Lexical Variability and Reusability

Even within one conversation, it turned out that in a less than a quarter of the cases, two individuals use the same symbolic reference for a concept, and hence the freedom to use synonyms should

be accommodated [8]. To engender creativity, domain experts should initially be allowed to use their own vocabularies, instead of being harshly restricted by an unfamiliar controlled taxonomy dictated by a central authorship. Gradually, this variability will converge towards one or more vocabularies that are commonly accepted.

For example, thousands of shared vocabularies or so-called folksonomies emerge, are sold and advertised, prosper or wither in a self-organizing manner on Web 2.0, through reuse and adaptation of natural language labels for tagging their resources. Natural language labels for concepts and relationships bring along their inherent ambiguity

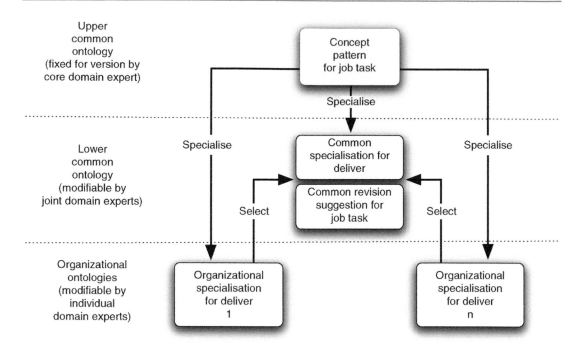

Ontology Elicitation, Fig. 3 A model for collaborative ontology elicitation

and variability in interpretation. Folksonomies provide on the one hand an unbounded reusability potential for specific reference in a given application context, which is important for scalable ontology elicitation. On the other hand, however, an analysis of multiple contexts is generally needed to disambiguate successfully [1, 5].

Semantic Versus Pragmatic Level

The meaning articulation mappings and the concept definition service respectively provide unambiguous reference and semantic explication of terms, independent of the preferred vocabulary. However, the meaning of these concepts should be further formalized for appropriately serving application purposes, by combining and linking them with other concepts, and axiomatizing them with semantic constraints and rules. In the section on applications, there is an overview of typical applications, including process logic and (legacy) information system interoperability, web service orchestration, and competency model gap analysis in human resources (HR).

The relevant properties and values for the concepts to be agreed on, depend on the application requirements. For the sake of scalability,

as in any realistic system or knowledge engineering scenario, in ontology elicitation, (parts of) existing semantic resources are reused and adopted as much as possible for new application purposes. This asks for a methodological trade-off between the reuse of relevant consensus from existing application contexts as much as possible, while allowing specific variations for new application requirements at stake to be collaboratively negotiated, based on (parts of) existing consensus.

Figure 3 illustrates this with a model for collaborative ontology elicitation, as introduced by [7], and inspired by the Delphi method (http://en.wikipedia.org/wiki/Delphi_method.). For bootstrapping the elicitation of a concept, knowledge workers are given a *pattern* that defines the current insights and interests of the community for that type of concepts. The example here concerns the elicitation of a concept "Deliver," which is a type of "Job Task." Each of the stakeholding organizations elicit the relevant properties and values of "Deliver" by specializing the pattern abstracted from "Job Task." If no pattern exists, it is bootstrapped by the core domain experts overlooking the domain.

Divergence and Conflict

Divergence is the point in collaborative ontology elicitation where domain experts disagree or have a conflict about the meaning of some concept in such a way that consequently their ontologies evolve in widely varying directions. Although they share common goals for doing business, divergent knowledge positions appear as a natural consequence when people collaborate in order to come to a unique common understanding.

Rather than considering it to be a problem, conflicts should be seen as an opportunity to *negotiate* about the subtle differences in interpretation of the domain, which will ultimately converge to a shared understanding disposed of any subjectivity. However, meaning conflicts and ambiguities should only be resolved when relevant. It is possible that people have alternative conceptualizations in mind for business or knowledge they do not wish to share. Therefore, in building the shared ontology, the individual ontologies of the various partners only need to be aligned insofar necessary, in order to avoid wasting valuable modeling time and effort.

Basically they only need to agree on a common specialization of the current properties present in the pattern. However, it could be the case that a considerable part of the stakeholders identifies new relevant properties, or see other properties to be obsolescent. This provides a feedback suggestion to revise the patterns for a next version of the ontology.

Relevant techniques for collaborative ontology negotiation and argumentation include [14, 17].

Convergence and Patterns

Once, a common specialization is agreed on, it is lifted up in the upper common levels of the ontology. Gradually, this would result in the emergence of increasingly stable generally deployable ontology patterns that are key for enabling future business interoperability needs in a scalable manner [2, 6, 9, 7].

Key Applications

Ontologies have become an integral part of many academic and industrial applications in various domains, including Semantic Web services, regulatory compliance, and human resources.

Semantic Web Services

Service-oriented (SOA) is an architecture that relies on *service-orientation* as its fundamental design principle. In a SOA environment, independent services can be accessed without knowledge of their underlying platform implementation. Within this paradigm, the creation of automation logic is specified in the form of services. Service orientation is another design paradigm that provides a means for achieving a separation of concerns, which obviously increases the potential for software reusability.

The Semantic Web aims to make data accessible and understandable to intelligent machine processing. Semantic Web services additionally aim to do the same for services available on the Semantic Web, targeting automation of service discovery, composition and invocation. For describing Semantic Web services, it is required to elicit the so-called "domain ontologies" or that formalize the knowledge necessary for capturing the meaning of services and exchanged data. In other words, given a particular business goal, the domain ontologies enable the weaving of the relevant concerns that are separated in relevant services.

A key challenge here is to overcome the *ontology-perspicuity bottleneck* [11] that constrains the use of ontologies, by finding a compromise between top-down imposed formal semantics expressed in expert language and bottom-up emerging real-world semantics expressed in layman user language.

For more on infrastructure, theory, business aspects, and experiences on ontology elicitation and management for Semantic Web applications, see [12].

Regulatory Compliance

Businesses and government must be able to show compliance of their outputs, and often also of their systems and processes, to specific regulations. Demonstrable evidence of this compliance is increasingly an auditable consideration and required in many instances to meet acceptable criteria for good corporate governance. Moreover the number and the complexity of applicable regulations in Europe and elsewhere is increasing. This includes mandatory compliance audits and assessments against numerous regulations and best practice guidelines over many disciplines and against many specific criteria. The implementation of information communications technology also means that previous manual business processes are now being performed electronically and the degree of compliance to applicable regulations depends on how the systems have been designed, implemented and maintained. Keeping up with the rate of new regulations for a major corporation and small business alike is a never ending task. What is the answer to all of this regulatory complexity? First one should simplify regulations where possible and then apply automatic tools to assist.

The automated data demands of networked economies and an increasingly holistic view on regulatory issues are driving and yet partially frustrating attempts to simplify regulations and statutes. In an ideal world companies and other organizations would have the tools and online services to check and measure their regulatory compliance; and governmental organizations would be able to electronically monitor the results. This requires a more systemic shared approach to regulatory assurance assessment and compliance certification.

Lessig [15] has a simple yet profound thesis "Code is law." The application of this concept taken in conjunction with the emergence of regulatory ontologies opens up a new way of assessing whether burgeoning systems are compliant with regulations they seek and claim to embody. First specific regulations (e.g., data privacy, digital rights management) are converted into and expressed as "Regulatory Ontologies." These ontologies are then used as the base platform for a "Trusted Regulatory Compliance Certification Service."

Over time the resulting ontology describing and managing the areas analyzed can literally re-place the regulations and compliance criteria. So much so, it is envisaged that an eventual outcome could be that the formal writing (codification) of future laws will start with the derived ontologies and use intelligent agents to help propose specific legal text which ensures that the policy objectives are correctly coded in law. In addition automatic generation of networked computer applications that are perfectly compliant with the wide variety of directives and laws in any country is one of the ultimate goals of this type of ontology based work.

For an overview of ontology-grounded trusted regulatory compliance, see [18].

Human Resources

Competencies describe the skills and knowledge individuals should have in order to be fit for particular jobs. Especially in the domain of vocational education, having a central shared and commonly used competency model is becoming crucial in order to achieve the necessary level of interoperability and exchange of information, and in order to integrate and align the existing information systems of competency stakeholders like schools or public employment agencies. Only few organizations however, have successfully implemented a company-wide "competency initiative," let alone a strategy for inter-organizational exchange of competency related information.

Several projects (See, e.g., the EU-funded CoDrive project.) aim at contributing to a competency-driven vocational education by using state-of-the-art ontology methodology and infrastructure in order to collaboratively develop a conceptual, shared and formal KR of competence domains.

For a business case study on vocational competency ontology elicitation, see [4].

Future Directions

The ever-changing interoperability requirements between the stakeholding communication partners (See, e.g., diverse (legacy) systems in the open extended enterprise.) requires ontologies to continuously evolve. Usually the domain is too large and complex to be explicated in one single effort, and the knowledge workers understanding of the domain is in continuously changing, requiring timely renegotiation of existing consensus. Therefore, one should not merely focus on the practice of eliciting ontologies in a project-like context, but consider it as a real-time collaborative and continuous process that is integrated with and in the operational processes of the community itself. The shared background of communication partners is continuously negotiated as are the characteristics or values of the concepts that are agreed upon.

There are many additional complexities that should be considered. As investigated in FP6 integrated projects (See, e.g., http://ecolead.vtt.fi/.) on collaborative networked organizations, the different professional, social, and cultural backgrounds among communities and organizations can lead to misconceptions, resulting in costly ambiguities and misunderstandings if not aligned properly. This is especially the case in inter-organizational settings, where there may be many pre-existing organizational sub-ontologies, inflexible data schemas interfacing to legacy data, and ill-defined, rapidly evolving collaborative requirements. Furthermore, participating stakeholders usually have strong individual interests, inherent business rules, and work practices. These may be tacit, or externalized in workflows that are strongly interdependent, hence further complicate the conceptual alignment. Finally this also involves ontology elicitation cost estimation. Simperl and Sure (chapter 7, [12]) propose a parametric cost estimation model for ontologies by identifying relevant cost drivers having a direct impact on the effort invested in ontology elicitation.

For an overview of future directions towards community-driven ontology elicitation and management, see [6].

Cross-References

▶ Emergent Semantics
▶ Ontology
▶ Ontology Engineering

Recommended Reading

1. Bachimont B, Troncy R, Isaac A. Semantic commitment for designing ontologies: a proposal. In: Proceedings of the 13th International Conference on Knowledge Engineering and Knowledge Management: Ontologies and the Semantic Web; 2002. p. 114–21.
2. Blomqvist E. OntoCase – a pattern-based ontology construction approach. In: Proceedings of the OTM Confederated International Conferences, CoopIS, DOA, GADA, and ODBASE; 2007. p. 971–88.
3. Buitelaar P, Cimiano P, Magnini B. Ontology learning from text: methods, evaluation and applications, vol. 123 of Frontiers in Artificial Intelligence and Applications, IOS, Amsterdam. 2005.
4. Christiaens S, De Leenheer P, de Moor A, Robert Meersman R. Ontologising competencies in an interorganisational setting. In: Ontology management. Semantic web and beyond computing for human experience, vol. 7. Berlin: Springer; 2008. p. 265–88.
5. De Leenheer P, de Moor A, Meersman R. Context dependency management in ontology engineering: a formal approach. J Data Semant. 2006;8:26–56.
6. De Leenheer P, Meersman R. Towards community-based evolution of knowledge-intensive systems. In: Proceedings of the OTM Confederated International Conferences, CoopIS, DOA, GADA, and ODBASE; 2007. p. 989–1006.
7. de Moor A, De Leenheer P, Meersman R. DOGMA-MESS: a meaning evolution support system for interorganizational ontology engineering. In: Proceedings of the 14th International Conference on Conceptual Structures; 2006. p. 189–203.
8. Furnas G, Landauer T, Dumais S. The vocabulary problem in human-system communication. Commun ACM. 1987;30(11):964–71.
9. Gangemi A. Ontology design patterns for semantic web content. In: Proceedings of the 4th International Semantic Web Conference; 2005. p. 262–76.
10. Ganter B, Stumme G, Wille R, editors. Formal concept analysis, foundations and applications, LNCS, vol. 3626. Berlin: Springer; 2005.
11. Hepp M. Possible ontologies: how reality constrains the development of relevant ontologies. IEEE Internet Comput. 2007;11(1):90–6.
12. Hepp M, De Leenheer P, de Moor A, Sure Y, editors. Ontology management, semantic web, semantic web services, and business applications, vol. 7 of semantic web and beyond computing for human experience. Berlin: Springer; 2008.
13. Jarrar M, Demey J, Meersman R. On reusing conceptual data modeling for ontology engineering. J. Data Semant. 2003;1(1):185–207.
14. Kotis K, Vouros G. Human-centered ontology engineering: the Hcome methodology. Knowl Inf Syst. 2005;10(1):109–31.
15. Lessig L. Ontology management, semantic web, semantic web services, and business applications. Basic Books. 1999.

16. Milton N. Knowledge acquisition in practice: a step-by-step guide. London: Springer; 2007.

17. Pinto H, Staab S, Tempich C. DILIGENT: towards a fine-grained methodology for DIstributed, Loosely-controlled and evolvInG Engineering of oNTologies. In: Proceedings of the 16th European Conference on AI; 2004.

18. Ryan H, Spyns P, De Leenheer P, Leary R. Ontology-based platform for trusted regulatory compliance services. In OTM workshops, LNCS. vol. 2889. Berlin: Springer: 2003. p. 675–89.

19. Siorpaes K, Hepp M. Games with a purpose for the semantic web. IEEE Intell Syst. 2008;23(3):50–60.

20. Spyns P, Meersman R, Jarrar M. Data modelling versus ontology engineering. ACM SIGMOD Rec. 2002;31(4):12–7.

21. Stamper R. Information in business and administrative systems. New York: Wiley; 1973.

22. Van Damme C, Hepp M, Siorpaes K. Folksontology: an integrated approach for turning folksonomies into ontologies. In: Proceedings of the ESWC Workshop Bridging the Gap between Semantic Web and Web 2.0; 2007.

Ontology Engineering

Avigdor Gal
Faculty of Industrial Engineering and
Management, Technion–Israel Institute of
Technology, Haifa, Israel

Synonyms

Ontological engineering

Definition

Ontology engineering is "the set of activities that concern the ontology development process, the ontology life cycle, and the methodologies, tools and languages for building ontologies" [1]. It provides "a basis of building models of all things in which computer science is interested" [4]. Ontology engineering aims at providing standard components for building knowledge models. Ontologies play a similar role to design ratio-nale in mechanical design. It allows the reuse of knowledge in a knowledge base by providing conceptualization, reflecting assumptions and requirements made in the problem-solving using the knowledge base. Ontology engineering pro-vides the means to build and use ontologies for building models.

Ontology Engineering: Details

Mizoguchi and Ikeda [4] define eight levels (from shallow to deep) of using ontologies. At level 1, ontologies are used as a common vocabulary for communication. At level 2, it is used as a conceptual schema of a relational database. In the third level, ontologies are used as backbone information for using a knowledge base. The remaining five levels are the levels were ontology engineering comes into play. Ontologies in the fourth level are used to answer competence questions, and then they are used for standardization (of terminology or of tasks) in level 5. In level 6, ontologies are used for structural and semantic transformation of schemas. Reusing knowledge is done at the seventh level, and knowledge reorga-nization is considered the eigth and highest level of using ontologies.

Ontology engineering makes use of ontologies (in the sense of levels 4–8) to generate standard tools for knowledge representation. This does not imply that knowledge is standardized. Using ontology engineering, one can design knowledge for specific applications, similarly to production based on engineering tools in other engineering fields.

We now illustrate the use of ontology engi-neering in two key applications, namely, func-tional design and schema matching. For the for-mer, Kitamura et al. [3] describe a seventh level of using ontologies in a real-world application of plant and production systems. There, an ontology that describes two types of functional models, two types of organization of generic knowledge, and two ontologies of functionality were put into use in sharing functional design knowledge on production systems. The users (engineers) of the system said that this framework enabled them to make implicit knowledge possessed by each designer explicit and to share it among team members.

Ontologies are used in schema matching in many ways. One of which, corresponding to the sixth level of use, was presented in the Onto-Builder toolcase [2]. Special ontological constructs were identified for the matching of Web form data. Ontologies were built using these constructs, and dedicated matching algorithms were constructed to determine the amount of certainty to assign with the matching of attribute pairs. An example of an ontological construct, unique to OntoBuilder, is *precedence*. This construct determines the order in which attributes are presented to the user on a Web form and generate a partial order on attributes. Attribute similarity is then measured based on their relative positioning in their own ontologies.

Cross-References

▶ Ontology
▶ Semantic Matching

Recommended Reading

1. G'omez-P'erez A, Fern'andez-L'opez M, Corcho O. Ontological engineering. London: Springer; 2004.
2. Gal A, Modica G, Jamil H, Eyal A. Automatic ontology matching using application semantics. AI Mag. 2005;26(1):21–32.
3. Kitamura Y, Kashiwase M, Fuse M, Mizoguchi R. Deployment of an ontological framework of functional design knowledge. Adv Eng Inform. 2004;18(2): 115–27.
4. Mizoguchi R, Ikeda M. Towards ontology engineering. Technical report AI-TR-96-1, I.S.I.R., Osaka University. 1996.
5. Paslaru Bontas E, Tempich C. Ontology engineering: a reality check. In: Proceedings of the 5th International Conference on Ontologies, DataBases, and Applications of Semantics; 2006. p. 836–54.
6. Sure Y, Tempich C, Vrandecic D. Ontology engineering methodologies. In: Semantic Web Technologies: Trends and Research in Ontology-based Systems. Springer; 2006.

Ontology Visual Querying

Sean Bechhofer and Norman W. Paton
University of Manchester, Manchester, UK

Definition

An *ontology definition language* provides constructs that can be used to describe concepts and the relationships in which they participate. Because such languages define the properties concepts can exhibit, they can be used to restrict the questions that can meaningfully be asked about the concepts. Given a specification of the questions that can legitimately be asked, a user interface can direct query construction tasks towards meaningful requests, which in turn are expected to yield non-empty answers. Thus *ontology visual querying* is the use of an ontology to direct interactive query construction. A related topic is *faceted browsing*, in which the incremental description of concepts of interest is closely integrated with retrieval, thereby providing information about the results of a request as it is being constructed.

Historical Background

The history of visual query languages is almost as long as that of textual query languages, with Query-by-Example [14] developed in parallel with SQL. Query-by-Example contained two features that recur in almost all visual query languages: (i) a representation of the model over which the query is to be expressed; and (ii) a notation for incrementally constructing queries from the collections, relationships and value ranges of interest, with reference to the representation in (i). The evolution of visual query languages [4] has tracked the evolution of data models and interactive paradigms, and proposals have been made that support querying over many different data models (relational, object-oriented, temporal, etc) using a variety of interaction objects (forms, graphs, icons, etc).

An orthogonal aspect is the closeness of the relationship between query construction and answer presentation; for example, in *dynamic queries* the answer to a request is constructed automatically and incrementally as a query is refined [12]. This provides immediate feedback to users on the size and nature of the result, but may require specialised storage structures to support incremental result computation.

Ontology visual querying has been developed so that the knowledge expressed in an ontology can be used to direct query construction; query answers may then be constructed from an instance store that is closely integrated with the ontology definition language, or by evaluating requests over external data sources. The latter is quite common, as ontologies are widely used to provide conceptual models for web (e.g., [1]) or data (e.g., [3]) resources.

Foundations

As in other visual query languages, ontology visual querying requires a visual representation of the concepts over which a request is to be constructed. Visual query formulation allows users to explore the domain of interest by *recognition* rather than *recall*: that is, it should not be necessary to remember (or even be fully aware of) the ontology in order to express a query over it.

Ontologies are represented visually other than for querying; for example, ontology design tools typically support both form and graph-based views of concepts and their relationships (e.g., [10]). As concept definition and query formulation may have significant common ground, representations that are useful for ontology browsing and concept definition may also be relevant for querying. For example, in the TAMBIS ontology-based data integration system [2], a query is a concept definition in a Description Logic (DL) [8], so writing a query is essentially the same as defining a new concept.

Although expressive ontology languages may present challenges for navigation and thus query construction, they also present certain opportunities for query interface designers, as various

forms of reasoning may be useful for guiding query construction. For example, an interface can prevent the submission of queries that are unsatisfiable (i.e., that are known from the definitions in the ontology to return no results) either by making it impossible for the user to construct such queries or by detecting when such requests have been created (e.g., [5, 7]). Such feedback can be seen as *intensional*, with the constraints or knowledge in the ontology determining the behaviour of the interface. Feedback may also be *extensional*, for example, with the number of results to be returned being shown to the user. This is common in faceted browsing systems, and such direct result construction is generally supported in combination with closely integrated stores.

In addition to the feedback described above, systems may also provide alternative renderings of the query being constructed – for example a natural language description of the query. This can be of use in helping naïve or inexperienced users in forming appropriate queries. Note that this involves the rendering of the query in natural language, rather than translating a query posed using natural language.

A common approach is to specify queries through an iterative refinement process, in which the content of the ontology and current context of the query impact on the options presented. The principle of intensional navigation uses the vocabulary to guide the user during query formulation, employing constraints in the ontology to either flag to the user that the query is in some way violating the constraints, or preventing the user from forming queries that would be unsatisfiable, and thus return no results. Operations available for query manipulation in SEWASIE [5] include the addition of a new role/property with an associated filler, or the replacement of a filler value. In the latter case, a classification or super/sub class taxonomy is used to support the manipulation, with value fillers being specialised or generalised. Although ontology languages differ in the constructors and expressivity offered, some notion of hierarchical classification is nearly always present, and so can be exploited in visual query interfaces. An additional operation offered by the TAMBIS system is to *refocus* the query, which takes a sub node of the query and reorganises the query to promote that node to the root. This operation introduces an additional requirement on the ontology language, namely that properties or relations have *inverses*.

Various of the notions discussed above are illustrated in Figs. 1 and 2 for SQoogle, which was developed in the SEWASIE project. Figure 1 shows the composition phase, with the graphical depiction of the query. In Fig. 2, the user is being offered generalisations or specialisations of a particular node in the query.

Overall, ontology visual query systems can be characterized by a number of features:

- *Identification of starting points.* The construction of a query has to start from some place in the ontology; systems may offer predetermined entry points, user defined bookmarks, or a search mechanism across the concepts in the ontology.
- *Query language.* More expressive query languages support more precise question

answering, but may contain constructs that require explanation for users or that are challenging to represent using certain visual paradigms. Ontology visual query languages rarely support features such as aggregation or grouping.
- *Query modification operations.* Query construction involves manipulation of query expressions, for example, to include additional relationships or to specialise a concept named in the query.
- *The ontology definition language.* Richer ontology definition languages are generally more complex to display and navigate, but may provide more options for generalizing and specializing query components, and can express constraints that are useful for directing query construction.
- *Relationship between query language and ontology language.* Proposals implement different relationships between the query language and the ontology definition language. For example, in SEWASIE, the query language and ontology language are

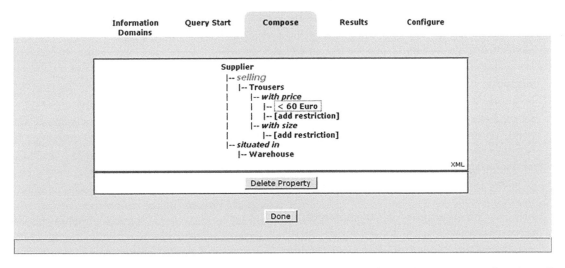

Ontology Visual Querying, Fig. 1 Visual query expression in SQoogle. The query is to select suppliers that sell trousers that cost less than 60 euros, where the supplier is situated in a warehouse

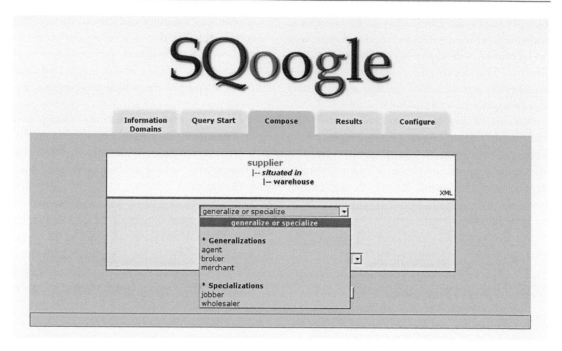

Ontology Visual Querying, Fig. 2 Visual query refinement in SQoogle. The query can be revised by replacing the concept supplier with a more general (e.g., agent, broker) or specialized (e.g., wholesaler) concept

explicitly separated. The ontology language supports reasoning services that are used to guide the intensional navigation process described above. In contrast, in TAMBIS, queries *are* concept descriptions, thus the interface is tied more closely to the ontology language.

- *Feedback mechanisms.* Feedback can inform the user about the results of the query or the state of the query with respect to the underlying ontology. Feedback may be intensional (in terms of the ontology), or extensional (in terms of the result set).
- *Query presentation.* Queries may be presented solely using the visual query, or may also offer, for example, natural language renderings of the query.
- *Domain specificity.* Visual interfaces provide the opportunity to represent models or results using general-purpose or domain-specific representations. Most ontology visual query languages are general purpose, but faceted browsing interfaces are often designed to support specific applications.

Table 1 describes a collection of representative visual query systems using a selection of the above criteria: the TAMBIS and SEWASI visual query languages, and the Flamenco and /facet faceted browsing systems. In all these proposals, the ontology directs query construction, and thus the design of the ontology has a significant influence on the utility of the interface.

Key Applications

Ontology visual query systems have most commonly been deployed in areas of science and culture where there are rich data resources to be explored. For example, TAMBIS provided access to multiple biological information sources. Faceted browsing and querying has been widely used to browse image collections and in the cultural heritage domain, for example to support access to Finnish Museums [9] and galleries in the Netherlands [11]. The faceted approach is also common in on-line shopping sites such as eBay.

Ontology Visual Querying, Table 1 Representative examples of ontology visual query systems

Proposal	TAMBIS	SEWASIE	Flamenco	Facet
Reference	[7]	[5]	[13]	[8]
Query language	Concept definition	Conjunctive queries	Path expressions	Path expressions
Query modification operations	Property add/remove; filler specialization or generalization; refocus	Property add/remove; filler specialization or generalization	Property add/remove; filler specialization or generalization	Property add/remove; filler specialization or generalization
Ontology definition language	Description logic	Description logic	Hierarchical categories	RDFS
Feedback mechanism	Intensional	Intensional	Extensional	Extensional
Query presentation	Visual	Visual plus NL rendering	Path expression	Path expression
Domain specificity	Generic	Generic	Specific (image repositories)	Generic

Future Directions

Large amounts of a data are beginning to emerge using representation languages like OWL. Current work in ontology languages should see standardisation of the SPARQL query language finalised in the near future. Query interfaces that sit on top these standardized languages will then be required in order to support access to this data – interfaces that support naïve or non-expert users will clearly be required.

Cross-References

▶ OWL: Web Ontology Language
▶ Visual Query Language

Recommended Reading

1. Antoniou G, van Harmelen F. A semantic web primer. Cambridge, MA: MIT Press; 2004.
2. Baader F, Calvanese D, McGuinness D, Nardi D, Patel-Schneider P, editors. The description logic handbook. Cambridge: Cambridge University Press; 2003.
3. Calvanese D, De Giacomo G, Lenzerini M, Nardi D, Rosati R. Data integration in data warehousing. Int J Coop Inf Syst. 2001;10(3):237–71.
4. Catarci T, Costabile MF, Levialdi S, Batín C. Visual query systems for databases: a survey. J Vis Lang Comput. 1997;8(2):215–60.
5. Catarci T, Dongilli P, Di Mascio T, Franconi E, Santucci G, Tessaris S. An ontology based visual tool for query formulation support. In: Proceedings of the 16th European Conference on AI; 2004. p. 308–12.
6. Colucci S, Noia TD, Sciascio ED, Donini FM, Ragone A, Rizzi R. A semantic-based fully visual application for matchmaking and query refinement in B2C e-marketplaces. In: Proceedings of the 8th ACM International Conference on Electronic Commerce; 2006. p. 174–84.
7. Goble CA, Stevens R, Ng G, Bechhofer S, Paton NW, Baker PG, Peim M, Brass A. Transparent access to multiple bioinformatics information sources. IBM Syst J. 2001;40(2):532–51.
8. Hildebrand M, van Ossenbruggen J, Hardman L. /facet: a browser for heterogeneous semantic web repositories. In: Proceedings of the 5th International Semantic Web Conference; 2006. p. 272–85.
9. Hyvyonen E, Myakelya E, Salminen M, Valo A, Viljanen K, Saarela S, Junnila M, Kettula S. Museum Finland – Finnish museums on the semantic web. J Web Semant. 2005;3(2):224–41.
10. Knublauch H, Fergerson RW, Noy NF, Musen MA. The protégé OWL plugin: an open development environment for semantic web applications. In: Proceedings of the 3rd International Semantic Web Conference; 2004. p. 229–43.
11. Schreiber G, et al. MultimediaN E-culture demonstrator. In: Proceedings of the 5th International Semantic Web Conference; 2006. p. 951–8.
12. Shneiderman B. Dynamic queries for visual information seeking. IEEE Softw. 1994;11(6):70–7.
13. Yee K-P, Swearingen K, Li K, Hearst MA. Faceted metadata for image search and browsing. In: Proceedings of the SIGCHI Conference on Human Factors in Computing Systems; 2003. p. 401–8.
14. Zloof MM. Query-by-example: the invocation and definition of tables and forms. In: Proceedings of the 1st International Conference on Very Large Data Bases; 1975. p. 1–24.

Ontology-Based Data Access and Integration

Diego Calvanese[1], Giuseppe De Giacomo[2],
Domenico Lembo[2], Maurizio Lenzerini[2], and
Riccardo Rosati[2]
[1]Research Centre for Knowledge and Data
(KRDB), Free University of Bozen-Bolzano,
Bolzano, Italy
[2]Dip. di Ingegneria Informatica Automatica e
Gestionale Antonio Ruberti, Sapienza Università
di Roma, Rome, Italy

Definition

An *ontology-based data integration* (OBDI) system is an information management system consisting of three components: an ontology, a set of data sources, and the mapping between the two. The ontology is a conceptual, formal description of the domain of interest to a given organization (or a community of users), expressed in terms of relevant concepts, attributes of concepts, relationships between concepts, and logical assertions characterizing the domain knowledge. The data sources are the repositories accessible by the organization where data concerning the domain are stored. In the general case, such repositories are numerous, heterogeneous, each one managed and maintained independently from the others. The mapping is a precise specification of the correspondence between the data contained in the data sources and the elements of the ontology. The main purpose of an OBDI system is to allow information consumers to query the data using the elements in the ontology as predicates.

In the special case where the organization manages a single data source, the term *ontology-based data access* (ODBA) system is used.

Historical Background

The notions of ODBA and ODBI were introduced in [3, 14] and originated from several disciplines, in particular, information integration, knowledge representation and reasoning, and incomplete and deductive databases.

OBDI can be seen as a sophisticated form of information integration [11], where the usual global schema is replaced by an ontology describing the domain of interest. The main difference between OBDI and traditional data integration is that in the OBDI approach, the integrated view that the system provides to information consumers is not merely a data structure accommodating the various data at the sources but a semantically rich description of the relevant concepts in the domain of interest, as well as the relationships between such concepts. Also, the distinction between the ontology and the data sources reflects the separation between the conceptual level, the one presented to the client, and the logical/physical level of the information system, the one stored in the sources, with the mapping acting as the reconciling structure between the two levels.

The central notion of OBDI is therefore the ontology, and reasoning over the ontology is at the basis of all the tasks that an OBDI system has to carry out. In particular, the axioms of the ontology allow one to derive new facts from the source data, and these inferred facts greatly influence the set of answers that the system should compute during query processing. In the last decades, research on ontology languages and ontology inferencing has been very active in the area of knowledge representation and reasoning. Description logics [1] (DLs) are widely recognized as appropriate logics for expressing ontologies and are at the basis of the W3C standard ontology language OWL. (http://www.w3.org/TR/owl2-overview/) These logics permit the specification of a domain by providing the definition of classes and by structuring the knowledge about the classes using a rich set of logical operators. They are decidable fragments of mathematical logic, resulting from extensive investigations on the trade-off between expressive power of knowledge representation languages and computational complexity of reasoning tasks. Indeed, the constructs appearing in the DLs used in OBDI are carefully chosen taking into account such a trade-off [3, 12]. We observe that many research papers on reasoning in DLs for ODBA and OBDI actually concentrate on query answering in a setting where the data are assumed to reside in ad hoc repositories (often in RDF format), rather than in independent sources. Since such ad hoc repositories are designed for storing the instances of the elements of the ontology, mappings are not present in this simplified setting, which is often called *ontology-based query answering* (OBQA).

As we said before, the axioms in the ontology can be seen as semantic rules that are used to complete the knowledge given by the raw facts determined by the data in the sources. In this sense, the source data of an OBDI system can be seen as an incomplete database, and query answering can be seen as the process of computing the answers logically deriving from the combination of such incomplete knowledge and the ontology axioms. Therefore, at least conceptually, there is a connection between OBDI and the two areas of incomplete information [8] and deductive databases [4]. The new aspect of OBDI is related to the kind of incomplete knowledge represented in the ontology, which differs both from the formalisms typically used in databases under incomplete information (e.g., Codd tables) and from the rules expressible in deductive database languages (e.g., logic programming rules).

Scientific Fundamentals

We deal here with the semantic and computational aspects related to the use of an ontology and of mappings to data sources in query processing in OBDI. Thus, we do not address the problems that specifically pertain to accessing and querying *multiple, heterogeneous* data sources in an integrated way, such as wrapping non-relational data, distributed query evaluation, and entity resolution. For those problems, we refer to the "Information Integration" entry. The distinction between OBDI and OBDA is therefore not significant for our treatment of the scientific fundamentals, and we refer to OBDA only. Coherently, we assume to deal with a single rela-

tional data source, whose schema might represent the federated schema of multiple, heterogeneous data sources, wrapped as relational databases.

OBDA Framework

We distinguish between the specification of an OBDA system and the OBDA system itself (cf. Fig. 1). An *OBDA specification* $\mathcal{J}$ determines the intensional level of the system and is expressed as a triple $\langle \mathcal{O}, S, \mathcal{M} \rangle$, where $\mathcal{O}$ is an ontology, S is the schema of the data source, and $\mathcal{M}$ is the mapping between S and $\mathcal{O}$. Specifically, $\mathcal{M}$ consists of a set of mapping assertions, each one relating a query over the source schema to a query over the ontology. An *OBDA system* $\langle \mathcal{J}, D \rangle$ is obtained by adding to $\mathcal{J}$ an extensional level, which is given in terms of a database D, representing the data at the source, structured according to the schema S.

The formal semantics of $\langle \mathcal{J}, D \rangle$ is specified by the set $Mod_D(\mathcal{J})$ of its models, which is the set of (logical) interpretations $\mathcal{I}$ for $\mathcal{O}$ such that $\mathcal{I}$ is a model of $\mathcal{O}$ and $\langle D, \mathcal{I} \rangle$ satisfies all the assertions in $\mathcal{M}$. The satisfaction of a mapping assertion depends on its form, which is meant to represent semantic assumptions about the completeness of the source data with respect to the intended ontology models. Specifically, *sound* (resp., *complete*, *exact*) mappings capture sources containing a subset (resp., a superset, exactly the set) of the expected data.

In OBDA, the main service to be provided by the system is *query answering*. The user poses queries by referring only to the ontology and is therefore masked from the implementation details and the idiosyncrasies of the data source. The fact that the semantics of $\langle \mathcal{J}, D \rangle$ is defined in terms of a set of models makes the task of query answering involved. Indeed, query answering cannot be simply based on evaluating the query expression over a single interpretation, like in traditional databases. Rather, it amounts to compute the so-called *certain answers*, i.e., the tuples that satisfy the query in all interpretations in $Mod_D(\mathcal{J})$ and has therefore the characteristic of a logical inference task. Obviously, the computation of certain answers must take into account the semantics of the ontology, the knowledge expressed in the mapping, and the content of the

Ontology-Based Data Access and Integration, Fig. 1 OBDI/OBDA specification and system

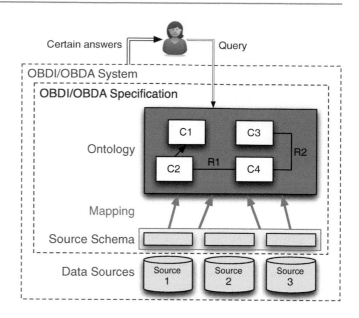

data source. Designing efficient query processing algorithms is one of the main challenges of OBDA.

We discuss now computational issues connected to query answering in OBDA, with the aim of showing which are the sources of computational complexity. An OBDA framework is characterized by three formalisms: (1) the language used to express the ontology, (2) the language used for queries, and (3) the language used to specify the mapping. The choices made for each of the three formalisms affect semantic and computational properties of the system.

The axioms of the ontology allow one to enrich the information coming from the source with domain knowledge and hence to infer additional answers to queries. The language used for the ontology deeply affects the computational characteristics of query answering. For this reason, instead of expressing the ontology in first-order logic (FOL), one adopts tailored languages, typically based on description logics (DLs), which ensure decidability and possibly efficiency of reasoning.

Also, the use of FOL (i.e., SQL), as a query language, immediately leads to undecidability of query answering, even when the ontology consists only of an alphabet (i.e., it is a flat

schema) and when the mapping is of the simplest possible form, i.e., it specifies a one-to-one correspondence between ontology elements and database tables. Hence, the language typically adopted is union of conjunctive queries (UCQs), i.e., FOL queries expressed as a union of select-project-join SQL queries.

With respect to mapping specification, the incompleteness of the source data is captured correctly by mappings that are sound. Moreover, allowing to mix sound mapping assertions with complete or exact ones leads to undecidability of query answering, even when only CQs are used in queries and mapping assertions and the ontology is simply a flat schema. As a consequence, all proposals for OBDA frameworks so far assume that mappings are sound. In addition, the concern above on the use of FOL applies also for the ontology queries in the mapping. Note instead that the source queries in the mapping are directly evaluated over the source database and hence are typically allowed to be arbitrary (efficiently) computable queries.

A Tractable OBDA Framework

Considering the discussion above, we present now a specific OBDA framework tailored towards efficiency and tractability. The framework makes use of a family of DLs, called *DL-Lite*,

Ontology-Based Data Access and Integration, Table 1 $DL\text{-}Lite_A$ assertions. Symbols in square brackets may or may not be present, and $R^-(x, y)$ stands for $R(y, x)$

Type of assertion	DL syntax		FOL semantics
ISA/disjointness between concepts	C_1	$\sqsubseteq$ $[\neg]C_2$	$\forall x.C_1(x) \to [\neg]C_2(x)$
Domain/range of a role	$\exists R^{[-]}$	$\sqsubseteq$ C	$\forall x, y.R^{[-]}(x, y) \to C(x)$
Participation to a role	C	$\sqsubseteq$ $\exists R^{[-]}$	$\forall x.C(x) \to \exists y.R^{[-]}(x, y)$
ISA/disjointness between roles	$R_1^{[-]}$	$\sqsubseteq$ $[\neg]R_2^{[-]}$	$\forall x, y.R_1^{[-]}(x, y) \to [\neg]R_2^{[-]}(x, y)$
Functionality of roles	(funct $R^{[-]}$)		$\forall x, y, z.R^{[-]}(x, y) \wedge R^{[-]}(x, z) \to y = z$

which has also given rise to the OWL 2 QL profile (http://www.w3.org/TR/owl2-profiles/) of the Web Ontology Language OWL standardized by the W3C.

DLs are class-based formalisms that represent the domain of interest in terms of classes, or *concepts*, and binary relationships, or *roles*, between classes. Here, we consider $DL\text{-}Lite_A$, which is able to capture essentially all features of entity-relationship diagrams and UML Class Diagrams, except for completeness of hierarchies. In $DL\text{-}Lite_A$, a concept is either an atomic concept C (i.e., a unary predicate) or the projection $\exists R$ or $\exists R^-$ of a role R on its first or second component, respectively. A role can be either an atomic role R or an inverse role R^-, allowing for a complete symmetry between the two directions. $DL\text{-}Lite_A$ includes also *value attributes* relating objects in classes to domain values (such as strings or integers), but we do not discuss this aspect here. The *ontology* is modeled by means of axioms that can express *inclusion* and *disjointness* between concepts or roles and (global) *functionality* of roles (with some restrictions on the interaction between functionality and role inclusions to ensure tractability). In Table 1, we illustrate the conceptual modeling constructs captured by $DL\text{-}Lite_A$ assertions and provide also their semantics expressed in FOL.

Mapping assertions are of type GAV, i.e., have the form $\varphi(\vec{x}) \to \psi(\vec{x})$, where $\varphi(\vec{x})$ is a domain-independent FOL query over the source schema, e.g., expressed in SQL, with answer variables $\vec{x}$, and $\psi(\vec{x})$ is a conjunction of atoms over the concepts and roles of the ontology, whose only variables are those in $\vec{x}$. However, we need to take into account the impedance mismatch between values in the data source and objects that populate

classes in the ontology. To do so, the arguments of the atoms in $\psi(\vec{x})$ might be not only constants or variables but also terms constructed by applying *functors* to them. Such functors act as object constructors, like in object-oriented approaches. The meaning of a mapping assertion $\varphi(\vec{x}) \to \psi(\vec{x})$ is to extract the tuples satisfying $\varphi(\vec{x})$ and to use them to (partially) populate according to $\psi(\vec{x})$ the concepts and roles, constructing suitable objects through the functors. Actually, such extraction is typically only virtually performed during query answering.

The DLs of the $DL\text{-}Lite$ family, including $DL\text{-}Lite_A$, combined with the GAV mapping assertions above, have been designed so as to enjoy the *FO-rewritability* property: given a UCQ q and an OBDA specification $\mathcal{J} = \langle \mathcal{O}, S, \mathcal{M} \rangle$, it is possible to compile q, $\mathcal{O}$, and $\mathcal{M}$ into a new FO query q' formulated over S. Such query q' has the property that when evaluated over a database D for S, it returns exactly the certain answers for q over the OBDA system $\langle \mathcal{J}, D \rangle$, for every data source D. Each such q' is called an (FO-) *perfect rewriting* of q w.r.t. $\mathcal{J}$. FO-rewritability immediately implies that the data complexity of computing certain answers is in AC^0, which is the same complexity as that of FOL query evaluation in relational databases.

Techniques for Query Answering via Rewriting

In the tractable OBDA framework previously described, one can think of a simple technique for query answering, which first retrieves an initial set of concept and role instances from the data source through the mapping and then, using the ontology axioms, "expands" such a set of instances deriving and materializing all the logically entailed concept and role assertions; finally, queries can be evaluated on such an expanded set of instances. Unfortunately, the instance materialization step of the above technique is not feasible in general, because the set of entailed instance assertions starting from even very simple OBDA specifications and small data sources may be infinite.

As an alternative to the above materialization strategy, most of the approaches to query answering in OBDA are based on query rewriting, where the aim is to first compute the perfect rewriting q' of a query q w.r.t. an OBDA specification $\mathcal{J}$ and then evaluate q' over the source database.

The above described OBDA framework allows for modularizing query rewriting. Indeed, the current techniques for OBDA consist of a phase of *query rewriting w.r.t. the ontology* followed by a phase of *query rewriting w.r.t. the mapping*. In the first phase, the initial query q is rewritten with respect to the ontology, producing a new query q_o, still over the ontology signature: intuitively, q_o "encodes" the knowledge expressed by the ontology that is relevant for answering the query q. In the second phase, the query q_o is rewritten with respect to the mapping $\mathcal{M}$, using the mapping assertions as rules for reformulating the query with respect to the source schema signature. We illustrate now the two phases more in detail.

Query Rewriting w.r.t. the Ontology

Most of the proposed techniques [3, 5, 13] start from a CQ or a UCQ and end up producing a UCQ (i.e., a set of CQs) expanding the initial query. They are based on variants of clausal resolution [10]: every rewriting step essentially corresponds to the application of clausal resolution between a CQ among the ones already generated and a concept or role inclusion axiom of the ontology. Each such step produces a new conjunctive query that is added to the resulting UCQ. The rewriting process terminates when a fixed point is reached, i.e., no new CQ can be generated.

A potential bottleneck of the rewriting approach is caused by the size of the rewritten query, and several research works aim at optimization techniques addressing this issue. For

example, the first algorithm for query rewriting w.r.t. a *DL-Lite* ontology [3] has been improved in [5, 13] by refining and optimizing the way in which term unification is handled by the above resolution step. Notice that the sentences corresponding to the ontology axioms may be Skolemized (e.g., due to the presence of existentially quantified variables in the right-hand side of a concept inclusion): to compute perfect rewritings, the unification of Skolem terms during resolution can actually be constrained in various ways with respect to standard resolution.

Some recent proposals for optimizing query rewriting w.r.t. the ontology (e.g., [5, 7, 15]) are based on the use of Datalog queries besides CQs and UCQs, to express either intermediate results or the final rewritten query. The same idea has also been used to extend query rewriting to more expressive, not necessarily FO-rewritable ontology languages [2, 5, 6, 13]. Other approaches take a more radical view and propose strategies based on partial materialization of instance assertions [9].

Query Rewriting w.r.t. the Mapping

It is well known by the studies on data integration that rewriting a query w.r.t. GAV mappings boils down to a simple unfolding strategy, which essentially means substituting every predicate of the input query with the queries that the mapping associates to that predicate [11]. In OBDA, however, query rewriting w.r.t. mappings is complicated by the following two aspects: (1) OBDA mappings allow for constructing objects that are instances of the ontology predicates from the values stored in the data source, in order to deal with the mentioned impedance mismatch problem; (2) the source queries in the mapping are expressed using the full expressive power of SQL, which is needed to bridge the large cognitive distance that may exist between the ontology and the source schema.

Solutions to the first problem depend on the strategy adopted to construct objects from values. When functors applied to values are used, as in the tractable framework for OBDA we presented above, logic terms constructed through such functors can be treated in the standard way in the unifications at the basis of the unfolding procedure: see, e.g., the algorithm proposed in [14], which relies on techniques from partial evaluation of logic programs. In the R2RML standard, (http://www.w3.org/TR/r2rml/) functors are realized through templates that construct W3C compliant URIs for objects from the values returned by the SQL query in the mapping assertion.

The second problem heavily affects the performance of the query-answering algorithm. Indeed, current SQL engines have hard times in optimizing the execution of queries expressed over virtual views, like those introduced by the unfolding that use complex SQL features such as union, nesting, or aggregation. Performance problems are of course amplified when there are several SQL queries mapping the same ontology predicate. Due to the abovementioned limitations, it is not realistic to group all such queries within a single mapping assertion for each predicate. However, without such grouping, the mapping associates several queries to the same predicate, and therefore, the size of the query obtained by rewriting w.r.t. the mapping may be exponential in the size of the input query. Indeed, in real-world applications, it may very well happen that the size of the produced rewriting is too large to be handled by current SQL engines. Techniques to avoid or mitigate these issues are currently under investigation.

Key Applications

The applications of OBDA and OBDI include all the real-world settings in which an organization needs a unified and transparent access to its data, based on a domain model. Examples are:

- Enterprise information systems, where data governance and data access can be greatly enhanced by the use of the ontology
- Scientific data management, at least in those fields where ontologies are available as unified representations of relevant meta-data

- Public administration and government data management, where the OBDI paradigm can be the enabling technology for information sharing and semantic interoperability
- Open data publishing, where the ontology can help determining what to publish and which strategy to follow in order to enrich the data with useful meta-data

Future Directions

OBDA and OBDI are young paradigms, and many problems related to them are still open. Here is an incomplete list:

- Although query processing in OBDI has been the main subject of investigation, there is still much room for optimization techniques, especially those aiming at making query processing feasible in the "big data" setting.
- Real-world applications often demand more expressive languages in the various components of the OBDI system. An interesting direction that is currently under investigation is to extend the query processing techniques to the case where both the ontology language and the mapping language go beyond the expressive power considered above and to the case of more powerful languages for specifying user queries.
- Although we concentrated on description logics, other types of ontology languages have been considered and studied, notably those based on extensions of Datalog [2].
- Since the ontology should reflect the conceptual model of the domain, and not the information at the sources, it is likely that source data are not fully coherent with the axioms in the ontology. How to design inconsistency-tolerant query answering methods is an important challenge in OBDI.
- In various applications of OBDI, e.g., enterprise information systems, there is the need to provide the user with update facilities. Obviously, the updates should be expressed at the level of the ontology, and the main challenge

is to design techniques for translating the update requests into appropriate updates on the source data, similarly to the notorious problem of view update.

– Finally, interesting research developments aim at going beyond query processing and exploring the power of OBDI in more general data governance tasks, including data quality checking, data cleaning, data profiling, and data provenance.

Cross-References

► Description Logics
► Incomplete Information
► Inconsistent Databases
► Information Integration
► Ontology
► View-Based Data Integration

Recommended Reading

1. Baader F, Calvanese D, McGuinness D, Nardi D, Patel-Schneider PF, editors. The description logic handbook: theory, implementation and applications. 2nd ed. Cambridge: Cambridge University Press; 2007.
2. Calì A, Gottlob G, Lukasiewicz T. A general datalog-based framework for tractable query answering over ontologies. J Web Semant. 2012;14(July):57–83.
3. Calvanese D, De Giacomo G, Lembo D, Lenzerini M, Rosati R. Tractable reasoning and efficient query answering in description logics: the DL-Lite family. J Autom Reason. 2007;39(3):385–429.
4. Ceri S, Gottlob G, Tanca L. Logic programming and databases. Berlin: Springer; 1990.
5. Chortaras A, Trivela D, Stamou GB. Optimized query rewriting for OWL 2 QL. In: Proceedings of the 23rd International Conference on Automated Deduction; 2011. p. 192–206.
6. Eiter T, Ortiz M, Simkus M, Tran T-K, Xiao G. Query rewriting for Horn-SHIQ plus rules. In: Proceedings of the 26th National Conference on Artificial Intelligence; 2012.
7. Gottlob G, Kikot S, Kontchakov R, Podolskii VV, Schwentick T, Zakharyaschev M. The price of query rewriting in ontology-based data access. Artif Intell. 2014;213(Aug):42–59.
8. Imielinski T, Lipski W Jr. Incomplete information in relational databases. J ACM. 1984;31(4):761–91.
9. Kontchakov R, Lutz C, Toman D, Wolter F, Zakharyaschev M. The combined approach to ontology-based data access. In: Proceedings of the 22nd International Joint Conference on AI; 2011. p. 2656–61.
10. Leitsch A. The resolution calculus. Berlin: Springer; 1997.
11. Lenzerini M. Data integration: a theoretical perspective. In: Proceedings of the 21st ACM SIGACT-SIGMOD-SIGART Symposium on Principles of Database Systems; 2002. p. 233–46.
12. Levy AY, Rousset M-C. Combining Horn rules and description logics in CARIN. Artif Intell. 1998;104(1–2):165–209.
13. Pérez-Urbina H, Horrocks I, Motik B. Efficient query answering for OWL 2. In: Proceedings of the 8th International Semantic Web Conference; 2009. p. 489–504.
14. Poggi A, Lembo D, Calvanese D, De Giacomo G, Lenzerini M, Rosati R. Linking data to ontologies. J Data Semant. 2008;X(3):133–73.
15. Rosati R, Almatelli A. Improving query answering over *DL-Lite* ontologies. In: Proceedings of the 12th International Conference on Principles of Knowledge Representation and Reasoning; 2010. p. 290–300.

Open Database Connectivity

Changqing Li
Duke University, Durham, NC, USA

Synonyms

ODBC

Definition

Open Database Connectivity (ODBC) [1] is an Application Programming Interface (API) specification to use database management systems (DBMS). The ODBC API is a library of functions for the ODBC-enabled applications to connect any ODBC-driver-available database, execute Structured Query Language (SQL) statements, and retrieve results. ODBC is independent of programming languages, database systems and operating systems.

Key Points

ODBC (pronounced as separate letters), is a standard database access method developed by the SQL Access Group in 1992. Its objective is to make any application to access any data regardless of the database management systems. To achieve this objective, ODBC inserts a database driver as a middle layer between an application and the DBMS, the purpose of which is to translate the application queries to commands understood by the DBMS. In practice, both the application and the DBMS must be ODBC-compliant; that is, the application must be capable of issuing ODBC commands and the DBMS must be capable of responding to them.

A procedural API is offered by the ODBC specification for using SQL queries to access data. One or more applications will be contained in an implementation of ODBC, a core ODBC library, and one or more "database drivers." Independent of the applications and DBMS, the core library acts as an "interpreter" between the applications and the database drivers, whereas the database drivers contain the DBMS-specific details. Thus applications can be written to use standard types and features without concerning the specifics of each DBMS that the applications may encounter. Similarly, database driver implementors only need to know how to attach to the core library. This makes ODBC modular.

ODBC operates with a variety of operating systems and drivers existing for relational database as well as non-relational data such as spreadsheets, text and XML files.

Cross-References

- ▶ .NET Remoting
- ▶ Database Adapter and Connector
- ▶ Interface
- ▶ Java Database Connectivity
- ▶ Web 2.0/3.0
- ▶ Web Services

Recommended Reading

1. Geiger K. Inside ODBC. Redmond: Microsoft Press; 1995.

Open Nested Transaction Models

Alejandro Buchmann
Darmstadt University of Technology, Darmstadt, Germany

Synonyms

Advanced transaction models; Extended Transaction Models and the ACTA Framework

Definition

Open nested transactions are hierarchically structured transactions with relaxed ACID properties. Individual subtransactions may commit independently before the complete top level transaction commits. Therefore, conventional rollback is not possible and the effects of a commited subtransaction have to be compensated if the top level transaction aborts. Depending on the particular open nested transaction model, subtransactions may be vital or non-vital and may have alternative or contingency subtransactions. Open nested transaction models are characterized through relaxed visibility rules, abort and commit dependencies.

Historical Background

Open nested transaction models evolved in the 1980s in response to two major sets of requirements: the needs of federated multidatabase systems integrating autonomous legacy database systems, and the demands of long running, cooperative processes for higher levels of concurrency while maintaining some

O

of the main benefits of transactional processes. Extended transaction models influenced the tightly coupled transaction model of distributed object systems and the activity model of the OMG, as well as the transaction and coordination models of Web Services.

Foundations

Transaction models are characterized by the structure of its transactions, the commit and abort dependencies and the visibility rules among transactions.

Nested transactions are hierarchically structured transactions consisting of a top level transaction and subtransactions that may themselves be tree-structured. Typical of the execution model of nested transactions is the fact that higher level transactions do not execute while their subtransactions are active. The order of execution of subtransactions can be either sequential or parallel. The closed nested transaction model proposed by Moss does not specify an execution order and preserves the atomicity, consistency, isolation and durability properties of traditional flat transactions. The commit dependencies of closed nested transactions specify that all subtransactions must commit to their immediate ancestor and the top level transaction can only commit after all the subtransactions terminated. The abort dependencies specify that the whole transaction tree must be aborted if the top level transaction aborts while the abort of a subtransaction can be handled by the immediate ancestor transaction. The visibility rules among transactions and subtransactions specify that subtransactions may see changes of other subtransactions only once they are committed to the common ancestor. Changes of a nested transaction become visible to the outside world only after the top level transaction commits. Since changes of committed subtransactions are made visible only after the top level transaction commits, they can be undone through conventional roll-back.

Open nested transaction models are also based on hierarchically structured transactions. However, depending on the particular transaction model, the subtransactions may be of different types. Different types of subtransactions require different commit and/or abort dependencies. The visibility rules are also relaxed with respect to those of closed nested transaction models.

Open nested transactions may consist of the following component transactions:

- One top level transaction that has mostly a coordination function
- Vital subtransactions that must all commit for the top level transaction to be allowed to commit
- Non-vital subtransactions of which one or more may abort without causing the top level transaction to fail
- Contingency transactions are alternative subtransactions that often are executed by different autonomous systems or service providers
- Compensating transactions that must be defined to undo the changes of subtransactions that may have committed but must be undone
- Triggers that are executed as subtransactions and may execute either immediately, deferred before the commit of the triggering transaction or as detached transactions.

Open nested transaction models were defined for long running transactions and for transactions executing on federated autonomous (legacy) systems. Therefore, the degree of control of the coordinating top level transaction over the execution of the subtransactions is reduced compared to closed nested transactions running on a single database management system. Distributed short lived transactions and closed nested transactions are typically implemented through a two-phase commit protocol. In a two-phase commit protocol the first phase serves to reach agreement among the participants whether to commit or to abort, and once consensus to commit has been reached, the commit is carried out in the second phase. This protocol requires holding all the resources required by all the subtransactions through the negotiation phase until the global commit. This approach is not feasible for long running transactions because of performance reasons and in federated multidatabase systems because of the autonomy of the participating database systems.

The coordinating top level transaction in an open nested transaction model cannot secure the resources of the participating systems that execute subtransactions as autonomous individual short transactions until all participating systems have executed their subtransactions and are ready to commit. Therefore, subtransactions may commit immediately upon completion and their results are thus visible to the outside world before the top level transaction commits. Compensating transactions execute the semantically inverse operation of the committed subtransaction but do not guarantee that the exact initial state can be restored since other transactions may have executed in the interim.

The commit dependencies of open nested transaction models depend on whether they distinguish between vital and non-vital subtransactions or not. A commit dependency exists between the top level transaction and all vital subtransactions, i.e., the coordinating top level transaction may only commit if all the vital subtransactions committed. This is the default case if non-vital subtransactions are not provided. The abort of a non-vital transaction does not affect the possibility to commit the top level transaction.

The abort dependencies of open nested transaction models are the same as for closed nested transactions: the abort of the parent transaction causes the abort and undo (roll-back or compensation) of the subtransaction. The abort of a non-vital subtransaction is handled by the parent transaction, the abort of a vital transaction causes the abort of the parent and all siblings.

Contingency transactions represent alternative actions. For example, if the top-level transaction represents the booking of a trip consisting of a roundtrip flight, a hotel and a rental car booking, two different flights on different airlines may be defined as a subtransaction and a contingency transaction. Contingency transactions may only commit if the primary subtransaction aborts. Open nested transaction models that do not provide contingency transactions must implement this functionality as application-specific code in the corresponding parent transaction or the top-level transaction.

Triggers have become commonplace mechanisms in commercial relational databases and execute as subtransactions according to the semantics of closed nested transactions. The order of execution may be immediate or deferred, meaning that the trigger executes either at the point of occurrence of the triggering event or at the end of the triggering transaction, respectively. Some extended transaction models allow the triggering of subtransactions as detached or autonomous transactions, i.e., following the semantics of open nested transactions. This, however, implies all the problems resulting from the violation of the isolation and atomicity properties and requires the definition of compensating transactions for those triggers.

The concepts developed as part of extended transaction models resulted in the definition of the CORBA Activity Service Framework. The Activity Service is a general purpose event signalling mechanism that can be used to program activities to coordinate themselves according to the transaction model under consideration. The Activity Service has also been incorporated in the J2EE framework but is not widely used.

Key Applications

The main application areas for open nested transaction models are multidatabase systems in which autonomous systems are loosely coupled. Web Services with their loose coupling, distribution and often long running interactions prompted renewed interest in extended transaction models and open nested transactions. Applications based on Web Services span the whole range of e-business applications. Mobile commerce is another key application domain for open nested transactions. Proposed mobile transaction models are instances of open nested transaction models, e.g., the model proposed by Chrysanthis and extended in Kangaroo Transactions.

Future Directions

Several proposals for long running activities based on Web Services have been advanced by different consortia. The OASIS Business Transaction Protocol was proposed in 2001 by a consortium consisting of HP, BEA, and Oracle. Their model provides for the execution of business logic between the two phases of a 2 phase commit protocol. Arjuna, Oracle, Sun Microsystems, IONA Technologies and Fujitsu in 2003 founded the OASIS Web Services Composite Application Framework that defines three transaction protocols, each aimed at a specific use case. The WS-TX builds on and extends the web Services Coordination specification and provides two kinds of transaction models. These are not meant to cover all possible use cases and can be extended with the semantics of other extended transaction models as required by emerging applications.

Atomic Transactions are meant for short lived transactional interactions within trusted domains and provide full isolation (no dirty reads and repeatable reads), atomicity (well-formedness and two-phase commit protocol), and durability. For loosely coupled, long running interactions Web Services Transactions can use the Business Activity protocol, a more flexible transaction and coordination protocol that relaxes the ACID properties and draws heavily on previous research on open nested transaction models.

Business Activities are designed for long-lived transactions. They are based on the original Sagas open nested transaction model. Services are treated as Sagas and if there is the appropriate compensation behavior defined, they can execute the undo behavior if so instructed by the Business Activity. The responsibility of writing correct compensation services to ensure consistency rests with the developer of a service.

Business Activities consist of (possibly nested) Saga-like service invocations. Such scopes can handle errors, i.e., the abort of a task, through application logic and continue processing without globally aborting. Upon completion a child subtransaction can either leave the scope of the Business Activity or it can signal the parent that its work can be compensated later. In any event, the visibility rules are such that the results of child tasks can be seen by the outside world. Business Activities must record application state and keep a record of all sent and received messages, all request messages must be acknowledged, and requests and responses are decoupled. Two different protocols exist for Business Activities: Business Agreement With Coordinator Complete and Business Agreement With Participant Complete. The main difference between these two protocols is that in the former a participating task may not leave the scope of the Business Activity unilaterally and must wait for it to terminate. In case of abort, the subtask must compensate. In the latter a task may leave the scope of the Business Activity unilaterally.

Cross-References

▶ Compensating Transactions
▶ ConTract
▶ CORBA
▶ Distributed Database Systems
▶ Distributed Transaction Management
▶ Extended Transaction Models and the ACTA Framework
▶ Flex transactions
▶ Loose Coupling
▶ Multilevel Transactions and Object-Model Transactions
▶ Nested Transaction Models
▶ Orchestration
▶ Sagas
▶ Transaction
▶ Web Transactions
▶ Workflow Transactions

Recommended Reading

1. Buchmann A, Özsu MT, Hornick M, Georgakopoulos D, Manola F. A transaction model for active distributed object systems. In: Elmagarmid AK, editor. Database transaction models for advanced applications. Los Altos: Morgan Kaufmann Publishers; 1992.
2. Cabrera LF, Copeland G, Feingold M, et al. Web services atomic transaction (WS-AtomicTransaction). Version 1.0. 2005. Available at: http://download. boulder.ibm.com/ibmdl/pub/software/dw/specs/ws-tx/WS-AtomicTransaction.pdf.

3. Cabrera LF., Copeland G, Feingold M, et al. Web services business activity framework (WS BusinessActivity). Version 1.0. 2005. Available at: http://download.boulder.ibm.com/ibmdl/pub/software/dw/specs/ws-tx/WS-BusinessActivity.pdf.

4. Cabrera LF, Copeland G, Feingold M, et al. Web services coordination (WS-Coordination). Version 1.0. 2005. Available at: http://download.boulder.ibm.com/ibmdl/pub/software/dw/specs/ws-tx/WS-Coordination.pdf.

5. Chrysanthis PK. Transaction processing in a mobile environment. In: Proceedings of the IEEE Workshop on Advances in Parallel and Distributed Systems; 1993. p. 77–82.

6. Chrysantis P, Ramamritham K. ACTA: the saga continues. In: Elmagarmid AK, editor. Database transaction models for advanced applications. Los Altos: Morgan Kaufmann Publishers; 1992.

7. Dunham MH, Helal A, Balakrishnan S. A mobile transaction model that captures both data and movement behavior. Mob Netw Appl. 1997;2(2):149–62.

8. Elmagarmid AK. Database transaction models for advanced applications. Los Altos: Morgan Kaufmann Publishers; 1992.

9. Garcia-Molina H, Salem K. SAGAS. In: Proceedings of the ACM SIGMOD International Conference on Management of Data; 1987. p. 249–59.

10. Houston I, Little M, Robinson I, Shrivastava SK, Wheater SM. The CORBA activity service framework for supporting extended transactions. In: Proceedings of the IFIP/ACM International Conference on Distributed Systems Platforms; 2001. p. 197–215

11. Little M. A history of extended transactions. Available at: http://www.infoq.com/articles/History-of-Extended-Transactions.

12. Moss E. Nested transactions. Cambridge: MIT Press; 1985.

13. Weikum G, Schek HJ. Concepts and applications of multilevel transactions and open nested transactions. In: Elmagarmid AK, editor. Database transaction models for advanced applications. Los Altos: Morgan Kaufmann Publishers; 1992.

Operator-Level Parallelism

Nikos Hardavellas[1] and Ippokratis Pandis[1,2]
[1]Carnegie Mellon University, Pittsburgh, PA, USA
[2]Amazon Web Services, Seattle, WA, USA

Synonyms

Inter-operator parallelism

Definition

Operator-level parallelism (or inter-operator parallelism) is a form of intra-query parallelism obtained by executing concurrently several operators of the same query. By contrast, intra-operator parallelism is obtained by executing the same operator on multiple processors, with each instance working on a different subset of data.

Historical Background

Parallelism has been a key focus of database research since the 1970s. For example, as early as 1978 Teradata was building highly-parallel database systems and quietly pioneered many of the ideas on parallel query execution [5]. However, the intra-query parallelism employed by these early systems was mostly intra-operator or independent parallelism (see Classes of Parallelism below). Gamma [4] was one of the first database systems that allowed operator-level parallelism through pipelining.

Foundations

Parallel processing uses multiple processors cooperatively to improve the performance of application programs. With relations growing larger and queries becoming more complex, parallel processing is an increasingly attractive option for improving the performance of database management systems. The widespread adoption of the relational database model has enabled the parallel execution of relational queries, as these queries are composed of uniform operators applied to uniform streams of data. Each operator produces a new relation, so the operators can be composed into highly parallel dataflow graphs. At the same time, multiprocessor systems and high-speed interconnection networks have become mainstream, providing an excellent basis for parallel execution.

Classes of Parallelism

Parallelism in the evaluation of database queries is classified into two main categories: inter-query parallelism (see Inter-query Parallelism), in which different queries execute on different processors to improve the overall throughput of the system, and intra-query parallelism, in which several processors cooperate for the faster execution of a single query. Intra-query parallelism is further classified into intra-operator and inter-operator parallelism. Intra-operator parallelism (see Intra-operator Parallelism) is obtained by executing the same operator on multiple processors, with each instance working on a different subset of data. Operator-level parallelism (or inter-operator parallelism), is obtained by executing concurrently several operators of the same query. This latter form of parallelism is the subject of this chapter.

Operator-level parallelism is in two forms: independent parallelism and pipelined parallelism. Independent parallelism (or bushy parallelism) is achieved when there is no dependency between the operators executed in parallel. For example, consider a simple query plan with two select operators and a join, that it is not nested-loops. The select operators are independent of each other and can execute concurrently, thereby exhibiting independent parallelism. Algebraically, independent parallelism can be expressed by a relation of the form $f(g(X), h(Y))$, where X and Y are relations and f, g, and h are relational operators. In this example, g and h exhibit independent parallelism.

Because the operators participating in bushy parallelism are independent, they do not directly affect the execution of one another. Interference is only indirect, e.g., due to the concurrent use of shared resources like disks, caches, or main memory bandwidth. Thus, independent parallelism is simpler to employ as it is easier to schedule the execution of the participating independent operators, and it has the potential to deliver high performance improvements.

Alternatively, operator-level parallelism can take the form of pipelined parallelism, also called dataflow parallelism. Pipelined parallelism can be achieved when the concurrent operators form producer/consumer pairs in which the consumer can start executing without requiring its entire input to be available. For example, consider the aforementioned simple query that consists of two select operators and a join. The select operator can execute in parallel with the join operator. However, they are not independent, because the intermediate results produced by the select are consumed by the subsequent join. Thus, the tuples output by the select can be pipelined to the join operator to be consumed immediately. This example illustrates a significant advantage of pipelined parallelism: intermediate results are used immediately and are not materialized, saving memory and disk accesses. Algebraically, pipelined parallelism can be expressed by a relation of the form $f(g(X))$, where X is a relation and f and g are relational operators. The operators that cannot produce tuples unless they have processed their entire input are called Stop-&-Go operators.

Effect of Query Plan Selection on Operator-Level Parallelism

The query plan determines the execution sequence of a query's operators. The selection of a query plan greatly affects the degree of attainable operator-level parallelism. To illustrate this point, and without loss of generality, let's assume a multi-way hash-join query with four joins: $A \times B \times C \times D \times E$ where A, B, C, D, E are relations and $\times$ is the join operator. The query plan is typically depicted graphically as a tree with vertices representing relations. Because every operator in the relational model defines a new relation, the operators in the internal vertices denote the relation they represent. If an operator Y takes relation X as one of its inputs, then a directed edge connects X to Y in the tree representation.

Three forms of query execution trees are explored in the literature: left-deep trees, right-deep trees, and bushy trees. Fig. 1 shows the query execution trees for the example multi-way join query used above. Left-deep trees and right-deep trees represent the two extreme strategies of query execution, while bushy trees claim a middle ground.

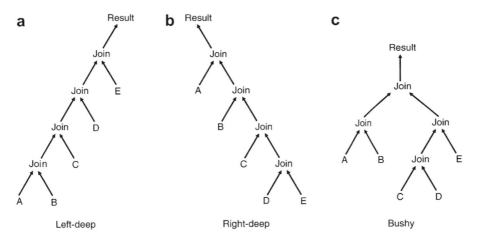

Operator-Level Parallelism, Fig. 1 (**a**) Left-deep, (**b**) right-deep, and (**c**) bushy query plans

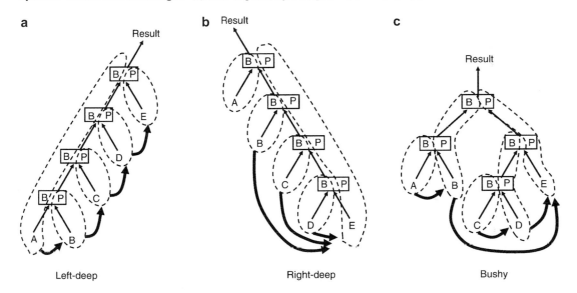

Operator-Level Parallelism, Fig. 2 Operator dependencies for (**a**) left-deep, (**b**) right-deep, and (**c**) bushy plans

To compare the trade-offs between the alternative query plans, Fig. 2 shows the execution dependencies between the operators of each execution strategy in Fig. 1. The execution dependencies are shown using operator dependency graphs [10]. The dotted lines encircle the operators amenable to pipelined parallelism. The bold directed arcs between subgraphs show which sets of operators must be executed before other sets of operators are executed, thereby determining the maximum level of parallelism and resource requirements (e.g., memory) for the query. As discussed in [10] hash joins have two distinct

phases, the build and the probe phase. Since the first phase must completely precede the second, the hash joins in Fig. 2 can be viewed as if consisting of two operators, the build and the probe operator.

The operator dependency graph of the left-deep query plan shows that only a scan, the build phase of a join, and the probe phase of a join can execute in parallel. Thus, although the left-deep query plan has low memory requirements (it needs enough memory to fit the hash tables of two joins) it offers only limited pipelined parallelism and no independent parallelism. In contrast, the operator dependency graph for the right-deep query plan shows that significant operator-level parallelism is available: all scans but one have a producer/consumer relationship with the build phase of the subsequent hash join, thereby exhibiting pipelined parallelism, while the scan/build pairs are independent of one another so they exhibit independent parallelism. However, the high degree of parallelism comes at the expense of high shared resource pressure. The hash tables for all joins should fit in main memory simultaneously, or risk spilling to disk.

Finally, the operator dependency graph for the bushy query plan has characteristics that are between the left-deep and the right-deep query plans. The bushy plan enables independent parallelism, albeit at a lower degree than the right-deep plan, as about half the scans can proceed in parallel. However, the bushy plan imposes lower pressure on shared resources than the right-deep plan, because fewer operators execute in parallel. Bushy query plans allow the formation of deeper pipelines, some of which extend all the way from a leaf to the root of the tree. Thus, bushy trees enable pipelined parallelism as well, but it may be harder to balance the load within their deeper pipelines due to execution skew.

Because bushy plans achieve a balance between pipelined parallelism, independent parallelism, and resource utilization, researchers further investigated their applicability in improving query execution. For example, segmented right-deep trees [3] (bushy trees of right-deep subtrees) have been shown to outperform their left-deep and right-deep counterparts.

Other Factors Limiting Operator-Level Parallelism

The selection of the query plan determines the degree of available operator-level parallelism. This section discusses factors that limit the effectiveness of operator-level parallelism, given a query plan. Among other things, the discussion in this section touches on issues of load balancing and processor allocation.

The operators participating in independent parallelism interfere only indirectly through the concurrent use of shared resources. The factors limiting the benefit of independent parallelism are the constraints imposed by the hardware resources. All relations that execute in parallel produce intermediate results which increase the data footprint of the application, resulting in higher cache miss rates and higher memory pressure. The larger data footprint, in turn, may oversubscribe memory bandwidth or induce more spills to disk if the relations do not fit in main memory.

Resource contention affects pipelined parallelism as well, but to a lesser degree because the intermediate data in pipelined parallelism are short lived as they are consumed immediately after their production. The benefits of pipelined parallelism are generally limited by three factors [5]: (i) relational pipelines are rarely very long - a chain of length ten is unusual. (ii) some relational operators are blocking operators, i.e., they do not emit their first output until they have consumed all their inputs. Sort and the partitioning phase of hash join are examples of blocking relational operators. Such operators cannot be pipelined, and (iii) there are dependencies between the operators participating in a pipeline. Often, the execution cost of one operator is much greater than the others, a phenomenon referred to as execution skew. In this case, the performance of the pipelined execution is dominated by the slowest operator, which significantly limits parallelism.

The execution skew also gives rise to startup/teardown execution delays: processors assigned to operators at the end of a pipeline are idle at the beginning of the computation, whereas processors assigned to operators at the beginning of a pipeline are idle towards the end of the computation. It is important to note here that data skew may induce execution skew in some cases. For example, in a sort-merge join with data skew, some sort partitions may be much larger than others, creating execution skew.

A potential solution to execution skew is to predict the execution load for each operator and schedule them accordingly across the parallel processors. However, the predictions may fail as the costs are estimated in the query optimization phase using typically inaccurate cost models and statistics.

The assignment of processors to operators and their scheduling is an important and hard problem that affects all forms of operator-level parallelism. It is an optimization problem that attempts to utilize all the available processors efficiently to minimize the execution time of a query. Sometimes operators may need to be scheduled as a team (e.g., producer/consumer pairs), while other times gang scheduling should be avoided (e.g., scheduling together the first and the last operator of a deep pipeline would leave the last operator mostly idle). Scheduling is easier when there are no dependencies between the operators executing in parallel, in which case load balancing is of primary concern.

The processor allocation is based on the selection of the query plan and estimates on the execution cost of each operator. If the execution cost of some operators is much higher than others, the system may be subject to fragmentation: after a sequence of processor allocations and releases there may be a few processors left idle and rebalancing the workload dynamically is not always possible or beneficial. These cases may benefit from the concurrent employment of multiple forms of parallelism (see next section). However, the application of multiple forms of parallelism adds an extra dimension to the processor allocation problem, making it harder to solve.

Relation to Inter-query and Intra-operator Parallelism

Operator-level parallelism is orthogonal to inter-query and intra-operator parallelism and can work synergistically with them to improve performance even further. For example, if there is imbalance in the execution times of a query's operators and there are free processors, intra-operator parallelism can be applied to split a long-running operator into multiple ones, each executing on a smaller subset of data. This will allow for faster execution of the expensive operators and may balance the execution times of operators participating in a pipeline, avoiding execution skew.

For a more comprehensive treatment of operator-level parallelism, the interested reader is referred to [5, 10, 11].

Key Applications

Several parallel database systems have been developed that utilize operator-level parallelism to improve performance. Systems built in academic institutions include GAMMA [4], BUBBA [2], Volcano [6], MonetDB/X100 [1], and StagedDB [7]. Commercial systems that support operator-level parallelism include Oracle [9] and IBM DB2 [8].

Cross-References

▶ Data Skew
▶ Execution Skew
▶ Inter-query Parallelism
▶ Intra-operator Parallelism
▶ Parallel Hash Join, Parallel Merge Join, Parallel Nested Loops Join
▶ Parallel Query Processing
▶ Pipelining
▶ Query Plan
▶ Stop-&-Go Operator

Recommended Reading

1. Boncz P, Zukowski M, Nes N. MonetDB/X100: hyper-pipelining query execution. In: Proceedings of the 2nd Biennial Conference on Innovative Data Systems Research; 2005. p. 225–37.
2. Boral H. Prototyping bubba: a highly parallel database system. IEEE Trans Knowl Data Eng. 1990;2(1):4.
3. Chen MS, Lo M, Yu PS, Young HC. Using segmented right-deep trees for the execution of pipelined hash joins. In: Proceedings of the 18th International Conference on Very Large Data Bases; 1992. p. 15–26.
4. DeWitt DJ, Gray J. Parallel database systems: the future of high-performance database computing. Commun ACM. 1992;35(6):85–98.
5. DeWitt DJ., Gerber RH, Graefe G, Heytens ML, Kumar KB, Muralikrishna M. GAMMA - a high performance dataflow database machine. In: Proceedings of the 12th International Conference on Very Large Data Bases; 1986. p. 228–37.
6. Graefe G. Volcano - an extensible and parallel query evaluation system. IEEE Trans Knowl Data Eng. 1994;6(1):120–35.
7. Harizopoulos S, Ailamaki A. Staged D.B.: designing database servers for modern hardware. IEEE Data Eng Bull. 2005;28(2):11–6.

8. IBM Corp. DB2 Version 9 Performance Guide. Part no. SC10-4222-00. 2006.
9. Oracle Corp. Oracle Database Data Warehousing Guide. 10 g Release 1 (10.1). Part no. B10736–01. 2003.
10. Schneider DA. DeWitt DJ. Tradeoffs in processing complex join queries via hashing in multiprocessor database machines. In: Proceedings of the 12th International Conference on Very Large Data Bases; 1986. p. 469–80.
11. Yu PS, Chen MS, Wolf JL, and Turek JJ. Parallel query processing. In: Adam N, Bhargava B, editors. Advanced database systems. LNCS, vol. 759. Berlin: Springer; 1993. p. 239–58.

Opinion Mining

Bing Liu
University of Illinois at Chicago, Chicago, IL, USA

Synonyms

Sentiment analysis

Definition

Given a set of evaluative text documents D that contain opinions (or sentiments) about an object, opinion mining aims to extract attributes and components of the object that have been commented on in each document $d \in D$ and to determine whether the comments are positive, negative or neutral.

Historical Background

Textual information in the world can be broadly classified into two main categories, *facts* and *opinions*. Facts are objective statements about entities and events in the world. Opinions are subjective statements that reflect people's sentiments or perceptions about the entities and events. Much of the existing research on text information processing has been (almost exclusively) focused on mining and retrieval of factual information, e.g., information retrieval, Web search, and many other text mining and natural language processing tasks. Little work has been done on the processing of opinions until only recently. Yet, opinions are so important that whenever one needs to make a decision one wants to hear others' opinions. This is not only true for individuals but also true for organizations.

One of the main reasons for the lack of study on opinions is that there was little opinionated text before the World Wide Web. Before the Web, when an individual needs to make a decision, he/she typically asks for opinions from friends and families. When an organization needs to find opinions of the general public about its products and services, it conducts surveys and focused groups. With the Web, especially with the explosive growth of the user generated content on the Web, the world has changed. One can post reviews of products at merchant sites and express views on almost anything in Internet forums, discussion groups, and blogs, which are collectively called the *user generated content*. Now if one wants to buy a product, it is no longer necessary to ask one's friends and families because there are plenty of product reviews on the Web that give the opinions of the existing users of the product. For a company, it may no longer need to conduct surveys, to organize focused groups or to employ external consultants in order to find consumer opinions or sentiments about its products and those of its competitors.

Finding opinion sources and monitoring them on the Web, however, can still be a formidable task because a large number of diverse sources exist on the Web and each source also contains a huge volume of information. In many cases, opinions are hidden in long forum posts and blogs. It is very difficult for a human reader to find relevant sources, extract pertinent sentences, read them, summarize them and organize them into usable forms. An automated opinion mining and summarization system is thus needed. *Opinion mining*, also known as *sentiment analysis*, grows out of this need.

Research on opinion mining started with identifying *opinion* (or *sentiment*) *bearing words*, e.g., great, amazing, wonderful, bad, and poor. Many researchers have worked on mining such words and identifying their *semantic orientations* (i.e., positive or negative). In [5], the authors identified several linguistic rules that can be exploited to identify opinion words and their orientations from a large corpus. This method has been applied, extended and improved in [3, 8, 12]. In [6, 9], a bootstrapping approach is proposed, which uses a small set of given seed opinion words to find their synonyms and antonyms in WordNet (http://wordnet.princeton.edu/). The next major development is sentiment classification of product reviews at the document level [2, 11, 13]. The objective of this task is to classify each review document as expressing a positive or a negative sentiment about an object (e.g., a movie, a camera, or a car). Several researchers also studied sentence-level sentiment classification [9, 14, 15], i.e., classifying each sentence as expressing a positive or a negative opinion. The model of feature-based opinion mining and summarization is proposed in [6, 10]. This model gives a more complete formulation of the opinion mining problem. It identifies the key pieces of information that should be mined and describes how a structured opinion summary can be produced from unstructured texts. The problem of mining opinions from comparative sentences is introduced in [4, 7].

Foundations

Model of Opinion Mining

In general, opinions can be expressed on anything, e.g., a product, a service, a topic, an individual, an organization, or an event. The general term *object* is used to denote the entity that has been commented on. An object has a set of *components* (or *parts*) and a set of *attributes*. Each component may also have its sub-components and its set of attributes, and so on. Thus, the object can be hierarchically decomposed based on the *part-of* relationship.

Definition (object): An *object O* is an entity which can be a product, topic, person, event, or organization. It is associated with a pair, (T, A), where T is a hierarchy or taxonomy of *components* (or *parts*) and *sub-components* of O, and A is a set of *attributes* of O. Each component has its own set of sub-components and attributes.

In this hierarchy or tree, the root is the object itself. Each non-root node is a component or sub-component of the object. Each link is a part-of relationship. Each node is associated with a set of attributes. An opinion can be expressed on any node and any attribute of the node.

However, for an ordinary user, it is probably too complex to use a hierarchical representation. To simplify it, the tree is flattened. The word "*features*" is used to represent both components and attributes. Using features for objects (especially products) is quite common in practice. Note that in this definition the object itself is also a feature, which is the root of the tree.

Let an evaluative document be d, which can be a product review, a forum post or a blog that evaluates a particular object O. In the most general case, d consists of a sequence of sentences $d = \langle s_1, s_2, ..., s_m \rangle$.

Definition (opinion passage on a feature): The *opinion passage* on a feature f of the object O evaluated in d is a group of consecutive sentences in d that expresses a positive or negative opinion on f.

This means that it is possible that a sequence of sentences (at least one) together expresses an opinion on an object or a feature of the object. It is also possible that a single sentence expresses opinions on more than one feature, e.g., "The picture quality of this camera is good, but the battery life is short."

Definition (opinion holder): The *holder* of a particular opinion is a person or an organization that holds the opinion.

In the case of product reviews, forum postings and blogs, opinion holders are usually the authors of the posts. Opinion holders are important in news articles because they often explicitly state the person or organization that holds a particular opinion [9]. For example, the opinion holder in

O

the sentence "John expressed his disagreement on the treaty" is "John."

Definition (semantic orientation of an opinion): The *semantic orientation* of an opinion on a feature *f* states whether the opinion is positive, negative or neutral.

Putting things together, a *model* for an object and a set of opinions on the features of the object can be defined, which is called the *feature-based opinion mining model*.

Model of Feature-Based Opinion Mining: An object O is represented with a finite set of features, $F = \{f_1, f_2,...,f_n\}$, which includes the object itself. Each feature $f_i \in F$ can be expressed with a finite set of words or phrases W_i, which are *synonyms*. That is, there is a set of corresponding synonym sets $W = \{W_1, W_2,...,W_n\}$ for the *n* features. In an evaluative document *d* which evaluates object O, an opinion holder *j* comments on a subset of the features $S_j \subseteq F$. For each feature $f_k \in S_j$ that opinion holder *j* comments on, he/she chooses a word or phrase from W_k to describe the feature, and then expresses a positive, negative or neutral opinion on f_k. The opinion mining task is to discover all these hidden pieces of information from a given evaluative document *d*.

Mining output Given an evaluative document *d*, the mining result is a set of quadruples. Each quadruple is denoted by (*H, O, f, SO*), where *H* is the opinion holder, *O* is the object, *f* is a feature of the object and *SO* is the semantic orientation of the opinion expressed on feature *f* in a sentence of *d*. Neutral opinions are ignored in the output as they are not usually useful.

Given a collection of evaluative documents *D* containing opinions on an object, three main technical problems can be identified (clearly there are more):

Problem 1: Extracting object features that have been commented on in each document $d \in D$.

Problem 2: Determining whether the opinions on the features are positive, negative or neutral.

Problem 3: Grouping synonyms of features (as different opinion holders may use different words or phrase to express the same feature).

*Digital_camera*_1:

```
Camera:
    Positive:  125        <individual review sentences>
    Negative:  7          <individual review sentences>
Feature: picture quality
    Positive:  123        <individual review sentences>
    Negative:  6          <individual review sentences>
Feature: size
    Positive:  82         <individual review sentences>
    Negative:  10         <individual review sentences>
...
```

Opinion Mining, Fig. 1 An example of a feature-based summary of opinions

Opinion Summary There are many ways to use the mining results. One simple way is to produce a *feature-based summary* of opinions on the object [6]. An example is used to illustrate what that means.

Figure 1 summarizes the opinions in a set of reviews of a particular digital camera, *digital_camera*_1. The opinion holders are omitted. In the figure, "CAMERA" represents the camera itself (the root node of the object hierarchy). One hundred and twenty-five reviews expressed positive opinions on the camera and seven reviews expressed negative opinions on the camera. "picture quality" and "size" are two product features. One hundred and twenty-three reviews expressed positive opinions on the picture quality, and only six reviews expressed negative opinions. The ⟨individual review sentences⟩ points to the specific sentences and/or the whole reviews that give the positive or negative comments about the feature. With such a summary, the user can easily see how existing customers feel about the digital camera. If he/she is very interested in a particular feature, he/she can drill down by following the ⟨individual review sentences⟩ link to see why existing customers like it and/or dislike it.

The summary in Fig. 1 can be easily visualized using a bar chart [10]. Figure 2a shows such a chart. In the figure, each bar above the *X*-axis gives the number of positive opinions on a feature (listed at the top), and the bar below the *X*-axis gives the number of negative opinions on the

Opinion Mining, Fig. 2
(**a, b**) Visualization of
feature-based opinion
summary and comparison

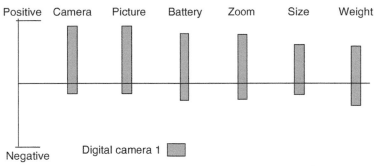

Feature-based summary of opinions on a digital camera

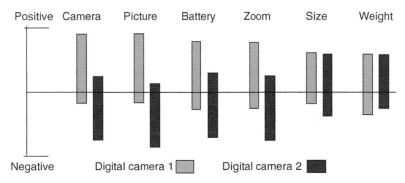

Opinion comparison of two digital cameras

same feature. Obviously, other visualizations are also possible. For example, one may only show the percentage of positive (or negative) opinions on each feature. Comparing opinion summaries of a few competing objects is even more interesting [10]. Figure 2b shows a visual comparison of consumer opinions on two competing digital cameras. One can clearly see how consumers view different features of each camera.

Sentiment Classification

Sentiment classification has been widely studied in the natural language processing (NLP) community [e.g., 2, 11, 13]. It is defined as follows: Given a set of evaluative documents D, it determines whether each document $d \in D$ expresses a positive or negative opinion (or sentiment) on an object. For example, given a set of movie reviews, the system classifies them into positive reviews and negative reviews.

This is clearly a classification learning problem. It is similar but also different from the classic topic-based text classification, which classifies documents into predefined topic classes, e.g., politics, sciences, and sports. In topic-based classification, topic related words are important. However, in sentiment classification, topic-related words are unimportant. Instead, opinion words that indicate positive or negative opinions are important, e.g., great, excellent, amazing, horrible, bad, worst, etc. There are many existing techniques. Most of them apply some forms of machine learning techniques for classification (e.g., [11]). Custom-designed algorithms specifically for sentiment classification also exist, which exploit opinion words and phrases together with some scoring functions [2, 13].

This classification is said to be at the document level as it treats each document as the basic information unit. Sentiment classification thus makes the following assumption: Each evaluative document (e.g., a review) focuses on a single object O and contains opinions of a single opinion holder. Since in the above opinion mining model an object O itself is also a feature (the root node of the object hierarchy), sentiment classification basically determines the semantic orientation of the opinion expressed on O in each evaluative document that satisfies the above assumption.

Apart from the document-level sentiment classification, researchers have also studied classification at the *sentence-level*, i.e., classifying each sentence as a subjective or objective sentence and/or as expressing a positive or negative opinion [9, 14, 15]. Like the document-level classification, the sentence-level sentiment classification does not consider object features that have been commented on in a sentence. Compound sentences are also an issue. Such a sentence often express more than one opinion, e.g., "The picture quality of this camera is amazing and so is the battery life, but the viewfinder is too small."

Feature-Based Opinion Mining

Classifying evaluative texts at the document level or the sentence level does not tell what the opinion holder likes and dislikes. A positive document on an object does not mean that the opinion holder has positive opinions on all aspects or features of the object. Likewise, a negative document does not mean that the opinion holder dislikes everything about the object. In an evaluative document (e.g., a product review), the opinion holder typically writes both positive and negative aspects of the object, although the general sentiment on the object may be positive or negative. To obtain such detailed aspects, going to the feature level is needed. Based on the model presented earlier, three key mining tasks are:

1. Identifying *object features*: For instance, in the sentence "The picture quality of this camera is amazing," the object feature is "picture quality." In [10], a supervised pattern mining method is proposed. In [6, 12], an unsupervised method is used. The technique basically finds frequent nouns and noun phrases as features, which are usually genuine features. Clearly, many information extraction techniques are also applicable, e.g., condi-

tional random fields (CRF), hidden Markov models (HMM), and many others.

2. Determining opinion orientations: This task determines whether the opinions on the features are positive, negative or neutral. In the above sentence, the opinion on the feature "picture quality" is positive. Again, many approaches are possible. A lexicon-based approach has been shown to perform quite well in [3, 6]. The lexicon-based approach basically uses opinion words and phrases in a sentence to determine the orientation of an opinion on a feature. A relaxation labeling based approach is given in [12]. Clearly, various types of supervised learning are possible approaches as well.

3. Grouping synonyms: As the same object features can be expressed with different words or phrases, this task groups those synonyms together. Not much research has been done on this topic. See [1] for an attempt on this problem.

Mining Comparative and Superlative Sentences

Directly expressing positive or negative opinions on an object or its features is only one form of evaluation. Comparing the object with some other similar objects is another. Comparisons are related to but are also different from direct opinions. For example, a typical opinion sentence is "The picture quality of camera x is great." A typical comparison sentence is "The picture quality of camera x is better than that of camera y." In general, a comparative sentence expresses a relation based on similarities or differences of more than one object. In English, comparisons are usually conveyed using the *comparative* or the *superlative* forms of adjectives or adverbs. The structure of a comparative normally consists of the stem of an adjective or adverb, plus the suffix *-er*, or the modifier "more" or "less" before the adjective or adverb. The structure of a superlative normally consists of the stem of an adjective or adverb, plus the suffix *-est.*, or the modifier "most" or "least" before the adjective or adverb. Mining of comparative sentences basically consists of identifying what features and objects are compared and which objected are preferred by their authors (opinion holders). Details can be found in [4, 7].

Key Applications

Opinions are so important that whenever one needs to make a decision, one wants to hear others' opinions. This is true for both individuals and organizations. The technology of opinion mining thus has a tremendous scope for practical applications.

Individual consumers If an individual wants to purchase a product, it is useful to see a summary of opinions of existing users so that he/she can make an informed decision. This is better than reading a large number of reviews to form a mental picture of the strengths and weaknesses of the product. He/she can also compare the summaries of opinions of competing products, which is even more useful.

Organizations and businesses Opinion mining is equally, if not even more, important to businesses and organizations. For example, it is critical for a product manufacturer to know how consumers perceive its products and those of its competitors. This information is not only useful for marketing and product benchmarking but also useful for product design and product developments.

Cross-References

▶ Text Mining

Recommended Reading

1. Carenini G, Ng R, Zwart E. Extracting knowledge from evaluative text. In: Proceedings of the 3rd International Conference on Knowledge Capture; 2005.
2. Dave D, Lawrence A, Pennock D. Mining the peanut gallery: opinion extraction and semantic classification of product reviews. In: Proceedings of the 12th International World Wide Web Conference; 2003.
3. Ding X, Liu B, Yu P. A holistic lexicon-based approach to opinion mining. In: Proceedings of the 1st ACM International Conference on Web Search and Data Mining; 2008.
4. Ganapathibhotla G, Liu B. Identifying preferred entities in comparative sentences. In: Proceedings of the 22nd International Conference on Computational Linguistics; 2008.
5. Hatzivassiloglou V, McKeown K. Predicting the semantic orientation of adjectives. In: Proceedings of the 8th Conference of the European Chapter of the Association Computational Linguistics; 1997.
6. Hu M, Liu B. Mining and summarizing customer reviews. In: Proceedings of the 10th ACM SIGKDD International Conference on Knowledge Discovery and Data Mining; 2004.
7. Jindal N, Liu B. Mining comparative sentences and relations. In: Proceedings of the 21st National Conference on Artificial Intelligence and 18th Innovative Applications of Artificial Intelligence Conference; 2006.
8. Kanayama H, Nasukawa T. Fully automatic lexicon expansion for domain-oriented sentiment analysis. In: Proceedings of the 2006 Conference on Empirical Methods in Natural Language Processing; 2006.
9. Kim S, Hovy E. Determining the sentiment of opinions. In: Proceedings of the 20th International Conference on Computational Linguistics; 2004.
10. Liu B, Hu M, Cheng J. Opinion observer: analyzing and comparing opinions on the web. In: Proceedings of the 14th International World Wide Web Conference; 2005.
11. Pang B, Lee L, Vaithyanathan S. Thumbs up? Sentiment classification using machine learning techniques. In: Proceedings of the 2002 Conference on Empirical Methods in Natural Language Processing; 2002.
12. Popescu A-M, Etzioni O. Extracting product features and opinions from reviews. In: Proceedings of the 2005 Conference on Empirical Methods in Natural Language Processing; 2005.
13. Turney P. Thumbs up or thumbs down? Semantic orientation applied to unsupervised classification of reviews. In: Proceedings of the 40th Annual Manufacturing of Association Computational Linguistics; 2002.
14. Wiebe J, Riloff E. Creating subjective and objective sentence classifiers from unannotated texts. In: Proceedings international conference on intelligent text processing and computational linguistics. 2005.
15. Wilson T, Wiebe J, Hwa R. Just how mad are you? Finding strong and weak opinion clauses. In: Proceedings of the 19th National Conference on Artificial Intelligence and 16th Innovative Applications of Artificial Intelligence Conference; 2004.

O

Optimistic Replication and Resolution

Marc Shapiro
Inria Paris, Paris, France
Sorbonne-Universités-UPMC-LIP6, Paris,
France

Synonyms

Asynchronous Replication; Lazy replication;
Optimistic replication; Reconciliation-based data
replication

The term "optimistic replication" is prevalent in
the distributed systems and distributed algorithms
literature. The database literature prefers "lazy
replication."

Definition

Data replication places physical copies of
a shared logical item onto different sites.
Optimistic replication (OR) [17] allows a
program at some site to read or update the
local replica at any time. An update is *tentative*
because it may conflict with a remote update.
Such conflicts are resolved after the fact, in the
background. Replicas may *diverge* occasionally
but are expected to converge eventually (see
"▸ Eventual Consistency").

OR avoids the need for distributed coordination prior to using an item. It allows a site to
execute even when remote sites have crashed,
when network connectivity is poor or expensive,
or while disconnected from the network.

The defining characteristic of OR is that any
communication between sites occurs in the background, after local commitment, i.e., off the critical path of the application.

OR enables parallelism, and updates occur
and propagate quickly. The OR approach is well
adapted to distributed databases over slow or
failure-prone networks, and OR is essential to be
able to access remote data with high availability.

Prominent examples include *geo-replication* (see
"▸ Multi-datacenter Consistency Properties")
and mobile computing scenarios. Indeed, the
"▸ CAP Theorem" states that, in a network that
is prone to disconnection, it is not possible to
ensure *both* strong consistency and availability.
When availability is paramount, for instance in e-
commerce applications, this leads to the choice of
the weak consistency levels (such as "▸ Eventual
Consistency") supported by OR.

Disconnected operation, the capability to
compute while disconnected from a data source,
e.g., in mobile computing, requires OR. In
computer-supported cooperative work, OR
enables a user to temporarily insulate himself
from other users. In *cloud computing* OR
enables the system to remain available for reads
and writes even when the network is slow or
partitioned away.

Historical Background

(The vocabulary used in this history is defined in
Section "Foundations.")

The first historical instance of OR is Johnson's and Thomas's Last-Writer-Wins replicated
database (1976).

Usenet News (1979) supports a large-scale
ever-growing database of (read-only) items,
posted by users all over the world. A Usenet
site connects infrequently (e.g., daily) with its
peers. New items are flooded to other sites
and are delivered in arbitrary order. Users
occasionally observe ordering anomalies, but this
is not considered a problem. However, system
administrators must deal manually with conflicts
over administrative operations.

In 1984, Wuu and Bernstein's replicated mutable key-value-pair database uses an operation
log, transmitted by an *anti-entropy* protocol: site
A sends to site B only the tail of A's log that B has
not yet seen [23]. Concurrent operations either
commute or have a natural semantic order; non-
concurrent operations execute in happens-before
order.

The Lotus Notes system (1988) supports cooperative work between mobile enterprise users.

It replicates a database of discrete items in a peer-to-peer manner. Notes is state-based and uses a Last-Writer Wins policy. A deleted item is replaced by a *tombstone*.

Several file systems, designed in the early 1990s to support disconnected work, e.g., Coda [9], are state based and use version vectors for conflict detection. Conflicts over some specific object types (e.g., directories or mailboxes) cause automatic resolver programs to run. The others must be resolved manually.

The Computer-Supported Cooperative Work (CSCW) community invented (1989) a form of OR called Operational Transformation (OT). Conflicting operations are transformed, by modifying their arguments, in order to execute in arbitrary but causal order [20].

Golding (1992) [5] studies a replicated database of mutable key-value pairs. This system purges an operation from the log when it can prove that it was delivered to all sites. Consistency is ensured by defining a total order of operations.

Bayou (1994–1997) is a seminal general-purpose database for mobile users [13]. Bayou is operation based and uses an anti-entropy protocol. Each site executes transactions in arbitrary order; transactions remain tentative. The eventual serialization order is the order of execution at a designated *primary* site. Other sites roll back their tentative state, and re-execute committed transactions in commit order.

In 1996, Gray et al. argued that OR databases cannot scale [7], because conflict reconciliation is expensive, conflict probability rises as the third power of the number of nodes, and the wait probability further increases quadratically with disconnection time.

In 1999, Breitbart et al. [2] describe a partially replicated database that uses a form of OR. Each item has a designated primary site and may be replicated at any number of secondary sites. A read may occur at a secondary site but a write must occur on the primary. It follows that write transactions update a single site. If transactions are serializable at each site, and update propagation is restricted to avoid ordering anomalies, then transactions are serializable despite lazy propagation.

Cloud computing has sparked a new interest in OR. In order to avoid synchronization, which is bad for performance and for fault-tolerance, AP (Available under Partition) databases are designed in an OR style, supporting only weakly-consistent key-value storage, such as Last-Writer Wins (Cassandra) or Multi-Value Register (Dynamo).

Geo-replication (see "▶ Multi-datacenter Consistency Properties") places database replicas at several data centers around the globe, for improved responsiveness and fault tolerance. Although a replica may be strongly consistent internally, geo-replication typically uses OR between data centers to ensure availability. Examples include Walter [19], Eiger [11], or Riak.

Around 2010, several researchers proposed the concept of a Replicated Data Type (RDT) [3, 15, 16, 18]. An RDT is similar to an ordinary data type; for instance, read-write register, set, map, graph, etc., may constitute RDT types. Abstractly, an RDT is similar to the corresponding ordinary abstract data type; for instance, the interface to a register RDT might have *read* and *assign* methods, whereas a set RDT would have methods for testing whether an element is a member of the set and for adding and removing elements to/from the set. Internally, an RDT is replicated, to provide reliability, availability, and responsiveness. Encapsulation hides the details of replication and conflict resolution.

Foundations

Figure 1 depicts a logical item *x*, concretely replicated at three different sites. In OR, any site may *submit* or *initiate* a transaction reading or writing the local replica. If the transaction succeeds locally, the system *propagates* it to other sites and *replays* the transaction on the remote sites, in a *lazy* manner, in the background. Local execution is *tentative* and may be resolved against a *conflict* with a concurrent remote transaction. (The happens-before and concurrency relations

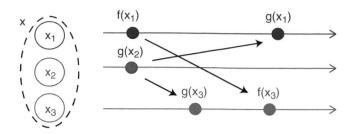

Optimistic Replication and Resolution, Fig. 1 Three sites with replicas of logical item x. Site 1 initiates transaction f; Site 2 initiates g. The system propagates and replays on remote sites. Site 3 executes in the order $g;f$, whereas Site 1 replays f before g. Eventually, Site 2 will also execute f

are defined formally by Lamport [10]. Transaction A happens-before B, if OR replayed B was initiated on some site after A executed at that site. Two transactions are concurrent if neither happens-before the other.)

OR is opposed to *pessimistic* (or *eager*) replication, where a local transaction terminates only when it commits globally. Pessimistic replication logically establishes a total order for committed transactions, at the latest when each transaction terminates. In contrast, OR generally relaxes the ordering requirements and/or converges to a common order a posteriori. The effects of a tentative transaction can be observed; thus OR protocols may violate the *isolation* property and allow cascading aborts and retries to occur.

Transmitting and Replaying Updates

In OR, updates are propagated lazily, in the background, after the transaction has terminated locally. Transmission usually uses peer-to-peer epidemic or anti-entropy techniques (see entry on "▶ Peer-to-Peer Content Distribution").

There are two main approaches to update transmission and delivery. In the *state-based* approach, a sender transmits the updated (after-values) of the object; a receiver *merges* the received value into its local state. In the *operation-based* approach, the sender transmits the *program* of the update transaction itself; a receiver *replays* the code of the transaction on its local replica.

The state-based approach is often perceived as being the simpler of the two. In the common case of last-writer-wins, state-based merge often reduces to overwriting the local replica with the received value; this is guaranteed to be deterministic. In the more general case, the merge procedure must be carefully designed to ensure convergence. The state-based approach tolerates unreliable and out-of-order delivery. However, if the replicated object is large, then state-based transmission is expensive, and replay is subject to false conflicts.

Conversely, the cost of transmitting an operation is often very small, similar to a remote procedure call.

Furthermore, logical operations are more likely to commute than writes; thus operation-based replay typically causes fewer aborts. However, the operation-based approach assumes communication layer, that ensures reliable, exactly-once delivery in happened-before order.

Conflicts

Each transaction taken individually is assumed correct (the C of the ACID properties), i.e., it maintains semantic invariants. For example, ensuring that a bank account remains positive, or that a person is not scheduled in two different meetings at the same time.

As is clear from Fig. 1, concurrent transactions may be delivered to different sites in different orders (see section "Scheduling Transactions Content and Ordering"). However, requires that local schedules be equivalent. In this respect, one may classify pairs of concurrent transactions as commuting, non-commuting, and antagonistic. Transactions *conflict* if they are mutually non-commuting or mutually antagonistic.

The relative execution order of *commuting* transactions is immaterial; they require no remote synchronization. Formally, two transactions T_1 and T_2 commute if execution order $T_1;T_2$ returns the same results to the user and leaves the database in the same state as the order $T_2;T_1$. For instance, depositing €10 in a bank account commutes with a depositing €20 into the same account and also commutes with withdrawing €100 from an independent account.

If running concurrent transactions together would violate an invariant, they are said *antagonistic*. Safety requires aborting one or the other (or both). For instance, if T_1 schedules me in a meeting from 10:00 to 12:00, and T_2 schedules a different meeting from 11:00 to 13:00, they are antagonistic since no combination of both T_1 and T_2 can be correct.

If two transactions are *non-commuting* and neither is aborted, then their relative execution order must be the same at all sites. Consider for instance T_1 = "transfer balance to savings" and T_2 = "deposit €100." Both orders $T_1;T_2$ and $T_2;T_1$ make sense, but the result is clearly different. There must be a system-wide consensus on the order $T_1;T_2$ or $T_2;T_1$.

Conflict Resolution and Reconciliation

Conflict resolution rewrites or aborts transactions to remove conflicts. Conflict resolution can be either manual or automatic. Manual conflict resolution simply allows conflicting transactions to proceed, thereby creating conflicting versions; it is up to the user to create a new, merged version.

Reconciliation detects and repairs conflicts and combines non-conflicting updates. Thus transactions are *tentative*, i.e., a tentatively successful transaction may have to roll back for reconciliation purposes. OR resolves conflicts a posteriori (whereas pessimistic approaches avoid them a priori).

In many systems, data invariants are either unknown or not communicated to the system. In this case, the system designer conservatively assumes that, if concurrent transactions access the same item, and one (or both) writes the item, then they are antagonistic. Then, one of them must abort, or both.

A few systems, such as Bayou [22] IceCube [14] or CISE [6] support an application-specific check of invariants. Bailis et al. [1] shows that application-level enforcement of invariants is error prone.

Last Writer Wins

When transactions consist only of reads and assignments, a common approach is to ensure a global precedence order.

For instance, many replicated file systems follow the "Last Writer Wins" (LWW) approach. Files have timestamps that increase with successive versions. When the file system encounters two concurrent versions of the same file, it overwrites the one with the smallest timestamp with the "younger" one (highest timestamp). The write with the smallest timestamp is lost; this approach violates the Durability property of ACID.

Semantic Resolvers

A resolver is an application-specific conflict resolution program that automatically merges two conflicting versions of an item into a new one. For example, the Amazon online bookstore resolves problems with a user's "shopping cart" by taking the union of any concurrent instances. This maximizes availability despite network outages, crashes, and the user opening multiple sessions.

A resolver should ensure that the conflicting transactions are made to commute. In a state-based approach, a resolver generally parses the item's state into small, independent sub-items. Then it applies an LWW policy to updated and tombstoned sub-items and a union policy to newly created sub-items.

The most elaborate example exists in Bayou. A Bayou transaction has three components: the dependency check, the write, and the merge procedure. The former is a database query that checks for conflicts when replaying. The write (a SQL update) executes only if the consistency check succeeds. If it fails, the merge procedure (an arbitrary but deterministic program) provides a chance to fix the conflict. However, it is very difficult to write merge procedures in the general case.

Operational Transformation

In Operational Transformation (OT), conflicting operations are *transformed* [20]. Consider two users editing the shared text "abc." User 1 initiates *insert("X",2)* resulting in "aXbc" and User 2 initiates *delete(3)*, resulting in "ab." When User 2 replays the insert, the result is "aXb" as expected. However for User 1 to observe the same result, the delete must be transformed to *delete(2)*.

In essence, the operations were specified in a non-commuting way, but transformation makes them commute. OT assumes that transformation is always possible. The OT literature focuses on a simple, linear, shared edit buffer data type, for which numerous transformation algorithms have been proposed.

OT requires two correctness conditions, often called TP1 and TP2. TP1 requires that, for any two concurrent operations A and B, running *"A followed by {B transformed in the context of A}"* yield the same result as *"B followed by {A transformed in the context of B}."* TP1 is relatively easy to satisfy and is sufficient if replay is somehow serialized.

TP2 requires that transformation functions themselves commute. TP2 is necessary if replay is in arbitrary order, e.g., in a peer-to-peer system. The vast majority of published non-serialized OT algorithms have been shown to violate TP2 [12].

Conflict-Free Replicated Data Types (CRDTs)

The common memory-cell data model is not well suited to an OR system, since concurrent assignments do not commute. OR will benefit from a data model where concurrent updates can be merged, ensuring that replicas converge without requiring synchronisation or consensus. For instance, concurrent increment and decrement operations to a shared counter can be naturally merged, because they commute.

Conflict-free Replicated Data Types (CRDTs) generalise this approach [18]. A CRDT is an abstract data type that extends some sequential type, and encapsulates algorithms ensuring that concurrent updates are merged deterministically and are guaranteed to converge. Thanks to this property, replicas of a CRDT can be updated in parallel without synchronisation. CRDT types include registers, counters, sets, maps, graphs and sequences.

When used in a sequential way, a CRDT type behaves just like its sequential counterpart. Furthermore, if two updates commute in the sequential specification, then executing the same two updates concurrently will converge to the same state. For instance, concurrently adding different elements to some CRDT set has the same result as adding them in any order. This means that a CRDT type is plug-in replacement for the corresponding sequential data type.

The key challenge in CRDT design is providing sensible concurrency semantics for updates that do not commute in the sequential specification. Thus, the concurrent specification of concurrently adding and removing the same element to a set might be "add wins," i.e., it appears in the set; but, it could equally be "remove wins" or "highest timestamp wins," depending on application requirements.

Scheduling Transactions Content and Ordering

In order to capture any causal dependencies, transactions execute in happens-before order i.e., causal consistency. As explained in Section "Conflicts," antagonistic transactions cause aborts, and non-commuting transactions must be mutually ordered. This so-called *serialization* requires a consensus, violating the availability requirement.

Whereas pessimistic approaches serialize *a priori*, most OR systems execute transactions tentatively in arbitrary order and serialize *a posteriori*. Some executions are rolled back; cascading aborts may occur.

A prime example is the Bayou system [22]. Each site executes transactions in the order received. Eventually, the transactions reach a distinguished *primary* site. If the dependency check of a transaction fails at the primary, then it aborts everywhere. Transactions that succeed commit and are serialized in the execution order of the primary.

The IceCube system showed that it is possible to improve the user experience by scheduling operations intelligently [14]. IceCube is a middleware that relieves the application programmer from many of the complexities of reconciliation. Multiple applications may coexist on top of Ice-Cube. Applications expose semantic annotations, indicating which operation pairs commute or not, are antagonistic, dependent, or have an inherent semantic order. The user may create atomic groups of operations from different applications. The IceCube scheduler performs an optimization procedure over a batch of operations, minimizing the number of aborted operations. The user commits any of the alternative schedules proposed by the system.

Freshness of Replicas

Applications may benefit from *freshness* or quality-of-service guarantees, e.g., that no replica diverges by more than a known amount from the ideal, strongly consistent state. Such guarantees come at the expense of decreased availability.

The Bayou system proposes qualitative "session guarantees" on the relative ordering of operations [21]. For instance, Read-Your-Writes (RYW) guarantees that a read observes the effect of a write by the same user, even if initiated at a different site. RYW ensures, that immediately after changing his password, a user can log in with the new password. Other similar guarantees are Monotonic-Reads, Writes-Follow-Reads, and Monotonic-Writes. The conjunction of their guarantees is equivalent to causal consistency.

Systems such as TACT control replica divergence quantitatively [8]. TACT provides a time-based guarantee, allowing an item to remain stale for only a bounded amount of time. TACT implements this by pushing an update operation to remote replicas before the time limit elapses. TACT also provides "order bounding," i.e., limiting the number of uncommitted operations: when a site reaches a user-defined bound on the number of uncommitted operations, it stops accepting new ones.

Finally, TACT can bound the difference between numeric values. For this, each replica is allocated a quota. Each site estimates the progress of other sites, using vector clock techniques. The site stops initiating operations once its cumulative modifications, or the estimated remote updates to the item, reach the quota. At that point, the site pushes its updates and pulls remote operations. For example, a bank account might be replicated at ten sites. To guarantee that the balance observed is within €50 of the truth, each site's quota is €50/10 = €5. Whenever the difference estimated by a site reaches €5, it synchronizes with the others.

Optimistic Replication Versus Optimistic Concurrency Control

The word "optimistic" has different, but related, meanings when used in the context of replication and of concurrency control.

Optimistic replication (OR) means that updates propagate lazily. There is no a priori total order of transactions. There is no point in time where different sites are guaranteed to have the same (or equivalent) state. Cascading aborts are possible.

Optimistic concurrency control (OCC) means that conflicting transactions are allowed to proceed concurrently. However, in most OCC implementations, a transaction validates before terminating. A transaction is serialized with respect to concurrent transactions, at the latest when it terminates, and cascading aborts do not occur.

Key Applications

Usenet News pioneered the OR concept, allowing to share write-only information over a slow, but cheap network using dial-up modems over telephone lines.

Mobile users want to be able to work as usual, even when disconnected from the network. Thus, mobile computing is a key driver for OR applications. Systems designed for disconnected work that use OR include the Coda file system [9], the Bayou shared database [22], or the Lotus Notes collaborative suite.

Another important application area is Computer-Supported Collaborative Work. In this domain, users must be able to update shared artefacts in complex ways without interfering with one another. OR allows a user to insulate himself temporarily from other users. A key example is the Concurrent Versioning System (CVS), which enables collaborative authoring of computer programs [4]. Bayou and Lotus Notes, just cited, are also designed for collaborative work.

OR is used for high performance and high availability in large-scale web sites. A well-known example is Amazon's "shopping cart," which is designed to be highly available, even if the same user connects to several instances of the Amazon store discussed earlier. For this reason, many NoSQL databases embrace the Available under Partition (AP) option (cf. CAS Theorem) which is OR.

Cross-References

▶ CAP Theorem
▶ Eventual Consistency
▶ Multi-datacenter Consistency Properties
▶ NoSQL Stores
▶ Peer-to-Peer Content Distribution
▶ Replicated Data Types

Recommended Reading

1. Bailis P, Fekete A, Franklin MJ, Ghodsi A, Hellerstein JM, Stoica I. Feral concurrency control: an empirical investigation of modern application integrity. In: Proceedings of the ACM SIGMOD International Conference on Management of Data; 2015. p. 1327–42. http://doi.acm.org/10.1145/2723372.2737784
2. Breitbart Y, Komondoor R, Rastogi R, Seshadril S. Update propagation protocols for replicated databases. In: Proceedings of the ACM SIGMOD International Conference on Management of Data; 1999. p. 97–108.
3. Burckhardt S, Leijen D. Semantics of concurrent revisions. In: Proceedings of the 20th European Conference on Programming Languages and Systems; 2011. Vol. 6602. p. 116–135. http://dx.doi.org/10.1007/978-3-642-19718-5_7
4. Cederqvist P, et al. Version Management with CVS. Bristol: Network Theory; 2006.
5. Golding RA. Weak-consistency group communication and membership. Ph.D. thesis, University of California, Santa Cruz. 1992. Technical Report no. UCSC-CRL-92-52. Available at: ftp://ftp.cse.ucsc.edu/pub/tr/ucsc-crl-92-52.ps.Z
6. Gotsman A, Yang H, Ferreira C, Najafzadeh M, Shapiro M. Cause I'm strong enough: reasoning about consistency choices in distributed systems. In:

POPL, St. Petersburg. 2016. p. 371–84. http://dx.doi.org/10.1145/2837614.2837625
7. Gray J, Helland P, O'Neil P, Shasha D. The dangers of replication and a solution. In: Proceedings of the ACM SIGMOD International Conference on Management of Data; 1996. p. 173–82.
8. Haifeng Yu, Amin V. Combining generality and practicality in a conit-based continuous consistency model for wide-area replication. In: Proceedings of the 21st International Conference on Distributed Computing Systems; 2001.
9. Kistler JJ, Satyanarayanan M. Disconnected operation in the Coda file system. ACM Trans Comp Syst. 1992;10(5):3–25.
10. Lamport L. Time, clocks, and the ordering of events in a distributed system. Commun ACM. 1978;21(7):558–65.
11. Lloyd W, Freedman MJ, Kaminsky M, Andersen DG. Don't settle for eventual: scalable causal consistency for wide-area storage with COPS. In: Proceedings of the 23rd ACM Symposium on Operating Systems Principles; 2011. p. 401–16. http://doi.acm.org/10.1145/2043556.2043593
12. Oster G, Urso P, Molli P, Imine A. Proving correctness of transformation functions in collaborative editing systems. Rapport de recherche RR-5795, LORIA – INRIA Lorraine. 2005. Available at: http://hal.inria.fr/inria-00071213/
13. Petersen K Spreitzer MJ, Terry DB, Theimer MM, Demers AJ. Flexible update propagation for weakly consistent replication. In: Proceedings of the 16th ACM Symposium on Operating System Principles; 1997. p. 288–301.
14. Preguiça N, Shapiro M, Matheson C. Semantics-based reconciliation for collaborative and mobile environments. In: Proceedings of the International Conference on Cooperative Information Systems; 2003. p. 38–55.
15. Preguiça N, Marquès JM, Shapiro M, Leţia M. A commutative replicated data type for cooperative editing. In: Proceedings of the 29th IEEE International Conference on Distributed Computing Systems; 2009. p. 395–403. http://doi.ieeecomputersociety.org/10.1109/ICDCS.2009.20
16. Roh H-G, Jeon M, Kim J-S, Lee J. Replicated abstract data types: building blocks for collaborative applications. J Parallel Distrib Comput. 2011;71(3):354–68. http://dx.doi.org/10.1016/j.jpdc.2010.12.006
17. Saito Y, Shapiro M. Optimistic replication. ACM Comput Surv. 2005;37(1):42–81.
18. Shapiro M, Preguiça N, Baquero C, Zawirski M. Conflict-free replicated data types. In: Proceedings of the 13th International Symposium on Stabilization, Safety, and Security of Distributed Systems; 2011. p. 386–400.
19. Sovran Y, Power R, Aguilera MK, Li J. Transactional storage for geo-replicated systems. In: Proceedings of the 23rd ACM Symposium on Operating System Principles; 2011. p. 385–400. http://doi.acm.org/10.1145/2043556.2043592

20. Sun C, Ellis C. Operational transformation in real-time group editors: issues, algorithms, and achievements. In: Proceedings of the International Conference on Computer-Supported Cooperative Work; 1998. p. 59.

21. Terry DB, Demers AJ, Petersen K, Spreitzer MJ, Theimer MM, Welch BB. Session guarantees for weakly consistent replicated data. In: Proceedings of the 3rd International Conference on Parallel and Distributed Information Systems; 1994. p. 140–9.

22. Terry DB, Theimer MM, Petersen K, Demers AJ, Spreitzer MJ, Hauser CH. Managing update conficts in Bayou, a weakly connected replicated storage system. In: Proceedings of the 15th ACM Symposium on Operating System Principles; 1995. p. 172–82.

23. Wuu GTJ, Bernstein AJ. Efficient solutions to the replicated log and dictionary problems. In: Proceedings of the ACM SIGACT-SIGOPS 3rd Symposium on the Principles of Distributed Computing; 1984. p. 233–42.

Optimization and Tuning in Data Warehouses

Ladjel Bellatreche
LIAS/ISAE-ENSMA, Poitiers University, Futuroscope, France

Synonyms

Physical design

Definition

Optimization and tuning in data warehouses ($\mathcal{DW}$) are the processes of selecting and managing adequate and optimal techniques in order to make queries and updates run faster and to maintain their performance by maximizing the use of $\mathcal{DW}$ system resources and satisfying specific constraints. A $\mathcal{DW}$ is usually accessed by complex queries for key business operations. They must be completed in seconds not days to satisfy the decision-makers' requirements. To continuously improve the query performance, two main phases are required: *physical design* and *tuning*. In the physical design phase, a $\mathcal{DW}$ administrator selects the best techniques such as *materialized views, advanced indexes, data compression, horizontal partitioning*, and *parallel processing* by exploiting advanced high-performance computing (HPC) and emerging hardware. Generally, this selection is based on most frequently asked queries and typical updates. The physical design generates a configuration Δ containing a number of optimization techniques. This configuration should evolve, since the $\mathcal{DW}$ dynamically changes during its lifetime. These changes necessitate a tuning phase to keep the performance of the $\mathcal{DW}$ from degrading. Changes may be related to the content of entities of the logical model of the $\mathcal{DW}$, sizes of optimization techniques, frequencies of queries/updates, addition/deletion of queries/updates, etc. The role of the tuning phase is to monitor and to diagnose the use of configuration Δ and different resources assigned to Δ (e.g., buffer, storage space, etc.). For instance, if an optimization technique, like an index, is not used by the whole workload, it will be dropped by a tuning tool and might be replaced by another index. A generic formalization of the physical design problem ($\mathbb{PDP}$) is given as follows:

Inputs: (**i**) a $\mathcal{DW}$; (**ii**) a workload of l queries $\overline{\mathcal{W}} = \{q_1, q_2, \ldots, q_l\}$, being each query q_i ($1 \leq i \leq l$) characterized by an access frequency f_i; (**iii**) the classes of optimization techniques $\mathcal{COT} = \{cot_1, cot_2, \ldots, cot_t\}$, supported by the target DBMS hosted the $\mathcal{DW}$; (**iv**) a set of constraints $C = \{c_1, c_2, \ldots, c_t\}$ being each constraint c_t associated to each type of optimization technique; and (**v**) a set of nonfunctional requirements $\mathcal{NFR} = \{nfr_1, nfr_2, \ldots, nfr_m\}$ that the $\mathbb{PDP}$ has to satisfy.

The objective of the $\mathbb{PDP}$ is to select a set of optimization techniques (belonging to one or several classes) that satisfies the $\mathcal{NFR}$s to respect the considered constraints.

Usually, all the inputs of the $\mathbb{PDP}$ are given, except the workload, where sometimes it is not given (in this case, we talk about *dynamic* $\mathbb{PDP}$). In the case, where it is given, we refer to *static* (workload-aware) $\mathbb{PDP}$.

Historical Background

The importance of the physical optimizations has been amplified as query optimizers became sophisticated to cope with *complex decision support queries* [8]. In the first generation of the $\mathcal{DW}$, a couple of instances of the generic $\mathbb{PDP}$ have been studied considering the traditional $\mathcal{NFR}$s such as the performance of the queries, the maintenance cost of the $\mathcal{DW}$, and the computation time of the selected optimization technique(s). These instances concern essentially three classes of optimization techniques: (1) *materialized views*, (2) *indexes*, and (3) *data partitioning*. With the explosion of the Cloud (as a deployment platform for the $\mathcal{DW}$) and the sustainable energy initiative advances to integrate the energy dimension in the process of designing $\mathcal{DW}$ applications, new $\mathcal{NFR}$s have been added to the $\mathbb{PDP}$ such as the pricing [27] and the electrical energy consumption [30].

It should be noticed that not only the $\mathcal{NFR}$ that have evolved, but the DBMS and their platforms (e.g., Cloud, database clusters, parallel database machines, etc.) have evolved dramatically. Concerning the DBMSs, their traditional row-storage has been competed with a new storage called column-storage, in which a $\mathcal{DW}$ is completely vertically partitioned into a collection of individual columns that are physically stored separately. Several academic and commercial systems built on this storage (e.g., Sybase IQ [20], Vertica [18], and MonetDB [16]). By late 2013, traditional DBMS editors (IBM, Microsoft, Oracle, SAP) have adopted this storage in their systems [1].

In the context of $\mathcal{DW}$, the physical phase is the *funnel* of all phases of the life cycle that includes (a) the collection of user requirements, (b) conceptual modeling, (c) logical modeling, (d) ETL (extract, transform, load), (e) deployment, (f) physical design, and (g) exploitation and exploration. This is because the physical phase retrieves several data from other phases, which augments its dependencies with these phases. In addition to the strong interaction that exists between physical design and deployment phases, another dependency exists between the physi-

cal and the logical phases. More concretely, the queries of the workload are expressed according to a query language conformed to the logical model of the $\mathbb{PDP}$. The properties of a query language are usually exploited by the approaches solving the $\mathbb{PDP}$. For instance, the unified query plan, obtained by merging all individual plans of the relational queries of a given workload has been widely used as a data structure of several algorithms statically and dynamically selecting materialized views [28, 35].

Based on the strong interaction between physical, logical, and deployment phases, the inputs of the above formalization of the $\mathbb{PDP}$ have to be more explicit. This explicitation can be illustrated as follows:

a *workload* $\mathcal{W}$ of queries, expressed in: a *query language* ($\mathcal{QL}$) related to: a *logical model* ($\mathcal{LM}$) translated to: a *physical model* ($\mathcal{PM}$) associated to: a set of *physical optimizations* ($\mathcal{PO}$) deployed in: a *platform* ($\mathcal{P}$).

The objective of the $\mathbb{PDP}$ is to optimize the workload $\mathcal{W}$ by *exploiting as much as possible* the characteristics of each input. Figure 1 presents an UML model detailing each input of the $\mathbb{PDP}$.

Scientific Fundamentals

The various optimization techniques selected during the physical design process may be classified into two main categories: redundant techniques and nonredundant techniques (Fig. 2).

Redundant Techniques

This category includes four main techniques: materialized views, advanced indexing schemes, denormalization, and vertical partitioning.

(1) *Materialized views.* Once materialized views are selected, all queries will be rewritten using materialized views (this process is known as *query rewriting*). A rewriting of a query q_i using views is a query expression q_j referencing to these views. The query rewriting is done *transparently* by the query optimizer. Two major problems related to materialized

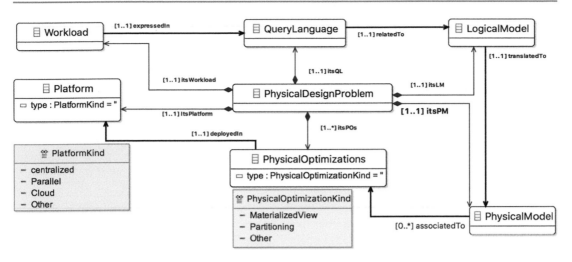

Optimization and Tuning in Data Warehouses, Fig. 1   An UML model the physical design problem

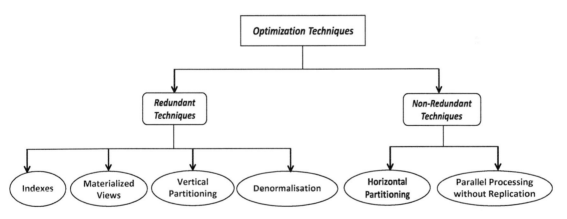

Optimization and Tuning in Data Warehouses, Fig. 2   A classification of optimization techniques

views are (a) the view selection problem and (b) the view maintenance problem.

(a) *View selection problem* ($\mathcal{VSP}$). The administrator cannot materialize all possible views, as he/she is constrained by some resources like the *disk space* (the storage of all views explodes in size), the *computation time*, the *maintenance overhead*, the time required for *query rewriting* process, etc. Hence, he/she needs to select an appropriate set of views to materialize under some resource constraints. This gave rise the $\mathcal{VSP}$, which is instance of the generic formalization described above. Several studies on the $\mathcal{VSP}$, covering various logical models of the $\mathcal{DW}$ associated to specific query languages, MOLAP (MDX)

[13], ROLAP (SQL) [35], XML (Xquery) [33], and semantic (SPARQL) [15] and Hive (HiveQL) [34], have been conducted. With regard to the input corresponding to the constraints, chronologically, the first studies related to the $\mathcal{VSP}$ assume the absence of the constraints [35]. After that and due to the large size of the stored selected materialized views, the storage space becomes the main constraint that the selection process has to integrated [36]. Then, maintenance cost becomes an important constraint, since the selected materialized views need to be updated once the base entities of its logical schema change. Some studies considered both constraints [14].

The variation of $\mathcal{NFR}$ also impacts the $\mathcal{VSP}$ formalizations. Five main variations of the $\mathcal{VSP}$, based on the used objective function to satisfied, are distinguished: (1) the minimization of the query response time [29], (2) the satisfaction of the maintenance cost [13], (3) the reduction of the energy consumption when executing a workload [30], (4) the minimization of the price due to materializing views in the Cloud [27], and (5) the number of selected views, as in [12], where the authors deal with the problem of speeding up query processing on big data. These objectives may be combined to give rise a multi-objective formalization of the $\mathcal{VSP}$ such as in [35] (where the response time and the maintenance cost were considered), [30] (where the response time and the consuming energy were used) and in [12] (where the total *MapReduce* query cost processing, the *MapReduce* cost for maintaining and the number of views were used).

The $\mathcal{VSP}$ in its different variants is known as an NP-hard problem [14]. A plethora of algorithms to solve the $\mathcal{VSP}$ has been proposed in the literature. A nice classification of these algorithms is given in [22], where four main classes have been distinguished: *deterministic algorithms*, *randomized algorithms*, *evolutionary algorithms*, and *hybrid algorithms*.

The explicitation of the inputs of the $\mathcal{VSP}$ contributes in proposing a hypercube representation allowing instancing any $\mathcal{VSP}$ work. Figure 3 gives a cube representation considering three dimensions representing, respectively, three inputs of the $\mathcal{VSP}$: the $\mathcal{NFR}$, the used algorithms, and the $\mathcal{NFR}$s [30].

The traditional formalization selects statically views. To incorporate the dynamic nature of decision support analysis, a dynamic selection has been proposed [17, 28].

(b) *View maintenance problem*. Note that materialized views store data from base entities of the logical schema of the $\mathcal{DW}$. In order to keep the views in the $\mathcal{DW}$ up to date, it is necessary to maintain the materialized views in response to the changes at the base entities of the considered logical model. This process of updating views is called view maintenance

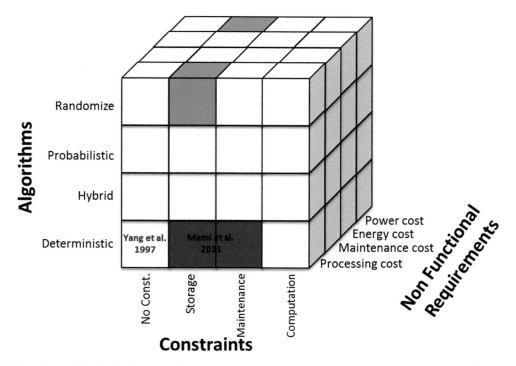

Optimization and Tuning in Data Warehouses, Fig. 3 Views selection problem

which has generated a great deal of interest in the past years. Views can be either recomputed from scratch, or incrementally maintained by propagating the base data changes onto the views. As recomputing the views can be prohibitively expensive, the incremental maintenance of views is of significant value [14].

(2) *Indexing* has been the foundation of performance tuning for databases for many years. It creates access structures that provide faster access to the base data relevant to the restriction criteria of queries. The size of the index structure should be manageable so that benefits can be accrued by traversing such a structure. The traditional indexing strategies used in database systems do not work well in $\mathcal{DW}$ environments. Most OLTP transactions typically access a few rows; most OLTP queries are point queries. B-trees, which are used in most common relational database systems, are geared toward such point queries. They are well suited for accessing a few rows. An OLAP query typically accesses many records for summarizing information. For example, an OLTP transaction would typically query for a customer who booked a flight on AF1234 on say June 11th; on the other hand, an OLAP query would be more like "Give me the number of customers who booked flight on AF1234 in say one month?." The second query would access more records and they are of type of range queries. B-tree indexing scheme is not suited to answer OLAP queries efficiently. An index can be defined on one table (or view) or many tables using a join index. In the $\mathcal{DW}$ context, when we talk about index, we refer to two different things: (1) indexing techniques and (2) index selection problem. A number of indexing strategies have been suggested for $\mathcal{DW}$: value-list index, projection index, bitmap index, bit-sliced index, data index, join index, and star join index [23].

The bitmap index is probably the most important result obtained in the $\mathcal{DW}$ physical optimization field. It is more suitable for *low* and *high* cardinality attributes since its size strictly depends on the *number of distinct values of the column on which it is built* [10]. Besides disk space saving (due to their binary representation and potential compression [7]), such index speeds up queries having Boolean operations (such as AND, OR, and NOT) and COUNT operations. A bitmap join index is proposed to speed up join operations. In its simplest form, it can be defined as a bitmap index on a fact table on one or several columns of other dimension tables.

(b) Index selection problem ($\mathcal{ISP}$). It should be noticed that materialized views and indexes are quite similar – they compete for the same resource representing storage and incur maintenance overhead in the presence of updates. Due to this similarity and the fact that the $\mathcal{ISP}$ is an instance of the generic formalization of the $\mathbb{PDP}$, the whole description related to $\mathcal{VSP}$, in terms of constraints, $\mathcal{NFR}$, used algorithms, logical schemes, etc. can be easily reproduced for the $\mathcal{VSP}$. An interesting specificity of the $\mathcal{ISP}$ is the usage of the data mining-driven approaches to select bitmap join indexes [5]. As for the $\mathcal{VSP}$, the $\mathcal{ISP}$ has been treated for statically and dynamically. Recently, index cracking techniques have been developed, where indexes are created adaptively and incrementally as a side-product of query processing [32].

(3) *Vertical partitioning* can be viewed as a redundant structure even if it results in little storage overhead. Initially it has been considered as a logical optimization structure, then a physical structure and actually a physical storage structure (column-store). Although column-store systems employ concepts that are at a high level similar to those in early research proposals for vertical partitioning.

The vertical partitioning of a table T splits it into two or more tables, called sub-tables or vertical fragments, each of which contains a subset of the columns in T. Note that the key columns are duplicated in each vertical fragment, to allow "reconstruction" of an original row in T. Since many queries access only a small subset of the columns in a table, vertical partitioning can reduce the amount of data that needs to be scanned to answer the query.

To vertically partition a table with m non-primary key attributes, the number of possible fragments is equal to $B(m)$, which is the mth Bell number [25]. For large values of m, $B(m) \cong m^m$. For example, for m = 10; $B(10) \cong$ 115 975. These values indicate that it is futile to attempt to obtain optimal solutions to the vertical partitioning problem. Many algorithms were proposed and classified into two categories: grouping and splitting [25]. Grouping starts by assigning each attribute to one fragment, and at each step, joins some fragments until some criteria is satisfied. Splitting starts with a table and decides on beneficial partitioning based on the query frequencies.

The vertical partitioning can be combined with other optimization techniques such as materialized views [11].

(4) *Denormalization* is the process of attempting to optimize the performance of a database by adding redundant data to save join operations. Denormalization is usually promoted in a $\mathcal{DW}$ environment.

Nonredundant Techniques

In this category, we focus on two main techniques which are horizontal partitioning and parallel processing.

(1) *Horizontal partitioning* represents an important aspect of the physical design phase. It allows tables, indexes, and materialized views to be partitioned into *disjoint* sets of rows that are physically stored and accessed separately [31] or in parallel. Horizontal partitioning may have a significant impact on performance of queries and manageability of very large $\mathcal{DW}$. Not only do data partitions reduce the time it takes to perform database maintenance and management tasks, by eliminating nonrelevant partition(s), they also have a positive effect on the performance of applications. Another characteristic of horizontal partitioning is its ability to be combined with other optimization structures like indexes and materialized views. Splitting a table, a materialized view, or an index into smaller pieces makes all operations on individual pieces much faster. Contrary to materialized views and indexes, data partitioning does not replicate data, thereby reducing space requirements and minimizing update overhead [26].

A native database definition language support is available for horizontal partitioning, where several fragmentation modes are available [24]: range, list, and hash. In the range partitioning, an access path (table, view, and index) is decomposed according to a range of values of a given set of columns. The hash mode decomposes the data according to a hash function (provided by the system) applied to the values of the partitioning columns. The list partitioning splits a table according to the listed values of a column. These methods can be combined to generate composite partitioning (List-List, Range-Range, Hash-Hash, Range-List,...). Recently, a new mode of horizontal partitioning became available in Oracle11g [24], called *virtual column-based partitioning*. It is defined by one of the abovementioned partition techniques, and the partitioning key is based on a virtual column. Virtual columns are not stored on disk and only exist as metadata.

Two versions of horizontal partitioning are available [25]: *primary* and *derived* horizontal partitioning. Primary horizontal partitioning of a table is performed using predicates defined on that relation. It can be performed using the different fragmentation modes above cited. Derived horizontal partitioning is the partitioning of a table that results from predicates defined in other table(s). The derived partitioning of a table R according to a fragmentation schema of table S is feasible if and only if there is a join link between R and S.

In the context of relational $\mathcal{DW}$, derived horizontal partitioning is well adapted. In other words, to partition a $\mathcal{DW}$, the best way is to *partition some/all dimension tables using their predicates, and then partition the fact table* based on the fragmentation schemas of dimension tables. This fragmentation takes into consideration star join queries requirements (these queries impose restrictions on the dimension values that are used for selecting specific facts; these facts are further grouped and aggregated according to the user demands). To illustrate this fragmentation, let us suppose a relational $\mathcal{D}\mathcal{W}$ modeled by a star schema with d dimension tables and a fact table F. Among these dimension tables, g tables are fragmented ($g \leq d$). Each dimension table D_i ($1 \leq i \leq g$) is partitioned into m_i fragments: $\{D_{i1}, D_{i2}, \ldots, D_{im_i}\}$, where each fragment D_{ij} is defined as:

$$D_{ij} = \sigma_{cl_j^i}(D_i), \text{ where } cl_j^i \text{ and } \sigma \ (1 \leq i \leq g, 1 \leq j \leq m_i)$$ represent a conjunction of simple predicates and the selection operator, respectively. Thus, the fragmentation schema of the fact table F is defined as follows: $F_i = F \ltimes D_{1j} \ltimes D_{2k} \ltimes \ldots \ltimes D_{gl}, (1 \leq i \leq m_i)$, where $\ltimes$ represents the semi join operation.

Derived horizontal partitioning has two main advantages in relational $\mathcal{D}\mathcal{W}$, in addition to classical benefits of data partitioning: (1) precomputing joins between fact table and dimension tables participating in the fragmentation process of the fact table [2] and (2) optimizing selections defined on dimension tables. Similar advantages hold for bitmap join indexes.

Horizontal Partitioning Schema Selection Problem $\mathcal{H}\mathcal{P}\mathcal{S}\mathcal{P}$. The $\mathcal{H}\mathcal{P}\mathcal{S}\mathcal{P}$ is also an instance of our generic problem. In its general formalization, it aims at providing a new schema of the $\mathcal{D}\mathcal{W}$ in which dimension tables are decomposed into dimension fragments using the primary horizontal partitioning and the fact table into N fragments by the means of the derived horizontal partitioning. The obtained partitioning schema shall minimize the execution cost of the workload $\mathcal{W}$, and the number of fact fragments N must not exceed the threshold W ($N \leq W$). This problem has been shown as NP-hard [3]. Large panoply of algorithms is proposed in different $\mathcal{D}\mathcal{W}$ contexts (Relational, XML, MOLAP) [3, 21] including hill climbing, genetic, simulated annealing, and data mining-driven algorithms.

(3) *Parallel processing.* Today volumes of data are increasing more and more due to the explosion of data providers such as sensors, social networks (e.g., Facebook, Twitter, and LinkedIn), network traffic, cybersecurity, etc. New requirements related to data exploration and exploitation have emerged to predict the behavior of users in order to improve their services via analyzing so-collected large data volumes. As a consequence, traditional $\mathcal{D}\mathcal{W}$ have become obsolete, and parallel $\mathcal{D}\mathcal{W}$, instead, have been proposed as a robust and scalable platform for storing, processing and analyzing large volumes of data within the layers of modern analytics infrastructures. Designing a parallel $\mathcal{D}\mathcal{W}$ passes through a lifecycle composed by a set of phases [4]: (i) choosing the hardware architecture (e.g., shared nothing, database clusters, etc.), (ii) fragmenting the $\mathcal{D}\mathcal{W}$ schema, (iii) allocating the generated fragments, (iv) replicating fragments in order to ensure high performance, and (v) defining the strategies for load balancing and query processing. A comprehensive approach including the above steps for designing a relational $\mathcal{D}\mathcal{W}$ in the context of database clusters is well described in [6].

Due to the complexity of the $\mathbb{PDP}$, commercial and academic tools were proposed to assist

Optimization and Tuning in Data Warehouses, Table 1 Commercial advisor

Advisor name	Supported techniques
SQL database tuning advisor (DTA)	Partitioning, materialized views, indexes
Oracle SQL access advisor	Partitioning, materialized views, indexes
DB2 index advisor tool	Partitioning, materialized views, indexes, clustering
Oracle database 11g automatic storage management	Storage management
DB2 9	Storage management

administrators in their tasks in choosing their relevant optimization techniques. Table 1 describes the commercial advisors and their supported optimization techniques.

Tuning

The major challenge for the physical design phase is the efficient utilization of different design approaches for more volatile workload requirements. This is because the physical design is usually static. It should be noticed that physical redesign is not applicable for a single query (or small set of) as well as it is a too costly (manual) task to perform in short intervals; thus, such approach is not sufficient to react on changing workloads. Consequently, automatic tuning techniques associated with tools (advisors) have been proposed. Two types of advisors exist: those that propose physical design improvements based on user's request. They alert user to redesign whenever a benefit above a given threshold is estimated for redesign. The second type of advisors aims at ensuring a zero-administration of the $\mathcal{DW}$ and consequently reducing the user interaction with the physical design module.

Two main self-tuning approaches are distinguished based on how the tuning is triggered [19]. In the first approaches, the profit of the tuning calculates the advantage for an estimated (new) configuration compared with the existing configuration. This calculation has to satisfy side conditions for system environment (e.g., available size) and (i.e., a threshold). In the second approaches, the self-tuning is based on a general framework issued from adaptive database systems [9]. This framework is composed of three main modules: *observe*, *predict*, and *react*. The *observe* module monitors the specified workload characteristics by collecting statistics of the $\mathcal{DW}$ components (e.g., collects statistics for an index candidates for each query). The *predict* component is used to assess performance of the current configuration in the near-future and to compare it with the possible configurations (e.g., continuously controls index statistics). The *react* module is engaged when the current configuration is found unfit and a new configuration should be selected (e.g., create/delete indexes to an appropriate moment).

Key Applications

The spectacular growth of big data has become a fact, and it is expected to get even bigger due to the development of applications related to the Internet of Things (IoT). Application areas of sensors and IoT technology includes various application domains such as smart cities, traffics monitoring, smart parking, smart lighting, environmental monitoring, manufacturing, supply chain management, smart agricultural irrigation, smart health monitoring, and smart homes and vehicles. A study by Dell EMC and IDC (https://www.emc.com/leadership/digital-universe/index.htm) revealed that the Digital Universe will grow ten times by 2020 and of this, data generated by Sensor-Enabled "Things" will represent 10%. Faced to the data volume and variety of deployment platforms, the $\mathbb{PDP}$ shall be revisited.

Future Directions

As mentioned in the previous section, physical design and tuning are very crucial decisions for the performance of $\mathcal{DW}$ applications. In this section, some interesting open issues for physical design and tuning are highlighted:

(1) *The development of mathematical cost models:* estimating the cost of $\mathcal{NFR}$. Different algorithms used in the process of selecting optimization techniques are based on cost models (usually called cost model-driven algorithms). The quality of the selected techniques strongly depends on used cost models. The development of cost models related to no traditional $\mathcal{NFR}$ have to use machine learning techniques to identify the values of some parameters like the energy consumption of a CPU [30] and to calibrate them.

(2) *Database cracking:* has been mainly focused on indexes. It incrementally builds an index structure over a table based on access patterns of queries. Due to the similarity that may exist between indexes, materialized views and data partitioning the reproduction of index

cracking effort to other optimization techniques such as materialized views and data partitioning by exploiting their similarities could be an interesting issue.

(3) *In-main memory processing:* traditional $\mathcal{DW}$s cannot satisfy the needs of big data applications. This is because they spent a lot of time to prepare data for analytical processing. This preparation is related to the fact that a traditional $\mathcal{DW}$ and its redundant optimization techniques are disk resident. Faced to this situation, several research efforts have been proposed offering systems and services to store huge amounts in RAM to process analytical operations (e.g., SAPA HANA, IBM DB2 with BLU Acceleration, etc.). This situation impacts the selection and the usage of optimization techniques.

(4) *Consideration of multi-query optimization (MQO) during the selection process of optimization techniques:* we have mentioned the existence of a strong interdependency between $\mathbb{PDP}$ and MQO (that captures the interaction between queries). This interdependency has to be exploited by algorithms selecting optimization techniques such as indexes, partitioning, etc. and tuning the $\mathcal{DW}$.

(5) *Platform selection problem:* certainly, several industrial and academic tools (advisors) exist, but they consider that DBMS hosting the $\mathcal{DW}$ is already deployed in a given platform (centralized, distributed, parallel, Cloud, etc.). Due to the large diversity of the deployment platforms and emerging hardware, it would be interesting to simulate the performance of $\mathcal{DW}$ applications in various platforms and based on that simulation. Based on its results, the $\mathcal{DW}$ may choose her/his favorite platform.

(6) *Reproduction of the test chain of $\mathbb{PDP}$ solutions:* researchers on the $\mathbb{PDP}$ consecrated important efforts in evaluating, benchmarking and testing their products in terms of selection algorithms, optimization techniques, storage systems, etc. These efforts are usually performed through simulation (by the means of mathematical cost models) and real validation. This expertise in validating proposals has to be stored in repositories to facilitate its reproduction by researchers, students, industries, etc.

Experimental Results

The experimental results related to this chapter are available in [4].

Data Sets

The datasets used in the experimental results related to this chapter are available in [4].

Cross-References

▶ Data Warehousing in Cloud Environments
▶ Parallel and Distributed Data Warehouses
▶ Query Processing (in Relational Databases)
▶ Rewriting Queries Using Views

Recommended Reading

1. Abadi D, Boncz PA, Harizopoulos S, Idreos S, Madden S. The design and implementation of modern column-oriented database systems. Found Trends Databases. 2013;5(3):197–280.
2. Bellatreche L, Boukhalfa K, Mohania MK. Pruning search space of physical database design. In: Proceedings of the 18th International Conference on Database and Expert Systems Applications; 2007. p. 479–88.
3. Bellatreche L, Boukhalfa K, Richard P, Woameno KY. Referential horizontal partitioning selection problem in data warehouses: hardness study and selection algorithms. Int J Data Warehouse Min. 2009;5(4):1–23.
4. Bellatreche L, Cuzzocrea A, Benkrid S. Effectively and efficiently designing and querying parallel relational data warehouses on heterogeneous database clusters: the F&A approach. J Database Manag. 2012;23(4):17–51.
5. Bellatreche L, Missaoui R, Necir H, Drias H. Selection and pruning algorithms for bitmap index selection problem using data mining. In: Proceedings of the 9th International Conference on Data Warehousing and Knowledge Discovery; 2007. p. 221–30.
6. Benkrid S, Bellatreche L, Cuzzocrea A. Designing parallel relational data warehouses: a global, comprehensive approach. In: Proceedings of the 17th East

European Conference on Advances in Databases and Information Systems; 2013. p. 141–50.

7. Chambi S, Lemire D, Kaser O, Godin R. Better bitmap performance with roaring bitmaps. Softw Pract Exper. 2016;46(5):709–19.

8. Chaudhuri S, Narasayya V. Self-tuning database systems: a decade of progress. In: Proceedings of the 33rd International Conference on Very Large Databases; 2007. p. 3–14.

9. Chaudhuri S, Weikum G. Self-management technology in databases. In: Encyclopedia of Database Systems; 2009. p. 2550–55.

10. Deliège F, Pedersen TB. Position list word aligned hybrid: optimizing space and performance for compressed bitmaps. In: Proceedings of the 13th International Conference on Extending Database Technology; 2010. p. 228–39.

11. Du J, Miller RJ, Glavic B, Tan W. Deepsea: progressive workload-aware partitioning of materialized views in scalable data analytics. In: Proceedings of the 20th International Conference on Extending Database Technology; 2017. p. 198–209.

12. Goswami R, Bhattacharyya DK, Dutta M. Materialized view selection using evolutionary algorithm for speeding up big data query processing. J Intell Inf Syst. 2017;49(3):407–33.

13. Gupta H. Selection of views to materialize in a data warehouse. In: Proceedings of the 6th International Conference on Database Theory; 1997. p. 98–112.

14. Gupta H. Selection and maintenance of views in a data warehouse. Ph.D. thesis, Stanford University; 1999.

15. Ibragimov D, Hose K, Pedersen TB, Zimányi E. Optimizing aggregate SPARQL queries using materialized RDF views. In: Proceedings of the 15th International Semantic Web Conference; 2016. p. 341–59.

16. Idreos S, Groffen F, Nes N, Manegold S, Sjoerd Mullender K, Kersten ML. Monetdb: two decades of research in column-oriented database architectures. IEEE Data Eng Bull. 2012;35(1):40–45.

17. Kotidis Y, Roussopoulos N. Dynamat: a dynamic view management system for data warehouses. In: Proceedings of the ACM SIGMOD International Conference on Management of Data; 1999. p. 371–82.

18. Lamb A, Fuller M, Varadarajan R, Tran N, Vandier B, Doshi L, Bear C. The vertica analytic database: C-store 7 years later. Proc VLDB Endow. 2012;5(12):1790–801.

19. Lübcke A. Automated query interface for hybrid relational architectures. Ph.D. thesis, University of Magdeburg; 2017.

20. MacNicol R, French B. Sybase IQ multiplex – designed for analytics. In: Proceedings of the 30th International Conference on Very Large Data Bases; 2004. p. 1227–30.

21. Mahboubi H, Darmont J. Data mining-based fragmentation of xml data warehouses. In: Proceedings of the ACM 11th International Workshop on Data Warehousing and OLAP; 2008. p. 9–16.

22. Mami I, Bellahsene Z. A survey of view selection methods. SIGMOD Rec. 2012;41(1):20–29.

23. O'Neil PE, Quass D. Improved query performance with variant indexes. In: Proceedings of the ACM SIGMOD International Conference on Management of Data; 1997. p. 38–49.

24. Oracle Data Sheet. Oracle partitioning. White Paper: http://www.oracle.com/technology/products/bi/db/11g/; 2007.

25. Özsu MT, Valduriez P. Principles of distributed database systems. 2nd ed. Upper Saddle River: Prentice Hall; 1999.

26. Papadomanolakis S, Ailamaki A. Autopart: automating schema design for large scientific databases using data partitioning. In: Proceedings of the 16th International Conference on Scientific and Statistical Database Management; 2004. p. 383–92.

27. Perriot R, Pfeifer J, d'Orazio L, Bachelet B, Bimonte S, Darmont J. Cost models for selecting materialized views in public clouds. Int J Data Warehouse Min. 2014;10(4):1–25.

28. Phan T, Li W. Dynamic materialization of query views for data warehouse workloads. In: Proceedings of the 24th International Conference on Data Engineering; 2008. p. 436–45.

29. Ross KA, Srivastava D, Sudarshan S. Materialized view maintenance and integrity constraint checking: trading space for time. In: Proceedings of the ACM SIGMOD International Conference on Management of Data; 1996. p. 447–458.

30. Roukh A, Bellatreche L, Bouarar S, Boukorca A. Eco-physic: eco-physical design initiative for very large databases. Inf Syst. 2017;68(Aug):44–63.

31. Sanjay A, Narasayya VR, Yang B. Integrating vertical and horizontal partitioning into automated physical database design. In: Proceedings of the ACM SIGMOD International Conference on Management of Data; 2004. p. 359–70.

32. Schuhknecht FM, Jindal A, Dittrich J. An experimental evaluation and analysis of database cracking. VLDB J. 2016;25(1):27–52.

33. Tang N, Xu Yu J, Tang H, Tamer Özsu M, Boncz PA. Materialized view selection in XML databases. In: Proceedings of the 14th International Conference on Database Systems for Advanced Applications; 2009. p. 616–30.

34. Thusoo A, Sen Sarma J, Jain N, Shao Z, Chakka P, Zhang N, Anthony S, Liu H, Murthy R. Hive – a petabyte scale data warehouse using hadoop. In: Proceedings of the 26th International Conference on Data Engineering; 2010. p. 996–1005.

35. Yang J, Karlapalem K, Li Q. Algorithms for materialized view design in data warehousing environment. In: Proceedings of the 23th International Conference on Very Large Data Bases; 1997. p. 136–45.

36. Zhang C, Yang J. Genetic algorithm for materialized view selection in data warehouse environments. In: Proceedings of the 1st International Conference on Data Warehousing and Knowledge Discovery; 1999. p. 116–25.

OQL

Peter M. D. Gray
University of Aberdeen, Aberdeen, UK

Synonyms

Object query language

Definition

OQL was developed to play the role of SQL for *Object-Oriented Databases*, especially those adhering to the *ODMG Standard* [4] where the language is defined. Unlike SQL, OQL is a functional language, and its operators can be composed to an arbitrary level of nesting within a query provided the query remains type-correct. Fegaras and Maier [8] have shown how OQL expressions have a direct translation into monoid *Comprehensions*.

Optimisation techniques for OQL that exploit its inherent functional nature are discussed in [5, 6, 8]. OQL has been influential in the development of the SQL3 standard and also the functional core of the XQuery language for XML. Thus optimisation techniques developed for OQL are also applicable to these languages.

Key Points

The fundamental modelling concept of *object identifiers* for entity instances was accepted into the database mainstream in the late 1980s, and the move to using SQL-like syntax for querying such data models followed soon after. Early influential systems were OSQL [2] and *AMOSQL* (q.v.). This resulted in query language proposals for object-oriented databases such as the very influential O2 query language [1] and its successor OQL, which was included in the ODMG Standard [4]. For example, the DAPLEX query.

```
FOR EACH S IN STUDENT
SUCH THAT name(S)="Fred Jones"
```

```
PRINT name(S), age(S);
```

is expressed as follows in OQL, basically by syntactic reordering of the query clauses and using path expressions rather than function application:

```
SELECT S.name, S.age FROM STUDENT S
WHERE S.name="Fred Jones"
```

When restricted to sets, monoid comprehensions are equivalent to set monad comprehensions [3], which capture precisely the nested relational algebra [8]. Most OQL expressions have a direct translation into the monoid calculus. For example, the OQL query

```
SELECT DISTINCT HOTEL.price
FROM HOTEL IN (
SELECT h
FROM c IN CITIES, h IN c.hotels
WHERE c.name="Arlington")
WHERE EXIST r IN HOTEL.rooms:r.bed_num
= 3
AND HOTEL.name IN (
SELECT t.name
FROM s IN STATES, t IN s.attractions
WHERE s.name = "Texas" );
```

finds the prices of hotels in Arlington that have rooms with three beds and are also are named after a tourist attraction in Texas. This query is translated into the following monoid comprehension [7]:

```
fold(Union,Empty,
[price(h)|c <- Cities; h <- hotels(c);
name(c) = ''Arlington'';
fold(Or,False,
[bednum(r)=3 | r <- rooms(h) ]),
fold(Or,False,
[ name(h)=name(t) | s <- States; t <-
attractions(s); name(s)=''
Texas'']])])
```

Here, as in Functional Programming

```
fold(Or,False,[x1,x2, ... xn]) = x1
Or x2 Or ... xn Or False
```

computes the logical *Or* of a list of boolean values, so it is true only if *some* of them are true. Likewise *fold(Union,Empty,L)* copies the list L into a set without duplicates. Mathematically *fold* implements *monoid* operations with a given *merge* operation and a *zero*.

Cross-References

▶ AMOSQL
▶ Comprehensions
▶ Functional Query Language

Recommended Reading

1. Bancilhon F, Delobel C, Kanellakis PC. Building an object-oriented database system, the story of O2. Los Altos: Morgan Kaufmann; 1992.
2. Beech D. A foundation of evolution from relational to object databases. In: Advances in Database Technology. Proceedings of the 1st International Conference on Extending Database Technology; 1988. p. 251–70.
3. Buneman P, Libkin L, Suciu D, Tannen V, Wong L. Comprehension syntax. ACM SIGMOD Rec. 1994;23(1):87–96.
4. Cattell RGG, editor. The object data standard: ODMG 3.0. Los Altos: Morgan Kaufmann; 2000.
5. Cluet S, Delobel C. A general framework for the optimization of object-oriented queries. In: Proceedings of the ACM SIGMOD International Conference on Management of Data; 1992. p. 383–92.
6. Fegaras L. Query unnesting in object-oriented databases. In: Proceedings of the ACM SIGMOD International Conference on Management of Data; 1998. p. 49–60.
7. Fegaras L. Query processing and optimization in λ-DB, Chapter 13. In: Gray PMD, Kerschberg L, King PJH, Poulovassilis A, editors. The functional approach to data management. Berlin: Springer; 2004.
8. Fegaras L, Maier D. Towards an effective calculus for Object Query Languages. In: Proceedings of the ACM SIGMOD International Conference on Management of Data; 1995. p. 47–58.

Orchestration

W. M. P. van der Aalst
Eindhoven University of Technology,
Eindhoven, The Netherlands

Definition

In a Service Oriented Architecture (SOA) services are interacting by exchanging messages, i.e., by combining services more complex services are created. Orchestration is concerned with the composition of such services seen from the viewpoint of single service.

Key Points

The terms "orchestration" and "choreography" describe two aspects of integrating services to create business processes [2, 3]. The two terms overlap somewhat and the distinction is subject to discussion. Orchestration and choreography can be seen as different "perspectives." Choreography is concerned with the exchange of messages between those services and is often be characterized by analogy "Dancers dance following a global scenario without a single point of control." Orchestration is concerned with the interactions of a single service with its environment. Here an analogy can also be used. In orchestration, there is someone, "the conductor", who tells everybody in the orchestra what to do and makes sure they all play in sync.

Figure 1 illustrates the notion of orchestration. Service A is interacting with other services to create a more complex service. The dashed area shows the focal point of orchestration, i.e., the control-flow related to message exchanges of a single party. Languages such a BPEL are pro-

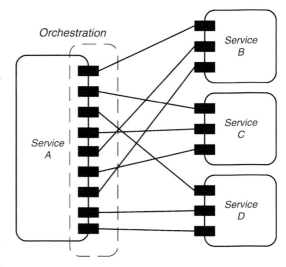

Orchestration, Fig. 1 Orchestration

posed to model and enact such orchestrations [1]. Note that languages like BPEL are very close to traditional workflow languages, i.e., the same types of control-flow patterns need to be supported.

Orchestration often assumes that services have a "buy side" and a "sell side," i.e., services can be used by other services ("sell side") and at the same time use services ("buy side"). Orchestration is mainly concerned with the "buy side." Unlike choreography, there is a single party coordinating the process.

Cross-References

▶ Business Process Execution Language
▶ Business Process Management
▶ Orchestration
▶ Web Services
▶ Workflow Management

Recommended Reading

1. Alves A, Arkin A, Askary S, Barreto C, Bloch B, Curbera F, Ford M, Goland Y, Guzar A, Kartha N, Liu CK, Khalaf R, Koenig D, Marin M, Mehta V, Thatte S, Rijn D, Yendluri P, Yiu A. Web services business process execution language version 2.0 (OASIS Standard). WS-BPEL TC OASIS; 2007. http://docs.oasis-open.org/wsbpel/2.0/wsbpel-v2.0.html
2. Dumas M, van der WMP, Aalst ter Hofstede AHM. Process-aware information systems: bridging people and software through process technology. New York: Wiley; 2005.
3. Weske M. Business process management: concepts, languages, architectures. Berlin: Springer; 2007.

Order Dependency

Jaroslaw Szlichta
University of Ontario Institute of Technology, Oshawa, ON, Canada

Synonyms

OD

Definition

Order dependencies (ODs) describe the relationship among lexicographical orderings of sets of tuples. An *order specification* is a list of attributes *marked* as *asc* (ascending) or *desc* (descending). This is the notion of order used in SQL and within query optimization, as by the order-by operator (nested sort). Let $\mathbf{X} = [\mathtt{A}|\mathbf{T}]$ be a list of marked attributes, the marked attribute $\mathtt{A}$ is the *head* of the list, and the marked list $\mathbf{T}$ is the tail. For two tuples s and t, $s \preceq_{\mathbf{X}} t$ *iff* $\mathbf{X} = [\ ]$; or $\mathbf{X} = [\mathtt{A}\ asc|\mathbf{T}]$ and $s_{\mathtt{A}} < t_{\mathtt{A}}$; or $\mathbf{X} = [\mathtt{A}\ desc|\mathbf{T}]$ and $s_{\mathtt{A}} > t_{\mathtt{A}}$; or $\mathbf{X} = [\mathtt{A}\ asc|\mathbf{T}]$ or $\mathbf{X} = [\mathtt{A}\ desc|\mathbf{T}]$, $s_{\mathtt{A}} = t_{\mathtt{A}}$, and $s \preceq_{\mathbf{T}} t$. Let $\mathbf{r}$ be a table over a relation $\mathbf{R}$ that contains the attributes that appear in $\mathbf{X}$ and $\mathbf{Y}$. Given two order specifications $\mathbf{X}$ and $\mathbf{Y}$ a table $\mathbf{r}$ satisfies an OD $\mathbf{X} \mapsto \mathbf{Y}$, read as $\mathbf{X}$ *orders* $\mathbf{Y}$, *iff* for all $s, t \in \mathbf{r}$, $s \preceq_{\mathbf{X}} t$ implies $s \preceq_{\mathbf{Y}} t$ [8]. An OD is *unidirectional* when attributes within it are all marked as *asc* or *desc* [4,6,7,9]. Unidirectional ODs are subsumed by ODs by definition.

Historical Background

Functional dependencies (FDs) cannot capture relationships among ordered attributes, such as between timestamps and numbers. ODs can express such semantics. ODs properly subsume FDs, as they can additionally express business rules involving order; e.g., an employee never has a higher salary while paying lower taxes compared with another employee. ODs subsume FDs as any FD can be mapped to an equivalent OD by prefixing the left-hand-side onto the right-hand-side attributes [8]. For example, if `salary` functionally determines `tax`, then `salary` *orders* `salary` and `tax` and vice versa. A *pointwise* OD $X \rightsquigarrow Y$ holds if order over the values of *each* attribute of set X implies an order over the values of *each* attribute of set Y [2]. Pointwise ODs properly subsume ODs [8]. *Sequential dependencies* (SDs) specify that when tuples have consecutive antecedent values, their consequents must be within a prescribed range specified by a *min* and *max* value [3].

Scientific Fundamentals

The OD inference problem is to answer whether an OD is logically entailed by a set of ODs. The inference problem for both ODs and unidirectional ODs is co-NP-complete by reduction to 3-SAT [8]. The inference problem of inferring FDs from ODs is also co-NP-complete; however, it is linear for the case of FDs over unidirectional ODs. The complexity of sound and complete chase procedure for unidirectional ODs [6] and ODs [8] is exponential in the number of attributes. A domain is restricted if an additional order property is guaranteed over the schema. The inference problem for ODs over the restricted transitive domain is polynomial [8]. A sound and complete axiomatization for unidirectional ODs is characterized by six axioms [7]. While formulating dependencies can be done manually, it requires domain expertise, is prone to human errors, and may be excessively time consuming. The automatic discovery of ODs has exponential worst-case time complexity in the number of attributes and linear complexity in the number of tuples [9].

Key Applications

The ODs can be used for improving data quality, where ODs can describe intended semantics and business rules, and their violations can point out possible data errors [4, 9]. Query optimizers can use ODs to eliminate costly operators such as sorts and to identify *interesting orders*: ordered streams between query operators that exploit available indices, enable pipelining, and eliminate intermediate sorts and partitioning steps [6, 7]. Sorting and interesting orders are integral parts of relational query optimizers for SQL order-by and group-by and for sort-merge joins. ODs enable optimization techniques for *online analytical processing* (OLAP): eliminating expensive joins from query plans in data warehouse environments [8] and improving the performance of queries with SQL functions and algebraic expressions (e.g., `date` *orders* *year*(`date`) and `date` *orders* *year*(`date`)*100 + *month*(`date`)) [5,8]. ODs can be also used for improved design to reduce the indexing space [1].

Recommended Readings

1. Dong J, Hull R. Applying approximate order dependency to reduce indexing space. In: Proceedings of the ACM SIGMOD International Conference on Management of Data; 1982. p. 119–27.
2. Ginsburg S, Hull R. Order dependency in the relational model. Theor Comput Sci. 1983;26(1):149–95.
3. Golab L, Karloff H, Korn F, Saha A, Srivastava D. Sequential dependencies. Proc VLDB Endow. 2009;2(1):547–85.
4. Langer P, Naumann F. Efficient order dependency detection. VLDB J. 2016;25(2):223–41.
5. Malkemus T, Padmanabhan S, Bhattacharjee B, Cranston L. Predicate derivation and monotonicity detection in DB2 UDB. In: Proceedings of the 21st International Conference on Data Engineering; 2005. p. 939–47.
6. Ng W. An extension of the relational data model to incorporate ordered domains. ACM Trans Database Syst. 2001;26(3):344–83.
7. Szlichta J, Godfrey P, Gryz J. Fundamentals of order dependencies. Proc VLDB Endow. 2012;5(11):1220–31.
8. Szlichta J, Godfrey P, Gryz J, Zuzarte C. Expressiveness and complexity of order dependencies. Proc VLDB Endow. 2013;5(11):1858–69.
9. Szlichta J, Godfrey P, Golab L, Kargar M, Srivastava D. Effective and complete discovery of order dependencies via set-based axiomatization. Proc VLDB Endow. 2017;10(7):12.

OR-Join

Nathaniel Palmer
Workflow Management Coalition, Hingham, MA, USA

Synonyms

Synchronous join

Definition

The point of convergence within a workflow following alternative, mutually exclusive paths.

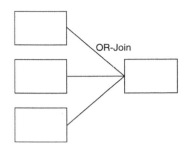

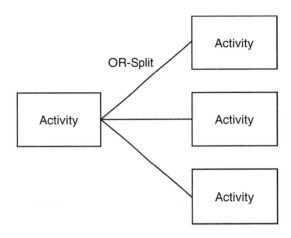

OR-Join, Fig. 1 OR-Join

OR-Split, Fig. 1 OR-split

Key Points

An OR-Join (Fig. 1) represents a point within a workflow where two or more alternative workflow branches re-converge following an OR-Split into a single common activity as the next step within the workflow. In contrast with an AND-Join, no parallel activity execution has occurred at the join point, therefore no synchronization is required. With an OR-Join a thread of control may arrive at the specific activity via any of several alternative preceding activities.

Definition

A point within the workflow where a single thread of control makes a decision upon which branch to take when encountered with multiple alternative workflow branches.

Key Points

An OR-Split (Fig. 1) establishes alternative and mutually exclusive workflow branches. For example, in a mortgage application, different paths may represent different branches based on conditional logic, such as credit risk or the amount to be barrowed. As paths are mutually exclusive, no parallel execution of activities occur and thus no synchronization is required, so the workflow branching converges with an OR-Join rather than an AND-Join. An OR-Split is conditional and the (single) specific transition to next activity is selected according to the outcome of the Transition Condition(s).

Cross-References

- ► Join
- ► OR-Split
- ► Process Life Cycle
- ► Workflow Management and Workflow Management System

OR-Split

Nathaniel Palmer
Workflow Management Coalition, Hingham, MA, USA

Synonyms

Branch; Conditional branching; Conditional routing; Switch

Cross-References

- ► OR-Join
- ► Process Life Cycle
- ► Split
- ► Workflow Management and Workflow Management System

OSQL

Tore Risch
Department of Information Technology, Uppsala
University, Uppsala, Sweden

Definition

OSQL [1, 2] is an functional query language
and data model similar to Daplex, first implemented in the Iris DBMS [4]. The data model
of OSQL is object oriented with three kinds of
system entities: objects, types, and functions. A
database consists of a set of objects, the objects
are classified into types, and functions define the
semantics of types. The data model is similar to
an ER model with the difference that both entity
relationships and attributes are represented as
functions and that (multiple) inheritance among
entity types is supported. OSQL provide object
identifiers (OIDs) as first class objects, and, unlike Daplex, queries can return OIDs in results.
Queries are expressed using a SELECT syntax
similar to SQL. Derived functions are also defined using select statements similar to functions
in SQL-2003.

Key Points

With the OSQL data model a database consists
of a set of objects. The objects are classified
into subsets by types and each type has an extent consisting of the objects belonging to that
type. Type inheritance is supported with the type
named OBJECT as most general type. The extent
of a type is a subset of the extents of its super-
type(s). For example if entity type STUDENT is
a subtype of type PERSON then the extent of
type STUDENT is also a subset of the extent
PERSON. Types are defined dynamically using
a CREATE TYPE statement, e.g.,:

CREATE TYPE STUDENT SUBTYPE OF
PERSON;

Functions define relationships among entities
and properties of entities. Functions can be stored

in the databases, derived in terms of other functions, or be defined as foreign functions implemented in some conventional programming language. Stored functions correspond to tables in
relational databases, and derived functions are parameterized views similar to function definitions
in SQL-2003.

Queries and derived functions are defined
declaratively using a SELECT statement, e.g.,:

SELECT NAME(P)
FOR EACH PERSON P
WHERE   AGE(P)>20   AND   SEX(P)   =
"Female";
CREATE   FUNCTION   GRANDPARENTS
(PERSON P)-> PERSON
AS SELECT PARENT(PARENT(P));

Queries are expressed as constraints over
extents. Functions composition allows easy
traversal of relationships between entity types.
As in Daplex, if a function returns a set of objects
(e.g., PARENT) functions applied on it iterate
over the elements of the set. This is a form
of extended path expressions through function
composition.

OSQL was implemented in the Iris DBMS
[4] and HP's OpenODB product. The Amos II
DBMS [3] uses a modified OSQL language,
AmosQL.

Cross-References

▶ AmosQL
▶ Daplex
▶ Functional Data Model

Recommended Reading

1. Beech D. A foundation of evolution from relational to
   object databases. In: Advances in Database Technology, Proceedings of the 1st International Conference
   on Extending Database Technology; 1988. p. 251–70.
2. Fishman DH, Beech D, Cate HP, Chow EC, Connors
   T, Davis JW, Derrett N, Hoch CG, Kent W, Lyngbaek
   P, Mahbod B, Neimat MA, Ryan TA, Shan Iris MC. An
   object-oriented database management system. ACM
   Trans Off Inf Syst. 1987;5(1):48–69.
3. Risch T, Josifovski V, Katchaounov T. Functional data
   integration in a distributed mediator system. In: Gray

P, Kerschberg L, King P, Poulovassilis A, editors. Functional approach to data management – modeling, analyzing and integrating heterogeneous data. Berlin: Springer; 2003.

4. Wilkinson K, Lyngbaek P, Hasan W. The iris architecture and implementation. IEEE Trans Knowl Data Eng. 1990;2(1):63–75.

Outlier Detection

Arthur Zimek[1,2] and Erich Schubert[3]
[1]Ludwig-Maximilians-Universität München, Munich, Germany
[2]Department of Mathematics and Computer Science, University of Southern Denmark, Odense, Denmark
[3]Heidelberg University, Heidelberg, Germany

Synonyms

Anomaly detection; Fraud detection; Identification of outliers; Rejection of outliers

Definition

Outlier detection aims at identifying those objects in a database that are unusual, i.e., different than the majority of the data and therefore suspicious resulting from a contamination, error, or fraud. In a statistical modeling, the assessment of "being unusual" is typically based on a parametric model of the data, identifying those objects that do not fit well to the modeled distribution as outliers. In the database context, the statistical intuition of "being unusual" is typically modeled in an approximate but more efficient, nonparametric way by (local) density estimates and comparison to some reference set.

Historical Background

Filtering out those observations that look suspiciously different than the majority of observations is a procedure probably tacitly practiced since people studied data collections and tried to make sense out of observations. In the eighteenth century, Bernoulli criticizes this practice among astronomers as he does not see sufficient reason to separate those out that do not fit to the model, pointing out that those rejected observations could have possibly served best to supply corrections to the model.

Statistical reasoning typically tackles the problem with parametric approaches, that is, one assumes the data follows some specific distribution and fits a model of such distribution to the data at hand. Outliers would then be those data objects that do not fit well to the fitted distribution model. Accordingly, there is an abundance of formulations of statistical tests, for different assumptions of distributions by type of parameters, number of distributions, number of variables (dimensions), etc. [1–3].

Scientific Fundamentals

Database-Oriented Efficient Outlier Models

Distance-Based Outlier Detection
The work of Knorr and Ng on the "distance-based" notion of outliers (DB-outlier) simplified statistical distribution-based approaches in order to enable efficient computation for large data sets [4]. Their method requires the specification of a distance threshold (ε) and a percentage (π). Those database objects that have less neighbors within the ε-range than specified by the percentage threshold π are considered outliers (cf. Fig. 1).

A method in the same spirit uses the distances to the k-nearest neighbors (kNN) of each object [5]. As a variant, the sum of distances to all points within the set of k-nearest neighbors (called the "weight") has been used as an outlier degree [6]. As a result, these methods do not deliver a binary decision (outlier vs. inlier) but a ranking of the objects w.r.t. the outlier degree or outlier score of the objects (as estimated by the outlier detection model), where outliers should be top ranked (cf. Fig. 2).

Outlier Detection, Fig. 1 The DB-outlier model: for a fixed neighborhood radius (ε-range query), neighbors are counted. While the amount of neighbors does not exceed $\pi = 10\%$ for point o, for point p it does. As a consequence, for the given ε and π, o will be labeled outlier; p will be labeled inlier

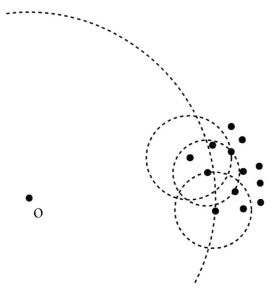

Outlier Detection, Fig. 2 The kNN outlier model: in areas of low density, larger radii are required to capture k neighbors than in areas of higher density (here: $k = 3$). The required radius (i.e., the k-distance) is used as outlier score. This allows to rank the data points according to their outlier characteristic. Clearly, o is more prominently an outlier than the other data points

The kNN-based method could be considered Curried (or Schönfinkeled) form of the DB-outlier model as its result could be seen as a function that takes an ε parameter to map to a DB-outlier solution: the decision inherent to the DB-outlier model could be derived from the kNN ranking by specifying some ε threshold as cutoff value.

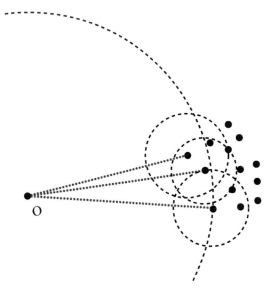

Outlier Detection, Fig. 3 The local outlier model: the density estimate for some point o is compared not to the density estimates of all the other points in the database but to the density estimates of o's local neighborhood only. This allows a flexible adaptation to local variations of typical density levels

From a statistical point of view, both models, DB-outlier and kNN, relate to density estimation using a uniform/kNN kernel.

Local Outlier Factor

To derive rankings of outlierness instead of binary decisions remained the method of choice for the design of subsequent methods. The often so-called density-based but more accurately called "local" approaches consider ratios between the local density around an object and the local density around its neighboring objects, starting with the seminal LOF ("local outlier factor") [7] algorithm (cf. Fig. 3).

Nevertheless, the method for density estimation is also a bit more refined in LOF compared to the previous models that were using the kNN distance directly. Instead, LOF uses an asymmetric *reachability distance*:

$$\text{reach-dist}_k (o \leftarrow p) = \max\{k\text{-dist}(p), d(o, p)\}$$

where k-dist(p) is the distance from p to its kth-nearest neighbor. This distance definition has been used in hierarchical density-based clustering. Based on the reachability distances, the *local reachability density* (lrd) is estimated by

$$\text{lrd}(o) := 1 \left/ \frac{\sum_{p \in k\text{NN}(o)} \text{reach-dist}_k(o \leftarrow p)}{|k\text{NN}(o)|} \right.$$

where $k\text{NN}(o)$ are the k-nearest neighbors of o.

The LOF score is a comparison of the local density estimate (lrd) for a given point o to the local density estimates (lrds) of o's neighborhood:

$$\text{LOF}(o) := \text{avg}_{n \in k\text{NN}(o)} \frac{\text{lrd}(n)}{\text{lrd}(o)}.$$

Many variants have adapted the original LOF idea in different aspects [8]. Major differences among the variants are in the method of density estimation and in the definition of "neighboring objects." In many variants of LOF, a simplified version ("simplified LOF" [8]) is used, substituting (presumably unintentionally) the regular distance $d(o, p)$ for the reachability distance.

Categories and Variants

Global and Local Outlier Detection
An important differentiation is whether an outlier score is "global" or "local." However, this is not a clear cut of methods but rather a characteristic of a "degree of locality." The differentiation of methods as "global" or "local" reflects the scale of computation and comparison of outlier score estimates for individual objects: are individual scores computed based on a local neighborhood (such as the typical so-called distance-based and the so-called "density-based" approaches) or based on the global data set (as it is typically the case in parametric statistical modeling)? Are the individual scores compared to the global database (as it is the case in the "distance-based" models DB-outlier and $k\text{NN}$ scores) or to a local reference set only (which was the genuine novelty of the "local" outlier factor (LOF))? Different variants of outlier detection in the literature are "local" in different meaning and to different degrees [8].

Distance-Based and Density-Based
The differentiation of methods into the categories "distance-based" vs. "density-based" is traditional yet meaningless. Approaches that are termed "distance-based" or "density-based" both consider more or less simplistic density estimates as base for the computation of outlier scores.

"Distance-based" methods could be seen, however, as a subset of "density-based" methods. The virtue of methods in this subset would be the simplicity of the density estimate that allows the use of indexing-related techniques for acceleration of the computation of the outlier score.

Variants for Improved Efficiency
When computing a ranking of outlier scores, one is typically only interested in the top-ranked objects (i.e., in the most likely outliers). Many efficient variants of basic models have been proposed, typically as filter-refinement processing, refining only the outlier scores of those objects that still have a chance to be among the top-n outliers. Filters are typically based on approximate neighborhood computation, pruning of partitions (during neighborhood search or during outlier search) and ranking of (neighborhood or outlier) candidates. Orair et al. [9] analyze and categorize such variants and the techniques used.

Application-Specific Methods and Generalized Models
The concept of "local density" can be abstracted to handle more complex data types: while density – or some other measure – is computed usually on numerical attributes, the neighborhood can be based on a different data source or a non-Euclidean semantic of closeness, and standard methods can thus be applied to complex data. Examples of such adaptations include the application to geo-spatial/temporal data, graph data, video streams, or text analysis [8].

Recent Developments
The challenge of high-dimensional data has triggered development of specialized solutions for outlier detection. Many database methods for outlier detection rely on nearest neighbor retrieval.

Since nearest neighbor search is negatively impacted by high dimensionality, there are two main categories of approaches to outlier detection in high-dimensional data: (i) methods to improve efficiency or effectiveness of outlier detection in a high-dimensional data space or (ii) methods identifying outliers in subspaces of a high-dimensional data space. For both directions, there are some first approaches, but there remain also many open challenges [10].

Another recent direction is to apply ensemble learning techniques, well known and studied in the area of classification or clustering, to the outlier detection problem. This transfer is nontrivial. For building effective ensembles, minimally accurate yet diverse base methods are required and their predictions need to be combined. All three requirements, accuracy, diversity, and combination procedures, have found only heuristic solutions in outlier detection so far [11].

Also methods for nonnumerical data types, such as symbolic sequences or graph data, find increasing attention in the research community [12, 13].

Key Applications

Traditionally, outlier detection was mostly used for data cleaning, as it can improve the results of other analysis methods that are sensitive to extreme deviations. With manually collected data, contamination would often stem from unintentional errors. However, not every unusual observation is an error – and non-erroneous, unusual data can often provide new insights. Furthermore, in electronic communication and automated data, an increasing amount of manipulation can be observed, such as insurance fraud. An outlier in network activity can arise from an attack. An unusual trading activity may be caused by manipulation or a software error. Outlier detection can be used when the exact nature of the effect is unknown beforehand: while an engineer may expect a device to fail eventually, it is hard to predict the exact way it will fail.

With the increasing volume of data, outliers become more and more common, and thus many

applications have to deal with thousands of outliers. For example, a LiDAR sensor will yield erroneous height measurements due to reflections and refraction that need to be quickly identified and filtered. At 150,000 laser pulses a second, a 1 in a million error will occur several times per minute.

Application scenarios thus include detecting human error, or fraud, data contamination, electronic attacks, machine failures, and interviewer fabrication, finding contradicting examples for a theory, detecting rare medical conditions, and cleaning data for further analysis with other methods.

Future Directions

Since ranking methods typically abstain from deciding on a good cutting point in the ranking (i.e., they do not by default deliver a binary decision based on the ranking), an open problem is to define a good value of n for the top-n outliers or to define a meaningful threshold for outlier scores. Also, improving the interpretability of outlier scores and the explanation of outlierness is a topic that just starts to gain attention in the research community [14]. However, these decisions will always remain highly application dependent. After all, the decision of some method about some database object being an outlier would only mean that this object is suspicious, i.e., a domain expert needs to examine the case.

Applying ensemble techniques to outlier detection is a novel direction of great potential where many open challenges remain [11].

Tackling high-dimensional data remains challenging as well. Existing methods are pointing out the challenges rather than truly tackling them [10].

URL to Code

Many standard methods for outlier detection are available in the open-source data mining framework ELKI [15] at http://elki.dbs.ifi.lmu.de/.

Cross-References

▶ Anomaly Detection on Streams
▶ Data Mining
▶ Density-Based Clustering
▶ Ensemble
▶ Indexing and Similarity Search
▶ Range Query

Recommended Reading

1. Hawkins D. Identification of outliers. London: Chapman and Hall; 1980.
2. Barnett V, Lewis T. Outliers in statistical data. 3rd ed. Chichester: Wiley; 1994.
3. Rousseeuw PJ, Hubert M. Robust statistics for outlier detection. Wiley Interdiscip Rev Data Min Knowl Discov. 2011;1(1):73–9.
4. Knorr EM, Ng RT, Tucanov V. Distance-based outliers: algorithms and applications. VLDB J. 2000;8(3–4):237–53.
5. Ramaswamy S, Rastogi R, Shim K. Efficient algorithms for mining outliers from large data sets. In: Proceedings of the ACM SIGMOD International Conference on Management of Data; 2000. p. 427–38.
6. Angiulli F, Pizzuti C. Outlier mining in large high-dimensional data sets. IEEE Trans Knowl Data Eng. 2005;17(2):203–15.
7. Breunig MM, Kriegel HP, Ng RT, Sander J. LOF: identifying density-based local outliers. In: Proceedings of the ACM SIGMOD International Conference on Management of Data; 2000. p. 93–104.
8. Schubert E, Zimek A, Kriegel HP. Local outlier detection reconsidered: a generalized view on locality with applications to spatial, video, and network outlier detection. Data Min Knowl Disc. 2014;28(1):190–237.
9. Orair GH, Teixeira C, Wang Y, Meira Jr W, Parthasarathy S. Distance-based outlier detection: consolidation and renewed bearing. Proc VLDB Endow. 2010;3(2):1469–80.
10. Zimek A, Schubert E, Kriegel HP. A survey on unsupervised outlier detection in high-dimensional numerical data. Stat Anal Data Min. 2012;5(5):363–87.
11. Zimek A, Campello RJGB, Sander J. Ensembles for unsupervised outlier detection: challenges and research questions. ACM SIGKDD Explor. 2013;15(1):11–22.
12. Chandola V, Banerjee A, Kumar V. Anomaly detection for discrete sequences: a survey. IEEE Trans Knowl Data Eng. 2012;24(5):823–39.
13. Akoglu L, Tong H, Koutra D. Graph-based anomaly detection and description: a survey. Data Min Knowl Disc. 2014; https://doi.org/10.1007/s10618-014-0365-y.
14. Kriegel HP, Kröger P, Schubert E, Zimek A. Interpreting and unifying outlier scores. In: Proceedings of the 11th SIAM International Conference on Data Mining; 2011. p. 13–24.
15. Achtert E, Kriegel HP, Schubert E, Zimek A. Interactive data mining with 3D-parallel-coordinate-trees. In: Proceedings of the ACM SIGMOD International Conference on Management of Data; 2013. p. 1009–12.

Overlay Network

Wojciech Galuba and Sarunas Girdzijauskas
EPFL, Lausanne, Switzerland

Definition

An *overlay network* is a communication network constructed on top of an another communication network. The nodes of the overlay network are interconnected with logical connections, which form an *overlay topology*. Two overlay nodes may be connected with a logical connection despite being several hops apart in the underlying network. Overlay networks may define their own *overlay address space* which is used for efficient message routing in the overlay topology.

Key Points

When a distributed application is deployed in a computer network, the individual nodes on which the application is running need to be able to discover and communicate with one another. A solution to this problem is the overlay network. The overlay network interconnects all the application nodes and provides the basic communication primitives such as flooding, random walks or point-to-point overlay message routing and multicast.

Overlay networks are typically deployed on top of the Internet and by far the most common usage is in peer-to-peer systems. For example, Gnutella, an early peer-to-peer file sharing system connects all the peers in an overlay network, each peer shares its files, and files are searched for using query flooding in the overlay network.

Overlay network topologies can be divided into two broad classes: unstructured and structured. Unstructured overlay networks do not construct a globally consistent topology, instead peers choose their neighbor sets independently and in a largely ad-hoc way. In unstructured overlay networks nodes reach the other nodes by message flooding or random walks. Structured overlays define an address space and each of the overlay nodes has a unique address. The addresses are used to construct an overlay topology that enables efficient and scalable messages passing between the overlay nodes. In most of the modern structured overlays the expected number of routing hops scales as $O(\log N)$ with the network size. Distributed Hash Tables are a specific case of structured overlay networks. Apart from structured and unstructured there also exist hybrid overlays.

Overlay networks are designed to be robust to *churn*, i.e., arrivals and departures of the overlay network nodes to and from the network. As overlay network nodes loose their overlay topology connections, new connections have to be added in their place. In structured overlay networks the additional challenge is to maintain the overlay topology such that the overlay routing remains efficient, i.e., the routing paths are kept short.

Cross-References

▶ Distributed Hash Table
▶ Peer-to-Peer Overlay Networks: Structure, Routing and Maintenance
▶ Peer-to-Peer System

OWL: Web Ontology Language

Sean Bechhofer
University of Manchester, Manchester, UK

Synonyms

Web ontology language

Definition

The Web Ontology Language OWL is a language for defining ontologies on the Web. An OWL Ontology describes a domain in terms of classes, properties and individuals and may include rich descriptions of the characteristics of those objects. OWL ontologies can be used to describe the properties of Web resources. Where earlier representation languages have been used to develop tools and ontologies for specific user-communities in areas such as sciences, health and e-commerce, they were not necessarily designed to be compatible with the World Wide Web, or more specifically the Semantic Web, as is the case with OWL.

Features of OWL are a collection of expressive operators for concept description including boolean operators (intersection, union and complement), plus explicit quantifiers for properties and relationships; the ability to specify characteristics of properties, such as transitivity or domains and ranges; a well defined semantics facilitating the use of inference and automated reasoning; the use of URIs for naming concepts and ontologies; a mechanism for importing external ontologies; and compatability with the architecture of the World Wide Web, in particular other representation languages such as RDF and RDF Schema.

OWL consists of a suite of World Wide Web Consortium (W3C) Recommendations - six documents published in February 2004 describe Use Cases and Requirements, an Overview of the language, a Guide, Reference, OWL Semantics and a collection of Test Cases [3].

Key Points

Ontology languages allow the representation of ontologies. An ontology defines a set of representational primitives with which to model a domain of knowledge or discourse (see Ontology). The definition of an ontology can encompass a wide range of artefacts, from simple word lists, through taxonomies, thesauri and rich logic-based models and there are a corresponding range of languages for their representation.

Standardization of representation languages is a cornerstone of the Semantic Web effort. A standard representation facilitates interoperation - in particular, well-defined, unambiguous *semantics* ensure that applications can agree on the meaning of expressions. OWL is intended to provide that standard representation.

OWL builds on RDF and RDF Schema and adds more vocabulary for describing properties and classes. The design of the language was influenced by a number of factors. Description Logics, Frame-based modeling paradigms, and Web languages RDF and RDF Schema were key inputs, as was earlier work on languages such as OIL and DAML+OIL.

Knowledge Representation in a Web setting introduces particular requirements such as the distribution across many systems; scalability to Web size; compatibility with Web standards for accessibility and internationalization; and openness and extensibility. OWL uses URIs for naming and extends the description framework for the Web provided by RDF to address some of the issues above.

OWL defines three sublanguages: OWL Lite, OWL DL and OWL Full. OWL Full is essentially RDF extended with additional vocabulary, with no restrictions on the way in which that vocabulary is used. OWL DL places restrictions on the way in which the vocabulary can be used in order to define a language for which a number of key reasoning tasks (for example concept satisfiability or subsumption) are decidable. OWL Lite further restricts the expressivity allowed - for example, explicit union or complement arc disallowed in OWL Lite. OWL DL and OWL Lite have a model theoretic semantics that corresponds to a Description Logic (DL) [1] and thus facilitate the use of DL reasoners to provide reasoning support for the language [2].

The design of representation languages often involves trade-offs, and there are limitations on what can be expressed using OWL, in particular in OWL-DL. These limitations have been selected primarily to ensure that these language subsets are well-behaved computationally, with decidable procedures for concept satisfiability. For example, OWL does not provide support for general purpose rules, which are seen as an important paradigm in knowledge representation, for example in expert systems or deductive databases. Extensions to OWL are being proposed to cover, among others, rules, query, additional expressivity, metamodeling and fuzzy reasoning.

Cross-References

▶ Description Logics
▶ Ontology
▶ Resource Description Framework
▶ Resource Description Framework (RDF) Schema (RDFS)
▶ Semantic Web

Recommended Reading

1. Baader F, Calvanese D, McGuinness DL, Nardi D, Patel-Schneider PF, editors. The description logic handbook: theory, implementation, and applications. Cambridge: Cambridge University Press; 2003.
2. Horrocks I, Patel-Schneider PF, van Harmelen F. From *SHIQ* and RDF to OWL: the making of a web ontology language. J Web Semant. 2003;1(1):7–26.
3. World Wide Web Consortium. Web Ontology Language (OWL). W3C Recommendation. Available at: http://www.w3.org/2004/OWL/.

P

P/FDM

Peter M. D. Gray
University of Aberdeen, Aberdeen, UK

Definition

P/FDM [5–7] integrated a functional data model with the logic programming language Prolog for general-purpose computation. The data model can be seen as an Entity-Relationship diagram with sub-types, much like a UML Class Diagram. The idea was for the user to be able to define a computation over objects in the diagram, instead of just using it as a schema design aid. Later versions of P/FDM included a graphic interface [2, 4] to build queries in DAPLEX syntax by clicking on the diagram and filling in values from menus.

P/FDM was subsequently extended with constraints [3] and with alternative back-ends to remote databases [6], in the spirit of the original MULTIBASE system.

P/FDM is a vehicle to test a system designed on the principle of Data Independence, whereby Functions represent computations that are expressed in a way that is completely independent of data storage (arrays, lists of objects, indexed files etc.). Functions can be sent across the internet and applied to data in a different form, which was very useful for federated data.

Key Points

In P/FDM, the DAPLEX language was deliberately altered from its original specification so that its semantics could be defined by equivalent *Comprehensions* [1]. In particular, simple assignment operations, if present, could only take place within the innermost loop. Based on this, a very successful early optimizer was written in Prolog by Paton, and used in several bioinformatics applications [6, 8].

In P/FDM, data independence is ensured by using a small sparse set of built-in predicates *getentity(entity-type, key, instance-variable)* and *getfunctionval(function-name, argument-variable, value-variable)*. The first of these predicates requires that abstract entities be identifiable by unique, possibly compound, scalar keys [7]. The keys help with object identity and are very important when accessing or loading bulk data. This is a feature first used in ADAPLEX and EFDM, though P/FDM extends it to allow composed (single-valued) functions, which aids in forming hierarchical keys.

P/FDM queries are written in DAPLEX and translated into Prolog for evaluation. Backtracking in Prolog is used to give the effect of lazy evaluation in a functional programming language, accessing tuples or objects on demand. For example, assuming the following P/FDM declarations:

© Springer Science+Business Media, LLC, part of Springer Nature 2018
L. Liu, M. T. Özsu (eds.), *Encyclopedia of Database Systems*,
https://doi.org/10.1007/978-1-4614-8265-9

```
declare student ->> entity
declare name(student) -> string
key_of student is name
declare course ->> entity
declare cname(course) -> string
key_of course is cname.
declare attends(student) ->> course
```
to print all the courses attended
by Fred Jones, the DAPLEX query is:
```
for each F in student such that
name(F) = "Fred Jones"
for each C in attends(F)
print(cname(C));
```

which generates the following Prolog:

```
getentity(student,&apos;Fred Jones
&apos;,F),getfunctionval(attends,F,C),
getfunctionval(cname,C,N), write(N),
fail; true.
```

Cross-References

▶ Comprehensions
▶ Functional Query Language
▶ Query Languages and Evaluation Techniques for Biological Sequence Data

Recommended Reading

1. Embury SM. User manual for P/FDM V.9.1. Technical report, Department of Computing Science, University of Aberdeen; 1995.
2. Gil I, Gray PMD, Kemp GJL. A Visual interface and navigator for the P/FDM object database. In: Proceedings of the User Interfaces to Data Intensive Systems; 1999. p. 54–63.
3. Gray PMD, Embury SM, Hui KY, Kemp GJL. The evolving role of constraints in the functional data model. J Intell Inf Syst. 1999;12(2–3):113–37.
4. Gray PMD, Kemp GJL. Capturing quantified constraints in FOL, through interaction with a relationship graph. In: Proceedings of the 15th International Conference on Knowledge Engineering and Knowledge Management: Ontologies and the Semantic Web; 2006. p. 19–26.
5. Gray PMD, Moffat DS, Paton NW. A Prolog interface to a functional data model database. In: Advances in Database Technology, Proceedings of the 1st International Conference on Extending Database Technology; 1988. p. 34–48.
6. Kemp GJL, Dupont J, Gray PMD. Using the functional data model to integrate distributed biological data sources. In: Proceedings of the 8th International Conference on Scientific and Statistical Database Management; 1996. p. 176–85.
7. Paton NW, Gray PMD. Identification of database objects by key. In: Proceedings of the 2nd International Workshop on Object-Oriented Database Systems; 1988. p. 280–5.
8. Paton NW, Gray PMD. Optimising and executing daplex queries using prolog. Comput J. 1990;33(6):547–55.

Parallel and Distributed Data Warehouses

Ladjel Bellatreche[1], Todd Davis[2], and Belayadi Djahida[3]
[1]LIAS/ISAE-ENSMA, Poitiers University, Futuroscope, France
[2]Department of Computer Science and Software Engineering, Concordia University, Montreal, QC, Canada
[3]National High School for Computer Science (ESI), Algiers, Algeria

Synonyms

High performance data warehousing; Scalable decision support systems

Definition

With the era of Big Data, we are facing a data deluge (http://www.economist.com/node/15579717). Multiple data providers are contributing to this deluge. We can cite three main examples : (i) the massive use of sensors (e.g. 10 Terabyte of data are generated by planes every 30 min), (ii) the massive use of social networks (e.g., 340 million tweets per day), (iii) transactions (Walmart handles more than one million customer transactions every hour, which is imported into databases estimated to contain more than 2.5 petabytes of data). The decision makers need fast response time to their requests in order to predict in *real time* the behavior of users, so they can offer them services via analyzing large volumes of data. The data warehouse ($\mathcal{DW}$) technology deployed on conventional platforms (e.g. centralized) has become obsolete, even with the spectacular progress in terms of advanced optimization structures (e.g., materialized views, indexes, storage layouts, etc.). Despite this, the sole use of these structures is not sufficient to gain efficiency during the evaluation complex OLAP queries over relational $\mathcal{DW}$. To deal with this data deluge and simultaneously satisfy the requirements of companies decision makers, distributed and parallel platforms have been proposed as a robust and scalable solution to store, process and analyze data, with the layers of modern analytic infrastructures.

Historical Background

The terms "parallel $\mathcal{DW}$" and "distributed $\mathcal{DW}$" are very often used *interchangeably*. In practice, however, the distinction between the two has historically been quite significant. Distributed $\mathcal{DW}$s, much like distributed databases, grew out of a need to place processing logic and data in close proximity to the users who might be utilizing them. In general, multi-location organizations (motivated by globalization) consist of a small number of distinct sites, each typically associated with a subset of the information contained in the global data pool. In the $\mathcal{DW}$ context, this has traditionally led to the development of some form of *federated* architecture. In contrast to monolithic, centralized $\mathcal{DW}$s, federated models are usually constructed as a cooperative coalition of departmental or process specific *data marts* [8]. A simple example is illustrated in Fig. 1. For the most part, design and implementation issues in such environments are similar to those of distributed operational DBMSs [20]. For example, it is important to provide a *single transparent conceptual model* for the distinct sites and to divide data so as to reduce the effects of *network latency* and *bandwidth limitations*. In this context three main issues shall be addressed; *two* inherited from traditional distributed databases which are the data partitioning [19] and fragment allocation and the third one concerns the process of ETL (Extract, Load, Transform).

With respect to parallelism, the research efforts have again been influenced by parallel DBMS projects such as Gamma [12] that were initiated from the middle to late 1980s. By the 1990s, it had become clear that commodity-based "shared nothing" databases provided significant advantages over the earlier SMP (Symmetric Multi-processor) architectures in terms of cost and scalability [13]. Subsequent research therefore focused on partitioning and replication models for the tables of the parallelized DBMS [24]. Note that the data partitioning is a fundamental phase of any parallel database/$\mathcal{DW}$

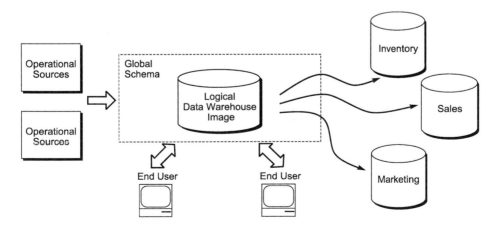

Parallel and Distributed Data Warehouses, Fig. 1 A simple federated model illustrating the mapping of a logical global schema on to a series of physically independent data marts

design methodology, and can also be considered as a *pre-condition* of this design.

Inside parallel solutions dedicated to $\mathcal{DW}$ we find two main tendencies. First, *turnkey parallel platforms* such as Teradata (www.teradata.com) offer analytic data platforms, applications and related services. In line with these major trends, Small and Medium-sized Enterprises cannot afford license fees and costs of installation and maintenance (stirred-up by the current economic crisis). As a consequence, clusters of workstations are very often used as a valid alternative to turnkey platforms. This parallel platform is essentially constructed as a series of commodity DBMS systems, "glued together" with a thin partition-aware wrapper layer. The most used parallel DW systems are Oracle Exadata, Teradata, IBM Netezza, Microsoft SQL server Parallel datawarehouse, Greenplum.

In addition to the continuation of the traditional partitioning work, a second important theme has been the parallelization of state-of-the-art sequential *data cube* generation methods. The data cube has become the primary abstraction for the multi-dimensional analysis that is central to modern $\mathcal{DW}$ systems. Cube parallelization efforts have in fact taken two forms, one based upon data that is physically represented as array-based storage [15], the other based upon relational storage [11, 16]. In both cases, the complexity of the algorithm and implementation issues has led to the development of relatively complete DBMS prototypes.

Scientific Fundamentals

Central to $\mathcal{DW}$ systems is a de-normalized logical multi-dimensional model known as the Star Schema (the normalized version is referred to as a Snowflake). A Star Schema consists of one or more extremely large table(s), called *fact* table(s) housing the measurement records associated with a given organizational process and several dimension tables with reasonable size that define specific business entities (e.g., customer, product, store). Queries defined on a star schema are called star join queries. Evaluating a star-join query

is *expensive* because the fact table participates in every join operation. Therefore, it is worth developing an efficient method for processing star-join queries. Typically, the basic Star Schema is augmented with compact, pre-computed aggregates (called *group-bys* or *cuboids*) that can be queried much more efficiently at run-time. This collection of aggregates and base data is known as the data cube. Specifically, for a d-dimensional space, $\{A_1, A_2, \ldots, A_d\}$, the cube defines the aggregation of the 2^d unique dimension combinations across one or more relevant measure attributes. In practice, the generation and manipulation of the data cube is often performed by a dedicated OLAP (on-line analytical processing) server that runs on top of the underlying relational data warehouse. While the OLAP server may utilize either array-based (MOLAP) or table-based (ROLAP) storage, both provide the same, intuitive multi-dimensional representation for the end user.

Designing a parallel $\mathcal{DW}$ goes through a well-identified life cycle including six main steps (Fig. 2) [6]: (1) choosing the hardware architecture, (2) partitioning the target $\mathcal{DW}$, (3) allocating the so-generated fragments over available nodes, (4) replicating fragments for efficiency purposes, (5) defining efficient query processing strategies and (6) defining efficient load balancing strategies. Each one of these steps will be detailed in the following paragraphs.

The choice of the platform: The choice of the deployment platform is a pre-condition to achieve the scalability and availability of data. Hardware architectures are classified into five main categories: (i) shared-memory, (ii) shared-disk, (iii) shared-nothing, (iv) shared-something and (v) shared-everything. As we said before, the **Shared-everything** architecture has been adopted by $\mathcal{DW}$ for parallelization purpose. **Shared memory and shared disk architectures** have the advantage that they are relatively easy to administer as the hardware transparently performs much of the "magic". This architecture has been recommended by the leaders of Gamma project as the reference architecture for supporting high-performance data warehouses modeled in terms of relational

Parallel and Distributed Data Warehouses, Fig. 2 The steps of the life cycle of parallel $\mathcal{DW}$ design

star schemes. That being said, shared everything designs also tend to be quite expensive and have limited scalability in terms of both the CPU count and the number of available disk heads. In Petabyte-scale $\mathcal{DW}$ environments, either or both of these constraints might represent a serious performance limitation. As the choice of the hardware architecture is influenced by price, high-performance features, extensibility and data availability, clusters of workstations are very often used as a valid alternative to Shared-Nothing architectures (e.g., [18]). Figure 3 proposes a model instantiating each architecture, by describing its components.

Data partitioning can be done horizontally, vertically and mixed [20]. The Horizontal Partitioning ($\mathcal{HP}$) is essentially used to design distributed/parallel relational $\mathcal{DW}$. $\mathcal{HP}$ is the process of splitting access objects (tables, materialized views, indexes) into set of disjoint rows. It was first introduced at the end of 1970s and beginning of the 1980s [10] for logically designing databases in order to improve the query performance by eliminating unnecessary accesses to non-relevant data. It was welcomed with a large success in designing homogeneous distributed databases [10, 20] and in (in the beginning of the 1990s) in designing parallel databases [13]. $\mathcal{HP}$ contributes in speeding query performance in the shared-nothing architecture, by offering inter-query and intra-query parallelisms [13]. In this case, an individual query Q_i is decomposed and simultaneously executed across the $\mathcal{P}$ nodes of the system, with each *partial query* running against approximately $(1/\mathcal{P})$ records of the partitioned fact table (the small dimension tables are typically replicated on each node). Figure 4b provides a simple example of this technique, contrasting with the non-partitioned model typically utilized on a single node server (Fig. 4a). Though merging of results may be necessary, it is important to note that while the input parti-

tions may be massive, the output of user-directed analytical $\mathcal{DW}$ queries is typically quite small. In the context of $\mathcal{DW}$ deployed in a centralized platform, $\mathcal{HP}$ has been considered as a part of the physical design [1], where in most of today's commercial and academic database systems propose native Data Definition language (DDL) support for defining $\mathcal{HP}$ for a table by the means of various partitioning modes: *range*, *list*, *hash* and *composite*.

There are two versions of $\mathcal{HP}$ [20]: *primary* and *derived* (Fig. 5). Primary $\mathcal{HP}$ of a relation R is performed using selection predicates that are defined on that relation. Note that a simple predicate (selection predicate) has the following form: $< A\ \theta\ value >$, where A is an attribute of the table R, θ is one of six comparison operators $\{=, <, >, \leq, \geq, \neq\}$ and value is the predicate constant belonging to the domain of the attribute A. When a partitioning schema (A partitioning schema is the result of the partitioning process of a given table.) of a table R is propagated to another table S, the result of the $\mathcal{HP}$ of the table S is called derived $\mathcal{HP}$. The type of partitioning is *feasible* if and only if a join link between the relations R and S exists [10]. This partitioning has been implemented in Oracle 11 G under the name of *Referential Partitioning* in 2008.

The formalization of $\mathcal{HP}$ problem in the context of distributed/parallel $\mathcal{DW}$ is quite different in the context of traditional databases. The main difference concerns the unit of partitioning (table vs. schema).

In the traditional distributed databases, the general formulation of the $\mathcal{HP}$ problem is the following: *given a relation R_i of a given database schema and a set of most frequently asked queries* $\mathcal{Q} = \{Q_1, Q_2, \ldots, Q_L\}$, the problem of $\mathcal{HP}$ consists in fragmenting R into a set of fragments $F^{R_i} = \{F_1^{R_i}, F_2^{R_i}, \ldots, F_N^{R_i}\}$, where every fragment $F_j^{R_i}$ $(1 \leq i \leq N)$ stores a subset of the tu-

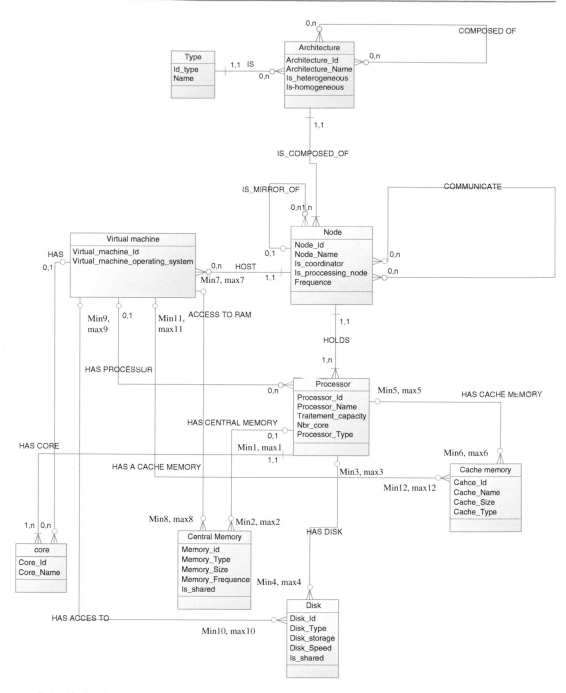

Parallel and Distributed Data Warehouses, Fig. 3 Model of deployment architectures

ples of a relation R_i such as the sum of the query costs when executed on top of the partitioned schema is minimized. The number of generated fragments is not known in advance; although it has a great impact on query processing (since a global query will be rewritten on the generated fragments) and data allocation (since one of the inputs of data allocation is the fragments). Several $\mathcal{HP}$ algorithms have been proposed in this context, that we can classify into three categories:

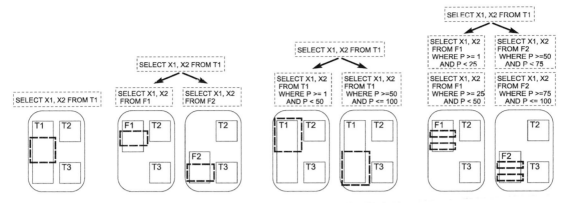

Parallel and Distributed Data Warehouses, Fig. 4 Fundamental $\mathcal{DW}$ partitioning schemes. (**a**) Query executed on a single node server that houses a large fact table (*T1*) and two small dimension tables (*T2, T3*). (**b**) Physical partitioning, creating fact table *fragments* (*F1, F2*). (**c**) Virtual partitioning on the *clustering attribute* **P** of the replicated fact table. (**d**) Adaptive virtual partitioning, using fragments and multiple sub-queries

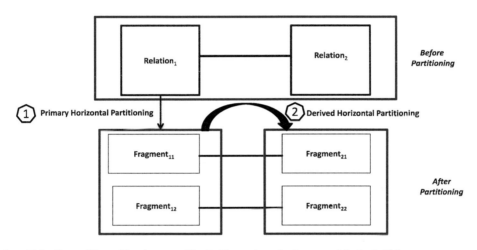

Parallel and Distributed Data Warehouses, Fig. 5 Illustration of primary and derived $\mathcal{HP}$

predicate-based algorithms [20], *affinity-based algorithms* [17] and c*ost-based algorithms* [9].

In the distributed and parallel $\mathcal{DW}$, the $\mathcal{HP}$ problem is formalized as follows: let a $\mathcal{DW}$, usually modeled using a $-+$ with one fact table and k dimension tables, a set of most frequently asked queries $\mathcal{Q} = \{Q_1, Q_2, \ldots, Q_L\}$ and a partitioning threshold $\mathcal{W}$ fixed by the designer representing the number of the final fragments that she/he wants to have [25]. Then the problem of $\mathcal{HP}$ consists in fragmenting some/all dimension tables using the primary $\mathcal{HP}$, and use them to derive partitioning the fact table such as the query processing cost is minimized. This formalization takes into consideration the queries requirements and the relationship between the fact table and dimension tables. Most of the existing algorithms in this context are based on a cost model. The main role of a cost model is to guide de partitioning algorithms by ensuring a partitioning satisfying the non functional requirements of the designer such as query performance, $\mathcal{DW}$, etc.

A large panoply of projects involving $\mathcal{HP}$ has been set in the parallel $\mathcal{HP}$ context. Essentially, there are *three forms* of $\mathcal{HP}$ partitioning. In the first case, the fact tables are physically partitioned using the derived partitioning. While the performance with this approach can be impressive [25], it is also true that imbalances caused by inherent data skew make effective a priori data striping quite challenging. In response, the *virtual* partition model was proposed [2]. Here, as illustrated in Fig. 4c, the fact tables are replicated in full across the nodes of the DBMS cluster. Queries are then decomposed into sub-queries $(1/P)$ records per node that are run against virtual partitions mapped on top of the fully replicated tables. The advantage is that sub-queries may be (i) run against any cluster node and (ii) dynamically migrated to under-utilized nodes. The downside is (i) the storage requirement associated with p copies of the primary fact table and (ii) the fact that the commodity DBMS will invoke a full table scan if the partitions are too large. The third partitioning method attempts to combine the best features of the previous methods. In *adaptive virtual partitioning* (AVP) [14], small virtual partitions are layered on top of larger physical partitions. The AVP algorithm dynamically determines the maximum sub-query size that does not invoke a table scan by iteratively probing the commodity DBMS with successively larger queries. The result is a set of re-locatable sub-queries that can generally be answered efficiently without access to the code base of the commodity DBMS systems. A simple illustration is provided in Fig. 4d. The use of partitioning throughout the database generations is illustrated in Fig. 6.

Data allocation consists in placing generated fragments over nodes of a reference parallel machine. This allocation may be either redundant (with replication) or non-redundant (without replication). Once fragments are placed, global queries are executed over the processing nodes according to parallel computing paradigms. In more detail, parallel query processing on top of a parallel DW (a critical task within the family of parallel computing tasks) includes the following phases: (i) rewriting the global query according to the fixed DW partitioning schema; (ii) scheduling the evaluation of so-generated sub-queries over the parallel machine according to a suitable allocation schema. Generating and evaluating sub-queries such that the query workload is evenly balanced across all the processing nodes is the most difficult task in the parallel processing above.

Data replication is related to two major problems: (1) replica creation and (2) maintenance of materialized replicas. Our study focuses on the problem of creating replicas which is strongly influenced by the number of replicas and their placement. Indeed, the reference Replica Placement Problem (RPP) consists in choosing the best replica placement on the distributed system in order to optimize given performance criteria. The optimal replica placement problem has been

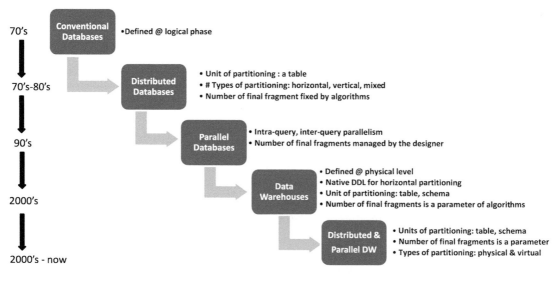

Parallel and Distributed Data Warehouses, Fig. 6 The use of partitioning throughout the database evolution

shown to be an NP-Hard problem [26]. Therefore, it is often solved by means of approximate solutions in a feasible time, and a relevant amount of work has been devoted to this paradigm in the literature [18].

Load imbalance can be caused by one or both of the following two phenomena: (i) *data skew*, which refers to the situation where data are unevenly distributed across the local memories of different processors. This usually occurs when the data partitioning function uses attributes whose data value distributions are non-uniform; (ii) *processing skew*, which is caused by the unpredictable nature of the processing itself and it may be propagated by the data skew at the beginning. It refers to the situation where a significant part of the workload is executed by a few processors while other processors are relatively idle.

Load balancing is usually performed by means of the so-called multi-reordering process. According to this process, multiple processors that have small average loads are selected in order to participate to the load balancing. Then, each free processor is moved as to become adjacent (according to the node network topology) to a high-loaded processor, the load of which is then shared with the (newly-introduced) free processor. This so-determined data migration task may cause high communication costs, which overall lower the global throughput of the parallel $\mathcal{DW}$ architecture. It is well-understood that communication cost is a factor that must be mastered depending on the available infrastructure, and that most of data access must be local (for efficiency purposes). Therefore, data replication has become a strict requirement of parallel $\mathcal{DW}$ architectures in order to guarantee avoiding bottlenecks and reducing communication costs. To this end, replication aims at (i) ensuring data availability and fault tolerance, (ii) improving data locality by following the criterion of placing a job at the same node where its data are located, and (iii) achieving load balancing by distributing work across data replicas.

Based on the above discussion, we assert that parallel $\mathcal{DW}$ design can be modeled by the following tuple [7]: $< Arch, DP, DA, DR, LB >$,

where $Arch$ denotes the parallel architecture, DP the data partitioning scheme, DA the data allocation scheme, DR the data replication scheme and LB the load balancing scheme, respectively. Note that the problem corresponding to each component of the above tuple is NP-hard [3,5,26].

Two main methodologies exist to deal with the problem of parallel $\mathcal{DW}$ design: *iterative design* and *conjoint design*. Iterative design methodologies have been proposed in the context of traditional distributed and parallel database design [14,21,23,25]. The idea underlying this class of methodologies consists in first fragmenting the relational $\mathcal{DW}$ using any partitioning algorithm, and then allocating the so-generated fragments by means of any allocation algorithm. In the most general case, each partitioning and allocation algorithm has its *own cost model*. The main limitation of iterative design is represented by the fact that they neglect the *inter-dependency* between the data partitioning and the fragment allocation phase, respectively. Figure 7a summarizes the steps of iterative design methodologies. To overcome these limitations, the combined design methodologies were proposed in [6]. They consist in merging some phases of the life cycle of the distributed/parallel life cycle. The main idea of this merging is that a phase is performed knowing the requirements of the next step. Three main variants of this methodology are distinguished (Fig. 7b–d).

Figure 7b depicts an architecture where the basic idea consists in first partitioning the $\mathcal{DW}$ using any partitioning algorithm and then determining how fragments are allocated to the nodes by also determining replication of fragments. The main advantage of this architecture is that it takes into account the inter-dependency between allocation and replication, which are closely related [6]. The main limitation is the fact that it neglects the inter-dependency between the data partitioning and fragment allocation. Figure 7c shows a second variant of the conjoint design, in which the $\mathcal{DW}$ is first horizontally partitioned into fragments, and then fragments are allocated to nodes within the *same* phase as in [23]. After that, a replication algorithm is used

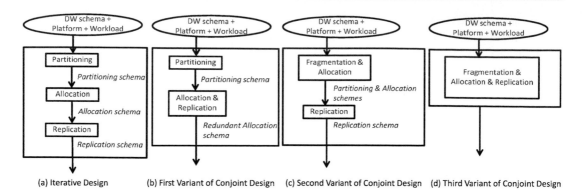

Parallel and Distributed Data Warehouses, Fig. 7 Iterative and variants of conjoint design methodologies

in order to determine how to allocate replicated fragments. The advantage of this variant consists in performing the allocation phase at the partitioning time, in a simultaneous manner. However, the drawback of the architecture is that it neglects the closely-related dependency between the allocation and the replication process. Finally, Fig. 7d depicts an architecture such that partitioning, allocation and replication are combined into a *unified process*.

The basic difference between the four reference architectures depicted in Fig. 7 is represented by the selection of partitioning attributes. The iterative approach determines the partitioning attributes using a cost model that neglects the inter-dependency between partitioning, allocation and replication.

Formally, our parallel $\mathcal{D}W$ design problem that considers the different phases of the life cycle can be formalized as the following *Constraint Optimization Problem*. Given:

(_) a parallel platform (PP) with $\mathcal{P}$ nodes $\mathcal{N} = \{N_1, N_2, \ldots, N_P\}$; (_) a relational $\mathcal{D}W$ modeled according to a star schema and composed of one fact table $\mathcal{F}$ and d dimension tables $\mathcal{D} = \{D_1, D_2, \ldots, D_d\}$ – similarly to [18], we suppose that all dimension tables are replicated over the nodes of the parallel platform and can fit in main memory; (_) a set of star join queries $Q = \{Q_1, Q_2, \ldots, Q_L\}$ to be executed over PP, each query Q_l characterized by an access frequency f_l; (_) the *maintenance constraint*

$\mathcal{W}$ representing the number of fragments that the designer considers relevant for his/her target allocation process (note that this number must be greater than the number of nodes, i.e. $\mathcal{W} \gg P$); (_) the *replication constraint* $\mathcal{R}$, such that $\mathcal{R} \leq \mathcal{P}$, representing the number of fragment copies that the designer considers relevant for his/her parallel query processing; (_) the *attribute skewness constraint* Ω representing the degree of non-uniform value distributions of the attribute sub-domain chosen by the designer for the selection of the partitioning attributes; (_) the *data placement constraint* α representing the degree of data placement skew that the designer allows for the placement of data; (_) the *load balancing constraint* δ representing the data processing skew that the designer considers relevant for his/her target query processing.

The problem of designing a $\mathcal{D}W$ over the parallel platform (PP) consists in *fragmenting the fact table $\mathcal{F}$ based on the partitioning schemes of some/all dimension tables into $\mathcal{N}, N \gg P$ fragments and allocating them and the replicated fragments over different nodes of the parallel platform such that the total cost of executing all the queries in Q can be minimized while all constraints of the problem are satisfied.*

Due to the complexity of this problem, in [6, 7], the authors proposed a stepwise approach in which the second and the third variants of the conjoint design approach are considered (Fig. 7). Several classes of algorithms (genetic, hill climbing, etc.) guided by a mathematical cost model have been proposed. Test batteries were conducted to evaluate the efficiency of our proposal in terms of our cost model and a real validation on *Teradata machine*.

Future Directions

In the short term, one would expect to see continued interest in the exploitation of commodity DBMS systems within loosely federated parallel DBMS clusters. This approach makes a great deal of sense given the considerable maturity of such platforms. Beyond this, OLAP query performance might be further improved by investigating the parallelization of more recent sequential cube methods.

In the longer term, one possible area for further investigation is the incorporation of other non functional requirements such as energy consumption when executing OLAP queries [22], providing mathematical models to bound the consumed energy and the confrontation of theoretical and real results.

Experimental Results

To evaluate the effectiveness and efficiency of the conjoint methodologies of designing parallel data warehouses, the variant 3 of the design [6,7], in which the partitioning and the allocation phases are combined on a Teradata machine is considered. Recall that Teradata is a massively parallel processing system running shared nothing architecture. The Teradata DBMS is linearly and predictably scalable in all dimensions of a database system workload (data volume, breadth, number of users, complexity of queries). The basic unit of parallelism in Teradata is a virtual processor (called Access Module Processor or AMP) which is assigned a data portion. Each *AMP* executes DBMS functions on its own data. This allows locks and buffers do not have to be shared which ensures scalability. A node is a multi-core system with disks and memory. It provides a pool of resources (disk, memory, etc.) for the AMPS in that node. *BYNET* is the network

used to link different AMPs within a node and across different nodes as well.

The experiments were designed around the star schema benchmark (SSB) (http://www.cs.umb.edu/~poneil/StarSchemaB.PDF) derived from the TPC-H benchmark schema. Like TPC-H the SSB benchmark comes with a data generation utility (dbgen) which is scalable. For our experiments we used 10 scale factor (10 GB). We implement the and iterative conjoint approaches and assess their respective effects in Teradata. The obtained results show that the conjoint method performed 38% better for the overall workload which validates the results since the workload performance is the objective of both algorithms [4].

Cross-References

Data warehousing systems:

▶ Cube Implementations
▶ Data Partitioning
▶ Data Warehousing Systems: Foundations and Architectures
▶ Multidimensional Modeling
▶ Online Analytical Processing
▶ Optimization and Tuning in Data Warehouses
▶ Query Processing in Data Warehouses
▶ Replication
▶ Star Schema

Recommended Reading

1. Agrawal S, Narasayya VR, Yang B. Integrating vertical and horizontal partitioning into automated physical database design. In: Proceedings of the ACM SIGMOD International Conference on Management of Data; 2004. p. 359–70.
2. Akal F, Böhm K, Schek HJ. OLAP query evaluation in a database cluster: a performance study on intra-query parallelism. In: Proceedings of the 6th East European Conference on Advances in Databases and Information Systems; 2002. p. 218–31.
3. Apers PMG. Data allocation in distributed database systems. ACM Trans Database Syst. 1988;13(3): 263–304.
4. Bellatreche L, Benkrid S, Ghazal A, Crolotte A, Cuzzocrea A. Verification of partitioning and

allocation techniques on teradata DBMS. In: Proceedings of the 11th International Conference on Algorithms and Architectures for Parallel Processing; 2011. p. 158–69.

5. Bellatreche L, Boukhalfa K, Richard P. Referential horizontal partitioning selection problem in data warehouses: hardness study and selection algorithms. Int J Data Warehouse Min. 2009;5(4):1–23.

6. Bellatreche L, Cuzzocrea A, Benkrid S. Effectively and efficiently designing and querying parallel relational data warehouses on heterogeneous database clusters: the F&A approach. J Database Manag. 2012;23(4):17–51.

7. Bellatreche L, Cuzzocrea A, Benkrid S. A global paradigm for designing parallel relational data warehouses in distributed environments. Trans Large-Scale Data-Knowl-Cent Syst J. 2014;XV:1–38 (To appear).

8. Bellatreche L, Karlapalem K, Mohania MK, Schneider M. What can partitioning do for your data warehouses and data marts? In: Proceedings of the International Symposium on Database Engineering and Applications; 2000. p. 437–46.

9. Bellatreche L, Karlapalem K, Simonet A. Algorithms and support for horizontal class partitioning in object-oriented databases. Distrib Parallel Databases J. 2000;8(2):155–79.

10. Ceri S, Negri M, Pelagatti G. Horizontal data partitioning in database design. In: Proceedings of the ACM SIGMOD International Conference on Management of Data; 1982. p. 128–36.

11. Dehne F, Eavis T, Rau-Chaplin A. The cgmCUBE Project: optimizing parallel data cube generation for ROLAP. J Distrib Parallel Databases. 2006;19(1):29–62.

12. DeWitt D, Ghandeharizadeh S, Schneider D, Bricker A, Hsaio H, Rasmussen R. The Gamma database machine project. Trans Knowl Data Eng. 1990;2(1):44–62.

13. DeWitt D, Gray J. Parallel database systems: the future of high performance database systems. Commun ACM. 1992;35(6):85–98.

14. Furtado C, Lima A, Pacitti E, Valduriez P, Mattoso M. Physical and virtual partitioning in OLAP database clusters. In: Proceedings of the International Symposium on Computer Architecture and High Performance Computing; 2005. p. 143–50.

15. Goil S, Choudhary A. High performance multidimensional analysis of large datasets. In: Proceedings of the 1st ACM International Workshop on Data Warehousing and OLAP; 1998. p. 34–9.

16. Jin R, Vaidyanathan K, Yang G, Agrawal G. Communication and memory optimal parallel data cube construction. Trans Parallel Distrib Syst. 2005;16(12):1105–19.

17. Karlapalem K, Li Q. A framework for class partitioning in object-oriented databases. Distrib Parallel Databases J. 2000;8(3):333–66.

18. Lima AB, Furtado C, Valduriez P, Mattoso M. Parallel OLAP query processing in database clusters with data replication. distributed and parallel databases. Distrib Parallel Database J. 2009;25(1–2):97–123.

19. Noaman AY, Barker K. A horizontal fragmentation algorithm for the fact relation in a distributed data warehouse. In: Proceedings of the 8th International Conference on Information and Knowledge Management; 1999. p. 154–61.

20. Özsu MT, Valduriez P. Principles of distributed database systems, 2nd ed. Upper Saddle River: Prentice-Hall; 1999.

21. Rao J, Zhang C, Lohman G, Megiddo N. Automating physical database design in a parallel database. In: Proceedings of the ACM SIGMOD International Conference on Management of Data; 2002. p. 558–69.

22. Roukh A, Bellatreche L, Boukorca A, Bouarar S. Eco-dmw: Eco-design methodology for data warehouses. In: Proceedings of the ACM 18th International Workshop on Data Warehousing and OLAP, DOLAP; 2015. p. 1–10.

23. Saccà D, Wiederhold G. Database partitioning in a cluster of processors. ACM Trans Database Syst. 1985;10(1):29–56.

24. Scheuermann P, Weikum G, Zabback P. Data partitioning and load balancing in parallel disk systems. VLDB J. 1998;7(1):48–66.

25. Stohr T, Märtens H, Rahm E. Multi-dimensional database allocation for parallel data warehouses. In: Proceedings of the 26th International Conference on Very Large Data Bases; 2000. p. 273–84.

26. Wolfson O, Jajodia S. Distributed algorithms for dynamic replication of data. In: Proceedings of the 11th ACM SIGACT-SIGMOD-SIGART Symposium on Principles of Database Systems; 1992. p. 149–63.

Parallel Coordinates

Alfred Inselberg
Tel Aviv University, Tel Aviv, Israel

Synonyms

∥-coords; Multidimensional visualization; Parallel axes; Parallel coordinates plot (PCP); Parallel coordinates system (PCS)

Definition

In the plane with xy-Cartesian coordinates N copies of the real line labeled $\bar{X}_1, \bar{X}_2, \ldots, \bar{X}_N$ are placed equidistant and perpendicular to the x-

axis. They are the axes of the multidimensional system of *Parallel Coordinates* all having the same positive orientation as the y-axis. An N-tuple, N-dimensional point, $C = (c_1, c_2, \ldots, c_N)$ is represented by the polygonal line C whose N vertices are at the c_i values on each $\bar{X}_i$-axis as shown in Fig. 1. In this way, a 1-1 correspondence between points in N-dimensional space and polygonal lines with vertices on the parallel axes is established. In principle, a large number of axes can be placed and be seen parallel to each other. The representation of points is deceptively simple and much development with additional ideas is needed to enable the visualization of *multivariate relations* or equivalently multidimensional objects.

A dataset with M items has two M subsets anyone of which may be the one wanted. With a good data display the fantastic human pattern-recognition can not only cut great swaths searching through this combinatorial explosion but also extract insights from the visual patterns. These are the core reasons for data visualization. With Parallel Coordinates (abbr. ∥-coords) the search for multivariate relations in high dimensional datasets is transformed into a 2-D pattern recognition problem. A geometric classification algorithm based on ∥-coords is presented and applied to a complex dataset. It has low computational complexity providing the classification rule explicitly and *visually*. The minimal set of variables required to state the rule is found and ordered *optimally* by their predictive value. A visual economic model of a real country is constructed and analyzed to illustrate how multivariate relations can be modeled by means of hypersurfaces. Recent results like viewing *convexity in any dimension* and non-orientability (as in the Möbius strip) provide a prelude of what is on the way.

Historical Background

Legend has it that while constructing a proof, Archimedes was absorbed in a diagram when he was killed by a Roman soldier. "Do not disturb my circles" he pleaded as he was being struck by the sword. Visualization flourished in Geometry and Archimedes' is the first recorded death in defense of visualization. Visual *interaction* with diagrams is interwoven with testing of conjectures and construction of proofs. The tremendous human *pattern recognition* enables interaction for the extraction of insight from images. This essence of visualization is abstracted and adapted into the general problem-solving process to the extent that, a *mental image* of a problem is formed and at times one says *see* when it is meant *understand*.

Is there a way to make accurate pictures of multidimensional problems analogous to Descartes coordinate system? What is "sacred" about *orthogonal* axes which use up the plane very fast? After all, in Geometry parallelism rather than orthogonality is the fundamental concept and they are not equivalent for orthogonality requires a prior concept of angle." By 1959, while studying Mathematics at the University of Illinois, these thoughts lead to the *multidimensional* coordinate system based on *Parallel Coordinates*. With the encouragement of Professors Cairns and Bourgin (both topologists), basic properties like the *point* ↔ *line* duality were derived. It was not until 1977 when, while teaching a Linear Algebra course, the "challenge" was raised to *show* spaces of dimension higher than 3; parallel coordinates were recalled and their systematic development

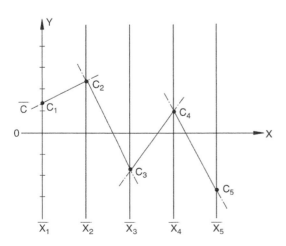

Parallel Coordinates, Fig. 1 The polygonal line $\bar{C}$ represents the N-tuple $C = (c_1, c_2, c_3, c_4, c_5)$

began. There was early interest and acceptance [6]. The comprehensive report [7] and later [9] laid the foundations of what became the ∥-coords methodology. Collaboration with Dimsdale (A long-time associate of John von Neumann.), Hurwitz, Boz and Addison ushered to five USA patents (Collision Avoidance Algorithms for Air-Traffic Control, Data Mining and Computer Vision) and applications to Optimization, Process Control [4] and elsewhere. Hurwitz, Adams and later Chomut experimented with multi-query interactive software written in APL for exploratory data analysis (EDA) with ∥-coords [5]. There followed [10] on the discovery of structure in data, Gennings et al. [4] on response surfaces based on ∥-coords for statistical applications, Wegman using the aforementioned duality promoted the EDA application, Fiorini on Robotics and Hinterberger [14] on comparative multivariate visualization (and earlier on data density analysis). The results of Eickemeyer [2], Hung and Inselberg [5] and Chatterjee were seminal. More recently the work of Ward et al. [15] (Hierarchical ∥-coords and more), Jones [13] (Optimization), Yang (Association Rules – Databases), Hauser (Categorical Variables and more) and Choi and Heejo Lee (on Intrusion Detection), increased the versatility and sophistication of ∥-coords. This list is by no means exhaustive, Shneiderman, Grinstein, Keim, Mihalisin, and others have made significant contributions to data and information visualization.

Foundations

"A picture is worth a thousand words" – But how does one say this with a picture?

The value of data visualization is not seeing "zillions" of objects but rather recognizing *relations* among them. Wonderful successes like Minard's "Napoleon's March to Moscow," Snow's "dot map," and others (see Friendly et al. [3]) are ad hoc (i.e., one-of-a-kind) and exceptional. Succinct multivariate relations are rarely apparent from *static* displays. With *interactivity* the visual clues in a good data display (*Parallax*, MDG's proprietary Data Mining software is used by permission.) can masterfully guide knowledge discovery. Follow up on anything that catches the eyes, gaps, regularities, holes, twists, peaks and valleys, density contrasts like the ones which reveal the water regions in Fig. 2. For the financial data in Fig. 3 (left) *multidimensional contouring* is applied to the axis with *SP500* index values

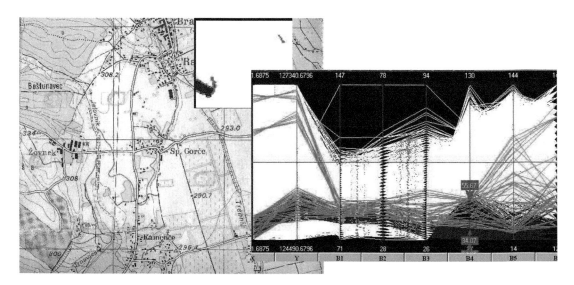

Parallel Coordinates, Fig. 2 Ground emissions measured by satellite on a region of Slovenia (*left*) are displayed on the right. The water and lake's edge are discovered with two queries

a

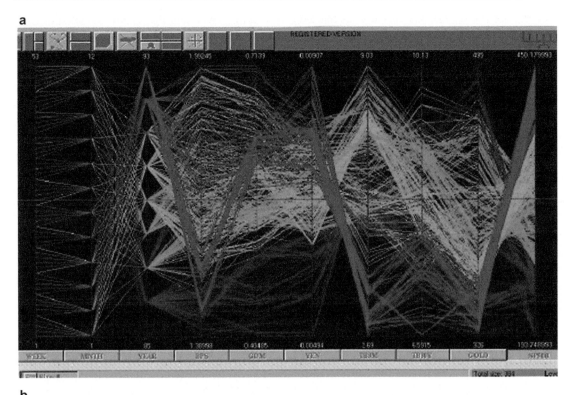

b

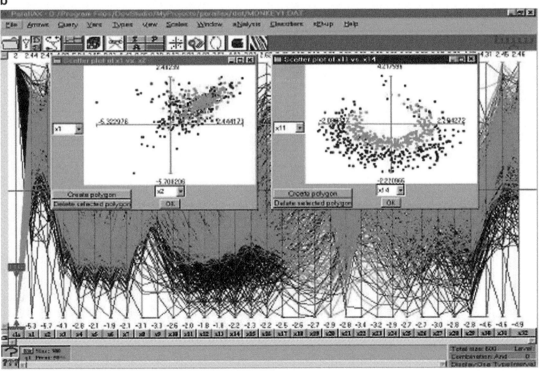

Parallel Coordinates, Fig. 3 (**a**) The *multidimensional contouring* on financial data reveals multiple interrelations. (**b**) Classification – dataset with 32 parameters and two categories (right) uncovering multivariate relations like high *SP500* low *Gold* and *Interests* correlate with high *Yen* and more. These are examples of two queries and they are others. With Boolean operators compound queries are formed to perform complex tasks. Classification is an important operation in data mining. Powerful geometrical classifiers based on ∥-coords [12] have been constructed. An example is shown Fig. 3 (right). The mixing of the two categories is seen on the left plot of the first two parameters. The classifier found the 9 (out of 32) parameters needed to state the rule with 4% error and ordered them according to their predictive value. The two best predictors are plotted on the right showing the separation achieved [12].

Multivariate relations can be modeled in terms of hypersurfaces – just as a relation between two variables can be represented by a planar region. From a dataset consisting of the outputs of various economic sectors of a real country a visual model of its economy is constructed and shown in Fig. 4 represented by the upper and lower curves. An interior (i.e., a combination of sector outputs) point satisfies all the constraints simultaneously therefore represents a feasible economic policy for that country. Such points can constructed by sequentially choosing variable values within their allowable range. Once a value of the first variable is chosen (in this case the *Agricultural* output) within its range, the dimensionality of the region is reduced by one. In fact, the upper and

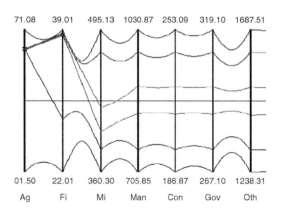
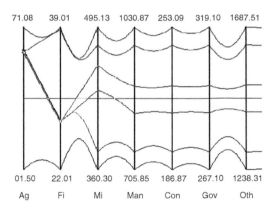

Parallel Coordinates, Fig. 4 Hypersurface modeling a country's economy. An interior point represented by a polygonal line depicts a feasible economic policy

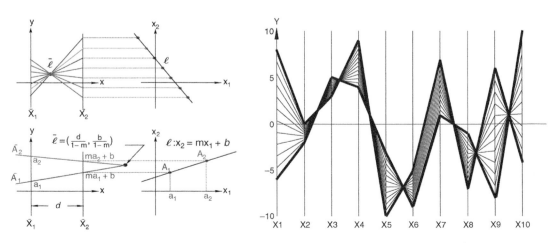

Parallel Coordinates, Fig. 5 (*Left*) *point* ↔ *line* duality in 2-D. (*Right*) A N-dimensional line ℓ is represented by $N-1$ points – the intersections of polygonal representing points of ℓ

lower curves between the second and third axes show the reduced available range of the second variable *Fishing* and similarly for the remaining variables. Hence the impact of a decision can be seen downstream. In this way it was found that a high value from the available range of *Fishing* corresponds to very low values of the *Mining* sector – seen in the right of Fig. 4 and vice versa. This inverse correlation was investigated and found that the two sectors compete for the *same* group of migrating workers. When fishing is doing well, most of them leave the mountains where the mines are located to work on the

fishing boats and the reverse. This is an example of ‖-coords used for Decision Support and Trade-Off Analysis.

Key Applications

There are numerous others like GIS, Process Control, Trading & Financial Analysis the visualization and analysis of multivariate/multidimensional problems in general.

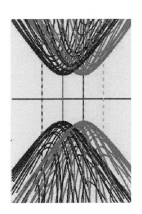
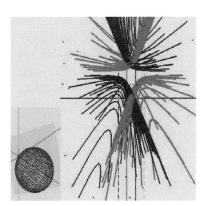
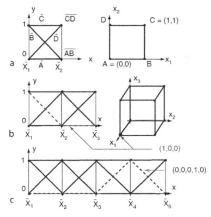

Parallel Coordinates, Fig. 6 Convex surfaces represented by $N-1$ hyperbola-like regions. (*left*) Sphere in 3-D centered at origin. (*Center*) Translated sphere *translation* $\leftrightarrow$ *rotation* duality (*right*) cube and hypercube in 5-D repeating pattern

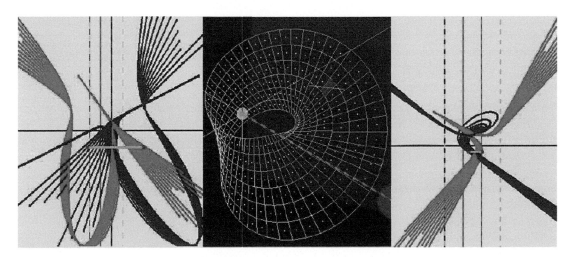

Parallel Coordinates, Fig. 7 Möbius strip and its representation for two orientations

Future Directions

Though some refer to ∥-coords as a "plot" it is actually *a coordinate system* where the goal, since its inception, is to concentrate N-dimensional relational information into *patterns* the earliest being that N-dimensional lines are represented by $N-1$ points with *two indices* Fig. 5. Hyperplanes are represented by $N-1$ points with N indices, and $N-1$ regions with N indices represent a surface – i.e., the tangent planes enveloping it. Figures 6 and 7 show recent breakthroughs where *convexity in any dimension* and *non-orientability* can be seen. Work is progressing on transforming multivariate relations in datasets into planar patterns completely eliminating display clutter [11].

Cross-References

▶ Abstraction
▶ Association Rule Mining on Streams
▶ Business Intelligence
▶ Classification
▶ Comparative Visualization
▶ Curse of Dimensionality
▶ Data Mining
▶ Data Visualization
▶ Data Warehouse
▶ Dimension
▶ Dimension Reduction Techniques for Clustering
▶ Dimensionality Reduction
▶ Dynamic Graphics
▶ Event Detection
▶ Exploratory Data Analysis
▶ Feature Selection for Clustering
▶ Geographic Information System
▶ High-Dimensional Indexing
▶ Individually Identifiable Data
▶ Information Extraction
▶ Interface
▶ Machine Learning in Computational Biology
▶ Mining of Chemical Data
▶ Model Management
▶ Multidimensional Data Formats
▶ Multidimensional Modeling
▶ Multidimensional Scaling

▶ Multivariate Visualization Methods
▶ Principal Component Analysis
▶ Process Optimization
▶ Query Evaluation Techniques for Multidimensional Data
▶ Range Query
▶ Scientific Visualization
▶ Spatial Data Analysis
▶ Spatial Data Mining
▶ Spatiotemporal Data Mining
▶ Temporal Data Mining
▶ Temporal Dependencies
▶ Visual Analytics
▶ Visual Association Rules
▶ Visual Classification
▶ Visual Clustering
▶ Visualizing Clustering Results
▶ Visual Interfaces
▶ Visual Online Analytical Processing (OLAP)
▶ Visualizing Quantitative Data
▶ What-if Analysis

Recommended Reading

A. Inselberg, Visual Multidimensional Geometry and its Applications, 550 pages, Springer, 2009

1. Chomut T. Exploratory data analysis in parallel coordinates. M.Sc. Thesis, Department of Computer Science, UCLA; 1987.
2. Eickemeyer J. Visualizing p-flats in N-space using parallel coordinates. Ph.D. Thesis, Department of Computer Science, UCLA; 1992.
3. Friendly M, et al. Milestones in thematic cartography. (2005) www.math.yorku.ca/scs/SCS/Gallery/milestones/.
4. Gennings C, Dawson KS, Carter WH, Myers RH. Interpreting plots of a multidimensional dose-response surface in parallel coordinates. Biometrics. 1990;46(3):719–35.
5. Hung CK, Inselberg A. Parallel coordinate representation of smooth hypersurfaces. USC Tech. Report # CS-92-531, Los Angeles; 1992.
6. Inselberg A. N-dimensional coordinates. In: Proceedings of the IEEE Conference on Picture Data Description; 1980.
7. Inselberg A. N-dimensional graphics, LASC Tech. Rep. G320-2711, IBM, 1981.
8. Inselberg A. Intelligent instrumentation and process control. In: Proceedings of the 2nd IEEE Conference on AI Application; 1985. p. 302–7.

9. Inselberg A. The plane with parallel coordinates. Vis Comp. 1985;1(2):69–97.
10. Inselberg A. Discovering multi-dimensional structure with parallel coordinates (*invited paper*). In: Proceedings of the American Statistical Association, Section on Statistical Graphics; 1989. p. 1–16.
11. Inselberg A. Parallel coordinates : *VISUAL* multidimensional geometry and its applications. Springer; 2009.
12. Inselberg A, Avidan T. The automated multidimensional detective. In: Proceedings of the IEEE Information Visualization; 1999. p. 112–9.
13. Jones C. Visualization and optimization. Boston: Kluwer Academic; 1996.
14. Schmid C, Hinterberger H. Comparative multivariate visualization across conceptually different graphic displays. In: Proceedings of the 6th International Conference on Scientific and Statistical Database Management; 1994.
15. Ward MO. XmdvTool: integrating multiple methods for visualizing multivariate data. In: Proceedings of the IEEE Conference on Visualization; 1994. p. 326–33.

Parallel Data Placement

Patrick Valduriez
INRIA, LINA, Nantes, France

Synonyms

Multiprocessor data placement

Definition

Parallel data placement refers to the physical placement of the data in a multiprocessor computer in order to favor parallel data access and yield high-performance. Most of the work on data placement has been done in the context of the shared-nothing architecture. Data placement in a parallel database system exhibits similarities with data fragmentation in distributed databases since fragmentation yields parallelism. However, there is no need to maximize local processing (at each node) since users are not associated with particular nodes and load balancing is much more difficult to achieve in the presence of a large number of nodes.

The main solution for parallel data placement is a variation of horizontal fragmentation, called *partitioning*, which divides database relations into partitions, each stored at a different disk node. There are three basic partitioning strategies: round-robin, hashing, and interval. Furthermore, to improve availability, replication of partitions is needed.

Historical Background

Parallel data placement was proposed in the early 1980s for the first software-oriented parallel database systems. Instead of proposing expensive changes to the disk technology as in database machines with filtering devices, the main idea was to distribute the data onto multiple (smaller) disks and parallelize the I/O bandwidth which could be better consumed by multiple processors.

Data partitioning was first proposed for shared-nothing architectures which are a natural evolution of distributed database architectures. The same basic idea has been also used (with different partitioning strategies) for the Redundant Arrays of Inexpensive Disks (RAID) to build powerful disks out of many smaller ones.

Most of the work in parallel data placement has been done in the context of the relational model with relation partitioning as the main strategy. Extensions have been proposed for object-oriented or semi-structured (XML) data by partitioning collections of objects or elements.

Foundations

Most of the work on data placement has been done in the context of the shared-nothing architecture which is the most general architecture. However, data placement is also important in shared-memory and shared-disk architectures since there can be multiple disks. Thus, the data placement techniques designed for shared-nothing can also be used, sometimes in a simplified form, to other architectures.

Data placement in a parallel database system exhibits similarities with data fragmentation in distributed databases. An obvious similarity

P

is that fragmentation can be used to increase parallelism. In what follows, the terms partitioning and partition are used instead of horizontal fragmentation and horizontal fragment, respectively, to contrast with the alternative approach which clusters all data of each relation at one node. Vertical fragmentation can also be used to increase parallelism and load balancing much as in distributed databases. Another similarity is that queries should be executed as much as possible where the data reside. However, there are two important differences with the distributed database approach. First, there is no need to maximize local processing (at each node) since users are not associated with particular nodes. Second, load balancing is much more difficult to achieve in the presence of a large number of nodes. The main problem is to avoid resource contention, which may result in the entire system thrashing (e.g., one node ends up doing all the work while the others remain idle). Since queries are executed where the data reside, data placement is a critical performance issue.

Data placement must be done to maximize system performance, which can be measured by combining the total amount of work done by the system and the response time of individual queries. Maximizing response time (through intra-query parallelism) may result in increased total work due to communication overhead. For the same reason, inter-query parallelism results in increased total work. On the other hand, clustering all the data necessary to a query minimizes communication and thus the total work done by the system in executing that query. In terms of data placement, the following trade-off exists: maximizing response time or inter-query parallelism leads to partitioning, whereas minimizing the total amount of work leads to clustering. This problem is addressed in distributed databases in a rather static manner. The database administrator is in charge of periodically examining fragment access frequencies, and when necessary, moving and reorganizing fragments.

A solution to data placement is full partitioning, whereby each relation is horizontally fragmented across all the nodes in the system. There are three basic strategies for data partitioning: round-robin, hash, and range partitioning.

1. *Round-robin partitioning* is the simplest strategy, it ensures uniform data distribution. With n partitions, the ith tuple in insertion order is assigned to partition (i mod n). This strategy enables the sequential access to a relation to be done in parallel. However, the direct access to individual tuples, based on a predicate, requires accessing the entire relation.

2. *Hash partitioning* applies a hash function to some attribute which yields the partition number. This strategy allows exact-match queries on the selection attribute to be processed by exactly one node and all other queries to be processed by all the nodes in parallel.

3. *Range partitioning* distributes tuples based on the value intervals (ranges) of some attribute. In addition to supporting exact-match queries as with hashing, it is well-suited for range queries. For instance, a query with a predicate A between a_1 and a_2 may be processed by the only node(s) containing tuples whose A value is in the range [a_1 and a_2]. However, range partitioning can result in high variation in partition size.

Full partitioning generally yields better performance than clustering relations on a single (possibly very large) disk. Although full partitioning has obvious performance advantages, highly parallel execution may cause a serious performance overhead for complex queries involving joins. Furthermore, full partitioning is not appropriate for small relations that span a few disk blocks. These drawbacks suggest that a compromise between clustering and full partitioning needs to be found.

A solution is to do data placement by variable partitioning, where the degree of partitioning, in other words, the number of nodes over which a relation is partitioned, is a function of the size and access frequency of the relation. This strategy is much more involved than either clustering or full partitioning because changes in data distribution may result in reorganization. For example, a relation initially placed across eight nodes may have its cardinality doubled by subsequent insertions, in which case it should be placed across 16 nodes.

In a highly parallel system with variable partitioning, periodic reorganizations for load balancing are essential and should be frequent unless the workload is fairly static and experiences only a few updates. Such reorganizations should remain transparent to compiled queries that run on the database server. In particular, queries should not be recompiled because of reorganization. Therefore, the compiled queries should remain independent of data location, which may change rapidly. Such independence can be achieved if the run-time system supports associative access to distributed data. This is different from a distributed DBMS, where associative access is achieved at compile time by the query processor using the data directory.

A serious problem in data placement is dealing with skewed data distributions which may lead to non-uniform partitioning and hurt load balancing. Range partitioning is more sensitive to skew than either round-robin or hash partitioning. A solution is to treat non-uniform partitions appropriately, e.g., by further fragmenting large partitions.

Another important function is data replication for high availability. The simple solution is to maintain two copies of the same data, a primary and a backup copy, on two separate nodes. This is the mirrored disks architecture promoted by many computer manufacturers. However, in case of a node failure, the load of the node having the copy may double, thereby hurting load balancing. To avoid this problem, an interesting solution is Teradata's interleaved partitioning which partitions the backup copy on a number of nodes. In failure mode, the load of the primary copy gets balanced among the backup copy nodes. But if two nodes fail, then the relation cannot be accessed thereby hurting availability. Reconstructing the primary copy from its separate backup copies may be costly. In normal mode, maintaining copy consistency may also be costly. A solution to this problem is Gamma's chained partitioning which stores the primary and backup copy on two adjacent nodes. The main idea is that the probability that two adjacent nodes fail is much less than the probability that any two nodes fail. In failure mode, the load of the failed node and the backup nodes are balanced among all remaining nodes by using both primary and backup copy nodes. In addition, maintaining copy consistency is cheaper.

Key Applications

Parallel data placement has been primarily used by parallel database systems on multiprocessor computers. Data partitioning is also used as a basic technique for dealing with very large volumes of data in different environments such as database clusters, data grids, search engines, document management systems, and P2P systems.

Recommended Reading

1. Copeland GP, Alexander W, Boughter EE, Keller, TW. Data placement in bubba. In: Proceedings of the ACM SIGMOD International Conference on Management of Data; 1988. p. 99–108.
2. DeWitt DJ, Gray J. Parallel database systems: the future of high performance database systems. Commun ACM. 1992;35(6):85–98.
3. Mehta M, DeWitt DJ. Data placement in shared-nothing parallel database systems. VLDB J. 1997;6(1):53–72.
4. Özsu MT, Valduriez P. Principles of distributed database systems. 3rd ed. New York: Prentice Hall; 2009.
5. Valduriez P, Pacitti E. Parallel database systems. In: Hammer J, Scheider M, editors. Handbook of database technology. Boca Raton: CRC Press; 2007.

Parallel Database Management

Patrick Valduriez
INRIA, LINA, Nantes, France

Synonyms

Multiprocessor database management

Definition

Parallel database management refers to the management of data in a multiprocessor computer and is done by a parallel database system, i.e., a full-fledge DBMS implemented on a multiprocessor computer. The basic principle employed by parallel DBMS is to partition the data across multiprocessor nodes, in order to increase performance through parallelism and availability through replication. This enables supporting very large databases with very high query or transaction loads.

Parallel database systems can exploit distributed database techniques. In particular, database partitioning is somewhat similar to database fragmentation. Essentially, the solutions for transaction management, i.e., distributed concurrency control, reliability, atomicity, and replication, can be reused. However, the critical issues for parallel database systems are data placement, parallel query processing (including optimization and execution) and load balancing. These issues are much more difficult than in distributed DBMS because the number of nodes may be much higher. Furthermore, the interconnection network of a multiprocessor system provides better opportunities for improving performance than a general-purpose network.

Historical Background

The performance objective of parallel database systems was also that of the database machines in the 1980s. The problem faced by conventional database management on a monoprocessor computer has long been known as "I/O bottleneck," induced by high disk access time. Database machine designers tackled this problem through special-purpose hardware, e.g., by introducing data filtering devices within the disk heads. However, this approach did not succeed because of a poor cost/performance ratio compared to the software solution which can easily benefit from progress in standard hardware technology (processor, RAM, disk). However, the idea of pushing database functions closer to the disk has received renewed interest with the introduction of general-purpose microprocessors in disk controllers, thus leading to intelligent disks.

An important result of database machine research, however, is in the general solution to the I/O bottleneck which can be summarized as increasing the I/O bandwidth through parallelism.

If a database is partitioned, e.g., by horizontal fragmentation of the relational tables, across a number d of disks, then disk throughput can be increased by accessing the d disks in parallel. The same principle has been applied to Redundant Arrays of Inexpensive Disks (RAID) which build a powerful disk with many small disks.

The main memory database solution which maintains the entire database in main memory to avoid disk accesses, is complementary rather than alternative to parallel database systems. In particular, the "memory access bottleneck" induced by high memory access time relative to processor speed can also be tackled using parallel database techniques. In other words, a parallel database system can use main memory database nodes.

Foundations

Parallel database system designers strive to develop software-oriented solutions in order to exploit the various kinds of multiprocessor systems which are now available. This can be achieved by extending distributed database technology, for example, by partitioning the database across multiple disks so that much inter- and intra-query parallelism can be obtained. This can lead to significant improvements in both response time and throughput (number of transactions per time unit).

A parallel database system can be loosely defined as a DBMS implemented on a multiprocessor computer. This definition includes many alternatives ranging from the porting of an existing DBMS, which may require only rewriting the operating system interface routines, to a sophisticated combination of parallel processing and database system functions into a new hardware/software architecture. Thus, there is a trade-off between portability (to several platforms) and efficiency. The sophisticated approach is better able to fully exploit the opportunities offered by a multiprocessor at the expense of portability. Interestingly, this gives different advantages to computer manufacturers and software vendors.

The main objectives of a parallel database system are the following:

High-Performance This can be obtained through several complementary solutions: data-based parallelism, query optimization, and load balancing. Parallelism can increase throughput, using inter-query parallelism, and decrease transaction response times, using intra-query parallelism. Load balancing is the ability of the system to divide a given workload equally among all processors. Depending on the parallel system architecture, it can be achieved by static physical database design or dynamically at run-time.

High-Availability Because a parallel database system consists of many redundant components, it can well increase data availability and fault-tolerance. Replicating data at several nodes is useful to support failover, a fault-tolerance technique which enables automatic redirection of requests from a failed node to another node which stores a copy of the data, thereby providing nonstop operation.

Extensibility In a parallel system, accommodating increasing database sizes or increasing performance demands (e.g., throughput) should be easier. Extensibility is the ability of smooth expansion of the system by adding processing and storage power to the system. Ideally, the increase in the configuration of the parallel database system (i.e., more nodes) should speed up existing workloads or scale up to larger databases.

The functions supported by a parallel database system are the same as in a typical DBMS. The differences, though, have to do with implementation of these functions which must now deal with parallelism, data partitioning and replication, and distributed transactions. Depending on the architecture, a processor node can support all (or a subset) of these functions. In the early database machines for instance, some nodes were specialized with hardware to implement some functions, e.g., data filtering.

The way the main hardware elements, i.e., processors, main memory, and disks, are connected (through some interconnection network) has a major impact on the design of a parallel database system. The term processor is used in the general sense of central processing unit (CPU). However, the processor itself can be made of several core processors integrated in a single chip, i.e., a multi-core processor, and perform instruction-level multithread parallelism. But the parallelism being discussed here is much higher-level (query- or operator-level) so processors (e.g., multi-core processors) are simply considered as black-box components accessing other resources (main memory, disk, network). Depending on how main memory or disk is shared, three basic architectures are obtained: shared-memory, shared-disk and shared-nothing.

In the shared-memory architecture, any processor has access to the entire main memory, and thus to all the disks, through the interconnection network. The network is typically very fast (e.g., a high-speed bus or a cross-bar switch) and redundant to avoid any unavailability. All the processors are under the control of a single operating system. Shared-memory makes it easy to build a parallel database system because the operating system deals with load balancing. It is also very efficient since processors can communicate via the main memory. However, the complexity and cost of the interconnection network hurt extensibility which is limited to a few tens of processors. Most parallel database system vendors provide support for shared-memory.

In the shared-disk architecture, any processor has access to any disk unit through the interconnection network but exclusive (non-shared) access to its main memory. Each processor-memory node is under the control of its own copy of the operating system. An important function that is needed is cache coherency which allows different nodes to cache a consistent disk page. This function is hard to support and requires some form of distributed lock management. Shared-disk has better extensibility than shared-memory since the main memory is distributed. However, scalability depends on the efficiency of the cache coherency mechanism. Migrating from a centralized system to shared-disk is relatively straightforward since the data on disk need not be reorganized. Shared-disk is the main architecture used by Oracle for its parallel database system.

The shared-nothing architecture is fully-distributed: each node is made of processor, main memory and disk and communicates with other nodes through the interconnection network.

Similar to shared-disk, each processor-memory-disk node is under the control of its own copy of the operating system. Then, each node can be viewed as a local site (with its own database and software) in a distributed database system. Therefore, most solutions designed for distributed databases such as database fragmentation (called partitioning in parallel databases), distributed transaction management and distributed query processing may be reused. Using a fast interconnection network, it is possible to accommodate large numbers of nodes. The major advantages of shared-nothing over shared-memory or shared-disk are those of distributed systems: relatively low cost, extensibility and availability. However, it is much more complex to manage as physical data portioning is necessary. Shared-nothing has been adopted by the major DBMS vendors (except Oracle) for their high-end parallel database systems.

Various possible combinations of these three basic architectures are possible to obtain different trade-offs between cost, performance, extensibility, availability, etc. Hierarchical architectures typically combine the three architectures using a shared-nothing design where each node can be shared-memory or shared-disk and communicates with other nodes through the interconnection network.

Key Applications

Parallel database systems are used to support very large databases (e.g., tens or hundreds of terabytes). Examples of applications which deal with very large databases are e-commerce, data warehousing, and data mining. Very large databases are typically accessed through high numbers of concurrent transactions (e.g.,

performing on-line orders on an electronic store) or complex queries (e.g., decision-support queries). The first kind of access is representative of On Line Transaction Processing (OLTP) applications while the second is representative of On Line Analytical Processing (OLAP) applications. Although both OLTP and OLAP can be supported by the same parallel database system (on the same multiprocessor), they are typically separated and supported by different systems to avoid any interference and ease database operation.

Recommended Reading

1. DeWitt DJ, Gray J. Parallel database systems: the future of high performance database systems. Commun ACM. 1992;35(6):85–98.
2. Özsu T, Valduriez P. Distributed and parallel database systems – technology and current state-of-the-art. ACM Comput Surv. 1996;28(1):125–8.
3. Özsu MT, Valduriez P. Principles of distributed database systems. 3rd ed. New York: Prentice Hall; 2009.
4. Valduriez P. Parallel database systems: the case for shared-something. In: Proceedings of the 9th International Conference on Data Engineering; 1993. p. 460–5.
5. Valduriez P, Pacitti E. Parallel Database Systems. In: Hammer J, Scheider M. Handbook of database technology. Boca Raton: CRC Press; 2007.

Parallel Hash Join, Parallel Merge Join, Parallel Nested Loops Join

Goetz Graefe
Google, Inc., Mountain View,
CA, USA

Synonyms

Parallel join algorithms

Definition

These join algorithms are parallel versions of the traditional serial join algorithms. They are designed to exploit multiple processors on a network, within a machine, or even within a single chip.

Key Points

Parallel join algorithms are based on the traditional serial join algorithms, namely (index) nested loops join, merge join, and (hybrid) hash join. The goal of parallel execution is to reduce the input sizes in each processing element, thus reducing the time for query completion even at the expense of increasing overall query execution effort due to data movement. Ideally, parallel join algorithms exhibit linear speed-up and linear scale-up.

Parallel join algorithms are orthogonal to pipelining among join operations in a complex query execution plan. Even a query with a single join can benefit from a parallel join algorithm. The essence of pipelining (and also of "bushy parallelism" in appropriate query execution plans) is to execute different operations with different predicates concurrently, whereas the essence of parallel join algorithms is that a single logical operation exploits multiple processors or processing cores. The conceptual foundations of parallel join algorithms are that database query processing manipulates sets of records, that sets can be divided into disjoint subsets, and that query results based on one record can be computed independently of results based on other records.

Parallel join algorithms apply not only to inner joins but also to semijoins, outer joins, and set operations such as intersection, union, and difference. Thus, they apply to query execution techniques such as index intersection for conjunctive predicates and the star join techniques that are generalizations traditional binary joins.

For a parallel join algorithm, input records are partitioned or replicated such that each result record is produced exactly once. The usual choices are to partition both inputs or to partition one input and to replicate the other. For parallel joins without any equality predicate, the processing elements can be organized in a rectangular grid, one input partitioned over rows of the grid and replicated within each row, and the other join input partitioned over columns of the grid and replicated within each column. These choices mirror schemes in which database records may be stored partitioned or replicated.

Partitioning schemes include range partitioning, hash partitioning, and hybrid schemes such as range partitioning of hash values or hashing (identifiers of) key ranges. For intermediate query results, hash partitioning is simple yet reasonably robust against skewed key distributions.

All of the join algorithms can reduce data transfer efforts using semijoin reduction or its heuristic approximation, bit vector filtering or Bloom filters. Hash join can benefit most obviously due to its separate build and probe phases, with the bit vector filter populated during the build phase and exploited during the probe phase. Merge join can exploit bit vector filtering if the filter can be populated prior to a stop-and-go operation such as sort, and can exploit it with the most gain if the other input also must be processed by an expensive operation such as a sort.

Symmetric partitioning means that both inputs are partitioned. Ideally, the inputs are stored partitioned in such a way that join processing does not incur any cost or delay for partitioning. This is often called "local indexing" or "table partitioning" when it refers to indexes of the same table and "aligned partitioning" when it refers to separate tables. Aligned partitioning often coincides with frequent join operations, foreign key constraints, and complex objects on the conceptual level. The opposites are "global indexing," "index partitioning," and "non-aligned partitioning."

If one of the join inputs is very large, and if the degree of parallelism is only moderate, it might be less expensive to replicate the small input than to partition the large input. An appropriate cost calculation needs to consider record counts, record sizes, communication costs per message or per byte, the available memory allocation for each thread, etc. A first approximation compares the quotient of the two input sizes with the degree of parallelism.

Cost calculation during compile-time query optimization can focus either on the average cost in each of the parallel processing elements or on the maximal cost. The former metric focuses on overall system throughput in multi-user systems, whereas the latter metric focuses on the query elapsed time as perceived by a single user or application.

Cross-References

▶ Parallel Query Execution Algorithms

Recommended Reading

1. Graefe G. Query evaluation techniques for large databases. ACM Comput Surv. 1993;25(2):73–170.
2. Mishra P, Eich MH. Join processing in relational databases. ACM Comput Surv. 1992;24(1):63–113.

Parallel Query Execution Algorithms

Goetz Graefe
Google, Inc., Mountain View, CA, USA

Synonyms

Partitioned query execution

Definition

Parallel query execution algorithms employ multiple threads that together execute a single query execution plan, with the goal of increased processing bandwidth and decreased response time. Multiple forms of parallelism may be employed in isolation or in combination. Most parallel query execution algorithms are serial algorithms executed in multiple threads, usually with no or only minor adaptations to function correctly or efficiently in the multi-threaded environment.

Historical Background

Set-oriented query languages and the relational algebra have given rise to query optimization and parallel database processing almost from their beginning in the 1970s. They represented a new complexity or liability in database software design as non-procedural languages and physical data independence prevent users from specifying query processing steps and thus require an automatic system component to provide plans for data access and for query execution. At the same time, they created a new opportunity as to prior database systems and their interfaces did permit query rewrite techniques or parallel execution.

The last 30 years have seen many successes and many failures in parallel query execution algorithms. Parallel versions of naïve nested loops join were a failure, because even hardware acceleration cannot compete with well-chosen indexes, whether those are persistent indexes supporting index nested loops join or temporary indexes in form hash tables in hash join. Parallel set operations, including non-indexed selection, join, grouping, and sorting, have been great successes, because they can provide both good speed-up and scale-up. Parallelism is sometimes associated with specific set operations, e.g., hash join instead of all joins, but almost all join algorithms benefit quite similarly if designed and implemented with similar care.

Foundations

In principle, parallel query execution algorithms are traditional sequential algorithms with multiple threads executing different operations or executing the same operation on different data. The relational algebra of sets permits pipelining between operations, partitioning of stored data and of intermediate results, and bushy execution in complex query execution plans. Individual parallel algorithms exploit partitioning but also overcome the issues introduced by partitioning and parallelism.

The value and success of parallel query execution is often expressed as speed-up and scale-up. Speed-up compares execution times of serial and parallel execution for a constant data volume. Linear speed-up is considered ideal; it is achieved if the ratio of execution times is equal to the degree of parallelism. Scale-up compares execution times for a constant ratio of data volume and degree of parallelism. Linear scale-up is considered ideal; it is achieved if the execution times remain constant. There are, however, some effects that permit super-linear speed-up. For example, adding nodes to a parallel database machine increases not only processing bandwidth but also available memory, and might thus turn an external sort into an in-memory sort at a high degree of parallelism.

Pipelining

The pipeline between a producer operation and a consumer operation may pass individual records, value packets, pages, or even runs. Value packets within a sorted stream are defined by equal values in the sort columns. As an example for iterators passing run information, consider a sort operation split into run generation and merging; the pipeline between those two components may pass entire runs at-a-time. In addition to data, the pipeline contents may include control records, e.g., artificially created keys design to advance the merge logic further up in the sequence of operations.

More importantly for multi-threaded query execution, the pipeline between threads may be demand-driven or data-driven. Within each thread, iterators commonly implement demand-driven dataflow, which lend themselves well to joins and other binary operations but not to sharing intermediate results or common sub-expressions, which occur much less frequently than joins in query execution plans. Between threads, data-driven dataflow is commonly regarded superior.

On the other hand, if flow control is implemented, the slower one among producer and consumer dictates the overall pipeline flow, and it might not truly be useful to attempt distinguishing demand- and data-driven dataflow. The important aspect is that some "slack" is created that permits the producer to get ahead of the consumer to some reasonable amount without requiring an excessive amount of buffer space. In general, the quotient of buffer space and bandwidth is equal to the interval that can be smoothed out, i.e., the amount of time the producer may get ahead of the consumer or the consumer may fall behind the producer.

Partitioning

Since pipelining with flow control often leads to consumers throttling producers or the other way around, and since pipelining alone permits only a fairly limited degree of parallelism, it is usually combined with partitioning. Partitioning means that the input into a relational algebra operation is divided into disjoint subsets, which are processed by separate threads or processors.

Many partitioning methods are in common use. The basic partitioning methods are random, round-robin, range, and hash. Both random and round-robin partitioning permit perfect load balancing but do not exploit key values and their importance in relational algebra operations such as joins and duplicate elimination. When random partitioning or round-robin partitioning is applied to stored data, neither queries nor updates can be directed to a single location and instead must be processed in all partitions. Range partitioning assigns key ranges (of the partitioning column) to storage or processing sites. For perfect load balancing, accurate quantile information is required. For intermediate results, quantiles must be estimated during query optimization or during an earlier query execution phase. Hash partitioning applies a hash function to determine storage or processing sites. A suitable hash function can achieve very good load balancing, which is why hash partitioning is sometimes associated with linear speed-up and scale-up. On the other hand, even hash partitioning fails if the data distribution suffers from duplicate skew, i.e., the data contain many duplicate key values.

There are many ways to combine these basic partitioning methods. For example, it may be desirable that small range queries are processed at a one or two sites with little overhead and

P

that large range queries are processed by many or all sites with maximum throughput. To achieve both these effects, many small key ranges can be hashed to storage sites. In other cases, one column may be used to hash data items to sites and another column to hash data items to storage locations within each site. This is just one of many multi-dimensional partitioning methods. Note that partitioning among sites, partitioning within sites, and local organization are orthogonal. The possibilities are endless, and almost each vendor claims to have found a new and superior "secret sauce." The most sophisticated systems even offer partitioning on computed columns or expressions.

Partitioning of stored data usually determines where initial query operations execute, including selection and projection as well as preliminary stages of sorting, duplicate elimination, and "group by" operations. Partitioning of stored data may also interact with high availability consideration, e.g., each data item is hashed to two sites using two hash functions, carefully designed to ensure that no data item is hashed to the same site by both hash functions.

For binary operations, "co-location" of the two inputs permits local execution. These binary operations include inner and outer joins, semi-joins including "exists" nested queries, and set operations such as intersection, union, and difference. Set operations after scans of non-clustered indexes are common in query execution plans for complex queries, for ad-hoc queries, and for star joins in relational data warehouses using a star schema. If all indexes for a table or view are partitioned in the same way, both set operations and fetch operations can remain local. On the other hand, there are many situations in which it is advantageous to partition each index on its search columns.

If co-location and local execution are not possible, join input data need to be re-partitioned. In most cases, both inputs are partitioned on the columns participating in the join operation's equality predicate. If one input is much smaller than the other, it may be more efficient to broadcast the small input and not move the large input. In some case, in particular non-equality joins, a combination can be applied. In this technique, the processing sites can be thought of as a rectangular matrix, with one input partitioned among rows and broadcast within rows, and the other input partitioned among columns and broadcast within columns. The essence of all partitioning strategies for binary operations such as joins is to ensure that each pair of input rows that may contribute to the join result "meets" at precisely one processing site.

Bushy Execution

In addition to pipelining and partitioning, a third form of parallel query execution is enabled by bushy query execution plans, as opposed to linear query execution plans. In a linear query execution plan, one input of each binary operation always is a stored table. In a bushy query execution plan, both inputs may be intermediate query results. For example, two sort operations may provide the inputs for a merge join. During the sort operations' input phases, the two sort operations can run in parallel. In more complex query execution plans, numerous operations and plan branches may be active concurrently and independently.

Even the simple example with two sort operations and a merge join suffices to demonstrate the substantial difficulties, however. One problem is finding an appropriate degree of parallelism for each operation or branch. A harder problem is assigning appropriate resources such as memory, disk space and bandwidth, etc. The hardest problem is to ensure that concurrent branches produce their results at the right time. In the example, resources may be wasted if one sort operation finishes before the other, because the former will likely hold on to resources such as memory while waiting for the merge join to need its input and thus for the other sort operation to finish.

Specifics of Parallel Algorithms

In order to keep parallel query execution engines modular, traditional serial query execution algorithms are combined with mechanisms for parallel execution, specifically pipelining and partitioning. The interface between traditional serial query execution operations usually takes the form of iterator. Iterator methods support forward-only scans of intermediate results, rewinding, binding parameters as needed for nested iteration and general predicate evaluation, etc. Iterator instances may recursively invoke their input iterators to implement those methods, and by doing so can execute a very complex query execution plan within a single thread.

Mechanisms for parallel query execution can be encapsulated in a special iterator. This design has been used in a number of research prototypes and commercial database systems. Commonly called the "exchange" operation but also known as "river" or "asynchronous table queue," this iterator encapsulates data transfer and flow control, often provides initialization and tear-down of threads and communication paths, and batches records in order to reduce inter-process communication. If parallel execution of nested iteration is desired, both data transfer and flow control must work both from producer to consumer for intermediate query results and from consumer to producer for bindings of correlation columns from outer to inner inputs of nested iteration operations.

As the exchange operation or its equivalent encapsulate the basic mechanisms for parallel query execution, the following only indicates how specific parallel algorithms differ from their serial equivalents.

For parallel sorting, after the input data is partitioned as appropriate, the individual threads of a parallel sort operation usually can work entirely independently. If, however, the query optimizer relies on the sort operation not only for the correct sequence of records but also to avoid the Halloween problem by means of phase separation, the individual threads must coordinate their transition from consuming unsorted input records to producing sorted output records. If one of them produces output too early, it might affect the input set of another thread still consuming input and thus create the Halloween problem.

A similar coordination point exists in parallel hash join. A serial hash join might choose between Grace join, which devotes its first phase entirely to partitioning, and hybrid hash join, which performs some join logic even during its first phase. Multiple threads in a parallel hash join might choose differently, e.g., due to different data volumes. Scheduling other operations within the query execution plan might require, however, that all hash join threads follow the same choice. As for parallel sorting with built-in Halloween protection, such a parallel hash join requires a very brief communication and decision phase.

For parallel hash aggregation and duplicate elimination, a common technique is to run the operation twice, once before and once after data partitioning, in cases in which the input data are not yet partitioned on the grouping columns. The goal is to reduce the number of data records that need to be shipped between threads, processes, and processors. Assuming uniform random distribution of input rows prior to re-partitioning, this goal can be achieved if the average group size, i.e., the ratio of row counts in the input and in the output, exceeds the degree of parallelism. A common name for this technique is "local-global aggregation." The initial, local aggregation may be opportunistic, i.e., it performs as much aggregation as can readily be achieved with the available memory, or it may be complete, i.e., it will spill overflow partitions to disk if required. If network transfer is faster than local disk I/O, overflow partitions may be send to their final destination rather than to a local disk.

Sort-based aggregation and duplicate elimination can benefit from the same ideas. The opportunistic variant performs merely run generation and early duplicate elimination prior to data transfer. The full variant performs complete local sort operations and all aggregation and duplicate elimination prior to the data transfer.

Order-preserving and merging re-partitioning operations may experience a deadlock, assuming flow control or bounded buffers in the data exchange operation. The essence of this deadlock is that producers need to send data in the order produced, whereas consumers need to consume data in the appropriate sort order. For example, if producer 1 has data only for consumer 1, and producer 2 has data only for consumer 2, then consumer 1 waits for data from producer 2 in order to advance its merge logic, yet producer 2 waits for consumer 2 to release flow control, etc. There are several possible solutions for this problem. A typical deadlock avoidance strategy is to let producers send dummy records to all consumers, such that the consumers can advance their merge logic. A typical deadlock resolution strategy is to spill data to disk, effectively going beyond bounded buffers and relaxing flow control.

Parallel nested loops join, index nested loops join, and general nested iteration add further complexity to the data exchange operation. One issue is the number of threads: if the nested operation should run in 4 threads (due to four on-disk partitions, for example) yet the outer operation runs in three threads, will in inner operation actually run in 12 threads? If not, communication and synchronization need careful design, including avoidance or resolution of deadlocks. Moreover, invoking the inner operation for batches of outer correlation values rather than for individual outer rows may reduce communication costs as well as create optimization opportunities in the inner query execution plan.

Another technique to reduce communication costs in all matching operations is bit vector filtering. Sometimes thought to apply only to hash join, it also applies to merge join and nested iteration. If the two inputs for the matching operation are scanned or computed in two separate phases, the first such phase can populate a bit vector filter that the second phase can exploit to eliminate some items without further predicate evaluation and in particular without communication in parallel query execution. The bit vector filter might permit hash collisions or hash collisions might be prevented by some means, typically involving dynamic growth of the bit vector filter. In addition to the actual bit vector filter, it can prove useful to retain information about the minimal and maximal values.

In addition to expensive join and grouping operations, parallel execution and bit vector filtering can benefit set operations such as intersecting lists of row identifiers when exploiting multiple indexes on the same table for a query with a conjunctive predicate. Other index operations include union, difference, and join. Index join sometimes refers to joining indexes of two tables prior to fetching complete rows from either table and sometimes refers to joining indexes from a single table in order to obtain all columns required in a query and to avoid fetching rows one-by-one. Both kinds of index joins can benefit from parallel execution and bit vector filtering.

Some operations, however, cannot exploit parallel execution. The most notable example is a "top" operation, e.g., a query to find the three most productive sales people in the organization. While the "top" operation permits local-global computation with potentially tremendous reduction in serial computation and in communication, the final computation cannot benefit from multiple concurrent threads. An exception to this exception are "top" operations that apply to groups of records, e.g., a query looking for the three most productive sales people in each region.

Some algorithm implementations also exploit local parallelism, independent of parallel execution of the overall query execution plan. Asynchronous I/O (sequential read-ahead, single-page prefetch, and write-behind) are forms of local concurrent execution. Expensive operations, e.g., sort operations, may benefit from multiple threads, e.g., fitting records into a sort operation's workspace including initialization of a pointer array, sorting and rearranging the pointer array to represent the desired sort order, and forming on-disk runs from pointer arrays already sorted.

Parallel Execution Beyond Queries

In addition to query execution, parallelism is also needed for large update operations and, perhaps most importantly, in utilities that scan or modify entire indexes, tables, or databases.

Parallel update operations are required in shared-nothing databases, simply because updates like filters are processed locally at each data site. Otherwise, if all update operations are relatively small, parallel update operations are not required for performance. However, update operations can be large, e.g., in bulk insertion and bulk deletion, also known as load, import, or roll-in and as roll-out.

A parallel update algorithm can follow the partitioning of stored data, it can assign disjoint key ranges within B-tree indexes to separate update threads, or it can assign threads to individual indexes. The latter approach is a special form of index-by-index updates (as opposed to row-by-row updates) as it permits the changes for each index to be sorted like the index for fast modification with read-ahead, etc.

Not only the actual database modifications but also other activities associated with updates can be parallel, including verification of integrity constraints, update propagation to materialized and indexed views, etc. Of course, verification of integrity constraints can be parallel also during definition of new constraints. Definition of new materialized and indexed views creates an opportunity for parallelism both while computing the query defining the view and while loading the query result into the new data structures. Refreshing a view by re-computation is very similar.

In practice, however, the most important parallel operation is probably index creation. High performance index creation is particularly important if online index creation is not supported, i.e., an entire table is read-locked while the index utility scans, sorts, and inserts data. Fortunately, parallel versions of all three of these steps are readily available using standard mechanisms for parallel query execution.

Key Applications

Parallel algorithms for database query execution are used for two reasons. First, if there are more processors than active users, the only means by which these processors can do useful work on behalf of the users are parallel algorithms. Pure pipelining often leads to poor load balancing; thus, partitioning stored data and intermediate results, in particular using hash functions, often permits higher and more reliable speed-up. In the future, many-core processors may well increase the importance of parallel algorithms in database query execution.

Second, the hardware might require parallel algorithm because the database and its data are partitioned across multiple nodes each with its own memory, operating system, etc. If communication is more expensive than immediate data reduction (using selection, projection, and local aggregation), parallel algorithm are required. The future of improvements in processing bandwidth and in communication bandwidth will probably continue to be unbalanced; the current trend towards networked storage over direct-attached storage may reverse, in particular for database systems that support a single-system image over many direct-attached storage devices including management of that storage.

Future Directions

While processor speed kept increasing, the value of parallel query execution has been doubted at times. With hardware development now focusing on many-core processors rather than ever higher clock speeds, there should be no doubt that parallel query execution is required for relational data warehousing and business intelligence. It may be worth noting, however, that parallel utilities such as index creation are more urgently needed by customers with growing data volumes than parallel query execution. Many parallel utilities are implemented using mechanisms shared with parallel query execution, however, such that parallel query technology is needed in any case.

With the emergence of XML and semi-structured data in database type systems, there probably will be growing interest in parallel algorithms for relational algebra extended for unstructured data and graph manipulation. In addition, more data cleaning, data mining, and scientific applications are integrated with databases, and they, too, require parallel execution very similar to parallel query execution. Perhaps the future of parallel query execution is limited by advances in extensible query optimization, such that both together enable deep integration of those parallel algorithm into the query processing framework.

P

Cross-References

► Parallel Query Optimization
► Relational Algebra
► Storage Resource Management

Recommended Reading

1. Bratbergsengen K. Hashing methods and relational algebra operations. In: Proceedings of the 10th International Conference on Very Large Data Bases; 1984. p. 323–33.
2. DeWitt DJ, Gerber RH, Graefe G, Heytens ML, Kumar KB, Muralikrishna M. GAMMA – a high performance dataflow database machine. In: Proceedings of the 12th International Conference on Very Large Data Bases; 1986. p. 228–37.
3. Fushimi S, Kitsuregawa M, Tanaka H. An overview of the system software of a parallel relational database machine GRACE. In: Proceedings of the 12th International Conference on Very Large Data Bases; 1986. p. 209–19.
4. Graefe G. Query evaluation techniques for large databases. ACM Comput Surv. 1993;25(2):73–170.

Parallel Query Optimization

Hans Zeller[1] and Goetz Graefe[2]
[1]Hewlett-Packard Laboratories, Palo Alto, CA, USA
[2]Google, Inc., Mountain View, CA, USA

Synonyms

Optimization of parallel query plans

Definition

Parallel query optimization is the process of finding a plan for database queries that employs parallel hardware effectively. The details of this process depend on the types of parallelism supported by the underlying hardware, but the most common method is partitioning of the data across multiple processors.

Historical Background

Most parallel database systems today can trace part of their heritage back to the Gamma project at the University of Wisconsin, Madison in the 1980s – with the exception of the Teradata system, which predates Gamma by several years. Also influential were the GRACE database machine, developed at the University of Tokyo, and work at the Norwegian Institute of Technology, University of Trondheim. These projects did not publish a description of their parallel query optimization algorithms, however. Later projects, like XPRS (University of California, Berkeley) and IBM DB2 Parallel Edition describe that process in more detail.

A simple way of generating parallel query plans is to take the optimal plan produced by a conventional query optimizer and to parallelize it where possible. This may not result in the optimal parallel plan. For example, the chosen join order may require repartitioning of the data, while a slightly different join order might have avoided this costly step. Therefore, most of today's systems integrate parallelism into the optimization phase itself. Different parallel and serial plans are compared on the basis of their estimated cost, and the best plan is selected.

Most optimizer designers of parallel systems chose a very natural extension of non-parallel optimizers. This approach is described in the next section.

Foundations

Most of the parallel database systems use an operator-based model for parallel queries: Traditional relational operators such as join and groupby are employed in a parallel framework, allowing individual CPUs of a parallel system to work on a partition of the data. The task of the query optimizer is to find the right partitioning and degree of parallelism for each step.

With such an operator-based approach, optimizing parallel queries is similar to conventional query optimization, except that a new dimension is added – partitioning.

Extending a non-parallel optimizer to handle parallelism consists of four basic steps: First, the space of semantically correct parallel plans is defined. Second, the optimizer's search algorithm needs to be extended to enumerate a good part of the search space that was defined in the first step. Third, the cost model of the optimizer needs to be able to cost parallel plans such that the best one can be selected. Finally, all these extensions will likely increase the complexity of the optimization algorithm significantly. Heuristics are required to avoid an explosion of the cost of the optimization itself.

The following explores each of these four steps briefly.

Extending Query Plans to Execute in Parallel on Partitioned Data

The optimizer is responsible for generating semantically valid parallel plans. It therefore needs to have a good understanding of and model for parallel operator semantics. This is typically done by assigning a "partitioning" property to each relational operator and by using these properties to constrain the search space to semantically correct query plans. The partitioning property specifies a partitioning scheme (usually hash-based), the partitioning key, and the number of partitions.

In the following, the basic idea behind partitioning properties of a few key relational operators will be discussed.

When a file scan operator is executed in parallel, each parallel instance simply reads a partition of the data. On a shared-nothing system, the partitioning property describes how the data is physically partitioned. On a shared-everything architecture, another way is to let parallel scan operators compete for each block as it is read from disk, achieving a natural load balancing between the parallel operators, with a partitioning property that specifies only the number of partitions, not a partitioning key.

For joins, there are basically two different parallelization strategies. The first is to let each parallel instance join one partition of the first table R with a matching (co-located) partition of the second table S. Both R and S must have a matching partitioning property on the equi-join columns for this case. This is so that for a given row from a partition of R, all its matching rows from S are found in the corresponding partition of S.

The second parallel join algorithm is to join each partition of R with every partition of S or vice-versa. Such joins combine a partitioned table R with a replicated table S or vice versa. This type of parallel join is applicable in nearly all cases, except for full outer joins.

Terms like "co-located nested loop join," "broadcast hash join" or "repartitioned merge join" are used to distinguish these different cases [13].

Semijoins and outer joins can be parallelized as well, with the exception that only the operand requiring special handling can be broadcast. Parallelizing the UNION operator puts no special constraints on the partitioning properties, except what is needed for duplicate elimination.

To increase its choices, the execution engine typically is capable to produce partitioning artificially. This is in most systems implemented by a separate operator, called Exchange [11], Table Queue (DB2), or River [2]. Generally, the exchange operator is the only one that sends data through messages, all other operators process local data and pass their results on in local memory or in a local disk file.

Extending the Search Algorithm to Include Parallel Plans

Almost all query optimizers operate under a principle borrowed from dynamic programming that combines optimal sub-solutions to a larger optimal solution. Sub-solutions are optimal with respect to relevant properties, as described first in [Selinger]. Probably the key aspect of query optimization for parallel queries is that this system of properties can be very elegantly extended to include partitioning properties. Initially only consisting of ordering, with a sort operator to create artificially sorted results, partitioning properties and the exchange operator are now included to create artificially partitioned results.

Extending the properties of query execution plans with partitioning solves the key part of parallel query optimization, namely the problem of enumerating the parallel plans that are possible for a query. The driver for this enumeration are the different ways the other operators can be executed in parallel.

The exchange operator acts as the "enforcer" [11] or "glue" [14] for partitioning.

This extension applies to classical Selinger-style optimizers (e.g., DB2) as well as to rule-based optimizers derived from the Exodus, Volcano and Cascades approaches (e.g., Microsoft SQL Server).

Costing Parallel Query Plans

Executing queries in parallel usually increases the amount of resources consumed, due to communication and synchronization overhead. Parallel execution is therefore not well-suited to minimize the resource cost of queries. Most parallel query optimizers try to minimize the time a query would take if executed in single-user mode on the system. When costing a parallel operator, they therefore compute the cost for a single partition only (assuming that the data is equally distributed over the partitions without skew) and add the cost for needed synchronization. With the goal of keeping the overall query response time low, it is usually not beneficial to reduce the degree of parallelism, therefore the optimizers consider only plans that use all the processing nodes of the system and compare them with a serial plan. The explosion of choices comes with different methods of partitioning and/or replicating data to allow for different types of join execution plans.

Heuristics to Reduce the Number of Parallel Plans Considered

One heuristic, the use of "interesting orders" [Selinger], can be used for partitioning as well. Optimizers with top-down exploration achieve this filtering effect naturally. Without some heuristics, the number of possible partitioning schemes could easily explode. For example, consider a co-located parallel merge join for the query "select * from R join S on R.a = S.a and R.b = S.b and R.c = S.c." Any ordering on a subset of columns (a,b,c), combined with a partitioning on some or all of the columns in this

subset would constitute a valid set of properties for the join operator – assuming it is present for both tables R and S. Most implementations of optimizers will heuristically try only a small number of possibilities in this case.

Key Applications

Parallel computers allow database applications to scale, and optimization of parallel queries allows users to write applications without expending effort on parallelizing database queries. Parallel databases are the main application today where automatic parallelization of generic requests is performed successfully.

Future Directions

Challenges for future systems include adaptive systems that learn from poorly parallelized queries and that avoid data skew automatically, as well as automatic advisors and optimizers of physical designs for parallel databases.

Cross-References

▶ Parallel Query Execution Algorithms
▶ Query Optimization

Recommended Reading

1. Ballinger C, Fryer R. Born to be parallel. why parallel origins give teradata an enduring performance edge. IEEE Data Eng Bull. 1997;20(2):3–12.
2. Barclay T, Barnes R, Gray J, Sundaresan P. Loading databases using dataflow parallelism. ACM SIGMOD Rec. 1994;23(4):72–83.
3. Baru CK, Fecteau G, Goyal A, Hsiao H-I, Jhingran A, Padmanabhan S, Wilson WG. An overview of DB2 parallel edition. In: Proceedings of the ACM SIGMOD International Conference on Management of Data; 1995. p. 460–2.
4. Boral H, Alexander W, Clay L, Copeland GP, Danforth S, Franklin MJ, Hart BE, Smith MG, Valduriez P. Prototyping Bubba, a highly parallel database system. IEEE Trans Knowl Data Eng. 1990;2(1):4–24.

5. Bratbergsengen K. Algebra operations on a parallel computer – performance evaluation. In: Proceedings of the 5th International Workshop on Data Machines; 1987. p. 415–28.
6. Chen A, Kao Y-F, Pong M, Shak D, Sharma S, Vaishnav J, Zeller H. Query processing in nonstop SQL. IEEE Data Eng Bull. 1993;16(4):29–41.
7. DeWitt DJ, Gerber RH, Graefe G, Heytens ML, Kumar KB, Muralikrishna M. GAMMA – a high performance dataflow database machine. In: Proceedings of the 12th International Conference on Very Large Data Bases; 1986. p. 228–37.
8. DeWitt DJ, Gray J. Parallel database systems: the future of high performance database systems. Commun ACM. 1992;35(6):85–98.
9. DeWitt DJ, Smith M, Boral H. A single-user performance evaluation of the Teradata database machine. In: Proceedings of the 2nd International Workshop on High Performance Transaction Systems; 1987.
10. Fushimi S, Kitsuregawa M, Tanaka H. An overview of the system software of a parallel relational database machine GRACE. In: Proceedings of the 12th International Conference on Very Large Data Bases; 1986. p. 209–19.
11. Graefe G. Encapsulation of parallelism in the volcano query processing system. In: Proceedings of the ACM SIGMOD International Conference on Management of Data; 1990. p. 102–11.
12. Graefe G, Davison DL. Encapsulation of parallelism and architecture-independence in extensible database query execution. IEEE Trans Softw Eng. 1993;19(8):749–64.
13. Jhingran A, Malkemus T, Padmanabhan S. Query optimization in DB2 parallel edition. IEEE Data Eng. Bull. 1997;20(2):27–34.
14. Lee MK, Freytag JC, Lohman GM. Implementing an interpreter for functional rules in a query optimizer. In: Proceedings of the 14th International Conference on Very Large Data Bases; 1988. p. 218–29.
15. Mohan C, Pirahesh H, Tang WG, Wang Y. Parallelism in relational database management systems. IBM Sys J. 1994;33(2):349–71.
16. Neches PM. The anatomy of a database computer system. In: Digest of papers – COMPCON; 1985. p. 252–4.
17. Selinger P, Astrahan M, Chamberlin D, Lorie R, Price T. Access path selection in a relational database management system. In: Proceedings of the ACM SIGMOD International Conference on Management of Data; 1979. p. 23–34.
18. Stonebraker M, Katz RH, Patterson DA, Ousterhout JK. The design of XPRS. In: Proceedings of the 14th International Conference on Very Large Data Bases; 1988. p. 318–30.
19. von Bueltzingsloewen G. Optimizing SQL queries for parallel execution. In: Proceedings of the ODBF Workshop on Database Query Optimization; 1989.

Parallel Query Processing

Esther Pacitti
INRIA and LINA, University of Nantes, Nantes, France

Synonyms

Multiprocessor query processing

Definition

Parallel query processing designates the transformation of high-level queries into execution plans that can be efficiently executed in parallel, on a multiprocessor computer. This is achieved by exploiting the way data is placed in parallel and the various execution techniques offered by the parallel database system. As in query processing, the transformation from the query into the execution plan to be executed must be both correct and yield efficient execution. Correctness is obtained by using well-defined mappings in some algebra, e.g., relational algebra, which provides a good abstraction of the execution system. Efficient execution is crucial for high-performance, e.g., good query response time or high query throughput. It is obtained by exploiting efficient parallel execution techniques and query optimization which selects the most efficient parallel execution plan among all equivalent plans.

The main forms of parallelism which can be inferred by high-level queries are inter-query parallelism (several queries executed in parallel) and intra-query parallelism (each query executed in parallel), and within a query, inter- and intra-operator parallelism. These various forms of parallelism are obtained based on how data is placed in the parallel system, i.e., physically on disk or memory. Hence, the term data-based parallelism is communally used. These forms of parallelism can be combined and are exploited by parallel execution techniques.

Historical Background

Parallel query processing had been initiated in the context of database machines in the late 1970s. At that time, parallel processing techniques were already used successfully in scientific computing to improve the response time of numerical applications. However, these techniques are very complex as they typically involve parallelizing compilation which must infer parallelism from programs written in a procedural language, e.g., Fortran or C. Since these programs may be sequential, inferring parallelism to obtain good performance is hard.

With a high-level language like SQL, parallelism is relatively easier to infer. SQL enables programmers to specify what data to access (with predicates) without any detail on how. Thus, this provides the parallel database system much leverage to decide how to access data.

In the context of database machines, parallel query processing was very simple and concentrated on select-project-join queries using brute-force algorithms and special-purpose hardware. As parallel database systems evolved to exploit general-purpose multiprocessors and software-oriented solutions, parallel query processing has become much more complex. Also, the emergence of new applications such as OLAP and data mining translated into new features in SQL which have made parallel query processing more complex.

Compared to distributed query processing (which can also exploit parallelism by executing a query using multiple sites), the main difference is that the parallel system can have many more nodes and a very fast interconnection network. Thus, the execution techniques and execution plans are fairly different.

Foundations

Parallel query processing is the major solution to high-performance management of very large databases. The basic principle is to partition the database across multiple disks or memory nodes so that much inter- and intra-query parallelism can be gained. This can lead to significant im-

provements in both response time (the time the query takes to be executed and return results) and throughput (number of queries or transactions per time unit). However, decreasing the response time of a complex query through large-scale parallelism may well increase its total time (by additional communication) and hurt throughput as a side-effect. Therefore, it is crucial to optimize and parallelize queries in order to minimize the overhead of parallelism, e.g., by constraining the degree of parallelism for the query.

The performance of a parallel database system should ideally demonstrate two advantages: linear speedup and linear scaleup. *Linear speedup* refers to a linear increase in performance for a constant database size and linear increase in the number of nodes (i.e., processing and storage power). Thus, the addition of computing power should yield proportional increase in performance. *Linear scaleup* refers to a sustained performance for a linear increase in both database size and workload and number of nodes. Furthermore, extending the system should require minimal reorganization of the existing database. Thus, an increase of computing power proportional to the increase in database size and load should yield the same performance. These advantages are ideal since, in practice, increasing the configuration tends to increase the overhead of parallelism (e.g., more communications between nodes, more interference when accessing shared resources, etc.).

Partitioned data placement is the basis for the parallel execution of database queries. Given a parallel system with n nodes (each node may have one or more processors accessing a main memory and disks), each database table T is typically partitioned onto a number of nodes (less or equal to n) so that each node stores a subset of the tuples of T. This corresponds to horizontal fragmentation in distributed databases and similar techniques can be used. However, because there can be many nodes, more scalable techniques are also often used such as round robin or hashing on the placement attribute(s). Furthermore, to improve availability and performance, some partitions (typically those which are accessed more than others) can be replicated.

Given a partitioned database, parallel query execution can exploit two forms of parallelism: inter- and intra-query. *Inter-query parallelism* enables the parallel execution of multiple queries, each at a different node, in order to increase query throughput. This form of parallelism, reminiscent of that used in transaction processing systems, is quite simple since queries need not be parallelized. Thus, incoming queries can simply be dispatched (by a load balancer of the parallel database system) to the nodes which store the data corresponding to the queries. However, this can only work if all the data accessed by a query are stored at the node, or accessible at the node. For instance, in the case of a shared-nothing parallel database system, if a query Q involves data that is partitioned at two different nodes, then either node cannot entirely execute the query. But in the case of a shared disk parallel database system, since all the disks can be accessed from each node, the same query Q could be executed by either node.

A query can be decomposed in a tree of relational operators, where each operator takes data as input (either base data or temporary data) and produces temporary data, with the root operator producing the final result. This decomposition allows for *intra-query parallelism*) which has two forms: inter-operator and intra-operator parallelism. *Inter-operator* parallelism is obtained by executing in parallel several operators of the query on several nodes while with intra-operator parallelism, the same operator is executed by many nodes, each one working on a different partition of the data. For inter-query parallelism, much attention has been devoted to pipelined (or consumer-producer) parallelism which enables consumer operators to start execution as soon they get input data from producer operators. Pipelined executions do not require temporary relations to be materialized, i.e., a tree node corresponding to an operator executed in pipeline is not stored. These two forms of parallelism complement each other well since inter-operator parallelism gets beneficial as the query gets complex (with many operators) while intra-operator parallelism gets beneficial for heavy operators (which access large parts of the data).

To exploit inter- and intra-query parallelism, using database partitioning, parallel algorithms for the execution of relational operators are necessary. Such algorithms must yield a good trade-off between parallelism and communication cost since increasing parallelism involves more communication among nodes. Parallel algorithms for unary operators (e.g., select) are relatively simple as they typically exploit data placement or secondary indices to restrict access to base data. Parallel algorithms for binary operators such as join are much more involved since join can incur much communication. The main algorithms exploit the placement of one of the two relation to reduce communication or artificially create a good data placement using hashing. When there is sufficient main memory to hold one of the two relations (which is a very practical case), the execution of hash-based join can be pipelined across multiple operator nodes, thus increasing performance. Since the intermediate results of the operators can be skewed, parallel algorithms must also deal with load unbalancing, e.g., by redistributing a heavy work at one node to multiple nodes using hashing.

Parallel query optimization is the process of selecting the best parallel execution plan for a query. Compared to distributed query optimization, it focuses on taking advantage of both intra-operator parallelism and inter-operator parallelism. As any query optimizer, a parallel query optimizer can be seen as three components: a search space, a cost model, and a search strategy. Parallel execution plans are abstracted by means of operator trees, which define the order in which the operators are executed. Operator trees are enriched with annotations, which indicate additional execution aspects, such as the algorithm of each operator. An important aspect to be reflected by annotations is that operators are executed in pipeline. The cost model provides the cost functions for parallel operator algorithms.

Key Applications

Parallel query processing has been developed in the context of parallel database systems, with a focus on OLAP applications, where good response time is crucial. Most of the work has been done in the context of the relational model. In order to support new OLAP applications which may access all kinds of data, including unstructured and semi-structured (XML) data, major extensions to parallel query processing are necessary. In particular, techniques from parallel database and information retrieval need to be combined.

Cross-References

▶ Parallel Data Placement
▶ Parallel Query Execution Algorithms
▶ Query Optimization

Recommended Reading

1. Graefe G. Query evaluation techniques for large databases. ACM Comput Surv. 1993;25(2):73–170.
2. Kossmann D. The state of the art in distributed query processing. ACM Comput Surv. 2000;32(4):422–69.
3. Lanzelotte R, Valduriez P, Zait M, Ziane M. Industrial-strenght parallel query optimization: issues and lessons. Inf Syst. 1994;19(4):311–30.
4. Özsu MT, Valduriez P. Principles of distributed database systems. 2nd ed. Upper Saddle River: Prentice-Hall; 1999.
5. Valduriez P, Pacitti E. Parallel database systems. In: Hammer J, Scheider M, editors. Handbook of database technology. CRC; 2007.

Parameterized Complexity of Queries

Christoph Koch
Cornell University, Ithaca, New York, NY, USA
EPFL, Lausanne, Switzerland

Definition

Parameterized complexity theory is the study of the interaction between the fixing of parameters of input problems and their computational complexity. A central parameterized complexity concept is that of a fixed-parameter tractable (FPT) problem, which captures a strong notion of well-behavedness of a problem under the assumption that parameter values do not grow with input sizes. There is also a solid theory of fixed-parameter intractability, which gives strong evidence that for certain parameterizations of problems, no FPT algorithms can be found.

Historical Background

Fixed-parameter complexity theory is strongly associated with R. Downey and M. Fellows, who did much seminal work in the area (cf. [3, 5]). The first fixed-parameter complexity result in the context of database query evaluation was the linear-time query processing algorithm for acyclic conjunctive queries by Yannakakis in 1981 [12], which preceded the development of parameterized complexity theory (cf. also [13]).

Foundations

Fixed-Parameter Tractability

A *parameterized (decision) problem* is a set of pairs (x, k), where x is called the input and k the parameter (an integer). As a convention, $n = ||x||$ will be used to denote the size of the input. A parameterized problem is called strongly uniformly *fixed-parameter tractable* (subsequently, just fixed-parameter tractable, or FPT), if there is a computable integer function f, a constant c, and an algorithm that, given a parameterized problem instance (x, k), decides x in time $O(f(k) \cdot n^c)$.

It is important to note the difference between an algorithm that runs in time $O(n^k)$ and an FPT algorithm. If $k = 10$ and $c = 1$, an algorithm that runs in time $O(f(10) \cdot n)$ is linear (with possibly a large constant), while an $O(n^{10})$ algorithm will not even scale to very moderately sized problem instances.

Consider the *data complexity* of queries [11], i.e., the complexity of evaluating a query on a database if only the database is considered part of the input, while the query is fixed and part of the problem specification. It is well known that quite naive query evaluation techniques can evaluate an arbitrary relational algebra query in time $O(n^k)$, where k is the size of the query (e.g., the number of algebra operators involved). By fixing k, the query evaluation time becomes polynomial, but the $O(n^k)$ time query evaluation technique does not yield a fixed-parameter tractability result. Indeed, such an FPT result is considered unlikely to exist.

For another example, consider the Set Cover problem, a well known combinatorial problem that appears in database problems such as subspace clustering or finding pipelined query plans in stream processors.

A set C is called a *cover* of a set of sets $\mathbf{S}$ if for each $S \in \mathbf{S}$, $S \cap C \neq \emptyset$. The Set Cover problem is defined as follows.

Set Cover

Input: An integer k, a finite set V, and a set $\mathbf{S} \subseteq 2^V$ of subsets of V.

Question: Is there a set C with $|C| \leq k$ such that C is a cover for $\mathbf{S}$?

The Set Cover problem is NP-complete. A naive brute-force algorithm would check for each set $C \subseteq V$ with $|C| \leq k$ whether C is a cover. This algorithm runs in time $O(|V|^k \cdot n)$, which is polynomial in n if k is fixed. However, this is not an FPT algorithm with parameter k because $|V|$ may get arbitrarily large with the input.

Consider the following refined algorithm for Set Cover (cf. [5, 8]).

```
C₀ : ={∅};
for i = 1 to k do
Cᵢ = {C ∪ {a}|C ∈ Cᵢ₋₁,a ∈ S(C)};
if at least one element of Cₖ is a
cover for S then output true;
else output false
```

Here, $S(C)$ is a function that returns, in a deterministic way, an element S of $\mathbf{S}$ such that, if this condition can be satisfied by any element of $\mathbf{S}$, $S \cap C \neq \emptyset$. One way of defining $S(C)$ is as

$$S(C) = \begin{cases} \min \mathbf{S}(C) \dots \mathbf{S}(C) \neq \theta \\ \min \mathbf{S} \dots \text{otherwise.} \end{cases}$$

where $\mathbf{S}(C) = \{S \in \mathbf{S} \mid S \cap C \neq \emptyset\}$ and min returns the smallest element of a subset of $\mathbf{S}$ with respect to some arbitrary fixed order among the elements of $\mathbf{S}$ (e.g., the order in which the elements of $\mathbf{S}$ are stored in memory).

Consider for example $\mathbf{S} = \{\{a, b\},\{a, c\},\{b, d\},\{c, e\},\{c, f\}\}$ with the elements of $\mathbf{S}$ ordered as just enumerated. Then, $S(\emptyset) = \{a, b\}$, $\mathbf{C}_1 = \{\{a\},\{b\}\}$, $S(\{a\}) = \{b, d\}$, $S(\{b\}) = \{a, c\}$, and $\mathbf{C}_2 = \{\{a, b\},\{a, d\},\{b, c\}\}$. Since $\{b, c\}$ is a cover, the algorithm returns true.

It is easy to check that this algorithm runs in time $O(s^k \cdot n)$ where s is the maximum cardinality among the elements of $\mathbf{S}$, i.e., $s = \max\{|S|: S \in \mathbf{S}\}$. Thus, Set Cover with parameter $k + s$ (which is not the standard parameterization for Set Cover, which would be k) is FPT. If s is small, this may be a useful algorithm.

Fixed-Parameter Intractability

For a number of parameterized versions of NP-hard problems it can be shown that they arc either provably not FPT or not FPT unless P=NP. An example for the former scenario would be EXPTIME Turing machine acceptance (which is EXPTIME-complete) with a dummy parameter that is unrelated to the input. An example of the latter scenario is the satisfiability of CNF formulae whose clauses have no more than k literals, with parameter k. For $k = 3$, this is the 3SAT problem, which is NP-complete. But there are also parameterized problems for which apparently no FPT algorithm exists and for which more subtle hardness arguments based on special complexity classes for parameterized problems have to be developed.

A *fixed-parameter many-one reduction* is a reduction that maps each instance (x, k) of a parameterized problem to an instance (x', k') of another parameterized problem using an FPT algorithm (i.e., in time $O(f(k) \cdot x^c)$), with the additional condition that the size of k' must only depend on k but not on x, such that (x, k) is a yes-instance of the first problem if and only if (x', k') is a yes-instance of the second problem. Closure under such reductions yields robust complexity classes. The class of parameterized problems that are fixed-parameter many-one reducible to problem Π shall be denoted by $[\Pi]^{fpt}$. A problem in $[\Pi]^{fpt}$ is called *complete* for $[\Pi]^{fpt}$ if Π is fixed-parameter many-one reducible to it.

Let $\Gamma_{0,d} = \Delta_{0,d}$ be the class of all propositional formulae constructible from propositional variables using negation and the *binary* operations conjunction $\wedge$ and disjunction $\vee$ such that the maximum expression depth, not taking into account negations, is d. In other words, for any such formula, the maximum number of disjunctions or conjunctions occurring on paths from the root of the expression tree to a leaf is d. Now, $\Gamma_{t,d}$ consists of the formulae of the form $\wedge \Phi$, where Φ is a finite subset of $\Delta_{t-1,d}$, and $\Delta_{t,d}$ consists of the formulae of the form $\vee \Psi$, where Ψ is a finite subset of $\Gamma_{t-1,d}$.

A usual way of defining the W hierarchy is in terms of a weighted version of the propositional satisfiability problem. The weight of a truth assignment for the variables of a propositional formula is the number of variables set to true in that truth assignment. The weighted satisfiability problem $\mathrm{WSAT}(\Gamma_{t,d})$ for the class of formulae $\Gamma_{t,d}$ is as follows: Given a formula $\varphi \in \Gamma_{t,d}$ and parameter k, does φ have a satisfying assignment of weight k?

The W-hierarchy, for each $t \geq 1$, is defined as

$$W[t] := \underset{d \geq 0}{U}\left[\mathrm{WSAT}\left(\Gamma_{t,d}\right)\right]^{fpt}$$

with $W[t] \subseteq W[t+1]$ and is believed to be strict ($W[t] \subsetneq W[t+1]$). For a few examples, natural parameterized versions of Clique and Independent Set are $W[1]$-complete, while parameterized versions of Hitting Set, Dominating Set, and Kernel are $W[2]$-complete (see [5]). These results depend on the choice of parameter, and the standard parameterization is the size of the structure (set) whose existence is to be guessed and verified. The question whether FPT $\neq W[1]$ is the parameterized complexity analogue of the question whether P $\neq$ NP. It remains unproven, but is strongly suspected.

Returning to the evaluation complexity of relational algebra queries with the size of queries as the parameter,

Parameterized Query Evaluation
Input: A Boolean relational calculus query Q and a relational database.
Parameter: The size k of a reasonable representation of Q in bits.
Question: Does Q return true on the input database?

which is known to be $W[1]$-complete [9] already for conjunctive queries (i.e., select-project-join queries) and AW[*]-complete for full relational calculus [4]. AW[*] is a complexity class that subsumes the $W[t]$ classes, for all t, but may contain additional problems.

Further Positive Results on Query Evaluation Complexity

While relational query languages such as relational algebra, calculus, and datalog (a generalization of conjunctive queries) are $W[1]$-hard and thus unlikely to be fixed-parameter tractable, there are important fragments of these languages that are FPT. Moreover, there are other logics and query languages, specifically on tree- and graph-structured data models, which are FPT with the size of the query as the parameter.

The first FPT result in the context of database query processing was on the evaluation of *acyclic* conjunctive queries by Yannakakis. Consider a Boolean conjunctive query in datalog notation, $q \leftarrow R_1\left(\overrightarrow{y}_1\right), ..., R_k\left(\overrightarrow{y}_k\right)$. The acyclicity of queries refers to a notion of acyclicity of associated hypergraphs, whose nodes are query variables and whose hyperedges are the sets of variables $\overrightarrow{y}_i$ occurring together in an atomic formula of the query. Acyclicity can be defined in a number of ways, such as by a low-complexity algorithm or using guarded logic. An exact definition is beyond the scope of this article, but in the case that all input relations are binary and no atom is of the form $R(y, y)$, the hypergraph is an undirected graph and hypergraph acyclicity coincides with the standard graph-theoretic notion of acyclicity. By exploiting the tree structure of this graph, which describes the necessary joins, and projecting away columns that will not be involved in further joins as early as possible, it is always possible to find a query plan that can be evaluated in time $O(f(k) \cdot n)$ where k is the size of the query – i.e., the problem is fixed-parameter linear.

A Boolean acyclic conjunctive query can be written using just selections and semijoin operations, with a $\pi_\emptyset$ operation on top. If Q_1 and Q_2 are select-semijoin-queries, then $Q_1(x) \ltimes Q_2(x) = \pi_{sch(Q1)}(Q_1(x) \bowtie Q_2(x))$ is a subset of $Q_1(x)$ and can be computed in linear time in $|Q_1(x)| + |Q_2(x)|$, where x is the input database. By induction it follows that overall query evaluation is FPT.

Consider the acyclic conjunctive query $\pi_\emptyset(R \bowtie S \bowtie T \bowtie U \bowtie V)$ for database schema $R(A,B),S(B,C),T(C,D),U(C,E),V(A,F)$. The hypergraph (here, graph) is acyclic:

The query plan $\pi_\emptyset((R \ltimes V) \ltimes ((S \ltimes \mathbf{T}) \ltimes U))$ admits efficient evaluation.

For nonboolean conjunctive queries, the running time is polynomial of degree c where c is the arity of the query result (which must be considered a true constant rather that a parameter if this is to be considered an FPT result), and clearly, this is optimal because the output size of a query that simply computes the product of the input relations is $\Omega(n^c)$.

The notion of hypergraph acyclicity has been generalized to queries of bounded hypertree-width, which is a measure of how tree-like the query hypergraph is: hypertree-width 1 coincides with hypergraph acyclicity. It was shown in [7] that conjunctive query evaluation with the hypertree-width of the query as the parameter is FPT and that this generalizes in a natural way to relational calculus queries. The fixed-parameter tractability of queries of bounded hypertree-width >1 (in fact, 2) has been used to explain the polynomial-time complexity of XPath queries [1].

A classical result by Courcelle [2] shows that very powerful queries – in monadic second-order logic, a language that strictly subsumes relational calculus – over databases of bounded tree-width (the exact definition is technical, but in the case that all relations are binary, the condition is bounded tree-width of the (undirected) graph obtained by unioning the relations together) are fixed-parameter linear. Note that in this result the structure of the data is restricted but the structure

of queries is not. The special case where the database is a tree with node labels from a finite alphabet is due to Doner, Thatcher and Wright [10]. The algorithm is based on compiling the query into a tree automaton which, once obtained, can be evaluated on the data tree in linear time. Thus, the running time is $O(f(k) + n)$, where k is the size of the query. This is a famous case where f is much worse than singly exponential: f is nonelementary – a tower of twoes $2^{2^{2^{.2}}}$ whose height grows with k. This is apparently necessarily so: Frick and Grohe [6] show that unless $P = NP$, any FPT algorithm for the problem must have a nonelementary f.

Key Applications

Fixed-parameter complexity results can assist designers of data management algorithms in two ways. There is now a large set of positive results for a variety of parameterized versions of NP-hard problems, which may surface in contexts such as query optimization and data mining. For a particular data management problem, it may be known that a particular parameter is always bounded in the problem instances that arise. If the problem is FPT for that parameter, there is an efficient algorithm for solving the problem. Furthermore, FPT results exist for fragments of the relational query languages such as relational algebra as well as for query languages for tree- and graph-structured data.

Conversely, if it is known that a parameterized problem is W[1]-hard, then it is quite hopeless to try to develop an efficient algorithm for solving the parameterized problem; in that case one may look for different acceptable parameterizations or for efficient approximation techniques.

References

1. Benedikt M, Koch C. XPath Leashed. ACM Comput Surv. 2008;4(1):1.
2. Courcelle B. Graph rewriting: an algebraic and logic approach. In: van Leeuwen J, editor. Handbook of theoretical computer science, vol. 2. Amsterdam: Elsevier B.V.; 1990. p. 193–242 .chap. 5

3. Downey RG, Fellows MR. Parameterized complexity. Berlin: Springer; 1999.
4. Downey RG, Fellows MR, Taylor U. The parameterized complexity of relational database queries and an improved characterization of W[1]. In: Proceedings of the First Conference of the Centre for Discrete Mathematics and Theoretical Computer Science Combinatorics, Complexity, and Logic; 1996. p. 194–213.
5. Flum J, Grohe M. Parameterized complexity theory. Berlin: Springer; 2006.
6. Frick M, Grohe M. The complexity of first-order and monadic second-order logic revisited. In: Proceedings of the 17th Annual IEEE Symposium on Logic in Computer Science; 2002. p. 215–24.
7. Gottlob G, Leone N, Scarcello F. Hypertree decompositions and tractable queries. J Comput Syst Sci. 2002;64(3):579–627.
8. Grohe M. Parameterized complexity for the database theorist. ACM SIGMOD Rec. 2002;31(4):86.
9. Papadimitriou CH, Yannakakis M. On the complexity of database queries. In: Proceedings of the 16th ACM SIGACT-SIGMOD-SIGART Symposium on Principles of Database Systems; 1997.
10. Thatcher J, Wright J. Generalized finite automata theory with an application to a decision problem of second-order logic. Math Syst Theory. 1968;2(1): 57–81.
11. Vardi MY. The complexity of relational query languages. In: Proceedings of the 14th Annual ACM Symposium on Theory of Computing; 1982. p. 137–46.
12. Yannakakis M. Algorithms for acyclic database schemes. In: Proceedings of the 7th International Conference on Very Data Bases; 1981. p. 82–94.
13. Yannakakis M. Perspectives on database theory. In: Proceedings of the 36th IEEE Symposium on Foundations of Computer Science; 1995. p. 224–46.

Parametric Data Reduction Techniques

Rui Zhang
University of Melbourne, Melbourne, VIC, Australia
Dataware Ventures, Tucson, AZ, USA
Dataware Ventures, Redondo Beach, CA, USA

Definition

A parametric data reduction technique is a data reduction technique that assumes a certain model for the data. The model contains some parameters and the technique fits the data into the model to determine the parameters. Then data reduction can be performed.

Key Points

Parametric data reduction (PDR) techniques is opposite to nonparametric data reduction (NDR) techniques. A model with parameters is used in a PDR technique and therefore some computation is required to determine these parameters, which may be costly. However, if a PDR technique is well-chosen, it may result in much more data reduction than NDR techniques. A representative example is linear regression [3]. Linear regression assumes that the data fall on a straight line, expressed by the following formula

$$Y = a + bX \qquad (1)$$

Given a set of points (Assuming two dimensions.) $\{\langle x_1, y_1 \rangle, \langle x_2, y_2 \rangle, \ldots\}$, parameters a and b in Eq. (1) are determined from the points using the least squares criteria. The result is

$$b = c \frac{\sum (X - \overline{X})(Y - \overline{Y})}{\sum (X - \overline{X})^2}$$

$$a = \overline{Y} - b\overline{X}$$

where $\overline{X}$ and $\overline{Y}$ are the average values of $x_1, x_2, \ldots$ and $y_1, y_2, \ldots$, respectively. If the data are actually distributed on almost a line, linear regression is a very efficient data reduction technique. Besides the values of a and b, only one dimension of every point is needed to represent the data set. The data volume is reduced by half. However, if the data are not distributed on a line, linear regression will result in large errors.

Other popular PDR techniques include Singular Value Decomposition (SVD) [2] and Discrete Wavelet Transform (DWT). SVD assumes that a matrix is decomposed to the form of $\mathbf{A} = \mathbf{USV}^t$ and $\mathbf{U}, \mathbf{S}, \mathbf{V}$ need to be determined; DWT assumes that a signal is projected to a set of orthogonal basis vectors and the wavelet coefficients need to be determined. A summary of data reduction techniques including PDR techniques can be found in [1].

Cross-References

▶ Discrete Wavelet Transform and Wavelet Synopses
▶ Linear Regression
▶ Nonparametric Data Reduction Techniques
▶ Singular Value Decomposition

Recommended Reading

1. Barbará D, DuMouchel W, Faloutsos C, Haas PJ, Hellerstein JM, Ioannidis YE, Jagadish HV, Johnson T, Ng RT, Poosala V, Ross KA, Sevcik KC. The New Jersey data reduction report. IEEE Data Eng Bull. 1997;20(4):3–45.
2. Jolliffe IT. Principal component analysis. Berlin: Springer; 1986.
3. Wonnacott RJ, Wonnacott TH. Introductory statistics. New York: Wiley; 1985.

Partial Replication

Bettina Kemme
School of Computer Science, McGill University, Montreal, QC, Canada

Definition

A replicated database consists of a set of nodes $\mathcal{N}$ (database servers) and each logical data item x has a physical copy on a subset $\mathcal{N}_x$ of the nodes in $\mathcal{N}$. The replication degree $r_x = |\mathcal{N}_x|$ of a data item is the number of copies it has. Using *full replication*, each logical data item has a copy on each of the nodes, i.e., for each data item x of the database, $\mathcal{N}_x = \mathcal{N}$. Whenever there is at least one data item that does not have copies at all nodes, one refers to a partial replication architecture.

Main Text

Full replication is expensive in update intensive environments as the updates have to be typically executed on all replicas limiting scalability and requiring costly coordination. Thus, the lower the replication factor, the better the potential for scalability and fast update operations. On the other hand, more replicas offer better scalability for read operations, replicas at strategic positions in a wide-area setting offer fast local access to clients, and more replicas increase fault-tolerance and availability.

Scalability

In order to achieve scalability, data can be replicated across nodes, typically residing in a single cluster. Read access to data items can then be distributed across the existing copies. Write operations, however, have to be performed on all copies. The potential for scalability can be best understood by a simple analytical model (derived from [4]). It assumes each transaction accesses one data item and has execution cost normalized to 1 unit. A fraction of wr are update transactions. Communication costs and concurrency control issues are ignored. A read-only transaction is executed at any node with a copy of the data item while an update transaction is executed at all nodes. The system consists of n nodes each having the capacity to process C execution units per time unit. There are m data items each having the same replication factor r. The copies are equally distributed across the nodes ($\frac{m*r}{n}$ copies per node). All data items are accessed with the same frequency and requests are equally distributed across all nodes.

If the replicated system can execute l transactions per unit then $(1 - wr) * l$ are read-only transactions executed at one node and $wr * l$ are update transactions executed at r nodes. Thus, the total capacity of the system is used as follows: $n * C = (1 - wr) * l + r * wr * l$. The scale-out is the number of transactions per unit l the replicated system can handle divided by the number of transactions per unit a non-replicated system can handle (which is C). Therefore, the scale-out of an n-node system is $\frac{n}{1+(r-1)*wr}$.

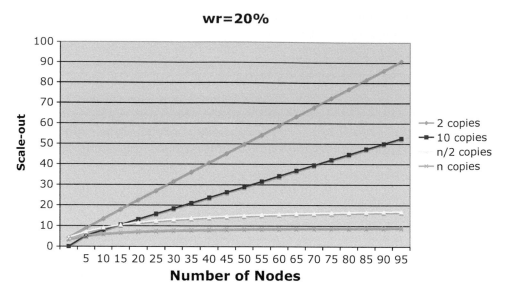

Partial Replication, Fig. 1 Scale-out of partial replication

Figure 1 shows the scale-out for systems up to 100 nodes with an update load of 10 %, and each data item has $2, 10, n/2$ or n copies. While a constant replication factor provides linear scalability (and the smaller the better), a replication factor that increases with the number of nodes leads to a scalability cealing. Once the saturation point is reached, adding more nodes will not increase the throughput because applying update transactions consumes most of the available resources.

Fast Local Access

Replicating data items at different geographical locations is mainly used to provide fast local access to clients of the different regions. Having a data item replicated at a specific location decreases communication costs for read operations but increases communication and processing costs for update transactions, and has additional storage costs. Thus, placement algorithms have to decide where it is worth to put replicas in order to find a trade-off between the different factors [5].

Fault-Tolerance and Available

In order to offer fault-tolerance and high availability it is not necessary to have full replication. In most solutions, the replication factor r for all data items is a small uneven number such as 3 or 5. It is common to use a Paxos [1] protocol to keep copies consistent where an update operation succeeds when a majority of replicas have applied the update (while the remaining updates can execute in the background). If a transparent solution is requested in which reads always should read the latest state, reads either also have to access a majority of replicas, or they always have to read the leader replica (the replica that coordinates the Paxos protocol upon an update request) as the leader always has the latest state. In order to achieve higher availability and reduce costs for read operations, in particular in wide-area settings (also referred to as geo-replication), reads could also be allowed to execute at any replica with the trade-off that this replica might have stale data.

Challenges

When a transaction executes at a specific node but the node does not have a copy of a requested data item, remote access is necessary. This poses several challenges. Firstly, a location mechanism must be implemented that finds nodes with appropriate data copies. Secondly, for complex SQL queries, query execution becomes distributed potentially requiring data shipping and advanced operators. Thirdly, concurrency control becomes more complicated. As no node has a full view of the entire database and only copies of some of the data items, it becomes more difficult to provide a globally unique serialization order of all update transactions and provide read transactions with a globally consistent snapshot of the database [2, 3]. A genuine solution for partial replication should only require coordination protocols for update transactions but not for read-only transactions, and this coordination should only involve nodes that have copies of the data items access by the transaction. Furthermore, it should keep the auxiliary information, such as clocks and timestamps, that needs to be maintained at each node and exchanged in coordination as small as possible.

Therefore, in practice, data that is often accessed together (within a transaction or query) is co-located in a partition, and then this partition is replicated several times. Thus, execution of transactions that only access data of a single partition is simplified, and only rare global transactions, spanning more than one partition, require global coordination. This, however, reduces the flexibility in terms of optimizing load-balancing and update processing.

Cross-References

▶ Replica Control
▶ Replication for Scalability

Recommended Reading

1. Lamport L. The part-time parliament. ACM Trans Comput Syst. 1998;16(2):133–69.
2. Peluso S, Romano P, Quaglia F. Score: a scalable one-copy serializable partial replication protocol. In: Proceedings of the ACM/IFIP/USENIX 13th International Middleware Conference; 2012. p. 456–75.
3. Schiper N, Sutra P, Pedone F. P-store: genuine partial replication in wide area networks. In: Proceedings of the 29th Symposium on Reliable Distributed Systems; 2010. p. 214–24.
4. Serrano D, Patiño-Martínez M, Jiménez-Peris R, Kemme B. Boosting database replication scalabil-
ity through partial replication and 1-copy-snapshot-isolation. In: Proceedings of the IEEE Pacific Rim Dependable Computing Conference; 2007.
5. Wolfson O, Jajodia S, Huang Y. An adaptive data replication algorithm. ACM Trans Database Syst. 1997;22(2):255–314.

Path Query

Yuqing Wu
Indiana University, Bloomington, IN, USA

Synonyms

Document path query

Definition

Given a semi-structured data set D, a *path query* identifies nodes of interest by specifying the *path* lead to the nodes and the predicates associated with nodes along the *path*. The *path* is identified by specifying the labels of the nodes to be navigated and structural relationship (parent-child or ancestor-descendant) among the nodes. A predicate can be a path query itself, relative to the node that it is associated with.

Historical Background

Using path information in query processing has been studied in the object-oriented database systems, in which most queries require the traversing from one object to another following object identifiers, in the mid 1990s. The notion of path query, in which the path and predicates along the path are specified as the core of the query, became popular with the growth of the information on the web and the introduction of semi-structured data, especially XML.

Most of the popular query languages for querying XML data, such as XPath [6] and

XQuery [4], employ path query as the approach to identify nodes of interest. However, some techniques, such as schema-free XQuery, relax the requirements of specifying the exact path, but rely more heavily on the database management systems to detect the least common ancestors of the keywords specified in the query.

Algebraic research has been fruitful in studying, comparing and characterizing fragments of path queries [2, 9], as well as methods and techniques for optimizing and evaluating path queries.

Foundations

Semi-structured data, for example, XML, consists of data entries and containment relationships among the data entries. The data, usually referred to as *a document*, is frequently represented by an ordered node-labeled tree or graph, depending on whether only the containment relationships are treated as first-class relationship, or the id-reference relationships among the data entries are also treated as first-class relationships. The tree representation is more popular:

A *document* D is a 3-tuple (V, E d, λ), with V the finite set of nodes, $Ed \subseteq V \times V$ a set of parent-child edges, and $\lambda : V \to \mathcal{L}$ a node-labeling function into a countably infinite set of labels $\mathcal{L}$.

In addition, even though not always substantial, the ordering among siblings is usually an important feature for nodes in a *document*. The pre-order among the data entries is called the *document order*. Among others, the pre-order is a dominant approach used to identify data entries in a document, while the document is stored in a database system, relational or native.

The aim of the path query is to express the precise requirement of retrieving a set of data entries that satisfy certain value and structural requirements.

The algebraic form of a path query consists of the primitives of $\emptyset$, ε, $\widehat{l}$ ($l \in \mathcal{L}$) for token matching, $\uparrow$ and $\downarrow$ for upward and downward navigation, and operations $\diamond$ for composition of

two algebraic expression $E_1 \diamond E_2$, Π_1 and Π_2 for the first and second projection of an algebraic expression, and set operations $\cap$, $\cup$, and $-$. Given a document $D = (N, Ed, \lambda)$ and a path query E, the path semantics of $E(D)$ is a binary relation:

$$
\begin{aligned}
\emptyset(D) &= \emptyset \\
\epsilon(D) &= \{(n, n) \mid n \in N\} \\
\widehat{l}(D) &= \{(n, n) \mid n \in N \ and \ \lambda(n) = l\} \\
\downarrow(D) &= Ed \\
\uparrow(D) &= Ed^{-1} \\
\Pi_1(E)(D) &= \pi_1(E(D)) \\
\Pi_2(E)(D) &= \pi_2(E(D)) \\
E_1 \diamond E_2(D) &= \pi_{1,4}\sigma_{2=3}(E_1(D) \times E_2(D)) \\
E_1 \cap E_2(D) &= E_1(D) \cap E_2(D) \\
E_1 \cup E_2(D) &= E_1(D) \cup E_2(D) \\
E_1 - E_2(D) &= E_1(D) - E_2(D)
\end{aligned}
$$

These primitives and operations identify the path of navigation in a document to reach the resultant data entries.

The $\uparrow^*$ and $\downarrow^*$ are frequently used to identify the ancestor-descendant relationship among data entries along a path:

$$
\downarrow^*(D) = \bigcup_{i=0..height(D)} \underbrace{\downarrow \ldots \downarrow}_{i}(D)
$$

$$
\uparrow^*(D) = \bigcup_{i=0..height(D)} \underbrace{\uparrow \ldots \uparrow}_{i}(D)
$$

The result of a path query against document D under the *path semantics* is a set of node pairs whose neighborhood data entries and structures satisfy the query expression.

A localized semantics, also called the *node semantics* of a path query Q is to apply the path expression to a document D and a specific node n_0 in the document, such that $Q(D, n_0) = \{n \mid (n_0, n) \in Q(D)\}$. Usually, the results are presented in a list that honors the *document order*.

For example, assuming document D is represented as a tree structure as shown in Fig. 1, some sample path queries are evaluated as follows:

Path query	Sample result
$Q_1 = \downarrow \diamond B$	$Q_1(D) = \{(A_1, B_1), (A_1, B_4), (A_2, B_2), (A_2, B_3), (B_4, B_5)\}$
	$Q_1(D, A_1) = \{B_1, B_4\}$
$pcQ_2 = \Pi_1(\downarrow \diamond \downarrow \diamond C)$	$Q_2(D) = \{(A_1, A_1), (A_2, A_2), (B_4, B_4)\}$
	$Q_2(D, A_2) = \{A_2\}$
	$Q_2(D, B_4) = \{B_4\}$
$Q_3 = \epsilon \diamond \Pi_1(\downarrow \diamond \downarrow \diamond C) \diamond \downarrow \diamond B \diamond \downarrow$	$Q_3(D) = \{(A_1, C_1), (A_1, B_5), (A_2, C_2), (A_2, D_1), (A_2, C_3),$
	$\qquad\qquad (B_4, C_4)\}$
	$Q_3(D, A_1) = \{C_1, B_5\}$
	$Q_3(D, B_4) = \{C_4\}$
	$Q_3(D, B_2) = \emptyset$
$Q_4 = \uparrow \diamond \Pi_1(\downarrow^* \diamond D) \diamond \Pi_1(\downarrow \diamond \Pi_1(\downarrow \diamond B))$	$Q_4(D) = \{(B_1, A_1), (A_2, A_1), (B_4, A_1)\}$
	$Q_4(D, B_1) = \{A_1\}$
$Q_5 = \epsilon \diamond \Pi_1((\downarrow \diamond B) \cup (\uparrow \diamond A)) \diamond \downarrow \diamond \downarrow$	$Q_5(D) = \{(A_2, C_2), (A_2, D_1), (A_2, C_3)\}$
	$Q_5(D, A_2) = \{C_2, D_1, C_3\}$

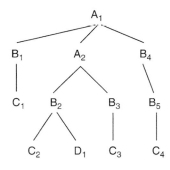

Path Query, Fig. 1 A sample semi-structured document presented in a tree structure (subscripts are used to distinguish data entries with the same label)

Path Query and Pattern Tree Matching

Path queries are frequently represented as tree structure, called *pattern trees*, in which nodes (optionally labeled) represents the query requirement on the data entries along the path, and edges (labeled with parent-child or ancestor-descendant relationship) represent the structural requirements among the data entries.

The evaluation of a path query is also called the process of *pattern matching*, which is to find the *witness trees* that consist of data entries that satisfy both the node and structural constraint expressed by the pattern tree.

Path Query Languages

Path query is the core of various query languages for semi-structured data.

XPath [6] is a W3C standard for addressing part of an XML document and for matching and testing whether a node satisfies a pattern. The primary syntactic construct in XPath is the expression, which is evaluated to yield a node set, a boolean value, a number, or a string. The core of the XPath query language is the path query, which is called *location paths* in XPath. Location path consists of relative location paths and absolute location path. A relative location path consists of a sequence of one or more location steps separated by "/". A step is composed from left to right and each step selects a set of nodes relative to a context node, which will in turn serve as the context node for the following step. An absolute location path consists of a leading "/", followed optionally by relative location paths.

XPath, in turn, is the core of other XML query languages and transformation languages, such as XSLT [5] and XQuery [4].

Path Query Evaluation

Even though the way in which the document is stored has great impact on how a path query can be evaluated, some common challenges exist in the evaluation of path queries, comparing to their relational peers, and numerous new operators, optimization techniques and index structures have been proposed to facilitate efficient path query evaluation.

One major characteristic of path query is that the query requirements are expressed not only on data entries, such as the tag (label), attributes, and text values of the entries, but also on the structural relationship among the data entries, which are highlighted by the primitives $\uparrow$ (parent axis), $\downarrow$ (child axis), $\uparrow^*$ (ancestor axis), and $\downarrow^*$ (descendant axis).

Navigation is a natural approach for evaluating a path query. The nagivational approach scans the whole document to verify the query requirement on data entries and the structural relationship among data entries. This approach works more naturally while the document is stored as file, in object-oriented database, or in a native data store. In addition, the nagivational approach is potentially very expensive in cost, especially when the ancestor/descendent axis is involved.

The invention of structural join operator [1, 13] and multiple algorithms for efficient computing of structural join advanced the evaluation technique of path queries dramatically. A structural join operation takes two sets of data entries as input and returns pairs of data entries that satisfy the desired structural relationship. The basis of the structural join operation is the pre-order and post-order numbering encoding of data entries in the document. The parent-child relationship is a special case of the ancestor-descendant relationship, in which the nodes that satisfies the desired structural relationship have to be exactly one level apart from each other.

$$(a, b) \in \downarrow^*(D) \iff a_{pre} < b_{pre} \wedge a_{post} > b_{post} \ (a, b) \in \downarrow (D) \iff a_{pre} < b_{pre} \wedge a_{post} b_{post} \wedge a_{level} + 1 = b_{level}$$

The structure join operation and various algorithms that support efficient evaluation of the operation enables the evaluation of path queries by decomposition. This approach partitions a path query into twigs that consists of either a node or a pair of nodes with a desired parent-child or ancestor-descendant relationship. The matching of the nodes can be easily evaluated via indices that are similar to value indices in RDB. The structural relationships between node pairs are evaluated by the structural join. Holistic approach has been proposed and widely adopted in answering path queries. This approach uses a chain of linked stacks to compute and represent intermediate results of a path in a compactly fashion.

Path Query Optimization

As any database management system, optimization is a critical step in evaluating path queries.

Syntax based optimization rewrites a path query into a path query in normal format, in pursuit of minimum expression length, minimum number of predicates, and no redundant value and structural requirements. The rewrite may also aim at decomposing a path query into subqueries that belong to some fragments of path queries that are simpler to answer.

The cost-based optimization relies on the data statistics of the document to be queried, and the cost-model of the physical operators to be employed, to enumerate various physical evaluation plans and choose one with minimum or acceptable estimate performance. This process involves the study of access method selection, query decomposition, cost estimation, etc.

Indices for Path Query Evaluation

Indexing, being one of the most important ingredients in efficient query evaluation, has seen its importance in the context of XML. Over 20 different types of indices have been proposed and have led to significant improvements in the performance of XML query evaluation.

Indices similar to the ones used in RDBs, namely value indices on element tags, attribute names and text values, are first used, together with the structural join algorithms [1, 3, 10, 13], in XML query evaluation. This approach turns out to be simple and efficient, but is not capable of capturing the structural containment relationships native to the XML data.

To directly capture the structural information of XML data, a family of structural indices has been introduced. DataGuide [7] was the first to be proposed, followed by the 1-index [11], which is based on the notion of bi-simulation among nodes in an XML document. These indices can be used to evaluate some path expressions accurately without accessing the original data graph. Milo and Suciu [11] also introduced the 2-index and T-index, based on similarity of pairs (vectors) of nodes. Unfortunately, these and other early structural indices tend to be too large for practical use because they typically maintain too fine-grained structural information about the document. To remedy this, Kaushik et al. introduced the $A(k)$ index which uses a notion of bi-similarity on nodes relativized to paths of length k [8]. This captures localized structural information of a document, and can support path expressions of length up to k. Focusing just on local similarity, the $A(k)$-index can be substantially smaller than the 1-index and others. Several works have investigated maintenance and tuning of the $A(k)$ indices. The $D(k)$-index and $M(k)$-index extend the $A(k)$-index to adapt to query workload. The integrated use of structural and value indices has been explored, and there have also been investigations on covering indices and index selection.

Other directions of XML indexing techniques proposed by researchers include indexing frequent sub-patterns, indexing XML tree and queries as sequences, forward and backward index, HOPI index, XR-tree, and encoding-based indices.

Key Applications

Path query is the core concept behind the query languages for semi-structured data. It is also the foundation of path algebra that are used to represent and reason about queries expressed against semi-structured data, especially those focusing on retrieving fragments of the document that satisfy certain value and structural constraints.

Future Directions

Even though the basic idea of the path query has been around for decades, its popularity exploded since XML prevails.

Path query has been adopted as the core of expressing queries again semi-structural data, especially XML. Even though XPath query language has been very stable, new language and language features keep emerging. As to the path query itself, there are on-going study of the sub-languages and the characteristics of such languages, their relationship to each other, the decidability of the queries in these languages, and the complexity of answer the queries.

On the practical side, systems have been developed to answer path queries. Various query evaluation, query optimization and indexing techniques have been proposed to facilitate efficient evaluation of path queries. However, the level of maturity of these techniques are not at the same level as those of relational queries.

In the relational world, the use cases are well understood and various benchmarks have been developed to measure the performance of relational DBMSs. Those of the queries on semi-structured documents are less understood, despite the existence of a few XML benchmarks, such as XMark [12]. In depth research on the usage of path queries and the design and development of benchmarks is yet another promising direction.

Data Sets

Example semi-structured data and queries can be found in benchmarks such as XMark (http://

www.xml-benchmark.org), XMach (http://sdbs.
uni-leipzig.de/en/projekte/XML/XmlBenchmark
ing.html), and MBench (http://www.eecs.umich.
edu/db/mbench).

Cross-References

▶ Semi-Structured Data Model
▶ XPath/XQuery
▶ XSL/XSLT

Recommended Reading

1. Al-Khalifa S, Jagadish HV, Patel JM, Koudas N,
   Srivastava D, Wu Y. Structural joins: a primitive for
   efficient XML query pattern matching. In: Proceedings of the 18th International Conference on Data
   Engineering; 2002.
2. Benedikt M, Fan W, Kuper GM. Structural properties of XPath fragments. Theor Comput Sci.
   2005;226(1):3–31.
3. Bruno N, Koudas N, Srivastava D. Holistic twig
   joins: optimal XML pattern matching. In: Proceedings of the ACM SIGMOD International Conference
   on Management of Data; 2002. p. 310–21.
4. Chamberlin D, Clark J, Florescu D, Robie J, Simeon
   J, Stefanescu M. XQuery 1.0: an XML Query Language, May 2003.
5. Clark J. XSL transformations (XSLT) version 1.0.
   http://www.w3.org/TR/XSLT.
6. Clark J, DeRose D. XML path language (XPath)
   version 1.0. http://www.w3.org/TR/XPATH.
7. Goldman R, Widom J. Data guides: enabling
   query formulation and optimization in semistructured databases. In: Proceedings of the 23th International Conference on Very Large Data Bases; 1997.
   p. 436–45.
8. Kaushik R, Shenoy P, Bohannon P, Gudes E. Exploiting local similarity for efficient indexing of paths
   in graph structured data. In: Proceedings of the 18th
   International Conference on Data Engineering; 2002.
9. Koch C. Processing queries on tree-structured data efficiently. In: Proceedings of the 25th ACM SIGACT-
   SIGMOD-SIGART Symposium on Principles of
   Database Systems; 2006. p. 213–24.
10. McHugh J, Widom J. Query optimization for XML.
    In: Proceedings of the 25th International Conference
    on Very Large Data Bases; 1999. p. 315–26.
11. Milo T, Suciu D. Index structures for path expres-
    sions. In: Proceedings of the 7th International Con-
    ference on Database Theory; 1999. p. 277–295.
12. Schmidt A, Waas F, Kersten ML, Carey MJ,
    Manolescu I, Busse R. XMark: a benchmark for
    XML data management. In: Proceedings of the 28th
    International Conference on Very Large Data Bases;
    2002. p. 974–85.
13. Zhang C, Naughton JF, DeWitt DJ, Luo Q, Lohman
    GM. On supporting containment queries in relational
    database management systems. In: Proceedings of the
    ACM SIGMOD International Conference on Man-
    agement of Data; 2001.

Pattern-Growth Methods

Hong Cheng[1] and Jiawei Han[2]
[1]Department of Systems Engineering and
Engineering Management, The Chinese
University of Hong Kong, Hong Kong, China
[2]University of Illinois at Urbana-Champaign,
Urbana, IL, USA

Definition

Pattern-growth is one of several influential frequent pattern mining methodologies, where a
pattern (e.g., an itemset, a subsequence, a subtree,
or a substructure) is *frequent* if its occurrence
frequency in a database is no less than a specified *minimum_support* threshold. The (frequent)
pattern-growth method mines the data set in a
divide-and-conquer way: It first derives the set of
size-1 frequent patterns, and for each pattern p,
it derives p's projected (or conditional) database
by data set partitioning and mines the projected
database recursively. Since the data set is decomposed progressively into a set of much smaller,
pattern-related projected data sets, the pattern-
growth method effectively reduces the search
space and leads to high efficiency and scalability.

Historical Background

Frequent itemset mining was first introduced as
an essential subtask of association rule mining by
Agrawal et al. [1]. A candidate set generation-
and-test approach, represented by the Apriori
algorithm, was proposed by Agrawal and Srikant
[2]. The approach effectively reduces the search

space by exploring the *downward closure* property of frequent patterns, i.e., *any subpattern of a frequent pattern is frequent*. Various kinds of refinements of and extensions to this approach were proposed afterwards. However, since the Apriori-like candidate set generation-and-test approach repeatedly scans the whole database and checks the candidates by pattern matching, it is still rather costly.

The pattern-growth approach, represented by the *FP-growth* algorithm, was first proposed by Han, Pei, and Yin [8] for mining frequent itemsets. Since then, the method has been developed in several directions: (i) further enhancement of mining efficiency using refined data structures, such as FP-growth* [6], which uses an array-based implementation of prefix-tree-structure of FP-growth; (ii) mining closed and max patterns [6, 13], where a pattern p is *closed* if there exists no super-pattern with the same support, and p is a *max pattern* if there exists no super-pattern that is frequent; (iii) mining sequential patterns [11] and frequent substructures [14]; and (iv) mining high-dimensional data set [7] and colossal patterns [15], and pattern-based classification [4] and clustering [12]. A comprehensive overview of such extensions is presented in [7].

Foundations

Frequent Itemset Mining

To illustrate the pattern-growth method, the FP-growth method is briefly introduced here that exploits pattern-growth in frequent itemset mining. FP-growth works in a divide-and-conquer way. The first scan of the database derives a list of frequent items in which items are ordered by frequency-descending order (Notice that this particular ordering is not essential, and different ordering schemes can be explored). According to this ordering, the database is compressed into a frequent-pattern tree, or *FP-tree*, which retains the itemset association information.

The FP-tree is mined by starting from each frequent length-1 pattern (as an initial suffix pattern), constructing its *conditional pattern base* (a "subdatabase", which consists of the set of prefix paths in the FP-tree co-occurring with the suffix pattern), then constructing its conditional FP-tree, and performing mining recursively on such a tree. The pattern growth is achieved by the concatenation of the suffix pattern with the frequent patterns generated from a conditional FP-tree. Figure 1 shows an example of a global FP-tree as well as a set of conditional trees and the recursive mining process on top of them. Therefore, the FP-growth algorithm transforms the problem of finding long frequent patterns to searching for shorter ones recursively and then concatenating the suffix.

Sequential Pattern Mining

The pattern-growth philosophy has been extended to sequential pattern mining, where a sequential pattern is a set of (gapped) subsequences that occur frequently in a set of sequences. PrefixSpan, developed by Pei et

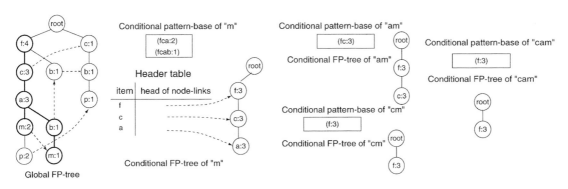

Pattern-Growth Methods, Fig. 1 FP-growth mining on conditional FP-trees

al. [11], is a typical sequential pattern mining algorithm based on the pattern-growth approach. It works in a divide-and-conquer way, by first scanning the sequence database to derive the set of length-1 sequential patterns. Then each sequential pattern is treated as a prefix and the complete set of sequential patterns can be partitioned into different subsets according to different prefixes. To mine the subsets of sequential patterns, corresponding projected databases are constructed and mined recursively.

Frequent Subgraph Mining

Inspired by the pattern-growth-based frequent itemset and sequential pattern mining algorithms, there are quite a few pattern-growth-based graph pattern mining algorithms developed. Here gSpan [14] is used an example to explain the ideas.

The pattern-growth graph mining algorithm extends a frequent graph by adding a new edge, in every possible position. A potential problem with the edge extension is that the same graph can be discovered many times. gSpan solves this problem by introducing a new lexicographic order among graphs, and maps each graph to a unique minimum DFS code as its canonical label. Based on this lexicographic order, gSpan adopts the pattern-growth philosophy with a *right-most extension* technique, where the only extensions take place on the *right-most path*. A right-most path is the straight path from the starting vertex v_0 to the last vertex v_n, according to a depth-first

search on the graph. In Fig. 2, the graph shown in 2a has several potential children with one edge growth, which are shown in 2b–f (assume the darkened vertices constitute the rightmost path). Among them, 2b–d grow from the rightmost vertex while 2e and 2f grow from other vertices on the rightmost path. 2(b.0)-(b.3) are children of 2b, and 2(e.0)-(e.2) are children of 2e. Backward edges can only grow from the rightmost vertex while forward edges can grow from vertices on the rightmost path. The enumeration order of these children is enhanced by the DFS lexicographic order, i.e., it should be in the order of 2b–f.

With the DFS lexicographic order definition, the frequent subgraph mining is performed in a *DFS Code Tree*. In a DFS Code Tree, each node represents a DFS code. The relation between parent and child node complies with the parent-child relation; the relation among siblings is consistent with the DFS lexicographic order. Figure 3 shows a DFS Code Tree, the nth level nodes contain DFS codes of $(n - 1)$-edge graphs. Through depth-first search of the code tree, all the minimum DFS codes of frequent subgraphs can be discovered. That is, all the frequent subgraphs can be discovered in this way. One should note that if in Fig. 3, the dark nodes contain the same graph but different DFS codes, then one of them must not be the minimum code. Therefore, the search space of that sub-branch can be pruned since it does not correspond to a minimum DFS code.

Pattern-Growth Methods, Fig. 2 Graph growth with DFS lexicographic order

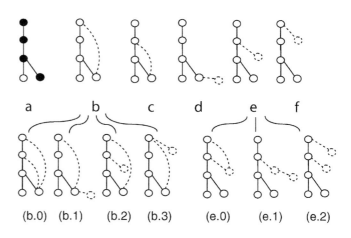

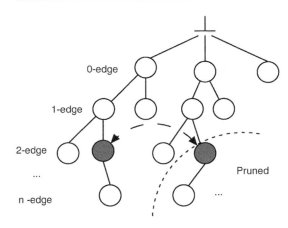

Pattern-Growth Methods, Fig. 3 A search space: DFS code tree

Key Applications

Pattern-growth methods have been used in many tasks that need the mining of frequent patterns, including the discovery of association and correlation relationships among large sets of transactions, event sequences, or complex structures, the discovery of discriminative frequent features for classification and clustering, as well as many other data mining and pattern recognition applications, such as biological data mining.

Future Directions

Mining Approximate or Noise-Tolerant Patterns

Pattern-growth method has been widely applied for efficient mining of precise and complete set of frequent patterns. However, in some real applications, the capacity to accommodate approximation in the mining process has become critical due to inherent noise and imprecision in complex data sets, for example, gene mutations in genomic DNA sequences, and protein-protein interaction networks. Approximate or noise-tolerant frequent patterns could be the natural choice to handle noise or variations in many applications. Some recent studies proposed new algorithms for mining approximate itemsets [9] and subgraphs [3]. Among them, [9] adopts a candidate generation-and-test approach with the noisy mining model

while [3] takes a pattern-growth approach. It is interesting to explore whether pattern-growth method could be naturally applied into a noisy mining model to efficiently discover the approximate or noise-tolerant patterns.

Pattern-Based Classification and Clustering

Frequent patterns have been demonstrated useful in other data mining tasks such as classification [4] and clustering [12], where frequent patterns are used as discriminative classification features and clustering subspaces, respectively. Instead of using the complete set of patterns, usually only a small number of frequent patterns are used in classification and clustering tasks, e.g., the subset of highly discriminative patterns in classification and the subset of frequent patterns with dense regions in clustering. It would be desirable to directly mine the subset of patterns of interests without generating the complete set, for efficiency consideration. To achieve this goal, pruning strategies need to be designed and integrated into the pattern-growth mining methodology to effectively prune the search space which does not yield high-quality patterns. This is a nontrivial task since many quality measures, such as information gain, density, or entropy, are not *antimonotonic* which is the essential pruning strategy in frequent pattern mining.

Experimental Results

In general, for every proposed method, there is an accompanying experimental evaluation in the corresponding reference. In addition, for frequent itemset mining methods, [5] (The FIMI workshop) provided a detailed and comprehensive experimental evaluation on a large set of benchmark data.

Data Sets

Synthetic Data

A synthetic tree generator can be found at http://www.cs.rpi.edu/~zaki/software

Real Data

A large collection of real transaction datasets can be found at http://fimi.cs.helsinki.fi/data/. Commonly used real graph data includes AIDS antiviral screening datasets at http://dtp.nci.nih.gov, and NCI anti-cancer screening datasets at http://pubchem.ncbi.nlm.nih.gov.

URL to Code

The binary codes for FP-growth, PrefixSpan and gSpan are provided by the IlliMine project at http://illimine.cs.uiuc.edu/.

The source codes of FP-growth*, FPClose, and FPMax* [6] are provided by Grahne and Zhu at http://fimi.cs.helsinki.fi/src/fimi06.tgz

Cross-References

▶ Apriori property and Breadth-First Search Algorithms
▶ Frequent Graph Patterns
▶ Frequent Itemsets and Association Rules
▶ Sequential Patterns

Recommended Reading

1. Agrawal R, Imielinski T, Swami A. Mining association rules between sets of items in large databases. In: Proceedings of the ACM SIGMOD International Conference on Management of Data; 1993. p. 207–16.
2. Agrawal R, Srikant R Fast algorithms for mining association rules. In: Proceedings of the 20th International Conference on Very Large Data Bases; 1994. p. 487–99.
3. Chen C, Yan X, Zhu F, Han J. gApprox: mining frequent approximate patterns from a massive network. In: Proceedings of the 7th IEEE International Conference on Data Mining; 2007. p. 445–50.
4. Cheng H, Yan X, Han J, Yu PS. Direct discriminative pattern mining for effective classification. In: Proceedings of the 24th International Conference on Data Engineering; 2008.
5. Goethals B, Zaki M. An introduction to workshop on frequent itemset mining implementations. In: Proceedings of the ICDM International Workshop on Frequent Itemset Mining Implementations; 2003. p. 1–13.
6. Grahne G, Zhu J. Efficiently using prefix-trees in mining frequent itemsets. In: Proceedings of the ICDM International Workshop on Frequent Itemset Mining Implementations; 2003.
7. Han J, Cheng H, Xin D, Yan X. Frequent pattern mining: current status and future directions. Data Mining Knowl Discov. 2007;15(1):55–86.
8. Han J, Pei J, Yin Y. Mining frequent patterns without candidate generation. In: Proceedings of the ACM SIGMOD International Conference on Management of Data; 2000. p. 1–12.
9. Liu J, Paulsen S, Sun X, Wang W, Nobel A, Prins J. Mining approximate frequent itemsets in the presence of noise: Algorithm and analysis. In: Proceedings of the SIAM International Conference on Data Mining; 2006. p. 405–16.
10. Pan F, Cong G, Tung AKH, Yang J, Zaki M. CARPENTER: Finding closed patterns in long biological datasets. In: Proceedings of the 9th ACM SIGKDD International Conference on Knowledge Discovery and Data Mining; 2003. p. 637–42.
11. Pei J, Han J, Mortazavi-Asl B, Wang J, Pinto H, Chen Q, Dayal U, Hsu M-C. Mining sequential patterns by pattern-growth: the prefixspan approach. IEEE Trans Knowl Data Eng. 2004;16(11):1424–40.
12. Pei J, Zhang X, Cho M, Wang H, Yu PS. Maple: a fast algorithm for maximal pattern-based clustering. In: Proceedings of the 1st IEEE International Conference on Data Mining; 2001. p. 259–66.
13. Wang J, Han J, Pei J. CLOSET+: Searching for the best strategies for mining frequent closed itemsets. In: Proceedings of the 9th ACM SIGKDD International Conference on Knowledge Discovery and Data Mining; 2003. p. 236–45.
14. Yan X, Han J. gSpan: Graph-based substructure pattern mining. In: Proceedings of the 2nd IEEE International Conference on Data Mining; 2002. p. 721–24.
15. Zhu F, Yan X, Han J, Yu PS, Cheng H. Mining colossal frequent patterns by core pattern fusion. In: Proceedings of the 23rd International Conference on Data Engineering; 2007. p. 706–15.

Peer Data Management System

Philippe Cudré-Mauroux
Massachusetts Institute of Technology, Cambridge, MA, USA

Synonyms

Decentralized data integration system; PDMS

Definition

A Peer Data Management System (PDMS) is a triple $S = \langle P, S, M \rangle$ where P is a set of autonomous peers, S a set of heterogeneous schemas used by the peers to represent their data, and M a set of schema mappings, each enabling the reformulation of queries between a given pair of schemas.

Key Points

A Peer Data Management System (PDMS) is a distributed data integration system providing transparent access to heterogeneous databases without resorting to a centralized logical schema. Instead of imposing a uniform query interface over a mediated schema, PDMSs let the peers define their own schemas and allow for the reformulation of queries through mappings relating pairs of schemas (see Fig. 1). PDMSs typically exploit the schema mappings transitively in order to retrieve results from the entire network.

Compared to centralized data integration systems, PDMSs suggest a scalable, decentralized and easily extensible integration architecture where any peer can contribute data, schemas, and mappings. Peers with new schemas simply need to provide a mapping between their schema and any other schema already used in the system to be part of the network.

The languages used to define the mappings in PDMSs may vary, but are typically derived from GLAV formulae with extensions to support both inclusion and equality mappings. The Piazza system [3] proposes new algorithms to retrieve certain answers in this context. Hyperion [1] is a system focusing on relating data not only at the schema level, but also at the instance level through mapping tables. GridVine [2] provides distributed probabilistic analyses in order to automatically detect mapping inconsistencies in PDMS settings.

PDMSs generally use Peer-to-Peer overlay networks to support their distributed operations. Some use unstructured overlay networks [1, 3] to organize the peers into a random graph and use flooding or random walks to contact distant peers. Other PDMSs maintain a decentralized yet structured Peer-to-Peer network [2] to allow any peer to contact any other peer by taking advantage of a distributed index.

The lack of global coordination has raised several questions as to the global properties of such systems in the large. In particular, new complex systems perspectives - such as the emergent semantics approach - have been proposed to characterize the global semantics of PDMSs.

a   **b**

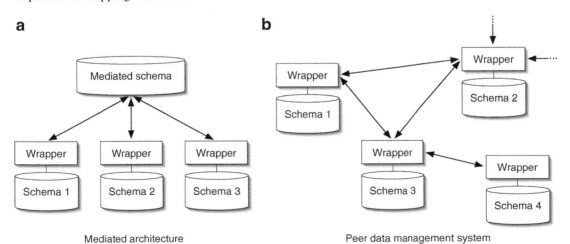

Mediated architecture   Peer data management system

Peer Data Management System, Fig. 1 Contrary to the mediated architecture (**a**), Peer Data Management Systems (**b**) do not impose any form of centralization but consider instead networks of heterogeneous data sources related to each other through pairwise schema mappings

Cross-References

► Emergent Semantics
► Peer-to-Peer Data Integration
► Peer-to-Peer Overlay Networks: Structure, Routing and Maintenance

Recommended Reading

1. Arenas M, Kantere V, Kementsietsidis A, Kiringa I, Miller RJ, Mylopoulos J. The hyperion project: from data integration to data coordination. ACM SIGMOD Rec. 2003;32(3):53–8.
2. Cudré-Mauroux P. Emergent semantics. EPFL Press; 2008.
3. Halevy A, Ives Z, Madhavan J, Mork P, Suciu D, Tatarinov I. The piazza peer data management system. IEEE Trans Knowl Data Eng. 2004;16(7): 787–98.

Peer-to-Peer Content Distribution

Ernst Biersack[1] and Pascal Felber[2]
[1]Eurecom, Sophia Antipolis, France
[2]University of Neuchatel, Neuchatel, Switzerland

Synonyms

Cooperative content distribution; Peer-to-peer file sharing

Definition

Peer-to-peer content distribution is an approach for cost-effective distribution of bandwidth-intensive content to large numbers of clients based on *swarming*. Content is split into blocks that are sent to different clients. Thereafter, the clients can directly exchange blocks with one another, in a peer-to-peer manner, without the help of the original server.

Historical Background

Peer-to-peer systems are application-layer networks that directly interconnect users. They have enjoyed a phenomenal success in the last 10 years together with the democratization of broadband access. The first disruptive peer-to-peer applications were designed for file sharing (e.g., Napster, Gnutella, KaZaA): they provided users with the means to (i) *search* for content, and (ii) *distribute* content. In the research community, most of the initial focus was on structured peer-to-peer overlays, also known as distributed hash tables (e.g., CAN, Chord). These systems have proved very efficient for *looking up* specific content.

Once users have started exchanging larger files, notably full-length movies, the classical client-server techniques used to transfer files were no longer sufficient and the focus has shifted on developing more scalable mechanisms for content distribution. The first and most emblematic example of peer-to-peer content distribution application is BitTorrent [5]. Its development started in 2001 and it reached widespread popularity in 2003.

BitTorrent is a peer-to-peer application that capitalizes the *bandwidth* of peer nodes to quickly replicate a single large file to a set of clients. The challenge is thus to maximize the speed of replication. The clients involved in a torrent cooperate to replicate the file among each other using *swarming* techniques: the file is broken into fixed size blocks and the clients in a peer set exchange blocks in parallel with one another. BitTorrent only deals with distribution and does not provide any mechanisms for locating content. Several alternatives to BitTorrent have been proposed, but none of them has reached the same level of popularity nor demonstrated sufficient benefits to supersede it.

More recently, some peer-to-peer content distribution systems have been developed specifically for streaming media, such as live television (e.g., PPLive). For such applications, blocks have to be received in order and in time, and clients are only interested in the part of the stream that is being broadcast while they are online.

Foundations

The principle underlying peer-to-peer content distribution is to capitalize the upstream bandwidth of the clients. A user that downloads some content will, at the same time, send part of the content to other peers.

Peer-to-peer content distribution networks are inherently *self-scalable*, in that the bandwidth capacity of the system increases as more peers arrive: each new peer requests service from, but also provides service to, the other peers. The peer-to-peer system can thus spontaneously adapt to the demand by taking advantage of the resources provided by every peer.

The behavior of each peer in a peer-to-peer content distribution is determined by two factors: (i) the *peer selection strategy* that determines the set of peers a given peer will exchange blocks with and (ii) the *block selection strategy* that determines which blocks will be exchanged. The choice of the peer and the block selection strategy has an important impact on the architecture of the content distribution system. One can distinguish between *structured* and *unstructured* architectures.

Structured architectures organize the peers in a directed acyclic graph such as linear chain or a tree (see Fig. 1). The peer selection is done once for the whole duration of the content distribution. The block selection becomes trivial as each peer simply forwards all the blocks it receives to its children. While structured architectures are easy to understand and model, they are best suitable for content distribution over networks where the bandwidth is homogeneous and nodes stay connected during the whole duration of the content distribution. If these conditions are not met, which is typically the case for the Internet, complex mechanisms are required to make structured approaches work.

Self-organization is the ability of a peer-to-peer network to dynamically determine how to best manage the block exchange as peers join, leave, or fail. Unstructured architectures are self-organizing: they use an underlying mesh structure and build directed graphs through which data is forwarded along several possible paths from the source to each peer. Meshes adapt well to bandwidth fluctuations/heterogeneity and nodes leaving and joining during the transfer at the price of more complex peer and block selection strategies. Since there exist multiple paths for receiving data, each peer must coordinate with its neighbors to avoid receiving the same block multiple times. One such system that uses an unstructured mesh-based approach is BitTorrent, a very popular peer-to-peer system for file distribution.

File Distribution

The BitTorrent protocol makes sensible choices for peer and block selection that are based on a few simple principles. *Peer selection:* first, every peer maintains a limited number of active connections to the other peers that offer the best upload and download rates, thus optimizing bandwidth utilization; second, a peer preferentially sends data to another peer that reciprocally sent data to it, which enforces fairness; third, every peer periodically sends some data to newcomers, so that peers can have an active role in the torrent independent of their arrival time. *Block selection:* a peer requests the block from its neighbors for which the least number of copies exist (rarest first), as rare blocks have a high trading value and can potentially increase the lifetime of the torrent. Maintaining a good diversity of the blocks available in the systems assures that each peer can fully use its upload capacity since it has blocks that its neighbors are interested in.

Despite the simplicity of these design principles and the lack of theoretical foundations to back them when they were introduced, studies [8] have shown they perform exceptionally well in practice. The uplink utilization at the peers is remarkably high, sharing incentives are very effective and resilient to freeriders, and one can observe that peers spontaneously cluster according to their capacity [10].

The choice of the peer and block selection strategy is very important for the good performance of mesh-based systems, because peers can fully use their upload capacity only if the *diversity* of blocks is high, i.e., if they own blocks their neighbors are interested in. Another way to assure that a peer has useful information to transmit to its neighbors is to use *network coding*. In this case, every block that is transmitted is a linear combination of all or a subset of the blocks available at the peer. Since with high probability every such block is unique and contains useful information for the receiving peer, the block selection strategy becomes trivial. Avalanche [6] is a peer-to-peer system for file distribution that uses network coding.

Video Streaming

File distribution systems such as BitTorrent are often used to download files that contain audio or video content. Since the blocks can be downloaded in any order, the consumption of the content cannot start before the download is complete.

In recent years, some peer-to-peer content distribution systems have been developed to support live streaming, where the playback of the content occurs simultaneously with its production. Supporting video streaming is more challenging than file distribution: the users expect that once viewing has started it will be continuous, which means that data must be received in sequence and the rate of data reception must be equal the rate at which data are injected by the source. The architectural variants available for live streaming are the same as for file distribution, namely structured and unstructured approaches.

SplitStream [4] uses a structured architecture and constructs parallel trees, with every peer belonging to all trees. Content is split in multiple layers, each sent along a different tree. As was discussed for the case of file distribution, structured architectures lack flexibility. For this reason, the systems that have been used most widely for video streaming are mesh-based. One can notably mention PPlive [7] that has been used to stream video to tens of thousands of peers.

A robust video streaming system must be able to cope with the heterogeneity of the peers, which can have widely varying download bandwidth. To assure real-time data reception for all clients despite heterogeneity, one can use layered coding: the video is encoded into a base layer and several enhancement layers. While the video cannot be viewed if the base layer is not completely received, the enhancement layer only improve the viewing experience of the video and can be omitted if the download bandwidth is not sufficient.

Many peers are connected via ADSL links that provide much higher download rates than upload rates. A peer-to-peer streaming system is only stable if the aggregate upload rate of the participating peers is at least as high as the aggregate playout rate [9]. In many cases this can only be achieved if there are some peers ("super-peers") that have an upload capacity much higher than the video playout rate [7].

Besides participating in a live streaming event, users may want to watch a video at any point of time, which is referred to as video on demand. While there exist designs for peer-to-peer systems that support video on demand [1], none of these systems has been deployed on a large scale.

Performance Analysis of Distribution Architectures

To evaluate the potential of peer-to-peer file distribution, one can consider very simple distribution models [3] where a server S distributes a file to N peers. The server splits a file into C blocks and serves the file sequentially and infinitely at rate b. The time needed to download the complete file from the server to a *single* peer at rate b is referred to as *one round*.

Consider first a linear chain architecture where the peers are organized in a chain with the server uploading the blocks to the first peer, which in turn uploads the blocks to the second peer and so on (Fig. 1, left). Peers disconnect once they have uploaded the whole file once. The number of peers served in a given number of rounds grows linearly with the number of blocks C, because one can faster engage all peers in the distribution process, and quadratically with the number of rounds t, because the source forks additional chains. Interestingly, when $N/C \ll 1$, the time necessary to serve N peers converges asymptotically to 1.

Consider now a tree architecture where the peers are organized in a tree with an outdegree k (Fig. 1, center). The server serves k peers in parallel, each at rate b/k, and all the peers that are

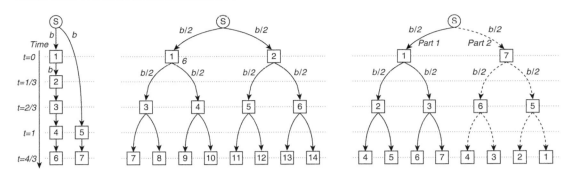

Peer-to-Peer Content Distribution, Fig. 1 Evolution of three simple distribution models with time: linear chains (*left*), trees (*center*), and parallel trees (*right*). The nodes in each graph are labeled with the peer number and the edges with the bandwidth allocated for the block transmission. In this example, $C = 3$ and $k = 2$

Peer-to-Peer Content Distribution, Table 1 Performance comparison of three structured distribution models

Architecture	Clients served	Service time	Copies served	Download rate	Upload rate
Linear chain	$C \cdot t^2$	$\frac{1}{2} + \sqrt{\frac{1}{4} + \frac{2 \cdot N}{C}}$	1	b	b (except leaves)
Tree	$k^{(t-k)\frac{C}{k}+1}$	$k + (l\log_k Nm - 1) \cdot \frac{k}{C}$	k	$1/b$	b (except leaves)
Parallel trees	$k^{(t-1)\frac{C}{k}}$	$1 + l\log_k Nm \frac{k}{C}$	1	b	b

not leaves in the tree in turn upload the blocks to k other peers at rate b/k. This means that it takes k rounds for a peer to download the file and interior nodes upload an amount of data equivalent to k times the size of the file, while leaf nodes do not upload the file at all. The performance of the tree architecture depends on the node degree, and it turns out that for $N/C \leq 1$ the best value is $k = 1$, i.e., a linear chain. For $N/C > 1$ a binary tree is more efficient. Trees with higher degrees are penalized by the higher number of non-contributing leaf nodes.

Finally, consider a forest of k parallel trees, each of which contains all the peers (Fig. 1, right). The server partitions the file into k parts and constructs k spanning trees to distribute each part along a separate tree to all peers. The trees can be built such that each peer is an interior node in at most one tree and a leaf in the remaining $k - 1$ trees. The parallel tree architecture is as efficient as the linear chain when $N/C \ll 1$ and significantly outperforms the other two architectures in other scenarios. The optimal performance is obtained for trees with an outdegree $k = e \approx 3$.

Table 1 summarizes the scaling behavior of the different approaches. One can see that for both the tree and parallel trees architectures, the number of clients served increases exponentially in time and in the number of blocks C.

These performance results are remarkable: with the same amount of effort at the server, distributing the content to N peers using a peer-to-peer architecture can be done in a little more than one round, i.e., it takes just slightly longer than to distribute the file from the server to a single peer. While structured architectures are neither robust nor practical, these performance models still give valuable insights on the asymptotically performance of peer-to-peer content distribution. See [2, 11] for studies that take into account other factors such as bandwidth heterogeneity.

Developing performance models for unstructured architectures is much more difficult than for structured ones. Yet, it turns out that the following unstructured architecture, called *Interleave*, is analytically tractable [12]. The performance model for Interleave assumes that all blocks are numbered in increasing order, all peers have the same bandwidth, and the transmission of a block takes one slot. Peer selection is random, while block selection is as follows: in every odd time slot the source and the other peers push the highest numbered block they own to a random peer; in every even time slot each peer requests from a random peer the lowest numbered block that the peer does not yet have. Note that Interleave does not require any peer to exchange information with other peers about the block it already has. As a consequence the protocol will experience some inefficiency as a peer p may ask another peer q for a block that q does not have, or may send to q a block that q already has. It can be shown analytically that with high probability the entire file that consists of C blocks can be distributed to all N peers in a time of $3.2 + \frac{\log N}{C}$. Using simulation the result obtained was $2 + \frac{\log N}{C}$, which is similar to the time required by a structured binary tree (see Table 1). Such good performance is quite astonishing as is indicates that there exist robust mesh-based file distribution schemes that can achieve performance as good as structured architectures.

Key Applications

Peer-to-peer content distribution is a technique to transfer large contents simultaneously to many users. Key applications include file sharing systems, live TV, video on demand, or distribution of critical updates.

It also represents a cost-effective alternative to scaling up server architectures. It has notably been used as a way to protect servers from unexpected surges in request traffic, called flash crowds, by replicating popular content on multiple peers.

Future Directions

Peer-to-peer content distribution networks are overlays that typically ignore the underlay, i.e., the underlying IP connectivity. Therefore, such networks generate much unnecessary traffic between Internet service providers.

Another important problem that plagues content distribution networks is the fact that the validity of content cannot be verified until after it is downloaded. The media industry contributes to content pollution in existing networks to deter exchange of copyrighted files.

One can expect video on demand to be an important application domain for peer-to-peer content distribution technology. A major challenge is that users must be able to start watching a movie at any time, as well as suspend it, rewind, skip chapters, etc. (VCR functionality). In these scenarios, there is no such synchronization between the peers as with live streaming.

Cross-References

▶ Distributed Hash Table
▶ Peer-to-Peer System

Recommended Reading

1. Annapureddy S, Guha S, Gkantsidis C, Gunawardena D, Rodriguez P. Is high quality VoD feasible using P2P swarming. In: Proceedings of the 16th International World Wide Web Conference; 2007.
2. Biersack EW, Carra D, Cigno RL, Rodriguez P, Felber P. Overlay architectures for file distribution: fundamental performance analysis for homogeneous and heterogeneous cases. Comput Netw. 2007;51(3):901–17.
3. Biersack E, Rodriguez P, Felber P. Performance analysis of peer-to-peer networks for file distribution. In: Proceedings of the 5th International Workshop on Quality of Future Internet services; 2004. p. 1–10.
4. Castro M, Druschel P, Kermarrec AM, Nandi A, Rowstron A, Singh A. SplitStream: high-bandwidth multicast in a cooperative environment. In: Proceedings of the 19th ACM Symposium on Operating System Principles; 2003.
5. Cohen B. Incentives to Build Robustness in BitTorrent. Tech. rep., http://www.bittorrent.org/. 2003.
6. Gkantsidis C, Miller J. Rodriguez P. Anatomy of a P2P content distribution system with network coding. In: Proceedings of the 5th International Workshop Peer-to-Peer Systems; 2006.
7. Hei X, Liang C, Liang J, Liu Y, Ross K. A measurement study of a large-scale P2P IPTV system. IEEE Trans Multimed. 2007;9(8):1672–87.
8. Izal M, Urvoy-Keller G, Biersack E, Felber P, Al Hamra A, Garces-Erice L. Dissecting BitTorrent: five months in a torrent's lifetime. In: Proceedings of the 5th Passive and Active Measurement Workshop; 2004.
9. Kumar R, Liu Y, Ross KW. Stochastic fluid theory for P2P streaming systems. In: Proceedings of the 26th Annual Joint Conferences of the IEEE Computer and Communications Societies; 2007.

10. Legout A, Liogkas N, Kohler E, Zhang L. Cluster-
    ing and sharing incentives in bittorrent systems. In:
    Proceedings of the 2007 ACM SIGMETRICS Inter-
    national Conference on Measurement and Modeling
    of Computer Systems; 2007.
11. Mundinger J, Weber R, Weiss G. Optimal schedul-
    ing of peer-to-peer file Dissemination. J Sched.
    2007;11(2):105.
12. Sanghavi S, Hajek B, Massoulie L. Gossiping
    with multiple messages. IEEE Trans Inf Theory.
    2007;53(12):4640–54.

Peer-to-Peer Data Integration

Anastasios Kementsietsidis
IBM T.J. Watson Research Center, Hawthorne,
NY, USA

Definition

Peer-to-Peer data integration lies in the intersection of two popular research topics, namely, Peer-
to-Peer systems and Data Integration, and is one
of the key topics in the area of *Peer-to-Peer Data
Management*. A Peer-to-Peer data integration setting involves a set P of autonomous, heterogeneous, independently evolving (peer) sources
whose pairwise schema or data-level mappings,
collectively denoted by M, induce a peer-to-peer
network. In this setting, each (peer) source in
the network can be queried and act as an *access
point* to the data residing in the other network
sources. Research in this area focuses on studying
the specification and expressiveness of the peer
mappings; the corresponding query languages
used; algorithms for rewriting queries between
peer source schemas; and, to some extent, topics
that concern the propagation of updates between
peer sources. Key characteristics of the peer-to-
peer data integration setting that differentiate it
from *traditional* data integration settings and typical Peer-to-Peer systems include (i) the fact that
each peer can be full-fledged database; (ii) the
lack of centralized control and global schemas;
(iii) the need for more diverse set of mapping
specifications; and (iv) the need for integration of
data across diverse domains.

Historical Background

Data integration [11] has been characterized as
one of the longest-standing research problems
faced by the data management community. A
typical data integration setting involves a set S
of sources, a global schema G, and a set M
of mappings between the sources in S and the
global schema G. Figure 1 illustrates such a
setting along with the logical steps during query
evaluation. Central to the evaluation of queries
are the mappings between the global and local
schemas (indicated as *Metadata* in the figure).
The mappings are used during the rewriting of a
user query over the global schema G to a set of
queries over the sources. Two basic approaches
have been used to specify the mappings. In a
nutshell, in the *Global-as-View* (GAV) approach,
the global schema G is expressed as a view of
the set of local schemas S. Main advantage of the
GAV approach is the simplicity of the rewriting
algorithm. However, a drawback of the approach
is that the addition of source in S results in a
revision of the mapping and, in the worst case, a
complete re-design of the global schema. In the

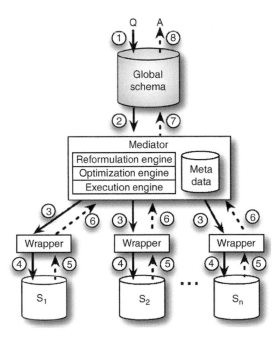

Peer-to-Peer Data Integration, Fig. 1 Typical data integration setting

second approach, called *Local-as-View* (LAV), each source $S_i \in S$ is expressed as a view over the global schema G. Query rewriting is substantially more involved here, but the approach handles source additions more gracefully.

Peer-to-peer computing involves an open-ended network of computational peers, where each peer exchanges data and services with a set of other peers, called its *acquaintances*. The peer-to-peer paradigm, initially popularized by file-sharing systems such as Napster [12] and Gnutella [5], offers an alternative to traditional architectures found in distributed systems and the web. Distributed systems are rich in services, but require considerable overhead to launch and have a relatively static, controlled architecture. In contrast, the web offers a dynamic, anyone-to-anyone architecture with minimum startup costs but limited services. Combining the advantages of both architectures, peer-to-peer offers an evolving architecture where peers come and go, choose with whom they interact, and enjoy some traditional distributed services with less startup cost. Figure 2 shows an example of such a system, often referred to in the literature as an *unstructured* peer-to-peer system, to differentiate it from the complementary class of *structured* peer-to-peer systems whose architecture is based on *Distributed Hash Tables* (DHT's). As shown in the figure, mappings in unstructured peer-

to-peer system (represented by solid lines) exist between any pair of peers while queries (represented by various line arrows) can be initiated at any peer in the system. Furthermore, different queries might involve a different set of peers.

Probably the first works to consider the interaction between database and peer-to-peer systems were the ones by Gribble et al. [6] and Bernstein et al. [3]. The former work focuses on the problem of *data placement*, that is, to find an optimal placement of a set of objects on a peer-to-peer system, under a pre-specified query workload, so as to minimize the cost of evaluating the queries. In more detail, the authors consider graph whose nodes correspond to the peers in a peer-to-peer network. Each node is associated with a storage capacity and a query workload. Furthermore, each pair of nodes connected by an edge has an associated data transfer cost, over this edge. Queries over this network are object lookups and each query has an associated frequency and cost (with the cost being zero, if the query can be served locally in a peer, and a function of the object size and edge transfer cost, if served remotely). In this setting, the main result of this work is that the problem of data placement with optimal cost is NP-complete. The work by Gribble et al. [6] initiated the Piazza System [17] (described in the next section) although

Peer-to-Peer Data Integration, Fig. 2 A Peer-to-Peer system

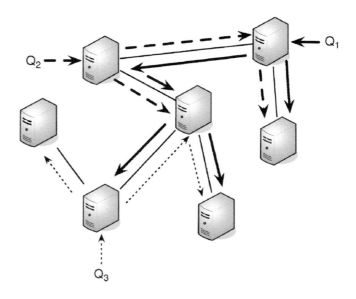

the follow-up work in Piazza addresses a setting much closer to data integration. As presented here, the setting in Gribble et al. [6] is more closely related to the one found on the work on DHT's like CAN [15] or Chord [16].

Bernstein et al. [3] is the first work to actually discuss data integration in a peer-to-peer context. The focus in this work is the introduction of the Local Relational Mode (LRM), a model designed specifically for the peer-to-peer setting. Key notions in the model are that of *coordination formulas* and *domain relations*. Much like a mapping in a typical data integration setting, a coordination formula establishes a relationship between the data items stored in acquainted peers. This relationship can be used to express a constraint, like for example, that a particular data item must be stored in either one of three different acquainted peer databases, or it can be used during query answering by expressing, for example, a GAV (LAV) type mapping where the data items returned by query Q_1 in a peer P_1 are contained in those returned by query Q_2 in peer P_2. Since different peers might use different vocabularies to express the same real-world notion, it is the responsibility of domain relations to express the mapping between these vocabularies. For example, a domain relation might be used to map prices expressed in US dollars, in a peer, to those in euro in another peer. The work by Bernstein et al. [3] initiated the Hyperion project [2] the details of which are described in the next section.

Foundations

Important aspects in every peer-to-peer data integration system include (i) the peer joining process and especially the type of mappings used by the system to resolve the heterogeneity of the acquainted peer; (ii) the supported query language and the processing of queries; and (iii) how the system deals with the lack of centralization and the inherent dynamic and unreliable nature of the peer-to-peer setting. Not every system addresses all these aspects and different systems put more emphasis on a different system aspect.

The following paragraphs provide an overview of some of the main systems in this area.

In Piazza [17], each peer in the system *exports* a schema which can be one of two kinds: (i) a virtual schema, used for querying and mapping the schemas of its acquainted peers, or (ii) a schema which internally is mapped to actual stored data in the peer. In either case, a peer joining the system establishes GAV or LAV mappings between its export schema and the export schemas of peers that are already part of the system. The creation of such mappings is a complicated process which cannot be fully automated. *Schema matching* techniques are used, both in Piazza and elsewhere, to facilitate their creation. Both the created mappings and the supported query language are expressed in a fragment of XQuery. User queries in this fragment are expressed over the schema of a single peer. The query is answered locally by the peer, if the peer actually stores data, and it is also reformulated to a set of queries over the acquaintances of the current peer. Briefly, the reformulation algorithm combines view unfolding, if a GAV mapping exists between acquainted peers, and an answering queries using views algorithm [7], in the case of LAV mappings. The reformulation algorithm terminates once all the queries that result in from the reformulation refer only to relations that correspond to stored data in peers (and no virtual schemas).

Influenced by the work in the LRM model, in Hyperion [2] two types of mappings are used to support the sharing of information between relational database peers, namely, *mapping expressions* and *mapping tables*. Similar to the GAV/LAV mappings in Piazza (and elsewhere), mapping expressions are schema-level mappings used during query answering for the reformulation of queries between peers. A distinguishing feature of Hyperion, mapping tables [10] are *data-level* mappings that associate the values of acquainted peers (inspired by the domain relations in the LRM model). A mapping table T is a relation over the set of attributes $X \cup Y$, where X and Y are non-empty sets of attributes belonging to two acquainted peers. Each tuple $t \in T$ (whose values might include constants and/or variables) associates the set of values in $t[X]$ to those in $t[Y]$.

While schema matching can still be employed for the creation of mapping expressions, the creation of mapping tables requires the development of specialized techniques. Indeed, the work in [10] formalizes mapping tables as data-level constraints over the sharing of data between peers and illustrates how new mapping tables can be inferred automatically (from existing tables) while establishing an acquaintance between two peers. Due to the dynamic nature of peer-to-peer system, peers are expected to continuously evolve and change their schema (less frequently) and their data (very frequently). This change influences both the existing mapping expressions and the existing mapping tables. As a result, work in the Hyperion project is concerned with how to maintain the existing mappings with emphasis in the frequently updated mapping tables.

In terms of query answering, for the case of mapping expressions, Hyperion relies on techniques similar to the ones developed for Piazza. For the case of mapping tables, Hyperion uses a specialized algorithm to rewrite select-project-join (SPJ) queries between peers, relying only on the use of mapping tables for the rewriting [9]. No algorithm is provided in Hyperion to rewrite queries by combining mapping expressions and tables. However, mapping expressions are used in conjunction with mapping tables in Hyperion for *data coordination* [8]. Indeed, another distinguishing characteristic of Hyperion is that it supports the creation and distributed execution of Event-Condition-Action (ECA) rules over multiple acquainted peer.

The PeerDB [14] system is built on top of a generic peer-to-peer platform, called BestPeer [13]. The BestPeer platform supports two types of peers, namely, a large number of normal (data) peers and a smaller number of *location independent global names lookup (LIGLO)* peers. A LIGLO peer acts as a name server with which every peer in the system registers, when joining, in order to acquire a unique identifier.

Similar to Piazza and Hyperion, in PeerDB, data sharing between heterogeneous peers is achieved in two steps. In the first step, mappings are established between the schemas of the peers. In the second step, the mappings are used to

rewrite a query over the schema of one peer to a query over the schema of another. PeerDB has a number of distinguishing characteristics over the previous approaches. First, it differs from the previous systems in that it uses an information retrieval-based technique as a basis for its schema mappings. In more detail, each peer relation and attribute name is associated with a set of descriptive keywords. A mapping between two relations (attributes) that reside on different peers is established if their corresponding descriptive keywords overlap *significantly*. Once a mapping is established, it is used to rewrite a query that refers to one relation into a query that refers to its mapped counterpart. The second distinguishing characteristic of PeerDB is that mappings are established dynamically, at query time. In more detail, PeerDB employs an agent-based technology both during the discovery of schema mappings and during the rewriting of queries. After a user initiates a query over a local peer schema, software agents are responsible for crawling the peer-to-peer network, looking for peers whose schemas can be mapped to the schema where the user query is posed. The agents carry all the necessary functionality to perform the rewriting once such schemas are discovered.

Similar to PeerDB, the GridVine [4] system is built on top its own generic peer-to-peer platform, called P-Grid [1]. However, while the architecture of BestPeer follows the peer/super-peer paradigm, the architecture of P-Grid is that of a structured overlay network. Capitalizing on the efficiency of P-Grid in terms of indexing, routing and load balancing, GridVine builds a semantic mediation layer on top of P-Grid. In this semantic layer, schemas and data are represented as RDF triples, while queries over these triples are expressed as *triple patterns*. A triple pattern query issued at any peer is routed to the peer that can answer the query by using the services of the overlay network (by hashing the constants appearing in the query). In terms of heterogeneity, GridVine employs OWL statements to relate semantically similar schema elements that belong to pairs of schemas. During query evaluation, these mappings are used for the rewriting of queries which are then executed using the overlay network, as described earlier.

Key Applications

Scientific databases, Health informatics databases, Business-to-Business (B2B).

Cross-References

▶ Distributed Database Systems
▶ Peer Data Management System
▶ Schema Matching

Recommended Reading

1. Aberer K, Cudré-Mauroux P, Datta A, Despotovic Z, Hauswirth M, Punceva M, Schmidt R. P-Grid: a self-organizing structured P2P system. ACM SIGMOD Rec. 2003;32(3):29–33.
2. Arenas M, Kantere V, Kementsietsidis A, Kiringa I, Miller RJ, Mylopoulos J. The hyperion project: from data integration to data coordination. ACM SIGMOD Rec. 2003;32(3):53–8.
3. Bernstein P, Giunchiglia F, Kementsietsidis A, Mylopoulos J, Serafini L, Zaihrayeu I. Data management for peer-to-peer computing: a vision. In: Proceedings of the 5th International Workshop on the World Wide Web and Databases; 2002.
4. Cudre-Mauroux P, Agarwal S, Aberer K. GridVine: an infrastructure for peer information management. IEEE Internet Comput. 2007;11(5):36–44.
5. Gnutella Protocol Specification. World Wide Web. http://gnet-specs.gnufu.net/
6. Gribble S, Halevy A, Ives Z, Rodrig M, Suciu D. What can databases do for peer-to-peer? In: Proceedings of the 4th International Workshop on the World Wide Web and Databases; 2001.
7. Halevy AY. Answering queries using views: a survey. VLDB J. 2001;10(4):270–94.
8. Kantere V, Kiringa I, Mylopoulos J, Kementsietsidis A, Arenas M. Coordinating peer databases using ECA rules. In: Proceedings of the International Workshop on Databases, Information Systems and Peer-to-Peer Computing; 2003. p. 108–22.
9. Kementsietsidis A, Arenas M. Data sharing through query translation in autonomous sources. In: Proceedings of the 30th International Conference on Very Large Data Bases; 2004. p. 468–79.
10. Kementsietsidis A, Arenas M, Miller RJ. Data mapping in peer-to-peer systems: semantics and algorithmic issues. In: Proceedings of the ACM SIGMOD International Conference on Management of Data; 2003. p. 325–36.
11. Lenzerini M. Data integration: a theoretical perspective. In: Proceedings of the 21st ACM SIGACT-SIGMOD-SIGART Symposium on Principles of Database Systems; 2002. p. 233–46.
12. Napster. World Wide Web. http://www.napster.com/
13. Ng WS, Ooi BC, Tan K-L. BestPeer: a self-configurable peer-to-peer system. In: Proceedings of the 18th International Conference on Data Engineering; 2002. p. 272.
14. Ng WS, Ooi BC, Tan K-L, Zhou A. PeerDB: a P2P-based system for distributed data sharing. In: Proceedings of the 19th International Conference on Data Engineering; 2003. p. 633–44.
15. Ratnasamy S, Francis P, Handley M, Karp R, Shenker S. A scalable content addressable network. In: Proceedings of the ACM International Conference on Data Communication; 2001. p. 161–72.
16. Stoica I, Morris R, Karger D, Kaashoek F, Balakrishnan H. Chord: a scalable peer-to-peer lookup service for internet applications. In: Proceedings of the ACM International Conference on Data Communication; 2001. p. 149–60.
17. Tatarinov I, Ives Z, Madhavan J, Halevy A, Suciu D, Dalvi N, Dong XL, Kadiyska Y, Miklau G, Mork P. The piazza peer data management project. ACM SIGMOD Rec. 2003;32(3):47–52.

Peer-to-Peer Overlay Networks: Structure, Routing and Maintenance

Wojciech Galuba and Sarunas Girdzijauskas
EPFL, Lausanne, Switzerland

Definition

A peer-to-peer overlay network is a computer network built on top of an existing network, usually the Internet. Peer-to-peer overlay networks enable participating peers to find the other peers not by the IP addresses but by the specific logical identifiers known to all peers. Usually, peer-to-peer overlays have the advantage over the traditional client-server systems because of their scalability and lack of single-point-of-failure. Peer-to-peer overlays are commonly used for file sharing and real time data streaming.

Historical Background

The rise of the Internet brought the first instances of peer-to-peer overlays like the Domain Name System (DNS), the Simple Mail Transfer Protocol (SMTP), USENET and more recently IPv6, which were needed to facilitate the operation of the Internet itself. These peer-to-peer overlays were intrinsically decentralized and represented symmetric nature of the Internet, where every node in the overlay had equal status and assumed cooperative behavior of the participating peers. The beginning of the file-sharing era and the rise and fall of the first file-sharing peer-to-peer system Napster [9] (2000–2001) paved the way for the second generation of peer-to-peer overlays like Gnutella [5] (2000) and Freenet [4] (2001). The simple protocols and unstructured nature made these networks robust and lacking Napster's drawbacks like single-point-of-failure. Since 2001, these peer-to-peer overlays became extensively popular and accounted for the majority of the Internet traffic. Soon after it was evident that the unstructured nature of Gnutella-like systems is embarrassingly wasteful in bandwidth, more efficient structured overlays appeared, like the Distributed Hash Tables (DHTs), which used the existing resources more effectively (e.g., Chord [13]). Currently, unstructured peer-to-peer overlays are sparsely used, as the most popular peer-to-peer applications for file-sharing and data-streaming (e.g., Skype [12], Kademlia [8], KaZaA [6]) are implemented using structured or hybrid overlay concepts.

Foundations

Taxonomy

There are many features of peer-to-peer overlays, by which they can be characterized and classified [2, 11]. However, strict classification is not easy since many features have mutual dependencies on each other, making it difficult to identify the distinct overlay characteristics (e.g., overlay topologies versus routing in overlays). Although every peer-to-peer overlay can differ by many parameters, but each of them will have to have

certain network structure with distinctive routing and maintenance algorithms allowing the peer-to-peer application to achieve its purpose. Thus, most commonly, peer-to-peer overlays can be classified by:

1. Purpose of use;
2. Overlay structure;
3. Employed routing mechanisms;
4. Maintenance strategies.

Purpose of Use

Peer-to-peer overlays are used for an efficient and scalable sharing of individual peers' resources among the participating peers. Depending on the type of the resources which are shared, the peer-to-peer overlays can be identified as oriented for:

1. Data-sharing (data storage and retrieval);
2. Bandwidth-sharing (streaming);
3. CPU-sharing (distributed computing).

Data-sharing peer-to-peer overlays can be further categorized by their purpose to perform one or more specific tasks like file-sharing (by-far the most common use of the peer-to-peer overlays), information retrieval (peer-to-peer web search), publish/subscribe services and semantic web applications. The examples of such networks are BitTorrent [3] (file-sharing), YaCy-Peer [14] (web search), etc.

Bandwidth-sharing peer-to-peer overlays to some extent are similar to the data-sharing ones, however, are mainly aimed at the efficient streaming of real-time data over the network. Overlay's ability to find several disjoint paths from source to destination can significantly boost the performance of the data streaming applications. Bandwidth-sharing peer-to-peer overlays are mostly found in peer-to-peer telephony, peer-to-peer video/TV, sensor networks and peer-to-peer publish/subscribe services. Currently Skype [12] is arguably the most prominent peer-to-peer streaming overlay application.

For the computationally intensive tasks, when the CPU resources of a single peer cannot fulfill

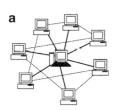
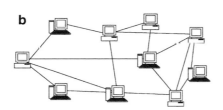
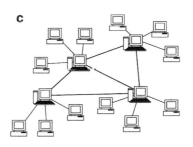

Centralized overlay.
Central peer facilitates
the interactions among
the leaf-peers.

Decentralized overlay.
No central authority, all
peers treated equally.

Hybrid overlay.
Hierarchical topology,
interconnected super-
peers locally serve the
subsets of leaf-peers.

Peer-to-Peer Overlay Networks: Structure, Routing and Maintenance, Fig. 1 Examples of peer-to-peer overlays

its needs, a *CPU-sharing* peer-to-peer overlays can provide plenty of CPU resources from the participating idle overlay peers. Currently, only a major scientific experiments employ such strategy for the tasks like simulation of protein folding or analysis of an astronomic radio signals. Although *not* being a pure peer-to-peer overlay, Berkeley Open Infrastructure for Network Computing (BOINC) is very popular among such networks, supporting such distributed computing projects as SETI@home, folding@home, AFRICA@home, etc.

Overlay Structure

Peer-to-peer overlays significantly differ by the topology of the networks which they form. There exist a wide scope of possible overlay instances, ranging from centralized to purely decentralized ones, however, most commonly, three classes of network topology are identified:

1. Centralized overlays;
2. Decentralized overlays;
3. Hybrid overlays.

Depending on the routing techniques and whether the overlay network was created by some specific rules (deterministically) or in ad hoc fashion (nondeterministically), overlay networks can be also classified into *structured* and *unstructured* peer-to-peer overlays.

Centralized Overlays

Peer-to-peer overlays based on *centralized* topologies are pretty efficient since the interaction between peers is facilitated by a central server which stores the global index, deals with the updates in the system, distributes tasks among the peers or quickly responds to the queries and give complete answers to them (Fig. 1a). However, not all the purposes of use fit the centralized network overlay model. Centralized overlays usually fail to scale with the increase of the number of participating peers. The centralized component rapidly becomes the performance bottleneck. The existence of a single-point-of-failure (e.g., Napster [9]) also prevents from using centralized overlays for many potential data-sharing applications.

Decentralized Overlays

Because of the aforementioned drawbacks, *decentralized* structured and unstructured overlays emerged, which use purely decentralized network model, and do not differ peers as servers or clients, but treat all of them equally - as they were both servers and clients at the same time (Fig. 1b). Thus, such peer-to-peer overlays successfully deal with the scalability and can exist without any governing authority.

The simplest decentralized overlays usually are *unstructured*. Unstructured overlay networks typically have arbitrary topology and use flooding based routing among the peers. The distribution of the resources among the peers is completely unrelated to the network topology. Because of their simplicity, the unstructured overlays are pretty robust to network and peer failures, although are rather inefficient in bandwidth consumption and have poor querying performance.

Structured overlay networks, however, use more efficient routing techniques and the topology of the structured overlays is not arbitrary but typically exhibit Small-World properties, specifically high clusterization and low network diameter. The link establishment among the peers is usually strictly defined by the specific protocols. The topologies can result in various structures like rings, toruses, hypercubes and de-Bruin networks or more loose randomized networks, which do have properties of Small-World networks. A particular instance of structured peer-to-peer overlays is a Distributed Hash Table (DHT) enabling an efficient lookup service, by using a predefined hashing algorithms to assign an ownership for a particular resource (e.g., Chord [13], P-Grid [1], Symphony [7], etc.). In contrast to the traditional Hash Table, the DHTs share the global hashing information among all the participating peers equally and the DHT protocols ensure that any part of a global hash table is easily reachable (usually in logarithmic steps) and there is enough replication to sustain the consistency in the system.

Hybrid Overlays

There also exist many *hybrid* peer-to-peer overlays (super-peer systems) which trade-off between different degree of topology centralization and structure flexibility. Hybrid overlays usually use hierarchical network topology consisting of regular peers and super-peers, which act as local servers for the subsets of regular peers (Fig. 1c). For example, a hybrid overlay might consist of the super-peers forming a structured network which serves as a backbone for the whole overlay, enabling an efficient communication among the super-peers themselves. Hybrid overlays have advantage over simple centralized networks since the super-peers can be dynam-

ically replaced by regular peers, hence do not constitute single points of failure, but have the benefits of centralized overlays.

Routing

Peer-to-peer overlay networks enable the peers to communicate with one another even if the communicating peers do not know their addresses in the underlying network. For example, in an overlay deployed on the Internet, a peer can communicate with another peer without knowing its IP address. The way it is achieved in the overlays is by *routing* overlay messages. Each overlay message originates at a source and is forwarded by the peers in the overlay until the message reaches one or more destinations. A number of routing schemes have been proposed.

Routing in Unstructured Overlays

Unstructured overlay networks use mainly two mechanisms to deliver routed messages: flooding and random walks (Figs. 2 and 3). When some peer v receives a flood message from one of its overlay neighbors w, then v forwards the flood message to all of its neighbors except w. When v receives the same flood message again, it is ignored. Eventually the flood reaches all of the destinations.

For example, in Gnutella [10], a file-sharing peer-to-peer system, a peer s that wants to download a file floods the network with queries. If some peer d that has the file desired by s is reached by the query flood, then d sends a response back to s. Flooding consumes a significant amount of network bandwidth. To reduce it, the flooded messages typically contain a Time-To-Live (TTL) counter included in every message that is decremented whenever the message is forwarded. This limits how far the flood can spread from the source but at the same time lowers the chance of reaching the peer that holds the searched file.

The high bandwidth usage of flooding has led to the design of an alternative routing scheme for unstructured overlay networks: *random walks*. Instead of forwarding a message to all of the neighbors, it is only forwarded to a randomly chosen one. Depending on the network topology

a

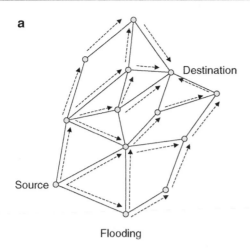

Flooding

b

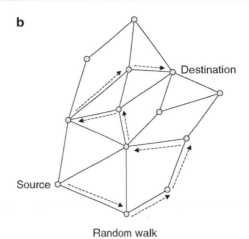

Random walk

Peer-to-Peer Overlay Networks: Structure, Routing and Maintenance, Fig. 2 *Routing in unstructured overlay networks.* The *circles* and *solid lines* represent the overlay topology. The *dashed arrows* illustrate the flow of messages. Peers routing in an unstructured network do not know the exact location of the destinations so they have to either look in all possible directions via flooding or randomly walk to find the destination peer

random walks provide different guarantees of locating the destination peer(s), however all of the random walk approaches share one disadvantage: a significant and in most cases intolerable delivery latency.

Routing in Structured Overlays

As more peers join the overlay network and there are more messages that need to be routed, flooding and random walks quickly reach their scalability limits. This problem has prompted the research on structured overlays.

In structured overlays each peer has a unique and unchanging identifier picked when a peer joins the overlay. The peer identifiers enable efficient routing in the structured overlays. Each routed message has a destination identifier selected from the peer identifier set. Instead of blindly forwarding the message to all neighbors as in the unstructured overlays, a peer in a structured overlay uses the destination identifier to forward the message only to one neighbor. The next hop neighbor is selected to minimize the number of hops to the destination, i.e., the routing is greedy. This selection is made using the *peer identifier distance*.

Most of the modern structured overlays define the notion of distance between any two peer identifiers. For example, in Chord [13] identifiers are selected from the set of integers $[0.2^m - 1]$ and are ordered in a modulo 2^m circle. The distance $d(x, y)$ between two identifiers x and y is defined as the difference between x and y on that identifier circle, i.e., $d(x, y) = (y - x) \bmod 2^m$ In another overlay, Kademlia [8], the identifiers are 160-bit integers and the distance between two identifiers x and y is defined as their exclusive bitwise OR (XOR) interpreted as an integer, i.e., $d(x, y) = x \oplus y$.

Although the modern structured overlays differ in the details of how they make use of the peer identifiers for routing efficiency, they are all based on the same general *greedy routing* principle. When some peer v receives a message with a given destination identifier it forwards the message to that next hop whose identifier is the closest to the destination identifier. In other words, in every hop the message gets as close as possible to the destination. Routing terminates when TTL is exhausted or one of the peers decides it is the destination for the message. The latter decision is application dependent. For example, in a Distributed Hash Table each peer knows for which hash table keys it is responsible. The key space is mapped onto the peer identifier space in DHTs and the destination identifier of each DHT lookup message specifies the hash table key K the lookup is querying for. The

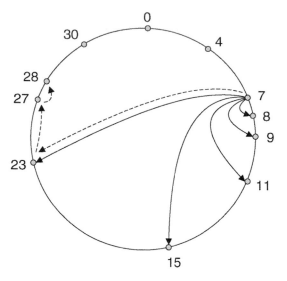

Peer-to-Peer Overlay Networks: Structure, Routing and Maintenance, Fig. 3 *Routing in chord.* The *big circle* represents the peer identifier space with IDs in the interval $[0.2^5]$. The *small circles* are the peers and the number beside them is their ID. Peer 7 is connected (*solid arrows*) to peers with exponentially increasing distance from 7: $7 + 1 = 8, 7 + 2 = 9, 7 + 4 = 11, 7 + 8 = 15, 7 + 16 = 23$. Assume that peer 7 wants to route a message to peer 28. *Dashed arrows* represent the routing path. The peer 23 is the neighbor of 7 that is closest on the ring to the destination 28 and that neighbor is chosen as the first hop for the message. One hop is not enough to reach 28, but the peer 23 brings the message closer to the destination and peer 27 finally delivers it. The greedy routing rule of always selecting such next hop that brings the message as close as possible to its destination is the main building block of all structured overlay networks

lookup message is greedily routed hop by hop from the origin until the message reaches a peer responsible for K. Routing then terminates and the responsible peer sends a lookup response to the origin. The lookup response contains the hash table value stored under the key K.

Maintenance

Peer-to-peer systems are commonly deployed in environments characterized by high dynamicity, peers can depart or join the system at any time. These continuous joins and departures are commonly referred to as *churn*. Instead of gracefully departing from the network peers can also abruptly fail or the network connection with some of its neighbors may be closed. In all of

these cases the changes in the routing tables may adversely affect the performance of the system. The overlay topology needs to be *maintained* to guarantee message delivery and routing efficiency.

There are two main approaches to overlay maintenance: proactive and reactive. In *proactive maintenance* peers periodically update their routing tables such that they satisfy the overlay topology invariants. For example, Chord periodically runs a "stabilization" protocol to ensure that every peer is linked to other peers at exponentially increasing distance. This ensures routing efficiency. To ensure message delivery each Chord peer maintains connections to its immediate predecessor and successor on the Chord ring.

In contrast to proactive maintenance, *reactive maintenance* is triggered immediately after the detection of a peer failure or peer departure. The missing entry in the routing table is replaced with a new one by sending a connect request to an appropriate peer.

Failures and departures of peers are detected in two ways: by probing or through usage. In probe-based failure detection each peer continuously runs a ping-response protocol with each of its neighbors. When ping timeouts occur repeatedly the neighbor is considered to be down and is removed from the routing table. In usage-based failure detection when a message is sent to a neighbor but not acknowledged within a timeout, the neighbor is considered to have failed.

The more neighbors a peer must maintain the higher the bandwidth overhead incurred by the maintenance protocol. In modern structured overlays maintenance bandwidth typically scales as $O(\log(N))$ in terms of the network size.

Key Applications

The key applications of peer-to-peer overlay networks include:

1. File-sharing systems, e.g., BitTorrent [3]

2. VoIP (Voice over IP) and VoD (Video on Demand) systems, e.g., Skype [12]
3. Information retrieval, e.g., YaCy-Peer [14]

Cross-References

▶ Load Balancing in Peer-to-Peer Overlay Networks
▶ Peer-to-Peer Content Distribution
▶ Peer-to-Peer Storage
▶ Peer-to-Peer System
▶ Peer-to-Peer Web Search

Recommended Reading

1. Aberer K. P-Grid: a self-organizing access structure for P2P information systems. In: Proceedings of the International Conference on Cooperative Information Systems; 2001.
2. Androutsellis-Theotokis S, Spinellis D. A survey of peer-to-peer content distribution technologies. ACM Comput Surv. 2004;36(4):335–71.
3. Bittorrent. http://www.bittorrent.com/.
4. Clarke I, Sandberg O, Wiley B, Hong TW. Freenet: a distributed anonymous information storage and retrieval system. In: Designing Privacy Enhancing Technologies: Proceedings of the International Workshop on Design Issues in Anonymity and Unobservability; 2001.
5. Gnutella Homepage. http://www.gnutella.wego.com/.
6. Kazaa Homepage. http://www.kazaa.com/.
7. Manku GS, Bawa M, Raghavan P. Symphony: distributed hashing in a small world. In: Proceedings of the 4th USENIX Symposium on Internet Technologies and Systems; 2003.
8. Maymounkov P, Mazières D. Kademlia: a peer-to-peer information system based on the XOR metric. In: Proceedings of the 1st International Workshop on Peer-to-Peer Systems; 2002. p. 53–65.
9. Napster. http://www.napster.com/.
10. Ripeanu M, Foster I, Iamnitchi A. Mapping the Gnutella network: properties of large-scale peer-to-peer systems and implications for system design. IEEE Internet Comput J. 2002;6(1).
11. Risson J, Moors T. Survey of research towards robust peer-to-peer networks: search methods. Comput Netw. 2006;50(17):3485–521.
12. Skype Homepage. http://www.skype.com/.
13. Stoica I, Morris R, Karger DR, Kaashoek MF, Balakrishnan H. Chord: a scalable peer-to-peer lookup service for internet applications. In: Proceedings of the ACM International Conference on Data Communication; 2001. p. 149–60.
14. YaCyPeer. http://www.yacyweb.de/.

Peer-to-Peer Publish-Subscribe Systems

Ioannis Aekaterinidis and Peter Triantafillou
University of Patras, Rio, Patras, Greece

Definition

Publish/Subscribe (a.k.a. pub/sub) software systems constitute a facility for asynchronous filtering of information. Users, consumers of information, present the system with continuous queries, coined *subscriptions*. Sources of data generation (producers) present the system with data-carrying *publication events*. The pub/sub system infrastructure is responsible for (asynchronously) matching the publication events to all relevant subscriptions. Hence, in essence, this infrastructure filters all available information for every user and presents to each user only the information units (s)he has defined as relevant. As such, a pub/sub infrastructure can play a vital role in large-scale data systems, with huge volumes of data, shielding users from the burden of always actively searching for and retrieving relevant information units.

Peer-to-Peer (P2P) systems are software systems, which in fact constitute *overlay* networks, which are built over physical networks, such as the Internet. Their key discriminative feature is the complete decentralized algorithmic and system design, which, in turn, leads to guarantees with respect to system *scalability* in terms of the network size and stored data. In addition, P2P systems are characterized by *self-organization*, being able to withstand high dynamics with respect to network nodes joining and leaving the system while continuing to offer efficient operation.

Pub/sub P2P systems are an attempt to combine these two important technologies. The central aim is to offer facilities for asynchronous matching of continuous user queries to publication events, at very large scales. This entails dealing with large numbers of producers and consumers of information, which are geographically distributed and, also, supporting large publication-event and subscription arrival rates. Coupling pub/sub technology with the P2P paradigm achieves this aim, while introducing the additional benefits of scalability and self-organization into the pub/sub realm.

Historical Background

Publish/subscribe systems have evolved significantly over time, differing on a number of fundamental characteristics. They are classified into two major categories, according to the way subscribers express their interests; the *topic-based* [6, 10, 16] and the *content-based* [4, 5, 7, 8] systems. Historically, the first systems were topic-based, in which users subscribe to specific topics. All incoming publication events associated with a particular topic are sent to all user subscribers of the topic. This paradigm mimics the way news groups operate. *Content-based* pub/sub systems emerged subsequently, offering users the much-needed ability to express their interests on specific publication events, carrying specific content. Principally, users issue subscriptions which specify predicates over the *values* of a number of well defined attributes. The matching of publication events to subscriptions is performed based on the content (i.e., the values of attributes) being carried by the publication events.

At the turn of the century, related research efforts were maturing and a number of content-based pub/sub systems were already available. Influential examples, with respect to content-based pub/sub systems research, include SIENA [5], Gryphon [4], Le Subscribe [8] and JEDI, [7]. Already, some of this work targeted the challenging issues arising in a distributed system, where publishers and subscribers are geographically distributed [4, 5]. Around the same time and in parallel, research related to P2P networks and systems was also maturing. Worthy of special mention are *structured P2P overlay networks*, like Chord [14], Pastry [13], CAN [12], and P-Grid [1]. Several of these networks influenced work on pub/sub systems, especially endeavors

aiming to leverage existing P2P networks for providing pub/sub functionality with scalability, efficiency, and self-organization. The Scribe system [6] was a first attempt at providing topic-based pub/sub functionality over the Pastry P2P network. This effort was followed by endeavors to build content-based publish/subscribe systems over P2P networks, [2, 3, 9, 11, 15, 17].

Foundations

Pub/sub data management represents a significant point of departure compared to traditional data management. In the latter, data items are stored and the system is responsible for appropriate indexing and processing queries and updates issued by users against these data items. In the pub/sub model, it is the user's continuous queries (subscriptions) that are stored and indexed appropriately and the system is responsible for processing data-carrying publication events, matching them to all relevant stored subscriptions.

The fundamental functionality offered by a publish/subscribe system rests on the pillars of subscription processing and publication-event processing and matching. In a P2P environment, all this functionality is based on distributed algorithms that appropriately leverage the P2P network capabilities in order to ensure scalability and efficiency of operation, free of concern for network topology dynamics. The core functionality exported by a P2P network is the so-called *lookup*() function, which receives as input a *key* and returns the network address of the peer node where the data item associated with the key is located. Structured P2P networks, such as those built using *Distributed Hash Tables* (*DHTs*) (such as [12–14]) can ensure that *lookup*() executes in $O(logN)$ messages, in a network of N nodes, a feature that in essence guarantees scalability and efficiency.

In topic-based publish/subscribe systems, events and subscriptions are associated with specific topics (a.k.a. groups/subjects). A straightforward approach, on which Scribe [6] for example is based for supporting topic-based pub/sub functionality in a P2P environment is to further associate each topic with a *multicast tree*. Topic (and thus multicast tree) creation, involves identifying the node responsible for the tree root, which is determined simply by issuing a *lookup(hash(topic_name))*, using the DHT's hash function. The DHT hash function introduces a randomizing effect when selecting root nodes for different multicast trees. Subscription processing then involves storing the subscription locally at the origin peer node and having that node join the multicast tree. This is basically accomplished by also issuing a *lookup(hash(topic_name))* and creating a path in the multicast tree consisting of all the DHT nodes visited during the execution of the *lookup()* call. Finally, publication-event processing is performed by locating the tree root node (again, using the *lookup()* function) and sending the publication to it, which it subsequently distributes to the multicast tree. Figure 1 illustrates this process.

The content-based pub/sub model has dominated the area, since it allows for greater user-query expressiveness, resulting in more efficient information filtering. However, this model incurs a much higher complexity. The publication event and subscription schema adopted in this model, defines a set of A attributes (a_i, $1 \leq i \leq A$). Each attribute a_i consists of a name, a type (usually string or numerical), and a value $v(a_i)$. A publication event is defined to be a set of $k <$ *attribute,value* > pairs ($k \leq A$), while a subscription is defined through an appropriate set of predicates on attribute's values over a subset of the A attributes of the schema. The complexity emerges from the need to support subscriptions with complex predicates. For numerical-typed attributes, the allowable predicates may involve equality, inequality, $\leq, \geq$, and ranges of values. For string-typed attributes, subscribers may define prefix, suffix, sub-string and equality predicates. Fundamentally, solutions in a P2P environment can be classified according to whether they require knowledge of the internal DHT routing state, which is maintained by each DHT node, and its association with additional state that is needed for subscription and publication processing [3, 11, 15]. A key characteristic in several of these approaches [11, 15] is that each node n_1 associates with each other node in its routing table, say n_2, additional state that consists of all subscriptions that n_1 has received from n_2. When a publication event arrives at n_1, it is forwarded to every node n_2 in its routing table only if the publication event matches one of the subscriptions sent to n_1 by n_2.

Approaches that do not belong in this category avoid the maintenance costs associated with the extra, per-node state and enjoy wider applicability as they can be easily integrated with a number of DHTs. Hereafter, the focus is on these approaches, where the content (i.e., the values defined by the predicates) will determine at which peer nodes the subscriptions will be stored. Similarly, the values carried by a publication event will determine which route must be followed within the network so to reach every possible peer node storing a relevant subscription.

Peer-to-Peer Publish-Subscribe Systems, Fig. 1 Multicast tree construction for event dissemination in topic-based publish/subscribe

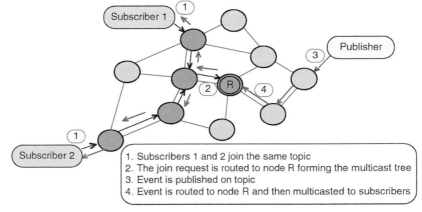

1. Subscribers 1 and 2 join the same topic
2. The join request is routed to node R forming the multicast tree
3. Event is published on topic
4. Event is routed to node R and then multicasted to subscribers

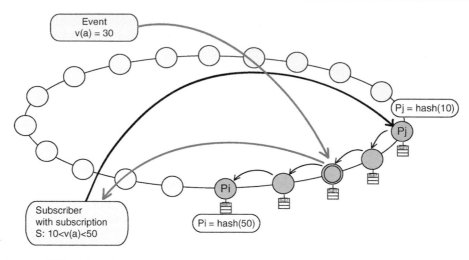

Peer-to-Peer Publish-Subscribe Systems, Fig. 2   Processing range queries in a content-based pub/sub system

Equality predicates can be handled straightforwardly – the basic idea being the following. A subscription s associating with attribute a the value $v(a)$, can be stored simply at the peer node with identifier p, where $p = hash(v(a))$. A publication event e carrying $a = v(a)$, is processed by visiting the peer node $p = hash(v(a))$ and retrieving all locally-stored subscriptions, such as s.

However, for more complex predicates, this is inadequate. At an abstract, high-level view, proposed solutions, albeit very different, share the following characteristics. The node ID namespace is typically associated in solution-specific manners with the value domain of an attribute. A subscription with a complex predicate on an attribute is stored at several nodes, whose IDs depend on the values defined by the subscription's predicate. In other words, a subscription is mapped to a subspace of the node ID namespace. Publications, associating an attribute with a value correspond to a point in the namespace. The node associated with this point, by construction, will be storing all subscriptions whose predicates on this attribute include this event's value. In this way, the *subscription to publication event rendezvous* occurs and in this rendezvous node the event can be matched to all relevant subscriptions.

For concreteness, the basics of the approach adopted in [17] – one of the first to leverage an existing P2P network on top of which to build scalable content-based pub/sub systems – for processing subscriptions with range predicates, is outlined. The approach is simple, requires no additional state maintenance, and no knowledge of the internals of the underlying DHT state. This approach utilizes the Chord DHT [14] in which nodes are arranged in a circular list according to their IDs. However, in [17] specific order-preserving hashing is employed to store subscriptions in the Chord network. That is, when processing subscriptions with values $v(a)_i$ and $v(a)_j$, they are stored at nodes p_i and p_j whose IDs are given by $hash(v(a)_i)$ and $hash(v(a)_j)$, respectively. If $v(a)_j < v(a)_i$, then $p_j < p_i$. An incoming subscription s, identifying a range of values $[v(a)_j, v(a)_i]$ is stored on all nodes in the arc of the ring starting at node $hash(v(a)_j)$ and ending at node $hash(v(a)_i)$. Given this, a publication carrying value $v(a)_k$, with $v(a)_j < v(a)_k < v(a)_i$ will be directed at node $hash(v(a)_k)$ and this node by construction falls on the arc of the ring where the subscription s has been stored. Thus, the matching can be locally performed at this node. Figure 2 illustrates this process.

Finally, the aforementioned discussion has focused on single-attribute (one-dimensional) sub-

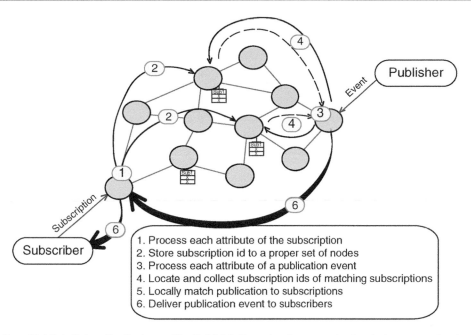

Peer-to-Peer Publish-Subscribe Systems, Fig. 3 Multi-dimensional event and subscription processing in content-based P2P publish/subscribe systems

scriptions and publications. In general, pub/sub systems involve subscriptions and publications specifying multi-dimensional content. Processing multi-dimensional publication events and subscriptions is a source of additional complexity. In this setting, a publication event e will match a subscription s if and only if all attributes named by s are also named by e and the predicates defined by s on its attributes are satisfied by the values for these attributes carried by the publication event. A straightforward approach dealing with multi-dimensional events and subscriptions is the following. First, perform the processing as discussed above for each named attribute in the publication event and subscription. Second, for each attribute collect a candidate result set, consisting of the subscriptions being matched only for that attribute of the publication. Lastly, merge and filter the per-attribute candidate result sets into a single result set, by removing those subscriptions in the candidate result sets which have at least one attribute predicate not satisfied by the publication event. Figure 3 illustrates this process.

As these candidate result sets are distributed and can contain very large numbers of subscriptions, the tasks of merging them and filtering subscriptions from them can introduce large overheads. Therefore, multi-dimensional event processing is open to a number of crucial performance optimizations which reveal key trade-offs with respect to network bandwidth overheads, number of required messages, and event-matching latencies [2].

Key Applications

In recent years one notices a proliferation of data-intensive and compute-intensive networked applications aiming for efficiency, scalability, and self-organization. A key characteristic in many such applications is the massive amounts of data of various types and characteristics being generated continuously, from many different sources, at different times, and possibly at very high rates. Users can thus be inundated by the sheer volume of this data and are confronted with severe difficulties in accessing only the typically very small fraction of this data that is of interest to them. Hence, what is very much needed for such applications is an infrastructure that can offer decentralized, scalable, self-organizing asynchronous filtering of this massive information base.

Content-based publish subscribe systems operating in the distributed environment of a P2P overlay network appear as the appropriate infrastructure, and are used to design and implement such applications. A few representative example applications, which also indicate open research and development challenges, are listed below:

1. Information Feeds

   A classic example for publish/subscribe application infrastructures are data applications based on information feeds. In these applications publishers can be, for example, news agencies that publish news articles. Human subscribers declare their interests with proper content-based subscriptions involving numerical-attribute predicates such as date ranges, and string-attribute predicates such as words defining article title prefixes. In general, any RSS feed-like information dissemination system like stock exchange reports falls in this category. The distributed aspects occur when considering large, multinational news agencies whose computers form a large, geographically-dispersed network. One can even further imagine alliances of such multi-nationals, increasing dramatically the application's scale.

2. Grid Computing

   Consider a farm of computing clusters, each one including a number of computing nodes with heterogeneous characteristics involving, for example, various types of operating systems, hardware specifications, etc. and forming a peer-to-peer computing grid. Each node publishes its characteristics while possible users that wish to execute their programs are interested in specific node attributes (subscriptions) defining for example acceptable CPU speed ranges, minimum memory requirements, desirable operating system versions, etc.

3. Pub/sub and the Web

   The web is of course a massive distributed data repository. Web information systems can be significantly enriched by adopting pub/sub technology. Overlay networks can be created consisting of nodes representing web pages. A pub/sub system over this overlay network can be used, for instance, to inform existing web users of interesting new pages being created, of updates to pages already marked as relevant, etc.

4. Bio-Informatics Applications

   In the area of bioinformatics research, one might imagine a network of research data bases, belonging to specific governmental or private, non-for-profit research institutions, storing results or matching protein sequences and specific sub-sequences. The publishers in this case are the institutions 'researchers publishing specific findings and subscribers are researchers inquiring for specific sequences' matching, as defined in their subscriptions.

Experimental Results

Typically, all proposed research solutions are accompanied by independent experimental results. So far the community has not produced widely acceptable general benchmarks, standard data sets, and workloads.

Data Sets

As mentioned, there are no generally-agreed upon real data sets to be used for testing research and development results. Some appropriate data sets are available, however, for some champion pub/sub applications, such as those based on RSS feeds. One example is data made available by the New York Stock Exchange (NYSE) (http://www.nysedata.com/nysedata/default.aspx).

Cross-References

▶ Channel-Based Publish/Subscribe
▶ Content-Based Publish/Subscribe

- ▶ Peer-to-Peer Content Distribution
- ▶ Peer-to-Peer Overlay Networks: Structure, Routing and Maintenance
- ▶ Peer-to-Peer System
- ▶ Publish/subscribe
- ▶ Publish/Subscribe over Streams
- ▶ State-Based Publish/Subscribe
- ▶ Topic-Based Publish/Subscribe
- ▶ Type-Based Publish/Subscribe

Recommended Reading

1. Aberer K. P-Grid: a self-organizing access structure for P2P information systems. In: Proceedings of the International Conference on Cooperative Information Systems; 2001.
2. Aekaterinidis I, Triantafillou P. Internet scale string attribute publish/subscribe data networks. In: Proceedings of the International Conference on Information and Knowledge Management; 2005.
3. Aekaterinidis I, Triantafillou P. PastryStrings: a comprehensive content-based publish/subscribe DHT Network. In: Proceedings of the 23rd International Conference on Distributed Computing Systems; 2006.
4. Banavar G, Chandra T, Mukherjee B, Nagarajarao J, Strom J, Sturman D. An efficient multicast protocol for content-based publish-subscribe systems. In: Proceedings of the 19th International Conference on Distributed Computing Systems; 1999.
5. Carzaniga A, Rosenblum DS, Wolf AL. Design and evaluation of a wide-area event notification service. ACM Trans Comput Syst. 2001; 19(3):332–383.
6. Castro M, Druschel P, Kermarrec A, Rowstron A. Scribe: a large-scale and decentralized application-level multicast infrastructure. J Sel Areas Commun. 2002;20(8):1489–99.
7. Cugola G, Nitto ED, Fuggetta A. The JEDI event-based infrastructure and its application to the development of the OPSS WFMS. In: Proceeding of the 23rd International Conference on Software Engineering; 2001.
8. Fabret F, Jacobsen A, Llirbat F, Pereira J, Ross K, Shasha D. Filtering algorithms and implementation for very fast publish/subscribe. In: Proceedings of the ACM SIGMOD International Conference on Management of Data; 2001.
9. Gupta A, Sahin OD, Agrawal D, Abbadi AE. Meghdoot: content-based publish subscribe over p2p networks. In: Proceedings of the ACM/IFIP/USENIX 5th International Middleware Conference; 2004.
10. Lehman T, Laughry S, Wyckoff P. Tspaces: the next wave. In: Proceedings of the 32nd Annual Hawaii International Conference on System Sciences; 1999.
11. Pietzuch PR, Bacon J. Hermes: a distributed event-based middleware architecture. In: Proceedings of the 1st International Workshop Distributed Event-Based Systems; 2002.
12. Ratnasamy S, Francis P, Handley M, Karp R, Shenker S. A scalable content addressable network. In: Proceedings of the ACM International Conference on Data Communication; 2001.
13. Rowstron A, Druschel P. Pastry: scalable and distributed object location and routing for large-scale peer-to-peer systems. In: Proceedings of the IFIP/ACM International Conference on Distributed Systems Platforms; 2001.
14. Stoica I, Morris R, Karger D, Kaashoek F, Balakrishnan H. Chord: a scalable peer-to-peer lookup service for internet applications. In: Proceedings of the ACM International Conference on Data Communication; 2001.
15. Terpstra WW, Behnel S, Fiege L, Zeidler A, Buchmann AP. A peer-to-peer approach to content-based publish/subscribe. In: Proceedings of the 2nd International Workshop Distributed Event-Based Systems; 2003.
16. TIBCO TIB/Rendezvous. Tech. rep., White paper, Palo Alto. http://www.tibco.com. 1999.
17. Triantafillou P, Aekaterinidis I. Publish-subscribe over structured P2P networks. In: Proceedings of the 3rd International Workshop Distributed Event-Based Systems; 2004.

Peer-to-Peer Storage

Anwitaman Datta
Nanyang Technological University, Singapore, Singapore

Synonyms

Cooperative storage systems; Distributed storage systems; Wide-area storage systems

Definition

Peer-to-peer (P2P) storage is a paradigm to leverage the combined storage capacity of a network of storage devices (peers) contributed typically by autonomous end-users as a common pool of storage space to store and share content, and is designed to provide persistence and availability of the stored content despite unreliability of the individual autonomous peers in a decentralized environment.

Historical Background

For diverse reasons including fault-tolerance, load-balance or response time, or geographic distribution of end users, distributed data stores have been around for a long while. This includes distributed databases, distributed file systems and *Usenet* servers among others. *Usenet* servers communicated among each other in a peer-to-peer manner, and replicated content.

While some redundancy is necessary for fault tolerance, replicating all content at all peers is a very special case of a peer-to-peer storage system, and is very inefficient. In general, an object is replicated at fewer locations. This leads to the problem of locating the object in the network. Plaxton et al.'s work on accessing nearby copies in a distributed environment [9] using key based routing is one of the seminal works, which subsequently influenced the design of many peer-to-peer storage systems. Advances in *structured overlay* networks address the problem of object location in a decentralized manner, that is "If an object is in the network, how to find it?" This is important in realizing an efficient decentralized peer-to-peer storage system.

Systems like *OceanStore* [7] aimed at archival storage, *Freenet* [3] for anonymous file sharing and distributed hash table based *DHash* storage layer of *CFS* [5] for a cooperative file system all store data at peers based on key associations determined by the structured overlay. These systems thus have the functionalities of object location and storage coupled together. However, note that locating and storing objects in a peer-to-peer system are in principle independent of each other. One can imagine the routing as a distributed index structure, which can store a pointer to the actual storage location, instead of the object itself. In the file sharing network *Napster*, the storage was peer-to-peer,

however the indexing was in fact fully centralized, and very well illustrates the orthogonality of object location and storage issues. So in the rest of this entry the focus will be only on storage.

A critical requirement in a peer-to-peer storage system is to ensure that once a user (application) stores an object, this object should be available and persist in the network, notwithstanding the unreliability of individual peers and membership dynamics (churn) in the peer-to-peer network. This throws open a host of interesting design issues. For resilience, redundancy is essential. Redundancy can be achieved either by replication or using coding techniques, using schemes similar to the *RAID* technique [8]. While coding is in principle storage space efficient for achieving a certain level of resilience, it leads to various kinds of overheads, including computational overhead, and in the context of peer-to-peer systems, communication overhead and even storage overhead to keep track of encoded object fragments, thus making coding mechanisms worthwhile only for relatively larger or rarely accessed objects and applications like archival storage.

Redundancy is lost unless replenished because of departure of peers from the system. Trade-off considerations of redundancy maintenance effort and resilience led to the design of different maintenance strategies [2, 6, 10].

Note that file sharing systems – early ones like *Napster* and *Gnutella* as well as more recent ones like *Kazaa* and *BitTorrent* – also store and provide access to stored objects. However they are not reliable storage systems. In file sharing networks, users store locally only files they are interested in, and allow others to download the same. However there is no explicit intention to provide a highly available and persistent storage, and hence there are no mechanisms for redundancy management. The design of these systems is often focused on improving the efficiency of search and data transfer, while the content available in the network is considered ephemeral. Nevertheless, popular content may get so widely replicated that it is coincidentally (but not by design) always available.

Foundations

Distributed storage systems have traditionally been managed in a centralized manner, for instance in distributed databases and distributed file systems. The advent of decentralized file sharing networks demonstrated the potential as well as feasibility of decentralized peer-to-peer storage systems composed of autonomous peers. Peer-to-peer storage systems use the combined capacity of the peers to provide storage functionality to end users.

There are several reasons to have such a distributed storage. Multiple users may share and access some stored objects – data, files, etc. Individual users may not have the capacity to store all the objects they wish to access. Furthermore, by storing the same objects at other peers (and reciprocating by storing other users' objects), all peers benefit from an automatic back up service. Against these advantages, the main drawbacks include the unreliability of individual peers, who may leave and re-join autonomously, or even leave permanently, as well as transient communication failures, and delay in accessing objects stored only at a remote peer. Thus the main functionality of peer-to-peer storage systems is to make the distribution of stored objects transparent to the end users, even while users benefit from this distribution.

File sharing networks can be viewed as a special case of a peer-to-peer storage system. However they are not designed to guarantee availability or persistence of stored content, which are essential for a storage system. Resilience and performance in terms of access cost and latency pose some of the crucial challenges in realizing peer-to-peer storage systems.

Resilience of storage systems is measured in terms of two metrics – availability and durability (persistence). Availability of an object within a period of time is essentially the time averaged probability that it is accessible at any random time within that period of time. Durability of an object is the probability that it persists in the system indefinitely (or long enough, depending on the application requirements) under an assumption of a worst case failure scenario.

Consequently, crucial to peer-to-peer storage systems research are some of the following questions: What kind of redundancy is most efficient from various perspectives including storage overhead as well as access and maintenance costs and latency, and implementation complexity? What minimal redundancy is necessary to meet a desired level of resilience? Which maintenance strategy to apply to maintain the necessary redundancy? Which peers to store the redundant blocks in, based possibly on issues like reliability of individual peers, locality and load? The following delves into these. Figure 1 summarizes some of the important design factors.

In a decentralized setting, it is not only essential to ensure that stored objects stay available and persist over time despite changes in the network membership – churn – caused by peers leaving and (re-)joining the network, but also it is important to locate the stored objects efficiently. Some of the early storage systems like *OceanStore* [7] use one of the precursors [9] of contemporary structured overlays to locate objects. The basic idea is to assign to each peer an unique identifier, and also to each object an unique identifier from a same key-space, for instance using a hashing function. Objects are stored at peers which have identifiers closest to the object's identifier. When looking for an object, or the peer(s) at which an object is to be stored, the network of peers is typically searched in a greedy manner, by approaching peers with closer identifiers to the object's key, that is, using *key based routing*. Similar ideas are also used in *Freenet* [3], which aims at providing anonymity to its users. Other subsequent storage systems have also followed this approach [5].

This traditional dual use of structured overlays for both routing (indexing) and storage tends to blur the difference between the two, and many people consider the structured overlay (e.g., distributed hash tables) to be the storage system also. In order to identify a full spectrum of design space of peer-to-peer storage systems it is crucial to understand that storage and indexing are actually two different but necessary ingredients for them. Combining these two sometimes simplifies the system architecture and implementation.

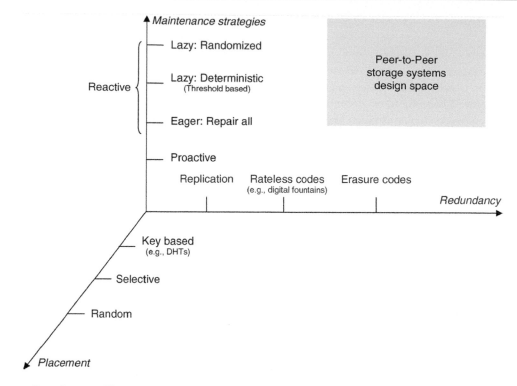

Peer-to-Peer Storage, Fig. 1 Some of the important design decisions that need to be taken into consideration when designing and deploying a peer-to-peer storage system

However, the structured overlay may be used to store only pointers to the actual objects, which are placed using any independent criterion. One may also employ any other kind of directory (for example, *Napster* has a centralized directory) or indexing mechanism instead.

Resilience from Redundancy

Redundancy is essential to achieve resilience. Storage systems realize redundancy by using either replication or error correcting codes (e.g., erasure codes) as discussed next.

Replication Replication, i.e., mirroring, stores the same object on multiple nodes. It is normally efficient when the size of the object is small, or the object is frequently accessed. Note that a large object may still be split in multiple fragments, and each of these fragments will then be replicated.

Erasure codes Erasure codes are a class of error-correcting codes, which can transform a

M-fragment object into N ($N > M$) encoded fragments, such that the original object can be reconstructed from any *M out of the N* encoded fragments. This typically leads to a storage overhead slightly more than N/M. The rate of erasure codes r is defined as the fraction of fragments required for decoding, i.e., $r = \frac{M}{N}$. Based on the coding rate, erasure codes can be categorized into two types, fixed-rate codes in which N is limited and r has a fixed value, and rateless codes (for example *Digital Fountains* [4]) in which N is potentially unlimited and r approaches zero (therefore called rateless).

Although replication is sometimes regarded as a special case of erasure codes, a subtle difference exists between them, which has its implications on the maintenance of redundancy. For replication, every reintegrated replica can enhance the object availability. In contrast, for a fixed-rate erasure code, if the reintegrated fragment is identical to one of the existing fragments, it does not improve the availability of the whole object.

Even with M fragments, one may still not be able to reconstruct the original object, unless the fragments are distinct. The above problem does not occur for rateless codes, in which unique fragments are generated, and thus every reintegrated fragment is useful. In this aspect, replication is more similar to rateless codes, instead of fixed-rate codes.

When accessing an object, if it had been stored in coded fragments, enough fragments need to be accessed first, and then the object needs to be decoded, causing computational and possibly communication overheads. The benefit of coding is that for the same targeted resilience, using coding techniques drastically reduces the storage overhead, alternatively said, for same storage overhead, coding provides a much higher resilience. However, depending on the implementation details, using coding techniques may also lead to additional storage overheads to keep track of the encoded fragments. For frequently accessed objects or small objects, the overheads associated with coding may thus outweigh the benefits.

Because of these considerations, a rule of thumb is small or frequently accessed objects are typically replicated, while large or rarely accessed objects erasure coded. Another option is to use a hybrid approach [11]. Erasure coded fragments are used to achieve high persistence at a low storage overhead. Replication is used for performance.

Maintaining Redundancy

Individual peers may be occasionally unavailable because users go offline, machines fail, or the network gets disconnected. Typically, if there is sufficient redundancy, such *temporary churn* has marginal effect on the availability of a stored object. Furthermore temporary churn does not directly affect long term persistence, since the peers join back, bringing back in the network whatever is stored in them. However, over a period of time, some participating peers may leave the system permanently, in turn leading to permanent loss of redundancy. *Permanent churn* thus makes the system more vulnerable to temporary churn, and unless mitigated, leads to permanent

loss of stored objects, thus adversely affecting persistence/durability. So while redundancy provides resilience against temporary churn, maintenance strategies are necessary to restore the lost redundancy and make the system resilient against permanent churn.

Henceforth both coded fragments or replicas will be referred as fragments. The simplest maintenance mechanism is to probe periodically all the peers which are supposed to store redundant fragments of an object, and whenever a probe fails, reactively replenish the redundancy be reintegrating a new suitable redundant fragment. Example storage systems employing this strategy include CFS [5]. This strategy is referred to as an *eager maintenance* strategy. Such an eager reactive (The simultaneous use of the adjectives eager and reactive is admittedly oxymoronic, and is a vestige of the historical development and nomenclature of various maintenance mechanisms.) maintenance mechanism means the system always operates in a state where redundancy level remains constant apart from temporary reduction between repair periods. It has been empirically observed [2] that eager (reactive) maintenance strategy is bandwidth expensive.

A periodic probing and reintegration of all unavailable fragments ignores the fact that many of the fragments are only temporarily unavailable, and will be restored back in the system automatically once the corresponding peers join back. Consequently, a lot of overhead in regenerating and transferring new redundant fragments is avoidable if only the system can wait long enough for some of the absent peers to rejoin. This observation is key to the design of the first lazy repair (reactive) strategy, used in the *TotalRecall* system [2]. All nodes are probed periodically, and repairs are initiated only when less than a certain threshold $T_a > M$ of nodes (and corresponding data) are available. Thus to say, when an object has no more than T_a fragments available in the system, then a repair process for the object is initiated so that at the end of the repair process all N fragments are again available. As long as an object has a redundancy more than the parameter T_a, there is no maintenance.

This lazy maintenance mechanism saves overheads caused by transient failures. However if the redundancy reduces below this threshold, then it is considered imprudent to further delay repair operations. So once the threshold is breached for an object, *TotalRecall* regenerates all the corresponding unavailable fragments. Thus, this reactive lazy mechanism has a deterministic trigger for initiating repairs.

While saving on bandwidth in comparison to the above mentioned eager reactive maintenance strategy, the threshold based lazy repair strategy suffers from several undesirable effects. First of all, by waiting for the redundancy to fall below a threshold, the system is allowed to degenerate and become more susceptible as compared to other systems which have the same maximum redundancy N. Secondly, while most of the time this approach does not use the bandwidth even if it were available, once the threshold is breached, this approach tries to replenish all the missing fragments at once, thus causing bandwidth spikes. Finally, between the maintenance spikes, as redundancy falls, the available fragments are accessed more frequently, causing access load imbalance, particularly overloading the available peers. Two different works try to address these shortcomings.

A randomized variant of the lazy repair strategy [6] is to probe only a fraction of the stored fragments randomly (uniformly), until a minimal $T_b \geq M$ number of live fragments are detected. Thus a random number $T_b + X$ of probes (determined according to a probability distribution which in turn depends on the actual number of live fragments) will be required to locate T_b live fragments. Then X fragments which were detected to be unavailable are replaced by the system. The beauty of this randomized approach is that the expected value of X adapts with the number of live fragments. If there are fewer live fragments, then X will typically be large, and vice versa. Normally X can be typically much smaller than the total number of unavailable fragments at that instant. As a consequence, the repair process is continuous – thus available bandwidth is used more judiciously, avoiding spikes. The randomized repair process is naturally more aggressive

when more redundant fragments are missing, while if very few fragments are missing then there are very few repairs.

While the randomized lazy repair strategy [6] tries to strike a balance between the periodic repair and the threshold based deterministic lazy repair strategies, and as a consequence the bandwidth usage is continuous and smoother, another independent approach [10] makes it an explicit goal to not exceed a bandwidth budget per unit time. Subject to the bandwidth budget per unit time it proactively creates new replicas. In contrast to the previously mentioned approaches all of which react to lost redundancy, the proactive approach does not aim to reduce overall bandwidth usage, but instead tries to ensure that by not exceeding the per unit time maintenance bandwidth budget, a better bandwidth provisioning can be achieved, and thus the maintenance operation does not interfere with applications. Another consequence of this approach is that typically enough redundancy is available to ensure equitable load distribution. However, a maximum redundancy threshold needs to be defined in using such a proactive replication approach to make a judicious use of the storage capacity.

Placement Strategies

A final system design issue is determining the placement strategy to be used to store the objects. One of the most widely used approach is to determine the placement of objects based on keys. In this scheme, all the peers as well as objects are assigned unique keys (e.g., by hashing), and then objects are stored at peers with closest or similar keys. The peers themselves communicate among each other by forming a structured overlay network, and messages are routed based on key similarity (key based routing). Such an object placement strategy can be seen in systems like *CFS* [5], *Oceanstore* [7], *Freenet* [3] and *Tempo* [10] to name a few. This approach essentially combines the storage of the object with the search mechanism.

Alternatively, object placement may be decoupled from the search mechanism, thus giving the systems designer, or even the applications using the storage system much more flexibility in choosing the storage location. This choice may be random (for example, in *TotalRecall* [2]), or based on reliability prediction derived from the history of peers' availability, or based on proximity (locality) from the end users accessing the object, load at peers, or other considerations like storing the object within a specific domain. Furthermore, this choice may be made by the peer-to-peer storage system designer or its administrator, or even independently by the applications and end users using the storage system.

Against this flexibility, the main disadvantage of decoupling storage from search is that one then needs some kind of directory service (potentially realized with a structured overlay) to perform the search functionality. The search provides pointer(s) to the stored object fragments. This creates additional storage and maintenance (of the pointers) overheads. Thus, in contrast to the key based storage approach, there is an additional level of indirection, and the storage system needs to explicitly keep track of any changes, including network level changes like change of peers' network address, unlike the structured overlay (key based storage) approach, where the structured overlay maintenance mechanism takes care of such changes and simplify storage systems design.

Analysis Techniques

There are several approaches to analyze and study the behavior of peer-to-peer storage systems to better understand and refine its algorithms and design, make better parameter choices and validate its implementation.

Static resilience Given a certain amount of redundancy, if a certain fraction of the peers are not accessible, either because they left the network, or the machines temporarily crashed, or because of communication problems, one can determine the probability that any specific object will become inaccessible or permanently lost. For example, for pure replication, if there are ρ replicas, and each peer storing a replica is available only p_{on} fraction of the time (randomly and independently from the other peers replicating the object), then the probability of the object becoming

unavailable is $(1 - p_{on})^\rho$. Such a static resilience analysis gives a system designer a reference for determining an adequate redundancy to tolerate a certain degree of membership dynamics.

Time-evolution Static resilience does not take into account the combined effect of membership dynamics, which leads to loss of redundancy over time, and repairing strategy, which regenerates redundancy, thus improving the health of the storage system. Given a particular level of network dynamics, the choice of maintenance strategy affects the resilience of the storage system, as well as the overheads incurred. A pragmatic system design thus needs to take into account the combined effect of individual peers' unreliability as well as the specifics of the deployed redundancy maintenance strategy.

Of particular interest is the actual probability distribution of object redundancy under the combined effect of churn and maintenance. This is in contrast to the maximum or average redundancy of objects in the system, based on which static resilience is typically estimated. Objects with smaller actual redundancy at a time instant are more vulnerable to become temporarily unavailable or even permanently lost. Also access to these corresponding objects causes higher load at the peers storing the object fragments or replicas. Finally, more effort and bandwidth is required to restore redundancy for the objects with fewer fragments, thus the maintenance operation witnesses spikes in bandwidth usage. The probability distribution provides a more fine-grained state of the storage system's health, by showing what fraction of all the stored objects are expected to have what level of actual redundancy. Such information can be obtained by studying the time evolution of storage systems [6, 12].

For example, Fig. 2 shows the probability density function of the actual redundancy of objects when the deterministic or the randomized lazy maintenance algorithms are used for an otherwise identical specific scenario, that is, same redundancy for objects, same churn level, and algorithm parameters chosen such that total bandwidth usage in both cases are comparable. There

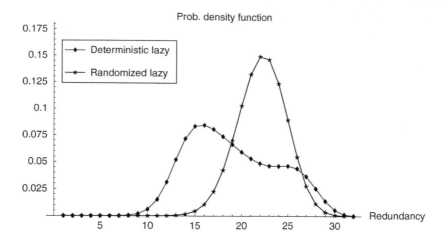

Peer-to-Peer Storage, Fig. 2 A snapshot of the probability distribution of the available number of object fragments in the storage system when using (i) Deterministic lazy (reactive) maintenance, (ii) Randomized lazy (reactive) maintenance. An any *8 out-of 32* fragments (rate 0.25) erasure code was used to store the objects. Churn was simulated synthetically, such that online peers could go offline with a probability of 0.2 and offline peers could come back online with a probability of 0.1, such that on an average one third of the peers were online, i.e., p_{on} = 1/3. The thresholds T_a and T_b for the two variants of the lazy maintenance strategies were chosen such that the aggregate bandwidth usage for maintenance was the same in both experiments. This figure has been (Obtained from Datta and Aberer [6])

is a greater area under the curve corresponding to low redundancy if the deterministic lazy maintenance mechanism is used, in comparison to the case when the randomized variation is used.

Key Applications

Peer-to-peer storage systems have been used to realize traditional applications like file systems [2, 5], backup [1] and archival storage [7]. A peer-to-peer backup system has several advantages in comparison to traditional offline backup systems based on secondary storage. It is easier to verify that the backup has indeed worked correctly (in comparison to physical medium like magnetic tapes), and backup as well as maintenance of the backed up data and its restoration whenever necessary can all be completely automated. Furthermore, peer-to-peer storage is not vulnerable to geographically localized catastrophes. This gives large corporations with geographically dispersed offices strong economic incentives to use their employees' desktop computers as a private corporate peer-to-peer storage infrastructure, instead of managing a separate secondary storage based backup. Similarly, individuals can back-up their data by relying on the resource pooled in a public peer-to-peer storage network.

Peer-to-peer storage systems also find use in web proxies for caching, content distribution and file sharing [3] applications. Peer data management systems too rely on an underlying reliable storage service.

Cross-References

▶ Peer Data Management System
▶ Peer-to-Peer Overlay Networks: Structure, Routing and Maintenance
▶ Peer-to-Peer Content Distribution
▶ Updates and Transactions in Peer-to-Peer Systems

Recommended Reading

1. http://www.cleversafe.org/dispersed-storage.
2. Bhagwan R, Tati K, Cheng Y, Savage S, Voelker GM. TotalRecall: systems support for automated availability management. In: Proceedings of the 1st

USENIX Symposium on Networked Systems Design & Implementation; 2004.

3. Clarke I, Miller SG, Sandberg O, Wiley B. Protecting free expression online using Freenet. IEEE Internet Comput. 2002;6(1):40–9.

4. Shokrollahi A. Raptor codes. IEEE Trans Inf Theory. 2006;52(6):2551–67.

5. Dabek F, Kaashoek F, Karger D, Morris R, Stoica I. Wide-area cooperative storage with CFS. In: Proceedings of the 18th ACM Symposium on Operating System Principles; 2001.

6. Datta A, Aberer K. Internet-scale storage systems under churn – a study of the steady state using Markov models. In: Proceedings of the 6th IEEE International Conference on Peer-to-Peer Computing; 2006.

7. Kubiatowicz J, Bindel D, Chen Y, Czerwinski S, Eaton P, Geels D, Gummadi R, Rhea S, Weatherspoon H, Weimer W, Wells C, Zhao B. OceanStore: an architecture for global-scale persistent storage. In: Proceedings of the 9th International Conference on Architectural Support for Programming Languages and Operating Systems; 2000.

8. Patterson D, Gibson GA, Katz R. A case for redundant arrays of inexpensive disks (RAID). In: Proceedings of the ACM SIGMOD International Conference on Management of Data; 1988.

9. Plaxton CG, Rajaraman R, Richa AW. Accessing nearby copies of replicated objects in a distributed environment. In: Proceedings of the ACM Symposium on Parallel Algorithms and Architectures; 1997.

10. Sit E, Haeberlen A, Dabek F, Chun BG, Weatherspoon H, Morris R, Frans Kaashoek M, Kubiatowicz J. Proactive replication for data durability. In: Proceedings of the 5th International Workshop Peer-to-Peer Systems; 2006.

11. Williams C, Huibonhoa P, Holliday J, Hospodor A, Schwarz T. Redundancy management for P2P storage. In: Proceedings of the 7th IEEE International Symposium on Cluster Computing and the Grid; 2007.

12. Wu D, Tian Y, Ng K-W, Datta A. Stochastic analysis of the interplay between object maintenance and churn. Elsevier Journal of Computer Communications, Special Issue on Foundations of Peer-to-Peer Computing. Elsevier; 2007.

Peer-to-Peer System

Wojciech Galuba and Sarunas Girdzijauskas
EPFL, Lausanne, Switzerland

Synonyms

Peer-to-peer network; Peer-to-peer overlay

Definition

A peer-to-peer system is a computer network which enables peers to share the network resources, computational power and data storage, without relying on a central authority. Most commonly, peer-to-peer systems form *overlay networks* deployed in the Internet and are used for file sharing, realtime data streaming and computationally intensive tasks.

Key Points

In contrast to client-server systems, peer-to-peer systems consist of interconnected peers of similar capabilities and responsibilities, where the peers can act as both servers and clients. Most commonly, the architecture of the peer-to-peer systems is flat and all peers are assumed to be functionally equal. However, a number of peer-to-peer systems employ hierarchical architecture where some peers (superpeers) act as local servers for the subsets of regular peers. It is also widely accepted in peer-to-peer systems, where some services can be provided by a centralized component, in particular system bootstrapping, authentication etc.

Peer-to-peer systems are designed to be self-organizing and to scale well with the number of participating peers. Each peer contributes a different amount of resources and peer-to-peer systems employ various methods to distribute the load evenly across these resources. Good load balancing is crucial to the scalability of peer-to-peer systems. To account for frequent peer departures and arrivals and to maintain high availability peer-to-peer systems replicate the services and data across several peers for redundancy.

Peer-to-peer systems have a wide range of applications in such areas as the Internet infrastructure (e.g., DNS), file sharing (e.g., Kazaa [3]), communication and data streaming (e.g., Skype [4]), distributed computing (e.g., BOINC [1]), and can even make use of human presence at the participating peers (e.g., Galaxy Zoo project [2]). In addition, the decentralized nature of peer-to-peer systems allows the peer-to-peer applications to be highly resistant to censorship.

High robustness, scalability and lack of the single-point-of-failure make peer-to-peer systems a viable alternative to large client-server systems.

Cross-References

► Distributed Hash Table

Recommended Reading

1. Berkeley open infrastructure for network computing homepage. http://boinc.berkeley.edu/.
2. Galaxy zoo homepage. http://galaxyzoo.org/.
3. Kazaa homepage. http://www.kazaa.com/.
4. Skype homepage. http://www.skype.com/.

Peer-to-Peer Web Search

Gerhard Weikum
Department 5: Databases and Information
Systems, Max-Planck-Institut für Informatik,
Saarbrücken, Germany

Definition

The peer-to-peer (P2P) computing paradigm is an intriguing alternative to centralized search engines for querying and ranking Web content. In a P2P network with many thousands or millions of computers, every peer can have locally compiled content such as recently visited or thematically gathered Web pages, and can employ its own, potentially personalized search engine on the locally indexed data. In addition, queries can be forwarded to judiciously chosen other peers for collaborative evaluation. Such P2P architectures enable keyword search and result ranking on the network-wide global content. Thus, all peers together form a P2P search engine. Conversely, such P2P architectures could be utilized for scalable, distributed implementations of Web indexing with appropriate partitioning across the nodes of a large server farm.

Historical Background

Peer-to-peer (*P2P*) *systems* aim to provide scalable and self-organizing ways of loosely coupling thousands or millions of computers in order to jointly achieve some global functionality [14]. In the last decade, very successful systems of this kind have been built. Most notably, these include *file-sharing networks* such as Gnutella or BitTorrent, and IP telephony like Skype and other collaborative messaging services. They organize peers in so-called *overlay networks* on top of the standard Internet infrastructure, and use various forms of *epidemic dissemination* or distributed data structures like *distributed hash tables* (*DHTs*) such as Chord or Pastry. The basic functionality that underlies many of these systems is the distributed and dynamic maintenance of a dictionary with efficient support for exact-match key lookups.

Web search requires much richer functionality than exact-match lookups: keyword queries combine multiple dimensions of a very-high-dimensional data space in an ad hoc manner so that standard multi-dimensional indexes are not applicable, and they require ranking of query results based on statistics about local and global keyword frequencies. On the other hand, it seems very natural to build Web search in a P2P manner as the producers and owners of Web pages are widely distributed and autonomous. In fact, the standard approach of crawling the Web for centralized search engines can be seen as rather artificial, but has advantages regarding system management and commercial services such as query-related advertisements.

Approaches to P2P Web search are reminiscent of earlier work on *distributed information retrieval* (*IR*), most notably, various kinds of *metasearch engines* where queries are routed to judiciously chosen search providers [15]. However, the P2P setting is much more challenging regarding the enormous scale of the underlying data sources, the dynamics of the system, and the autonomy of the individual peers.

Foundations

Peer-to-peer (P2P) Web search has been studied from two perspectives: *global computing* (*GC*) and *social computing* (*SC*). The GC approaches consider peers that dedicate all their storage and computing resources to the P2P network. The network as a whole emulates a traditional search engine architecture by partitioning a global Web index and assigning partitions and query load to peers. The SC approaches, on the other hand, consider architectures where each peer corresponds to one user, and emphasizes the autonomy of peers, with every peer having full control over its local content and extent of sharing it with other peers. Both GC and SC can be embedded in a server-farm environment with a high-speed physical network or in a geographically distributed environment over the wide-area Internet.

P2P Global Computing for Scalable Search-Engine Functionality

For *indexing the Web*, commercial search engines use large data centers consisting of thousands of low-end computers, carefully designed parallelism and redundancy, and customized software with very low overhead [2]. The key technique to sustain a peak throughput of thousands of queries per second with sub-second response times is *data partitioning*. Index entries – postings that consist of a keyword id, a page id, and a score – are hashed on page id, and each resulting hash partition is assigned to one computer in a very large *cluster* (with hundreds or thousands of nodes). A query with one or more keywords is simply sent to all nodes for parallel evaluation. The load is perfectly balanced across the nodes of the cluster, so that the approach scales up extremely well. For failure resilience, high availability, and higher throughput, an entire cluster is often replicated sufficiently.

The above architecture is distributed, but it is not a P2P approach, for it assumes a fixed, reasonably time-invariant system configuration with a high-speed homogeneous network between nodes. If one tried to carry this design over to a setting with a dynamically varying number of peers that communicate over a high-latency wide-area network, the fine-grained parallelism would probably result in a poor cost/performance ratio. Thus, a geographically distributed architecture for Web indexing needs more sophisticated techniques and is still viewed as a research challenge, but promising work is on its way [1].

The current approaches differ in their ways of partitioning the data across peers, the overlay networks that are employed, their load balancing mechanisms, and their methods for caching and replication, which in turn influences query processing strategies. As for *data partitioning*, three common ways are: hash-partitioning on page ids (documents), hash- partitioning on keywords (terms), or thematic clustering based on page contents or query logs. *Load balancing* needs to counter skewed distributions of both index-list lengths and query popularities for different keywords; many techniques from the distributed computing literature are applicable. *Caching* can consider different granularities like index lists for individual keywords or entire query results, and needs to employ appropriate strategies for cache refreshing and replacement. Finally, efficient algorithms for *distributed top-k query processing* are called for.

P2P Social Computing with Autonomous Peers

In the SC line of work, an architecture is pursued where every peer has a powerful local search engine, with its own crawler, indexer, and query processor. Such a peer can compile its own content from thematically focused crawls and other sources, and make this content available in a P2P overlay network. Search requests issued by a peer can first be executed locally, on the peer's locally indexed content. When the recall of the local search result is unsatisfactory, the query can be forwarded to a small set of other peers that are expected to provide thematically relevant, high-quality and previously unseen results. Deciding on this set of target peers is the *query routing* problem in a P2P search network, also known as *collection selection*. Subsequently, the actual search on the chosen target peers requires efficient algorithms for *top-k query processing*. Search results are then returned by different peers and need to be meaningfully merged, which entails specific problems for *result ranking*. Both query routing and result ranking can build on various forms of *distributed statistics*, computed in a decentralized way and aggregated and disseminated in a scalable P2P manner (using compact synopses such as Bloom filters or hash sketches and leveraging the DHT infrastructure).

Query Routing

A practically viable query routing strategy needs to consider the similarity of peers in terms of their thematic profiles, the overlap of their contents, the potential relevance of a target peer's collection for the given query, and also the costs of network communication and peer-specific processing loads.

Traditionally, the most important measure for assessing the benefit of a candidate target peer for a given query is the estimated relevance of the peer's overall content for the query. This standard IR measure can be estimated frequency statistics over the query's keywords (terms in IR jargon). In conventional, document-oriented IR, these would be term frequencies (tf) within a document and the so-called inverse document frequencies (idf), the reciprocal of the global number of documents that contain a given term. In P2P IR, the estimation is based on the overall content of a peer as a whole. Instead of tf, the *document frequency* (*df*) of a peer is considered, which is the total number of documents that contain the term and are in the peer's collection; and instead of idf, the *inverse collection frequency*

(*icf*) is considered, which is the reciprocal of the total number of peers that contain (at least one document with) the term. These basic measures are combined into a relevance or query-specific quality score for each candidate peer. There are various models for the combined scores; among the most cited and best performing models are *CORI* [4], based on probabilistic IR, and statistical language models adapted to the setting of P2P collection selection [9]. The Decision-Theoretic Framework (DTF) [11] provides a unified model for incorporating quality measures of this kind as well as various kinds of cost measures.

Selecting peers solely by query relevance, like CORI routing, potentially wastes resources when executing a query on multiple peers with highly overlapping contents. To counter this problem, methods for estimating the overlap of two peers' contents have been developed. These estimates are then factored into an *overlap-aware query routing* [10] method by using a weighted combination of peer quality and overlap (or novelty) as ranking and decision criterion.

Query routing decisions are typically made at query run-time, when the query is issued at some peer. But the above methods involve directory lookups, statistical computations, and multi-hop messages; so it is desirable to precompute preferred routing targets and amortize this information over many queries. A technique for doing this is to encode a similarity-based precomputed binary relation among peers into a *Semantic Overlay Network* (*SON*) [5]. The routing strategy would then select target peers only or preferably from the SON neighbors of the query initiator.

Search Result Ranking

When a query returns results that have been obtained from different peers, the scores that the peers assign to them are usually not comparable. The reason is that different peers may use different statistics, for example, for estimating the idf value of a term which is crucial for weighting the importance of different query terms, or they may even use completely different IR models. This situation leads to the problem of *result merging*. It is addressed by re-normalizing scores from different peers to make results meaningfully comparable [15]. A variety of such methods exist in the literature, some using only the peer-specific scores and some aggregated measures about peers (e.g., the total number of documents per peer), some using sampling-based techniques, and some using approaches that first reconstruct the necessary global statistics (e.g., global document frequencies for each term) for optimal re-normalization of scores.

Web search ranking usually also considers the query-independent authority of pages as derived from link analysis, and a P2P network is a natural habitat for such "social ratings." *Link analysis* algorithms such as PageRank are centralized algorithms with very high memory demand. Executing them in a distributed manner would allow scaling up to even larger link graphs, by utilizing the aggregated memory of a P2P system. Various decentralized methods have been developed to this end, including a general solution to the spectral analysis of graphs and matrices [8], which underlies the PageRank computation.

Most of these methods assume that the underlying Web graph can be nicely partitioned among peers. In contrast, a P2P system with autonomous peers faces a situation where the Web pages and links that are known to the individual peers are not necessarily disjoint. The JXP algorithm [12] computes global authority measures such as PageRank in a decentralized and scalable P2P manner, when the Web graph is spread across autonomous peers and peers' local graph fragments overlap arbitrarily, and peers are (a priori) unaware of other peers' fragments. The scores computed by JXP provably converge to the same values that would be obtained by a centralized PageRank computation on the full Web graph. These kinds of algorithms seem to be highly relevant also for analyzing authority and reputation measures in large-scale social networks.

Key Applications

The technology for P2P Web search has many potential applications, ranging from keyword search on the Web or in blogs, possibly in a personalized manner and exploiting social-network affinities among users, all the way to more "semantic" search capabilities in large enterprises, scholarly communities, federations of digital libraries, and distributed Internet archives. The latter may include searching XML, RDF, or multimedia data as well as providing temporal querying and other forms of enhanced search capabilities. With richer functionality, centralized systems face scalability bottlenecks, whereas decentralized approaches can leverage the fact that the underlying information is naturally distributed at a large scale.

Future Directions

Today, there is still no P2P Web search system that would scale anywhere near the sizes of the major commercial search engines. But as the Web continues to grow, search functionality is becoming richer (e.g., by personalization), and workloads are becoming much more demanding, the P2P paradigm is likely to gain more momentum for Web applications. Currently, there is a variety of research prototypes (e.g., [3, 6, 7, 13]), including some that offer open-source software.

Cross-References

▶ Distributed Hash Table
▶ Peer Data Management System
▶ Social Networks

Recommended Reading

1. Baeza-Yates RA, Castillo C, Junqueira F, Plachouras V, Silvestri F. Challenges on distributed web retrieval. In: Proceedings of the 23rd International Conference on Data Engineering; 2007. p. 6–20.

P

2. Barroso LA, Dean J, Hölzle U. Web search for a planet: the Google cluster architecture. IEEE Micro. 2003;23(2):22–8.

3. Bender M, Michel S, Parreira JX, Crecelius T. P2P web search: make it light, make it fly. In: Proceedings of the 3rd Biennial Conference on Innovative Data Systems Research; 2007. p. 164–8.

4. Callan JP, Lu Z, Croft WB. Searching distributed collections with inference networks. In: Proceedings of the 18th Annual Interntional ACM SIGIR Conference on Research and Development in Information Retrieval; 1995. p. 21–8.

5. Crespo A, Garcia-Molina H. Semantic overlay networks for P2P systems. In: Proceedings of the 3rd International Workshop Agents and Peer-to-Peer Computing; 2004. p. 1–13.

6. Cuenca-Acuna FM, Peery C, Martin RP, Nguyen TD. PlanetP: using gossiping to build content addressable peer-to-peer information sharing communities. In: Proceedings of the 12th IEEE International Symposium High Performance Distributed Computing; 2003. p. 236–49.

7. Kalnis P, Ng WS, Ooi BC, Tan K-L. Answering similarity queries in peer-to-peer networks. Inf Syst. 2006;31(1):57–72.

8. Kempe D, McSherry F. A decentralized algorithm for spectral analysis. In: Proceedings of the 36th Annual ACM Symposium on Theory of Computing; 2004. p. 561–8.

9. Lu J, Callan JP. Content-based retrieval in hybrid peer-to-peer networks. In: Proceeings of the International Conference on Information and Knowledge Management; 2003. p. 199–206.

10. Michel S, Bender M, Triantafillou P, Weikum G. IQN routing: integrating quality and novelty in P2P querying and ranking. In: Advances in Database Technology, Proceedings of the 10th International Conference on Extending Database Technology; 2006. p. 149–66.

11. Nottelmann H, Fuhr N. Comparing different architectures for query routing in peer-to-peer networks. In: Proceedings of the 28th European Conference on IR Research; 2006. p. 253–64.

12. Parreira JX, Donato D, Michel S, Weikum G. Efficient and decentralized pageRank approximation in a peer-to-peer web search network. In: Proceedings of the 32nd International Conference on Very Large Data Bases; 2006. p. 415–26.

13. Podnar I, Rajman M, Luu T, Klemm F, Aberer K. Scalable peer-to-peer web retrieval with highly discriminative keys. In: Proceedings of the 23rd International Conference on Data Engineering; 2007. p. 1096–105.

14. Steinmetz R, Wehrle K. Peer-to-peer systems and applications. Springer; 2005.

15. Weiyi M, Yu CT, Liu K-L. Building efficient and effective metasearch engines. ACM Comput Surv. 2002;34(1):48–89.

Performance Analysis of Transaction Processing Systems

Alexander Thomasian
Thomasian and Associates, Pleasantville, NY, USA

Synonyms

Cache performance; Concurrency control; Probabilistic analysis; Queueing analysis; Storage systems

Definition

The performance of *transaction (txn) processing (TP)* systems and more generally *database management systems (DBMSs)* is measured on operational systems, prototypes, and benchmarks. Probabilistic and queueing analyses have been used to gain insight into TP system performance, but also to develop capacity planning tools. The following is considered: (i) queueing analysis of processors and disks, (ii) *queueing network models (QNMs)* of computer systems, (iii) techniques to estimate the database buffers miss rate, (iv) factors affecting RAID performance, (v) *concurrency control (CC)* methods for high data contention TP systems and their analyses.

Historical Background

Early performance studies of TP were concerned with processor or *central processing unit (CPU)* scheduling. *Queueing network models – QNMs* were developed in the 1970s to estimate delays at active computer system resources, i.e., CPU and disks. The effect of passive resources, such as the memory size constraint, was later incorporated into capacity planning tools for TP systems [5]. The *Transaction Processing Council's (TPC's)* debit-credit benchmark in 1985 compares the performance of TP systems using their throughput at which a certain txn response time is reached (http://www.tpc.org). Buffer/cache management policies in processors, databases, and disk controllers have been studied since the 1970s. Coherence issues of CPU caches in multiprocessors and database buffers in shared disk systems gained importance in 1980s. Magnetic disks invented in 1950s have a significantly improved capacity and transfer rate, but not random access time. This led to the proposal and analysis of numerous disk arm scheduling policies. The 1988 *Redundant Array of Independent Disks (RAID)* classification provided renewed impetus to improve the performance and reliability of storage systems [1]. Shared-everything, −disk, and -nothing computer organizations have limitations for high-performance TP and DBMS applications, i.e., the first two pose cache coherence problems and shared-nothing systems are susceptible to processing time skew. (For example, there is a twofold increase with exponentially distributed processing times at four nodes, since $\sum_{i=1}^{4} 1/i = 2.08$.) CC has limited effect on TP performance in modern DBMSs, but this was not so in early relational DBMSs with coarse granularity locking, therefore many CC methods were proposed and evaluated [12].

Foundations

CPU Scheduling. Processing of txns at the CPU has been modeled by an M/G/1 queueing model, which consists of a *First-Come, First-Served (FCFS)* queue and a single server [4]. M stands for Poisson arrivals with rate λ, G stands for a general service time distribution with $\overline{x^i}$ as the ith moment. The server utilization is $\rho = \lambda \overline{x} < 1$, the mean waiting time at the queue is $W_{M/G/1} = \lambda \overline{x^2} / (2(1-\rho))$, and the mean response time $R = W_{M/G/1} + \overline{x}$ [4]. M/M/1 is a special case of M/G/1 with an exponential service time distribution with $\overline{x^2} = 2(\overline{x})^2$, so that the mean of the exponentially distributed response time is $R_1 = \overline{x}/(1-\rho)$. M/M/m has m servers, hence $\rho = \lambda \overline{x}/m$. For $m = 2$ there is $R_2 = \overline{x}/\left(1-\rho^2\right)$ (for $m > 2$ see [4]). For a single-server which is twice as fast as the previous servers: $R_3 = (\overline{x}/2)/(1-\rho)$.

The same ρ is maintained in the three systems with R_i, $1 \leq i \leq 3$, by setting the arrival rate to 2λ, but utilizing two M/M/1 queues with uniform routing in the first case, so that the total processing capacity is $2\,\mu$ in all three cases. There is $R_1 > R_2 > R_3$, where the first inequality reflects the resource sharing advantage of a shared queue, while the second inequality reflects the advantage associated with a single fast server when service times are not highly variable [4]. For K fork-join requests initiated at K M/M/1 queues the expected value of the maximum of K response times is: $R_k^{max} = H_K R$, where $H_K = \sum_{k=1}^{K} 1/k$ is the Harmonic sum. The components of a fork-join request are correlated, so that $R_2^{F/J} < R_k^{max}$, e.g., $R_2^{F/J} = (1.5 - \rho/8)R < R_2^{max}$. An approximate expression for $R_k^{F/J}$ for $K > 2$ has been used in the analysis of RAID5 disk arrays. With two txn classes the response time for the more urgent (class 1) txns can be improved by processing them at a higher priority than less urgent (class 2) txns. The arrival rates are λ_1 and λ_2 and the ith moments of service time are $\overline{x_1^i}$ and $\overline{x_2^i}$, respectively. With preemptive priorities and a negligible preemption overhead, class 1 txns are processed without being affected by class 2 txns. With nonpreemptive priorities $W_1 = W_0/(1-\rho_1)$ and $W_2 = W_0/((1-\rho_1)(1-\rho))$, where $\rho_j = \lambda_j \overline{x}_j$, $j = 1, 2$ and $W = \frac{1}{2} \sum_{j=1}^{2} \lambda_j \overline{x_j^2}$ is the remaining processing time of txns at the CPU [4]. Note that W_1 is only affected by ρ_1, but not $\rho = \rho_1 + \rho_2$. Kleinrock's *conservation law* states that the improvement in waiting time of one txn class is at the cost of the other, so that the weighted sum of waiting times remains constant: $\sum_{j=1}^{2} \rho_j W_j = W/(1-\rho)$. Preemptive (resp. nonpreemptive) priorities are applicable to CPU (resp. disk) scheduling.

Queueing Network Models: QNMs

In *open QNMs* there are txn arrivals and departures, while in a *closed QNM* a completed txn is immediately replaced by a new txn, so that the number of txns remains fixed at N. Txns are processed at $K > 1$ nodes, where each node is a single or multiserver queueing system. The QNM of a computer system may be organized as the *Central Server Model (CSM)*, with the CPU the central server and the disks as peripheral servers. Txns arrive at the CPU (node 1), access one of the $K - 1$ disks with probability p_k, $2 \leq k \leq K$, return to the CPU for additional processing, and leave the system after CPU processing with probability p_1, so that $\sum_{k=1}^{K} p_k = 1$. The transition probability matrix yields the mean number of visits to the nodes: i.e., $v_1 = 1/p_1$ and $v_k = p_k/p_1, 2 \leq k \leq K$. QNMs satisfying the BCMP theorem allow four types of nodes, most notably exponential service times at single- or multi-server queues with FCFS scheduling and nodes with an infinite number of servers [4, 5]. The steady-state probability $Pr[n_1, n_2, \ldots, n_K]$ in such QNMs can be expressed in product-form and their performance metrics computed efficiently. Approximation techniques or simulation can be applied otherwise.

The mean residence times at the nodes of an open product-form QNM can be obtained separately. With an external arrival rate Λ, the arrival rate to node k is $\lambda_k = v_k \Lambda$, its utilization $\rho_k = \lambda_k \overline{x}_k / m_k$, where $\overline{x}_k$ is its mean service time and m_k is the number of its servers. The service demand or the total txn processing time at a node k is given as $D_k = v_k \overline{x}_k$, $1 \leq k \leq K$, so that alternatively $\rho_k = \Lambda D_k$. The mean txn residence time at node k is $r_k = \overline{x}_k / \left(1 - \rho_k^{m_k}\right), 1 \leq k \leq K$ with $m_k \leq 2$. The mean txn response time is $R = \sum_{j=1}^{K} v_k r_k$ and the mean number of txns at the system following Little's result is the product of the arrival rate and the mean response time: $\overline{N} = \Lambda R$ [4, 5]. $\overline{N}$ can be used to estimate the degree of txn concurrency or *Multiprogramming Level (MPL)* and the memory size requirements for TP.

Closed QNMs can be analyzed via the convolution or the *mean value analysis – MVA* algorithm [4, 5]. The MVA *arrival theorem* states that in a closed QNM with n txns the mean queue-length at node k at an arrival instant, is that of a closed system with $n - 1$ txns, i.e., $A_k (n) = Q_k (n - 1)$ [5]. The mean queue-length includes the request at the server, whose mean residual service time is the same as its service time due to the memoryless property of the exponential distribution: $\overline{x}'_k =$

$\overline{x^2}_k / (2\overline{x}_k) = 2(\overline{x}_k)^2 / (2\overline{x}_k) = \overline{x}_k$ [4]. The mean residence time at node k for v_k visits is $R_k(n) = v_k r_k(n) = v_k \overline{x}_k [1 + Q_k (n - 1)] = D_k [1 + Q_k (n - 1)]$.

- *MVA Algorithm for Closed QNM with Single Servers.*
- Input parameters: N (MPL), K (number of nodes), D_k, $1 \leq k \leq K$ (mean service demands).
- for $k = 1$ to K do $Q_k (0) = 0$ (initialize queuelengths at $n = 0$)
- for $n = 1$ to N do {(vary the number of txns)
- for $k = 1$ to K do {(vary the node index)
  - if delay server $R_k (n) = D_k$ (no queueing at "infinite servers")
  - if single server $R_k (n) = D_k [1 + Q_k (n - 1)]$}
- $R(N) = \sum_{k=1}^{K} R_k (n)$ (txn residence time)
- $X(n) = n/R(n)$ (txn throughput using Little's result)
- for $k = 1$ to K do $Q_k (n) = X(n)R_k (n)$ (mean queuelength at the nodes, Little's result)}

The convolution algorithm applied to single-server queues computes $G(0: N)$, after the initialization $G(0) = 1$ and $G(n) = 0$, $n \neq 0$, as: $G(n) = G(n) + D_k G(n - 1)$, $1 \leq n \leq N$, $1 \leq k \leq K$, so that $X(N) = G(N - 1)/G(N)$. Compared to MVA the order of the iterations on n and k are reversed. Given that $\sum_{k=1}^{K} D_k = $ constant, the throughput of a closed QNM is maximized when the service demands at all the nodes are equal. Due to symmetry $Q_k (N) = (N - 1)/K$, so that $R(N) = KD(1 + (N - 1)/K) = (N + K - 1)D$. Next, consider Poisson arrivals to a QNM with a maximum degree of concurrency M due to limited memory size. For $n < M$ arriving txns are activated immediately and otherwise enqueued at a FCFS memory queue. A hierarchical solution method is applicable in this case [5]. The lower level model yields the throughput characteristic of the closed QNM: $X(n)$, $1 \leq n \leq M$. Additional txns waiting in the memory queue do not contribute to throughput, so that $X(n) = X(M)$, $n > M$. A birth-death model is used at the higher level, with the state n denoting the number of txns at the system. The arrival rate at all states is Λ and the completion rate at state n is $X(n)$, $n \geq 1$. The probability that there are n txns in the system is: $P_n = \Lambda P_{n-1}/X(n)$, $n \geq 1$. The normalization condition $\sum_{n \geq 0} P_n = 1$ yields P_0 and hence $P_n, \forall n > 0$. The mean number of txns at the computer system, mean number of txns at the memory queue, and the mean number of activated txns is: $\overline{N} = \sum_{n>0} n P_n$, $\overline{N}_B = \sum_{n>M}(n - M) P_n$, and $\overline{N}_A = \sum_{n>0} \min(n, M) P_n$. The mean txn response time, mean waiting time at the memory queue, and the mean residence time at the computer system are, respectively: $R = \overline{N}/\Lambda$, $W_M = \overline{N}_B/\Lambda$, and $R_A = \overline{N}_A/\Lambda$. $\overline{N} = \overline{N}_B + \overline{N}_A$, so $R = W_M + R_A$. The throughput characteristic $X(n)$, $n \geq 1$ vs. n (without an MPL constraint) is a convex function, which reaches an asymptotic value $\min\{m_k/D_k, \forall k\}$. The node with the smallest throughput is the bottleneck resource, since it determines the maximum system throughput Λ_{max}. The throughput may drop due to thrashing in an overloaded virtual memory system [5] or due to excessive lock contention with the standard locking method.

Buffer Miss Rate (BMR)

BMR is applicable to the various levels of the memory hierarchy. Misses at the database buffer and disk controller cache result in disk accesses, which significantly degrade the performance of TP systems. BMR is affected by the buffer size, the replacement policy, and the workload. Analyses of replacement policies, such as FIFO, LRU, and CLOCK, using Markov chain modeling are based on the *Independent Reference Model (IRM)*. (http://www.informatik. uni-trier.de/~ley/db/dbimpl/buffer.html.) Data streams can also be characterized by stack depth and the hierarchical reuse or fractal models. Trace-driven simulation is a straightforward technique for evaluating the performance of buffer management policies, which can be used for calibrating empirical formulas. A recent top-down analysis derives an equation for the number of page faults (n) vs. memory size (M) [9]. M_0 and M^* are the min and max M that are sufficient and necessary. n^* is the number of cold memory misses, i.e., misses starting with an empty buffer.

The parameters vary with program behavior and replacement policy.

$$n = \frac{1}{2}\left[H + \sqrt{H^2 - 4}\right](n^* + n_0) - n_0,$$

$$H = 1 + \frac{M^* - M_0}{M - M_0}, M \leq M^*.$$

Performance Analysis of Storage Systems

RAID performance is improved by striping and caching. Striping eliminates disk access skew by partitioning large datasets into fixed size *stripe units (SUs)*, which are allocated in round-robin manner across all disks [1]. The RAID controller cache complementing the database buffer in main memory is partially *nonvolatile storage (NVS)* and is used for *fast writes*, which are as reliable as writing to disk. Disk writes can be deferred and processed at a lower priority than reads, since read response times affect txn response time. Dirty data blocks in NVS are possibly overwritten several times, before they are destaged or written to disk. This eliminates unnecessary disk accesses. Applying disk scheduling to batched destaging of multiple blocks leads to higher disk access efficiency. Both effects can be captured by trace analysis [15].

RAID1 or disk mirroring, which predates the RAID classification, replicates data on two disks to achieve fault-tolerance. A data block can be read from either disk, but both copies should be eventually updated. Disk bandwidth can be saved by writing modified blocks to NVS first. The doubling of data access bandwidth is beneficial in view of increasing disk capacities, but an even further improvement in bandwidth can be attained by judicious routing of requests. When one of mirrored disks fails the read load on the surviving disk is doubled, but a smaller load increase can be attained with *interleaved declustering*, which replicates the contents of one disk on the other $n - 1$ disks of the cluster, so that the read load increase is $n/(n - 1)$ [13].

Erasure coding in RAID5 and RAID6 uses one and two check disks to protect against single and double disk failures, respectively [1]. RAID5 uses parity coding, so that the check SU is the *exclusive-OR (XOR)* of the remaining SUs in the same stripe or row. RAID6 uses Reed-Solomon or specialized parity codes to protect two disks out of N. The updating of a data block requires the corresponding check block(s) to be computed and updated. If the old copies of data and check blocks are not cached, four (resp. six) disk accesses are required for RAID5 (resp. RAID6), this is therefore referred to as the *small write penalty* [1]. Check SUs are placed in repeating left-to-right diagonals to balance the load for updating the check blocks. When one of the N disks fails, RAID5 continues its operation in degraded mode, by reconstructing missing blocks of the broken disk on demand as follows. The controller issues a fork-join request to the $N - 1$ corresponding blocks on surviving disks and these blocks are XORed to reconstruct the missing block. The read load on surviving disks is doubled, but the load increase for write requests is smaller. The *clustered RAID* paradigm reduces the disk load in degraded mode by using a parity group size $G < N$, so that the increase in read load is $\alpha = (G - 1)/(N - 1) < 1$ [1]. The RAID5 rebuild process reconstructs the contents of a failed disk track-by-track on a spare disk, by reading successive tracks from surviving disks, XORing all corresponding tracks to reconstruct a missing track, and writing it on the spare disk [15]. Distributed sparing distributes the spare areas among $N + 1$ disks. so that the bandwidth of the spare disk is not wasted [15]. The *vacationing server method (VSM)* starts reading tracks from disks when they become idle, so the rebuild process does not affect the disk load. No new tracks are read after there is an external arrival, but the reading of the current track is completed. The increase in mean disk waiting time over M/G/1 is the mean residual time to read a track: $W_{VSM} = W_{M/G/1} + \overline{x}'_{track}$ [4]. RAID6 deals with the possibility of unsuccessful rebuild, when unreadable sectors or *latent sector failures – LSFs* are encountered during the rebuild process or to cope with rare two disk failures.

Mean disk response time with FCFS disk scheduling can be obtained using the M/G/1 queuing model [4]. Random disk accesses incur a seek, latency or rotational delay (half a rotation time for small blocks), and transfer time (negligible for small blocks). One has $x_{disk} = x_{seek} + x_{latency} + x_{xfer}$. The moments of disk access time can be obtained under various assumptions, e.g., all disk cylinders are accessed uniformly. A significant reduction in disk response time is possible via the *Shortest Access Time First (SATF)* policy [14]. Given n random disk requests, the disk access time drops as $n^{1/5}$, e.g., $n = 32$ halves the access time with respect to FCFS.

Performance Analysis of Concurrency Control Methods

Performance degradation due to CC is in the form of txn blocking and restarts. The *standard locking* method is based on the strict *two-phase locking (2PL)* method [3], i.e., locks are requested on demand and held to completion, at which point the txn commits, releasing all of its locks. Locks are also released when a txn is aborted to resolve deadlocks or other reasons [3]. Restart-oriented locking methods and *optimistic CC (OCC)* methods are also discussed.

Standard Locking

Consider an abstract model of standard locking [11]. There are M txns in a closed system. A txn of size k has $k + 1$ steps with mean duration s, so that the mean txn residence time with no lock contention is: $r_k = (k + 1)s$. A lock is requested at the end of the first k steps and all locks are released at the end of the $k + 1$st step, i.e., strict 2PL, so that lock requests are uniformly distributed over the lifetime of a txn. The mean number of locks held by a txn is the ratio of the time-space of the number of locks held by a txn and r_k, which is an approximation to R_k at lower levels of lock contention, which yields: $\overline{L}_k \approx k/2$. Assuming that all lock requests are exclusive and are uniformly distributed over the D objects in the database, the probability of lock conflict is: $P_c(k) \approx (M - 1)\overline{L}_k/D$. When the fraction of shared lock requests is f_S, the analysis can proceed as if all the lock requests are exclusive with an effective database size $D_{eff} = D/(1 - f_S^2)$ [8]. The probability that a txn encounters a lock conflict for $P_c(k) \ll 1$

can be used to obtain the probability of a two-way deadlock [3]:

$$P_w(k) = 1 - (1 - P_c(k))^k$$

$$\approx k P_c(k) \approx \frac{(M-1)k^2}{2D}.$$

$$P_{D(2)}(k) = \frac{1}{M-1} \Pr[T_1 \to T_2] \Pr[T_2 \to T_1]$$

$$= \frac{(P_w(k))^2}{M-1} \approx \frac{(M-1)k^4}{4D^2}.$$

A two-way deadlock can only occur if txn T_A requests a lock held by txn T_B, which is blocked because of its previous lock conflict with T_A. Multiway deadlocks are much less common and are ignored [3]. We obtain the mean blocking delay of a txn with respect to an active txn holding the requested lock ($W_k^{(1)}$) by noting that the probability of lock conflict increases with the number of locks held by the txn (j) and that there are ($k - j$) remaining steps each with mean duration ($s + u$) [10, 12]:

$$W_k^{(1)} \approx \sum_{j=1}^{k} \frac{2j}{k(k+1)}[(k-j)(s+u)]$$

$$+ s' = \frac{k-1}{3}[r+u] + s'.$$

$u = P_c(k)W_k$ is the mean blocking time per step, where W_k is the mean waiting time per lock conflict, and s' is the remaining processing time of a step. Since deadlocks are rare, the mean txn response time only takes into account txn blocking time due to lock conflicts: $R_k \approx (k+1)s + kP_c(k)W_k$. The fraction of txn blocking time per lock conflict with an active txn is $A = W_k^{(1)}/R_k \approx 1/3$. A more accurate expression for two-way deadlocks as verified by simulation is: $P'_{D(2)}(k) \approx AP_{D(2)}(k) = P_{D(2)}(k)/3$ [16]. Next consider K txn classes in a closed system, so that a completed txn is replaced by a class k txn with frequency f_k, $\sum_{k=1}^{K} f_k = 1$. The mean response time

over all txn classes is: $R = (K_1 + 1)s + K_1 P_c W$, where $K_i = \sum_{k \geq 0} k^i f_i$ is the ith moment of txn size. The mean number of txns in class k is $\overline{M}_k = MR_k f_k / R$ (the summation $\sum_{k=1}^{K}$ yields M on both sides). Given that $\overline{M}_k = M(k+1)f_k/(K_1+1) \approx Mkf_k/K_1$, the mean number of locks held per txn is then:

$$\overline{L} = \frac{1}{M} \sum_{k=1}^{K} \overline{L}_k \overline{M}_k \approx \frac{1}{M} \sum_{k=1}^{K} \frac{k}{2} \overline{M}_k$$

$$\approx \frac{1}{2K_1} \sum_{k=1}^{K} k^2 f_k = \frac{K_2}{2K_1}.$$

Similarly to fixed size txns $P_c \approx (M-1)\overline{L}/D$ and the expression for $P_{D(2)}$ is given in [16, 11]. Both P_c and $P_{D(2)}$ are affected by the variability of txn sizes, e.g., for a truncated geometric distribution there is a twofold and tenfold increase with respect to fixed txn sizes, respectively [16]. W_1 for variable txn sizes can be obtained as the expected value of $W_1(k)$ over all txn sizes. The normalized value for W_1 is [10, 12]:

$$A = \frac{W_1}{R} \approx \frac{K_3 - K_1}{3K_1(K_2 + K_1)}.$$

Denote the fraction of blocked txns with β and ignore the difference in the number of locks held by active and blocked txns. The probability that a txn has a lock conflict with an active txn at level $i = 0$ (resp. has a lock conflict with a txn blocked at level $i = 1$) is $1 - \beta$ (resp. β) and the mean blocking time is W_1 (resp. $W_2 = 1.5 W_1$). The expression for W_2 relies on the fact that at the random instant when the lock conflict occurs the active txn in the waits-for-graph is halfway to its completion. Generally, the probability that a txn is blocked at level i is: $P_b(i) \approx \beta^{i-1}, i > 1$ and $P_b(1) = 1 - \beta - \beta^2 - \beta^3 \ldots$. The mean waiting time at level $i > 1$ is $W_i \approx (i - 0.5)W_1$, which is motivated by the expression for W_2. The mean blocking time is a weighted sum of delays incurred by txns blocked at different levels:

$$W = \sum_{i \geq 1} P_b(i) W_i$$

$$= W_1 \left[1 - \sum_{i \geq 1} \beta^i + \sum_{i > 1} (i - 0.5) \beta^{i-1} \right].$$

Let $\overline{M}_A$ and $\overline{M}_B$ denote the mean number of active and blocked txns, so that $\overline{M}_A + \overline{M}_B = M$. The fraction of blocked txns is $\beta = \overline{M}_B/M$. Dividing $\overline{M}_B$ and M by the system throughput yields $\beta = K_1 P_c W/R$. Multiplying both sides of the above equation by $K_1 P_c /R$ yields β on the left hand side and $\alpha = K_1 P_c W_1/R = A K_1 P_c$ on the right hand side. Note that α is the product of the mean number of lock conflicts per txn and the normalized waiting time with respect to active txns. It follows: $\beta = \alpha(1 + 0.5\beta + 1.5\beta^2 + 2.5\beta^3 + \ldots)$. A closed-form expression for β can be obtained by noting that $\beta < 1$ and assuming that the series is infinite:

$$f(\beta) = \beta^3 - (1.5\alpha + 2)\beta^2$$
$$+ (1.5\alpha + 1)\beta - \alpha = 0.$$

Plotting $f(\beta)$ vs. β it is observed that the cubic equation has three roots $0 \leq \beta_1 < \beta_2 \leq 1$ and $\beta_3 \geq 1$ for $\alpha \leq \alpha^* = 0.226$, which makes its discriminant $f(\alpha) = \alpha^3 + \frac{4}{5}\alpha^2 + \frac{28}{15}\alpha - \frac{64}{135} < 0$ (http://en. wikipedia.org/wiki/ Cubic_equation.), and a single root $\beta_3 > 1$ for $\alpha > \alpha^*$. *The single parameter α determines the level of lock contention for the standard locking and its critical value α^* determines the onset of thrashing.* The smallest root β_1 for $\alpha \leq \alpha^*$ determines $\overline{M}_A = M(1 - \beta_1)$, which increases with M, but drops after a maximum value is achieved. This is because of a snowball effect that blocked txns cause further txn blocking. Given the throughput characteristic $X(n)$, $n \geq 1$ with no lock contention, the throughput of a TP system when $\overline{M}_A$ is non-integer is: $X(\overline{M}_A) = (j\overline{M}_A k - \overline{M}_A) X(l\overline{M}_A m) + (\overline{M}_A - l\overline{M}_A m) X(j\overline{M}_A k)$. The analysis for txns with variable step durations uses iteration to estimate ρ, which is the ratio of the number locks held by blocked and active txns. It is shown

in [11] that the critical value of ρ matches the *conflict ratio* parameter in [17], which is the ratio of the number of locks held by all and active txns.

Restart-Oriented Locking Methods

Performance degradation in standard locking is mainly caused by txn blocking. Txn aborts and restarts to resolve deadlocks have little effect on performance, because deadlocks are rare. Restart-oriented locking methods follow the strict 2PL paradigm, but txns encountering lock conflicts are restarted to increase the degree of txn concurrency, so that a higher txn throughput can be attained. This is beneficial for high lock contention TP systems with excess processing capacity, but in hardware resource bound systems this may result in performance degradation with respect to standard locking. Since active txns are not guaranteed to complete successfully and commit, txn throughput cannot be expressed as $X(\overline{M}_A)$, as in the case of standard locking. The *Running Priority* (RP) method aborts txn T_B in the waits-for-graph $T_A \to T_B \to T_C$, when T_A has a lock conflict with T_B [2]. Similarly T_B which is already blocking T_A is aborted when it is blocked requesting a lock held by T_C. In effect RP increases the degree of txn concurrency even if the restart of an aborted txn is delayed. The no-waiting [8] and cautious waiting methods abort and restart a txn encountering a lock conflict with an active and blocked txn, respectively. There is no benefit, since the restarted txn will encounter the same lock conflict as before. The *Wait Depth Limited* (WDL) method measures txn progress by its length (L), which in [2] is approximated by number of locks that it holds in a lock contention bound system, while attained CPU time would be a concern otherwise. WDL differs from RP in that it attempts to minimize wasted processing by not restarting txns holding many locks or txns nearing completion. This requires a priori knowledge of txn's locking requirements. For example, if T_A which is blocking another txn has a lock conflict with T_B, then if $L(T_A) < L(T_B)$ then abort T_A, else abort T_B [2]. The superior performance of WDL over RP and especially standard locking has been shown via random-number-driven (Monte-Carlo) simulation in [2], but also trace-driven simulation in [17]. RP and WDL attempt to attain *essential blocking*, i.e., allowing a txn to be blocked only by active txns and not requesting a lock until it is required. Immediately restarting an aborted txn may result in *cyclic restarts* or *livelocks*, which should be prevented to reduce wasted processing and to ensure timely txn completion. *Restart waiting* delays the restart of an aborted txn until *all* conflicting txns are completed [2]. Conflict avoidance delays with random duration are a less reliable method to prevent cyclic restarts.

Restart-oriented methods are analyzed using a Markov chain model in [10], while flow-diagrams are utilized in the analysis of the no-waiting method in [8]. Active (resp. blocked) states correspond to $S_{2j}, 0 \leq j \leq k$ (resp. $S_{2j+1}, 0 \leq j \leq k-1$). For example, in the case of the no-waiting policy there is a transition forward when there is no lock conflict: $S_{2j} \rightarrow S_{2j+2}, 0 \leq j \leq k-1$, otherwise the txn is aborted: $S_{2j} \rightarrow S_0$. The state equilibrium equations for the Markov chain yield the steady-state probabilities $\pi_i, 0 \leq i \leq 2k$. The mean number of visits v_i to S_i can be similarly obtained by noting that $v_{2k} = 1$, which is the state at which the txn commits. Given that h_i denotes the mean holding time at S_i, then $\pi_i = v_j h_i / \sum_{j=0}^{2k} v_j h_j$, $0 \leq i \leq 2k$. The mean number of txn restarts is given by v_0-1. A shortcoming of the analytic approach is that requested locks are implicitly resampled.

Optimistic Concurrency Control

A txn starting its execution first copies all objects required for its processing into its private workspace [6]. To ensure serializability after completing its processing, the txn undergoes validation to ascertain that none of the copied objects was modified by another txn after it was read. If validation is successful the txn commits and otherwise it is immediately restarted. Given that the k objects updated by txns are uniformly distributed over a database of size D, the probability of data conflict between two txns of size k is: $\psi = 1 - (1 - k/D)^k \approx k^2/D$. The completion rate of successfully validated txns is $p \mu s(M)$, where p is the probability of successful validation, μ is the completion rate of txns, and

$s(M)$ takes into account the hardware resource contention due to multiprogramming. A txn with processing time x will require $xM/s(M)$ time units to complete. It observes the commits of other txns as a Poisson process with rate $\lambda = (1-1/M)p\mu s(M)$. Txns encountering data conflict with rate $\gamma = \lambda \psi$, fail their validation with probability $q = 1 - e^{-\gamma xM/s(M)}$. The number of txn executions in the system follows a geometric distribution $P_j = q(1-q)^{j-1}, j \geq 1$ with a mean $\overline{J} = 1/q = e^{\gamma xM/s(M)}$. The system efficiency p can be expressed as the ratio of the mean execution time of a txn without and with data contention, i.e., with the possibility of restarts due to failed validation:

$$p = \frac{E[xM/s(M)]}{E[(xM/s(M)) e^{\gamma xM/s(M)}]}$$

$$= \frac{\int_0^\infty x e^{-\mu x} dx}{\int_0^\infty x e^{(\gamma M/s(M)-\mu)x} dx}$$

$$= [1 - (M-1)\psi p]^2.$$

The equation yields one acceptable root $p = [1 + 2(M-1)\psi - \sqrt{1 + 4(M-1)\psi}] / [2(M-1)^2 \psi^2] < 1$, provided $\gamma M/s(M) < \mu$ and the integral in the denominator converges. The txn execution time is not resampled in this analysis, because resampling processing times results in a mix of completed txns which is shorter than original. Shorter txns have a higher probability of successful validation and resampling results in overestimating performance. This analysis in [6] is that of the OCC *silent* [7] or *die* method [2], which is inefficient in that a conflicted txn is allowed to execute to the end, at which point it is restarted. The analysis in [6] also considers *static* data access, i.e., all objects are accessed at the beginning of execution, while *dynamic* or *on-demand* data access is more realistic. The OCC *broadcast* or *kill* method aborts a txn as soon as a conflict is detected. The analysis of the static/silent method in [6] is extended to the other three cases in [7]. The distribution of txn processing time affects performance and in a system with multiple txn sizes the wasted processing is mainly due to lengthy txns [7]. This is due the *quadratic effect* that the probability that a txn encounters a data conflict increases with the number of objects it accesses (k) and its processing time, which is also proportional to k [2]. While the optimistic die method seems to be less efficient than the optimistic kill method, this is not the case when due to buffer misses txns make disk accesses. It is advantageous then to allow a conflicted txn to run to its completion to prime the database buffer with all the objects required for its re-execution in a second processing phase. In *two-phase processing* txns in the second execution phase have a much shorter processing time, since disk accesses are not required, provided *access invariance* prevails, i.e., a restarted txn accesses the same set of objects as it did in its first execution phase [2]. The second execution phase may use the optimistic kill method or even lock preclaiming which ensures that the txn will not be restarted. In general we have multiphase processing methods with different CC methods with increasing strengths, e.g., locking vs. OCC. WDL and RP restart-oriented locking methods outperform two-phase methods [2]. The analysis in [7] can be extended to take into account the variability in txn processing times and the use of different CC methods across txn phases.

Conclusion

There is a need for methodologies combining analytic models, simulation, and trace analysis to explore the effectiveness of new computer organizations for TP, since it is expensive to build realistic prototypes for experimentation. An abstract model of a standard locking system is analyzed here to provide an understanding of the thrashing phenomenon in standard locking. It is shown that a single parameter α determines the level of lock contention with a critical value, which determines the onset of thrashing. Restart-oriented locking methods selectively abort txns to increase the degree of txn concurrency and reduce

the level of lock contention, by disallowing txns to wait for the completion of already blocked txns. Two-phase processing methods reduce the data contention level in TP systems with access invariance, by lowering the holding time of database objects. Both methods require excess processing capacity to cope with the additional processing when txns are restarted.

Cross-References

▶ Redundant Arrays of Independent Disks
▶ Two-Phase Locking

Recommended Reading

1. Chen PM, Lee EK, Gibson GA, Katz RH, Patterson DA. RAID: high-performance, reliable secondary storage. ACM Comput Surv. 1994;26(2):145–85.
2. Franaszek P, Robinson JT, Thomasian A. Concurrency control for high contention environments. ACM Trans Database Syst. 1992;17(2):304–45.
3. Gray JN, Reuter A. Transaction processing: concepts and facilities. Los Altos: Morgan Kauffmann; 1992.
4. Kleinrock L. Queueing systems, Theory/computer applications, vol. 1/2. New York: Wiley; 1975/1976.
5. Lazowska ED, Zahorjan J, Graham GS, Sevcik KC. Quantitative system performance. Englewood Cliffs: Prentice-Hall; 1984.
6. Morris RJT, Wong WS. Performance analysis of locking and optimistic concurrency control algorithms. Perform Eval. 1985;5(2):105–18.
7. Ryu IK, Thomasian A. Performance evaluation of centralized databases with optimistic concurrency control. Perform Eval. 1987;7(3):195–211.
8. Tay YC. Locking performance in centralized databases. New York: Academic; 1987.
9. Tay YC, Zou M. A page fault equation for modeling the effect of memory size. Perform Eval. 2006;63(2):99–130.
10. Thomasian A. Two-phase locking and its thrashing behavior. ACM Trans Database Syst. 1993;18(4):579–625.
11. Thomasian A. Concurrency control: methods, performance, and analysis. ACM Comput Surv. 1998;30(1):70–119.
12. Thomasian A. Performance analysis of locking policies with limited wait-depth. Perform Eval. 1998;33(1):1–21.
13. Thomasian A, Blaum M. Mirrored disk reliability and performance. IEEE Trans Comput. 2006;55(12):1640–4.

14. Thomasian A, Fu G, Han C. Performance evaluation of two-disk failure tolerant arrays. IEEE Trans Comput. 2007;56(6):799–814.
15. Thomasian A, Menon J. RAID5 performance with distributed sparing. IEEE Trans Parallel Distr Syst. 1997;8(6):640–57.
16. Thomasian A, Ryu IK. Performance analysis of two-phase locking. IEEE Trans Softw Eng. 1991;17(5):386–402.
17. Weikum G, Hasse C, Moenkeberg A, Zabback P. The COMFORT automatic tuning project. Inf Syst. 1994;19(5):381–432.

Performance Monitoring Tools

Philippe Bonnet[1] and Dennis Shasha[2]
[1]Department of Computer Science, IT University of Copenhagen, Copenhagen, Denmark
[2]Department of Computer Science, New York University, New York, NY, USA

Definition

Performance monitoring tools denote the utilities and programs that give access to database server internals.

Historical Background

Relational database systems have had to prove their performance from the outset [1]. Tools to measure their performance have therefore been present in early versions of most major systems.

Foundations

Performance monitoring tools are useful in finding out how queries are being serviced and how the underlying resources are being used. In this entry, the characteristics of the most relevant types of tools are described.

Event Monitor

Event monitors capture the aggregate resources associated to a given event (e.g., query execution, deadlock, session) and report the collected data when the event completes. Event monitors should have low overhead as the collected data is usually accumulated in performance counters that are updated as a side effect of the operations monitored (e.g., CPU usage, IO issued, locks collected during query execution) and only accessed once the event has completed.

Event monitors are useful to identify the queries that consume most resources and should require the attention of the database tuner.

Query Plan Explainer

Query plan explainers display the execution plan chosen by the query optimizer for a given query. Explainers represent a query plan, either in textual form or graphically, as a tree whose leaves are base tables and internal nodes are access methods and relational operators. The nodes are possibly annotated with relevant properties (e.g., cardinality, estimated cost).

Query plan explainers are useful to check that the optimizer relies on updated statistics to generate a query execution plan, to check that indexes are used as intended, or to find out whether costly operators are inserted against the programmer's better judgment.

Profiler

There are two types of profilers:

1. Time-based profilers address the question: *how is the time spent*? A time-based profiler logs the time spent in the different components of the database system and presents the time spent processing as well as the time spent waiting for resources.
2. Counter-based profilers address the question: *how are resources used*? A counter-based profiler uses counters to monitor the resources used during execution and presents either database-wide or system-wide aggregate counters.

Profilers might incur high running cost because of the overhead of logging a potentially large number of actions. Profilers are most useful to understand the behavior of critical queries.

Key Applications

Database management system developers need performance monitoring tools to make sure that their system is performing well. Database administrators need performance monitoring tools to make sure that their database instance is performing as well as possible.

Cross-References

▶ Query Optimization

Recommended Reading

1. McJones P. The 1995 SQL reunion: people, projects, and politics. Technical report: SRC-TN-1997-018.
2. Millsap C, Holt J. Optimizing oracle performance. Sebastopol: O'Reilly; 2003.
3. Shasha D, Bonnet P. Database tuning: principles, experiments and troubleshooting techniques. San Francisco: Morgan Kaufmann; 2002.

Period-Stamped Temporal Models

Nikos A. Lorentzos
Informatics Laboratory, Department of
Agricultural Economics and Rural Development,
Agricultural University of Athens, Athens,
Greece

Synonyms

Interval-based temporal models

Definition

A period-stamped temporal model is a *temporal data model* for the management of data in which time has the form of a *time period*.

Historical Background

There are applications that require the recording and management not only of data but also of the time during which this data is valid (*valid time*). A typical example is the data maintained by pension and life insurance organizations in order to determine the benefits for which a person qualifies. Similarly, such organizations have to record their financial obligations at various times in the future.

There are also sensitive applications, in which it is important to record not only the data but also the time at which this data was either recorded in the database or deleted or updated (*transaction time*). This time is recorded automatically by the database management system (DBMS) and it cannot be modified by the user.

Finally, there are applications in which data, valid time, and transaction time need be recorded.

A data model for the management of data and also of either valid time or transaction time is a *temporal data model*. If time has the form of a *time period*, the model is a *period-stamped data model*.

In the case of the relational model, data and time are recorded in a relation. A relation to record data and valid time is a *valid time* relation. A relation to record data and transaction time is a *transaction time* relation. Finally, a relation to record data, valid time, and transaction time is a bitemporal relation.

In spite of the interest for the management of temporal data, there are many practical problems, which cannot be supported directly by the use of a conventional DBMS. As a consequence, much programming is required in order to support such applications. Hence, the satisfaction of actual user requirements necessitated the definition of a period-stamped temporal model. The bulk of research, on the definition of such a model, appeared in the 1980s, at a time when computers became powerful enough to process large volumes of data.

Scientific Fundamentals

Problems related to the management of temporal data are depicted next. The illustration is restricted to the relational model, in particular to the management of valid time relations. Note, however, that relevant problems can be identified in the management of *transaction time* relations whereas the management of *bitemporal relations* is much more complicated.

Examples Illustrating Inadequacy of Conventional Models to Support Time Period

Figure 1 shows SALARY, a valid time relation that records salary histories. The notation "d_number," for values recorded in attributes Begin and End, represents a point in time, which is of a date-time data type. For example, d100 could represent "January 1, 2007." The interpretation of the first tuple is that Alex earned 100 on each of the dates in the *time period* [d100, d299], i.e., on each of the dates d100, d101, . . . ,d299. Next, the amount earned by Alex became 150 for each of the dates in [d300, d499].

Contrary to the above simple representation of a valid time relation, a conventional DBMS lacks the functionality required for the management of such relations. This is illustrated below by a number of examples.

Data insertion: If the tuple (Alex, 150, d400, d799) is inserted into SALARY, it will be recorded in addition to the other tuples. In this case, however, SALARY will contain duplicate data for Alex's payment on the dates d400, d401,...,d499, since this data is already recorded in the second tuple of the relation. Moreover, the query, to return *the time at which Alex's salary became 150*, if not formulated carefully, will return three dates, d300, d400 and d800, whereas the correct answer should be d300 and d800. To avoid such problems, it would be appropriate for the newly inserted tuple to be combined with the second and third tuple for Alex into a single one (Alex, 150, d300, d999), i.e., tuples with identical data for attributes Name and Amount *(value equivalence)*, which have either *overlapping* or *adjacent* time periods, to coalesce into a single time period *(temporal coalescing)*.

Key: Declaring the primary key of SALARY, to be a multi-attribute key of Name and Amount, will not disallow the recording of the tuple (Alex, 200, d400, d799). As a result, SALARY will contain conflicting data, since its second tuple will be showing that Alex's payment is 150 for each of the dates d400, d401,...,d499 whereas, from the newly inserted tuple, it will also be shown that, for these dates, this payment is 200. Hence, in the case of a conventional DBMS, special integrity constraints have to be defined for every relation like SALARY.

Data deletion: Assuming that John's payment for the dates d400, d401,...,d799 was recorded by mistake and has to be deleted, the last tuple of SALARY has to be replaced by two tuples, (John, 100, d200, d399) and (John, 100, d800, d899).

Data update: Assuming that John's payment for dates d400, d401,...,d799 has to be corrected to be 150, the last tuple of SALARY has to be replaced by three tuples, (John, 100, d200, d399), (John, 150, d400, d799) and (John, 100, d800, d899).

Data projection: The query "for every employee, list the time during which he is paid," requires a projection on attributes Name, Begin, and End. This will generate four tuples. In prac-

Period-Stamped Temporal Models, Fig. 1 Example of a valid time relation

SALARY

Name	Amount	Begin	End
Alex	100	d100	d299
Alex	150	d300	d499
Alex	150	d800	d999
John	100	d200	d899

tice, however, a *temporal coalescing* of the time periods recorded in SALARY is also required in order to obtain the correct result, consisting of only three tuples, (Alex, d100, d499), (Alex, d800, d999), and (John, d200, d899).

Since a conventional DBMS lacks the functionality illustrated above, a special code has to be written to handle correctly all the cases described. This limitation gave rise to research on the definition of a temporal data model. Many of the research efforts led to proposals for the definition of various period-stamped data models.

Common Characteristics of Period-Stamped Data Models

The majority of researchers adopted the principle that a temporal data model should be a minimal extension of the conventional relational data model. The most common characteristics of these approaches are outlined below.

A period-stamped, valid time relation always has two types of attributes: The first type consists of one or more ordinary attributes of a conventional relation. These attributes are termed *explicit* by some authors. Valid time itself is not considered to be data; it is considered as being *orthogonal* to data. Hence, the second type consists of special attributes to record valid time. These attributes are often termed *implicit* and may be system assigned, by default. Some approaches consider one pair of such attributes, to record the Begin and End of valid time periods. The valid time recorded in them is usually assumed to represent a time period of the form [Begin, End], i.e., a period closed on both sides. It is said that the data recorded in the explicit attributes is *stamped* by the time recorded in the implicit attributes.

According to the concepts discussed above, the explicit attributes of the relation in Fig. 1 are Name and Amount and the implicit attributes are Begin and End. The first tuple of the relation is (Alex, 100), and it is *stamped* by the time period [d100, d299].

Some approaches define a *time period* data type; therefore, they consider only one implicit attribute. In such approaches, the equivalent scheme for the relation in Fig. 1 is SALARY (Name, Amount, Period).

In many approaches, valid time is not considered to be data; hence, this time may not be part of the primary key of a relation. Therefore, Name is designated as the primary key of SALARY, i.e., it matches the primary key of an ordinary relation.

In all the approaches, valid time is considered to be discrete. Hence, the time period [d100, d299] is interpreted as consisting of the dates d100, d101,...,d299. Various *temporal granularities* can be declared by the user, depending on the application.

Temporal algebras, temporal query languages, or some temporal calculus are proposed in various approaches. The functionality of the operations of the conventional relational model is revised accordingly, so as to overcome the problems illustrated earlier. Appropriate predicates are also defined, applicable to time periods, which can be used in selection operations. In some approaches additional operations are defined to capture temporal functionality that cannot be achieved by simply extending existing languages. In general, an implicit *temporal coalescing* takes place in all the temporal operations introduced.

Most of the proposed temporal models support the valid time property over past, present, and future time periods. Some of the proposed temporal models also support the transaction time property over past and present time periods.

Desired Behavior of Period-Stamped Data Models

Given the limitations outlined earlier of the conventional relational model for the management of temporal data, new problems had to be faced by the proposed period-stamped models. The most interesting of them are illustrated next by making use of relation SALARY in Fig. 1.

Data projection: The query "list all the employees ever paid" requires a projection of SALARY on Name. Such an operation normally yields a relation R with tuples (Alex) and (John). Given however that R lacks implicit attributes, it gives rise to the question whether it is an

**Period-Stamped
Temporal Models, Fig. 2**
Another valid time relation

ASSIGNMENT

Name	Department	Begin	End
Alex	Shoe	d100	d199
Alex	Food	d300	d800
John	Food	d200	d899

R

Name	Amount	Department	Begin	End
Alex	100	Shoe	d100	d199
Alex	150	Food	d300	d499
Alex	150	Food	d800	d800
John	100	Food	d200	d899

Period-Stamped Temporal Models, Fig. 3 Result of a temporal join operation

appropriate response. To overcome this problem, in some approaches projection is defined to yield, after a *temporal coalescing*, the rows (Alex, d100, d499), (Alex, d800, d999), and (John, d200, d899). Notice however that, in this case, the result matches exactly that of another query, "for every employee, list the time during which he is paid."

Stamp projection: Since time is not data, some special operation, to project only on time, may be necessary. For example, the answer to the query "give the time during which at least one employee was paid" should consist of one row, (d100, d999). Given however that this result lacks explicit attributes, it is necessary to determine whether it is appropriate to allow time stamp projection without explicit attributes. Similar questions arise in the case of projections on only one of the two implicit attributes.

Association of data in distinct relations: Such an association requires the use of the Cartesian product operation. For an illustration, consider also the relation in Fig. 2, which is used to record the history of employee assignments to departments. Then an ordinary *Cartesian product* of SALARY with ASSIGNMENT yields a relation with two pairs of Begin and End time stamps, one pair from each relation. In most *temporal data models* such a result is undesirable. The same is also true for the ordinary *join* operation. Due to this, some researchers define instead only a *temporal join* operation, an example of which is given in Fig. 3.

Literature Overview

The characteristics of individual approaches are outlined below. Unless otherwise specified, a period-stamped valid time relation, in the approach under discussion, is that of Fig. 1.

Jones and Mason [3] define LEGOL, a language for the management of period-stamped, valid time relations. It is an early, yet incomplete piece of work.

Ben-Zvi defines a model for the management of period-stamped, bitemporal relations ([15], chapter 8). One more attribute is used, to record the time at which a tuple is deleted. Sets of time periods are also considered, consisting of mutually disjoint periods. Some SQL extension is provided, too. It is also an early and incomplete piece of work.

Snodgrass defines a QUEL extension, for the management of valid time, transaction time, and bitemporal relations [12]. Data may alternatively be stamped by time points (*point-stamped temporal models*). Time may not be specified for the primary key. Formal semantics are provided. A *relational calculus* and a *relational algebra* are defined in [15] (chapter 6).

Navathe and Ahmed propose an SQL extension for period-stamped valid time relations ([15], chapter 4, [8, 9]). In this approach, the primary key of the relation in Fig. 1 is <Name, Begin>. The definition of the primary key of a relation, in conjunction with the definition of a *time normal form*, enables temporal coalescing. One variation of the select operation takes

P

as argument a time period value. A *moving window* operation enables a temporary *split* of the time stamps of all the tuples into time periods of a fixed size, and the subsequent application of aggregate functions on the data that are associated with these fixed size periods.

Lorentzos and Johnson define a relational algebra for the management of either period-stamped or point-stamped (*time instant*) *valid time* [5, 6]. It is also shown that there are practical considerations for having relations with more than one *valid time*. A *time period* data type is later defined in [15] (chapter 3). A generic *period data type* is defined in [15], chapter 3, and in [7], enabling the use of periods of numbers and strings. Such periods can be used for the management of nonvalid time relations which, however, require a functionality identical with that of period-stamped, *valid time* relations. The operations of the conventional relational model are not revised. Instead, two new algebraic operations are defined, which enable coalescing on multiple attributes. Time is treated as data. As such, it may participate in the primary key [7]. In [15], chapter 3, and in [7], it is shown that there are applications in which a *temporal coalescing* should be disallowed. An SQL extension defined in [7] proposes among others the inclusion of a *PORTION* clause in the ISO:SQL DELETE and UPDATE operations. Finally, the design of temporal databases is investigated in much detail in [1].

Sarda defines a relational algebra in [10] and an SQL extension for valid time relations in [15], chapter 5, and in [11]. A time period data type is also defined in [10]. Data may alternatively be stamped by time points. Time may not be specified for the primary key. The operations of the conventional relational model are not revised, but, as reported in [15], if time is projected out, the result relation is not a correct valid time relation. Similarly, the result returned by the Cartesian product operation is considered not to be a correct period-stamped relation. Two additional relational algebra operations are also defined, functionally equivalent with those defined by Lorentzos.

TSQL2 is a consensus temporal SQL2 extension for the management of either period or point-stamped valid time relations, transaction time relations, and bitemporal relations. It is complemented by the definition of a relational algebra.

The ISO SQL:2011 standard [2] provides some primitive temporal support, which is outlined in [4]. It supports *valid time, transaction time,* and *bitemporal relations*. The data representation has many similarities with that in [5, 6]. It also includes a *PORTION* clause, originally proposed in [7]. Curiously enough, *temporal coalescing* is not supported.

All the previous approaches incorporate time at the tuple level (*tuple time-stamping*). Contrary to these, Tansel incorporates time at the attribute level (*attribute time-stamping*) [13]. Hence, only one relation scheme EMPLOYEE (Name, Salary, Department) is needed in this approach to record the history of every employee's salary and of his assignment to departments. Four types of attributes are considered: *Atomic*, to record only one piece of data (e.g., Alex); *set valued*, to record a set of atomic data (e.g., {Alex, Tom}; *triplet valued*, to record the Begin and End of a period during which an associated piece of data is valid (e.g., a triplet <[d100, d300), 100>) under attribute Salary denotes that some salary was 100 during the period [d100, d300); and *set triplet valued*, to record a set of triplet valued data (e.g., {<[d100, d200), Shoe>, <[d300, now], Food>}under attribute Department denotes that some employee was assigned to the shoe department during [d100, d200) and to the food department during [d300, now]). As can be seen from this last example, time periods are in general open for the End time. One exception is the case where the end point of a time period equals *now*, in which case the period is closed. Hence, data valid in the future is not supported. The operations of the relevant conventional model are revised accordingly. Four additional operations are defined, which enable transformations between relations with different types of attributes. A QUEL extension is defined in [14]. The approach in [13] is extended in [17], chapter 7, and in [15], to a nested model, incorporating the well-known nest and unnest operations. A calculus and algebra are also defined. The approach in [15] is extended in [16], in order to support nested bitemporal relations, in which valid and transaction time are incorporated at the attribute level.

Key Applications

There are many applications that necessitate the management of period-stamped and, more generally, temporal data. Some examples are as follows.

Period-stamped, valid time relations can be used by pension and life insurance organizations in order to determine the benefits a person qualifies for. Such relations are also needed for these organizations to record their future financial obligations.

Similarly, period-stamped, transaction time relations can be used in sensitive applications, to enable tracing the content of the database and evaluate some decision taken in the past, with respect to the content of the database at that time.

Future Directions

Given the major need for the management of period-stamped data, the *ISO SQL* Committee should urgently include *temporal coalescing* in the SQL standard.

Further research is necessary for the definition of a nested period-stamped data model and for the management of period-stamped geographical data.

Cross-References

▸ Point-Stamped Temporal Models
▸ Supporting Transaction Time Databases
▸ Temporal Algebras
▸ Temporal Data Models
▸ Temporal Database
▸ Temporal Logical Models
▸ Temporal Object-Oriented Databases
▸ Temporal Query Languages
▸ Temporal Relational Calculus
▸ Temporal Extensions in the SQL Standard

Recommended Reading

1. Date CJ, Darwen H, Lorentzos NA. Time and relational theory, temporal databases in the relational model and SQL. Morgan Kaufmann Publishers; 2013.
2. International Organization for Standardization (ISO). Database language SQL. Document ISO/IEC 9075:2011. 2011.
3. Jones S, Mason PS. Handling the time dimension in a database. In: Proceedings of the International Conference Data Bases; 1980. p. 65–83.
4. Kulkarni K, Michels J-E. Temporal features in SQL:2011. ACM SIGMOD Rec. 2012;41(3):34–43.
5. Lorentzos NA, Johnson RG. Extending relational algebra to manipulate temporal data. Inf Syst. 1988;13(3):289–96.
6. Lorentzos NA, Johnson RG. TRA a model for a temporal relational algebra. In: Rolland C, Bodart F, Leonard M, editors. Temporal aspects in information systems. North-Holland; 1988. p. 203–15.
7. Lorentzos NA, Mitsopoulos YG. SQL extension for interval data. IEEE Trans Knowl Data Eng. 1997;9(3):480–99.
8. Navathe SB, Ahmed R. TSQL: a language interface for history databases. In: Rolland C, Bodart F, Leonard M, editors. Temporal aspects in information systems. North-Holland; 1988. p. 109–22.
9. Navathe SB, Ahmed R. A temporal relational model and a query language. Inf Sci Int J. 1989;47(2):147–75.
10. Sarda NL. Algebra and query language for a historical data model. Comput J. 1990;33(1):11–8.
11. Sarda NL. Extensions to SQL for historical databases. IEEE Trans Knowl Data Eng. 1990;2(2):220–30.
12. Snodgrass S. The temporal query language TQUEL. ACM Trans Database Syst. 1987;12(2):247–98.
13. Tansel AU. Adding time dimension to relational model and extending relational algebra. Inf Syst. 1986;11(4):343–55.
14. Tansel AU. A historical query language. Inf Sci. 1991;53(1–2):101–33.
15. Tansel AU. Temporal relational data model. IEEE Trans Knowl Data Eng. 1997;9(3):464–79.
16. Tansel AU, Canan EA. Nested bitemporal relational algebra. In: Proceedings of the 21st International Symposium on Computer and Information Sciences; 2006. p. 622–33.
17. Tansel A, Clifford J, Gadia J, Segev A, Snodgrass R, editors. Temporal databases: theory, design and implementation. Benjamin/Cummings; 1993.

P

Personalized Web Search

Ji-Rong Wen[1], Zhicheng Dou[2], and Ruihua
Song[1]
[1]Microsoft Research Asia, Beijing, China
[2]Nankai University, Tianjin, China

Synonyms

Personalized search

Definition

For a given query, a personalized Web search
can provide different search results for different
users or organize search results differently for
each user, based upon their interests, preferences,
and information needs. Personalized web search
differs from generic web search, which returns
identical research results to all users for identical
queries, regardless of varied user interests and
information needs.

Historical Background

Web search engines have made enormous contributions to the web and society. They make finding
information on the web quick and easy. However,
they are far from optimal. A major deficiency
of generic search engines is that they follow the
"one size fits all" model and are not adaptable to
individual users. This is typically shown in cases
such as these:

1. Different users have different backgrounds
   and interests. They may have completely
   different information needs and goals when
   providing exactly the same query. For
   example, a biologist may issue "mouse"
   to get information about rodents, while
   programmers may use the same query to
   find information about computer peripherals.
   When such a query is issued, generic search
   engines will return a list of documents on
   different topics. It takes time for a user
   to choose which information he/she really
   wants, and this makes the user feel less
   satisfied. Queries like "mouse" are usually
   called ambiguous queries. Statistics has shown
   that the vast majority of queries are short
   and ambiguous. Generic web search usually
   fails to provide optimal results for ambiguous
   queries.
2. Users are not static. User information needs
   may change over time. Indeed, users will have
   different needs at different times based on
   current circumstances. For example, a user
   may use "mouse" to find information about
   rodents when the user is viewing television
   news about a plague, but would want to find
   information about computer mouse products
   when purchasing a new computer. Generic
   search engines are unable to distinguish between such cases.

Personalized web search is considered a
promising solution to address these problems,
since it can provide different search results
based upon the preferences and information
needs of users. It exploits user information
and search context in learning to which sense
a query refers. Consider the query "mouse"
mentioned above: Personalized web search
can disambiguate the query by gathering the
following user information:

1. The user is a computer programmer, not a
   biologist.
2. The user has just input a query "keyboard," but
   not "biology" or "genome." Before entering
   this query, the user had just viewed a web
   page with many words related to computer
   mouse, such as "computing," "input device,"
   and "keyboard."

Foundations

User Profiling

To provide personalized search results to users,
personalized web search maintains a user profile for each individual. A user profile stores

approximations of user tastes, interests and preferences. It is generated and updated by exploiting user-related information. Such information may include:

1. Demographic and geographical information, including age, gender, education, language, country, address, interest areas, and other information;
2. Search history, including previous queries and clicked documents. User browsing behavior when viewing a page, such as dwelling time, mouse click, mouse movement, scrolling, printing, and bookmarking, is another important element of user interest.
3. Other user documents, such as bookmarks, favorite web sites, visited pages, and emails. Teevan et al. [15] and Chirita et al. [2] demonstrate that external user data stored in a user client is useful to personalize individual search results.

User information can be specified by the user (explicitly collecting) or can be automatically learnt from a user's historical activities (implicitly collecting). As the vast majority of users are reluctant to provide any explicit feedback on search results and their interests, many works on personalized web search focus on how to automatically learn user preferences without involving any direct user efforts [6, 8, 9, 10, 13]. Collected user information is processed and organized as a user profile in a certain structure, depending on the need of personalization algorithm. This can be completed by creating vectors of URLs/domains, keywords, topic categories, tensors, or the like.

A user profile can usually aggregate a user's history information and represent the user's *long-term* interests (information needs). Some work has investigated whether such a long-term user profile is ineffective in some cases. Consider the second case that was described in the historical background section: a user will have different needs at different times based on circumstances. In such situations, personalization based on a user's long-term interests may not provide a satisfying performance, because similar results could be returned. Some work [10] has considered the use of a user's active context to represent *short-term* information needs. Search context is incorporated into the user profile, or is constructed as a separate short-term user model/profile and is used in helping infer a user's information needs.

Personalized Search Based on Content Analysis

Personalized web search can be achieved by checking content similarity between web pages and user profiles.

Some work has represented user interests with topical categories. User's topical interests are either explicitly specified by users themselves, or can be automatically learned by classifying implicit user data. Search results are filtered or re-ranked by checking the similarity of topics between search results and user profiles. In some work [1, 8], a user profile is structured as a concept/topic hierarchy. User-issued queries and user-selected snippets/documents are categorized into concept hierarchies that are accumulated to generate a user profile. When the user issues a query, each returned snippet/document is also classified. The documents are re-ranked based upon how well the document categories match user interest profiles. Chirita et al. [1] use the ODP (Open Directory Project, http://www.dmoz.org/) hierarchy to implement personalized search. User favorite topics nodes are manually specified in the ODP hierarchy. Each document is categorized into one or several topic nodes in the same ODP hierarchy. The distances between the user topic nodes and the document topic nodes are then used to re-rank search results.

Some other work uses lists of keywords (bags of words) to represent user interests. In [13], a user profile is built as a vector of distinct terms and is constructed by aggregating past user click history. The cosine similarity between the user profile vector and the feature vector of returned web pages are used to re-rank results. Shen et al. [10] first use language modeling to mine immediate search contextual and implicit feedback information. The approach selects appropriate terms from related preceding queries and corresponding search results to expand the current query. In a query session, the viewed document summaries are used to immediately re-rank documents that have not yet been seen by the user. Teevan et al. [15] and Chirita et al. [2] exploit rich models of user interests, built from both search-related information, and other information about the user. This includes documents and emails the user has read and created. In [6], keywords are associated with categories and thus user profiles are represented by a hierarchical category tree based on keywords categories.

Personalized Web Search Based on Hyperlink Analysis

Most generic web search approaches rank importance of documents based on the linkage structure of the web. An intuitive approach of personalized web search is to adapt these algorithms to compute personalized importance of documents. A large group of these works focuses on personalized PageRank. PageRank, proposed by Page and Brin [7], is a popular link analysis algorithm used in web search. The fundamental motivation underlying PageRank is the recursive notion that important pages are those linked-to by many important pages. This recursive notion can be formalized by the "random surfer" model [7] on the directed web graph G. A directed edge $<p, q>$ exists in G if page p has a hyperlink to page q. Let $O(p)$ be the outdegree of web page p in G. $O(p)$ is equivalent to number of web pages that linked by page p. Let $\mathbf{A}$ be the matrix corresponding to the web graph G, where $A_{ij} = 1/O(j)$ if page j links to page i, and $A_{ij} = 0$ otherwise. In the random surfer model, when a surfer visits page p, he/she keeps clicking outlinks at random with probability $(1-c)$, and jumps to a random web page with probability c. c is called teleportation constraint or damping factor. The PageRank of a page p is defined as the probability that the surfer visited page p. Iterative computation of PageRank is done as the following equation:

$$v^{k+1} = (1 - c)\, A v^k + c\mathbf{u} \qquad (1)$$

Here, $\mathbf{u}$ is defined as a preference vector, where $|\mathbf{u}| = 1$ and $u(i)$ denotes the amount of preference for page i when the surfer jumps to a random web page i. The global PageRank vector is computed when there is no particular preference on any pages, i.e., $\mathbf{u} = [1/n,...,1/n]^T$. By setting variant preference to web pages, a PageRank vector with personalized views of web page importance is generated. It recursively favors pages with high preference, and pages linked by high-preference page. This PageRank vector is called a *personalized PageRank vector (PPV)*. To accomplish personalized web search, a personalized PageRank is computed for each user based upon the user's preference. For example, web pages in the user's bookmarks are set higher preferences in $\mathbf{u}$. Rankings of the user's search results can be biased according to the user's Personalized PageRank vector instead of the global PageRank.

Unfortunately, computing a PageRank vector usually requires multiple scans of the web graph [7], which makes it impossible to carry out online in response to a user query. Furthermore, when a large number of users employ a search engine, it is impossible to compute and store so many personalized PageRank vectors offline. Many later works [4, 5] make efforts to reduce the computation and storage cost of personalized PageRank vectors. Jeh and Widom [5] support the concept that a user's preference set is a subset of a set of hub pages H, selected as those of greater interest for personalization. For each hub page p in H, setting the preference to 1 for page p and 0 for other pages, the corresponding personalized PageRank vector is called a basis hub vector. The authors decompose each basis hub vector in two parts: hub skeleton vector and partial vector. Hub skeleton vector represents common interrelationships between hub vectors, and is computed offline. Each partial vector for a hub page p represents the part of p's hub vector unique to itself. Partial vector can be computed at construction-time efficiently. Finally, a personalized PageRank vector can be expressed as a linear combination of a set of basis hub vectors, and is computed at query time efficiently. Experiments show that the approach is feasible when size of hub set $>10^4$.

Haveliwala [4] use personalized PageRank to enable "topic-sensitive" web search. The approach precomputes k personalized PageRank vectors using k topics, e.g., the 16 top level topics of the Open Directory. For each topic i, a preference vector u_i is generated. $(u_i)_j$ represents the confidence that web page j is classified into topic i. A PPV is computed base upon preference vector u_i. The k personalized PageRank vectors are combined at query time, using the context of the query to compute the appropriate topic weights. The experiments concluded that the use of personalized PageRank scores can improve web search, but the number of personalized PageRank vectors used was limited due to the computational requirements. In fact, this approach modulates the rankings based on the topic of the query and query context, rather than for truly "personalizing" the rankings to a specific individual. Qiu and Cho [9] develop a method to automatically estimate a user's topic preferences based on Topic-Sensitive PageRank scores of the user's past clicked pages. The topic preferences are then used to bias future search results.

Community-Based Personalized Web Search

In most of the above personalized search strategies, each user has a distinct profile and the profile is used to personalize search results for the user. There are also some approaches that personalize search results for the preferences of a community of like-minded users. These approaches are called community-based personalized web search or collaborative web search. In a community-based personalized web search, when a user issues a query, search histories of users who have similar interests to the user are used to filter or re-rank search results. For example, documents that have been selected for the target query or similar queries by the community are re-ranked higher in the results list. Sugiyama et al. [13] use a modified collaborative

filtering algorithm to constructed user profiles to accomplish personalized search. Sun et al. [14] proposed a novel method named CubeSVD to apply personalized web search by analyzing correlations among users, queries, and web pages in clickthrough data. Smyth et al. [12] show that collaborative web search can be efficient in many search scenarios when natural communities of searchers can be identified.

Server-Side and Client-Side Implement

Personalized web search can be implemented on either server side (in the search engine) or client side (in the user's computer or a personalization agent).

For server-side personalization, user profiles are built, updated, and stored on the search engine side. User information is directly incorporated into the ranking process, or is used to help process initial search results. The advantage of this architecture is that the search engine can use all of its resources, for example link structure of the whole web, in its personalization algorithm. Also, the personalization algorithm can be easily adapted without any client efforts. This architecture is adopted by some general search engines such as Google Personalized Search. The disadvantage of this architecture is that it brings high storage and computation costs when millions of users are using the search engine, and it also raises privacy concerns when information about users is stored on the server.

For client-side personalization, user information is collected and stored on the client side (in the user's computer or a personalization agent), usually by installing a client software or plug-in on a user's computer. In client side, not only the user's search behavior but also his contextual activities (e.g., web pages viewed before) and personal information (e.g., emails, documents, and bookmarks) could be incorporated into the user profile. This allows the construction of a much richer user model for personalization. Privacy concerns are also reduced since the user profile is strictly stored and used on the client side. Another benefit is that the overhead in computation and storage for personalization can be distributed among the clients. A main drawback of personalization on the client side is that the personalization algorithm cannot use some knowledge that is only available on the server side (e.g., PageRank score of a result document). Furthermore, due to the limits of network bandwidth, the client can usually only process limited top results.

Challenges of Personalized Search

Despite the attractiveness of personalized search, there is no large-scale use of personalized search services currently. Personalized web search faces several challenges that retard its real-world large-scale applications:

1. Privacy is an issue. Personalized web search, especially server-side implement, requires collecting and aggregating a lot of user information including query and clickthrough history. A user profile can reveal a large amount of private user information, such as hobbies, vocation, income level, and political inclination, which is clearly a serious concern for users [11]. This could make many people nervous and feel afraid to use personalized search engines. A personalized web search will be not well-received until it handles the privacy problem well.
2. It is really hard to infer user information needs accurately. Users are not static. They may randomly search for something which they are not interested in. They even search for other people sometimes. User search histories inevitably contain noise that is irrelevant or even harmful to current search. This may make personalization strategies unstable.
3. Queries should not be handled in the same manner with regard to personalization. Personalized search may have little effect on some queries. Some work [1, 2, 3] investigates whether current web search ranking might be sufficient for clear/unambiguous queries and thus personalization is unnecessary. Dou et al. [3] reveal that personalized search has little effect on queries with high user

selection consistency. A specific personalized search also has different effectiveness for different queries. It even hurts search accuracy under some situations. For example, topical interest-based personalization, which leads to better performance for the query "mouse," is ineffective for the query "free mp3 download." Actually, relevant documents for query "free mp3 download" are mostly classified into the same topic categories and topical interest-based personalization has no way to filter out desired documents. Dou et al. [3] also reveal that topical interest-based personalized search methods are difficult to deploy in a real world search engine. They improve search performance for some queries, but they may hurt search performance for additional queries.

Key Applications

Personalized web search is considered a promising solution to improve the performance of generic web search. Currently, Google and other web search engines are trying to do personalized search.

Experimental Results

Experimental results have shown that personalized web search can indeed improve performance of web search. Detailed experimental results can be found in the corresponding reference for each presented method. Dou et al. [3] propose a personalized web search evaluation framework based upon large-scale query logs.

Cross-References

▶ Information Retrieval
▶ Privacy
▶ WEB Information Retrieval Models
▶ Web Search Relevance Feedback

Recommended Reading

1. Chirita PA, Nejdl W, Paiu R, Kohlschütter C. Using ODP metadata to personalize search. In: Proceedings of the 31st Annual International ACM SIGIR Conference on Research and Development in Information Retrieval; 2008. p. 178–85.
2. Chirita PA, Firan C, Nejdl W. Summarizing local context to personalize global web search. In: Proceedings of the 15th ACM International Conference on Information and Knowledge Management; 2006.
3. Dou Z, Song R, Wen J. A large-scale evaluation and analysis of personalized search strategies. In: Proceedings of the 33rd Annual International ACM SIGIR Conference on Research and Development in Information Retrieval; 2010.
4. Haveliwala TH. Topic-sensitive PageRank. In: Proceedings of the 11th International World Wide Web Conference; 2002.
5. Jeh G, Widom J. Scaling personalized web search. In: Proceedings of the 12th International World Wide Web Conference; 2003. p. 271–79.
6. Liu F, Yu C, Meng W. Personalized web search by mapping user queries to categories. In: Proceedings of the 11th International Conference on Information and Knowledge Management; 2002. p. 558–65.
7. Page L, Brin S, Motwani R, Winograd T. The PageRank citation ranking: bringing order to the web. Technical report, Computer Science Department, Stanford University. 1998.
8. Pretschner A, Gauch S. Ontology based personalized search. In: Proceedings of the 11th IEEE International Conference on Tools with AI; 1999. p. 391–98.
9. Qiu F, Cho J. Automatic identification of user interest for personalized search. In: Proceedings of the 15th International World Wide Web Conference; 2006. p. 727–36.
10. Shen X, Tan B, Zhai C. Implicit user modeling for personalized search. In: Proceedings of the 14th ACM International Conference on Information and Knowledge Management; 2005. p. 824–31.
11. Shen X, Tan B, Zhai C. Privacy protection in personalized search. SIGIR Forum. 2007;41(1):4–17.
12. Smyth B, Coyle M, Boydell O, Briggs P, Balfe E, Freyne J, Bradley K. A live-user evaluation of collaborative web search. In: Proceedings of the 19th International Joint Conference on AI; 2005.
13. Sugiyama K, Hatano K, Yoshikawa M. Adaptive web search based on user profile constructed without any effort from users. In: Proceedings of 12th International World Wide Web Conference; 2003. p. 675–84.
14. Sun J-T, Zeng H-J, Liu H, Lu Y, Chen Z. CubeSVD: a novel approach to personalized web search. In: Proceedings of the 14th International World Wide Web Conference; 2005. p. 382–90.
15. Teevan J, Dumais ST, Horvitz E. Personalizing search via automated analysis of interests and activities. In: Proceedings of the 31st Annual International ACM SIGIR Conference on Research and Development in Information Retrieval; 2008. p. 449–56.

Petri Nets

W. M. P. van der Aalst
Eindhoven University of Technology,
Eindhoven, The Netherlands

Synonyms

Colored nets; Condition event nets; Place transition nets

Definition

The Petri net formalism provides a graphical but also formal language which is appropriate for modeling systems and processes with concurrency and resource sharing. It was introduced in the beginning of the 1960s by Carl Adam Petri and was the first formalism to adequately describe concurrency. The classical Petri net is a directed bipartite graph with two node types called *places* and *transitions*. The nodes are connected via directed *arcs*. Places are represented by circles and transitions by rectangles. The network structure of the Petri net is static. However, places may contain tokens, and the distribution of tokens of places may change as described in the *firing rule*. Petri nets have formal semantics and allow for all kinds of analysis. Moreover, due to the strong theoretical foundation, much is known about the properties of different subclasses of Petri nets. Petri nets have been extended in many different application domains, e.g., workflow management systems, flexible manufacturing, embedded systems, communication protocols, web services, asynchronous circuits, etc. Moreover, many modeling languages have been influenced by Petri nets (e.g., UML) and have been mapped onto Petri nets for analysis purposes (e.g., BPEL, BPMN, etc.). The classical Petri net has been extended to allow for the modeling of complex systems, e.g., so-called colors have been added to model data, time stamps have been added to model time, and hierarchy concepts have been proposed to structure large models. The resulting models are called *high-level Petri nets*.

Historical Background

Petri started his scientific career with his dissertation "Communication with Automata" [7], which he submitted to the Science Faculty of Darmstadt Technical University in July 1961. He defended his thesis there in June 1962 [8]. Petri's dissertation has been quoted frequently, but the Petri net formalism as it is known today emerged later. However, the fundamental idea that asynchronous systems are more powerful than synchronous systems was already present. Petri nets as they are used today first appeared in Petri's talk "Fundamentals on the description of discrete processes" at the Third Colloquium on Automata Theory in Hannover in 1965. Using a simple graphical representation consisting of two types of nodes (places and transitions), Petri was able to describe processes exhibiting concurrency and possibly infinitely many states. Figure 1 shows a simple Petri net consisting of six transitions and seven places. The black dots in places *p1* and *p3* denote tokens and are used to represent the initial state. Despite the addition of concurrency, various properties are decidable for Petri nets which are not decidable for Turing machines (e.g., reachability, liveness, boundedness, etc.).

Initially the focus was on understanding concurrency through Petri nets, and little emphasis was put on practical applications. This changed over time and Petri nets were used more and more in all kinds of applications. A nice example is the work on office information systems in the late 1970s and early 1980s. People like Skip Ellis, Anatol Holt, and Michael Zisman worked on information systems which were driven by explicit process models. It is interesting to see that the three pioneers in this area independently used Petri net variants to model office procedures. The ideas of these people resulted in all kinds of workflow management systems, and today most of the workflow management systems use a notation close to Petri nets [1]. Petri nets have been used in many other application domains ranging from embedded systems and asynchronous circuits to flexible manufacturing, communication protocols, and web services. Petri nets also influenced the development of many languages ranging from process calculi to more application-oriented languages such as UML, EPCs, etc.

The classical Petri net allows for the modeling of states, events, conditions, synchronization, parallelism, choice, and iteration. However, Petri nets describing real processes tend to be complex and extremely large. Moreover, the classical Petri net does not allow for the modeling of data and time. To solve these problems, many extensions have been proposed. Three well-known extensions of the basic Petri net model are (1) the extension with *color* to model data, (2) the extension with *time*, and (3) the extension with *hierarchy* to structure large models. A Petri net extended with color, time, and hierarchy is called a *high-level Petri net* [2]. Using high-level Petri nets, it is much easier to describe complex systems and processes. However, analysis of high-level Petri net other than simulation is often intractable.

Foundations

The classical Petri net is a directed bipartite graph consisting of *places* and *transitions* connected via directed *arcs*. Connections between two nodes of the same type are not allowed. Places are repre-

Petri Nets, Fig. 1 A Petri net consisting of six transitions and seven places

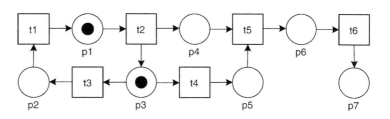

sented by circles and transitions by rectangles as shown in Fig. 1.

A *Petri net* is formally defined by a triple (P, T, F):

1. P is a finite set of *places*.
2. T is a finite set of *transitions* $(P \cap T = \varnothing)$.
3. $F \subseteq (P \times T) \cup (T \times P)$ is a set of *arcs* (flow relation).

A place p is called an *input place* of a transition t if there exists a directed arc from p to t. Place p is called an *output place* of transition t if there exists a directed arc from t to p. The notation •t is used to denote the set of input places for a transition t. The notations t•, •p, and p• have similar meanings, e.g., p• is the set of transitions sharing p as an input place. In Fig. 1 transition $t2$ has one input place ($p1$) and two output places ($p3$ and $p4$).

At any time a place contains zero or more *tokens*, drawn as black dots. The *state*, often referred to as marking, is the distribution of tokens over places, i.e., $M \in P \to IN$. A state is represented as follows: $1p_1 + 2p_2 + 1p_3 + 0p_4$ is the state with one token in place p_1, two tokens in p_2, one token in p_3, and no tokens in p_4. The notation $p_1 + 2p_2 + p_3$ can also be used to represent this state. To compare states a partial ordering is defined. For any two states M_1 and M_2, $M_1 \leq M_2$ if for all $p \in P$: $M_1(p) \leq M_2(p)$.

The number of tokens may change during the execution of the net. Transitions are the active components in a Petri net: they change the state of the net according to the following *firing rule*:

1. A transition t is said to be *enabled* if each input place p of t contains at least one token.
2. An enabled transition may *fire*. If transition t fires, then t *consumes* one token from each input place p of t and *produces* one token for each output place p of t.

The initial marking shown in Fig. 1 is $p1 + p3$. In this marking, $t2$, $t3$, and $t4$ are enabled. Firing $t2$ results in the state $2p3 + p4$. Firing $t3$ results in the state $p1 + p2$. Firing $t4$ results in the state $p1 + p5$. Note that firing $t3$ disables $t4$ and vice versa, i.e., there is a nondeterministic choice between $t3$ and $t4$. However, $t2$ can fire independently from $t3$ and $t4$ because it is in parallel.

Given a Petri net (P, T, F) and a state M_1, the following notations are used:

1. $M_1 \xrightarrow{t} M_2$: transition t is enabled in state M_1, and firing t in M_1 results in state M_2.
2. $M_1 \to M_2$: there is a transition t such that $M_1 \xrightarrow{t} M_2$.
3. $M_1 \xrightarrow{\sigma} M_n$: the firing sequence $\sigma = t_1\ t_2\ t_3 \ldots t_{n-1}$ leads from state M_1 to state M_n via a (possibly empty) set of intermediate states $M_2, \ldots M_{n-1}$, i.e., $M_1 \xrightarrow{t_1} M_2 \xrightarrow{t_2} \ldots \xrightarrow{t_{n-1}} M_n$.

A state M_n is called *reachable* from M_1 (notation $M_1 \xrightarrow{*} M_n$) if there is a firing sequence σ such that $M_1 \xrightarrow{\sigma} M_n$. Note that the empty firing sequence is also allowed, i.e., $M_1 \xrightarrow{*} M_1$.

(PN, M) is used to denote a Petri net PN with an initial state M. A state M' is a *reachable state* of (PN, M) if $M \xrightarrow{*} M'$.

The marked Petri net shown in Fig. 1 has an unbounded number of states. Note that the sequence $t2\ t3\ t1$ can be repeated over and over again. In each cycle an additional token is put in place $p4$.

In the remainder, some standard properties for Petri nets are defined [3, 4]. First, properties related to the dynamics of a Petri net are defined. Then some structural properties are given.

A Petri net (PN, M) is *live* if for every reachable state M' and every transition t, there is a state M'' reachable from M' which enables t. Note that the marked Petri net shown in Fig. 1 is not live. After firing $t4$ twice, it becomes impossible to execute any of the first three transitions. A Petri net is *structurally live* if there exists an initial state such that the net is live. Figure 1 is also not structurally live.

A Petri net (PN, M) is *bounded* if for each place p, there is a natural number n such that for every reachable state the number of tokens in p is less than n. The net is safe if for each place the maximum number of tokens does not exceed 1. As indicated before, the marked Petri net shown in Fig. 1 is not bounded because the number of tokens in *p4* can exceed any number. A Petri net is *structurally bounded* if the net is bounded for any initially state.

Figure 2 shows a more meaningful example of a Petri net. In fact it models the workflow related to some ordering process, and the transitions correspond to tasks. The token in *start* corresponds to an order. Task *register* is represented by a transition bearing the same name. From a routing point of view, it acts as a so-called AND-split (two outgoing arcs) and is enabled in the state shown. If a person executes this task, the token is removed from place *start* and two tokens are produced: one for *c0* and one for *c2*. Then, in parallel, two tasks are enabled: *check_availability* and *send_bill*. Depending on the eagerness of the workers executing these two tasks either *check_availability* or *send_bill* is executed first. Suppose *check_availability* is executed first. Based on the outcome of this task, a choice is made. This is reflected by the fact that three arcs are leaving *c1*. If the ordered goods are available, they can be shipped, i.e., firing *in_stock* enables task *ship_goods*. If they are not available, either a replenishment order is issued or not. Firing *out_of_stock_repl* enables task *replenish*. Firing *out_of_stock_no_repl* skips task *replenish*. Note that *check_availability*, place *c1* and the three transitions *in_stock*, *out_of_stock_repl*, and *out_of_ stock_no_repl* together form a so-called OR-split: As a result of this construct, one token is produced for either *c3*, *c4*, or *c5*. Suppose that not all ordered goods are available, but the appropriate replenishment orders were already issued. A token is produced for *c3* and task *update* becomes enabled. Suppose that at this point in time task *send_bill* is executed, resulting in the state with a token in *c3* and *c6*. The token in *c6* is input for two tasks. However, only one of these tasks can be executed and in this state only *receive_payment* is enabled. Task *receive_payment* can be executed the moment the payment is received. Task *reminder* is an AND-join/AND-split and is blocked until the bill is sent and the goods have been shipped.

However, it is only possible to send a reminder if the goods have actually been shipped. Assume that in the state with a token in *c3* and *c6* task *update* is executed. This task does not require human involvement and is triggered by a message of the warehouse indicating that relevant goods have arrived. Again *check_availability* is enabled. Suppose that this task is executed and the result is positive, i.e., the path via *in_stock* is taken. In the resulting state, *ship_goods* can be executed. Now there is a token in *c6* and *c7*, thus enabling task *reminder*. Executing task *reminder* enables the task *send_bill* for the second time. A new copy of the bill is sent with the appropriate text. It is possible to send several reminders by alternating *reminder* and *send_bill*. However, assume that after the first loop the customer pays resulting in a state with a token in *c7* and *c8*. In this state, the AND-join *archive* is enabled and executing this task results in the final state with a token in *end*.

The marked Petri net shown in Fig. 2 is bounded but not live. In fact the model is safe because there are never two tokens in the same place.

Reachability, liveness, and boundedness can be decided for any classical Petri net. For example, it is possible to construct the so-called coverability graph. However, this may be quite inefficient. Fortunately, more efficient techniques are available for different subclasses of Petri nets. Some of these subclasses are listed below.

A Petri net is a *free-choice Petri net* [5], if, for every two transitions t_1 and t_2, $\bullet t_1 \cap \bullet t_2 \neq \emptyset$ implies $\bullet t_1 = \bullet t_2$. Figure 1 is free choice but Fig. 2 is not. Figure 2 is not free choice because of the two arcs between *c7* and *reminder*. Many properties of free-choice nets can be exploited in their analysis. For example, the combination of liveness and boundedness can be decided in polynomial time.

A Petri net is *state machine* if each transition has exactly one input and one output place. A Petri net is *marked graph* if each place has exactly one input and one output transition. Both state machines and marked graphs are examples of free-choice nets and can be analyzed efficiently.

A Petri net $PN = (P, T, F)$ is a WF-net (workflow net [1, 6]) if and only if:

1. There is one source place $i \in P$ such that $\bullet i = \emptyset$.
2. There is one sink place $o \in P$ such that $o\bullet = \emptyset$.
3. Every node $x \in P \cup T$ is on a path from i to o.

The class of WF-net has been extensively studied because of its applications in workflow management processes and other case-driven processes. Clearly, Fig. 2 is a WF-net while Fig. 1 is not.

For a WF-net with one token on the source place i, it is interesting to know whether the net will always terminate properly with a token in the sink place o. This corresponds to the so-called soundness property. Soundness can be expressed in terms of liveness and boundedness. If the sink place o is connected to the source place i using some transition t^*, the so-called short-circuited net is obtained. The original WF-net is sound if and only if the corresponding short-circuited net is live and bounded. This result can be exploited in the analysis of complex workflows.

Key Applications

Workflow Management and BPM

Petri nets are used for the modeling, analysis, and enactment of workflows. Many workflow languages have a graphical representation close to Petri nets. Moreover, today's workflow engines use mechanisms comparable to using tokens and the firing rule. Also in business process management (BPM), Petri nets are often used for process modeling and analysis.

Petri nets are widely used in process mining [9]. Process mining bridges the gap between traditional model-based process analysis (e.g., simulation and other business process management techniques) and data-centric analysis techniques such as machine learning and data mining. Process mining seeks the confrontation between event data (i.e., observed behavior) and process models (hand-made or discovered automatically). Example applications include: analyzing treatment processes in hospitals, improving customer service processes in a multinational, understanding the browsing behavior of customers using booking site, analyzing failures

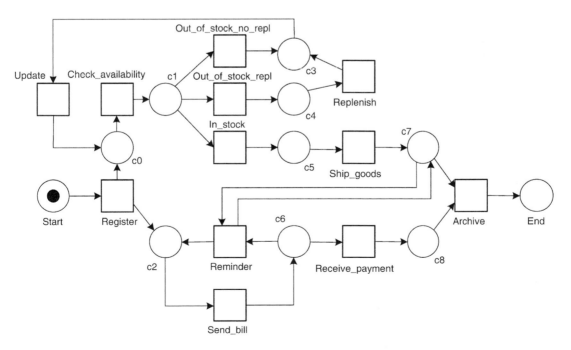

Petri Nets, Fig. 2 A workflow expressed in terms of a Petri net

of a baggage handling system, and improving the user interface of an X-ray machine. All of these applications have in common that dynamic behavior needs to be related to process models.

Discrete/Flexible Manufacturing

Traditionally, there have been many applications of Petri nets in manufacturing. The models are used to analyze the manufacturing process in detail, e.g., to find deadlocks or to analyze the performance.

Communication Protocols

Different protocols have been modeled using Petri nets. Various errors have been discovered by verifying established protocols using Petri-net-based analysis techniques.

Embedded Systems

The interactions between hardware and software in embedded systems (e.g., copiers, cars, mobile phones, etc.) may result in all kinds of problems. Therefore, Petri nets are used to model and analyze such systems.

Cross-References

▶ Business Process Management
▶ Process Mining
▶ Web Services
▶ Workflow Model Analysis
▶ Workflow Patterns

Recommended Reading

1. Brauer W, Reisig W. Carl Adam Petri and Petri Nets. Informatik-Spektrum. 1996;29(5):369–74.
2. Desel J, Esparza J. Free choice Petri nets, Cambridge tracts in theoretical computer science, vol. 40. Cambridge, UK: Cambridge University Press; 1995.
3. Jensen K, Kristensen LM, Wells L. Coloured Petri nets and CPN tools for modelling and validation of concurrent systems. Int J Softw Tools Technol Trans. 2007;9(3–4):213–54.
4. Murata T. Petri nets: properties, analysis and applications. Proc IEEE. 1989;77(4):541–80.
5. Petri CA. Kommunikation mit Automaten. PhD Thesis, Fakultät für Mathematik und Physik, Technische Hochschule Darmstadt, Darmstadt, 1962.
6. Reisig W, Rozenberg G, editors. Lectures on Petri nets I: basic models. Berlin/Heidelberg/New York: Springer; 1998.
7. van der Aalst WMP. The application of Petri nets to workflow management. J Circ Syst Comput. 1998;8(1):21–66.
8. van der Aalst WMP, van Hee KM. Workflow management: models, methods, and systems. Cambridge, MA: MIT Press; 2004.
9. van der Aalst WMP. Process Mining: Data Science in Action. Springer-Verlag, Berlin, 2016.

Physical Clock

Curtis E. Dyreson
Utah State University, Logan, UT, USA

Synonyms

Clock

Definition

A *physical clock* is a physical process coupled with a method of measuring that process to record the passage of time. For instance, the rotation of the Earth measured in *solar days* is a physical clock. Most physical clocks are based on cyclic processes (such as a celestial rotation). One or more physical clocks are used to establish a *timeline clock* for a temporal database.

Key Points

Every concrete time is a measurement of some physical clock. For instance a windup watch provides the time "now" by measuring the rate at which a coiled, wound spring unwinds. A physical clock is limited by the durability and regularity of the underlying physical process that it measures, so to provide a clock for every instance in a time line for a temporal database, several physical clocks might be needed. For instance, a clock based on the rotation of the Earth will (likely) be limited as current models predict the Earth will eventually be gravitationally drawn into the Sun. The modifier "physical" distinguishes this kind of clock from other kinds of clocks in an application, in particular a "timeline" clock. Atomic clocks, which are a kind of physical clock, provide the measurements for International Atomic Time (TAI).

Cross-References

▶ Chronon
▶ Time-Line Clock
▶ Time Instant

Recommended Reading

1. Dyreson CE, Snodgrass RT. Timestamp semantics and representation. Inf Syst. 1993;18(3):143–66.
2. Dyreson CE, Snodgrass RT. The baseline clock. The TSQL2 temporal query language. Kluwer; 1987. p. 73–92.
3. Fraser JT. Time: the familiar stranger. Amherst: University of Massachusetts Press; 1987.

Physical Database Design for Relational Databases

Sam S. Lightstone
IBM Canada Ltd, Markham, ON, Canada

Synonyms

Clustering; Database design; Database implementation; Database materialization; Indexing; Materialized query tables; Materialized views; Multidimensional clustering; Range partitioning; Table design; Table normalization

Definition

Physical database design represents the materialization of a database into an actual system. While logical design can be performed independently of the eventual database platform, many physical database attributes depend on the specifics and semantics of the target DBMS. Physical design is performed in two stages:

1. Conversion of the logical design into table definitions (often performed by an application developer): includes pre-deployment design, table definitions, normalization, primary and foreign key relationships, and basic indexing.
2. Post deployment physical database design (often performed by a database administrator): includes improving performance, reducing I/O, and streamlining administration tasks.

Generally speaking, physical database design covers those aspects of database design that impact the actual structure of the database on disk. Stage 1 is variably referred to in the industry as an aspect of *logical database design* or *physical database design*. It is known as *logical database design* in the sense that it can be designed independent of the data server or the particular DBMS used. It is also often performed by the same people who perform the early requirements building and entity relationship modeling. Conversely, it is also called *physical database design* in the sense that it affects the physical structure of the database and its implementation. For the sake of this entry, the latter assumption is used, and it is therefore included as part of *physical database design*.

Although it is possible to perform logical design independently of the physical platform and knowledge of database software that will eventually be used, many physical database design tasks depend on the specifics and semantics of the target DBMS (software and hardware). The major tasks of physical database design include: table normalization, table denomalization, indexing, clustering, multidimensional clustering

(MDC), range partitioning, hash partitioning, shared nothing partitioning (hash), materialized views (MVs), memory allocation, database storage topology, and database storage object allocation.

Historical Background

Physical database design is as old as database research. Specifically, if this question is limited to relational databases, the first systems were prototyped in the early 1970s. The most elementary problems of physical database design are those of table normalization and index selection. The relational model for databases was first proposed in 1970 by E.F Codd at IBM. The first relational databases using SQL and B+-tree was IBM's System R, in 1976 and the INGRES database at the University of California, Berkeley. The B+-tree, the most commonly-used indexing storage structure for user designed indexes, was first described in the paper *Organization and Maintenance of Large Ordered Indices* by Rudolf Bayer and Edward M. McCreight. While System R was a research prototype, commercialization followed with products by IBM (DB2 – DB2 and Informix are trademarks or registered trademark of International Business Machines Corporation in the United States, other countries, or both. Oracle is a trademarks or registered trademark of Oracle Corporation in the United States, other countries, or both. Other company, product, or service names may be trademarks or service marks of others.) and Oracle around 1980. INGRES also became commercial and was followed by POSTGRES which was incorporated into Informix. Prior to these relational systems, other data models for databases had been used, such as IMS which uses a hierarchical storage model and dominated industrial practice between 1967 and 1980. These pre-relational systems also had unique physical database design challenges.

As relational database systems advanced, new techniques were introduced to help improve operational efficiency, and reduce I/O by partitioning, distributing, and/or better indexing data.

All of these innovations, while improving the capabilities and potential for databases, expanded the scope of physical database design increasing the number of design choices and therefore the complexity of optimizing database structures. Though the 1980s and 1990s were dominated by the introduction of new physical database design capabilities, it can also be said that the years since have been dominated by efforts to simplify the process through automation, best practices, or deploying pre-configured "database appliances".

Foundations

The vast majority of physical database design tasks have the primary goal of reducing I/O consumption at runtime for query processing. However, to a lesser degree there are physical design aspects that help improve administrative efficiency by reducing the granularity of backup and restore operations (as in the case of range partitioning), or by improving the efficiency of mass insertion or deletion of data (so called "roll-in" and "roll-out" of data) as in the case of range partitioning. Some features can help reduce CPU or network consumption as in the case of memory tuning for heaps associated with these resources (such as the compiled SQL statement cache found in many commercial database server products).

Infrastructure Mechanics

Indexing schemes typically exploit tree-based structures or hash methods to provide fast access to data, with B+-trees being the dominant strategy. Range partitioning and clustering methods are based on grouping strategies to physically cluster like or similar data nearby on disk in order to reduce I/O. Hash partitioning methods hash data horizontally across storage objects so that processing can be performed on records associated with the different hash partitioning by different CPUs, providing parallel processing efficiencies. MVs are based on the strategy of pre-computation, so that results are available either in full or in part when needed without scanning and calculation. Considerable literature has been published on strategies to match the information within the MV at query compilation time to the queries requiring them, which is non trivial except in cases where the incoming query is an exact match to the MV.

The full breadth of physical database design is beyond the scope of this entry, so this discussion touches only on major aspects.

Design Choices

Physical database design, as performed by end users, has not yet been codified into a scientifically based strategy, and remains as much an art as a science. Typical strategies include cost-benefit analysis of applying various design features. Trial-and-error is still a significant part of the design process. For query processing, indexing is the first design feature considered, and the design process can generally explore the columns used in query predicates within the workload. Range partitioning is generally designed around date fields (by month or by quarter) for roll-in and roll-out processing, and backup granularity. Hash partitioning typically focuses on a unique key or near-unique key in order to efficiently distribute records across CPUs without skew. Data clustering strategies typically focus on columns used in range predicates, or in equality clauses that have high cardinality results sets in order to place similar data that are likely to be jointly required for query resolution physically nearby one another on disk.

Combining Physical Design Choices

Figure 1 shows a conceptual illustration of data stored in one large physical object without any specific treatment of the data, so that the data is unsorted, unclustered, unpartitioned, and not indexed. Data of different characteristics is illustrated by the different colored shapes that appear in the set. One can consider a query that was interested in data represented by the blue triangles in the image, which appear scattered throughout the object. A full data scan of the set would be required to find and fetch these data.

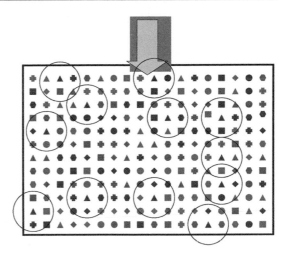

Physical Database Design for Relational Databases, Fig. 1 Data of various attributes, stored in a single monolithic table (Image courtesy of K. Beck at IBM)

Figure 2 shows a database with the same data as Fig. 1, which has been carefully designed to exploit a number of physical database design features in combination, such as indexing, range partitioning, MVs, MDC, and hash partitioning. In this new configuration a query for the data represented by the blue triangles is not only clustered in order to perform minimal IO, but the clearly indexed and can be distributed across several CPUs due to hash partitioning. Aggregation on the data can be precomputed and stored in the MV as well.

Other Physical Database Design Techniques

1. Other indexing mechanisms: hash index, bitmap index
   - These indexing strategies use hash functions, or bitmaps to index data that is typically slow changing. Bitmap indexes are favors for data with low cardinality (i.e., few distinct values).
2. Physical design for XML data
   - Semi-structured databases, like XML databases, require alternate indexing strategies.

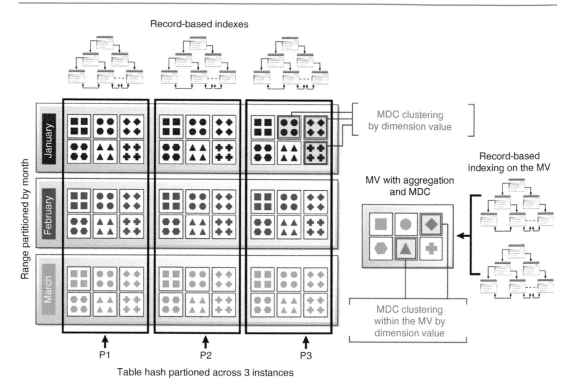

Physical Database Design for Relational Databases, Fig. 2 Data stored using a number of physical design extensions concurrently

3. Indexing for Information Retrieval
   - Text indexing is often based on an "inverted tree" index method.
4. Storage and object layout
   - Data placement on disk is affects the efficiency of storage and retrieval. Particularly important when some data has very different frequency or urgency of access than others.
5. Page and block size for I/O efficiency
   - Size of the storage pages, and the blocking levels (number of pages) used during I/O dramatically affect I/O efficiency, where random access favors smaller blocking and while sequential access favors large blocking.
6. Memory configuration and management
   - Complex design choice in the distribution of memory for caching, sorting, hashing, locking, tat typically has large impact on the efficiency of a database, especially if any key memory consumer is constrained.

Since memory is a fixed resource, the distribution of memory is a zero sum game (i.e., adding memory for one purpose requires reducing it for another)

7. Distributed database design
   - Distributed databases require complex design choices for data placement and choice of access (particularly where the distributed database includes redundancy, providing choices for which node to access).
8. Shared nothing partitioning
   - Shared nothing partitioning, common in data warehousing, hash partitions data across CPUs (and/or server) to achieve parallel processing over large volumes of data. The choice of hashing key is non trivial. However, in a shared nothing architecture, queries over each hash bucket later need to be shipped and merged since each node (or partition) is assumed to be distinct.

9. Hash partitioned tables
   - Hash partitioning within a server, can divide process of a query across CPUs, while still allowing the efficiencies of shared memory processing within a server.

Key Applications

Lifecycle Differences

It is important to distinguish between physical database design tasks performed during application development, versus those performed following deployment by an administrator.

During application development, the application designer performs logical database design and derives from it a physical design of the database. The physical design assigns typing to the elements of the entities in the database. This phase also designs the database tables, including normalization. Indices are defined, most commonly on primary and foreign keys. Views (non-materialized) are created to abstract the physical implementation away from the application as much as possible and to improve security of the system by controlling access to the data by the logical definition of the views, rather than by the physical structures of the storage.

Following deployment of the database, physical design continues, usually conducted by a database administrator. The design tasks here focus on performance tuning, or performance problem alleviation. Common tasks include the selection of secondary indexes, clustering, MDC, and range partitioning.

A common problem in database application development is that the application designer is forced to develop the database application using small samples of data. DBAs face design challenges when trying to optimize data access for applications that operate over large data sets, which the application designers themselves did not have access to when performing the initial application and database design.

Another key difference between the database design tasks performed at application development time and those following deployment relate to the physical resources of the database server.

In most cases the properties of the database server – number and speed of CPUs, CPU-cores, disks, memory, bus bandwidth, etc. – are unknown during application development. For applications that are deployed to possibly many sites or many customers there can be many different eventual target servers. Thus physical design qualities of the database that depend on the server's characteristics are inherently the role of the DBA to design. These include data placement and storage topology, memory allocation, and to a lesser degree clustering choices.

The design tasks performed by the DBA occur frequently with the deployment of new applications, application upgrades, or due to the evolution of the data set.

Application Domains

Different application domains require different physical design techniques. The section below highlights the two largest categories of application domains, namely Online Transaction processing (OLTP) and Decision Support Systems (DSS).

OLTP/ERP. Such applications deal with small numbers of records per transaction, and are commonly subject to extremely high concurrency rates. Key physical design characteristics include: (i) Small storage page sizes; (ii) Highly normalized data; (iii) Use of clustering indices or single-dimension MDC to enforce clustering of data; (iv) Memory allocation is heavily used for data caching (bufferpools), and to a lesser degree for caching compiled SQL statements. Memory work area for sorting, hash joins, etc, has considerably reduced demand in this space; (v) Due to the small selectivity of the transactions, the use of range partitioning, MDC with high dimensionality, and view materialization are not heavily used.

Decision Support, Data Warehousing and OLAP. These applications are characterized by large numbers of records accessed for aggregation, cubing etc. Table scans are common, and the workloads often include a high percentage of ad-hoc unpredictable queries, which can not be explicitly optimized for. Key physical design considerations here include: (i) Larger page size and storage blocking to increase the efficiency of scans and prefetching; (ii) Data may be stored in 3NF or Star Schema; (iii) Use of clustering is extremely common, especially long date and geography dimensions; (iv) Range partitioning is heavily used for roll-in and roll-out of data within the warehouse. This is almost always performed along a date dimension (such as by month or by quarter); (v) Memory allocation is complex in these environments. Data caching remains dominant, however, the common occurrence of sorting and hashing in these workloads places increased demand on work memory for these operations. The low concurrency characteristics of these workloads (compared to OLTP systems) place a reduced demand on memory for locking and caching of compiled statements.

Future Directions

The powerful capabilities of modern RDBMSs add complexity and have given rise to an entire professional domain of experts, known as Database Administrators (DBAs), to manage these systems. The problem is not unique to RDBMSs, but is ubiquitous in Information Technology today. Some have called this explosive growth of complexity "creeping featurism". In fact many companies now spend more money recruiting administrators to manage their technology than the money they have spent on the technology itself. One of the key roles of such DBAs is in the development and refinement of physical database design, to achieve "tuning" of the database.

The only real viable long term strategic solution is to develop systems that manage themselves. Such systems are called "self-managing" or "autonomic" systems. Several research efforts have focused on such self-designing database systems, working on problems such as:

1. What-if analysis reusing a database's native query optimizer to explore hypothetical design choices over a real or synthetic workload.
2. Data sampling and statistical analysis to explore the impact of design choices on data

distribution and density, or conversely to assess the suitability of target data for design attributes.

Cross-References

▶ Database Design
▶ Database Management System
▶ Database Middleware
▶ Database Security
▶ Distributed Database Systems
▶ Self-Management Technology in Databases
▶ Workflow Management

Recommended Reading

1. Astrahan MM, Blasgen MW, Chamberlin DD, Jim Gray W, King FII, Lindsay BG, Lorie RA, Mehl JW, Price TG, Putzolu GR, Schkolnick M, Selinger PG, Slutz DR, Strong HR, Tiberio P, Traiger IL, Bradford W, Yost WRA. System R: a relational data base management system. IEEE Comput. 1979;12(5): 42–8.
2. Bayer R, McCreight EM. Organization and maintenance of large ordered indexes. SIGFIDET Workshop; 1970. p. 107–41.
3. Finkelstein S, Schikolnick M, Tiberio P. Physical database design for relational databases. ACM Trans Database Syst. 1988;13(1):91–128.
4. Jun R, Chun Zhang, Megiddo N, Lohman GM. Automating physical database design in a parallel database. In: Proceedings of the ACM SIGMOD International Conference on Management of Data; 2002. p. 558–69.
5. Lightstone S, Bishwaranjan B. Automated design of multi-dimensional clustering tables for relational databases. In: Proceedings of the 30th International Conference on Very Large Data Bases; 2004. p. 1170–181.
6. Lightstone S, Teory T, Nadeau T. Physical database design: the database professional's guide to exploiting indexes, views, storage, and more. Morgan Kaufmann; 2007.
7. Markos Z, Cochrane R, Lapis G, Hamid P, Monica U. Answering complex SQL queries using automatic summary tables. In: Proceedings of the ACM SIGMOD International Conference on Management of Data; 2000. p. 105–16.
8. Sanjay A, Surajit C, Narasayya VR. Automated selection of materialized views and indexes in SQL databases. In: Proceedings of the 26th International Conference on Very Large Data Bases; 2000. p. 496–505.

9. Sriram P, Bishwaranjan B, Malkemus T, Cranston L, Huras M. Multi-dimensional clustering: a new data layout scheme in DB2. In: Proceedings of the ACM SIGMOD International Conference on Management of Data. p. 637–41.

10. Surajit C, Narasayya VR. AutoAdmin "What-if" index analysis utility. In: Proceedings of the ACM SIGMOD International Conference on Management of Data; 1998. p. 367–78.

11. Surajit C, Narasayya VR. Microsoft index tuning wizard for SQL Server 7.0. In: Proceedings of the ACM SIGMOD International Conference on Management of Data; 1998. p. 553–4.

12. Teorey T, Lightstone S, Nadeau T. Database modeling & design: logical design. 4th ed. Morgan Kaufmann; 2005.

13. Zilio DC, Jun R, Lightstone S, Lohman GM, Storm AJ, Garcia-Arellano Christian, Fadden Scott. DB2 design advisor: integrated automatic physical database design. In: Proceedings of the 30th International Conference on Very Large Data Bases; 2004. p. 1087–97.

14. Zilio DC, Zuzarte C, Lightstone S, Wenbin Ma, Lohman GM, Cochrane R, Hamid P, Latha SC, Gryz J, Alton E, Liang D, Valentin G. Recommending materialized views and indexes with IBM DB2 design advisor. In: Proceedings of the 1st International Conference on Autonomic Computing; 2004. p. 180–8.

Physical Layer Tuning

Philippe Bonnet[1] and Dennis Shasha[2]
[1]Department of Computer Science, IT
University of Copenhagen, Copenhagen,
Denmark
[2]Department of Computer Science, New York
University, New York, NY, USA

Definition

Tuning the physical layer entails choosing and configuring the underlying hardware and operating system to improve the performance of a database system running a given set of applications.

Historical Background

Ultimately, a database management system runs on processor and memory chips as well as secondary devices such as flash memory and disks. Trying to use those physical resources in the best way possible has been a problem since the first databases were available. Many researchers have noticed that disk capacity has been less of an issue than disk speed (or aggregate bandwidth from an entire disk array). Main memory size has always been a critical factor, because accesses to main memory are so much faster than accesses to disk. Processor speed has been important in applications that mix sophisticated computation with data access (for example in financial analysis), such computations are often handled outside the database server. New technologies have improved the capacity and speed of processors, memory, and disks, but memory is still vastly faster than disk (and somewhat faster than flash memory), so certain tuning principles will remain true.

Foundations

Hardware Tuning

Hardware tuning essentially consists of picking the hardware components on which the database management server is run. The goal is to design a balanced system [2].

Each processing unit consists of one or more processors, one or more disks, and some memory. Assuming a n-GIPS (billion instructions per second) processor, disks will be the bottleneck for on-line transaction-processing applications until the processor is attached to around 30n disks (counting 500,000 instructions per random IO issued by the database system and 70 random IOs per second). Each transaction spends far more time waiting for head movement on disk than in the processor.

Decision-support queries, by contrast, often entail massive scans of a table. In theory, an n-GIPS processor is saturated when connected to n/4 disks (counting 500,000 instruction per sequential IO and 8000 IO per second, considering 50 MB/second per disk and 64 KB per IO). In practice, the system bus might become the bottleneck. Thus, decision-support sites may need fewer disks per processor than transaction-processing sites for the purposes of matching aggregate disk bandwidth to processor speed. The numbers change when disks are implemented using flash memory, but the principle remains the same.

Operating System Tuning

Tuning the operating system primarily consists of providing an efficient input/output (IO) subsystem to the database server. A database server can issue buffered or unbuffered IO using the file system, or it can issue IO on raw disks thus bypassing the file system. Raw disks offer control and performance to the database system at the cost of a very low level of abstraction. The file system, on the other hand, provides a high level of abstraction, at the cost of decreased performance and double buffering. Further, IO is either synchronous or asynchronous. A synchronous IO forces the thread that issued the issuing thread to block until the IO is complete. Asynchronous IO allows the database server to issue concurrent IO requests on multiple files. Modern operating system such as Linux or Windows, provide database servers with direct asynchronous IO, that provide file abstraction and efficient IO scheduling at a low overhead. Note that out-of-the-box, a database system will most likely be configured to use standard buffered IO and thus needs to be tuned to use direct asynchronous IO.

As noted above, if one could eliminate the overhead caused by seeks and rotational delay, the aggregate bandwidth could increase by a factor of 10–100. Making this possible requires laying out the data to be read sequentially along disk tracks. Recognizing the advantage of sequential reads on properly laid-out data, most database systems encourage administrators to lay out tables in relatively large *extents* (consecutive portions of disk). Having a few large extents is a good idea for tables that are scanned frequently or (like database recovery logs or history files) are written sequentially. Large extents, then, are a necessary condition for good performance, but not sufficient, particularly for history files. Consider, for example, the scenario in which a database log is laid out on a disk in a few large extents, but another hot table is also on that disk.

The accesses to the hot table may entail a seek from the last page of the log; the next access to the log will entail another seek. So, much of the gain of large extents will be lost. For this reason, each log or history file should be the only hot file on its disk, unless the disk makes use of a large RAM cache to buffer the updates to each history file.

When accesses to a file are entirely random (as is the case in on-line transaction processing), seeks cannot be avoided. But placement can still minimize their cost, since seek time is roughly proportional to a constant plus the square root of the seek distance.

The operating system also allows to control the thread model used by the database server and to assign priorities to the different database server threads (and processes). The goal is to favor low latency requests (log writes), ensure fairness among database clients while avoiding starvation. This has several pitfalls however as discussed in [3] below.

Key Applications

A database system depends on the underlying hardware and operating system. Tuning the physical layer is thus a necessary step when deploying any database application.

Cross-References

▶ Database Management System
▶ Disk
▶ Main Memory
▶ Processor Cache

Recommended Reading

1. Gray J, Kukol P. Sequential disk IO tests for GBps land speed record. MS Research Technical Report MSR-TR-2004-62.
2. Gray J, Shanoy PJ. Rules of thumb in data engineering. In: Proceedings of the 18th International Conference on Data Engineering; 2000.
3. Hall C, Bonnet Ph. Getting priorities straight: improving Linux support for database I/O. In: Proceedings of

the 31st International Conference on Very Large Data Bases; 2005.

4. Shasha D, Bonnet P. Database tuning: principles, experiments and troubleshooting techniques. San Francisco: Morgan Kaufmann; 2002.

Pipeline

Ryan Johnson
Carnegie Mellon University, Pittsburg, PA, USA

Definition

Pipelining is a performance-enhancing technique which breaks a job into multiple overlapping pieces of work assigned to *pipeline stages*. Each pipeline stage operates in parallel, receiving inputs from a previous stage and passing intermediate results to the next. A pipeline with N stages can accommodate up to N in-progress jobs at a time. A well-designed pipeline requires only a fraction of the resources needed to achieve the same throughput with non-pipelined execution. A *balanced* pipeline, which distributes processing time evenly between all stages, can produce N results during the time it would have taken to process a single job all at once; unbalanced pipelines do not perform nearly as well. Pipelining also provides a secondary benefit of allowing each pipeline stage to specialize for its particular task, potentially making the combined stages both faster and less expensive than the non-pipelined server. Factory assembly lines leverage pipelining to great effect, allowing N specialized workers to achieve far higher throughput than N general purpose workers, while requiring less training per worker and fewer resources overall. The impressive throughput and efficiency pipelining provides make it a useful tool in both hardware and software design. Database systems often improve performance of query processing by executing query plans in pipelined fashion. Pipelining is a form of parallel dataflow execution.

Key Points

Suppose a non-pipelined server requires T time units and R resources to process a job all at once. The server's *latency* (time to process a job start-to-finish) and *occupancy* (time until it can accept more work) are both T. In contrast, a pipelined server with N stages ideally provides latency and occupancy of T' $\leq$ T and T'/N time units, respectively, while consuming R + ε resources. Because peak throughput is the reciprocal of the server's occupancy, pipelined execution potentially provides an N-fold increase in throughput over non-pipelined execution. When plenty of work is available, an initially empty pipeline *fills* or *ramps up* over a period of T' time units to reach peak operating capacity with N in-progress jobs. The pipeline must *stall* any time the first pipeline stage becomes ready and no new jobs are available. Stalls propagate through the pipeline as *bubbles* and reduce throughput by leaving stages idle.

Ideal pipelines assume that (i) the job can be broken into N pieces requiring no more than T/N time units each (possibly less due to gains from specialization), and that (ii) passing work between stages imposes no extra overhead. In practice, pipelines usually have non-negligible stage overhead and tend to be unbalanced, with a *bottleneck stage* that runs slower than the others. Real pipelines therefore have occupancy and

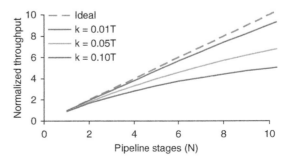

Pipeline, Fig. 1 Impact of unbalanced pipelines and pipelining overhead for as pipeline depth varies. Normalized throughput is equal to $1/(T/N + k)/T$. Note that just 1% overhead (and/or imbalance) costs 10% ideal throughput in a 10-stage pipeline

latency of T/N + k and T + kN, where k is the total extra delay imposed by imbalance and overhead. Pipeline bubbles, unbalanced stages, and overhead all become more severe for *deep* pipelines containing many stages, and limit the useful depth of any pipeline. Fig. 1 illustrates how even small values of k reduce peak throughput from the ideal as a pipeline gets deeper.

Pipelining

Evaggelia Pitoura
Department of Computer Science and Engineering, University of Ioannina, Ioannina, Greece

Definition

The query execution engine operates on an execution plan produced during query processing which typically is a physical operator graph whose edges specify the dataflow among the operators. There are two basic alternatives for evaluating the execution plan: materialization and pipelining. With materialized evaluation, the results of each operator are created (materialized) and stored in disk and then used as input for the evaluation of the next operator. With pipelining, the results of an operator are passed along as input to the next operator, before the first operator completes its execution. In general, pipelining improves the efficiency of query evaluation by reducing the number of temporary files that are produced.

Key Points

The output of query optimization is an execution plan or operator tree. The execution plan can be thought of as a dataflow graph where the nodes correspond to the physical operators and the edges represent the data flow among these physical operators. The query execution engine is responsible for the execution of this plan. The engine provides generic implementations of all physical operators that take as input one or more data streams and produce one output data stream.

In most database systems, each operator supports an *iterator* interface that allows a parent operator to get the results from its children one tuple at a time. The iterator interface supports *pipelining*, since tuples produced as output by a child operator can be fed as input to their parent in the operator tree, even before the child operator has finished its execution. The alternative would be *materialized evaluation*, in which each operator would fully complete its operation and write its results to disk before the next operator starts execution. By reducing the number of intermediate results written to disk, pipelining improves the efficiency of query evaluation. However, not all physical operators can use pipelining efficiently. For instance, merge-join cannot exploit pipelining in general, since it requires both its input relations to be sorted and it is not possible to sort a relation until all its tuples are available.

Pipelines can be executed in one of two ways: demand-driven or producer-driven. In a *demand-driven* or *demand-pull* pipeline, an operator computes the next tuple or tuples to be returned after receiving a corresponding request from its parent operator. In a *producer-driven* or *push* pipeline, operators do not wait for requests to produce tuples, instead, they generate output tuples eagerly and put them in their output buffers until the buffers become full. Demand-driven pipelining is more common, since it is easier to implement. It also allows for the whole plan to be executed by a single process or thread.

Cross-References

▶ Evaluation of Relational Operators
▶ Iterator
▶ Parallel Query Execution Algorithms
▶ Query Processing

Recommended Reading

1. Graefe G. Query evaluation techniques for large databases. ACM Comput Surv. 1993;25(2):73–170.
2. Silberschatz A, Korth HF, Sudarshan S. Database system concepts. 5th ed. New York: McGraw Hill; 2005.

Platform-as-a-Service (PaaS)

Chandra Krintz
Department of Computer Science, University of
California, Santa Barbara, CA, USA

Synonyms

Application platform-as-a-service (APaaS); Integration platform-as-a-service (APaaS)

Definition

Platform-as-a-service (PaaS) describes a distributed software system that exports a wide variety of scalable application services via application programming interfaces (APIs). These services include program deployment and execution support, data storage and management, authentication and user management, and background tasking, among others. PaaS typically targets Internet-accessible software written in high-level programming languages.

Overview

Platform-as-a-service (PaaS) characterizes a software architecture that significantly simplifies and automates the deployment and management of Internet-accessible applications by providing access to and management of scalable application infrastructure services. These services facilitate application execution (via load balancers, application servers, and background tasking) and provide scalable implementations of components that are commonly used by web-facing software. *Polyglot* PaaS systems support applications written in multiple high-level programming languages through the integration of different application servers. The platform automatically deploys, configures, and manages these services and can add and remove services according to load/usage or as directed by users.

PaaS systems make application components available via application programming interfaces (APIs) to network-attached services. Most PaaS systems provide APIs for (un-)structured data storage and caching, queuing systems, as well as user authentication and management. Some PaaS systems also offer system-wide services for platform customization, deployment governance, and business process management. Services are shared by multiple applications and isolated via system-level virtualization or other operating system and programmatic mechanisms (e.g., processes, containers, and namespaces).

Users of *public* PaaS systems rent resources on a pay-per-use basis, using quotas and pricing models (e.g., per access, by volume, by duration of use, etc.) in exchange for performance guarantees specified via service-level agreements (SLAs). Some PaaS systems restrict application functionality (e.g., communication protocols, language and library use, request-response and service access rates, etc.) to protect system stability and to meet SLAs and scaling requirements. Examples of public PaaS systems include Google App Engine and Microsoft Azure. Examples of open source PaaS software for private use via on-premise clusters or IaaS systems include AppScale [1], CloudFoundry, and RedHat OpenShift.

Cross-References

▶ Cloud Computing
▶ Infrastructure-as-a-Service (IaaS)
▶ Software-as-a-Service (SaaS)

References

1. Krintz C. The appscale cloud platform: enabling portable, scalable web application deployment. IEEE Internet Comput. 2013;17(2):72–5.

Point-in-Time Copy

Kazuhisa Fujimoto
Hitachi Ltd., Tokyo, Japan

Synonyms

PiT copy; Snapshot

Definition

A point-in-time copy is a copy of original data as it appeared at a point in time. In a conventional backup operation, users often create a PiT Copy, while an application is in quiescing, to make the PiT Copy a consistent copy of original data.

Key Points

There are two popular implementation techniques for creating PiT Copies inside a storage system: split mirror and copy on write.

Split mirror is a technique for replicating the original data at a point in time. In some implementations, a storage system replicates the original data, and when users create the PiT Copy, a storage system splits the replication. Copy on Write (CoW) is a technique for capturing data changes to storage and creating a PiT Copy after specifying the point in time. In some implementations, when users create the PiT Copy, a storage system creates its image with both the original data and the modified data.

Techniques for creating PiT Copies have a trade-off between occupied storage capacity and performance. A PiT Copy created by split mirror occupies as much storage as the original data, but a PiT Copy by CoW usually occupies less storage. On the other hand, CoW usually has more time for accessing the copy source and creating PiT Copies than split mirror.

Cross-References

▶ Backup and Restore
▶ Logging and Recovery

Recommended Reading

1. Alain Azagury, et al. Point-in-time copy: yesterday, today, and tomorrow. In: Proceedings of the IEEE Symposium Conference on Mass Storage Systems and Technologies; 2002.

Point-Stamped Temporal Models

David Toman
University of Waterloo, Waterloo, ON, Canada

Synonyms

Point-based temporal models

Definition

Point-stamped temporal data models associate database objects with time instants, *indivisible* elements drawn from an underlying time domain. Such an association usually indicates that the information represented by the database object in question is *valid* (i.e., believed to be true by the database) at that particular time instant. The time instant is then called the *timestamp* of the object.

Historical Background

Associating time-dependent data with time instants has been used in sciences, in particular in physics, at least since Isaac Newton's development of classical mechanics: time is commonly modeled as a two-way infinite and continuous linear order of indivisible time instants (often with distance defined as well). Similarly, in many areas of computer science, ranging from protocol specifications to program verification to model checking, discrete point-based timestamps play an essential role. In database systems (and in AI), however, the requirement of *finite* and compact representation has often lead to the use of more complex timestamps, such as intervals. This is particularly common when information about *durations* is stored in a database (in AI, such timestamps are called *fluents*). However, Chomicki [2] has shown that many of these approaches are simply compact representations of large and potentially infinite sets of time instants associated with a particular fact and that, at the conceptual level, the underlying information is often better understood as being timestamped by time instants.

Foundations

Temporal data models are typically defined as extensions of standard non-temporal data models that provide means for representation and manipulation of time-dependent data. Temporal extensions of the relational model accomplish this by relativizing *truth*, i.e., the sets of facts the database believes to be true in the modeled reality, with respect to elements of the time domain. These elements are then called the timestamps of the facts and, intuitively, specify *at which times* the associated facts are true. In the case of point-stamped temporal models, these elements are indivisible time instants drawn from an appropriate time domain.

The main ideas are outlined in a setting in which the time domain is an unbounded countably infinite linear order of *time instants*. Moreover, while the main focus is on *temporal extensions of the relational model*, most of the issues arise in other data models as well, and admit similar solutions.

Timestamps and Database Objects

The choice of timestamps is orthogonal to other choices that define the flavor and properties of the resulting temporal data model, in particular:

1. To the decision of *what database objects the timestamps are attached to*, and
2. To the decision of *how many timestamps* (per such object) are used.

First to explore is the choice of objects to be timestamped. Intuitively, timestamps should only be attached to objects that actually *capture* information in the underlying data model, indicating that the particular (piece of) information is valid for that timestamp. In the relational model these are *tuples* and their membership in *relations*. Thus, the two principal choices are as follows:

The Snapshot Model

Temporal databases in the *snapshot model* are formally defined as mappings from the time domain T to (the class of) standard relational databases with a particular fixed schema ρ. In the case of a linearly ordered time domain, such a temporal database can be viewed as a time-indexed sequence of standard relational databases, commonly called *snapshots*. Note that such a sequence is not necessarily discrete or finite: that depends on the choice of the underlying time domain. Intuitively, to capture the fact that, in a snapshot temporal database D, a relationship r holds among uninterpreted constants $a_1, \ldots, a_k$ at time t it must be the case that $r(a_1, \ldots, a_k)$ holds in $D(t)$, where $D(t)$ is a standard relational instance, the snapshot of D at time t; this statement is denoted by writing $r^{D(t)}(a_1, \ldots, a_k)$.

These structures, in the area of *modal logics*, are called *Kripke structures* in which worlds are described by relational databases and where the time domain serves as the accessibility relation.

The Timestamp Model

Temporal databases in the *timestamp model* are defined in terms of temporal relations, relations whose schemas are extended with an additional attribute ranging over the time domain. This attribute is commonly referred to as the *timestamp attribute* or simply *timestamp*.

More formally, a relational symbol R is a *timestamped extension* of a symbol $r \in \rho$ if it contains all attributes of r and a single additional attribute t of the temporal sort (without loss of generality, assuming that it is always the first attribute). A *timestamp temporal database* D is then a first-order structure $D = \{R_1^D, \ldots, R_k^D\}$ consisting of the interpretations (instances) for all the temporal extensions R_i of r_i in ρ. The instances R_i^D are called *temporal relations*. Similarly to the snapshot case, there is no restriction on the number of timestamps (i.e., the cardinality of the set of timestamps) in such instances; issues connected with the actual finite representation of these relations are addressed below. However, the relation

P

$$\left\{ a_1, \ldots, a_k : (t, a_1, \ldots, a_k) \in R_i^D \right\},$$

a snapshot of R at t, must be finite for every timestamp $t \in T$.

It is easy to see that snapshot and timestamp temporal models are simply different views of the same (isomorphic) sets of facts and thus represent the same class of temporal databases. Formally, a snapshot temporal database D corresponds to a timestamp temporal database D' (and vice versa) as follows:

$$\forall t. \forall x_1, \ldots, x_k \, r^{D(t)} (x_1, \ldots, x_k)$$

$$\Longleftrightarrow R^{D'} (t, x_1, \ldots, x_k),$$

where r and R are a relation in the schema of D and its temporal extension in the schema of D', respectively. This correspondence makes the two models interchangeable. Hence, for temporal queries formulated in query languages such as Temporal Relational Calculus (TRC) or First-order Temporal Logic (FOTL), these two models of point-stamped temporal databases can be used interchangeably.

The Parametric Model

Several temporal data models propose to use time-dependent *unary functions* as attribute values of tuples in temporal relations. These models are called *parametric models* or *attribute timestamped models*. As unary functions can be represented by binary relations (e.g., with the first attribute being the timestamp and the second attribute the value of the function for that timestamp), these models are examples of nested, non-first normal form models [4, 5]. Moreover, if the implicit grouping of tuples in such relations is accidental or in the presence of appropriate keys, a first-normal form representation can be obtained by unfolding, similarly to the case of the nested relational model. The transformation is then defined by:

$$R := \{ (t, f_1(t), \ldots, f_k(t)) \mid$$

$$(f_1, \ldots, f_k) \in R^P, t \in T \},$$

where R is a timestamp relation corresponding to the parametric relation R^P. Note also that if a varying number of tuples is to be modeled in this model, *partial functions* must be used in R^P; then, however, tuples can become only partially defined for a given time instant, and therefore additional conditions must be enforced. This makes the model quite cumbersome to use, in particular when compared to the snapshot and timestamp models. Many of these difficulties originate from associating timestamps with *uninterpreted constants* that, in the relational model, *do not* carry any information on their own.

Multiple Atomic Timestamps

The second issue that arises is the question of *how many* timestamps are attached to database objects. So far single-dimensional temporal databases were considered, i.e., where each database object was associated with a single timestamp. The intuition behind such models is that *validity* of data is determined by a single time instant. In the case of the timestamp model, this translates to the fact that temporal relations were allowed only a single (often *distinguished*) temporal attribute. However, there are two natural reasons to relax this requirement and to allow multiple timestamps to be associated with database objects (i.e., multiple timestamp attributes to appear in schemas of relations, either explicitly or implicitly). There are two cases to consider:

Models with a Fixed Number of Timestamp Attributes

In many cases data modeling requires attaching several timestamps to database objects. Intuitively, a particular piece of information can be associated with, e.g., a timestamp for which the information is *valid* in the modeled world and another timestamp that states when the information is *recorded* in the database. Hence, every object has two timestamps. The resulting data model is called the bitemporal model [6] and the timestamps are called valid time and transaction time, respectively. Similarly, one can envision temporal models with three (or any fixed number) of *distinguished temporal attributes* – attributes with a predetermined interpretation – that are *common to all temporal relation schemas.* Temporal data models with an apriori *fixed* number of timestamp attributes can still be equivalently represented using the snapshot and the timestamp approaches. In the snapshot case, database instances are indexed by fixed tuples of time instants. Hence the bitemporal model can be seen as a two-dimensional plane of relational instances. The timestamp model simply adds an appropriate number of (distinguished) attributes to the timestamp extensions of relational schemes.

Models with a Varying Number of Timestamp Attributes

While most of the data modeling techniques require only a fixed number of timestamp attributes in schemas of temporal relations, it is often convenient to allow arbitrary number of timestamp attributes to be associated with a database object. This leads to allowing temporal data models without limits on the number of timestamp attributes in schemas of temporal relations. In such models, time dependencies are captured by explicit (user-defined) attributes ranging over the time domain. For a varying number of timestamps, only the *timestamp* view of the temporal data model makes sense.

Fixed-dimensional temporal data models are appealing as they commonly provide an additional built-in *interpretation* for timestamps, e.g., the valid time timestamp always states that the information is *true* in the world at that particular time. Temporal data models with varying, user-defined timestamps do not posses this additional interpretation, and the exact meaning of the timestamps depends on how the world is modeled by the associated attributes (similarly to the standard relational case). However, there is another need for models with varying and unbounded number of timestamp attributes: for temporal query languages based on temporal relational calculus, there cannot be an equivalent temporal relational algebra defined over any of the fixed-dimensional temporal data models (see

the entry on Temporal Logic in Database Query Languages or [7, 8] for details).

Sets of Timestamps: Compact Representation

The point-stamped temporal data models and the associated temporal query languages provide an excellent vehicle for defining a precise semantics of queries and for studying their properties. However, in *practical applications* an additional hurdle has to be overcome: a naive storage of point-stamped temporal relations, either in the timestamp or in the snapshot model, is often not possible (as the instances of the relations can be infinite, e.g., sets of time instants that represent bounded durations are infinite when a *dense time domain* is used) or impractical (the number of instants is very large). For these and other reasons, many temporal data models associate database objects with *sets* of time instants rather than with single individual time instants.

Temporal models that use complex timestamps, such as intervals, no longer appear to be point-stamped – or in the first normal form (1NF). However, Chomicki [2] has shown that in many cases these complex objects are merely compact representations of a possibly large or even infinite number of time instants associated with a particular fact. The most common approach along these lines is to attach timestamps in the form of *intervals* to facts that persist over time. For example, the temporal relation **WorksFor** in Fig. 1 can be compactly (and finitely) represented by the relation:

$$\mathbf{WorksFor'} = \Big\{ ([2001, 2002], \text{John}, \text{IBM}), \\ ([2004, \infty], \text{John}, \text{Microsoft}) \Big\}.$$

Note that such an *interval encoding* is not unique and multiple snapshot equivalent representations can exist. For single-dimensional temporal relations, uniqueness of the representation can be achieved using temporal coalescing. However, it is not clear that meaningful, canonical ways of defining coalescing for multi-dimensional temporal relations, including bitemporal relations, exist [3].

P

Year	Database snapshot (as a set of facts)
...	
1997	{**Degree**(John,BS,MIT)}
1998	{ }
1999	{**Degree**(John,MS,UofT)}
2000	{ }
2001	{**WorksFor**(John,IBM)}
2002	{**WorksFor**(John,IBM)}
2003	{ }
2004	{**Degree**(John,PhD,UW), **WorksFor**(John,Microsoft)}
2005	{**WorksFor**(John,Microsoft)}
...	{**WorksFor**(John,Microsoft)

WorksFor

Year	Name	Company
200	John	IBM
200	John	IBM
200	John	Microsoft
200	John	Microsoft
..	John	Microsoft

Degree

Year	Name	Degree	School
1997	John	BS	MIT
1999	John	MS	UofT
2004	John	PhD	UW

Point-Stamped Temporal Models, Fig. 1 Instances of matching snapshot and timestamp temporal database

On the other hand, the use of temporal coalescing and other set-oriented operations on interval timestamps, such as intersection of timestamps in temporal joins, indicates that the interval timestamps are indeed solely a representation tool for point-stamped relations: these operations cannot be used in a model that associates truth with the intervals themselves as there is no clear way to do so for the intervals that result from such operations, e.g., it is unclear – from a conceptual point of view – what facts should be associated with an intersection of two intervals *without* implicitly assuming that facts associated with the original two intervals hold *for all time points contained in those two intervals*, or at least for certain subintervals.

The choice of *intervals* to compactly represent adjacent timestamps originates from the structure of the temporal domain: intervals are the only (convex) one-dimensional sets that can be defined by (first-order) formulas in the language of linear order. For more structured time domains, however, the repertoire of encodings for sets of timestamps can be richer, e.g., using formulas describing *periodic sets*, etc., as compact timestamps.

Query Languages and Integrity Constraints

Point-stamped data models support point-based temporal query languages that are commonly based on extensions of the relational calculus (first-order logic). The two principal extensions are as follows:

1. First-order Temporal Logic: a language with an implicit access to timestamps using temporal connectives and
2. Temporal Relational Calculus: a two-sorted logic with variables and quantifiers explicitly ranging over the time domain.

These languages are the counterparts of the snapshot and timestamp models. However, unlike the data models that are equivalent in their expressiveness, the second language is strictly more expressive than the first [1, 8]. Temporal integrity constraints can also be conveniently expressed in these languages [3].

Key Applications

Point-based temporal data models serve as the underlying data model for most abstract temporal query languages.

Cross-References

▶ Bitemporal Relation
▶ Constraint Databases
▶ Key
▶ Nested Transaction Models

Recommended Reading

1. Abiteboul S, Herr L, Van den Bussche J. Temporal versus first-order logic to query temporal databases. In: Proceedings of the 15th ACM SIGACT-SIGMOD-SIGART Symposium on Principles of Database Systems; 1996. p. 49–57.
2. Chomicki J. Temporal query languages: a survey. In: Proceedings of the 1st International Conference on Temporal Logic; 1994. p. 506–34.
3. Chomicki J, Toman D. Temporal databases. In: Gabbay D, Villa L, Fischer M, editors. Handbook of temporal reasoning in artificial intelligence. London: Elsevier; 2005. p. 429–67.
4. Clifford J, Croker A, Tuzhilin A. On the completeness of query languages for grouped and ungrouped historical data models. In: Tansel A, Clifford J, Gadia S, Jajodia S, Segev A, Snodgrass RT, editors. Temporal databases: theory, design, and implementation. Redwood City: Benjamin/Cummings; 1993. p. 496–533.
5. Clifford J, Croker A, Tuzhilin A. On completeness of historical relational query languages. ACM Trans Database Syst. 1994;19(1):64–116.
6. Jensen CS, Soo MD, Snodgrass RT. Unification of temporal data models. In: Proceedings of the 9th International Conference on Data Engineering; 1993.
7. Toman D. On incompleteness of multi-dimensional first-order temporal logics. In: Proceedings of the 10th International Symposium on Temporal Representation and Reasoning/4th International Conference on Temporal Logic; 2003. p. 99–106.
8. Toman D, Niwinski D. First-order queries over temporal databases inexpressible in temporal logic. In: Advances in Database Technology, Proceedings of the 5th International Conference on Extending Database Technology; 1996.

Polytransactions

George Karabatis
University of Maryland, Baltimore Country (UMBC), Baltimore, MD, USA

Definition

A polytransaction T^+ is a transitive closure of a transaction T submitted to an Interdependent Data Management System (a type of a multidatabase system which enforces dependencies among related data objects). The transitive closure is computed with respect to an Interdatabase Dependency Schema (IDS) consisting of a collection of data dependency descriptors (D^3s), which specify the dependencies between data objects.

Key Points

Data objects in multiple (possibly distributed and/or heterogeneous) systems which form dependencies among them are called *interdependent data*. These dependencies are identified through *data dependency descriptors* (D^3s). A D^3 specifies the relationships between source and target objects including how much inconsistency can be tolerated between them before it is necessary to restore it, and the actions to take to maintain consistency when it reaches intolerable levels. All these D^3s together form an *Interdatabase Dependency Schema* (IDS) [3].

When a source object in a D^3 is updated, the consistency between itself and the target object in the same D^3 may be violated beyond allowable levels. Then, a system-generated transaction updates the target object to maintain consistency. The updated target object may participate as source in another D^3, therefore a series of related transactions may execute to maintain consistency. These related transactions form a transaction tree called a *polytransaction* [1, 2, 3].

A polytransaction tree contains nodes (corresponding to its component transactions), and edges (identifying the "execution mode" between the parent and children transactions) and is determined as follows. For each D^3_ε IDS containing a source object updated by T, and in need to restore consistency with its target object, a new node is added corresponding to a new transaction T' (child of T). The operations of T' which restore consistency are the actions specified in the D^3.

Execution modes: A child transaction is *coupled* if the parent transaction must wait until the child transaction completes before proceeding further. It is *decoupled* otherwise. Also, a coupled transaction is *vital* (the parent transaction must fail if the child fails), or *nonvital* (the parent transaction survives the failure of a child).

Cross-References

▶ Database Dependencies
▶ Database Management System
▶ Distributed Transaction Management
▶ Extended Transaction Models and the ACTA Framework
▶ Inconsistent Databases
▶ Transaction
▶ Transaction Management
▶ Transaction Manager

Recommended Reading

1. Karabatis G, Rusinkiewicz M, Sheth A. Correctness and enforcement of multidatabase interdependencies. In: Advanced database systems LNCS, vol 759. London: Springer; 1993. p. 337–58.
2. Karabatis G, Rusinkiewicz M, Sheth A. Interdependent database systems. In: Elmagarmid A, Rusinkieuicz M, Sheth A, editors. Management of heterogeneous and autonomous database systems. San Francisco: Morgan-Kaufmann; 1999. p. 217–52.
3. Rusinkiewicz M, Sheth A, Karabatis G. Specifying interdatabase dependencies in a multidatabase environment. IEEE Comput. 1991;24(12):46–53.

Positive Relational Algebra

Cristina Sirangelo
IRIF, Paris Diderot University, Paris, France

Synonyms

SPJRU-algebra; SPCU-algebra

Definition

Positive relational algebra is the fragment of relational algebra which excludes the difference operator. Relational algebra queries are expressions defining mappings from database instances of an input database schema to an output relation. In particular a positive relational algebra query over a database schema τ is one of the following expressions, each one with an associated set of attributes:

- A constant relation over a set of attributes U is a positive relational algebra query with associated set of attributes U.
- If $R(U)$ is a relation schema in τ, the relation symbol R is a positive relational algebra query with associated set of attributes U.
- If Q and Q' are positive relational algebra queries with sets of attributes U and U', respectively, the following are positive relational algebra queries:
  - The *selection* $\sigma_{A=B}(Q)$ or $\sigma_{A=c}(Q)$, with set of attributes U, where A and B are attributes in U, and c is a constant value
  - The *projection* $\pi_X(Q)$, with set of attributes X, where $X \subseteq U$
  - The *natural join* $Q \bowtie Q'$, with set of attributes $U \cup U'$
  - The *renaming* $\rho_{U \to W}(Q)$, with set of attributes W
  - The *union* $Q \cup Q'$, in the case that $U = U'$, with associated set of attributes U

The semantics of a positive relational algebra query on a database instance I of schema τ is a relation instance defined as follows: the semantics of a constant relation is the relation itself; the semantics of a relation symbol R of τ is the value of R in I; the semantics of selection, projection, natural join, renaming, and union expressions above is defined according to the semantics of the corresponding operator on inputs given by the semantics of Q and Q'.

Key Points

The basic operators of the positive relational algebra are a nonredundant set: by removing any of these operators, the set of expressible query mappings would be reduced. Moreover they allow the simulation of other operators. Among these, the intersection $R_1 \cap R_2$ of two relations over the same set of attributes can be simulated as $R_1 \bowtie R_2$; the generalized selection whose condition is a positive Boolean combination of equality atoms can be simulated using composition and union of primitive selection operators; consequently also the theta-join, with the same restriction on the join condition, and the equijoin can be simulated.

An example of positive relational algebra query over a database schema consisting of relation schemas *Students(student-number, student-name)* and *Exams(course-number, student-number, grade)* is the expression $\pi_{student-name} (Students \bowtie \sigma_{grade=A} (Exams))$. It returns the names of the students that have passed at least one exam with grade A.

The positive relational algebra is equivalent (in that it expresses the same query mappings) to other query languages such as *unions of conjunctive queries* in safe relational calculus and *nonrecursive datalog* with one target predicate.

In the absence of attribute names, the basic operators of positive relational algebra are $\{\sigma, \pi, \times, \cup\}$. The positive relational algebra without names turns out to be equivalent to the positive relational algebra with names and thus equivalent to the other above-mentioned query paradigms.

Cross-References

▶ Difference
▶ Join
▶ Datalog
▶ Projection
▶ Relational Algebra
▶ Relational Calculus
▶ Selection
▶ Union

Recommended Reading

1. Abiteboul S, Hull R, Vianu V. Foundations of databases. Reading: Addison-Wesley; 1995.

Possible Answers

Gösta Grahne
Concordia University, Montreal, QC, Canada

Synonyms

Consistent facts; Credulous reasoning; Maybe answer

Definition

In an incomplete database, which is a set of complete databases (possible worlds), the possible answer to a query is the set of tuples that are in the answer to the query, when posed on *some* possible world. The dual of the possible answer is the *certain* answer, which consists of all tuples true in the answer to the query when posed simultaneously on *all* possible worlds.

Key Points

For more information on the certain and possible answers, as well as on various (partial) orders on incomplete databases, see [1, 2].

Cross-References

▶ Certain (and Possible) Answers
▶ Incomplete Information
▶ Naive Tables

Recommended Reading

1. Grahne G. The problem of incomplete information in relational databases. Berlin: Springer; 1991.
2. Libkin L. Aspects of partial information in databases. PhD Thesis, University of Pennsylvania. 1994.

PRAM

Josep Domingo-Ferrer
Universitat Rovira i Virgili, Tarragona, Catalonia, Spain

Synonyms

Post-randomization method

Definition

The Post-RAndomization Method (PRAM) is a probabilistic, perturbative masking method for disclosure protection of categorical microdata. In the masked file, the scores on some categorical attributes for certain records in the original file are changed to a different score according to a prescribed probability mechanism, namely a Markov matrix. The Markov approach makes PRAM very general, because it encompasses noise addition, data suppression and data recoding.

Key Points

The PRAM matrix contains a row for each possible value of each attribute to be protected. This rules out using the method for continuous data. PRAM was invented by Gouweleeuw et al. [1].

The information loss and disclosure risk in data masked with PRAM largely depend on the choice of the Markov matrix and are still (open) research topics [2].

Cross-References

▶ Inference Control in Statistical Databases
▶ Microdata
▶ SDC Score

Recommended Reading

1. Gouweleeuw JM, Kooiman P, Willenborg LCRJ, DeWolf P-P. Post randomisation for statistical disclosure control: theory and implementation, 1997. Statistics Netherlands. Voorburg: Research Paper No. 9731; 1997.
2. de Wolf P-P. Risk, utility and PRAM. In: Domingo-Ferrer J, Franconi L, editors. Privacy in statistical databases LNCS, vol. 4302. 2006. p. 189–204.

Precision

Ethan Zhang[1] and Yi Zhang[2]
[1] University of California, Santa Cruz, CA, USA
[2] Yahoo! Inc., Santa Clara, CA, USA

Definition

Precision measures the accuracy of an information retrieval (IR) system. More precisely, precision is the fraction of retrieved documents that are relevant. Consider a test document collection and an information need Q. Let R be the set of documents in the collection that are relevant to Q. Assume an IR system processes the information need Q and retrieves a document set A. Let $|R_a|$ denote the number of documents that are in both R and A. Further let $|R|$ and $|A|$ be the numbers of documents in R and A, respectively. The *precision* of the IR system for Q is defined as $P = |R_a|/|A|$.

Key Points

Precision and recall are the most frequently used and basic retrieval performance measures. Many other standard performance metrics are based on the two concepts.

Cross-References

▶ Eleven Point Precision-Recall Curve
▶ Average Precision
▶ F-Measure
▶ Precision-Oriented Effectiveness Measures
▶ Recall
▶ Standard Effectiveness Measures

Precision and Recall

Ben Carterette
University of Massachusetts Amherst, Amherst, MA, USA

Synonyms

False negative rate; Positive predictive value; Sensitivity

Definition

Recall measures the ability of a search engine or retrieval system to locate relevant material in its index. *Precision* measures its ability to not rank nonrelevant material. With everything above rank cut-off n considered "retrieved" and everything below considered "not retrieved," precision and recall can be stated mathematically as:

$$\text{precision} = \frac{|\text{retrieved \& relevant at rank } n|}{|\text{retrieved at rank } n|}$$

$$\text{recall} = \frac{|\text{retrieved \& relevant at rank } n|}{|\text{relevant}|}$$

Key Points

Precision and recall are the traditional metrics for retrieval system performance evaluation, and nearly all other performance measures can be seen as either precision-based, recall-based, or a combination of the two.

		Condition	
		True	False
Prediction	True	True positive	False positive
	False	False negative	True negative

There is a trade-off between precision and recall: as more is done to bring more relevant results into a ranking, increasing recall, inevitably nonrelevavnt results will be captured as well, decreasing precision. This can be seen in precision-recall curves, which plot precision against recall while varying rank cut-off n.

Precision and recall are not limited to information retrieval; they can be defined more formally on any 2×2 contingency table. For the table precision is True Positives divided by True Positives + False Positives and recall is True Positives divided by True Positives + False Negatives. In medical literature, precision is known as *positive predictive value* and recall as *sensitivity*; in the statistics literature, recall is equivalent to the *power* (or $1 - $ Type II error rate) of a hypothesis test (precision has no analogue).

Cross-References

▶ Average Precision
▶ Precision at n
▶ Recall

Recommended Reading

1. van Rijsbergen CJ. Information retrieval. London: Butterworths; 1979.

Precision at n

Nick Craswell
Microsoft Research Cambridge, Cambridge, UK

Synonyms

P@n

Definition

In an information retrieval system that retrieves a ranked list, the top-n documents are the first n in the ranking. Precision at n is the proportion of the top-n documents that are relevant.

Key Points

If r relevant documents have been retrieved at rank n, then:

$$\text{Precision at n} = \frac{r}{n}$$

The value of n can be chosen based on an assumption about how many documents the user will view. In Web search a results page typically contains ten results, so $n = 10$ is a natural choice. However, not all users will use the scrollbar and look at the full top ten list. In a typical setup the user may only see the first five results before scrolling, suggesting Precision at 5 as a measure of the initial set seen by users. It is the document at rank 1 that gets most user attention, because this is the document that users view first, suggesting the use of Precision at 1 (which is equivalent to Success at 1).

It is possible to calculate precision at a later cutoff, although precision gives equal weight to every result in the list. For example, when calculating precision at 1,000 the 1,000th document is as important as the 1st, whereas users of a ranked retrieval system are likely to consider the 1st document most important. Other information retrieval measures place a greater emphasis on early ranks, such as Mean Average Precision and Mean Reciprocal Rank.

In set-based retrieval, where there is no ranking, the precision of a retrieved set of size n can still be calculated, as r/n.

In order to calculate precision at n it is only necessary to obtain relevance judgments for the top-n documents, unlike recall, which can only be measured if the complete set of relevant documents has been identified.

Cross-References

▶ Average Precision
▶ Precision-Oriented Effectiveness Measures
▶ Recall
▶ Success at n

Precision-Oriented Effectiveness Measures

Nick Craswell
Microsoft Research Cambridge, Cambridge, UK

Definition

Precision-oriented evaluation in information retrieval considers the relevance of the top n search results, for small n and using a set of relevance judgments that need not be complete. Such "shallow" evaluation is consistent with a user who only cares about the top-ranked documents. Relaxing the requirement of identifying all relevant documents for every query means that certain measures, such as recall at n, cannot be applied. However, it also allows evaluation on a very large corpus, where employing human relevance assessors to find the complete relevant set for each query would be too expensive. Both aspects of precision-oriented evaluation, the shallow viewing of results and the large corpus, are associated with Web search, where search results are typically a top-10 and the corpus may contain tens of billions of documents.

Historical Background

The Cranfield II experiments in 1963 were a landmark effort in information retrieval evaluation [3]. A test collection comprising 1,440 documents, 225 queries and relevance judgments was created. Because the document set was small, it was possible to judge the relevance of every document for each query. Using complete relevance judgments it is possible to evaluate precision, the proportion of retrieved documents that are relevant, and recall, the proportion of relevant documents that are retrieved. The test collection is reusable, in that it is possible to run a new retrieval experiment using existing queries and relevance judgments, without encountering any unjudged document in the search results.

Judging all documents for every query does not scale well with corpus size, and becomes untenable even with tens of thousands of documents. The best-known solution to this problem is the pooling method, as introduced in the Text Retrieval Conference (TREC) in 1991 [5]. Under pooling, the top-n documents are merged from a large number of systems. This pool of documents does not necessarily contain all relevant documents. However, if a sufficient variety of top-n lists are merged, for a large enough n, the pool will contain the majority of the relevant documents. After judging the pool, the set of relevance judgments can be thought of as "sufficiently complete" to be reusable. In other words, a new retrieval experiment may retrieve unjudged documents, but these can be assumed to be irrelevant since almost all the relevant documents have already been identified.

Most TREC experiments have resulted in test collections where the pools are sufficiently complete for reuse. However, for the largest TREC corpora, it is unlikely that the full relevant set has been identified, or could be identified due to practical limitations in the judging resources [4]. In such cases it is possible to judge with shallow pools, for example judging the top 20 documents and using a precision-oriented measure. Since the full relevant set has not been discovered, it is not straightforward to measure recall, or use any measure that relies on knowing the size of the relevant set. The test collection may not be reusable.

Foundations

To understand precision-oriented evaluation, it is useful to revisit the fundamental measures of information retrieval, precision and recall. If there are R known relevant documents, and r of them have been retrieved at rank n, then:

$$\text{Precision at } n = \frac{r}{n}$$

$$\text{Recall at } n = \frac{r}{R}$$

For a given cutoff n, these are just different ways of normalizing r. If r increases, because a new retrieval system retrieves a better set of n documents, both precision and recall improve.

One aspect of precision-oriented evaluation is where not all relevant documents have been identified, so R is unknown. In such a case it is not straightforward to estimate recall, which depends on knowledge of R. However, if the top n documents have been judged, it is possible to accurately measure precision.

Another aspect of precision-oriented evaluation is shallow evaluation. At an early cutoff (small n) few relevant documents will have been retrieved (small r), but precision may still be quite high. By contrast, recall might be low, particularly if R is much higher than r and n. In a standard precision-recall curve, which shows precision and recall at multiple cutoffs, precision-oriented evaluation considers the left-hand end of the curve. Precision-oriented evaluation models a user who does not need to see all relevant documents and is impatient, unwilling to look deep into the ranking.

Some experiments have both shallow evaluation and incomplete judgments, for example the TREC Very Large Collection Track [4] judged with a pool depth of 20, to measure Precision at 20. Metrics such as Precision at n, Success at n, Mean Reciprocal Rank and Average Precision at n have been applied in such settings. More details about these metrics can be found from the corresponding entries in this encyclopedia.

In other experiments, judgments may be incomplete or evaluation shallow, but not both. In those cases, it is not clear that the evaluation should be called precision-oriented. It is possible to evaluate with a recall-oriented measure using incomplete judgments, via measures such as bpref and inferred AP [1, 6]. It is also possible to perform shallow evaluation with complete judgments, particularly in cases with very few relevant documents such as known item search. In a setting where R tends to be between 1 and 10, a metric such as R-Precision could be thought of as precision-oriented.

Key Applications

Web search is probably the most well-known application where precision-oriented evaluation is used. Users do not often look deep into the results list, so evaluation that concentrates on shallow ranks is appropriate. In addition, the document collection is so large, that it is difficult to be sure that all good answers for a query have been identified. The exception is where a user has a very focused need, for example navigational search, where only a single document is required. Even then the web is volatile, with documents being created, modified and deleted without warning, so it is necessary to perform careful maintenance of a query set. For queries with multiple correct answers, it is usual to calculate precision-oriented measures, and very difficult to estimate recall.

Data Sets

Data sets for precision-oriented evaluation, and Information Retrieval experimentation in general, are available from the National Institute of Standards and Technology, in the Text Retrieval Conference (TREC) series [5]. TREC experiments typically involve hundreds of thousands of documents or more, and 50 or more query topics.

Cross-References

▶ Average Precision at n
▶ Bpref
▶ Mean Reciprocal Rank
▶ Precision
▶ Precision at n
▶ R-Precision
▶ Standard Effectiveness Measures
▶ Success at n

Recommended Reading

1. Buckley C, Voorhees EM. Retrieval evaluation with incomplete information. In: Proceedings of the 30th Annual International ACM SIGIR Conference on Research and Development in Information Retrieval; 2007. p. 25–32.
2. Clarke CLA, Scholer F, Soboroff I. The TREC 2005 terabyte track. In: Proceedings of the 5th Text Retrieval Conference; 2005.
3. Cleverdon C. The Cranfield tests on index language devices. Readings in information retrieval. San Fransisco: Morgan Kaufmann; 1997. p. 47–59.
4. Hawking D, Craswell N. The very large collection and web tracks. In: Voorhees E, Harman D, editors. TREC experiment and evaluation in information retrieval. Cambridge, MA: MIT Press; 2005. p. 199–231.
5. Voorhees EM, Harman DK. TREC: experiment and evaluation in information retrieval. Cambridge, MA: MIT Press; 2005.
6. Yilmaz E, Aslam JA. Estimating average precision with incomplete and imperfect judgments. In: Proceedings of the 15th ACM International Conference on Information and Knowledge Management; 2006. p. 102–11.

Predictive Analytics

Ugur Cetintemel
Department of Computer Science, Brown University, Providence, RI, USA

Synonyms

Predictive analytics

Definition

Predictive analytics refers to the practice of using a class of analytical techniques that involve data-driven modeling, mining, and learning over historical data to make predictions about missing, incomplete, or future data values, events, or patterns.

Main Text

Predictive analytics combines historical data with predictive models to produce additional information not readily available within the data.

Predictive analytics is often cited as a major analytics category along with descriptive analytics, which focuses on the analysis of historical data for postmortem insight, and prescriptive analytics that uses predicted data to help with antemortem decision making.

Predictive models that underlie predictive analytics are diverse yet they all commonly describe the relationships present in the data. The most popular models are based on regression (e.g., linear and logistic) and machine learning techniques (e.g., support vector machines and k-nearest neighbors). Regression models commonly attempt to identify a relationship between known and unknown variables in the form of a mathematical equation. Machine learning techniques often rely on less formal representations to capture, which can sometimes be highly complex, difficult to formalize relationships in the data.

Predictive analytics has broad applications in many science, engineering, and business domains. For example, predictive analytics is used (i) in banking for credit scoring, default probability analysis, and customer targeting; (ii) in manufacturing for root cause analysis of defects, and yield and warranty analysis; (iii) in retail for customer segmentation, response modeling, and product recommendation; (iv) in utilities for predicting equipment and power line failures, product bundling, and fraud detection; and (v) in health care for patient procedure recommendations, treatment outcome prediction, and doctors' note analysis.

Software support for predictive analysis has significantly improved in recent years. In addition to specialized packages (e.g., SAS Enterprise Miner, IBM SPSS Modeler, Tableau, Weka) and language extensions (R, Mahout), database management systems (e.g., Oracle Data Miner, SQL Server Data Mining, and DB2 Intelligent Miner) have also incorporated predictive features. Recent research efforts have primarily focused on scaling and simplifying data-intensive predictive modeling and analysis [1–5].

Cross-References

▶ Business Intelligence
▶ Prescriptive Analytics
▶ Text Mining
▶ Visual Data Mining
▶ What-If Analysis

Recommended Readings

1. Mert Akdere, Ugur Çetintemel, Matteo Riondato, Eli Upfal, Stanley B. Zdonik. The case for predictive database systems: opportunities and challenges. In: Proceedings of the 5th Biennial Conference on Innovative Data Systems Research; 2011. p. 167–74.
2. Amol Ghoting, Rajasekar Krishnamurthy, Edwin P. D. Pednault, Berthold Reinwald, Vikas Sindhwani, Shirish Tatikonda, Yuanyuan Tian, Shivakumar Vaithyanathan. SystemML: declarative machine learning on MapReduce. In: Proceedings of the 27th International Conference on Data Engineering; 2011. p. 231–42.
3. Hellerstein JM, Ré C, Schoppmann F, Wang DZ, Fratkin E, Gorajek A, Ng KS, Welton C, Feng X, Li K, Kumar A. The MADlib analytics library or MAD skills, the SQL. Proc VLDB Endow. 2012;5(12): 1700–11.
4. Tim Kraska, Ameet Talwalkar, John C. Duchi, Rean Griffith, Michael J. Franklin, Michael I. Jordan. MLbase: a distributed machine-learning system. In: Proceedings of the 6th Biennial Conference on Innovative Data Systems Research; 2013.
5. Xixuan Feng, Arun Kumar, Benjamin Recht, Christopher Ré. Towards a unified architecture for in-RDBMS analytics. In: Proceedings of the ACM SIGMOD International Conference on Management of Data; 2012. p. 325–36.

P

Preference Queries

Jan Chomicki
Department of Computer Science and
Engineering, State University of New York at
Buffalo, Buffalo, NY, USA

Synonyms

Ranking queries

Definition

A preference query is a query obtained by augmenting a database query with user-defined preferences (*preference specification*), so that the preference query returns not all the answers but only the best, most preferred answers. For example, given a binary preference relation, a winnow preference query may return all the undominated tuples belonging to the input (all the tuples to which no tuples in the input are preferred). As another example, given a numeric scoring function a preference *Top-K query* returns K answers, if available, with the highest scores. Note: *Top-K queries* are not considered in this article because they are covered in a separate article.

Historical Background

An early paper [1] viewed preference queries in the context of relational query relaxation. The answers satisfying the query condition would be the best; the answers obtained by progressively weakening the query condition would be less and less preferred. So in the absence of the best answers, the less preferred answers would be the *best available*. This is an instance of *implicit* preference. The *explicit preference specification* in queries was first proposed in logic-programming-based approaches [2, 3].

Scientific Fundamentals

The most common kind of relational preference query has been formalized as the *winnow* operator ω [4], also called *BMO* [5] and *Best* [6]. The operator retrieves all the best (undominated) tuples from a given input relation. Note that the tuples in the result of winnow do not have to dominate all the tuples in the given relation, they just cannot be dominated by any other tuple in the relation. A winnow tuple may, in fact, dominate no tuples in the input relation.

Formally, given a preference relation $\succ$ and a database relation r:

$$\omega_\succ(r) = \{t \in r \mid \neg \exists t' \in r.\, t' \succ t\}.$$

If a preference relation $\succ^C$ is defined using a formula C, then we write $\omega_C(r)$, instead of $\omega_{\succ^C}(r)$. Winnow is syntactically a unary operator, akin to selection, but its semantics is that of a *self-anti-join*.

It is clear that a *skyline* is the result of computing winnow under the skyline preference relation (also called Pareto dominance). This simple observation has important consequences: the techniques for evaluating and optimizing winnow apply immediately to skylines.

Winnow seamlessly integrates with the operators of the relational algebra. In fact, winnow can be expressed in relational algebra and provides the nonmonotonic functionality equivalent to that of set difference [4].

Consider the preference relation $\succ^{C_1}$ that for each make minimizes the car price, given the input relation $Car(Make, Age, Price)$:

$$t_1 \succ^{C_1} t_2 \equiv t_1[Make]$$
$$= t_2[Make] \wedge t_1[Price] < t_2[Price].$$

and the following relation instance

$$r_1 = \{w_1 : (mazda, 4, 20\,K),$$
$$w_2 : (ford, 4, 17\,K),$$
$$w_3 : (mazda, 6, 17\,K)\}.$$

Winnow $\omega_{C_1}(r_1)$ returns the tuples w_2 and w_3, The tuple w_1 is dominated by w_3.

There are two algorithms that compute winnow for arbitrary strict partial order preference relations: BNL [7] and SFS [8]. Those algorithms were first proposed in the context of skylines but they only require irreflexivity and transitivity of preferences [4].

A major advantage of the logic-based approach to preference queries is that *rewrite-based query optimization* is done in a natural and clean way. For example, the algebraic laws involving winnow are formulated analogously to the well-known laws of relational algebra [4, 5, 9]. However, often the laws do not hold unconditionally. Consider commuting winnow and selection. It is the case that $\sigma_C(\omega_\gamma(r)) = \omega_\gamma(\sigma_C(r))$ for every relation instance r if the formula $\forall t_1, t_2.[C(t_2) \land \gamma(t_1, t_2) \Rightarrow C(t_1)]$ is valid. Under the preference relation $\succ^{C_1}$, the selection $\sigma_{Price<20K}$ commutes with ω_C but $\sigma_{Price>20K}$ does not. Other laws involving winnow were studied in [4, 9].

Semantic query optimization (query optimization using integrity constraints) also fits in very well here. As shown in [10], the information about integrity constraints can be used to eliminate redundant occurrences of winnow and make more efficient computation of winnow possible. We say that ω_C is *redundant w.r.t.* a set of integrity constraints F if $w_C(r) = r$ for all r satisfying F. Now ω_C is *redundant w.r.t.* F iff F implies the formula

$$\forall t_1, t_2. \ R(t_1) \land R(t_2) \Rightarrow t_1 \not\succ^C t_2 \land t_2 \not\succ^C t_1.$$

Different *preference specification* frameworks provide different ways in which the preference relation $\succ$, which is the parameter of winnow, can be specified. For instance, one could use algebraic operators of Pareto or prioritized accumulation. Preference constructors of [11] are integrated with the Preference SQL query language. If the PREFERRING clause is present in the query, then the BMO semantics (retrieving the most preferred tuples) is implicitly assumed. In the running example, the winnow ω_{C_1} is formulated as:

```
SELECT * FROM Car
GROUP BY Make
PREFERRING LOWEST(Price)
```

In the approach of [12], the preferences are specified as pairs (*Condition, Score*), where *Condition* is a conjunction of selection and join conditions. Given a conjunctive query, the approach identifies the preferences the most relevant to the query and uses them to personalize the query. Specifically, the top K preferences are used to modify the query and to generate results that satisfy at least L of the K preferences. Suitable query rewriting algorithms are proposed. Clearly, query results in this approach depend on how condition scores are defined and how relevant preferences are determined. Arvanitis and Koutrika [13] introduces a new operator λ which uses a preference to assign scores to the tuples in a relation. In this case, *preference specification* assumes the form (*Condition, Scoring Function, Confidence*). Such preferences cannot capture Pareto dominance, and λ cannot in general represent skyline queries.

It may be too difficult for database users to directly specify the preferences they have in mind. Jiang et al. [14] propose to use *superior* and *inferior* examples to determine missing attribute orderings in an incomplete Pareto dominance specification. The superior examples are the tuples that have to be in the skyline, and the inferior examples, those that should not be in the skyline because of being dominated by a skyline element. Mindolin and Chomicki [15] also use superior and inferior examples, but for a different purpose. Assuming that attribute orderings are given, their approach seeks to find a maximal prioritized dominance relation that satisfies the conditions on superior and inferior result tuples.

Key Applications

Several preference-aware database management systems have been developed and deployed. Preference SQL [11] extends SQL queries with several new clauses, including PREFERRING for preference and CASCADE for prioritized accumulation. The system has been in development and use for over 12 years, and has supported e-commerce applications. FLEXPREF [16] is a prototype system, supporting user-defined preference query classes through a simple, general API. Example query classes include *skyline* and *Top-k queries*. PrefDB [13] supports preferences in a framework generalizing [12] and embeds them into relational algebra using an operator λ that scores individual tuples. Arvanitis and Koutrika [13] studies algebraic properties of λ and shows how they can be used in query optimization.

Applications of preference queries include query personalization, comparison shopping, and e-commerce. Further applications are discussed in *preference specification*.

Future Directions

In the basic preference query frameworks discussed above preferences are loosely coupled with queries: the former serve as parameters of the latter. But tighter coupling is also possible: preferences can be defined not only over database tables but also over views. Among others, this makes it possible to formulate some extrinsic preferences.

Let *Rebate(Make,Amount)* be another database relation. Then joining *Car* and *Rebate* enables calculating the final prices of cars:

```
CREATE VIEW CarFinal(Make,Age,
   FinalPrice) AS
SELECT Make, Age, Car.Price -
   Rebate.Amount
FROM Car, Rebate
WHERE Car.Make=Rebate.Make
```

Now preferences can be defined over the view *CarFinal*, instead of *Car*. The computational challenge is to calculate the skyline over a view without materializing the view. This problem occurs in the context of *creating competitive products* [17].

A research direction which is in some sense opposite to tighter coupling of queries and preferences is *decoupling* them. Such a need arises, for example, when user preferences are not known at query time. Nanongkai et al. [18] proposes a *regret-minimizing* approach, under the assumption that user preferences are captured using an unknown scoring function. A preference query returns a small set of tuples that makes every user at least $x\%$ happy. A user is $x\%$ happy with a list of tuples if the utility it obtains from the best tuple in the list is at least $x\%$ of the utility she obtains from the best tuple in the database. Nanongkai et al. [18] shows that the size of the query result is a small number which can be bounded independently of the database. The same paper presents various algorithms for computing the query answer. A follow-up paper [19] shows how the approach can be enhanced with user interaction.

Still another potential research direction involves computing *provenance* of preference queries. Certainly, there are scenarios in which the user would like to know why a specific answer was (or was not) among the best answers.

Cross-References

▶ Preference Specification
▶ Probabilistic Skylines
▶ Provenance in Databases
▶ Similarity and Ranking Operations
▶ Skyline Queries and Pareto Optimality

Recommended Reading

1. Lacroix M, Lavency P. Preferences: putting more knowledge into queries. In: Proceedings of the 13th International Conference on Very Large Data Bases; 1987. p. 217–25.
2. Govindarajan K, Jayaraman B, Mantha S. Preference logic programming. In: Proceedings of the 12th International Conference on Logic Programming; 1995. p. 731–45.
3. Kießling W, Güntzer U. Database reasoning – a deductive framework for solving large and complex problems by means of subsumption. In:

Proceedings of the 3rd Workshop on Information Systems and Artificial Intelligence; 1994. p. 118–38.

4. Chomicki J. Preference formulas in relational queries. ACM Trans Database Syst. 2003;28(4):427–66.

5. Kießling W. Foundations of preferences in database systems. In: Proceedings of the 28th International Conference on Very Large Data Bases; 2002. p. 311–22.

6. Torlone R, Ciaccia P. Which are my preferred items? In: Proceedings of the AH'2002 Workshop on Recommendation and Personalization in eCommerce; 2002.

7. Börzsönyi S, Kossmann D, Stocker K. The skyline operator. In: Proceedings of the 17th International Conference on Data Engineering; 2001. p. 421–30.

8. Chomicki J, Godfrey P, Gryz J, Liang D. Skyline with presorting. In: Proceedings of the 19th International Conference on Data Engineering; 2003. p. 717–19.

9. Hafenrichter B, Kießling W. Optimization of relational preference queries. In: Proceedings of the 16th Australasian Database Conference; 2005. p. 175–84.

10. Chomicki J. Semantic optimization techniques for preference queries. Inf Syst. 2007;32(5):660–74.

11. Kießling W, Köstler G. Preference SQL – design, implementation, experience. In: Proceedings of the 28th International Conference on Very Large Data Bases; 2002. p. 990–1001.

12. Koutrika G, Ioannidis Y. Personalization of queries in database systems. In: Proceedings of the 20th International Conference on Data Engineering; 2004. p. 597–608.

13. Arvanitis A, Koutrika G. PrefDB: supporting preferences as first-class citizens in relational databases. IEEE Trans Knowl Data Eng. 2014;26(6):1430–46.

14. Jiang B, Pei J, Lin X, Cheung DW, Han J. Mining preferences from superior and inferior examples. In: Proceedings of the 14th ACM SIGKDD International Conference on Knowledge Discovery and Data Mining; 2008. p. 390–98.

15. Mindolin D, Chomicki J. Preference elicitation in prioritized skyline queries. VLDB J. 2011;20(2):157–82. Special issue: selected papers from VLDB 2009.

16. Levandoski JJ, Eldawy A, Mokbel MF, Khalefa ME. Flexible and extensible preference evaluation in database systems. ACM Trans Database Syst. 2013;38(3):17.

17. Wan Q, Wong RCW, Ilyas IF, Özsu MT, Peng Y. Creating competitive products. Proc VLDB Endow. 2009;2(1):898–909.

18. Nanongkai D, Sarma AD, Lall A, Lipton RJ, Xu J. Regret-minimizing representative databases. Proc VLDB Endow. 2010;3(1):1114–24.

19. Nanongkai D, Lall A, Sarma AD, Makino K. Interactive regret minimization. In: Proceedings of the ACM SIGMOD International Conference on Management of Data; 2012. p. 109–20.

Preference Specification

Jan Chomicki
Department of Computer Science and Engineering, State University of New York at Buffalo, Buffalo, NY, USA

Synonyms

Preference Definition

Definition

Usually, preference relations are formalized as *strict partial orders (SPO)*, which means that a preference relation $\succ$ satisfies the following properties:

- *irreflexivity*: $\forall x\ (x \not\succ x)$,
- *transitivity*: $\forall x, y, z\ (x \succ y \wedge y \succ z \rightarrow x \succ z)$.

Additionally, an SPO $\succ$ is a *weak order* if $\forall x, y, z\ (x \succ y \rightarrow x \succ z \vee z \succ y)$, and a *total order* if $\forall x, y\ (x \succ y \vee y \succ x \vee x = y)$.

The formal properties of orders capture the nature of preferences in an abstract, application-independent way. It is obvious that irreflexivity should hold: preferring an object over itself seems to violate the basic intuitions behind preference. But transitivity is debatable. On one hand, it captures the *rationality* of preferences [1, 2]. On the other, transitivity is sometimes violated by preference aggregation in voting scenarios [3]. The weak order property implies that the set ordered by the preference relation can be viewed as consisting of disjoint layers each of which contains equivalent elements. Preferences can also be formalized using non-strict orders that mean "preferred or equivalent to."

Historical Background

The study of preferences originated in philosophy [4] and originally focused on the *logics of preference* that axiomatize natural and desirable properties of preferences. Later, the use of preferences in *multi-dimensional decision making* became an important topic in economics and decision theory [1]. In economics, preferences are usually expressed using numeric utility functions. Recently, there has been a surge of interest in preferences in artificial intelligence [5], primarily in the context of autonomous agents. Independently, *preference queries* have been studied in the database research community [6]. In AI, the dominant topic is *reasoning* with and about preferences, typically limited to propositional languages. Relations are first-order structures, and thus preferences in databases are captured by first-order formulas, and the emphasis is on database querying. Preferences are embedded into queries to form *preference queries*.

Scientific Fundamentals

In many application scenarios, the database user is interested in retrieving only the *best* objects in the database, according to some criterion. In order to be able to delegate this task to a database system, she has to explicitly state her *preferences:* what it means for an object to be *better* in her view than another object.

Finite preference relations are specified by enumerating their elements. To specify *infinite preference relations*, which is common in the presence of infinite domains, several approaches have been pursued: logical or algebraic specification, new query language constructs, or scoring functions.

A preference relation is denoted by $\succ$, possibly with a subscript or a superscript. The relationship $x \succ y$ is variously described as: x is *is preferred to y, x is better than y*, or *x dominates y*. Similarly, *preference relation* and *dominance* are treated as synonyms.

Logical specification. The running example is the database relation *CarPrices(Age, Price)* over which several preference relations are defined:

$$t_1 \succ_{Age} t_2 \equiv t_1[Age] < t_2[Age]$$

$$t_1 \succ_{Price} t_2 \equiv t_1[Price] < t_2[Price]$$

$$t_1 \succ_{Age,Price} t_2 \equiv (t_1[Age] < t_2[Age] \wedge t_1[Price] \le t_2[Price]$$

$$\vee\, t_1[Age] \le t_2[Age] \wedge t_1[Price] < t_2[Price]).$$

The preference relation $\succ_{Age}$ (resp. $\succ_{Price}$) captures the (atomic) preference for newer (resp. cheaper) cars. In contrast, the preference relation $\succ_{Age,Price}$ is complex: it captures the preference for cars that are newer but not more expensive or cheaper but not older. This is an example of a 2-dimensional *Pareto dominance*. The concept of Pareto dominance, well known in economics, serves as a foundation of *skyline queries*. Let $\mathcal{A} = \{A_1, \ldots, A_d\}$ be a finite set of attributes. The general definition of Pareto dominance is as follows:

$$t_1 \succ_{\mathcal{A}}^{pto} t_2 \equiv \bigvee_{A_i \in \mathcal{A}} t_1 \succ_{A_i} t_2 \wedge \bigwedge_{A_i \in \mathcal{A}} t_1 \succeq_{A_i} t_2.$$

where the number d is the *dimension* of Pareto dominance.

Typically, logic formulas defining preferences (*preference formulas*) contain constants, variables, comparison operators (like $<$), and Boolean connectives. Thus, it is possible to perform a dominance check (whether one tuple is preferred to another) by substituting tuple attribute values into the preference formula and computing the truth value of the resulting ground formula. The preferences which are based only on the contents of the tuples being compared are called *intrinsic* [7], in contrast to *extrinsic* preferences which may also refer to the contents of database relations. Also, intrinsic preference formulas usually admit quantifier elimination, and thus quantifiers are not needed in preference specification. The generality of the logic notation has a disadvantage: preference relations defined using logic formulas do not have to be strict partial orders. However, since the order properties of preference relations, for example transitivity, are also defined logically using axioms, one can check such properties of a given intrinsic preference relation by substituting its defining formula into the appropriate axioms and then applying quantifier elimination to determine the validity of the resulting formula.

Extrinsic preferences are less well-behaved. Assuming a preference relation is stored as a finite relation $Pref$, the following is an extrinsic preference relation:

$$x \succ^{Pref} y \equiv Pref(x, y).$$

The dominance test $u \succ^{Pref} v$ for some u and v and the satisfaction of the order properties by $\succ^{Pref}$ clearly depend on the contents of the relation $Pref$, and thus require querying the database.

Algebraic specification. Another approach to preference specification uses *algebraic operators* to define complex preference relations [8]. From an algebraic perspective, Pareto dominance can be defined using a binary operator $\otimes$ (called *Pareto accumulation*):

$$\succ^{pto}_{\mathcal{A}} = \succ_{A_1} \otimes \succ_{A_2} \otimes \cdots \otimes \succ_{A_d},$$

where $\succ_{XY} = \succ_X \otimes \succ_Y$ and

$$t_1 \succ_{XY} t_2 \equiv t_1 \succ_X t_2 \wedge t_1[Y] = t_2[Y]$$
$$\vee \, t_1 \succ_Y t_2 \wedge t_1[X] = t_2[X]$$
$$\vee \, t_1 \succ_X t_2 \wedge t_1 \succ_Y t_2$$

for $XY \subseteq \mathcal{A}$ and $X \cap Y = \emptyset$. The preference relations are subscripted by the sets of attributes used in their definition. The preference $\succ_{Age,Price}$ in the running example can now be defined as

$$\succ_{Age,Price} = \succ_{Age} \otimes \succ_{Price}.$$

Similarly to Pareto accumulation one can define *prioritized accumulation* [8] in a logical or algebraic fashion. Prioritized accumulation may be used to capture differences in the importance of atomic preferences. For instance, in the car-buying scenario minimizing Age may be more important than minimizing $Price$. Complex preferences are defined by nesting accumulation operators. Since the algebraic operators are defined logically, the logical approach is more general and flexible than the algebraic one. But the algebraic properties of the operators, for example commutativity and associativity of Pareto accumulation, have a potential of influencing preference query optimization.

Preference constructors. The approach adopted in PreferenceSQL [9] relies on *preference constructors*. The constructors are used to specify both atomic and complex preferences. The instances of the constructors are added to SQL in a new PREFERRING clause. For example, the preference for lower price is expressed as:

PREFERRING LOWEST(Price).

Preference SQL provides atomic constructors to capture positive (POS) and negative (NEG) preferences for specific attribute values and positive preferences for values that are closer to a target value. Complex preferences involve Pareto and prioritized accumulation. The preferences in the running examples can be captured as follows:

PREFERRING LOWEST(Age) AND
    LOWEST(Price).

Preference SQL preferences can all be expressed logically. For example, the POS preference for Mazda and the NEG preference for Fiat can be expressed as

$$x \succ^{PN} y \equiv x \neq \text{Fiat} \wedge y = \text{Fiat} \vee x$$
$$= \text{Mazda} \wedge y \neq \text{Mazda}.$$

Through careful design, preference relations defined in Preference SQL are all strict partial *orders*.

Scoring functions. In the context of *ranking queries* preferences are captured using numeric-valued *scoring functions*. Such functions constitute special cases of preference relations. Indeed, a scoring function f represents a preference relation $\succ_f$ such that

$$x \succ^f y \equiv f(x) > f(y).$$

It is easy to see that preference relations represented by scoring functions are *weak orders*. The weak order property has an important consequence: preference relations that are strict partial orders but not weak orders *cannot be represented using scoring functions*. Nevertheless, such preferences are very common: every skyline preference relation of dimension at least two, e.g., $\succ_{Age,Price}$ in the running example, is not a weak order. Moreover, numeric scores do not have an intuitive interpretation (what does it mean that a tuple has the score of 0.8 as opposed to 0.9?) and lack natural language formulation.

Often the objects to be compared with respect to the given preferences are more complex than flat tuples. For example, some applications require finding the best sets [10–12]. Aggregate properties of sets can be succinctly represented as tuples of function values (*profiles*). Examples include cardinality and minimum value. The universe of sets may consist of all k-element subsets of a given set of tuples [10–12], or all the tuple groups in a set of tuples [13]. Now a set T dominates another set S if the profile of T dominates the profile of S. In the literature set preference is often restricted to Pareto dominance of profiles.

The approach to preference specification in [14] combines logical specification with scoring functions. In that approach scores are assigned to sets of tuples, which are defined using selection and join conditions. A preference is then specified as a pair (*Condition, Score*). This means that the tuples satisfying *Condition* have the score equal to *Score*. For example, assuming the score is 0.5 for all 3-years-old cars with price less than $10 K, the preference is specified as ($Age = 3 \land Price < 10$ K$, 0.5$). Note that the same tuple may satisfy the conditions of multiple pairs that have different scores, so there is a need to specify how such scores are aggregated. This approach is generalized in [15], where a preference is a triple (*Condition, ScoringFunction, Confidence*). Now *ScoringFunction* assigns scores, not necessarily the same, to the tuples satisfying *Condition*, with confidence equal to *Confidence*. Like other approaches based on scoring, [14, 15] cannot represent Pareto dominance.

Key Applications

Preferences are essential ingredients of preference queries. They are used in query personalization [14]. Two different users querying a database using the same query may expect different answers because they may use different, implicit criteria of which answers are the *best*. For example, when purchasing a car one customer may be primarily concerned about the price, while another may be looking for a newer car. The preference specification frameworks presented above allow users to state such (and more complicated) criteria and see them reflected in queries.

The ability to compose preferences is essential in multi-agent scenarios. For example, if vendor preferences are added to the car buying example, it is natural to talk about composing customer and vendor preferences using some form of accumulation [16].

Preferences are also used to address the problems of *query answer inadequacy*. To deal with the problem of *too few query answers* (perhaps none), hard constraints are replaced by soft constraints that are satisfied only to the extent possible. The problem of *too many answers* can be alleviated by having the user specify which query answers are preferred.

Future Directions

Preference strength. Most of the recent research on preferences has assumed the *ordinal*

view of preferences: which objects are preferred
to which but not by how much. Nevertheless,
there are situations where the *cardinal* view is
more appropriate. In the car buying example
augmented with a *Make* attribute a user may say
that she prefers Mazda over Fiat much more than
Alfa Romeo over Nissan. In this view, binary
relations between objects are no longer sufficient
to capture preferences [17].

Preference elicitation. Capturing user preferences exactly is difficult. A user may lack the
skills to formulate them. Automated preference
elicitation may be time consuming and error
prone. The popularity of skyline queries is partly
due to the fact that they require only rudimentary
user input in the form of single-attribute preferences $\succ_{A_i}$. However, Pareto dominance is not
always semantically adequate, and thus the study
of preference elicitation [18] needs to continue.
Perhaps such elicitation should proceed in stages,
starting from a rough-draft version and ending
with the fine-tuning of details.

Preference dynamics. Oftentimes preferences
change, which may be due to a variety of factors:
context, external influences, or change in user
status or availability of different options. It is a
challenge how to capture such changes (which
may be quite subtle) and appropriately revise the
affected preference relations.

Cross-References

▶ Preference Queries
▶ Probabilistic Skylines
▶ Similarity and Ranking Operations
▶ Skyline Queries and Pareto Optimality

Recommended Reading

1. Fishburn PC. Utility theory for decision making. New
   York: Wiley; 1970.
2. Fishburn PC. Preference structures and their
   numerical representations. Theor Comput Sci.
   1999;217(2):359–83.
3. Sen AK, Pattanaik PK. Necessary and sufficient conditions for rational choice under majority decision. J
   Econ Theory. 1969;1(2):178–202.
4. Hansson SO. Preference logic. In: Gabbay D, editor.
   Handbook of philosophical logic. vol 4. Dordrecht/-
   Boston: Kluwer; 2001.
5. Brafman RI, Domshlak C. Preference handling
   – an introductory tutorial. AI Mag. 2009;30(1):
   58–86.
6. Stefanidis K, Koutrika G, Pitoura E. A survey on
   representation, composition and application of preferences in database systems. ACM Trans Database
   Syst. 2011;36(4):1–45.
7. Chomicki J. Preference formulas in relational
   queries. ACM Trans Database Syst. 2003;28(4):
   427–66.
8. Kießling W. Foundations of preferences in database
   systems. In: Proceedings of the 28th International Conference on Very Large Data Bases; 2002.
   p. 311–22.
9. Kießling W, Köstler G. Preference SQL – design,
   implementation, experience. In: Proceedings of the
   28th International Conference on Very Large Data
   Bases; 2002. p. 990–1001.
10. Im H, Park S. Group skyline computation. Inf Sci.
    2012;188(0):151–69.
11. Li C, Zhang N, Hassan N, Rajasekaran S, Das
    G. On skyline groups. In: Proceedings
    of the 21st ACM International Conference on
    Information and Knowledge Management; 2012.
    p. 2119–23.
12. Zhang X, Chomicki J. Preference queries over sets.
    In: Proceedings of the 27th International Conference
    on Data Engineering; 2011. p. 1019–30.
13. Antony S, Wu P, Agrawal D, El Abbadi A. Aggregate
    skyline: analysis for online users. In: Proceedings of
    the 9th Annual International Symposium on Applica-
    tions and the Internet; 2009. p. 50–56.
14. Koutrika G, Ioannidis Y. Personalization of queries
    in database systems. In: Proceedings of the 20th
    International Conference on Data Engineering; 2004.
    p. 597–608.
15. Arvanitis A, Koutrika G. PrefDB: support-
    ing preferences as first-class citizens in relational
    databases. IEEE Trans Knowl Data Eng. 2014;26(6):
    1430–46.
16. Kießling W, Fischer S, Döring S. COSIMA ^{B2B}
    – sales automation for E-procurement. IEEE Press;
    2004. p. 59–68.
17. Köbberling V. Strength of preference and cardinal
    utility. Econ Theory. 2006;27(2):375–91.
18. Mindolin D, Chomicki J. Preference elicitation in
    prioritized skyline queries. VLDB J. 2011;20(2):157–
    82. Special issue: selected papers from VLDB
    2009.

P

Prescriptive Analytics

Laurynas Šikšnys and Torben Bach Pedersen
Department of Computer Science, Aalborg
University, Aalborg, Denmark

Synonyms

Prescriptive business intelligence; Third phase of
business analytics

Definition

Prescriptive analytics is the third and final stage
of *business analytics* dedicated to finding and
suggesting (i.e., prescribing) the best decision options for a given situation. Prescriptive analytics
encompasses the activities of (1) *data collection
and consolidation*, (2) *information extraction*, (3)
forecasting, (4) *optimization*, (5) *visualization*,
and (6) *what-if analysis* for first making predictions and then, based on these predictions, (a)
suggesting the most appropriate time-dependent
decisions (i.e., *prescriptions*) and (b) illustrating
the implications of each decision option.

Descriptive, predictive, and prescriptive analytics are the three stages of business analytics
[1], characterized by different levels of difficulty,
value, and intelligence (see Fig. 1).

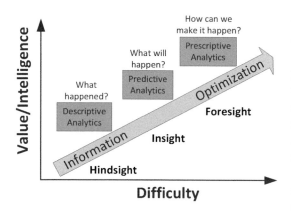

Prescriptive Analytics, Fig. 1 Three stages of business
analytics (Source "Gartner Inc.")

Descriptive analytics gives insights into what
has happened in the past or what is happening now. *Predictive analytics* forecasts what will
most likely happen in the future. *Prescriptive
analytics* seeks to determine the best course of
action among various choices to achieve a desired
future goal, given a set of known parameters,
objectives, requirements, and constraints. In comparison to descriptive and predictive analytics,
prescriptive analytics allows decision-makers not
only to see issues and opportunities (by looking
into past, presence, or future) but also actively
suggests (prescribes) the best decision options to
take advantage of the predicted future and illustrates the implications of each decision option.
Hence, prescriptive analytics is the most sophisticated type of business analytics and can bring the
greatest intelligence and value to businesses (see
Fig. 1).

Main Text

In practice, prescriptive analytics is a continuous and data-intensive process, during which
data is automatically collected and processed,
predictions are updated, potential decisions are
generated, interactions between decisions are analyzed, and the optimal course of actions is ultimately suggested (i.e., *prescribed*). This optimal
course of actions, termed a *prescription*, is a
time-dependent plan of what to do next in the
present situation. To derive such prescriptions, as
well as to explore factors, constraints, and the
interactions between them, various *optimization*,
statistical, and *simulation* models are employed
[2]. The effectiveness of the prescriptions depends on how well these models represent the
domain under study and capture impacts of decisions being analyzed. To improve the effectiveness of prescriptions, the use of *hybrid* data, a
combination of structured and *unstructured* data,
is encouraged [3].

Tool support for prescriptive analytics is still
not mature, with different tools being used for
each of steps (1–6) above. Typically, Hadoop
or RDBMSes such as PostgreSQL or MySQL
are used for data collection and consolidation;

purpose-specific data mining/analytics tools such WEKA or MLLIB are used for information extraction; numerical computing environments such as MatLab or R are used for forecasting; optimization solvers such as CPLEX or Gurobi are used for optimization; dashboard tools such as Cyfe or ClicData are used for visualization; what-if analysis is done in Excel or specialized tools. This results in complex, labor-intensive, and error-prone workflows. Commercial prescriptive analytics suites from vendors such as Ayata, IBM, FICO, and Tableau provide better integration across the prescriptive analytics steps, but are still tied to specific business domains, data types, and techniques, and use several proprietary models and languages. The most integrated type of tools are database systems with integrated analytics and optimization problem solving. These tools offer a common language and environment for *all* prescriptive analytics steps; examples include LogicBlox, based on the Datalog-like LogiQL language, and SolveDB [4], based on an extension of SQL.

Cross-References

▶ Business Intelligence
▶ Information Extraction
▶ Visualization for Information Retrieval
▶ Predictive Analytics
▶ What-If Analysis

Recommended Reading

1. Holsapple C, Post AL, Pakath R. A unified foundation for business analytics. Decis Support Syst. 2014;64:130–41.
2. Haas PJ, Maglio PP, Selinger PG, Tan WC. Data is dead ... without what-if models. PVLDB. 2011;4(12):1486–9.
3. Basu A. Five pillars of prescriptive analytics success. Analytics Magazine; 2013. p. 812.
4. Šikšnys L, Pedersen TB, Solve DB. Integrating Optimization Problem Solvers Into SQL Databases. In: Proceedings of the 28th International Conference on Scientific and Statistical Database Management; 2016.

Presenting Structured Text Retrieval Results

Jaap Kamps
University of Amsterdam, Amsterdam, The Netherlands

Synonyms

Structured text retrieval tasks

Definition

Presenting structured text retrieval results refers to the fact that, in structured text retrieval, results are not independent and a judgment on their relevance needs to take their presentation into account. For example, HTML/XML/SGML documents contain a range of nested sub-trees that are fully contained in their ancestor elements. As a result, structured text retrieval should make explicit the assumptions on how the retrieval results are to be presented. Four of the main assumptions to be addressed are the following. First, the *unit of retrieval* assumption: is there a designated retrieval unit (such as the document or root node of the structured document) or can every sub-tree be retrieved in principle? Second, the *overlap* assumption: may retrieval results contain text or content already part of other retrieval results (such as a full article and one of its individual paragraphs)? Third, the *context* assumption: can results from the same structured document be interleaved with results from other structured documents? Fourth, the *display* assumption: is a retrieval result (say a document sub-tree corresponding to a paragraph) presented as an autonomous unit of text, or as an entry-point within a structured document?

Historical Background

Although similar considerations play an important role in the design of user interfaces (see, for example, [6]), this entry will focus on the underlying principles of the different structured text retrieval tasks. Structured text retrieval dates back, at least, to the early days of passage retrieval [14]. Early passage retrieval approaches have been using either the document structure (sentences, paragraphs, sections, etc.), or arbitrary text windows of fixed length. The early experimental results primarily confirmed the effectiveness of passage-level evidence for boosting document retrieval. The use of document structure derived from SGML mark-up was pioneered by [20], studying adhoc SGML element retrieval. Probabilistic indexing approaches for databases have been studied even earlier [4], allowing to rank results based on vague queries. Adhoc XML element retrieval and best entry point retrieval was studied in the Focus project [3, 10].

The main thrust in recent years is the initiative for the evaluation of XML retrieval [7]. The retrieval task descriptions heavily evolved during the different years. Initially, in 2002, INEX studied adhoc XML element retrieval for keyword (Content-Only) and structured (Content-And-Structure) queries with the goal to "retrieve the most specific relevant document components" [5, p. 2]. This generic adhoc XML element retrieval task was continued at INEX 2003 [9, p. 200] and at INEX 2004 [12, p. 237], asking for "components that are most specific and most exhaustive with respect to the topic of request." Ongoing discussion, and vivid disagreement, on the interpretation of generic adhoc XML element retrieval task prompted the introduction of three different retrieval strategies at INEX 2005 [11, pp. 385–386]: *Thorough* aims to find all highly exhaustive and specific elements (roughly corresponding to the earlier INEX task); *Focussed* aims to find the most specific and exhaustive element in path (no overlapping results); and *Fetch and browse* aims to first identify relevant articles, and then to find the most specific and exhaustive elements within the fetched articles (results grouped by article). These different adhoc XML element retrieval tasks have been continued and further explicated at INEX 2006 [1], with the Fetch and browse task refined to: *Relevant in Context* aims to retrieve a set of non-overlapping

relevant elements per article; and *Best in Context* aims to retrieve, per article, a single best entry point to read its relevant content. At INEX 2007 three tasks are continued: Focused, Relevant in Context, and Best in Context, but liberalized to arbitrary passages [2].

Foundations

The way in which retrieval results are presented to users, is always a crucial factor determining the success or failure of an operational retrieval system. However, within the Cranfield/TREC tradition of evaluating document retrieval systems it is unproblematic to abstract away from presentation issues and analyze retrieval effectiveness by regarding retrieved documents as atomic and independent results. In structured text retrieval, the situation is different, and there is a need to make explicit some of the assumptions underlying the retrieval task since these have an impact on what is regarded as a "relevant" retrieval result.

First, the *unit of retrieval* assumption: is there a designated retrieval unit (such as the document or root node of the structured document) or can every sub-tree be retrieved in principle? Rather than treating documents as atomic, structured documents have internal document structure that allows any logical unit of them to be retrieved. For example, in case of a textual document where the layout structure is marked up, it is possible to retrieve sections, paragraphs, or still the whole article if its completely devoted to the topic of request. Figure 1 contains a screen-shot of a XML element retrieval system that retrieves a ranked list of XML elements.

Second, the *overlap* assumption: may retrieval results contain text or content already part of other retrieval results (such as a full article and one of its individual paragraphs)? Interactive experiments at INEX 2004 [17] clearly revealed that test persons disliked a ranked list of element results that overlap in whole or part in their content. Hence, if the retrieval tasks should reflect a scenario in which the ranked elements are directly displayed to an end-user, retrieval results should be disjoint.

dbdk_training in **Baseline System**

[   ] Search

query was: text classification naive bayes
Results 1 - 10 of 100.
Result pages: **1** 2 3 4 5 6 7 8 9 10 next

Search Result

1: (0.247) **Scalable Feature Mining for Sequential Data**
   Neal Lesh Mitsubishi Electric Research Lab Mohammed J. Zaki Rensselaer Polytechnic Institute Mitsunori Ogihara University of Rochester
   Result path: /article[1]/bdy[4]/sec[5]

2: (0.204) **Probability and Agents**
   Marco G. Valtorta University of South Carolina mgv@cse.sc.edu Michael N. Huhns University of South Carolina huhns@sc.edu
   Result path: /article[1]/bdy[4]/sec[3]

3: (0.176) **Combining Image Compression and Classification Using Vector Quantization**
   Karen L. Oehler Member IEEE Robert M. Gray Fellow IEEE
   Result path: /article[1]/bdy[4]/sec[4]/ss1[2]/ss2[4]

4: (0.175) **Text-Learning and Related Intelligent Agents: A Survey**
   Dunja Mladenic J. Stefan Institute
   Result path: /article[1]/bm[5]/app[4]/sec[5]

5: (0.175) **Detecting Faces in Images: A Survey**
   Ming-Hsuan Yang Member IEEE David J. Kriegman Senior Member IEEE Narendra Ahuja Fellow IEEE
   Result path: /article[1]/bdy[4]/sec[2]/ss1[9]/ss2[10]

Presenting Structured Text Retrieval Results, Fig. 1 Displaying structured text retrieval results as a ranked list of elements (Reproduced from Malik et al. [13])

Third, the *context* assumption: can results from the same structured document be interleaved with results from other structured documents? A further finding of the interactive experiments at INEX 2004 [17] is that test persons prefer results from the same document be grouped together. Figure 2 contains a screen-shot of a XML element retrieval system that retrieves XML elements displayed in document order in their article's context.

Fourth, the *display* assumption: is a retrieval result (say a document sub-tree corresponding to a paragraph) presented as an autonomous unit of text, or as an entry-point within a structured document? A decision on the relevance of a particular document component crucially depends on whether it will be presented as an isolated excerpt, or within its original context. In the first case, the component should be fully self-contained: it should not only contain the relevant information (say, for example, a description of an algorithm) but also establish that this information is, indeed, satisfying the topic of request (for example, that the algorithm is the fastest way to lexicographically sort a list, if that were the topic of request). This is related to linguistics,

where there is a common distinction between the context (or topic/theme: that what is being talked about), and the information (or comment/rheme/focus: that what is being said). If results are to be presented in their document context, the link to the topic of request can be taken for granted and only the sought information can be regarded as relevant. If results are to be presented out of context, both the information and its relation to the topic of request are needed to establish the relevance of a document component.

Table 1 shows how the different structured text retrieval tasks are based on different underlying assumptions. For traditional document or article retrieval, there is a fixed unit of retrieval and assumptions on overlap, context, or display do not apply. For the generic adhoc element retrieval task (INEX 2002–2004) or Thorough (INEX 2005–2006), any document component can be retrieved, and there are no restrictions on overlap, context, or display. Basically, the task is system-biased, reflecting the ability of the retrieval engine to estimate the relevance of individual document components, for example for further processing methods. For Focussed/Focused (INEX 2005–2007), a ranked list of

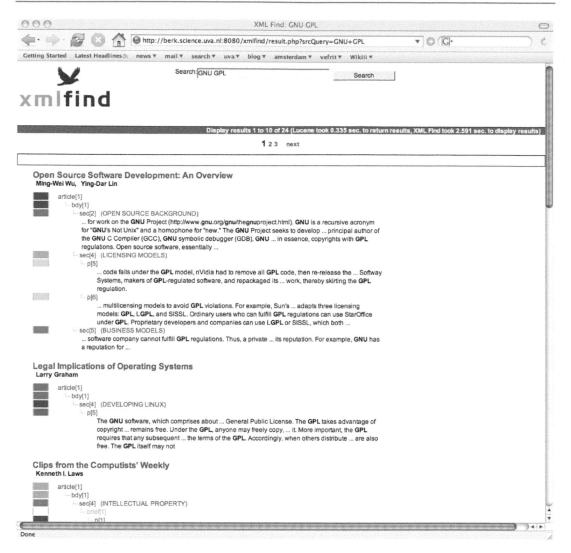

Presenting Structured Text Retrieval Results, Fig. 2 Displaying structured text retrieval results within article context (Reproduced from Sigurbjörnsson [15])

non-overlapping document components is asked for, with no restrictions on context or display. This task reflects a scenario where a ranked-list of document components is directly presented to the searcher. For Fetch and browse (INEX 2005), retrieval results from the same structured document need to be returned consecutive, with no restriction on overlap or display. This results in a tasks resembling on the one hand traditional document retrieval, whilst on the other hand providing deep-linking to relevant document components. The same holds for Relevant in context (INEX 2006–2007), where there is an unranked

set of now non-overlapping elements per article, reflecting results to be presented in document order. Finally, Best in context (INEX 2006–2007) explicitly asks for a single best entry point into the article (so non-overlapping and non-scattered articles by definition). This scenario captures a "relative" notion of relevance, where users desire access to the best information, rather than all relevant information.

These different retrieval tasks lead to different evaluations of what systems and techniques are effective for structured text retrieval. Although these tasks are not unrelated, for example, the

Presenting Structured Text Retrieval Results, Table 1 Structured text retrieval tasks

	Unit of retrieval	Overlap	Context	Display
Article retrieval	Whole article	–	–	–
Thorough	Arbitrary element	Allowed	Scattered	Elements
Focussed/Focused	Arbitrary element	Non-overlapping	Scattered	Elements/Passages
Fetch and browse	Arbitrary element	Allowed	List per article	Elements
Relevant in context	Arbitrary element	Non-overlapping	Set per article	Elements/Passages
Best in context	Arbitrary element	Non-overlapping	One result per article	Starting point

generic Thorough task (capturing the ability of a system to estimate the relevance of an element) can be use as input for further processing for the other tasks, each of these different retrieval tasks is capturing a different aspect of structured text. The retrieval tasks bring in elements from the task context in which they are to be applied, either in a end-user setting or system setting. As a result, the richer descriptions of the task's context and underlying assumptions are resonating more closely with actual real-world applications [19]. Bringing task-specific elements into information retrieval benchmark testing has been identified as one of the main research directions for further enhancing information access in general [18].

Key Applications

Structured text retrieval has the potential to improve information access by giving more direct access to the relevant information inside documents. As [14, p. 49] put it:

Large collections of full-text documents are now commonly used in automated information retrieval. When the stored document texts are long, the retrieval of complete documents may not be in the users' best interest. In such circumstances, efficient and effective retrieval results may be obtained by using passage retrieval strategies designed to retrieve text excerpts of varying size in response to statements of user interest.

Structured document retrieval is becoming increasingly important in all areas of information retrieval, the application to full-text book searching is obvious and such commercial systems already exist [19].

Future Directions

Improving information access by formulating retrieval tasks that capture interesting aspects of real-world structured text searching is an ongoing open problem. There has been a series of workshops addressing open problems, including real-world applications, the unit of retrieval, tasks and measures, and the problem of overlap [18].

The traditional picture of IR takes as input a document collection and a query, and gives as output a ranked list of documents. In the retrieval task, there no distinction between the hit list (communicating the ranked list) and the actual result documents. Where structured document retrieval is going beyond a linear ranked list of results, at least conceptually, interesting new research questions present themselves. By presenting related results from the same article, like in Fig. 2, the hit-list becomes a query-biased summary of the discourse structure of the retrieved article. [16] conduct experiments on the level of detail desired by searchers. Evaluation of such a system seems to require taking both retrieval effectiveness and document summarization aspects into account.

Data Sets

Notable data-sets are:

1. The *Shakespeare test collection* used in the Focus project 2000–2001 [3].
2. The *IEEE Computer Society collection* used at INEX 2002–2004 [7].

3. The expanded *IEEE Computer Society collection* used at INEX 2005 [7].
4. The *Wikipedia XML Corpus* used at INEX 2006–2007 [7].

Cross-References

▸ Evaluation Metrics for Structured Text Retrieval
▸ INitiative for the Evaluation of XML Retrieval
▸ XML Retrieval

Recommended Reading

1. Clarke C, Kamps J, Lalmas M. INEX 2006 retrieval task and result submission specification. In: Proceedings of the 5th International Workshop of the Initiative for the Evaluation of XML Retrieval; 2006. p. 381–8.
2. Clarke CLA, Kamps J, Lalmas M. INEX 2007 retrieval task and result submission format. In: Proceedings of the 6th International Workshop of the Initiative for the Evaluation of XML Retrieval; 2007. p. 445–53.
3. Focus. Focussed retrieval of structured documents – A Large Experimental Study. 2001.
4. Fuhr N. A probabilistic framework for vague queries and imprecise information in databases. In: Proceedings of the 16th International Conference on Very Large Data Bases; 1990. p. 696–707.
5. Gövert N, Kazai G. Overview of the INitiative for the Evaluation of XML retrieval (INEX) 2002. In: Proceedings of the 2nd International Workshop of the Initiative for the Evaluation of XML Retrieval; 2003. p. 1–17.
6. Hearst MA. User interfaces and visualization, Chapter X. In: Modern information retrieval. New York: ACM. 1999. p. 257–324.
7. INEX. INitiative for the evaluation of XML retrieval. 2007. http://inex.is.informatik.uni-duisburg.de/.
8. Jones KS. What's the value of TREC – is there a gap to jump or a chasm to bridge? SIGIR Forum. 2006;40(1):10–20.
9. Kazai G, Lalmas M, Gövert N, Malik S. INEX'03 retrieval task and result submission specification. In: Proceedings of the 2nd International Workshop of the Initiative for the Evaluation of XML Retrieval; 2003. p. 200–3.
10. Kazai G, Lalmas M, Reid J. Construction of a test collection for the focussed retrieval of structured documents. In: Proceedings of the 25th European Conference on IR Research; 2003. p. 88–103.
11. Lalmas M, Kazai G. INEX 2005 retrieval task and result submission specification. In: Proceedings of the 4th International Workshop of the Initiative for the Evaluation of XML Retrieval; 2005. p. 385–90.
12. Lalmas M, Malik S. INEX 2004 retrieval task and result submission specification. In: Proceedings of the 3rd International Workshop of the Initiative for the Evaluation of XML Retrieval; 2004. p. 237–40.
13. Malik S, Klas C-P, Fuhr N, Larsen B, Tombros A. Designing a user interface for interactive retrieval of structured documents – lessons learned from the INEX interactive track. In: Proceedings of the 10th European Conference on Research and Advanced Technology for Digital Libraries; 2006. p. 291–302.
14. Salton G, Allan J, Buckley C. Approaches to passage retrieval in full text information systems. In: Proceedings of the 16th Annual International ACM SIGIR Conference on Research and Development in Information Retrieval; 1993. p. 49–58.
15. Sigurbjörnsson B. Focused information access using XML element retrieval. SIKS dissertation series 2006-2028, University of Amsterdam. 2006.
16. Szlávik Z, Tombros A, Lalmas M. Feature- and query-based table of contents generation for XML documents. In: Proceedings of the 29th European Conference on IR Research; 2007. p. 456–67.
17. Tombros A, Larsen B, Malik S. The interactive track at INEX 2004. In: Proceedings of the 3rd International Workshop of the Initiative for the Evaluation of XML Retrieval; 2004. p. 410–23.
18. Trotman A, Geva S, Kamps J, editors. In: Proceedings of the SIGIR 2007 Workshop on Focused Retrieval; 2007.
19. Trotman A, Pharo N, Lehtonen M. XML-IR users and use cases. In: Proceedings of the 5th International Workshop of the Initiative for the Evaluation of XML Retrieval; 2006. p. 400–12.
20. Wilkinson R. Effective retrieval of structured documents. In: Proceedings of the 17th Annual International ACM SIGIR Conference on Research and Development in Information Retrieval; 1994. p. 311–7.

Primary Index

Yannis Manolopoulos[1], Yannis Theodoridis[2], and Vassilis J. Tsotras[3]
[1]Aristotle University of Thessaloniki, Thessaloniki, Greece
[2]University of Piraeus, Piraeus, Greece
[3]University of California-Riverside, Riverside, CA, USA

Synonyms

Clustering index; Sparse index

Definition

A tree-based index is called a *primary index* if its order is the same as the order of the file which it indexes Consider a relation R with some numeric attribute A taking values over an (ordered) domain D. Assume that relation R has been physically stored as an *ordered* file, following the order of the values of attribute A. Furthermore, assume that a tree-based index (e.g., B +-tree) has been created on attribute A. Then this index is primary.

Key Points

Tree-based indices are built on numeric attributes and maintain an order among the indexed search-key values. They are further categorized by whether their search-key ordering is the same with the file's physical order (if any). Note that a file may or may not be ordered. Ordered is a file whose records are stored in pages according to the order of the values of an attribute. Obviously, a file can have at most a single such order since it is physically stored once. For example, if the *Employee* relation is ordered according to the *name* attribute, the values in the other attributes will not be in order. A file stored without any order is called an unordered file or heap. If the search-key of a tree-based index is the same as the ordering attribute of a (ordered) file then the index is called *primary*. An index built on any non-ordering attribute of a file is called *secondary*.

Note that a primary index also controls the physical placement of the data records in a file. Since such an index also clusters the values of the file, the term *clustering* index has alternatively been used. In contrast, the order in the leaf pages of a secondary index is not the same as the order of the records in the file. Hence, a secondary index (also called *non-clustering*) needs an extra level of indirection, namely, a pointer to the actual position of a record with a given value in the relation file. In other words, a secondary index only clusters references to records (in the form of <value,pointer> fields), but *not* the records themselves. This extra indirection from a leaf page of a secondary index to the actual position of a record in a file has important subsequences on optimization. Consider for example a secondary index (B+ -tree) on the *ssn* attribute of the *Employee* relation (which assume is ordered by the *name* attribute). A query that asks for the salaries of employees with *ssn* in the range *[x,y]* can facilitate the B+ -tree on *ssn* to retrieve references to all records in the query range. Assume there are 1,000 such *ssn* values in the *Employee* file. Since the actual *Employee* records must be retrieved (so as to report their salaries), each such reference needs to be materialized by possibly a separate page I/O (since the actual records can be in different pages of the *Employee* file). If instead the file was ordered on the *ssn* attribute, the B+ -tree on *ssn* would have clustered (as primary index) the *Employee* records on *ssn*, and thus the 1,000 records that need to be retrieved would be located within few pages.

In practice, the search-key of a primary index is usually the file's primary key, however this is not necessary. That is, primary indexes need not be on the primary keys of relations. (In the above example, it was first assumed that the *Employee* file is ordered according to the *name* attribute and not according to the primary key *ssn* attribute). A relation can have several indices, on different search-keys; among them, at most one is primary (clustering) index and the rest are secondary ones.

Cross-References

▶ Access Path
▶ Index Creation and File Structures
▶ Indexed Sequential Access Method
▶ Secondary Index

Recommended Reading

1. Elmasri R, Navathe SB. Fundamentals of database systems. 5th ed. Boston: Addisson-Wesley; 2007.
2. Manolopoulos Y, Theodoridis Y, Tsotras VJ. Advanced database indexing. Dordrecht: Kluwer; 1999.
3. Ramakrishnan R, Gehrke J. Database management systems. 3rd ed. New York: McGraw-Hill; 2003.

Principal Component Analysis

Heng Tao Shen
School of Information Technology and Electrical
Engineering, The University of Queensland,
Brisbane, QLD, Australia
University of Electronic Science and Technology
of China, Chengdu, Sichuan, Sheng, China

Synonyms

PCA

Definition

Principal components analysis (PCA) is a linear technique used to reduce a high-dimensional dataset to a lower dimensional representations for analysis and indexing. For a dataset P in D-dimensional space with its principal component set Φ, given a point $p \in P$, its projection on the lower d-dimensional subspace can be defined as: p. Φ_d, where Φ_d represents the matrix containing 1st to d^{th} largest principal components in Φ and $d < D$.

Key Points

PCA finds a low-dimensional embedding of the data points that best preserves their variance as measured in the high-dimensional input space [1–3]. It identifies the directions that best preserve the associated variances of the data points while minimize "least-squares" (Euclidean) error measured by analyzing data covariance matrix. The first principal component is the eigenvector corresponding to the largest eigenvalue of the dataset's co-variance matrix, the second component corresponds to the eigenvector with the second largest eigenvalue and so on. The chosen principal components form the lower dimensional subspace of interest. Typically, the top d principal components are selected since they carry the most information of original data. The

lower dimensional representation for a point is then generated by multiplying the data point with the selected principal components. PCA makes a stringent assumption of orthogonality to make it amenable to linearity.

PCA has been widely used in many applications for exploratory data analysis and compression, making predictive models, indexing, information retrieval, etc.

Cross-References

▶ Dimensionality Reduction
▶ Discrete Wavelet Transform and Wavelet Synopses
▶ Multidimensional Scaling
▶ Singular Value Decomposition

Recommended Reading

1. Jolliffe IT. Principal component analysis. 2nd ed. New-York: Springer; 2002.
2. Huang Z, Shen HT, Shao J, Rüger SM, Zhou X. Locality condensation: a new dimensionality reduction method for image retrieval. In: Proceedings of the 16th ACM International Conference on Multimedia; 2008. p. 219–28.
3. Shen HT, Zhou X, Zhou A. An adaptive and dynamic dimensionality reduction method for high-dimensional indexing. VLDB J. 2007;16(2):219–34.

Privacy

Patrick C. K. Hung[1] and Vivying S. Y. Cheng[2]
[1]University of Ontario Institute of Technology,
Oshawa, ON, Canada
[2]Hong Kong University of Science and
Technology, Hong Kong, China

Definition

Privacy helps to establish personal autonomy and create individualism. Privacy is a state or condition of limited access to a person (e.g., client). In particular, information privacy relates to an individual's right to determine how, when, and to what extent information about the self will be released to another person or to an organization.

Key Points

With the rising occurrence of information privacy violations, people have begun to take interest in how, when, and where their personal information is being used. People usually interchange the concepts of security with privacy. In fact, they are essentially two different concepts used to protect data. In general, security involves the use of cryptographic tools to protect information in terms of confidentiality, availability, and access control and integrity enforcement. Privacy mainly focuses on the ability of keeping data away from public access, and the way to protect it according to the individual rights.

There are various definitions of privacy in the literature. Some researchers describe confidentiality as privacy, while some regard privacy as an aspect different from using cryptographic tools. Anderson [1] defined privacy as secrecy for benefit of the individual, and confidentiality as secrecy for the benefit of the organization. Alternatively, privacy can also be described by the ability to have control over the collection, storage, access, communication, manipulation and disposition of data. Privacy is also referred as the right for individuals to determine for themselves when, how, and to what extent information about them is communicated to others. As Westin [2] notes,

> No definition [of privacy] ... is possible, because [those] issues are fundamentally matters of values, interests and power.

It can be said that privacy is a much broader concept than security; privacy protection is based on security protection. Security may enable privacy protection from authorized access, but security alone cannot provide privacy.

To enhance the privacy protection of personal information, legislative schemas and practice guidelines have been proposed by different organizations such as the Health Insurance Portability and Accountability Act (HIPAA) in the United States of America (USA) and the European Union (EU) Data Protection Act. These legislations and requirements can be abstracted as access control policies or rules to serve as a privacy foundation to control access to personal information.

Cross-References

▶ Data Privacy and Patient Consent
▶ Privacy-Enhancing Technologies
▶ Privacy Metrics
▶ Privacy Policies and Preferences
▶ Privacy-Preserving Data Mining

Recommended Reading

1. Anderson RJ. A security policy model for clinical information systems. In: Proceedings of the 1996 IEEE Symposium on Security and Privacy; the 1996.
2. Westin A. Privacy and freedom. New York: Atheneum; 1967.

Privacy Metrics

Chris Clifton
Department of Computer Science, Purdue University, West Lafayette, IN, USA

Synonyms

Privacy measures

Definition

Measures to determine the susceptibility of data or a dataset to revealing private information. Measures include ability to link private data to an individual, the level of detail or correctness of sensitive information, background information needed to determine private information, etc.

Historical Background

Legal definitions of privacy are generally based on the concept of *Individually Identifiable Data*. Unfortunately, this concept does not have a clear meaning in the context of many database privacy technologies. The official statistics (census) community has long been concerned with measures for privacy, particularly in the contexts of microdata sets (datasets that represent real data, but obscured in ways to protect privacy) and tabular datasets. Measures have largely been based on the probability that a specific value belongs to a given individual, given the disclosed data. As technologies have been developed to anonymize and analyze private data, measures have been developed to quantify the potential for disclosure of private information by those techniques. In 2001, Samarati and Sweeney published papers on *k*-anonymity as a measure of individual identifiability, leading to considerable research in anonymity measures. At the same time, the rise in privacy-preserving data mining techniques lead to measures based on the ability to estimate values from disclosed data. As of this writing, this field is still open, with many competing measures, often tied to specific anonymization, analysis, or privacy protection techniques.

Foundations

There have been two distinct approaches to measuring privacy, or more specifically probability of disclosure. The first is measuring anonymity. *k*-anonymity states that any released data item must be linked to at least k individuals with equal probability. It builds on a concept of *Quasi-Identifier QI*: a quasi-identifier is information that an adversary can use to link a released data record to an individual. Formally,

k-Anonymity [8] A given table T^* is said to satisfy *k*-anonymity if and only if each sequence of values in $T*[QI_T*]$ appears at least k times in $T*$.

While simple to compute, *k*-anonymity does not by itself guarantee that sensitive data is protected. As first pointed out in [7], if all k values

with the same quasi-identifiers also have the same value for a sensitive attribute, *k*-anonymity can be met while still revealing the sensitive value for an attribute to an adversary. *k*-anonymity can be extended to deal with this issue:

ℓ-Diversity Principle [4] A q^*-block is $ℓ$-diverse if contains at least $ℓ$ "well-represented" values for the sensitive attribute S. A table is $ℓ$-diverse if every q^*-block is $ℓ$-diverse.

This still does not resolve the issue, as real values may have skewed distributions; a particular rare value may be "well-represented" by occurring in only one q^*-block, identifying the individuals represented in that block as having an unusually high probability of possessing that rare value. A further refinement is:

The t-closeness Principle [3] An equivalence class is said to have *t*-closeness if the distance between the distribution of a sensitive attribute in this class and the distribution of the attribute in the whole table is no more than a threshold t. A table is said to have *t*-closeness if all equivalence classes have t-closeness.

A challenge with any of these metrics is choosing appropriate values for k. Achieving a suitably low risk of disclosure may require a high value of k to protect particularly easy to identify individuals, resulting little specificity (and thus low value) in the anonymized dataset. This leads to a second set of metrics for anonymity, measuring not privacy but the fidelity of the disclosed data. These are typically based on the levels of generalization required to achieve a given level of anonymization, but as yet it isn't clear how well they relate to actual data utility (see [6]).

An alternative is to measure specifically the risk of identifying individuals, as with

δ- Presence [6] Given an external public table P, and a private table T, *δ-presence* holds for a generalization T^* of T, with $δ = (δ_{min}, δ_{max})$ if

$$\delta_{min} \leq \mathcal{P}(t \in T \mid T^*) \leq \delta_{max} \qquad \forall \ t \in P$$

The above approaches measure the ability of an adversary to link a data item to a specific individual. An alternative is to measure the ability to estimate sensitive data for an individual, once such linkage is performed. In [2], the assumption was that individuals provided sensitive data directly to the data analyst; the link between data and individual was thus known. Instead, the sensitive data is distorted to prevent the analyst from knowing values for an individual. This leads to a metric based on bounding the knowledge gained by an adversary from seeing the (distorted) sensitive value. The metric in [2] was based on two things: an interval and confidence level. If the analyst can determine that a value lies within the range $[x_1, x_2]$ with $c\%$ confidence, then the privacy at that confidence level is $x_2 - x_1$.

This measure raises two issues. First (as with ℓ-diversity), it relies on two numbers. This increases the difficulty of using it in practice. Second, as pointed out in [1], knowledge of the distribution of the original data (acquired by the analyst from the distorted dataset) may allow tightening of the bounds. They propose a measure based on the *differential entropy* of a random variable. The differential entropy $h(A)$ is a measure of the uncertainty inherent in A. Their metric for privacy is $2^{h(A)}$. Specifically, when adding noise from a random variable A, the privacy is:

$$\Pi(A) = 2^{-\int_{\Omega_A} f_A(a) log_2 f_A(a) da}$$

where Ω_A is the domain of A. This metric has several nice features. It is intuitively satisfying for simple cases. For noise generated from A, a uniformly distributed random variable between 0 and a, $\Pi(A) = a$. Thus this privacy metric is exactly the width of the unknown region. Furthermore, if a sequence of random variables A_i converges to B, then $\Pi(A_i)$ converges to $\Pi(B)$. For most random variables, e.g., a gaussian, the notion of width of the unknown region does not make sense. However, by calculating Π for such random variables, the above properties can be used to make the case that the privacy is equivalent to having no knowledge of the value except that it is within a region of width Π. This gives an intuitively satisfying way of comparing the privacy of different methods of adding random noise.

The authors extend this definition to *conditional privacy*, capturing the possibility that the

inherent privacy from obscuring data may be reduced by what can be learned from a collection. The conditional privacy $\Pi(A|B)$ follows from the definition of conditional entropy:

$$\Pi(A|B)$$
$$= 2^{-\int_{\Omega_{A,B}} f_{A,B}(a,b) log_2 f_{A|B=b}(a) da db}$$

They show how this can be applied to measure the actual privacy after reconstructing distributions of the original data to improve the accuracy of decision trees build on the obscured data [1, 2].

Another use of this metric is to evaluate the inherent loss of privacy caused by data mining results. The use of conditional privacy enables estimating how much privacy is lost by knowing the data mining results, even with a "perfect" privacy-preserving technique such as secure multiparty computation. The literature has not yet addressed this issue; the assumption has generally been that the data mining results do not of themselves violate privacy.

Numerous other metrics have been proposed; the above give a representative view of the approaches and challenges faced in measuring privacy.

Key Applications

Primary applications are in aggregate analysis of privacy-sensitive data, such as individual healthcare, personal preference, or financial information. Direct access to individual data is typically a confidentiality and access control issue; access to data is either allowed or it is not. Privacy measures come into play when individual data is analyzed as part of a broader study, where the end result is not solely for the benefit of the individuals whose data was used. (e.g., Studies of demographics or medical research.) Privacy laws typically control the use of data when not specifically to complete a transaction on behalf of the individual; measures are thus needed to show that privacy is maintained.

Future Directions

One approach is measures that correlate with legal and regulatory standards (e.g., the HIPAA safe harbor rules). Perhaps more promising as a research direction is risk-based measures such as δ-presence [5]. Location data (as from mobile devices) also poses new issues; European Community rules specifically discuss protection of location, but it is not clear to what extent existing measures can be applied. Another need is measures that adjust for differences in individual privacy needs, e.g., protecting outliers.

Cross-References

▸ Individually Identifiable Data
▸ Privacy
▸ Privacy-Preserving Data Mining
▸ Randomization Methods to Ensure Data Privacy
▸ Statistical Disclosure Limitation for Data Access

Recommended Reading

1. Agrawal D, Aggarwal CC. On the design and quantification of privacy preserving data mining algorithms. In: Proceedings of the 20th ACM SIGACT-SIGMOD-SIGART Symposium on Principles of Database Systems; 2001. p. 247–55.
2. Agrawal R, Srikant R. Privacy-preserving data mining. In: Proceedings of the ACM SIGMOD International Conference on Management of Data; 2000. p. 439–50.
3. Li N, Li T. T-closeness: privacy beyond k-anonymity and l-diversity. In: Proceedings of the 23rd International Conference on Data Engineering; 2007.
4. Machanavajjhala A, Gehrke J, Kifer D, Venkitasubramaniam M. l-diversity: privacy beyond k-anonymity. ACM Trans Knowl Discov Data. 2007;1(1): No.3.
5. Nergiz M, Atzori M, Clifton C. Hiding the presence of individuals from shared databases. In: Proceedings of the ACM SIGMOD International Conference on Management of Data; 2007. p. 665–76.
6. Nergiz ME, Clifton C. Thoughts on k-anonymization. Data Knowl Eng. 2007;63(3):622–45.
7. Øhrn A, Ohno-Machado L. Using boolean reasoning to anonymize databases. Artif Intell Med. 1999;15(3):235–54.
8. Sweeney L. Achieving k-anonymity privacy protection using generalization and suppression. Int J Uncertainty Fuzziness Knowledge Based Syst. 2002;10(5):557–70.

Privacy Policies and Preferences

Patrick C. K. Hung, Yi Zheng, and Stephanie Chow
University of Ontario Institute of Technology, Oshawa, ON, Canada

Definition

Privacy policies describe an enterprise's data practices, what information is collected from individuals (subjects), what the information (objects) will be used for, whether the enterprise provides access to the information, who the recipients are of any result generated from the information, how long the information will be retained, and who will be informed in the event of a dispute. A subject releases his or her data to the custody of an enterprise while consenting to the set of purposes for which the data may be used. The subject can express his or her preferences in a set of preference rules to make decisions regarding the acceptability of privacy policies.

Historical Background

People have been concerned with privacy policies and preferences for more than 200 years. For example, the Hippocratic oath was written as a guideline of medical ethics for doctors in respect to a patient's health condition, and states as follows: "Whatsoever things I see or hear concerning the life of men, in my attendance on the sick or even apart there from, which ought not to be noised abroad, I will keep silence thereon, counting such things to be sacred secrets." Privacy policies and preferences are often expressed in natural language to specify or regulate how a system or an organization is to preserve privacy. In response to privacy policy and preference violations, many countries have enacted legislation to protect privacy for individuals. There are also different legislations for different industry sectors. For example, in the financial sector the Gramm-Leach-Bliley Act (GLB Act) of the U.S. requires banks to implement security programs to protect customer information; in the healthcare sector the Health Insurance Portability and Accountability Act (HIPAA) of the U.S. sets an American national standard to protect and enhance the rights of consumers, clients, and patients to control how their health information is used and disclosed. Similarly, the Personal Information Protection and Electronic Documents Act (PIPEDA) of Canada sets out ground rules for how private sector organizations may collect, use, or disclose personal information in the course of commercial activities. Failing to comply with these legislations, in their respective countries, may lead to civil and/or criminal penalties and/or imprisonment. In addition to the penalties, organizations may even suffer the loss of reputation and goodwill when the non-compliance of legislation is publicized. Looking at the above factors, it is evident that the privacy legislations have forced a widespread impact on the privacy policies and preferences.

Foundations

One can imagine that information privacy is usually concerned with the confidentiality of information. Threats to information privacy can come from insiders and from the outsiders in each enterprise. Privacy control is usually not concerned with individual subjects. Privacy policies are often expressed in natural language to specify or regulate how a system or an enterprise is to preserve privacy. Here are the principles of information privacy protection:

- Principle 1: Data-level security protection principle
  - Principle 1.1: Personal information shall be protected by security safeguards in a secure way such that data confidentiality, integrity, and availability can be achieved.
  - Principle 1.2: In needs to be accurate, complete, and up-to-date. The correction of inaccurate information should be allowed for maintaining data quality.

- Principle 2: Communication-level security protection principle
  - Principle 2.1: Information shall be transported in a secure way that data confidentiality and authentication of users/systems/services are achieved.
  - Principle 2.2: Data integrity must be maintained during the transmission in the communication channel.
- Principle 3: Consent requirement principle
  - Principle 3.1: Consent must be given by the owner of the information for the collection, disclosure, use, and retention of his/her information.
  - Principle 3.2: Owners of information should be allowed to review, control, and setup restrictions to their information on collection, usage, disclosure, and retention.

Principle 3 includes four limitations to specify how personal information can be released upon request:

- *Limitation on Collection*: The collection of personal information shall be limited to specific legitimate purposes of collection only.
- *Limitation on Disclosure*: The owner of information should be able to make special restrictions on the disclosure of his/her own information.
- *Limitation on Use*: The use of personal information shall be identified as legitimate use by the services provider and/or the owner of information.
- *Limitation on Retention*: Personal information shall be retained for only as long as the purpose for which it is used.

A subject releases his or her data to the custody of an enterprise while consenting to the set of purposes for which the data may be used. The subject can express his or her preferences in a set of preference-rules to make decisions regarding the acceptability of privacy policies. Preference assumes a real or imagined choice between alternatives and the possibility of rank ordering of these alternatives in a privacy policy. Privacy preferences are formally expressed by a set of rules and should preferably be captured through an interface. Subjects who are aware of how their personally identifiable information is being used, collected or disclosed can better understand how to protect it. Personally Identifiable Information (PII) includes information that identifies and individual and information that an organization may be able to identify. This includes, but is not limited to a subject's name, address, telephone number, social insurance number and credit card number(s). Among the most sensitive PII are medical and financial records.

Key Applications

In many instances, subjects do not have the knowledge or understanding of how his or her PII is being used, collected or disclosed, nor what privacy preferences are available. The adoption of information technology and the Internet have further added to this complexity. Even the most common daily transactions such as Internet browsing, grocery shopping and online banking increase the exposure and vulnerability of threats to information privacy. In response to these threats, and concerns surrounding data integrity, security, online privacy and confidentiality, professional services firms have begun to expand their service lines to include third-party enforcement programs. Seals, or other easily distinguishable symbols, are issued to enterprises whose privacy policies and procedures have been concluded to be in compliance with its governing board(s). The purposes of third-party enforcement programs are threefold: (i) to build consumer trust; (ii) to educate subjects of their privacy preferences; and (iii) to develop a complaint resolution mechanism. Here are the three major procedures of enforcing privacy policies in an organization:

- Building consumer trust
  - There exists a natural conflict of interest between organization and subject.

  - Subject is absent from the development of privacy policy.
  - Independent review of an organization's compliance with governing board(s) adds reliability and credibility.
- Educate subjects of their privacy preferences
  - Promote awareness of privacy issues and how to get in contact with privacy coordinator.
  - If for any reason a subject's PII is required for another purpose from when it was collected, it is the enterprise's responsibility to obtain additional consent.
  - At a minimum, the enterprise is to inform the subject of the circumstance and provide an opportunity for the subject to opt out of such a use.
- Complaint resolution mechanism
  - Organizations rewarded a seal of validation are required to provide subjects a method to resolve any problems or discuss any complaints.
  - Complaint resolution process should be easily accessible and comprehendible.

In order to stand apart from industry rivals, companies strive to obtain a competitive advantage. As an organization, it is advantageous to be knowledgeable of the external risks associated affecting the protection of a subject's privacy. Participants of third-party enforcement programs are continuously monitored for adaptability to legislative frameworks and threats to privacy compliance.

Privacy technologies have been researched for a period of time. For example, the Platform for Privacy Preferences Project (P3P) working group at World Wide Web Consortium (W3C) develops the P3P specification for enabling Web sites to express their privacy practices in a standard and machine-readable XML format. P3P user agents allow users to automatically be informed of site practices and to automate decision-making based on the Web sites' privacy practices. In addition, P3P also provides a language called P3P Preference Exchange Language 1.0 (APPEL1.0) that is used to express the user's preferences for making automated or semi-automated decisions regarding the acceptability of machine-readable privacy policies from P3P enabled Web sites. It provides a base schema for the data collected and a vocabulary to express purposes, the recipients, and the retention policy. Although it captures common elements of privacy policies, sites may have to provide further explanations in human-readable policies.

Furthermore, WS-Privacy has been mentioned in industry for a period of time for defining subject privacy preferences and organizational privacy practice statements for Web services. At this minute, the WS-Privacy specification has not been released to public yet. Then, the Enterprise Privacy Authorization Language (EPAL) technical specification is used to formalize privacy authorizations for actual enforcement within an intra- or inter-enterprise for business-to-business privacy control. On the other hand, the XACML is a general-purpose access control policy language used to describe policy and access control decision request/response [11]. Though XACML has drafted a privacy policy profile document [12], the current XACML framework can not handle the privacy enforcement.

One of the most significant privacy technologies is the IBM Tivoli Privacy Manager for e-business. This privacy middleware technology converts privacy policy and data-handling rules from applications and IT systems to P3P format. The future development will include the integration of privacy technologies into some specific Web-based applications like in healthcare sector (e.g., Microsoft Healthvault and Google Health Portal), especially in light of recent changes in health privacy legislative environment.

Cross-References

▶ Data Privacy and Patient Consent
▶ Privacy
▶ Privacy-Enhancing Technologies
▶ Privacy Metrics
▶ Privacy-Preserving Data Mining

Recommended Reading

1. Cheng VSY, Hung PCK. Health Insurance Portability and Accountability Act (HIPAA) compliant access control model for web services. Int J Health Inf Syst Inform. 2005;1(1):22–39.
2. Fischer-Hubner S. IT-security and privacy. Lecture notes in computer science. Berlin/Heidelberg/New York: Springer; 2001.
3. Online Privacy Alliance. Effective enforcement of self regulation. Online: http://www.privacyalliance.org/resources/enforcement.shtml
4. Powers CS, Ashley P, Schunter M. Privacy promises, access control, and privacy management – enforcing privacy throughout an enterprise by extending access control. In: Proceedings of the 3rd International Symposium on Electronic Commerce; 2002. p. 13–21.

Privacy Through Accountability

Anupam Datta
Computer Science Department and Electrical and Computer Engineering Department, Carnegie Mellon University, Pittsburgh, PA, USA

Definition

Privacy through accountability refers to the principle that entities that hold personal information about individuals are accountable for adopting measures that protect the privacy of the data subjects [1]. This article focuses on computational treatments of this principle. This research area has produced precise definitions of privacy properties and computational accountability mechanisms to aid in their enforcement.

Formally, privacy properties impose restrictions on personal information flows. Information flow types encompass context-specific direct flows (e.g., transfer of health information from a hospital to an insurance company) [2–4], implicit flows (e.g., the use of users' location in a web advertising system) [5], and flows of noisy statistics from databases of personal information (e.g., the use of customers' ratings to recommend movies) [6]. The restrictions on these types of information flow include role-based restrictions (e.g., permitting certain types of flows between a patient and a doctor) [2–4], temporal restrictions (e.g., permitting flows if the data subject consents or has been given notice) [2, 3], and purpose restrictions (e.g., requiring that information be used only for certain purposes like treatment) [7].

Computational accountability mechanisms aid in demonstrating that programs and people are compliant with privacy properties, detecting and explaining violations, and designing interventions (e.g., fixing program bugs or punishing human violators). Significant results include audit algorithms that check logs for compliance with temporal [2, 8, 9] and purpose restrictions [7] on flows and provide explanations for detected violations [10]. Other major results include information flow analysis of black-box software systems [11, 12] and program analysis methods that check privacy compliance of the source code of software systems [5, 13].

Historical Background

The principle of privacy through accountability goes back at least to the 1970s and early 1980s when it was embodied into several privacy guidelines including an influential one developed by the OECD [1]. Computational treatments of this principle are much more recent. They do not trace back to a single source. However, several early works that emphasize this viewpoint for direct flows include work on privacy and contextual integrity (which relies on monitors for privacy protection) [2], privacy APIs (which relies also on monitors) [4], and information accountability (a position paper) [14].

Significant scientific results in this area include audit algorithms that check logs for compliance with temporal restrictions on direct flows [8, 9] and purpose restrictions on direct and implicit flows [7] and provide explanations for detected violations [10]. Other major results include statistical information flow analysis of black-box software systems [11, 12] and program analysis

methods that check privacy compliance of the source code of software systems encompassing direct and implicit flows [5] and statistical information flow properties [13].

Key applications include applications to the HIPAA Privacy Rule for healthcare privacy in the USA [3, 8], checking programs in Microsoft Bing's advertising pipeline for compliance with its privacy policies [5], and black-box analysis of personal data use for online advertising by several web service providers [12].

Scientific Fundamentals

This section summarizes some significant results in the area of privacy through accountability.

Checking Audit Logs for Temporal Restrictions on Direct Flows

Contextual integrity is a philosophical theory of privacy. The descriptive component of this theory has been formalized in computational terms as imposing role-based and temporal restrictions on personal information flow [2, 3]. The building blocks of this theory are social contexts and context-relative informational norms. A context captures the idea that people act and transact in society not simply as individuals in an undifferentiated social world, but as individuals in certain capacities (roles) in distinctive social contexts, such as healthcare, education, friendship, and employment.

Norms prescribe the flow of personal information in a given context, e.g., in a healthcare context, a norm might prescribe flow of personal health information from a patient to a doctor and proscribe flows from the doctor to other parties who are not involved in providing treatment. Norms are a function of the following parameters: the respective roles of the sender, the subject, and the recipient of the information, the type of information, and the principle under which the information is sent to the recipient. Examples of transmission principles include confidentiality (prohibiting agents receiving the information from sharing it with others), reciprocity (requiring bidirectional information flow, e.g., in a friendship context), consent (requiring permission from the information subject before transmission), and notice (informing the information subject that a transmission has occurred). When norms are contravened, people experience a violation of privacy.

This theory has been used to explain why a number of technology-based systems and practices threaten privacy by violating entrenched informational norms. The theory is now well known in the privacy community and has influenced privacy policy in the USA (e.g., "respect for context" was included as an important principle in the Consumer Privacy Bill of Rights released by the White House in 2012).

The idea that privacy expectations can be stated using context-relative informational norms is formalized in a semantic model and logic of privacy [2] and developed further in follow-up work [3]. At a high level, the model consists of a set of interacting agents in roles that perform actions involving personal information in a given context. For example, Alice (a patient) may send her personal health information to Bob (her doctor). Following the structure of context-relative informational norms, each transmission action is characterized by the roles of the sender, subject, recipient, and the type of the information sent. Interactions among agents give rise to traces where each trace is an alternating sequence of states (capturing roles and knowledge of agents) and actions performed by agents that update state (e.g., an agent's knowledge may increase upon receiving a message or his role might change). Transmission principles prescribe which traces respect privacy and which traces don't.

While contextual integrity talks about transmission principles in the abstract, a precise logic enables specification in a form that information processing systems can check for violations of such principles. Two considerations are particularly important in designing the logic: (a) expressivity, the logic should be able to represent practical privacy policies, and (b) enforceability, it should be possible to provide automated support for checking whether traces satisfy policies expressed in the logic. A *logic of privacy* that meets these goals is presented in recent work [8]. This privacy logic is an expressive fragment of first-order logic. It has been used to develop the first complete formalization of two US privacy laws – the HIPAA Privacy Rule for healthcare organizations and the Gramm-Leach-Bliley Act for financial institutions [3].

These comprehensive case studies shed light on common concepts that arise in transmission principles in practice – data attributes, dynamic roles, notice and consent (formalized as temporal properties), purposes of uses and disclosures, and principals' beliefs – as well as how individual transmission principles are composed in privacy policies.

The *reduce* algorithm is an accountability mechanism that checks audit logs for compliance with privacy properties specified in this logic [8]. This algorithm addresses two fundamental challenges in compliance checking that arise in practice. First, in order to be applicable to realistic policies, reduce operates on policies expressed in a first-order logic that allows restricted quantification over infinite domains. It builds on ideas from logic programming to identify the restricted form of quantified formulas. The logic can, in particular, express all 84 disclosure-related clauses of the HIPAA Privacy Rule, which involve quantification over the infinite set of messages containing personal information. Second, since audit logs are inherently incomplete (they may not contain sufficient information to determine whether a policy is violated or not), reduce proceeds iteratively: in each iteration, it provably checks as much of the policy as possible over the current log and outputs a residual policy that can only be checked when the log is extended with additional information. The algorithm comes with proofs of correctness, termination, time, and space complexity results. It has also been implemented and experimentally evaluated to check simulated audit logs for compliance with the HIPAA Privacy Rule. The algorithm has also been augmented to provide explanations when a violation is detected [10].

Checking Audit Logs for Purpose Restrictions on Direct and Implicit Flows

Purpose restrictions occupy a central place in numerous influential privacy guidelines and regulations, including OECD's Privacy Guidelines, the EU Privacy Directive, US privacy laws and organizational privacy policies in sectors as diverse as healthcare, finance, web services, insurance, education, and government. For example, the HIPAA Privacy Rule requires that hospital employees use personal health information *only for* certain purposes (e.g., treatment) and not for other purposes (e.g., gossip). Thus, it is important to define what purpose restrictions mean and design algorithms for their enforcement using information processing systems.

One body of work assumes that the auditor has a method for determining which behaviors are for a purpose. An example of this approach associates purposes with resources and roles; only agents in roles (e.g., a doctor) are permitted to access resources (e.g., medical records) if the user's role matches the resources purpose (e.g., treatment). Related approaches associate groups, sequences of actions, or programs with purposes and similarly mediate access or information flow. These approaches are useful in settings where the auditor has a way of labeling roles, sequences of actions, or programs with their purposes. However, these approaches do not provide a systematic basis for such labeling.

An alternative body of work provides a semantic basis for purpose restrictions [7]. The central thesis is that an action is for a purpose if it is part of a *plan* for achieving that purpose and information is used for a purpose if it affects the planning process. Planning is modeled using (partially observable) Markov decision processes. Audit algorithms check audit logs for compliance with purpose restriction policies. The algorithms compare logged actions to a model of how an agent attempting to achieve the allowed purpose would plan to do so. If the logged actions differ from the model, the algorithm reports a potential violation. When Markov decision processes (MDPs) are used to model planning, the algorithms effectively check whether the logged actions could have been produced by an optimal plan for the MDP; if not, it declares that the actions were not only for the purpose modeled by the reward function of the MDP.

Checking Black-Box Software Systems for Statistical Privacy Properties

Web advertisers, such as Google's DoubleClick, the Microsoft Media Network, or the Yahoo Ad Exchange, use complex behavioral profiles to select the ads that they show to website visitors. These profiles and their uses raise privacy concerns about the data used. An important set of scientific questions arise in studying these systems like *black boxes*, i.e., without access to the source code and data models internally used by the systems. This is a useful setting since web users, privacy advocacy groups, or government regulators are interested in understanding how these systems use personal information to serve advertisements, so that they can hold the companies accountable for inappropriate uses.

For example, a user may be interested in understanding whether gender has a significant effect on job-related online advertisements (a finding of discrimination) or whether browsing substance-abuse-related websites results in related advertisements (a finding suggestive of a healthcare privacy concern). These questions are all instances of detecting statistical personal information flow in a black-box setting. Recent work on *information flow experiments* [11] provides a precise definition of this problem, proves that it is equivalent to a problem of causal inference, and presents a methodology based on experimental science and statistical analysis for detecting such flows. A central insight from that work is that permutation testing (a particular nonparametric statistical test) is well suited for sound analysis in this setting. In contrast, assumptions underlying a number of other commonly used tests, such as independence of ads and absence of cross-unit effects, don't hold in the online advertising setting.

There are a number of other significant empirical studies in this emerging area (e.g., [12]).

Checking Source Code of Software Systems for Privacy Properties

A complementary setting is one in which the data holder would like to use a computational tool to check that the *source code* of its software systems uses personal information in ways that respect privacy expectations. This problem reduces to program analysis for various kinds of information flow properties [5, 13]. We mention two significant results in this space.

Recent work [5] develops a methodology and tool chain for checking software systems written in big data programming languages (e.g., Scope, Hive, Dremel) for compliance with a class of privacy policies. The privacy policies restrict direct and implicit information flows based on role, purpose, and other considerations. The tool chain has been applied to check over a million lines of source code in Microsoft Bing's data analytics pipeline for compliance with its privacy policies. This work addresses two central challenges in making privacy compliance tools practical.

First, privacy policies are often crafted by legal teams, while software that has to respect these policies is written by developers. An important challenge is thus to design privacy policy languages that are usable by legal privacy teams, yet have precise operational meaning (semantics) that software developers can use to restrict how their code operates on personal information of users. The *Legalese* language designed in this work has these characteristics. It supports nested allow-deny information flow rules (inspired by [3]) in a natural language-like format that restrict direct and implicit flows in software systems based on role, purpose, and other constraints. A user study at Microsoft indicates that the language is usable by the target group consisting of lawyers and privacy champions (who sit between legal privacy teams and software developers).

Second, software systems that perform data analytics over personal information of users are often written without a technical connection to the privacy policies that they are meant to respect. Tens of millions of lines of such code are already in place in companies like Facebook, Google, and Microsoft. An important challenge is thus to bootstrap existing software for privacy compliance. This work includes a tool called Grok to annotate software written in programming languages that support the map-reduce programming model (e.g., Dremel, Hive, Scope) with Legalese's policy datatypes. A simple way to conduct the bootstrapping process would be to ask developers to manually annotate all code with policy datatypes (e.g., labeling variables as IPAddress, programs as being for the purpose of Advertising, etc.). However, this process is too labor-intensive to scale. Instead, they develop a set of techniques to automate the bootstrapping process.

Another work [13] presents a programming language method for checking software systems for compliance with differential privacy – a statistical privacy property [6]. They present a type system that guarantees that well-typed programs satisfy differential property. Other researchers have developed related methods for checking models of software systems for compliance with differential privacy based on unwinding relations and relational logic. These are significant scientific results although their application to large-scale software systems remains future work.

Key Applications

This article interleaves the discussion of scientific fundamentals with the applications that motivate their development. Most of the significant reported applications are in healthcare privacy [2–4, 7, 8, 10] and web privacy [5, 11, 12], although the methods are general enough to be applicable to many other domains.

Recommended Reading

1. OECD. Fair information practices principles. http://www.oecd.org/internet/ieconomy/oecdguidelinesontheprotectionofprivacyandtransborderflowsofpersonaldata.htm

2. Barth A, Datta A, Mitchell JC, Nissenbaum H. Privacy and contextual integrity: framework and applications. In: Proceedings of the 2006 IEEE Symposium on Security and Privacy; 2006. p. 184–98.

3. DeYoung H, Garg D, Jia L, Kaynar DK, Datta A. Experiences in the logical specification of the HIPAA and GLBA privacy laws. In: Proceedings of the 2010 ACM Workshop on Privacy in the Electronic Society; 2010. p. 73–82.

4. May MJ, Gunter CA, Lee I. Privacy APIs: access control techniques to analyze and verify legal privacy policies. In: Proceedings of the 19th IEEE Computer Security Foundations Workshop; 2006. p. 85–97.

5. Sen S, Guha S, Datta A, Rajamani S, Tsai J, Wing JM. Bootstrapping privacy compliance in big data systems. In: Proceedings of the 2010 IEEE Symposium on Security and Privacy; 2014.

6. Dwork C. Differential privacy. In: Proceedings of the 33rd International Colloquium on Automata, Languages, and Programming; 2006. p. 1–12.

7. Tschantz MC. Formalizing and enforcing purpose restrictions. PhD thesis, Computer Science Department, Carnegie Mellon University, Technical Report CMU-CS-12-117, May 2012.

8. Garg D, Jia L, Datta A. Policy auditing over incomplete logs: theory, implementation and applications. In: Proceedings of the 18th ACM Conference on Computer and Communication Security; 2011. p. 151–62.

9. Basin DA, Klaedtke F, Muller S, Pfitzmann B. Runtime monitoring of metric first-order temporal properties. In: Proceedings of the 28th International Conference on Foundations of Software Technology and Theoretical Computer Science; 2008. p. 49–60.

10. Oh SE, Chun JY, Jia L, Garg D, Gunter CA, Datta A. Privacy-preserving audit for broker-based health information exchange. In: Proceedings of the 4th ACM Conference on Data and Application Security and Privacy; 2014. p. 313–20.

11. Tschantz MC, Datta A, Datta A, Wing JM. A methodology for information flow experiments. CoRR abs/1405.2376. 2014.

12. Lecuyer M, Ducoffe G, Lan F, Papancea A, Petsios T, Spahn R, Chaintreau A, Geambasu R. XRay: increasing the web's transparency with differential correlation. In: Proceedings of the 23rd USENIX Security Symposium; 2014.

13. Reed J, Pierce BC. Distance makes the types grow stronger: a calculus for differential privacy. In: Proceeding of the 15th ACM SIGPLAN International Conference on Functional Programming; 2010. p. 157–68.

14. Weitzner DJ, Abelson H, Berners-Lee T, Feigenbaum J, Hendler JA, Sussman GJ. Information accountability. Commun ACM. 2008;51(6):82–7.

15. Kagal L, Pato J. Preserving privacy based on semantic policy tools. IEEE Secur Priv. 2010;8(4): 25–30.

Privacy-Enhancing Technologies

Simone Fischer-Hübner
Karlstad University, Karlstad, Sweden

Synonyms

PETs

Definition

Privacy-enhancing technologies (PETs) can be defined as technologies that are enforcing privacy principles in order to protect and enhance the privacy of users of information technology (IT) and/or of individuals about whom personal data are processed (the so-called data subjects). Privacy principles that PETs are enforcing can be derived from internationally acknowledged privacy guidelines or legislation, such as the OECD Privacy Guidelines, the EU Data Protection Directive 95/46/EC and the EU General Data Protection Regulation (GDPR), and the proposed General EU Data Protection Regulation. One fundamental privacy principle that serves as the foundation for the privacy-enhancing technologies that are aiming at providing *anonymity*, *pseudonymity*, or *unobservability* for users and/or other data subjects is the privacy principles of data minimization. It requires that the collection of personally identifiable data should be minimized (and if possible avoided), because obviously privacy is best protected if no personal data at all (or at least as little data as possible) is collected or processed. Further important privacy principles addressed by other PETs are purpose specification and limitation, the informed consent by data subjects (as a prerequisite for making data processing legitimate), transparency (i.e., openness) of data processing, and appropriate technical security means for protecting the confidentiality, integrity, and availability of personal data.

Historical Background

IT security technologies for protecting the confidentiality, integrity, and availability of (personal) data, such as *access control*, have been developed and researched since the beginning of computing – *data encryption* as a technique for protecting the confidentiality of data has been used since ancient times. *Inference controls* for protecting the data of individuals stored in statistical databases (e.g., medical databases) have been elaborated since the 1970s.

Most of the fundamental anonymity technology concepts were introduced by David Chaum in the 1980s (see: [1–3]). The term "privacy-enhancing technologies" was first introduced in 1995 in a report on data minimization technologies, which was jointly published by the Dutch Registratiekamer and the Information and Privacy Commissioner in Ontario, Canada, with the title "Privacy-Enhancing Technologies: The path to Anonymity" [4]. Since then, further research and development has been done in data minimization technologies (e.g., [5], [7], [9], [11], [14], [15]) and areas such as privacy policies and privacy-enhancing access control [10], [12], [13], [16] and privacy-enhancing identity management [6].

The related concept of Privacy by Design, which has been marketed by the Information and Privacy Commissioner of Ontario since the late 1990s,is a system engineering approach which requires that privacy protection, particularly by the implementation of PETs, should be incorporated into the overall system design, i.e., it should be embedded throughout the entire system life cycle.

Privacy as an expression of the rights of self-determination and human dignity is considered a core value in democratic societies and is recognized either explicitly or implicitly as a fundamental human right by most constitutions of democratic societies. However, in the network society, individuals are increasingly at risk that all their communications, transactions, and movements can be monitored and profiled. Profiles collected at various sites can be easily combined, aggregated, and retained without time limitations and without the individuals' knowledge or con-

sent. Research and development of PETs have been motivated by the vision to provide technical means allowing individuals to retain control over their personal spheres, i.e., to protect their privacy in the electronic information age.

Scientific Fundamentals

Privacy-enhancing technologies can basically be divided into:

1. The class of PETs for minimizing or avoiding identifiable data for users and/or other data subjects
2. The class of PETs for safeguarding lawful and privacy-friendly personal data processing
3. PETs that are a combination of the two aforementioned PET classes

PETs for Minimizing or Avoiding Personally Identifiable Data

PETs that are minimizing/avoiding identifiable data of users, and that are thereby providing *anonymity*, *pseudonymity*, and/or *unobservability*, can be divided dependent on whether data is minimized on communication level or on application level.

Data Minimization at Communication Level

Examples for anonymous communication schemes for achieving sender anonymity are DC-nets [3], Mixnets [1], Crowds [9], or Tor [7], where Mixnets will be presented in more detail below.

With DC-nets, the fact that someone is sending a message is hidden by a one-time pad encryption, which means that DC-nets can offer *perfect sender anonymity* (in the information-theoretic sense). By the use of message broadcast and implicit addresses (i.e., by the use of an attribute which allows only the addressee to recognize that the message is addressed to him, e.g., the message is public-key encrypted and the addressee is the

P

only one who can successfully decrypt it with his secret key), it also provides *receiver and relationship anonymity*.

Mix Nets The technique of a Mix network, which was originally introduced by David Chaum [1], realizes unlinkability of a sender and recipient (*relationship anonymity*), as well as *sender anonymity* against the recipient, and optionally *recipient anonymity* (via so-called anonymous return addresses, which is however not elaborated further here).

A mix is a special network station, which collects and stores incoming messages, discards repeats, changes their appearance by encryption, and outputs them in a different order and by this hides the relation between incoming and outgoing messages. If a sender of a message uses one mix for forwarding a message to a recipient on his behalf, the relation between sender and recipient is hidden from everybody but the mix and the sender of the message. This means also that the recipient only learns that the message was sent to him by the mix, but he does not know the identity of the real sender. However, if only one mix is used, this mix can in detail learn and potentially profile who is communicating with whom. To improve security, a message is sent over a mix net, which consists of a chain of independent mixes. The sender must perform cryptographic

operations inverse to those of the mix, because the recipient must be able to read the message. A global attacker, who can monitor all communication lines, should in principle only be able to trace a message through the mix network, if he has the cooperation of all mix nodes on the path or if he can break the cryptographic operations. Thus, in order to ensure unlinkability of sender and recipient, at least one mix in the chain has to be trustworthy.

Assume that Alice wants to send anonymously a message (msg) to recipient Bob (which could be encrypted with Bob's public key for providing also message secrecy). Alice chooses a mix sequence $Mix_1, Mix_2, \ldots, Mix_m$ (in Fig. 1, Alice chooses m = 3 Mixes). Let for simplicity Mix_{m+1} denote the recipient (Bob). Each Mix_i with address A_i has initially chosen a key pair (c_i, d_i), where c_i is a public key of Mix_i and d_i is its private key. Let z_i be a random string.

Alice recursively creates the following encrypted messages, where M_i is the message that Mix_i will receive:

```
M_{m+1} = msg.
M_i = E_{ci}(z_i, A_{i+1}, M_{i+1})  for i = 1, ..., m,
```

i.e., she adds one layer of public key encryption to the message in reverse order to the sequence of mixes in the chain. She then sends the result M_1 to Mix_1.

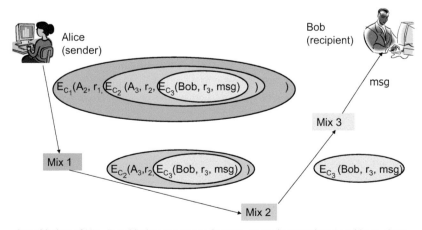

C_i: public key of Mix_i, E: public key encryption function, r_i: random number, A_i: address of Mix_i

Privacy-Enhancing Technologies, Fig. 1 Mixnetwork with a chain of three mixes

Each Mix_i decrypts the incoming message $M_i = E_{ci}(z_i, A_{i+1}, M_{i+1})$ with its private key d_i, discards the random string z_i, and finds the address A_{i+1} of the next mix Mix_{i+1} in the chain and the message M_{i+1}, which it forwards to Mix_{i+1}. The random string z_i is needed, because otherwise an attacker could again, by encrypting an outgoing message from a mix with the public key of the mix and comparing the result with the former inputs to the mix, succeed to relate an input to its output (i.e., he would be able to trace the message flow through the mix). Besides *anonymity* protection, mix nets can also provide *unobservability*, if dummy messages (i.e., meaningless messages) are sent out by the participants in order to hide information about if and when meaningful messages were sent.

Mixnets have first been developed for anonymous email applications. Later the Mixnet concept has been used for establishing anonymous bidirectional real-time communication. One example for distributed overlay networks designed to anonymize real-time, bidirectional TCP-based communications is onion routing, in which the sender's proxy constructs an onion which encapsulates a route to the responder (e.g., Web server). An onion is an object with layers of public key encryption, which is used to produce anonymous bidirectional virtual circuit between communication partners (in Mix nets layered public key encryption are also used to build up a path) and to distribute symmetric keys to the nodes on the path. The sender's proxy constructs an onion which encapsulates a route to the responder. Faster symmetric encryption with the symmetric keys that were distributed to the nodes on the path with the help of the onion is used for data communication via the virtual path once it has been established.

Tor [7], the secondgeneration of onion routing, has added several improvements. In particular, it provides forward secrecy, i.e., once the session keys are deleted, they cannot be obtained any longer, even if all communication has been wiretapped the long-term secret keys of the onion routers ("Mixes") become compromised. Therefore, instead of using hybrid encryption for distributing symmetric session keys, the Diffie-Hellman key negotiation protocol is used, which provides forward secrecy.

Data Minimization at Application Level

To the class PETs that are minimizing or eliminating personally identifiable data of data subjects (that are not acting as users) belong *inference controls* in statistical databases. In general-purpose database systems (e.g., medical databases), some users (e.g., physicians) need to access personal data attributes, while other users (e.g., researchers) only need to access a personal database by statistical queries. Thus, access control mechanisms should allow certain users to access data in anonymous form by performing only statistical queries. However, by correlating different statistics (i.e., by launching so-called tracker attacks), a user may succeed in deducing confidential information about some individual. *Inference controls* in statistical databases shall ensure that the statistics released by the database do not lead to the disclosure of any confidential personal data. While it has been shown that it is impossible to guarantee that nothing about an individual should be learnable from a database, *differential privacy* can be achieved by data modification, which means that the risk of one's privacy does not significantly increase as a result of participating in a statistical database [11].

Another example for a protocol providing receiver anonymity is private information retrieval [14], which assures that a powerful attacker is not able to find out what information another user has requested (i.e., who has received certain information). The goal is to request exactly one datum that is stored in a remote memory cell of a server without revealing which datum is requested. Oblivious transfer is private information retrieval, where in addition the user may not learn any item other than the one that he requested.

PETs for anonymous or pseudonymous applications/transactions can be implemented on top of anonymous communication protocols. Examples are anonymous E-cash protocols based on *blind signatures* [15].

Another example is **anonymous credential systems**, which were first introduced by Chaum [2] and later enhanced by Brands [5] and by Camenisch and Lysyanskaya (IdeMix credentials, see [15]) and can be used to implement, for instance, anonymous or pseudonymous access control, e-government, and e-commerce. Anonymous credentials have stronger privacy properties than traditional credentials that require that all attributes are disclosed together if the user wants to prove certain properties, so that the verifier can check the issuer's signature. Besides, with traditional credentials, the verifier and issuer can link the different uses of the user's credential to the issuing of the credential.

Anonymous credentials allow the user to essentially "transform" the certificate into a new one that contains only a subset of attributes of the original certificate, i.e., it allows proving only a subset of its attributes to a verifier (selective disclosure property). In terms of cryptography, with the IdeMix protocol by Camenisch et al., the user is basically using a zero-knowledge proof to convince the verifier of possessing a signature generated by the issuer on a statement containing the subset of attributes. Instead of revealing the exact value of an attribute, anonymous credential

systems also enable the user in the transformation to apply any mathematical function to the (original) attribute value, allowing him to prove only attribute properties without revealing the attribute itself. Besides with IdeMix, the issuer's signature is also transformed in such a way that the signature in the new certificate cannot be linked to the original signature of the issuer. Hence, different credential uses cannot be linked by the verifier and/or issuer.

Figure 2 provides an example scenario how anonymous credentials can enforce data minimization for online transactions: First, user Bob obtains an anonymous identity card credential from an identity provider including personal attributes such as his birth date. Later, he would like to rent a video from an online video rental site, which can only be rented by adult users. After Bob sent a service request, the online video rental site will answer with a data request for a proof that he is older than 18. Bob can now take advantage of the selective disclosure feature of the anonymous credential protocol to prove just the fact that he is older than 18 without revealing his birth date or any other attributes of his credential. If Bob later wants to rent another video which is only permitted for adults at the same

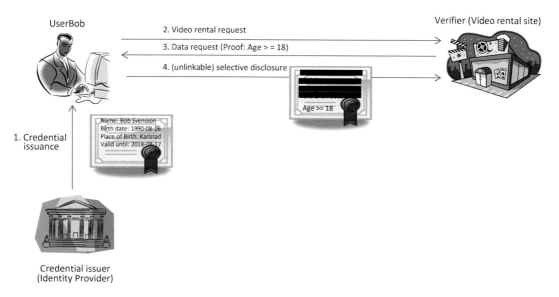

Privacy-Enhancing Technologies, Fig. 2 Example for achieving data minimization (selective disclosure) for online transactions by anonymous credential proofs

video rental site, he can use the same anonymous credential for a proof that he is over 18, and if the IdeMix protocol is used, the video rental site will even be unable to recognize that the two proofs are based on the same credential. Hence, the two rental transactions cannot be linked to the same person (Bob), which means that Bob cannot be profiled.

PETs for Safeguarding Lawful and Privacy-Friendly Personal Data Processing

It is not always possible to avoid the processing of personal data. Public authorities, healthcare providers, and employers are example of organizations that still need to process personal data about their citizens, patients, and employers for various legitimate reasons. If personal data are collected and processed, legal privacy requirements need to be fulfilled. The class of PETs described in this section comprises technologies that enforce legal privacy requirements in order to safeguard the lawful processing of personal data. The driving principles behind these types of PETs can also be found in data protection legislation, such as the EU Data Protection Directive 95/46/EC and the GDPR which require that controllers implement security measures which are appropriate to the risks presented for personal data in storage or transmission, with a view to protecting personal data against accidental loss, alteration, unauthorized access, and against all other unlawful forms of processing.

Examples of technologies belonging to this class of PETs include classical security technologies, such as *data encryption* or *access control*, which protect the confidentiality and integrity of personal data. Other examples are technologies for stating or enforcing privacy policies: The Platform for Privacy Preferences Protocol (P3P) [10] increases transparency for the end users by informing them about the privacy policies of web sites and can hence be used to enforce the legal principles and requirement (e.g., according to Art.10 EU Data Protection Directive

95/46/EC and Art. 13 GDPR) to inform data subjects about the purpose of data processing, identity of the controller, retention period, etc. Privacy policy models (such as the privacy model presented in [12]) can technically enforce privacy requirements, such as purpose binding, and privacy policy languages, such as the Enterprise Privacy Authorization Language EPAL [16], can be used to encode and to enforce more complex enterprise privacy policies within and across organizations.

A more advanced policy language, the PrimeLife Policy Language (PPL), was developed in the EU FP7 project PrimeLife [13]. PPL is a language to specify privacy policies of data controllers and also of third parties (so-called downstream data controllers) to whom personal data are further forwarded as well as privacy preferences of the users in machine-readable form (in XACML). The data controller and downstream data controllersdefine policies basically specifying which data are requested from the user, for which purposes and obligations (e.g., the obligation to delete the data after 1 year or to notify the data subject if the data will be forwarded to third parties). PPL allows specifying both uncertified data requests as well as certified data requests based on proofs of the possession of (anonymous IdeMix or traditional X.509) credentials that fulfil certain properties. The user's preferences allow expressing for each data item to which data controllers the data can be released (e.g., data may only be disclosed to service providers that have received a EuroPrise privacy seal) and how the user expectsher data to be treated (e.g., the data may be used for marketing purposes, but should be deleted after 2 years). These statements may include requirements that a downstream data controller has to fulfil in order to obtain the data from the (primary) data controller. The PPL engine conducts an automated matching of the data controller's policy and the user's preferences, which can result in a mutual agreement concerning the usage of data in form of a so-called sticky policy, which should be enforced by the access control systems at the backend sides and will "travel" with the data that are transferred to downstream data controllers.

PETs that Are a Combination of the Two Aforementioned PET Classes

The third class of PETs comprises technologies that are combining PETs of class 1 and class 2. An example is provided by privacy-enhancing identity management technologies as the ones that have been developed within the EU project PrimeLife. According to the PrimeLife architecture, all data communication is by default anonymized (e.g., through the use of Tor). For services that request personal data, PPL in combination with anonymous credential is used to negotiate (sticky) privacy policies and to achieve data minimization on application level. Transparency-enhancing tools, such as the PrimeLife Data Track, allow users to keep track of what data they have disclosed to whom under which agreed-upon policies and enable them to exercise their data subject rights to access their data online (see [6] for more details).

Key Applications

As illustrated above, PETs can be applied in all kinds of application areas of IT, in which personal data is or can be collected or processed.

Cross-References

▶ Access Control
▶ Anonymity
▶ Blind Signatures
▶ Data Encryption
▶ Database Security
▶ Inference Control in Statistical Databases
▶ Privacy Policies and Preferences
▶ Pseudonymity
▶ Unobservability

Recommended Reading

1. Chaum DL. Untraceable electronic mail, return addresses, and digital pseudonyms. Commun ACM. 1981;24(2):84–8.
2. Chaum DL. Security without identification: card computers to make big brother obsolete. Informatik-Spektrum. 1987;10:262–77.
3. Chaum DL. The dining cryptographers problem: unconditional sender and recipient untraceability. J Cryptol. 1988;1(1):65–75.
4. Registratiekamer, Information and Privacy Commissioner/Ontario. Privacy-enhancing technologies: the path to anonymity. Achtergrondstudies en Verkenningen 5B, vols. I and II, Rijswijk. 1995.
5. Brands S. Rethinking public key infrastructure and digital certificates – building in privacy. PhD thesis. Eindhoven: Institute of Technology; 1999.
6. Camenisch J, Fischer-Hübner S, & Rannenberg K (Eds.). Privacy and identity management for life. Springer Science & Business Media; 2011.
7. Dingledine R, Mathewson N, Syverson P. Tor: the second-generation onion router. In: Proceedings of the 13th USENIX Security Symposium; 2004.
8. Reed MG, Syverson PF, Goldschlag DM. Anonymous connections and onion routing. IEEE J Select Areas Commn. 1998;16(4):482–94.
9. Reiter M, Rubin A. Anonymous web transactions with crowds. Commun ACM. 1999;42(2):32–48.
10. Cranor L. Web privacy with P3P. Sebastopol: O'Reilly; 2002.
11. Dwork C. Differential privacy. Automata, languages and programming. Berlin/Heidelberg: Springer; 2006. p. 1–12.
12. Fischer-Hübner S. IT-security and privacy: design and use of privacy enhancing security mechanisms, LNCS, vol. 1958. Berlin: Springer LNCS; 2001. ISBN:3-540-42142-4.
13. PrimeLife. Privacy and identity management in Europe for life – policy languages. http://primelife.ercim.eu/results/primer/133-policy-languages. Accessed 11 Aug 2014.
14. Cooper DA, Birman KP. Preserving privacy in a network of mobile computers. In: Proceedings of the 1995 IEEE Symposium on Security and Privacy; 1995. p. 26–83.
15. Camenisch J, van Herreweghen E. Design and implementation of the idemix anonymous credential system. In: Proceedings of the 9th ACM Conference on Computer and Communications Security; 2002. p. 21–30.
16. Karjoth G, Schunter M, Waidner M. Platform for enterprise privacy practices: privacy-enabled management of customer data. In: Proceedings of the 2nd Workshop on Privacy Enhancing Technologies; 2002. p. 69–84.

Privacy-Preserving Data Mining

Chris Clifton
Department of Computer Science, Purdue
University, West Lafayette, IN, USA

Definition

Data Mining techniques that use specialized approaches to protect against the disclosure of private information may involve anonymizing private data, distorting sensitive values, encrypting data, or other means to ensure that sensitive data is protected.

Historical Background

The field of privacy-preserving data mining began in 2000 with two papers of that name [1, 4]. Both papers addressed construction of decision trees, approximating the ID3 algorithm while limiting disclosure of data. While the problems appeared similar on the surface, the fundamental difference in privacy constraints shows the complexity of this field. In [1], the assumption was that individuals were providing their own data to a common server, and added noise to sensitive values to protect privacy. The key to the technique was to discover the original distribution of the data, enabling successful construction of the decision tree. In [4], the data was presumed to be divided between two (or a small number) of parties, and cryptographic *Secure Multiparty Computation* techniques are used to construct the decision tree without any site disclosing the actual data values it holds.

A third approach to privacy-preserving data mining, *Data Transformation*, was introduced in [5]. The idea behind this approach is to transform the data space (e.g., using techniques such as Random Projection) so that data items can no longer be tied to individuals, but sufficient information is preserved to enable data mining.

Research in the field has largely been directed toward:

- Supporting more data mining tasks and algorithms (e.g., clustering, classification, outlier detection)
- Different privacy constraints (e.g., *vertically partitioned* and *horizontally partitioned* data)

Techniques provide varying levels of privacy guarantees, data mining result accuracy, and runtime performance.

Foundations

There are several factors that must be identified in developing or choosing a privacy-preserving data mining approach. The first is the source of and privacy constraints on the data. In some cases, data is held by a relatively small number of trusted parties, such as credit reporting bureaus, insurance companies, or government agencies. The challenge is that while these parties are allowed to use the data for their own purposes, privacy constraints prevent them from disclosing the data to others.

Secure Multiparty Computation Methods address this problem, allowing a data mining model to be constructed from the distributed data using cryptographic approaches that prevent disclosure of the individual data items between the parties. The key components of such a method are the algorithm, which uses cryptographic computations to duplicate a distributed data mining algorithm without disclosing data from any site, and a proof that disclosure is controlled. Typically such proofs use a simulation argument, showing that given the final result, a party can simulate the messages received in running the protocol (thus showing that no valuable information was disclosed other than that inherent in the result.)

When data is more widely distributed, such as individuals providing their data to a server for use in data mining, *Data Transformation*

Methods and *Data Perturbation* come into play. Data transformation methods modify the data so that the original values are lost, but important information (e.g., distances) is preserved. A typical example of transformation is dimensionality reduction, for example random projection [3]. Typically the providers of data must agree on a transformation, then transform their data and send it to a central data collector / data miner. Details of the transformation are kept secret from the data miner. In many cases, the transformation is individualized or constructed collaboratively in a way that protects against collusion. Such techniques require specialized versions of data mining algorithms to mine the transformed data. The techniques also involve a proof; in this case, showing the difficulty/impossibility of reversing the transformation, or the amount of background knowledge required to do so.

Data Perturbation is similar in that individuals modify their data before providing it to a central data warehouse. The differences is that data perturbation techniques add noise, while preserving the structure of the original data. The key to these techniques is specialized data mining algorithms that can utilize knowledge of the distribution of the noise, along with the noisy data, to obtain better results than mining on the noisy data alone. These approaches typically reconstruct the data *distribution*, then use this to guide the data mining, rather than mining the data directly. As the goal of most data mining is to obtain models that generalize well, models consistent with the distribution of the data are likely to have good accuracy. In addition to the reconstruction algorithm, data perturbation methods need a measure of the privacy provided; as data is not completely hidden, it is important to be able to measure the tradeoff between privacy and model accuracy.

Once the data privacy constraints are understood and a general approach is chosen, the next step is to develop an algorithm for the data mining model to be built. Privacy-preserving data mining algorithms typically replicate the results of traditional data mining algorithms, and have been developed for many of the standard machine learning approaches.

Key Applications

Privacy-preserving data mining has application wherever data mining is applied to data about individuals. Examples include recommender systems, medical studies, intelligence, and social

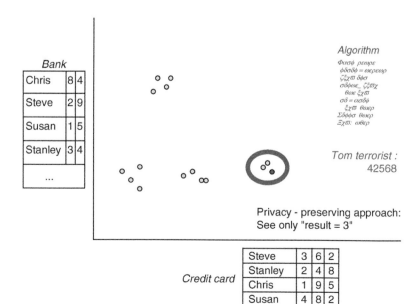

Privacy-Preserving Data Mining, Fig. 1 Using a privacy-preserving outlier detection algorithm to determine if an algorithm isolates terrorist behavior

network analysis. In addition, this technology can be useful for sharing sensitive corporate information, for example in supply chain management [2].

As an example, Fig. 1 shows a scenario evaluating an Algorithm developed to identify terrorist organization financial patterns. To evaluate this, it is necessary to compare real financial records to see how many are identified by the algorithm. An outlier detection algorithm could be used to determine if the space identified by the algorithm reflects a cluster of "normal" financial patterns or just a few outliers, but accessing real financial records is problematic. Using the secure multiparty computation technique in [6], it is possible to identify the number of items in the vicinity of "Tom Terrorist" without revealing either the identity or specific values of any of the items. This gives us the result sought without exposing private data.

Future Directions

The main factors limiting adoption of this technology are:

1. The difficulty of doing exploratory data analysis. Privacy-preserving data mining techniques require that the data mining task to be performed be known without seeing the data.
2. Privacy standards and definitions. Current privacy regulations do not make it clear if this technology is either necessary or sufficient. Further work is needed to establish technically meaningful definitions of privacy.
3. Cost/performance. Many of these techniques are expensive in either computational power, their effect on result quality, and perhaps most important the cost of implementing a solution for a particular task.

Ongoing work is addressing these issues. Applications in corporate collaboration using corporate-sensitive data is likely to lead commercialization, as cost/benefit tradeoffs are easier to evaluate than with personal private data.

The field is closely related to *inference control in statistical databases* and *statistical disclosure limitation*; these fields have extensively studied the privacy risks inherent in releasing data under various types of perturbation and transformation. The key difference with privacy-preserving data mining is that the specific use of the data is assumed to be known. This allows methods that drastically distort the data, keeping only specific information intact rather than trying to preserve general statistical properties. The success of statistical disclosure limitation provides a good historical framework for the adoption of privacy-preserving data mining.

Cross-References

▶ Inference Control in Statistical Databases
▶ Privacy Metrics
▶ Randomization Methods to Ensure Data Privacy
▶ Secure multiparty Computation Methods
▶ Statistical Disclosure Limitation for Data Access

Recommended Reading

1. Agrawal R, Srikant R. Privacy-preserving data mining. In: Proceedings of the ACM SIGMOD International Conference on Management of Data; 2000. p. 439–50.
2. Atallah MJ, Elmongui HG, Deshpande V, Schwarz LB. Secure supply-chain protocols. In: Proceedings of the IEEE International Conference on E-commerce; 2003. p. 293–302.
3. Kaski S. Dimensionality reduction by random mapping. In: Proceedings of the International Joint Conference on Neural Networks; 1999. p. 413–8.
4. Lindell Y, Pinkas B. Privacy preserving data mining. In: Advances in cryptology – CRYPTO 2000. Heidelberg: Springer; 2000. p. 36–54.
5. Oliveira SRM, Zaïane OR. Privacy preserving clustering by data transformation. In: Proceedings of the 18th Brazilian Symposium on Databases; 2003.
6. Vaidya J, Clifton C. Privacy-preserving outlier detection. In: Proceedings of the 4th IEEE International Conference on Data Mining; 2004. p. 233–40.
7. Vaidya J, Clifton C, Zhu M. Privacy preserving data mining. Berlin: Springer; 2006.

P

Privacy-Preserving DBMSs

Tyrone Grandison
Proficiency Labs, Ashland, OR, USA

Introduction

The concept of a privacy-preserving database management system (PP-DBMS) is a relatively recent one – dating back to the 2000s [1, 8]. Such a system assumes that privacy is a fundamental property of the data in the DBMS and that the database management system automatically and seamlessly adheres to the privacy dictates of the data owners. As a first step, we must understand the notion of privacy.

Privacy Fundamentals

Privacy is a complex and multifaceted topic that is steeped in history and rich with subtleties. The task of understanding the fundamental underpinnings, semantics, and nuisances of the concept of privacy has been underway in the legal profession for many decades. In 1928, US Supreme Court Justice Louis Brandeis stated that privacy was "the right to be left alone" [31]. Brandeis postulated that privacy is one of the "conditions favorable to the pursuit of happiness" [31].

Over the years, other legal scholars have established that privacy is one of the necessary conditions for the advance of individual identity, for the creation of intimacy, and for the proper operation of a democracy. In the legal community, privacy is either considered a fundamental right or a necessary construction to protect other fundamental rights ([28, 26).

In 1967, Alan Westin provided the definition of privacy that has emerged as the de facto definition for computer scientists [32]. Westin defines privacy as an individual's right "to control, edit, manage, and delete information about them[selves] and decide when, how, and to what extent information is communicated to others" [32]. Current privacy legislature around the globe, e.g., OECD Data Protection Guidelines [22], HIPAA [30], and PIPEDA [23], leverages this perspective as their guiding premise. Additionally, a large majority of privacy-enhancing computer technologies, across industries, tend to utilize Westin's definition as their starting point [11, 14, 29].

With this knowledge, we are properly positioned to discuss the features of a privacy-preserving database management system.

Fundamentals of a PP-DBMS

The concept of a privacy-preserving database management system was first introduced by the team at the IBM Almaden Research Center, in their paper on a Hippocratic database [1]. The vision of a Hippocratic database is founded on ten principles, based on OECD Data Protection Guidelines [22]:

1. **Purpose Specification**. The purposes for which personal information has been collected shall be associated with that information in the database.
2. **Consent**. The purposes associated with personal information shall have the consent of the individual who is the subject of the information.
3. **Limited Collection**. The personal information collected shall be limited to the minimum necessary for accomplishing the specified purposes.
4. **Limited Use**. The database shall run only those queries and operations that are consistent with the purposes for which the information has been collected.
5. **Limited Disclosure**. Personal information stored in the database shall not be communicated outside of the database for purposes other than those to which the individual consented.
6. **Limited Retention**. Personal information shall be retained only as long as necessary to fulfill the purposes for which it was collected.
7. **Accuracy**. All personal information in the database shall be accurate and current.

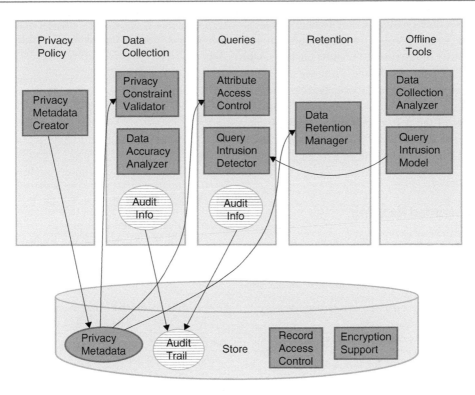

Privacy-Preserving DBMSs, Fig. 1 Douglas-Peucker heuristic

8. **Safety**. Personal information shall be protected by security safeguards against theft and other misappropriation.
9. **Openness**. An individual shall be able to access all information about him or her stored in the database.
10. **Compliance**. An individual shall be able to verify compliance with the above principles, and the database be capable of responding to individual challenges.

The initial strawman architecture of a Hippocratic vision can be seen in Figure 1. More detail on the individual components of the strawman is available in the original paper [1]. However, it is important to note that Agrawal et al. stated that Hippocratic databases should be architected to natively regulate the use and disclosure of personal information in strict accordance with privacy and security laws, enterprise policies, and individual choices. They also postulated that these systems should be designed to safeguard the information and protect individual privacy without impeding legitimate and beneficial uses of information.

Since this initial articulation, and its subsequent proof points [2–5, 7, 13, 16, 18, 25], the idea of a privacy-preserving database management system has been applied to multiple contexts, applications, and domains, e.g., Web services [10], hierarchical databases [19], XML databases [17], PostgreSQL [24], location-based databases [21], eLearning [6], log files [25], etc.

Though a lot of work has occurred in this space, there are still areas of research that require attention [9, 12, 15, 20].

Future Directions

There is still need for improved privacy policy specification and enforcement. Enforcement of privacy policy after data is extracted from the DBMS is still a hard problem. Purpose specification and automatic assignment of purpose to programs is still difficult. Limited retention is a thorny issue that involves technological and legal concerns. Limited collection is counter to the standard way of operating for most businesses in today's economy. Real-time intrusion detection of databases is a desirable goal that researchers are still trying to achieve. Guarantees on the soundness of the data contained in a DBMS are still being developed. All in all, the space has a lot of opportunity and is still ripe for innovation.

Recommended Reading

1. Agrawal R, Kiernan J, Srikant R, Xu Y. Hippocratic databases. In: Proceedings of the 28th International Conference on Very Large Data Bases; 2002.
2. Agrawal R, Evfimievski A, Srikant R. Information sharing across private databases. In: Proceedings of the ACM SIGMOD International Conference on Management of Data; 2003.
3. Agrawal R, Bayardo R, Faloutsos C, Kiernan J, Rantzau R, Srikant R. Auditing compliance with a hippocratic database. In: Proceedings of the 30th International Conference on Very Large Data Bases Endowment; 2004. p. 516–27.
4. Agrawal R, Kiernan J, Srikant R, Xu Y. Order-preserving encryption for numeric data. In: Proceedings of the ACM SIGMOD International Conference on Management of Data; 2004.
5. Agrawal R, Bird P, Grandison T, Kiernan J, Logan S, Rjaibi W. Extending relational database systems to automatically enforce privacy policies. In: Proceedings of the 21st International Conference on Data Engineering; 2005. p. 1013–22.
6. Azemović J. Privacy aware eLearning environments based on hippocratic database principles. In: Proceedings of the 5th Balkan Conference in Informatics; 2012. p. 142–9.
7. Bayardo RJ, Agrawal R. Data privacy through optimal k-anonymization. In: Proceedings of the 21st International Conference on Data Engineering; 2005.
8. Bertino E, Byun JW, Li N. Privacy-preserving database systems. In: Foundations of security analysis and design III. Berlin/Heidelberg: Springer; 2005. p. 178–206.
9. Bottcher S, Hartel R, Kirschner M. Detecting suspicious relational database queries. In: Proceedings of the 3rd International Conference on Availability, Reliability and Security; 2008. p. 71–778.
10. Cheng VS, Hung PC. Towards an integrated privacy framework for HIPAA-compliant web services. In: Proceedings of the 7th IEEE International Conference on E-commerce Technology; 2005. p. 480–3.
11. Goldberg I. Privacy-enhancing technologies for the Internet, II: five years later. Berlin/Heidelberg: Springer; 2003. p. 1–2.
12. Grandison T, Johnson C, Kiernan J. Hippocratic databases: current capabilities and future trends. In: Handbook of database security. New York: Springer; 2008. p. 409–29.
13. Johnson CM, Grandison TWA. Compliance with data protection laws using hippocratic database active enforcement and auditing. IBM Syst J. 2007;46(2): 255–64.
14. Juels A. RFID security and privacy: a research survey. IEEE J Sel Areas Commun. 2006;24(2): 381–94.
15. Kirchberg M, Link S. Hippocratic databases: extending current transaction processing approaches to satisfy the limited retention principle. In: Proceedings of the 43rd Annual Hawaii International Conference on System Sciences; 2010. p. 1–10.
16. Laura-Silva Y, Aref WG. Realizing privacy-preserving features in Hippocratic databases. In: Proceedings of the IEEE 23rd International Conference on Data Engineering Workshop; 2007. p. 198–206.
17. Lee JG, Whang KY, Han W, Song I. Hippocratic XML databases: a model and an access control mechanism. Comput Syst Sci Eng. 2006;21(6)
18. LeFevre K, Agrawal R, Ercegovac V, Ramakrishnan R, Xu Y, DeWitt D. Limiting disclosure in hippocratic databases. In: Proceedings of the 30th International Conference on Very Large Data Bases Endowment; 2004. p. 108–19.
19. Massacci F, Mylopoulos J, Zannone N. Hierarchical hippocratic databases with minimal disclosure for virtual organizations. VLDB J. 2006;15(4):370–87.
20. Mohamed Sidek Z, Abdul Ghani N. Utilizing hippocratic database for personal information privacy protection. Jurnal Teknologi Maklumat. 2008;20(3): 54–64.
21. Mokbel FM. Towards privacy-aware location-based database servers. In: Proceedings of the 22nd International Conference on Data Engineering Workshops; 2006.
22. OECD. Guidelines on the protection of privacy and transborder flows of personal data. 1980. http://www.oecd.org/internet/ieconomy/ oecdguidelinesontheprotectionofprivacyandtransborde rflowsofpersonaldata.htm. Accessed 27 Aug 2014.
23. Office of the Privacy Commissioner of Canada. The personal information protection and electronic documents act (PIPEDA). 4 1, 2011. http://laws-lois. justice.gc.ca/PDF/P-8.6.pdf. Accessed 27 Aug 2014.
24. Padma J, Silva YN, Arshad MU, Aref WG. Hippocratic PostgreSQL. In: Proceedings of the 25th International Conference on Data Engineering; 2009. p. 1555–8.
25. Rutherford A, Botha R, Olivier M. Towards Hippocratic log files. In: Proceedings of the 4th Annual Information Security South Africa Conference; 2004. p. 1–10.
26. Solove DJ. Understanding privacy: Harvard University Press; 2010.

27. Solove DJ. Nothing to hide: the false tradeoff between privacy and security. J Value Inq. 2012;46(1):107–112.
28. Solove DJ, Schwartz PM. Privacy law fundamentals, Second Edition. Aspen Publishers, 2013.
29. Such JM, Espinosa A, García-Fornes A. A survey of privacy in multi-agent systems. Knowl Eng Rev. 2014;29(03):314–44.
30. U.S. Department of Health and Human Services Office for Civil Rights. HIPAA Administrative Simplification. 3 26, 2013. http://www.hhs.gov/ocr/privacy/hipaa/administrative/combined/hipaa-simplification- 201303.pdf. Accessed 27 Aug 2014.
31. U.S. Supreme Court. Osmalt v. U.S. Government (1928). 1928. Accessed 27 Aug 2014.
32. Westin AF. Privacy and freedom. New York: Athenum; 1967.
33. Zhu H, Lü K. Fine-grained access control for database management systems. In: Richard C, Jessie K, editors. Data management. Data, data everywhere. Berlin/Heidelberg: Springer; 2007. p. 215–23.

Private Information Retrieval

Xun Yi
Computer Science and Info Tech, RMIT University, Melbourne, VIC, Australia

Definition

Private information retrieval (PIR) protocol allows a user to retrieve the i-th bit of an n-bit database, without revealing to the database server the value of i. A trivial solution is for the user to retrieve the entire database, but this approach may incur enormous communication cost. A good PIR protocol is expected to have considerably lower communication complexity. Private block retrieval (PBR) is a natural and more practical extension of PIR in which, instead of retrieving only a single bit, the user retrieves a block of bits from the database.

Historical Background

PIR was first introduced by Chor, Goldreich, Kushilevitz, and Sudan [4] in 1995 in a multi-server setting, where the user retrieves information from multiple database servers, each of which has a copy of the same database. To ensure user privacy in the multi-server setting, the servers must be trusted not to collude. In [4], Chor et al. have shown that if only a single database is used, n bits must be communicated in the information-theoretic sense, that is, the user's query gives absolutely no information about i. They have also shown that any PIR protocol can be converted to a PBR protocol.

In 1997, using the quadratic residue computational assumption, Kushilevitz and Ostrovsky [14] constructed a single-database PIR with communication complexity of $O(2^{\sqrt{\log n \log \log n}})$ which is less than $O(n^\epsilon)$ for any $\epsilon > 0$, where N is the composite modulus. Their basic idea is viewing the database as a matrix $M = (x_{ij})_{s \times t}$ of bits. To retrieve the (a, b) entry of the matrix, the user sends to the database server a composite (hard-to-factor) modulus N and t randomly chosen integers $y_1, y_2, \cdots, y_t$ such that only y_a is not a quadratic residue modulo N, that is, $y_i \neq \alpha^2 (mod\ N)$ for any integer α. The server sends back $z_i = \prod_{j=1}^{t} y_j^{2-x_{ij}} (mod\ N)$ for $1 \leq i \leq s$. The user concludes that $x_{ij} = 0$ if z_a is a quadratic residuosity modulo N and $x_{ij} = 1$ otherwise.

In 1999, Cachin, Micali, and Stadler [2] constructed the first single-database PIR with poly-logarithmic communication complexity $O(\log^8 n)$. The security of their protocol is based on the Φ-hiding number-theoretic assumption, that is, it is hard to distinguish which of the two primes divides $\phi(N)$ for the composite modulus N. Their basic idea is mapping each index i to a distinct prime p_i. To retrieve bit b_i from a database $B = b_1 b_2 \cdots b_n$, the user sends to the database server a composite (hard-to-factor) modulus N such that p_i divides $\phi(N)$ and a generator g with order divisible by p_i. The server sends back $r = g^P (mod\ N)$ where $P = \prod_j p_j^{b_j}$. The user concludes that $b_i = 1$ if r is a p_i-residue modulo N; otherwise, $b_i = 0$.

In 2000, Kushilevitz and Ostrovsky [15] constructed a PIR protocol with total communication complexity $n - \frac{cn}{2k} + O(k^2)$, where k is a security parameter and c is a constant. Their protocol is built on the Naor-Yung one-way 2-to-1 trapdoor permutations [20] and the Goldreich-Levin hardcore predicates [12]. Their basic idea is dividing an n-bit database into k-bit blocks and organizing the database into pairs of blocks, denoted by $z_{i,L}$ and $z_{i,R}$, where $i = 1, 2, \cdots, \frac{n}{2k}$. Suppose that the user wishes to retrieve $z_{s,L}$. The user sends to the database server the descriptions of one-way trapdoor permutations f_L and f_R, to which the user has the trapdoors. The server computes $f_L(z_{i,L})$ and $f_R(z_{i,R})$ for all i and returns these values to the user. With trapdoors, the user computes two possible pairs of pre-images $(z_{s,L}, z'_{s,L})$ and $(z_{s,R}, z'_{s,R})$. Next, the user sends the server two hardcore predicates r_L and r_R such that $r_L(z_{s,L}) \neq r_L(z'_{s,L})$ but $r_R(z_{s,R}) = r_R(z'_{s,R})$. The server responds with $r_L(z_{i,L}) \oplus r_R(z_{i,R})$ for all i. At the end, the user learns which pre-image is $z_{s,L}$ from $r_L(z_{s,L})$.

In 2005, Gentry and Ramzan [11] extended the single-database PIR protocol of Cachin et al. [2] to a PBR with communication complexity $O(\log^2 n)$, the current best bound for communication complexity. The security of their protocol is also based on the Φ-hiding assumption. Assuming that an n-bit database B is partitioned into m blocks, each has ℓ bits, denoted as $B = C_1 \| C_2 \| \cdots \| C_m$. Their basic idea is associating C_i with a distinct small prime p_i, rather than associating a (largish) prime with each bit. The database server uses the Chinese remainder theorem to determine an integer e such that $e = C_i (mod\ p_i^{c_i})$ for $1 \leq i \leq m$, where c_i is the smallest integer such that $p_i^{c_i} \geq 2^\ell$. To retrieve block C_i, the user sends to the database server a composite (hard-to-factor) modulus N such that $p_i^{c_i}$ divides $\phi(N)$ and a generator g with order divisible by $p_i^{c_i}$. The server sends back $r = g^e (mod\ N)$. Let $q = order(g)/p_i^{c_i}$ and then $order(g^q) = p_i^{c_i}$. Since p_i is a small prime, the user can compute the discrete logarithm $\log_{(g^q)}(r^q) = e(mod\ p_i^{c_i}) = C_i$ using the Pohlig-Hellman algorithm [27].

Like the original single-database PIR of Kushilevitz and Ostrovsky [14], Chang [3] and Lipmaa [16, 17] also constructed PIR protocols with communication complexity of $O(\log^2 n)$. The difference is that the former is based on the Goldwasser-Micali homomorphic encryption [13], but the latter is built on the Damgard-Jurik homomorphic encryption [6], a variant of the Paillier homomorphic encryption [24].

Since PIR was first introduced by Chor, Goldreich, Kushilevitz, and Sudan [4] in 1995, lots of PIR protocols have been proposed in the literature [8, 22]. Generally, PIR protocols can be classified into two categories: information-theoretical PIR protocols such as [1,4,5,7] (assume that there are multiple non-cooperating servers, each having a copy of the database) and computational PIR protocols such as [2,3,11,14–17] (assume that the server is computationally bounded). There also exist some PIR protocols based on the trusted hardware [18,28,29] (assume that a trusted hardware exists in the server to response the user's query without revealing to the server any query information).

Foundations

Informally, a single-database PIR protocol is a two-party protocol, where a user retrieves the i-th bit from an n-bit database $DB = b_1 b_2 \cdots b_n$, without revealing to the database server the value of i. Formally, a single-database PIR protocol consists of three algorithms as in [2,11]:

(1) Query Generation (QG): Takes as input a security parameter k, the size n of the database, and the index i of a bit in the database and output a query Q and a secret s, denoted as $(Q, s) = \mathsf{QG}(n, i, 1^k)$.
(2) Response Generation (RG): Takes as input the security parameter k, the query Q, and the database DB and output a response R, denoted as $R = \mathsf{RG}(DB, Q, 1^k)$.
(3) Response Retrieval (RR): Takes as input the security parameter k, the response R, the index i of the bit, the size n of the database, the query Q, and the secret s and output a bit b', denoted as $b' = \mathsf{RR}(n, i, (Q, s), R, 1^k)$.

The security of single-database PIR protocol can be defined with a game as follows.

Give an n-bit database, $DB = b_1 b_2 \cdots b_n$. Consider the following game between an adversary (the database server) $\mathcal{A}$ and a challenger C. The game consists of the following steps:

(1) The adversary $\mathcal{A}$ chooses two different indices $1 \le i, j \le n$ and sends them to C.
(2) Let $\lambda_0 = i$ and $\lambda_1 = j$. The challenger C chooses a random bit $b \in \{0, 1\}$, executes $\mathsf{QG}(n, \lambda_b, 1^k)$ to obtain (Q_b, s), and then sends Q_b back to $\mathcal{A}$.
(3) The adversary $\mathcal{A}$ can experiment with the code of Q_b in an arbitrary non-black-box way and finally outputs $b' \in \{0, 1\}$.

The adversary wins the game if $b' = b$ and loses otherwise. We define the adversary $\mathcal{A}$'s advantage in this game to be

$$\mathsf{Adv}_{\mathcal{A}}(k) = |\mathsf{Pr}(b' = b) - 1/2|.$$

Security Definition. A single-database PIR protocol is semantically secure if for any probabilistic polynomial time (PPT) adversary $\mathcal{A}$, we have that $\mathsf{Adv}_{\mathcal{A}}(k)$ is a negligible function, where the probability is taken over coin tosses of the challenger and the adversary.

Similarly, a single-database private block retrieval (PBR) protocol can be defined by viewing the n-bit database as $DB = B_1 \| B_2 \| \cdots \| B_m$, where B_i is a block with n/m bits.

Correctness. A single-database PIR protocol is correct if, for any security parameter k, any database DB with any size n, and any index i for $1 \le i \le n$, $b_i = \mathsf{RR}(n, i, (Q, s), R, 1^k)$ holds, where $(Q, s) = \mathsf{QG}(n, i, 1^k)$ and $R = \mathsf{RG}(DB, Q, 1^k)$.

In simple terms, the correctness definition means that for every query Q, the correct bit/block is retrieved, while the security definition means that for any two queries Q_1, Q_2, with indices i, j, respectively, an adversary cannot distinguish them from one another with probability greater than $\frac{1}{2}$.

Homomorphic encryption techniques are often very natural ways to construct PIR. For examples, the PIR approach by Kushilevitz and Ostrovsky is based on the Goldwasser-Micali homomorphic encryption. A generic method to construct single-database PIR from homomorphic encryption scheme was given by Ostrovsky and Skeith [22]. These underlying encryption schemes allow homomorphic computation of only one operation (either addition or multiplication) on plaintexts. In 2009, Gentry [9, 10] constructed the first fully homomorphic encryption (FHE) scheme using lattice-based cryptography. FHE allows homomorphic computation of two operations (both addition and multiplication) of plaintexts. Motivated by recent breakthrough in FHE, Yi, Kaosar, Paulet, and Bertino [30] proposed single-database PIR and private block retrieval (PBR) protocols from FHE. Their solution is conceptually simpler than any existing PIR and more efficient in terms of computation complexity.

Key Applications

Usually, to retrieve a record from a database, a user will send a request pointing out which record he wants to retrieve, and the database server will send back the requested record. Which record a user is interested in may be an information he would like to keep secret, even for the database server. For example, the database may be

- An electronic library, and which books we are interested in may provide information about our politic or religious beliefs, or other details about our personality it may be desirable to keep confidential;
- Stock exchange share prices, and the clients may be investors reluctant to divulge which share they are interested in;
- A pharmaceutical database, and some client laboratories may wish that nobody may learn which are the active principles they may want to use;

Private Information Retrieval, Fig. 1 High level overview of the protocol

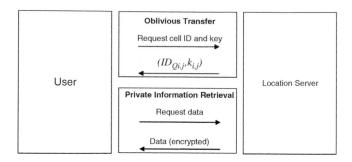

- A location-based database, and some mobile users may wish to keep his location private

To protect his privacy, a user accessing a database may therefore want to retrieve a record without revealing which record he is interested in. PIR can be applied to all of these cases.

As an example [25, 26], let us see how PIR is applied to privacy-preserving and content-protecting location-based queries as follows.

In this example we describe a solution to one of the location-based query problems. This problem is defined as follows: (1) a user wants to query a database of location data, known as points of interest (POIs), and does not want to reveal his/her location to the server due to privacy concerns and (2) the owner of the location data, that is, the location server, does not want to simply distribute its data to all users. The location server desires to have some control over its data, since the data is its asset.

We can achieve this by applying a two-stage approach shown in Fig. 1. The first stage is based on a two-dimensional oblivious transfer [19], and the second stage is based on a communicationally efficient PIR [11]. The oblivious transfer-based protocol is used by the user to obtain the cell ID, where the user is located, and the corresponding symmetric key. The knowledge of the cell ID and the symmetric key is then used in the PIR-based protocol to obtain and decrypt the location data.

The user determines his/her location within a publicly generated grid P by using his/her GPS coordinates and forms an oblivious transfer query. (An oblivious transfer query is such that a server cannot learn the user's query, while the user cannot gain more than they are entitled. This

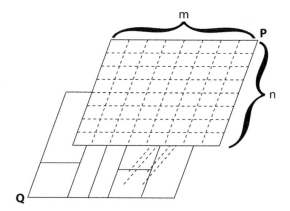

Private Information Retrieval, Fig. 2 The public grid superimposed over the private grid

is similar to PIR, but oblivious transfer requires protection for the user and server. PIR only requires that the user is protected.) The minimum dimensions of the public grid are defined by the server and are made available to all users of the system. This public grid superimposes over the privately partitioned grid generated by the location server's POI records, such that for each cell $Q_{i,j}$ in the server's partition, there is at least one $P_{i,j}$ cell from the public grid. This is illustrated in Fig. 2.

Since PIR does not require that a user is constrained to obtain only one bit/block, the location server needs to implement some protection for its records. This is achieved by encrypting each record in the POI database with a key using a symmetric key algorithm, where the key for encryption is the same key used for decryption. This key is augmented with the cell info data retrieved by the oblivious transfer query. Hence, even if the user uses PIR to obtain more than one record, the data will be meaningless resulting in improved security for the server's database.

Future Directions

Recently, we have seen a significant evolution from single-database PIR to private searching on streaming data. The problem of private searching on streaming data was first put forward by Ostrovsky and Skeith [21]. It was motivated by one of the tasks of the intelligence community, to collect potentially useful information from huge volumes of streaming data flowing through a public server. What is potentially useful and raises a red flag is often classified and satisfies secret criteria. The challenge is how to keep the criteria classified even if the program residing in the public server falls into enemy's hands. This work has many applications for the purposes of intelligence gathering. For example, in airports one can use this technique to find if any of hundreds of passenger lists has a name from the possible terrorists list and if so his/hers itinerary without revealing the secret terrorists list.

Private searching on steaming data and single-database PIR have the same aim, that is, retrieving information from a source of data without revealing user's query to the owner of the data. Both of them intend to protect the privacy of the user. The first difference between them is that the data in PIR is fixed, while the data in private searching is streaming data. The second difference is that the user in PIR knows the index of the data he wishes to retrieve, while the user in private searching only knows the searching criteria. The third difference is that the outcome in PIR is immediately returned to the user, while the searching results in private searching are stored in a buffer which is returned to the user after certain period.

Compared with PIR, private searching on streaming data is more complicated and difficult to design and implement. The first solution for private searching on streaming data was given by Ostrovsky and Skeith in 2005 [21, 23]. It was built on the concept of public-key program obfuscation, where an obfuscator compiles a given program f from a complexity class C into a pair of algorithms (F, Dec), such that $Dec(F(x)) = f(x)$ for any input x and it is impossible to distinguish for any polynomial time adversary which f from C was used to produce a given code for F.

Private searching on streaming data allows to search for streaming documents containing one or more of classified keywords $K = \{k_1, k_2, \cdots, k_{|K|}\} \subset D$, where D is the dictionary, and store the findings in a buffet and send it to the user after a certain period. Ostrovsky and Skeith [21, 23] suggested to throw the findings into randomly chosen boxes in the buffet. If two different matching documents are ever added to the same buffer box, a collision will occur, and both copies will be lost. To avoid the loss of matching documents, the buffer size has to be sufficiently large so that each matching document can survive in at least one buffer box with overwhelming probability.

Cross-References

▶ Privacy-Enhancing Technologies
▶ Privacy Metrics
▶ Privacy Policies and Preferences
▶ Privacy-Preserving Data Mining

Recommended Reading

1. Beimel A, Ishai Y. Information-theoretic private information retrieval: a unified construction. In: Proceedings of the 28th International Colloquium on Automata, Languages and Programming; 2001. p. 912–26.
2. Cachin C, Micali S, Stadler M. Computationally private information retrieval with polylogarithmic communication. In: Proceedings of the International Conference on the Theory and Application of Cryptographic Techniques; 1999. p. 402–14.
3. Chang Y. Single database private information retrieval with logarithmic communication. In: Proceedings of the 9th Australasian Conference on Information Security and Privacy; 2004. p. 50–61.
4. Chor B, Goldreich O, Kushilevitz E, Sudan M. Private information retrieval. In: Proceedings of the 36th IEEE Symposium on Foundations of Computer Science; 1995. p. 41–51.

5. Chor B, Kushilevitz E, Goldreich O, Sudan M. Private information retrieval. J ACM. 1998;45(6) L965–81.

6. Damgard I, Jurik M. A generalisation. A simplification and some applications of Paillier's probabilistic public-key system. In: Proceedings of the 4th International Workshop on Practice and Theory in Public Key Cryptography; 2001. p. 119–36.

7. Devet C, Goldberg I, Heninger N. Optimally robust private information retrieval. In: Proceedings of the 12th USENIX Security Symposium; 2012. p. 13–13.

8. Gasarch, W. A survey on private information retrieval. Bull EATCS. 2004; 82:72–107.

9. Gentry C. Fully homomorphic encryption using ideal lattices. In: Proceedings of the 41st Annual ACM Symposium on Theory of Computing; 2009. p. 169–78.

10. Gentry C. Toward basing fully homomorphic encryption on worst-case hardness. In: Proceedings of the 30th Annual Conference on Advances in Cryptology; 2010. p. 116–37.

11. Gentry C, Ramzan Z. Single database private information retrieval with constant communication rate. In: Proceedings of the 32nd International Colloquium on Automata, Languages, and Programming; 2005. p. 803–15.

12. Goldreich, O. and Levin, LA. A hard-core predicate for all one-way functions. In: Proceedings of the 21st Annual ACM Symposium on Theory of Computing; 1989. p. 25–32.

13. Goldwasser S, Micali S. Probabilistic encryption. J Comput Syst Sci. 1984;28(2):270–99.

14. Kushilevitz E, Ostrovsky R. Replication is not needed: single database, computationally-private information retrieval. In: Proceedings of the 38th Annual IEEE Symposium on the Foundations of Computer Science; 1997. p. 364–73.

15. Kushilevitz E, Ostrovsky R. One-way trapdoor permutations are sufficient for non-trivial single-server private information retrieval. In: Proceedings of the 19th International Conference on Theory and Application of Cryptographic Techniques; 2000. p. 104–21.

16. Lipmaa H. An oblivious transfer protocol with log-squared communication. In: Proceedings of the 8th Information Security Conference; 2005. p. 314–28.

17. Lipmaa H. First CPIR protocol with data-dependent computation. In: Proceedings of the 12th International Conference on Information Security and Cryptology; 2009 p. 193–210.

18. Lliev A, Smith SW. Protecting client privacy with trusted computing at the server. IEEE Secur Priv. 2005;3(2):20–28.

19. Naor M, Pinkas B. Oblivious transfer with adaptive queries. In: Proceedings of the 19th Annual International Cryptology Conference on Advances in Cryptology; 1999. p. 791–91.

20. Naor M, Yung M. Universal one-way hash functions and their cryptographic applications. In: Proceedings of the 21st Annual ACM Symposium on Theory of Computing; 1989. p. 33–43.

21. Ostrovsky R, Skeith WE III. Private searching on streaming data. In: Proceedings of the 25th Annual International Conference on Advances in Cryptology; 2005. p. 223–40.

22. Ostovsky R, Skeith WE III. A survey of single-database PIR: Techniques and applications. In: Proceedings of the 10th International Conference on Practice and Theory in Public-Key Cryptography; 2007. p. 393–411.

23. Ostrovsky R, Skeith WE III. Private searching on streaming data. J Cryptol. 2007;20(4):397–430.

24. Paillier P. Public key cryptosystems based on composite degree residue classes. In: Proceedings of the 17th International Conference on the Theory and Application of Cryptographic Techniques; 1999. p. 223–38.

25. Paulet R, Kaosar MG, Yi X, Bertino E. Privacy-preserving and content-protecting location based queries. In: Proceedings of the 28th International Conference on Data Engineering; 2012. p. 44–53.

26. Paulet R, Kaosar MG, Yi X, Bertino E. Privacy-preserving and content-protecting location based queries. IEEE transactions on knowledge and data engineering, 2014; 26(5):1200–10.

27. Pohlig S, Hellman M. An improved algorithm for computing logarithms over GF(p) and its cryptographic significance. IEEE Trans Inf Theory. 1978;24(1):106–10.

28. Sion R, Carbunar B. On the computational practicality of private information retrieval. In: Proceedings of the Network and Distributed Systems Security Symposium; 2007.

29. Wang S, Ding X, Deng RH, Bao F. Private information retrieval using trusted hardware. In: Proceedings of the 11th European Symposium on Research in Computer Security; 2006. p. 49–64.

30. Yi X, Kaosar M, Paulet R, Bertino E. Single-database private information retrieval from fully homomorphic encryption. IEEE Trans Knowl Data Eng. 2013;25(5):1125–34.

Probabilistic Databases

Dan Suciu
University of Washington, Seattle, WA, USA

Synonyms

Probabilistic databases extend standard databases with probabilities, in order to model uncertainties in data; Query evaluation becomes probabilistic inference

Definition

A probabilistic database is a database in which every tuple t belongs to the database with some probability $\mathbf{P}(t)$; when $\mathbf{P}(t) = 1$ then the tuple is certain to belong to the database; when $0 < \mathbf{P}(t) < 1$ then it belongs to the database only with some probability; when $\mathbf{P}(t) = 0$ then the tuple is certain not to belong to the database, and we usually don't even bother representing it. A traditional (deterministic) database corresponds to the case when $\mathbf{P}(t) = 1$ for all tuples t. Tuples with $\mathbf{P}(t) > 0$ are called *possible tuples*. In addition to indicating the probabilities for all tuples, a probabilistic database must also indicate somehow how the tuples are correlated. In the simplest cases the tuples are declared to be either independent (when $\mathbf{P}(t_1 t_2) = \mathbf{P}(t_1 t_2)$), or exclusive (or disjoint, when $\mathbf{P}(t_1 t_2) = 0$).

Historical Background

Extensions of databases to handle probabilistic data have been considered since the 1980s. Barbara, Garcia-Molina, and Porter [1] developed in 1992 a probabilistic data model that is an extension of the relational data model, in which relations have deterministic keys, and non-key attributes that are stochastic. They also introduced probabilistic analogs to the relational algebraic operators, later called *extensional operators*. Queries were restricted to return all deterministic keys of relations mentioned in the query body. Efforts to remove this restriction on queries lead to several extensions. Lakshmanan, Leone, Ross, and Subrahmanian in 1997 [13] to modify the query semantics and propose that probabilities be combined by user-defined *strategies*. Fuhr and Roellke in 1997 [9] propose an alternative way to compute the correct semantics for all queries, using *intensional semantics*, by which every tuple in the query answer is annotated with a propositional formula called *event expression*: a general-purpose probabilistic inference algorithm can be used to compute the tuple's probability from the event expression, but this is known

to be expensive in general. Dalvi and Suciu in 2004 [3] show that some queries can be computed efficiently using extensional operators if these are ordered carefully, to ensure that they are applied correctly: a plan consisting of correctly applied operators is called a *safe plan*. Not all queries admit a safe plan, but those that don't admit one can be shown to be #P-hard, thus they cannot be evaluated efficiently unless $P = NP$. Systems for managing probabilistic or incomplete databases include Trio, MystiQ, and MayBMS. The extensional query evaluation in probabilistic databases corresponds to lifted inference in statistical relational models [5]. Connections between probabilistic databases with independent tuples, and Markov Logic Networks [6], where tuples are correlated in complex ways, have recently been described [12].

Scientific Fundamentals

For an illustration, consider the probabilistic database in Fig. 1. It has three tables, of which Product^p and Customer^p are probabilistic and Order^p is deterministic. Product^p contains three products; their names and their prices are known, but we are unsure about their color and shape. For example Gizmo may be have color = red and shape = oval with probability $p_1 = 0.25$, or may have color = blue and shape = square with probability $p_2 = 0.75$ respectively. Camera has three possible combinations of color and shape, and IPod has two. In other words, for every product in the database we don't know the values for color and shape, but instead we have a probability space over their values. We want each color-shape combination to exclude the others, so we must have $p_1 + p_2 \leq 1$, $p_3 + p_4 + p_5 \leq 1$ and $p_6 + p_7 \leq 1$, which indeed holds for our table. When the sum is strictly less than one then that product may not occur in the table at all: for example Camera may be missing from the table with probability $1 - p_3 - p_4 - p_5$.

Representation and semantics. A popular representation of a probabilistic database consists of

Probabilistic Databases, Fig. 1 A probabilistic database

$Product^p$

prod	price	color	shape	**P**
Gizmo	20	red	oval	$p_1 = 0.25$
		blue	square	$p_2 = 0.75$
Camera	80	green	oval	$p_3 = 0.3$
		red	round	$p_4 = 0.3$
		blue	oval	$p_5 = 0.2$
IPod	300	white	square	$p_6 = 0.8$
		black	square	$p_7 = 0.2$

Order

prod	price	cust
Gizmo	20	Sue
Gizmo	80	Fred
IPod	300	Fred

$Customer^p$

cust	city	**P**
Sue	New York	$q_1 = 0.5$
	Boston	$q_2 = 0.2$
	Seattle	$q_3 = 0.3$
Fred	Boston	$q_4 = 0.4$
	Seattle	$q_5 = 0.3$

a regular database with two additional pieces of information: for each probabilistic table a distinguished attribute **P** is designated as the probability attribute, and a set of attributes is designated as *possible worlds key*. The key is interpreted as follows. If two tuples have the same values of the key attributes, then they are exclusive, i.e. only one of them may actually occur in the database. Otherwise, if the values of their key attributes differ then the tuples are independent. Tuples from different tables are always assumed to be independent.

Tuples with the same key are sometimes called *xor-tuples* or *x-tuples*. The entire set of such tuples is called a *maybe tuple*. The three Camera tuples in Fig. 1 are x-tuples, and together they form one maybe-tuple.

The semantics of a probabilistic database is a probability space over *possible worlds*. $Product^p$ has 16 possible worlds, since there are two choices for Gizmo, four for Camera (including removing Camera altogether) and two for IPod. Figure 2 illustrate three of these sixteen possible worlds and their probabilities.

Query Evaluation. The semantics of a query q on a probabilistic database is defined as follows. Consider all possible worlds of the probabilistic database: each such world is a standard, deterministic database. Evaluate q on each possible

$Product_1$

prod	price	color	shape
Gizmo	20	red	oval
Camera	80	blue	oval
IPod	300	white	square

$\mathbf{P}(Product_1) = p_1 p_5 p_6$

$Product_2$

prod	price	color	shape
Gizmo	20	blue	square
Camera	80	blue	oval
IPod	300	white	square

$\mathbf{P}(Product_2) = p_2 p_5 p_6$

$Product_3$

prod	price	color	shape
Gizmo	20	red	oval
IPod	300	white	square

$\mathbf{P}(Product_3) = p_1 (1\text{-}p_3\text{-}p_4\text{-}p_5) p_6$

Probabilistic Databases, Fig. 2 Three possible worlds

world, using the standard semantics of a query on a deterministic database. Any tuple that is an answer to the query in some possible world is called a *possible answer tuple*. The *probability of a possible answer tuple* is defined as the sum of the probabilities of all worlds where the tuple is an answer. The semantics of the query is the set of possible answer tuples together with their probabilities.

A central problem in probabilistic databases is query evaluation. It is infeasible to enumerate all possible worlds, because their number is exponential is the size of the database. There are two approaches to query evaluation: extending relational algebra operators to manipulate probabilities explicitly, or using a general-purpose probabilistic inference algorithm. The first approach is rather similar to standard query processing, and is quite efficient: the declarative SQL query is translated into a relational algebra plan, called an *extensional plan*, then this plan is executed. We illustrate this below. However not all queries admit correct extensional plans. Those that do not admit such plans require a more expensive, general-purpose probabilistic inference, which have been studied in the context of knowledge representation systems and are not be discussed here.

We illustrate the three main extensional operators on the three queries in Fig. 3, then show how to combine them for more complex queries. The left column in the Figure shows the queries in SQL syntax, the right column shows the same queries in datalog notation. In datalog we underline the variables that occur in the key positions. The first query, Q_1, asks for all the oval products in the database, and it returns two possible answer tuples:

prod	price	**P**
Gizmo	20	p_1
Camera	80	$p_3 + p_5$

For example the probability of the possible answer tuple Gizmo is the sum of the probabilities of the 8 possible worlds for Product (out of 16) where Gizmo appears as oval, and this turns out (after simplifications) to be p_1. These probabilities can be computed without enumerating all possible worlds, directly from the table in Fig. 1 by the following process: (1) Select all rows with shape='oval', (2) project on prod, price, and **P** (the probability), (3) eliminate duplicates, by replacing their probabilities with the sum, because they are disjoint events. The latter two steps are called a disjoint project:

Disjoint Project, $\pi_{\bar{A}}^{pD}$ If k tuples with probabilities $p_1, \cdots, p_k$ have the same value, $\bar{a}$, for their $\bar{A}$ attributes, then the disjoint project will associated the tuple $\bar{a}$ with the probability $p_1 + \cdots + p_k$. The disjoint project is correctly applied if any two tuples that share the same values of the $\bar{A}$ attributes are disjoint events.

Q_1 can therefore be computed by the following plan:

$$Q_1 = \pi_{\text{prod,price}}^{pD}(\sigma_{\text{shape='oval'}}(\text{Product}^P))$$

$\pi_{\text{prod,price}}^{pD}$ is correctly applied, because any two tuples in Product^P that have the same prod and price are disjoint events.

Query Q_2 in Fig. 3 asks for all cities in the Customer table, and its answer is:

city	**P**
New York	q_1
Boston	$1-(1-q_2)(1-q_4)$
Seattle	$1-(1-q_3)(1-q_5)$

This answer can also be obtained by a projection with a duplicate elimination, but now the probabilities $p_1, p_2, p_3, \ldots$ of duplicate values are replaced with $1-(1-p_1)(1-p_2)(1-p_3)\ldots$, since in this case all duplicate occurrences of the same city are independent. This is called an independent project:

Independent Project, $\pi_{\bar{A}}^{pI}$ If k tuples with probabilities $p_1, \cdots, p_k$ have the same value, $\bar{a}$, for their $\bar{A}$ attributes, then the independent project will associated the tuple $\bar{a}$ with the probability $1-(1-p_1)(1-p_2)\cdots(1-p_k)$. The independent project is correctly applied if any two tuples that share the same values of the $\bar{A}$ attributes are independent events.

Thus, the disjoint project and the independent project compute the same set of tuples, but with different probabilities: the former assumes disjoint probabilistic events, where $\mathbf{P}(t \vee t') = \mathbf{P}(t) + \mathbf{P}(t')$, while the second assumes independent probabilistic events, where $\mathbf{P}(t \vee t') =$

```
Q₁:   SELECT DISTINCT prod, price
      FROM Productᴾ
      WHERE shape='oval'
```
$$Q_1(x,y) \;:- \;\; \text{Product}^P(\underline{x,y},'oval',z)$$

```
Q₂:   SELECT DISTINCT city
      FROM Customerᴾ
```
$$Q_2(y) \;:- \;\; \text{Customer}^P(\underline{x},y)$$

```
Q₃:   SELECT DISTINCT *
      FROM Productᴾ, Order, Customerᴾ
      WHERE Productᴾ.prod = Order.prod
      and Productᴾ.price = Order.price
      and Order.cust = Customerᴾ.cust
```
$$Q_3(*) \;:- \;\; \text{Product}^P(\underline{x,y,z}),$$
$$\text{Order}(\underline{x,y,u})$$
$$\text{Customer}^P(\underline{u},v)$$

Probabilistic Databases, Fig. 3 Three simple queries, expressed in SQL and in datalog

$1 - (1 - \mathbf{P}(t))(1 - \mathbf{P}(t'))$. Continuing our example, the following plan computes Q_2: $Q_2 = \pi^{pI}_{\text{city}}(\text{Customer}^P)$. Here π^{pI}_{city} is correctly applied because any two tuples in Customer^P that have the same city are independent events.

Query Q_3 in Fig. 3 illustrates the use of a join, and its answer is:

prod	price	color	shape	cust	city	**P**
Gizmo	20	Red	Oval	Sue	New York	p_1q_1
Gizmo	20	Red	Oval	Sue	Boston	p_1q_2
Gizmo	20	Red	Oval	Sue	Seattle	p_1q_3
Gizmo	20	Blue	Square	Sue	New York	p_2q_1
...	...					

It can be computed by modifying the join operator to multiply the probabilities of the input tables:

Join, $\bowtie^P$ Whenever it joins two tuples with probabilities p_1 and p_2, it sets the probability of the resulting tuple to be p_1p_2.

A plan for Q_3 is: $Q_3 = \text{Product} \bowtie^P \text{Order} \bowtie^P \text{Customer}$.

Extensional operators can be combined to form an extensional plan. A plan is called *safe* if each operator is applied correctly. We illustrate with the following probabilistic database schema:

$\text{Product}^P(\underline{\text{prod, price}},\text{color},\text{shape})$
$\text{Order}^P(\underline{\text{prod, price, cust}})$

Both tables are probabilistic. The following query finds all colors of products ordered by the customer named Sue:

$$Q_4(c) : -\text{Product}^P(x, y, c, s),$$
$$\text{Order}^P(\underline{x, y, Sue})$$

Q_4 admits the following safe plan:

$$Q_4 = \Pi^{pI}_{\text{color}}(\Pi^{pD}_{\text{prod},\text{price},\text{color}}(\text{Product}^P)$$
$$\bowtie^P \sigma_{\text{cust}='Sue'}(\text{Order}^P))$$

This is safe because $\Pi^{pD}_{\text{prod},\text{price},\text{color}}$ combines only disjoint tuples, while Π^{pI}_{color} combines only independent tuples. Similarly, the join operators combines independent tuples, since the left operand consists of tuples combined from the Product^P table while the right operand consists of tuples from the order^P table.

In contrast, the following plan is not safe:

$$\Pi^{pI}_{\text{color}}(\text{Product}^P \bowtie^P \sigma_{\text{cust}='Sue'}(\text{Order}^P))$$

because the outermost projection Π^I_{color} will combine tuples that are not independent: there may be two or more distinct tuples in the join Product^P $\bowtie^P$ Order^P that have the same Product^P tuple, hence they are not independent.

Some queries do not admit any safe plans at all. Figure 4 illustrates three queries that are known not to admit any safe plan. All three are boolean queries, i.e. they return either 'true' or nothing, but they still have a probabilistic semantics, and we have to compute the probability of the answer 'true'. No combination of extensional operators result in a safe plan for any of these

Probabilistic Databases, Fig. 4 Three queries that do not admit safe plans, and that are known to be #P-complete

```
Schema:  R(A),S(A,B),T(B)
H₁:  SELECT DISTINCT 'true' AS A
     FROM R, S, T
     WHERE R.A=S.A and S.B=T.B
```

$$H_1 \; :- \; R(\underline{x}), S(\underline{x}, \underline{y}), T(\underline{y})$$

```
Schema:  R(A,B),S(B)
H₂:  SELECT DISTINCT 'true' AS A
     FROM R, S
     WHERE R.B=S.B
```

$$H_2 \; :- \; R(\underline{x}, y), S(\underline{y})$$

```
Schema:  R(A,B),S(A, B)
H₃:  SELECT DISTINCT 'true' AS A
     FROM R, S
     WHERE R.A = S.A and R.B=S.B
```

$$H_3 \; :- \; R(\underline{x}, y), S(x, \underline{y})$$

queries. For example $\pi_\emptyset^{pI}$ (R ⋈ S ⋈ T) is an incorrect plan for H_1, because two distinct rows in R ⋈ S ⋈ T may share the same tuple in R, hence they are not necessarily independent events. In fact it has been shown that the complexity of each of these three queries is #P-hard, hence there is no polynomial time algorithm for computing their probabilities unless $P = NP$.

Probabilistic Databases and Markov Logic Networks

Markov Logic Networks (MLN) [6] have emerged as the most popular formalism for describing both Machine Learning and probabilistic inference tasks in statistical relational models. An MLN consists of probabilistic relations, plus several *soft constraints*. For example, the soft rule

$$2.6 \quad \text{Friends}(x, y) \wedge \text{Smoker}(x) \Rightarrow \text{Smoker}(y) \tag{1}$$

says that, typically, friends of smokers tend to be smokers. The number $w = 2.6$ represents the weight associated to the rule; the higher the weight, the stronger the rule is. An MLN consists of several such rules, and describes a probability distribution on possible worlds where the tuples are correlated in complex ways. Probabilistic inference is usually done using variants of MCMC. It has been shown that every MLN can be converted into a distribution where all tuples are independent, such that the probability of an event

in the MLN is equal to a conditional probability in the new space [12]. For example, for the MLN in Eq. (2) we introduce a new relational symbol $R(x, y)$ where all tuples have weight 2.6 and add the hard constraint:

$$\forall x \forall y R(x, y) \Leftrightarrow (\text{Friends}(x, y) \wedge \text{Smoker}(x)$$
$$\Rightarrow \text{Smoker}(y)) \tag{2}$$

Exploiting the connection between MLN's and probabilistic databases is being actively researched.

Key Applications

Several large *Knowledge Bases* have been developed recently, which include Yago [10], Nell [2], DeepDive [11], Reverb [8], Microsoft's Probase [15] or Google's Knowledge Vault [7]. These knowledge bases have millions to billions of triples of the form (subject, predicate, objetct). They are constructed using a large variety of automated and semi-automated techniques, like automatic extraction from large corpora of text, data integration, path prediction, crowd curation, or simply acquisition from external sources. As a consequence most of their triples are uncertain, and their degree of uncertainty is expressed as a probability. These knowledge bases are the largest probabilistic databases to date.

In *Information Extraction* the goal is to extract structured data from a collection of unstructured text documents. Examples include address segmentation, citation extraction, extractions for comparison shopping, hotels, restaurant guides, etc. The schema is given in advance by the user, and the extractor is tailored to that specific schema. All approaches to extraction are imprecise, and most often can associate a probability score to each item extracted. A probabilistic database allows these probability scores to be stored natively.

In *Fuzzy Object Matching* the problem is to reconcile objects from two collections that have used different naming conventions. This is a central problem in data cleaning and data integration, and has also been called record linkage, or deduplication. The basic approach is to compute some similarity score between pairs of objects, usually by starting from string similarity scores on their attributes then combining these scores into a global score for the pair of objects. Thus, the result of fuzzy object matching is inherently probabilistic. In traditional data cleaning it needs to be determinized somehow, e.g. by comparing the similarity score with a threshold and classifying each pair of object into a *match*, a *non-match*, and a *maybe-match*. A probabilistic database allows users to keep the similarity scores natively.

Other applications include: management of sensor data and of RFID data, probabilistic data integration, probabilistic schema matches, and flexible query interfaces.

Cross-References

▶ Data Uncertainty Management in Sensor Networks
▶ Incomplete Information
▶ Inconsistent Databases
▶ Uncertainty Management in Scientific Database Systems

Recommended Reading

The book [14] is the most comprehensive coverage of the topic up to 2011. The complexity of query evaluation is established in [4]. Markov Logic Networks are covered in [6].

1. Barbara D, Garcia-Molina H, Porter D. The management of probabilistic data. IEEE Trans Knowl Data Eng. 1992;4(5):487–502.
2. Carlson A, Betteridge J, Kisiel B, Settles B, Hruschka ER Jr, Mitchell TM. Toward an architecture for never-ending language learning. In: Proceedings of the 24th National Conference on Artificial Intelligence; 2010.
3. Dalvi N, Suciu D. Efficient query evaluation on probabilistic databases. In: Proceedings of the 30th International Conference on Very Large Data Bases; 2004.
4. Dalvi NN, Suciu D. The dichotomy of probabilistic inference for unions of conjunctive queries. J ACM. 2012;59(6):30.
5. Van den Broeck G, Suciu D. Tutorial: Lifted probabilistic inference in relational models. In: Proceedings of the 25th International Joint Conference on AI; 2016.
6. Domingos P, Lowd D. Markov logic: an interface layer for artificial intelligence. Synthesis lectures on artificial intelligence and machine learning. San Rafael: Morgan & Claypool Publishers; 2009.
7. Dong X, Gabrilovich E, Heitz G, Horn W, Lao N, Murphy K, Strohmann T, Sun S, Zhang W. Knowledge vault: a web-scale approach to probabilistic knowledge fusion. In: Proceedings of the 20th ACM SIGKDD International Conference on Knowledge Discovery and Data Mining; 2014.
8. Fader A, Soderland S, Etzioni O. Identifying relations for open information extraction. In: Proceedings of the Conference on Empirical Methods in Natural Language Processing. A meeting of SIGDAT, a special interest group of the ACL; 2011. p. 1535–45.
9. Fuhr N, Roelleke T. A probabilistic relational algebra for the integration of information retrieval and database systems. ACM Trans Inf Syst. 1997;15(1):32–66.
10. Hoffart J, Suchanek FM, Berberich K, Weikum G. Yago2: a spatially and temporally enhanced knowledge base from wikipedia. Artif Intell. 2013;194(Jan):28–61.
11. http://deepdive.stanford.edu/.
12. Jha AK, Suciu D. Probabilistic databases with markoviews. Proc VLDB Endow. 2012;5(11):1160–71.
13. Lakshmanan L, Leone N, Ross R, Subrahmanian VS. Probview: a flexible probabilistic database system. ACM Trans Database Syst. 1997;22(3):419–69.
14. Suciu D, Olteanu D, Ré C, Koch C. Probabilistic databases. Synthesis lectures on data management. San Rafael: Morgan & Claypool Publishers; 2011.
15. Wu W, Li H, Wang H, Zhu KQ. Probase: a probabilistic taxonomy for text understanding. In: Proceedings of the ACM SIGMOD International Conference on Management of Data; 2012. p. 481–92

Probabilistic Entity Resolution

Ekaterini Ioannou
Faculty of Pure and Applied Sciences, Open
University of Cyprus, Nicosia, Cyprus

Synonyms

Deduplication; Linkage; Matching

Definition

Entity Resolution is the task of analyzing a collection of data (e.g., database, data set) in order to create *entities* by merging the data instances that describe the same *real-world objects*. Uncertain entity resolution is a group of resolution methodologies focusing on handling the uncertainties that are present either in the data or are generated during the resolution process.

Historical Background

The fundamental component of resolution techniques is an instance that provides some characteristic of a real-world object. An instance is a tuple with k attributes $\langle v_1, \ldots, v_k \rangle$, with each attribute being one characteristic of the corresponding object. Consider now a collection of **instances**. The goal of resolution is to detect the instances that describe the same real-world objects and merge them into entities, i.e., create entity e for representing instances r_1, r_2, and r_3.

The initial resolution approaches focused on handling the problem on structured data, such as in relational databases. Each table record was considered as an instance, and due to misspellings, abbreviations, acronyms, etc., the table contained more than one instance for the same object. Addressing the problem was achieved by computing the similarities between the data of two instances, for example, by measuring the numbers of operations we would need for converting the attributes of the first instance to the ones of the second instance. The processing can identify additional information that could assist resolution and use this for improving the accuracy of results. A major source for this additional resolution information is the relationships between instances, e.g., knowing that instance r_1 is the same with r_1', instance r_1 is a co-author of instance r_2, and instance r_1' is a co-author of instance r_2', we can increase our belief that instance r_2 describes the same author as r_2'. More information on these approaches is available in [4,6].

Although these approaches provide sufficient results for structured and static data, they have various limitations when applied to data of modern systems, such as Web applications. A major limitation arises because these approaches assume that the data is relatively static, and thus the processing is performed offline. Once performed, the instances found to describe the same objects are merged, and the created entities are used to replace the original instances. Query answering is then performed on the entities. When the original data is highly volatile, this process needs to be continuously repeated, and it becomes inefficient. The data and schema heterogeneity is one more of the data from the sources that may be extracted using information extraction techniques and so may yield erroneous data. Another limitation is that none of these techniques gives certain results. They typically merge instances that are found to have a belief above some specific threshold value. The selection of the threshold value is not easy and is often done experimentally. In addition, it causes recall-precision dilemma: opt for high precision (by using a high threshold) that will return the most certain results but typically a subset of the complete result set or high recall (by using a low threshold) that will return many results but with a higher percentage of errors.

Probabilistic entity resolution appeared for handling the probabilities that appeared either in the original data or during the data processing. In both cases, the main source of probabilities is primarily the data and schema heterogeneity and the evolving nature of the data.

P

Scientific Fundamentals

Exclusive Instance Alternatives

During the traditional resolution process (explained in *Historical Background*), the instances are compared, and this results in a probability encoding the process's belief that the specific two instances should be matched, i.e., represent the same real-world object. A number of approaches explain that handling these probabilities during query processing is more beneficial than the traditional approach of merging instances based on a probability threshold. Thus, these approaches maintain and keep all potential matches between instances and use them during query processing for retrieving all possible answers.

One such approach is [7]. It focuses on the integration of two data collections and considers a probabilistic database composed by the probabilistic matches between the instances of the first collection with the instances of the second collection. Query processing follows the "possible world" semantics (see Reference) for computing the answers (i.e., entities) and their probabilities. For example, consider that instance r_{A1} from the first collection matches instance r_{B1} from the second collection with probability p_1 and also matches instance r_{B2} from the second collection with probability p_2. Query processing must consider that: (1) None of the instances is an answer; (2) r_{A1} is an answer and does not match other instances; (3) r_{B1} is an answer and does not match other instances; (4) r_{B2} is an answer and does not match other instances; (5) r_{A1} matches r_{B1} and is a valid answer; (6) r_{A1} matches r_{B2} and is a valid answer; The paper introduces a query processing mechanism for computation of the k answers with the highest probabilities. The processing runs in parallel with several Monte-Carlo simulations (i.e., one for each possible answer) and approximates each probability only to the extent needed to compute correctly the top-k answers.

Probabilistic Schema Mappings

Schema heterogeneity is another reason causing probabilities that the resolution process needs to handle. This situation typically appears when considering the resolution of data from sources that have different schemata. For example, a data integration system would use mappings for specifying the mappings (i.e., relationships) between the instance attributes contained in the given collections with the ones used by a mediated schema. This means that we can have attribute v_1 of the instances mapping to attribute v_α of the mediated schema S with a probability p_1 and to attribute v_β of S with a probability p_2.

The approach introduced in [3] focuses on this issue, i.e., investigates the use of the probabilistic mappings between the attributes of the contributing sources with a mediated schema. Interpreting these mappings can be done using the following semantics: (i) Create a possible world for each mapping; (ii) Create possible worlds by combining mappings. The authors introduce algorithms for using the provided mapping to reformulate a given query into different queries, explain the issues for processing some of the semantics, and provide restriction on the query syntax that can be executed.

Querying Probabilistic Matches

Andritsos et al. [1] do not focus on the schema information, as the approach presented in [7], but on the actual data. The authors assume that the duplicate tuples for each representation are given, for example, as the results computed by a technique from [7]. Thus, all tuples describing alternative representations of the same representation have the same identifier. The tuples of the alternative representations are considered as disjoined, which means that only one tuple for each identifier can be part of the final resulted representation [2].

The approach in [5] addresses more challenges of heterogeneous data. In particular, this approach does not assume that the alternative representations of representations are known but that a representation collection comes with a set of possible linkages between representations. Each linkage represents a possible match between two representations and is accompanied with a probability that indicates the belief we have that the specific representations are for the same real-world object. Representations are compiled on the fly, by effectively processing the incoming query over representations and linkages, and thus, query answers reflect the most probable solution for the specific query.

Key Applications

Probabilistic entity resolution is a task required when integrating and cleaning data in highly heterogeneous environments. As an example, consider data from modern social Web applications, such as Blogosphere, Facebook, and Twitter. These applications need to integrate heterogeneous, typically unstructured data, with frequent additions and updates since users are allowed, and even encouraged, to interact with the data.

Cross-References

▶ Probabilistic Databases

Recommended Reading

1. Andritsos P, Fuxman A, Miller R. Clean answers over dirty databases: a probabilistic approach. In: Proceedings of the 22nd International Conference on Data Engineering; 2006.
2. Beskales G, Soliman M, Ilyas I, Ben-David S. Modeling and querying possible repairs in duplicate detection. Proc VLDB Endow. 2009;2(1):598–609.
3. Dong XL, Halevy A, Yu C. Data integration with uncertainty. In: Proceedings of the 33rd International Conference on Very Large Data Bases; 2007. p. 687–98.
4. Elmagarmid A, Ipeirotis P, Verykios V. Duplicate record detection: a survey. IEEE Trans Knowl Data Eng. 2007;19(1):1–16.
5. Ioannou E, Nejdl W, Niederée C, Velegrakis Y. On-the-fly entity-aware query processing in the presence of linkage. Proc VLDB Endow. 2010;3(1):429–38.
6. Ioannou E, Staworko S. Management of inconsistencies in data integration. In: Data exchange, integration, and streams. 2013. p. 217–25.
7. Re C, Dalvi N, Suciu D. Efficient top-k query evaluation on probabilistic data. In: Proceedings of the 23rd International Conference on Data Engineering; 2007. p. 886–95.

Probabilistic Retrieval Models and Binary Independence Retrieval (BIR) Model

Thomas Roelleke[1], Jun Wang[1], and Stephen Robertson[2]
[1]Queen Mary University of London, London, UK
[2]Microsoft Research Cambridge, Cambridge, UK

Synonyms

BIR model; Probabilistic model; RSJ model

Definition

Information retrieval (IR) systems aim to retrieve relevant documents while not retrieving non-relevant ones. This can be viewed as the foundation and justification of the binary independence retrieval (BIR) model, which proposes to base the ranking of documents on the division of the probability of relevance and non-relevance.

For a set r of relevant documents, and a set $\overline{r}$ of non-relevant documents, the BIR model defines the following term weight and retrieval status value (RSV) for a document-query pair "d, q":

$$\mathrm{birw}(t, r, \overline{r}) := \frac{P(t|r)\cdot P(\overline{t}|\overline{r})}{P(t|\overline{r})\cdot P(\overline{t}|r)} \quad (1)$$

$$\mathrm{RSV}_{\mathrm{BIR}}(d, q, r, \overline{r}) := \sum_{t \in d \cap q} \log \mathrm{birw}(t, r, \overline{r}) \quad (2)$$

Here, $P(t|r)$ is the probability that term t occurs in the relevant documents, and $P(t|\overline{r})$ is the respective probability for term t in non-relevant documents.

There are two ways to estimate the term probabilities: $(I_1) P(t|r) := \frac{r_t}{R}$ and $P(t|\overline{r}) := \frac{n_t}{N}$, or $(I_2) P(t|r) := \frac{r_t}{R}$ and $P(t|\overline{r}) := \frac{n_t - r_t}{N - R}$. R denotes the number of relevant documents, and N denotes the number of all documents; r_t denotes the number of relevant documents containing term t, and n_t denotes the respective number of all documents. I_1 assumes term independence in the relevant documents and in *all* documents; I_2 assumes term independence in the relevant documents and in the *non-relevant* documents.

Given the two independence assumptions, and given the option to consider term presence only, there are four forms/variations of the BIR term weight. Further, the BIR model proposes a technique for dealing with missing relevance, where a zero probability problem needs to be avoided. The definitions given next show the case referred to as F4 where I_2 is assumed and term absence is considered. The constant 0.5 caters for missing relevance data.

$$\text{birw}_{+0.5}(t, r, \overline{r}) :=$$
$$\frac{(r_t + 0.5) / (R - r_t + 0.5)}{(n_t - r_t + 0.5) / (N - R - (n_t - r_t) + 0.5)} \tag{3}$$

$$\text{RSV}_{\text{BIR,F4}}(d, q, r, \overline{r}) := \sum_{t \in d \cap q} \log \text{birw}_{+0.5}(t, r, \overline{r}) \tag{4}$$

The derivation of the term weight and the retrieval status value is to be found in the scientific fundamentals of this entry.

Historical Background

The paper [10] on "Relevance Weighting of Search Terms" proposed what became known as the probabilistic retrieval model, binary independence retrieval model, or RSJ (initials of the authors) model.

Closely related is the "The Probability Ranking Principle in IR" [8] to justify that a ranking based on the probability of relevance is an optimal ranking, since the ranking minimises the costs for browsing the ranking.

Reference [2] on "Using Probabilistic Models of Document Retrieval without Relevance Information" investigated the generalisation for missing relevance, and [4] contributed "An Evaluation of Feedback in Document Retrieval using Co-occurrence Data."

Foundations

Derivation of the BIR Model

Derivations of the BIR model can be found in [6], [3] (pp. 21–31), and [5] (pp. 167–175).

The BIR model is derived from the odds of the probability of relevance:

$$O(r|d, q) := \frac{P(r|d, q)}{P(\overline{r}|d, q)} = \frac{P(r|d, q)}{1 - P(r|d, q)} \tag{5}$$

Here, $P(r|d, q)$ is the probability that the event relevant (r) occurs for a given document-query pair. $P(\overline{r}|d, q)$ is the respective probability for non-relevant.

The derivation of the BIR model can be represented in six steps:

1. Bayes theorem and reduction of $O(r|d, q)$.
2. Representation of d as a vector $\vec{x}$ of binary term features (the binary term features are "term does occur" and "term does not occur").
3. Independence assumption for features.
4. Split product over features into two products: a product for term presence, and a product for term absence.
5. Non-query term assumption: if non-query terms occur with the same frequency in relevant and non-relevant documents, then they do not affect the ranking.
6. Rewrite the expression to obtain a compact form.

Bayes Theorem

To estimate the unknown probability $P(r|d, q)$, the theorem of Bayes is applied:

$$P(r|d, q) := \frac{P(r|d, q)}{P(d, q)}$$
$$= \frac{P(d|r, q) \cdot P(r|q) \cdot P(q)}{P(d, q)} \tag{6}$$

For $O(r|d, q)$, the probabilities $P(q)$ and $P(d, q)$ drop out. With respect to the ranking of documents, the factor $\frac{P(r|q)}{P(\overline{r}|q)}$ has no effect, since it is document-independent. Thus, the ranking of documents is based on the following ranking equivalence:

$$O(r|d,q) \stackrel{rank}{=} \frac{P(d|r,q)}{P(d|\overline{r},q)} \quad (7)$$

Binary Feature Vector $\vec{x}$

The next step concerns the representation and decomposition of the document d. For this, the BIR model assumes that d is a vector of binary features; this vector is denoted $\vec{x} = (x_1, ..., x_n)$. Each x_i corresponds to a term t_i, and for all x_i, $x_i = 1$ if term is present, and $x_i = 0$ if term is absent.

Independence Assumption

Next, the BIR model assumes that the binary features are independent. This leads to the following equation for $P(d|r,q)$:

$$P(d|r,q) = P(\vec{x}|r,q) = \prod_{x_i} P(x_i|r,q) \quad (8)$$

The equation for $P(d|\overline{r},q)$ is accordingly.

Product Split

The product over all features is split into two products: one product for term presence ($x_i = 1$), and one for absence ($x_i = 0$).

$$P(d|r,q) = \left[\prod_{x_i=1} P(x_i|r,q) \right]$$
$$\cdot \left[\prod_{x_i=0} P(x_i|r,q) \right] \quad (9)$$

Since $x_i = 1$ corresponds to "term occurs," and $x_i = 0$ corresponds to "term does not occur," the next equation replaces x_i by the term event t, where t means "term occurs."

$$P(d|r,q) = \left[\prod_{t \in d} P(t|r,q) \right] \cdot \left[\prod_{t \notin d} P(\overline{t}|r,q) \right] \quad (10)$$

In classical literature [10, 6], the symbols p_i and q_i are employed instead of $P(t|r, q)$ and $P(\overline{t}|r,q)$; however, since the probabilities are

more explicit, and avoid a confusion of query (q) and absence of term (q_i), and facilitate to relate the BIR model to other retrieval modes, the probabilistic notion is chosen here, while keeping in mind that $P(t|r, q)$ refers to the binary feature of term t.

Non-query Term Assumption

For all non-query terms, it is assumed that their frequency in relevant documents is equal to their frequency in non-relevant documents, i.e., $\forall t \notin q : P(t|r,q) = P(t|\overline{r},q)$. This leads to a reduction of the product, i.e., the non-query terms are dropped.

$$\frac{P(d|r,q)}{P(d|\overline{r},q)} = \left[\prod_{t \in d \cap q} \frac{P(t|r,q)}{P(t|\overline{r},q)} \right]$$
$$\cdot \left[\prod_{t \in q \setminus d} \frac{P(\overline{t}|r,q)}{P(\overline{t}|\overline{r},q)} \right] \quad (11)$$

This assumption corresponds to $\frac{P(t|r,q)}{P(t|\overline{r},q)} = 1$, and the more general assumption is $\frac{P(t|r,q)}{P(t|\overline{r},q)} = c$, where c is a constant.

Rewriting to Achieve Compact Form

Finally, a rewriting leads to a compact form. The rewriting is based on the following equation:

$$1.0 = \prod_{t \in d \cap q} \left[\frac{P(\overline{t}|\overline{r},q)}{P(\overline{t}|r,q)} \cdot \frac{P(\overline{t}|r,q)}{P(\overline{t}|\overline{r},q)} \right] \quad (12)$$

Equation 11 is multiplied with this 1.0. This fills up the right product over query-only terms to a product over all query terms; then, this product is document-independent, and it can be dropped for ranking purpose, as the next equations illustrate.

$$O(r|d,q) = \left[\prod_{t \in d \cap q} \frac{P(t|r,q)}{P(t|\overline{r},q)} \right]$$
$$\cdot \left[\prod_{t \in q \setminus d} \frac{P(\overline{t}|r,q)}{P(\overline{t}|\overline{r},q)} \right]$$

$$\cdot \left[\prod_{t \in d \cap q} \frac{P(\bar{t}|\bar{r},q)}{P(\bar{t}|r,q)} \cdot \frac{P(\bar{t}|r,q)}{P(\bar{t}|\bar{r},q)} \right]$$

$$\overset{rank}{=} \prod_{t \in d \cap q} \frac{P(t|r,q) \cdot P(\bar{t}|\bar{r},q)}{P(t|\bar{r},q) \cdot P(\bar{t}|r,q)}$$

The fractional expression is referred to as the BIR term weight, where either the above form or the logarithm of it may be viewed as the term weight. The following definition follows the first approach, and the logarithm is applied to the term weight in the definition of RSV_{BIR}.

$$\text{birw}(t,r,\bar{r}) := \frac{P(t|r,q) \cdot P(\bar{t}|\bar{r},q)}{P(t|\bar{r},q) \cdot P(\bar{t}|r,q)} \qquad (14)$$

$$\text{RSV}_{\text{BIR}}(d,q,r,\bar{r}) := \sum_{t \in d \cap q} \log \text{birw}(t,r,\bar{r}) \qquad (15)$$

The BIR weight is greater than 1.0 for *good* terms, i.e., good terms occur more frequently in relevant documents than in non-relevant documents. For $P(t|r,q) = P(t|\bar{r},q)$, the BIR weight is 1.0; thus, such terms have no effect on the ranking. For *poor terms*, the BIR weight is less than 1.0; thus, poor terms have a negative effect on the RSV_{BIR}.

Alternative Derivation

An alternative, shorter derivation is shown next. The abbreviations $p_i := P(x_i = 1|r)$ and $q_i := P(x_i = 1|\bar{r})$ are applied, and the document probabilities are expressed as follows:

$$\frac{P(d|r)}{P(d|\bar{r})} = \prod_i \frac{p_i^{x_i} \cdot (1-p_i)^{1-x_i}}{q_i^{x_i} \cdot (1-q_i)^{1-x_i}} \qquad (16)$$

The exponent $x_i = 1$ selects the probability p_i that the term occurs, and $x_i = 0$ selects the probability $(1 - p_i)$ that the term does not occur. Resolving the exponent $1 - x_i$ leads to the next equation and ranking equivalence.

$$\frac{P(d|r)}{P(d|\bar{r})} = \left[\prod_i \left(\frac{p_i}{1-p_i} \cdot \frac{1-q_i}{q_i} \right)^{x_i} \right] \cdot \prod_i \frac{1-p_i}{1-q_i} \qquad (17)$$

$$\overset{rank}{=} \prod_{t \in d \cap q} \text{birw}(t,r,\bar{r}) \qquad (18)$$

The right product $\left(\prod_i \frac{1-p_i}{1-q_i} \text{ in Eq. 17} \right)$ is constant, and therefore does not affect the ranking. Since $x_i = 0$ for non-document terms, the first product reduces to $x_i = 1$ in d. With $p_i = q_i$ for non-query terms, the product reduces to $x_i = 1$ in d and q, i.e., the product is over $t \in d \cap q$.

Estimation of Term Probabilities

For describing the estimation of term probabilities, the following notation is employed:

Traditional notation	Event-based notation	Comment
r_t	$n_D(t,r)$	Number of relevant documents with term t
R	$N_D(r)$	Number of relevant documents
n_t	$n_D(t,c)$	Number of documents with term t in the collection c
N	$N_D(c)$	Number of documents in the collection c
p_t	$P_D(t|r)$	Term probability in relevant documents
q_t	$P_D(t|\bar{r})$	term probability in non-relevant documents

The notation comprises the traditional symbols, and it shows an alternative, namely an event-based notation. The event-based notation is explicit regarding the event-space (D for documents). Therefore, it is applicable in a dual way to document-based and location-based event spaces. Also, events such as a document, a collection, and any set of documents (relevant, non-relevant, retrieved, all) can be referred to systematically.

The estimates of the term probabilities are given next:

The I_2 assumption and the consideration of term absence lead to the following equation for the BIR term weight:

Traditional notation	Event-based notation	Comment
$p_t = \frac{r_t}{R}$	$P_D(t\|r) = \frac{n_D(t,r)}{N_D(r)}$	Probability of term in relevant
$q_t = \begin{cases} \frac{n_t}{N}I_1 \\ \frac{n_t - r_t}{N - R}I_2 \end{cases}$	$P_D(t\|\bar{r}) = \begin{cases} \frac{n_D(t,c)}{N_D(c)} & I_1 \\ \frac{n_D(t,c) - n_D(t,r)}{N_D(c) - N_D(r)} & I_2 \end{cases}$	Probability of term in non-relevant assumptions I_1 and I_2

$$\text{birw}(t,r,\bar{r}) = \frac{r_t / (R - r_t)}{(n_t - r_t) / (N - R - (n_t - r_t))} \quad (19)$$

The next section groups the variations of the term weight.

Variations of the BIR Term Weight

[10] introduced four variations (forms) of the term weights: the four variations (referred to as F1, F2, F3, and F4) follow from the two estimates for $P(t|\bar{r})$, and whether or not term absence is considered.

Assumption	Term absence	Variation	Term weight
I_1	no	F1	$\frac{r_t / R}{n_t / N}$
I_2	no	F2	$\frac{r_t / R}{(n_t - r_t)/(N - R)}$
I_1	yes	F3	$\frac{r_t / (R - r_t)}{n_t / (N - n_t)}$
I_2	yes	F4	$\frac{r_t / (R - r_t)}{(n_t - r_t)/(N - R - (n_t - r_t))}$

For further reading, the BIR term weights are summarised in [3], page 27. The variations form a comprehensive mathematical coverage; F4 is the prime choice.

Solving the Zero Probability Problem

The BIR term weight is not defined for missing relevance; $R = r_t = 0$ leads to a division by zero. Therefore, [10] proposes to add constants to the frequency counts:

$$\text{birw}_{+0.5}(t,r,\bar{r}) :=$$

$$\frac{(r_t + 0.5) / (R - r_t + 0.5)}{(n_t - r_t + 0.5) / (N - R - (n_t - r_t) + 0.5)} \quad (20)$$

The subscript $+0.5$ in $\text{birw}_{+0.5}(t,r,\bar{r})$ marks this term weight to be different from the bare term weight $birw(t,r,\bar{r})$ in Eq. 19; $\text{birw}_{+0.5}(t,r,\bar{r})$ is also referred to as $rsj(t,r,\bar{r})$. What is the explanation underlying this zero-probability "fix"?

To reach an explanation, insert $r_t + 0.5$ for each r_t, and insert $R + 1$ for each R. This corresponds to assuming that there is a *virtual* document that is relevant, and each term occurs in *half* of the relevant documents, i.e., $p_t = 0.5$. The virtual document is retrieved, i.e., $n_t + 1$. Finally, this requires to assume $N + 2$ documents, which corresponds to assuming that there are two virtual documents, one of which is relevant. The next equation illustrates the explanation via virtual documents:

$$\text{birw}_{+0.5}(t,r,\bar{r}) :=$$

$$\frac{(r_t + 0.5) / (R + 1 - (r_t + 0.5))}{((n_t + 1) - (r_t + 0.5)) / \left((N + 2) - (R + 1) - ((n_t + 1) - (r_t + 0.5)) \right)} \quad (21)$$

This expanded form reduces to Eq. 20.

From a more general point of view, the estimate $P(t|r) := \frac{r_t + k}{R + K}$ is applied to cover the case for missing relevance, where $\frac{k}{K} = 0.5$, and $k = 0.5$, and $K = 1$, for $\text{birw}_{+0.5}$. This formulation reminds of the Laplace law of succession. Since the notion of "half of a relevant document" has no intuition, the number of virtual documents could be scaled to four rather than two [13]. This leads to: $N + 4$, $n_t + 2$, $R + 2$, $r_t + 1$. Then, the BIR weight is as follows:

$$\text{birw}_{+1}(t,r,\bar{r}) :=$$

$$\frac{(r_t + 1) / (R - r_t + 1)}{(n_t - r_t + 1) / (N - R - (n_t - r_t) + 1)} \quad (22)$$

Relationship Between the BIR Model, IDF, and BM25

The BIR model can be viewed as a theoretical argument to support IDF-based retrieval.

[2] proposes for missing relevance to assume that for all term, $P(t|r)$ is the same. This leads to a co-ordination level component in the retrieval function:

P

$$\sum_{t \in d \cap q} \log \frac{P(t|r)}{1 - P(t|r)} + \sum_{t \in d \cap q} - \log \frac{P(t|\bar{r})}{1 - P(t|\bar{r})}$$

The left component expresses the co-ordination level match. For large N and $N \gg n_t$, the right component is similar to the IDF component. i.e., $\sum_{t \in d \cap q} - \log \frac{n_t}{N - n_t} \approx \sum_{t \in d \cap q} idf(t, c)$. [11] investigates how these assumptions are affected in the case of little relevance information.

[9] reviews the relationship of the BIR model and IDF, and theoretical arguments for IDF. The relationship between BIR and IDF is based on the definition $idf(t, c) := -\log P_D(t|c)$ and the estimation of the BIR probability $P(t|\bar{r})$. By assuming $P(t|\bar{r}) = P_D(t|c)$, the relationship is established, i.e., $idf(t, c) = -\log P(t|\bar{r})$.

[13] builds on this relationship and shows a fully idf-based formulation of the BIR term weight:

$$\log birw(t, r, \bar{r}) = \log \frac{P(t|r).P(\bar{t}|\bar{r})}{P(t|\bar{r}).P(\bar{t}|r)} \quad (23)$$

$$= idf(t, \bar{r}) - idf(t, r) + idf(\bar{t}, r) - idf(\bar{t}, \bar{r}) \quad (24)$$

$$\approx idf(t, c) - idf(t, r) + idf(\bar{t}, r) - idf(\bar{t}, c) \quad (25)$$

This linear combination of idf values under-lines the issue that the collection c is much larger than the set r of relevant documents ($|c| \gg |r|$), and therefore the maximum-likelihood estimates may be not comparable, and may need to be normalised [13].

Ignoring the negated term events (dropping the term absence, see section 1, variations of BIR term weight) leads to the following simplified term weight:

$$\log \frac{P(t|r)}{P(t|\bar{r})} = idf(t, \bar{r}) - idf(t, r)$$
$$\approx idf(t, c) - idf(t, r) \quad (26)$$

The formulation shows that the term weight is $idf(t, c)$, if $idf(t, r) = 0$. The latter is the case for terms that occur in *all* relevant documents, i.e.,

$P(t|r) = 1$. Therefore, from this simplified form of the BIR model, idf-based retrieval assumes $P(t|r) = 1$. On the other hand, the full formulation Eq. (25) assumes $idf(t, r) = idf(\bar{t}, r)$, i.e., $P(t|r) = 0.5$, for missing relevance.

Regarding BM25, the BIR term weight is in BM25 what IDF is in TF-IDF, i.e., in BM25, a TF-component is multiplied with the BIR F4 term weight, whereas in basic TF-IDF, a TF-component is multiplied with the IDF term weight.

Key Applications

The BIR model is applied to incorporate rele-vance feedback data into document ranking; this model has become a key foundation of retrieval models. In 2004, the BIR model served as a foundation for a probabilistic ranking of tuples for database queries [1].

Cross-References

▶ BM25
▶ Language Models
▶ TF*IDF

Recommended Reading

1. Chaudhuri S, Das G, Hristidis V, Weikum G. Proba-bilistic ranking of database query results. In: Proceed-ings of the 30th International Conference on Very Large Data Bases; 2004. p. 888–99.
2. Croft WB, Harper DJ. Using probabilistic models of document retrieval without relevance information. J Doc. 1979;35(4):285–95.
3. Grossman DA, Frieder O. Information retrieval. Al-gorithms and heuristics. 2nd ed. The information retrieval series, vol. 15. Berlin:Springer; 2004.
4. Harper DJ, van Rijsbergen CJ. An evaluation of feedback in document retrieval using cooccurrence data. J Doc. 1978;34(3):189–216.
5. Belew RK. Finding out about: Cambridge University Press; 2000.
6. van Rijsbergen CJ. Information Retrieval. 2nd ed. London: Butterworths; 1979. http://www.dcs. glasgow.ac.uk/Keith/Preface.html
7. Robertson S. On event spaces and probabilistic models in information retrieval. Inform Retr J. 2005;8(2):319–29.

8. Robertson SE. The probability ranking principle in IR. J Doc. 1977;33(4):294–304.
9. Robertson SE. Understanding inverse document frequency: On theoretical arguments for idf. J Doc. 2004;60(5):503–20.
10. Robertson SE, Sparck JK. Relevance weighting of search terms. J Am Soc Inf Sci. 1976;27(3):129–46.
11. Robertson SE, Walker S. On relevance weights with little relevance information. In: Proceedings of the 20th Annual International ACM SIGIR Conference on Research and Development in Information Retrieval; 1997. p. 16–24.
12. Roelleke T, Wang J. A parallel derivation of probabilistic information retrieval models. In: Proceedings of the 32nd Annual International ACM SIGIR Conference on Research and Development in Information Retrieval; 2009. p. 107–14.
13. de Vries A, Roelleke T. Relevance information: a loss of entropy but a gain for IDF? In: Proceedings of the 31st Annual International ACM SIGIR Conference on Research and Development in Information Retrieval; 2008. p. 282–9.

Probabilistic Skylines

Niccolò Meneghetti
Computer Science and Engineering Department, University at Buffalo, Buffalo, NY, USA

Synonyms

Uncertain skylines; Stochastic skylines

Definition

Given an arbitrary set P of points in the positive orthant $\mathbb{R}^d_+$, we say that a point p *Pareto-dominates* a point q ($p \succ q$) if p is no worse than q in all the d dimensions and strictly better in at least one dimension. Without lack of generality, we can assume that a point p is better than a point q over the dimension i if p is smaller than q when both are projected on the i-coordinate. The *skyline* of P is the subset of points that are not dominated by any other point in P. The goal of *probabilistic skylines* is to compute the skyline over uncertain data, i.e. when there is no perfect information about the location of each point in P. There are several ways to define probabilistic skylines; the most appropriate definition usually depends on the task at hand. The first, original definition is due to Jiang et al. [1]: given a probability threshold τ, the probabilistic skyline is the set of points p such that the probability of p being dominated is less than $(1 - \tau)$. The other alternative definitions will be discussed in the following sections.

Historical Background

The concept of Pareto dominance has been studied for decades within several disciplines, including economics, game theory and, more recently, computer science. The first efficient algorithms for identifying non-dominated points have been proposed in the context of computational geometry, as a solution to the *maximal vector problem*. Later, the seminal of work of Börzsönyi et al. [2] introduced the problem to the attention of the database community, in the form of *skyline queries*. More recently, as the research on probabilistic databases gained momentum [3, 4], many well-established query paradigms, such as range and top-k queries, have been extended in order to deal with uncertain data. Skyline queries made no exception: several generalizations of the Pareto dominance criterion have been proposed, as a tool for supporting decisions under imperfect information.

Scientific Fundamentals

In the context of probabilistic skylines, uncertain data is usually modeled under the *possible worlds* semantics: a probabilistic relation R^P is defined by a pair (R, p), where R is a relation instance in the deterministic sense (a set of tuples), and p is a probability distribution over the power-set of R (i.e. the set of possible worlds). The probability for a tuple t to belong to R^P is simply the sum of the probabilities of all the possible worlds containing t. This general model is often simplified, by setting some restrictions on the probability distribution p used to define R^P. One popular approach, adopted by the original work by Jiang et al. [1], is the *x-relation* model [4]: an *uncertain tuple* T is a set of mutually exclusive tuple instances ($\{t_1, t_2, \ldots, t_n\}$), together with a local probability distribution p_t; each possible world can contain only one instance for each uncertain tuple, that is selected according to the local probability distribution p_t. The local probability distributions are assumed to be independent on each other. Under this model, $T \succ T'$ denotes the event that the tuple instance selected for T' is dominated by the one selected for T. It is important to notice that, given three arbitrary uncertain tuples T, T' and T'', the event $T \succ T'$ is not necessarily independent from the event $T'' \succ T'$, even under the simplifying assumptions of the x-relation model (see Fig. 1, for a simple example). Hence, computing the exact probability for an uncertain tuple T to belong to the skyline usually requires to take into consideration a large number of possible worlds.

A point p in P belongs to the deterministic skyline of P if and only if it belongs to the maxima w.r.t. some monotone decreasing scoring function, defined over all the d attributes. In other words, skylines contains all the top-1 results for a large family of intuitive scoring functions, making them appealing for information filtering applications. Several probabilistic skyline semantics have been proposed in order to reproduce this property in the context of uncertain data. The first step in this direction is due to Lin

et al. [5]. The authors proposed to define the dominance between two uncertain tuples using the *lower orthant order*. Given an arbitrary point p in $\mathbb{R}_+^d$, its lower orthant is defined as the minimum bounding box containing both p and the origin. According to the semantics introduced in [5], $T \succ T'$ holds if and only if for each point p in $\mathbb{R}_+^d$ the cumulative probability of T being instantiated somewhere in the lower orthant of p is not smaller than the same probability for T', and strictly larger for at least one specific point p^* in $\mathbb{R}_+^d$. The probabilistic skyline consists of all uncertain tuples that are not dominated by any other element of the probabilistic relation R^P. The results obtained this way are those that maximize the expectation for some multiplicative utility function $E[u(T)] = E[\Pi_{i=1}^d u_i(T[i])]$.

Future Directions

The models discussed above have been extended in several ways. Böhm et al. [6] extended the original semantics by [1], modeling the probabilistic relation R^P in parametric form, using mixtures of Gaussians, while [7] provided an efficient method for computing all skyline probabilities. The semantics of [1] was used also in [8] for computing probabilistic reverse skylines, and in [9] for computing skylines over uncertain data streams. The idea of defining probabilistic skylines as the union of the results of probabilistic top-1 queries, introduced in [5], has been expanded by Bartolini et al. [10], where the semantics used for answering top-k queries over uncertain data determines the final content of the probabilistic skyline. More recently, [11] addressed the problem of computing the skyline of a deterministic relation under the assumption the preferences are uncertain.

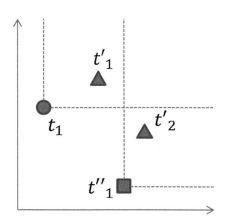

Probabilistic Skylines, Fig. 1 A probabilistic relation with two certain tuples, t and t'', and one uncertain tuple, t'. Notice how the Pareto dominance $t \succ t'$ is not probabilistically independent from the dominance $t'' \succ t'$

Key Applications

Multi-criteria Decision Making
Information Filtering
Sensors networks

Cross-References

- ► Preference Queries
- ► Preference Specification
- ► Probabilistic Databases
- ► Query Processing Over Uncertain Data
- ► Skyline Queries and Pareto Optimality
- ► Uncertain Top-k Queries

Recommended Reading

1. Jiang B, Pei J, Lin X, Yuan Y. Probabilistic sky-lines on uncertain data: model and bounding-pruning-refining methods. J Intell Inf Syst. 2012;38(1): 1–39.
2. Börzsönyi S, Kossmann D, Stocker K. The sky-line operator. In: Proceedings of the 17th International Conference on Data Engineering; 2001. p. 421–30.
3. Dalvi NN, Suciu D. Efficient query evaluation on probabilistic databases. In: Proceedings of the 30th International Conference on Very Large Data Bases; 2004. p. 864–75.
4. Agrawal P, Benjelloun O, Sarma AD, Hayworth C, Nabar SU, Sugihara T, et al. Trio: a system for data, uncertainty, and lineage. In: Proceedings of the 32nd International Conference on Very Large Data Bases; 2006. p. 1151–54.
5. Lin X, Zhang Y, Zhang W, Cheema MA. Stochastic skyline operator. In: Proceedings of the 27th International Conference on Data Engineering; 2011. p. 721–32.
6. Böhm C, Fiedler F, Oswald A, Plant C, Wacker-sreuther B. Probabilistic skyline queries. In: Proceedings of the 18th ACM International Conference on Information and Knowledge Management; 2009. p. 651–60.
7. Atallah MJ, Qi Y, Yuan H. Asymptotically efficient algorithms for skyline probabilities of uncertain data. ACM Trans Database Syst. 2011;36(2):12.
8. Lian X, Chen L. Monochromatic and bichromatic reverse skyline search over uncertain databases. In: Proceedings of the ACM SIGMOD International Conference on Management of Data; 2008. p. 213–26.
9. Zhang W, Lin X, Zhang Y, Wang W, Yu JX. Prob-abilistic skyline operator over sliding windows. In: Proceedings of the 25th International Conference on Data Engineering; 2009. p. 1060–71.
10. Bartolini I, Ciaccia P, Patella M. The skyline of a probabilistic relation. IEEE Trans Knowl Data Eng. 2013;25(7):1656–69.
11. Zhang Q, Ye P, Lin X, Zhang Y. Skyline probability over uncertain preferences. In: Proceedings of the 16th International Conference on Extending Database Technology; 2013. p. 395–405.

Probabilistic Spatial Queries

Reynold Cheng[1] and Jinchuan Chen[2]
[1]Computer Science, The University of Hong Kong, Hong Kong, China
[2]Key Laboratory of Data Engineering and Knowledge Engineering, Ministry of Education, Renmin University of China, Beijing

Synonyms

Imprecise spatial queries

Definition

An uncertain item is defined as a range-limited probability density function (pdf) in a multi-dimensional space, which can model the uncertainty of location, sensor and biological data. Given a set of uncertain items, a probabilistic spatial query returns results augmented with probabilistic guarantees for the validity of answers. The impreciseness of query answers is an inherent property of these applications due to data uncertainty, unlike the techniques for approximate processing that trade accuracy for performance. New query definitions, processing and indexing techniques are required to handle these queries.

Historical Background

Data uncertainty is an inherent property in a number of important and emerging applications. Consider, for example, a habitat monitoring system used in scientific applications, where data such as temperature, humidity, and wind speed are acquired from a sensor network. Due to physical imperfection of the sensor hardware, the data obtained are often inaccurate [9]. Moreover, a sensor cannot report its value at every point in time, and so the system can only obtain data samples at discrete time instants. As another example, in the Global-Positioning System (GPS), the location collected from the GPS-enabled devices (e.g., PDAs) also has measurement and sampling error [14, 16]. The location data transmitted to the system may further encounter some network delay. In biometric databases, the attribute values of the feature vectors stored are not exact [1]. Hence, the data collected in these applications are often imprecise, inaccurate, and stale.

Services or queries that base their decisions on these data can produce erroneous results. There is thus a need to manage these data errors more carefully. In particular, the idea of *probabilistic spatial queries* (PSQ in short), which is a variant of spatial queries that handle data uncertainty, has been recently proposed. The main idea of a PSQ is to consider the models of the data uncertainty (instead of just the data value reported), and augment probabilistic guarantees to the query results. For example, a traditional query asking who is the nearest neighbor of a given point q can tell the user that John is the answer, while a PSQ informs the user that John has a probability of 0.8 of being the closest to q. The probabilities reflect the degree of correctness of query results, thereby facilitating the system to produce a more confident decision.

In this entry, the recent research efforts on PSQ will be summarized. Specifically, the details of how a PSQ can be classified will be discussed. Then, the issues of evaluating and indexing different types of PSQ in a large database will be addressed.

Foundations

Spatial Uncertainty Models To understand a PSQ, it is important to first discuss a commonly-used model of data uncertainty. This model assumes that the actual data value is located within a closed region, called the *uncertainty region*.

In this region, a non-zero probability density function (*pdf*) of the value is defined, where the integration of pdf inside the region is equal to one. The cumulative density function (*cdf*) of the item is also provided. In an LBS, a normalized Gaussian pdf is used to model the measurement error of a location stored in a database [14, 16] (Fig. 1). The uncertainty region is a circular area, with a radius called the "distance threshold"; the newest location is reported to the system when it deviates from the old one by more than this threshold (Fig. 1). Gaussian distributions are also used to model values of a feature vector in biometric databases [1]. Figure 1 shows the histogram of temperature values in a geographical area observed in a week. The pdf, represented as a histogram, is an arbitrary distribution between 10 and 20 °C.

A logical formulation of queries for this kind of uncertainty model has been recently studied in [12, 15]. Other variants have also been proposed. In [11], piecewise linear functions are used to approximate the cdf of an uncertain item. Sometimes, point samples are derived from an item's pdf [10, 13]. In the *existential uncertainty model*, every object is represented by the value in the space, as well as the probability that this object exists [8]. With these modified models, it is possible to develop fast processing techniques for PSQs.

Query Classification Given the spatial uncertainty model, the semantics of PSQs can be defined. Cheng et al. proposed a classification scheme for different types of PSQ [4]. In that scheme, a PSQ is classified according to the forms of answers. An *entity-based query*

Probabilistic Spatial Queries, Fig. 1 Location and sensor uncertainty

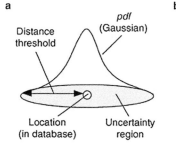

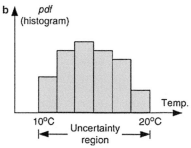

is one that returns a set of objects (e.g., list of objects that satisfy a range query or join conditions), whereas a *value-based query* returns a single numeric value (e.g., value of a particular sensor). Another criterion is based on whether an *aggregate* operator is used to produce results. An aggregate query is one where there is interplay between objects that determines the results (e.g., a nearest-neighbor query). Based on these two criteria, four different types of probabilistic queries are defined. Each query type has its own methods for computing answer probabilities. In [4], the notion of *quality* has also been defined for each query type, which provides a metric for measuring the ambiguity of an answer to the PSQ.

In the rest of this section, two important PSQs, namely the probabilistic range queries and the probabilistic nearest-neighbor queries, will be studied.

Probabilistic Range Queries A well-studied PSQ is the *probabilistic range query* (PRQ). Figure 2 illustrates this query, which shows the shape of the uncertainty regions, as well as the user-specified range, R. The task of the range query is to return the each object that can be inside R, as well as its probability. The PRQ can be used in location-based services, where queries like: "return the suspect vehicles in a crime scene" can be asked. It is also used in sensor network monitoring, where sensor IDs whose physical values (e.g., temperature, humidity) are returned to the user. Figure 2 shows the probability values of the items (A, B, and C) that are located inside R. A PRQ is an *entity-based* query, since it returns a list of objects. It is also

a *non-aggregate* query, because the probability of each object is independent of the existence of other objects [4].

To compute an item's probability for satisfying the PRQ, one can first find out the overlapping area of each item's region within R (shaded in Fig. 2), and perform an integration of the item's pdf inside the overlapping area. Unfortunately, this solution may not be very efficient, since expensive numerical integration may need to be performed if the item's pdf is arbitrary [7]. Even if an R-tree is used to prune items that do not overlap R, the probability of each item that are non-zero still needs to computed. A more efficient solution was developed in [7], where the authors proposed a user-defined constraint, called the *probability threshold P*, with $P \in (0,1]$. An item is only returned if its probability of satisfying the PRQ is not less than P. In Fig. 2, if $P = 0.6$, then only A and C will be returned to the user. Under this new requirement, it is possible to incorporate the uncertainty information of items into a spatial index (such as R-tree). The main idea is to precompute the *p-bounds* of an item. A p-bound of an uncertain item is essentially a function of p, where $p \in [0,0.5]$. In a 2D space, it is composed of four line segments, as illustrated by the hatched region in Fig. 2. The requirement of right p-bound (illustrated by the thick solid line) is that the probability of the location of the item on the right of the line has to be exactly equal to p (the shaded area). Similarly, the probability of the item on the left of the left p-bound is exactly equal to p. The remaining line segments (top and bottom p-bounds) are defined analogously. Once these p-bounds are known, it is possible to know immediately whether an item satisfies

Probabilistic Spatial Queries, Fig. 2
Probabilistic range queries over uncertain items, showing (**a**) the probability of each item, and (**b**) the p-bound of an uncertain item

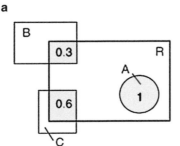
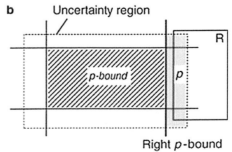

the PRQ. Figure 2 shows that a range query R overlaps the item's uncertainty region, but does not cut the right p-bound, where p is less than P. Since the integration of the item's pdf inside the overlapping area of R and the uncertainty region cannot be larger than P, the item is pruned without doing the actual probability computation.

By precomputing a finite number of p-bounds, it is possible to store them in a modified version of the R-tree. Called *Probability Threshold Index* (PTI), this index can facilitate the pruning of uncertain items in the index level [7]. Compared with the R-tree which uses the MBR of the object for pruning, the use of p-bounds in the PTI provides more pruning power. In Fig. 2, for example, although the range R overlaps with the MBR of the object (dashed rectangle), it does not cut the p-bounds, and so it can be pruned by the PTI but not by the R-tree. [7] also examined special cases of pdf (uniform and Gaussian distributions) and proposed an indexing scheme where p-bounds can be computed on-the-fly without being stored in the index. Since storing p-bounds for high-dimensional uncertain items can be expensive, Tao et al. [17, 18] proposed a variant of PTI called *U-tree*, which only stores approximate information of p-bounds in the index. With these improvements, it is possible to index uncertain items in the high-dimensional space. The p-bound techniques were also used in [6] to facilitate the processing of join queries over uncertain spatial data.

Another PRQ evaluation technique was recently proposed by Ljosa et al. [7], who used piecewise linear functions to approximate the cdf of an uncertain item in order to avoid expensive integration. They also described an index that stored these piecewise linear functions, so that a PRQ can be evaluated more efficiently. More recently, the problem of evaluating *imprecise location dependent range queries* is studied [2, 18]. This is a variant of PRQ, where the range query is defined with reference to the (imprecise) position of the query issuer. For instance, if the query issuer looks for his friends within 2 miles of his current position, and his position is uncertain, then the actual query range (a circle with a 2-mile radius) cannot be known precisely. The

authors of [2] proposed several approaches to efficiently evaluate these queries, by (i) using the Minkowski Sum (a computational geometry technique), (ii) switching the role of query issuer and data being queried, and (iii) using p-bounds.

Nearest-Neighbor Queries Another important PSQ for uncertain items is the probabilistic nearest-neighbor queries (PNNQ in short). This query returns the non-zero probability of each object for being the nearest neighbor of a given point q [4]. A PNNQ can be used in a sensor network, where sensors collect the temperature values in a natural habitat. For data analysis and clustering purposes, a PNNQ can find out the district(s) whose temperature values is (are) the closest to a given centroid. Another example is to find the IDs of sensor(s) that yield the minimum or maximum wind-speed from a given set of sensors [4, 9]. A minimum (maximum) query is essentially a special case of PNNQ, since it can be characterized as a PNNQ by setting q to a value of $-\infty$ (∞).

Evaluating a PNNQ is not trivial. In particular, since the exact value of a data item is not known, one needs to consider the item's possible values in its uncertainty region. Moreover, since the PNNQ is an entity-based aggregate query [4], an item's probability depends not just on its own value, but also on the relative values of other objects. If the uncertainty regions of the objects overlap, then their pdfs must be considered in order to derive their corresponding probabilities. This is unlike the evaluation of PRQ, where each item's probability can be computed independent of others. To evaluate PNNQ, one method is to derive the pdf and cdf of each item's distance from q. The probability of an item for satisfying the PNNQ is then computed by integrating over a function of distance pdfs and cdfs [4, 5, 9]. In [5], an R-tree-based solution for PNNQ was presented. The main idea is to prune items with zero probabilities, using the fact that these items' uncertainty regions must not overlap with that of an item whose maximum distance from q is the minimum in the database. The *probabilistic verifiers*, recently proposed in [3], are algorithms for efficiently computing the lower and upper bounds of each object's probability for satisfying a PNNQ. These algorithms, when used together with the probability threshold defined by the user, avoid the exact probability values to be calculated. In this way, a PNNQ can be evaluated more efficiently.

There are two other solutions for PNNQ that base on a different representation of uncertain items. Kriegel et al. [10] used the Monte-Carlo method, where the pdf of each object was sampled as a set of points. The probability was evaluated by considering the portion of points that could be the nearest neighbor. In [11], Ljosa et al. used piecewise linear representation of the cdf for an uncertain item to propose efficient evaluation and indexing techniques.

Another important entity-based aggregate query over uncertain items, namely the probabilistic skyline queries, has been studied in [13]. A skyline query returns a set of items that are not dominated by other items in all dimensions. In that paper, the issues of defining and computing the probability that an uncertain item was in the skyline were addressed. Two bounding-pruning-refining based algorithms were developed: the bottom-up algorithm used selected instances of uncertain items to prune other instances of uncertain items, while the top-down algorithm recursively partitions the instances of uncertain items into subsets. The authors showed that both techniques enable probabilistic skyline queries to be efficiently computed.

Key Applications

A PSQ can be used in applications that require the processing of uncertain spatial data. These applications include location-based services, road traffic monitoring, wireless sensor network applications, and biometric feature matching, where the data collected from the physical environments (e.g., location, temperature, humidity, images) cannot be obtained with a full accuracy. Recent works that propose to inject uncertainty to a user's location for location privacy protection also requires the use of a PSQ [2].

Future Directions

A lot of work remains to be done in the area of uncertain spatial data processing. An important future work will be the definition and evaluation of important spatial queries, such as reverse nearest-neighbor queries. It will also be interesting to study the development of data mining algorithms for spatial uncertainty. Another direction is to study spatio-temporal queries over historical spatial data (e.g., trajectories of moving objects). Other works include revisiting query cost estimation, query plan evaluation, and user interface design that allows users to visualize uncertain data. A long term goal is to consolidate these research ideas and develop a comprehensive spatio-temporal database system with uncertainty management facilities.

URL to Code

The following URL contains source codes of the ORION system, which is a database system that provides querying facilities for uncertain spatial data: http://orion.cs.purdue.edu

Cross-References

▶ Nearest Neighbor Query
▶ R-Tree (and Family)
▶ Spatial Anonymity
▶ Spatial Indexing Techniques

Recommended Reading

1. Böhm C, Pryakhin A, Schubert M. The Gauss-Tree: efficient object identification in databases of probabilistic feature vectors. In: Proceedings of the 22nd International Conference on Data Engineering; 2006.
2. Chen J, Cheng R. Efficient evaluation of imprecise location-dependent queries. In: Proceedings of the 23rd International Conference on Data Engineering; 2007.
3. Cheng R, Chen J, Mokbel M, Chow C. Probabilistic verifiers: evaluating constrained nearest-neighbor queries over uncertain data. In: Proceedings of the 24th International Conference on Data Engineering; 2008.

P

4. Cheng R, Kalashnikov D, Prabhakar S. Evaluating probabilistic queries over imprecise data. In: Proceedings of the ACM SIGMOD International Conference on Management of Data; 2003. p. 551–62.

5. Cheng R, Kalashnikov DV, Prabhakar S. Querying imprecise data in moving object environments. IEEE Trans Knowl Data Eng. 2004;16(9):1112–27.

6. Cheng R, Singh S, Prabhakar S, Shah R, Vitter J, Xia Y. Efficient join processing over uncertain data. In: Proceedings of the 15th ACM International Conference on Information and Knowledge Management; 2006.

7. Cheng R, Xia Y, Prabhakar S, Shah R, Vitter JS. Efficient indexing methods for probabilistic threshold queries over uncertain data. In: Proceedings of the 30th International Conference on Very Large Data Bases; 2004. p. 876–87.

8. Dai X, Yiu ML, Mamoulis N, Tao Y, Vaitis M. Probabilistic spatial queries on existentially uncertain data. In: Proceedings of the 9th International Symposium on Advances in Spatial and Temporal Databases; 2005. p. 400–17.

9. Deshpande A, Guestrin C, Madden S, Hellerstein J, Hong W. Model-driven data acquisition in sensor networks. In: Proceedings of the 30th International Conference on Very Large Data Bases; 2004.

10. Kriegel H, Kunath P, Renz M. Probabilistic nearest-neighbor query on uncertain objects. In: Proceedings of the 12th International Conference on Database Systems for Advanced Applications; 2007. p. 337–48.

11. Ljosa V, Singh A. APLA: indexing arbitrary probability distributions. In: Proceedings of the 23rd International Conference on Data Engineering; 2007. p. 946–55.

12. Parker A, Subrahmanian V, Grant J. A logical formulation of probabilistic spatial databases. IEEE Trans Knowl Data Eng. 2007;19(11):1541–56.

13. Pei J, Jiang B, Lin X, Yuan Y. Probabilistic skylines on uncertain data. In: Proceedings of the 33rd International Conference on Very Large Data Bases; 2007.

14. Pfoser D, Jensen C. Capturing the uncertainty of moving-objects representations. In: Proceedings of the 11th International Conference on Scientific and Statistical Database Management; 1999.

15. Singh S, Mayfield C, Shah R, Prabhakar S, Hambrusch S, Neville J, Cheng R. Database support for probabilistic attributes and tuples. In: Proceedings of the 24th International Conference on Data Engineering; 2008.

16. Sistla PA, Wolfson O, Chamberlain S, Dao S. Querying the uncertain position of moving objects. In: Etzion O, Jajodia S, Sripada S, editors. Temporal databases: research and practice. Berlin/New York: Springer; 1998.

17. Tao Y, Cheng R, Xiao X, Ngai WK, Kao B, Prabhakar S. Indexing multi-dimensional uncertain data with arbitrary probability density functions. In: Proceedings of the 31st International Conference on Very Large Data Bases; 2005. p. 922–33.

18. Tao Y, Xiao X, Cheng R. Range search on multidimensional uncertain data. ACM Trans Database Syst. 2007;32(3):15.

Probabilistic Temporal Databases

V. S. Subrahmanian
University of Maryland, College Park, MD, USA

Synonyms

Temporally indeterminate databases; Temporally uncertain databases

Definition

There are many applications where the fact that a given event occurred is known, but where there is uncertainty about exactly when that event occurred. Such events are called temporally indeterminate events. Probabilistic temporal databases attempt to store information about events that are both temporally determinate and temporally indeterminate. For the latter, they specify a set of time points (often an interval) – it is known that the event occurred at some time point in this set. The probability that the event occurred at a specific time point is given by a probability distribution or by one of a set of probability distributions.

Historical Background

There is no shortage of events that are known to have certainly occurred, but where the exact dates are not known with certainty. For instance, the exact date of the extinction of dinosaurs is unknown – nor does the historical record show the precise date when Cyrus the Great of Persia was born. The latter, estimated by scholars to be between 590 and 576 BC, is an excellent example of a temporally indeterminate event.

Vehicle ID	Vehicle type	Location	Time	PDF
V1	T72	a	$11 \leq t \leq 20$	U
V1	T72	b	$18 \leq t \leq 25$	$\gamma_{0.5}$
V1	T72	c	$24 \leq t \leq 27$	$\gamma_{0.2}$

Vehicle ID	Vehicle type	Location	Time	PDF
V1	T72	b	$20 < t < 23$	$\gamma_{0.5}$

Vehicle ID	Vehicle type	Location	Time	Time 2	PDF
V1	T72	a	$11 \leq t \leq 20$	$11 \leq t \leq 20$	U
V1	T72	b	$18 \leq t \leq 25$	$18 \leq t \leq 25$	$G_{0.5}$
V1	T72	c	$24 \leq t \leq 27$	$24 \leq t \leq 27$	$G_{0.2}$

Vehicle ID	Vehicle type	Location	Time	Time 2	PDF
V1	T72	B	$20 < t < 23$	$18 \leq t \leq 25$	$G_{0.5}$

Techniques to study such temporally indeterminate events in databases started with information about *partial null values* [9, 13]. Meanwhile, in a largely separate community, researchers focused on the problem of incorporating probabilistic information within a relational database. One of the earliest efforts was due to Cavallo and Pittarelli [4] who proposed an extension of the relational data model to include probabilities – they proposed a partial algebra consisting of projection and join operators. Much early work in this field, such as that of [2, 7] made important advances under some restrictions. Lakshmanan et al.'s ProbView system [11] proposed two representations of probabilistic data: a *probabilistic tuple* has the form $((t_1, V_1), \ldots, (t_n, V_n))$ where each t_i is a set of possible values and V_i is a probability distribution over the set. They then showed that each probabilistic tuple could be "flattened" into an "annotated" representation. Annotated tuples look like ordinary tuples except that a probability is attached to the tuple as a whole (and a special "path" field was introduced to handle tuples with identical data). ProbView introduced the important concept of conjunction and disjunction strategies that allowed users to specify – in their

query – their knowledge about the dependencies between events when posing the query. Thus, the barrier of the independence assumption in previous works was overcome. They proposed a query algebra for annotated tuples, developed query rewrite rules, and developed view maintenance algorithms.

Much work has subsequently been done on temporal probabilistic data. Despite a long history of reasoning about time and uncertainty in AI [10], a major advance in temporal probabilistic databases occurred when Dyreson and Snodgrass [8] proposed the notion of an indeterminate instant. This is just like one of the (t_i, V_i) pairs in the annotated representation of [11] with the exception that the t_i component represents a set of time points. They then extended the SQL query language in several ways. First, they added constructs to indicate that a temporal attribute in indeterminate. Second, they added the concept of "correlation credibility" which allows a query to modify indeterminate temporal data (e.g., by choosing a max temporal value or an expected temporal value). Third, they introduced the concept of "ordering plausibility" which specifies an ordering about the plausibility levels of different conditions in the WHERE clause of an SQL query. The next major breakthrough in temporal probabilistic databases came when Dekhtyar et al. [6] proposed a temporal probabilistic relational database model that added "tp-cases" to ordinary relational tuples. A tp-case contained two constraints one of which was used to denote valid time points as solutions of the constraints, and the other was used in conjunction with a distribution function to infer a probability distribution on the solutions of the first constraint. They developed an extension of the relational algebra that directly manipulated tp-tuples and used the conjunction and disjunction strategies of ProbView [11] to avoid making independence assumptions. TP-databases were one of the first temporal probabilistic DBMSs to report real world applications for the US Navy [12]. It was later used to build similar applications for the US Army as well. Later, Biazzo et al. [3] extended this work to the case of object bases containing temporal indeterminacy. The Trio system [1] extends temporal

P

indeterminacy models to include lineage information as well. More recently, SPOT databases [14] allow reasoning in the presence of space, time and uncertainty.

Foundations

Consider a relation $R(A_1, \ldots, A_n)$ which describes events whose temporal validity is not precisely known. In order to identify when the tuples in such a relation are valid, *constraints* can be to describe the period of validity of the tuple. The table below (ignoring the shaded column for now) is a temporally indeterminate database about vehicle locations.

The first row in this table says that the event "Vehicle V1 was at location a" occurred at some time during the closed interval [11, 20]. This is a form of temporal indeterminacy with no probabilities. Suppose there is a probability distribution over the last column. A common temptation is to add an additional column specifying the "name" of the distribution (e.g., "u" might be the uniform distribution, g_r may be a geometric distribution with a parameter r, and so forth). Such a table might now look like this:

Now consider the table above with the shaded column included. The first row in the table now says that there is a 10 % probability that vehicle V1 will be at location a at time 11 (and the same for times 12, 13, and so on till time 20) and that the probability is uniformly distributed (probability distribution function or pdf "u"). In contrast, the second row says something different because the pdf γ treats time intervals differently. It says, implicitly, that the probability that vehicle V1 will be at location b at time 18 is 0.15, at time 19 is 0.25, at time 20 is 0.125, and so on.

How should queries over this representation of the database be answered? To see this, consider a temporal query which says "Select all tuples where $20 < Time < 23$." The only tuple that has any chance of satisfying this constraint is the second one. However, there is no *guarantee* that the second tuple actually is valid during this interval. Fortunately, there is a 9.375 % chance that this is

the case – so the system could return the probabilistically valid response saying that the second tuple is valid with probability 9.375 %. Unfortunately, there is no way of returning this answer to the user *unless the implicit assumption in all relational databases that the output schema for selection queries should match the input schema is sacrificed.* It is clearly inappropriate to just return the table:

The *Time* field here is computed merely by solving the constraint in the second tuple of the input relation in conjunction with the constraint in the query. Unfortunately, the distribution in the *PDF* field is incorrect, and would need to be recomputed. This is further complicated by the fact that the shape of the original geometric distribution does not look the same when restricted to a portion (of interest) of the original distribution.

There have been two attempts to solve the problem of dealing with what happens when a distribution is manipulated. Dyreson and Snodgrass [8] develop a "rod and point" method to store distributions and infer new distributions when selection operations of the kind above are performed. They approximate a probability mass by splitting it into chunks called "rods." However, each chunk can have a different length. An approximate representation of the original distribution is obtained through this rod mechanism.

Dekhtyar et al. [6] solve this problem by using some extra space. They require that the "base relation" contains *two constraints*, both of which are identical initially. They would represent the original relation as follows:

In base relations, the *Time* and *Time2* fields have exactly the same constraints in them. When queries are executed, the *Time* field ends up denoting valid time and changes based on the query – however, the *Time2* field rarely changes. When the selection query mentioned above is execution, they would return the answer:

This is very subtle. The *Time* attribute here specifies the valid time (in this case, time points 21 and 22). The *Time2* attribute is a system-maintained attribute that need not be shown to the user which says apply the PDF mentioned in the *PDF* field to the solutions of the *Time2* constraint – but only show the probability values for the solutions of the *Time* constraint). Thus, *Time2* is used to derive the probabilities for each valid time point. In the above case, the valid time points are 21 and 22, and their probabilities are derived – using the distribution in the *PDF* column applied to the constraint in the *Time2* column – to get probabilities of 0.0625 and 0.03125, respectively. Dekhtyar et al. [6] goes on and specifies how to add additional "low" and "high" probability fields to such relations and provides normalization methods.

Cartesian products between two relations are more complex. When concatenating tuple s*t1* and *t2* from two different relations, it is important to consider the probability that both tuples will be valid at a given time point. This requires knowing the relationship between the events being denoted by these two tuples: are they independent? Are they correlated somehow? Is there no information about the relationship? Thus, a *conjunction strategy* must be specified in the query when a Cartesian product (and hence a join operation) is being performed. Dekhtyar et al. [6] show how Cartesian products and joins can be computed under any assumption specified in a user query.

Another recent effort is the one on Trio [1] which attempts to deal with time, uncertainty, and lineage. They address the fact that in many applications involving uncertainty (such as crime-fighting applications), any "final" answer needs to be explainable. As a consequence, they introduce "lineage" parameters when answering a query – informally speaking, the lineage parameter associated with a tuple in an answer (similar to the "path" parameter in [11]) associates a "justification" for each tuple. This justification references the set of base tuples that caused the derived tuple to be placed in an answer.

Key Applications

Temporal probabilistic databases have already been used in defense applications. For instance, [12] describes work in which temporal probabilistic databases are used to store the results of predictions about where enemy submarines will be in the future, when they will be there, and with what probability. In fact, this raises a large set of possible applications based on reasoning about moving objects. For defense applications, cell phone applications, logistics applications, and many other application domains, there is interest in knowing where a moving object will be in the future, when it is expected to be there, and with what probability. Cell phone companies can use such data to understand and better handle load on cell towers.

Moreover, as the world is becoming increasing "geo-location aware" through the use of devices like RFID tags and GPS locators, it is clear that reasoning about where vehicles will be in the future and with what probability will be important in a wide range of applications such as traffic light settings to ease road congestion, recommending detours on highway signs, and more effectively directing 911 traffic in congested situations. These would not be possible without a good estimate of when and where and with what probabilities vehicles will be in the future.

Logistics applications are another important class of applications where companies need to plan activities in the presence of uncertainty about when various supply items will arrive. Corporations today use complex prediction models to learn about suppliers' performance.

Financial applications are another major source of temporal uncertainty. Banks need to have a good idea of their incoming funds. For instance, a credit card provider deals with constant uncertainty about when people will pay their credit card bills, how much of the bills they will pay, and how much they will carry forward as debt. Such applications embody a mix of data uncertainty and temporal uncertainty.

Future Directions

Three major areas of expansion include:

1. *Temporal probabilistic aggregates.* To date, there are almost no techniques to manage aggregates efficiently in temporal probabilis-

tic databases. Most methods would compute aggregates by first answering a non-aggregate query and then deriving aggregates from there: however, techniques to scalably answer aggregate queries are required.

2. *Probabilistic spatiotemporal reasoning*. Moving objects clearly have a spatial component – hence, reasoning about them involves a neat fusion of temporal reasoning, spatial reasoning, and probabilistic reasoning. Some work on indexing in such domains has recently been proposed. However, much future work is needed, especially in understanding the correlations that exist between the presence of a vehicle at time *t* and its presence at another location at time *t + 1*.

3. *Query optimization*. Recent work [5] has made a good start on query optimization in probabilistic databases – however, query optimization in databases involving time and uncertainty has a ways to go. Such methods are critical for scaling applications.

Cross-References

▶ Qualitative Temporal Reasoning
▶ Temporal Constraints

Recommended Reading

1. Agrawal P, Benjelloun O, Sarma AD, Hayworth C, Nabar SU, Sugihara T, Widom J. Trio: a system for data, uncertainty, and lineage. In: Proceedings of the 32nd International Conference on Very Large Data Bases; 2006. p. 1151–4.
2. Barbará D, Garcia-Molina H, Porter D. The management of probabilistic data. IEEE Trans Knowl Data Eng. 1992;4(5):487–502.
3. Biazzo V, Giugno R, Lukasiewicz T, Subrahmanian VS. Temporal probabilistic object bases. IEEE Trans Knowl Data Eng. 2003;15(4):921–39.
4. Cavallo R, Pittarelli M. The theory of probabilistic databases. In: Proceedings of the 13th International Conference on Very Large Data Bases; 1987. p. 71–81.
5. Dalvi N, Suciu D. Answering queries from statistics and probabilistic views. In: Proceedings of the 31st International Conference on Very Large Data Bases; 2005. p. 805–16.
6. Dekhtyar A, Ross R, Subrahmanian VS. Probabilistic temporal databases, I: algebra. ACM Trans Database Syst. 2001;26(1):41–95.
7. Dey D, Sarkar S. A probabilistic relational model and algebra. ACM Trans Database Syst. 1996;21(3): 339–69.
8. Dyreson CE, Snodgrass RT. Supporting valid-time indeterminacy. ACM Trans Database Syst. 1998;23(1):1–57.
9. Grant J. Partial values in a tabular database model. Inf Process Lett. 1979;9(2):97–9.
10. Kraus S, Subrahmanian VS. Multiagent reasoning with probability, time and beliefs. Int J Intell Syst. 1994;10(5):459–99.
11. Lakshmanan LVS, Leone N, Ross RB, Subrahmanian VS. ProbView: a flexible probabilistic database system. ACM Trans Database Syst. 1997;22(3):419–69.
12. Mittu R, Ross R. Building upon the coalitions agent experiment (COAX) – integration of multimedia information in GCCS-M using IMPACT. In: Proceedings of the 9th International Workshop on Multimedia Information Systems; 2003. p. 35–44.
13. Ola A. Relational databases with exclusive disjunctions. In: Proceedings of the 8th International Conference on Data Engineering; 1992. p. 328–36.
14. Parker A, Subrahmanian VS, Grant J. A logical formulation of probabilistic spatial databases. IEEE Trans Knowl Data Eng. 2007;19(11):1541–56.

Probability Ranking Principle

Ben He
University of Glasgow, Glasgow, UK

Synonyms

PRP

Definition

The probability ranking principle asserts that relevance has a probabilistic interpretation. According to this principle documents are ranked by a probability $p(Rel|\, d, q)$, where *Rel* denotes the event of a document *d* being relevant to a query *q*. Robertson called this principle the *probability ranking principle* [1].

Key Points

By assuming independence between query terms, Robertson and Sparck-Jones proposed for the probability $p(Rel|d, q)$ the following model (the RSJ model [2]):

$$\log\left(p\left(Rel|d, q\right)\right) \propto \sum_{t \in q} \log \frac{p\left(t|Rel\right) \cdot p\left(\bar{t}|\overline{Rel}\right)}{p\left(t|\overline{Rel}\right) \cdot p\left(\bar{t}|Rel\right)} \tag{1}$$

where $\overline{Rel}$ indicates the event of non-relevance; t and $\bar{t}$ indicate the events that the term t occurs in document d or does not, respectively. For each query term t, the probability $p(Rel|d, t)$ is given by the sum of two log-odds, $\log \dfrac{p\left(t|Rel\right)}{p\left(t|\overline{Rel}\right)}$ and

$\log \dfrac{p\left(\bar{t}|\overline{Rel}\right)}{p\left(\bar{t}|Rel\right)}$.

If N is the number of documents in the whole collection, R is the number of relevant documents, r is the number of relevant documents containing t, N_t is the document frequency, i.e., the number of documents containing t, [3] instantiated the RSJ model as follows:

$$w^{(1)} = \log \frac{(r + 0.5)\,(N - N_t - R + r + 0.5)}{(R - r + 0.5)\,(N_t - r + 0.5)} \tag{2}$$

where $w^{(1)}$ is the raw weight of a term t in a document d. The number 0.5 is used to avoid assigning negative weights. The formula is called the "point-5" formula.

If relevance information is not available, i.e., $R = r = 0$, the point-5 formula can be written as:

$$w^{(1)} = \log \frac{N - N_t + 0.5}{N_t + 0.5} \tag{3}$$

As one of the most well-established IR systems, Okapi uses a weighting model that is based on the RSJ model introduced above, and takes also term frequency (tf) and query term frequency (qtf) into consideration.

Cross-References

▶ Information Retrieval
▶ Information Retrieval Models
▶ Term Weighting

Recommended Reading

1. Robertson SE. The probability ranking principle in IR. J Doc. 1977;33(4):294–304.
2. Robertson SE, Sparck-Jones K. Relevance weighting of search terms. J Am Soc Inf Sci. 1977;27(3):129–46.
3. Robertson SE, Walker S. On relevance weights with little relevance information. In: Proceedings of the 20th Annual International ACM SIGIR Conference on Research and Development in Information Retrieval; 1997. p. 16–24.

Probability Smoothing

Djoerd Hiemstra
University of Twente, Enschede,
The Netherlands

Definition

P

Probability smoothing is a language modeling technique that assigns some nonzero probability to events that were unseen in the training data. This has the effect that the probability mass is divided over more events; hence, the probability distribution becomes more *smooth*.

Key Points

Smoothing overcomes the so-called sparse data problem, that is, many events that are plausible in reality are not found in the data used to estimate probabilities. When using maximum likelihood estimates, unseen events are assigned a zero probability. In case of information retrieval, most events are unseen in the data, even if simple unigram language models are used documents that are relatively short (say on average several hundreds of words), whereas the vocabulary is typically big (maybe millions of words), so the vast majority of words does not occur in the document. A small document about "information retrieval" might not mention the word "search," but that does not mean it is not relevant to the query "text search." The sparse data problem is the reason that it is hard for information retrieval systems to obtain high recall values without degrading values for precision, and smoothing is a means to increase recall (possibly degrading precision in the process). Many approaches to smoothing are proposed in the field of automatic speech recognition [1]. A smoothing method may be as simple as the so-called Laplace smoothing, which adds an extra count to every possible word. The following equations show, respectively, (1) the unsmoothed or maximum likelihood estimate, (2) Laplace smoothing, (3) linear interpolation smoothing, and (4) Dirichlet smoothing [3]:

$$P_{ML}(T = t | D = d) = \text{tf}(t, d) \Big/ \sum_{t'} \text{tf}(t', d) \tag{1}$$

$$P_{LP}(T = t | D = d) = (\text{tf}(t, d) + 1)$$
$$\Big/ \sum_{t'} (\text{tf}(t', d) + 1) \tag{2}$$

$$P_{LI}(T = t | D = d) = \lambda P_{ML}(T = t | D = d)$$
$$+ (1 - \lambda) P_{ML}(T = t | C) \tag{3}$$

$$P_{Di}(T = t | D = d) = (\text{tf}(t, d) + \mu P(T = t | C))$$
$$\Big/ \left(\left(\sum_{t'} \text{tf}(t', d) \right) + \mu \right) \tag{4}$$

Here, tf(t, d) is the frequency of occurrence of the term t in the document d, and $P_{ML}(T|C)$ is the probability of a term occurring in the entire collection C. Both linear interpolation smoothing and Dirichlet smoothing assign a probability proportional to the term occurrence in the collection to unseen terms. Here, λ ($0 < \lambda < 1$) and μ ($\mu > 0$) are unknown parameters that should be

tuned to optimize retrieval effectiveness. Linear interpolation smoothing has the same effect on all documents, whereas Dirichlet smoothing has a relatively big effect on small documents but a relatively small effect on bigger documents. Many smoothed estimators used for language models in information retrieval (including Laplace and Dirichlet smoothing) are approximations to the *Bayesian predictive distribution* [2].

Cross-References

▶ Language Models
▶ N-Gram Models

Recommended Reading

1. Chen SF, Goodman J. An empirical study of smoothing techniques for language modeling. Technical report TR-10-98, Center for Research in Computing Technology, Harvard University, August 1998.
2. Zaragoza H, Hiemstra D, Tipping M, Robertson S. Bayesian extension to the language model for ad hoc information retrieval. In: Proceedings of the 26th Annual International ACM SIGIR Conference on Research and Development in Information Retrieval; 2003. p. 4–9.
3. Zhai C, Lafferty J. A study of smoothing methods for language models applied to information retrieval. ACM Trans Inf Syst. 2004;22(2):179–214.

Process Life Cycle

Nathaniel Palmer
Workflow Management Coalition, Hingham, MA, USA

Synonyms

Process state model; Thread lifecycle; Workflow lifecycle

Definition

The stages of life from the start to the end of a process instance within the context of workflow management.

Key Points

The Process Life Cycle represents the stages of a process instance as it evolves from instantiation to termination. This life cycle is most closely related to the life cycle of a thread, and is distinct from the life cycle approach to Business Process Management initiatives, involving an iterative or recursive evolution through the five stages of design, modeling, execution, monitoring, and optimization.

The latter notion of Business Process Management Life cycle is associated with the discipline of continuous process improvement, whereby processes are never deemed "complete" and thus no longer subject to change, but rather are continuously improved through multiple instances of execution, examination, and modification. In contrast, the Process Life Cycle as defined herein refers to the "life span" of a process instance and has definitive start and end points. Thus the Process Life Cycle of the individual process instance is more aptly described as linear as opposed to cyclical, although at various steps in the process it may cycle between running and suspended states.

The steps of the Process Life Cycle are "Instantiate" representing the creation of a new instance (making it live but not necessarily running); "activate" representing the activation of the process instance (now live and running); "passivate" which refers to temporarily suspending the instance (live but not running); "terminate" which represents the end of life of the process instances, through either abortion, cancellation, or completion after running through the full process as defined.

Cross-References

► Business Process Modeling
► Workflow Model

Process Mining

W. M. P. van der Aalst
Eindhoven University of Technology,
Eindhoven, The Netherlands

Synonyms

Workflow mining

Definition

Process mining techniques allow for the analysis of business processes based on event logs. For example, the audit trails of a workflow management system, the transaction logs of an enterprise resource planning system, and the electronic patient records in a hospital can be used to discover models describing processes, organizations, and products. Moreover, such event logs can also be used to compare event logs with some a priori model to see whether the observed reality conforms to some prescriptive or descriptive model.

The basic idea of *process mining* is to discover, monitor, and improve *real* processes (i.e., not assumed processes) by extracting knowledge from event logs [1]. Today many of the activities occurring in processes are either supported or monitored by information systems. Consider, for example, ERP, WFM, CRM, SCM, and PDM systems to support a wide variety of business processes while recording well-structured and detailed event logs. However, process mining is not limited to information systems and can also be used to monitor other operational processes or systems. For example, process mining has been applied to complex X-ray machines, high-end copiers, web services, careflows in hospitals, etc. All of these applications have in common that *there is a notion of a process* and that *the occurrences of activities are recorded in so-called event logs*. Assuming that the supporting systems log events, a wide range of *process mining techniques* comes into reach. The basic idea of process mining is to learn from observed executions of a process and can be used to (i) *discover* new models (e.g., constructing a Petri net that is able to reproduce the observed behavior), (ii) check the *conformance* of a model by checking whether the modeled behavior matches the observed behavior, and (iii) *extend* an existing model by projecting information extracted from the logs onto some initial model (e.g., show bottlenecks in a process model by analyzing the event log). All three types of analysis have in common that they assume the existence of some *event log*.

Key Points

The goal of process mining is to discover, monitor, and improve real processes by extracting knowledge from event logs. Clearly, process mining is relevant in a setting where much flexibility is allowed or needed, because the more ways in which people and organizations can deviate, the more variability and the more interesting it is to observe and analyze processes as they are executed. Three basic types of process mining can be identified (Fig. 1):

1. *Discovery*. There is no a priori model, i.e., based on an event log some model is constructed. For example, using the α-algorithm [2] a process model can be discovered based on low-level events.
2. *Conformance*. There is an a priori model. This model is used to check if reality conforms to the model. For example, there may be a process model indicating that purchase orders of more than one million Euro require two checks. Another example is the so-called four-

Process Mining, Fig. 1
Three types of process mining: (i) discovery, (ii) conformance, and (ii) extension

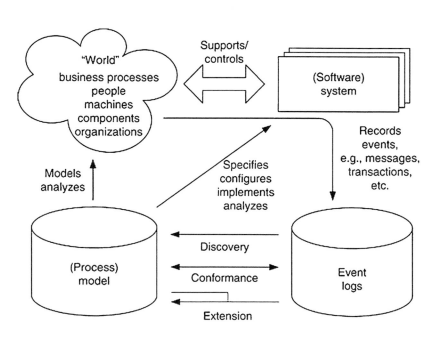

eye principle where two activities need to be executed by different people. Conformance checking may be used to detect deviations, to locate and explain these deviations, and to measure the severity of these deviations.

3. *Extension.* There is an a priori model. This model is extended with a new aspect or perspective, i.e., the goal is not to check conformance but to enrich the model. An example is the extension of a process model with performance data, i.e., some a priori process model dynamically annotated with performance data (e.g., bottlenecks are shown by coloring parts of the process model). Figure 1 shows that a log and a model are used to create a new model.

Traditionally, process mining has been focusing on *discovery*, i.e., deriving information about the original process model, the organizational context, and execution properties from enactment logs. An example of a technique addressing the control flow perspective is the α-algorithm, which constructs a Petri net model describing the behavior observed in the event log. However, process mining is not limited to process models (i.e., control flow), and recent process mining techniques are more and more focusing on other perspectives, e.g., the organizational perspective or the data perspective. For example, there are approaches to extract social networks from event logs and analyze them using social network analysis [1]. This allows organizations to monitor how people and groups are working together.

Conformance checking compares an a priori model with the observed behavior as recorded in the log. In [4] it is shown how a process model (e.g., a Petri net) can be evaluated in the context of a log using metrics such as "fitness" (Is the observed behavior possible according to the model?) and "appropriateness" (Is the model "typical" for the observed behavior?). However, it is also possible to check conformance based on organizational models, predefined business rules, temporal formulas, quality of service (QoS) definitions, etc.

There are different ways to *extend* a given process model with additional perspectives based on event logs, e.g., decision mining. Decision mining, also referred to as decision point analysis, aims at the detection of data dependencies that affect the routing of a case. Starting from a process model, one can analyze how data attributes influence the choices made in the process based on past process executions. Classical data mining techniques such as decision trees can be leveraged for this purpose. Similarly, the process model can be extended with timing information (e.g., bottleneck analysis).

In recent years new types of process mining emerged, for example, the analysis of concept drift: This is the situation where the process is changing while being analyzed. Another example is the prediction of remaining processing times for concrete process instances.

Process mining is strongly related to classical data mining approaches. However, the focus is not on data but on process-related information (e.g., the ordering of activities). Process mining is also related to monitoring and business intelligence [3].

Cross-References

▶ Association Rule Mining on Streams
▶ Data Mining
▶ Workflow Management

Recommended Reading

1. van der Aalst WMP. Process mining: discovery, conformance and enhancement of business processes. Berlin: Springer; 2011.
2. van der Aalst WMP, Weijters AJMM, Maruster L. Workflow mining: discovering process models from event logs. IEEE Trans Knowl Data Eng. 2004;16(9):1128–42.
3. Grigori D, Casati F, Castellanos M, Dayal U, Sayal M, Shan MC. Business process intelligence. Comput Ind. 2004;53(3):321–43.
4. Rozinat A, van der Aalst WMP. Conformance checking of processes based on monitoring real behavior. Inf Syst. 2007;33(1):64–95.

Process Optimization

Danilo Ardagna
Politechnico di Milano University, Milan, Italy

Synonyms

Business process optimization; QoS-based web
services composition

Definition

With the development of the service oriented
architecture (SOA), complex applications can be
composed as business processes invoking a variety of available Web services (WSs) with different characteristics. Advanced SOA systems [1,
12, 15] allow the development of applications
by specifying component WSs in a process only
through their required functional characteristics,
and to select WSs during process execution from
the ones included in a registry of available services. Service descriptions are stored and retrieved from enhanced UDDI registries, which
also provide information about quality of service
(QoS) on the provider side. Usually, a set of functionally equivalent services can be selected, i.e.,
services which implement the same functionality
but differ for the quality parameters.

Process optimization identifies the best set of
services available at run time, taking into consideration end-user preferences and constraints on
the QoS properties.

Historical Background

Process optimization has its roots in workflow
scheduling problems. Scheduling of workflows
[13] is the problem of finding a correct execution
sequence of workflow tasks such that some
temporal constraints or resource constraints (i.e.,
agents which can support tasks executions) are
met. In workflow management systems, agents
could be human beings or software applications,

in Process optimization the available resources
are component WSs. Process optimization
is related also to workflow/process planning
[3] where the problem of synthesizing a
complex behavior from an explicit goal and a
set of candidates which contribute to a partial
reaching of this goal is investigated. In Process
optimization, vice versa, the *process schema*,
i.e., the sequence of activities, is given and the
optimum mapping of activities to component
WSs candidate for their execution is identified.

Process optimization has been applied in
context-aware business processes and e-science
research fields. The literature has provided
three generations of solutions. First generation
solutions implemented *local* approaches [3,
14, 15] which select WSs one at the time
by associating the running abstract activity to
the best candidate service which supports its
execution. Local approaches can guarantee only
local QoS constraints, i.e., candidate WSs are
selected according to a desired characteristic,
e.g., the price of *a single WS* is lower than a
given threshold.

Second generation solutions proposed *global*
approaches [5, 8, 10, 12, 15]. The set of services
which satisfy the process constraints and user
preferences for the whole application are identified before executing the process. In this way,
QoS constraints can predicate at a global level,
i.e., constraints posing restrictions over the *whole
composed service execution* can be introduced. In
order to guarantee the fulfilment of global QoS
constraints, second generation optimization techniques consider the worst case execution scenario
for the composed service. For cyclic processes,
loops are unfolded, i.e., unrolled according to
their maximum number of iterations [5, 15]. This
approach could be very conservative and constitutes the main limitation of second generation
techniques. Furthermore, global approaches introduce an increased complexity with respect to
local solutions. The main issue for the fulfill-
ment of global constraints is WSs performance
variability. Indeed, the QoS of a WS may evolve
relatively frequently, either because of internal
changes or because of workload fluctuations [15].
If a business process has a long duration, the set

of services identified by the optimization may change their QoS properties during the process execution or some services can become unavailable or others may emerge. In order to guarantee global constraints, WS selection and execution are interleaved: optimization is performed when the business process is instantiated and its execution is started, and is iterated during the process execution performing *re-optimization* at run time.

To reduce optimization/re-optimization complexity, a number of solution have been proposed which guarantee global constraints only for the critical path [15] (i.e., the path which corresponds to the highest execution time), or reduce loops to a single task [5], satisfying global constraints only statistically, by applying the reduction formula proposed in [7].

Another drawback of second generation solutions is that, if the end-user introduces severe QoS constraints for the composed service execution, i.e., limited resources which set the problem close to un-feasibility conditions (e.g., limited budget or stringent execution time limit), no solutions could be identified and the composed service execution fails [5].

Third generation techniques [3] overcome the limits of the previous approaches and focus on the execution of processes under severe QoS constraints. Severe constraints are very relevant whenever processes have to be performed with stringently limited resources. Third generation solutions are based on loops peeling, which significantly improves the solutions based on loops unfolding. Furthermore, negotiation is exploited if a feasible solution cannot be identified, to bargain QoS parameters with service providers offering services, reducing process invocation failures.

Foundations

Process optimization allows the specification of complex applications as business processes composed by *abstract services* which act as *place holders* of WS components invoked at run time. The best set of services, selected by solving an optimization problem, is invoked at run time

by implementing a *dynamic/late binding* mechanism.

Process optimization is usually formalized as a multi-objective optimization problem since several quality criteria can be associated with WS execution. Past approaches [5, 12, 15] focussed on *execution time* (the expected delay, between the time instant when a request is sent and the time when the result is obtained), *availability* (the probability that the service is accessible), *price* (the fee that a service requester has to pay to the Service Provider for the service invocation), and *reputation* (a measure of the service trustworthiness). Furthermore, the optimization is performed statistically, i.e., by considering the probability of execution of the execution paths of the business process (i.e., any possible sequence of invocations of abstract services). For this reason, some annotations are added to the BPEL specification in order to identify: (i) the maximum [15] or the probability distribution [3] of the number of iterations of loops; (ii) the expected frequency of execution of conditional branches; (iii) global and local constraints on quality dimensions.

Figure 1 shows an example of composed process which implements a virtual travel agency, and the corresponding annotations which specify constraints. The BPEL specification includes invocation to abstract WSs which can be supported at run time by *concrete* WS components.

The probability distribution of the number of iterations of loops and the frequency of execution of conditional branches can be evaluated from past executions by inspecting system logs or can be specified by the composite service designer. If an upper bound for loops execution cannot be determined, then the optimization cannot guarantee that global constraints are satisfied [15]. Prior to perform process optimization, loops are unfolded [15] or peeled [3] (see Fig. 2) and are modeled as directed acyclic graphs (DAGs). Loops peeling is a form of loops unrolling where loop iterations are represented as a sequence of branches and each branch condition evaluates if the loop has to continue with the next ith iteration or it has to exit with probability p_i.

The objective function to be optimized is, usually [3, 14, 15], the aggregated value of QoS for

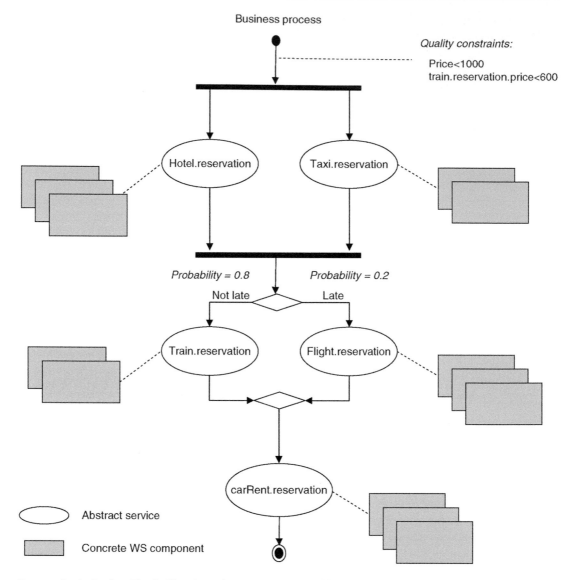

Process Optimization, Fig. 1 Virtual travel agency process specification

the end user which can be obtained by applying the simple additive weighting (SAW) technique. SAW is one of the most widely used techniques to obtain a *score* from a list of dimensions. Since the quality dimensions have different units of measure, the SAW method first normalizes the raw values for each quality dimension. Each quality dimension is also associated with a weight which expresses the user preferences among multiple quality parameters. The overall value of QoS is calculated as a weighted sum of the normalized values of quality dimension. The SAW method originates a linear objective function, other proposal introduces more general utility functions (i.e., functions which map each possible configuration of the business process to a scalar value) which can be non-linear [5].

First generation solutions considered only *local constraints*. In that case, the process optimization is very simple and the optimum solution can be identified by a greedy algorithm which selects the best candidate service suitable for

Process Optimization, Fig. 2 Loops unfolding and peeling

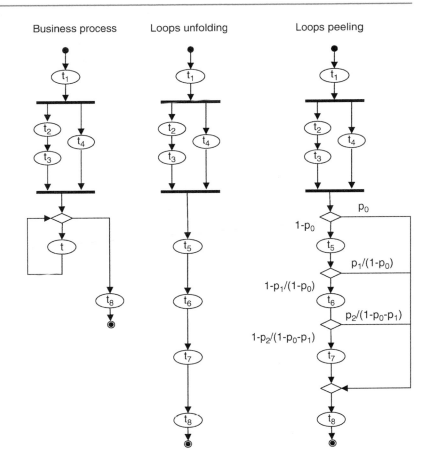

the execution. An example of first generation technique can be found in [11], where Web agents can migrate to invoke services locally in order to minimize also the network bandwidth.

Second generation solutions support *global constraints* and introuce NP-hard optimization problems. In [4] the complexity of some variants of the global process optimization problem is analyzed, while an overview of heuristic techniques, which hence identify only suboptimal solutions, can be found in [10]. In [14] global process optimization has been modeled as a multiple choice multiple dimension knapsack problem (MMKP) and as a graph constrained optimum path problem. A MMKP is one kind of knapsack problem where the resources are multidimensional, that is, there are multiple resource constrains for the knapsack (e.g., weight and volume) and items are classied in groups. Each item of the group has a particular value

and it requires resources. The objective of the MMKP is to pick exactly one item from each group for maximum total value of the collected items, subject to the resource constraints of the knapsack. Process optimization can be reduced to a MMKP since each abstract service corresponds to a group, each concrete WS is an item in a group and each QoS constraint corresponds dimension of the knapsack. Authors in [14] has implemented ad hoc efficient techniques to identify sub-optimal solutions of the MMKP.

Global approaches have been proposed for the first time in [15], where the Process optimization problem has been formalized as a mixed integer linear programming problem, solved by integer linear programming solvers. The authors separately optimize each execution path and obtain the abstract services to concrete services mapping by composing separate solutions according to the frequency of execution. This approach has some limitations (e.g., availability and response time constraints are guaranteed only for the critical path, and global constraints cannot always be fulfilled) which have been solved by following research proposals [3, 14].

Some recent proposals face the Process optimization problem by implementing genetic algorithms [5, 8]. In Canfora et al. [5] the reduction formulas presented in [7] are adopted, the re-optimization is considered but abstract services specified in loops are always assigned to the same concrete WS component. Furthermore, by applying reduction formulas the solution guarantees global constraints only statistically. At run time, if low probability paths are taken (see [5]), then the solution could become infeasible and re-optimization must be triggered. In [8], the multi-objective evolutionary approach NSGA-II (non-dominated sorting genetic algorithm) is implemented, which identifies a set of Pareto optimal solutions without introducing a ranking among different quality dimensions. Every identified solution is characterized by the fact that no other plans exist such that a quality dimension is improved without worsening the other ones. Genetic algorithms are more flexible than mixed integer linear approaches, since they allow considering also non-linear composition rules for composed WSs, but are less computationally efficient. In current implementations, some execution time is wasted by generating also non-feasible solutions and, sometimes, no solution can be identified even when the problem is feasible in case the global constraints are stringent.

The work presented in [3] proposes a third generation solution which poses the basis for the execution of processes under severe QoS constraints. The solution is based on loops peeling which significantly improves the solutions based on loops unfolding (up to 40%). Furtheremore, negotiation techniques are exploited to identify a feasible solution of the problem, if one does not exist, reducing process invocation failures. The joint optimization and negotiation approach has been proved to be effective for large processes (including up to 10,000 abstract services), when QoS constraints are severe and reduces also the re-optimization overhead.

Key Applications

Context-Aware Business Process

Dynamic WS selection for composed WSs focused in particular on context aware business processes. Context awareness may be needed both when considering WS personalization, where a generic process is personalized choosing services according to user preferences, and in mobile composed services, to provide ubiquitous services where selection and execution depend on the available services and their QoS [1].

E-Science and Grid Computing

In e-science complex processes, defined as workflows enacted in grid environments, are being developed reaching the dimension of thousands of tasks in "in silico" experiments [9]. Each task is performed selecting and invoking a service. In this case, the Process optimization problem is more challenging, since the concrete resources have to be modeled with a more fine grain and are represented by the physical machines which can support tasks execution instead of abstract WSs. Current solutions propose to create a hierarchy of processes or distributing the workflow over a number of engines, partitioning in this way the optimization problem but leading to sub-optimal solutions.

Future Directions

All of the above approaches consider the optimization of a single process instance and assume a constant QoS profile. Cardellini et al. [6] tackled the problem of optimization of multiple process instances in order to reduce optimization overhead. The work presented in [2] considered variable (periodic) quality of service profiles of component WSs and explicitly addressed long term process execution. The execution of multiple instances, variable QoS profiles, and long term process execution make the optimization problem more cumbersome. Only heuristic approaches have been proposed so far, more efficient solutions, both in term of optimization time and quality of the final solution, are needed.

Experimental Results

For every presented approach, there is an experimental evaluation in the corresponding reference. Zeng et al. [15] discuss local and global approaches. Tao Yu et al. [14] present a comparison among linear integer programming approaches and heuristic solutions. The work presented in [3] analyzes loops peeling and second and third generation solutions.

Cross-References

► Grid and Workflows
► Workflow Management and Workflow Management System

Recommended Reading

1. Ardagna D, Comuzzi M, Mussi E, Pernici B, Plebani P. PAWS: a framework for executing adaptive web-service processes. IEEE Softw. 2007;24(6):39–46.
2. Ardagna D, Giunta G, Ingraffia N, Mirandola R, Pernici B. QoS-driven Web services selection in autonomic grid environments. In: Proceedings of the OTM Confederated International Conferences, CoopIS, DOA, GADA, and ODBASE; 2006. p. 1273–89.
3. Ardagna D, Pernici B. Adaptive service composition in flexible processes. IEEE Trans Softw Eng. 2007;33(6):369–84.
4. Bonatti PA, Festa P. On optimal service selection. In: Proceedings of the 14th International World Wide Web Conference; 2005. p. 530–8.
5. Canfora G, Penta M, Esposito R, Villani ML. QoS-aware replanning of composite web services. In: Proceedings of the IEEE International Conference on Web Services; 2005.
6. Cardellini V, Casalicchio E, Grassi V, Mirandola R. A framework for optimal service selection in broker-based architectures with multiple QoS classes. In: Proceedings of the IEEE Services Computer Workshops; 2006. p. 105–12.
7. Cardoso J Quality of service and semantic composition of workflows, Ph. D. Thesis, University of Georgia. 2002.
8. Claro DB, Albers P, Hao JK. Selecting web services for optimal composition. In: Proceedings of the IEEE International Conference on Web Services; 2005.
9. Fox GC, Gannon D. Workflow in grid systems. Concurr Comp Pract Exper. 2006;18(10):1009–19.
10. Jaeger MC, Muhl G, Golze S. QoS-aware composition of web services: an evaluation of selection algorithms. In: Proceedings of the International Conference on Cooperative Information Systems; 2005.
11. Maamar Z, Sheng QZ, Benatallah B. Interleaving web services composition and execution using software agents and delegation. In: Proceedings of the Web Services and Agent-Based Engineering; 2003.
12. Patil AA, Oundhakar SA, Sheth AP, Verma K. METEOR-S web service annotation framework. In: Proceedings of the 12th International World Wide Web Conference; 2004. p. 553–62.
13. Senkul P, Toroslu IH. An architecture for workflow scheduling under resource allocation constraints. Inf Syst. 2005;30(5):399–422.
14. Yu T, Zhang Y, Lin KJ. Efficient algorithms for web services selection with end-to-end QoS constraints. ACM Trans Web. 2007;1(1):1–26.
15. Zeng L, Benatallah B, Dumas M, Kalagnamam J, Chang H. QoS-aware middleware for web services composition. IEEE Trans Softw Eng. 2004;30(5):311–27.

Process Structure of a DBMS

Pat Helland
Microsoft Corporation, Redmond, WA, USA

Synonyms

Cluster databases; Scale-out databases; Scale-up databases; Shared-disk databases; Shared-everything databases; Shared-nothing databases

Definition

Database Management Systems are typically implemented on top of operating systems which allow execution within processes. Different systems have chosen different process structures as they map their computation onto the operating system. This section surveys some of these choices.

Historical Background

The first database management systems were simple libraries that ran inside the process of the application. While the use of these libraries offered leverage to the applications by providing essential functionality, they did not offer protection for the data in the presence of application errors.

To provide protection, DBMSs were initially moved into higher security rings accessible by hardware protected transitions to memory and code which was more secure than the application but less secure than the operating system kernel. Running the DBMS in shared (but secured) memory allowed access by multiple applications (in separate processes). The shared memory within the DBMS allowed for efficient cross application management of data (See Fig. 1).

Two trends caused gradual retreat from the implementation of the DBMS within a trusted security ring. First, there is the emergence of DBMSs that were designed to be ported across operating systems. Second, the emergence of distributed computing drove the need to run different portions of the computing stack (both application and database) across different machines. Both of these trends grew throughout the 1980s and led to a constellation of process and processor architectures for both applications and database management systems.

The process structure of database management systems has evolved and today can be seen in many forms. As mentioned above, initial imple-

mentations of DBMS systems in the 1960s and 1970s were embedded in the same process (but soon with protection for the DBMS within a security ring).

In the 1980s, a number of distributed databases emerged, exemplified by Tandem's NonStop SQL. In these, the application interface remained an in-memory call to portions of the database system but behind the scenes there were cross-process and cross-processor calls to other portions of the DBMS. This was implemented transparently to the application except, of course, with some performance implications which could be both positive and negative.

In the 1990s, the client-server computing first arrived with the separation of the client and server-side database in what became known as *two-tier client-server* applications. This necessitated the creation of the *database connection*, exemplified by ODBC (Open Database Connectivity). With a database connection, both DML (Data Manipulation Language) and the resultant data sets were returned across process boundaries, allowing the application process (and processor) to be different than the database process and processor. Subsequently, the application itself began to break across processes and processors resulting in *three-tier client-server* or even later *N-tier client-server systems*. Within each of these architectures, the application program still

Process Structure of a DBMS, Fig. 1 Early implementations of DBMS systems used shared memory as a technique to allow the DBMS to run in process with the application. This minimized app to DBMS communication costs

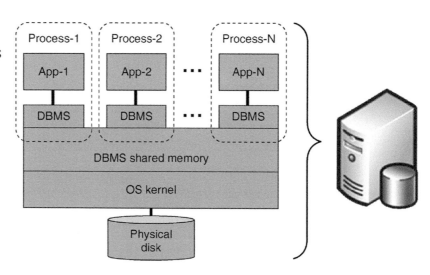

perceived the notion of a single database even though the work was potentially spread across multiple systems.

Shortly after the year 2000, the industry began to recognize the importance of the relationships of applications and databases running independently and speak about what is today called *SOA (Service Oriented Architecture)*. This is delineated from N-tier client-server by the absence of a common DBMS; SOA services each have the own DBMS whereas a client-server system shares a common DBMS, even if the common DBMS is distributed across processes and/or processors.

Foundations

To understand the process and processor architecture of DBMS and applications, the reader first looks at a sketch of their high-level architecture independent of the processes and processors implementing the components of the architecture, followed by some common patterns for breaking this work up.

Layers of the DBMS and Application

Figure 2, depicts a breakdown of both the 3-tier application and the underlying DBMS. The dotted lines show the classic 3-tier architecture. The *presentation tier* is connected to the *logic tier* with an application specific call or RPC. The *logic tier* is connected with the *database tier* using a database connection.

Inside the database, more layers of abstraction are utilized to examine different process and processor architectures. Closest to the application is the front-end to the *Query and DML Processor*: it accepts a database connection from the application and is responsible for the processing DML and queries and returning the results across the connection. The front-end of this function will interact with a back-end which is intimate with the various access methods which hold the database records and/or alternate keys. The back-

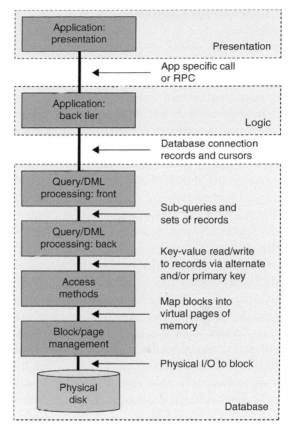

Process Structure of a DBMS, Fig. 2 The architecture of a 3-Tier application and its supporting database management system. The notations beside the arrows denote the formats of the requests and data flowing between the layers

end of the Query and DML Processor will interact with the *Access Methods* using keys and records. The Access Methods, in turn, use blocks (which are mapped into in-memory pages) by the *Block and Page Manager*, which issues physical I/Os to the disks themselves.

In the simplest DBMS architectures, the *Shared-Everything* design, all DBMS components are resident on the same computer system and they interact through memory. In the *Shared Disk* architecture, the interactions between the *Block and Page Management* and the *Physical Disk* are spread apart. Finally, the *Shared Nothing* architecture separates the front and back ends of the Query and DML Processor.

Client-Server Computing

When client-server computing first arrived on the scene, its hallmark was the separation of the application from the DBMS itself leaving the database on its own system. Initially, this was done with two tiers, the client and DBMS-server. The client interacted with the server using a database connection such as ODBC (See Fig. 3).

The transition between two-tier client-server systems and three-tier (or even N-tier) client-server systems lies in the architecture of the application itself. It is simply the splitting up of the application tiers that differentiate these (See Fig. 4).

As soon as the 3-tier architecture was introduced, new challenges arose in the management of transactions as they propagated through the system. In 2-tier client-server architectures, the transactional scope was bounded by a time interval on the database connection. Now, in a 3-tier client-server scheme, the work initiated by the Presentation layer may need to be atomic as it propagates through different servers implementing the Logic layer. For the moment, consider this

architecture with a single process DBMS at the back-end (See Fig. 5).

New notions in the management of transaction identifiers needed to be created for this to work. The identity of the atomic transaction needed to be propagated with the RPC or other app-specific call from the client to the middle-tier server. Also, the same transaction-id could now arrive at the database from *different* database connections. A window of time using the database-connection could no longer be a surrogate for the transaction. These challenges came as the application that related to the database underwent changes in its process architecture.

Multi-Database Computing

As distributed systems progressed, there were occasions in which a 2-tier client wished to do transactional work across multiple database servers. This was only possible with the arrival of BOTH distributed transactions (implemented with two-phase commit) AND the ability for the client application to create a transaction independent of

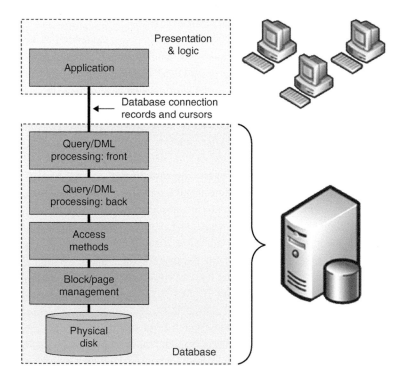

Process Structure of a DBMS, Fig. 3 Two-tier client-server architecture. The application is separated out from the database. The interaction is maintained with a database connection

Process Structure of a DBMS, Fig. 4
Three-tier client-server architecture. The presentation, logic, and database are distributed across different systems

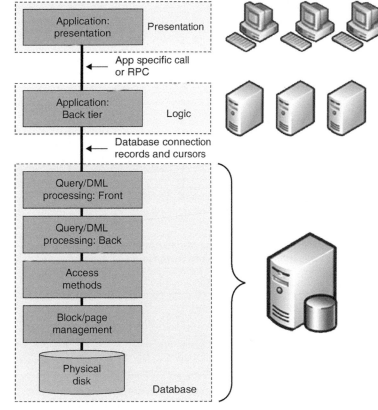

Process Structure of a DBMS, Fig. 5
Building a 3-tier client-server application necessitates both the management of the same transaction coming into the DBMS on different database connections AND the propagation of transactions from the client to the middle-tier server

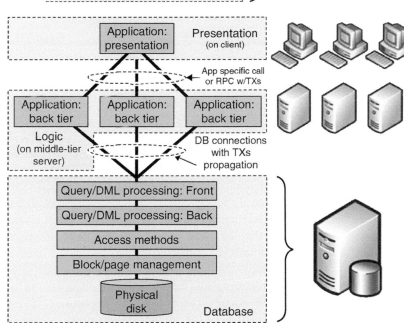

the database connection and manage the association of the transaction to the connections to the separate databases. This involved both extensions to the client libraries and the creation of DBMS server to DBMS server distributed transaction management.

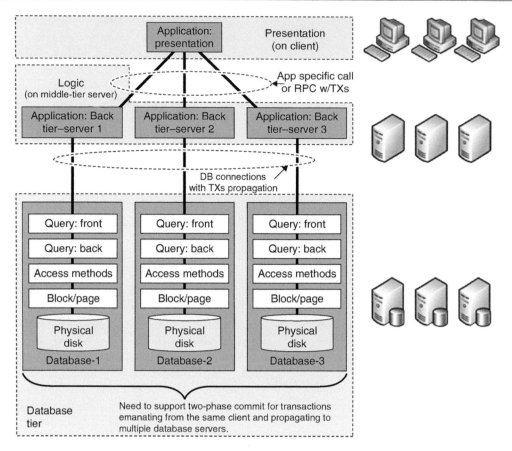

Process Structure of a DBMS, Fig. 6 A 3-tier client-server application with DIFFERENT databases not only has to manage the propagation of transactions from client to app-server AND app-server to DBMS, it must in addition manage the atomic two-phase commit of the transactions across database servers

Most systems did not implement transactional support for multiple databases in 2-tier applications until after this was accomplished for 3-tier applications. The need to have a client call different middle-tier application servers with different databases was the pressure that led to the implementation of two phase commit for distributed transactions across the multiple databases. Only later were the facilities for sharing transactions across database connections directly connected to a single client implemented. See Fig. 6 for a depiction of a 3-tier client-server application where some of the middle-tier servers use different databases all associated with a single common transactional scope.

Service-Oriented Architectures (SOA)

A recently popular application architecture is the service-oriented architecture. It is distinguished from the client-server application architecture in that there are no transactions shared across the service boundaries and, indeed, no direct visibility to the partner's database. All access to the database is indirect and mitigated by the application. Because there are no shared transactions across the service boundaries and there are no database connections or semantics across these boundaries, the database management system does not see the SOA architecture.

SOA implementations often use an internal client-server architecture to build out a single service. Still, from the standpoint of the process architecture of a DBMS, the use of an application combined with its database in a larger SOA system is not of any impact on the DBMS process structure.

Shared-Memory (Single Database) DBMS Architectures

In a shared-memory (or sometimes called shared-everything) system, the DBMS runs in a single process or, at least, in a fashion where the processes are able to share their memory. Figure 1 is an example of a shared-memory system, one where multiple processes have a special mechanism to access special (protected) memory. The architecture depicted in Fig. 1 is not a client-server system.

Most shared-nothing systems are a little less exotic and look like Figs. 3, 4, and 5. There is a single database and it runs on a large process (perhaps with multiple threads within a shared-memory multiprocessor).

Figure 6 is considered to be a shared-everything system *precisely because each database is a separate database*, and there cannot be queries issued across them. There is no way to view the data from these multiple databases except through the assistance of application logic. There cannot exist a single database connection to that pool of databases – they are separate databases

The presence of distributed transactions spanning multiple databases does not make them a single database.

Shared-Disk (Single Database) DBMS Architectures

Shared-Disk systems comprise a cluster in which multiple independent machines have access to a pool of disks. In addition to some mechanism for sending messages across the machines, each of the computers in the cluster can read and write pages of disk across a collection of many disk spindles. Typically, this is implemented with some form of disk controller managing access by the computers to the disks. Figure 7 shows a sketch of the hardware architecture of a shared-disk cluster.

The distinction between the database process architecture versus the application process architecture is sometimes confusing. The first shared-disk systems used block mode terminals and there was no notion of a smart client at that time. As an example, consider Fig. 8 which is a slight modification of Fig. 7. Figure 8 depicts a system in which the business logic (the middle tier of a three tier system) runs on the same processors (or potentially the same processes in the same spirit as shown in Fig. 1). In this example, the business logic of the application as well as the entire processing of the DBMS can run in the same process (or at least the same processor) while still scaling across a multi-processor cluster with tremendous efficiency if the load characteristics are appropriate for the architecture.

Shared-Disk DBMS systems have the advantage that, once all of the blocks are brought into memory, the entire query can be processed within a single process. This style of distributed implementation delivers high performance when the workload splits into easily partitioned sets of blocks but works less efficiently when the workload has conflicts over the blocks needed by the different processes.

Shared-Nothing (Single Database) DBMS Architectures

Share-Nothing DBMS architectures are implemented by splitting the query processing engine into a front-end and a back-end. Messaging is used to pass portions of the query or update from the front-end to the back-end. Resulting sets of records are passed from the back-end to the front-end where the completion of the query is performed. The optimizations used in these architectures have been the source of significant research and engineering and can result in fascinating performance gains. Figure 9 shows a Shared-Nothing DBMS in which the applica-

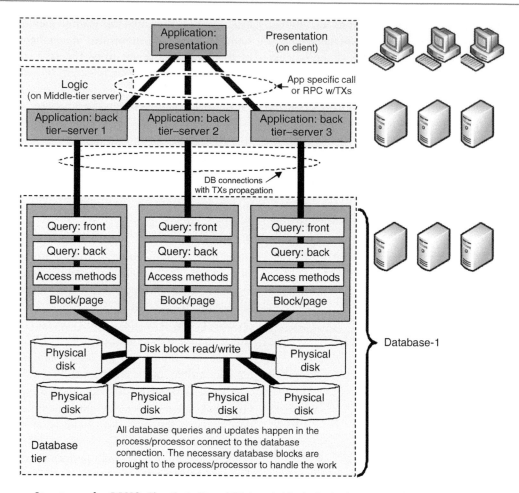

Process Structure of a DBMS, Fig. 7 A Shared-Disk Database Management System. For the first time in our figures, the single database SPANS multiple processes and, indeed, multiple processors. The single database (with single database semantics presented to the application) runs on many different computers but sharing the access to the physical disks. Special locking infrastructure on the contents of the blocks of disk must be maintained

tion's Business Logic is running on the same scale out cluster as the DBMS. Just like the Shared-Disk DBMS, it is important to realize that the application's process/processor architecture may take different forms. Running the Business Logic close to the Front-End of the DBMS is one configuration of Shared-Nothing.

Implications of Shared-Nothing, Shared-Disk, and Shared-Everything Architectures

For a number of years, there have been debates in the industry about the strengths and weaknesses of different process architectures for a DBMS system. Before even engaging in these, it is important to remember the delineation of the DBMS architecture from that of a Service Oriented Architecture and, also, from the application architecture of an N-tier system. The term DBMS is used to refer to a single collection of records across which relational operations may occur. Service Oriented Architectures (SOAs) offer an aggregation of computation connected by business logic without the presence of spanning relational operations or transactions. N-tier application environments frequently offer atomic transactions across different databases but do not offer relational operations across the contents

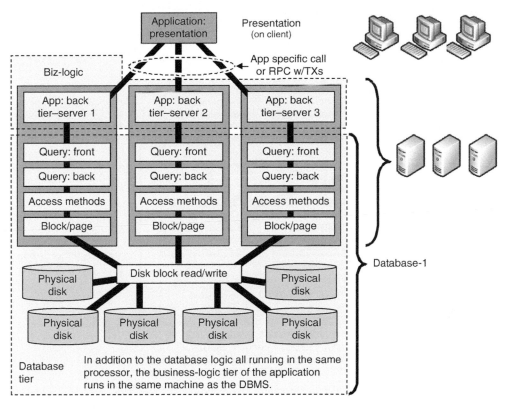

Process Structure of a DBMS, Fig. 8 Shared-Disk DBMS within a Two-Tier application architecture

of the databases. So, the taxonomy of DBMS process structures refers only to the portion of the system providing a single relational database semantic.

The *Shared-Everything* DBMS architecture is by far the highest performing architecture in that it offers the most throughput for a single computer (it is not necessarily the most scalable one; scalability is discussed below). Shared-Everything DBMSs do not need to perform any messaging or other cross process communication because they don't cross processes in their implementation of the DBMS. All of the work is done within a single machine. Included in this architecture are various single memory multi-processor implementations. There are few concerns in a Shared-Everything DBMS about the types of queries and usage patterns because everything coexists in a single shared process. Again, Shared-Everything works wonderfully until the DBMS grows too large to fit into one system.

A *Shared-Disk* DBMS architecture allows for additional sharing by letting multiple DBMS processes and processors to access the same physical disks. When the usage pattern of the application has a low probability of conflict in its updates of a shared page, Shared-Disk DBMS systems are very efficient. The pages needed for a query are brought into the processor requesting them and the work of the application's transaction is handled inside one processor of the cluster in what is (hopefully) a very efficient fashion. When the data in question has low update rates, this typically works well. When a single data item is rapidly updated (called a "hot-spot"), this can lead to performance conflicts which cause the block of the database to be pulled back-and-forth across processors. Another challenge occasionally presented by Shared-Disk DBMS systems lies in block (or page) mode locking. Distinct records in the same block may observe lock conflicts causing performance challenges that would not be present in other architectures.

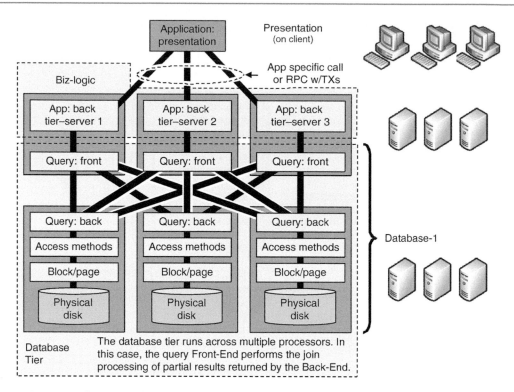

Process Structure of a DBMS, Fig. 9 A Shared-Nothing DBMS running with an application configuration which places the business logic tier on the same cluster as the Shared-Nothing DBMS. Also, in this configuration, the Query Front-End of the DBMS runs on the same processors as the Business-Logic

Still, Shared-Disk DBMS systems have been wildly successful for many applications and allow many databases to scale beyond what a single shared-memory system could offer.

Shared-Nothing DBMS architectures typically carry a heavier cost to set up a complex transaction but can, in some cases, offer greater throughput over a particular amount of data. It is typical for a Shared-Nothing system to offer record locking. For systems with very high throughput "hotspots," the data for the "hot-spot" does not move around the cluster and it is possible for the system to get more transactions over the same piece of data in a fixed period of time. This is sometimes referred to as "moving the operation to the data" (Shared-Nothing) rather than "moving the data to the operation" (Shared-Disk).

When a database can fit into a single shared-memory multiprocessor, Shared-Everything offers distinct advantages (at least until the application's demands exceed the single machine and

you have a problem). For databases that exceed this size and yet want full database semantics, there is a lively debate within the community about which architecture is better and different applications offer different performance characteristics.

Key Applications

Database process structures are an essential part of scaling past a single system. As discussed, the distinction between an application multiprocessor architecture and a DBMS multiprocessor architecture has many nuances.

Cross-References

► Application Server
► Client-Server Architecture
► Distributed Concurrency Control

- ▶ Distributed Database Systems
- ▶ Distributed DBMS
- ▶ Distributed Query Processing
- ▶ Distributed Transaction Management
- ▶ Multitier Architecture
- ▶ Open Database Connectivity
- ▶ Service-Oriented Architecture
- ▶ Shared-Disk Architecture
- ▶ Shared-Memory Architecture
- ▶ Shared-Nothing Architecture
- ▶ Two-Phase Commit

Recommended Reading

1. Gray J, Reuter A. Transaction processing: concepts and techniques. San Mateo: Morgan Kaufmann; 1992.
2. Michael S. (UC Berkeley). The case for shared nothing architecture. Database Eng. 1986;9(1):4–9.
3. Oracle RAC (Real Application Clusters). http://www.oracle.com/database/rac_home.html.
4. Susanne E, Jim G, Terrye K, Praful S. A Benchmark of NonStop SQL Release 2 Demonstrating Near Linear Speedup and Scaleup on Large Databases. In: Proceedings of the 2000 ACM SIGMETRICS International Conference on Measurement and Modeling of Computer Systems; 1990. p. 24–35.
5. The Tandem Database Group. NonStop SQL: a distributed high performance, high availability implementation of SQL. In: Proceedings of the 2nd High Performance Transaction Processing Workshop; 1989.

Processing Overlaps in Structured Text Retrieval

Georgina Ramírez
Yahoo! Research Barcelona, Barcelona, Spain

Synonyms

Controlling overlap; Removing overlap

Definition

In semi-structured text retrieval, processing overlap techniques are used to reduce the amount of overlapping (thus redundant) information returned to the user. The existence of redundant information in result lists is caused by the nested structure of semi-structured documents, where the same text fragment may appear in several of the marked up elements (see Fig. 1). In consequence, when retrieval systems perform a focused search on this type of document and use the marked up elements as retrieval objects, very often result lists contain overlapping elements. In retrieval applications where it is assumed that the user does not want to see the same information twice, it may be necessary to reduce or completely remove this overlap and return a ranked list of no overlapping elements. Thus, depending on the underlying user model and retrieval application, different processing overlap techniques are used in order to decide, given a set of relevant but overlapping elements, what are the most appropriate elements to return to the user.

Historical Background

Although the problem of overlap in semi-structured text retrieval is as old as the semi-structured documents themselves, not much work has been published on processing techniques for reducing or removing overlap. Some related work can be found in the area of passage retrieval, where approaches that use a varying window size for passage selection might produce result lists with overlapping passages. However, most of this work is performed on unstructured documents and the approaches taken for processing overlap tend to be simpler.

In the domain of semi-structured documents, there are several areas where different overlap issues are studied. For example, there is quite some work in the area of evaluation of XML systems that addresses the so called overlap problem (e.g., [3]). A different overlap problem is created by the possibility that standards like SGML provide of having multiple annotations (markups) on the same document (a.k.a. multiple hierarchies). In this case the overlap is produced by the structure of the different annotations. Since the multiple hierarchies complicate the use of standard retrieval techniques on this type of documents, work on this area is still focusing on addressing other indexing and retrieval issues. It is only recently that in the domain of XML documents, several approaches have been presented that address the problem of processing overlap from result lists containing overlapping elements. The next section summarizes some of them.

Foundations

One of the advantages of semi-structured documents is that retrieval systems can perform focused search by simply using the marked up divisions of the documents (elements) and retrieving those instead of the whole documents. However, since elements overlap with each other (see Fig. 1), when using traditional ranking techniques to independently rank these elements, result lists often contain many overlapping elements. This is due to the nested structure of semi-structured documents, where the same text fragment is usually contained in several of the marked up elements. Thus, when a specific element is estimated relevant to the query, all the elements containing this element (a.k.a. ancestors) will also be estimated to some degree relevant to the query.

Furthermore, this element is probably estimated relevant because it contains several relevant elements (a.k.a. descendants). For example, if a section of a document is estimated highly relevant, it probably contains several highly relevant paragraphs and it is contained in a relevant article. If all of these elements are returned to the user, the amount of redundant information contained in the result list will be considerable. In retrieval scenarios where users do not like to see the same information twice, retrieval systems need to decide which of these relevant but overlapping elements is the most appropriate piece of information to return to the user. The final decision on which elements the system should return depends on the search application and the underlying user model but a common goal is to reduce redundancy in the result lists. Although this can be done at indexing time (e.g., by selecting a subset of non-overlapping elements as potential retrievable objects), commonly this is done by removing overlapping elements from the result set, after retrieval systems have produced an initial ranking of all elements. Processing overlap techniques have recently been widely discussed in the domain of XML retrieval, where different approaches have been presented. The rest of this section presents and discusses some of them.

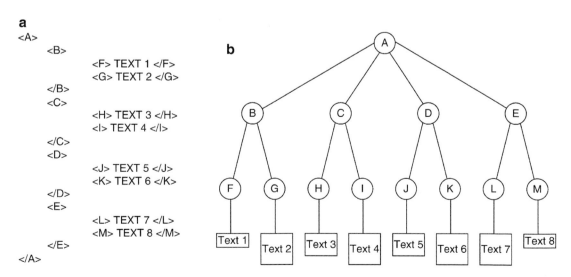

Processing Overlaps in Structured Text Retrieval, Fig. 1 Example of a semi-structured document and its tree structure representation. Note that each fragment of the document is contained in three different elements (nodes in the tree)

Using Element Types

A simple way to reduce overlap is to select a subset of element types and consider only these for retrieval. For example, if sections of documents are considered to be the most appropriate pieces of information, retrieval systems might want to use only these as retrievable objects and ignore all the other element types. This can be done at indexing time or by post-filtering the result lists. Depending on the number and element types selected, overlap is reduced at different degrees. Note that it might not always be possible to completely remove overlap; even if a single element type is selected as a unique retrievable object, it is still possible that elements of the same type overlap each other. The main drawback of this approach is that, since it is desirable to select element types that are likely to be relevant and useful to the user, it requires knowledge of the structure of the documents and its common usage.

Using Paths

A common approach for processing overlaps keeps the highest ranked element on each path and removes its ancestors and descendants from the result list, i.e., all the elements in the result list that contain or are contained within it (e.g., [2, 6]). It is important to notice that depending on the order in which the different paths are processed different outputs might be produced.

In [6] the authors present a two steps algorithm to remove the overlap. The first step is used to select the highest scored element from each relevant path. Since their algorithm selects the elements from the different paths simultaneously, the output may still contain overlapping elements. That is why a second step is needed, to completely remove overlap. This is done by selecting again (this time from the output of the first step) the highest scored element from each path:

Algorithm 1

1. Select highest scored element from each relevant path.

2. Select highest scored element from each relevant path in output of step 1.

Another common way to remove overlap using paths [5] is to recursively process the result list by selecting the highest ranked element and removing any element from lower ranks that belongs to the same path (it contains the selected element or it is contained within it):

Algorithm 2

1. Return highest ranked element from result list.
2. Remove from result list all the elements belonging to the same path.
3. Repeat step 1 and 2 until result list is empty.

The underlying assumption of this type of approaches is that the most appropriate piece of information in each path has been assigned a higher score than the rest and therefore, removing overlap is simply a presentation issue. These approaches rely completely on the underlying retrieval models to produce the best ranking. This could indeed be the case if the retrieval model would consider, when ranking, not only the estimated relevance of the element itself but also its *appropriateness* compared to other elements in the same path. However, since many retrieval models rank elements independently, the highest scored element may not be the most appropriate one, i.e., the one the user prefers to see.

To illustrate the different outputs of the previous algorithms, have another look at the example document from Fig. 1. Imagine now that, given a query, the retrieval model estimates the relevance of each element in the document. Figure 2 shows the retrieval scores obtained by each of the elements and the outputs produced when removing overlap with the algorithms described above. Both algorithms produce a result list of non-overlapping elements. However, there are substantial differences. The main drawback of the first algorithm is that it might miss some relevant information. For example, one could argue that element K should also be contained in the output list. To be able to do that, the algorithm should consider structural relationships between

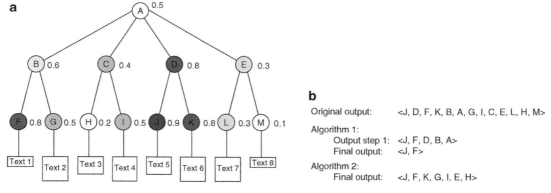

Processing Overlaps in Structured Text Retrieval, Fig. 2 Example of a retrieval run and the resulting outputs when removing overlapping elements with Algorithms 1 and 2. The numbers on the tree indicate the initial retrieval scores obtained by each of the elements

elements and re-add element K to the result list when it decides to remove element D in the second step.

Although the second algorithm produces a more complete list, someone could argue that the output produced is not the most desirable. For example, imagine that elements F and G are, respectively, the title and the abstract of the document. In this case, even if the title has been ranked high (it may contain most or all of the query terms), users might prefer to see a result item containing both, title and abstract (i.e., element B) instead of seeing both elements independently. In general, it can be argued that retrieval systems should only return those elements that have enough content information to be useful and can stand alone as independent objects. A similar example can be seen for elements D, J, and K. Even if J is ranked higher, element D contains mostly relevant information and therefore, it might be more desirable (from a user perspective) to read element D, than J, and K independently. Furthermore, for elements E, L, and M it could be argued that it is better to return L than E because the relevance estimated for element E is due to the content of L and no extra benefit is obtained when returning E.

In these, cases a better output might be produced if the algorithms would consider structural relationships between elements when deciding which elements to return to the user.

Using Structural Relationships for Re-Ranking

The following approaches present more advanced techniques that exploit the structural relationships within a document to decide which elements should be removed or pushed down the ranked list and which ones should be returned to the user (e.g., [1, 4, 5]). These techniques often modify the initial ranking predicted by the retrieval model.

In [1] elements are re-ranked by adjusting the element scores of the lower ranked elements according to their containment relationship with other higher ranked elements. The assumption underlying this approach is that the score of the elements that are contained or contain other elements that have been already shown to the user (i.e., are ranked higher) should be adjusted in order to reflect that the information they contain might be redundant. They do that by reducing the importance of terms occurring in already reported elements. The basic algorithm is similar to Algorithm 2 but instead of removing the ancestors and descendants of the reported elements, their scores get adjusted:

Algorithm 3

1. Report the highest ranked element.
2. Adjust the scores of the unreported elements.
3. Repeat steps 1 and 2 until m elements are reported.

The author also presents an extended version of the algorithm where different weighting values are used for ancestors and descendants and where the number of times an element is contained within others is considered. For example, a paragraph contained in an already seen section and in an already seen article is further punished because the user has already seen this information twice. Note that this algorithm is not designed to remove the overlap but to push down the result list those elements that contain redundant information.

In [4] a completely different approach is taken. The authors present a two step re-ranking algorithm for removing overlap. The first step identifies clusters of highly ranked results and picks the most relevant element from each cluster. The second round is used to remove any remaining overlap between the selected elements. The selection criteria for the first step is based on three different cases (illustrated in Fig. 3): (i) if an element N has a descendant that is substantially more relevant, the element N is removed from the result list, (ii) if case 1 does not hold and the element N has a child that contains most of the relevant information (the relevant elements are concentrated under this child), the element N is also removed from the result list, and (iii) if none of the previous cases hold and the results are evenly distributed under the element N, then the element N is kept and all its descendants removed from the result list. In the rest of the cases they do not do anything and leave the final overlap removal for the second phase. In the second step, they remove overlap by comparing the score of each element with the ones of its descendants. If the score of the element is bigger, all the descendants are removed and the element is kept. Otherwise, the element is removed.

In [5] the authors present an approach that makes use of an *utility* function that captures the amount of *useful* information contained in each element. They argue that to model the *usefulness* of a node three important aspects need to be considered: (i) the relevance score estimated by the retrieval model, (ii) the size of the element, and (iii) the amount of irrelevant information the element contains. They present an algorithm

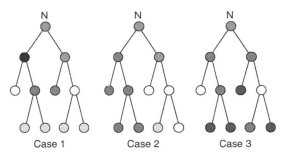

Processing Overlaps in Structured Text Retrieval, Fig. 3 Illustration of the three cases considered in the approach presented in [4]. The grayer the node, the higher the relevancy estimated for that node

that selects elements according to the estimated *usefulness* of each element. If an element has an estimated *usefulness* value higher than the sum of the *usefulness* values of its children, then the element is selected and the children are removed. Otherwise, the children elements whose *usefulness* value exceeds some threshold are selected and the element is removed.

Key Applications

Processing overlap techniques are needed in any retrieval application where users need to be directed to specific parts of documents and there are not predefined retrieval units.

Experimental Results

For every presented approach, there is an accompanying experimental evaluation in the corresponding reference. Commonly, the referred article provides an overview of the performance of the approach and of its variations. However, it is difficult to use the reported evaluations to compare approaches between articles. The main reason is that all approaches make use of different retrieval models for their initial runs, thus it is not clear whether the performance obtained after removing overlap is due to the underlying retrieval model or to the approach used to remove the overlap. Besides, not all the presented approaches experiment on the same dataset or use the same evaluation measures to report their numbers.

In [5] the authors present an experimental comparison between several of the approaches described above. They show that the best performing approach is the one that returns only paragraphs. As a general trend for the first type of approach (the ones that select a specific element type), the longer the element type selected, the worse the performance. The authors also show that their approach of estimating the *usefulness* of an element can help to improve retrieval performance (in terms of precision at low recall levels) when compared to the approach of using paths (Algorithm 2).

Data Sets

Since 2005 the *INitiative for the Evaluation of XML Retrieval* (INEX) provides a data-set that can be used to test processing overlap strategies (see http://inex.is.informatik.uni-duisburg.de/).

Cross-References

▶ Structured Document Retrieval
▶ XML Retrieval

Recommended Reading

1. Clarke CLA. Controlling overlap in content-oriented XML retrieval. In: Proceedings of the 31st Annual International ACM SIGIR Conference on Research and Development in Information Retrieval; 2008. p. 314–21.
2. Geva S. GPX – gardens point XML IR at INEX 2005. In: Proceedings of the 4th International Workshop of the Initiative for the Evaluation of XML Retrieval; 2006. p. 240–53.
3. Kazai G, Lalmas M, de Vries AP. The overlap problem in content-oriented XML retrieval evaluation. In: Proceedings of the 30th Annual International ACM SIGIR Conference on Research and Development in Information Retrieval; 2007. p. 72–9.
4. Mass Y, Mandelbrod M. Using the INEX environment as a test bed for various user models for XML retrieval. In: Proceedings of the 4th International Workshop of the Initiative for the Evaluation of XML Retrieval; 2006. p. 187–95.
5. Mihajlovi V, Ramírez G, Westerveld T, Hiemstra D, Blok HE, de Vries AP. TIJAH scratches INEX 2005: vague element selection, image search, overlap and relevance feedback. 2006. p. 72–87.
6. Sauvagnat K, Hlaoua L, Boughanem M. XFIRM at INEX 2005: ad-hoc and relevance feedback tracks. In: Proceedings of the 4th International Workshop of the Initiative for the Evaluation of XML Retrieval; 2006. p. 88–103.

Processing Structural Constraints

Andrew Trotman
University of Otago, Dunedin, New Zealand

Definition

When searching unstructured plain text, the user is limited in the expressive power of their query – they can only ask for documents that are about something. When structure is present in the document, and with a query language that supports its use, the user is able to write far more precise queries. For example, searching for "smith" in a document is not necessarily equivalent to searching for "smith" as an author of a document. This increase in expressive power should lead to an increase in precision with no loss in recall. By specifying that "smith" should be the author, all those instances where "smith" was the profession will be dropped (increasing precision), while all those in which "smith" is the author will still be found (maintaining recall).

Historical Background

With the proliferation of structured and semi-structured markup languages such as SGML and XML came the possibility of unifying database and information retrieval technologies. The Evaluation of XML Retrieval (INEX) was founded in 2002 to examine the use of semi-structured data for both technologies. It was expected that the use of structure would not only unify the two technologies but would also improve the performance of both.

Foundations

User Querying Behavior

When using an information retrieval search engine, the user typically has some information need. This information need is expressed by the user as a keyword query. There are many different queries that could be drawn from the same information need. Some might contain only keywords, others phrases, and others a combination of the two. It is the task of the search engine to satisfy the information need given the query. Because the task is not to satisfy the query, the terms in the query can be considered nothing more than hints by the user on how to identify relevant documents. It is likely that some relevant documents will not contain the user's keywords, while others that do might not be relevant.

If the user does not immediately find an answer, they will often change their query, perhaps by adding different keywords or by removing keywords. If the query syntax permits, they might add emphasis markers (plus and minus) to some terms.

With a few exceptions, semi-structured search engines remain experimental, so user behavior cannot be studied in a natural environment. Instead the user behavior is expected to mirror that of other search engines or search engines that include some structural restriction.

The model used at INEX is that a user will give a query containing only search terms; then, if they are dissatisfied with the results, they might add structural constraints to their query. The keyword searches are known as content only (CO) and when structure is added as content-only + structure (CO + S) or content-and-structure (CAS) queries. Just as the keywords are hints, so too are the structural constraints. For this reason they are commonly referred to as structural hints.

The addition of structure to an otherwise content-only search leads to a direct comparison of the performance of a search engine before and after the structural constraint has been added. The two queries are instantiations of the same information need, so the same documents or document components are relevant to each query making a direct comparison meaningful.

The analysis of runs submitted to INEX 2005 (against the IEEE document collection) showed no statistical difference in performance between the top CO and top CO + S runs – having structural hints in the query did not improve performance [1]. Even at low levels of recall (1 and 10%), no significant improvement was seen. About half the systems showed a performance gain and the other half no gain.

There are several reasons why improvements are not seen: first it could be a consequence of the structure present in the IEEE collection; second (and more likely) it could be that users are not proficient at providing structural hints.

The result was backed up by a user study [2] in which users were presented with three ways of querying the document collection: keywords, natural language (including structure), and bricks [3] (a graphical user interface). Sixteen users each performed six simulated work tasks, two with each interface. The same conclusion is drawn, that is, no significant improvement was seen when structure was used in querying.

INEX subsequently reexamined this problem, first using the Wikipedia, and then in 2010 the Data Centric Track [4] examined queries over IMDb (people and movies). No improvements were seen when structure was used on the Wikipedia [5]. It was not until the second evaluation over the IMDb (in 2011) that evidence started to emerge that structure may help early precision, but may negatively impact recall [6, 7]. INEX has not run this kind of experiment since 2011.

Structural Constraints

There are two reasons a user might add structural constraints to a query. The first is to constrain the size of the result. When searching a collection of textbooks, it is, perhaps, of little practical use to identify a book that satisfies the user need. A better result might be a chapter from the book, or a section from the chapter, or even a single paragraph. One way to identify the best granularity of result is to allow the user to specify this as part of the query. These elements are known as target elements.

The user may also wish to narrow the search to just those parts of a document he or she knows to be appropriate. In this case the user might search for "smith" as an author in order to disambiguate the use from that as any of: an author, a profession, a street, or a food manufacturer. Restricting a query to a given element does not affect the granularity of the result; instead it lends support on where to look so such elements are known as support elements. Both target elements and support elements can appear in the same query.

It is not at all obvious from a query whether or not the user expects the constraint to be interpreted precisely (strictly) or imprecisely (vaguely). In the case of "smith" as an author, it is likely that "smith" as a profession is inappropriate, but "smith" as an editor might be appropriate. If the target element is a paragraph, then a document abstract (about the size of a paragraph) is likely to be appropriate, but a book not so.

The four possible interpretations of a query were examined at INEX 2005 [8]. Runs that perform well with one interpretation of the target elements do so regardless of the interpretation of the support elements. The interpretation of the target element does, however, matter. The consequence is that the search engine needs to know, as part of the query, whether a strict or vague interpretation of the target element is expected by the user.

Processing Structural Constraints

Given a search engine, the strict interpretation of target elements can be satisfied by a simple post-process eliminating all results that do not match. As just discussed above, strictly processing support elements has been shown to be unnecessary.

Several techniques for vaguely satisfying target element constraints have been examined including ignoring them, pre-generating a set of tag equivalences, boosting the score of elements that match the target element, and propagating scores up the document tree.

Ignoring Structural Constraints

Structural constraints might be removed from the query altogether and a content-only search engine used to identify the correct granularity of result.

Tag Equivalence

A straightforward method for vaguely processing structural constraints is tag equivalence. A set of informational groups are chosen a priori, and all tags in the DTD (DTD is the document type definition, specifying the format of the XML documents forming the collection) are mapped to these groups. If, for example, <p> is used for paragraphs and <ip> is used for initial paragraphs, these would be grouped into a single paragraph group.

Mass and Mandelbrod [9] a priori choose appropriate retrieval units (target elements) for the document collection and build a separate index for each. The decision about which units these are is made by a human before indexing. A separate index is built for each unit, and the search is run in parallel on each index. Within each index the traditional vector space model is used for ranking. The lexicon of their inverted index contains term and context (path) information making strict evaluation possible. Vague evaluation of paths is done by matching lexicon term contexts against a tag equivalence list.

Mihajlović et al. [10] build their tag equivalence lists using two methods, both based on prior knowledge of relevance. For INEX 2005 they build the first list by taking the results from INEX 2004 and selecting the most frequent highly and fairly relevant elements and adding the most frequently seen elements from the queries. In the second method, they take the relevant elements from previous queries targeting the same element and normalize a weight by the frequency of the element in the previous result set (the training data, in this case INEX 2004). Using this second method, they automatically construct many different tag equivalence sets using the different levels of relevance seen in the training data.

In a heterogeneous environment in which many different tags from many different DTDs are semantically but not syntactically identical, techniques from research into schema matching

[11] might be used to automatically identify tag equivalence lists.

Structure Boosting

Van Zwol [12] generates a set of results ignoring structural constraints and then boosts the score of those that do match the constraints by linearly scaling by some tag-specific constant. The consequence is to boost the score of elements that match the structural constraints while not removing those that do not. A score penalty is also used for deep and frequent tags in the expectation of lowering the score of highly frequent (and short) tags. A similar technique is used by Theobald et al. [13] who use it with score propagation.

Score Propagation

Scores for elements at the leaves of the document tree (i.e., the text) are computed from their content. Scores for nodes internal to the document tree are computed from the leaves by propagating scores up the tree until finally a score for the root is computed. Typically as the score propagates further up the tree, its contribution to the score of an ancestor node is reduced (see the entry on "▶ Propagation-Based Structured Text Retrieval" and for details).

Figure 1 illustrates score propagation. A search term is found to occur two times in the p element and four times in the (left) sec element. With a decay factor of 0.5, the score of the bdy element is computed as 4 * 0.5 + 2 * 0.5 * 0.5 = 2.5. The score for the article element is computed from that score likewise. If the target element is bdy and the score, for example, is boosted by $K = 5$, then the element with the highest score is that element. The score for K and the propagation value are chosen here for illustrative purposes only and should be computed appropriately for a given document collection.

Hurbert [14] uses score propagation with structure reduction – if a node in the tree does not match a constraint in the query, then the score there is reduced by some factor. In this way all nodes in the tree obtain scores, but those matching the constraints are over-selected for. Sauvagnat et al. [15] use score propagation

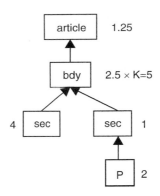

Processing Structural Constraints, Fig. 1 Score propagation, each time a score is propagated, the score is weakened (in the example: halved), but at target nodes it is boosted (in the example: $K = 2$)

in a similar way but in combination with tag equivalence.

Key Applications

Retrieval of document components from structured document collections.

Future Directions

The best performing search engines that interpret structural constraints were shown at INEX 2011 to outperform those that ignore them; however this result is shown on only one document collection, with a relatively small number of topics and a small number of participants – it should be considered preliminary. Several reasons have been suggested for the long string of negative results.

There is evidence to suggest specifying a structural constraint is difficult for a user. Studies into the use of structure in INEX queries suggest that even expert users, when asked to give structured queries, give simple queries [1]. This is in line with studies that show virtually no use of advanced search facilities on the web.

Structure-aware search engines are not as mature as web search engines, and as yet the best way to use structural constraints (when present in

P

a query) is unknown. The annual INEX workshop provides a forum for testing and presenting new methods.

Improvements were not seen when the IEEE document collection or Wikipedia was used, but were seen when IMDb was used. At the very least, this suggests that the result is collection specific. Alternative collections including newspapers, radio broadcast, and television have been suggested [16, 17]. It is not known that characteristic of a document collection indicates that a structured approach will be effective.

Relevance feedback including structural constraints has been examined. Users might provide feedback on both the desired content and the preferred target element or just one of these. Evidence suggests that including structure in relevance feedback does improve precision.

Experimental Results

Evidence that the use of structure increases precision is tentative. In the XML search engine of Kamps et al. [18], no significant difference is seen overall; however significant differences are seen at early recall points (the first few tens of documents). Wang et al. [7] observe an early precision increases but loss in recall. The search engine of Geva [19] performs better without structure than with. That of Trotman [20], often used to generate baselines at INEX, also performs well by ignoring structure.

At INEX 2005, a comparative analysis of performance with and without structural constraints on the same set of information needs was performed [5]. The best structure run was compared to the best non-structure run, and no significant difference was found. Not even a significant difference at early recall points was found. On a system-by-system basis, about half the search engines show a performance increase. However, at INEX 2011, and using the IMDb collection, the top runs used structural hints, and analysis suggests that structure does help early precision but negatively affects recall.

Cross-References

▶ Average Precision
▶ Content-and-Structure Query
▶ Content-Only Query
▶ INitiative for the Evaluation of XML Retrieval
▶ Narrowed Extended XPath I
▶ Propagation-Based Structured Text Retrieval
▶ XML Retrieval

Recommended Reading

1. Trotman A, Lalmas M. Why structural hints in queries do not help XML retrieval. In: Proceedings of the 32nd Annual International ACM SIGIR Conference on Research and Development in Information Retrieval; 2009. p. 711–2.
2. Woodley A, Geva S, Edwards SL. Comparing XML-IR query formation interfaces. Aust J Intell Inf Proc Syst. 2007;9(2):64–71.
3. van Zwol R, Baas J, van Oostendorp H, Wiering F. Bricks: the building blocks to tackle query formulation in structured document retrieval. In: Proceedings of the 28th European Conference on IR Research; 2006. p. 314–25.
4. Trotman A, Wang Q. Overview of the INEX 2010 data centric track. In: Proceedings of the 9th International Workshop of the Initiative for the Evaluation of XML Retrieval; 2010. p. 171–81.
5. Arvola P, Geva S, Kamps J, Schenkel R, Trotman A, Vainio J. Overview of the INEX 2010 ad hoc track. In: Proceedings of the 9th International Workshop of the Initiative for the Evaluation of XML Retrieval; 2010. p. 1–32.
6. Schuth A, Marx M. University of Amsterdam data centric ad hoc and faceted search runs. In: Proceedings of the 10th International Workshop of the Initiative for the Evaluation of XML Retrieval; 2011. p. 155–60.
7. Wang Q, Gan Y, Sun Y. RUC @ INEX 2011 datacentric track. In: Proceedings of the 10th International Workshop of the Initiative for the Evaluation of XML Retrieval; 2011. p. 167–79.
8. Trotman A, Lalmas M. Strict and vague interpretation of XML-retrieval queries. In: Proceedings of the 32nd Annual International ACM SIGIR Conference on Research and Development in Information Retrieval; 2009. p. 709–10.
9. Mass Y, Mandelbrod M. Using the INEX environment as a test bed for various user models for XML retrieval. In: Proceedings of the 4th International Workshop of the Initiative for the Evaluation of XML Retrieval; 2005. p. 187–95.
10. Mihajlovic V, Ramírez G, Westerveld T, Hiemstra D, Blok HE, de Vries AP. Vtijah scratches INEX 2005: vague element selection, image search, overlap, and

relevance feedback. In: Proceedings of the 4th International Workshop of the Initiative for the Evaluation of XML Retrieval; 2005. p. 72–87.

11. Doan A, Halevy AY. Semantic integration research in the database community: a brief survey. AI Mag. 2005;26(1):83–94.

12. van Zwol R. B³-sdr and effective use of structural hints. In: Proceedings of the 4th International Workshop of the Initiative for the Evaluation of XML Retrieval; 2005. p. 146–60.

13. Theobald M, Schenkel R, Weikum G. Topx and xxl at INEX 2005. In: Proceedings of the 4th International Workshop of the Initiative for the Evaluation of XML Retrieval; 2005. p. 282–95.

14. Hubert G. XML retrieval based on direct contribution of query components. In: Proceedings of the 4th International Workshop of the Initiative for the Evaluation of XML Retrieval; 2005. p. 172–86.

15. Sauvagnat K, Hlaoua L, Boughanem M. Xfirm at INEX 2005: ad-hoc and relevance feedback tracks. In: Proceedings of the 4th International Workshop of the Initiative for the Evaluation of XML Retrieval; 2005. p. 88–103.

16. O'Keefe RA If INEX is the answer, what is the question? In: Proceedings of the 3rd International Workshop of the Initiative for the Evaluation of XML Retrieval; 2004. p. 54–9.

17. Trotman A. Wanted: element retrieval users. In: Proceedings of the 4th International Workshop of the Initiative for the Evaluation of XML Retrieval; 2005. p. 63–9.

18. Kamps J, Marx M, Rijke MD, Sigurbjörnsson B. Articulating information needs in XML query languages. Trans Inf Sys. 2006;24(4):407–36.

19. Geva S. GPX – gardens point XML IR at INEX 2006. In: Proceedings of the 5th International Workshop of the Initiative for the Evaluation of XML Retrieval; 2006. p. 137–50.

20. Trotman A, Jia X-F, Crane M. Towards an efficient and effective search engine. In: Proceedings of the SIGIR 2012 Workshop on Open Source Information Retrieval; 2012. p. 40–7.

Processor Cache

Peter Boncz
CWI, Amsterdam, The Netherlands

Synonyms

CPU cache; Data cache; Instruction cache; L1 cache; L2 cache; L3 cache; Translation Lookaside Buffer (TLB)

Definition

To hide the high latencies of DRAM access, modern computer architecture now features a memory hierarchy that besides DRAM also includes SRAM cache memories, typically located on the CPU chip. Memory access first check these caches, which takes only a few cycles. Only if the needed data is not found, an expensive memory access is needed.

Key Points

CPU caches are SRAM memories located on the CPU chip, intended to hide the high latency of accessing off-chip DRAM memory. Caches are organized in cache lines (typically 64 bytes). In a fully-associative cache, each memory line can be stored in any location of the cache. To make checking the cache fast, however, CPU caches tend to have limited associativity, such that storage of a particular cache line is possible in only two or FOUR locations. Thus only two or four locations need to be checked during lookup (these are called 2- resp. 4-way associative caches). The cache hit ratio is determined by the spatial and temporal locality of the memory accesses generated by the running program(s).

Cache misses can either be compulsory misses (getting the cache lines of all used memory once), capacity misses (caused by the cache being too small to keep all multiply used lines in cache), or conflict misses (due to the limited associativity of the cache).

Most modern CPUs have at least three independent caches: an instruction cache to speed up executable instruction fetch, a data cache to speed up data fetch and store, and a Translation Lookaside Buffer (TLB) used to speed up virtual-to-physical address translation for both executable instructions and data. The TLB is not organized in cache lines, it simply holds pairs of (virtual, logical) page mappings, typically a fairly limited amount (e.g., 64). In practice, this mean that algorithms that repeatedly touch memory in more than 64 pages (whose size is often 4 KB) shortly after each other, run into

P

TLB thrashing. This problem can sometimes be mitigated by setting a large virtual memory page size, or by using special large OS pages (sometimes supported in the CPU with a separate, smaller, TLB for large pages).

Another issue is the tradeoff between latency and hit rate. Larger caches have better hit rates but longer latency. To address this tradeoff, many computers use multiple levels of cache, with small fast caches backed up by larger slower caches. Multi-level caches generally operate by checking the smallest Level 1 (L1) cache first; if it hits, the processor proceeds at high speed. If the smaller cache misses, the next larger cache (L2) is checked, and so on, before external memory is checked. As the latency difference between main memory and the fastest cache has become larger, some processors have begun to utilize as many as three levels of on-chip cache.

For multi-CPU and multi-core systems, the fact that some of the higher levels of cache are not shared, yet provide coherent access to shared memory, causes additional cache-coherency inter-core communication to invalid stale copies of cache lines on other cores when one core modifies it. In multi-core CPUs, an important issue is which cache level is shared among all cores – this cache level is on the one hand a potential hot-spot for cache conflicts, on the other hand provides an opportunity for very fast inter-core data exchange.

In case of sequential data processing, the memory controller or memory chipset in modern computers often detect this access pattern and start requesting the subsequent cache lines in advance. This is called hardware prefetching. Prefetching effectively allows to hide compulsory cache misses. Without prefetching, the effective memory bandwidth would equate cache line size divided my memory latency (e.g., 64/50 ns = 1.2 GB/s). Thanks to hardware prefetching, modern computer architectures reach four times that on sequential access. Modern CPUs also offer explicit prefetching instructions, which a software writer can exploit to perform (non-sequential) memory accesses in advance, hiding their latency. In database systems, such software prefetching has successfully been used in making hash-table lookup faster (e.g., in hash-join and hash-aggregation).

In database systems, a series of cache-conscious data storage layouts (e.g., DSM and PAX) have been proposed to improve cache line usage. Also, a number of cache-conscious query processing algorithms, such as cache-partitioned hash join and hash-join using memory prefetching, have been studied. In the area of data structures and theoretical computer science, there has recently been interest in cache-oblivious algorithms, that regardless the exact parameters of the memory hierarchy (number of levels, cache size, cache line sizes and latencies) perform well.

Cross-References

▶ Architecture-Conscious Database System
▶ Cache-Conscious Query Processing
▶ Disk
▶ Main Memory
▶ Main Memory DBMS

Profiles and Context for Structured Text Retrieval

Marijn Koolen[1] and Toine Bogers[2]
[1]Research and Development, Huygens ING, Royal Netherlands Academy of Arts and Sciences, Amsterdam, The Netherlands
[2]Department of Communication and Psychology, Aalborg University Copenhagen, Copenhagen, Denmark

Definition

The combination of structured information retrieval with user profile information represents the scenario where systems search with an explicit statement of the information need – a search query – as well as a profile of a user, which can contain information about previous interactions, search history, user demographics, or other relevant information about the user's preferences. The relation between the profile and the information need is implicit and may contain many irrelevant signals. The task of the system then is to model both the current information need and the background user preferences to derive notions of topical relevance as well as user relevance and to find the right balance between these notions to determine the optimal ranking of search results.

Historical Background

Information retrieval research has traditionally focused on locating documents that are relevant to a user's search query – an explicit statement of that user's underlying information need. However, it is well known that a query is only a limited and compromised representation of the underlying need [15]. Which documents are relevant to the user is related to the user's context, which includes among other things her background knowledge, interests, and what the requested information will be used for. A user's previous interactions with the system can be captured in a user profile to build a model of her preferences and interests. Such a profile can consist of, among others, a list of viewed or selected documents, bookmarks, ratings, reviews, tags, and other signals of interaction.

Early work on combining user profiles with search queries was based on literature search, where a user's profile consisted of papers previously collected by that user [3, 12]. This model was also used for the TREC (TExt Retrieval Conference, see also http://trec.nist.gov) Filtering Track [14], where the system has to filter an incoming stream of documents and track the topic(s) of interest to a user based on their previously selected documents. The TREC Contextual Suggestion Track [5], starting in 2012, addresses the related problem of suggesting activities for a user in a specific location, where there is no query but only a profile of user preferences. Just as the information retrieval community has investigated how to integrate user profile information into retrieval algorithms, the recommender

systems community has long attempted to integrate content-based information and user profile information [1, 2, 4, 13].

Due to the rise of social media, an increasing amount of information about the user can be exploited by search engines and recommender systems. Users have rich profiles on social network sites such as Facebook, LinkedIn, and Google+ that reveal their interests, activities, and their social circles. Other platforms allow users to rate, review, and tag books, movies, or songs and recommend new items based on these interactions. These are domains with highly structured information in the form of metadata and user-generated content. Searching in such collaboratively created catalogs provides an interesting scenario to study the value of user profiles for retrieval.

One such platform is the social cataloging site LibraryThing (LT) (http://www.librarything.com), where users can build up a personal catalog of books and rate, review, and tag these books or discuss them with other members. LT has a recommendation engine to support users in discovering new books, but users can also use directed search, which is the scenario that is investigated in the Social Book Search (SBS) tracks (http://social-book-search.humanities.uva.nl/) at INEX and CLEF [11]. The SBS track investigates book search in a large collection of book descriptions containing professional metadata and user-generated content, with an elaborate statement of user information needs and an extensive profile of those users, all of which have rich structure that can be used to support retrieval [10, 11].

Scientific Fundamentals

User profiles are typically used in recommender systems, where users rate items and the system derives implicit interests from these ratings to suggest other items to the user. These signals of interest may also help to contextualize the information need for directed search. However, few retrieval systems make use of such profiles to improve the ranking of retrieval results.

P

Exploiting user profiles for directed search poses a substantial challenge. The profile contains signals of implicit interest and background knowledge, but how should these signals be combined with the explicit statement of the specific information need expressed by the search query? How is a user profile related to the search topic and which of a user's interests are relevant to that topic?

Formally, a user u has an information need represented by a query q and related interests and background knowledge represented by a user profile p_u, and the challenge is to rank a collection D of documents d by relevance $R(d|q, p_u)$ such that the most relevant documents are ranked highest. Documents and profiles can have a rich metadata structure, allowing for complex retrieval and ranking models that carefully balance the evidence from q, p, and d [9].

The SBS evaluation campaign studies this problem in the context of the LT discussion forums, which are used to discuss a broad range of book-related topics. Many LT members turn to these forums asking for book suggestions with other members replying and providing suggestions. The LT discussion forums provide

a unique opportunity to unobtrusively investigate complex, realistic search scenarios with highly structured data in the form book metadata and user profiles. At the same time, the data from the forums is used to construct test collections for developing and evaluating search systems that combine structured retrieval with user profile information. Examples of book metadata that could be exploited for structured retrieval are book titles, author names, subject descriptors, product descriptions, and user tags and reviews. User profiles also contain personal tags, ratings, and reviews. The result of so many structural elements is a variety of perspectives on a book from different types of agents (authors, publishers, library catalogers, and readers) and for a broad range of information needs – combining aspects of topic, genre, style, mood, time, and interest. The challenge is then to identify and weight the strongest signals of relevance in this abundance of evidence.

The request in Fig. 1 is highly complex, providing requirements about the content, examples of books and authors that the poster of the request is already familiar with, and contextual cues on usage. The username links to their

Profiles and Context for Structured Text Retrieval, Fig. 1   Book request on the LibraryThing forum

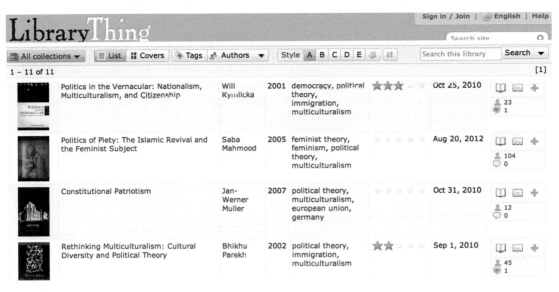

Profiles and Context for Structured Text Retrieval, Fig. 2 User profile on LibraryThing

profile, which provides contextual information on the user (Fig. 2), including the books they cataloged and how they tagged and rated them. The example books introduce a form of querying by example that could also be seen as a recommendation task. Other forum members reply in the thread with book suggestions relevant to the request. These suggestions represent human relevance judgments and are used, in combination with the user profile of the topic creator, to evaluate retrieval systems. The user profile of the requester contains information on when they added each book to their catalog, and the forum thread has timestamps on when the request was posted and when suggestions were made. Book suggestions in the forum thread that are cataloged by the requester after posting the request are considered more relevant than suggestions that the requester ignores or already cataloged before the request. This provides a complex picture of relevance that goes beyond topical relevance, touching on situational and user aspects of relevance.

Key Applications

There are several domains where search engines could make use of rich document structure and contextual information from user profiles, such as travel, online shopping, and scientific literature.

Travel Users who are organizing a trip and are searching options for travel, accommodation, and activities. This is a complex search task where specific requirements and personal preferences need to be balanced, and the available data is often highly structured (names, locations, dates, prices, ratings, reviews). Information about past travel plans could help direct the search process toward the most optimal arrangements.

Online shopping Shopping on e-commerce websites such as Amazon.com often involves complex information needs that require multiple searches to locate the most appropriate products. User profile information could help to personalize the search results and provide more relevant product suggestions.

Literature search There is rich structured information that could be exploited when a researcher searches for relevant literature to reference in their paper. The researcher's interest and background knowledge can be represented by their own publications and the publications they reference. Each individual publication has metadata such as journal or proceedings title, year, keywords, and references [2] .

Experimental Results

A well-established result in book search is that user-generated content is very effective for retrieving relevant suggestions for a range of tasks [8, 9]. The most effective approaches use a mixture of models for different types of metadata, essentially exploiting the principle of *polyrepresentation* (representing the same information by data from multiple sources).

Forum suggestions are different from traditional topical relevance judgments [9]. Forum members take into account writing style, engagement, humor, or how comprehensively a topic is discussed. Members have read most of the books they suggest in response to a request and mention books mostly positively [10].

The best performing systems use all available evidence for the information need (represented by different fields such as thread title, discussion group, and text of the first post), as they all provide different perspectives on the information need, again exploiting the principle of polyrepresentation.

The user profiles have been effectively used for combining retrieval results with collaborative filtering recommendations, where books are recommended if they are prominent in the catalogs of users similar to the requester [7]. More recently, Hafsi et al. [6] rank books from similar users and find that precision is low, but that recall goes up linearly after rank 500 whereas runs based on textual queries gain almost no recall that far down the results list.

Data Sets

Data sets in IR are typically test collections constructed in evaluation campaigns like TREC, CLEF, NTCIR, and INEX. Test collections with structured document collections and user profiles are available from the CLEF Social Book Search Lab (Available at http://social-book-search.humanities.uva.nl/.). The SBS test collection consists of a set of documents, user profiles, search topics, and relevance judgments.

The document set contains 2.8 million book descriptions with professional metadata from Amazon – such as title, author, publisher, and subject descriptors – and user-generated content from Amazon (user reviews and ratings) and LT (user tags). The search topics are made up of LT forum topics that are focused on a request for book suggestions and consist of the thread title, the initial post, and discussion group name. The relevance judgments are based on the forum suggestions by other LT users. In addition, there is a set of 94,610 user profiles containing the books that each user cataloged, the cataloging date, and user ratings and tags. Another relevant test collection containing structured document collections and user profiles is that of the TREC Contextual Suggestion track (Available at https://sites.google.com/site/treccontext/.).

Recommended Reading

1. Basilico J, Hofmann T. Unifying collaborative and content-based filtering. In: Proceedings of the 21st International Conference on Machine Learning; 2004. p. 9–16.
2. Bogers T, Van den Bosch A. Collaborative and content-based filtering for item recommendation on social bookmarking websites. In: Jannach D, Geyer W, Freyne J, Anand SS, Dugan C, Mobasher B, Kobsa A, editors, Proceedings of the ACM RecSys '09 Workshop on Recommender Systems and the Social Web; 2009. p. 9–16.
3. Chavarria-Garza H. A user profile-query model for document retrieval. PhD thesis, Southern Methodist University, 1982.
4. Claypool M, Gokhale A, Miranda T, Murnikov P, Netes D, Sartin M. Combining content-based and collaborative filters in an online newspaper. In: Proceedings of the ACM SIGIR Workshop on Recommender Systems; 1999.
5. Dean-Hall A, Clarke CLA, Kamps J, Thomas P, Simon N, Voorhees EM. Overview of the TREC 2013 contextual suggestion track. In: Voorhees EM, editor, The Twenty-Second Text REtrieval Conference (TREC 2013) Notebook. National Institute for Standards and Technology. NIST Special Publication; 2014. p. 302–500.
6. Hafsi M, Géry M, Beigbeder M. LaHC at INEX 2014: social book search track. In: Working Notes of the Conference and Labs of the Evaluation Forum; 2014.
7. Huurdeman HC, Kamps J, Koolen M, van Wees J. Using collaborative filtering in social book search. In:

CLEF 2012 Evaluation Labs and Workshop, Online Working Notes; 2012. p. 17–20. http://ceur-ws.org/Vol-1178/CLEF2012wn-INEX-HuurdemanEt2012.pdf

8. Koolen M. User reviews in the search index? That'll never work! In: Proceedings of the 36th European Conference on IR Research; 2014. p. 323–34. http://dx.doi.org/10.1007/978-3-319-06028-6_27.

9. Koolen M, Kamps J, Kazai G. Social book search: comparing topical relevance judgements and book suggestions for evaluation. In: Proceedings of the 21st ACM International Conference on Information and Knowledge Management; 2012. p. 185–94.

10. Koolen M, Kazai G, Preminger M, Doucet A. Overview of the INEX 2013 social book search track. In: Proceedings of the CLEF 2013 Evaluation Labs and Workshop; 2013.

11. Koolen M, Bogers T, Kamps J, Kazai G, Preminger M. Overview of the INEX 2014 social book search track. In: Working Notes for CLEF 2014 Conference. CEUR Workshop Proceedings, vol. 1180; 2014. p. 462–79. www.CEUR-WS.org

12. Korfhage RR, Chavarria-Garza H. Retrieval improvement by the interaction of queries and user profiles. Department of Computer Science and Engineering, Southern Methodist University, 1982.

13. Mahoui M, Bhargava B, Mohania M. User content mining supporting usage content for web personalization. CERIAS technical report 2001-21, Department of Computer Science, Purdue University, 2001.

14. Robertson S, Hull DA. The TREC-9 filtering track final report. 2000; p. 25–40.

15. Voorhees EM. The philosophy of information retrieval evaluation. In: Revised Papers from the Second Workshop of the Cross-Language Evaluation Forum on Evaluation of Cross-Language Information Retrieval Systems, CLEF'01. London: Springer; 2002. p. 355–70. ISBN:3-540-44042-9. http://dl.acm.org/citation.cfm?id=648264.753539

Projection

Cristina Sirangelo
IRIF, Paris Diderot University, Paris, France

Synonyms

Projection (Relational Algebra)

Definition

Given a relation instance R over set of attributes U, and given a subset X of U, the projection of R on X – denoted by $\pi_X(R)$ – is defined as a relation over set of attributes X whose tuples are the restriction of tuples of R to attributes X. That is, $t \in \pi_X(R)$ if and only if $t = t'(X)$ for some tuple t' of R (here $t'(X)$ denotes the restriction of t' to attributes X).

Key Points

The projection is one of the basic operators of the relational algebra. It operates by "restricting" the input relation to some of its columns.

The arity of the output relation is bounded by the arity of the input relation. Moreover the number of tuples in $\pi_X(R)$ is bounded by the number of tuples in R. In particular, the size of $\pi_X(R)$ can be strictly smaller than the size of R since different tuples of R may have the same values on attributes X.

As an example, consider a relation *Goods* over attributes (*code, price, quantity*), containing tuples {(001, 5.00, 10), (002, 5.00, 10), (003, 25.00, 3)}. Then $\pi_{price,quantity}$ (*Goods*) is a relation over attributes (*price, quantity*) with tuples {(5.00, 10), (25.00, 3)}.

In the case that attribute names are not present in the relation schema, the projection is specified by an expression of the form $\pi_{i_1,\ldots,i_n}(R)$. Here $i_1,\ldots,i_n$ is a sequence of positive integers – where each i_j is bounded by the arity of R – identifying attributes of R. The output is a relation of arity n with tuples $(t(i_1),\ldots,t(i_n))$, for each tuple t in R. In this case the arity n of the output relation can be possibly larger than the arity of R, since integers $i_1,\ldots,i_n$ need not be distinct.

Cross-References

► Relational Algebra

Propagation-Based Structured Text Retrieval

Karen Pinel-Sauvagnat
IRIT laboratory, University of Toulouse,
Toulouse, France

Synonyms

Relevance propagation; Score propagation

Definition

Propagation-based structured retrieval is a structure-based retrieval paradigm based on the logical structure of documents. Other structural retrieval paradigms include contextualization and content aggregation. Approaches using propagation view the logical structure of a structured document as a tree whose nodes are components of the document and whose edges represent the relationships between the connected nodes. A document component can be either a leaf or an inner node. Leaf nodes are document components that correspond to the last elements of hierarchical relationship chains and that contain raw data (textual information). With propagation, relevance scores are first calculated for leaf components. They are then propagated upward in the document tree structure to evaluate relevance scores for the inner components.

Historical Background

With appropriate query languages, users may want to exploit the structure of structured text documents to perform fine-grained and flexible retrieval. Instead of processing documents as atomic units, structured text retrieval systems aim at retrieving document components that answer a given information need in the most *specific* way. Classical information retrieval (IR) models (e.g., the vector space model, the probabilistic model, language models) have been extended in order to take into account the structural dimension of documents. IR research has shown that document term weighting is a crucial concept for effective retrieval; thus, classical formalisms were adapted with additional weighting parameters: component type, number of descendant components, and frequency of the component type, to name a few. However, most of these adaptations do not use (or make little use of) the tree representation of structured documents to identify the most specific components.

The idea behind relevance propagation method is to follow the way *relevance* changes in a document tree to estimate the relevance of the document components. Indeed *relevance* in structured text documents has been expressed in terms of *specificity* and exhaustivity: the specificity (its coverage of the topic and nothing else) of components in the tree typically decreases as one moves up the tree, and when a component has multiple relevant descendants, its exhaustivity (coverage of the topic) usually increases compared to that of each of the descendant elements.

In propagation methods, relevance scores are thus first computed at leaf level and then propagated up the document tree: to allow the identification of the most specific components, the relevance score of leaf components should be somehow decreased while being propagated upward in the document tree, and to allow the identification of the most exhaustive components, relevance scores may be aggregated (a parent score should be evaluated using its children scores). A naive solution would be to just sum the relevance scores of each inner component relevant child. However, this would ultimately result in root components being returned at top ranks although they may not be the most specific components for the given information need.

Few relevance propagation approaches were proposed before 2002 (when *INEX*, the Initiative for the Evaluation of XML Retrieval, was set up). One can however cite the method presented in [6] using inference nets. The retrieval process is applied to SGML documents but can be extended to any type of structured documents. The basic retrieval strategy is to calculate the degree to which a component at any level of the document hierarchy satisfies the query by considering what components are contained in that component. This strategy makes it possible to systematically calculate the expected relevance of a component at any level, taking into account its relationship with other components in the hierarchy. Consider the following net made of two section components S_1 and S_2 that are the parent node of a term node T and the children nodes of the component C (see Fig. 1).

The retrieval process is performed in a bottom-up fashion, because of the way documents are represented in the inference net: no components other than leaf components contain actual text, and the retrieval process must start from leaf components whose text contains a query term.

The degree of belief of T given the network topology is computed with a simplified formula as follows (the simplification comes from considering only positive events):

$$B(T|C) = P(T|S_1) \times P(S_1) + P(T|S_2) \times P(S_2)$$
$$= P(T|S_1) \times P(S_1|C) \times P(C) + P(T|S_2)$$
$$\times P(S_1|C) \times P(C). \qquad (1)$$

Children components S_i and their parent C are represented as $S_i = <s_{i1}, s_{i2}, \ldots, s_{in}>$ and $C = <c_1, c_2, \ldots, c_n>$, respectively, where n is the number of index terms and s_{ij} and c_i are calculated using standard term frequency (TF) and inverse document frequency (IDF) statistics. The probability $P(S_i|C)$ of observing S_i given C (or the degree of belief that C supports S_i) is calculated as a similarity between the two vectors:

$$P(S_i|C) = \lambda_i \times (S_i \cdot C), \qquad (2)$$

where λ_i is a weight associated to the component type, representing its overall importance relative to other types of components sharing the same parent. This computation incorporates both the content and the type of components.

The probability $P(T|S_i)$ of observing a term T given a component S_i (or the degree of belief that S_i supports T) is estimated with:

$$P(T|S_i) \cong IDF_T \times TF_{i,S_i}. \qquad (3)$$

$P(C)$ is the probability of observing C assuming that C is the root node (i.e., document) in the inference net. In the implementation, it is set to 1. This approach was however not evaluated, since no suitable test collection was available.

Other approaches presented in the rest of the entry were developed and evaluated in the context of the *INEX* evaluation campaign, which is concerned with the evaluation of XML retrieval. In this case, document components correspond to XML elements.

Foundations

As opposed to other structured ranking strategies such as contextualization or content aggregation, propagation-based approaches need an indexing strategy based on leaf nodes.

Indexing

Approaches using relevance propagation consider *indexing units* as disjoint units: the text of each element is the union of one or more of its disjoint parts. Thus, as textual information is only present in leaf elements, inverted index only concerns leaf elements. Propagation accounts then for the fact that the text in a given leaf element is also contained in its ancestors. The document structure is generally stored in a separate index and used to build the document trees.

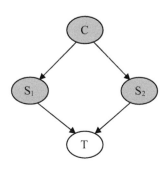

Propagation-Based Structured Text Retrieval, Fig. 1
A simple network

Relevance Score Evaluation for Content-Only Queries

In the rest of the entry, the following notations are used. Let q be a query composed of k terms $t_1, \ldots t_k$. In the document tree, le is a leaf element and e an inner element having n children. RSV (q, le) is the relevance score of the leaf element le with respect to query q, and $RSV(q, e)$ is the final score of element e with respect to query q.

Leaf elements are first scored, and their relevance scores are then used to evaluate the inner nodes relevance.

Scoring Leaf Elements

To evaluate leaf element scores ($RSV(q, le)$), approaches found in the literature have used the following parameters:

1. Frequency of term t_i in query q or leaf element le
2. Frequency of term t_i in the whole collection
3. Number of leaf elements containing t_i
4. Length of leaf element
5. Average length of leaf elements
6. Inverse element frequency *ief*, which is similar to *idf* (inverse document frequency) but that takes into account the collection of leaf elements instead of the collection of documents.

In [1], the weighting scheme used to compute the relevance score of leaf elements also uses the cross-structural importance of t_i relative to le and q, which allows to increase or decrease the importance of a term depending on its location in the query (some terms may be more important than others in a content-and-structure query). In [3], the frequency of terms in the leaf elements and in the whole collection is used with a parameter scaling up the score of elements having multiple query terms. The formula penalizes elements with frequently occurring query terms (frequent in the collection) and rewards elements with more unique query terms within a result element. One can also cite the leaf element weighting scheme presented in [9], where term frequency and inverse element frequency *ief* are applied together with *idf*. This allows to take into account the importance of terms in both the

collection of leaf elements and the collection of documents.

Propagating Relevance Scores

Once the relevance score of the leaf elements have been calculated, they are used to evaluate the relevance score of the inner elements. The main issue here is how to combine these scores. As already said, a naive solution that simply sums the relevance sores of leaf elements will result in root elements being ranked at the top of the result lists, although they are likely to not constitute the most specific elements to the query.

In [1] the relevance of inner elements is, for instance, evaluated using the maximum of their leaf element scores. Other approaches use a weighted sum of leaf or children element scores and make some assumptions related to the document tree structure. They are described below.

For example, in the GPX approach [3], a heuristically derived formula is proposed to evaluate the scores of inner elements that account for specificity and exhaustivity:

$$\text{RSV}(q, e) = D(n) \sum_{l=1}^{n} \text{RSV}(q, l), \quad (4)$$

where n is the number of children elements, $D(n) = 0.49$ if $n = 1$, 0.99 otherwise, and $RSV(q, l)$ is the relevance score of the lth child element.

The value of the decay factor D depends on the *number of relevant children* that the inner element has. If the element has one relevant child, then the decay constant is 0.49. An element with only one relevant child will be ranked lower than its child. If the element has multiple relevant children, the decay factor is 0.99. An element with many relevant children will be ranked higher than its descendants. Thus, a section with a single relevant paragraph would be judged less relevant than the paragraph itself, but a section with several relevant paragraphs will be ranked higher than any of the paragraphs.

"$t_2 t_3 t_4$." The method is illustrated in Fig. 2, with the content-only query "$t_2 t_3 t_4$." To simplify, the score of a leaf element is equal to the number

Propagation-Based Structured Text Retrieval, Fig. 2 Relevance propagation according to [3]

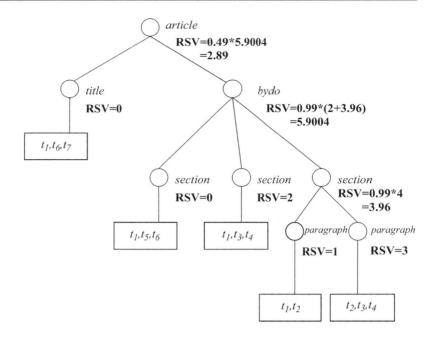

of query terms it contains. For instance, the /article[1]/body[1]/section[3]/paragraph[2] element that contains three query terms has thus a score of 3. Its parent (/article[1]/body[1]/section[3]) contains two relevant children with scores 1 and 3. Its score is then calculated with a decay factor equal to 0.99 as follows: 0.99 × (1 + 3). To evaluate the score of the root element (which only contains one relevant child), the decay factor used is 0.49, and the root element has consequently a lower score than its child /article[1]/body[1]. As a result of the propagation, the /article[1]/body[1] element has the highest score and will be ranked first by the GPX system.

The *distance between an element and its descendant leaf elements* has also been used as a weight in the weighted sum that calculates the relevance score of inner elements. Terms that occur close to the root of a given subtree are more significant to the root element than ones at deeper levels of the subtree. It seems therefore that the greater the distance of an element from its ancestor, the less it should contribute to the relevance of its ancestor. This can be modeled with the $d(x, y)$ parameter, which is the distance between elements x and y in the document tree, i.e., the number of arcs joining x and y.

In [4], the relevance score of an inner element takes into account this distance and also the distance separating the root element and leaf elements. The latter is used as a normalization factor. The relevance score of an inner element e is calculated as follows:

$$
\mathrm{RSV}(q, e)
$$
$$
= \sum_{le_j \in L_e} \left(1 - 2\lambda\lambda \frac{d(e, le_j)}{d(e, le_j) + d(root, le_j)}\right)^2
$$
$$
\times \mathrm{RSV}(q, le_j). \tag{5}
$$

λ is a constant coefficient ≥ 0, le_j are leaf elements being descendant of e, and L_e is the set of leaf elements being descendant of e. This process tends to consider an element having a relevant descendant element less relevant than the descendant element itself, which is comparable to the approach followed in [3], as above described.

Finally, in [9], the distance between elements and the number of relevant descendant leaf elements are used together with an additional parameter, β, in the weighted sum. The β factor captures the assumption that small elements may be used by authors to highlight important information (*title* elements, *bold* elements, etc.). *Small elements can therefore give important indications on the relevance of their ancestors,* and their importance should be increased during propagation. The relevance value of an element *e* is thus computed according to the following formula:

$$\text{RSV}(q, e) = |L_e^r| . \sum_{le_j \in L_e} \alpha^{d(e, le_j) - 1} \times \beta(le_j)$$

$$\times \text{RSV}(q, le_j), \qquad (6)$$

where α is a constant coefficient $\in]0,1]$ used to tune the importance of the distance $d(e, le_j)$ parameter and $|L_e^r|$ is the number of leaf elements being descendant of *e* and having a nonzero relevance value. $\beta(le_j)$ is experimentally fixed and allows to increase the role of elements smaller than the average leaf element size in the propagation function.

Although upward propagation is the most used in state-of-the-art approaches, backward propagation has also been used in Sauvagnat et al. [9] or Laitang et al. [5] in order to account for the whole document relevance (and thus for the element context) in the calculation of the relevance score of inner elements. Laitang et al. also use siblings propagation to better represent the context.

Content-and-Structure Query Processing

In most of the approaches using relevance propagation, result elements are simply filtered to satisfy the structural constraints of *content-and-structure queries* [3, 4]. The approach described in [10] however uses relevance propagation to process structural constraints. Queries may have several constraints, and each constraint is composed of both a content and a structure condition, one of them indicating which type of elements should be returned (target elements). For each constraint, propagation starts from leaf nodes answering the content condition and goes until an element that matches the structure constraint is found. The score of result elements of each constraint is then propagated again to elements belonging to the set of targeted structures.

Key Applications

The access to structured text documents such as SGML or XML documents is the main application of the relevance propagation approach described in this entry. Relevance propagation from a component to another can also be used to access HTML documents in the context of web retrieval, for example, to determine among a set of related web pages which one corresponds to the best entry in the set. In [8], link information is used as a source of evidence to find relevant web pages (a relevant page may be linked to other relevant pages). Moreover, relevance propagation has also been used in distributed IR [2], expert finding [12], or entity ranking [11].

Experimental Results

Good performances were achieved with relevance propagation methods in the *INEX* evaluation campaign. For example, the approach in [3] was ranked in the top 5 for the focused retrieval task (which aims at targeting the appropriate level of granularity of relevant content that should be returned to the user for a given topic), and comparable results were achieved by the approach described in [9] using content-and-structure queries.

Data Sets

Test collections have been released by the INitiative for the Evaluation on XML retrieval (INEX) since 2002 (https://inex.mmci.uni-saarland.de).

Cross-References

▶ Aggregation-Based Structured Text Retrieval
▶ Content-and-Structure Query
▶ Content-Only Query
▶ Initiative for the Evaluation of XML Retrieval
▶ Relevance
▶ Specificity
▶ Structured Document Retrieval
▶ Term Statistics for Structured Text Retrieval

Recommended Reading

1. Anh VN, Moffat A. Compression and an IR approach to XML Retrieval. In: Proceedings of the 1st International Workshop of the Initiative for the Evaluation of XML Retrieval; 2002.
2. Baumgarten C. A probabilistic model for distributed information retrieval. In: Proceedings of the 20th Annual International ACM SIGIR Conference on Research and Development in Information Retrieval; 1997. p. 258–66.
3. Geva S. GPX – gardens point XML IR at INEX 2005. In: Proceedings of the 4th International Workshop of the Initiative for the Evaluation of XML Retrieval; 2005. p. 240–53.
4. Hubert G. XML retrieval based on direct contribution of query components. In: Proceedings of the 4th International Workshop of the Initiative for the Evaluation of XML Retrieval; 2005. p. 172–86.
5. Laitang C, Pinel-Sauvagnat K, Boughanem M. DTD-based costs for tree-edit distance in structured information retrieval. In: Proceedings of the 35th European conference on IR research; 2013. p. 158–70.
6. Myaeng S-H, Jang D-H, Kim M-S, Zhoo Z-C. A flexible model for retrieval of SGML documents. In: Proceedings of the 21st Annual International ACM SIGIR Conference on Research and Development in Information Retrieval; 1998. p. 138–45.
7. Ogilvie P, Callan J. Parameter estimation for a simple hierarchical generative model from XML retrieval. In: Proceedings of the 4th International Workshop of the Initiative for the Evaluation of XML Retrieval; 2005. p. 211–24.
8. Qin T, Liu T-Y, Zhang X-D, Chen Z, Ma W-Y. A study of relevance propagation for web search. In: Proceedings of the 31st Annual International ACM SIGIR Conference on Research and Development in Information Retrieval; 2008. p. 408–15.
9. Sauvagnat K, Boughanem M, Chrisment C. Answering content-and-structure-based queries on XML documents using relevance propagation. Inf Syst. 2006;31(7):621–35.
10. Sauvagnat K, Hlaoua L, Boughanem M. XFIRM at INEX 2005: adhoc and relevance feedback tracks. In: Proceedings of the 4th International Workshop of the Initiative for the Evaluation of XML Retrieval; 2005. p. 88–103.
11. Serdyukov P, Rode H, Hiemstra D. Modeling multi-step relevance propagation for expert finding. In: Proceedings of the 17th ACM International Conference on Information and Knowledge Management; 2008. p. 1133–42.
12. Tsikrika T, Serdyukov P, Rode H, Westerveld T, Aly R, Hiemstra D, de Vries AP. Structured document retrieval, multimedia retrieval, and entity ranking using PF/Tijah. In: Proceedings of the 6th International Workshop of the Initiative for the Evaluation of XML Retrieval; 2007. p. 306–20.

Protection from Insider Threats

William R. Claycomb and John McCloud
CERT Insider Threat Center, Software
Engineering Institute, Carnegie Mellon
University, Pittsburgh, PA, USA

Synonyms

Insider; Spy

Definition

Insider threats can be either *malicious* or *unintentional*.

A *malicious insider threat* is any current or former employee, contractor, or other business partner who has or had authorized access to an organization's network, system, or data and intentionally exceeded or misused that access in a manner that negatively affected the confidentiality, integrity, or availability of the organization's information or information systems.

An *unintentional insider threat* is a current or former employee, contractor, or other business partner who has or had authorized access to an organization's network, system, or data and who, through action/inaction and without malicious intent, causes harm or substantially increases the probability of future harm to the confidentiality, integrity, or availability of the organization's information or information systems.

The content of this entry will describe malicious insider threats only.

Malicious insiders pose a substantial threat by virtue of their knowledge of, and access to, their employers' systems and/or databases and can bypass existing physical and electronic security measures through legitimate measures. The target of a malicious insider's behavior is most often the critical assets of an organization: people, technology, data, money, and/or facilities. Impacts include damage to the confidentiality, availability, and/or integrity of these critical assets.

Through empirical analysis [1], insider activities have been categorized broadly into three distinct categories:

- Sabotage of information technology (IT): an insider's use of IT to direct specific harm at an organization or an individual
- Theft of intellectual property: removal of proprietary or sensitive information by an insider
- Fraud: an insider's use of IT for the unauthorized modification, addition, or deletion of an organization's data (not programs or systems) for personal gain or theft of information which leads to fraud

Most insider incidents fall into one or more of the three categories listed above. However, other types of malicious insider activities exist, including:

- Unauthorized disclosure, such as broadly releasing sensitive information to the public domain
- Unauthorized use of IT for personal satisfaction, such as inappropriately accessing medical records of high-profile individuals

References

1. Cappelli D, Moore AP, Trzeciak RF. The CERT guide to insider threats. 1st ed. New York: Addison-Wesley Professional; 2012.

Provenance and Reproducibility

Fernando Chirigati[1] and Juliana Freire[1,2,3]
[1]NYU Tandon School of Engineering, Brooklyn, NY, USA
[2]NYU Center for Data Science, New York, NY, USA
[3]New York University, New York, NY, USA

Synonyms

Computational reproducibility

Definition

A computational experiment composed by a sequence of steps S created at time T, on environment (hardware and operating system) E, using data D is *reproducible* if it can be executed with a sequence of steps S' (modified from or equal to S) at time $T' > T$, on environment E' (potentially different than E), using data D' that is similar to (or the same as) D with consistent results [5]. *Replication* is a special case of reproducibility where $S' = S$ and $D' = D$. While there is substantial disagreement on how to define reproducibility [1], in particular across different domains, in this entry, we focus on *computational reproducibility*, i.e., reproducibility for computational experiments or processes.

The information needed to reproduce an experiment can be obtained from its provenance: the details of how the experiment was carried out and the results it derived. For computational experiments, provenance can be systematically and transparently captured. In addition to enabling reproducibility, provenance allows results to be *verified*, i.e., one can determine whether the experiment solves the problem it claims to solve, which helps identify common issues and bias in research such as p-hacking. Provenance also describes the chain of reasoning used to derive the results, which provides additional insights when the findings are reproduced or extended for other experiments.

Historical Background

The concept of reproducible research for computational experiments is similar in nature to the ideas on literate programming introduced by Knuth [6]. Claerbout and his colleagues coined the term reproducible research [3] and pioneered the adoption of reproducibility in geophysics research.

The ability to reproduce experiments is a requirement in the scientific process and the benefits of reproducibility have long been known. Readers (consumers) can validate and compare methods published in the literature, as well as build upon previous work. In fact, reproducibility may increase the value of subsequent work: if a better method is developed, and the original paper is reproducible, it is possible for researchers to quantify the validity of their own ideas in comparison to statements made in the original publication. Authors (producers) also benefit in multiple ways. Making an experiment reproducible forces the researcher to document execution pathways. This in turn enables the pathways to be analyzed (and audited). It also helps newcomers to get acquainted with the problem and tools used. Furthermore, reproducibility forces portability which simplifies the dissemination of the results. Last, preliminary evidence exists that reproducibility increases impact, visibility, and research quality [10, 11] and helps defeat self-deception [9].

Although a standard in natural science and in Math, where results are accompanied by formal proofs, the task of reproducibility has not been widely applied for results backed by computational experiments. Scientific papers published in conferences and journals present a large number of tables, plots, and beautiful pictures that summarize the obtained results but that loosely describe the steps taken to derive them. Not only can the methods and the implementation be complex, but their configuration may require setting many parameters. Consequently, reproducing the results from scratch is both time-consuming and error-prone and sometimes impossible [8]. As a matter of fact, studies have shown that the number of scientific publications that can be effectively reproduced is far from ideal [2, 4, 7, 11].

The credibility crisis in computational science has led to many efforts aimed at establishing the publication of reproducible results the norm. Funding agencies in the USA, top-tier conferences and journals, and academic institutions have started to encourage and sometimes require authors to create reproducible experiments, i.e., to publish data and code needed to reproduce results presented in publications. Many different techniques and open-source tools have been developed to facilitate the practice of reproducible research.

Scientific Fundamentals

Provenance Components for Reproducibility. To enable reproducibility, provenance should be executable, and include *input data, specification of experiment*, and *specification of environment*.

The **input data** includes the data D used in the experiment, parameters, and additional variables, which must be included in extension (e.g., a text file) or in intension (e.g., a script that generates the data). It may be also useful to include intermediate and output data, in case some steps of the experiment cannot be fully reproduced (e.g., data generated by a third-party software that cannot be openly distributed). **Specification of experiment** relates to a description of the experiment and the set of computational steps S executed. Examples of specifications include scripts that glue together the different pieces of an experiment and workflows created by authors using scientific workflow systems. The **specification of environment** consists of information about the computational environment E where the experiment was originally created and executed. It includes the operating system (OS), hardware architecture, and library dependencies.

Gathering and managing this provenance is challenging because experiments tend to use different software systems, library dependencies, programming languages, and platforms. In addition, the specification of environment may not be trivial to capture, as operating systems and dependencies are often complex and have many stateful components. Tools are available that simplify this process (see below).

Granularity. Provenance can be captured at different levels of granularity. *OS-based capture* works at the OS level, i.e., capturing system calls, kernel information, and computational processes. Therefore, a detailed description with respect to the specification of environment can be captured. However, because this provenance is fine-grained, it may be hard to reconcile it with the different computational steps of the experiment and to provide a high-level, human-readable representation. *Workflow-based*

P

capture uses a scientific workflow system to track and store the computational steps (wrapped in modules) and their data dependencies, but information about the environment is rarely gathered. *Code-instrumented capture* entails instrumenting the source code, automatically or manually through annotations, to capture the provenance components. Similar to workflows, the input data and specification of experiment are captured, but not the environment specification.

Axes of Reproducibility. Computational reproducibility may come in different levels depending on how much provenance information is captured. Full reproducibility, although desirable, may be hard or impossible to attain. To characterize these different levels, three axes of reproducibility were introduced [5]: *transparency level*, *portability*, and *coverage*.

The *transparency level* considers how much information with respect to data and computational steps is available for an experiment. This relates to the capture of input data and specification of experiment. The default level today is to provide a document with a set of figures (and captions) that represent the computational results. The level increases by providing (i) the input data D and intermediate results derived by the experiment, (ii) the binaries used to generate the figures included in the document, (iii) a high-level, human-readable description of the experiment (e.g., a workflow), and (iv) the source code, which allows further exploration of the experiment.

Portability is related to the ability to run an experiment in different environments and depends on the available specification of the environment. Computational results may be reproduced (i) on the original environment E, (ii) on similar environments (i.e., compatible operating system but different machines), or (iii) on different environments (i.e., different operating systems and machines).

Coverage takes into account how much of the experiment can be reproduced, i.e., if the experiment can be (1) partially or (2) fully reproduced. Many experiments cannot be fully reproduced, including experiments that rely on data

derived by third-party Web services or special hardware or that require nondeterministic computational processes. But such experiments can, sometimes, be partially reproduced: a sequence of steps $S' \subset S$ is often available. For example, if the data derived by a special hardware is made available, although the data derivation step cannot be reproduced, the subsequent analysis that depends on this data can still be performed. The level of coverage is higher if more detailed provenance (for all the components) is captured.

Note that these axes, although related, are independent from each other. An experiment may have a high coverage (e.g., data for all the steps is available) but a low transparency (e.g., no source code or binaries are available) or vice versa. Besides, even with high coverage and transparency, the experiment may still not be portable (e.g., if only binaries are made available, it is hard to run in other operating systems).

Reproducibility Modes. One can *plan for reproducibility* and capture provenance as an experiment is carried out, or attain it *after the fact* – after the experiment is completed. The former enables the capture of the entire history of the experiment, which helps understand the chain of reasoning behind all the steps. In the latter, the nature of the experiment determines how to capture the provenance for reproducibility purposes. If the source code is available, it is possible to instrument the code to capture the required information. On the other hand, if only binaries are provided, provenance can be captured at the OS level.

Key Applications

Scientific Workflow Systems (SWS). In these systems, provenance capture is tied to the workflow definition (workflow-based capture). They capture and store the *input data* as well as the *specification of experiment* as a workflow, together with a description of the modules that compose this workflow. However, the *specification of environment* is rarely captured: workflows usually do not include in their bundles precise information about the computational environment (e.g., library dependencies), which hampers portability. Coverage will depend on how much of the experiment is represented in the workflow. The level of transparency often depends on how the underlying computational steps are wrapped for integration with SWS. Since modules are black boxes, the more information is abstracted within a module, the lower the transparency is. Examples of workflow systems include VisTrails, Taverna, Kepler, and Galaxy.

Programming Tools and Software Packages. Tools are available for standard programming environments that can gather provenance without requiring researchers to port (or wrap) their experiment to other systems such as workflow systems. Some of them involve instrumenting the source code for the provenance capture (code-instrumented capture). The captured provenance includes *input data* and *specification of experiment*, but it often lacks full details for the *specification of environment*. Examples include Sumatra, noWorkflow, and ProvenanceCurious.

Literate programming tools encompass another class of tools relevant to this category: they interleave source code and text in natural language, which allows users to integrate code fragments with graphical, numerical, and textual output. Similar to the previous tools, they often capture *input data* and *specification of experiment*, but portability cannot be entirely guaranteed. Sweave and Dexy are examples of such tools. Last but not least, interactive notebooks are closely related to literate programming tools, but instead of requiring a separate compilation step, they are based on interactive worksheets. Examples include Jupyter and Sage.

Packaging Applications. These tools create executable compendiums that allow experiments to be reproduced in environments E' potentially different from the original environment E, usually having high coverage, transparency, and portability. Linux-based packing tools, such as CDE and ReproZip, use OS-based capture to obtain all the three provenance components related to an experiment. Using the captured information, these tools create an executable package for the experiment that can be used to reproduce the original results.

Executable Documents. A number of tools have been developed to aid researchers in creating *executable documents*, which can be defined as provenance-rich files that, in addition to text, also include the computational objects (e.g., code and data) used to generate the experimental results of a publication. They can blend static and dynamic content, and are often represented by HTML pages or enhanced PDF files, allowing researchers to interact and reproduce the findings described by authors. The creation of executable documents is a feature of many tools, such as VisTrails, Galaxy, and most literate programming tools and interactive notebooks. Other tools (e.g., Janiform) are entirely focused on producing these documents: they do not capture provenance but they encapsulate the provenance within documents.

Repositories. Repositories have been created to host, store, and share computational experiments. Although provenance capture is often not supported, these repositories aim at preserving the provenance components to allow experiments to be maintained and reproduced long after they were created. Repositories such as myExperiment, Dataverse, DataONE, and figshare support the archival of experiments that can be downloaded at any time by researchers, while others such as crowdLabs also allow results to be reproduced through a Web browser.

URL to Code

- CDE: http://www.pgbovine.net/cde.html
- crowdLabs: http://www.crowdlabs.org/
- DataONE: https://www.dataone.org/
- Dataverse: https://dataverse.harvard.edu/
- Dexy: http://www.dexy.it/
- figshare: https://figshare.com/
- Galaxy: https://galaxyproject.org/

- Janiform: https://github.com/uds-datalab/PDBF
- Jupyter: http://jupyter.org/
- Kepler: https://kepler-project.org/
- myExperiment: http://www.myexperiment.org/
- noWorkflow: https://github.com/gems-uff/noworkflow
- ProvenanceCurious: https://github.com/rezwan4438/ProvenanceCurious
- ReproZip: https://vida-nyu.github.io/reprozip/
- Sage: https://cloud.sagemath.com/
- Sumatra: https://pythonhosted.org/Sumatra/
- Sweave: https://www.statistik.lmu.de/~leisch/Sweave/
- Taverna: http://www.taverna.org.uk/
- VisTrails: http://www.vistrails.org/

Cross-References

► Data Provenance
► Provenance in Databases
► Provenance in Scientific Databases
► Provenance in Workflows
► Provenance Standards

Recommended Reading

1. Baker M. Muddled meanings hamper efforts to fix reproducibility crisis. Nature News & Comment. 14 Jun 2006 (2016).
2. Bonnet P, Manegold S, Bjørling M, Cao W, Gonzalez J, Granados J, Hall N, Idreos S, Ivanova M, Johnson R, Koop D, Kraska T, Müller R, Olteanu D, Papotti P, Reilly C, Tsirogiannis D, Yu C, Freire J, Shasha D. Repeatability and workability evaluation of SIGMOD'2011. ACM SIGMOD Rec. 2011;40(2):45–8.
3. Claerbout J, Karrenbach M. Electronic documents give reproducible research a new meaning. In: Proceedings of the 62nd Annual International Meeting of the Society of Exploration Geophysics; 1992. p. 601–4.
4. Collberg C, Proebsting T, Warren AM. Repeatability and benefaction in computer systems research. Technical report. TR 14-04, University of Arizona; 2015.
5. Freire J, Bonnet P, Shasha D. Computational reproducibility: state-of-the-art, challenges, and database research opportunities. In: Proceedings of the 2012 ACM SIGMOD International Conference on Management of Data; 2012. p. 593–6.
6. Knuth DE. Literate programming. Comput J. 1984;27(2):97–111.
7. Kovacevic J. How to encourage and publish reproducible research. In: Proceedings of the IEEE International Conference on Acoustics, Speech and Signal Processing; 2007. p. IV-1273–6.
8. LeVeque R. Python tools for reproducible research on hyperbolic problems. Comput Sci Eng. 2009;11(1):19–27.
9. Nuzzo R. How scientists fool themselves, and how they can stop. Nature. 2015;526(7572):182–5.
10. Piwowar HA, Day RS, Fridsma DB. Sharing detailed research data is associated with increased citation rate. PLoS One. 2007;2(3):e308.
11. Vandewalle P, Kovacevic J, Vetterli M. Reproducible research in signal processing – what, why, and how. IEEE Signal Process Mag. 2009;26(3):37–7.

Provenance in Databases

James Cheney[1] and Wang-Chiew Tan[2]
[1]University of Edinburgh, Edinburgh, UK
[2]University of California-Santa Cruz, Santa Cruz, CA, USA

Synonyms

History; Lineage; Origin; Pedigree; Source

Definition

Let t be a data element in the result of a query Q applied to a dataset D. The *provenance* of t is the set of all proofs for t according to Q and D. A proof for t according to Q and D is a subset D' of data elements in D so that t is in the result of applying Q on D'. In some cases, a proof also details the process by which t is derived from Q and D'.

Most work on provenance in databases focused on finding minimal subsets of D that witness the existence of t in the result, as well as which parts of D are t copied from. More general forms of provenance based on annotations (e.g., elements of algebraic structures such as semirings) have also been investigated. Provenance is also important for understanding how data in databases has evolved as a result of updates over time, particularly in *curated* scientific databases.

Historical Background

Data provenance (or fine-grained provenance) is an account of the derivation of a piece of data in a dataset that is typically the result of executing a database query against a source database [8]. This type of provenance determines the parts of source data that were used to generate a piece of data in the resulting dataset. Typically, data provenance is obtained by carefully reasoning about the algebraic form of the database query and the underlying data model of the source and resulting databases. In contrast, transformations occurring in scientific workflows are often external processes (e.g., Perl scripts), and the log files of scientific workflows provide only object identifiers to pieces of data involved in the transformation at best. Often, such external processes and data involved do not possess good properties for detailed analysis. Hence, a fine-grained analysis of an external process is not always possible, and the provenance recorded for such transformation is usually more coarse grained.

Scientific Fundamentals

Here, a (small) biological database (Fig. 1) is used as an example for illustrating data provenance. An example SQL query is shown below.

```
SELECT swissprot, pir, s.desc
  FROM SWISS-PROT s, PIR p,
       Mapping-Table m
  WHERE s.id = m.swissprot AND
        p.id = m.pir
```

The result of executing the SQL query on the example data shown in Fig. 1 consists of two tuples (a231, p445, AB) and (w872, p267, CD). Work on data provenance [6, 11] has provided explanations for why a tuple, such as (a231, p445, AB), is in the query result. The reason is because the first tuples in SWISS-PROT, PIR, and Mapping-Table, respectively, joined according to the query to produce the output tuple. This type of provenance is termed *why-provenance* in [6]. In contrast, the *where-provenance* of "AB" in the output tuple (i.e., where "AB" is copied from) is the desc attribute of the first tuple in SWISS-PROT. In particular, no other source tuples are involved in the explanation of where-provenance. Subsequent work by Green et al. [16] is also able to explain *how* a tuple is derived in the result of a query using annotations based on algebraic structures, such as semirings. Extensions of this model to consider negation, aggregation, and XML data have all been considered. Cheney, Chiticariu, and Tan [9] formulate, compare, and survey research on why-, where-, and how-provenance, and Karvounarakis and Green [17] survey subsequent developments on the semiring-annotated data model.

Another line of work captures provenance by propagating annotations of source data to the output, based on provenance, along query transformations. Wang and Madnick [19] first articulated the idea of using propagated annotations to analyze source attribution. Subsequently, Buneman et al. [7] studied the *annotation placement problem* using a variation of the propagation scheme of [19]. The propagation scheme of [7, 19] is essentially based on where data is copied from. This scheme forms the default propagation scheme of the DBNotes system (Bhagwat et al. [4]). A serious drawback of the default scheme is that two equivalent SQL queries (select-project-join-union queries) may not propagate annotations in the same way. DBNotes proposed an alternative default-all scheme to overcome this limitation.

Provenance in Databases, Fig. 1 Example databases

Swiss-Prot

id	desc
a231	AB
w872	CD
u812	DD

PIR

id	desc
p445	AB
p267	CD
p547	ED

Mapping-Table

mid	swissprot	pir
1	a231	p445
2	w872	p267

In the default-all propagation scheme, equivalent queries always propagate annotations in the same way. Additionally, DBNotes also supports the custom propagation scheme, where one is allowed to customize a propagation scheme as desired. So far, all systems [4, 7, 19] only allow annotations on attribute values.

Subsequent research efforts have proposed techniques to relax this restriction. Representative systems exploring different points in this design space include Trio, Orchestra, Perm/GProM, and ProQL. Trio [3] is a database system implementing the *databases with uncertainty and lineage* model. In Trio, source data is annotated with probabilities, and derived data is accompanied by lineage explaining how it was derived. The lineage is also used to compute probabilities associated with derived results. In the Perm [14] system, data and its provenance are represented together in a single relation, and provenance queries are rewritten to standard SQL queries, leveraging existing query optimization techniques. The GProM [1] system builds upon Perm's query rewriting approach to support updates, transactions, and operation-spanning transactions. In the ProQL [18] system, SQL is extended to support semiring provenance queries, and processing and indexing techniques are investigated for evaluating them.

The database provenance techniques discussed so far provide little help for managing provenance in curated databases. This is because curated databases are, by definition, not constructed from the result of executing a query but rather, created manually by scientists through the analysis of information from several sources. Buneman et al. [5] proposed a copy-and-paste model that captures a scientist's manual curation efforts as provenance-aware transactions. These transactions are logged and various hierarchical compression techniques were proposed and validated experimentally. The LIVE system [12] combines the ULDB model with support for updates to the data, which are fully versioned; lineage is also used to maintain derived data incrementally. Archer et al.'s *multi-granularity, multi-provenance* (MMP) model [2]

is a conceptual model for curated relational data that tracks provenance for database updates at multiple levels (e.g., field vs. row) and allows users to express degrees of confidence or doubt about data.

Key Applications

Data quality: By providing the source data and transformations, provenance may help to estimate data quality and data reliability.

Probabilistic databases: Provenance has been used to compute the confidence in the result of a probabilistic query correctly, as well as to correctly reason about the set of possible instances in the result of the query [3].

Data sharing and data integration: Provenance has also been used to describe trust policies in the data sharing system Orchestra and to prioritize updates, as well as to understand and debug a data exchange or data integration specification [10].

View maintenance and updates: Provenance or lineage can be used as a framework for view maintenance [12, 15] and view updates [13].

URL to Code

DBNotes project: http://www.soe.ucsc.edu/~wctan/Projects/dbnotes

Orchestra project: http://www.cis.upenn.edu/~zives/orchestra

SPIDER project: http://www.soe.ucsc.edu/wctan/~Projects/debugger/index.html

Trio project: http://infolab.stanford.edu/trio/

PERM project: http://www.cs.iit.edu/~dbgroup/research/perm.php

GProM project: http://www.cs.iit.edu/~dbgroup/research/gprom.php

Cross-References

▸ Data Provenance
▸ Provenance in Workflows
▸ Provenance Standards

Recommended Reading

1. Arab B, Gawlick D, Radhakrishnan V, Guo H, Glavic B. A generic provenance middleware for database queries, updates, and transactions. In: Proceedings of the 6th USENIX Workshop on the Theory and Practice of Provenance; 2014.
2. Archer DW, Delcambre LML, Maier D. User trust and judgments in a curated database with explicit provenance. In: In search of elegance in the theory and practice of computation. Lecture notes in computer science, vol. 8000. Heidelberg: Springer; 2013. p. 89–111.
3. Benjelloun O, Sarma AD, Halevy AY, Theobald M, Widom J. Databases with uncertainty and lineage. VLDB J. 2008;17(2):243–64.
4. Bhagwat D, Chiticariu L, Tan W-C, Vijayvargiya G. An annotation management system for relational databases. Very Large Data Bases (VLDB) J. 2005;14(4):373–96.
5. Buneman P, Chapman A, Cheney J. Provenance management in curated databases. In: Proceedings of the ACM SIGMOD International Conference on Management of Data; 2006. p. 539–50.
6. Buneman P, Khanna S, Tan W-C. Why and where: a characterization of data provenance. In: Proceedings of the 8th International Conference on Database Theory; 2001. p. 316–30.
7. Buneman P, Khanna S, Tan W-C. On propagation of deletions and annotations through views. In: Proceedings of the 21st ACM SIGACT-SIGMOD-SIGART Symposium on Principles of Database Systems; 2002. p. 150–8.
8. Buneman P, Tan W-C. Provenance in databases. In: Proceedings of the ACM SIGMOD International Conference on Management of Data; 2007. p. 1171–3. (Tutorial Track).
9. Cheney J, Chiticariu L, Tan WC. Provenance in databases: why, how, and where. Found Trends Databases. 2009;1(4):379–474.
10. Chiticariu L, Tan W-C. Debugging schema mappings with routes. In: Proceedings of the 32nd International Conference on Very Large Data Bases; 2006. p. 79–90.
11. Cui Y, Widom J, Wiener JL. Tracing the lineage of view data in a warehousing environment. ACM Trans Database Syst. 2000;25(2):179–227.
12. Das Sarma A, Theobald M, Widom J. LIVE: a lineage-supported versioned DBMS. In: Proceedings of the 22nd International Conference on. Scientific and Statistical Database Management; 2010.
13. Fegaras L. Propagating updates through XML views using lineage tracing. In: Proceedings of the 26th International Conference on Data Engineering; 2010. p. 309–20.
14. Glavic B, Alonso G. Perm: processing provenance and data on the same data model through query rewriting. In: Proceedings of the 25th International Conference on Data Engineering; 2009.
15. Green TJ, Ives ZG, Tannen V. Reconcilable differences. Theory Comput Syst. 2011;49(2):460–88.
16. Green TJ, Karvounarakis G, Tannen V. Provenance semirings. In: Proceedings of the 26th ACM SIGACT-SIGMOD-SIGART Symposium on Principles of Database Systems; 2007.
17. Karvounarakis G, Green TJ. Semiring-annotated data: queries and provenance. ACM SIGMOD Rec. 2012;41(3):5–14.
18. Karvounarakis G, Ives ZG, Tannen V. Querying data provenance. In: Proceedings of the ACM SIGMOD International Conference on Management of Data; 2010.
19. Wang Y, Madnick SE. A polygen model for heterogeneous database systems: the source tagging perspective. In: Proceedings of the 16th International Conference on Very Large Data Bases; 1990. p. 519–38.

Provenance in Scientific Databases

Sarah Cohen-Boulakia[1] and Wang-Chiew Tan[2]
[1]University Paris-Sud, Orsay Cedex, France
[2]University of California-Santa Cruz, Santa Cruz, CA, USA

Synonyms

History; Lineage; Origin; Pedigree; Source

Definition

Scientific databases contain data which may have been produced as answer to a query posed over other resources, or generated by in silico experiments (or scientific workflow) involving various softwares, or manually curated by domain experts based on analysis of several other resources. The provenance of a piece of data in scientific databases typically includes information of where this piece of data originates from, as well as details of the scientific process (e.g., parameters used in the experiments, software versions, etc.) by which it arrived in the scientific database.

Historical Background

Provenance of scientific databases has been studied in two granularities: *workflow provenance* and *data provenance*.

Workflow provenance (or *coarse-grained provenance*) refers to the record of the history (or workflow) of the derivation of some dataset in a scientific workflow [1–3]. The amount of information recorded for workflow provenance varies, depending on application needs. For example, in some cases, a workflow provenance may record the type and model of external devices used, such as sensors, cameras, or other data collecting equipments, as well as the associated versions of software used for processing data; in other cases, these information may be deemed unnecessary.

More recently, workflow systems developed within the scientific community to conduct and manage experiments have started recording information about the processes used to derive intermediate and final data objects from raw data. Survey papers dedicated to workflow provenance approaches have been proposed [1, 3], and "provenance challenges" [2] have been held to encourage system designers to learn about the capabilities and expressiveness of each other's systems and work toward interoperable solutions.

Data provenance (or *fine-grained provenance*) is an account of the derivation of a piece of data in a dataset that is typically the result of executing a database query against a source database [4]. This type of provenance determines the parts of source data that were used to generate a piece of data in the resulting dataset. Typically, data provenance is obtained by carefully reasoning about the algebraic form of the database query and the underlying data model of the source and resulting databases. In contrast, transformations occurring in scientific workflows are often external processes (e.g., Perl scripts), and the log files of scientific workflows provide only object identifiers to pieces of data involved in the transformation at best. Often, such external processes and data involved do not possess good properties for detailed analysis. Hence, a fine-grained analysis

of an external process is not always possible, and the provenance recorded for such transformation is usually more coarse-grained.

Foundations

Workflow Provenance: The notion of workflow provenance is illustrated using the workflow of the first provenance challenge, which aimed to establish an understanding of the capabilities of available provenance-related systems. This simple workflow is inspired from a real experiment in the area of functional magnetic resonance imaging (fMRI) and formed the basis of the challenge. It is represented in Fig. 1 as a graph where oval nodes represent steps and rectangle nodes represent the kind of data exchanged between them. Typical provenance queries ask for the history of some data item, e.g., "What caused Atlas X Graphic (one final output) to be as it is?". Depending on whether the complete history of the data is asked or only the previous step and its inputs, queries are qualified as *deep* (or *recursive*) or *immediate*. For example, the immediate provenance of *Atlas X Graphic* is given by the step *13.convert* and its input, *Atlas X slice*, while its deep provenance is given by all the data and steps used to compute the step *9.soft-mean* and *Atlas Image*, *Atlas Header*, *10.slicer*, *Atlas X slice*, and *13.convert*. Other provenance queries may include *annotation queries* finding data annotated with some specific metadata, such as "Find all invocations of procedure align_warp using a twelfth order nonlinear 1365 parameter model?" (metadata on the procedure used should be recorded) and "Find the outputs of align_warp where the inputs are annotated with center = UChicago" (metadata on the data used should be recorded).

Various workflow provenance approaches have been designed and applied to very diverse scientific domains (e.g., biology, ecology, astronomy, meteorology). As workflow provenance is by nature associated with actual scientific experiments, most of these approaches have been implemented into prototypes or systems that have typically participated in such challenges [2].

Approaches differ in various aspects. In particular, provenance can be tracked at various degrees of granularity: from OS or instrument level to input/output data and metadata associated to them. The graphs formed by the workflow provenance information are also represented using various data models, from relational to tree-based data models (e.g., XML, such as in the Kepler and ES3 projects) or graph-based data models (e.g., Web semantic approaches based on RDF, such as in the myGrid or Wings/Pegasus projects). As a consequence, a plethora of languages have been used to query workflow provenance information including SQL, XQuery, and SPARQL, while several dedicated graph query languages have been designed.

More precisely, Foster et al. [2] tackle with the problem of reproducing experiments by providing a *virtual data system* approach offering the ability to not only track the data consumed and produced during computations but also rederive deleted intermediate results of an experiment (virtual data).

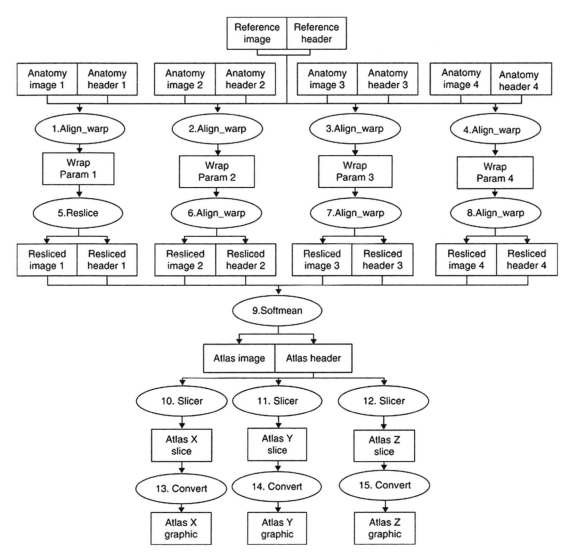

Provenance in Scientific Databases, Fig. 1 Workflow of the first provenance challenge

Callahan et al. [2] focus on the problem of *workflow evolution*: *VisTrails* provides techniques to compute workflow differences and similarities.

In an orthogonal way, Miles et al. define an *open architecture for provenance* [2] designed to be domain and technology independent. In this architecture, *process documentations* describe what actually occurred at execution time and are created and stored by provenance-aware applications.

Finally, Biton et al. provide abstraction mechanisms to help users face the overwhelming amount of provenance information in workflow environments, offering the possibility of focusing on the most relevant provenance information. The technique pursued in *ZOOM*UserViews* is that of "user views" [5]: Since bioinformatics tasks may themselves be complex sub-workflows, a user view determines what level of sub-workflow the user can see and thus what data and steps are visible in provenance queries. Algorithms to compute relevant user views have been proposed in [6].

SWISS-PROT

id	desc
a231	AB
w872	CD
u812	DD

PIR

id	desc
p445	AB
p267	CD
p547	ED

Mapping-Table

mid	swissprot	pir
1	a231	p445
2	w872	p267

Data Provenance: Here, biological databases are used as example scientific databases for illustrating data provenance. The relations and SQL query are shown below.

SELECT swissprot, pir, s.desc

FROM SWISS-PROT s, PIR p, Mapping Table m

WHERE s.id = m.swissprot AND p.id = m. pir

The result of executing the SQL query consists of two tuples (a231, p445, AB) and (w872, p267, CD). Work on data provenance [7, 8] has provided explanations for why a tuple, such as (a231, p445, AB), is in the query result. The reason is because the first tuples in SWISS-PROT, PIR, and Mapping-Table, respectively, joined according to the query to produce the output tuple. This type of provenance is termed *why-provenance* in [7]. In contrast, the *where-provenance* of "AB" in the output tuple (i.e., where "AB" is copied from) is the desc attribute of the first tuple in SWISS-PROT. In particular, no other source tuples are involved in the explanation of where-provenance. Subsequent work by Green et al. [9] is also able to explain *how* a tuple is derived in the result of a query.

Another line of work captures provenance by propagating annotations of source data to the output, based on provenance, along query transformations. Wang and Madnick [10] first articulated the idea of using propagated annotations to analyze source attribution. Subsequently, Buneman et al. [11] studied the *annotation placement problem* using a variation of the propagation scheme of [10]. The propagation scheme of [10, 11] is essentially based on where data is copied from. This scheme forms the *default* propagation scheme of DBNotes [12]. A serious drawback of the default scheme is that two equivalent SQL queries (select-project-join-union queries) may not propagate annotations in the same way. In DBNotes, the authors proposed an alternative *default-all* scheme to overcome this limitation. In the default-all propagation scheme, equivalent queries always propagate annotations in the same way. Additionally, DBNotes also support the *custom* propagation scheme, where one is allowed to customize a propagation scheme as desired. So far, all systems [10–12] only allow annotations on attribute values. Subsequent research efforts have proposed techniques to relax this restriction.

Most database and workflow techniques provide little help for managing provenance in curated databases. This is because curated databases are, by definition, not constructed from the result of executing a query but, rather, created manually by scientists through the analysis of information from several sources. In [13], the authors proposed a copy-and-paste model that captures a scientist's manual curation efforts as provenance-aware transactions. These transactions are logged, and various hierarchical compression techniques were proposed and validated experimentally.

Key Applications

There are many application domains that can benefit from a provenance system in science. Several uses of provenance information are considered here, as described in the literature [1–3].

Informational: Interpreting and understanding the result of a scientific experiment necessitate to know the provenance or context in which the data has been produced.

Data Quality: By providing the source data and transformations, provenance may help to estimate data quality and data reliability.

Error Detection: Provenance can be used to detect errors in data generation.

Reuse: Reproducing a local experiment or an experiment described in a research paper may be possible if precise provenance information is provided.

Attribution: The copyright and ownership of data can be determined using provenance information. It enables its citation and may determine liability in case of erroneous data.

Probabilistic Databases: Provenance has been used to compute the confidences in the result of a probabilistic query correctly, as well as to correctly reason about the set of possible instances in the result of the query [14].

Data Sharing and Data Integration: Provenance has also been used to describe trust policies in the data sharing system Orchestra and to prioritize updates, as well as to understand and debug a data exchange or data integration specification [15].

URL to Code

Chimera Project: http://www.ci.uchicago.edu/wiki/bin/view/VDS/VDSWeb/WebMain

DBNotes Project: http://www.soe.ucsc.edu/~wctan/Projects/dbnotes

ES3 Project: http://eil.bren.ucsb.edu

GriPhyN Virtual Data System Project: http://www.ci.uchicago.edu/wiki/bin/view/VDS/VDSWeb/WebMain

Kepler Project: http://kepler-project.org

myGrid Project: http://www.mygrid.org.uk

Orchestra Project: http://www.cis.upenn.edu/~zives/orchestra

Pasoa Project: http://users.ecs.soton.ac.uk/lavm/projects/pasoa.html

Provenance Challenges: http://twiki.ipaw.info/bin/view/Challenge

SPIDER Project: http://www.soe.ucsc.edu/wctan/Projects/debugger/index.html

VisTrails Project: http://vistrails.sci.utah.edu/index.php/Main_Page

Wings/Pegasus Project: http://www.isi.edu/ikcap/wings

ZOOM*UserViews Project: http://zoomuserviews.db.cis.upenn.edu"

Cross-References

▶ Data Provenance
▶ Provenance in Databases
▶ Provenance in Workflows
▶ Scientific Databases
▶ Scientific Workflows

Recommended Reading

1. Bose R, Frew J. Lineage retrieval for scientific data processing: a survey. ACM Comput Surv. 2005;37(1):1–28.

2. Moreau L, Ludäscher B, Altintas I, Barga RS, Bowers S, Callahan S, Chin G Jr, Clifford B, Cohen S, Cohen-Boulakia S, Davidson S, Deelman E, Digiampietri L, Foster I, Freire J, Frew J, Futrelle J, Gibson T, Gil Y, Goble C, Golbeck J, Groth P, Holland DA, Jiang S, Kim J, Koop D, Krenek A, McPhillips T, Mehta G, Miles S, Metzger D, Munroe S, Myers J, Plale B, Podhorszki N, Ratnakar V, Santos E, Scheidegger C, Schuchardt K, Seltzer M, Simmhan YL, Silva C, Slaughter P, Stephan E, Stevens R, Turi D, Vo H, Wilde M, Zhao J, Zhao Y. The first provenance challenge. Concurrency Comput Pract Exp. 2007;20(5):409–18. Special issue on the First Provenance Challenge.

3. Simmhan Y, Plale B, Gannon D. A survey of data provenance in e-science. ACM SIGMOD Rec. 2005;34:31–6.

4. Buneman P, Tan WC. Provenance in databases. In: Proceedings of the ACM SIGMOD International Conference on Management of Data; 2007. p. 1171–3.

5. Biton O, Cohen-Boulakia S, Davidson S. Zoom* UserViews: querying relevant provenance in workflow systems. In: Proceedings of the 33rd International Conference on Very Large Data Bases; 2007. p. 1366–9.

6. Biton O, Cohen-Boulakia S, Davidson S, Hara CS. Querying and managing provenance through user views in scientific workflows. In: Proceedings of the 24th International Conference on Data Engineering; 2008.

7. Buneman P, Khanna S, Tan WC. Why and where: a characterization of data provenance. In: Proceedings of the 8th International Conference on Database Theory; 2001. p. 316–30.

8. Cui Y, Widom J, Wiener JL. Tracing the lineage of view data in a warehousing environment. ACM Trans Database Syst. 2000;25(2):179–227.

9. Green TJ, Karvounarakis G, Tannen V. Provenance semirings. In: Proceedings of the 26th ACM SIGACT-SIGMOD-SIGART Symposium on Principles of Database Systems; 2007. p. 31–40.

10. Wang YR, Madnick SE. A polygen model for heterogeneous database systems: the source tagging perspective. In: Proceedings of the 16th International Conference on Very Large Data Bases; 1990. p. 519–38.

11. Buneman P, Khanna S, Tan WC. On propagation of deletions and annotations through views. In: Proceedings of the 21st ACM SIGACT-SIGMOD-SIGART Symposium on Principles of Database Systems; 2002. p. 150–8.

12. Bhagwat D, Chiticariu L, Tan WC, Vijayvargiya G. An annotation management system for relational databases. VLDB J. 2005;14(4):373–96.

13. Buneman P, Chapman A, Cheney J. Provenance management in curated databases. In: Proceedings of the ACM SIGMOD International Conference on Management of Data; 2006. p. 539–50.

14. Benjelloun O, Sarma AD, Halevy AY, Widom J. ULDBs: databases with uncertainty and lineage. In: Proceedings of the 32nd International Conference on Very Large Data Bases; 2006. p. 953–64.

15. Chiticariu L, Tan WC. Debugging schema mappings with routes. In: Proceedings of the 32nd International Conference on Very Large Data Bases; 2006. p. 79–90.

Provenance in Workflows

David Koop[1], Marta Mattoso[2], and Juliana Freire[3,4,5]

[1]University of Massachusetts Dartmouth, Dartmouth, MA, USA

[2]Federal University of Rio de Janeiro, Rio de Janeiro, Brazil

[3]NYU Tandon School of Engineering, Brooklyn, NY, USA

[4]NYU Center for Data Science, New York, NY, USA

[5]New York University, New York, NY, USA

Synonyms

Computational provenance; Lineage; Origin; Source; History

Definition

Data and compute-intensive science require the ability to orchestrate computational steps and integrate distinct tools. Scientific workflow systems have been developed to structure such computations. A *scientific workflow* is a directed graph where a set of computational steps are linked together. Each computational module/actor/processor contains a set of input and output ports; a link/edge/channel/connection between an output of one module and the input of another indicates a data dependency. Modules may also have settable parameters that influence their computations. *Workflow provenance* may then include information about the specification of the workflow, the evolution of that specification, and executions of the workflow.

Historical Background

Workflows have been used to model business processes [14]. Business workflows, scripts, coordination languages, and dataflow systems are precursors of today's scientific workflow systems [16], and besides the ability to orchestrate disparate tools, workflows also help abstract lower-level computations, giving users an understandable view of complex tasks. In addition, many workflow systems support execution on different types of computing resources, hiding the complexities of cloud or HPC resources, and some, such as Kepler, provide different execution modes.

In the course of data-driven analyses, a given workflow may be executed multiple times generating a large number of data products. Manually keeping track of different runs, parameters values are used, and derived data is time-consuming and error prone. Because a workflow can encapsulate an entire experiment, a scientific workflow system is able to systematically and uniformly track all computational steps and trace the lineage of data it derives. Furthermore to enable reproducibility, it is also important to track detailed information about the computing resources and environment utilized. Workflow provenance seeks to capture all of this information, not only for reproducibility, archival, and verification purposes but also as a source of information for debugging and future extensions [3, 6, 8]. The volume and type of provenance captures varies based on the system used, the domain, and what the provenance information will be used for. While database provenance is fine grained and captures transformations at the tuple level, workflow provenance allows some latitude with respect to the granularity and abstraction.

Scientific Fundamentals

Capture and storage. There are many facets of provenance that may be captured from workflows. A baseline for workflow provenance includes, for any execution, the original inputs, set of steps followed, resulting outputs, and any errors that occurred during execution. There are other types of provenance information such as manual annotations, execution environment, and evolution of the workflow specification [5, 8].

The workflow definition documents the computation it represents (prospective provenance). Provenance must also be captured for the actual computational events (retrospective provenance). Many workflow systems capture high-level information when modules are executed, while others rely on system-level mechanisms to capture the computational steps. The high-level information is usually easier to understand, but the system-level information can capture accesses or interactions that have been abstracted. For each module, provenance information may include the identity of that module, any parameters, input and output data, the start/end time of execution, the executing machine, and any computation-specific annotations (e.g., the seed used for random number generation). The high-level mechanisms for capturing provenance usually involve wrapping each computational step so that the start, end, and status of the step are noted. Lower-level mechanisms often involve system event monitors and require some reconciliation with the abstract representation. To reduce redundant information in multiple provenance traces [4], some systems (e.g., VisTrails) normalize execution provenance so that only information that differs between executions is stored (e.g., runtime errors). Provenance information can be stored in different ways, including textual files and structured files, and in a database.

Granularity. The different abstractions workflows use permit different levels of granularity in their provenance. Some workflows do this by supporting *subworkflows* that encapsulate pieces of a workflow. If the provenance of the subworkflows is captured in addition to the top-level workflow provenance, users may wish to drill down to the finer provenance in subworkflows like they would with system-call-level provenance. Even without the structure of subworkflows, systems like ZOOM provide the ability for users to define and navigate different views of the provenance [2].

Workflow evolution. While the provenance of workflow execution focuses on the steps taken to compute a particular set of results, the provenance of workflow evolution focuses on the steps taken to build a particular set of workflows. The change-based provenance model captures workflow evolution by treating the workflow specification as a first-class data product ; it keeps track of the steps involved in building a workflow or transforming one workflow into another [9]. By storing this information in a tree data structure, it is possible to concisely represent a set of related workflows because only the changes between them need to be stored. The tree can also be used to support reflective reasoning, allowing users to follow chains of reasoning backward and forward by navigating over different workflow versions, modifying them, and comparing the workflow definitions as well as their results.

Dataflow vs. control-flow. In contrast to business workflows which are control driven, scientific workflows are often dataflows. Some scientific workflow systems, however, do offer support for control-driven structures, including loops and conditional statements. While for dataflows the provenance graph mirrors that of the workflow definition, this is not the case for workflows with control structures. For example, loops are unrolled in the execution graph. To make the connection between the specification and execution steps clearer, some workflow systems structure provenance or store annotations to indicate that a computation takes place as part of a loop or the result of some condition. In some cases, the interior of loops or the condition-dependent subworkflow are grouped like subworkflow provenance.

Distributed computation. Because workflows abstract the details of how a computation is executed, the same workflow can be executed sequentially on a desktop or in parallel on a distributed cluster. Provenance of distributed computations is important to track potential discrepancies and troubleshoot the performance and configuration of distributed applications [17]. Different architectures are possible: the provenance of an entire run can be collected or it can

remain distributed. This ties into concerns about compressing provenance for transmission and the accessibility and maintenance of distributed provenance.

Key Applications

Domains. As scientific domains become increasingly dependent on data-intensive analyses, the ability of workflows to orchestrate different tools has encouraged their adaptation in many fields, including bioinformatics (e.g., Taverna, Galaxy), climate science (e.g., UVCDAT), quantum physics (e.g., ALPS), and ecological modeling (e.g., SAHM). These adaptations are driven by the need to connect multiple tools and also by the integrated, understandable provenance scientific workflow systems provide. In addition, there has been work to link ontologies with workflow provenance so that queries over the provenance can be posed in domain-specific languages.

Uses of workflow provenance. Provenance is necessary to reproduce a computation and the results it derives. However, the provenance captured by scientific workflow systems is often limited to the workflow steps. This information can insufficient to reproduce the results at a later time or on a system different from the one where the workflow was created. To ensure reproducibility as well as portability, provenance must be captured about data input, code, software, library dependencies, and information about the source computational environment (e.g., operating system and hardware architecture).

A key goal of provenance is to establish the lineage of data products, allowing users to follow the trail of computations and transformations to the original inputs. In practice, users often wish to query provenance collections to determine, for example, where a particular dataset was used or which results are affected by a bug in a particular computational module. More complicated queries might include finding all executions where a particular set of modules and parameter ranges were tested. Queries may be posed textually, using graph-based, datalog, or traditional languages, or interactively via query-by-example techniques where users build the sub-workflow they are interested in [12]. In addition to querying, workflow provenance can be distilled into summaries [1, 10]. Workflow evolution provenance can also be reused to repeat similar transformations via analogies [12].

Provenance can also be used for teaching. It can help instructors be more effective and improve the learning experience for students. Silva et al. developed a provenance-rich teaching methodology where students both receive and submit work with provenance information [13]. Lecture notes are fully documented, allowing students to not only reproduce the examples shown in class but also follow the instructor's reasoning. The provenance of the assignments allows instructors to understand how students approached problems.

Parameter sweeps. Due to the exploratory analysis of experiments, users often need to run parameter sweep [15] workflows, which are workflows that are invoked repeatedly using different input data combinations. These workflows generate a large amount of tasks that are naturally submitted to parallel execution environments. Manual analysis of the results of a large-scale many-task scientific computation is generally unfeasible. This involves, for instance, checking inputs and outputs of each component application of the workflow, verifying if jobs failed on remote computational resources, and checking all processes that contributed to the creation of a particular data set. Provenance data registers the data-flow associated to each parameter combination. Analytical queries on this database help on the evaluation of the parameters associated with input and output relationships between data sets and processes [7].

Iterative methods. Unlike the direct methods that calculate the exact answer to a given problem, iterative methods provide an approximate answer based on a sequence of solutions that converges to the exact answer. Usually, itera-

tive methods are halted when the approximate solution meets a preset user supplied criterion, typically a tolerance on the ratio of vector norms. The analysis of the iteration steps can be improved by querying the workflow provenance. Going further, the decision whether to interfere on the dataflow or to reduce, increase, or even stop the iterative loop becomes more interesting when they can be steered by users. These decisions can be made based on partial results within the execution context, which provides a dynamic execution behavior for the workflow. Provenance data helps find and analyze this context [7].

User steering. Scientific workflow systems often provide results, and provenance data only after the workflow finishes executing all steps. By monitoring workflow as it runs, a user can verify the status of the execution at specific points to discover if anything out of the norm has happened (or is happening). Based on this monitoring, the user is able to decide if she should analyze partial results, stop the execution, or reexecute some of the activities, possibly with adjustments such as filtering boundaries, error precisions, or the maximum number of iterations on a loop [11].

URL to Code

- ALPS: http://alps.comp-phys.org/
- Galaxy: http://galaxyproject.org
- Kepler: http://kepler-project.org
- Pegasus: http://pegasus.isi.edu
- Provenance Challenges: http://twiki.ipaw.info/bin/view/Challenge
- ReproZip: https://github.com/ViDA-NYU/reprozip
- SAHM: https://www.fort.usgs.gov/products/sb/5090
- Taverna: http://www.taverna.org.uk
- UVCDAT: http://uvcdat.llnl.gov/index.html
- VisTrails: http://vistrails.org
- W3C PROV: https://www.w3.org/TR/prov-overview/
- Wings: http://www.wings-workflows.org
- ZOOM*UserViews: http://zoomuserviews.db.cis.upenn.edu

Cross-References

Recommended Reading

1. Alper P, Belhajjame K, Goble C, Karagoz P. Small is beautiful: summarizing scientific workflows using semantic annotations. In: Proceedings of the 2013 IEEE International Congress on Big Data; 2013. p. 18–25.
2. Biton O, Cohen-Boulakia S, Davidson SB. Zoom* userviews: querying relevant provenance in workflow systems. In: Proceedings of the 33rd International Conference on Very Large Data Bases. VLDB Endowment; 2007. p. 1366–69.
3. Bose R, Frew J. Lineage retrieval for scientific data processing: a survey. ACM Comput Surv. 2005;37(1):1–28.
4. Chapman AP, Jagadish HV, Ramanan P. Efficient provenance storage. In: Proceedings of the 2008 ACM SIGMOD International Conference on Management of Data; 2008. p. 993–1006.
5. Chirigati FS, Shasha D, Freire J. Packing experiments for sharing and publication. In: Proceedings of the ACM SIGMOD International Conference on Management of Data; 2013. p. 977–80.
6. Davidson SB, Boulakia SC, Eyal A, Ludäscher B, McPhillips TM, Bowers S, Anand MK, Freire J. Provenance in scientific workflow systems. IEEE Data Eng Bull. 2007;30(4):44–50.
7. Dias J, Guerra G, Rochinha F, Coutinho ALGA, Valduriez P, Mattoso M. Data-centric iteration in dynamic workflows. Futur Gener Comput Syst. 2015;46:114–26. http://dx.doi.org/10.1016/j.future.2014.10.021.
8. Freire J, Koop D, Santos E, Silva C. Provenance for computational tasks: a survey. Comput Sci Eng. 2008;10(3):11–21.
9. Freire J, Silva C, Callahan S, Santos E, Scheidegger C, Vo H. Managing rapidly-evolving scientific workflows. In: International Provenance and Annotation Workshop (IPAW), LNCS, vol. 4145. Springer; 2006. p. 10–8.
10. Koop D, Freire J, Silva CT. Visual summaries for graph collections. In: Visualization Symposium (PacificVis), 2013 IEEE Pacific; 2013. p. 57–64.
11. Mattoso M, Dias J, Ocaña KACS, Ogasawara E, Costa F, Horta F, Silva V, de Oliveira D. Dynamic steering of HPC scientific workflows: a survey. Futur Gener Comput Syst. 2015;46(May):100–13.
12. Scheidegger CE, Vo HT, Koop D, Freire J, Silva CT. Querying and creating visualizations by analogy. IEEE Trans Vis Comput Graph. 2007;13(6):1560–67.
13. Silva CT, Anderson E, Santos E, Freire J. Using VisTrails and provenance for teaching scientific visualization. Comput Graphics Forum. 2011;30(1):75–84.
14. Van Der Aalst WMP, Ter Hofstede AHM, Weske M. Business process management: a survey. In: Business Process Management. Springer; 2003. p. 1–2.
15. Walker E, Guiang C. Challenges in executing large parameter sweep studies across widely distributed computing environments. In: Proceedings of the 5th IEEE Workshop on Challenges of Large Applications in Distributed Environments; 2007. p. 11–8.
16. Zhao Y, Foster I. Scientific workflow systems for 21st century, new bottle or new wine. In: IEEE Workshop on Scientific Workflows; 2008.
17. Zhou W, Mapara S, Ren Y, Li Y, Haeberlen A, Ives Z, Loo BT, Sherr M. Distributed time-aware provenance. Proc VLDB Endow. 2012;6(2):49–60.

Provenance Standards

Paolo Missier
School of Computing Science, Newcastle University, Newcastle upon Tyne, UK

Synonyms

PROV

Definition

PROV, the Provenance standard, is a family of specifications released in 2013 by the Provenance Working Group, as a contribution to the Semantic Web suite of technologies at the World Wide Web Consortium. The specifications define a data model along with a number of serializations for representing aspects of provenance. The term provenance, as understood in these specifications, refers to *information about entities, activities, and people involved in producing a piece of data or thing, which can be used to form assessments about its quality, reliability, or trustworthiness* (PROV-Overview [1]). The specifications include a combination of W3C *Recommenda-*

tion and *Note* documents. Recommendation documents include:

(i) The main PROV data model specification (PROV-DM [2]), with an associated set of constraints and inference rules (PROV-CONSTRAINTS [3])
(ii) An OWL ontology that allows a mapping of the data model to RDF (PROV-O [4])
(iii) A notation for PROV with a relational-like syntax, aimed at human consumption (PROV-N [5])

All other documents are Notes. These include PROV-XML, which defines a XSD schema for XML serialization (http://www.w3.org/TR/prov-xml/); PROV-AQ, the Provenance Access and Query document (http://www.w3.org/TR/prov-aq/), which defines a Web-compliant mechanism to associate a dataset to its provenance; PROV-DICTIONARY (http://www.w3.org/TR/prov-dictionary/), for expressing the provenance of data collections defined as sets of key-entity pairs; and PROV-DC (http://www.w3.org/TR/prov-dc/), which provides a mapping between PROV-O and Dublin Core terms.

Historical Background

The idea of a community-grown data model for describing the provenance of data originated around 2006, when consensus began to emerge on the benefits of having a uniform representation for "data provenance, process documentation, data derivation, and data annotation", as stated in [6]. The first Provenance Challenge [7] was then launched, to test the hypothesis that heterogeneous systems (mostly in the e-science/cyberinfrastructure space), each individually capable of producing provenance data by observing the execution of data-intensive processes, could successfully exchange such provenance observations with each other, without loss of information. The Open Provenance Model (OPM) [6] was proposed as a common data model for the experiment. Other Provenance

Challenges followed, to further test the ability of the OPM to support interoperable provenance.

Central to the OPM is the notion of *causal relationships*, or dependencies, involving *artifacts* (e.g., data items), *processes*, and *agents*. Using the OPM, one can assert that an artifact A was produced or consumed by a process P, e.g., "the cake C was produced by a baking process B, which used eggs E and flour F." Here C, E, and F are artifacts, and B is a process. One can also assert a derivation dependency between two artifacts, A1 and A2, without mentioning any mediating process, i.e., "A2 was derived from A1." Agents, including humans, software systems, etc., can be mentioned in OPM as process controllers, i.e., "the baking was controlled by Bob (the cook)."

OPM statements attempt to explain the existence of artifacts. Since such statements may reflect an incomplete view of the world, obtained from a specific perspective, the OPM adopts an open world assumption, whereby dependencies are interpreted as correct but possibly incomplete knowledge: "A2 was derived from A1" asserts a certain derivation, but does not exclude that other, possibly unknown artifacts, in addition to A1, may have contributed to explaining the existence of A2. Other features of the OPM, including built-in rules for inference of new provenance facts, are described in detail in [6].

In September, 2009, the W3C Provenance Incubator Group was created. Its mission, as stated in the charter (http://www.w3.org/2005/Incubator/prov/charter), was to "provide a state-of-the art understanding and develop a roadmap in the area of provenance for Semantic Web technologies, development, and possible standardization." W3C Incubator Groups produce recommendations on whether a standardization effort is worth undertaking. Led by Yolanda Gil at USC/ISI, the group produced its final report in December 2010 (http://http://www.w3.org/2005/Incubator/prov/XGR-prov-20101214). The report highlighted the importance of provenance for multiple application domains, outlined typical scenarios that would benefit from a rich provenance description, and summarized the state of the art from the literature, as well as in the

P

Web technology available to support tools that exploit a future standard provenance model. As a result, the W3C Provenance Working Group was created in 2011, chaired by Luc Moreau (University of Southampton) and Paul Groth (VU University Amsterdam). The group released its final recommendations for PROV in June 2013.

Scientific Fundamentals

While PROV builds upon the prior experience gained from the OPM, and therefore it echoes some of the notions presented above (see "Historical Background"), its design emerges from a more disciplined community effort, governed by standard W3C Working Group policy. PROV is the result of 2 years of work and incorporates the expectations of group members representing over 50 organizations from a broad range of application domains, each bringing different sets of requirements.

The brief account of PROV that follows cannot possibly cover all the features of the family of specifications. The reader is referred to the overview document [1], which provides the main entry point and a roadmap to the other documents, including the nonnormative Notes. What follows is a summary of the main principles that informed the design, following mainly the PROV-DM document [2] (please note, all sentences in italics below are quotes from that document).

The scenario depicted in Fig. 1 will be used to illustrate those principles. The primer document [8] also provides a complete running example.

In this scenario, two coauthors, Alice and Bob, are responsible for editing a document. After Alice has edited a first version draft, Bob comments on the draft, and then Alice edits a second version, based upon the first draft and Bob's comments.

Entities, Activities, and Agents

At the core of the PROV data model are the notions of entities, activities, and agents. *Provenance describes the use and production of entities by activities, which may be influenced in various ways by agents.*

Entities may represent data, but they are more generally defined as *physical, digital, conceptual, or other kind of thing with some fixed aspects.* Entities may be real or even imaginary. Examples of entities are a particular version of a document, the output produced by some algorithm, a record in a database, a car at a particular stage of its lifetime, etc. In practice, anything that may have a provenance is an entity.

Importantly, the "fixed aspects" mentioned informally above refer to the characteristics that are relevant to describing the provenance of the entity. For instance, a document as in Fig. 1 may be characterized by a filename, version number, and current content. Some of these properties may persist over time (e.g., the filename), while others may change. A document entity, _aDoc.v1, is a document with values specified for each of those properties. When any of those values change, a different document entity, for instance, _aDoc.v2, is defined, with a different provenance. In our example, this is the document with updated content and a new version number. The editing activity *editing2* accounts for the change relative to _aDoc.v1.

Unlike entities, activities such as *drafting1*, *commenting1*, and *editing2* have a duration: *An activity is something that occurs over a period of time and acts upon or with entities; it may include consuming, processing, transforming, [...,] using, or generating entities.* Typically, activities may *use* existing entities (*editing2* used entity _aDoc.comments) or *generate* new ones (*editing2* generated _aDoc.v2).

Agents, on the other hand, *bear some form of responsibility for an activity taking place, for the existence of an entity, or for another agent's activity.* Agents may be humans, as in the case of *Alice* and *Bob* in the example, or, for instance, software systems. Note that one may want to describe the provenance of an agent – for instance, to help explain their behavior vis a vis carrying out an activity (for instance, what knowledge did *Alice* have during her *drafting1* activity?) Therefore, in PROV, agents are viewed as a particular type of entity.

All entities, activities, and agents are given a unique ID, drawn from a specific namespace,

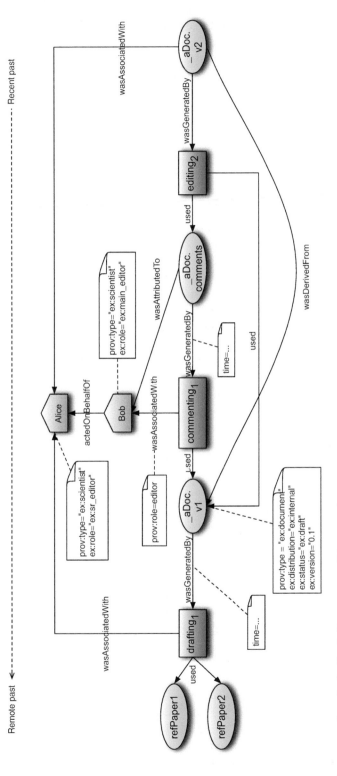

Provenance Standards, Fig. 1 Document coediting scenario

which is valid within a given scope, i.e., a *provenance document*. Furthermore, they can be annotated with sets of properties, i.e., of name-value pairs. PROV reserves certain properties, using the PROV namespace. Thus, for instance, the following statement in PROV-N notation:

```
entity(ex:_aDoc.v1; [prov:type
= "ex:document", ex:distribution
="ex:internal",...])
```

defines a new entity with unique name *_aDoc.v1* in a custom namespace denoted by prefix *ex* and annotated with two properties, one of which is the standard *prov:type* property (but with a value in the *ex* namespace).

Core Provenance Relationships

The provenance of entities is expressed by means of a small set of core concepts, namely, generation, usage, derivation, communication, attribution, association, and delegation. In [2], these are defined independently of any formalism and are then manifested as relationships in a UML data model specification, reproduced in Fig. 2. The PROV-N notation [5] is recommended as a syntactic rendering of relationship instances and will be used in these examples.

A key principle is that entities have a lifetime, which begins with generation, defined as *the completion of production of a new entity by an activity. The entity did not exist before generation and becomes available for usage after this generation*. Associated with generation is a *generation event*, which can be thought of as a point in time. (However, PROV avoids explicit notions of time, which are difficult to manage when provenance is recorded by multiple distributed and autonomous systems, each possibly using a different clock.) Note that the production of an entity, for instance, a file produced incrementally by a program execution, may extend over time. In this case, the generation event marks the completion of the production. Symmetrically, usage may also extend over time (i.e., the reading of the file). Thus, usage is defined as *the beginning of utilizing an entity by an activity*. Note how these definitions are suitable to model typical producer/consumer patterns in data processing.

Here are examples of generation and usage in the document editing scenario:

```
used(drafting1, refPaper1),
used(drafting1, refPaper2),
wasGeneratedBy(_aDoc.v2, drafting1)
```

as well as

```
used(editing2, _aDoc.comments)
wasGeneratedBy(_aDoc.v2, editing2)
```

In PROV, all relationships may optionally be given an explicit, unique ID, just like entities etc. Using IDs, the example above could also have been written as:

```
used(ex:u1; editing2, _aDoc.comments)
wasGeneratedBy(ex:g1; _aDoc.v2,
editing2)
```

where *ex:u1* and *ex:g1* are the new IDs. This design principle is useful when introducing derivations.

A derivation is a data dependency between two entities, e1 and e2, where e2 (the derived entity) is the result of some transformation that occurred to e1. The nature of such transformation may be implicit, for example:

```
wasDerivedFrom(_aDoc.v2, _aDoc.v1)
```

However, it may also be expressed in terms of a mediating activity *a* that "explains" the derivation in terms of usage of *e1* and generation of *e2*. More specifically, in abstract one could have the following statements, involving the IDs for generation and usage relationships:

```
used(u; a,e1)
wasGeneratedBy(g,e2,a)
wasDerivedFrom(e2, e1, a, g, u)
```

This is an example of a binary relationship, derivation, which admits additional arguments. Such optional arguments are common in PROV.

Similar to derivation, communication relates two activities, *a1* and *a2*, such that *a2* is dependent upon *a1, by way of some unspecified entity that is generated by a1 and used by a2*.

Constraints and Inferences

The previous example suggests, intuitively, possible connections among some of the relationships, e.g., derivation, usage, and generation. Such connections are indeed formalized, as

**Provenance Standards,
Fig. 2** UML diagram for
core PROV entities and
relationships (From
[PROV-DM])

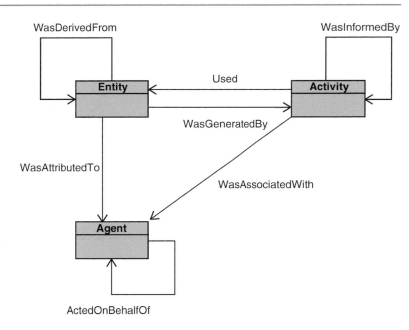

part of a comprehensive normative PROV-CONSTRAINTS document [3], which specifies *definitions of some provenance statements in terms of others, inferences over PROV instances that applications may employ,* and *a class of valid PROV instances by specifying constraints that valid PROV instances must satisfy.* In this context, the term *valid* refers to *a consistent history of objects and interactions to which logical reasoning can be safely applied.* As an example, returning to the derivation/usage/generation statements, consider the following inference rule, from [3#inf 11]:

```
IF wasDerivedFrom(_id; e2,e1,a,gen2,
use1,_attrs),
THEN there exists _t1 and _t2 such that
used(use1; a,e1,_t1,[])
and
wasGeneratedBy(gen2; e2,a,_t2,[]).
```

Here *_t1* and *_t2* indicate timestamps, a detail that can be overlooked at this stage. Informally the rule states that if derivation of *e2* from *e1* involves an activity *a*, then *a* must be involved in the usage of *e1* and the generation of *e2*.

The simple usage/generation provenance pattern shown earlier is helpful to illustrate *event ordering constraints.* Intuitively, if an activity *a* generates (resp. uses) entities *e2* (resp. *e1*), then it must be the case that the generation (resp.

usage) event lies within the lifetime of *a*. The start and end events of an activity can indeed be expressed, i.e.:

```
wasStartedBy(start; a,_e1,_a1,_t1,
_attrs1)
wasEndedBy(end; a,_e2,_a2,_t2,_attrs2)
```

denote the start and end events, resp., for *a*.

Constraints 33 and 34 in [3] state that a pre-order relation must exist, whereby the start event (resp. end event) must precede (resp. not precede) any usage and generation event involving *a*. Formally:

```
IF wasStartedBy(start; a,_e1,_a1,_t1,
_attrs1) and used(use; a,_e2,_t2,
_attrs2) THEN start precedes use.
IF used(use; a,_e1,_t1,_attrs1) and
wasEndedBy(end; a,_e2,_a2,_t2,_attrs2)
THEN use precedes end.
IF wasStartedBy(start; a,_e1,_a1,_t1,
_attrs1) and wasGeneratedBy(gen; _e2,a,
_t2,_attrs2) THEN start precedes gen.
IF wasGeneratedBy(gen; _e,a,_t,_attrs)
and wasEndedBy(end; a,_e1,_a1,_t1,
_attrs1) THEN gen precedes end.
```

These exemplar rules and constraints are representative of a much larger collection. PROV-CONSTRAINTS include 21 inference rules and 55 constraints.

Actors and Their Relationships

The core elements of PROV also include three relationships involving agents, namely, attribution, association, and delegation. Their use is illustrated in the scenario of Fig. 1:

```
wasAttributedTo(_aDoc.comments, Bob)
```

denotes that entity _aDoc.comments_ is ascribed to Bob, without specifying any associated activity. One can also explicitly ascribe responsibility for an activity to an agent using the association relation, for example:

```
wasAssociatedWith(commenting1, Bob).
```

Inference rules are defined in [3], which formalize the relationship between attribution and association when they are both present.

Finally, one can specify chains of responsibility by means of the delegation relationship, as follows:

```
actedOnBehalfOf(Bob, Alice)
```

PROV-O

The PROV-O document [PROV-O] specifies the PROV data model as an OWL ontology. This makes it natural to express provenance statements, of the kind shown here using the PROV-N, as RDF triples. The PROV-PRIMER [8] document provides examples in both notations, as well as in PROV-XML (http://www.w3.org/TR/prov-xml/).

Extensibility

PROV is designed as _upper level_ model, agnostic to any application domain. Two main mechanisms exist for extending the model. Firstly, one can introduce custom properties, as mentioned, as well as custom values for standard properties, such as _prov:type_. Secondly, one can extend the PROV-O ontology using the standard extension mechanisms available in the Semantic Web framework (e.g., subClass, subProperty).

Further Reading

A survey of foundations for Provenance on the Web, predating PROV, is available [9], as well

as an introduction to PROV [10]. Research on Database Provenance has been developing alongside the more general model for provenance described here; however, it is not primarily within the scope of PROV. An account of Database Provenance can be found in [11]. Several tutorials on PROV are also available, including one that describes the data model, the constraints model, and a number of known implementations and extensions as of 2013 [12].

Key Applications

1. Attribution of user-authored Web pages (e.g., blogs) and social media content, trust in Web content.
2. Attribution of published science data, derivation history of scientific dataset that represents the outcome of experiments. In particular, workflow management systems have been early generators of provenance traces, which can be used to help validate published datasets in the eyes of potential new users. These include, among others, the VisTrails, Taverna, Kepler, Pegasus, and more. A recent analysis of workflow provenance can be found in [13].
3. Provenance, when it is sufficiently detailed, can be instrumental in some cases, to facilitate the reproducibility of scientific experiments [14–15].

Future Directions

Many implementations of PROV along with a variety of tools are currently being developed. An initial list, which however evolves rapidly, can be found in the PROV Implementation Report (http://www.w3.org/TR/prov-implementations/). Useful annual or biannual conferences to monitor include TAPP and IPAW.

URL to CODE

http://lucmoreau.github.io/ProvToolbox/

Cross-References

▶ Data Provenance
▶ Provenance in Scientific Databases

Recommended Reading

1. Groth P, Moreau L. PROV-Overview: an overview of the PROV Family of Documents [Internet]. 2012. Available from: http://www.w3.org/TR/prov-overview/
2. Moreau L, Missier P, Belhajjame K, B'Far R, Cheney J, Coppens S, et al. PROV-DM: the PROV Data Model [Internet]. In: Moreau L, Missier P, editors. 2012. Available from: http://www.w3.org/TR/prov-dm/
3. Cheney J, Missier P, Moreau L. Constraints of the provenance data model [Internet]. 2012. Available from: http://www.w3.org/TR/prov-constraints/
4. Lebo T, Sahoo S, McGuinness D, Belhajjame K, Cheney J, Corsar D, et al. PROV-O: the PROV ontology [Internet]. In: Lebo T, Sahoo S, McGuinness D, editors. 2012. Available from: http://www.w3.org/TR/prov-o/
5. Moreau L, Missier P, Cheney J, Soiland-Reyes S. PROV-N: the provenance notation [Internet]. In: Moreau L, Missier P, editors. 2012. Available from: http://www.w3.org/TR/prov-n/
6. Moreau L, Clifford B, Freire J, Futrelle J, Gil Y, Groth P, et al. The open provenance model – core specification (v1.1). Futur Gener Comput Syst Elsevier. 2011;7(21):743–56.
7. Moreau L, Ludäscher B, Altintas I, Barga RS. The first provenance challenge. Concurr Comput Pract Exp [Internet]. 2008;20:409–18. Available from: http://www3.interscience.wiley.com/journal/116837632/abstract
8. Gil Y, Miles S, Belhajjame K, Deus H, Garijo D, Klyne G, et al. PROV model primer [Internet]. In: Gil Y, Miles S, editors. 2012. Available from: http://www.w3.org/TR/prov-primer/
9. Moreau L. The foundations for provenance on the web. Found Trends Web Sci [Internet]. Citeseer; 2009 [cited 2011 Oct 18];131. Available from: http://citeseerx.ist.psu.edu/viewdoc/download?doi=10.1.1.155.784&rep=rep1&type=pdf
10. Moreau L, Groth P. Provenance: an introduction to PROV. Synth Lect Semant Web Theory Technol [Internet]. Morgan & Claypool Publishers; 2013 Sep 15 [cited 2014 Aug 25];3(4):10–129. Available from: http://www.morganclaypool.com/doi/abs/10.2200/S00528ED1V01Y201308WBE007
11. Cheney J, Chiticariu L, Tan W-C. Provenance in databases: why, how, and where. Found Trends{\textregistered} Databases. 2009;1: 379–474.
12. Missier P, Belhajjame K, Cheney J. The W3C PROV family of specifications for modelling provenance metadata. In: Proceedings of the 16th International Conference on Extending Database Technology (Tutorial) [Internet]. 2013. Available from: http://www.edbt.org/Proceedings/2013-Genova/papers/edbt/a80-missier.pdf
13. Bowers S. Scientific workflow, provenance, and data modeling challenges and approaches. J Data Semant [Internet]. 2012 Apr 11 [cited 2013 Jun 8];1(1):19–30. Available from: http://link.springer.com/10.1007/s13740-012-0004-y
14. Missier P, Woodman S, Hiden H, Watson P. Provenance and data differencing for workflow reproducibility analysis. Concurr Comput Pract Exp [Internet]. 2013;n/a–n/a. Available from: http://dx.doi.org/10.1002/cpe.3035
15. Peng R. Reproducible research in computational science. Science. 2011;334(6060):1226–127.

Provenance Storage

Thomas Heinis[1] and Adriane Chapman[2]
[1]Imperial College London, London, UK
[2]University of Southampton, Southampton, UK

Synonyms

History storage; Lineage storage; Pedigree organization; Provenance organization

Definition

Given the provenance of data processing or manipulation (e.g., through ad hoc manipulations, workflows, or database operators), provenance storage defines how the provenance information is stored on disk. Provenance information essentially captures all information describing the history, creation, and modification of a data product. In the context of workflows, for example, relevant information includes but is not limited to the parameters used in each step of the workflow recursively, software versions used, etc. Provenance storage defines where and how this information is stored and organized on disk.

Historical Background

The original academic works on digital provenance focused on provenance within relational databases [1–3]. However, workflow systems also found a use for provenance and quickly began capturing and storing provenance information [4–8]. Throughout the history of digital provenance, it has been stored in the following methods: relational [9–13], flat file [14], bound to the data itself [15], and graph-based [16–19].

It should be noted that storage data model does not necessarily equate to interoperable format. There have been many provenance standards through the years, Dublin Core [20], OPM [21], W3C PROV [22, 23], as well as many domain-specific ones such as CDISC (clinical research) and ISO-19115 (geospatial). All of these standards create specifications on how to represent provenance during exchange, but do not need to be adhered to (nor should they) for efficient and usable storage.

There has also been research in how to architecturally utilize the storage solutions. While most storage systems assume a centralized provenance store, several options exist for a decentralized provenance store, including [24–27].

Scientific Fundamentals

Provenance information typically captures the recursive relationship between data transformations and the data products used as input of transformations and produced as output. Critical for storing and organizing data provenance is establishing identity between data products. If identity cannot be determined, a data product dp generated by a data transformation cannot be connected to the next transformation where dp is used as input, thereby severely limiting change propagation queries or recursive queries about the origin of data.

The PROV data model [22, 23] uses URLs/URIs as identifiers to identify data products and algorithms: if two entities have the same identifier, then they are considered equal.

Imposing an artifact naming or URI convention on provenance capture, however, will not easily work in wide collaborations, and essentially no assumptions can be placed on the storage or transmission method of the data. To establish identity in this context, content- or context-bound identifiers need to be adopted. A content-bound identifier (CBI) is any identifier computed based on the content of a data item. Content-bound identifiers permit two independent observers to identify the item the same way, even if they are ignorant of the existence of other observers. Provenance reporting systems can use content-bound identifiers as a way of synchronizing multiple observers/reporting clients and de-duplicating what would otherwise become redundant and disconnected.

There are several available methods to compute content-bound identifiers based on cryptographic hash functions. Hash functions for provenance identity need to be evaluated in terms of three aspects: performance (data volume hashed in a given period of time), resistance to collision (likelihood that two different data items have the same digest), and size (size of the digest). In terms of these trade-offs, SHA-256 is larger, more robust/resistant to collisions, but slower than MD5.

By establishing identity between data products and between transformations, provenance information is captured in a directed acyclic graph that expresses what data product depends on what other data and transformations and what transformations and data influence a particular data product. Provenance queries consequently usually are also graph queries asking, for example, for the shortest paths between two data products, the dependencies of a data product, etc.

Once identity between data products and data transformations is established, a directed acyclic graph representing the provenance information can be assembled and then needs to be organized and stored. Provenance information can either be stored with the data products or without.

Storing the provenance with the data products makes it very easily portable, i.e., easy to share and interpret by other systems, but limits efficiency in the absence of indexes and other auxiliary data structures. It also limits the ability to query and process across the entire provenance store.

XML and other hierarchical document formats such as JSON and BSON are workable solutions to store provenance with the data. Despite the data model behind XML and JSON fundamentally is a tree and not a directed graph (as is required for provenance information), XML languages that support directed graphs (i.e., GraphML) can help. XML is well-suited for expressing a subset of provenance graphs and data structures, but to express the full range of directed graphs, implementers will either fall back on the use of pointers (e.g., XML ID/IDREF) or data duplication within the document to express directed graphs without tree assumptions. XML is therefore particularly useful as an interchange format, but not ideal as a storage format because it complicates query.

Storing the data provenance separately enables more flexibility and supports the optimization of provenance information for queries. At the same time, it limits portability.

Relational databases are attractive because of their wide adoption and mature tooling. RDMBS, however, make path-associative query extremely difficult. Storing provenance in an RDBMS typically involves a table of nodes and a table of edges. These designs are excellent for bulk queries that do not require edge traversal ("Fetch all provenance owned by Bob"), but tend to be very poor at path-associative queries ("Fetch all provenance that is between 2 and 5 steps downstream of X"). Path-associative queries typically end up being translated as dynamically constructed, variably recursive SQL queries that join nodes to edges. Graph queries, however, are not well supported in RDBMS developers that often have to re-implement basic graph techniques the RDBMS does not provide (e.g., shortest path algorithms) rather than exploiting known good implementations. By using auxiliary data structures and indexing approaches designed for directed graphs, however, RDBMS become

a viable approach to enable graph queries over provenance information.

Better support for graph queries is provided in graph databases (such as Neo4J or RDF triple stores). They are a better fit for provenance for two reasons:

1. The graph model under the hood of a graph database is fundamentally a match for the core of provenance (a directed graph).
2. Graph databases will typically provide graph-oriented query languages (such as Cypher within Neo4J or SPARQL within RDF triple stores) which greatly facilitate provenance queries.

The negative aspect of graph databases is that they are "naturally indexed" by relationships/edges and do not perform as well on bulk queries (mentioned before). While such bulk queries do have important uses, the most interesting and powerful provenance queries typically are path-associative. This style of query emphasizes the strengths of graph query languages, an emphasis which plays to many of the weaknesses of other languages.

Key Applications

There are a few basic provenance storage solutions that are generic and can be used straight out of the box [28–30]. These storage solutions provide their own API to report and query provenance and maintain that provenance within their own provenance store.

Cross-References

▶ Data Provenance
▶ Digital Archives and Preservation
▶ Provenance in Scientific Databases

Recommended Reading

1. Buneman P, Khanna S, Tan W-C Why and where: a characterization of data provenance. In: Proceedings

P

of the 8th International Conference on Database Theory; p. 316–30.

2. Cui Y, Widom J. Practical lineage tracing in data
warehouses. In: Proceedings of the 16th International
Conference on Data Engineering; p. 367–78.

3. Woodruff A, Stonebraker M. Supporting fine-grained
data Lineage in a database visualization environment.
In: Proceedings of the 13th International Conference
on Data Engineering; p. 97–102.

4. Altintas I, Barney O, Jaeger-Frank E. Provenance
collection support in the Kepler scientific workflow system. In: Proceedings of the International
Provenance and Annotation Workshop; 2006. p.
118–32.

5. Foster I, Vockler J, Eilde M, Zhao Y. Chimera:
a virtual data system for representing, querying,
and automating data derivation. In: Proceedings
of the 14th International Conference on Scientific and Statistical Database Management; 2002. p.
37–46.

6. Freire J, Silva CT, et al. Managing rapidly-evolving
scientific workflows, managing rapidly-evolving scientific workflows. 2006.

7. Simmhan Y, Plale B, Gannon D. A framework for
collecting provenance in data-centric scientific workflows. In: Proceedings of the IEEE International Conference on Web Services; 2006.

8. Wong SC, Miles S, Fang W, Groth P, Moreau L.
Provenance-based validation of E-Science experiments. In: Proceedings of the 4th International Semantic Web Conference, Lecture Notes in Computer
Science. 2005. p. 801–15.

9. Anand MK, Bowers S, McPhillips T, Ludascher
B. Efficient provenance storage over nested data
collections. In: Advances in Database Technology, Proceedings of the 12th International Conference on Extending Database Technology; 2009.
p. 958–69.

10. Artem Chebotko SL, Fei X, Fotouhi F. RDFPROV:
a relational RDF store for querying and managing
scientific workflow provenance. Data Knowl Eng.
2010;69(8):836–65.

11. Buneman P, Chapman A, Cheney J. Provenance
management in curated databases. In: Proceedings
of the ACM SIGMOD International Conference on
Management of Data; 2006. p. 539–50.

12. Heinis T, Alonso G. Efficient lineage tracking for
scientific workflows. In: Proceedings of the ACM
SIGMOD International Conference on Management
of Data; 2008. p. 1007–18.

13. Xiey Y, Muniswamy-Reddy K-K, Fengy D, Liz Y,
Longz DDE, Tany Z, Chen L. A hybrid approach for
efficient provenance storage. In: Proceedings of the
21st ACM International Conference on Information
and Knowledge Management; 2012.

14. Park H, Ikeda R, Widom J. RAMP: a system for
capturing and tracing provenance in mapreduce

workflows. In: Proceedings of the 37th International
Conference on Very Large Data Bases; 2011.

15. Mason C. Cryptographic binding of metadata, The
National Security Agency's Review of Emerging
Technologies, vol. 18. 2009.

16. Allen MD, Chapman A, Blaustein B. Engineering
choices for open world provenance. In: Proceedings
of the 6th International Provenance and Annotation
Workshop; 2014.

17. Dey S, Agun M, Wang M, Ludäscher B, Bowers S,
Missier P. A provenance repository for storing and
retrieving data lineage information, Technical Report,
DataONE Provenance & Workflow Working Group.
2011.

18. Missier P, Chen Z. Extracting PROV provenance
traces from Wikipedia history pages. In: Proceedings
of the 16th International Conference on Extending
Database Technology; 2013.

19. Robinson I, Webber J, E. Eifrem. Graph databases.
O'Reilly Media, Inc.; 2013.

20. Dublin Core Metadata Initiative Usage Board. DCMI
Metadata Terms: A complete historical record.
Dublin Core Metadata Initiative (DCMI), Online,
2014.

21. Moreau L, Clifford B, Freire J, Futrelle J, Gil Y,
Groth P, Kwasnikowska N, Miles S, Missier P, Myers
J, Plale B, Simmhan Y, Stephan E, Van den Bussche J. The Open Provenance Model core specification (v1.1), Future Generation Computer Systems
2011;27:6, 743–756.

22. Moreau L, Groth P. Provenance an introduction to
PROV. Morgan & Claypool Publishers; 2013.

23. Groth P, Moreau L. PROV-Overview. World Wide
Web Consortium (W3C), Online, 2013.

24. Abawajy JH, Jami SI, Shaikh ZA, Hammad SA. A
framework for scalable distributed provenance storage system. Comput Stand Interfaces. 2013;35(1):
179–86.

25. Allen MD, Chapman A, Blaustein B, Seligman
L. Getting it together: enabling multi-organization
provenance exchange. In: Proceedings of the 3rd
USENIX Workshop on the Theory and Practice of
Provenance; 2011.

26. Groth P, Jiang S, Miles S, Munroe S, Tan V, Tsasakou
S, Moreau L. An architecture for provenance systems,
Technical Report. ECS, University of Southampton.
2006.

27. Zhao D, Shou C, Malik T, Raicu I. Distributed data
provenance for large-scale data-intensive computing.
IEEE Cluster. 2013.

28. Groth P, Miles S, Moreau L. PReServ: provenance
recording for services, UK OST e-Science second
AHM. 2005.

29. PLUS. https://github.com/plus-provenance/plus

30. Simmhan Y, Plale B, Gannon D. Karma2: provenance
management for data driven workflows. J Web Ser
Res. 2008;5(2):1–22.

Provenance: Privacy and Security

Susan B. Davidson[1] and Sudeepa Roy[2]
[1]Department of Computer and Information Science, University of Pennsylvania, Philadelphia, PA, USA
[2]Department of Computer Science, Duke University, Durham, NC, USA

Synonyms

Confidentiality; Integrity; Lineage; Origin

Definition

Data provenance is information about the origins of data and its movement between databases and processes. It can be used to understand and debug the process by which data was obtained and transformed, to ensure reproducibility of results, and to establish trust. Provenance therefore has implications for both the security and privacy of the associated data. As metadata, there are also security and privacy concerns associated with provenance itself, including the integrity, confidentiality, and availability of provenance information.

Historical Background

Tracking the provenance of data within a system includes (i) capturing metadata associated with raw data that is input to the system and (ii) details of computations that transform the raw data to create new information, e.g., the sequence of steps or processes, parameter settings (in a program), and inputs and outputs of each step. Queries over provenance typically answer a question of the form *What are the input data and/or processing steps that led to the creation or modification of a given data item?*.

Since provenance is itself data (metadata), at a first glance, it might seem that the security and privacy of provenance records are already addressed by the vast amount of work that has been done in these domains. However, the application of these techniques for provenance is not straightforward due to the interaction between the entities, agents and activities involved in the data generation process, and the types of questions that are asked of provenance data.

As an example, *differential privacy* [9] is a widely accepted notion of privacy which focuses on aggregate queries and obscures sensitive information about individuals in the aggregates by adding random noise. How to use these ideas in the context of data provenance is not clear: (i) queries over provenance information are typically *not* aggregate queries, and (ii) adding random noise to provenance makes it of little use for purposes of reproducibility and trust.

Another subtlety is that the privacy levels of different components of provenance may vary both within and across applications [3]. For example, the decision of whether to accept a paper for a refereed journal is based on a set of reviews. While the reviews can be accessed by the authors, the reviewers themselves should not be, making the provenance information more sensitive than data. Similarly, the decision of whether to accept a student to a graduate program is based on a set of letters of recommendation. While the identity of the recommenders is known to the student since they provide the names, the recommendations should not be revealed to the student. In this case, the data is more sensitive than provenance.

Note that in both of these applications, the processes by which the decision is made, including the activity nodes and connections between nodes, are not confidential. In other applications, however, an activity may represent proprietary code or the connections between activities (e.g., the sequence of steps taken in a biological/clinical experiment) may be required to remain confidential. Privacy and security concerns in provenance therefore require careful study at the level of data, agents, activities, and the connections between activities.

P

Scientific Fundamentals

Provenance is typically studied in the context of *data flow* applications, such as databases, workflows, or storage systems. Such applications can be modeled as a directed acyclic graph in which the nodes represent atomic processes (e.g., a program or a procedure), and the edges represent the data flow between processes. As an example, Fig. 1a shows the *specification* (i.e., the template) of a simplified phylogenetic tree construction workflow. Here the input sequence data flows into a process called `Split_Entries`, which splits the input into annotation and sequence information. The annotations are sent to process `Curate_Annotations` and sequences to `Align_Sequences`. The curated annotations, aligned sequences, as well as input functional data are then formatted and combined in `Construct_Tree` to create the final output tree.

In an *execution* of this data flow, data (e.g., in the form of files) will be passed from one process to others when the processes are executed. Provenance in such an execution can be modeled using PROV, a W3C standard that has been proposed to unify provenance representations and enable their interchanges on the Web (PROV

Model Primer: http://www.w3.org/TR/prov-primer/.) [14]. There are three types of nodes in PROV: (i) *entity nodes*, denoting individual data items; (ii) *activity nodes*, denoting processes that consume one or more data items and produce one or more data items; and (iii) *agent nodes* that are *attributed to* one or more entity nodes or are *associated with* one or more activity nodes. Edges in this model include `used` (from an activity to an entity), `wasGeneratedBy` (from an entity to an activity), `wasAttributed to` (from an entity to an agent), and `wasAssociatedWith` (from an activity to an agent). Each data item is either an *initial input* (i.e., it was not generated by any activity, e.g., `Input_Sequence_Data`) or was generated by exactly one activity (e.g., `Curated_Annotation`). Similarly, if a data item is a final output (e.g., `Output_Tree`), it is not used in any activity; otherwise, it can be used in one or more activities.

A PROV graph for an execution of the workflow specification in Fig. 1a is shown in Fig. 1b. Its structure is similar to the specification, but additional nodes are present: ovals are used to represent data items as entities (e.g., `Annotation`), and pentagons represent agents (e.g., `Institute_A`). Edges also have different meanings and use the reverse orientation. Note

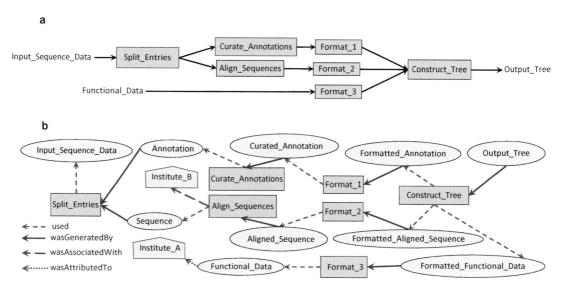

Provenance: Privacy and Security, Fig. 1 (a) A workflow specification, (b) the provenance of an execution of the workflow using PROV

that each execution of the workflow would generate a new provenance graph.

The provenance for a data item is the subgraph that is reachable from it and includes *provenance records* (nodes in the PROV graph) as well as information about *connectivity* between records (edges in the PROV graph). For example, Fig. 1b is the provenance of `Output_Tree`, and the provenance of `Annotation` is the path `wasGeneratedBy`, `Split_Entries`, `Used`, and `Input_Sequence_Data`.

Desiderata of Provenance Privacy and Security

Secure and privacy preserving provenance must ensure the following [5, 10]:

- **Confidentiality:** Unauthorized parties should not have access to provenance records or connectivity information.
- **Availability:** Authorized parties should have access to provenance records and connectivity information.
- **Integrity:** Adversaries cannot tamper with individual provenance records or connectivity information without being detected.

Provenance privacy requires ensuring confidentiality while maintaining availability, whereas provenance security requires ensuring availability while maintaining integrity.

Provenance Privacy

While capturing and publishing provenance enable transparency, trust, and reproducibility of results, it may expose sensitive information contained in the provenance graph. Provenance privacy therefore concerns data (entities and agents), activities (e.g., modules or processes), and structure (relationships between entities, agents, and activities) [7].

Data privacy. The first and foremost concern is ensuring the confidentiality of data items while publishing provenance. For example, in the refereed journal paper review example, the author should not be able to see information about agents who are reviewers, but should be able to see that of the agent who is the editor, the reviews, and the review process itself. All of the provenance information should be visible to the editor and his/her superiors. Assuming a double-blind review process, reviewers should be able to see the reviews, information about the editor and reviewer agents, as well as the review process, but not information about the author agent. Note that it is not necessary to disallow individuals from seeing the entire provenance graph, which would violate availability, as long as they are prevented from accessing confidential components. Most of such privacy concerns (involving entities and agents) can be managed using standard role-based access control techniques, i.e., by specifying categories of users and access privileges against the application specification.

Module privacy. Ensuring the confidentiality of certain activities contained within provenance is trickier. For example, if an activity node represents the execution of a proprietary software, it is not enough to disallow access to the provenance record for that activity in all executions. If a user is allowed to see provenance records of input (`used`) and output (`wasGeneratedBy`) entities to that activity, and the software is run over a large fraction of possible inputs, then the user might be able to emulate the proprietary software. To avoid this, some of the inputs and/or outputs to that activity across all executions must also be hidden, compromising availability to guarantee confidentiality of the activity.

Structural privacy. Publishing provenance information can also reveal confidential connectivity information among entities, agents, and activities. For example, suppose it is known that a set of modules is typically used to perform a particular analysis in a workflow (e.g., for functional annotation [7]). These modules can be executed in different orders, sequentially or in parallel, and some of them may be skipped under certain conditions. The order of processing may significantly influence the efficiency of the overall analysis, as well as the overall cost. In the published provenance information, the workflow owner may be willing to reveal which modules were used, but the order in which they were executed could be confidential.

Provenance Security

Provenance security involves ensuring the integrity and availability of provenance records and connectivity. The threat is that adversaries may attempt to subvert provenance by altering, adding, or deleting either data, provenance records, or connectivity. Ultimately, this means that provenance must be captured and stored using trusted hardware.

However, provenance is frequently captured by database or workflow systems without such support or may travel across software domains. The focus therefore becomes one of ensuring that provenance records and connectivity are *tamper evident* and that a *trusted auditor* can verify their authenticity. Note that if the auditor does not have access privileges for the provenance records, they must be able verify authenticity without actually seeing the contents.

Mechanisms for Provenance Privacy and Security

As mentioned earlier, role-based access control is widely used to ensure the privacy of entities and agents, in mechanisms for addressing module and structural privacy, as well as in ensuring the security of provenance records themselves.

A formal model for module privacy based on l-diversity was given in [6]. Given a desired privacy level l, they considered the problem of determining what input data to hide to ensure that for every input x to the confidential module, the output is be indistinguishable from at least $l - 1$ other possible outputs. However, the problem becomes much more complex in the setting considered with provenance graphs, in particular when the module is part of a workflow and interacts arbitrarily with other modules, some of which may be public with known functionality (e.g., a sorting or reformatting module).

A common technique for addressing structural privacy is to use *composite processes*. In contrast to an atomic process, such as Split_Entries and Align_Sequences in Fig. 1a, a compos-

ite process is one which itself has structure. For example, the entire workflow in Fig. 1a could be a single composite process (say, Construct_All), which takes sequence and functional data as input and generates a tree as output. Composite processes form a basis for *views* of provenance information [1, 4] in which details of the composite process execution (a subgraph) are grouped within a single, composite node. A user without sufficient privileges would not be allowed to expand the composite node and see details contained within it.

Although confidentiality of structural information can be addressed using composite nodes, availability may be compromised, and the user may be able to infer incorrect connectivity. For example, if the Curate_Annotations and Align_Sequences in the PROV graph are put in a composite module, the user could infer a path from Curated_Annotation to Sequence, which is not present in the fully expanded PROV graph. Initial ideas of how to use composite node expansion and contraction are being developed for PROV [14], but more work remains to be done. Other approaches for addressing structural privacy include using a declarative framework that allows owners to specify which parts of the provenance graph are to be hidden or abstracted and checking user requests against the specification (PROPUB [8]) and using *surrogate* nodes and edges to protect sensitive graph components while maximizing graph connectivity [2].

Security mechanisms used to secure data can also be used for provenance, e.g., signatures, checksums, and signed hashes. These techniques can be enforced at the kernel, file system, or application layers and at different stages in a workflow as the data is curated, annotated, and queried [11, 12, 15]. However, the problem becomes much more difficult when provenance can cross multiple domain boundaries and can be passed through untrusted domain owners.

Key Applications

In addition to data flow applications like document management and scientific workflows, provenance privacy and security are important for a wide range of applications. Three of these are discussed below.

1. **Cloud computing and service-oriented architectures:** In recent years, cloud services have gained popularity due to their affordability and ease of use. In a cloud platform, resources (storage and computation) are managed by the cloud service provider, accessed by multiple users, and may be used to process and store sensitive information about individuals. In addition to ensuring the confidentiality of sensitive information, the identity of users may be confidential. Data-intensive collaborations on cloud computing therefore require ensuring the privacy and security of provenance.

2. **Networks:** In networked applications, the provenance of exchanged information may be used to judge its credibility. For example, in a military command and control application, information about location status, intelligence reports, and operational plans may be communicated between different units, and provenance may be used to judge its trustworthiness. In a network routing application, provenance may be used to identify faulty or misbehaving nodes and to assess damage such nodes may cause to the network. In such systems, care must be taken to ensure that provenance information is secure, i.e., that a compromised node cannot forge or tamper with provenance, or reveal the identities of other nodes in the network to unauthorized agents.

3. **Healthcare:** Privacy is a critical concern in healthcare applications, where patients, practitioners, and other people involved in the system of care must have authorized access to patient or treatment information. Enabling provenance in this domain can significantly increase trust in the data and be used to improve the quality of service. However, since provenance may leak information about patients and their treatment to unauthorized agents, stringent measures must be taken to ensure its privacy and security.

Future Directions *

There are several important directions for future research in provenance privacy and security. First, there is a gap between theory and practice in this area. For example, PROV has been proposed as a standard for provenance representation and exchange and is starting to be used in provenance-enabled systems. However, it lacks certain abstractions, such as composite modules, which are important for mechanisms which enable provenance privacy and security. On the other hand, the practical effectiveness of provenance privacy techniques has not been successfully evaluated due to the lack of real datasets and well-defined measures on availability and confidentiality requirements. It is important to narrow this gap in future research. Second, existing research on provenance security has focused on identifying requirements and has proposed solutions using standard techniques used for securing data. It would be interesting to understand whether there are novel security problems and solutions arising from the interplay between data and provenance, as there are with privacy. Third, privacy notions currently being used for provenance are fairly weak and make strong assumptions. For example, module privacy uses ℓ-diversity [13] and assumes no background knowledge of the adversary; it is implemented by hiding portions of the provenance information using access control techniques. However, there are much stronger notions of privacy such as differential privacy [9], which rely on adding random noise to the output of aggregate queries. It is important to understand how these techniques can be used in the context of provenance, where queries are typically not aggregates and reproducibility is essential.

Cross-References

► Data Provenance
► Provenance in Databases
► Provenance in Workflows
► Provenance Standards

► Randomization Methods to Ensure Data Privacy

► Role-Based Access Control

Recommended Reading*

1. Bao Z, Davidson SB, Milo T. Labeling workflow views with fine-grained dependencies. Proc VLDB Endow. 2012;5(11):1208–19.
2. Blaustein B, Chapman A, Seligman L, Allen MD, Rosenthal A. Surrogate parenthood: protected and informative graphs. Proc VLDB Endow. 2011;4(8):518–25.
3. Braun U, Shinnar A, Seltzer M. Securing provenance. In: Proceedings of the 3rd USENIX Workshop on Hot Topics in Security; 2008. p. 4:1–5.
4. Chebotko A, Chang S, Lu S, Fotouhi F, Yang P. Scientific workflow provenance querying with security views. In: Proceedings of the 9th International Conference on Web-Age Information Management; 2008. p. 349–56.
5. Cheney J. A formal framework for provenance security. In: Proceedings of the 2011 IEEE 24th Computer Security Foundations Symposium; 2011. p. 281–93.
6. Davidson SB, Khanna S, Milo T, Panigrahi D, Roy S. Provenance views for module privacy. In: Proceedings of the 30th ACM SIGACT-SIGMOD-SIGART Symposium on Principles of Database Systems; 2011. p. 175–86.
7. Davidson SB, Khanna S, Roy S, Stoyanovich J, Tannen V, Chen Y. On provenance and privacy. In: Proceedings of the 14th International Conference on Database Theory; 2011. p. 3–10.
8. Dey SC, Zinn D, Ludäscher B. PROPUB: Towards a declarative approach for publishing customized, policy-aware provenance. In: Proceedings of the 23rd International Conference on Scientific and Statistical Database Management; 2011. p. 225–43.
9. Dwork C. Differential privacy: a survey of results. In: Proceedings of the 5th Annual Conference on Theory and Applications of Models of Computation; 2008. p. 1–9.
10. Hasan R, Sion R, Winslett M. Introducing secure provenance: problems and challenges. In: Proceedings of the 2007 ACM Workshop on Storage Security and Survivability; 2007. p. 13–8.
11. Hasan R, Sion R, Winslett M. Preventing history forgery with secure provenance. Trans Storage 2009;5(4):12:1–43.
12. Lu R, Lin X, Liang X, Shen XS. Secure provenance: the essential of bread and butter of data forensics in cloud computing. In: Proceedings of the 5th ACM Symposium on Information, Computer and Communications Security; 2010. p. 282–92.
13. Machanavajjhala A, Kifer D, Gehrke J, Venkitasubramaniam M. L-diversity: privacy beyond k-anonymity. ACM Trans Knowl Discov Data 2007;1(1):3.
14. Missier P, Bryans J, Gamble C, Curcin V, Dánger R. ProvAbs: model, policy, and tooling for abstracting PROV graphs. In: IPAW; 2014. p. 3–15.
15. Zhang J, Chapman A, LeFevre K. Do you know where your data's been? – tamper-evident database provenance. In: SDM; 2009. p. 17–32.

Pseudonymity

Simone Fischer-Hübner
Karlstad University, Karlstad, Sweden

Synonyms

Nymity

Definition

The term pseudonymous originates from the Greek word *pseudonymos* meaning "having a false name."

A pseudonym is an identifier of a subject other than one of the subject's real names, and pseudonymity is the use of pseudonyms as identifiers. Sender pseudonymity is defined as the sender being pseudonymous, recipient pseudonymity as the recipient being pseudonymous [1].

In the EU General Data Protection Regulation (GDPR) defines "pseudonymisation of data" as "the processing of personal data in such a manner that the personal data can no longer be attributed to a specific data subject without the use of additional information, provided that such additional information is kept separately and is subject to technical and organisational measures to ensure that the personal data are not attributed to an identified or identifiable natural person" [2].

Key Points

Pseudonymity resembles anonymity as both concepts aim at protecting the real identity of a subject. The use of pseudonyms, however, allows one to maintain a reference to the subject's real identity, e.g., for accountability purposes [3]. A trusted third party, adhering to agreed rules, could, for instance, reveal the real identities of misbehaving pseudonymous users. Pseudonymity also allows a user to link certain actions under one pseudonym. For instance, a user could reuse the same pseudonym for purchasing items at a certain online shop for building up a reputation or for collecting bonus points under this so-called relationship pseudonym.

The degree of anonymity protection provided by pseudonyms depends on the amount of personal data of the pseudonym holder that can be linked to the pseudonym and on how often the pseudonym is used in various contexts/for various transactions. The best protection can be achieved if for each transaction a new so-called transaction pseudonym is used that is unlinkable to any other transaction pseudonyms and at least initially unlinkable to any other personal data items of its holder (see also [1]).

Cross-References

▶ Anonymity
▶ Privacy
▶ Privacy-Enhancing Technologies

Recommended Reading

1. Pfitzmann A, Hansen M. A terminology for talking about privacy by data minimization: anonymity, unlinkability, unobservability, pseudonymity, and identity management. Version 0.34. http://dud.inf.tu-dresden.de/Anon_Terminology.shtml. 10 Aug 2010.
2. EU Commission. Regulation (EU) 2016/679 of the European Parliament and of the Council of 27 April 2016 on the protection of natural persons with regard to the processing of personal data and on the free movement of such data, and repealing Directive 95/46/EC (General Data Protection Regulation). Off J Eur Union. L 119/1. 4 May 2016.
3. Common Criteria for Information Technology Security Evaluation. Version 3.1, revision 4 Sept 2012. Part 2: Functional security components. http://www.commoncriteriaportal.org.

Publish/Subscribe

Hans-Arno Jacobsen
Department of Electrical and Computer Engineering, University of Toronto, Toronto, ON, Canada

Definition

Publish/Subscribe is an interaction pattern that characterizes the exchange of messages between publishing and subscribing clients. Subscribers express interest in receiving messages and publishers simply publish messages without specifying the recipients for a message. The publish/subscribe message exchange is decoupled and anonymous. That is, publishers neither know subscribers' identities nor whether any subscribers with matching interests exist at all. This supports a many-to-many style of communication, where data sources publish and data sinks subscribe. Different classes of publish/subscribe approaches have crystallized. Their main differences lie in the way subscribers express interest in messages, in the structure and format of messages, in the architecture of the system, and in the degrees of decoupling supported. Publish/Subscribe is widely used as middleware abstraction, applied to enterprise application integration, system and network monitoring, and selective information dissemination.

Historical Background

The definition in this entry aims to be general and characterizes publish/subscribe as an *interaction pattern* that governs the interaction between *many* publishing data sources and *many* subscribing data sinks. However, in practice, publish/subscribe is often interpreted to mean many slightly different concepts, such as an asynchronous communication style, a messaging paradigm, a message routing approach, an event filtering (matching) approach, or a design pattern. Moreover, research on publish/subscribe has been conducted in different communities, such as distributed systems, networking, programming languages, software engineering, and databases. The exact origins of publish/subscribe are therefore difficult to pinpoint exactly. Today, publish/subscribe-style abstractions can be found in many messaging standards, messaging products, databases, and even in special purpose hardware solutions.

The interpretation of publish/subscribe as asynchronous communication style emphasizes the data dissemination aspect and the associated qualities of service. The primary concern is the distribution of data from *many* sources to *many* sinks. Sometimes only a single source is considered and publish/subscribe is viewed as a *one-to-many* data dissemination paradigm, analogous to multicast. The channel-based publish/subscribe model best represents this interpretation. The indirect ancestry of publish/subscribe in this context are early reliable broadcast protocols [3] and primitives for process group management in operating systems [4], which inspired work on group communication as abstraction for reliable many-to-many communication, surveyed in [5]. It is this context that resulted in the subject-based publish/subscribe approach (a.k.a. topic-based model), first articulated by Oki et al. [16] and popularized by TIBCO with its TIBCO/RV product.

The interpretation of publish/subscribe as mesaging paradigm emphasizes the asynchronously decoupled nature of publishing data sources and subscribing data sinks. This view of publish/subscribe results from the inclusion of publish/subscribe functionality into standard messaging system specifications and products, such as MQ Series from IBM or the Java Message Service specification [9]. While these abstractions focus more on asynchronous one-to-one communication realized through message queues, they also include publication, subscription and filtering capabilities, often realized as subscriber-side filtering.

The interpretation of publish/subscribe as event filtering and matching approach emphasizes the selective filtering capabilities of the approach. Subscriptions represent filter expressions and publications represent observations about events in the environment that need to be selectively brought to the attention of subscribing entities. The ancestry of this work goes back to the formulation of the *many-to-many pattern matching problem* in the artificial intelligence domain by Forgy [7] who proposed the Rete algorithm for solving this problem. Rete is the basis of many rule-based expert systems. It efficiently evaluates observed facts provided as input against rules compiled into a memory resident graph (network), or an equivalent program [7]. This work was succeeded by approaches for efficient trigger management in database systems [2, 6] and research on active databases. In software engineering, similar concepts were applied for integrating software tools resulting in approaches such as YEAST [12]. Around the same time work on specifying events for event correlation and root cause analysis appeared in network management [14].

The interpretation of publish/subscribe as design pattern is based on the *Observer design pattern*, first articulated in the Gang of Four book [8], where the Observer pattern is also synonymously referred to as Publish/Subscribe. This reference is somewhat unfortunate, as the prescribed realization of the Observer pattern in the literature violates a key property of publish/subscribe. The Observer pattern is suggested for use in expressing one-to-many dependency between objects in a system. When the state of one object (the subject) changes state, all dependent objects are notified. This is achieve by having the subject know about its dependents by maintaining a list of them and requiring dependents to register, if they are interested in the subject's state changes. This violates the anonymous communication property of publish/subscribe, which requires that publishers and subscribers do not know each other.

There are a few other approach that have appeared independently in different contexts, such as tuple spaces in the programming languages context to model concurrency in the 1980's blackboard architectures in the context of artificial intelligence to model the interaction of agents, continuous query processing in the data management context, and stream processing, also in the data management context. All these approaches resemble publish/subscribe, but also differ in fundamental ways.

Foundations

A publish/subscribe system comprises *publishers*, *subscribers*, and *publish/subscribe message broker(s)*, also referred to as *message router(s)*.

Publishers and subscribers are roles held by applications built with the publish/subscribe abstraction. That is a client of the system could be publisher as well as subscriber at the same time. Subscribers express interest by registering subscriptions with the publish/subscribe system and publishers report on events by publishing messages to the publish/subscribe system. The system evaluates publications against registered subscriptions and determines which subscriptions match for a given publication. The complexity of matching varies among different publish/subscribe models. In the content-based publish/subscribe model, matching involves the evaluation of publication message content against expressive subscription filters. In the topic-based publish/subscribe model matching involves the evaluation of message topics against path-like subscription language expressions. In the channel-based publish/subscribe model, no explicit matching takes place; subscribers select among a set of channels and listen for messages broadcast on the channel.

Publications are transient and once match are not further stored or processed. Exceptions to this treatment are state-based publish/subscribe systems and the Subject spaces model [10, 11] where publications might be maintained as partial matching state and as persistent state per se, respectively.

Matching and notification are performed by the publish/subscribe message brokers. In centralized installations, there is a single broker to which subscribers and publishers connect. In distributed installations, there are multiple brokers to which subscribers and publishers connect.

Publish/Subscribe offers the following decoupling characteristics. Decoupling in space allows clients to be physically distributed. Decoupling in time allows clients to be independently available. Decoupling in location means that clients do not know each others identity. This last characteristics is also referred to as anonymous communication.

A publish/subscribe system is defined by the subscription language model, the publication data model, and the matching semantic. All these elements closely depend on each other and define the subscription, the notification, the advertisement, the publication and specify the matching of the former.

The subscription language model defines the language for expressing subscriptions. It determines the expressiveness of the publish/subscribe model. For example, in the content-based model a subscription is a Boolean function over Boolean predicates. Predicates test conditions, such as equality, binary relations, or string operators over attribute values in publications. Subscriptions are also referred to as *filters*, as they specify which publication to filter out from a flow of publications processed. Some systems distinguish among the *subscriber* and the *consumer*. The subscriber is the entity that specifies subscriptions and the *consumer* is the entity that receives notifications when certain subscriptions match. Subscriber and consumer entities must not be the same.

The publication data model defines the structure, the format, and the content type of publication messages. Publications are the messages emitted by publishers. They represent the *event* of interest about the state of the system or world in the context of the modeled application. An *event* is an asynchronous state transition of interest to subscribers. The publication is the message that conveys the occurrence of the event to any interested subscribers. In practice, the term publication and event are often used synonymously without distinguishing between the actual state transition and the message published about the event. The publication concept can be further refined by introducing *notifications*. A notification is the message sent from the publish/subscribe system to subscribers with matching subscriptions, whereas a publication is the message

P

Publish/Subscribe, Table 1 Comparison of models

Model	Filtering	Publication	Subscription
Channel-based	No filtering	Messages	Listening to channels
Topic-based	Topics &topic hierarchy	Messages tagged with topics	Expressions with wildcards over topics
Type-based	Type checking	Objects	–
Content-based	Message content	Messages	Content-based filters

published by publishers. A notification must not be identical to a publication that triggered it. Systems may define a notification semantic that specified which values of a publication to forward to subscribers. Also, more refined notification semantics that apply transformations to publications before notifying subscribers are imaginable. However, many authors do not distinguish between the publication and notification concept defined above and use both terms synonymously. Some publish/subscribe approaches rely on the concept of an *advertisement*. An *advertisement* is similar to a type in programming languages or schemas in databases and specifies the kind of information a publisher will publish. It is used by publish/subscribe systems to optimize matching and routing of publications. In symmetry to the difference between subscriber and publisher, a similar difference could be made among publisher and producer, where one entity publishes, while the other entity merely advertises. However, this difference does not seem to have been explored in practice so far. Not all publish/subscribe approaches use advertisements.

The matching semantic defines the conditions under which a publication matches a subscription. For example, in content-based publish/subscribe, subscriptions are often conjuncts of Boolean predicates. That is a publication matches a subscription, if each predicate in the subscription evaluates to true given the values specified in the publication. This is a *crisp* matching semantic; it requires that the publication matches the subscription exactly by either evaluating to true or false. In contrast, an approximate matching semantic weakens the matching condition and tolerates that certain predicates do not match or only match to a certain degree. A model based on fuzzy set theory

and possibility distribution is model realized in the Approximate *Toronto Publish/Subscribe* (A-ToPSS) project [13]. Similarly, probabilistic matching semantics that are defined by evaluating the probability that a publication matches a subscription are conceivable. Also, similarity-based semantics that measure the similarity between a publication and subscription based on some similarity measure or metric are conceivable.

Various publish/subscribe models have crystallized over time. These models are the channel-based, topic-based, type-based, content-based, state-based, and subject spaces. Table 1 compares these models with respect to publications, subscriptions, and filtering capabilities they support. More details about each model and additional subject spaces and state-based smodels is provided in a separate definition for each term.

Rule-based systems are intended to process *facts* against *rules* through logical inference, forward chaining or backward chaining algorithms, or by evaluating rules on events. Facts represent the state of the world and rules represent knowledge. Rule-based systems generally require the maintenance of state in the rule engine, as multiple facts processed over time may contribute to the evaluation of rules.

Key Applications

Applications of publish/subscribe include Information dissemination, information filtering, alerting and notification.

Standards that implement the publish/subscribe are the CORBA Event Service [18], the CORBA Notification Service [19], AMQP [1],

JMS [9], WS Topics, WS Notifications, WS Brokered Notifications, WS Eventing, OMG's Data Dissemination Service Specification [17], OGF's INFO-D [15].

Data Sets

The publish/subscribe research community has yet to produce benchmarks and collect data sets. Initial efforts are starting to emerge [20].

Cross-References

▶ Channel-Based Publish/Subscribe
▶ Content-Based Publish/Subscribe
▶ Database Trigger
▶ State-Based Publish/Subscribe
▶ Subject Spaces
▶ Topic-Based Publish/Subscribe
▶ Type-Based Publish/Subscribe

Recommended Reading

1. AMQP Consortium. Advanced message queuing protocol specification. version 0-10 edition. 2008.
2. Chakravarthy S, Mishra D. Snoop: an expressive event specification language for active databases. Data Knowl Eng. 1994;14(1):1–26.
3. Chang JM, Maxemchuk NF. Reliable broadcast protocols. ACM Trans Comput Syst. 1984;2(3):251–73.
4. Cheriton DR, Zwaenepoel W. Distributed process groups in the v kernel. ACM Trans Comput Syst. 1985;3(2):77–107.
5. Chockler GV, Keidar I, Vitenberg R. Group communication specifications: a comprehensive study. ACM Comput Surv. 2001;33(4):427–69.
6. Cohen D. Compiling complex database transition triggers. ACM SIGMOD Rec. 1989;18(2):225–34.
7. Forgy CL. Rete: a fast algorithm for the many pattern/many object pattern match problem. Artifi Intell. 1982;19(1):17–37.
8. Gamma E, Helm R, Johnson R, Vlissides J. Design patterns: elements of reusable object-oriented software. Reading: Addison-Wesley Longman Publishing Co.; 1995.
9. Hapner M, Burridge R, Sharma R. Java message service. sun microsystems. version 1.0.2 edition, 1999.
10. Ka Yau Leung H. Subject space: a state-persistent model for publish/subscribe systems. In: Proceedings

of conference of the centre for advanced studies on collaborative research. 2002. p. 7.
11. Ka Yau Leung H, Jacobsen HA. Efficient matching for state-persistent publish/subscribe systems. In: Proceedings of conf. of the centre for advanced studies on collaborative research. 2003. p. 182–96
12. Krishnamurthy B, Rosenblum DS. Ycast: a general purpose event-action system. IEEE Trans Softw Eng. 1995;21(10):845–57.
13. Liu H, Jacobsen HA. Modeling uncertainties in publish/subscribe system. In: Proceedings of the 20th International Conference on Data Engineering; 2004.
14. Mansouri-samani M, Sloman M. Gem: a generalized event monitoring language for distributed systems. Distrib Syst Eng. 1997;4(2):96–108.
15. OGF. Information dissemination in the grid environment base specifications. 2007.
16. Oki B, Pfluegl M, Siegel A, Skeen D. The information bus: an architecture for extensible distributed systems. In: Proceedings of the 14th ACM Symposium on Operating System Principles; 1993. p. 58–68.
17. OMG. Data distribution service for real-time systems. version 1.2, formal/07-01-01 edition. 2007.
18. OMG. Event service specification. version 1.2, formal/04-10-02 edition. 2004.
19. OMG. Notification service specification. version 1.1, formal/04-10-11 edition. 2004.
20. The PADRES Team. Publish/subscribe data sets. http://research.msrg.utoronto.ca/Padres/DataSets, 2008.

Publish/Subscribe Over Streams

Yanlei Diao[1] and Michael J. Franklin[2]
[1]University of Massachusetts Amherst, Amherst, MA, USA
[2]University of California-Berkeley, Berkeley, CA, USA

Definition

Publish/subscribe (pub/sub) is a many-to-many communication model that directs the flow of messages from senders to receivers based on receivers' data interests. In this model, publishers (i.e., senders) generate messages without knowing their receivers; subscribers (who are potential receivers) express their data interests, and are subsequently notified of the messages from a variety of publishers that match their interests.

P

Historical Background

Distributed information systems usually adopt a three-layer architecture: a presentation layer at the top, a resource management layer at the bottom, and a *middleware layer* in between that integrates disparate information systems. Traditional middleware infrastructures are tightly coupled. Publish/Subscribe [13] was proposed to overcome many problems of tight coupling:

- With respect to communication, tightly coupled systems use static point-to-point connections (e.g., remote procedure call) between senders and receivers. In particular, a sender needs to know all its receivers before sending a piece of data. Such communication does not scale to large, dynamic systems where senders and receivers join and leave frequently. Pub-/sub offers loose coupling of senders and receivers by allowing them to exchange data without knowing the operational status or even the existence of each other.
- With respect to content, tight coupling can occur in remote database access. To access a database, an application needs to have precise knowledge of the database *schema* (i.e., its structure and internal data types) and is at risk of breaking when the remote database schema changes. Extensible Markup Language (XML)-based pub/sub has emerged as a solution for loose coupling at the content level. Since XML is flexible, extensible, and self-describing, it is suitable for encoding data in a generic format that senders and receivers agree upon, hence allowing them to exchange data without knowing the data representation in individual systems.

In many pub/sub systems, message brokers serve as central exchange points for data sent between systems. Figure 1 illustrates a basic context in which a broker operates. Publishers provide information by creating streams of messages (Besides "messages," the words "events," "tuples," and "documents" are often used with similar meanings in various contexts in the database literature.) that each contain a header describing application-specific information and a payload capturing the content of the message. Subscribers register their data interests with a message broker in a subscription language that the broker supports. Inside the broker, arriving subscriptions are stored as *continuous queries* that will be applied to all incoming messages. These queries remain effective until they are explicitly deleted. Incoming messages are processed on-the-fly against all stored queries. For each message, the broker determines the set of queries matched by the message. A query result is created for each matched query and delivered to its subscriber in a timely fashion.

Figure 2 shows a design space for publish/-subscribe over data streams. In this diagram, pub/sub systems are first classified by the data model and the query language that these systems support. Roughly speaking, there are three main categories.

- *Subject-based*: Publishers label each message with a subject from a pre-defined set (e.g., "stock quote") or hierarchy (e.g., "sports/-golf"). Users subscribe to the messages in a particular subject. These queries can also contain a filter on the data fields of the message header to refine the set of relevant messages within a particular subject.

- *Complex predicate-based*: Some pub/sub systems model the message content (payload) as a set of attribute-value pairs, and allow

Publish/Subscribe Over Streams, Fig. 1 Overview of publish/subscribe

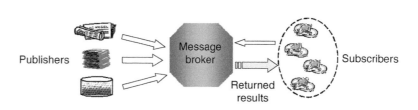

Publishers   Message broker   Subscribers

Returned results

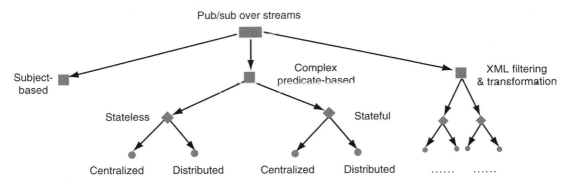

Publish/Subscribe Over Streams, Fig. 2 Design space for publish/subscribe over streams

user queries to contain predicates connected using "and" and "or" operators to specify constraints over values of the attributes. For example, a predicate-based query applied to the stock quotes can be "Symbol = 'ABC' and (Change >1 or Volume >50,000)."

- *XML filtering and transformation*: Recent pub/sub systems have started to exploit the richness of XML-encoded messages, in particular, the hierarchical, flexible XML structure. User queries can be written using an existing XML query language such as XQuery. The rich XML structure and use of an XML query language enable potentially more accurate filtering of messages and further restructuring of messages for customized result delivery.

Pub/sub systems can be further classified based on the style of query processing. In some systems, queries are applied only to individual messages, e.g., filtering messages, which does not involve any interaction across message boundaries. Such processing is referred to as *stateless*. Stateless processing is in contrast to stream query processing that maintains *state* over a long stream of messages, hence referred to as *stateful* processing. This distinction is illustrated for complex predicate-based systems in Fig. 2.

Finally, pub/sub systems can be distinguished based on the distribution of the architecture, as also shown in Fig. 2. In a coarse-grained fashion, this design space considers centralized and distributed processing. Distributed processing spreads the processing load for larger-scale pub/sub services; accordingly, it requires a more sophisticated routing functionality.

Foundations

As with stream processing, subscriptions, stored as continuous query inside a broker, need to be evaluated as data continuously arrives from other sources; that is, queries are evaluated every time when a new data item is received. Besides stream processing, pub/sub raises several additional challenges:

- *Scalability*. A key distinguishing requirement of pub/sub is scalability, in particular, in query population that pub/sub systems need to support. Such query populations can range from hundreds to millions in applications such as personalized content delivery. Given such populations, a salient issue is to efficiently search the huge set of queries to find those that can be matched by a message and to construct complete query results for them.
- *Robustness*. A second requirement of message brokers is the ability to perform in highly-dynamic environments where subscribers join and leave and their data interests change over time. Since message brokers see a constantly changing collection of queries, they must react quickly to query changes without adversely affecting the processing of incoming messages.

- *Distribution*. Due to the scale of message volume and query population, large-scale pub-/sub may require the use of a network of message brokers to distribute the query population and message processing load. In this case, an additional issue is how to efficiently route a message from its publishing site to the set of brokers hosting relevant queries for complete query processing.

Scope of this entry The rest of the entry focuses on complex predicate-based pub/sub systems. Pub/sub systems exploring XML filtering and transformation are described in detail in the entry "XML Publish/Subscribe."

Centralized, Stateless Publish/Subscribe

Le Subscribe [9] and *Xlyeme* [12] are predicate-based message filtering systems that use centralized processing. In these systems, a predicate is a comparison between an attribute and a constant using relational operators such as "=", ">", and "<". The main issue they address is how to efficiently match an incoming event, in the form of attribute value pairs, with the predicates of a large number of queries. The key idea is to *index* predicates as well as to *cluster* queries. In particular, *Le Subscribe* uses multi-attribute hash indexes to evaluate several predicates in a query with a single operation. In addition, it groups queries based on the number of contained predicates and the common conjunction of equality predicates, so many queries can be (partly) evaluated using a single operation. It further offers cost-based algorithms to find optimal clustering and to dynamically adjust it.

Centralized, Stateful Publish/Subscribe

NiagaraCQ [6] considers continuous queries with more complex predicates that can compare attributes of an input message to constants or to attributes of another message. To efficiently handle multiple queries, it groups query plans of continuous queries based on common *expression signatures*: an expression signature presents the same syntax structure, but possibly different constant values, in different queries. Consider

queries that are interested in stock quotes of different symbols. Traditional query processing involves repeated retrieval of the symbol attribute from input and evaluation of different predicates on this attribute for different queries. The group plan employs a constant table to store the constant values form different queries, and retrieves this attribute from the input once and then performs an equality join of the retrieved value and the constant table to find all matching queries. For robust processing, NiagaraCQ constructs group plans incrementally: Given a new query, it constructs the query plan, and merges the query plan bottom up with a group plan with the same signature, extending the group plan with the mismatched branch(es) if necessary. This process is incremental as the addition of the new query plan does not affect existing queries.

Recently, there has been a significant amount of activity on handling continuous and time-varying tuple streams, resulting in the development of multiple general-purpose stream management systems [1, 5, 11]. These systems support complex continuous queries that join multiple streams and/or compute aggregate values over a period of time called a *window* (hence, performing stateful processing). While this surge of research explores a broad set of issues such as adaptivity and approximation, shared processing of window-based queries [5, 11, 10] is of particular relevance to pub/sub.

Several special-purpose pub/sub systems have been recently proposed to handle temporal correlations among events in a stream. SASE [15] supports sequencing operators that integrate parameterized predicates (i.e., predicates that compare different events), negation, and windowing. It explores a new query processing abstraction that uses an automaton-based implementation for fast sequence operations and relational-style post-processing for other tasks such as negation and windowing. It also devises a set of optimizations in this automaton-based framework for efficiency and scalability. Cayuga [7] offers an algebra for expressing event sequences that may address a finite yet unbounded number of events with a similar property, and employs a more sophisticated automaton model to support this algebra. Its implementation focuses on multi-query optimization including merging states of automata for different queries and further indexing query predicates.

Distributed, Stateless Publish/Subscribe

In distributed pub/sub systems, messages are published and subscriptions are registered to different brokers. A key issue is to efficiently route each message from its publishing site to the subset of brokers hosting relevant queries for complete query processing. For complexity reasons, most distributed pub/sub systems restrict themselves to stateless services.

ONYX [8] presents an overview of a pub/sub network exploring *content-based routing*. In this paradigm, brokers are organized as an application-level overlay network with a particular topology. When a new message enters the broker network, the root broker as well as each intermediate broker routes the message to its neighboring brokers based on the correspondence between the content of the message and the subscriptions residing at and reachable from those brokers. Content-based routing is used as a key mechanism to avoid the flooding of messages to all brokers in the network, hence reducing bandwidth usage and broker processing load and rendering better scalability.

ONYX consists of two layers of functionality. The lower layer deals with the overlay network; in particular, for each broker, it constructs a broadcast tree that is rooted at that broker and reaches all other brokers in the network. On top of these broadcast trees, the higher layer performs content-based processing by dealing with subscriptions and messages. Two issues determine the effectiveness of content-based routing. The first is how to partition subscriptions and assign them to host brokers. Results of ONYX show that content-based routing is most effective if the clustering of subscriptions results in mutual exclusiveness in data interest among host brokers. The second issue is how to aggregate subscriptions from their host brokers and place such aggregations as routing specifications in the intermediate brokers for later directing the mes-

sage flow. Different degrees of generalization are possible depending on the precision-efficiency tradeoff suitable for each pub/sub system.

For content-based routing, Gryphon [2] and Siena [3] both aggregate subscriptions into compact, precise in-network data structures and use efficient algorithms to search these data structures to determine the routing of the messages to other brokers in the network.

XPORT [14] focuses on the construction, maintenance, and optimization of an overlay dissemination tree of the available message brokers in the system. Its tree-oriented optimization framework consists of a generic aggregation model that allows system cost to be expressed through various combinations of aggregation functions and local metrics, and distributed iterative optimization protocols for cost-based optimization of the overlay structure.

Distributed, Stateful Publish/Subscribe

For stateful publish/subscribe, Chandramouli et al. [4] adopt the following model: Users define subscriptions as SQL views over a conceptual (possibly distributed) database and messages are published as updates to the database; if a database update affects a subscription, the pub/sub system sends a notification to the subscriber containing the change to the content of the subscription view. The main idea of this work is to explore appropriate interfacing between the database and the pub/sub network so that existing stateless pub/sub networks can be leveraged for efficient dissemination. To do so, the key is to transform published messages into a semantic description of affected subscriptions (performed at the database side) and subscriptions into a predicate over the semantic description (evaluated in the stateless pub/sub network). Consider a selection-join subscription $\sigma_p(\sigma_{p\text{-}R} R \bowtie \sigma_{p\text{-}S} S)$. If a new message applies an update ΔR to table R, its effect on the subscription, $\sigma_p(\sigma_{p\text{-}R} \Delta R \bowtie \sigma_{p\text{-}S} S)$, requires access to table S that is not in the original update message (hence, stateful processing). The proposed solution reformulates each message ΔR into a series of messages containing the tuples in $\Delta R \bowtie S$ at the database side; to utilize a stateless pub/sub network, it transforms the select-join subscription into a simple condition that evaluates $\sigma_{p \wedge p\text{-}R \wedge p\text{-}S}$ over reformulated messages.

Key Applications

Personalized content delivery This class of applications provide personalized filtering and delivery of news feeds, web feeds (RSS), stock tickers, sport tickers, etc. to large numbers of online users.

Online auction and online procurement Through these applications, users can create their own feeds for their favorite searches, for example, on eBay.

Enterprise information management Pub/sub has been traditionally used to support application integration, which integrates disparate, independently-developed applications into new services via the loose coupling of senders and receivers based on receivers' data interest.

System and network monitoring Pub/sub has been recently used in system and network monitoring, where large-scale complex systems generate reports categorizing various aspects of system performance and resource utilization, and system administrators, end users, and visualization applications subscribe to receive updates on particular aspects of those reports.

Data Sets

Many data sources are available online, including RSS feeds (indicated by the orange button labeled with "RSS" or "XML" in many web pages) and financial feeds (e.g., Yahoo! Finance).

Cross-References

▶ Continuous Query
▶ Stream Processing
▶ XML
▶ XML Document
▶ XML Publish/Subscribe
▶ XPath/XQuery

Recommended Reading

1. Abadi D, Carney D, Cetintemel U, Cherniack M, Convey C, Lee S, Stonebraker M, Tatbul N, Zdonik S. Aurora: a new model and architecture for data stream management. VLDB J. 2003;12(2):120–39.
2. Aguilera MK, Strom RE, Sturman DC, Astley M, Chandra TD. Matching events in a content-based subscription system. In: Proceedings of the ACM SIGACT-SIGOPS 18th Symposium on the Principles of Distributed Computing; 1999.
3. Carzaniga A, Wolf AL. Forwarding in a content-based network. In: Proceedings of the ACM International Conference on Data Communication; 2003. p. 163–74.
4. Chandramouli B, Xie J, Yang J. On the database/network interface in large-scale publish/subscribe systems. In: Proceedings of the ACM SIGMOD International Conference on Management of Data; 2006. p. 587–98.
5. Chandrasekaran S, Cooper O, Deshpande A, Franklin MJ, Hellerstein JM, Hong W, Krishnamurthy S, Madden S, Raman V, Reiss F, Shah MA. TelegraphCQ: continuous dataflow processing for an uncertain world. In: Proceedings of the 1st Biennial Conference on Innovative Data Systems Research; 2003.
6. Chen J, Dewitt DJ, Tian F, Wang Y. NiagaraCQ: a scalable continuous query system for Internet databases. In: Proceedings of the ACM SIGMOD International Conference on Management of Data; 2000. p. 379–90.
7. Demers AJ, Gehrke J, Hong M, Riedewald M, White WM. Towards expressive publish/subscribe systems. In: Advances in Database Technology, Proceedings of the 10th International Conference on Extending Database Technology; 2006. p. 627–44.
8. Diao Y, Rizvi S, Franklin MJ. Towards an Internet-scale XML dissemination service. In: Proceedings of the 30th International Conference on Very Large Data Bases; 2004. p. 612–23.
9. Fabret F, Jacobsen HA, Llirbat F, Pereira J, Ross KA, Shasha D. Filtering algorithms and implementation for very fast publish/subscribe systems. In: Proceedings of the ACM SIGMOD International Conference on Management of Data; 2001. p. 115–26.
10. Krishnamurthy S. Shared query processing in data streaming systems. Ph.D. Dissertation, University of California, Berkeley.
11. Motwani R, Widom J, Arasu A, Babcock B, Babu S, Datar M, Manku G, Olston C, Rosenstein J, Varma R Query processing, resource management, and ap-

proximation in a data stream management system. In: Proceedings of the 1st Biennial Conference on Innovative Data Systems Research; 2003.

12. Nguyen B, Abiteboul S, Cobena G, Preda M. Monitoring XML data on the Web. In: Proceedings of the ACM SIGMOD International Conference on Management of Data; 2001. p. 437–48.

13. Oki B, Pfleugl M, Siegel A, Skeen D. The information bus: an architecture for extensible distributed system. In: Proceedings of the 14th ACM Symposium on Operating System Principles; 1993. p. 58–68.

14. Papaemmanouil O, Ahmad Y, Çetintemel U, Jannotti J, Yildirim Y. Extensible optimization in overlay dissemination trees. In: Proceedings of the ACM SIGMOD International Conference on Management of Data; 2006. p. 611–22.

15. Wu E, Diao Y, Rizvi S. High-performance complex event processing over streams. In: Proceedings of the ACM SIGMOD International Conference on Management of Data; 2006. p. 407–18.

Punctuations

David Maier[1] and Peter A. Tucker[2]
[1]Portland State University, Portland, OR, USA
[2]Whitworth University, Spokane, WA, USA

Definition

A *punctuation* is an item embedded in a data stream that specifies a subset of the domain of that stream. A data item that belongs to the subset specified by a punctuation is said to *match* that punctuation. A data stream is said to be *grammatical* if, for each punctuation, no data items will follow that match that punctuation. Consider a stream of bids for online auctions. When an auction closes, a punctuation p is embedded in the stream that matches all bids for that auction. In a grammatical stream, p indicates no more bids will arrive from that stream for that auction.

In the most common format, a punctuation is a tuple of *patterns*, where each pattern corresponds to an attribute of the data items in a stream. Typically, four patterns are used: a *wildcard* (denoted by '*') is matched by all values, a *literal* is matched by only the given value, a *list* (denoted by { }) is matched by any value in the given list, and a *range* (denoted by [ ]) is matched by any value that falls in the given range. If each value in a data item matches its corresponding pattern in a punctuation, then the data item matches the punctuation. For example, if auction bids have schema *<auction_id,bidder_id,price,timestamp>*, given the punctuation $p = P<\{105,106\},*,[0.00,100.00],*>$, all bids for auctions 105 and 106 with prices between 0 and 100 match p. Thus, the data item <105,95,90.00,10495> matches p, but the data items <104,95,90.00,10495> and <105,95,105.00,10495> do not.

Key Points

Punctuations have been shown to unblock query operators and reduce the amount of state required by query operators [2] when processing non-terminating data streams. In addition, punctuations have proven useful for dealing with disorder in streams [1] and in specifying window semantics [1, 3].

A *punctuation-aware operator* is a stream query operator that can take advantage of grammatical streams. Such an operator implements three different behaviors in the presence of punctuations [2]. A *pass behavior* allows operators to output data items due to punctuations. For example, when a punctuation arrives, an aggregate can output results for a particular group if all possible data items for that group match that punctuation. A *keep behavior* specifies which data items must remain in state when a punctuation arrives. For example, when a punctuation arrives, some data items that are held in state for symmetric hash join may be released. Finally, a *propagation behavior* defines which punctuations can be output when punctuation arrives. For example, union can output any punctuation that has arrived from both inputs.

Most punctuation-aware operators are counterparts of operators on finite relations. However, there can be more than one way to define pass, keep and propagation behavior when extending a relational operator to be punctuation-aware. An

P

important property here is for a stream operator s to be *faithful* to its analogous relational operator r, which means that on any finite prefix x of a stream, the output of s on x is consistent with the output of r on any finite extension of x.

Cross-References

▶ Continuous Query
▶ Data Stream
▶ Stream-Oriented Query Languages and Operators
▶ Stream Models
▶ Stream Processing
▶ Window-Based Query Processing

Recommended Reading

1. Li J, Maier D, Tufte K, Papadimos V, Tucker PA. Semantics and evaluation techniques for window aggregates in data streams. In: Proceedings of the ACM SIGMOD International Conference on Management of Data; 2005. p. 311–22.
2. Tucker PA, Maier D, Sheard T, Fegaras L. Exploiting punctuation semantics in continuous data streams. IEEE Trans Knowl Data Eng. 2003;15(3):555–68.
3. Tucker PA, Maier D, Sheard T, Stephens P. Using punctuation sschemes to characterize strategies for querying over data streams. IEEE Trans Knowl Data Eng. 2007;19(9):1227–40.

Printed by Printforce, the Netherlands